Graduate Texts in Contemporary Physics

Series Editors:

Joseph L. Birman
Jeffrey W. Lynn
Mark P. Silverman
H. Eugene Stanley
Mikhail Voloshin

Springer

New York
Berlin
Heidelberg
Barcelona
Budapest
Hong Kong
London
Milan
Paris
Santa Clara
Singapore
Tokyo

Graduate Texts in Contemporary Physics

Philippe Di Francesco
Pierre Mathieu
David Sénéchal

Conformal Field Theory

With 57 Illustrations

 Springer

Philippe Di Francesco
Commissariat l'Énergie Atomique
Centre d'Études de Saclay
Service de Physique Théorique
Gif-sur-Yvette, 91191 France

Pierre Mathieu
Département de Physique
Université Laval
Québec, QC G1K 7P4 Canada

David Sénéchal
Département de Physique
Université de Sherbrooke
Sherbrooke, QC J1K 2R1 Canada

Series Editors

Joseph L. Birman
Department of Physics
City College of CUNY
New York, NY 10031, USA

Jeffrey W. Lynn
Reactor Radiation Division
National Institute of Standards
and Technology
Gaithersburg, MD 20899, USA

Mark P. Silverman
Department of Physics
Trinity College
Hartford, CT 06106, USA

H. Eugene Stanley
Center for Polymer Studies
Physics Department
Boston University
Boston, MA 02215, USA

Mikhail Voloshin
Theoretical Physics Institute
Tate Laboratory of Physics
University of Minnesota
Minneapolis, MN 55455 USA

Library of Congress Cataloging-in-Publication Data
Di Francesco, Philippe.
    Conformal field theory / Philippe Di Francesco, Pierre Mathieu,
David Sénéchal.
        p.   cm. — (Graduate texts in contemporary physics)
    Includes bibliographical references and index.
    ISBN 0-387-94785-X (hrdcvr : alk. paper)
    1. Conformal invariants.   2. Quantum field theory.   I. Mathieu,
Pierre, 1957– .   II. Sénéchal, David.   III. Title.   IV. Series.
QC174.52.C66D5   1996
530.1′43—dc20   96-23155

Printed on acid-free paper.

Production managed by Robert Wexler; manufacturing supervised by Joe Quatela.
Photocomposed copy prepared from the authors' TeX files.
Printed and bound by Braun-Brumfield, Inc., Ann Arbor, MI.
Printed in the United States of America.

9 8 7 6 5 4 3 2 1

ISBN 0-387-94785-X Springer-Verlag New York Berlin Heidelberg   SPIN 10524551

This book is dedicated to our families

Preface

This is the first extensive textbook on conformal field theory, one of the most active areas of research in theoretical physics over the last decade. Although a number of review articles and lecture notes have been published on the subject, the need for a comprehensive text featuring background material, in-depth discussion, and exercises has not been satisfied. The authors hope that this work will efficiently fill this gap.

Conformal field theory has found applications in string theory, statistical physics, condensed matter physics, and has been an inspiration for developments in pure mathematics as well. Consequently, a reasonable text on the subject must be adapted to a wide spectrum of readers, mostly graduate students and researchers in the above-mentioned areas. Background chapters on quantum field theory, statistical mechanics, Lie algebras and affine Lie algebras have been included to provide help to those readers unfamiliar with some of these subjects (a knowledge of quantum mechanics is assumed). This textbook may be used profitably in many graduate courses dealing with special topics of quantum field theory or statistical physics, string theory, and mathematical physics. It may also be an instrument of choice for self-teaching. At the end of each chapter several exercises have been added, some with hints and/or answers. The reader is encouraged to try many of them, since passive learning can rapidly become inefficient.

It is impossible to encompass the whole of conformal field theory in a pedagogical manner within a single volume. Therefore, this book is intentionally limited in scope. It contains some necessary background material, a description of the fundamental formalism of conformal field theory, minimal models, modular invariance, finite geometries, Wess-Zumino-Witten models, and the coset construction of conformal field theories. Chapter 1 provides a general introduction to the subject and a more detailed description of the role played by each chapter. In building the list of references listed at the end of this volume, the authors have tried to be as complete as possible and hope to have given appropriate credit to all.

The authors intend to complete this work with a second volume, that would deal with the following subjects: Superconformal field theory ($N = 1, 2$), parafermionic models, W-algebras, critical integrable lattice models, perturbed conformal field theories, applications to condensed matter physics, and two-dimensional quantum gravity.

ACKNOWLEDGMENTS

Modern, computerized book production minimizes the number of trivial errors, but one still has to rely on friendly humans to detect what the authors themselves have overlooked! We are grateful to Dave Allen, Luc Bégin, Denis Bernard, François David, André-Marie Tremblay, Mark Walton, and Jean-Bernard Zuber for their useful reading of various parts of the manuscript and, in many cases, their much-appreciated counsel. In particular, we thank M. Walton for numerous discussions on the subjects covered in part C of this volume and his constant interest in this project. P.D.F. is especially indebted to J.-B. Zuber, who patiently introduced him to the conformal world, and to the late C. Itzykson, who guided his steps through modern mathematics with his extraordinary and communicative enthusiasm. P.M. and D.S. acknowledge the support of the Natural Sciences and Engineering Research Council of Canada (NSERC) and of "le Fonds pour la Formation de Chercheurs et l'Aide à la Recherche" (F.C.A.R.) of Québec.

Philippe Di Francesco
Pierre Mathieu
David Sénéchal
February 1996

Contents

7 Minimal Models I   **200**

  7.1 Verma Modules   200
      7.1.1 Highest-Weight Representations   201
      7.1.2 Virasoro Characters   203
      7.1.3 Singular vectors and Reducible Verma Modules   204
  7.2 The Kac Determinant   205
      7.2.1 Unitarity and the Kac Determinant   205
      7.2.2 Unitarity of $c \geq 1$ Representations   209
      7.2.3 Unitary $c < 1$ Representations   210
  7.3 Overview of Minimal Models   211
      7.3.1 A Simple Example   211
      7.3.2 Truncation of the Operator Algebra   214
      7.3.3 Minimal Models   215
      7.3.4 Unitary Minimal Models   218
  7.4 Examples   219
      7.4.1 The Yang-Lee Singularity   219
      7.4.2 The Ising Model   221
      7.4.3 The Tricritical Ising Model   222
      7.4.4 The Three-State Potts Model   225
      7.4.5 RSOS Models   227
      7.4.6 The $O(n)$ Model   229
      7.4.7 Effective Landau-Ginzburg Description of Unitary
            Minimal Models   231
            Exercises   235

8 Minimal Models II   **239**

  8.1 Irreducible Modules and Minimal Characters   240
      8.1.1 The Structure of Reducible Verma Modules for Minimal
            Models   240
      8.1.2 Characters   242
  8.2 Explicit Form of Singular Vectors   243
  8.3 Differential Equations for the Correlation Functions   247
      8.3.1 From Singular Vectors to Differential Equations   247
      8.3.2 Differential Equations for Two-Point Functions in
            Minimal Models   250
      8.3.3 Differential Equations for Four-Point Functions in
            Minimal Models   252
  8.4 Fusion Rules   255
      8.4.1 From Differential Equations to Fusion Rules   255
      8.4.2 Fusion Algebra   257
      8.4.3 Fusion Rules for the Minimal Models   259
  8.A General Singular Vectors from the Covariance of the OPE   265
      8.A.1 Fusion of Irreducible Modules and OPE Coefficients   266
      8.A.2 The Fusion Map $\mathcal{F}$: Transferring the Action of Operators 271

PART A

INTRODUCTION

CHAPTER 1

Introduction

> A vast similitude interlocks all,
> All distances of space however wide,
> All distances of time...
> – Walt Whitman

The æsthetic appeal of symmetry has been a guide—sometimes a tyrannic one—for philosophers of nature since the dawn of science. Ancient Greeks, in their belief that celestial bodies followed perfectly circular orbits, demonstrated an attachment to the circle as the most symmetric curve of all. In elaborating more complex systems involving scores of epicycles and eccentrics, they gave up the idea that celestial orbits should be explicitly symmetric, but invented unknowingly the concept of "hidden symmetry", for the circle remained the building block of their cosmology. Modern science, with Kepler, Galileo, and Newton, gave symmetry a deeper realm: that of the physical "laws." Circles gave way to ellipses and more complicated trajectories; the richness and variety of Nature became, in the Heavens like on Earth, compatible with symmetric laws, even without the exterior appearance of symmetry.

Twentieth-century physics has witnessed the triumph of symmetry and its precise formulation in theoretical language. The work of Lie and Cartan (among others) paved the way for the general application of symmetries in microscopic physics within quantum mechanics. Wigner, probably the most important figure in the application of group theory to physics, fitted the possible elementary particles into representations of the Lorentz and Poincaré groups. The principles of special and general relativity—the seeds of the other great revolution of twentieth-century physics—were also motivated by the appeal of symmetry. Modern theories of elementary particles (the so-called *standard model*) rest on the principle of local *gauge symmetry*. Our understanding of phase transitions and critical phenomena draws a great deal on the concept of *broken symmetry*. In particular, broken gauge symmetries are central to our understanding of weak interactions, superconductivity, and cosmology.

This book is about conformal symmetry in two-dimensional field theories. Conformal field theory plays a central role in the description of second- or higher-order phase transitions in two-dimensional systems, and in string theory, the (so far speculative) attempt at unifying all forces of Nature. To the practical man, this may seem a narrow field of application for a book of this size. However, two-dimensional conformal field theories are perfect examples of systems in which the symmetries are so powerful as to allow an exact solution of the problem. This feature, as well as the great variety of mathematical concepts needed in their solution and definition, have made conformal field theories one of the most active domains of research in mathematical physics.

In the context of a physical system with local interactions such as those studied in this work, conformal invariance is an immediate extension of scale invariance, a symmetry under dilations of space. This important fact was first pointed out by Polyakov [295]. Conformal transformations are nothing but dilations by a scaling factor that is a function of position (local dilations). It is entirely natural that a local theory (i.e., without action at a distance) that is symmetric under rigid (or global) dilations should also be symmetric under local dilations.

Even after being augmented to conformal invariance, the symmetry remains finite, in the sense that a finite number of parameters are needed to specify a conformal transformation in d spatial dimensions (specifically, $\frac{1}{2}(d+1)(d+2)$). The consequence of this finiteness is that conformal invariance can say relatively little about the form of correlations, in fact just slightly more than rotation or scale invariance. The exception is in two dimensions, where the above formula gives only the number of parameters specifying conformal transformations that are everywhere well-defined, whereas there is an infinite variety of local transformations (the conformal mappings of the complex plane) that, although not everywhere regular, are still equivalent to local dilations. The number of parameters specifying such local conformal transformations in two dimensions is infinite, because any locally analytic function provides a bona fide conformal mapping. This richness of conformal symmetry in two dimensions is the reason for the success of conformal invariance in the study of two-dimensional critical systems.

Scale invariance is by no means an exact symmetry of Nature, since our description of physical phenomena involves a number of *characteristic length scales* that indicate the typical distances over which the "action is taking place." These length scales are not invariant under dilations, and the latter result in a modification of the physical parameters of the system. The important exception occurs, of course, when these characteristic length scales are either zero or infinite. Let us illustrate this with some examples.

CRITICAL PHENOMENA

Consider first an infinite lattice of atoms in interaction, such as in a solid. Among the various forces involving ions and electrons, which are the source of so many interesting collective phenomena, consider for definiteness the magnetic (exchange) interaction that couples the spins of adjacent atoms. A very simplified version of this interaction is embodied in the *Ising model*, in which the spins σ_i at site i take only two definite values ($+1$ and -1) and the magnetic energy of the system is a sum over pairs of adjacent atoms:

$$E = \sum_{\langle ij \rangle} \sigma_i \sigma_j$$

An obvious characteristic length scale of this system is the lattice spacing a between adjacent atoms. Another, more important length scale is the so-called *correlation length* ξ, defined as the typical distance over which the spins are statistically correlated. More precisely, we write

$$\langle \sigma_i \sigma_j \rangle - \langle \sigma_i \rangle \langle \sigma_j \rangle \sim \exp - \frac{|i-j|}{\xi}$$

where $\langle \cdots \rangle$ denotes a thermal average at a temperature T and where $|i-j| \gg 1$ is the distance between the positions i and j. Since observable magnetic properties are derived from such correlations, they are quite affected by the value of ξ, which is a function of temperature.

For a generic value of the temperature, there is no symmetry of the model under scale transformations, because of the two length scales a and ξ. However, there are special circumstances, dictated by external parameters such as temperature, under which ξ grows without bounds.[1] Such values of the parameters of the model are called *critical points*, and the behaviors of systems at or near these critical points constitute what is called *critical phenomena*. When studying correlations over distances large compared to the lattice spacing, yet small compared to the correlation length, these two length scales lose their relevance, and scale invariance emerges.

The physical picture of a critical system one must keep in mind is that of an assembly of regions of $(+)$ spins (called *droplets*), within which smaller droplets of $(-)$ spins are included, and yet smaller droplets of $(+)$ spins are included within those, and so on.[2] This droplet structure is self-similar—in the sense that it has the same general appearance after zooming in or out a few times—as long as the droplet size ℓ satisfies $a \ll \ell \ll \xi$.

The Ising model is just one among an infinite variety of models that can provide an approximate description of complex systems with local interactions. One of the key ideas in our understanding of critical phenomena is that of *universality*: despite this continuous variety of models that possess critical points, their behaviors at (or near) the critical point belong to a discrete set of *universality classes*, corresponding to different realizations of scale invariance. One of the goals of conformal field

---

[1]   In real physical systems, the correlation length typically never grows beyond $\sim 10^3$ lattice spacings, because of the presence of impurities, defects, and inhomogeneities. But 10^3 is sufficiently close to infinity for scale invariance to have striking experimental consequences.

[2]   This is wonderfully illustrated by a computer simulation of the two-dimensional Ising model in the introductory paper by Zuber [370].

theory—so far only partially achieved—is a classification of all universality classes of two-dimensional critical systems.

CRITICAL QUANTUM SYSTEMS

For a special class of critical phenomena, the critical temperature vanishes or is small compared to other relevant energy scales. A quantum description of the system is then indispensable. Essentially, the statistical fluctuations giving rise to correlations are not thermal, but mainly quantum-mechanical in origin. An example of such a system is the so-called Heisenberg spin-$\frac{1}{2}$ chain, which represents an infinite chain of magnetic atoms, each carrying a spin one-half operator S_i and interacting with its immediate neighbors via the Heisenberg Hamiltonian:

$$H = \sum_{\langle ij \rangle} S_i \cdot S_j$$

One of the main characteristics of this model (in one spatial dimension) is the infinite correlation length, which means that the quantum correlations $\langle S_i S_j \rangle$ decay with distance according to a power law, not exponentially. This property is intimately related to the existence of *gapless excitations* in the system, namely, a continuum of excited states arbitrarily close in energy to the ground state. In any field theory (or any model involving an infinite number of degrees of freedom) the presence of gapless excitations is a signal of scale invariance, since the energy gap Δ between the ground state and the first excited state—the rest mass of the excitation—constitutes a characteristic length scale via the associated Compton wavelength $\lambda = \hbar/(v\Delta)$ (v being the characteristic velocity of the system, equal to the speed of light in relativistic field theories).

The mathematical formalism used in the description of quantum systems, and field theories in particular, bears a striking resemblance to the formalism of statistical mechanics describing finite-temperature critical phenomena. This similitude between the statistical and field-theoretical formalisms allows for a common treatment of both classes of phenomena. However, the field theory describing a statistical system (like the Ising model) lives in one spatial dimension less than the statistical system itself, since time constitutes an extra dimension inherently incorporated in the quantum description of the field theory. Critical quantum phenomena on which the methods of two-dimensional conformal field theory can be applied are thus one-dimensional, like the *spin chain* described above. Another example of a one-dimensional quantum system with scale invariance is constituted by the electrons moving on the *edge* of a microscopic layer of a semiconductor submitted to a large magnetic field of the appropriate strength. This is an aspect of the so-called *fractional quantum Hall effect*. It may also happen that a quantum system be only *formally* one-dimensional, after some simplifying treatment of its mathematical description. This is the case of the magnetic impurity problem (or *Kondo problem*), which has been successfully studied with the methods of conformal field theory.

DEEP INELASTIC SCATTERING

Another, very different area in which scale invariance has emerged[3] is the scattering of high-energy electrons from protons. Put very simply, scattering experiments failed to detect a characteristic length scale when probing the proton deeply with inelastically scattered electrons. This supported the idea that the proton is a composite object made of point-like constituents, the quarks.[4] This is quite reminiscent of Rutherford's study of the scattering of alpha particles off gold atoms, which revealed the absence of a length scale in the atom over five orders of magnitude, between the Bohr radius and the size of the nucleus.

Let us be more precise. Consider an electron (or any other lepton) of energy E scattered inelastically from a proton at an angle θ, with an energy $E' < E$. The quantity of experimental interest is the inclusive, inelastic cross-section, which gives the ratio of scattered flux to incident flux per unit solid angle and unit energy of the scattered particles:

$$\frac{d\sigma}{d\Omega' dE'} = \frac{\alpha^2}{4E^2 \sin^4(\theta/2)} \left[2W_1 \sin^2(\theta/2) + W_2 \cos^2(\theta/2) \right]$$

where α is the fine structure constant and $W_{1,2}$ are structure functions encapsulating the dynamics of the proton's interior. These structure functions depend on the kinematical parameters of the collision: the four-momentum q transferred from the lepton to the proton and the energy loss $(E - E') \equiv v/m$ (m is the lepton's mass). However, it turns out that the dimensionless quantities $2mW_1$ and vW_2/m depend only on the dimensionless ratio $x = 2v/(-q^2)$, if q^2 is negative enough (corresponding to large transferred spatial momentum). In other words, in this *deep-inelastic* range, the internal dynamics of the proton does not provide its own length scale ℓ that could justify a separate dependence of the structure functions on the dimensionless variables $\ell^2 v$ and $\ell^2 q^2$. In the context of quantum chromodynamics (QCD, the modern theory of strong interactions), this reflects the asymptotic freedom of the theory, namely, the quasi-free character of the quarks when probed at very small length scales.

Of course, the quark-gluon system underlying the scaling phenomena of deep inelastic scattering is thoroughly quantum-mechanical, just like systems undergoing quantum-critical phenomena. However, scale invariance manifests itself at *short* distances in QCD, whereas it emerges at *long* distances in quantum systems like the Heisenberg spin chain.

STRING THEORY

Whether statistical or quantum-mechanical, the physical systems enjoying scale invariance mentioned above were all in the same class, in the sense that they are

---

[3] It is interesting to note that scaling emerged as an important concept in the theory of critical phenomena *and* in high-energy physics at about the same time (the late 1960s) and over such widely different length scales! It was also at this time that a very fruitful interplay between high-energy theory and statistical mechanics started to develop, resting on the renormalization group theory.

[4] Although quarks had been hypothesized earlier from flavor symmetry considerations, prudent physicists initially called these constituents *partons*.

8

made of an infinite number of degrees of freedom (atoms, spins, etc.) fluctuating in space or space-time and characterized by a divergent correlation length or, equivalently, by power-law correlations. However, conformal invariance has appeared in other areas of theoretical physics. H. Weyl proposed in 1918 a generalization of general coordinate invariance (general relativity) in which local scale transformations would also be possible, in the hope of unifying electromagnetism and gravitation within the same formalism.[5] Since then, the hope of formulating a generalization of general relativity that would include the other known fundamental interactions has motivated an immense theoretical effort. Notable attempts in this direction come under the name of Kaluza-Klein theories and supergravity. In particular, theories of *conformal supergravity* are constructed to be invariant under conformal transformations of space-time.

Efforts toward unifying all forces of Nature in a single, comprehensive theory have culminated in what is known as *string theory*, in which two-dimensional scale invariance appears naturally. String theory originates from the *malaise* afflicting relativistic field theories in the 1960s, at a time when no consistent field theory could describe strong and weak interactions. An alternative to field theory, consisting of a set of prescriptions for scattering amplitudes between hadrons, was developed under the name of *dual models*. Curiously, the construction of dual models could follow from the assumption that mesons were in fact microscopic *strings*, or extended one-dimensional objects. The discovery of deep inelastic scattering and the subsequent development of QCD caused the demise of dual models, but some of their interesting features, such as finiteness in perturbation theory, inspired their transposition to the realm of quantum gravity, albeit at length scales much smaller (the Planck scale, 10^{-35} m). The great wave of activity in string theory occurred in the 1980s, after it was realized that consistent, finite first-quantized theories unifying gravitation and other interactions could be formulated.

We do not provide, in this work, an introduction to string theory; this can be found elsewhere (see the notes at the end of this introduction). Let us simply mention here some basic concepts. The time evolution of a one-dimensional extended object (i.e., a string) sweeps a two-dimensional manifold within space-time, which is called the *world-sheet* of the string. In a given classical configuration of the string, each point on this world-sheet corresponds to a point in space-time. The first-quantized formulation of string theory involves fields (representing the physical shape of the string) that reside on the world-sheet. From the point of view of field theory, this constitutes a two-dimensional system, endowed with *reparametrization invariance* on the world-sheet, meaning that the precise coordinate system used on the world-sheet has no physical consequence. This is particularly clear in Polyakov's formulation of string theory, and revives Weyl's idea of invariance under general coordinate transformations (this time on the world-sheet), augmented by local dilations. This reparametrization invariance is tantamount to conformal invariance. Conformal invariance of the world-sheet theory is essential for prevent-

---

[5] In Weyl's theory [353], the local dilations were called *gauge transformations*, a terminology that was recycled later for describing local group transformations.

ing the appearance of *ghosts* (states leading to negative probabilities in quantum mechanics). The various string models that have been elaborated basically differ in the specific content of this conformally invariant two-dimensional field theory (including boundary conditions). A classification of conformally invariant theories in two dimensions gives a perspective on the variety of consistent first-quantized string theories that can be constructed.

MODERN BREAKTHROUGHS

The modern study of conformal invariance in two dimensions was initiated by Belavin, Polyakov, and Zamolodchikov, in their fundamental 1984 paper [36]. These authors combined the representation theory of the Virasoro algebra— developed shortly before by Kac and by Feigin and Fuchs—with the idea of an algebra of local operators and showed how to construct completely solvable conformal theories: the so-called minimal models. An intense activity at the border of mathematical physics and statistical mechanics followed this initial envoi and the minimal models were identified with various two-dimensional statistical systems at their critical point. More solvable models were found by including additional symmetries or extensions of conformal symmetry in the construction of conformal theories.

A striking feature of the work of Belavin, Polyakov, and Zamolodchikov—and of previous work of Polyakov and other members of the Russian school—regarding conformal theories is the minor role played (if at all) by the Lagrangian or Hamiltonian formalism. Rather, the dynamical principle invoked in these studies is the associativity of the operator algebra, also known as the *bootstrap hypothesis*. This approach originates from the difficulty of describing strong interactions with quantum field theory. Instead of trying to solve the problem piecemeal with perturbative (or even nonperturbative) methods based on a local action, some physicists proposed a program designed to solve the whole problem at once—that is, to calculate all the correlations between all the fields—based only on criteria of self-consistency and symmetry.[6] The key ingredient of this approach is the assumption that the product of local quantum operators can always be expressed as a linear combination of well-defined local operators. Schematically,

$$\phi_i(x)\phi_j(y) = \sum_k C_{ij}^k(x-y)\phi_k(y) \tag{1.1}$$

where $C_{ij}^k(x-y)$ is a c-number function, not an operator. This is the *operator product expansion*, initially put forward by Wilson. This expansion constitutes an *algebra*—that is, a set of multiplication rules—for local fields. The dynamical principle of the bootstrap approach is the associativity of this algebra. In practice, a

---

[6] Put in an intuitive way, the strong interactions were thought to be mediated by a series of particles (the mesons), whose existence could in turn be inferred from a knowledge of the strong interaction. The term *bootstrap* is borrowed from the baron of Münchhausen, who made a similar-minded attempt at flying by pulling on his boot laces. A better analogy is found in the theory of communications, with Marshall McLuhan's famous phrase: "the medium is the message."

successful application of the bootstrap approach is hopeless, unless the number of local fields is finite. This is precisely the case in minimal conformal field theories.

By a fortunate coincidence, important progress in string theory was realized in the same year (1984) by Green and Schwarz [186] (see also [187]). In the years that followed, the development of conformal field theory and of string theory often went hand-in-hand. In particular, string scattering amplitudes were expressed in terms of correlation functions of a conformal field theory defined on the plane (tree amplitudes), on the torus (one-loop amplitudes), or on some higher-genus Riemann surface. Consistency requirements on the torus (modular invariance) turned out to be as fruitful in analyzing critical statistical models (e.g., the Potts model) as in constructing consistent string models in four space-time dimensions. The name of Cardy is associated with the early discovery of the importance of modular invariance in the context of critical statistical models.

Following the pioneering work of Belavin, Polyakov, and Zamolodchikov, conformal field theory has rapidly developed along many directions. The work of Zamolodchikov has strongly influenced many of these developments: conformal field theories with Lie algebra symmetry (with Knizhnik), theories with higher-spin fields—the W-algebras—or with fractional statistics—parafermions (with Fateev), vicinity of the critical point, etc. These developments, and their offspring, still constitute active fields of research today and make conformal field theory one of the most active areas of research in mathematical physics.

Contents of this Volume

This volume is divided into three parts of unequal lengths. Part A (Chapters 1 to 3) plays an introductory or preliminary role. Part B (Chapters 4 to 12) describes the core of conformal field theory and some of its immediate applications to classical statistical systems. Part C (Chapters 13 to 18) deals with conformal field theories with current algebras, essentially Wess-Zumino-Witten models.

Chapters 2 and 3 are preliminary chapters that do not deal with conformal symmetry, but provide a background essential to the comprehension of the remainder of the book. Readers with experience with quantum field theory and statistical mechanics will be able to start reading at Chapter 4. However, those readers might want to take a close look at Sections 2.4 and 2.5, dealing with continuous symmetries and the energy-momentum tensor, in which some conventions are set on the definition of symmetry operations. Chapter 3 provides a general background on critical phenomena as a theater of application of conformal invariance. An introduction to the renormalization group is provided, which helps in understanding the context in which conformal field theory is useful. We hope that mathematicians and entry-level physicists will find these two chapters instructive.

Part B starts with Chapter 4, which defines conformal transformations in arbitrary dimension and derives the basic consequences of conformal invariance on classical and quantum field theories, including the form of correlation functions and the Ward identities. Chapter 5 adapts these results to two dimensions and introduces the technique of complex (holomorphic and antiholomorphic) variables and components. The notion of operator product expansion is introduced and some free-field examples are worked out. Chapter 6 describes the "canonical" quantization of two-dimensional conformal field theories, including radial quantization, the Virasoro algebra, mode expansions, and their application to free bosons and fermions. The important notions of operator algebra and conformal bootstrap are introduced at the end of this chapter. Chapters 5 and 6 thus initiate the core of the subject.

Chapters 7 and 8 are devoted to minimal models, describing critical points of discrete two-dimensional statistical systems. Chapter 7 presents an overview of the subject and some examples, and Chapter 8, which is more technical, provides constructive proofs of many of the results presented in the previous chapter. Chapter 9 explains an alternate construction of minimal models, within the so-called Coulomb gas approach. This approach offers the simplest route to the calculation of four-point correlations.

Chapter 10 is devoted to conformal field theories defined on a torus and issues of modular invariance. The torus geometry brings an additional input in the construction of conformal field theories because it forces a consistent fusion of their holomorphic and antiholomorphic components.

Chapter 11 is a basic introduction to conformal field theories defined on finite geometries, in particular with boundaries. The two main issues are the influence of the size of the system on correlation functions and the interaction of the holomorphic and antiholomorphic components of the theory through the boundary. An application of these concepts to critical percolation is presented at the end of this chapter.

Chapter 12 is devoted entirely to the two-dimensional Ising model at its critical point. The goal is to calculate multipoint correlation functions of the various operators (energy and spin) in different schemes (bosonization and fermionization). Ample space is given to an extension of the techniques of previous chapters to the torus geometry in the particular case of the Ising model.

Part C of the book launches the analysis of conformal field theories with additional symmetries. New symmetries imply the existence of new conserved currents, apart from the energy-momentum tensor, the generator of the conformal algebra. The complete set of conserved currents span an *extended* conformal algebra. Part C is concerned with the most important class of extended conformal theories, those for which the additional currents generate an affine Lie algebra, the physicist's "current algebras."

Affine Lie algebras are introduced in Chapter 14. This is preceded by a detailed introduction to the theory of simple Lie algebras in Chapter 13. These two chapters are conceptually self-contained, and no background on the theory of Lie algebras is required. Chapters 13 and 14 may be safely skipped by readers familiar with these subjects. In order to facilitate this omission, we have presented our notation in an appendix at the end of each of these chapters. The few sections that are less standard are clearly identified in the introduction of each chapter.

The conformal-field theoretical study of models with Lie algebra invariance, called Wess-Zumino-Witten (WZW) models, starts with Chapter 15. Unlike many conformal field theories, these models may be defined in terms of an action functional, in addition to their algebraic formulation—heavily based on the theory of integrable representations of affine Lie algebra. A central concept is the Sugawara construction, which expresses the energy-momentum tensor in terms of the current algebra generators. An important part of our analysis of WZW models is devoted to their free-field representations.

The following two chapters are somewhat more technical. Chapter 16 is almost completely devoted to the analysis of fusion rules, which, roughly speaking, specify which three-point functions are nonzero. Chapter 17 explores techniques ensuring the compatibility between the field content of a theory with Lie algebra symmetry and modular invariance. The full classification of such Lie-symmetric modular invariant partition functions is a key step in the classification of all conformal field theories and, accordingly, of all string vacua. We stress that these two chapters are not essential in understanding most of Chapter 18 which, in contradistinction, is more fundamental.

Quotienting a WZW model, invariant under a Lie group G, by another WZW model, invariant with respect to a subgroup of G, produces what is called a *coset*. It is expected that any solvable conformal field theory can be described by some coset model. This makes the coset construction one of the very fundamental tools in conformal field theory. This is the subject of Chapter 18.

READING GUIDE

The size of this book might scare the reader willing to learn some aspects of conformal field theory without working through the 850 or so pages that follow. The figure on the next page illustrates (imperfectly) the logical flow of the book. We hope this short reading guide will propose useful paths through the book. A solid-line arrow indicates an essential logical dependence, meaning that the target chapter could not be well understood without the "mother" chapter. A dashed-line arrow indicates a weaker dependence, by which only parts of the target chapter necessitate previous reading. Of course, this diagram does no justice to the structure of each chapter. At the beginning of each chapter, a short introduction explains the purpose of the chapter and describes briefly its content. The chapters belonging to the central trunk of this diagram form the core of conformal field theory. Chapters located at the left of the diagram play an introductory role, physical or mathematical. Chapters located at the right of the diagram contain mostly applications of the formalism described in the core chapters, or provide additional information that is not essential for an understanding of the formalism of conformal field theory.

Notes

Introductory papers on conformal invariance for nonspecialists include that of Zuber [370] and Cardy [72]. Some texts already published in totality or partly to conformal field theory include those of Kaku [227], Christe and Henkel [76], and Ketov [235].

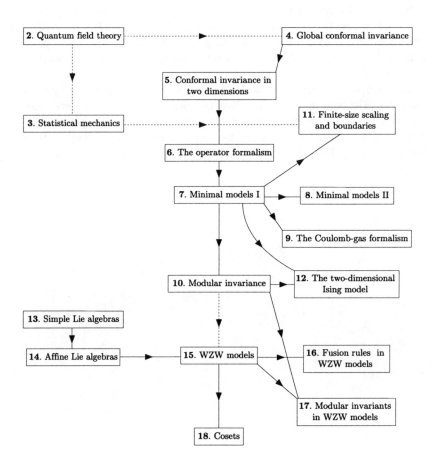

Figure 1.1. Logical flow of the book.

References on critical phenomena appear at the end of Chapter 3. A pedagogical review of some applications of conformal invariance to quantum critical phenomena can be found in Ref. [2]. Deep inelastic scattering is discussed in most texts on particle physics and in many texts on quantum field theory, including Ref. [205], in which further references can be found.

H. Weyl's extension of general relativity to include local scale invariance appeared in [353]. Conformal supergravity is reviewed in Ref. [134]. String theory is a vast subject, but the monograph of M. Green, J. Schwarz and E. Witten [187] is fairly comprehensive. Kaku's text on string theory [226] provides a more concise introduction to the subject. Polyakov's formulation of string theory appeared in Refs.[297, 298].

The operator product expansion (or operator algebra) was put forth by K. Wilson [356]. The bootstrap approach, based on operator algebra, was proposed by Polyakov [296]. The mathematical foundations of the algebraic representation of conformal invariance in two dimensions were found by Kac [213] and Feigin and Fuchs [127]. The work of Belavin, Polyakov, and Zamolodchikov appears in Ref. [36].

Quantum Field Theory

This chapter provides a quick—and therefore incomplete—introduction to quantum field theory. Those among our readers who know little about it will find here the basic material allowing them to appreciate and understand the remaining chapters of this book. Section 2.1 explains the canonical quantization of free fields, bosons and fermions, starting from a discrete formulation. It is appropriate for readers without any previous knowledge of quantum field theory; some experience with quantum mechanics remains an essential condition, however. Section 2.2 reviews the path-integral formalism of quantum mechanics for a single degree of freedom, and then for quantum fields, especially fermions. Section 2.3 introduces the central notion of a correlation function, both in the canonical and path-integral formalisms. The Wick rotation to imaginary time is performed, with the example of the free massive boson illustrating the exponential decay of correlations with distance. Section 2.4 explains the meaning of a symmetry transformation and the consequences of symmetries in classical and quantum field theories. This section deserves special attention—even from experienced readers—because the notion of a symmetry transformation and how it is implemented is fundamental to this work. Section 2.5 is devoted to the energy-momentum tensor, the conserved current associated with translation invariance, which plays a central role as the generator of conformal transformations when suitably modified.

§2.1. Quantum Fields

2.1.1. The Free Boson

The simplest system with an infinite number of degrees of freedom is a real scalar field $\varphi(x, t)$, a function of position and time. Its dynamics is specified by an action functional $S[\varphi]$, which explicitly depends on φ and its derivatives. For a generic action, the system is not soluble (by this we mean that the quantum stationary states cannot be written down). The simplest exception is the *free scalar field*, with the following action:

$$S[\varphi] = \int d\!x dt \; \mathcal{L}(\varphi, \dot{\varphi}, \nabla\varphi) \qquad\qquad \dot{\varphi} \equiv \frac{\partial\varphi}{\partial t}$$

$$\mathcal{L} = \frac{1}{2}\left\{ \frac{1}{c^2}\dot{\varphi}^2 - (\nabla\varphi)^2 - m^2\varphi^2 \right\} \tag{2.1}$$

$\mathcal{L}$ is the Lagrangian density (usually called *Lagrangian* by abuse of language) and m is the mass of the field (this terminology will be justified below). In a relativistic theory, the constant c stands for the speed of light, but in a different context (e.g., condensed matter physics) it stands for some characteristic velocity of the theory. We shall set c equal to 1, thus using the same units of measure for space and time. Our goal here is to solve this system within quantum mechanics, that is, to find the eigenstates of the associated Hamiltonian and provide some physical interpretation.

In order to simplify the notation we shall restrict ourselves to one spatial dimension. The conceptual difficulties associated with the continuum of degrees of freedom may be lifted by replacing space with a discrete lattice of points at positions $x_n = an$, where a is the lattice spacing and n is an integer. We shall assume that this one-dimensional lattice is finite in extent (with N sites) and that the variables defined on it obey periodic boundary conditions ($\varphi_N = \varphi_0$). The above Lagrangian $L = \int dx \; \mathcal{L}$ is then replaced by the following expression:

$$L = \sum_{n=0}^{N-1} \frac{1}{2} a \left\{ \dot{\varphi}_n^2 - \frac{1}{a^2}(\varphi_{n+1} - \varphi_n)^2 - m^2\varphi_n^2 \right\} \tag{2.2}$$

In the limit $a \to 0$ the action derived from (2.2) tends toward the continuum action (2.1).

The classical dynamics of such a system may be described in the canonical formalism, which first requires the introduction of the canonical momentum conjugate to the variable φ_n:

$$\pi_n = \frac{\partial L}{\partial \dot{\varphi}_n} = a\dot{\varphi}_n \tag{2.3}$$

The Hamiltonian function, or total energy, is then

$$H = \frac{1}{2}\sum_{n=0}^{N-1}\left\{ \frac{1}{a}\pi_n^2 + \frac{1}{a}(\varphi_{n+1} - \varphi_n)^2 - am^2\varphi_n^2 \right\} \tag{2.4}$$

If the mass m is set to zero, the above Hamiltonian describes the collective oscillations of atoms having their equilibrium positions on a regular lattice, with a potential energy varying as the square of the interatomic distance $|\varphi_{n+1} - \varphi_n|$.

The canonical quantization of such a system is done by replacing the classical variables φ_n and their conjugate momenta π_n by operators, and by imposing the following commutation relations at equal times:

$$[\varphi_n, \pi_m] = i\delta_{nm}$$
$$[\pi_n, \pi_m] = [\varphi_n, \varphi_m] = 0 \qquad (t_n = t_m) \qquad (2.5)$$

It is customary in quantum field theory to work in the Heisenberg picture, that is, to give operators a dependence upon time, while keeping the quantum states time-independent. Notice that we have set Planck's constant equal to 1, which amounts to using the same units for momentum and inverse distance, and similarly for energy and frequency.

The Hamiltonian (2.4) does not explicitly depend upon position: it is invariant under translations. This motivates the use of discrete Fourier transforms:

$$\tilde{\varphi}_k = \frac{1}{\sqrt{N}} \sum_{n=0}^{N-1} e^{-2\pi i k n/N} \varphi_n$$

$$\tilde{\pi}_k = \frac{1}{\sqrt{N}} \sum_{n=0}^{N-1} e^{-2\pi i k n/N} \pi_n \qquad (2.6)$$

where the index k takes integer values from 0 to $N-1$, since $\tilde{\varphi}_{k+N} = \tilde{\varphi}_k$. However, this range is arbitrary, the important point being to restrict summations over k to any range of N consecutive integers. Since φ_n and π_n are real, the Hermitian conjugates are

$$\tilde{\varphi}_k^\dagger = \tilde{\varphi}_{-k} \qquad\qquad \tilde{\pi}_k^\dagger = \tilde{\pi}_{-k} \qquad (2.7)$$

The Fourier modes $\tilde{\varphi}_k$ and $\tilde{\pi}_k$ obey the following commutation rules:

$$[\tilde{\varphi}_k, \tilde{\pi}_q^\dagger] = \frac{1}{N} \sum_{m,n=0}^{N-1} e^{-2\pi i(km-qn)/N} [\varphi_m, \pi_n]$$

$$= \frac{i}{N} \sum_{n=0}^{N-1} e^{-2\pi i n(k-q)/N} \qquad (2.8)$$

$$= i\delta_{kq}$$

In terms of these modes, the Hamiltonian (2.4) becomes

$$H = \frac{1}{2} \sum_{k=0}^{N-1} \left\{ \frac{1}{a} \tilde{\pi}_k \tilde{\pi}_k^\dagger + a\tilde{\varphi}_k \tilde{\varphi}_k^\dagger \left[m^2 + (2/a^2) \left(1 - \cos \frac{2\pi k}{N} \right) \right] \right\} \qquad (2.9)$$

Since $\tilde{\varphi}_k$ and $\tilde{\pi}_k$ obey canonical commutation relations, this is exactly the Hamiltonian for a system of uncoupled harmonic oscillators, with frequencies ω_k defined by

$$\omega_k^2 = m^2 + \frac{2}{a^2} \left(1 - \cos \frac{2\pi k}{N} \right) \qquad (2.10)$$

The inverse lattice spacing here plays the role of the harmonic oscillator's mass. Following the usual methods, we define raising and lowering operators

$$a_k = \frac{1}{\sqrt{2a\omega_k}}\left(a\omega_k\tilde{\varphi}_k + i\tilde{\pi}_k\right)$$

$$a_k^\dagger = \frac{1}{\sqrt{2a\omega_k}}\left(a\omega_k\tilde{\varphi}_k^\dagger - i\tilde{\pi}_k^\dagger\right)$$

(2.11)

obeying the commutation rules

$$[a_k, a_q^\dagger] = \delta_{kq} \tag{2.12}$$

When expressed in terms of these operators, the Hamiltonian takes the form

$$H = \frac{1}{2}\sum_{k=0}^{N-1}(a_k^\dagger a_k + a_k a_k^\dagger)\omega_k$$

$$= \sum_{k=0}^{N-1}(a_k^\dagger a_k + \frac{1}{2})\omega_k$$

(2.13)

The ground state $|0\rangle$ of the system is defined by the condition

$$a_k|0\rangle = 0 \qquad \forall\, k \tag{2.14}$$

and the complete set of energy eigenstates is obtained by applying on $|0\rangle$ all possible combinations of raising operators:

$$|k_1, k_2, \cdots, k_n\rangle = a_{k_1}^\dagger a_{k_2}^\dagger \dots a_{k_n}^\dagger |0\rangle \tag{2.15}$$

where the k_i are not necessarily different (as written, these states are not necessarily normalized). The energy of such a state is

$$E[k] = E_0 + \sum_i \omega_{k_i} \tag{2.16}$$

where E_0 is the ground state energy:

$$E_0 = \frac{1}{2}\sum_{k=0}^{N-1}\omega_k \tag{2.17}$$

When N is large and $ma \ll 1$, E_0 behaves like N/a.

The time evolution of the operators a_k is determined by the Heisenberg relation:

$$\dot{a}_k = i[H, a_k] = -i\omega_k a_k \tag{2.18}$$

whose solution is

$$a_k(t) = a_k(0)e^{-i\omega_k t} \tag{2.19}$$

From this, (2.6) and (2.11) follows the time dependence of the field itself:

$$\varphi_n(t) = \sum_{k=0}^{N-1}\sqrt{\frac{2}{Na\omega_k}}\left[e^{i(2\pi kn/N - \omega_k t)}a_k(0) + e^{-i(2\pi kn/N - \omega_k t)}a_k^\dagger(0)\right] \tag{2.20}$$

The continuum limit is obtained by sending the lattice spacing a to zero, and the number N of sites to ∞, while keeping the volume $V = Na$ constant. The infrared limit is taken in sending V to ∞, while keeping a constant. We now translate the relations found above in terms of continuous field operators. The continuum limits of the field and conjugate momentum are

$$\varphi_n \to \varphi(x) \qquad \frac{1}{a}\pi_n \to \pi(x) = \dot{\varphi}(x) \qquad (x = na) \qquad (2.21)$$

Sums over sites and Kronecker deltas become

$$a\sum_{n=0}^{N-1} \to \int dx \qquad \delta_{nn'} \to a\delta(x - x') \qquad (2.22)$$

Therefore, the canonical commutation relations of the field with its conjugate momentum become

$$[\varphi(x), \pi(x')] = i\delta(x - x') \qquad (2.23)$$

The discrete Fourier index k is replaced by the physical momentum $p = 2\pi k/V$. Sums over Fourier modes and Kronecker deltas in mode indices become

$$\frac{1}{V}\sum_{k=0}^{N-1} \to \int \frac{dp}{2\pi} \qquad \delta_{kk'} \to \frac{2\pi}{V}\delta(p - p') \qquad (2.24)$$

We define the continuum annihilation operator and the associated frequency as

$$a(p) = a_k\sqrt{V} \qquad \omega(p) = \sqrt{m^2 + p^2} \qquad (2.25)$$

whose commutation relations are therefore

$$[a(p), a^\dagger(p')] = (2\pi)\delta(p - p') \qquad (2.26)$$

The field $\varphi(x)$ admits the following expansion in terms of the continuum creation and annihilation operators:

$$\varphi(x) = \int \frac{dp}{2\pi}\left\{a(p)e^{i(px-\omega(p)t)} + a^\dagger(p)e^{-i(px-\omega(p)t)}\right\} \qquad (2.27)$$

The simplest excited states, the so-called *elementary excitations*, are of the form $a^\dagger(p)|0\rangle$ with energy

$$\omega(p) = \sqrt{m^2 + p^2} \qquad (2.28)$$

This dispersion relation (i.e., the functional relation between energy and momentum) is characteristic of relativistic particles. We thus interpret these elementary excitations as particles of mass m and momentum p. The states (2.15) physically represent a collection of independent particles. The momenta of these particles are conserved separately (they are "good quantum numbers"). Since the energy of an assembly of particles is simply the sum of the energies of the individual particles, we say that these particles do not interact: they are free. Furthermore, the states (2.15) are symmetric under the interchange of momenta; this follows from the commutation rules (2.12). Therefore these particles are bosons, hence the name *free boson* given to the field φ with action (2.1). We say that these particles are the "quanta" of the field φ. The ground state is also called the *vacuum*, since it contains no particles. The Hilbert space constructed from the action of all creation operators receives the special name of *Fock space*.

The vacuum energy E_0 poses a slight conceptual problem. We have seen that $E_0 \sim N/a = V/a^2$. This corresponds to a vacuum energy *density* of order $1/a^2$, which diverges in the continuum limit. This is the first instance of a "divergence" encountered in quantum field theory (it is, of course, due to the infinite number of degrees of freedom present in the system). This vacuum energy problem is circumvented by defining the energy of a state with respect to the vacuum, which is most easily implemented by introducing a "normal ordering" of operators (denoted by surrounding colons) which, in a given monomial, puts the operators annihilating the vacuum to the right. For instance,

$$:a(p)a^\dagger(p): \; = \; a^\dagger(p)a(p) \tag{2.29}$$

By definition the vacuum expectation value $\langle 0| : \mathcal{O} : |0\rangle$ of a normal-ordered operator vanishes. Since the ordering of classical quantities is immaterial, the canonical quantization procedure necessarily introduces ordering ambiguities in the definition of operators like the Hamiltonian. Some of these ambiguities may be lifted by requiring the vanishing of vacuum expectation values.

The expansion (2.27) splits the free Bose field φ into two parts: φ^+ and φ^-. The first one (the *positive frequency* part) contains only annihilation operators, whereas the second one (the *negative frequency* part) contains only creation operators. The positive frequency parts at different points commute, and likewise for the negative frequency parts, since the lack of commutativity comes solely from the relation (2.12). For instance, the normal-ordered product of $\varphi_1 = \varphi(x_1)$ with $\varphi_2 = \varphi(x_2)$ is

$$:\varphi_1\varphi_2: \; = \; \varphi_1^+\varphi_2^+ + \varphi_1^-\varphi_2^- + \varphi_1^-\varphi_2^+ + \varphi_2^-\varphi_1^+ \tag{2.30}$$

Finally, we briefly comment on interacting fields. As soon as we depart from the simple form (2.1), for instance by adding a term such as $g\varphi^4$, the system is no longer exactly soluble. If the *coupling constant* g is small, one may find approximate solutions using perturbation theory. By this we mean a calculation of the transition probability amplitude (S matrix) from a given initial state of free particles (with definite momenta) to another, final state of particles. The technique of Feynman diagrams is especially suited to this task. However, it is not the purpose of this introduction to explain standard perturbation theory, since it will not be used in the remainder of this book. The interested reader may consult one of the many texts on quantum field theory, which devote ample space to diagrammatic techniques.

Divergences encountered when calculating the vacuum energy density of the free field, and attributed to the continuum of degrees of freedom, are still present for interacting fields, and are the cause of more severe difficulties. These problems have stopped the development of quantum field theory for almost twenty years, and were formally resolved with the introduction of *renormalization*. The interpretation given to this procedure has evolved over the decades. In recent years, it has become customary to regard continuum field theories as approximations to more fundamental theories (a natural standpoint in condensed-matter applications of quantum field theory). This justifies the use of a cutoff: a lattice spacing, or some other kind of *regularization* that effectively suppresses the degrees of freedom associated with very small distances. It is thus necessary, in order to make sense of a field theory, to know not only its action functional, but also some regularization procedure, and an approximate estimate of the cut-off.

2.1.2. The Free Fermion

The defining property of fermions is the antisymmetry of many-particle states under the exchange of any two particles. In the context of a free-field theory, and in terms of mode operators $a(p)$ and $a^\dagger(p)$, this property follows from *anticommutation* relations:

$$\{a(p), a^\dagger(q)\} = (2\pi)2\omega_p\delta(p - q)$$
$$\{a(p), a(q)\} = \{a^\dagger(p), a^\dagger(q)\} = 0 \tag{2.31}$$

where $\{a, b\} = ab + ba$ is the anticommutator. However, the canonical quantization of a field taking its values in the set of real or complex numbers can lead only to *commutation* relations, as opposed to anticommutation relations.[1]

However, a classical description of Fermi fields can be given in terms of anti-commuting (or *Grassmann*) numbers. Appendix 2.B defines these entities, and the newcomer should read it through before proceeding. This description is especially suited for the extension to fermions of functional integrals (introduced in the next section), but it may also be used in the context of canonical quantization.

We apply to Grassmann variables the same canonical formalism as for real or complex variables, except that their anticommuting properties forbid the existence in the Lagrangian of terms quadratic in derivatives. Specifically, let us consider a discrete set $\{\psi_i\}$ of real Grassmann variables with the Lagrangian

$$L = \frac{i}{2}\psi_i T_{ij}\dot{\psi}_j - V(\psi) \tag{2.32}$$

(repeated indices are summed over). The time derivative $\dot{\psi}_j$ is still a Grassmann number:

$$\psi_i\dot{\psi}_j + \dot{\psi}_j\psi_i = 0 \tag{2.33}$$

It follows that only the symmetric part of the matrix T_{ij} is relevant. Indeed, its antisymmetric part couples to

$$\psi_i\dot{\psi}_j - \psi_j\dot{\psi}_i = \psi_i\dot{\psi}_j + \dot{\psi}_i\psi_j \tag{2.34}$$

---

[1] Viewed differently, a given fermionic mode cannot hold more than one particle and consequently a Fermi field cannot have a macroscopic value: its classical limit does not exist in terms of real or complex numbers.

which is a total derivative. The kinetic term of the Lagrangian (2.32) is real, as is easily seen by taking the complex conjugate. The Euler–Lagrange equations of motion are

$$\frac{d}{dt}\left\{-\frac{i}{2}\psi_i T_{ij}\right\} - \frac{i}{2}T_{ji}\dot{\psi}_i + \frac{\partial V}{\partial \psi_j} = 0 \tag{2.35}$$

or, in matrix notation,

$$\dot{\psi} = -i\, T^{-1}\frac{\partial V}{\partial \psi} \tag{2.36}$$

These equations are recovered in the quantum case from the Heisenberg time evolution equation $\dot{\psi} = i[H, \psi]$ provided we use the following Hamiltonian and anticommutation rules:

$$H = V(\psi) \qquad \{\psi_i, \psi_j\} = (T^{-1})_{ij} \tag{2.37}$$

wherein ψ_i is now an operator. The proof of this statement is straightforward, and is left as an exercise.

The closest analogue of (2.2) for a system of real Grassmann variables is

$$L = \frac{1}{2}i\sum_{n=0}^{N-1}\{a\psi_n\dot{\psi}_n + \psi_n\psi_{n+1}\} \tag{2.38}$$

Here a is the lattice spacing and we still assume periodic boundary conditions $(\psi_{n+N} = \psi_n)$. Notice that a term such as $(\psi_{n+1} - \psi_n)^2$ would automatically vanish, being the square of a Grassmann number. The above Lagrangian is real, but is not invariant under the parity transformation $\psi_n \to \psi_{-n}$ (the potential V changes sign). The Hamiltonian and anticommutation rules are

$$H = -\frac{i}{2}\sum_{n=0}^{N-1}\psi_n\psi_{n+1} \qquad \{\psi_n, \psi_m\} = \frac{1}{a}\delta_{mn} \tag{2.39}$$

Again, translation invariance motivates the use of Fourier transformed operators:

$$b_k = \sqrt{\frac{a}{N}}\sum_{n=0}^{N-1}\psi_n\, e^{-2\pi i k n/N} \qquad (k \in \mathbb{Z})$$

$$\psi_n = \frac{1}{\sqrt{aN}}\sum_{k=0}^{N-1}b_k\, e^{2\pi i k n/N} \tag{2.40}$$

where $b_{-k} = b_k^\dagger$. The mode operators b_k obey the anticommutation relation

$$\{b_k, b_q^\dagger\} = \delta_{kq} \tag{2.41}$$

The Hamiltonian $H = V$ is then a sum over modes:

$$H = \frac{1}{2} \sum_{k=0}^{N-1} \omega_k b_k^\dagger b_k \qquad \omega_k = \frac{1}{a} \sin \frac{2\pi k}{N}$$

$$= E_0 + \sum_{k>0}^{(N-1)/2} \omega_k b_k^\dagger b_k \qquad (2.42)$$

where for simplicity we have assumed N to be odd, and

$$E_0 = -\frac{1}{2} \sum_{k>0}^{(N-1)/2} \omega_k \qquad (2.43)$$

The time evolution of b_k follows from the Heisenberg equation:

$$\dot{b}_k = i[H, b_k] = -i\omega_k b_k \quad \Rightarrow \quad b_k(t) = e^{-i\omega_k t} b_k(0) \qquad (2.44)$$

The definition of the vacuum state for fermions is not exactly the same as for bosons. Since $b_k^\dagger = b_{-k}$, the condition $b_k|0\rangle = 0$ for all k leads to a Fock space made of only one state. This problem did not arise for bosons since $a_k^\dagger \neq a_{-k}$ or, more simply, because the classical Hamiltonian was a real number. This is no longer true for fermions: H takes classically its values in a Grassmann algebra (see App. 2.B) in which no ordering is defined a priori. The question of which classical configuration has the lowest energy is not well defined. The definition of the theory must be supplemented with a consistent definition of the vacuum, which we choose to be

$$b_k|0\rangle = 0 \qquad 0 < k \le N/2 \qquad (2.45)$$

(we shall treat later the zero mode b_0, which does not enter the Hamiltonian). The energy eigenstates are then

$$b_{k_1}^\dagger b_{k_2}^\dagger \ldots b_{k_n}^\dagger |0\rangle \qquad (0 < k_i \le N/2) \qquad (2.46)$$

with energy $E = E_0 + \sum_i \omega_{k_i}$. These states are, of course, antisymmetric under interchange of particles and are interpreted as free fermions, each with energy $\omega_k = \sin(2\pi k/N)/a$. In the continuum limit, this dispersion relation becomes

$$E(p) = p \qquad p = 2\pi k/(Na) \qquad (2.47)$$

These fermions are therefore massless.

The continuum limit is taken by introducing the continuum field $\psi(x) = \psi_n$ ($x = na$). The term $\psi_n \psi_{n+1}$ becomes $a\psi(x)\partial_x \psi(x)$ and

$$L = \frac{i}{2} \int dx\, \psi(\partial_t + \partial_x)\psi \qquad (2.48)$$

Had we used instead the potential $\sum_n \psi_{n+1}\psi_n$, the sign in front of ∂_x would have been the opposite. As noticed above, both choices lead to a violation of parity.

That symmetry can be restored by considering two Fermi fields ψ_1 and ψ_2 with opposite signs of the potential:

$$L = \frac{i}{2} \int dx \, \{\psi_1(\partial_t + \partial_x)\psi_1 + \psi_2(\partial_t - \partial_x)\psi_2\} \qquad (2.49)$$

Under a parity transformation the two fields are interchanged:

$$\psi_1(x) \to \psi_2(-x) \qquad\qquad \psi_2(x) \to \psi_1(-x) \qquad (2.50)$$

It is customary to write the above Lagrangian in terms of a two-component field $\Psi = (\psi_1, \psi_2)$:

$$\mathcal{L} = \frac{i}{2} \Psi^t \gamma^0 \gamma^\mu \partial_\mu \Psi \qquad (2.51)$$

where Ψ^t is the transpose of Ψ and

$$\gamma^0 = \begin{pmatrix} 0 & 1 \\ 1 & 0 \end{pmatrix} \qquad\qquad \gamma^1 = \begin{pmatrix} 0 & -1 \\ 1 & 0 \end{pmatrix} \qquad (2.52)$$

Since the zero mode b_0 does not enter the Hamiltonian, it commutes with H and therefore any two states $|\chi\rangle$ and $b_0|\chi\rangle$ are degenerate, including the vacuum $|0\rangle$: The whole spectrum is two-fold degenerate. This is no longer true if we impose antiperiodic boundary conditions on the lattice fermions:

$$\psi_{n+N} = -\psi_n \qquad (2.53)$$

The mode expansion (2.40) still applies, provided the indices k take their values among the half-integers $\frac{1}{2}, \frac{3}{2}, \cdots$. The remaining part of the argument is identical, except that the zero mode b_0 and the corresponding degeneracy no longer exist. The antiperiodic boundary conditions are called Neveu-Schwarz (NS) boundary conditions, whereas the periodic ones are called Ramond (R) boundary conditions.

Another remark is in order, concerning the so-called *fermion doubling* problem. The energy ω_k of a single fermion is minimum when $k \sim 1$ or $k \sim N/2$. When taking the continuum limit, the second minimum of the dispersion relation disappears, and the corresponding excitations are no longer admitted in the spectrum. Viewed the opposite way, additional low-energy excitations appear at the upper limit of the momentum range when a continuous theory of fermions is put on a lattice. These new excitations have the appearance of a new species of fermions, hence the expression "fermion doubling." In fact, there is a doubling of fermions for each dimension of space being discretized.

We treat systems described by complex Grassmann variables ψ_i and $\bar\psi_i$ ($i = 1, \cdots, n$) in a similar way. A generic Lagrangian is then

$$L = i\bar\psi_i T_{ij} \dot\psi_j - V(\psi) \qquad (2.54)$$

where T is Hermitian: $T^\dagger = T$. The Hamiltonian is still $H = V$, and the relevant anticommutation relations are

$$\{\psi_i, \psi_j\} = \{\psi_i^\dagger, \psi_j^\dagger\} = 0 \qquad \{\psi_i, \psi_j^\dagger\} = (T^{-1})_{ij} \qquad (2.55)$$

ψ_i and $\psi_i^\dagger$ being the quantum operators corresponding respectively to the classical variables ψ_i and $\bar\psi_i$. The vacuum state can now be defined without problem by the condition

$$\psi_i|0\rangle = 0 \qquad \forall\, i \qquad (2.56)$$

and the Hilbert space V is spanned by the following states:

$$\psi_{i_1}^\dagger \psi_{i_2}^\dagger \cdots \psi_{i_k}^\dagger |0\rangle \qquad k \in \mathbb{N} \qquad (2.57)$$

(the above states are not energy eigenstates, however). The dimension of the Hilbert space is

$$\sum_{k=0}^{n} \binom{n}{k} = 2^n \qquad (2.58)$$

§2.2. Path Integrals

The quantum description of a physical system may be done according to two equivalent methods, often complementary. The first one, older and better known, should be familiar to all our readers: canonical quantization. Classical quantities are replaced by operators acting on a vector space in which the states of the system reside. The second method, twenty years younger, is called *path integration* or *functional integration*. It has the advantage of being more intuitive, and of allowing formal manipulations, which, despite their lack of rigor, provide important results with the minimum of fuss. In practice, however, these advantages become apparent only for systems with an infinite number of degrees of freedom. In other cases, its interest is more or less academic and pedagogical. Another advantage of path integration resides in its formal analogy with statistical mechanics. This not only facilitates the formulation of quantum mechanics (or quantum field theory) at finite temperature, but also establishes a correspondence between many classical statistical systems and quantum field theories. This analogy will be exploited throughout many of the following chapters.

2.2.1. System with One Degree of Freedom

In this section we shall "derive" the path-integral method from the canonical quantization of a simple system: that of a point particle of mass m moving in an external potential $V(x)$. The Hamiltonian of this system is time-independent:

$$H = K + V(\hat{x}) \quad , \quad K = \frac{\hat{p}^2}{2m} \quad , \quad [\hat{x}, \hat{p}] = i \tag{2.59}$$

The hat (^) distinguishes the quantum operator from the corresponding classical quantity. To represent the dynamics we introduce an evolution operator $U(t)$, which brings a state $|\psi\rangle$ at time t_0 to the time $t_0 + t$:

$$U(t) = e^{-iHt} \tag{2.60}$$

First, we calculate the matrix elements of $U(\delta t)$ in the basis $\{|x\rangle\}$ of position eigenstates, where δt is an infinitesimal time interval. Calculations are done to first order in δt:

$$
\begin{aligned}
\langle x|e^{-i(K+V)\delta t}|x'\rangle &= \langle x|e^{-iK\delta t}e^{-iV\delta t}e^{O((\delta t)^2)}|x'\rangle \\
&\approx \int \frac{dp}{2\pi} \langle x|e^{-iK\delta t}|p\rangle \langle p|e^{-iV\delta t}|x'\rangle \\
&= \int \frac{dp}{2\pi} \exp\left\{-i\delta t\left[\frac{p^2}{2m} - p\frac{(x-x')}{\delta t} + V(x')\right]\right\} \\
&= \sqrt{\frac{m}{2\pi i \delta t}} \exp\left\{i\delta t\left[\frac{1}{2}m\frac{(x-x')^2}{\delta t^2} - V(x')\right]\right\}
\end{aligned} \tag{2.61}
$$

In the first step we have used the approximate relation

$$e^{\epsilon(A+B)} = e^{\epsilon A}e^{\epsilon B}e^{O(\epsilon^2)}$$

In the second step we have neglected the terms of order $(\delta t)^2$ and inserted a completeness relation

$$\int \frac{dp}{2\pi} |p\rangle\langle p| = 1$$

where $|p\rangle$ is an eigenstate of momentum, with $\langle x|p\rangle = e^{ipx}$. In the last step, we completed the square and performed a Gaussian integration, which is strictly valid only when the time interval δt has a small, negative imaginary part. This assumption will be implicit in what follows. The quantity in brackets on the last line of (2.61) is nothing but the infinitesimal action $S(x', x; \delta t)$ corresponding to the passage of the system from x' to x in a time δt. One may therefore write, to first order,

$$\langle x|U(\delta t)|x'\rangle = \sqrt{\frac{m}{2\pi i \delta t}} \exp iS(x', x; \delta t) \tag{2.62}$$

Second, we consider $\langle x_f|U(t)|x_i\rangle$, which is the probability amplitude for the system, initially at a well-defined position x_i, to evolve in a finite time t toward the position x_f. This amplitude is called *propagator* and may be obtained by dividing the interval of time t in N subintervals t/N and inserting completeness relations:

$$\langle x_f|U(t)|x_i\rangle = \left\{\frac{m}{2\pi i\delta t}\right\}^{N/2} \int \prod_{j=1}^{N-1} dx_j \, \langle x_f|U(t/N)|x_{N-1}\rangle$$

$$\times \langle x_{N-1}|U(t/N)|x_{N-2}\rangle \cdots \langle x_1|U(t/N)|x_i\rangle$$

The error made in using Eq. (2.62) for each factor is of order $1/N^2$, and the total error is of order $1/N$. Therefore, in the large N limit one may write

$$\langle x_f|U(t)|x_i\rangle = \lim_{N\to\infty} \left\{\frac{mN}{2\pi it}\right\}^{N/2} \int \prod_{j=1}^{N-1} dx_j \, \exp iS[x]$$

where $S[x]$ is the action associated with the discrete trajectory x_j, $j = 0, 1 \cdots N$ (we take $x_0 = x_i$ and $x_N = x_f$). If we define the following "functional integration measure":

$$[dx] = \lim_{N\to\infty} \prod_{j=1}^{N-1} \left\{\sqrt{\frac{mN}{2\pi it}} \, dx_j\right\} \tag{2.63}$$

we may then write our fundamental result as follows:

$$\boxed{\langle x_f|U(t)|x_i\rangle = \int_{(x_i,0)}^{(x_f,t)} [dx] \, \exp iS[x]} \tag{2.64}$$

where the action is, of course, given by

$$S[x] = \int dt \left(\frac{1}{2}m\dot{x}^2 - V(x)\right) \tag{2.65}$$

The interpretation of Eq. (2.64) is the following. Each possible trajectory going from x_i to x_f in a time t contributes to the amplitude $\langle x_f|U(t)|x_i\rangle$ with a weight equal to the exponential of i times its action. Within the set of possible trajectories, most are highly irregular, but they contribute little overall, since the kinetic term $\frac{1}{2}m\dot{x}^2$ drives up their action, and their contributions tend to cancel each other because of the oscillating exponential. The trajectories contributing most are those around which the phase of the exponential varies the least, that is, those with stationary action: the classical trajectories. In order to sharpen this remark, we restore the factors of $\hbar$, which have been suppressed so far. Planck's constant has the dimensions of action, and we simply have to replace every occurrence of the action S by $S/\hbar$. The classical limit is then valid when the action of the classical trajectory is much larger than $\hbar$: this is the correspondence principle. Otherwise, fluctuations about the classical trajectory are not sufficiently suppressed and a full quantum treatment is necessary (i.e., an exact use of Eq. (2.64)).

The propagator may also be used to express the probability amplitude for a state $|\psi_i\rangle$ to evolve, after a time t, toward another state $|\psi_f\rangle$. Indeed,

$$\langle \psi_f|U(t_f - t_i)|\psi_i\rangle = \int dx_i dx_f \, \psi_f^*(x_f)\psi_i(x_i)\langle x_f|U(t_f - t_i)|x_i\rangle \tag{2.66}$$

where $\psi_i(x) = \langle x|\psi_i\rangle$ is the wave function associated with $|\psi_i\rangle$, and similarly for $\psi_f(x) = \langle x|\psi_f\rangle$.

The amplitude (2.64) can be used as a starting point for all of quantum mechanics. It is fully equivalent to the Schrödinger equation (in the sense that it incorporates the dynamics of the system) and allows for the calculation of the same quantities, although in a different manner. We have derived it for a time-independent Hamiltonian, but only in order to keep the notation as simple as possible. The result is identical for a time-dependent Hamiltonian, and the derivation is almost identical, since it is the infinitesimal propagator (2.62) that matters.

2.2.2. Path Integration for Quantum Fields

The path-integral quantization of a bosonic field is not conceptually more difficult than that of a point particle. The integration measure may be defined by dividing time and space into infinitesimal intervals and integrating over each field variable $\varphi(x,t)$ at every point. Contrary to canonical quantization, path-integral quantization does not pick time as a special dimension at the outset. This contributes greatly to the apparent simplicity and beauty of the method. In particular, if a field theory is Lorentz invariant classically, this invariance is manifestly maintained by path-integral quantization. We may then write, without further ado, the probability amplitude for the transition between configurations $\varphi_i(x, t_i)$ and $\varphi_f(x, t_f)$ as

$$\boxed{\langle \varphi_f(x, t_f)|\varphi_i(x, t_i)\rangle = \int [d\varphi(x,t)]\, e^{iS[\varphi]}} \tag{2.67}$$

When dealing with fermions, we need to recast the demonstration of the preceding subsection into the language of Grassmann variables. For the sake of argument, let us consider the generic Lagrangian (2.54) involving complex Grassmann variables. The Hilbert space $\mathcal{V}$ is generated by the states (2.57) with complex coefficients. In order to formulate path integrals for fermions, we need eigenstates of the operators ψ_i, in analogy with the eigenstates $|x\rangle$ of position in ordinary quantum mechanics. This is impossible within $\mathcal{V}$ since it is a vector space over $\mathbb{C}$, whereas we need Grassmann eigenvalues. We must therefore work in an extended space $\mathcal{V} \otimes \Lambda$ (Λ is the Grassmann algebra) in which the coefficients can be Grassmann numbers.[2] In this extended space, we introduce an overcomplete basis of states $|\xi\rangle = |\xi_1, \cdots, \xi_n\rangle$ defined by

$$|\xi\rangle = e^{\psi^\dagger T\xi}|0\rangle \tag{2.68}$$

where ξ_i is a complex Grassmann number. These are called *coherent states* and satisfy the following three important properties:

$$\psi_i|\xi\rangle = \xi_i|\xi\rangle \tag{2.69a}$$

---

[2] This is not so different from ordinary quantum mechanics, since the states $|x\rangle$ are not bona fide members of the physical Hilbert space, not being properly normalizable.

$$1 = (\det T)^{-1} \int d\bar{\xi}d\xi \; |\xi\rangle \exp(-\xi^\dagger T\xi)\langle\xi| \qquad (2.69b)$$

$$\langle\xi|\xi'\rangle = \exp(-\xi^\dagger T\xi') \qquad (2.69c)$$

Given any state $|\Psi\rangle$, we define its wavefunction as $\Psi(\xi) = \langle\xi|\Psi\rangle$. The time evolution of the wavefunction is then given by

$$
\begin{aligned}
\Psi(\xi, t) &= \langle\xi|e^{-iHt}|\Psi\rangle \\
&= (\det T)^{-1} \int d\bar{\xi}'d\xi' \; \langle\xi|e^{-iHt}|\xi'\rangle \exp(-\xi'^\dagger T\xi')\langle\xi'|\Psi\rangle \qquad (2.70) \\
&= \int d\bar{\xi}'d\xi' \; K(t, \xi, \xi')\Psi(\xi', 0)
\end{aligned}
$$

where we have defined the propagator

$$K(t, \xi, \xi') = (\det T)^{-1}\langle\xi|e^{-iHt}|\xi'\rangle \; \exp(-\xi'^\dagger T\xi') \qquad (2.71)$$

which is the kernel of the evolution operator for wavefunctions.

In evaluating $\langle\xi|e^{-iHt}|\xi'\rangle$, we face the following difficulty: in the Hamiltonian $H = V(\psi^\dagger, \psi)$, the conjugate operators $\psi_i^\dagger$ sit at the left of the ψ_i. But this is not true of the exponential e^{-iHt}. Therefore we cannot use property (2.69a) to evaluate the propagator for arbitrary t. However, for t infinitesimal, we may expand the exponential to first order in δt and use (2.69a):

$$
\begin{aligned}
\langle\xi|e^{-iVt}|\xi'\rangle &\approx \langle\xi| \left(1 - i\delta t V(\psi^\dagger, \psi)\right) |\xi'\rangle \\
&= \left(1 - i\delta t V(\bar{\xi}, \xi')\right) \langle\xi|\xi'\rangle \qquad (2.72) \\
&\approx e^{-i\delta t V(\bar{\xi}, \xi')} e^{-\xi^\dagger T\xi'}
\end{aligned}
$$

In so doing we commit only an error of order $(\delta t)^2$. Therefore, to first order in δt, the propagator may be written as

$$
\begin{aligned}
K(\delta t, \xi, \xi') &= (\det T)^{-1} \exp\left[-\xi^\dagger T\xi' - \xi'^\dagger T\xi' - i\delta t V(\bar{\xi}, \xi)\right] \\
&= (\det T)^{-1} \exp\left\{(i\delta t)\left[-i\frac{(\xi - \xi')^\dagger}{\delta t}T\xi' - V(\bar{\xi}, \xi)\right]\right\} \qquad (2.73) \\
&= (\det T)^{-1} \exp iS(\bar{\xi}, \xi; \delta t)
\end{aligned}
$$

where, of course, $S(\bar{\xi}, \xi; \delta t)$ is the infinitesimal action for a trajectory in the classical (Grassmann) configuration space going from ξ to ξ' in a time δt. We used the property $i\bar{\psi}_i T_{ij}\dot{\psi}_j = -i\dot{\bar{\psi}}_i T_{ij}\psi_j$.

From this expression for the infinitesimal propagator, the finite time propagator follows exactly in the same way as for bosons. As the time slices δt and the lattice spacing go to zero, the path integration measure is written as

$$(\det T)^{-1} \prod_i d\bar{\xi}_i d\xi_i \;\; \rightarrow \;\; [d\bar{\xi}d\xi] \qquad (2.74)$$

wherein the index i distinguishes between not only the different fermionic degrees of freedom, but also the different time slices. From now on we will use the same symbol for the Grassmann variables appearing in the functional integral and the fermionic operators ($\xi \to \psi$). The transition amplitude between the classical field configurations $\psi_i(x, t_i)$ and $\psi_f(x, t_f)$ is then written as

$$\langle \psi_f(x, t_f) | \psi_i(x, t_i) \rangle = \int [d\bar{\psi} d\psi] \, e^{iS[\bar{\psi}, \psi]} \tag{2.75}$$

§2.3. Correlation Functions

Quantum field theory traditionally deals with scattering amplitudes between various asymptotic states (free particles). In practice these amplitudes are given by Green functions, or, by analogy with statistical mechanics, *correlation functions*.[3]

2.3.1. System with One Degree of Freedom

For a point particle, the n-point correlation function is defined as

$$\langle x(t_1) x(t_2) \cdots x(t_n) \rangle = \langle 0 | \mathcal{T} \left(\hat{x}(t_1) \cdots \hat{x}(t_n) \right) | 0 \rangle \tag{2.76}$$

where $|0\rangle$ is the ground state (or vacuum) and $\mathcal{T}$ is the time ordering operator, which sorts the factors that follow in chronological order from right to left:

$$\mathcal{T}(x(t_1) \cdots x(t_n)) = x(t_1) \cdots x(t_n) \quad \text{if} \quad t_1 > t_2 > \cdots > t_n \tag{2.77}$$

Correlation functions can be calculated by path integration as follows:

$$\langle x(t_1) x(t_2) \cdots x(t_n) \rangle = \lim_{\varepsilon \to 0} \frac{\int [dx] x(t_1) \cdots x(t_n) \exp iS_\varepsilon[x(t)]}{\int [dx] \exp iS_\varepsilon[x(t)]} \tag{2.78}$$

where S_ε is the action obtained by replacing t by $t(1 - i\varepsilon)$ (complex time) and where the functional integral is taken with bounds at $t \to \pm\infty$.

To prove this, we notice that[4]

$$\hat{x}(t) = e^{iHt} \hat{x} e^{-iHt} \tag{2.79}$$

($\hat{x}$ being taken at time $t = 0$). Therefore,

$$\langle x(t_1) x(t_2) \cdots x(t_n) \rangle = \frac{\langle 0 | \hat{x} e^{iH(t_2 - t_1)} \hat{x} e^{iH(t_3 - t_2)} \cdots \hat{x} | 0 \rangle}{\langle 0 | e^{iH(t_n - t_1)} | 0 \rangle} \tag{2.80}$$

---

[3] To be more precise, the relationship between scattering amplitudes and Green functions is given by the so-called *reduction formulas*.

[4] Again we consider a time-independent Hamiltonian for simplicity, although the result is quite general.

The outermost exponentials have been converted into a denominator, since $|0\rangle$ is an eigenstate of H (the normalization $\langle 0|0\rangle = 1$ is assumed). Now, let $|\psi_i\rangle$ and $|\psi_f\rangle$ be two arbitrary states with a component along the vacuum $|0\rangle$ (i.e., $\langle 0|\psi_{i,f}\rangle \neq 0$) and let us consider a general ratio of the type

$$\frac{\langle 0|\mathcal{O}_1|0\rangle}{\langle 0|\mathcal{O}_2|0\rangle}$$

where $\mathcal{O}_{1,2}$ are two generic operators. This ratio is equal to

$$\lim_{T_i,T_f \to \infty} \frac{\langle \psi_f | e^{-iT_f H(1-i\varepsilon)} \mathcal{O}_1 e^{-iT_i H(1-i\varepsilon)} |\psi_i\rangle}{\langle \psi_f | e^{-iT_f H(1-i\varepsilon)} \mathcal{O}_2 e^{-iT_i H(1-i\varepsilon)} |\psi_i\rangle} \tag{2.81}$$

Indeed, if $|n\rangle$ is the energy eigenstate with energy E_n, we have

$$\begin{aligned} e^{-iT_i H(1-i\varepsilon)} |\psi_i\rangle &= \sum_n e^{-iT_i H(1-i\varepsilon)} |n\rangle \langle n|\psi_i\rangle \\ &= \sum_n e^{-iT_i E_n(1-i\varepsilon)} |n\rangle \langle n|\psi_i\rangle \\ &\to e^{-iT_i E_0(1-i\varepsilon)} |0\rangle \langle 0|\psi_i\rangle \quad \text{if} \quad \varepsilon \to 0, \ T_i \to \infty \end{aligned} \tag{2.82}$$

Of course, this strictly holds only if the vacuum is nondegenerate and if there is an energy gap between the vacuum and the first excited state. The r.h.s. of Eq. (2.80) may now be written as

$$\lim_{\substack{T_i,T_f \to \infty \\ \varepsilon \to 0}} \frac{\langle \psi_f | e^{-iHT_f(1-i\varepsilon)} \hat{x} e^{-iH(t_1-t_2)(1-i\varepsilon)} \dots \hat{x} e^{-iHT_i(1-i\varepsilon)} |\psi_i\rangle}{\langle \psi_f | e^{-iH(T_f+T_i+t_1-t_n)(1-i\varepsilon)} |\psi_i\rangle} \tag{2.83}$$

By inserting completeness relations at each $\hat{x}$ and replacing each evolution operator by a path integral, we obtain

$$\int_{x_i}^{x_f} [dx(t)] \, \psi_f^*(x_f)\psi(x_i) \, x(t_1)\cdots x(t_n) \, e^{iS_\varepsilon[x(t)]} \tag{2.84}$$

for the numerator (x_i and x_f are taken at $t \to \mp\infty$, respectively). Each occurrence of $\hat{x}$ initially at time t_j has been replaced by the integration variable x_j corresponding to time t_j. Since the wavefunctions $\psi_{i,f}$ are arbitrary, one may choose $\psi_i(x_i) = \psi_f(x_f) = 1$, which concludes the demonstration of Eq. (2.78).

The time-ordering prescription may appear artificial within canonical quantization, but it is necessary to ensure convergence of the vacuum expectation values, assuming that a ground state exists with energy bounded from below. Notice, however, that this prescription is automatically satisfied (and hence completely natural) in the path-integral formalism.

2.3.2. The Euclidian Formalism

The ε prescription, that is, replacing t by $t(1-i\varepsilon)$, is crucial in the derivation of formula (2.78). It is customary in quantum field theory to "saturate" this prescription, that is, to define all correlation functions in imaginary time $t = -i\tau$ ($\tau \in \mathbb{R}$) and to integrate over time along the imaginary axis. The underlying assumption is, of course, that correlation functions may be analytically continued from imaginary time to real time. Since the space-time metric goes from the Minkowski to the Euclidian form when $t \to -i\tau$, we call this imaginary time method the *Euclidian formalism*. Formula (2.78) for the correlation functions then becomes (we redefine $x(-i\tau)$ as $x(\tau)$)

$$\langle x(\tau_1)x(\tau_2)\cdots x(\tau_n)\rangle = \frac{\int [dx] x(\tau_1) \cdots x(\tau_n) \exp -S_E[x(\tau)]}{\int [dx] \exp -S_E[x(\tau)]} \tag{2.85}$$

where S_E is the Euclidian action:

$$iS_E[x(\tau)] = S[x(t \to -i\tau)] \tag{2.86}$$

The Euclidian action is the integral over imaginary time of the Euclidian Lagrangian L_E:

$$L_E(x(\tau)) = -L(x(t \to -i\tau)) \tag{2.87}$$

We define likewise a Euclidian Lagrangian density $\mathcal{L}_E$. For instance, the Euclidian action of a point particle of mass m is

$$S_E[x(\tau)] = \int d\tau \left\{ \frac{1}{2m}\dot{x}^2 + V(x) \right\} \tag{2.88}$$

The Euclidian Lagrangian is then equal to the real-time Hamiltonian in this case (this is not true for fermions), hence the perfect analogy with classical statistical mechanics (see the next chapter). The other advantage of the Euclidian formalism is that path integrals are then much better defined than in Minkowski space-time. The oscillatory behavior that suppressed the contribution of large action trajectories is replaced by a simple exponential damping. Indeed, a more rigorous approach to path integration consists in defining path integrals and correlation functions in Euclidian space, and obtaining physical quantities through analytic continuation.

Important note: Unless otherwise indicated, we shall from now on work within the Euclidian formalism, and we shall drop the subscript E from the Euclidian action and replace τ by t.

Since the passage to Euclidian time affects the space-time metric, this is a good place to state our conventions in this respect. We denote by $\eta_{\mu\nu}$ the diagonal metric tensor of flat d-dimensional space-time:

$$\eta_{\mu\nu} = \begin{cases} \text{diag}(1,-1,\cdots,-1) & \text{(Minkowski)} \\ \text{diag}(1,1,\cdots,1) & \text{(Euclidian)} \end{cases} \tag{2.89}$$

The notation $\eta_{\mu\nu}$ is reserved for the metric tensor in a coordinate system that is not necessarily Cartesian. Boldface characters will denote points in Euclidian space-time (e.g., x, y, and so on). From here on the covariant notation will be used, with the summation convention for repeated (contracted) indices and the usual rules for converting between covariant and contravariant indices. Thus,

$$\eta_{\mu\nu}a^{\mu}b^{\nu} \quad \text{means} \quad \sum_{\mu,\nu=1}^{d} \eta_{\mu\nu}a^{\mu}b^{\nu} \tag{2.90}$$

and

$$a_{\mu} = \eta_{\mu\nu}a^{\nu} \qquad a^{\nu} = \eta^{\nu\mu}a_{\mu} \qquad \eta_{\mu\nu}\eta^{\nu\sigma} = \delta_{\mu}^{\sigma} \tag{2.91}$$

2.3.3. The Generating Functional

Correlation functions may be formally generated through the so-called *generating functional*:

$$Z[j] = \int [dx(t)] \exp - \left\{ S[x(t)] - \int dt\, j(t)x(t) \right\} \tag{2.92}$$

where $j(t)$ is an auxiliary "current" coupled linearly to the dynamical variable x. Formula (2.85) may be recast into

$$Z[j] = Z[0]\langle \exp \int dt j(t)x(t)\rangle$$
$$= Z[0] \sum_{n=0}^{\infty} \int dt_1 \cdots dt_n \frac{1}{n!} j(t_1)\cdots j(t_n)\langle x(t_1)\cdots x(t_n)\rangle \tag{2.93}$$

or, equivalently,

$$\langle x(t_1)\cdots x(t_n)\rangle = Z[0]^{-1} \frac{\delta}{\delta j(t_1)} \cdots \frac{\delta}{\delta j(t_n)} Z[j]\Big|_{j=0} \tag{2.94}$$

This definition is easily extended to a quantum field $\phi(x)$. The current is then a function $j(x)$ of Euclidian space-time:

$$Z[j] = Z[0]\langle \exp \int d^d x\, j(x)\phi(x)\rangle \tag{2.95}$$

If the field is fermionic, then the current j is a Grassmann number and care must be given to the ordering of the functional derivatives (2.94). By analogy with statistical mechanics, the generating functional at zero current $Z[0]$ is called the *partition function*.

2.3.4. Example: The Free Boson

In two dimensions, the free boson has the following Euclidian action:

$$S = \frac{1}{2}g \int d^2x \left\{ \partial_{\mu}\varphi\partial^{\mu}\varphi + m^2\varphi^2 \right\} \tag{2.96}$$

where g is some normalization parameter that we leave unspecified at the moment. We first calculate the two-point function, or propagator:

$$K(x,y) = \langle \varphi(x)\varphi(y)\rangle \tag{2.97}$$

If we write the action as

$$S = \frac{1}{2} \int d^2x\, d^2y\ \varphi(x) A(x,y) \varphi(y) \tag{2.98}$$

where $A(x,y) = g\delta(x-y)(-\partial^2 + m^2)$, the propagator is then $K(x,y) = A^{-1}(x,y)$, or

$$g(-\partial_x^2 + m^2) K(x,y) = \delta(x-y) \tag{2.99}$$

This follows from a continuous generalization of the results of App. 2.A on Gaussian integrals. This differential equation may also be derived from the quantum equivalent of the equations of motion, as done in Ex. (2.2). Because of rotation and translation invariance, the propagator $K(x,y)$ should depend only on the distance $r = |x-y|$ separating the two points, and we set $K(x,y) = K(r)$. Integrating (2.99) over x within a disk D of radius r centered around y, we find

$$\begin{aligned}
1 &= 2\pi g \int_0^r d\rho\, \rho \left(-\frac{1}{\rho} \frac{\partial}{\partial \rho} (\rho K'(\rho)) + m^2 K(\rho) \right) \\
&= 2\pi g \left\{ -rK'(r) + m^2 \int_0^r d\rho\, \rho K(\rho) \right\}
\end{aligned} \tag{2.100}$$

where $K'(r) = dK/dr$. The massless case ($m = 0$) can be solved immediately, the solution being, up to an additive constant,

$$K(r) = -\frac{1}{2\pi g} \ln r \tag{2.101}$$

or, in other words,

$$\langle \varphi(x)\varphi(y) \rangle = -\frac{1}{4\pi g} \ln (x-y)^2 \tag{2.102}$$

The massive case is solved by taking one more derivative with respect to r, which leads to the modified Bessel equation of order 0:

$$K'' + \frac{1}{r}K' - m^2 K = 0 \tag{2.103}$$

On physical grounds we are interested in solutions that decay at infinity, and therefore

$$K(r) = \frac{1}{2\pi g} K_0(mr) \tag{2.104}$$

where K_0 is the modified Bessel function of order 0:

$$K_0(x) = \int_0^\infty dt\, \frac{\cos(xt)}{\sqrt{t^2 + 1}} \qquad (x > 0) \tag{2.105}$$

The constant factor $1/2\pi g$ may be checked by taking the limit $r \to 0$. At large distances (i.e., when $mr \gg 1$) the modified Bessel function decays exponentially and

$$K(r) \sim e^{-mr} \tag{2.106}$$

This is also obvious from (2.103) when the second term is neglected. It is a generic feature of massive fields that correlation functions decay exponentially, with a characteristic length (the *correlation length*) equal to the inverse mass.

From the elementary Gaussian integral (2.209), it is a simple matter to argue that the generating functional (2.95) for the free boson is equal to

$$Z[j] = Z[0] \exp \left\{ \frac{1}{2} \int d^d x d^d y \, j(x) K(x,y) j(y) \right\} \qquad (2.107)$$

2.3.5. Wick's Theorem

We have defined two special orderings on field operators: normal ordering, which places all annihilation operators on the right, and time ordering, which sorts operators in chronological order. The first guarantees the vanishing of the vacuum expectation value, and the second expresses correlation functions in terms of a vacuum expectation value. Wick's theorem relates these two orderings in the case of free fields and will often be useful in subsequent chapters.

Before stating the theorem, we must define the *contraction* of two operators within a normal order. Given the product $:\phi_1 \cdots \phi_n:$, the contraction of ϕ_i with ϕ_j is simply the omission of these two operators from the normal order and their replacement by the two-point function $\langle \phi_1 \phi_2 \rangle$. We denote the contraction by brackets and write

$$:\overset{\frown}{\phi_1 \phi_2 \phi_3} \phi_4: \; = \; :\phi_1 \phi_3: \langle \phi_2 \phi_4 \rangle \qquad (2.108)$$

Now, the theorem itself: The time-ordered product is equal to the normal-ordered product, plus all possible ways of contracting pairs of fields within it. For instance,

$$\mathcal{T}(\phi_1 \phi_2 \phi_3 \phi_4) = \quad :\phi_1 \phi_2 \phi_3 \phi_4: + :\overset{\frown}{\phi_1 \phi_2} \phi_3 \phi_4: + :\overset{\frown}{\phi_1 \phi_2 \phi_3} \phi_4: +$$

$$:\overset{\frown}{\phi_1 \phi_2 \phi_3 \phi_4}: + :\phi_1 \overset{\frown}{\phi_2 \phi_3} \phi_4: + :\phi_1 \overset{\frown}{\phi_2 \phi_3 \phi_4}: +$$

$$:\phi_1 \phi_2 \overset{\frown}{\phi_3 \phi_4}: + :\overset{\frown}{\phi_1 \phi_2} \overset{\frown}{\phi_3 \phi_4}: + :\overset{\frown}{\phi_1 \phi_2 \phi_3 \phi_4}: +$$

$$:\overset{\frown}{\phi_1 \phi_2 \phi_3 \phi_4}: \qquad (2.109)$$

The simplest application of Wick's theorem is the following relation:

$$\mathcal{T}(\phi_1 \phi_2) \; = \; :\phi_1 \phi_2: + \langle \phi_1 \phi_2 \rangle \qquad (2.110)$$

This relation is rather obvious, since, for a Lagrangian quadratic in ϕ (a free field), the only difference between $\mathcal{T}(\phi_1 \phi_2)$ and $:\phi_1 \phi_2:$ comes from a rearrangement of the factors involving c-number commutators only. The difference can thus be evaluated by taking a vacuum expectation value, which leads directly to (2.110). The general form of Wick's theorem can be proven by recursion. The proof will not be given here, but can be found in standard texts on quantum field theory.

Wick's theorem also applies to free fermions, with the difference that a sign must be included in front of each term, according to the number of anticommutations required to bring the contracted fields next to each other. For instance, Eq. (2.109) applied to Fermi fields $\psi_{1,4}$ becomes

$$
\begin{aligned}
\mathcal{T}\left(\psi_1 \psi_2 \psi_3 \psi_4\right) = \quad & :\psi_1 \psi_2 \psi_3 \psi_4: + :\overline{\psi_1 \psi_2} \psi_3 \psi_4: - :\overline{\psi_1 \psi_2 \psi_3} \psi_4: + \\
& :\overline{\psi_1 \psi_2 \psi_3 \psi_4}: + :\psi_1 \overline{\psi_2 \psi_3} \psi_4: - :\psi_1 \overline{\psi_2 \psi_3 \psi_4}: + \\
& :\psi_1 \psi_2 \overline{\psi_3 \psi_4}: + :\overline{\psi_1 \psi_2} \overline{\psi_3 \psi_4}: + :\overline{\psi_1 \psi_2} \overline{\psi_3 \psi_4}: - \\
& :\overline{\psi_1 \psi_2} \overline{\psi_3 \psi_4}:
\end{aligned}
\tag{2.111}
$$

§2.4. Symmetries and Conservation Laws

One cannot overemphasize the importance of symmetries in physics. Indeed, this whole book is nothing but an analysis of the consequences of scale invariance for two-dimensional systems. In this section we give the precise meaning of symmetries in the context of a generic field theory and derive Noether's theorem, which states that to every continuous symmetry of a field theory corresponds a conserved current, and hence a conserved "charge."

2.4.1. Continuous Symmetry Transformations

Consider a collection of fields, which we collectively denote by Φ. The action functional will depend in general on Φ and its first derivatives:

$$
S = \int d^d x \, \mathcal{L}(\Phi, \partial_\mu \Phi)
\tag{2.112}
$$

In this section we study the effect, on the action functional, of a transformation affecting in general both the position and the fields:

$$
\begin{aligned}
x &\to x' \\
\Phi(x) &\to \Phi'(x')
\end{aligned}
\tag{2.113}
$$

In these transformations the new position x' is a function of x and the new field Φ' at x' is expressed as a function of the old field Φ at x:

$$
\Phi'(x') = \mathcal{F}(\Phi(x))
\tag{2.114}
$$

This is an important point: the field Φ, considered as a mapping from space-time to some target space $\mathcal{M}$ ($\Phi : \mathbb{R}^d \to \mathcal{M}$), is affected by the transformation (2.113) in two ways: first by the functional change $\Phi' = \mathcal{F}(\Phi)$, and second by the change of argument $x \to x'$. This way of looking at symmetry transformations is often called "active", in opposition to a "passive" point of view, in which the mapping

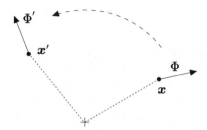

Figure 2.1. Pictorial representation of an active transformation, here a rotation. The arrows stand for a vector field that undergoes an internal rotation identical to that of the coordinate. Notice that this particular transformation is simpler to understand from a passive point of view, in which the observer rotates in the opposite direction.

$x \to x'$ is viewed simply as a coordinate transformation. The active point of view is illustrated in Fig. 2.1.

The change of the action functional under the transformation (2.113) is obtained by substituting the new function $\Phi'(x)$ for the function $\Phi(x)$ (we note that the argument x is the same in both cases). In other words, the new action is

$$
\begin{aligned}
S' &= \int d^d x \, L(\Phi'(x), \partial_\mu \Phi'(x)) \\
&= \int d^d x' \, L(\Phi'(x'), \partial'_\mu \Phi'(x')) \\
&= \int d^d x' \, L(\mathcal{F}(\Phi(x)), \partial'_\mu \mathcal{F}(\Phi(x))) \\
&= \int d^d x \left| \frac{\partial x'}{\partial x} \right| L(\mathcal{F}(\Phi(x)), (\partial x'^\nu / \partial x'^\mu) \partial_\nu \mathcal{F}(\Phi(x)))
\end{aligned}
\tag{2.115}
$$

In the second line, we have performed a change of integration variables $x \to x'$ according to the transformation (2.113), which allows us to express $\Phi'(x')$ in terms of $\Phi(x)$ in the third line. In the last line, we express x' in terms of x.

We now consider some examples, starting with a rather trivial one: a translation, defined as

$$
\begin{aligned}
x' &= x + a \\
\Phi'(x + a) &= \Phi(x)
\end{aligned}
\tag{2.116}
$$

Here $\partial x'^\nu / \partial x'^\mu = \delta^\nu_\mu$ and $\mathcal{F}$ is trivial. It follows that $S' = S$. The action is invariant under translations, unless it depends explicitly on position.

Next, we consider a Lorentz transformation. In general it takes the following form:

$$
\begin{aligned}
x'^\mu &= \Lambda^\mu{}_\nu x^\nu \\
\Phi'(\Lambda x) &= L_\Lambda \Phi(x)
\end{aligned}
\tag{2.117}
$$

where Λ is a matrix satisfying

$$\eta_{\mu\nu}\Lambda^{\mu}{}_{\rho}\Lambda^{\nu}{}_{\sigma} = \eta_{\rho\sigma} \tag{2.118}$$

and where L_{Λ} is another matrix, depending on Λ and acting on Φ if the latter has more than one component. The set of matrices Λ obeying the constraint (2.118) forms a group: the Lorentz group. The matrices L_{Λ} form a representation of the Lorentz group. In Euclidian space-time, Lorentz transformations are simply rotations. The difference between Minkowski and Euclidian space-time lies in the metric $\eta_{\mu\nu}$, and does not affect the rest of the present discussion. In d-dimensional Minkowski space-time, the Lorentz group is isomorphic to $SO(d-1, 1)$, the group of pseudo-orthogonal rotations. In two-dimensional Euclidian space-time, in which will be set the action of the near totality of this book, the rotation group is $SO(2)$, which is Abelian (commutative) and therefore admits only one-dimensional irreducible representations. The fields are then characterized by a (real) value of the planar spin.

Because of the condition (2.118), the Jacobian $|\partial x'/\partial x|$ is unity and the transformed action is

$$S' = \int d^d x \, \mathcal{L}(L_{\Lambda}\Phi, \Lambda^{-1} \cdot \partial(L_{\Lambda}\Phi)) \tag{2.119}$$

For a scalar field φ the representation is trivial ($L_{\Lambda} = 1$) and the action is invariant under Lorentz transformations ($S' = S$) if the derivatives ∂_{μ} appear in a Lorentz-invariant way. The most general Lorentz-invariant Lagrangian containing at most two derivatives is then

$$\mathcal{L}(\varphi, \partial_{\mu}\varphi) = f(\varphi) + g(\varphi)\partial_{\mu}\varphi\partial^{\mu}\varphi \tag{2.120}$$

where f and g are arbitrary functions (these functions are not arbitrary if further conditions, like renormalizability, are imposed).

Scale transformations will play a central part in this work. They are defined as

$$x' = \lambda x$$
$$\Phi'(\lambda x) = \lambda^{-\Delta}\Phi(x) \tag{2.121}$$

where λ is the dilation factor and where Δ is the scaling dimension of the field Φ. Since the Jacobian of this transformation is $|\partial x'/\partial x| = \lambda^d$, the transformed action is

$$S' = \lambda^d \int d^d x \, \mathcal{L}(\lambda^{-\Delta}\Phi, \lambda^{-1-\Delta}\partial_{\mu}\Phi) \tag{2.122}$$

We consider in particular the action of a massless scalar field φ in space-time dimension d:

$$S[\varphi] = \int d^d x \, \partial_{\mu}\varphi\partial^{\mu}\varphi \tag{2.123}$$

We check that this action is scale invariant provided we make the choice

$$\Delta = \frac{1}{2}d - 1 \tag{2.124}$$

A power φ^n may be added to the Lagrangian while preserving the scale invariance of the action provided $\Delta n = d$, or $n = 2d/(d-2)$. The only possibilities for n even (ensuring stability) are a φ^6 term in $d = 3$ and a φ^4 term in $d = 4$.

Finally, various transformations may be defined that affect only the field Φ and not the coordinates. The simplest example is that of a complex field with an action invariant under global phase transformations $\Phi'(x) = e^{i\theta}\Phi(x)$. A more complicated example is that of a multi-component field Φ transforming as $\Phi'(x) = R_\omega\Phi(x)$ where R_ω belongs to some representation of a Lie group parametrized by the group coordinate ω.

2.4.2. Infinitesimal Transformations and Noether's Theorem

We now study the effect of infinitesimal transformations on the action. Such transformations may in general be written as

$$x'^\mu = x^\mu + \omega_a \frac{\delta x^\mu}{\delta \omega_a}$$
$$\Phi'(x') = \Phi(x) + \omega_a \frac{\delta \mathcal{F}}{\delta \omega_a}(x) \qquad (2.125)$$

Here $\{\omega_a\}$ is a set of infinitesimal parameters, which we shall keep to first order only. It is customary to define the *generator* G_a of a symmetry transformation by the following expression for the infinitesimal transformation at a same point:

$$\delta_\omega \Phi(x) \equiv \Phi'(x) - \Phi(x) \equiv -i\omega_a G_a \Phi(x) \qquad (2.126)$$

We may relate this definition to Eq. (2.125) by noting that, to first order in ω_a,

$$\Phi'(x') = \Phi(x) + \omega_a \frac{\delta \mathcal{F}}{\delta \omega_a}(x)$$
$$= \Phi(x') - \omega_a \frac{\delta x^\mu}{\delta \omega_a} \partial_\mu \Phi(x') + \omega_a \frac{\delta \mathcal{F}}{\delta \omega_a}(x') \qquad (2.127)$$

The explicit expression for the generator is therefore

$$iG_a \Phi = \frac{\delta x^\mu}{\delta \omega_a} \partial_\mu \Phi - \frac{\delta \mathcal{F}}{\delta \omega_a} \qquad (2.128)$$

We consider here some examples. For an infinitesimal translation by a vector ω^μ (the index a becomes here a space-time index) one has $\delta x^\mu/\delta\omega^\nu = \delta^\mu_\nu$ and $\delta \mathcal{F}/\delta\omega^\nu = 0$. Therefore the generator of translations is simply

$$\boxed{P_\nu = -i\partial_\nu} \qquad (2.129)$$

An infinitesimal Lorentz transformation has the form

$$x'^\mu = x^\mu + \omega^\mu{}_\nu x^\nu$$
$$= x^\mu + \omega_{\rho\nu}\eta^{\rho\mu}x^\nu \qquad (2.130)$$

Substitution into the condition (2.118) yields the antisymmetry property $\omega_{\rho\nu} = -\omega_{\nu\rho}$. A general transformation has thus $\frac{1}{2}d(d-1)$ parameters. Using this antisymmetry, one may write the variation of the coordinate under an infinitesimal Lorentz transformation as

$$\frac{\delta x^\mu}{\delta \omega_{\rho\nu}} = \frac{1}{2}(\eta^{\rho\mu}x^\nu - \eta^{\nu\mu}x^\rho) \tag{2.131}$$

Its effect on the generic field Φ is

$$\mathcal{F}(\Phi) = L_\Lambda \Phi \qquad L_\Lambda \approx 1 - \frac{1}{2}i\omega_{\rho\nu}S^{\rho\nu} \tag{2.132}$$

where $S^{\rho\nu}$ is some Hermitian matrix obeying the Lorentz algebra. From (2.128), one therefore writes

$$\frac{1}{2}i\omega_{\rho\nu}L^{\rho\nu}\Phi = \frac{1}{2}\omega_{\rho\nu}(x^\nu\partial^\rho - x^\rho\partial^\nu)\Phi + \frac{1}{2}i\omega_{\rho\nu}S^{\rho\nu}\Phi \tag{2.133}$$

where $L^{\rho\nu}$ is the generator. The factor of $\frac{1}{2}$ preceding $\omega_{\rho\nu}$ in the definitions of $L^{\rho\nu}$ and $S^{\rho\nu}$ compensates for the double counting of transformation parameters caused by the full contraction of indices. The generators of Lorentz transformations are thus

$$\boxed{L^{\rho\nu} = i(x^\rho\partial^\nu - x^\nu\partial^\rho) + S^{\rho\nu}} \tag{2.134}$$

We now demonstrate Noether's theorem, which states that to every continuous symmetry of the action one may associate a current that is classically conserved. Given such a symmetry, the action is invariant under the transformation (2.125) only if the transformation is *rigid*, that is, if the parameters ω_a are independent of position. However, an especially elegant way to derive Noether's theorem is to suppose, as we will, that the infinitesimal transformation (2.125) is *not* rigid, with ω_a depending on the position.

From the last of Eqs. (2.115), we may write the effect on the action of the infinitesimal transformation (2.125). To first order, the Jacobian matrix is

$$\frac{\partial x'^\nu}{\partial x^\mu} = \delta^\nu_\mu + \partial_\mu\left(\omega_a \frac{\delta x^\nu}{\delta \omega_a}\right) \tag{2.135}$$

The determinant of this matrix may be calculated to first order from the formula

$$\det(1 + E) \approx 1 + \operatorname{Tr} E \qquad (E \text{ small}) \tag{2.136}$$

We obtain

$$\left|\frac{\partial x'}{\partial x}\right| \approx 1 + \partial_\mu\left(\omega_a \frac{\delta x^\mu}{\delta \omega_a}\right) \tag{2.137}$$

The inverse Jacobian matrix may be obtained to first order simply by reversing the sign of the transformation parameter:

$$\frac{\partial x^\nu}{\partial x'^\mu} = \delta^\nu_\mu - \partial_\mu\left(\omega_a \frac{\delta x^\nu}{\delta \omega_a}\right) \tag{2.138}$$

With the help of these preliminary steps, the transformed action S' may be written as

$$S' = \int d^d x \left(1 + \partial_\mu \left(\omega_a \frac{\delta x^\mu}{\delta \omega_a} \right) \right) \tag{2.139}$$

$$\times \mathcal{L} \left(\Phi + \omega_a \frac{\delta \mathcal{F}}{\delta \omega_a} \, , \, \left[\delta_\mu^\nu - \partial_\mu (\omega_a (\delta x^\nu / \delta \omega_a)) \right] (\partial_\nu \Phi + \partial_\nu \left[\omega_a (\delta \mathcal{F} / \delta \omega_a) \right]) \right)$$

The variation $\delta S = S' - S$ of the action contains terms with no derivatives of ω_a. These sum up to zero if the action is symmetric under rigid transformations. Then δS involves only the first derivatives of ω_a, obtained by expanding the Lagrangian. We write

$$\delta S = - \int dx \, j_a^\mu \, \partial_\mu \omega_a \tag{2.140}$$

where

$$j_a^\mu = \left\{ \frac{\partial \mathcal{L}}{\partial(\partial_\mu \Phi)} \partial_\nu \Phi - \delta_\nu^\mu \mathcal{L} \right\} \frac{\delta x^\nu}{\delta \omega_a} - \frac{\partial \mathcal{L}}{\partial(\partial_\mu \Phi)} \frac{\delta \mathcal{F}}{\delta \omega_a} \tag{2.141}$$

The quantity j_a^μ is called the *current* associated with the infinitesimal transformation (2.125). Integration by parts yields

$$\delta S = \int d^d x \, \partial_\mu j_a^\mu \, \omega_a \tag{2.142}$$

Now comes Noether's theorem: if the field configuration obeys the classical equations of motion, the action is stationary against any variation of the fields. In other words, δS should vanish for any position-dependent parameters $\omega_a(x)$. This implies the conservation law

$$\partial_\mu j_a^\mu = 0 \tag{2.143}$$

In words, every continuous symmetry implies the existence of a current given by (2.141), which is classically conserved.

The conserved charge associated with j_a^μ is

$$Q_a = \int d^{d-1} x \, j_a^0 \tag{2.144}$$

where j_a^0 is the time component of j_a^μ, and $d^{d-1} x$ stands for the purely spatial integration measure.[5] Its time derivative indeed vanishes:

$$\dot{Q}_a = \int d^{d-1} x \, \partial_0 j_a^0$$

$$= - \int d^{d-1} x \, \partial_i j^i \tag{2.145}$$

$$= - \int_\infty j^i d\sigma^i$$

---

5 In Euclidian space-time, the distinction between "time" and "space" is somewhat arbitrary.

where $d\sigma^i$ is a surface element at spatial infinity (Latin indices are summed over the "spatial" dimensions only). Therefore $\dot{Q}_a = 0$, provided the current j^i vanishes sufficiently rapidly as $x \to \infty$.

The expression (2.141) for the conserved current is termed "canonical", implying that there are other admissible expressions. In fact we may freely add to it the divergence of an antisymmetric tensor without affecting its conservation:

$$j_a^\mu \to j_a^\mu + \partial_\nu B_a^{\nu\mu} \quad , \quad B_a^{\nu\mu} = -B_a^{\mu\nu} \tag{2.146}$$

Indeed, $\partial_\mu \partial_\nu B_a^{\nu\mu} = 0$ by antisymmetry. The definition of j_a^μ is therefore ambiguous to some extent.

We stress here that Noether's theorem is a classical result that says little about the quantum realization of the symmetries. We shall see that classical symmetries imply constraints on correlation functions (the Ward identities). However, it may happen that the path integration measure does not possess the symmetry of the action, in which case that symmetry is said to be *anomalous*.

2.4.3. Transformation of the Correlation Functions

Classically, the invariance of the action under a continuous symmetry implies the existence of a conserved current. At the quantum level, correlation functions are the main object of study, and a continuous symmetry leads to constraints relating different correlation functions.

Consider again a theory involving a collection of fields Φ with an action $S[\Phi]$ invariant under a transformation of the type (2.113). Consider then the general correlation function

$$\langle \Phi(x_1) \cdots \Phi(x_n) \rangle = \frac{1}{Z} \int [d\Phi] \Phi(x_1) \cdots \Phi(x_n) \exp -S[\Phi] \tag{2.147}$$

where Z is the vacuum functional. The consequence of the symmetry of the action and of the invariance of the functional integration measure under the transformation (2.113) is the following identity:

$$\langle \Phi(x_1') \cdots \Phi(x_n') \rangle = \langle \mathcal{F}(\Phi(x_1)) \cdots \mathcal{F}(\Phi(x_1)) \rangle \tag{2.148}$$

where the mapping $\mathcal{F}$ describes the functional change of the field under the transformation, as in Eq. (2.114). The demonstration of this identity is straightforward:

$$
\begin{aligned}
\langle \Phi(x_1') \cdots \Phi(x_n') \rangle &= \frac{1}{Z} \int [d\Phi]\, \Phi(x_1') \cdots \Phi(x_n') \exp -S[\Phi] \\
&= \frac{1}{Z} \int [d\Phi']\, \Phi'(x_1') \cdots \Phi'(x_n') \exp -S[\Phi'] \\
&= \frac{1}{Z} \int [d\Phi]\, \mathcal{F}(\Phi(x_1)) \cdots \mathcal{F}(\Phi(x_n)) \exp -S[\Phi] \\
&= \langle \mathcal{F}(\Phi(x_1)) \cdots \mathcal{F}(\Phi(x_n)) \rangle
\end{aligned}
\tag{2.149}
$$

An explanation is in order. In going from the first to the second line of Eq. (2.149) we have just renamed the dummy integration variable $\Phi \rightarrow \Phi'$, without performing a real change of integration variables. In going from the second to the third line we have performed a change of functional integration variables, in which $\Phi'(x')$ is expressed in terms of $\Phi(x)$. We know by hypothesis that the action is invariant under such a change, which should be carried through as in Eq. (2.115). We need the further hypothesis that the Jacobian of this change of variable is trivial (i.e., does not depend on the field Φ). This is in fact the main obstacle to conformal invariance in a quantum symmetry: the action may well be scale invariant, but the measure is not because of the regularization procedure needed to define it properly.

For instance, invariance under translation $x' = x + a$ has the following consequence on the correlation functions:

$$\langle \Phi(x_1 + a) \cdots \Phi(x_n + a) \rangle = \langle \Phi(x_1) \cdots \Phi(x_n) \rangle \qquad (2.150)$$

In words, only the relative positions of the fields are important in a correlator. Likewise, Lorentz invariance has the following effect on correlators of scalar fields:

$$\langle \Phi(\Lambda^\mu{}_\nu x_1^\nu) \cdots \Phi(\Lambda^\mu{}_\nu x_n^\nu) \rangle = \langle \Phi(x_1^\mu) \cdots \Phi(x_n^\mu) \rangle \qquad (2.151)$$

Finally, scale invariance implies the following relation for correlators of a collection of fields ϕ_i with scaling dimensions Δ_i (cf. Eq. (2.121)):

$$\langle \phi_1(\lambda x_1) \cdots \phi_n(\lambda x_n) \rangle = \lambda^{-\Delta_1} \cdots \lambda^{-\Delta_n} \langle \phi_1(x_1) \cdots \phi_n(x_n) \rangle \qquad (2.152)$$

We shall come back to these relations in Chap. 4.

2.4.4. Ward Identities

The consequence of a symmetry of the action and the measure on correlation functions may also be expressed via the so-called Ward identities, which we shall now demonstrate. An infinitesimal transformation may be written in terms of the generators as

$$\Phi'(x) = \Phi(x) - i\omega_a G_a \Phi(x) \qquad (2.153)$$

where ω_a is a collection of infinitesimal, constant parameters. Note that the positions are the same on both sides of this expression. We make a change of functional integration variables in the correlation function (2.147), in the form of the above infinitesimal transformation with ω_a now a function of x. The action is not invariant under such a local transformation, its variation being given by (2.142). Denoting by X the collection $\Phi(x_1) \cdots \Phi(x_n)$ of fields in the correlation function and by $\delta_\omega X$ its variation under the transformation, we can write

$$\langle X \rangle = \frac{1}{Z} \int [d\Phi'] \, (X + \delta X) \, \exp - \left\{ S[\Phi] + \int dx \, \partial_\mu j_a^\mu \omega_a(x) \right\} \qquad (2.154)$$

We again assume that the functional integration measure is invariant under the local transformation (i.e., $[d\Phi'] = [d\Phi]$). When expanded to first order in $\omega_a(x)$, the above yields

$$\langle \delta X \rangle = \int dx \, \partial_\mu \langle j_a^\mu(x)X \rangle \omega_a(x) \qquad (2.155)$$

The variation δX is explicitly given by

$$\delta X = -i \sum_{i=1}^{n} \left(\Phi(x_1) \cdots G_a \Phi(x_i) \cdots \Phi(x_n) \right) \omega_a(x_i)$$

$$= -i \int dx \, \omega_a(x) \sum_{i=1}^{n} \left\{ \Phi(x_1) \cdots G_a \Phi(x_i) \cdots \Phi(x_n) \right\} \delta(x - x_i) \qquad (2.156)$$

Since (2.155) holds for any infinitesimal function $\omega_a(x)$, we may write the following local relation:

$$\boxed{\begin{aligned}
&\frac{\partial}{\partial x^\mu} \langle j_a^\mu(x)\Phi(x_1) \cdots \Phi(x_n) \rangle \\
&= -i \sum_{i=1}^{n} \delta(x - x_i) \langle \Phi(x_1) \cdots G_a \Phi(x_i) \cdots \Phi(x_n) \rangle
\end{aligned}} \qquad (2.157)$$

This is the Ward identity for the current j_a^μ. Note that the form of the current may be modified from the canonical definition (2.141) without affecting the Ward identity, if one adds to j_a^μ a quantity that is divergenceless identically (i.e., without using the equations of motion), such as in Eq. (2.146).

We integrate the Ward identity (2.157) over a region of space-time that includes all the points x_i. On the left-hand side (l.h.s.), we obtain a surface integral

$$\int_\Sigma ds_\mu \langle j_a^\mu(x)\Phi(x_1) \cdots \Phi(x_n) \rangle \qquad (2.158)$$

which vanishes, since the hypersurface Σ may be sent to infinity without affecting the integral: indeed, the divergence $\partial_\mu \langle j_a^\mu X \rangle$ vanishes away from the points x_i and the correlator $\langle j_a^\mu(x)X \rangle$ goes to zero sufficiently fast as $x \to \infty$, by hypothesis. For the right-hand side (r.h.s.) of Eq. (2.157), this implies

$$\delta_\omega \langle \Phi(x_1) \cdots \Phi(x_n) \rangle \equiv -i\omega_a \sum_{i=1}^{n} \langle \Phi(x_1)G_a \Phi(x_i) \cdots \Phi(x_n) \rangle = 0 \qquad (2.159)$$

In other words, the variation of the correlator under an infinitesimal transformation vanishes. This is simply the infinitesimal version of Eq. (2.149) (see also the definition (2.126)).

The Ward identity allows us to identify the conserved charge

$$Q_a = \int d^{d-1}x \, j_a^0(x) \qquad (2.160)$$

as the generator of the symmetry transformation in the Hilbert space of quantum states. Let $Y = \Phi(x_2) \cdots \Phi(x_n)$ and suppose that the time $t = x_1^0$ is different from all the times in Y. We integrate the Ward identity (2.157) in a very thin "pill box" bounded by $t_- < t$, by $t_+ > t$, and by spatial infinity, which excludes all the other points $x_2, \cdots, x_n$. The integral of the l.h.s. of (2.157) is converted into a surface integral and yields

$$\langle Q_a(t_+)\Phi(x_1)Y\rangle - \langle Q_a(t_-)\Phi(x_1)Y\rangle = -i\langle G_a\Phi(x_1)Y\rangle \qquad (2.161)$$

Remembering that a correlation function is the vacuum expectation value of a time-ordered product in the operator formalism, and assuming, for the sake of argument, that all other times x_i^0 are greater than t, we write, in the limit $t_- \to t_+$,

$$\langle 0|[Q_a, \Phi(x_1)]Y|0\rangle = -i\langle 0|G_a\Phi(x_1)Y|0\rangle \qquad (2.162)$$

This being true for an arbitrary set of fields Y, we conclude that

$$[Q_a, \Phi] = -iG_a\Phi \qquad (2.163)$$

In other words, the conserved charge Q_a is the generator of the infinitesimal symmetry transformations in the operator formalism. Of course, these identities are obtained in the Euclidian formalism. An easy way to go back to Minkowski space-time is to replace the charge Q by $-iQ$, since it is the outcome of an integration of the time-like component of a vector.

§2.5. The Energy-Momentum Tensor

Here we apply the general results of the previous section to the invariance of a theory with respect to translations and rotations (or Lorentz transformations). The conserved current associated with translation invariance is the *energy-momentum tensor*, whose components are the density and flux density of energy and momentum. In Chapters 4 and 5, the consequences of conformal symmetry will be expressed in terms of the Ward identities associated with the energy-momentum tensor; this section more or less paves the way for later discussions.

The infinitesimal translation $x'^\mu \to x^\mu + \epsilon^\mu$ induces the following variations in the coordinates and the fields (see Eq. (2.125)):

$$\frac{\delta x^\mu}{\delta \epsilon^\nu} = \delta^\mu_\nu \quad , \quad \frac{\delta \Phi}{\delta \epsilon^\nu} = 0 \qquad (2.164)$$

Consequently the corresponding canonical conserved current is

$$T_c^{\mu\nu} = -\eta^{\mu\nu}\mathcal{L} + \frac{\partial\mathcal{L}}{\partial(\partial_\mu\Phi)}\partial^\nu\Phi \qquad (2.165)$$

and the conservation law is $\partial_\mu T_c^{\mu\nu} = 0$. The conserved charge is the four-momentum

$$P^\nu = \int d^{d-1}x \, T_c^{0\nu} \qquad (2.166)$$

In particular, the energy is

$$P^0 = \int d^{d-1}x \left\{ \frac{\partial\mathcal{L}}{\partial\dot{\Phi}}\dot{\Phi} - \mathcal{L} \right\} \qquad (2.167)$$

which is the usual definition of the Hamiltonian. As an operator, the conserved charge P_μ has therefore the following effect in Euclidian time, according to Eq. (2.163):

$$[P_\mu, \Phi] = -\partial_\mu \Phi \tag{2.168}$$

In real time, this relation becomes $[P_\mu, \Phi] = -i\partial_\mu \Phi$, which is the well-known commutator of an x-dependent operator with momentum in ordinary quantum mechanics.

2.5.1. The Belinfante Tensor

In general, the canonical energy-momentum tensor $T_c^{\mu\nu}$ is not symmetric. However, we have the freedom to modify this tensor by adding the divergence of a tensor $B^{\rho\mu\nu}$ antisymmetric in the first two indices:

$$T_B^{\mu\nu} = T_c^{\mu\nu} + \partial_\rho B^{\rho\mu\nu} \quad , \quad B^{\rho\mu\nu} = -B^{\mu\rho\nu} \tag{2.169}$$

This addition does not affect the classical conservation law nor the Ward identity. Indeed, the variation of the action under a nonuniform translation with position-dependent parameter $\epsilon^\mu(x)$ is still given by

$$\delta S = -\int d^d x \, \partial_\mu T_B^{\mu\nu} \, \epsilon_\nu \tag{2.170}$$

since $\partial_\mu T_B^{\mu\nu} = \partial_\mu T_c^{\mu\nu}$ identically. If we succeed in finding $B^{\rho\mu\nu}$ such that the new tensor $T_B^{\mu\nu}$ is symmetric, then the latter is called the *Belinfante* energy-momentum tensor. In order to accomplish this, we consider the conserved currents associated with Lorentz transformations.

From (2.131) and (2.132), the variations of the coordinates and fields under an infinitesimal Lorentz transformation are

$$\frac{\delta x^\rho}{\delta\omega_{\mu\nu}} = \frac{1}{2}(\eta^{\rho\mu}x^\nu - \eta^{\rho\nu}x^\mu) \quad , \quad \frac{\delta\mathcal{F}}{\delta\omega_{\mu\nu}} = -i\frac{1}{2}S^{\mu\nu}\Phi \tag{2.171}$$

and the associated canonical conserved current is

$$j^{\mu\nu\rho} = T_c^{\mu\nu}x^\rho - T_c^{\mu\rho}x^\nu + \frac{1}{2}i\frac{\partial\mathcal{L}}{\partial(\partial_\mu\Phi)}S^{\nu\rho}\Phi \tag{2.172}$$

We look for $B^{\rho\mu\nu}$ such that this current may be expressed as

$$j^{\mu\nu\rho} = T_B^{\mu\nu}x^\rho - T_B^{\mu\rho}x^\nu \tag{2.173}$$

This relation ensures that $T_B^{\mu\nu} = T_B^{\nu\mu}$, as is easily seen by applying the conservation laws $\partial_\mu j^{\mu\nu\rho} = 0$ and $\partial_\mu T_B^{\mu\nu} = 0$. However, this implies only that $T_B^{\mu\nu}$ is symmetric classically (i.e., for field configurations obeying the equations of motions).

An explicit expression for $B^{\rho\mu\nu}$ can be found by inspection:

$$B^{\mu\rho\nu} = \frac{1}{4}i\left\{ \frac{\partial\mathcal{L}}{\partial(\partial_\mu\Phi)}S^{\nu\rho}\Phi + \frac{\partial\mathcal{L}}{\partial(\partial_\rho\Phi)}S^{\mu\nu}\Phi + \frac{\partial\mathcal{L}}{\partial(\partial_\nu\Phi)}S^{\mu\rho}\Phi \right\} \tag{2.174}$$

We check that this expression is indeed antisymmetric in the first two indices, since $S^{\mu\nu} = -S^{\nu\mu}$. In order to show that the above has the right form, we calculate its antisymmetric part in $(\rho\nu)$:

$$B^{\mu\rho\nu} - B^{\mu\nu\rho} = \frac{1}{2}i\frac{\partial\mathcal{L}}{\partial(\partial_\mu\Phi)}S^{\nu\rho}\Phi \tag{2.175}$$

On the other hand, the antisymmetric part of $T_c^{\rho\nu}$ in a classical configuration is obtained by applying the conservation laws to Eq. (2.172):

$$T_c^{\rho\nu} - T_c^{\nu\rho} = -\frac{1}{2}i\partial_\mu\left\{\frac{\partial\mathcal{L}}{\partial(\partial_\mu\Phi)}S^{\nu\rho}\Phi\right\} \tag{2.176}$$

We see that the antisymmetric part of $T_c^{\rho\nu} + \partial_\mu B^{\mu\rho\nu}$ vanishes, that is, $T_B^{\mu\nu}$ is indeed symmetric in a classical configuration. Note that the form given in Eq. (2.174) for $B^{\mu\rho\nu}$ is not unique; further modifications of the energy-momentum tensor are possible.

We can illustrate this with an example. Consider the following Lagrangian for a massive vector field A_μ (in Euclidian space-time):

$$\mathcal{L} = \frac{1}{4}F^{\alpha\beta}F_{\alpha\beta} + \frac{1}{2}m^2A^\alpha A_\alpha \tag{2.177}$$

wherein $F_{\alpha\beta} = \partial_\alpha A_\beta - \partial_\beta A_\alpha$. The canonical energy-momentum tensor is

$$T_c^{\mu\nu} = F^{\mu\alpha}\partial^\nu A_\alpha - \eta^{\mu\nu}\mathcal{L} \tag{2.178}$$

and it is not symmetric. We now calculate $T_B^{\mu\nu}$ as defined in Eq. (2.169), with

$$B^{\alpha\mu\nu} = F^{\alpha\mu}A^\nu \tag{2.179}$$

We end up with

$$T_B^{\mu\nu} = T_c^{\mu\nu} + F^{\alpha\mu}\partial_\alpha A^\nu + \partial_\alpha F^{\alpha\mu}A^\nu \tag{2.180}$$

This tensor is classically symmetric, as may be seen from the following: we define the identically symmetric tensor

$$\tilde{T}_B^{\mu\nu} = F^{\mu\alpha}F^\nu{}_\alpha - \frac{1}{4}\eta^{\mu\nu}F^{\alpha\beta}F_{\alpha\beta} + m^2\left[A^\mu A^\nu - \frac{1}{2}\eta^{\mu\nu}A^\alpha A_\alpha\right]$$

$$= T_B^{\mu\nu} - (\partial_\alpha F^{\alpha\mu} - m^2 A^\mu)A^\nu \tag{2.181}$$

The two tensors $\tilde{T}_B^{\mu\nu}$ and $T_B^{\mu\nu}$ coincide for classical configurations, since the equations of motion are

$$\partial_\alpha F^{\alpha\mu} - m^2 A^\mu = 0 \tag{2.182}$$

It is $\tilde{T}_B^{\mu\nu}$ which is written down in standard texts, whereas it is $T_B^{\mu\nu}$ which a priori appears in the Ward identity:

$$\partial_\mu\langle T_B^{\mu\nu}X\rangle = -\sum_j \delta(x - x_j)\langle\Phi(x_1)\cdots\partial^\nu\Phi(x_j)\cdots\Phi(x_n)\rangle \tag{2.183}$$

If we wish to use a symmetric tensor in the Ward identity, we must replace $T_B^{\mu\nu}$ by $\tilde{T}_B^{\mu\nu}$ therein, but this modifies the Ward identity. However, as we shall see presently, the modification to the Ward identity coming from this substitution has no effect and may be ignored in general.

Indeed, the Ward identity in terms of $\tilde{T}_B^{\mu\nu}$ is

$$\partial_\mu \langle \tilde{T}_B^{\mu\nu} X \rangle = - \sum_j \delta(x - x_j) \langle \Phi(x_1) \cdots \partial^\nu \Phi(x_j) \cdots \Phi(x_n) \rangle$$

$$- \partial_\mu \langle [\partial_\alpha F^{\alpha\mu}(x) - m^2 A^\mu(x)] A^\nu(x) X \rangle \qquad (2.184)$$

We wish to show that the last term is of no consequence. For this we need to use the following relation, written here in Euclidian time, which is a consequence of the equations of motion on correlation functions (see Ex. 2.2):

$$\left\langle \frac{\delta Y}{\delta \Phi(x)} \right\rangle = \left\langle Y \frac{\delta S}{\delta \Phi(x)} \right\rangle \qquad (2.185)$$

Here, Y is a product of local fields. We apply this relation to our system, with $Y = A_\nu(y) X$ (X is again a product of local fields) and $\Phi(x) \to A_\mu(x)$. We find

$$\frac{\delta S}{\delta A_\mu(x)} = -\partial_\alpha F^{\alpha\mu}(x) + m^2 A^\mu(x) \qquad (2.186)$$

Therefore,

$$\left\langle \frac{\delta X}{\delta A_\mu(x)} A_\nu(y) \right\rangle + \delta(x - y) \delta_{\mu\nu} \langle X \rangle$$

$$= \langle (-\partial_\alpha F^{\alpha\mu}(x) + m^2 A^\mu(x)) A^\nu(y) X \rangle \qquad (2.187)$$

We take the limit $x \to y$ and ignore the delta function $\delta(x - y)$, which is automatically subtracted if normal order is used for the product $[\partial_\alpha F^{\alpha\mu} - m^2 A^\mu] A^\nu$. We find

$$\langle [(-\partial_\alpha F^{\alpha\mu} + m^2 A^\mu) A^\nu]|_x X \rangle = \left\langle \frac{\delta X}{\delta A_\mu(x)} A_\nu(y) \right\rangle \Big|_{x \to y} \qquad (2.188)$$

This last expression will vanish for all x except at the isolated points x_j, the positions of the fields appearing in the product X. For instance, if $X = A_\rho(x_1) A_\sigma(x_2)$, then

$$\left\langle \frac{\delta X}{\delta A_\mu(x)} A_\nu(y) \right\rangle \Big|_{x \to y} = \delta_\rho^\mu \delta(x - x_1) \langle A_\sigma(x_2) A_\nu(x) \rangle$$

$$+ \delta_\sigma^\mu \delta(x - x_2) \langle A_\rho(x_1) A_\nu(x) \rangle \qquad (2.189)$$

In general, the additional contribution to the Ward identity will have the following form:

$$\partial_\mu \langle [\partial_\alpha F^{\alpha\mu}(x) - m^2 A^\mu(x)] A^\nu(x) X \rangle = \partial_\mu \sum_j \delta(x - x_j) f_j^\mu(x_1, \cdots, x_n) \qquad (2.190)$$

The reason such an addition is of no consequence is that the Ward identity, like any other expression involving delta functions, has a precise meaning only after integration through some arbitrary volume. The added term is a total divergence containing delta functions and can thus be converted into a surface integral, which receives no contribution from the delta functions.

In summary, provided the theory has rotation symmetry, we may define a new energy-momentum tensor $T_B^{\mu\nu}$, which is conserved, classically symmetric, and plays the same role in Ward identities as $T_c^{\mu\nu}$. In fact, one may use the equations of motion to bring $T_B^{\mu\nu}$ into another form (noted $\tilde{T}_B^{\mu\nu}$ above) which is now identically symmetric, still conserved, and still plays the same role as $T_c^{\mu\nu}$ in the Ward identity, except for terms that may be ignored. Consequently, we shall no longer distinguish between $T_B^{\mu\nu}$ and $\tilde{T}_B^{\mu\nu}$ (as far as Ward identities are concerned) in the remainder of this work.

2.5.2. Alternate Definition of the Energy-Momentum Tensor

We now consider a general infinitesimal transformation of the coordinates $x^\mu \to x'^\mu = x^\mu + \epsilon^\mu(x)$. This can be considered as a translation with an x-dependent parameter $\epsilon^\mu(x)$. According to (2.142) the induced change in the action is

$$
\delta S = \int d^d x \, T^{\mu\nu} \, \partial_\mu \epsilon_\nu
$$
$$
= \frac{1}{2} \int d^d x \, T^{\mu\nu} \, (\partial_\mu \epsilon_\nu + \partial_\nu \epsilon_\mu)
$$

(2.191)

where we have assumed that $T^{\mu\nu}$ is identically symmetric.[6] If the diffeomorphism $x' = x + \epsilon$ is considered as an infinitesimal change of coordinates, the corresponding change in the metric tensor $g_{\mu\nu}$ is (to first order in ϵ)

$$
g'_{\mu\nu} = \frac{\partial x^\alpha}{\partial x'^\mu} \frac{\partial x^\beta}{\partial x'^\nu} g_{\alpha\beta}
$$
$$
= (\delta_\mu^\alpha - \partial_\mu \epsilon^\alpha)(\delta_\nu^\beta - \partial_\nu \epsilon^\beta) g_{\alpha\beta}
$$
$$
= g_{\mu\nu} - (\partial_\mu \epsilon_\nu + \partial_\nu \epsilon_\mu)
$$

(2.192)

This prompts for an alternate definition of the energy-momentum tensor, as the functional derivative of the action with respect to the metric, evaluated in flat space:

$$
\boxed{\delta S = -\frac{1}{2} \int d^d x \, T^{\mu\nu} \delta g_{\mu\nu}}
$$

(2.193)

---

[6] As explained in the previous section, extra terms may appear in the above equations, but these terms vanish for classical configurations and are of no consequence on the Ward identities. We consequently ignore them.

For instance, on a general manifold, the action for a free scalar field φ is

$$
\begin{aligned}
S &= \int d^d x \sqrt{g}\, \mathcal{L} \\
&= \frac{1}{2} \int d^d x \sqrt{g} \left\{ g^{\mu\nu} \partial_\mu \varphi \partial_\nu \varphi + m^2 \varphi^2 \right\}
\end{aligned}
\tag{2.194}
$$

where $g \equiv \det g_{\mu\nu}$ and the factor $\sqrt{g}$ is required for the invariance of the space-time integration measure. Using the identities

$$
\det A = e^{\operatorname{Tr}\,\ln A} \quad \text{and} \quad \delta g^{\mu\nu} = -g^{\alpha\mu} g^{\beta\nu} \delta g_{\alpha\beta}
\tag{2.195}
$$

we find

$$
\delta \sqrt{g} = \frac{1}{2} \sqrt{g}\, g^{\mu\nu} \delta g_{\mu\nu}
\tag{2.196}
$$

and the definition (2.193) yields

$$
T^{\mu\nu} = -g^{\mu\nu} \mathcal{L} + \partial^\mu \varphi \partial^\nu \varphi
\tag{2.197}
$$

which coincides with the canonical definition (2.165). The advantage of the new definition (2.193) is that the energy-momentum tensor is identically symmetric. However, obtaining an explicit expression for $T^{\mu\nu}$ from (2.193) requires more involved calculations than going through the canonical definition, or its Belinfante generalization.

If a tetrad e^a_μ is used instead of a metric (see App. 2.C) then the energy-momentum tensor is endowed with a Lorentz index and an Einstein index: Since $g_{\mu\nu} = e^a_\mu e^a_\nu$, we easily find that

$$
\delta S = - \int d^d x\, e\, T^\mu_a \delta e^a_\mu
\tag{2.198}
$$

where $e = \det e^a_\mu$.

In the quantum theory, the alternate definition (2.193) of the energy-momentum tensor takes the following meaning. Let Φ represent the set of dynamical fields of the theory, and g the metric. On a general manifold the action is a functional $S[\Phi, g]$ of both quantities. The vacuum functional $Z[g]$ and the functional integration measure $[d\Phi]_g$ both depend on the metric:

$$
\begin{aligned}
Z[g] &= \int [d\Phi]_g \exp -S[\Phi, g] \\
&= \exp -W[g]
\end{aligned}
\tag{2.199}
$$

where we have defined the *connected functional* $W[g]$. Under an infinitesimal variation δg of the metric, the vacuum functional is modified:

$$
\begin{aligned}
Z[g + \delta g] &= \int [d\Phi]_{g+\delta g} \, \exp -S[\Phi, g + \delta g] \\
&= \int [d\Phi]_g \left\{ 1 + \frac{1}{2} \int d^d x \, \sqrt{g} \delta g_{\mu\nu} T^{\mu\nu} \right\} \exp -S[\Phi, g] \qquad (2.200) \\
&= Z[g] + \frac{1}{2} Z[g] \int d^d x \, \sqrt{g} \delta g_{\mu\nu} \langle T^{\mu\nu} \rangle
\end{aligned}
$$

In the second equation, we have assumed that the energy-momentum tensor takes care of the variation of the action *and* of the integration measure, if any. This is the essential difference between the classical and quantum definitions of the energy-momentum tensor. The variation of the connected functional $W[g]$ is then

$$
\delta W[g] = -\frac{\delta Z[g]}{Z[g]} = -\frac{1}{2} \int d^d x \, \sqrt{g} \delta g_{\mu\nu} \langle T^{\mu\nu} \rangle \qquad (2.201)
$$

or, in functional notation,

$$
\langle T^{\mu\nu}(x) \rangle = -\frac{2}{\sqrt{g}} \frac{\delta W[g]}{\delta g_{\mu\nu}(x)} \qquad (2.202)
$$

Again, if a tetrad is used instead of a metric, the above quantum definition becomes

$$
\delta W[e] = -\frac{1}{2} \int d^d x \, e \delta e^a_\mu \langle T^\mu_a \rangle \qquad (2.203)
$$

Appendix 2.A. Gaussian Integrals

In this appendix we consider integrals of the type

$$
I(A, b) = \int d^n x \, \exp \left\{ -\frac{1}{2} x^t A x + b^t x \right\} \qquad (2.204)
$$

where A is an $n \times n$ symmetric matrix whose eigenvalues have positive real parts, and where x and b are n-dimensional column vectors (the transpose of an object x is written x^t). We first evaluate the integral when $b = 0$. Since A is symmetric, it can be diagonalized by an orthogonal matrix: $A = O^t D O$ where D is diagonal with entries D_i and where $O^t O = 1$. By the change of variables $y = Ox$, for which the Jacobian is unity, the integral becomes

$$
\begin{aligned}
I(A, 0) &= \int d^n y \, \exp \left\{ -\frac{1}{2} \sum_i D_i y_i^2 \right\} \\
&= \left\{ \frac{(2\pi)^n}{\det A} \right\}^{\frac{1}{2}} \qquad \det A = \det D = \prod_j D_j
\end{aligned} \qquad (2.205)
$$

If $b \neq 0$, one simply has to complete the square of the exponent:

$$-\frac{1}{2}x^t A x + b^t x = \frac{1}{2}b^t A^{-1} b - \frac{1}{2}(x - A^{-1}b)^t A (x - A^{-1}b) \qquad (2.206)$$

and the change of variables $x \to x - A^{-1}b$ brings this case back to the above form, except for a prefactor:

$$I(A,b) = \left\{ \frac{(2\pi)^n}{\det A} \right\}^{\frac{1}{2}} \exp\left(\frac{1}{2}b^t A^{-1} b \right) \qquad (2.207)$$

We turn now to the evaluation of moments of order m:

$$\langle x_{i_1} x_{i_2} \cdots x_{i_m} \rangle = \frac{\int d^n x \, x_{i_1} x_{i_2} \cdots x_{i_m} \exp\left(-\frac{1}{2}x^t A x\right)}{\int d^n x \, \exp\left(-\frac{1}{2}x^t A x\right)} \qquad (2.208)$$

These are the discrete analog of the correlation functions. To this end we introduce the generating function

$$Z(b) = \int d^n x \, \exp\left(-\frac{1}{2}x^t A x + b^t x \right)$$

$$= Z(0) \exp\left(\frac{1}{2}b^t A^{-1} b \right) \qquad (2.209)$$

It then immediately follows that

$$\langle x_{i_1} x_{i_2} \cdots x_{i_m} \rangle = \frac{1}{Z(0)} \frac{\partial}{\partial b_{i_1}} \cdots \frac{\partial}{\partial b_{i_m}} Z(b) \Big|_{b=0} \qquad (2.210)$$

For instance, the second-order moment (or "propagator") is

$$\langle x_i x_j \rangle = (A^{-1})_{ij} \qquad (2.211)$$

It is straightforward to verify that the three-point moment $\langle x_i x_j x_k \rangle$ vanishes. This follows from the reflection symmetry $x_i \to -x_i$ of the exponential at $b = 0$. The four-point moment, along with all moments with an even number of points, can be expressed in terms of the two-point moment. Specifically,

$$\langle x_1 x_2 x_3 x_4 \rangle = \langle x_1 x_2 \rangle \langle x_3 x_4 \rangle + \langle x_1 x_3 \rangle \langle x_2 x_4 \rangle + \langle x_1 x_4 \rangle \langle x_2 x_3 \rangle \qquad (2.212)$$

This follows directly from (2.210). In general, the $2n$-point function is given by a sum over all ways of pairing the points, each pair being then replaced by the corresponding two-point moment. This constitutes a weak version of Wick's theorem (2.109).

Appendix 2.B. Grassmann Variables

We recall that an *algebra* is a vector space endowed with a product. A *Grassmann algebra* is a vector space constructed from a set of n generators θ_i on which an antisymmetric product is defined:

$$\theta_i \theta_j + \theta_j \theta_i = 0 \qquad (2.213)$$

A generic element of a Grassmann algebra is therefore a first-degree polynomial in the generators θ_i, namely

$$f(\theta_i) = \sum_{k=0}^{n} \sum_{i_1,\cdots,i_k}^{n} C_{i_1,\cdots,i_k}^{(k)} \theta_{i_1} \theta_{i_2} \cdots \theta_{i_k} \qquad (2.214)$$

where the complex coefficients $C_{i_1,\cdots,i_k}^{(k)}$ are defined only if all their indices are different, and where a standard ordering is defined on these indices. The dimension of the Grassmann algebra is then the number of distinct monomials that can be constructed from the θ_i, namely 2^n. For instance, generic elements of a Grassmann algebra with $n = 1$ and $n = 2$ are respectively

$$\begin{aligned} (n = 1) \quad & f(\theta) = c_0 + c_1\theta \\ (n = 2) \quad & f(\theta_1, \theta_2) = c_0 + c_1\theta_1 + c_2\theta_2 + c_{12}\theta_1\theta_2 \end{aligned} \qquad (2.215)$$

Any other term that we might add to these expressions is either redundant or zero because of the anticommutation properties.

The generators of a Grassmann algebra are often called *Grassmann variables*. Correspondingly, elements of the algebra, since they are polynomials in the generators, are called "functions" of Grassmann variables.

We define a differentiation on the Grassmann algebra in the obvious way, that is, by treating the generators θ_i like normal variables, except for their anticommuting properties. Consequently we must adopt a convention: The variable of differentiation must be brought to the left of every expression before taking the derivative:

$$df = \sum_i d\theta_i \frac{\partial f}{\partial \theta_i} \qquad (2.216)$$

For the function $f(\theta_1, \theta_2)$ defined above, we have

$$\frac{\partial f}{\partial \theta_2} = c_2 - c_{12}\theta_1 \qquad (2.217)$$

Since functions of Grassmann variables are at most linear in each variable, the differential operator $\partial/\partial\theta_i$ is nilpotent, that is, $(\partial/\partial\theta_i)^2 = 0$. In fact, these operators, together with the variables θ_i themselves, form a Clifford algebra:

$$\theta_i\theta_j + \theta_j\theta_i = 0$$

$$\frac{\partial}{\partial\theta_i}\frac{\partial}{\partial\theta_j} + \frac{\partial}{\partial\theta_j}\frac{\partial}{\partial\theta_i} = 0 \qquad (2.218)$$

$$\theta_i\frac{\partial}{\partial\theta_j} + \frac{\partial}{\partial\theta_j}\theta_i = \delta_{ij}$$

Integration over Grassmann variables is *defined* to be identical to differentiation:

$$\int d\theta_i f(\theta_1, \cdots, \theta_n) = \frac{\partial}{\partial\theta_i} f(\theta_1, \cdots, \theta_n) \qquad (2.219)$$

This definition may seem strange, but it should be kept in mind that we are defining *definite* integrals. Therefore the result of the integration does not depend on the integration variable any more, and its derivative vanishes. Conversely, the definite integral of a derivative vanishes if there are no boundary terms. Consequently, a natural definition of definite integration should have the properties

$$\frac{\partial}{\partial \theta_i} \int d\theta_i \, f(\theta) = \int d\theta_i \, \frac{\partial}{\partial \theta_i} f(\theta) = 0 \qquad (2.220)$$

which are satisfied by the definition (2.219) by virtue of the nilpotency of the derivative. The integral over several Grassmann variables of a generic function always yields the highest term of the expansion:

$$\int d\theta_n \cdots d\theta_1 \, f(\theta) = C^{(k)}_{i_1,\dots,i_n} \qquad (2.221)$$

Under a change of integration variables $\theta_i \rightarrow \theta_i'$, the integration measure $d\theta_1 \cdots d\theta_n$ changes according to

$$d\theta_1 \cdots d\theta_n = \left| \frac{\partial \theta'}{\partial \theta} \right| d\theta_1' \cdots d\theta_n' \qquad (2.222)$$

This is the opposite of ordinary integration, wherein the Jacobian occurs with the opposite power, and follows directly from the identification of integrals with derivatives.

Finally, we evaluate Gaussian integrals of Grassmann variables. We first consider the integral

$$I = \int d\theta_1 \cdots d\theta_n \, \exp -\frac{1}{2} \theta^t A \theta \qquad (2.223)$$

where θ is the column vector of the θ_i, θ^t is its transpose and A is an antisymmetric matrix (otherwise only its antisymmetric part contributes) of even dimension n. The series expansion of the exponential contains a finite number of terms (no summation over repeated indices here):

$$I = \int d\theta_n \cdots d\theta_1 \, \prod_{i<j} \exp -\theta_i A_{ij} \theta_j$$
$$= \int d\theta_n \cdots d\theta_1 \, \prod_{i<j} \left(1 - \theta_i A_{ij} \theta_j \right) \qquad (2.224)$$

Each factor commutes with the other, and thus we are free to order them according to increasing i. If we expand the product, the terms that survive the integration are those that contain each variable exactly once, and consequently contain $n/2$ matrix elements A_{ij}. Therefore, the result of the integration is

$$I = \sum_{p \in S_n} \varepsilon(p) A_{p(1)p(2)} A_{p(3)p(4)} \cdots A_{p(n-1)p(n)} \qquad (2.225)$$

$(\varepsilon(p)$ is the signature of the permutation p) with the constraints

$$p(1) < p(2) \,,\; p(3) < p(4) \,,\; p(5) < p(6) \,,\; \cdots$$
$$p(1) < p(3) < p(5) < p(7) < \cdots \tag{2.226}$$

The expression (2.225) is known as the *Pfaffian* of the matrix A and denoted $\mathrm{Pf}(A)$. The Pfaffian is defined for antisymmetric matrices of even dimension. It can be shown without difficulty (see Ex. 12.12) that

$$\mathrm{Pf}(A)^2 = \det A \tag{2.227}$$

The integral with a source

$$I(b) = \int d\theta_1 \cdots d\theta_n \; \exp\left\{-\frac{1}{2}\theta^t A\theta + b^t\theta\right\} \tag{2.228}$$

is done the same way as for the ordinary Gaussian integral. We proceed to a shift of integration variables: $\theta' = \theta - A^{-1}b$, and, the Jacobian being unity, the result is

$$I(b) = I(0)\exp\frac{1}{2}b^t A^{-1}b \tag{2.229}$$

The details of the calculation are slightly different from the ordinary Gaussian integral since b_i anticommutes with θ_j, but this is compensated by the antisymmetry of A. The moment $\langle\theta_i\theta_j\rangle$ is given by (notice the order of the derivatives)

$$\langle\theta_i\theta_j\rangle = I(0)^{-1}\frac{\partial}{\partial b_j}\frac{\partial}{\partial b_i}I(b)\Big|_{b=0} \tag{2.230}$$
$$= (A^{-1})_{ij}$$

Wick's theorem is also valid here, except that the two-point moments occur with the appropriate sign obtained by bringing together the members of the pair [cf. Eq. (2.111)].

We now turn to the integral

$$I_2 = \int d\bar{\theta}d\theta \; \exp-\bar{\theta}M\theta \tag{2.231}$$

where M is an $n \times n$ matrix and where $d\bar{\theta}d\theta$ stands for

$$d\bar{\theta}d\theta = \prod_{i=1}^{n} d\bar{\theta}_i d\theta_i \tag{2.232}$$

The variables θ_i and $\bar{\theta}_i$ may be thought of as conjugate to each other, although this is not necessary. Again, by expanding the exponential,

$$I_2 - \int d\bar{\theta}d\theta \; \prod_{ij}\left(1 - \bar{\theta}_i M_{ij}\theta_j\right)$$
$$= \sum_{p\in S_n} \varepsilon(p)\, M_{1p(1)}M_{2p(2)}\cdots M_{np(n)} \tag{2.233}$$
$$= \det M$$

Overall, we obtain results that are similar to those obtained for ordinary Gaussian integrals, except that the determinant occurs with the opposite power.

Appendix 2.C. Tetrads

This appendix offers a quick introduction to the concept of a tetrad, which is necessary in order to define spinor fields on a general curved manifold. In the usual formalism of Christoffel symbols, only the action for integer-spin fields can be written down in a covariant manner.

At each point of a manifold, coordinate differentials dx^μ span a local vector space (the cotangent space). Under a change of coordinate system $x \to x'$, the differentials transform as follows:

$$dx'^\mu = \frac{\partial x'^\mu}{\partial x^\nu} dx^\nu \qquad (2.234)$$

The only requirement imposed on the Jacobian matrix $\partial x'^\mu / \partial x^\nu$ is invertibility: It should be an element of the group $GL(d)$ of invertible d-dimensional matrices. We therefore say that an action with general covariance is endowed with a local $GL(d)$ symmetry. However, local fields have been defined according to their trans-formation properties in Euclidian (or Minkowski) space, where the corresponding symmetry group is $SO(d)$ (resp. $SO(d-1, 1)$). In order to carry over the Lorentz group formalism to a general manifold in a general coordinate system, we in-troduce at each point a local orthogonal frame of basis vectors for the cotangent space:

$$e^a = e^a_\mu dx^\mu \qquad a = 1, \cdots, d \qquad (2.235)$$

where the frame vectors e^a form a *tetrad*, or *vierbein*. These names are four-dimension specific, but will be used here in a general setting, rather than the imaginative "zweibein" and "vielbein" (Cartan's terminology of "repères mobiles" may also be used). A natural choice for the tetrad is determined by the conditions

$$e^a_\mu e^b_\nu g^{\mu\nu} = \eta^{ab} \qquad g_{\mu\nu} = \eta_{ab} e^a_\mu e^b_\nu \qquad (2.236)$$

which express the orthogonality of the tetrad. The lower (Greek) index of e^a_μ is called an Einstein index, while the upper (Latin) index is called a Lorentz index.

In order to compare vectors belonging to different (but nearby) cotangent spaces, we need to introduce a prescription for parallel transport, specified by the so-called *spin connection* ω^{ab}_μ:

$$V^a \to V^a - \omega^{ab}_\mu dx^\mu V^b \qquad (2.237)$$

where dx^μ is the amount of transport. The covariant derivative is defined as

$$(D_\mu V)^a = \partial_\mu V^a + \omega^{ab}_\mu V^b \qquad (2.238)$$

and results from the comparison of a vector at x with a vector parallel-transported from $x + dx$. Since parallel transport changes only the direction of a vector and not its length, the spin-connection is antisymmetric in its Lorentz indices: $\omega^{ab}_\mu = -\omega^{ba}_\mu$.

The tetrad e^a_μ may be used to convert between Lorentz and Einstein indices: $V^a = e^a_\mu V^\mu$. The Christoffel symbols $\Gamma^\mu_{\nu\lambda}$ are used to specify the parallel transport in a tetrad-free language:

$$V^\mu \to V^\mu - \Gamma^\mu_{\nu\lambda} dx^\lambda V^\nu \tag{2.239}$$

Since by definition the tetrad e^a_μ is invariant under parallel transport, we have the relation

$$\Gamma^\mu_{\nu\lambda} e^a_\mu = \partial_\lambda e^a_\nu + \omega^{ab}_\lambda e^b_\nu \tag{2.240}$$

The curvature of a manifold manifests itself when a vector is parallel-transported around a closed path. Around an infinitesimal "square" loop of sides dx and dy, the difference between the initial vector V^a and the transported vector V'^a is

$$\begin{aligned} V'^a - V^a &= -[D_\mu, D_\nu]^{ab} V^b dx^\mu dy^\nu \\ &= R^{ab}_{\mu\nu} V^b dx^\mu dy^\nu \end{aligned} \tag{2.241}$$

More explicitly, the curvature tensor $R^{ab}_{\mu\nu}$ is

$$R^{ab}_{\mu\nu} = \partial_\mu \omega^{ab}_\nu - \partial_\nu \omega^{ab}_\mu + \omega^{ac}_\mu \omega^{cb}_\nu - \omega^{ac}_\nu \omega^{cb}_\mu \tag{2.242}$$

This tensor is related to the usual Riemann tensor $R^{\rho\sigma}{}_{\mu\nu}$ by contraction with $e^a_\rho e^b_\sigma$.

The connection is determined by the metric $g_{\mu\nu}$, together with the torsion-free condition $\Gamma^\mu_{\nu\lambda} = \Gamma^\mu_{\lambda\nu}$. The latter condition is natural if we define the manifold as embedded in a higher-dimensional Euclidian space, as a hypersurface $\mathbf{X}(x)$. Then, the metric is given by

$$g_{\mu\nu} = \partial_\mu \mathbf{X} \cdot \partial_\nu \mathbf{X} \tag{2.243}$$

and the Christoffel symbols are easily derived to be

$$\begin{aligned} \Gamma^\mu_{\nu\lambda} &= \partial^\mu \mathbf{X} \cdot \partial_\nu \partial_\lambda \mathbf{X} \\ &= \frac{1}{2} g^{\mu\rho} (\partial_\nu g_{\rho\lambda} + \partial_\lambda g_{\rho\nu} - \partial_\rho g_{\nu\lambda}) \end{aligned} \tag{2.244}$$

On a two-dimensional manifold, the spin-connection can be expressed in terms of a single-covariant vector ω_μ:

$$\omega^{ab}_\mu = \epsilon^{ab} \omega_\mu \tag{2.245}$$

while the curvature tensor is

$$\begin{aligned} R^{ab}_{\mu\nu} &= \epsilon^{ab} (\partial_\mu \omega_\nu - \partial_\nu \omega_\mu) \\ &= \sqrt{g} \epsilon^{ab} \epsilon_{\mu\nu} R \end{aligned} \tag{2.246}$$

where R is the scalar curvature.

Exercises

2.1 *Expansion in eigenfunctions*
Consider a generalization of the Lagrangian (2.1):

$$\mathcal{L} = \frac{1}{2}(\dot{\varphi}^2 + \varphi D\varphi)$$

in which D is some Hermitian linear differential operator. For instance, $D = \partial_x^2 - m^2$ for the free scalar field. A possible generalization could be $D = \nabla^2 - V(x)$ (in d dimensions), in which case there is no translation invariance. In general, the above Lagrangian is not Lorentz invariant. The eigenfunctions of D are denoted $u_n(x)$, and form by assumption a discrete spectrum, with eigenvalues $-\omega_n^2$. We have the relations

$$\int d^d x\, u_n^* u_m = \delta_{mn}$$

Show that the quantum field may be expanded as

$$\varphi(x) = \sum_n \sqrt{\frac{1}{2\omega_n}} \left(a_n u_n(x) + a_n^\dagger u_n^*(x)\right)$$

where the a_n are annihilation operators, obeying the standard commutation relations. Show also that the Hamiltonian may be written as

$$H = \sum_n \omega_n(a_n^\dagger a_n + \frac{1}{2})$$

2.2 *Equations of motion for correlation functions*
Consider a generic action $S[\phi]$ involving some quantum field ϕ, and the correlation function

$$\langle X \rangle = \frac{1}{Z} \int [d\phi]\, X\, e^{iS[\phi]}$$

where X stands for an expression involving ϕ. By performing an infinitesimal change of functional integration variables $\phi \to \phi + \delta\phi$, demonstrate the following relation:

$$\left\langle \frac{\delta X}{\delta\phi(x)} \right\rangle = -i \left\langle X \frac{\delta S}{\delta\phi(x)} \right\rangle$$

Then take $X = \phi(y)$ and the Lagrangian (2.1), and show that the two-point function $\langle\phi(y)\phi(x)\rangle$ satisfies the equation

$$i\left(\frac{\partial}{\partial x^\mu}\frac{\partial}{\partial x_\mu} + m^2\right)\langle\phi(y)\phi(x)\rangle = \delta(x-y)$$

2.3 Demonstrate Eq. (2.37), i.e., that the equations of motion following from (2.32) are recovered in the Heisenberg equations of motion, provided the Hamiltonian and the commutation rules be as in (2.37). One may start with a simple quartic potential.

2.4 Prove the properties (2.69) of fermionic coherent states. For the second one (2.69b), it is useful to diagonalize the matrix T ($T = UDU^{-1}$) and to work with the rotated variables and operators Uz and $U\psi$.

2.5 From the expression (2.229) for the Gaussian Grassmann integral with a source, show how to recover the following special case of Wick's theorem:

$$\langle \theta_i \theta_j \theta_k \theta_l \rangle = \langle \theta_i \theta_j \rangle \langle \theta_k \theta_l \rangle - \langle \theta_i \theta_k \rangle \langle \theta_j \theta_l \rangle + \langle \theta_i \theta_l \rangle \langle \theta_j \theta_k \rangle$$

2.6 Demonstrate explicitly the relation (2.233).

Notes

There are many good texts on quantum field theory. However, most of this chapter does not follow any particular text. Some sections, in particular the treatment of the dynamics of Grassmann variables, are inspired by a graduate course given in 1986 at Cornell University by H. Kawai [233]. Among modern texts emphasizing the functional formulation of quantum field theory are those of Brown [60], Collins [79], Ramond [303], Weinberg [351] and Zinn-Justin [369]. More classic texts, such as Bjorken and Drell [48] and Itzykson and Zuber [205], are still very useful.

The method of path integrals was invented by R. Feynman [130]. Grassmann variables were applied to the functional description of fermions by F.A. Berezin [40]. The Belinfante energy-momentum tensor is discussed by Callan, Coleman, and Jackiw [62] and Jackiw [208].

CHAPTER 3

Statistical Mechanics

Most applications of conformal invariance pertain to statistical systems at criticality. A brief introduction to statistical mechanics is therefore required for those readers unfamiliar with the subject. The emphasis is put on the concepts underlying the hypothesis of conformal invariance in critical systems. Some parallels are to be drawn with the previous chapter, since quantum field theory and statistical mechanics walk hand in hand in the modern theory of critical phenomena. Section 3.1 reviews the notion of statistical ensemble of states and describes some basic models defined on the lattice or in the continuum. Section 3.2 explains the basic features of critical phenomena and how the scaling hypothesis provides a unified understanding of phenomena at or near the critical point. Section 3.3 justifies the scaling hypothesis with the idea of real-space renormalization. Section 3.4 applies the concepts of the renormalization group to continuum models and gives deeper meaning to the notion of scale invariance for Euclidian field theories. Finally, Sect. 3.5 briefly explains the transfer matrix method, a discrete analogue in statistical mechanics of the operator formalism of quantum theory.

§3.1. The Boltzmann Distribution

Statistical mechanics describes complex physical systems (i.e., systems made of a large number of atoms in interaction) whose exact states cannot be specified because of this complexity. Instead, macroscopic properties alone may be specified, and the role of the theory is to infer these properties from the microscopic Hamiltonian. Thus, statistical mechanics distinguishes *microscopic states* (or *microstates*) from *macroscopic states* (or *macrostates*). A microstate is specified by the quantum numbers of all the particles in the system or, classically, by the exact configuration (positions and momenta) of all the particles. It characterizes the system from a dynamical point of view in the sense that its future state is fixed by its present state through deterministic laws. A macrostate is specified by a finite number of macroscopic parameters, which characterize the system from the point of view of observation, such as pressure, temperature, magnetization, and so on. To a given macrostate corresponds a large number of microstates, each leading to the same macroscopic properties. Having no more information about an isolated system than that given by the macroscopic parameters, we assume that all the microstates associated with the observed macrostate have equal probabilities to be the actual state of the system.

The basic idea behind the statistical study of a complex system is that any physical property—like the energy, the magnetization, and so on—may be regarded as a statistical average, calculated over a suitable *ensemble* of microstates. Of course, at any instant, the system is in a specific (but unknown) microstate. The replacement of this microstate by a statistical ensemble needs some justification. It has long been customary to justify this replacement by invoking the so-called *ergodic hypothesis*, which states that the time average of a quantity over the time evolution of a specific microstate is equal to the average of the same quantity, at fixed time, over some statistical ensemble of microstates. If one accepts this hypothesis, then the use of a statistical ensemble is justified provided the time necessary for an efficient sweep of the ensemble by any of its microstates is short enough compared with the time of measurement of the physical quantity of interest. This is far from obvious. A better justification for the use of statistical ensembles follows from dividing the system into a very large number of mesoscopic parts, each of them large enough to display the complex properties of the whole system. At any instant, each of these mesoscopic subsystems is characterized by its own microstate, but the properties of the whole system are obtained by averaging over all subsystems. Thus, the ensemble averaging amounts more to a spatial averaging than to a time averaging.

Which ensemble of states is most appropriate for averaging depends on how isolated the system is. If it is completely isolated, with no exchange of energy or particles with its surroundings, the relevant ensemble of microstates is made of all states on a given energy "shell", occurring with equal probabilities. It is called the *microcanonical* ensemble.

If, on the other hand, a system S is in thermal contact with its surroundings and hence is free to exchange energy with it, then all microstates of S do not have equal probabilities. However, all microstates of the "universe" (S plus its surroundings) have equal probabilities. This, in turn, provides us with a distribution of probabilities for the microstates of S: The probability that a specific microstate of S be the actual state of the system depends only on its energy and is given by the *Boltzmann distribution*:

$$P_i = \frac{1}{Z} \exp{-\beta E_i} \qquad \beta = \frac{1}{T} \qquad (3.1)$$

where T is the absolute temperature[1] and Z is the normalization of the distribution, called the *partition function*:

---

[1] This definition of temperature includes the unit-dependent Boltzmann constant k_B. Thus T has the dimension of energy.

$$Z = \sum_i \exp -\beta E_i \qquad (3.2)$$

The ensemble of microstates defined by the Boltzmann distribution is the *canonical ensemble*.

The partition function (3.2) is of central importance in statistical mechanics since macroscopic quantities are generically related to derivatives of Z. For instance, the average energy within the canonical ensemble is obtained by lowering a factor of E_i in the sum of Boltzmann weights through differentiation with respect to β:

$$
\begin{aligned}
U &= \frac{1}{Z} \sum_i E_i \, \exp -\beta E_i \\
&= -\frac{1}{Z} \frac{\partial Z}{\partial \beta} \\
&= -T^2 \frac{\partial}{\partial T}(F/T)
\end{aligned}
\qquad (3.3)
$$

where we have introduced the *free energy*:

$$F = -T \ln Z \qquad (3.4)$$

Similarly, the heat capacity C at constant volume is

$$C = \left(\frac{\partial U}{\partial T}\right)_V = -T \frac{\partial^2 F}{\partial T^2} \qquad (3.5)$$

The *specific heat* is defined as the heat capacity per unit volume. Thus, the partition function is the generating function of all the thermodynamic functions of interest.

In practice, statistical mechanics studies systems composed of a large quantity of N identical components (atoms, molecules). The properties of each individual atom (e.g., energy, spin, etc.) fluctuate according to the Boltzmann distribution, but the physical quantities of interest are summed over all N components of the system. Because of the law of large numbers, their fluctuations vary as $1/\sqrt{N}$ and are completely negligible when N is large. The limit $N \to \infty$ is called the *thermodynamic limit* since then the variance of the macroscopic properties vanishes and their values cease to be random variables, becoming instead exact variables to be treated in the formalism of thermodynamics.

3.1.1. Classical Statistical Models

In practice the number of systems for which the partition function can be calculated, even in an approximate way, is very small. Confronted with the extreme complexity of most realistic systems one relies on simplified models to investigate finite-temperature properties. Some of these models are defined in terms of discrete, classical variables, which live on a lattice of sites. The best-known and simplest of these discrete models is the *Ising model*. It consists of a discrete lattice of spins σ_i, each taking the value -1 or 1. Unless otherwise indicated, a square lattice is used and i stands for a lattice site. For a lattice with N sites the number of different spin configurations $[\sigma]$ is 2^N, and the energy of a given configuration is

$$E[\sigma] = -J \sum_{\langle ij \rangle} \sigma_i \sigma_j - h \sum_i \sigma_i \qquad (3.6)$$

where the notation $\langle ij \rangle$ indicates that the summation is taken over pairs of nearest-neighbor lattice sites. The first term in the energy represents the interaction of neighboring spins through a ferromagnetic ($J > 0$) or antiferromagnetic ($J < 0$) coupling. The second term represents the interaction with an external magnetic field h. We shall not try to explain how such a simple model can arise from the microscopic quantum theory of magnetism but will be content in considering it for its own sake. We will assume that $J > 0$, although the case $J < 0$ is strictly equivalent at zero field ($h = 0$). In zero field, the lowest energy configuration is doubly degenerate: The spins can be either all up $(+1)$ or all down (-1). If the field h is nonzero, the lowest energy configuration will have all spins aligned with h (i.e., of the same sign as h).

The first thermodynamic quantity of interest is the magnetization M, the mean value of a single spin. By translation invariance, this is the same for all spins, and we can write:

$$M = \langle \sigma_j \rangle \qquad (\text{any } j)$$

$$= \frac{1}{NZ} \sum_{[\sigma]} \left\{ \sum_i \sigma_i \right\} \exp -\beta E[\sigma] \qquad (3.7)$$

$$= -\frac{1}{N} \frac{\partial F}{\partial h}$$

where the notation $\langle \dots \rangle$ denotes an ensemble average. Also of interest is the magnetic susceptibility, which indicates how the magnetization responds to a very small external field:

$$\chi = \frac{\partial M}{\partial h} \bigg|_{h=0}$$

$$= \frac{1}{N} \frac{\partial}{\partial h} \left\{ \frac{1}{Z} \sum_{[\sigma]} \left(\sum_i \sigma_i \right) \exp -\beta E[\sigma] \right\} \qquad (3.8)$$

$$= \frac{1}{NT} \left\{ \langle \sigma_{\text{tot.}}^2 \rangle - \langle \sigma_{\text{tot.}} \rangle^2 \right\}$$

where $\sigma_{\text{tot.}} = \sum_i \sigma_i$. The susceptibility is therefore proportional to the variance of the total spin, and measures its fluctuations.

The susceptibility is also related to the *pair correlation function* $\Gamma(i)$:

$$\Gamma(i - j) = \langle \sigma_i \sigma_j \rangle \qquad (3.9)$$

Because of translation invariance, the correlator Γ can depend only on the difference of lattice sites. Moreover, for large distances $|i - j|$, the lattice structure is less relevant, some rotation symmetry is restored and the correlators depend only on the distance $|i - j|$. The *connected* correlation function

$$\Gamma_c(i - j) = \langle\sigma_i\sigma_j\rangle_c = \langle\sigma_i\sigma_j\rangle - \langle\sigma_i\rangle\langle\sigma_j\rangle \tag{3.10}$$

is a measure of the mutual statistical dependence of the spins σ_i and σ_j, in terms of which the susceptibility may be rewritten as

$$\chi = \beta \sum_{i=0}^{\infty} \Gamma_c(i) \tag{3.11}$$

We therefore expect the susceptibility to be a measure of the statistical coherence of the system, increasing with the statistical dependence of all the spins.

The Boltzmann distribution is ,of course, invariant under a constant shift of the energy. This allows us to write the Hamiltonian of the Ising model in a slightly different way. Indeed, since $\sigma_i\sigma_j = 2\delta_{\sigma_i,\sigma_j} - 1$, the configuration energy is, up to a constant,

$$E[\sigma] = -2J \sum_{\langle ij\rangle} \delta_{\sigma_i,\sigma_j} - h \sum_i \sigma_i \tag{3.12}$$

This form lends itself to an immediate generalization of the Ising model, the so-called q-state Potts model, in which the spin σ_i takes q different integer values: $\sigma_i = 1, 2, \cdots, q$. To each possible value of σ we associate a unit vector $\boldsymbol{d}(\sigma)$ in $q-1$ dimensional space such that $\sum_\sigma^q \boldsymbol{d}(\sigma) = 0$. $\boldsymbol{d}(\sigma)$ plays the role of the magnetic dipole moment associated with the spin value σ. The configuration energy in an external field is

$$E[\sigma] = -\alpha \sum_{\langle ij\rangle} \delta_{\sigma_i,\sigma_j} - \boldsymbol{h} \cdot \sum_i \boldsymbol{d}(\sigma_i) \tag{3.13}$$

Other generalizations of the Ising model are possible, wherein for instance the spins are regarded as "flavors" of atoms interacting with their nearest neighbors with coupling constants depending on which flavors are paired (Ashkin-Teller models) and so on.

In Ising-type models, the variables (spins) reside on the sites of the lattice whereas the interaction energy resides on the links between nearest-neighbor pairs. In systems such as the eight-vertex model the opposite is true: The variables are arrows living on the links, each taking one of two possible directions along the link. The interaction energy resides on the sites and its value depends on how the four arrows come together at that point, with the constraint that the number of arrows coming into (and out of) a site must be even.

Other statistical models involve continuous degrees of freedom rather than discrete ones. For instance, a more realistic treatment of classical ferromagnetism is obtained by assuming the local spin to be a unit vector $\boldsymbol{n}$, with the configuration energy

$$E[\boldsymbol{n}] = J \sum_{\langle ij\rangle} \boldsymbol{n}_i \cdot \boldsymbol{n}_j - \sum_i \boldsymbol{h} \cdot \boldsymbol{n}_i \tag{3.14}$$

where h is some external magnetic field. This is the classical Heisenberg model, or the classical $O(n)$ model if the vector n is taken to have n components.

When discussing critical properties (in the next section) it is often more convenient to replace the lattice by a continuum, in which case the use of continuous degrees of freedom is mandatory. The above Hamiltonian is then equivalent to

$$E[n] = \int d^d x \ \{J\partial_k n \cdot \partial_k n - h \cdot n\} \tag{3.15}$$

wherein n_i and h_i are replaced by $n(x)$ and $h(x)$. The gradient term is the equivalent of the nearest-neighbor interaction of the discrete case.

Because the constraint $n^2(x) = 1$ at every position is difficult to implement in practical calculations, we may consider the simpler alternative in which it is replaced by the single constraint

$$\frac{1}{V} \int d^d x \ n^2 = 1 \tag{3.16}$$

where V is the volume of the system. One then obtains the *spherical model*, which differs from the $O(n)$ model by the constraint imposed. Another way to approximate the constraint $n^2(x) = 1$ is to make it energetically unfavorable for $n^2(x)$ to be different from 1. This may be done with the help of a quartic potential $V(|n|)$ having a minimum at $|n| = 1$. After rescaling the field n, the energy functional may be taken as

$$E[n] = \int d^d x \ \{\frac{1}{2}\partial_k n \cdot \partial_k n - \frac{1}{2}\mu^2 n^2 + \frac{1}{4}u(n^2)^2\} \tag{3.17}$$

The position of the minimum of energy as a function of $|n|$ depends on the relative values of μ and u. If n has a single component φ, this is termed the φ^4 model. The sign of the φ^2 term (positive or negative) determines whether the ground state value of φ vanishes or not. The case $u = 0$ is exactly solvable, and is called the Gaussian model since the partition function reduces to a product of Gaussian integrals. The associated configuration energy is

$$E[\varphi] = \int d^d x \ (\frac{1}{2}(\nabla\varphi)^2 + \frac{1}{2}\mu^2\varphi^2) \tag{3.18}$$

All of these models were extensively studied and are discussed in great detail in most texts devoted to critical phenomena.

For models defined on the continuum, the analogy between statistical mechanics and quantum field theory is manifest. The partition function of the φ^4 model is a sum over the possible configurations of the field φ (i.e., a functional integral):

$$\begin{aligned} Z &= \int [d\varphi] \ \exp -\beta E[\varphi] \\ &= \int [d\varphi] \ \exp \left\{ -\int d^d x \ \left[\frac{1}{2}(\nabla\varphi)^2 + \frac{1}{2}r\varphi^2 + \frac{1}{4}u\varphi^4 \right] \right\} \end{aligned} \tag{3.19}$$

Here we have rescaled the field φ by $\sqrt{\beta}$ and the φ^4 coupling u by $1/\beta$, so that the inverse temperature does not explicitly appear. The partition function of a

d-dimensional statistical model is thus entirely analogous to the generating functional of a quantum field in d space-time dimensions in the Euclidian formalism. Changing the temperature then amounts to scaling the field φ and modifying the φ^4 coupling.

3.1.2. Quantum Statistics

The statistical models described in the preceding subsection are all classical: All physical quantities have a definite value within each microstate of the statistical ensemble. In *quantum* statistical mechanics, we must deal with quantum indeterminacy as well as with thermal fluctuations. In that context, we define the *density operator*

$$\rho = \exp -\beta H \tag{3.20}$$

where H is the Hamiltonian of the system. The partition function may be expressed as a sum over the eigenstates of H:

$$Z = \sum_n e^{-\beta E_n} = \operatorname{Tr} \rho \tag{3.21}$$

The statistical average of an operator A is then

$$\langle A \rangle = \sum_n \langle n|e^{-\beta H}A|n\rangle = \operatorname{Tr}(\rho A) \tag{3.22}$$

The resemblance between the density operator $e^{-\beta H}$ and the evolution operator e^{-iHt} allows for the representation of the density operator as a functional integral. This introduces the Lagrangian formalism into statistical mechanics. Explicitly, consider the kernel of the density operator for a single degree of freedom:

$$\rho(x_f, x_i) = \langle x_f|e^{-\beta H}|x_i\rangle \tag{3.23}$$

The path integral is adapted to this kernel by substituting $t \rightarrow -i\tau$ (the Wick rotation), where τ is a real variable going from 0 to β. The action $S[x(t)]$ then becomes the Euclidian action $iS_E[x(\tau)]$. The kernel of the density operator ρ becomes then

$$\rho(x_f, x_i) = \int_{(x_i,0)}^{(x_f,\beta)} [dx] \exp -S_E[x] \tag{3.24}$$

The partition function may be expressed as

$$Z = \int dx \, \rho(x,x) = \int [dx] \exp -S_E[x] \tag{3.25}$$

This time, the integration limits are no longer specified: all "trajectories" such that $x(0) = x(\beta)$ contribute. Here the "time" τ is merely an auxiliary variable introduced to take advantage of the analogy with path integrals. The expectation value of an operator A is

$$\langle A \rangle = \frac{1}{Z} \int dx \, \langle x|\rho A|x\rangle$$

$$= \frac{1}{Z} \int dx dy \, \langle x|\rho|y\rangle\langle y|A|x\rangle$$

$$= \frac{1}{Z} \int dx dy \int_{(x,0)}^{(y,\beta)} [dx] \, \langle y|A|x\rangle \exp -S_E[x]$$

$$= \frac{1}{Z} \int dx dy \int_{(x,0)}^{(y,\beta)} [dx] \, A(x)\delta(x-y) \exp -S_E[x]$$

$$= \frac{1}{Z} \int [dx] \, A(x(0)) \, \exp -S_E[x] \tag{3.26}$$

where we have supposed that A is a function of x only, so that

$$\langle y|A|x\rangle = A(x)\delta(x-y) \tag{3.27}$$

Hence, the expectation value of A is calculated as in the path-integral method. Note, however, that the operator A is evaluated at $\tau = 0$.

The generalization to a system with a continuum of degrees of freedom and to multipoint correlation functions is straightforward. The key point here is that the partition function of a quantum system in the path integral formalism is obtained from the ordinary path integral by a Wick rotation and by restricting the Euclidian time to a finite domain of extent β. At zero temperature this domain is infinite in extent and we recover the usual generating functional in Euclidian time. At finite temperatures, the quantum partition function of a d-dimensional system resembles that of a $(d+1)$-dimensional classical system defined on a strip of width β.

§3.2. Critical Phenomena

3.2.1. Generalities

Phase transitions are arguably the most interesting feature of statistical systems. They are characterized by a sudden and qualitative change in the macroscopic properties of the system as the temperature (or some other control parameter) is varied. We distinguish *first-order* transitions from *continuous* transitions. First-order transitions are characterized by a finite jump in the energy U (the *latent heat*) at the transition temperature. This means that the system must absorb or deliver a finite amount of energy before leaving the transition temperature. Liquid-gas transitions and other structural transitions are generally of this type. On the other hand, continuous phase transitions do not involve any latent heat, nor any abrupt change in the average value of microscopic variables, such as the magnetization. However, the derivatives of such quantities, such as the specific heat or the susceptibility, are discontinuous or display some singular behavior at continuous phase transitions.

Strictly speaking, phase transitions exist only in the thermodynamic limit. The reason is clear: In systems such as the Ising model in zero field, where the energy of any configuration is an integer multiple of a fundamental energy scale ε, the partition function for a finite number of lattice sites is a polynomial in $z = \exp -\beta\varepsilon$.

For instance, in the Ising model, one can choose $\varepsilon = -J$, and the configuration of highest energy has $E = 2N\varepsilon$. Each configuration contributes a power of z to the partition function, with unit coefficient. Therefore Z is a polynomial of degree $2N$ in z, whose roots lie away from the positive real axis, and occur as complex conjugated pairs. Singularities of the free energy or of its derivatives can occur only at those roots, which all lie outside of the physical domain of interest as long as N is finite. As $N \to \infty$, the number of these roots becomes infinite, and they tend to form various arcs, some of them touching the real positive axis. It is at these locations on the positive real axis that the behavior of thermodynamic quantities becomes singular in the thermodynamic limit.

Continuous phase transitions will be of central interest to us because of their relation to conformal invariance. The two-dimensional Ising model, of which the exact solution is known, exhibits such a transition. Let us describe this transition before commenting on the general case: The critical temperature T_c is related to the coupling J by

$$\sinh(2J/T_c) = 1 \tag{3.28}$$

Above T_c, the magnetization at zero field (or *spontaneous magnetization*) vanishes, whereas below T_c it takes a nonzero value, tending toward 1 at $T = 0$ and toward 0 as $T \to T_c$ according to the power law

$$M \sim (T_c - T)^{1/8} \tag{3.29}$$

The system is then in its *ferromagnetic phase*. The two directions of spontaneous magnetization (up and down) are energetically equivalent, and which one is actually realized depends on how the external field h was brought to zero. Although the magnetization is continuous at T_c, its derivative with respect to the magnetic field—the susceptibility χ—diverges as $T \to T_c$, according to

$$\chi = \frac{\partial M}{\partial h} \sim (T - T_c)^{-7/4} \tag{3.30}$$

Away from T_c, the correlations $\Gamma_c(i)$ decay exponentially with distance, with a temperature-dependent characteristic length ξ called the *correlation length*, expressed here in units of the lattice spacing:

$$\langle s_i s_j \rangle_c \sim \exp -|i - j|/\xi(T) \qquad |i - j| \gg 1 \tag{3.31}$$

As T approaches its critical value, the correlation length increases toward infinity, like the inverse power of $T - T_c$:

$$\xi(T) \sim \frac{1}{|T - T_c|} \tag{3.32}$$

As we shall see, this divergence of the correlation length is the most fundamental characteristic of continuous phase transitions. Such transitions are termed *critical phenomena* and occur at so-called *critical points* of the phase diagram.

The importance of the correlation length in the behavior of thermodynamic quantities near the critical point is intuitively clear. Near a critical point, a spin system such as the Ising model is an aggregate of domains (or *droplets*) of different magnetizations. At first thought, the typical size of such droplets should be ξ, roughly the maximum scale over which the spins should be correlated. But in fact, droplets of all sizes up to the correlation length must be present, and droplets within droplets, etc. Otherwise the connected correlation functions $\Gamma_c(n)$ would have a peak near $n \sim \xi$ but would be small below that scale, which is not true: This can be seen from the observed divergence of the susceptibility χ as $T \to T_c$ and the expression (3.11) for χ. In other words, the spins fluctuate over all length scales between the lattice spacing and ξ. The free energy F will receive contributions from the domain walls separating spin droplets, integrated from the lattice spacing up to ξ, and it is plausible that its singular behavior (or, rather, that of its derivatives) be governed by the "upper integration bound", which is ξ.

At T_c or sufficiently close to it, the correlation length exceeds the physical dimension L of the system (we suppose, for the sake of argument, that the system lives in a square box of side L). At this point the free energy no longer depends on the correlation length but is limited by the box volume.[2] The pair correlation function does not have enough room to decay exponentially within the box, and its spatial dependence is algebraic (d is the dimension of space):

$$\Gamma(n) \sim \frac{1}{|n|^{d-2+\eta}} \tag{3.33}$$

The behavior of thermodynamic functions near or at the critical point is characterized by *critical exponents* defining power laws as $T \to T_c$. The most common exponents are defined in Table 3.1.

Table 3.1. Definitions of the most common critical exponents and their exact value within the two-dimensional Ising model. Here d is the dimension of space.

Exponent	Definition	Ising Value
α	$C \propto (T - T_c)^{-\alpha}$	0
β	$M \propto (T_c - T)^{\beta}$	1/8
γ	$\chi \propto (T - T_c)^{-\gamma}$	7/4
δ	$M \propto h^{1/\delta}$	15
ν	$\xi \propto (T - T_c)^{-\nu}$	1
η	$\Gamma(n) \propto |n|^{2-d-\eta}$	1/4

---

[2] In real systems, the correlation length is limited not by the physical size of the sample, but by the presence of sample inhomogeneities. It rarely goes beyond a thousand lattice sites, even in very pure samples.

We conclude this section by a remark on the relevance of classical statistical mechanics in a quantum world. Classical statistical mechanics is an approximation to quantum statistical mechanics, valid in the context of critical phenomena when the statistical coherence length ξ exceeds the characteristic de Broglie wavelength of the system. For a system with a characteristic velocity v (e.g., the speed of light, the Fermi velocity or the speed of some other excitation), the de Broglie wavelength at temperature T is $\lambda_T = v\hbar/k_B T \propto \beta$. Classical statistics takes over at large enough temperatures, or close to a finite-temperature critical point, where the classical correlation length ξ exceeds λ_T. This justifies the extensive use of classical models in a realistic study of critical phenomena. The exception to this rule occurs when $T_c = 0$, which happens in a large class of low-dimensional systems.

3.2.2. Scaling

The critical exponents of Table 3.1 can be related to each other by use of the *scaling hypothesis*, which stipulates that the free energy density (or the free energy per site, in the discrete case) near the critical point is a homogeneous function of its parameters, the external field h, and the *reduced temperature* $t = T/T_c - 1$. In other words, there should be exponents a and b such that

$$f(\lambda^a t, \lambda^b h) = \lambda f(t, h) \tag{3.34}$$

This hypothesis will be justified below, but for now let us derive its consequences on critical exponents.

First, the homogeneity relation (3.34) implies that the function $t^{-1/a} f$ is invariant under the scalings $t \to \lambda^a t$ and $h \to \lambda^b h$. Therefore it must depend only on the scale-invariant variable $y = h/t^{b/a}$, and the free energy density may be expressed as

$$f(t, h) = t^{1/a} g(y) \qquad y = h/t^{b/a} \tag{3.35}$$

where g is some function. The spontaneous magnetization near criticality is then

$$M = -\left.\frac{\partial f}{\partial h}\right|_{h=0} = t^{(1-b)/a} g'(0) \tag{3.36}$$

One more derivative yields the magnetic susceptibility:

$$\chi = \left.\frac{\partial^2 f}{\partial h^2}\right|_{h=0} = t^{(1-2b)/a} g''(0) \tag{3.37}$$

Similarly, the specific heat (heat capacity per unit volume) is

$$c = -T\left.\frac{\partial^2 f}{\partial T^2}\right|_{h=0} = -\frac{1}{T_c} t^{1/a-2} g''(0) \tag{3.38}$$

Finally, in the limit $t \to 0$, the behavior of M as a function of h is $M \sim h^{1/\delta}$, which implies the asymptotic behavior $g(y) \sim y^{1/\delta}$ as $y \to \infty$, and imposes the constraint $1 - b - b/\delta = 0$, if the limit $t \to 0$ is to be finite and nonzero. We have therefore obtained a set of four constraints on some of the critical exponents introduced in Table 3.1:

$$\alpha = 2 - 1/a$$
$$\beta = (1 - b)/a$$
$$\gamma = -(1 - 2b)/a$$
$$\delta = b/(1 - b)$$

(3.39)

We now justify the scaling hypothesis, and at the same time express a and b in terms of the remaining exponents ν and η, both pertaining to the pair correlation function. Following Kadanoff, we focus our attention on the Ising model on a hypercubic lattice, with the Hamiltonian

$$H = -J \sum_{\langle ij \rangle} \sigma_i \sigma_j - h \sum_i \sigma_i$$

(3.40)

We now reduce the number of degrees of freedom of the system by grouping spins into *blocks* of side r (in units of lattice spacings), as indicated in Fig. 3.1. If d is the dimension of space there are r^d elementary spins within a block and the sum of spins therein can take values ranging from $-r^d$ to r^d. Accordingly, we define a *block spin* variable Σ_I as

$$\Sigma_I = \frac{1}{R} \sum_{i \in I} \sigma_i$$

(3.41)

where the sum is taken over the sites i within the block I and where R is some normalization factor introduced so that Σ_I can effectively take the values ± 1. For instance, R would be equal to r^d if the spins within the block were always perfectly aligned (since this is not true, R will be lower than that).

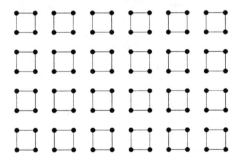

Figure 3.1. Block spins: an illustration of how four spins may be combined into a single site variable.

We will assume that the cooperative phenomena observed near the critical point can be accounted for equally well by a description in terms of block spins with a nearest-block Hamiltonian of the same form as the original Ising Hamiltonian,

$$H' = -J' \sum_{\langle IJ \rangle} \Sigma_I \Sigma_J - h' \sum_i \Sigma_I \qquad (3.42)$$

but with different parameters J' and h'. This is plausible since near criticality the correlation length ξ is much larger than the block side r. The correlation length of the blocks (the number of blocks over which the block spins are correlated) is, of course, ξ/r, which means that the effective reduced temperature t' is different from the original reduced temperature by a factor $r^{1/\nu}$:

$$t' = r^{1/\nu}t \qquad (3.43)$$

The two Hamiltonians H and H' should involve the same interaction energy with an external field, and therefore

$$h \sum_i \sigma_i = h' \sum_I \Sigma_I$$
$$= h'R^{-1} \sum_i \sigma_i \qquad (3.44)$$

which implies $h' = Rh$. Since our grouping procedure should in no way affect the total free energy of the system, the free energy per block should be r^d times the original free energy per site, and should moreover have the same functional dependence because H and H' have the same form:

$$f(t',h') = r^d f(t,h) \quad \text{or}$$
$$f(t,h) = r^{-d} f(r^{1/\nu}t, Rh) \qquad (3.45)$$

It remains to find R as a function of r in order to recover the scaling hypothesis (3.34). This is done by looking at the pair correlation function at criticality: The block-spin correlation function is then

$$\Gamma'(n) = \langle \Sigma_I \Sigma_J \rangle - \langle \Sigma_I \rangle \langle \Sigma_J \rangle$$
$$= R^{-2} \sum_{i \in I} \sum_{j \in J} \left\{ \langle \sigma_i \sigma_j \rangle - \langle \sigma_i \rangle \langle \sigma_j \rangle \right\}$$
$$= R^{-2} r^{2d} \Gamma(m)$$
$$= \frac{R^{-2} r^{2d}}{|rn|^{d-2+\eta}}$$
$$= \frac{R^{-2} r^{d+2-\eta}}{|n|^{d-2+\eta}} \qquad (3.46)$$

which implies

$$R = r^{(d+2-\eta)/2} \quad \text{so that} \quad h' = r^{(d+2-\eta)/2}h \qquad (3.47)$$

Looking back at the scaling hypothesis (3.34) and letting $r = \lambda^{1/d}$, we conclude that

$$a = 1/(\nu d) \quad \text{and} \quad b = (d+2-\eta)/(2d) \qquad (3.48)$$

The critical exponents α through δ can thus be expressed in terms of η and ν:

$$\alpha = 2 - \nu d$$

$$\beta = \frac{1}{2}\nu(d - 2 + \eta)$$

$$\gamma = \nu(2 - \eta)$$

$$\delta = (d + 2 - \eta)/(d - 2 + \eta)$$

(3.49)

We have succeeded in expressing all six critical exponents in terms of two of them (η and ν) pertaining more directly to the correlation functions. Of course, these relations can be written with a different set of "independent exponents." Table 3.2 gives the four scaling relations in their original form, with their accepted names.

Table 3.2. Summary of the scaling laws.

Rushbrooke's law	$\alpha + 2\beta + \gamma = 2$
Widom's law	$\gamma = \beta(\delta - 1)$
Fisher's law	$\gamma = \nu(2 - \eta)$
Josephson's law	$\nu d = 2 - \alpha$

3.2.3. Broken Symmetry

Phase transitions are generally associated with broken symmetries. By *broken symmetry*, we mean a symmetry of the configuration energy (or the action, in the quantum case) that is no longer reflected in the macrostate of the statistical system (or the ground state of the quantum system). For instance, the configuration energy of the two-dimensional Ising model at zero field is invariant with respect to the reversal of spins $\sigma_i \rightarrow -\sigma_i$. We say that this symmetry is broken if quantities that are not invariant under this symmetry operation have a nonvanishing expectation value. The magnetization $\langle \sigma_i \rangle$ is nonzero in the low temperature phase of the Ising model in the limit of zero external field, and the spin reversal symmetry is then broken. The simplest quantity that is not invariant under the symmetry considered and has a nonzero expectation value, such as the magnetization here, is called an *order parameter*. The phase with broken symmetry is often called the *ordered phase*. On the other hand, the high-temperature phase, in which the symmetry in unbroken, is often called the *symmetric phase*. We notice that in field theories, the analogue of temperature, after a rescaling of the fields, is some nonlinear coupling constant. Phase transitions in this case occur as a function of coupling; the interpretation is different, but the underlying physics is identical.

The spin-reversal symmetry of the Ising model has a discrete character. On the other hand, the $O(n)$ model (3.15) is endowed with a continuous symmetry: Its configuration energy is invariant under a rotation of its order parameter

n by a uniform $O(n)$ matrix. The average $\langle n \rangle$ would be nonzero in the ordered phase, except that a slow, continuous change of $\langle n \rangle$ throughout the system would cost very little energy. The consequence of this is the impossibility to break a continuous symmetry in a classical statistical system in one or two dimensions: this is the Mermin-Wagner-Coleman theorem. Simply put, long-wavelength thermal fluctuations of the order parameter take too much place in the phase space of low-dimensional systems (infrared divergence), and these fluctuations always succeed in destroying the order. The implications of this theorem to quantum statistical systems follow from the analogy between a quantum system in d spatial dimensions and a classical system in $d + 1$ dimensions, where the extra (imaginary time) dimension is limited in extent by the inverse temperature β. At any nonzero temperature, a certain class of fluctuations of the continuous order parameter occurs on a length scale greater than $v\beta$ (v is the characteristic velocity), and these long-wavelength fluctuations are thus governed by classical statistical mechanics. The Mermin-Wagner-Coleman theorem then implies that no continuous symmetry can be broken in two dimensions except at zero temperature. In a one-dimensional quantum system, such breaking is impossible even at zero temperature.

We point out that the Mermin-Wagner-Coleman theorem does not forbid all transitions implying a continuous order parameter. Such transitions are possible, provided they do not imply an expectation value of the order parameter. The best-known example is the Kosterlitz-Thouless transition in the $O(2)$ model defined on a plane (the two-dimensional *XY* model). In this model, the local order parameter is a planar, fixed-length vector n, and topological defects (vortices) play an important role. These vortices are bound in pairs below some critical temperature and are *deconfined* above that temperature. In both phases the average $\langle n \rangle$ vanishes.

§3.3. The Renormalization Group: Lattice Models

The scaling hypothesis of Sect. 3.2.2 has been motivated by the introduction of block spins with an effective Hamiltonian having the same form as the original Hamiltonian, albeit with different values of the couplings (this last step has not been demonstrated, but seems plausible; in fact it is only approximately valid). This procedure is called *block-spin renormalization* or *real-space renormalization* and defines a map between an original Hamiltonian H and a new *scaled* Hamiltonian H'. This map and its iterations form what we call the *renormalization group*, the most powerful tool at our disposal in the analysis of critical phenomena. In this section we present a survey of the basic concepts, along with a more detailed calculation within the Ising model on a triangular lattice. An exhaustive presentation of the renormalization group lies outside the scope of this review chapter and may be found in many good texts.

3.3.1. Generalities

We consider a general d-dimensional lattice model with N spins σ_i and Hamiltonian

$$H(\mathbf{J}, [s], N) = J_0 + J_1 \sum_i \sigma_i + J_2 \sum_{\langle ij \rangle}^{(1)} \sigma_i \sigma_j + J_3 \sum_{\langle ij \rangle}^{(2)} \sigma_i \sigma_j + \cdots \qquad (3.50)$$

$\mathbf{J}$ represents the collection of couplings $J_0, J_1, \cdots$ and the symbol $\sum_{\langle ij \rangle}^{(1)}$ means a summation over nearest neighbors, while $\sum_{\langle ij \rangle}^{(2)}$ means a summation over next-to-nearest neighbors, etc. Other couplings can possibly be included, with three-spin couplings and so on. We then define block spins Σ_I, along with a set of independent variables collectively denoted by ξ_I and describing the remaining degrees of freedom within each block. The Hamiltonian can in principle be rewritten in terms of these variables, and the partition function is

$$Z(\mathbf{J}, N) = \sum_{[\Sigma][\xi]} \exp -H(\mathbf{J}, [\Sigma], [\xi], N) \qquad (3.51)$$

The inverse temperature β has been absorbed in the couplings J_i. Each block is of size r in units of the lattice spacing, and the number of blocks is therefore Nr^{-d}. The block Hamiltonian $H'(\mathbf{J}', [\Sigma], Nr^{-d})$ is obtained by tracing over the internal variables ξ:

$$\exp -H'(\mathbf{J}', [\Sigma], Nr^{-d}) = \sum_{[\xi]} \exp -H(\mathbf{J}, [\Sigma], [\xi], N) \qquad (3.52)$$

We have assumed that H' has the same functional form as H, and this fixes the value of the effective coupling $\mathbf{J}'$. This assumption is only approximately valid, but the closer we are to the critical point, the better this approximation is. Its validity can also be improved with the inclusion of a more complete set of couplings in the theory. The partition function is then

$$Z(\mathbf{J}, N) = \sum_{[\Sigma]} \exp -H'(\mathbf{J}', [\Sigma], Nr^{-d})$$
$$= Z(\mathbf{J}', Nr^{-d}) \qquad (3.53)$$

The free energy per site is therefore mapped as

$$f(\mathbf{J}) = r^{-d} f(\mathbf{J}') \qquad (3.54)$$

The map $\mathbf{J} \to \mathbf{J}'$ from the original set of couplings to the set of effective block couplings generates the renormalization group.[3] We write

$$\mathbf{J}' = \mathbf{T}(\mathbf{J}) \qquad (3.55)$$

Iterations of this map generate a sequence of points in the space of couplings, which we call a renormalization group (RG) trajectory. Since the correlation length is

---

[3] In fact, some information is lost during the process of tracing over the internal variables σ. Thus the map $\mathbf{J} \to \mathbf{J}'$ is not reversible and the renormalization group is only a *semi-group*.

reduced by a factor r at each step, a typical renormalization-group trajectory tends to take the system away from criticality. Because the correlation length is infinite at the critical point, it takes an infinite number of iterations to leave that point. In general, a system is critical not only at a given point in coupling space but on a whole "hypersurface", which we call the *critical surface*, or sometimes the *critical line*. Under renormalization-group flow, a point on the critical surface stays on the critical surface. A point $\mathbf{J}_c$ on the critical surface that is stationary under renormalization-group flow is called a *fixed point* of the renormalization group:

$$\mathbf{J}_c = \mathbf{T}(\mathbf{J}_c) \tag{3.56}$$

In general, the map (3.55) is nonlinear and its exact analysis is difficult. What is most important, however, is its behavior near a fixed point, which can be obtained by *linearizing* the renormalization-group map around $\mathbf{J}_c$. This is done by defining the difference $\delta \mathbf{J} = \mathbf{J} - \mathbf{J}_c$ and expanding $\mathbf{T}$ to first order in a multivariable Taylor series. The resulting truncation is a linear map of the differences $\delta \mathbf{J}$:

$$\delta \mathbf{J}' = A\,\delta \mathbf{J} \qquad\qquad A_{ij} = \frac{\partial T_i}{\partial J_j} \tag{3.57}$$

The matrix A may be diagonalized, with eigenvalues λ_i and eigenvectors $\mathbf{u}_i$. These eigenvectors form a basis of coupling space, that is,

$$\mathbf{J} = \mathbf{J}_c + \sum_i t_i \mathbf{u}_i \tag{3.58}$$

with the t_i's playing the role of "proper couplings." In terms of these, the renormalization-group linearized action is diagonal:

$$\begin{aligned} t_i' &= \lambda_i t_i \\ &= r^{y_i} t_i \end{aligned} \tag{3.59}$$

The exponents y_i are precisely the scaling exponents[4] a and b (times d) of Eq. (3.34), since the singular part of the free energy density transforms like

$$f(t_1, t_2, \cdots) = r^{-d} f(r^{y_1} t_1, r^{y_2} t_2, \cdots) \tag{3.60}$$

Therefore all critical exponents can be obtained from the eigenvalues of the linearized renormalization-group transformation at the fixed point. To find these eigenvalues is the prime objective of renormalization-group calculations.

The character of a fixed point is determined by whether the eigenvalues λ_i are greater or smaller than 1, or equivalently whether the exponents y_i are positive or negative. A fixed point with positive and negative exponents is called *hyperbolic* because of the shape of renormalization-group trajectories near $\mathbf{J}_c$. A two-parameter example is illustrated in Fig. 3.2. The critical surface (which is a line on the figure)

---

[4] The reduced temperature may undergo a sign change, since $J_i \propto 1/T$, but this does not affect the critical exponents.

is the set of points in coupling space whose renormalization-group trajectories end up at the fixed point:

$$\lim_{n \to \infty} \mathbf{T}^n(\mathbf{J}) = \mathbf{J}_c \qquad (3.61)$$

The critical surface near $\mathbf{J}_c$ is a vector space spanned by the eigenvectors $\mathbf{u}_i$ such that $\lambda_i < 1$. Off the critical surface, the system is taken away from it by the renormalization-group flow.

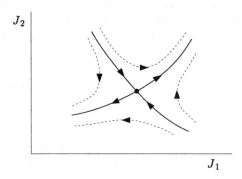

Figure 3.2. Schematic renormalization-group flow around a generic hyperbolic fixed point.

A parameter t_i associated with a positive scaling exponent ($\lambda_i > 1$) is called *relevant*, since it grows under renormalization-group flow (i.e., when the system is scaled away from criticality). If, on the contrary, $y_i < 0$ ($\lambda_i < 1$), t_i is said to be *irrelevant*, whereas if $y_i = 0$ ($\lambda_i = 1$) it is *marginal*. Marginal operators do not scale with a power law behavior near a critical point, but rather logarithmically; the linear approximation around the fixed point $\mathbf{J}_c$ is then invalid.

The existence of critical surfaces and fixed points is thought to explain the *universality* of critical exponents (i.e., that many different systems are characterized by the same critical exponents). In other words, statistical systems seem to fit into *universality classes* whose members share the same critical behavior. This can be understood if different systems live on submanifolds of one large coupling space, and if these submanifolds intersect the same critical surface. At criticality, all of these systems will be (presumably) driven toward the same fixed point, with the same scaling exponents.

3.3.2. The Ising Model on a Triangular Lattice

In order to illustrate some of the previous statements we will perform an explicit real-space renormalization-group calculation for the Ising model living on a triangular lattice.

The block structure is indicated on Fig. 3.3. The Ising Hamiltonian is written as

$$H(k,h) = -k \overset{(1)}{\sum_{\langle ij \rangle}} \sigma_i \sigma_j - h \sum_i \sigma_i \qquad (3.62)$$

Each lattice site has 6 nearest neighbors. A block I is made of three spins, which we call σ_1^I, σ_2^I and σ_3^I . We define the block spin Σ_I as

$$\Sigma_I = \text{sgn}\,(\sigma_1^I + \sigma_2^I + \sigma_3^I) \qquad (3.63)$$

In other words, Σ_I adopts the sign of the majority. The three spins within a block lead to $2^3 = 8$ different states, which makes four different states for the internal variable ξ_I and two for the block spin Σ_I. The four states are chosen to be

$$\xi_I : \quad (+,+,-)\,,\ (+,-,+)\,,\ (-,+,+)\,,\ (+,+,+) \qquad (3.64)$$

and the actual state of the spins σ_i is obtained by multiplying by $\Sigma_I = \pm 1$.

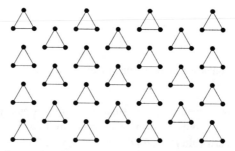

Figure 3.3. Block spins on the triangular lattice.

We decompose the Hamiltonian into the sum of a "free" part H_0 containing only the interaction within blocks, plus an "interaction" part V containing the interaction between blocks and with the external field:

$$H_0 = -k \sum_I \overset{(ij)}{\underset{(i,j \in I)}{\sum}} \sigma_i \sigma_j$$

$$V = -k \sum_{\langle IJ \rangle} \overset{(ij)}{\underset{i \in I, j \in J}{\sum}} \sigma_i \sigma_j - h \sum_I \sum_{i \in I} \sigma_i \qquad (3.65)$$

We also define the following expectation values in which only the variables internal to a block are summed:

$$\langle F[S] \rangle = Z_f^{-1} \sum_{[\xi]} F[\Sigma, \xi] \exp -H_0([\Sigma],[\xi]) \qquad (3.66)$$

$$Z_f = \sum_{[\xi]} \exp -H_0([\Sigma],[\xi]) \qquad (3.67)$$

According to (3.53), the block Hamiltonian $H(k', h')$ is defined by

$$\exp -H(k', h') = Z_f \langle e^V \rangle \tag{3.68}$$

The "free" partition function Z_f is easily calculated, since different blocks do not interact within H_0:

$$Z_f = Z_0^{N/3}$$

where Z_0 is the sum over states within a given block:

$$Z_0 = \sum_{\sigma_I} \exp \left\{ k(\Sigma_1^I \Sigma_2^I + \Sigma_2^I \Sigma_3^I + \Sigma_3^I \Sigma_1^I) \right\} \tag{3.69}$$

$$= 3e^{-k} + e^{3k}$$

This last step follows from Eq. (3.64), wherein three states have energy k and one state has energy $-3k$.

The expectation value $\langle e^V \rangle$ can be expressed as a *cumulant expansion*:

$$\langle e^V \rangle = \exp \left\{ \langle V \rangle + \frac{1}{2}(\langle V^2 \rangle - \langle V \rangle^2) + \cdots \right\} \tag{3.70}$$

At this point we will make the approximation of keeping only the first term of this expansion. This amounts to neglecting the fluctuations of the interaction term within each block. The expectation value $\langle V \rangle$ is relatively easy to calculate. We start with the block-block interaction V_{IJ}. There are two elementary links between a pair of nearest-neighbor blocks and, as shown in Fig. 3.4, the interaction V_{IJ} is

$$V_{IJ} = -k\Sigma_3^I(\Sigma_1^I + \Sigma_2^I) \tag{3.71}$$

Since the expectation value within different blocks factorizes, we have

$$\langle V_{IJ} \rangle = -2k\langle \Sigma_3^J \rangle \langle \Sigma_3^I \rangle \tag{3.72}$$

where $\langle \Sigma_i^I \rangle$ is the same for all $i = 1, 2, 3$. The expectation value $\langle \Sigma_3^I \rangle$ is readily calculated:

$$\langle \Sigma_3^I \rangle = Z_0^{-1} \sum_{\xi_I} \Sigma_3^I \exp -k(\Sigma_1^I \Sigma_2^I + \Sigma_2^I \Sigma_3^I + \Sigma_3^I \Sigma_1^I) \tag{3.73}$$

$$= Z_0^{-1}(e^{3k} + e^{-k})\Sigma_I$$

where we have used the definition (3.63) for the block spin Σ_I. Consequently, the mean interaction term between blocks is

$$\langle V_{IJ} \rangle = -2k \left(\frac{e^{3k} + e^{-k}}{e^{3k} + 3e^{-k}} \right)^2 \Sigma_I \Sigma_J \tag{3.74}$$

Since the average interaction with the external field involves only the expectation value $\langle \Sigma_3^I \rangle$, we find

$$\langle V \rangle = -2k \left(\frac{e^{3k} + e^{-k}}{e^{3k} + 3e^{-k}} \right)^2 \sum_{\langle IJ \rangle} \Sigma_I \Sigma_J - 3 \left(\frac{e^{3k} + e^{-k}}{e^{3k} + 3e^{-k}} \right) h \sum_I \Sigma_I \tag{3.75}$$

To first order in the cumulant expansion, the block-spin Hamiltonian is therefore

$$H(k',h') = 3 \ln Z_0 + \langle V \rangle \tag{3.76}$$

The first term is independent of Σ_I and may be ignored (except if one is interested in the value of the free energy F). We therefore end up with the following map between the block-spin couplings and the original ones:

$$k' = 2k \left(\frac{e^{3k} + e^{-k}}{e^{3k} + 3e^{-k}} \right)^2$$

$$h' = 3h \left(\frac{e^{3k} + e^{-k}}{e^{3k} + 3e^{-k}} \right) \tag{3.77}$$

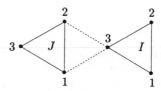

Figure 3.4. Interaction between block spins in the nearest-neighbor Ising model on the triangular lattice.

The renormalization-group (RG) flow associated with the above map is illustrated schematically in Fig. 3.5. There are 9 fixed points on this diagram, corresponding to the possible combinations of $h = 0, -\infty, \infty$ and $k = 0, k_c, \infty$, where k_c is determined by the equation

$$\frac{1}{2} = \left(\frac{e^{3k_c} + e^{-k_c}}{e^{3k_c} + 3e^{-k_c}} \right) \quad \Rightarrow \quad k_c = \frac{1}{4} \ln(1 + 2\sqrt{2}) \approx 0.336 \tag{3.78}$$

The fixed point $(k,h) = (k_c, 0)$ is unstable in both directions and corresponds to a continuous phase transition. Near this point, the RG flow admits the following linearization:

$$\begin{pmatrix} \delta k' \\ \delta h' \end{pmatrix} = \begin{pmatrix} 1.62 & 0 \\ 0 & 2.12 \end{pmatrix} \begin{pmatrix} \delta k \\ \delta h \end{pmatrix} \tag{3.79}$$

with the eigenvalues $\lambda_k = 1.62$ and $\lambda_h = 2.12$. Since the scale factor for the triangular matrix is $r = \sqrt{3}$, the free energy density scales as

$$f(k,h) = r^{-d} f(r^{0.88} k, r^{1.37} h) \tag{3.80}$$

The critical exponents can be calculated from (3.39) and from the scaling laws of Table 3.2. We list them here, together with the exponents obtained in the exact solution of the same model:

	α	β	γ	δ	ν	η
RG:	-0.27	0.72	0.84	2.17	1.13	1.26
exact:	0	$\frac{1}{8}$	$\frac{7}{4}$	15	1	$\frac{1}{4}$

Notice that the simplest RG calculation described here is not very successful at predicting the exponent η. The difference between its predictions and the exact exponents is attributed to the approximation made in neglecting higher-order terms in the cumulant expansion. If these terms were considered, more couplings would have to be included in order for the effective block Hamiltonian to have the same form as the original Hamiltonian, but a better agreement with the exact result would be found.

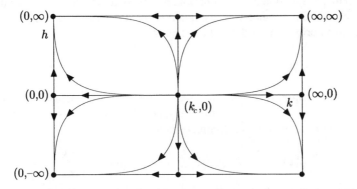

Figure 3.5. Schematic renormalization-group flow for the Ising model on a triangular lattice. The k and h axes have been contracted to display the points at infinity. The completely unstable fixed point $(k_c, 0)$ corresponds to the continuous phase transition, whereas the other fixed points are associated with phases (with or without an external field).

The other fixed points in Fig. 3.5 do not have the interpretation of phase transitions governed by temperature. Recall that the physical inverse temperature $\beta = 1/T$ is included in the definitions of the coupling k and of the field h. The "physical" field is rather $\tilde{h} = Th$. Thus, the fixed point $(k, h) = (0, 0)$ corresponds to infinite temperature and small field $\tilde{h}$ and describes a disordered phase. This point is unstable when an "infinite" field $\tilde{h}$ is turned on and a nonzero magnetization then appears, in one direction or the other. These ordered states are described by the points $(0, \pm\infty)$. At the other extreme, the fixed point $(k, h) = (\infty, 0)$ corresponds to zero temperature and describes an ordered phase in the absence of a field. It is unstable against an infinitesimal field $\tilde{h}$, which drives the system into a state of nonzero magnetization, described by the points $(\infty, \pm\infty)$. In general, stable fixed points describe stable phases of the system. This interpretation is natural since the correlation length decreases along the RG trajectory and the statistical mechanics of the system becomes simpler, since more and more degrees of freedom have been eliminated. The unstable fixed points located between the basins of attraction of stable fixed points are, on the contrary, associated with phase transitions governed by temperature (e.g., $(k, h) = (k_c, 0)$) or by other parameters (e.g., $(k, h) = (0, 0)$).

§3.4. The Renormalization Group: Continuum Models

Block-spin—or *real-space*—renormalization is an intuitive procedure designed for lattice models. If we want to apply renormalization ideas to continuum models, be it in the context of statistical mechanics or that of quantum field theory, a different procedure is needed, namely *momentum-space renormalization*. In what follows, the term *action* functional is used instead of *energy* functional, as it should be in statistical mechanics, since we have quantum field theory in mind and will refer to scale transformations as defined in Chap. 2.

3.4.1. Introduction

For the sake of introduction, we consider a statistical model defined in terms of a single scalar field $\varphi(x)$ in d-dimensional space (boldface letters denote vectors). The field $\varphi(x)$ may be Fourier decomposed as follows:

$$\varphi(x) = \int (dk)\, \tilde{\varphi}(k)\, e^{ik \cdot x} \qquad (dk) \equiv \frac{d^d k}{(2\pi)^d} \qquad (3.81)$$

The action functional $S[\varphi]$ may be expressed in terms of the Fourier components $\tilde{\varphi}(k)$. For instance, the action for the φ^4 theory in Eq. (3.19) becomes

$$S[\varphi; r, u] = \int (dk)\, \frac{1}{2}\tilde{\varphi}(-k)\tilde{\varphi}(k)(k^2 + r)$$

$$+ \frac{1}{4}u \int (dk_1)(dk_2)(dk_3)\, \tilde{\varphi}(-k_1 - k_2 - k_3)\tilde{\varphi}(k_1)\tilde{\varphi}(k_2)\tilde{\varphi}(k_3)$$

$$(3.82)$$

In general, we write the action as $S[\varphi; u_i]$, where u_i stands for the collection of parameters multiplying the various terms of the Lagrangian density.

Naturally, the continuum theory is defined only through some regularization procedure, which we take here as a cutoff Λ, meaning that the integration is restricted to the region of momentum space such that all arguments k of $\tilde{\varphi}(k)$ lie within the cutoff: $|k| < \Lambda$. The Fourier decomposition (3.81) amounts to a unitary transformation of the degrees of freedom, as could easily be seen in a discrete version of the Fourier transform for a finite lattice of points. Therefore, the functional integration measure may be formally written as

$$[d\varphi]_\Lambda = \prod_x d\varphi(x) = \prod_{|k| < \Lambda} d\tilde{\varphi}(k) \qquad (3.83)$$

since no Jacobian arises from the change of integration variables $\varphi(x) \to \tilde{\varphi}(k)$.

The first step of the renormalization procedure[5] consists in integrating out the Fourier components $\tilde{\varphi}(k)$ such that $\Lambda/s < |k| < \Lambda$ (the so-called *fast modes*), where s is some dilation factor ($s > 1$). The number of degrees of freedom is then effectively reduced, with a new cutoff equal to Λ/s. The remaining degrees of freedom (the *slow modes*) are governed by a modified action $S'[\varphi; u_i]$:

$$\exp{-S'[\varphi; u_i]} = \int \prod_{\Lambda/s < |k| < \Lambda} d\tilde{\varphi}(k) \, \exp{-S[\varphi; u_i]} \qquad (3.84)$$

As long as we are interested in correlation functions of slow modes only, the effective action S' is entirely equivalent to the original action S which includes fast modes.

The second step of the renormalization procedure is a scale transformation on the slow-mode action, as defined in Eq. (2.121):

$$k \rightarrow k' = sk \quad \text{or} \quad x \rightarrow x' = x/s \qquad (3.85)$$

Here the scaling factor λ is $1/s$. In general such a transformation also affects the field:

$$\varphi(x) \rightarrow \varphi'(x/s) = s^{\Delta}\varphi(x) \quad \text{or} \quad \tilde{\varphi}'(sk) = s^{\Delta-d}\tilde{\varphi}(k) \qquad (3.86)$$

The exponent Δ is the *scaling dimension* of the field φ and is related to the exponent η: $\Delta = \eta/2$. Such a transformation of the field affects the functional integration measure only through a multiplication factor. After this rescaling, the modified action S' can be rightfully compared with the initial action S, because they now have the same cutoff Λ, that is, the same set of degrees of freedom (this was not true before rescaling). As said above, the two actions S and S' are equivalent as far as the slow modes are concerned: they describe the same long-distance properties. However, the parameters u_i defining these two action functionals are different in general: $S'[\varphi] = S[\varphi; u_i']$.[6] We thus generate a curve $u_i(s)$ in parameter space (s is the rescaling factor), and each point on this curve defines an action functional with the same long-distance properties. The outcome of the renormalization procedure can be expressed in a set of coupled flow equations in parameter space:

$$\frac{du_i}{d\ln s} = \beta_i(u_j) \qquad (3.87)$$

where β_i is commonly referred to as the *beta function* associated with the parameter u_i. Like before, a *fixed point* u_i^* of the renormalization group is a point in parameter space that is unaffected by the renormalization procedure. In other words, it is characterized by a vanishing beta function:

$$\beta_i(u_j^*) = 0 \qquad (3.88)$$

---

[5] This procedure is known as the *Wilson-Kadanoff renormalization scheme*.

[6] This equation supposes that the scaling dimension Δ has been chosen appropriately; otherwise, the two actions are not equal, but differ by a multiplicative constant. Also, the number of parameters needed for the new action to be of the same form as the old action is in principle infinite. In practice, however, one keeps only a finite subset of parameters: relevant and marginal ones. Irrelevant parameters (in the RG sense) rapidly decrease under RG flow.

To summarize, a renormalization-group transformation amounts to a scale transformation applied both to the action and to the integration measure (i.e., the Fourier modes that would be scaled beyond the cutoff Λ are integrated out). A fixed point of the renormalization-group transformation thus defines a theory that has scale invariance at the quantum level.

THE GAUSSIAN MODEL

The simplest example of a continuum model for which the renormalization procedure can be carried out exactly is the free boson, or *Gaussian model*, obtained from Eq. (3.82) by setting $u = 0$:

$$S[\varphi; r] = \int_\Lambda (dk) \frac{1}{2} \tilde{\varphi}^*(k)\tilde{\varphi}(k)(k^2 + r) \tag{3.89}$$

In this model the fast and slow modes are decoupled, since different values of the wavevector do not mix in the action. Therefore, integrating the fast modes produces only an irrelevant multiplicative constant in front of the partition function. The effective slow-mode action is then

$$\begin{aligned}
S'[\varphi] &= \int_{\Lambda/s} (dk) \frac{1}{2} \tilde{\varphi}(-k)\tilde{\varphi}(k)(k^2 + r) \\
&= s^{-d} \int_\Lambda (dk') \frac{1}{2} \tilde{\varphi}(-k'/s)\tilde{\varphi}(k'/s)(k'^2/s^2 + r) \\
&= s^{d-2\Delta-2} \int_\Lambda (dk') \frac{1}{2} \tilde{\varphi}'(-k')\tilde{\varphi}'(k')(k'^2 + s^2 r)
\end{aligned} \tag{3.90}$$

We immediately see that S', in terms of φ', has the same form as $S[\varphi]$, provided $r = 0$ and $\Delta = \frac{1}{2}d - 1$. This we knew already from Eq. (2.124). In this particular case, the scale transformation on the path-integral measure brings nothing new and the scaling properties all follow from the action alone. Thus, the massless ($r = 0$) Gaussian model is a fixed point of the renormalization group—in fact, the simplest of all fixed points from the present point of view.

3.4.2. Dimensional Analysis

We consider a field ϕ (not necessarily a scalar field) governed by an action functional $S[\phi]$ and let us assume that there exists a fixed-point action $S_0[\phi]$ (not necessarily Gaussian) at some point in parameter space, which we take, for convenience, as the origin. In the vicinity of this fixed point, the generic action $S[\phi]$ may be expressed as

$$S[\phi] = S_0[\phi] + \sum_i u_i \int dx\, O_i(x) \tag{3.91}$$

where the $O_i(x)$'s are some local operators, expressible in terms of the field ϕ. The couplings u_i must be small if we are close to the fixed point. Under a renormalization-group (scale) transformation, the field ϕ transforms like

$\phi'(x) = s^\Delta \phi(sx)$ and only $S_0[\phi]$ is invariant. The other terms are modified through their couplings:

$$S'[\phi] = S_0[\phi] + \sum_i u_i'(s) \int dx \, O_i(x) \tag{3.92}$$

In principle, the series on the r.h.s. may be infinite, and the transformed couplings u_i' may depend on s in a complicated way, because of the functional integration of the fast modes. We assume, however, that the couplings u_i are so small that they have a negligible effect on the fast mode integration. In this approximation, the new couplings u_i' may be obtained simply from the behavior of the operators O_i under a scale transformation, which follows from the expression of O_i in terms of ϕ:

$$O_i'(x) = s^{\Delta_i} O_i(sx)$$
$$u_i \int dx \, O_i'(x) = u_i s^{\Delta_i - d} \int dx \, O_i(x) \tag{3.93}$$

Therefore

$$u_i' = u_i s^{d - \Delta_i} \tag{3.94}$$

In other words, in this zeroth-order approximation, the dimensions of couplings are obtained from the scaling dimension Δ of ϕ by applying *dimensional analysis*.

Adopting the terminology of the previous section, a coupling is said to be *relevant* if $\Delta_i < d$: It will grow as the fast modes are integrated. An *irrelevant coupling* is such that $\Delta_i > d$, and will shrink as the fast modes are integrated. At last, a *marginal* coupling will stay the same, or rather vary logarithmically near the fixed point.[7]

For instance, we now look at some operators within the Gaussian model. The first operator that comes to mind is the mass term $O_2 = \frac{1}{2}\varphi^2$, with coupling r. With Gaussian scaling (i.e., $\Delta = \frac{1}{2}d - 1$) we find that $\Delta_2 - d = -2$, and hence $r' = s^2 r$. This, of course, was already known from Eq. (3.90). Thus, the mass term is relevant at the Gaussian fixed point, in all dimensions. This is a trivial statement since we know from Chap. 2 that the mass is the inverse correlation length ($\mu \sim \xi^{-1}$) and that ξ decreases under scaling ($\xi' = \xi/s$). The quartic coupling of the φ^4 theory is associated with the operator $O_4 = \varphi^4$, with $\Delta_4 - d = d - 4$. Thus the quartic coupling u is relevant in dimensions smaller than four, irrelevant in $d > 4$, and marginal in $d = 4$ (still at zeroth order). At this order, it looks as if any (positive) value of u yields a fixed point in $d = 4$.

---

[7] It is important to keep in mind that the scaling dimensions of operators, or the relevance or irrelevance of couplings, depends not only on the form of these operators in terms of ϕ, but also on the fixed point considered.

3.4.3. Beyond Dimensional Analysis: The φ^4 Theory

To go beyond dimensional analysis, we generally use perturbation theory: We expands the exponential $\exp -S$ in powers of the perturbing coupling. The problem is then reduced to the calculation of Gaussian correlators, which can be done using Wick's theorem. Since we will make little use of perturbation theory in this work, this method is not reviewed in these introductory chapters; again we refer the reader to the standard texts on quantum field theory. Here we simply cite known results.

To first order in u and r, perturbation theory leads to the following renormalization-group transformation of the couplings:

$$r' = s^2(r + ub(1 - s^{2-d}))$$
$$u' = s^{4-d}u \tag{3.95}$$

with

$$b = K_d \Lambda^{d-2}/(2d - 4) \quad , \quad K_d^{-1} = (4\pi)^{d/2}\Gamma(d/2)/2 \tag{3.96}$$

In matrix form this becomes

$$\begin{pmatrix} r \\ u \end{pmatrix}' = \begin{pmatrix} s^2 & b(s^2 - s^{4-d}) \\ 0 & s^{4-d} \end{pmatrix} \begin{pmatrix} r \\ u \end{pmatrix} \tag{3.97}$$

We recall that the proper couplings t_i of Eq. (3.58) are obtained by diagonalizing this matrix. The eigenvalues and eigenvectors are

$$\begin{aligned} \lambda_1 &= s^2 & \mathbf{u}_1 &= (1,0) \\ \lambda_2 &= s^{4-d} & \mathbf{u}_2 &= (-b, 1) \end{aligned} \tag{3.98}$$

Since by definition $(r, u) = t_1 \mathbf{u}_1 + t_2 \mathbf{u}_2$, we have the proper couplings

$$t_1 = r + bu \qquad t_2 = u \tag{3.99}$$

At this order, there is a critical line in $d > 4$ specified by the equation $t_1 = 0$, or $r = -bu$. In $d = 4$, it still looks as if any value of $t_2 = u$ constitutes a fixed point.

However, this picture breaks down once we take into account higher orders of u in the perturbation expansion. At second order, we find that

$$r' = s^2\left[r + \frac{u}{16\pi^2}\left(\frac{1}{2}\Lambda^2(1 - s^{-2}) - r\ln s\right)\right]$$
$$u' = s^{4-d}\left[u - \frac{3u^2}{16\pi^2}\ln s\right] \tag{3.100}$$

The quartic coupling then receives logarithmic corrections in $d = 4$. This RG mapping is better expressed by the corresponding beta functions:

$$\frac{dr}{d\ln s} = 2r - \frac{ur}{16\pi^2} + \frac{u\Lambda^2}{16\pi^2}$$
$$\frac{du}{d\ln s} = (4-d)u - \frac{3}{16\pi^2}u^2 \tag{3.101}$$

This shows the emergence of a new (non-Gaussian) fixed point at $r, u \neq 0$, whose location is readily found from the above beta functions:

$$u^* = \frac{16\pi^2}{3}(4-d) \qquad r^* = \frac{d-4}{6}\Lambda^2 \qquad (3.102)$$

It is a straightforward exercise to linearize the flow (3.101) around this new fixed point and to find the critical exponents. For reasons that will not be explained here, the critical exponents of the φ^4 theory (and of other Gaussian-like models) are calculated in the form of a series in powers of $\varepsilon \equiv 4 - d$ (the so-called ε-expansion). Each additional order in perturbation theory leads to the correct evaluation of a new term of this expansion. To order ε^2, the exponents of the φ^4 theory are calculated to be

$$\nu = \frac{1}{2} + \frac{1}{12}\varepsilon + O(\varepsilon^2) \qquad \eta = 0 + O(\varepsilon^2) \qquad (3.103)$$

The φ^4 model in $d = 4$ clearly illustrates that scale invariance of the action (here on the line $r = 0$) does not guarantee scale invariance at the quantum level (i.e., a renormalization-group fixed point). This breakdown of dimensional analysis is due to interactions.

§3.5. The Transfer Matrix

A powerful way to solve the Ising model and other related statistical models is the *transfer matrix* method, which is the analogue in statistical mechanics of the operator formalism in quantum field theory. In this section we will describe this formalism and indicate how it can lead to an analogy between quantum field theories and statistical systems near criticality.

Again, we turn to the Ising model on a square lattice with m rows and n columns. A spin is here indexed by two integers[8] for the row number and column number, respectively, and we will impose periodic boundary conditions

$$\sigma_{i,j+n} = \sigma_{ij} \qquad \sigma_{i+m,j} = \sigma_{ij} \qquad (3.104)$$

thereby defining the lattice on a torus. Let us denote by μ_i the configuration of spins on the i-th row:

$$\mu_i = \{\sigma_{i1}, \sigma_{i2}, \cdots, \sigma_{in}\} \qquad (3.105)$$

There are 2^n such configurations. The row configuration μ_i has an energy of its own:

$$E[\mu_i] = \sum_{k=1}^{n} \sigma_{ik}\sigma_{i,k+1} \qquad (3.106)$$

---

[8] The two indices will be separated by a comma only when necessary to avoid confusion.

as well as an interaction energy with the neighboring rows:

$$E[\mu_i, \mu_j] = \sum_{k=1}^{n} \sigma_{ik}\sigma_{jk} \tag{3.107}$$

We next define a formal vector space V of row configurations spanned by the $|\mu_i\rangle$, for which we introduce a "bra-ket" notation in analogy with quantum mechanics. On this space, we define the action of the *transfer matrix* T by its matrix elements:

$$\langle\mu|T|\mu'\rangle = \exp -\beta\left(E[\mu,\mu'] + \frac{1}{2}E[\mu] + \frac{1}{2}E[\mu']\right) \tag{3.108}$$

In terms of the operator T, the partition function has the following simple form:

$$Z = \sum_{\mu_1,\cdots,\mu_m} \langle\mu_1|T|\mu_2\rangle\langle\mu_2|T|\mu_3\rangle \cdots \langle\mu_m|T|\mu_1\rangle \tag{3.109}$$

$$= \operatorname{Tr} T^m$$

The transfer matrix defined in (3.108) is manifestly symmetric, and therefore diagonalizable. The partition function may be expressed in terms of the 2^n eigenvalues Λ_k of T:

$$Z = \sum_{k=0}^{2^n-1} \Lambda_k^m \tag{3.110}$$

The thermodynamic limit is obtained when $m, n \rightarrow \infty$. In this limit, the free energy can be extracted by keeping only the largest eigenvalue of T, assuming, for the sake of argument, that it is nondegenerate. Indeed, the free energy per site f is given by

$$-f/T = \lim_{m,n\rightarrow\infty} \frac{1}{mn} \ln\left(\Lambda_0^m + \Lambda_1^m + \cdots\right)$$

$$= \lim_{m,n\rightarrow\infty} \frac{1}{mn} \left\{m \ln \Lambda_0 + \ln\left(1 + (\Lambda_1/\Lambda_0)^m + \cdots\right)\right\} \tag{3.111}$$

$$= \lim_{n\rightarrow\infty} \frac{\ln \Lambda_0}{n}$$

since $\Lambda_1/\Lambda_0 < 1$. The calculation of more complicated thermodynamic quantities requires the knowledge of more eigenvalues.

In order to express correlation functions in terms of the transfer matrix, we introduce a spin operator $\hat{\sigma}_i$ acting on V and giving the value of the spin on the i-th column when acting on basis vector $|\mu\rangle$:

$$\hat{\sigma}_i|\mu\rangle = \sigma_i|\mu\rangle \tag{3.112}$$

Then

$$\langle \sigma_{ij}\sigma_{i+r,k} \rangle = \frac{1}{Z} \sum_{\mu_1, \cdots, \mu_m} \langle \mu_1|T|\mu_2 \rangle \cdots \langle \mu_i|\hat{\sigma}_j T|\mu_{i+1} \rangle \cdots$$

$$\cdots \langle \mu_{i+r}|\hat{\sigma}_k T|\mu_{i+r+1} \rangle \cdots \langle \mu_m|T|\mu_1 \rangle \tag{3.113}$$

$$= \frac{\text{Tr} \left(T^{m-r} \, \hat{\sigma}_j \, T^r \, \hat{\sigma}_k \right)}{\text{Tr} \, T^m}$$

This should be reminiscent of the passage from the operator formalism to the path integral formalism in Euclidian quantum field theory. The transfer matrix here plays the role of the evolution operator $U(a)$ over a "distance of time" equal to the lattice spacing a. In other words, one can define a Hamiltonian operator $\hat{H}$ as

$$T = \exp -a\hat{H} \tag{3.114}$$

The eigenstates of T are the analogue of the energy eigenstates of quantum mechanics, the eigenvalues E_r of $\hat{H}$ (the energy levels) being expressed as

$$E_r = -\frac{1}{a} \ln \Lambda_r \tag{3.115}$$

in terms of the eigenvalues of T. Therefore, the free energy density f/a^2 is proportional to the vacuum energy per site, or the vacuum energy density in field theoretic language:

$$f/a^2 = \lim_{n \to \infty} \frac{E_0}{na} \tag{3.116}$$

The magnetization $\langle \sigma_{ij} \rangle$ in the thermodynamic limit is

$$\langle \sigma_{11} \rangle = \lim_{m \to \infty} (\text{Tr} \, T^m)^{-1} \text{Tr} \, (\hat{\sigma} \, T^m)$$

$$= \lim_{m \to \infty} e^{-ma(E_l - E_0)} \sum_l \langle 0|\hat{\sigma}_1|l \rangle \tag{3.117}$$

$$= \langle 0|\hat{\sigma}_1|0 \rangle$$

where we have inserted a complete set of T eigenstates, which reduces to $|0\rangle\langle 0|$ in the limit $m \to \infty$ because of the exponential factor. The statistical average of the spin is therefore given by the "vacuum expectation value" of the corresponding operator S. This applies to any local quantity and its operator.

Likewise, the pair correlation function can be expressed in the thermodynamic limit:

$$\langle s_{11}s_{1+r,1} \rangle = \lim_{m \to \infty} (\text{Tr} \, T^m)^{-1} \text{Tr} \, (T^{m-r} \, S_1 \, T^r \, S_1)$$

$$= \lim_{m \to \infty} e^{maE_0} \sum_l \langle 0|e^{(m-r)aE_0}S_1|l \rangle \langle l|e^{-raE_l}S_1|0 \rangle \tag{3.118}$$

$$= \langle s_{11} \rangle^2 + |\langle 0|S_1|1 \rangle|^2 \exp -ra(E_1 - E_0) + \cdots$$

The connected correlation function in the long distance limit ($r \gg 1$) is therefore

$$\langle s_{11} s_{1+r,1} \rangle \sim |\langle 0|S_1|1\rangle|^2 \exp -ra(E_1 - E_0) \tag{3.119}$$

The energy gap $E_1 - E_0$ is the mass m of the field quantum: It is the energy of a particle at rest. The relation between the correlation length and the mass of the associated Euclidian quantum field theory is therefore

$$\xi = \frac{1}{ma} \tag{3.120}$$

Near a critical point the correlation length grows without bounds and correspondingly the mass goes to zero (for fixed a). In other words, the largest eigenvalues of the transfer matrix coalesce at the critical point.

To summarize, we have shown how a lattice model can be described in an operator formalism, which makes clear the very close analogy with Euclidian quantum field theories. The free energy density is then the vacuum energy density, the pair correlation function is the field's propagator, and the correlation length is proportional to the inverse mass. A system at the critical point is therefore equivalent to a massless field theory, provided the lattice spacing a is not exactly zero.

Exercises

3.1 The binomial distribution
Consider a set of N particles moving almost freely in a box of volume V, with occasional collisions among themselves. The probability that a given particle be within the left half of the box at any moment is $\frac{1}{2}$. If we neglect the volume of the particles, i.e., if the density of the gas is not too large, then the fact that a particle is in the left half of the box is independent of the situation of other particles, and the number n of particles in the left half obeys a binomial probability distribution:

$$P(n) = \frac{N!}{n!(N-n)!} 2^{-N}$$

a) Compute the expectation value of the binomial distribution, namely the quantity $\langle n \rangle = \sum_{n=0}^{N} nP(n)$, which represents the average number of particles in the left half of the box.

b) Compute the standard deviation $\Delta n = \left\langle (n - \langle n \rangle)^2 \right\rangle^{1/2}$.

c) By expanding the probability $P(n)$ around the mean value $\langle n \rangle$, find the thermodynamic limit of the distribution $P(n)$.
Result: Writing $n = \frac{N}{2} + \varepsilon$, and using Stirling's formula

$$\ln x! = (x + \frac{1}{2})\ln x - x + \frac{1}{2}\ln 2\pi + O(1/x)$$

for large x, we find that

$$P(\frac{N}{2} + \varepsilon) \sim \sqrt{\frac{2}{\pi N}} e^{-2(2N-1)\varepsilon^2/N^2}$$

Hence, in terms of the scaling variable $x = 2\varepsilon/\sqrt{N}$, the thermodynamic distribution becomes the Gaussian distribution

$$P(x) = \frac{1}{\sqrt{\pi}} e^{-x^2}$$

3.2 *The one-dimensional Ising model*
We consider the one-dimensional Ising model, with energy (3.6). We introduce the scaled variables $K = -J/k_BT$ and $H = h/k_BT$.

a) Show that the partition function on a chain of N sites $i = 1, .., N$, with periodic boundary conditions $N + 1 \equiv 1$, can be expressed as the trace

$$Z_N(K,H) = \sum_{\substack{s_i=\pm 1 \\ s_{N+1}=s_1}} \exp\left\{K\sum_{(ij)} s_is_j + H\sum_i s_i\right\}$$

$$= \mathrm{Tr}\Big(T(K,H)^N\Big)$$

where $T(K,H)$ is the 2×2 transfer matrix of the model. Show that $T(K,H)$ is

$$T(\beta,H) = \begin{pmatrix} e^{K+H} & e^{-K} \\ e^{-K} & e^{K-H} \end{pmatrix}$$

in the basis $(+1,-1)$ for s.

b) Compute the thermodynamic free energy

$$f(K,H) = \lim_{N\to\infty} -(1/N)\ln Z_N(K,H)$$

Hint: $(Z_N)^{1/N}$ is dominated by the largest eigenvalue of the transfer matrix T, namely

$$\lambda_{\max} = e^K \cosh(H) + \sqrt{e^{-2K} + e^{2K}\sinh(H)}$$

c) Compute the magnetization $M = -\partial f/\partial K$. Show in particular that the magnetization is linear for h small ($M \sim he^{2K}$). Deduce that the magnetic susceptibility diverges at zero temperature. Show that there is no phase transition at finite temperature for the one-dimensional Ising model.

d) Compute the spin-spin correlation in the thermodynamic limit.

3.3 *Free energy of the one-dimensional Potts model*
In the q-state Potts model, the spin variable s_i takes q possible values, in the set $\{0, 1, ..., q - 1\}$. The energy of a configuration reads

$$E(s_1, ..., s_N) = -J\sum_{(ij)} \delta_{s_i,s_j}$$

and we use the scaled variable $K = J/k_BT$.

a) Write the transfer matrix T of the one-dimensional model with periodic boundary conditions in terms of the $q \times q$ matrix J, with all entries equal to 1.
Result: $T = (e^K - 1)\mathbb{I} + J$.

b) Compute the thermodynamic free energy of the one-dimensional q-state Potts model.
Hint: Note that $J^2 = qJ$, and use this fact to compute $\mathrm{Tr}(T^N)$.

3.4 *Transfer matrix for the two-dimensional Ising model*
The two-dimensional Ising model with spins s_{ij} sitting at the vertices (i, j) of a square lattice of size $N \times L$ in zero magnetic field has the energy

$$E[s] = -J \sum_{\langle (i,j)(k,l) \rangle} s_{ij} s_{kl}$$

where the sum extends over all the bonds of the lattice. We use the scaled variable $K = J/k_B T$.

Write the row-to-row transfer matrix for this model, namely the $2^L \times 2^L$ matrix $T_L(K)$, such that the partition function $Z_{N,L}$ with periodic boundary conditions reads

$$Z_{N,L}(K) = \mathrm{Tr}\left(T_L(K)^N\right)$$

3.5 *Numerical diagonalization of transfer matrices*

a) Given a symmetric indecomposable $r \times r$ matrix T, show that it has a unique maximal eigenvalue $\lambda_{\max}$. Let $v_{\max}$ denote the corresponding (normalized) eigenvector.

b) We define the sequence of vectors $v_0, v_1, v_2, \cdots$ where v_0 is arbitrary and the other members of the sequence are defined by recursion: $v_{n+1} = T v_n / |T v_n|$ ($|x|$ denotes the Euclidian norm of x). Show that if the scalar product $v_0 \cdot v_{\max}$ does not vanish, then the sequence v_n converges exponentially fast to $v_{\max}$.
Hint: Decompose v_0 in the orthonormal diagonalization basis of T.

c) Using the above, write a computer program to extract the largest eigenvalue of a symmetric matrix T.

d) Application: Evaluate numerically the thermodynamic free energy of the two-dimensional Ising model on an infinite strip of width L, at the known critical value of the coupling $K = K_c = -(1/2) \ln (\sqrt{2} - 1)$. (Use Ex. 3.4 above for the definition of the relevant transfer matrix.) Plot the results for various widths L. Fit the results with the ansatz

$$f_L = L f_0 - \frac{\pi}{6L} c + O(\frac{1}{L^2})$$

and evaluate the constants f_0 and c. The quantity c is the central charge of the corresponding conformal field theory. Its exact value for the two-dimensional Ising model is $c = 1/2$.

Notes

There are many excellent texts on statistical mechanics; we cannot list them all here. The very thorough and pedagogical text by Diu and collaborators [106] deserves special mention. Texts by Ma [261], Huang [194] and Pathria [292] are widely used. Among texts emphasizing critical phenomena are those of Amit [13], Binney et al. [47], Le Bellac [253], Ma [260] and Parisi [287].

Some discrete statistical models are described and solved using transfer matrix techniques in Baxter's text [31]. The scaling hypothesis for the free energy was introduced by Widom [355]. The idea of introducing block spins to calculate critical exponents is due to Kadanoff [222]. Applications of the renormalization group to critical phenomena were initiated by Wilson and are described in Ref. [357]. The real-space renormalization group treatment of the Ising model on a triangular lattice was done by Niemeijer and van Leeuwen [282]. The emergence of conformal invariance at critical points was shown by Polyakov [295].

PART B

FUNDAMENTALS

CHAPTER 4

Global Conformal Invariance

This relatively short chapter provides a general introduction to conformal symmetry in arbitrary dimension. Conformal transformations are introduced in Sect. 4.1, with their generators and commutation relations. The conformal group in dimension d is identified with the noncompact group $SO(d+1, 1)$. In Sect. 4.2 we study the action of a conformal transformation on fields, at the classical level. The notion of a quasi-primary field is defined. We relate scale invariance, conformal invariance, and the tracelessness of the energy momentum tensor. In Sect. 4.3 we look at the consequences of conformal invariance at the quantum level on the structure of correlation functions. The form of the two- and three-point functions is given, and the Ward identities implied by conformal invariance are derived. Aspects of conformal invariance that are specific to two dimensions, including local (not globally defined) conformal transformations, are studied in the next chapter. However, the proof that the trace $T^\mu{}_\mu$ vanishes for a two-dimensional theory with translation, rotation, and dilation invariance is given at the end of the present chapter.

§4.1. The Conformal Group

We denote by $g_{\mu\nu}$ the metric tensor in a space-time of dimension d. By definition, a conformal transformation of the coordinates is an invertible mapping $x \rightarrow x'$, which leaves the metric tensor invariant up to a scale:

$$g'_{\mu\nu}(x') = \Lambda(x)g_{\mu\nu}(x) \qquad (4.1)$$

In other words, a conformal transformation is locally equivalent to a (pseudo) rotation and a dilation. The set of conformal transformations manifestly forms a group, and it obviously has the Poincaré group as a subgroup, since the latter corresponds to the special case $\Lambda(x) \equiv 1$. The epithet *conformal* derives from the property that the transformation does not affect the angle between two arbitrary curves crossing each other at some point, despite a local dilation: the conformal group preserves angles. (This is of some importance in cartography applied to navigation, since the relative size of nations is then less important than aiming in the right direction!)

We investigate the consequences of the definition (4.1) on an infinitesimal transformation $x^\mu \rightarrow x'^\mu = x^\mu + \epsilon^\mu(x)$. The metric, at first order in ϵ, changes as follows (cf. Eq. (2.192)):

$$g_{\mu\nu} \rightarrow g_{\mu\nu} - (\partial_\mu \epsilon_\nu + \partial_\nu \epsilon_\mu) \tag{4.2}$$

The requirement that the transformation be conformal implies that[1]

$$\partial_\mu \epsilon_\nu + \partial_\nu \epsilon_\mu = f(x) g_{\mu\nu} \tag{4.3}$$

The factor $f(x)$ is determined by taking the trace on both sides:

$$f(x) = \frac{2}{d} \partial_\rho \epsilon^\rho \tag{4.4}$$

For simplicity, we assume that the conformal transformation is an infinitesimal deformation of the standard Cartesian metric $g_{\mu\nu} = \eta_{\mu\nu}$, where $\eta_{\mu\nu} = \mathrm{diag}(1, 1, \ldots, 1)$. (If the reader insists on living in Minkowski space, the treatment is identical, except for the explicit form of $\eta_{\mu\nu}$.) By applying an extra derivative ∂_ρ on Eq. (4.3), permuting the indices and taking a linear combination, we arrive at

$$2\partial_\mu \partial_\nu \epsilon_\rho = \eta_{\mu\rho} \partial_\nu f + \eta_{\nu\rho} \partial_\mu f - \eta_{\mu\nu} \partial_\rho f \tag{4.5}$$

Upon contracting with $\eta^{\mu\nu}$, this becomes

$$2\partial^2 \epsilon_\mu = (2 - d)\partial_\mu f \tag{4.6}$$

Applying ∂_ν on this expression and ∂^2 on Eq. (4.3), we find

$$(2 - d)\partial_\mu \partial_\nu f = \eta_{\mu\nu} \partial^2 f \tag{4.7}$$

Finally, contracting with $\eta^{\mu\nu}$, we end up with

$$(d - 1)\partial^2 f = 0 \tag{4.8}$$

From Eqs. (4.3)–(4.8), we can derive the explicit form of conformal transformations in d dimensions.

First, if $d = 1$, the above equations do not impose any constraint on the function f, and therefore any smooth transformation is conformal in one dimension. This is a trivial statement, since the notion of angle then does not exist. The case $d = 2$ will be studied in detail later. For the moment, we concentrate on the case $d \geq 3$. Equations (4.8) and (4.7) imply that $\partial_\mu \partial_\nu f = 0$ (i.e., that the function f is at most linear in the coordinates):

$$f(x) = A + B_\mu x^\mu \qquad (A, B_\mu \text{ constant}) \tag{4.9}$$

---

[1] The summation convention on repeated indices is used unless explicitly stated.

If we substitute this expression into Eq. (4.5), we see that $\partial_\mu \partial_\nu \epsilon_\rho$ is constant, which means that ϵ_μ is at most quadratic in the coordinates. We therefore write the general expression

$$\epsilon_\mu = a_\mu + b_{\mu\nu} x^\nu + c_{\mu\nu\rho} x^\nu x^\rho \qquad c_{\mu\nu\rho} = c_{\mu\rho\nu} \qquad (4.10)$$

Since the constraints (4.3)–(4.5) hold for all x, we may treat each power of the coordinate separately. It follows that the constant term a_μ is free of constraints. This term amounts to an infinitesimal translation. Substitution of the linear term into (4.3) yields

$$b_{\mu\nu} + b_{\nu\mu} = \frac{2}{d} b^\lambda_{\ \lambda} \eta_{\mu\nu} \qquad (4.11)$$

which implies that $b_{\mu\nu}$ is the sum of an antisymmetric part and a pure trace:

$$b_{\mu\nu} = \alpha \eta_{\mu\nu} + m_{\mu\nu} \qquad m_{\mu\nu} = -m_{\nu\mu} \qquad (4.12)$$

The pure trace represents an infinitesimal scale transformation, whereas the antisymmetric part is an infinitesimal rigid rotation. Substitution of the quadratic term of (4.10) into Eq. (4.5) yields

$$c_{\mu\nu\rho} = \eta_{\mu\rho} b_\nu + \eta_{\mu\nu} b_\rho - \eta_{\nu\rho} b_\mu \quad \text{where} \quad b_\mu \equiv \frac{1}{d} c^\sigma_{\ \sigma\mu} \qquad (4.13)$$

and the corresponding infinitesimal transformation is

$$x'^\mu = x^\mu + 2(x \cdot b)x^\mu - b^\mu x^2 \qquad (4.14)$$

which bears the name of *special conformal transformation* (SCT).

The finite transformations corresponding to the above are the following:

(translation)	$x'^\mu = x^\mu + a^\mu$	
(dilation)	$x'^\mu = \alpha x^\mu$	
(rigid rotation)	$x'^\mu = M^\mu_{\ \nu} x^\nu$	(4.15)
(SCT)	$x'^\mu = \dfrac{x^\mu - b^\mu x^2}{1 - 2b \cdot x + b^2 x^2}$	

The first three of the above "exponentiations" are fairly familiar, whereas the last one is not. We shall not demonstrate its validity here, but it is trivial to verify that its infinitesimal version is indeed (4.14), and straightforward to show that it is indeed conformal, with a scale factor $\Lambda(x)$ given by

$$\Lambda(x) = (1 - 2b \cdot x + b^2 x^2)^2 \qquad (4.16)$$

The SCT can also be expressed as

$$\frac{x'^\mu}{x'^2} = \frac{x^\mu}{x^2} - b^\mu \qquad (4.17)$$

Manifestly, the SCT is nothing but a translation, preceded and followed by an inversion $x^\mu \to x^\mu/x^2$.

We recall the definition (2.126) of the generator of an infinitesimal transformation. If we suppose for the moment that the fields are unaffected by the transformation (i.e., $\mathcal{F}(\Phi) = \Phi$), the generators of the conformal group are easily seen to be

$$
\begin{array}{ll}
\text{(translation)} & P_\mu = -i\partial_\mu \\
\text{(dilation)} & D = -ix^\mu \partial_\mu \\
\text{(rotation)} & L_{\mu\nu} = i(x_\mu \partial_\nu - x_\nu \partial_\mu) \\
\text{(SCT)} & K_\mu = -i(2x_\mu x^\nu \partial_\nu - x^2 \partial_\mu)
\end{array}
\tag{4.18}
$$

These generators obey the following commutation rules, which in fact define the conformal algebra:

$$
\begin{aligned}
&[D, P_\mu] = iP_\mu \\
&[D, K_\mu] = -iK_\mu \\
&[K_\mu, P_\nu] = 2i(\eta_{\mu\nu}D - L_{\mu\nu}) \\
&[K_\rho, L_{\mu\nu}] = i(\eta_{\rho\mu}K_\nu - \eta_{\rho\nu}K_\mu) \\
&[P_\rho, L_{\mu\nu}] = i(\eta_{\rho\mu}P_\nu - \eta_{\rho\nu}P_\mu) \\
&[L_{\mu\nu}, L_{\rho\sigma}] = i(\eta_{\nu\rho}L_{\mu\sigma} + \eta_{\mu\sigma}L_{\nu\rho} - \eta_{\mu\rho}L_{\nu\sigma} - \eta_{\nu\sigma}L_{\mu\rho})
\end{aligned}
\tag{4.19}
$$

In order to put the above commutation rules into a simpler form, we define the following generators:

$$
\begin{array}{ll}
J_{\mu\nu} = L_{\mu\nu} & J_{-1,\mu} = \dfrac{1}{2}(P_\mu - K_\mu) \\
J_{-1,0} = D & J_{0,\mu} = \dfrac{1}{2}(P_\mu + K_\mu)
\end{array}
\tag{4.20}
$$

where $J_{ab} = -J_{ba}$ and $a, b \in \{-1, 0, 1, \ldots, d\}$. These new generators obey the $SO(d+1, 1)$ commutation relations:

$$
[J_{ab}, J_{cd}] = i(\eta_{ad}J_{bc} + \eta_{bc}J_{ad} - \eta_{ac}J_{bd} - \eta_{bd}J_{ac})
\tag{4.21}
$$

where the diagonal metric η_{ab} is diag$(-1, 1, 1, \ldots, 1)$ if space-time is Euclidian (otherwise an additional component, say η_{dd}, is negative). This shows the isomorphism between the conformal group in d dimensions and the group $SO(d+1, 1)$, with $\frac{1}{2}(d+2)(d+1)$ parameters. Notice that the Poincaré group together with dilations forms a subgroup of the full conformal group. This means that a theory invariant under translations, rotations, and dilations is not necessarily invariant under special conformal transformations. Conditions under which it should be invariant are studied in the next section.

We end this section by constructing conformal invariants, that is, functions $\Gamma(x_i)$ of N points x_i that are left unchanged under all types of conformal transformations. Translation and rotation invariance imply that Γ can depend only on the distances

$|x_i - x_j|$ between pairs of distinct points. Scale invariance implies that only ratios of such distances, such as

$$\frac{|x_i - x_j|}{|x_k - x_l|}$$

will appear in Γ. Finally, under a special conformal transformation, the distance separating two points x_i and x_j becomes

$$|x'_i - x'_j| = \frac{|x_i - x_j|}{(1 - 2b \cdot x_i + b^2 x_i^2)^{1/2}(1 - 2b \cdot x_j + b^2 x_j^2)^{1/2}} \qquad (4.22)$$

It is therefore impossible to construct an invariant Γ with only 2 or 3 points. The simplest possibilities are the following functions of four points:

$$\frac{|x_1 - x_2||x_3 - x_4|}{|x_1 - x_3||x_2 - x_4|} \qquad \frac{|x_1 - x_2||x_3 - x_4|}{|x_2 - x_3||x_1 - x_4|} \qquad (4.23)$$

Such expressions are called *anharmonic ratios* or *cross-ratios*. With N distinct points, $N(N - 3)/2$ independent anharmonic ratios may be constructed.

§4.2. Conformal Invariance in Classical Field Theory

A field theory has conformal symmetry at the classical level if its action is invariant under conformal transformations. As a first step in the description of such theories we define the effect of conformal transformations on classical fields. We then show how, in certain theories, conformal invariance is a consequence of scale and Poincaré invariance. Again, it is important to realize that conformal invariance at the quantum level generally does not follow from conformal invariance at the classical level. A quantum field theory does not make sense without a regularization prescription that introduces a scale in the theory. This scale breaks the conformal symmetry, except at particular values of the parameters, which constitute a renormalization-group fixed point.

4.2.1. Representations of the Conformal Group in d Dimensions

We first show how classical fields are affected by conformal transformations. Given an infinitesimal conformal transformation parametrized by ω_g, we seek a matrix representation T_g such that a multicomponent field $\Phi(x)$ transforms as

$$\Phi'(x') = (1 - i\omega_g T_g)\Phi(x) \qquad (4.24)$$

The generator T_g must be added to the space-time part given in (4.18) to obtain the full generator of the symmetry, as in Eq. (2.128). In order to find out the allowed form of these generators, we shall use the same trick, which may be used for the smaller Poincaré algebra: We start by studying the subgroup of the Poincaré group that leaves the point $x = 0$ invariant, that is, the Lorentz group. We then introduce a matrix representation $S_{\mu\nu}$ to define the action of infinitesimal Lorentz transformations on the field $\Phi(0)$:

$$L_{\mu\nu}\Phi(0) = S_{\mu\nu}\Phi(0) \tag{4.25}$$

$S_{\mu\nu}$ is the spin operator associated with the field Φ. Next, by use of the commutation relations of the Poincaré group, we translate the generator $L_{\mu\nu}$ to a nonzero value of x:

$$e^{ix^\rho P_\rho}L_{\mu\nu}e^{-ix^\rho P_\rho} = S_{\mu\nu} - x_\mu P_\nu + x_\nu P_\mu \tag{4.26}$$

The above translation is explicitly calculated by use of the Hausdorff formula (A and B are two operators):

$$e^{-A}Be^A = B + [B,A] + \frac{1}{2!}[[B,A],A] + \frac{1}{3!}[[[B,A],A],A] + \cdots \tag{4.27}$$

This allows us to write the action of the generators:

$$P_\mu\Phi(x) = -i\partial_\mu\Phi(x)$$
$$L_{\mu\nu}\Phi(x) = i(x_\mu\partial_\nu - x_\nu\partial_\mu)\Phi(x) + S_{\mu\nu}\Phi(x) \tag{4.28}$$

We proceed in the same way for the full conformal group. The subgroup that leaves the origin $x = 0$ invariant is generated by rotations, dilations, and special conformal transformations. If we remove the translation generators from the algebra (4.19), we obtain something identical to the Poincaré algebra augmented by dilations, because of the similar roles played by P_μ and K_μ. We then denote by $S_{\mu\nu}$, $\tilde\Delta$, and κ_μ the respective values of the generators $L_{\mu\nu}$, D, and K_μ at $x = 0$. These must form a matrix representation of the reduced algebra

$$[\tilde\Delta, S_{\mu\nu}] = 0$$
$$[\tilde\Delta, \kappa_\mu] = -i\kappa_\mu$$
$$[\kappa_\nu, \kappa_\mu] = 0 \tag{4.29}$$
$$[\kappa_\rho, S_{\mu\nu}] = i(\eta_{\rho\mu}\kappa_\nu - \eta_{\rho\nu}\kappa_\mu)$$
$$[S_{\mu\nu}, S_{\rho\sigma}] = i(\eta_{\nu\rho}S_{\mu\sigma} + \eta_{\mu\sigma}S_{\nu\rho} - \eta_{\mu\rho}S_{\nu\sigma} - \eta_{\nu\sigma}S_{\mu\rho})$$

The commutations (4.19) then allow us to translate the generators, using the Hausdorff formula (4.27):

$$e^{ix^\rho P_\rho}De^{-ix^\rho P_\rho} = D + x^\nu P_\nu$$
$$e^{ix^\rho P_\rho}K_\mu e^{-ix^\rho P_\rho} = K_\mu + 2x_\mu D - 2x^\nu L_{\mu\nu} + 2x_\mu(x^\nu P_\nu) - x^2 P_\mu \tag{4.30}$$

from which we arrive finally at the following extra transformation rules:

$$D\Phi(x) = (-ix^\nu\partial_\nu + \tilde\Delta)\Phi(x)$$
$$K_\mu\Phi(x) = \left\{\kappa_\mu + 2x_\mu\tilde\Delta - x^\nu S_{\mu\nu} - 2ix_\mu x^\nu\partial_\nu + ix^2\partial_\mu\right\}\Phi(x) \tag{4.31}$$

If we demand that the field $\Phi(x)$ belong to an irreducible representation of the Lorentz group, then, by Schur's lemma, any matrix that commutes with all the generators $S_{\mu\nu}$ must be a multiple of the identity. Consequently, the matrix $\tilde{\Delta}$ is a multiple of the identity and the algebra (4.29) forces all the matrices κ_μ to vanish. $\tilde{\Delta}$ is then simply a number, manifestly equal to $-i\Delta$, where Δ is the scaling dimension of the field Φ, as defined in Eq. (2.121). That the eigenvalue of $\tilde{\Delta}$ is not real simply reflects the non-Hermiticity of the generator $\tilde{\Delta}$ (i.e., representations of the dilation group on classical fields are not unitary).

In principle, we can derive from the above the change in Φ under a finite conformal transformation. However, we shall give the result only for spinless fields ($S_{\mu\nu} = 0$). Under a conformal transformation $x \to x'$, a spinless field $\phi(x)$ transforms as

$$\phi(x) \to \phi'(x') = \left|\frac{\partial x'}{\partial x}\right|^{-\Delta/d} \phi(x) \tag{4.32}$$

where $\left|\partial x'/\partial x\right|$ is the Jacobian of the conformal transformation of the coordinates, related to the scale factor $\Lambda(x)$ of Eq. (4.1) by

$$\left|\frac{\partial x'}{\partial x}\right| = \Lambda(x)^{-d/2} \tag{4.33}$$

A field transforming like the above is called "quasi-primary."

4.2.2. The Energy-Momentum Tensor

Under an arbitrary transformation of the coordinates $x^\mu \to x^\mu + \epsilon^\mu$, the action changes as follows:

$$\delta S = \int d^d x \, T^{\mu\nu} \partial_\mu \epsilon_\nu$$
$$= \frac{1}{2} \int d^d x \, T^{\mu\nu} (\partial_\mu \epsilon_\nu + \partial_\nu \epsilon_\mu) \tag{4.34}$$

where $T^{\mu\nu}$ is the energy-momentum tensor, assumed to be symmetric.[2] This is valid even if the equations of motion are not satisfied (cf. Eq. (2.191)). The definition (4.3) of an infinitesimal conformal transformation implies that the corresponding variation of the action is

$$\delta S = \frac{1}{d} \int d^d x \, T^\mu_{\ \mu} \, \partial_\rho \epsilon^\rho \tag{4.35}$$

The tracelessness of the energy-momentum tensor then implies the invariance of the action under conformal transformations. The converse is not true, since $\partial_\rho \epsilon^\rho$ is not an arbitrary function.

---

[2] We have seen that in theories with rotation (or Lorentz) invariance, the energy-momentum tensor can be made symmetric, i.e., can be put in the Belinfante form.

Under certain conditions, the energy-momentum tensor of a theory with scale invariance can be made traceless, much in the same way as it can be made symmetric in a theory with rotation invariance. If this is possible, then it follows from the above that full conformal invariance is a consequence of scale invariance and Poincaré invariance.

We first consider a generic field theory with scale invariance in dimension $d > 2$. The conserved current associated with the infinitesimal dilation

$$x'^\mu = (1 + \alpha)x^\mu \qquad \mathcal{F}(\Phi) = (1 - \alpha\Delta)\Phi \tag{4.36}$$

is, according to (2.141),

$$
\begin{aligned}
j_D^\mu &= -\mathcal{L}x^\mu + \frac{\partial\mathcal{L}}{\partial(\partial_\mu\Phi)}x^\nu\partial_\nu\Phi + \frac{\partial\mathcal{L}}{\partial(\partial_\mu\Phi)}\Delta\Phi \\
&= T_{c\ \nu}^\mu x^\nu + \frac{\partial\mathcal{L}}{\partial(\partial_\mu\Phi)}\Delta\Phi
\end{aligned}
\tag{4.37}
$$

where $T_c^{\mu\nu}$ is the canonical energy-momentum tensor (2.165). Since by hypothesis this current is conserved, we have

$$
\begin{aligned}
\partial_\mu j_D^\mu &= T_{c\ \mu}^\mu + \Delta\,\partial_\mu\left(\frac{\partial\mathcal{L}}{\partial(\partial_\mu\Phi)}\Phi\right) \\
&= 0
\end{aligned}
\tag{4.38}
$$

We now define the *virial* of the field Φ:

$$V^\mu = \frac{\delta\mathcal{L}}{\delta(\partial^\rho\Phi)}\left(\eta^{\mu\rho}\Delta + iS^{\mu\rho}\right)\Phi \tag{4.39}$$

where $S^{\mu\rho}$ is the spin operator of the field Φ. We also assume that the virial is the divergence of another tensor $\sigma^{\alpha\mu}$:

$$V^\mu = \partial_\alpha\sigma^{\alpha\mu} \tag{4.40}$$

This last condition is obeyed in a large class of physical theories. Then we define

$$
\begin{aligned}
\sigma_+^{\mu\nu} &= \frac{1}{2}(\sigma^{\mu\nu} + \sigma^{\nu\mu}) \\
X^{\lambda\rho\mu\nu} &= \frac{2}{d-2}\Big\{\eta^{\lambda\rho}\sigma_+^{\mu\nu} - \eta^{\lambda\mu}\sigma_+^{\rho\nu} - \eta^{\lambda\mu}\sigma_+^{\nu\rho} + \eta^{\mu\nu}\sigma_+^{\lambda\rho} \\
&\qquad + \frac{1}{d-1}(\eta^{\lambda\rho}\eta^{\mu\nu} - \eta^{\lambda\mu}\eta^{\rho\nu})\sigma_{+\ \alpha}^\alpha\Big\}
\end{aligned}
\tag{4.41}
$$

and we consider the following modified energy-momentum tensor:

$$T^{\mu\nu} = T_c^{\mu\nu} + \partial_\rho B^{\rho\mu\nu} + \frac{1}{2}\partial_\lambda\partial_\rho X^{\lambda\rho\mu\nu} \tag{4.42}$$

The first two terms of the above expression constitute the Belinfante tensor (see Eq. (2.174)). The last term is an addition that will make $T^{\mu\nu}$ traceless. Because of the symmetry properties of $X^{\lambda\rho\mu\nu}$, this additional term does not spoil the conservation law:

$$\partial_\mu \partial_\lambda \partial_\rho X^{\lambda\rho\mu\nu} = 0 \qquad (4.43)$$

Indeed, the addition would not be conserved if $X^{\lambda\rho\mu\nu}$ had a part completely symmetric in the first three indices, but this is not the case. This new term does not spoil the symmetry of the Belinfante tensor either, since the part of $X^{\lambda\rho\mu\nu}$ antisymmetric in μ, ν is

$$X^{\lambda\rho\mu\nu} - X^{\lambda\rho\nu\mu} = \frac{2}{(d-2)(d-1)} \sigma^\alpha_{+\alpha}(\eta^{\lambda\mu}\eta^{\rho\nu} - \eta^{\lambda\nu}\eta^{\rho\mu})$$

and it vanishes upon contraction with $\partial_\lambda \partial_\rho$. Finally, the trace of the new term is

$$\frac{1}{2}\partial_\lambda \partial_\rho X^{\lambda\rho\mu}{}_\mu = \partial_\lambda \partial_\rho \sigma^{\lambda\rho}_+$$
$$= \partial_\mu V^\mu \qquad (4.44)$$

Since

$$\partial_\rho B^{\rho\mu}{}_\mu = \frac{1}{2}i\partial_\rho \left(\frac{\delta\mathcal{L}}{\delta(\partial^\mu \Phi)} S^{\mu\rho}\Phi \right)$$

it follows from (4.38) and (4.39) that

$$T^\mu{}_\mu = \partial_\mu j^\mu_D \qquad (4.45)$$

and therefore scale invariance implies that the modified energy-momentum tensor (4.42) is traceless, provided, of course, that the virial satisfies condition (4.40). This relation also means that the dilation current may be generally written as

$$j^\mu_D = T^\mu{}_\nu x^\nu \qquad (4.46)$$

This argument holds only in dimensions greater than two, since $X^{\mu\nu\lambda\sigma}$ is defined only for $d > 2$. However, the result still holds in dimension two. This is easily seen in particular cases. For instance, we know from Eq. (2.124) that the scaling dimension of the free scalar field vanishes if $d = 2$. Therefore, it follows from Eq. (4.38) that the canonical (or Belinfante) energy-momentum tensor is already traceless, and no modification thereof is necessary. The same is true of the free fermion action. We know of no general proof that the energy-momentum tensor of a two-dimensional field theory with scale invariance can be made traceless. However, we shall hold it to be true. To corroborate this hypothesis, we shall show in the next section, in a quantum context, that the vacuum expectation value of $(T^\mu{}_\mu)^2$ vanishes in dimension two if conformal invariance is present.

§4.3. Conformal Invariance in Quantum Field Theory

4.3.1. Correlation Functions

In this section we examine the consequences of conformal invariance on two- and three-point correlation functions of quasi-primary fields. Consider the two-point function

$$\langle \phi_1(x_1)\phi_2(x_2) \rangle = \frac{1}{Z} \int [d\Phi] \, \phi_1(x_1)\phi_2(x_2) \, \exp -S[\Phi] \qquad (4.47)$$

where ϕ_1 and ϕ_2 are quasi-primary fields (not necessarily distinct). Φ denotes the set of all functionally independent fields in the theory (to which ϕ_1 and ϕ_2 may belong), and $S[\Phi]$ is the action, which we assume to be conformally invariant.

We should remark here on an important detail that sometimes leaves newcomers puzzled. When one speaks of a *field* in conformal field theory, it does not necessarily mean that this field figures independently in the functional integral measure. For instance, a single boson ϕ, its derivative $\partial_\mu \phi$, and a composite quantity such as the energy-momentum tensor are all called *fields*, since they are local quantities, with a coordinate dependence. However, only some fields (such as the boson ϕ in this example) are integrated over in the functional integral. The richness of conformal invariance in two dimensions allows us to define theories based solely on the symmetry properties of the correlation functions, without reference (except in a few cases) to an action or a functional integral. The question "How many continuous, independent degrees of freedom are there?" is often an obscure one in this context, whereas the question "How many basic local operators are there that transform among themselves under conformal transformations?" is more relevant.

The assumed conformal invariance of the action and of the functional integration measure leads to the following transformation of the correlation function, according to Eq. (2.148) (we consider spinless fields for simplicity):

$$\langle \phi_1(x_1)\phi_2(x_2) \rangle = \left| \frac{\partial x'}{\partial x} \right|^{\Delta_1/d}_{x=x_1} \left| \frac{\partial x'}{\partial x} \right|^{\Delta_2/d}_{x=x_2} \langle \phi_1(x_1')\phi_2(x_2') \rangle \qquad (4.48)$$

If we specialize to a scale transformation $x \to \lambda x$ we obtain

$$\langle \phi_1(x_1)\phi_2(x_2) \rangle = \lambda^{\Delta_1 + \Delta_2} \langle \phi_1(\lambda x_1)\phi_2(\lambda x_2) \rangle \qquad (4.49)$$

Rotation and translation invariance require that

$$\langle \phi_1(x_1)\phi_2(x_2) \rangle = f(|x_1 - x_2|) \qquad (4.50)$$

where $f(x) = \lambda^{\Delta_1 + \Delta_2} f(\lambda x)$ by virtue of (4.49). In other words,

$$\langle \phi_1(x_1)\phi_2(x_2) \rangle = \frac{C_{12}}{|x_1 - x_2|^{\Delta_1 + \Delta_2}} \qquad (4.51)$$

where C_{12} is a constant coefficient. It remains to use the invariance under special conformal transformations. We recall that, for such a transformation,

$$\left| \frac{\partial x'}{\partial x} \right| = \frac{1}{(1 - 2b \cdot x + b^2 x^2)^d} \tag{4.52}$$

Given the transformation (4.22) for the distance $|x_1 - x_2|$, the covariance of the correlation function (4.51) implies

$$\frac{C_{12}}{|x_1 - x_2|^{\Delta_1 + \Delta_2}} = \frac{C_{12}}{\gamma_1^{\Delta_1} \gamma_2^{\Delta_2}} \frac{(\gamma_1 \gamma_2)^{(\Delta_1 + \Delta_2)/2}}{|x_1 - x_2|^{\Delta_1 + \Delta_2}} \tag{4.53}$$

with

$$\gamma_i = (1 - 2b \cdot x_i + b^2 x_i^2) \tag{4.54}$$

This constraint is identically satisfied only if $\Delta_1 = \Delta_2$. In other words, two quasi-primary fields are correlated only if they have the same scaling dimension:

$$\langle \phi_1(x_1) \phi_2(x_2) \rangle = \begin{cases} \dfrac{C_{12}}{|x_1 - x_2|^{2\Delta_1}} & \text{if} \quad \Delta_1 = \Delta_2 \\ 0 & \text{if} \quad \Delta_1 \neq \Delta_2 \end{cases} \tag{4.55}$$

Comparison with Table 3.1 shows that the exponent η is

$$\eta = 2\Delta + 2 - d \tag{4.56}$$

A similar analysis may be performed on three-point functions. Covariance under rotations, translations, and dilations forces a generic three-point function to have the following form:

$$\langle \phi_1(x_1) \phi_2(x_2) \phi_3(x_3) \rangle = \frac{C_{123}^{(abc)}}{x_{12}^a x_{23}^b x_{13}^c} \tag{4.57}$$

where $x_{ij} = |x_i - x_j|$ and with a, b, c such that

$$a + b + c = \Delta_1 + \Delta_2 + \Delta_3 \tag{4.58}$$

Actually, a sum (over a, b, c) of such terms is also acceptable, as long as the above equality is satisfied. Under special conformal transformations Eq. (4.57) becomes

$$\frac{C_{123}^{(abc)}}{\gamma_1^{\Delta_1} \gamma_2^{\Delta_2} \gamma_3^{\Delta_3}} \frac{(\gamma_1 \gamma_2)^{a/2} (\gamma_2 \gamma_3)^{b/2} (\gamma_1 \gamma_3)^{c/2}}{x_{12}^a x_{23}^b x_{13}^c}$$

For this expression to be of the same form as Eq. (4.57), all the factors involving the transformation parameter b^μ must disappear, which leads to the following set of constraints:

$$a + c = 2\Delta_1 \qquad a + b = 2\Delta_2 \qquad b + c = 2\Delta_3 \tag{4.59}$$

The solution to these constraints is unique:

$$a = \Delta_1 + \Delta_2 - \Delta_3$$
$$b = \Delta_2 + \Delta_3 - \Delta_1 \tag{4.60}$$
$$c = \Delta_3 + \Delta_1 - \Delta_2$$

Therefore, the correlator of three quasi-primary fields is made of a single term of the form (4.57), namely

$$\langle \phi_1(x_1)\phi_2(x_2)\phi_3(x_3) \rangle = \frac{C_{123}}{x_{12}^{\Delta_1+\Delta_2-\Delta_3} \, x_{23}^{\Delta_2+\Delta_3-\Delta_1} \, x_{13}^{\Delta_3+\Delta_1-\Delta_2}} \tag{4.61}$$

At this point the reader might feel encouraged by our success at calculating correlation functions (up to multiplicative constants, which only reflects a freedom in normalization for our fields). However, this impressive performance stops at three-point functions. Indeed, with four points (or more), it is possible to construct conformal invariants, the anharmonic ratios (4.23). The n-point function may have an arbitrary dependence (i.e., not fixed by conformal invariance) on these ratios. For instance, the four-point function may take the following form:

$$\langle \phi_1(x_1) \ldots \phi_4(x_4) \rangle = f\left(\frac{x_{12}x_{34}}{x_{13}x_{24}}, \frac{x_{12}x_{34}}{x_{23}x_{14}}\right) \prod_{i<j}^{4} x_{ij}^{\Delta/3-\Delta_i-\Delta_j} \tag{4.62}$$

where we have defined $\Delta = \sum_{i=1}^{4} \Delta_i$.

4.3.2. Ward Identities

We shall now write the Ward identities implied by conformal invariance, according to the general identity (2.157). The Ward identity associated with translation invariance appears in Eq. (2.183) and we reproduce it here:

$$\partial_\mu \langle T^\mu_{\ \nu} X \rangle = -\sum_i \delta(x - x_i) \frac{\partial}{\partial x_i^\nu} \langle X \rangle \tag{4.63}$$

This identity holds even after a modification of the energy-momentum tensor, as in Eq. (4.42). Recall that X stands for a product of n local fields, at coordinates x_i, $i = 1, \ldots, n$.

We consider now the Ward identity associated with Lorentz (or rotation) invariance. Once the energy-momentum has been made symmetric, the associated current $j^{\mu\nu\rho}$ has the form given in Eq. (2.172):

$$j^{\mu\nu\rho} = T^{\mu\nu}x^\rho - T^{\mu\rho}x^\nu \tag{4.64}$$

The generator of Lorentz transformations is given by Eq. (2.134). Consequently, the Ward identity is

$$\partial_\mu \langle (T^{\mu\nu}x^\rho - T^{\mu\rho}x^\nu)X \rangle = \sum_i \delta(x - x_i) \left[(x_i^\nu \partial_i^\rho - x_i^\rho \partial_i^\nu)\langle X \rangle - iS_i^{\nu\rho}\langle X \rangle \right] \quad (4.65)$$

where $S_i^{\nu\rho}$ is the spin generator appropriate for the i-th field of the set X. The derivative on the l.h.s. of the above equation may act either on the energy-momentum tensor or on the coordinates. Using the first Ward identity (4.63), we reduce the above to

$$\langle (T^{\rho\nu} - T^{\nu\rho})X \rangle = -i \sum_i \delta(x - x_i)S_i^{\nu\rho}\langle X \rangle \quad (4.66)$$

which is the Ward identity associated with Lorentz (rotation) invariance. It states that the energy-momentum tensor is symmetric within correlation functions, except at the position of the other fields of the correlator.

Finally, we consider the Ward identity associated with scale invariance. We shall assume that the dilation current j_D^μ may be written as in Eq. (4.46), which supposes that the energy-momentum tensor has been suitably modified (if needed) to be traceless. So far we have not shown how this can be done generally in two dimensions, although we hold that it can be done. In the next chapter we shall provide an alternate derivation of the Ward identity, which circumvents this problem. Since the generator of dilations is $D = -ix^\nu \partial_\nu - i\Delta$ for a field of scaling dimension Δ, the Ward identity is

$$\partial_\mu \langle T^\mu{}_\nu x^\nu X \rangle = -\sum_i \delta(x - x_i) \left\{ x_i^\nu \frac{\partial}{\partial x_i^\nu}\langle X \rangle + \Delta_i \langle X \rangle \right\} \quad (4.67)$$

Here again the derivative ∂_μ may act on $T^\mu{}_\nu$ and on the coordinate. Using Eq. (4.63), this identity reduces to

$$\langle T^\mu{}_\mu X \rangle = -\sum_i \delta(x - x_i)\Delta_i \langle X \rangle \quad (4.68)$$

Eqs. (4.63), (4.66), and (4.68) are the three Ward identities associated with conformal invariance.

4.3.3. Tracelessness of $T_{\mu\nu}$ in Two Dimensions

In this section we show that the vacuum expectation value of the trace of the energy-momentum tensor (or of its square) vanishes in two-dimensions if the theory has scale, rotation, and translation invariance. This implies that this trace is identically zero in the quantum theory and that conformal invariance follows from scale, rotation, and translation invariance in dimension two.

We consider the two-point function of the energy-momentum tensor (called the Schwinger function):

$$S_{\mu\nu\rho\sigma}(x) = \langle T_{\mu\nu}(x)T_{\rho\sigma}(0) \rangle \quad (4.69)$$

Since by assumption the theory is translation and rotation invariant, $T_{\mu\nu}$ is conserved and symmetric (or can be made symmetric). The symmetry of $T_{\mu\nu}$ implies that

$$S_{\mu\nu\rho\sigma} = S_{\nu\mu\rho\sigma} = S_{\mu\nu\sigma\rho} = S_{\nu\mu\sigma\rho} \tag{4.70}$$

Translation invariance implies that

$$\begin{aligned}
S_{\mu\nu\rho\sigma}(x) &= \langle T_{\mu\nu}(0)T_{\rho\sigma}(-x)\rangle \\
&= \langle T_{\rho\sigma}(-x)T_{\mu\nu}(0)\rangle \\
&= S_{\rho\sigma\mu\nu}(-x)
\end{aligned} \tag{4.71}$$

If the theory is invariant under parity, we conclude that

$$S_{\mu\nu\rho\sigma}(x) = S_{\rho\sigma\mu\nu}(x) \tag{4.72}$$

Finally, scale invariance implies that $T_{\mu\nu}$ transforms covariantly under scale transformations, with scaling dimension 2 since it is a density. This means that

$$S_{\mu\nu\rho\sigma}(\lambda x) = \lambda^{-4}S_{\mu\nu\rho\sigma}(x) \tag{4.73}$$

All these constraints restrict the most general form that $S_{\mu\nu\rho\sigma}$ can take:

$$\begin{aligned}
S_{\mu\nu\rho\sigma}(x) = (x^2)^{-4}\Big\{ &A_1 g_{\mu\nu}g_{\rho\sigma}(x^2)^2 + A_2(g_{\mu\rho}g_{\nu\sigma} + g_{\mu\sigma}g_{\nu\rho})(x^2)^2 \\
&+ A_3(g_{\mu\nu}x_\rho x_\sigma + g_{\rho\sigma}x_\mu x_\nu)x^2 + A_4 x_\mu x_\nu x_\rho x_\sigma \Big\}
\end{aligned} \tag{4.74}$$

(cf. Ex. 4.9). The constants A_1 to A_4 are not all arbitrary. Indeed, the conservation law $\partial^\mu T_{\mu\nu} = 0$ obviously extends to the Schwinger function. Taking the derivative, we find

$$\begin{aligned}
\partial^\mu S_{\mu\nu\rho\sigma}(x) = -(x^2)^{-4}\Big\{ &3(A_4 + 2A_3)x_\nu x_\rho x_\sigma + (4A_1 + 3A_3)g_{\rho\sigma}x_\nu x^2 \\
&+ (4A_2 - A_3)(g_{\rho\nu}x_\sigma + g_{\nu\sigma}x_\rho)x^2 \Big\}
\end{aligned} \tag{4.75}$$

This vanishes everywhere only if each combination of coefficients in parentheses vanishes. This leaves only one arbitrary constant:

$$A_1 = 3A \qquad A_2 = -A \qquad A_3 = -4A \qquad A_4 = 8A \tag{4.76}$$

Upon inserting these values into Eq. (4.74), we find

$$\begin{aligned}
S_{\mu\nu\rho\sigma}(x) = \frac{A}{(x^2)^4}\Big\{ &(3g_{\mu\nu}g_{\rho\sigma} - g_{\mu\rho}g_{\nu\sigma} - g_{\mu\sigma}g_{\nu\rho})(x^2)^2 \\
&- 4x^2(g_{\mu\nu}x_\rho x_\sigma + g_{\rho\sigma}x_\mu x_\nu) + 8x_\mu x_\nu x_\rho x_\sigma \Big\}
\end{aligned} \tag{4.77}$$

It is then straightforward to show that the trace

$$S^\mu{}_\mu{}^\sigma{}_\sigma(x) = \langle T^\mu{}_\mu(x)T^\sigma{}_\sigma(0)\rangle \tag{4.78}$$

vanishes everywhere. In particular $\langle T^\mu_{\ \mu}(0)^2\rangle = 0$, which implies that the operator $T^\mu_{\ \mu}$ has zero expectation value and zero standard deviation in the ground state. In fact, the general result is the Ward identity (4.68), which states that $T^\mu_{\ \mu}(x)$ vanishes within correlation functions, except when x coincides with the position of another field present in the correlator.

Exercises

4.1 Check Eqs. (4.3) and (4.5) explicitly.

4.2 Demonstrate that the metric scale factor produced by a special conformal transformation is given by Eq. (4.16).

4.3 Check Eq. (4.22) explicitly.

4.4

a) Show that the expression (4.62) for the four-point function is conformally covariant.

b) Show that there are only two independent cross-ratios of the form (4.23) that can be built out of four points, except in dimension two, where the two cross-ratios are related.

4.5 *Scale invariance in momentum space*
In momentum space, a correlation function of a set X of n fields $\phi_i(x_i)$ is represented by its Fourier transform $\Gamma_X(k_1,\cdots,k_n)$:

$$\langle\phi_1(x_1)\cdots\phi_n(x_n)\rangle = \int \frac{dk_1}{(2\pi)^d}\cdots\frac{dk_{n-1}}{(2\pi)^d}\Gamma_X(k_1,\cdots,k_n)e^{i(k_1\cdot x_1+\cdots+k_n\cdot x_n)} \tag{4.79}$$

where $-k_n = k_1+\cdots+k_{n-1}$ is fixed by momentum conservation (translation invariance).
a) Show that scale invariance imposes the following constraint on Γ_X:

$$\Gamma_X(k_1,\cdots,k_n) = s^{(n-1)d-\Delta_1-\cdots-\Delta_n}\Gamma_X(sk_1,\cdots,sk_n) \tag{4.80}$$

where Δ_i is the scaling dimension of the field ϕ_i.
b) Show that the two-point function $\Gamma_2(k)$ of a scale-invariant theory is of the form

$$\Gamma_2(k) \sim \frac{1}{k^{2-\eta}} \tag{4.81}$$

where η is the critical exponent defined in Table 3.1 and $k = |k|$.
c) In dimension two, show that the two-point function in coordinate space must accordingly be

$$G(r) = \int_{1/L}^\infty \frac{dk}{k^{1-\eta}}J_0(kr) \tag{4.82}$$

where $r = |x_1 - x_2|$, $k = |k|$, J_0 is the zeroth-order Bessel function and L^{-1} is a low-momentum (infrared) cutoff. Explain how this is compatible with the form (4.55).

4.6 Consider the Lagrangian of a free fermion in dimension two:

$$\mathcal{L} = \frac{i}{2}\Psi^t\gamma^0\gamma^\mu\partial_\mu\Psi$$

Obtain the precise form of the spin generator $S_{\mu\nu}$ that would ensure Lorentz invariance. Then, write down the canonical energy-momentum tensor, the Belinfante modification to the latter, and the dilation current.

4.7 *Traceless energy-momentum tensor*

a) Write down a modification of the energy-momentum tensor for the massless scalar field that is traceless in $d > 2$.

b) Repeat the exercise for the massless φ^4 theory in $d = 4$.

4.8 *Liouville field theory*

Consider the Liouville field theory in $d = 2$, with Lagrangian density

$$\mathcal{L} = \frac{1}{2}\partial_\mu\varphi\partial^\mu\varphi - \frac{1}{2}m^2 e^\varphi$$

Write down the canonical energy-momentum tensor and add a term that makes it traceless without affecting the conservation laws.

4.9 *The Schwinger function*

Eq. (4.74) gives the most general form of the Schwinger function compatible with translation, rotation, and scale invariance, as well as parity, in dimension two. The requirement of invariance under parity transformations is not essential in order to prove the tracelessness $S^\mu{}_\mu{}^\nu{}_\nu(x) = 0$, but simplifies the discussion. However, nothing in the form (4.74) is specific to two dimensions. The specificity comes from the possible introduction of the antisymmetric tensor in dimensions higher than two.

a) Show that a possible addition to (4.74) in two dimensions, compatible with all the symmetries, is

$$A_5(\epsilon_{\mu\sigma}\epsilon_{\nu\rho} + \epsilon_{\mu\rho}\epsilon_{\nu\sigma})(x^2)^2$$

and demonstrate that it reduces to a linear combination of the first two terms of (4.74).

b) Show that an admissible generalization of this addition in three dimensions is

$$A_5(\epsilon_{\mu\sigma\alpha}\epsilon_{\nu\rho\beta} + \epsilon_{\mu\rho\alpha}\epsilon_{\nu\sigma\beta})x^\alpha x^\beta(x^2)$$

Show that this addition is not equivalent to a combination of the other terms and that the imposition of the conservation law $\partial^\mu S_{\mu\nu\rho\sigma}(x) = 0$ does not lead to the tracelessness property $S^\mu{}_\mu{}^\nu{}_\nu(x) = 0$ in three dimensions.

Notes

The conformal group was studied early on by mathematicians, in particular by Lie [256]. The invariance of Maxwell's equation under the conformal group was noticed by Bateman [26] and Cunningham [85] at the beginning of the century. Even before, the tracelessness of the electrodynamic energy-momentum tensor had been noticed indirectly by Bartoli in 1876 and by Boltzmann [50], who wrote down the relation $P = \frac{1}{3}\mathcal{E}$ between the radiation pressure P and the energy density $\mathcal{E}$.

A detailed account of the applications of conformal invariance in four-dimensional quantum field theory and an extensive bibliography of early work on the subject are found in Todorov, Mintchev, and Petkova [335]. The representations of the conformal group acting on fields were studied by Mack and Salam [264] and Schroer and Swieca [324].

The form of the two-, three- and four-point functions in a conformally invariant theory was obtained by Polyakov [295]. The procedure followed to make the symmetric energy-momentum tensor traceless is borrowed from Ref. [312]. The proof that the energy-momentum tensor is traceless in dimension two if the theory has translation, rotation, and scale invariance is due to Lüscher and Mack [259].

CHAPTER 5

Conformal Invariance in Two Dimensions

Conformal invariance takes a new meaning in two dimensions. As already apparent in Section 4.1, the case $d = 2$ requires special attention. Indeed, there exists in two dimensions an infinite variety of coordinate transformations that, although not everywhere well-defined, are locally conformal: they are holomorphic mappings from the complex plane (or part of it) onto itself. Among this infinite set of mappings one must distinguish the 6-parameter *global* conformal group, made of one-to-one mappings of the complex plane into itself. The analysis of the previous chapter still holds when considering these transformations only. However, a local field theory should be sensitive to local symmetries, even if the related transformations are not globally defined. It is *local* conformal invariance that enables exact solutions of two-dimensional conformal field theories.

Section 5.1 introduces the essential language of holomorphic and antiholomorphic coordinates on the plane, used in the remaining chapters of this book. This section also clarifies the distinction between local and global transformations, introduces generators for local conformal transformations, defines the notion of a primary field, and translates the results of Sect. 4.3.1 on correlation functions in holomorphic language. Section 5.2 adapts the Ward identities of conformal invariance to complex coordinates and also provides an alternate derivation of the Ward identities, specific to two dimensions. Section 5.3 introduces the notion of a short-distance product of operators (operator product expansion) and applies this language to the Ward identities and to specific examples of free conformal fields: the boson, the fermion, and ghost systems. Section 5.4 describes the transformation properties of the energy-momentum tensor itself and introduces the central charge c. Throughout this chapter, no mention is made of the operator formalism (radial quantization and so on), which is introduced in the next chapter.

§5.1. The Conformal Group in Two Dimensions

5.1.1. Conformal Mappings

We consider the coordinates (z^0, z^1) on the plane. Under a change of coordinate system $z^\mu \to w^\mu(x)$ the contravariant metric tensor transforms as

$$g^{\mu\nu} \to \left(\frac{\partial w^\mu}{\partial z^\alpha}\right)\left(\frac{\partial w^\nu}{\partial z^\beta}\right) g^{\alpha\beta} \tag{5.1}$$

The condition (4.1) that defines a conformal transformation is $g'_{\mu\nu}(w) \propto g_{\mu\nu}(z)$ or, explicitly,

$$\left(\frac{\partial w^0}{\partial z^0}\right)^2 + \left(\frac{\partial w^0}{\partial z^1}\right)^2 = \left(\frac{\partial w^1}{\partial z^0}\right)^2 + \left(\frac{\partial w^1}{\partial z^1}\right)^2 \tag{5.2}$$

$$\frac{\partial w^0}{\partial z^0}\frac{\partial w^1}{\partial z^0} + \frac{\partial w^0}{\partial z^1}\frac{\partial w^1}{\partial z^1} = 0 \tag{5.3}$$

These conditions are equivalent either to

$$\frac{\partial w^1}{\partial z^0} = \frac{\partial w^0}{\partial z^1} \quad \text{and} \quad \frac{\partial w^0}{\partial z^0} = -\frac{\partial w^1}{\partial z^1} \tag{5.4}$$

or to

$$\frac{\partial w^1}{\partial z^0} = -\frac{\partial w^0}{\partial z^1} \quad \text{and} \quad \frac{\partial w^0}{\partial z^0} = \frac{\partial w^1}{\partial z^1} \tag{5.5}$$

In Eq. (5.4) we recognize the Cauchy-Riemann equations for holomorphic functions, whereas Eq. (5.5) defines antiholomorphic functions.

This motivates the use of complex coordinates z and $\bar{z}$, with the following translation rules:

$$\begin{array}{ll}
z = z^0 + iz^1 & z^0 = \dfrac{1}{2}(z + \bar{z}) \\[2mm]
\bar{z} = z^0 - iz^1 & \\[2mm]
\partial_z = \dfrac{1}{2}(\partial_0 - i\partial_1) & z^1 = \dfrac{1}{2i}(z - \bar{z}) \\[2mm]
 & \partial_0 = \partial_z + \partial_{\bar{z}} \\[2mm]
\partial_{\bar{z}} = \dfrac{1}{2}(\partial_0 + i\partial_1) & \partial_1 = i(\partial_z - \partial_{\bar{z}})
\end{array} \tag{5.6}$$

We shall sometimes write $\partial = \partial_z$ and $\bar{\partial} = \partial_{\bar{z}}$ when there is no ambiguity about the differentiation variable. In terms of the coordinates z and $\bar{z}$, the metric tensor is

$$g_{\mu\nu} = \begin{pmatrix} 0 & \frac{1}{2} \\ \frac{1}{2} & 0 \end{pmatrix} \qquad g^{\mu\nu} = \begin{pmatrix} 0 & 2 \\ 2 & 0 \end{pmatrix} \tag{5.7}$$

where the index μ takes the values z and $\bar{z}$, in that order. This metric tensor allows us to transform a covariant holomorphic index into a contravariant antiholomorphic index and vice versa. The antisymmetric tensor $\varepsilon_{\mu\nu}$ in holomorphic form is[1]

$$\varepsilon_{\mu\nu} = \begin{pmatrix} 0 & \frac{1}{2}i \\ -\frac{1}{2}i & 0 \end{pmatrix} \qquad \varepsilon^{\mu\nu} = \begin{pmatrix} 0 & -2i \\ 2i & 0 \end{pmatrix} \tag{5.8}$$

In this language, the holomorphic Cauchy-Riemann equations become simply

$$\partial_{\bar{z}} w(z, \bar{z}) = 0 \tag{5.9}$$

whose solution is any holomorphic mapping (no $\bar{z}$ dependence):

$$z \to w(z) \tag{5.10}$$

It is a well-known result that any analytic mapping of the complex plane onto itself is conformal (i.e., preserves angles). This is made plainly obvious by considering the differential

$$dw = \left(\frac{dw}{dz}\right) dz \tag{5.11}$$

The derivative dw/dz contains a dilation factor $|dw/dz|$, along with a phase Arg (dw/dz), which embodies a rotation. The conformal "group" in two dimensions is therefore the set of all analytic maps, wherein the group multiplication is the composition of maps. This set is infinite-dimensional, since an infinite number of parameters (the coefficients of a Laurent series) is needed to specify all functions analytic in some neighborhood. It is precisely this infinite dimensionality that allows so much to be known about conformally invariant field theories in two dimensions.

The first question that comes to mind regards the status of the variables z and $\bar{z}$, that is, whether they should be considered as independent. The proper approach is to extend the range of the Cartesian coordinates z^0 and z^1 to the complex plane. Then Eq. (5.6) is a mere change of independent variables, and $\bar{z}$ is not the complex conjugate of z, but rather a distinct complex coordinate. It should be kept in mind, however, that the physical space is the two-dimensional submanifold (called the *real surface*) defined by $z^* = \bar{z}$.

5.1.2. Global Conformal Transformations

All that we have inferred from Eq. (5.4) ff. is purely local, that is, we have not imposed the condition that conformal transformations be defined everywhere and be invertible. Strictly speaking, in order to form a group, the mappings must be invertible, and must map the whole plane into itself (more precisely the Riemann sphere, i.e., the complex plane plus the point at infinity). We must therefore distinguish *global* conformal transformations, which satisfy these requirements, from *local* conformal transformations, which are not everywhere well-defined. The set

---

[1] Usually, $\varepsilon_{\mu\nu}$ is taken as a pseudotensor, always given by its Cartesian components (0 or 1), but multiplied by the Jacobian of the coordinate transformation. Here we choose to include this Jacobian in the definition of $\varepsilon_{\mu\nu}$.

of *global* conformal transformations form what we call the *special conformal group*. It turns out that the complete set of such mappings is

$$f(z) = \frac{az+b}{cz+d} \quad \text{with} \quad ad - bc = 1 \tag{5.12}$$

where $a, b, c,$ and d are complex numbers. These mappings are called *projective transformations*, and to each of them we can associate the matrix

$$A = \begin{pmatrix} a & b \\ c & d \end{pmatrix} \tag{5.13}$$

We easily verify that the composition of two maps $f_1 \circ f_2$ corresponds to the matrix multiplication $A_2 A_1$. Therefore, what we call the *global* conformal group in two dimensions is isomorphic to the group of complex invertible 2×2 matrices with unit determinant, or $SL(2, \mathbb{C})$. It is known that $SL(2, \mathbb{C})$ is isomorphic to the Lorentz group in four dimensions, that is, to $SO(3, 1)$. Therefore, as far as the conformal group proper is concerned, we have learned nothing new since the previous chapter: the global conformal group is the 6-parameter (3 complex) pseudo-orthogonal group $SO(3, 1)$.

It is interesting to show explicitly why the transformations (5.12) are the only globally defined invertible holomorphic mappings. Consider such a mapping, say $f(z)$. It is clear that f should not have any branch point or any essential singularity. Indeed, around a branch point the map is not uniquely defined, whereas in any (however small) neighborhood of an essential singularity the function f sweeps the entire complex plane, and is therefore not invertible. Consequently, the only singularities deemed acceptable are poles, and the function f can be written as a ratio of polynomials (without common zeros):

$$f(z) = \frac{P(z)}{Q(z)} \tag{5.14}$$

If $P(z)$ has several distinct zeros, then the inverse image of zero is not uniquely defined and f is not invertible. If, moreover, $P(z)$ has a multiple zero z_0 of order $n > 1$, then the image of a small neighborhood of z_0 is wrapped n times around 0, and therefore f is not invertible. Thus $P(z)$ can be only a linear function: $P(z) = az + b$. The same argument applies for $Q(z)$ when looking at the behavior of $f(z)$ near the point at infinity. We therefore arrive at the form (5.12) with the proviso that the determinant $ad - bc$ be nonzero in order for the mapping to be invertible. Since an overall scaling of all coefficients a, b, c, d does not change f, the conventional normalization $ad - bc = 1$ has been adopted.

5.1.3. Conformal Generators

As is typical in physics, the local properties are more immediately useful than the global properties, and the local conformal group (the set of all, not necessarily invertible, holomorphic mappings) is of great importance. We now find the algebra

of its generators. Any holomorphic infinitesimal transformation may be expressed
as

$$z' = z + \epsilon(z) \qquad \epsilon(z) = \sum_{-\infty}^{\infty} c_n z^{n+1} \tag{5.15}$$

where, by hypothesis, the infinitesimal mapping admits a Laurent expansion around
$z = 0$. The effect of such a mapping (and of its antiholomorphic counterpart) on
a spinless and dimensionless field $\phi(z, \bar{z})$ living on the plane is

$$\begin{aligned}
\phi'(z', \bar{z}') &= \phi(z, \bar{z}) \\
&= \phi(z', \bar{z}') - \epsilon(z')\partial'\phi(z', \bar{z}') - \bar{\epsilon}(\bar{z}')\bar{\partial}'\phi(z', \bar{z}')
\end{aligned} \tag{5.16}$$

or

$$\begin{aligned}
\delta\phi &= -\epsilon(z)\partial\phi - \bar{\epsilon}(\bar{z})\bar{\partial}\phi \\
&= \sum_n \left\{ c_n \ell_n \phi(z, \bar{z}) + \bar{c}_n \bar{\ell}_n \phi(z, \bar{z}) \right\}
\end{aligned} \tag{5.17}$$

where we have introduced the generators

$$\ell_n = -z^{n+1}\partial_z \qquad \bar{\ell}_n = -\bar{z}^{n+1}\partial_{\bar{z}} \tag{5.18}$$

These generators obey the following commutation relations:

$$\begin{aligned}
[\ell_n, \ell_m] &= (n - m)\ell_{n+m} \\
[\bar{\ell}_n, \bar{\ell}_m] &= (n - m)\bar{\ell}_{n+m} \\
[\ell_n, \bar{\ell}_m] &= 0
\end{aligned} \tag{5.19}$$

Thus the conformal algebra is the direct sum of two isomorphic algebras, each
with very simple commutation relations. The algebra (5.19) is sometimes called
the Witt algebra.

Each of these two infinite-dimensional algebras contains a finite subalgebra
generated by ℓ_{-1}, ℓ_0, and ℓ_1. This is the subalgebra associated with the global
conformal group. Indeed, from the definition (5.18) it is manifest that $\ell_{-1} = -\partial_z$ generates translations on the complex plane, that $\ell_0 = -z\partial_z$ generates scale
transformations and rotations, and that $\ell_1 = -z^2\partial_z$ generates special conformal
transformations. The generators that preserve the real surface $z_0, z_1 \in \mathbb{R}$ are the
linear combinations

$$\ell_n + \bar{\ell}_n \quad \text{and} \quad i(\ell_n - \bar{\ell}_n) \tag{5.20}$$

In particular, $\ell_0 + \bar{\ell}_0$ generates dilations on the real surface, and $i(\ell_0 - \bar{\ell}_0)$ generates
rotations.

5.1.4. Primary Fields

In two dimensions the definition of quasi-primary fields applies also to fields with
spin. Indeed, given a field with scaling dimension Δ and planar spin s, we define

the *holomorphic conformal dimension* h and its antiholomorphic counterpart $\bar{h}$ as[2]

$$h = \frac{1}{2}(\Delta + s) \qquad \bar{h} = \frac{1}{2}(\Delta - s) \qquad (5.21)$$

Under a conformal map $z \to w(z)$, $\bar{z} \to \bar{w}(\bar{z})$, a quasi-primary field transforms as

$$\phi'(w, \bar{w}) = \left(\frac{dw}{dz}\right)^{-h} \left(\frac{d\bar{w}}{d\bar{z}}\right)^{-\bar{h}} \phi(z, \bar{z}) \qquad (5.22)$$

This constitutes a generalization of Eq. (4.32). The above shows that a quasi-primary field of conformal dimensions $(h, \bar{h})$ transforms like the component of a covariant tensor of rank $h + \bar{h}$ having h "z" indices and $\bar{h}$ "$\bar{z}$" indices.

If the map $z \to w$ is close to the identity—that is, if $w = z + \epsilon(z)$ and $\bar{w} = \bar{z} + \bar{\epsilon}(z)$ with ϵ and $\bar{\epsilon}$ small (at least in some neighborhood)—the variation of quasi-primary fields is

$$\begin{aligned}
\delta_{\epsilon, \bar{\epsilon}} \phi &\equiv \phi'(z, \bar{z}) - \phi(z, \bar{z}) \\
&= -(h \phi \partial_z \epsilon + \epsilon \partial_z \phi) - (\bar{h} \phi \partial_{\bar{z}} \bar{\epsilon} + \bar{\epsilon} \partial_{\bar{z}} \phi)
\end{aligned} \qquad (5.23)$$

In fact, a field whose variation under *any* local conformal transformation in two dimensions is given by (5.22) (or, equivalently, (5.23)) is called *primary*. All primary fields are also quasi-primary, but the reverse is not true: A field may transform according to (5.22) under an element of the global conformal group $SL(2, \mathbb{C})$, but for those conformal transformations only. As we shall see, an example of a quasi-primary field that is not primary is the energy-momentum tensor. A field which is not primary is generally called *secondary*. For instance, the derivative of a primary field of conformal dimension $h \neq 0$ is secondary.

5.1.5. Correlation Functions

Expressed in terms of holomorphic and antiholomorphic coordinates, the relation (2.149) for conformal transformations of n primary fields ϕ_i with conformal dimensions h_i and $\bar{h}_i$ becomes

$$\begin{aligned}
\langle \phi_1(w_1, \bar{w}_1) \ldots \phi_n(w_n, \bar{w}_n) \rangle = \\
\prod_{i=1}^{n} \left(\frac{dw}{dz}\right)^{-h_i}_{w=w_i} \left(\frac{d\bar{w}}{d\bar{z}}\right)^{-\bar{h}_i}_{\bar{w}=\bar{w}_i} \langle \phi_1(z_1, \bar{z}_1) \ldots \phi_n(z_n, \bar{z}_n) \rangle
\end{aligned} \qquad (5.24)$$

This relation fixes the form of two- and three-point functions. The novelty here is the possibility of nonzero spin, incorporated in the difference $h_i - \bar{h}_i$. The relations (4.55) and (4.61) are still valid in two dimensions. Let us express them in terms of

---

[2] One often uses the terminology *left* and *right*, instead of *holomorphic* and *antiholomorphic*, in that order.

complex coordinates, taking spin into account when imposing rotation invariance. The distance x_{ij} is equal to $(z_{ij}\bar{z}_{ij})^{1/2}$ and Eq. (4.55) becomes

$$\langle \phi_1(z_1,\bar{z}_1)\phi_2(z_2,\bar{z}_2)\rangle = \frac{C_{12}}{(z_1-z_2)^{2h}(\bar{z}_1-\bar{z}_2)^{2\bar{h}}} \quad \text{if} \quad \begin{cases} h_1 = h_2 = h \\ \bar{h}_1 = \bar{h}_2 = \bar{h} \end{cases} \tag{5.25}$$

The two-point function vanishes if the conformal dimensions of the two fields are different. The additional condition on the conformal dimensions comes from rotation invariance: the sum of the spins within a correlator should be zero.

Equation (4.61) for the three-point function becomes

$$\langle \phi_1(x_1)\phi_2(x_2)\phi_3(x_3)\rangle = C_{123}\frac{1}{z_{12}^{h_1+h_2-h_3} z_{23}^{h_2+h_3-h_1} z_{13}^{h_3+h_1-h_2}}$$

$$\times \frac{1}{\bar{z}_{12}^{\bar{h}_1+\bar{h}_2-\bar{h}_3} \bar{z}_{23}^{\bar{h}_2+\bar{h}_3-\bar{h}_1} \bar{z}_{13}^{\bar{h}_3+\bar{h}_1-\bar{h}_2}} \tag{5.26}$$

Again, the sum of the spins of the holomorphic part cancels that of the antiholomorphic part, thus ensuring rotation invariance.

The forms (5.25) and (5.26) of the simple correlators raises the question of multivaluedness and locality. Indeed, the two-point function (5.25) will have a branch cut at $z_1 = z_2$, $\bar{z}_1 = \bar{z}_2$ if the spin s of the two fields is not an integer or a half-integer. This is an aspect of the spin-statistics theorem. However, in two dimensions it is possible to bypass this theorem. The price to pay is the introduction of fields, called *parafermions*, which have a mutual long-ranged interaction. These fields will not be studied in this volume.

As before, global conformal invariance does not fix the precise form of the four-point function and beyond, because of the existence of anharmonic ratios. However, in two dimensions the number of independent anharmonic ratios is reduced, since the four points of the ratio are forced to lie in the same plane, which leads to an additional linear relation between them. Indeed, we have

$$\eta = \frac{z_{12}z_{34}}{z_{13}z_{24}} \qquad 1-\eta = \frac{z_{14}z_{23}}{z_{13}z_{24}} \qquad \frac{\eta}{1-\eta} = \frac{z_{12}z_{34}}{z_{14}z_{23}} \tag{5.27}$$

The four-point function may then depend on η and $\bar{\eta}$ in an arbitrary way—provided the result is real. The general expression (4.62) translates into

$$\langle \phi_1(x_1)\dots\phi_4(x_4)\rangle = f(\eta,\bar{\eta})\prod_{i<j}^4 z_{ij}^{h/3-h_i-h_j}\bar{z}_{ij}^{\bar{h}/3-\bar{h}_i-\bar{h}_j} \tag{5.28}$$

where $h = \sum_{i=1}^4 h_i$ and $\bar{h} = \sum_{i=1}^4 \bar{h}_i$. This form for the four-point function may also be understood as follows. Given three distinct points z_1 to z_3, it is always possible to find a global conformal transformation that maps these three points to three other points fixed in advance, for instance 0, 1, and the point at infinity. Indeed, the transformations (5.12) involve three independent complex parameters. Consider the anharmonic ratio η above. If we use a global conformal map to send

z_1 to 1, z_2 to ∞, and z_3 to 0, then $\eta = -z_4$ and a generic four-point function will depend on this last point.

The expression (5.28) may, of course, take different forms, since the product multiplying $f(\eta, \bar{\eta})$ may be modified by insertions of anharmonic ratios. Take, for instance, the four-point function of a single field ϕ of conformal dimension $h = \bar{h}$. Eq. (5.28) becomes

$$\langle \phi(x_1) \ldots \phi(x_4) \rangle = f(\eta, \bar{\eta})\left\{ \left(z_{12}z_{13}z_{14}z_{23}z_{24}z_{34} \right)^{-2h/3} \times \text{c.c.} \right\} \qquad (5.29)$$

(c.c. stands for "complex conjugate"). This may also be expressed as

$$f(\eta, \bar{\eta}) \left\{ \frac{(1-\eta)^{4h/3}}{\eta^{2h/3}} \frac{1}{(z_{14}z_{23})^{2h}} \times \text{c.c.} \right\} \qquad (5.30)$$

or as follows:

$$f(\eta, \bar{\eta}) \left\{ [\eta(1-\eta)]^{4h/3} \left(\frac{z_{13}z_{24}}{z_{12}z_{23}z_{14}z_{34}} \right)^{2h} \times \text{c.c.} \right\} \qquad (5.31)$$

§5.2. Ward Identities

5.2.1. Holomorphic Form of the Ward Identities

In Chap. 4 we have derived a set of Ward identities associated with translation, rotation, and scale invariance: Eqs. (4.63), (4.66), and (4.68), respectively. In so doing, we used the canonical definition of the energy-momentum tensor, with suitable modifications needed to make it symmetric and traceless.[3] Recall that the tracelessness of the energy-momentum tensor implies the conformal invariance of the action. Let us assemble these three Ward identities:[4]

$$\frac{\partial}{\partial x^\mu} \langle T^\mu{}_\nu(x)X \rangle = -\sum_{i=1}^n \delta(x-x_i) \frac{\partial}{\partial x_i^\nu} \langle X \rangle$$

$$\varepsilon_{\mu\nu} \langle T^{\mu\nu}(x)X \rangle = -i \sum_{i=1}^n s_i \delta(x-x_i)\langle X \rangle \qquad (5.32)$$

$$\langle T^\mu{}_\mu(x)X \rangle = -\sum_{i=1}^n \delta(x-x_i)\Delta_i \langle X \rangle$$

Here X stands for a string of n primary fields $\Phi(x_1) \cdots \Phi(x_n)$. In the second equation we have used the specific two-dimensional form $s_i \varepsilon_{\mu\nu}$ of the spin generators $S^i_{\mu\nu}$, where $\varepsilon_{\mu\nu}$ is the antisymmetric tensor and s_i is the spin of the field ϕ_i.

[3] We have not shown in general that the energy-momentum tensor of a two-dimensional scale-invariant theory can always be made traceless, but we know no example of the contrary.

[4] Ward identities, whether in Cartesian or holomorphic form, are valid only in the sense of distributions, that is, when integrated against suitable test functions.

We wish to rewrite these identities in terms of complex coordinates (cf. Eq. (5.6)) and complex components. We use expressions (5.7) and (5.8) for the metric tensor and the antisymmetric tensor, respectively. For the delta functions we use the identity

$$\delta(x) = \frac{1}{\pi}\partial_{\bar{z}}\frac{1}{z} = \frac{1}{\pi}\partial_{z}\frac{1}{\bar{z}} \tag{5.33}$$

This identity is justified as follows. We consider a vector F^{μ} whose divergence is integrated within a region M of the complex plane bounded by the contour ∂M. Gauss's theorem may be applied:

$$\int_{M} d^{2}x\, \partial_{\mu}F^{\mu} = \int_{\partial M} d\xi_{\mu}\, F^{\mu} \tag{5.34}$$

where $d\xi_{\mu}$ is an outward-directed differential of circumference, orthogonal to the boundary ∂M of the domain of integration. It is more convenient to use a counterclockwise differential ds^{ρ}, parallel to the contour ∂M: $d\xi_{\mu} = \varepsilon_{\mu\rho}ds^{\rho}$. In terms of complex coordinates, the above surface integral is nothing but a contour integral, where the (anti)holomorphic component of ds^{ρ} is dz $(d\bar{z})$:

$$\int_{M} d^{2}x\, \partial_{\mu}F^{\mu} = \int_{\partial M} \left\{ dz\, \varepsilon_{\bar{z}z}F^{\bar{z}} + d\bar{z}\, \varepsilon_{z\bar{z}}F^{z} \right\}$$
$$= \frac{1}{2}i \int_{\partial M} \left\{ -dz\, F^{\bar{z}} + d\bar{z}\, F^{z} \right\} \tag{5.35}$$

Here the contour ∂M circles counterclockwise. If $F^{\bar{z}}$ (F^{z}) is holomorphic (anti-holomorphic), then Cauchy's theorem may be applied; otherwise the contour ∂M must stay fixed. We consider then a holomorphic function $f(z)$ and check the correctness of the first representation in Eq. (5.33) by integrating it against $f(z)$ within a neighborhood M of the origin:

$$\int_{M} d^{2}x\, \delta(x)f(z) = \frac{1}{\pi} \int_{M} d^{2}x\, f(z)\partial_{\bar{z}}\frac{1}{z}$$
$$= \frac{1}{\pi} \int_{M} d^{2}x\, \partial_{\bar{z}} \left(\frac{f(z)}{z} \right) \tag{5.36}$$
$$= \frac{1}{2\pi i} \int_{\partial M} dz\frac{f(z)}{z}$$
$$= f(0)$$

In the second equation we have used the assumption that $f(z)$ is analytic within M, in the third equation we used the form (5.35) of Gauss's theorem with $F^{\bar{z}} = f(z)/\pi z$ and $F^{z} = 0$, and in the last equation we used Cauchy's theorem. A similar proof may be applied to the second representation in Eq. (5.33), this time with an antiholomorphic function $\bar{f}(\bar{z})$. Of course, one may in principle use either one of the two representations in Eq. (5.33), but the first one will be useful if the integrand is holomorphic and vice versa.

The Ward identities are then explicitly written as

$$2\pi\partial_z\langle T_{\bar{z}z}X\rangle + 2\pi\partial_{\bar{z}}\langle T_{zz}X\rangle = -\sum_{i=1}^{n}\partial_{\bar{z}}\frac{1}{z-w_i}\partial_{w_i}\langle X\rangle$$

$$2\pi\partial_z\langle T_{\bar{z}\bar{z}}X\rangle + 2\pi\partial_{\bar{z}}\langle T_{z\bar{z}}X\rangle = -\sum_{i=1}^{n}\partial_z\frac{1}{\bar{z}-\bar{w}_i}\partial_{\bar{w}_i}\langle X\rangle$$

$$2\langle T_{z\bar{z}}X\rangle + 2\langle T_{\bar{z}z}X\rangle = -\sum_{i=1}^{n}\delta(x-x_i)\Delta_i\langle X\rangle$$

$$-2\langle T_{z\bar{z}}X\rangle + 2\langle T_{\bar{z}z}X\rangle = -\sum_{i=1}^{n}\delta(x-x_i)s_i\langle X\rangle$$

$$(5.37)$$

The n points x_i are now described by the $2n$ complex coordinates $(w_i, \bar{w}_i)$, on which the set of primary fields X generally depends. If we add and subtract the last two equations of the above, we find

$$2\pi\langle T_{\bar{z}z}X\rangle = -\sum_{i=1}^{n}\partial_{\bar{z}}\frac{1}{z-w_i}h_i\langle X\rangle$$

$$2\pi\langle T_{z\bar{z}}X\rangle = -\sum_{i=1}^{n}\partial_z\frac{1}{\bar{z}-\bar{w}_i}\bar{h}_i\langle X\rangle$$

$$(5.38)$$

where we have chosen the representation (5.33) appropriate to each case and used the definition (5.21) of the holomorphic and antiholomorphic conformal dimensions. Inserting these relations into the first two equations of (5.37), we find

$$\partial_{\bar{z}}\left\{\langle T(z,\bar{z})X\rangle - \sum_{i=1}^{n}\left[\frac{1}{z-w_i}\partial_{w_i}\langle X\rangle + \frac{h_i}{(z-w_i)^2}\langle X\rangle\right]\right\} = 0$$

$$\partial_z\left\{\langle \bar{T}(z,\bar{z})X\rangle - \sum_{i=1}^{n}\left[\frac{1}{\bar{z}-\bar{w}_i}\partial_{\bar{w}_i}\langle X\rangle + \frac{\bar{h}_i}{(\bar{z}-\bar{w}_i)^2}\langle X\rangle\right]\right\} = 0$$

$$(5.39)$$

where we have introduced a renormalized energy-momentum tensor

$$T = -2\pi T_{zz} \qquad\qquad \bar{T} = -2\pi T_{\bar{z}\bar{z}} \qquad\qquad (5.40)$$

Thus the expressions between braces in (5.39) are respectively holomorphic and antiholomorphic: we may write

$$\langle T(z)X\rangle = \sum_{i=1}^{n}\left\{\frac{1}{z-w_i}\partial_{w_i}\langle X\rangle + \frac{h_i}{(z-w_i)^2}\langle X\rangle\right\} + \text{reg.} \qquad (5.41)$$

where "reg." stands for a holomorphic function of z, regular at $z = w_i$. There is a similar expression for the antiholomorphic counterpart.

5.2.2. The Conformal Ward Identity

It is possible to bring the three Ward identities (5.32) into a single relation as follows. Given an arbitrary conformal coordinate variation $\epsilon^\nu(x)$, we can write

$$\partial_\mu(\epsilon_\nu T^{\mu\nu}) = \epsilon_\nu \partial_\mu T^{\mu\nu} + \frac{1}{2}(\partial_\mu \epsilon_\nu + \partial_\nu \epsilon_\mu)T^{\mu\nu} + \frac{1}{2}(\partial_\mu \epsilon_\nu - \partial_\nu \epsilon_\mu)T^{\mu\nu}$$

$$= \epsilon_\nu \partial_\mu T^{\mu\nu} + \frac{1}{2}(\partial_\rho \epsilon^\rho)\eta_{\mu\nu}T^{\mu\nu} + \frac{1}{2}\varepsilon^{\alpha\beta}\partial_\alpha \epsilon_\beta \varepsilon_{\mu\nu}T^{\mu\nu}$$

(5.42)

where the relations

$$\frac{1}{2}(\partial_\mu \epsilon_\nu + \partial_\nu \epsilon_\mu) = \frac{1}{2}(\partial_\rho \epsilon^\rho)\eta_{\mu\nu}$$

$$\frac{1}{2}(\partial_\mu \epsilon_\nu - \partial_\nu \epsilon_\mu) = \frac{1}{2}\varepsilon^{\alpha\beta}\partial_\alpha \epsilon_\beta \varepsilon_{\mu\nu}$$

(5.43)

have been used. We note that $\frac{1}{2}\partial_\rho \epsilon^\rho$ is the local scale factor $f(x)$ of Eq. (4.3) and $\frac{1}{2}\varepsilon^{\alpha\beta}\partial_\alpha \epsilon_\beta$ is a local rotation angle. Integrating both sides of (5.42), the three Ward identities (5.32) derived in Sect. 4.3.2 may be encapsulated into

$$\delta_\epsilon \langle X \rangle = \int_M d^2x \, \partial_\mu \langle T^{\mu\nu}(x)\epsilon_\nu(x)X \rangle$$

(5.44)

where $\delta_\epsilon \langle X \rangle$ is the variation of X under a local conformal transformation. Here the integral is taken over a domain M containing the positions of all the fields in the string X.

Since the integrand is the divergence of a vector field F^μ, Gauss's theorem may be used. Applying (5.35) to $F^\mu = \langle T^{\mu\nu}(x)\epsilon_\nu(x)X \rangle$, one finds

$$\delta_{\epsilon,\bar\epsilon} \langle X \rangle = \frac{1}{2}i \int_C \{-dz\langle T^{\bar z \bar z}\epsilon_{\bar z}X \rangle + d\bar z \langle T^{zz}\epsilon_z X \rangle \}$$

(5.45)

We have defined $\epsilon = \epsilon^z$ and $\bar\epsilon = \epsilon^{\bar z}$, respectively holomorphic and antiholomorphic. Note that $\langle T_{\bar z z}X \rangle$ and $\langle T_{zz}X \rangle$ do not contribute to the contour integrals, since the contours do not exactly go through the positions contained in X, and since these expressions vanish outside these points, according to Eq. (4.68).[5] Finally, substituting the definition (5.40), we obtain the so-called *conformal Ward identity*:[6]

$$\boxed{\delta_{\epsilon,\bar\epsilon} \langle X \rangle = -\frac{1}{2\pi i}\oint_C dz \, \epsilon(z)\langle T(z)X \rangle + \frac{1}{2\pi i}\oint_C d\bar z \, \bar\epsilon(\bar z)\langle \bar T(\bar z)X \rangle}$$

(5.46)

Again, the counterclockwise contour C needs only to include all the positions $(w_i, \bar w_i)$ of the fields contained in X. The relative sign of the two terms on the

---

[5] Of course, Eq. (4.68) itself holds only for primary fields; however, this specific property is general: it only depends on the locality of the transformation properties of X.

[6] Some readers may be puzzled by the sign appearing in front this equation, since many review papers on the subject have it differently. This sign reflects our conventions on what is the variation of a field under a symmetry transformation, stated in Eq. (2.114) and its infinitesimal version (2.125).

r.h.s. reflects the use of a counterclockwise integration contour for the antiholo-morphic variable $\bar{z}$ or, said otherwise, that Cauchy's theorem has been complex conjugated (and $2\pi i \to -2\pi i$).

In deriving the identity (5.46), we have used the property that the fields in the set X are primary, through the Ward identities (5.32). However, the validity of Eq. (5.46) extends beyond primary fields, and may be taken as a definition of the effect of conformal transformations on an arbitrary local field within a correlation function. Indeed, the r.h.s. of the identities (5.32) needs not have this precise form in order for Eq. (5.46) to follow. However, the variation $\delta\Phi$ of the local field Φ under a conformal transformation should be local, ensuring the presence of delta functions $\delta(x - x_i)$ on the r.h.s. of Eq. (5.32).

If the fields in X are primary, the integral in the conformal Ward identity (5.46) may be done by the method of residues:

$$\delta_\epsilon \langle X \rangle = -\sum_i \left(\epsilon(w_i)\partial_{w_i} + \partial\epsilon(w_i)h_i \right) \langle X \rangle \tag{5.47}$$

We recover formula (5.23) for the variation of a primary field under an infinitesimal holomorphic conformal mapping:

$$\delta_\epsilon \phi = -\epsilon\partial\phi - h\phi\partial\epsilon \tag{5.48}$$

It is interesting to apply the conformal Ward identity to global conformal trans-formations (the $SL(2, \mathbb{C})$ mappings of Eq. (5.12)). According to the discussion surrounding Eq. (2.159), the variation $\delta_\epsilon \langle X \rangle$ must vanish for infinitesimal $SL(2, \mathbb{C})$ mappings, since they constitute a true symmetry of the theory. Such infinitesimal mappings have the form

$$f(z) = \frac{(1 + \alpha)z + \beta}{\gamma z + 1 - \alpha} \tag{5.49}$$

where α, β, and γ are infinitesimal. At first order, the coordinate variation $\epsilon(z)$ is

$$\epsilon(z) = \beta + 2\alpha z - \gamma z^2 \tag{5.50}$$

For α, β, and γ arbitrary, this implies the following three relations on correlators of primary fields:

$$\sum_i \partial_{w_i} \langle \phi_1(w_1) \cdots \phi_n(w_n) \rangle = 0$$

$$\sum_i (w_i\partial_{w_i} + h_i) \langle \phi_1(w_1) \cdots \phi_n(w_n) \rangle = 0 \tag{5.51}$$

$$\sum_i (w_i^2\partial_{w_i} + 2w_ih_i) \langle \phi_1(w_1) \cdots \phi_n(w_n) \rangle = 0$$

It is a simple matter to check that the two- and three-point functions (5.25) and (5.26) satisfy these constraints. In fact, it is possible to infer the forms (5.25) and (5.26) from the above relations. The relations (5.51) simply embody global confor-mal invariance. In the first of these relations we recognize the obvious consequence of translation invariance.

The Ward identity (5.46) sums up the consequences of local conformal symmetry on correlation functions, and is the main result of this section. It should be mentioned that its application rests on the assumption that the energy-momentum tensor is regular, meaning that it is everywhere well-defined. In particular, $T(0)$ should be finite (in the sense of correlation functions). This implies that $T(z)$ should decay as z^{-4} as $z \to \infty$. This may be seen as follows: Since the energy-momentum tensor is symmetric, traceless, and represents an energy density, it should have scaling dimension 2 and spin 2, leading to conformal dimensions $h = \bar{h} = 2$. Under the global conformal transformation $z \to w = 1/z$, it should transform as

$$T'(w) = \left(\frac{dw}{dz} \right)^{-2} T(z) = z^4 T(z) \tag{5.52}$$

Since the resulting tensor $T'(1/z)$ is just as regular as $T(z)$ the condition that $T'(0)$ be finite implies that $T(z)$ decay as z^{-4} as $z \to \infty$. This may be argued differently: The trivial correlator $\langle 1 \rangle$ must be invariant under an infinitesimal special conformal transformation. In other words,

$$\delta_\epsilon \langle 1 \rangle = -\frac{1}{2\pi i} \oint_C dz \, \epsilon(z) \langle T(z) \rangle = 0 \tag{5.53}$$

This must be true for any contour circling the point at infinity. Since $\epsilon(z)$ is quadratic in z for special conformal transformations, $T(z)$ must behave as z^{-4} near infinity if no residue is to be picked up around that point.

5.2.3. Alternate Derivation of the Ward Identities

This subsection provides an alternate derivation of the Ward identities (4.63), (4.66), and (4.68), based on the quantum definition of the energy-momentum tensor, given by Eqs. (2.202) or (2.203). The advantage of proceeding this way is to avoid the hypothesis that the canonical energy-momentum tensor can be made traceless in two dimensions. The following demonstration is not specific to two dimensions, except for scale invariance, where the aspects particular to two dimensions will be stressed. Accordingly, the formalism will be as general as possible, without holomorphic coordinates. The reader willing to accept the use of the Ward identity (4.68) in dimension two may skip this subsection, since nothing in the remainder of the text rests on it.

We shall assume that the action may be expressed on a Riemannian manifold in terms of a collection Φ of fields and of a tetrad e_μ^a (see App. 2.C for an introduction to tetrads):

$$S = \int d^2x \, e \, \mathcal{L}(\Phi, D_\mu \Phi, e_\mu^a) \tag{5.54}$$

The use of tetrads is necessary if the derivation is to apply to theories involving spinor fields (e.g., Dirac fermions). Here $e = \det(e_\mu^a)$ ensures that the measure $e \, d^2x$ is reparametrization invariant, and D_μ is the covariant derivative appropriate to the field Φ: it reduces to ∂_μ for a scalar field. For instance, the action for a simple scalar field ϕ is

$$S = \int d^2x\, e \left\{ e^{a\mu} \partial_\mu \phi\, e^{a\nu} \partial_\nu \phi - V(\phi) \right\} \qquad (5.55)$$

Recall that the greek (or Einstein) index of the tetrad is raised and lowered with the help of the metric tensor $g_{\mu\nu}$, whereas the Latin (or Lorentz) index is moved with the help of the Minkowski tensor η_{ab}.

Translation invariance—that is, the absence of *explicit* dependence of the Lagrangian density upon the coordinate of the local field—is generalized into *reparametrization invariance* on a Riemannian manifold. The action and the functional integration measure should be independent of the coordinate system used. Under a reparametrization $x \to x'(x)$ the tetrad e_μ^a and the fields transform as follows:

$$e_\mu^a \to e_\mu'^a = \frac{\partial x^\nu}{\partial x'^\mu} e_\nu^a$$

$$\Phi(x) \to \Phi'(x') = \Phi(x) \qquad (5.56)$$

Covariant derivatives transform like tensors of rank 1, like any quantity with one Einstein index. In the tetrad formalism the local fields $\Phi(x)$ do not carry Einstein indices, but they are affected by reparametrizations through their arguments and covariant derivatives.

In order to derive the Ward identity associated with reparametrization invariance, we first consider a generic correlation function $\langle X \rangle_e$ in some background tetrad e (as before, we denote by X a product $\phi_1(x_1)\dots\phi_n(x_n)$ of various fields taken at different positions):

$$Z_e \langle X \rangle_e = \int [d\Phi]_e\, X\, e^{-S[\Phi,e]} \qquad (5.57)$$

where Z_e is the vacuum functional. Implicit in this expression is the choice of a coordinate system. We then perform an infinitesimal reparametrization $x' = x + \xi(x)$. The variations of the tetrad and fields is then

$$\delta\Phi(x) = -\xi^\mu \partial_\mu \Phi(x)$$

$$\delta e_\mu^a = -\partial_\nu e_\mu^a \xi^\nu - \partial_\mu \xi^\nu e_\nu^a \qquad (5.58)$$

The above variations reflect a change in the *functional dependence* of the fields on the coordinates. We then assume that the action and the measure are invariant under such variations:

$$S[\Phi + \delta\Phi, e + \delta e] = S[\Phi, e]$$

$$[d\Phi + d\delta\Phi]_{e+\delta e} = [d\Phi]_e \qquad (5.59)$$

The effect of this infinitesimal reparametrization on the correlation function is

$$
\begin{aligned}
Z_{e+\delta_e}\langle X + \delta X\rangle_{e+\delta e} &= \int [d\Phi + d\delta\Phi]_{e+\delta e}(X + \delta X)e^{-S[\Phi+\delta\Phi,e+\delta e]}\\
&= \int [d\Phi]_e(X + \delta X)e^{-S[\Phi,e]} \qquad\qquad (5.60)\\
&= Z_e\langle X\rangle_e + Z_e\langle\delta X\rangle_e
\end{aligned}
$$

In particular, by taking $X = 1$ we conclude that $Z_{e+\delta e} = Z_e$: the vacuum functional is reparametrization invariant. Therefore, we may write

$$
\langle X + \delta X\rangle_{e+\delta e} = \langle X\rangle_e + \langle\delta X\rangle_e \qquad\qquad (5.61)
$$

On the other hand, a change of functional integration variables from $\Phi + \delta\Phi$ to Φ in the first of Eqs. (5.60) yields

$$
\begin{aligned}
Z_{e+\delta e}\langle X + \delta X\rangle_{e+\delta e} &= \int [d\Phi]_{e+\delta e} X\, e^{-S[\Phi,e+\delta e]}\\
&= \int [d\Phi]_e X\, e^{-S[\Phi,e]}\left\{1 + \int d^2x\, e\, \delta e^a_\mu T^\mu_a\right\} \qquad (5.62)\\
&= Z_e\langle X\rangle_e + Z_e\int d^2x\, e\, \delta e^a_\mu \langle T^\mu_a X\rangle
\end{aligned}
$$

where we have used the quantum definition (2.203) of the energy-momentum tensor. Comparing Eqs. (5.61) and (5.62), we conclude that

$$
\langle\delta X\rangle_e = \int d^2x\, e\, \delta e^a_\mu \langle T^\mu_a X\rangle_e \qquad\qquad (5.63)
$$

Strictly speaking, this identity is true only when δX and δe^a_μ are obtained through an infinitesimal reparametrization (5.58). Since these variations involve d parameters in d dimensions, the number of Ward identities implied is d, corresponding to the conservation of energy and momentum. If we substitute the variations (5.58) into (5.63) and restrict ourselves to flat space with $e^a_\mu = \delta^a_\mu$, we obtain

$$
\begin{aligned}
\langle\delta X\rangle_e &= -\sum_i \xi^\nu(x_i)\frac{\partial}{\partial x_i^\nu}\langle X\rangle\\
\int d^2x\, e\, \delta e^a_\mu \langle T^\mu_a X\rangle_e &= -\int d^2x\, \partial_\mu\xi^\nu\langle T^\mu_{\ \nu}X\rangle = \int d^2x\, \xi^\nu\partial_\mu\langle T^\mu_{\ \nu}X\rangle
\end{aligned} \qquad (5.64)
$$

Since the function $\xi^\nu(x)$ is arbitrary, this allows us to write our first Ward identity:

$$
\frac{\partial}{\partial x^\mu}\langle T^\mu_{\ \nu}(x)X\rangle = -\sum_{i=1}^n \delta(x - x_i)\frac{\partial}{\partial x_i^\nu}\langle X\rangle \qquad\qquad (5.65)
$$

This indeed coincides with Eq. (4.63).

In order to obtain the second Ward identity associated with rotation (or Lorentz) invariance, we must perform on the fields and tetrad an infinitesimal local rotation:

$$e^a_\mu \rightarrow e^a_\mu + \omega^{ab}(x)e_{b\mu}$$
$$\phi_i \rightarrow \phi_i - \frac{i}{2}\omega^{ab}(x)S_{i,ab}\phi_i \tag{5.66}$$

Here $S_{i,ab}$ is the spin generator for the field ϕ_i, and $\omega^{ab} = -\omega^{ba}$. The use of tetrads (or of a metric tensor in arbitrary coordinates) has promoted rotation invariance to the status of a local symmetry. The action and the integration measure are invariant under such local rotations, and consequently Eq. (5.59) still holds, except that the variations δe and $\delta \Phi$ are of the form above. The same argument applies and the identity (5.63) follows. If we substitute the explicit form of the variation, the flat space form of the tetrad, and if we use the arbitrariness of the antisymmetric function ω^{ab}, we obtain the following Ward identity:

$$\langle T_{\mu\nu}(x)X\rangle - \langle T_{\nu\mu}(x)X\rangle = -i\sum_{i=1}^{n}\delta(x-x_i)S_{i,\mu\nu}\langle X\rangle \tag{5.67}$$

associated with rotation invariance. This, apart from the covariant indices, coincides with Eq. (4.66).

Finally, we derive the Ward identity associated with scale invariance. We perform an infinitesimal, local scale transformation of the frames:

$$e^a_\mu \rightarrow e^a_\mu + \epsilon(x)e^a_\mu$$
$$\phi_i \rightarrow \phi_i - \epsilon(x)\Delta_i\phi_i \tag{5.68}$$

The scale factor $\Lambda(x)$ of Eq. (4.1) is here equal to $1 + 2\epsilon(x)$, and, according to Eqs. (4.32) and (4.33), the variation of a quasi-primary field is indeed given by the above in terms of its scaling dimension Δ_i. Since we are performing an arbitrary local scaling, only primary fields (as opposed to quasi-primary) will transform as above. It is here that we must distinguish the case of two dimensions from the others. In three or more dimensions an action cannot be invariant under a local scale transformation: The use of tetrads and covariant derivatives allows us to define actions invariant under local rotations of the frames, but not under local scalings. In contrast, the two-dimensional conformal group includes local scale transformations and we may proceed as before, and end up with the following Ward identity, the same as Eq. (4.68):

$$\langle T^\mu_{\ \mu}(x)X\rangle = -\sum_{i=1}^{n}\delta(x-x_i)\Delta_i\langle X\rangle \tag{5.69}$$

§5.3. Free Fields and the Operator Product Expansion

It is typical of correlation functions to have singularities when the positions of two or more fields coincide. This reflects the infinite fluctuations of a quantum field taken at a precise position. To be more precise, the average

$$\phi_{\text{av.}} \equiv \frac{1}{V} \int_V d^2x \, \phi(x) \tag{5.70}$$

of a quantum field within a volume V has a variance $\langle \phi_{\text{av.}} \phi_{\text{av.}} \rangle$ which diverges as $V \to 0$. The operator product expansion, or OPE, is the representation of a product of operators (at positions z and w, respectively) by a sum of terms, each being a single operator, well-defined as $z \to w$, multiplied by a c-number function of $z - w$, possibly diverging as $z \to w$, and which embodies the infinite fluctuations as the two positions tend toward each other.

The holomorphic version (5.41) of the Ward identity gives the singular behavior of the correlator of the field $T(z)$ with primary fields $\phi_i(w_i, \bar{w}_i)$ as z approaches the points w_i. The OPE of the energy-momentum tensor with primary fields is written simply by removing the brackets $\langle \ldots \rangle$, it being understood that the OPE is meaningful only within correlation functions. For a single primary field ϕ of conformal dimensions h and $\bar{h}$, we have

$$T(z)\phi(w,\bar{w}) \sim \frac{h}{(z-w)^2}\phi(w,\bar{w}) + \frac{1}{z-w}\partial_w\phi(w,\bar{w})$$
$$\bar{T}(\bar{z})\phi(w,\bar{w}) \sim \frac{\bar{h}}{(\bar{z}-\bar{w})^2}\phi(w,\bar{w}) + \frac{1}{\bar{z}-\bar{w}}\partial_{\bar{w}}\phi(w,\bar{w}) \tag{5.71}$$

Whenever appearing in OPEs, the symbol $\sim$ will mean equality modulo expressions regular as $w \to z$. Of course, the OPE contains also an infinite number of regular terms which, for the energy-momentum tensor, cannot be obtained from the conformal Ward identity. In general, we would write the OPE of two fields $A(z)$ and $B(w)$ as

$$A(z)B(w) = \sum_{n=-\infty}^{N} \frac{\{AB\}_n(w)}{(z-w)^n} \tag{5.72}$$

where the composite fields $\{AB\}_n(w)$ are nonsingular at $w = z$. For instance, $\{T\phi\}_1 = \partial_w\phi(w)$.

We stress that, so far, the quantities appearing in Eq. (5.71) are not operators but simply fields occurring within correlation functions. We shall now proceed with specific examples, in order to familiarize ourselves with basic techniques and with simple (but important) systems.

5.3.1. The Free Boson

From the point of view of the canonical or path integral formalism, the simplest conformal field theory is that of a free massless boson φ, with the following action:

$$S = \frac{1}{2}g \int d^2x \, \partial_\mu \varphi \partial^\mu \varphi \qquad (5.73)$$

where g is some normalization parameter that we leave unspecified at the moment. The two-point function, or propagator, has been calculated in Section 2.3:

$$\langle \varphi(x)\varphi(y) \rangle = -\frac{1}{4\pi g} \ln(x-y)^2 + \text{const.} \qquad (5.74)$$

In terms of complex coordinates, this is

$$\langle \varphi(z,\bar{z})\varphi(w,\bar{w}) \rangle = -\frac{1}{4\pi g}\left\{ \ln(z-w) + \ln(\bar{z}-\bar{w}) \right\} + \text{const.} \qquad (5.75)$$

The holomorphic and antiholomorphic components can be separated by taking the derivatives $\partial_z \varphi$ and $\partial_{\bar{z}} \varphi$:

$$\langle \partial_z\varphi(z,\bar{z})\partial_w\varphi(w,\bar{w}) \rangle = -\frac{1}{4\pi g}\frac{1}{(z-w)^2}$$
$$\langle \partial_{\bar{z}}\varphi(z,\bar{z})\partial_{\bar{w}}\varphi(w,\bar{w}) \rangle = -\frac{1}{4\pi g}\frac{1}{(\bar{z}-\bar{w})^2} \qquad (5.76)$$

In the following we shall concentrate on the holomorphic field $\partial\varphi \equiv \partial_z\varphi$. It is now clear that the OPE of this field with itself is

$$\boxed{\partial\varphi(z)\partial\varphi(w) \sim -\frac{1}{4\pi g}\frac{1}{(z-w)^2}} \qquad (5.77)$$

This OPE reflects the bosonic character of the field: exchanging the two factors does not affect the correlator.

The energy-momentum tensor associated with the free massless boson is

$$T_{\mu\nu} = g(\partial_\mu\varphi\partial_\nu\varphi - \frac{1}{2}\eta_{\mu\nu}\partial_\rho\varphi\partial^\rho\varphi) \qquad (5.78)$$

Its quantum version (5.40) in complex coordinates is

$$T(z) = -2\pi g : \partial\varphi\partial\varphi : \qquad (5.79)$$

Like all composite fields, the energy-momentum tensor has to be normal ordered, in order to ensure the vanishing of its vacuum expectation value. More explicitly, the exact meaning of the above expression is

$$T(z) = -2\pi g \lim_{w \to z} \left(\partial\varphi(z)\partial\varphi(w) - \langle \partial\varphi(z)\partial\varphi(w) \rangle \right) \qquad (5.80)$$

The OPE of $T(z)$ with $\partial\varphi$ may be calculated from Wick's theorem:

$$T(z)\partial\varphi(w) = -2\pi g : \partial\varphi(z)\partial\varphi(z): \partial\varphi(w)$$

$$\sim -4\pi g : \partial\varphi(z)\partial\,\overline{\varphi(z)} : \partial\varphi(w) \tag{5.81}$$

$$\sim \frac{\partial\varphi(z)}{(z-w)^2}$$

By expanding $\partial\varphi(z)$ around w, we arrive at the OPE

$$T(z)\partial\varphi(w) \sim \frac{\partial\varphi(w)}{(z-w)^2} + \frac{\partial_w^2\varphi(w)}{(z-w)} \tag{5.82}$$

This shows that $\partial\varphi$ is a primary field with conformal dimension $h = 1$. This was expected, since φ has no spin and no scaling dimension; hence its derivative has scaling dimension 1.

Wick's theorem also allows us to calculate the OPE of the energy-momentum tensor with itself:

$$T(z)T(w) = 4\pi^2 g^2 : \partial\varphi(z)\partial\varphi(z):: \partial\varphi(w)\partial\varphi(w):$$

$$\sim \frac{1/2}{(z-w)^4} - \frac{4\pi g : \partial\varphi(z)\partial\varphi(w):}{(z-w)^2} \tag{5.83}$$

$$\sim \frac{1/2}{(z-w)^4} + \frac{2T(w)}{(z-w)^2} + \frac{\partial T(w)}{(z-w)}$$

In the second equation the first term is the result of two double contractions, whereas the second term comes from four single contractions. We immediately see that the energy-momentum tensor is not strictly a primary field, because of the anomalous term $\frac{1}{2}/(z-w)^4$, which does not appear in Eq. (5.71).

5.3.2. The Free Fermion

In two dimensions, the Euclidian action of a free Majorana fermion is

$$S = \frac{1}{2}g \int d^2x \, \Psi^\dagger \gamma^0 \gamma^\mu \partial_\mu \Psi \tag{5.84}$$

where the Dirac matrices γ^μ satisfy the so-called Dirac algebra:

$$\gamma^\mu\gamma^\nu + \gamma^\nu\gamma^\mu = 2\eta^{\mu\nu} \tag{5.85}$$

If $\eta^{\mu\nu} = \text{diag}(1, 1)$, a representation thereof is[7]

$$\gamma^0 = \begin{pmatrix} 0 & 1 \\ 1 & 0 \end{pmatrix} \qquad \gamma^1 = i\begin{pmatrix} 0 & -1 \\ 1 & 0 \end{pmatrix} \tag{5.86}$$

---

[7] The factor of i in γ^1 was not present in Sect. 2.1.2 since we were then working in Minkowski space-time.

and therefore

$$\gamma^0(\gamma^0 \partial_0 + \gamma^1 \partial_1) = 2 \begin{pmatrix} \partial_{\bar{z}} & 0 \\ 0 & \partial_z \end{pmatrix} \tag{5.87}$$

Writing the two-component spinor Ψ as $(\psi, \bar{\psi})$, the action becomes

$$S = g \int d^2x \; (\bar{\psi} \partial \bar{\psi} + \psi \bar{\partial} \psi) \tag{5.88}$$

The classical equations of motion are $\partial \bar{\psi} = 0$ and $\bar{\partial} \psi = 0$, whose solutions are any holomorphic function $\psi(z)$ and any antiholomorphic function $\bar{\psi}(\bar{z})$.

Our first task is to calculate the propagator $\langle \Psi_i(x)\Psi_j(y) \rangle$ $(i, j = 1, 2)$. This is done by expressing the action as

$$S = \frac{1}{2} \int d^2x \, d^2y \; \Psi_i(x) A_{ij}(x, y) \Psi_j(y) \tag{5.89}$$

where we have defined the kernel

$$A_{ij}(x, y) = g\delta(x - y)(\gamma^0 \gamma^\mu)_{ij} \, \partial_\mu \tag{5.90}$$

From previous knowledge of Gaussian integrals of Grassmann variables, the two-point function is then $K_{ij}(x, y) = (A^{-1})_{ij}(x, y)$, or[8]

$$g\delta(x - y)(\gamma^0 \gamma^\mu)_{ik} \frac{\partial}{\partial x^\mu} K_{kj}(x, y) = \delta(x - y)\delta_{ij} \tag{5.91}$$

In terms of complex coordinates, this becomes

$$2g \begin{pmatrix} \partial_{\bar{z}} & 0 \\ 0 & \partial_z \end{pmatrix} \begin{pmatrix} \langle \psi(z, \bar{z})\psi(w, \bar{w}) \rangle & \langle \psi(z, \bar{z})\bar{\psi}(w, \bar{w}) \rangle \\ \langle \bar{\psi}(z, \bar{z})\psi(w, \bar{w}) \rangle & \langle \bar{\psi}(z, \bar{z})\bar{\psi}(w, \bar{w}) \rangle \end{pmatrix}$$

$$= \frac{1}{\pi} \begin{pmatrix} \partial_{\bar{z}} \dfrac{1}{z - w} & 0 \\ 0 & \partial_z \dfrac{1}{\bar{z} - \bar{w}} \end{pmatrix} \tag{5.92}$$

where we translated $x \to (z, \bar{z})$ and $y \to (w, \bar{w})$ and used the representations (5.33) for the delta function. The solution of the above matrix equation is easily read off:

$$\langle \psi(z, \bar{z})\psi(w, \bar{w}) \rangle = \frac{1}{2\pi g} \frac{1}{z - w}$$

$$\langle \bar{\psi}(z, \bar{z})\bar{\psi}(w, \bar{w}) \rangle = \frac{1}{2\pi g} \frac{1}{\bar{z} - \bar{w}} \tag{5.93}$$

$$\langle \psi(z, \bar{z})\bar{\psi}(w, \bar{w}) \rangle = 0$$

---

[8] This differential equation may also be derived from the equations of motion, as done in Ex. (2.2) for the boson.

These, after differentiation, imply

$$\langle \partial_z \psi(z,\bar z)\psi(w,\bar w)\rangle = -\frac{1}{2\pi g}\frac{1}{(z-w)^2}$$

$$\langle \partial_z \psi(z,\bar z)\partial_w \psi(w,\bar w)\rangle = -\frac{1}{\pi g}\frac{1}{(z-w)^3} \qquad (5.94)$$

and so on. The OPE of the fermion with itself (holomorphic components) is then

$$\boxed{\psi(z)\psi(w) \sim \frac{1}{2\pi g}\frac{1}{z-w}} \qquad (5.95)$$

Again, this OPE reflects the anticommuting character of the field: exchanging the two factors $\psi(z)$ and $\psi(w)$ produces a sign that is mirrored in the two-point function.

Second, we wish to calculate the OPE of the energy-momentum tensor with ψ and with itself. The canonical energy-momentum tensor for the above action may be found from the general expression (2.165) even if we use holomorphic coordinates, with the indices $\mu = 0, 1$ standing for z and $\bar z$, respectively, provided we start from the expression (5.88) for the action. We find

$$T^{\bar z z} = 2\frac{\partial \mathcal{L}}{\partial \bar\partial \Phi}\partial\Phi \qquad = 2g\psi\partial\psi$$

$$T^{zz} = 2\frac{\partial \mathcal{L}}{\partial \partial \Phi}\bar\partial\Phi \qquad = 2g\bar\psi\partial\bar\psi \qquad (5.96)$$

$$T^{z\bar z} = 2\frac{\partial \mathcal{L}}{\partial \partial \Phi}\partial\Phi - 2\mathcal{L} = -2g\psi\bar\partial\psi$$

We see that the energy-momentum tensor is not identically symmetric, since $T^{z\bar z} \neq 0$. However, $T^{z\bar z}$ vanishes if we use the classical equations of motion. According to the discussion of Section 2.5.1, we need not worry and may keep the energy-momentum tensor in its present form. The standard holomorphic component is then

$$T(z) = -2\pi T_{zz}$$

$$= -\frac{1}{2}\pi T^{\bar z \bar z} \qquad (5.97)$$

$$= -\pi g : \psi(z)\partial\psi(z):$$

where, as before, we have used the normal-ordered product:

$$:\psi\partial\psi:(z) = \lim_{w\to z}\left(\psi(z)\partial\psi(w) - \langle\psi(z)\partial\psi(w)\rangle\right) \qquad (5.98)$$

Again, the OPE between T and the fermion ψ is calculated using Wick's theorem:

$$
T(z)\psi(w) = -\pi g : \psi(z)\partial\psi(z): \psi(w)
$$

$$
\sim \frac{1}{2}\frac{\partial\psi(z)}{z-w} + \frac{1}{2}\frac{\psi(z)}{(z-w)^2} \tag{5.99}
$$

$$
\sim \frac{\frac{1}{2}\psi(w)}{(z-w)^2} + \frac{\partial\psi(w)}{z-w}
$$

In contracting $\psi(z)$ with $\psi(w)$ we have carried $\psi(w)$ over $\partial\psi(z)$, thus introducing a $(-)$ sign by Pauli's principle. We see from this OPE that the fermion ψ has a conformal dimension $h = \frac{1}{2}$.

The OPE of $T(z)$ with itself is calculated in the same way, with, however, a greater number of contractions:

$$
T(z)T(w) = \pi^2 g^2 : \psi(z)\partial\psi(z):: \psi(w)\partial\psi(w):
$$

$$
\sim \frac{1/4}{(z-w)^4} + \frac{2T(w)}{(z-w)^2} + \frac{\partial T(w)}{(z-w)} \tag{5.100}
$$

This OPE has the same form as Eq. (5.83) except for a numerical difference in the anomalous term.

5.3.3. The Ghost System

In string theory applications, there appears another simple system, with the following action:

$$
S = \frac{1}{2}g \int d^2x \, b_{\mu\nu}\partial^\mu c^\nu \tag{5.101}
$$

where the field $b_{\mu\nu}$ is a traceless symmetric tensor, and where both c^μ and $b_{\mu\nu}$ are fermions (i.e., anticommuting fields). These fields are called *ghosts* because they are not fundamental dynamical fields, but rather represent a Jacobian arising from a change of variables in some functional integrals. More precisely, they are known as *reparametrization ghosts*.

The equations of motion are

$$
\partial^\alpha b_{\alpha\mu} = 0 \qquad \partial^\alpha c^\beta + \partial^\beta c^\alpha = 0 \tag{5.102}
$$

In holomorphic form we write $c = c^z$ and $\bar{c} = c^{\bar{z}}$. The only nonzero components of the traceless symmetric tensor $b_{\mu\nu}$ are $b = b_{zz}$ and $\bar{b} = b_{\bar{z}\bar{z}}$. The equations of motion are then

$$
\bar{\partial}b = 0 \qquad\qquad \partial c = 0
$$

$$
\partial\bar{b} = 0 \qquad\qquad \partial\bar{c} = 0 \tag{5.103}
$$

$$
\partial c = -\bar{\partial}\bar{c}
$$

The propagator is calculated in the usual way, by writing the action as

$$S = \frac{1}{2} \int d^2x d^2y \, b_{\mu\nu}(x) A_\alpha^{\mu\nu}(x,y) c^\alpha(y)$$

$$A_\alpha^{\mu\nu}(x,y) = \frac{1}{2} g \, \delta_\alpha^\nu \, \delta(x-y) \, \partial^\mu \tag{5.104}$$

where we must consider (μ, ν) as a single composite index, symmetric under the exchange of μ and ν. The factor of $\frac{1}{2}$ in front of $A_\alpha^{\mu\nu}(x,y)$ compensates the double counting of each pair (μ, ν) in the sum, which should be avoided since $b^{\mu\nu}$ is the same degree of freedom as $b^{\nu\mu}$. Again, the propagator is $K = A^{-1}$, satisfying[9]

$$\frac{1}{2} g \, \delta_\alpha^\mu \, \partial^\nu K_{\mu\nu}^\beta(x,y) = \delta(x-y) \delta_{\alpha\beta} \tag{5.105}$$

or, in complex representation,

$$g \partial_{\bar{z}} K_{zz}^\beta = \frac{1}{\pi} \partial_{\bar{z}} \frac{1}{z-w} \delta_{\beta z} \tag{5.106}$$

which implies

$$\langle b(z)c(w) \rangle = K_{zz}^z(z,w) = \frac{1}{\pi g} \frac{1}{z-w} \tag{5.107}$$

In OPE form, this is

$$\boxed{b(z)c(w) \sim \frac{1}{\pi g} \frac{1}{z-w}} \tag{5.108}$$

from which we immediately derive the following:

$$\langle c(z)b(w) \rangle = \frac{1}{\pi g} \frac{1}{z-w}$$

$$\langle b(z)\partial c(w) \rangle = -\frac{1}{\pi g} \frac{1}{(z-w)^2} \tag{5.109}$$

$$\langle \partial b(z)c(w) \rangle = \frac{1}{\pi g} \frac{1}{(z-w)^2}$$

The canonical energy-momentum tensor for this system is

$$T_c^{\mu\nu} = \frac{1}{2} g \left(b^{\mu\alpha} \partial^\nu c_\alpha - \eta^{\mu\nu} b^{\alpha\beta} \partial_\alpha c_\beta \right) \tag{5.110}$$

Again this tensor is not identically symmetric, and should be put in the Belinfante form before proceeding: We add $\partial_\rho B^{\rho\mu\nu}$, where

$$B^{\rho\mu\nu} = -\frac{1}{2} g (b^{\nu\rho} c^\mu - b^{\nu\mu} c^\rho) \tag{5.111}$$

---

[9] Again, this differential equation could also be obtained from the equations of motion, as in Ex. (2.2).

The antisymmetric part of $T_c^{\mu\nu}$ is

$$\frac{1}{2}(T_c^{\mu\nu} - T_c^{\nu\mu}) = \frac{1}{4}g(b^{\mu\alpha}\partial^\nu c_\alpha - b^{\nu\alpha}\partial^\mu c_\alpha) \tag{5.112}$$

and we easily verify, with the help of the classical equations of motion, that this is compensated exactly by the antisymmetric part of $\partial_\rho B^{\rho\mu\nu}$. Therefore, the identically symmetric Belinfante tensor is, after using the equations of motion,

$$T_B^{\mu\nu} = \frac{1}{2}g\left\{b^{\mu\alpha}\partial^\nu c_\alpha + b^{\nu\alpha}\partial^\mu c_\alpha + \partial_\alpha b^{\mu\nu}c^\alpha - \eta^{\mu\nu}b^{\alpha\beta}\partial_\alpha c_\beta\right\} \tag{5.113}$$

This tensor is not only symmetric, but also identically traceless.

The normal-ordered holomorphic component is obtained from the above by setting $\mu = \nu = 1$, that is, by considering $T^{\bar{z}\bar{z}} = 4T_{zz}$:

$$T(z) = \pi g \;:(2\partial c\, b + c\partial b): \tag{5.114}$$

The OPE of the energy-momentum tensor with c is again calculated using Wick's theorem:

$$\begin{aligned}
T(z)c(w) &= \pi g \;:(2\partial c\, b + c\partial b): c(w) \\
&\sim -\frac{c(z)}{(z-w)^2} + 2\frac{\partial_z c(z)}{z-w} \\
&\sim -\frac{c(w)}{(z-w)^2} + \frac{\partial_w c(w)}{z-w}
\end{aligned} \tag{5.115}$$

Therefore c is a primary field with conformal dimension $h = -1$. On the other hand, b is a primary field with conformal dimension $h = 2$:

$$\begin{aligned}
T(z)b(w) &= \pi g \;:(2\partial c\, b + c\partial b): b(w) \\
&\sim 2\frac{b(z)}{(z-w)^2} - \frac{\partial_z b(z)}{z-w} \\
&\sim 2\frac{b(w)}{(z-w)^2} + \frac{\partial_w b(w)}{z-w}
\end{aligned} \tag{5.116}$$

We note that the anticommuting nature of b and c is crucial in order to obtain the above OPEs. The OPE of T with itself contains many more terms, which add up to the following:

$$\begin{aligned}
T(z)T(w) &= \pi g^2 \;:(2\partial c(z)b(z) + c(z)\partial b(z))::(2\partial c(w)b(w) + c(w)\partial b(w)): \\
&\sim \frac{-13}{(z-w)^4} + \frac{2T(w)}{(z-w)^2} + \frac{\partial T(w)}{(z-w)}
\end{aligned} \tag{5.117}$$

Again, but for a different coefficient of the anomalous term, this OPE has the same form as (5.83).

An alternate theory is obtained by modifying the action in such a way that the OPE of the fields c and b with themselves are not changed, but the energy-momentum tensor is modified, by subtracting a total derivative : $\partial(cb)$: as follows:

$$T(z) = \pi g \; : \partial c \; b: \tag{5.118}$$

We shall call this new theory the *simple ghost system*. The OPE of T with the fields c, b, and with itself is, of course, modified:

$$T(z)c(w) \sim \frac{\partial c(w)}{z - w}$$

$$T(z)b(w) \sim \frac{b(z)}{(z - w)^2} \tag{5.119}$$

$$\sim \frac{b(w)}{(z - w)^2} + \frac{\partial b(w)}{z - w}$$

In this new theory, c is therefore a primary field of conformal dimension $h = 0$, and b is a primary field of conformal dimension $h = 1$. The OPE of T with itself is

$$T(z)T(w) \sim \frac{-1}{(z - w)^4} + \frac{2T(w)}{(z - w)^2} + \frac{\partial T(w)}{(z - w)} \tag{5.120}$$

We still have the same form as above, albeit with a different coefficient in the anomalous term.

§5.4. The Central Charge

The specific models treated in the last section lead us naturally to the following general OPE of the energy-momentum tensor:

$$\boxed{T(z)T(w) \sim \frac{c/2}{(z - w)^4} + \frac{2T(w)}{(z - w)^2} + \frac{\partial T(w)}{(z - w)}} \tag{5.121}$$

where the constant c—not to be confused with the ghost field described above—depends on the specific model under study: it is equal to 1 for the free boson, $\frac{1}{2}$ for the free fermion, -26 for the reparametrization ghosts, and -2 for the simple ghost system. This model-dependent constant is called the *central charge*. Except for this anomalous term, the OPE (5.121) simply means that T is a quasi-primary field with conformal dimension $h = 2$. Bose symmetry and scale invariance make const./$(z - w)^4$ the only sensible addition to the standard OPE (5.71). Moreover, we already know from symmetry considerations that the Schwinger function $\langle T_{\mu\nu}(x)T_{\rho\sigma}(0)\rangle$ takes the form (4.77). This is, of course, compatible with the OPE (5.121), and further confirms that the latter is the most general form the OPE of T with itself can take. Indeed, if we convert Eq. (4.77) to holomorphic coordinates using Eqs. (5.6), (5.7), and (5.40), we find

$$\langle T(z)T(0)\rangle = \frac{c/2}{z^4} \qquad \langle \bar{T}(z)\bar{T}(0)\rangle = \frac{c/2}{\bar{z}^4} \tag{5.122}$$

All other components of the Schwinger function vanish. The constant A of Eq. (4.77) is proportional to the central charge: $A = c/(4\pi^2)$.

The central charge may not be determined from symmetry considerations: its value is determined by the short-distance behavior of the theory. For free fields, as seen in the previous section, it is determined by applying Wick's theorem on the normal-ordered energy-momentum tensor. When two decoupled systems (e.g., two free fields) are put together, the energy-momentum tensor of the total system is simply the sum of the energy-momentum tensors associated with each part, and the associated central charge is simply the sum of the central charges of the parts. Thus, the central charge is somehow an extensive measure of the number of degrees of freedom of the system.

5.4.1. Transformation of the Energy-Momentum Tensor

The departure of the OPE (5.121) from the general form (5.71) means that the energy-momentum tensor does not exactly transform like a primary field of dimension 2, contrary to what we expect classically. According to the conformal Ward identity (5.46) the variation of T under a local conformal transformation is

$$
\begin{aligned}
\delta_\epsilon T(w) &= -\frac{1}{2\pi i} \oint_C dz\, \epsilon(z) T(z) T(w) \\
&= -\frac{1}{12} c\, \partial_w^3 \epsilon(w) - 2T(w)\, \partial_w \epsilon(w) - \epsilon(w)\, \partial_w T(w)
\end{aligned}
\tag{5.123}
$$

The "exponentiation" of this infinitesimal variation to a finite transformation $z \to w(z)$ is

$$
T'(w) = \left(\frac{dw}{dz}\right)^{-2} \left[T(z) - \frac{c}{12}\{w; z\} \right]
\tag{5.124}
$$

where we have introduced the *Schwarzian derivative*:

$$
\{w; z\} = \frac{(d^3 w/dz^3)}{(dw/dz)} - \frac{3}{2}\left(\frac{d^2 w/dz^2}{dw/dz}\right)^2
\tag{5.125}
$$

This induction is far from obvious and we shall be content in verifying it for infinitesimal transformations. For an infinitesimal map $w(z) = z + \epsilon(z)$, the Schwarzian derivative becomes, at first order in ϵ,

$$
\{z + \epsilon; z\} = \frac{\partial_z^3 \epsilon}{1 + \partial_z \epsilon} - \frac{3}{2}\left(\frac{\partial_z^2 \epsilon}{1 + \partial_z \epsilon}\right)^2 \approx \partial_z^3 \epsilon
\tag{5.126}
$$

The infinitesimal version of Eq. (5.124) is therefore, at first order in ϵ,

$$
\begin{aligned}
T'(z + \epsilon) &= T'(z) + \epsilon(z)\partial T(z) \\
&= (1 - 2\partial\epsilon(z))(T(z) - \frac{1}{12}c\partial_z^3 \epsilon(z))
\end{aligned}
\tag{5.127}
$$

or

$$\delta_\epsilon T(z) = T'(z) - T(z)$$

$$= -\frac{1}{12}c\partial_z^3\epsilon(z) - 2\partial_z\epsilon(z)T(z) - \epsilon(z)\partial_z T(z) \qquad (5.128)$$

which indeed coincides with Eq. (5.123).

To confirm the validity of the transformation law (5.124), we must verify the following group property: The result of two successive transformations $z \to w \to u$ should coincide with what is obtained from the single transformation from $z \to u$, that is

$$T''(u) = \left(\frac{du}{dw}\right)^{-2}\left[T'(w) - \frac{c}{12}\{u; w\}\right]$$

$$= \left(\frac{du}{dw}\right)^{-2}\left[\left(\frac{dw}{dz}\right)^{-2}\left[T(z) - \frac{c}{12}\{w; z\}\right] - \frac{c}{12}\{u; w\}\right] \qquad (5.129)$$

$$= \left(\frac{du}{dz}\right)^{-2}\left[T(z) - \frac{c}{12}\{u; z\}\right]$$

The last equality requires the following relation between the Schwarzian derivatives:

$$\{u; z\} = \{w; z\} + \left(\frac{dw}{dz}\right)^2\{u; w\} \qquad (5.130)$$

It is a straightforward exercise to demonstrate that this condition is indeed satisfied. Moreover, if we set $u = z$, we find that

$$\{w; z\} = -\left(\frac{dw}{dz}\right)^2\{z; w\} \qquad (5.131)$$

and this relation allows us to rewrite the transformation law (5.124) as

$$T'(w) = \left(\frac{dw}{dz}\right)^{-2}T(z) + \frac{c}{12}\{z; w\} \qquad (5.132)$$

It is equally straightforward to verify that the Schwarzian derivative of the global conformal map

$$w(z) = \frac{az + b}{cz + d} \qquad (ad - bc = 1) \qquad (5.133)$$

vanishes. This needs to be so, for $T(z)$ is a quasi-primary field. In fact, it can be shown that the Schwarzian derivative in (5.124) is the only possible addition to the tensor transformation law that satisfies the group property (5.130) and vanishes for global conformal transformations.

Instead of providing a long and technical proof of this last statement, we shall derive Eq. (5.124) directly by means of the free boson representation. We write the free boson energy-momentum tensor (5.80) as

$$T(z) = -2\pi g \lim_{\delta \to 0} \left(\partial \varphi(z + \frac{1}{2}\delta) \partial \varphi(z - \frac{1}{2}\delta) + \frac{1}{4\pi g \delta^2} \right) \tag{5.134}$$

Consider the transformation $z \to w(z)$. Since φ has conformal dimension zero, $\partial \varphi$ transforms as

$$\partial_z \varphi(z) = w^{(1)} \partial_w \varphi'(w) \tag{5.135}$$

(here we denote the n-th derivative of w by $w^{(n)}$ in order to lighten the notation). Hence $T(z)$ transforms as

$$
\begin{aligned}
T(z) &= -2\pi g \lim_{\delta \to 0} \left\{ w^{(1)}(z + \frac{1}{2}\delta) w^{(1)}(z - \frac{1}{2}\delta) \partial_w \varphi'(w(z + \frac{1}{2}\delta)) \partial_w \varphi'(w(z - \frac{1}{2}\delta)) \right. \\
&\quad \left. + \frac{1}{4\pi g \delta^2} \right\} \\
&= \lim_{\delta \to 0} \left\{ w^{(1)}(z + \frac{1}{2}\delta) w^{(1)}(z - \frac{1}{2}\delta) \left[-2\pi g :\partial_w \varphi'(w) \partial_w \varphi'(w): \right. \right. \\
&\quad \left. \left. + \frac{1}{2(w(z + \frac{1}{2}\delta) - w(z - \frac{1}{2}\delta))^2} \right] - \frac{1}{2\delta^2} \right\} \\
&= \left(w^{(1)}(z) \right)^2 T'(w) + \lim_{\delta \to 0} \left\{ \frac{w^{(1)}(z + \frac{1}{2}\delta) w^{(1)}(z - \frac{1}{2}\delta)}{2(w(z + \frac{1}{2}\delta) - w(z - \frac{1}{2}\delta))^2} - \frac{1}{2\delta^2} \right\} \\
&= \left(w^{(1)}(z) \right)^2 T'(w) + \frac{1}{12} \left\{ \frac{w^{(3)}}{w^{(1)}} - \frac{3}{2} \left(\frac{w^{(2)}}{w^{(1)}} \right)^2 \right\}
\end{aligned}
\tag{5.136}
$$

Since $c = 1$ for a free boson, we recover (5.124) after isolating $T'(w)$.

5.4.2. Physical Meaning of c

The appearance of the central charge c, also known as the *conformal anomaly*, is related to a "soft" breaking of conformal symmetry by the introduction of a macroscopic scale into the system. In other words, c describes the way a specific system reacts to macroscopic length scales introduced, for instance, by boundary conditions. To make this statement more specific, we consider a generic conformal field theory living on the whole complex plane, and we map this theory on a cylinder of circumference L by way of the transformation

$$z \to w = \frac{L}{2\pi} \ln z \tag{5.137}$$

Then, $dw/dz = L/(2\pi z)$ and the Schwarzian derivative is $1/2z^2$. The energy-momentum tensor $T_{\text{cyl.}}(w)$ on the cylinder is related to the corresponding tensor

$T_{\text{pl.}}(z)$ on the plane by

$$T_{\text{cyl.}}(w) = \left(\frac{2\pi}{L}\right)^2 \left\{T_{\text{pl.}}(z)z^2 - \frac{c}{24}\right\} \tag{5.138}$$

If we assume that the vacuum energy density $\langle T_{\text{pl.}}\rangle$ vanishes on the plane, then taking the expectation value of the above equation yields a nonzero vacuum energy density on the cylinder:

$$\langle T_{\text{cyl.}}(w)\rangle = -\frac{c\pi^2}{6L^2} \tag{5.139}$$

The central charge is seen to be proportional to the *Casimir energy*, the change in the vacuum energy density brought about by the periodicity condition on the cylinder. The Casimir energy naturally goes to zero as the macroscopic scale L goes to infinity.

This remark allows us to relate the central charge to the free energy per unit length of a statistical system defined on a cylinder. The free energy F, which coincides with the connected functional W, varies in the following way when the metric tensor is changed:

$$\delta F = -\frac{1}{2} \int d^2x \sqrt{g}\,\delta g_{\mu\nu}\langle T^{\mu\nu}\rangle \tag{5.140}$$

In cylindrical geometry, we apply an infinitesimal scaling of the circumference: $L \to (1+\varepsilon)L$ or $\delta L = \varepsilon L$. This is realized by applying a coordinate transformation $w^0 \to (1+\varepsilon)w^0$, where w^0 is the coordinate running across the cylinder ($w \equiv w^0 + iw^1$). According to Eq. (2.192), the infinitesimal variation of the coordinate is $\epsilon^\mu = \varepsilon w^0 \delta_{\mu 0}$ and the corresponding variation of the metric is $\delta g_{\mu\nu} = -2\varepsilon \delta_{\mu 0}\delta_{\nu 0}$. Since

$$\langle T^{00}\rangle = \langle T_{zz}\rangle + \langle T_{\bar{z}\bar{z}}\rangle = -(1/\pi)\langle T\rangle = \frac{\pi c}{6L^2} \tag{5.141}$$

the variation of the free energy is

$$\delta F = \int dw^0 dw^1 \frac{\pi c}{6L^2}\frac{\delta L}{L}$$

This equation supposes that $\langle T^{00}\rangle$ vanishes in the $L \to \infty$ limit or, in other words, that $\langle T_{\text{pl.}}(z)\rangle = 0$. If, on the contrary, we suppose that there is a free energy f_0 per unit area in the $L \to \infty$ limit, then the above equation is replaced by

$$\delta F = \int dw^0 dw^1 \left(f_0 + \frac{\pi c}{6L^2}\right)\frac{\delta L}{L}$$

The integral over w^0 gives a trivial factor of L, and we can dispose of the integral over w^1 by defining a free energy F_L per unit length of the cylinder, in terms of which the variation is

$$\delta F_L = \left(f_0 + \frac{\pi c}{6L^2}\right)\delta L \tag{5.142}$$

After integration, it follows immediately that

$$F_L = f_0 L - \frac{\pi c}{6L}$$

(5.143)

This relation is important in the study of finite-size effects of statistical systems and numerical simulations; we shall come back to this in Chap. 11 (cf. also Ex. 3.5).

The central charge also arises when a conformal field theory is defined on a curved two-dimensional manifold. The curvature introduces a macroscopic scale in the system, and the expectation value of the trace of the energy-momentum tensor, instead of vanishing, is proportional to both the curvature R and the central charge c:

$$\langle T^\mu_{\ \mu}(x) \rangle_g = \frac{c}{24\pi} R(x)$$

(5.144)

This quantum breaking of scale invariance is called the *trace anomaly*. The proof of (5.144) is not simple, and is given in App. 5.A for the free boson, although the argument may be generalized to other systems.

Appendix 5.A. The Trace Anomaly

In this appendix we demonstrate Eq. (5.144) for the trace anomaly for a free boson. We consider the generating functional

$$Z[g] = \int [d\varphi]_g \, e^{-S[\varphi,g]}$$
$$= e^{-W[g]}$$

(5.145)

where $S[\varphi, g]$ is the action of a free scalar field in a background metric $g_{\mu\nu}$:

$$S[\varphi,g] = \int d^2x \, \sqrt{g} \, g^{\mu\nu} \partial_\mu \varphi \partial_\nu \varphi$$
$$= - \int d^2x \, \sqrt{g} \, \varphi \Delta \varphi$$

(5.146)

We have introduced the Laplacian operator Δ:

$$\Delta \varphi = \frac{1}{\sqrt{g}} \partial_\mu \left(\sqrt{g} g^{\mu\nu} \partial_\nu \varphi \right)$$

(5.147)

Under a local scale transformation of the metric $g_{\mu\nu} \rightarrow (1 + \sigma(x)) g_{\mu\nu}$, the action varies according to

$$\delta S[\varphi,g] = -\frac{1}{2} \int d^2x \, T^{\mu\nu} \delta g_{\mu\nu} = -\frac{1}{2} \int d^2x \, \sigma(x) T^\mu_{\ \mu}$$

(5.148)

where $\sigma(x)$ is infinitesimal. Consequently, the variation of the connected vacuum functional $W[g]$ is

$$\delta W[g] = -\frac{1}{2} \int d^2x \, \sigma(x) \langle T^\mu_\mu(x) \rangle \tag{5.149}$$

According to the Ward identities previously derived, this variation vanishes in flat space, since $\langle T^\mu_\mu(x) \rangle = 0$. This is no longer true on an arbitrary manifold.

To see this, we define the functional measure $[d\varphi]$ in a fashion more suited to an arbitrary metric. We proceed by analogy with integration on a general manifold of dimension d: the line element is then $ds^2 = g_{\mu\nu}dx^\mu dx^\nu$, and the volume element is $d\Omega = \sqrt{g} \, dx^1 \ldots dx^d$. If a coordinate system can be found such that $g_{\mu\nu} = \eta_{\mu\nu}$, then $\sqrt{g} = 1$ and the integration measure simplifies accordingly. In the space of field configurations, the analog of the metric is defined in a reparametrization invariant way:

$$(\varphi_1, \varphi_2) = \int d^2x \, \sqrt{g} \, \varphi_1^* \varphi_2 \tag{5.150}$$

and the line element is simply

$$||\delta\varphi||^2 = (\delta\varphi, \delta\varphi) \tag{5.151}$$

In order to diagonalize this "functional metric", we introduce a complete set of orthonormal functions $\{\varphi_n\}$ (i.e., such that $(\varphi_m, \varphi_n) = \delta_{mn}$) and express any general field configuration as $\varphi = \sum_n c_n \varphi_n$. The line element thus reduces to

$$||\delta\varphi||^2 = \sum_n (\delta c_n)^2 \tag{5.152}$$

which allows us to define the functional integration measure as

$$[d\varphi] = \prod_n dc_n \tag{5.153}$$

Of all possible complete sets $\{\varphi_n\}$, the most useful is the set of normalized eigenfunctions of the Laplacian, with eigenvalues $-\lambda_n$:

$$\Delta\varphi_n = -\lambda_n \varphi_n \tag{5.154}$$

The action of a configuration specified by the expansion coefficients c_n is then simply

$$S[\varphi, g] = \sum_n \lambda_n c_n^2 \tag{5.155}$$

which means that the modes φ_n decouple. However, all is not trivial since the eigenfunctions φ_n and the eigenvalues λ_n depend on the background metric $g_{\mu\nu}$.

The vacuum functional may be written as

$$Z[g] = \int \prod_n \left\{ dc_n \, e^{-\lambda_n c_n^2} \right\}$$

$$= \prod_n \sqrt{\frac{2\pi}{\lambda_n}} \tag{5.156}$$

We must be cautious here, since the Laplacian always has a zero-mode $\varphi_0 = const.$ with vanishing eigenvalue. Such a mode is a source of divergence in the vacuum functional. To fix this "infrared" problem, we "compactify" the field φ: We assume that φ takes its values on a circle, such that the values φ and $\varphi + a$ are equivalent. The circumference a can be chosen very large, and taken to infinity at the end of the calculation. Then the range of integration of c_0 is no longer the whole real axis, but the segment $[0, a\sqrt{A}]$, where A is the area of the manifold. This follows from the normalization condition $(\varphi_0, \varphi_0) = A\varphi_0^2 = 1$ and the condition $0 < c_0 \varphi_0 < a$. The above expression for the vacuum functional is then replaced by

$$Z[g] = a\sqrt{A} \prod_{n \neq 0} \sqrt{\frac{2\pi}{\lambda_n}} \tag{5.157}$$

The connected functional $W[g]$ is then

$$W[g] = -\ln a - \frac{1}{2} \ln A + \frac{1}{2} \text{Tr}' \ln \frac{-\Delta}{2\pi} \tag{5.158}$$

where Tr' indicates a trace taken over all nonzero modes. We then use the following representation of the logarithm:

$$\ln B = -\lim_{\varepsilon \to 0} \int_\varepsilon^\infty \frac{dt}{t} \left(e^{-Bt} - e^{-t} \right) \tag{5.159}$$

in order to write

$$W[g] = -\ln a - \frac{1}{2} \ln A - \frac{1}{2} \text{Tr}' \left\{ \int_\varepsilon^\infty \frac{dt}{t} \left(e^{t\Delta} - e^{-2\pi t} \right) \right\} \tag{5.160}$$

(we have scaled $t \to 2\pi t$). From now on we keep ε finite and shall send it to zero at the end of the calculation.

We now perform an infinitesimal local scale transformation. The variation of the metric is $\delta g_{\mu\nu} = \sigma g_{\mu\nu}$, and that of the Laplacian is $\delta \Delta = -\sigma \Delta$. The variation of the second term of (5.160) is

$$\delta\left(-\frac{1}{2} \ln A\right) = -\frac{\delta A}{2A} = -\frac{1}{2A} \int d^2 x \sqrt{g} \sigma \tag{5.161}$$

and that of the trace in Eq. (5.160) is

$$\frac{1}{2} \text{Tr}' \left\{ \int_\varepsilon^\infty dt \, \sigma \Delta e^{t\Delta} \right\} = \frac{1}{2} \text{Tr}' \left\{ \int_\varepsilon^\infty dt \, \sigma \frac{d}{dt} e^{t\Delta} \right\} = -\frac{1}{2} \text{Tr}'(\sigma e^{\varepsilon\Delta}) \tag{5.162}$$

In the second equality, we used the property that all nonzero eigenvalues of Δ are negative, so that only the lower-bound of the integral over t contributes. Since

$$-\frac{1}{2A}\int d^2x\,\sqrt{g}\sigma = -\frac{1}{2}(\varphi_0, \sigma\varphi_0) = -\frac{1}{2}(\varphi_0, \sigma e^{\varepsilon\Delta}\varphi_0) \qquad (5.163)$$

we may combine the two variations into a single expression:

$$\delta W[g] = \frac{1}{2}\,\mathrm{Tr}\,(\sigma e^{\varepsilon\Delta}) \qquad (5.164)$$

This expression contains the contribution of the zero-mode, hence $\mathrm{Tr}\,'$ has been replaced by Tr.

To proceed, we introduce the heat kernel

$$G(x,y;t) = \begin{cases} \langle x|e^{t\Delta}|y\rangle & (t \geq 0) \\ 0 & (t < 0) \end{cases} \qquad (5.165)$$

Since the eigenvalues of Δ can be arbitrarily negative, the expression $e^{t\Delta}$ has meaning only for $t \geq 0$. In terms of this kernel, the variation of $W[g]$ is

$$\delta W[g] = -\frac{1}{2}\int d^2x\,\sqrt{g}\,\sigma(x)G(x,x;\varepsilon) \qquad (5.166)$$

The crucial point here is the short-time behavior of the diagonal kernel, which can be shown to be

$$G(x,x;\varepsilon) = \frac{1}{4\pi\varepsilon} + \frac{1}{24\pi}R(x) + O(\varepsilon) \qquad (5.167)$$

(this result is proven in the App. 5.B). It follows that

$$\delta W[g] = -\frac{1}{8\pi\varepsilon}\int d^2x\,\sqrt{g}\,\sigma(x) - \frac{1}{48\pi}\int d^2x\,\sqrt{g}\,\sigma(x)R(x) \qquad (5.168)$$

In the limit $\varepsilon \to 0$, the first term seems problematic, being infinite. The origin of this divergence lies in the assumed finite size of the manifold and has nothing to do with curvature. To fix it, we add to the original action the following φ-independent counterterm:

$$S_1[g] = \mu\int d^2x\,\sqrt{g} \qquad (5.169)$$

which is simply equal to μA. Under a local scale transformation it undergoes the following variation:

$$\delta S_1[g] = \mu\int d^2x\,\sqrt{g}\,\sigma(x) \qquad (5.170)$$

By suitably choosing μ to be equal to $-1/8\pi\varepsilon$, the variation of the counterterm action S_1 cancels the divergent term in (5.168). The second term in (5.168) cannot be eliminated in the same way. Indeed, if we add a second counterterm of the form

$$S_2[g] = \int d^2x\,\sqrt{g}\,R(x) \qquad (5.171)$$

we find that it is proportional to the Euler characteristics χ, a topological invariant that depends only on the number of handles of the manifold. Therefore, it is invariant under a local scale transformation, and cannot cancel the rest of the variation $\delta W[g]$. Then, the equivalence of (5.149) and (5.168) implies that the trace of the energy-momentum tensor does not vanish, according to (5.144), with the value $c = 1$ appropriate for a free boson.

In order to relate the trace anomaly to the central charge figuring in the OPE of the energy-momentum tensor or, equivalently, in the two-point function $\langle T_{\mu\nu}(x)T_{\rho\lambda}(y)\rangle$, we proceed as follows. We use the "conformal gauge", a coordinate system in which the metric tensor is diagonal:

$$g_{\mu\nu} = \delta_{\mu\nu}e^{2\varphi(x)} \tag{5.172}$$

In two dimensions it is always possible to find such a system, at least locally. In terms of the field φ, the determinant $\sqrt{g}$ and the curvature are

$$\sqrt{g} = e^{2\varphi} \qquad \sqrt{g}R = \partial^2\varphi \tag{5.173}$$

Since a local scale transformation amounts to a local variation of the field φ, the corresponding variation of the connected functional $W[g]$ is

$$\delta W[g] = -\frac{c}{24\pi}\int d^2x\, \partial^2\varphi\, \delta\varphi \tag{5.174}$$

where c is some constant, equal to unity in the case of a free boson, as argued above. This implies that

$$W[g] = \frac{c}{48\pi}\int d^2x\, (\partial\varphi)^2 \tag{5.175}$$

up to terms independent of φ. In terms of the Green function $K(x,y)$ of the Laplacian, this is

$$W[g] = -\frac{c}{48\pi}\int d^2x\, d^2y\, \partial^2\varphi(x)K(x,y)\partial^2\varphi(y) \tag{5.176}$$

This follows from the defining property $\partial_x^2 K(x,y) = \delta(x-y)$ and integration by parts. The natural extension of the above to an arbitrary coordinate system is

$$W[g] = -\frac{c}{48\pi}\int d^2x\, d^2y\, \sqrt{g(x)}\sqrt{g(y)}\, R(x)K(x,y)R(y) \tag{5.177}$$

where $K(x,y)$ now satisfies

$$\sqrt{g(x)}\,\triangle_x K(x,y) = \delta(x-y) \tag{5.178}$$

The above expression for $W[g]$ can be used to calculate the two-point function of the energy-momentum tensor (the Schwinger function):

$$\langle T_{\mu\nu}(x)T_{\rho\lambda}(y)\rangle = \frac{\delta^2 W}{\delta g_{\mu\nu}(x)\delta g_{\rho\lambda}(y)} \tag{5.179}$$

Without a detailed calculation, it is by now clear that the Schwinger function will be proportional to c, which confirms that the central charge and the coefficient of the trace anomaly are one and the same thing.

Appendix 5.B. The Heat Kernel

In this appendix we show that the heat kernel $G(x,y;t)$ defined in (5.165) has the short-time behavior given in (5.167) for $x = y$.

From the definition of the heat kernel, we see that it satisfies the equations

$$\frac{\partial}{\partial t} G(x,y;t) = \Delta_x G(x,y;t) \qquad \Delta = \frac{1}{\sqrt{g}} \partial_\mu \sqrt{g} \partial^\mu$$

$$G(x,y;0) = \frac{1}{\sqrt{g}} \delta(x-y) \tag{5.180}$$

These two equations may be combined into

$$(\partial_t - \Delta_x) G(x,y;t) = \frac{1}{\sqrt{g}} \delta(x-y)\delta(t) \tag{5.181}$$

The equivalence of this single equation with Eq. (5.180) may be seen by first considering the case $t > 0$, and then by integrating the above equation over t from $-\varepsilon$ to ε, where ε is infinitesimal, remembering that $G(x,y;t) = 0$ if $t < 0$. The heat kernel is then the Green function for the diffusion equation:

$$G(x,y;t) = \langle x,t| (\partial_t - \Delta)^{-1} |y,0\rangle \tag{5.182}$$

We know the (normalized) solution to this equation in flat infinite space:

$$G_0(x,y,t) = \frac{1}{4\pi t} \exp -\frac{(x-y)^2}{4t} \tag{5.183}$$

We now wish to find the small t behavior of $G(x,x;t)$ on a general curved manifold. Physically, $G(x,x;t)$ is the probability that a random walker will diffuse from x back to x in a time t. If t is small the diffusion cannot go very far, and we can restrict our attention to the immediate neighborhood of x. To this end we write $y = x + \delta x$ and use a locally inertial frame at x, with $g_{\mu\nu}(x) = \eta_{\mu\nu}$ and $\partial_\lambda g_{\mu\nu}(x) = 0$:

$$g_{\mu\nu}(x+\delta x) \sim \eta_{\mu\nu} + \frac{1}{2} C_{\mu\nu\rho\lambda} \delta x^\rho \delta x^\lambda \tag{5.184}$$

where the constants $C_{\mu\nu\rho\lambda}$ are symmetric under the interchanges $\mu \leftrightarrow \nu$ and $\rho \leftrightarrow \lambda$. It is then a simple exercise to show that

$$\Delta(x) \sim \partial_\mu \partial^\mu + a^{\mu\nu} \partial_\mu \partial_\nu + b^\mu \partial_\mu \tag{5.185}$$

wherein

$$a^{\mu\nu} = -\frac{1}{2} C^{\mu\nu}{}_{\rho\lambda} \delta x^\rho \delta x^\lambda \quad \text{and} \quad b^\mu = \frac{1}{2}(C^\rho{}_\rho{}^\mu{}_\lambda - C^{\mu\rho}{}_{\rho\lambda}) \delta x^\lambda \tag{5.186}$$

The heat kernel then becomes

$$G(x,y;t) = \langle x,t| \frac{1}{A-B} |y,0 \rangle \tag{5.187}$$

where

$$A = \partial_t - \partial_\mu \partial^\mu$$
$$B = a^{\mu\nu} \partial_\mu \partial_\nu + b^\mu \partial_\mu \tag{5.188}$$

A perturbative solution for $G(x,y;t)$ is obtained by expanding

$$\frac{1}{A-B} = \frac{1}{A} + \frac{1}{A}B\frac{1}{A} + \frac{1}{A}B\frac{1}{A}B\frac{1}{A} + \cdots \tag{5.189}$$

To first order, this yields

$$G(x,y;t) = \langle x,t|A^{-1}|y,0 \rangle + \int d\tau dz \, \langle x,t|A^{-1}|z,\tau \rangle \langle z,\tau|BA^{-1}|y,0 \rangle$$

$$= G_0(x,y;t) + \int_0^t d\tau \int d^2z \, G_0(x,z;t-\tau) \tag{5.190}$$

$$\times \left\{ a^{\mu\nu}(z) \frac{\partial^2}{\partial z^\mu \partial z^\nu} + b^\mu(z) \frac{\partial}{\partial z^\mu} \right\} G_0(z,y;\tau)$$

The range of the τ integration follows from the vanishing of $G_0(x,y;t)$ for $t < 0$. One checks that the low t behavior of the n-th order contribution in perturbation theory is t^{n-1}. We are thus justified in keeping only the first-order contributions. Substitution of the explicit form (5.183) of $G_0(x,y;t)$ and (5.186) of $a^{\mu\nu}$ and b^μ yields

$$G(x,x;t) = \frac{1}{4\pi t} + \frac{1}{24\pi} \left(C^{\mu\lambda}{}_{\mu\lambda} - C^\mu{}_{\mu}{}^\lambda{}_\lambda \right) + \mathcal{O}(t) \tag{5.191}$$

On the other hand, it is straightforward to show that the scalar curvature is given by

$$R(x) = \left(C^{\mu\lambda}{}_{\mu\lambda} - C^\mu{}_{\mu}{}^\lambda{}_\lambda \right) \tag{5.192}$$

Therefore, the short-time behavior of the heat kernel on a curved manifold is given by

$$G(x,x;\varepsilon) = \frac{1}{4\pi\varepsilon} + \frac{1}{24\pi} R(x) + \mathcal{O}(\varepsilon) \tag{5.193}$$

Even if this result is obtained in a specific local inertial frame, the relation of the curvature with the short-time heat kernel is coordinate independent.

Exercises

5.1 *The group SL(2, C)*

a) Write down the explicit $SL(2, \mathbb{C})$ matrices corresponding to translations, rotations, dilations, and special conformal transformations.

b) Given three points z_1, z_2, and z_3, find the explicit $SL(2, \mathbb{C})$ transformation (5.12) that maps these three points respectively to 0, 1, and ∞.
We have seen in Chap. 4 that the global conformal group in Euclidian space is isomorphic to $SO(d+1, 1)$. For $d = 2$, this means that $SL(2, \mathbb{C})$ should be isomorphic to the Lorentz group $SO(3, 1)$ of Minkowski four-dimensional space-time. The Lorentz group is the set of linear transformations on a four-vector x^μ that leaves the interval $s^2 = (x^0)^2 - (x^1)^2 - (x^2)^2 - (x^3)^2$ invariant. To x^μ we may associate a 2×2 matrix $X = x^\mu \sigma^\mu$, where $\sigma^{1,2,3}$ are the usual Pauli matrices and σ^0 is the unit matrix.

c) Show that $s^2 = \det X$ and that any transformation $X \to S^\dagger X S$ leaves the interval invariant if S is a $SL(2, \mathbb{C})$ matrix, and vice versa. Conclude on the isomorphism of $SL(2, \mathbb{C})$ with the Lorentz group. What about the topology of these two groups? Hint: Changing the sign of the $SL(2, \mathbb{C})$ matrix should have no consequence on the Lorentz transformation.

5.2 *Cluster property of the four-point function*
Consider the expression (5.28) for a generic four-point function. Show how a product of two-point functions is recovered when the four points are paired in such a way that the two points in each pair are much closer to each other than the distance between the pairs. You must assume that the scaling dimensions are positive (i.e., that the correlations do not increase with distance).

5.3 *Four-point function for the free boson*
Calculate the four-point function $\langle \partial\varphi\partial\varphi\partial\varphi\partial\varphi \rangle$ for the free boson using Wick's theorem. Compare the result with the general expression (5.28). What is the function $f(\eta, \bar{\eta})$?

5.4 Verify the details of the calculation of the OPE of the energy-momentum tensor with itself, in Eqs. (5.83), (5.100), (5.117), and (5.120).

5.5 *Free complex fermion*
Given two real fermions ψ_1 and ψ_2, one may define a single *complex* fermion ψ and its Hermitian conjugate $\psi^\dagger$ this way (with holomorphic and antiholomorphic modes):

$$\psi = \frac{1}{\sqrt{2}}(\psi_1 + i\psi_2) \qquad \bar{\psi} = \frac{1}{\sqrt{2}}(\bar{\psi}_1 + i\bar{\psi}_2)$$

$$\psi^\dagger = \frac{1}{\sqrt{2}}(\psi_1 - i\psi_2) \qquad \bar{\psi}^\dagger = \frac{1}{\sqrt{2}}(\bar{\psi}_1 - i\bar{\psi}_2) \tag{5.194}$$

The real fermions ψ_1 and ψ_2 are governed by the action and energy-momentum tensor of Sect. 5.3.2.

a) Show that the OPE of the complex fermion with itself is

$$\psi^\dagger(z)\psi(w) \sim \frac{1}{z - w} \qquad \psi(z)\psi(w) \sim \psi(z)^\dagger \psi(w)^\dagger \sim 0 \tag{5.195}$$

b) Show that the energy-momentum tensor may be expressed as

$$T(z) = \frac{1}{2}(\partial\psi^\dagger \psi - \psi^\dagger \partial\psi) \tag{5.196}$$

and that the conformal dimension of ψ is $\frac{1}{2}$ and that the central charge is $c = 1$.

c) Show that the action describing the complex fermion system may be written as

$$S[\psi] = g \int d^2x \; \Psi^\dagger \gamma^0 \gamma^\mu \partial_\mu \Psi \tag{5.197}$$

where $\Psi = (\psi, \bar{\psi}^\dagger)$ is a two-component field.

5.6 *Generalized ghost system*

The ghost system may be generalized to a pair of fields $\tilde{b}(z)$ and $\tilde{c}(z)$, either both anticommuting ($\epsilon = 1$) or commuting ($\epsilon = -1$). Their OPE is defined to be

$$\tilde{c}(z)\tilde{b}(w) \sim \frac{1}{z-w} \qquad \tilde{b}(z)\tilde{c}(w) \sim \frac{\epsilon}{z-w} \tag{5.198}$$

and the associated energy-momentum tensor is defined as

$$T(z) = (1-\lambda)(\partial\tilde{b}\tilde{c})(z) - \lambda(\tilde{b})(\tilde{c})(z) \tag{5.199}$$

where λ is some constant.

a) Show that the ghosts $\tilde{b}(z)$ and $\tilde{c}(z)$ have, respectively, dimensions λ and $1 - \lambda$.

b) Calculate the central charge of this system. Answer:

$$c = -2\epsilon(6\lambda^2 - 6\lambda + 1). \tag{5.200}$$

What is the range of c if λ is real ?

5.7 Calculate explicitly the transformation property of the energy-momentum tensor of a free fermion using the point-splitting method, as has been done for the free boson. Check that the Schwarzian derivative appears there also, with the correct value of the central charge.

5.8 Express all components of the Schwinger function (4.77) in terms of holomorphic coordinates. What are the only nonzero (anti)holomorphic components of the Schwinger function?

5.9 *The Schwarzian derivative*

a) Demonstrate explicitly the group property (5.130) of the Schwarzian derivative.

b) Show that the Schwarzian derivative of the $SL(2,\mathbb{C})$ transformation (5.12) vanishes.

5.10 Demonstrate in detail the expressions (5.185) and (5.186) for the Laplacian in a locally inertial frame near the origin.

5.11 *Heat kernel on a sphere*

The Laplacian operator on a sphere of radius r embedded in three-dimensional space is $\Delta = -\mathbf{L}^2/r^2$, where $\mathbf{L}$ is the angular momentum operator of quantum mechanics.

a) Show that the heat kernel $G(x, x; t)$ is given by

$$G(x,x;t) = \frac{1}{r^2} \sum_{l,m} |Y_{l,m}(x)|^2 e^{-tl(l+1)/r^2} \tag{5.201}$$

where x stands for the angular coordinates (θ, φ). The *spherical harmonics* $Y_{l,m}(\theta, \varphi)$ are eigenfunctions of $\mathbf{L}^2$ and L_z:

$$\mathbf{L}^2 Y_{l,m} = l(l+1)Y_{l,m} \qquad L_z Y_{l,m} = mY_{l,m} \tag{5.202}$$

b) By setting $x = 0$ (the north pole $\theta = 0$) and using Euler's summation formula, show explicitly that

$$G(0,0;t) = \frac{1}{4\pi t} + \frac{1}{12\pi r^2} + \cdots \tag{5.203}$$

This result agrees with Eq. (5.167), since the scalar curvature R of a sphere of radius r is $R = 2/r^2$.

Notes

The seminal work of Belavin, Polyakov, and Zamolodchikov [36] (henceforth referred to as BPZ; see also [35]) had an immense influence on the developments of two-dimensional conformal field theory. Some of these developments are described in many review articles and lecture notes. Of note are those of Alvarez-Gaumé, Sierra, and Gómez [12], Cardy [68, 69], Christe and Henkel [76], Ginsparg [177], Saint-Aubin [312], and Zamolodchikov and Zamolodchikov [367]. A large chapter of the two-volume set by Itzykson and Drouffe [203] is devoted to conformal invariance in two dimensions. Recent books [235, 227] cover a great variety of subjects. The collection of reprints assembled by Itzykson, Saleur, and Zuber [204] is a handy reference and contains an extensive bibliography (up to 1989). The two-volume set by Green, Schwarz, and Witten [187] on superstring theory also contains a generous bibliography in which early work on free-field theories can be found.

The use of holomorphic and antiholomorphic coordinates in the context of string theory appears in Polyakov [297, 298] and in a lecture by Friedan [139]. The definition of a primary field appeared in BPZ [36]. The Ward identities were used extensively in BPZ, but the conformal Ward identity appears in the present form in Ref. [142]. The alternate derivation of the Ward identities on a Riemannian manifold follows the presentation of H. Kawai [233]. The operator product expansion was first introduced in field theory by Wilson [356] and Kadanoff [223]; it was used in string theory by Friedan [139] (see also [142]).

Bosons, fermions, and ghosts in dimension two were studied in the context of string theory. Fermions were introduced in string theory by Ramond [302] and Neveu and Schwarz [281]. Reparametrization ghosts were introduced in string theory by Polyakov [297, 298]; the extension of bosons, fermions, and ghosts to superstrings was studied in Friedan, Martinec, and Shenker [142], in which a detailed discussion of the relation between string theory and conformal field theory can also be found.

It was recognized in BPZ [36] that the central charge is a fundamental characteristic of a conformal field theory. Its deeper significance as a measure of the number of degrees of freedom in a theory is discussed by Zamolodchikov [363]. The behavior of the energy-momentum tensor under conformal transformation appeared in BPZ [36]; the argument given in this chapter for the free boson is due to Cardy [69]. The interpretation of the central charge as a Casimir energy is due to Affleck [1] and Blöte, Cardy, and Nightingale [49]. Our treatment of the trace anomaly follows H. Kawai [233], derived from the original work of Polyakov [297].

CHAPTER 6

The Operator Formalism

In the previous chapter, the consequences of conformal symmetry on two-dimensional field theories were embodied in constraints imposed on correlation functions known as the Ward identities. These Ward identities were most easily expressed in the form of an operator product expansion of the energy-momentum tensor with local fields. It was implicit, however, that operator product expansions were occurring within correlation functions and no use was made of any operator formalism or Hilbert space: The correlation functions could in principle be obtained in the path integral formalism.

Hilbert spaces and operators are nonetheless extremely useful in conformal field theory because of the power of algebraic and group-theoretical methods. The operator formalism of quantum mechanics implies a choice of reference frame, as it is not manifestly Lorentz invariant; this amounts to choosing a time axis in space-time. In a Euclidian theory, the time direction is somewhat arbitrary; in particular, it may be chosen as the radial direction from the origin. This is the object of radial quantization, described in Sect. 6.1. The use of complex coordinates then allows a representation of commutators in terms of contour integrals, making the operator product expansion a particularly useful computational tool. Section 6.2 expresses the conformal transformation of fields in terms of quantum generators, whose commutation relations define the Virasoro algebra. The general features of the Hilbert space and the notion of descendant states are also introduced. Section 6.3 discusses at length the quantization of the free boson on the cylinder with various boundary conditions. Some notions introduced here (e.g., vertex operators) will be of great importance later. Section 6.4 gives a comparable treatment of free fermions. Section 6.5 describes a new definition of normal ordering for interacting conformal fields. Section 6.6 introduces the notion of descendant fields, conformal families and operator algebra, and is of special importance for a good understanding of the structure of conformal field theories.

§6.1. The Operator Formalism of Conformal Field Theory

6.1.1. Radial Quantization

The operator formalism distinguishes a time direction from a space direction. This is natural in Minkowski space-time, but somewhat arbitrary in Euclidian space-time. In the context of statistical mechanics, choosing time and space directions amounts to selecting a direction in the lattice (e.g., rows) that we call "space", and defining a space of states spanned by all the possible spin configurations along that direction. The time direction is then orthogonal to space, and the transfer matrix makes the link between state spaces at different times. In the continuum limit the lattice spacing disappears and we are free to choose the space direction in more exotic ways, for instance along concentric circles centered at the origin. This choice of space and time leads to the so-called *radial quantization* of two-dimensional conformal field theories.

In order to make this choice more natural from a Minkowski space point of view (in particular in the context of string theory), we may initially define our theory on an infinite space-time cylinder, with time t going from $-\infty$ to $+\infty$ along the "flat" direction of the cylinder, and space being compactified with a coordinate x going from 0 to L, the points $(0, t)$ and (L, t) being identified. If we continue to Euclidian space, the cylinder is described by a single complex coordinate $\xi = t + ix$ (or equivalently $\xi = t - ix$). We then "explode" the cylinder onto the complex plane (or rather, the Riemann sphere) via the mapping illustrated on Fig. 6.1.

$$z = e^{2\pi\xi/L} \tag{6.1}$$

The remote past $(t \to -\infty)$ is situated at the origin $z = 0$, whereas the remote future $(t \to +\infty)$ lies on the point at infinity on the Riemann sphere.

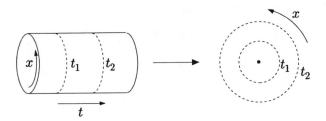

Figure 6.1. Mapping from the cylinder to the complex plane.

We must also assume the existence of a vacuum state $|0\rangle$ upon which a Hilbert space is constructed by application of creation operators (or their likes). In free-field theories, the vacuum may be defined as the state annihilated by the positive frequency part of the field (see Sect. 2.1). For an interacting field ϕ, we assume that the Hilbert space is the same as for a free field, except that the actual energy eigenstates are different. We suppose then that the interaction is attenuated as $t \to \pm\infty$ and that the asymptotic field

$$\phi_{\text{in}} \propto \lim_{t \to -\infty} \phi(x, t) \tag{6.2}$$

is free. Within radial quantization, this asymptotic field reduces to a single operator, which, upon acting on $|0\rangle$, creates a single asymptotic "in" state:

$$|\phi_{\text{in}}\rangle = \lim_{z, \bar{z} \to 0} \phi(z, \bar{z})|0\rangle \tag{6.3}$$

THE HERMITIAN PRODUCT

On this Hilbert space we must also define a bilinear product, which we do indirectly by defining an asymptotic "out" state, together with the action of Hermitian conjugation on conformal fields. In Minkowski space, Hermitian conjugation does not affect the space-time coordinates. Things are different in Euclidian space, since the Euclidian time $\tau = it$ must be reversed ($\tau \to -\tau$) upon Hermitian conjugation if t is to be left unchanged. In radial quantization this corresponds to the mapping $z \to 1/z^*$. This (almost) justifies the following definition of Hermitian conjugation on the real surface $\bar{z} = z^*$:

$$[\phi(z, \bar{z})]^\dagger = \bar{z}^{-2h} z^{-2\bar{h}} \phi(1/\bar{z}, 1/z) \tag{6.4}$$

where by assumption ϕ is a quasi-primary field of dimensions h and $\bar{h}$. The prefactors on the r.h.s. may be justified by demanding that the asymptotic "out" state

$$\langle \phi_{\text{out}}| = |\phi_{\text{in}}\rangle^\dagger \tag{6.5}$$

have a well-defined inner product with $|\phi_{\text{in}}\rangle$. Following the definition (6.4) of Hermitian conjugation, this inner product is

$$
\begin{aligned}
\langle \phi_{\text{out}}|\phi_{\text{in}}\rangle &= \lim_{z, \bar{z}, w, \bar{w} \to 0} \langle 0|\phi(z, \bar{z})^\dagger \phi(w, \bar{w})|0\rangle \\
&= \lim_{z, \bar{z}, w, \bar{w} \to 0} \bar{z}^{-2h} z^{-2\bar{h}} \langle 0|\phi(1/\bar{z}, 1/z)\phi(w, \bar{w})|0\rangle \\
&= \lim_{\xi, \bar{\xi} \to \infty} \bar{\xi}^{2h} \xi^{2\bar{h}} \langle 0|\phi(\bar{\xi}, \xi)\phi(0, 0)|0\rangle
\end{aligned}
\tag{6.6}
$$

According to the form (5.25) of conformally covariant two-point functions, the last expression is independent of ξ, and this justifies the prefactors appearing in Eq. (6.4): Had they been absent, the inner product $\langle \phi_{\text{out}}|\phi_{\text{in}}\rangle$ would not have been well-defined as $\xi \to \infty$. Notice that the passage from a vacuum expectation value to a correlator in the last equation is correct since the operators are already time-ordered within radial quantization: The first one is associated with $t \to \infty$ and the second one to $t \to -\infty$.

MODE EXPANSIONS

A conformal field $\phi(z, \bar{z})$ of dimensions $(h, \bar{h})$ may be mode expanded as follows:

$$\phi(z, \bar{z}) = \sum_{m \in \mathbb{Z}} \sum_{n \in \mathbb{Z}} z^{-m-h} \bar{z}^{-n-\bar{h}} \phi_{m,n}$$

$$\phi_{m,n} = \frac{1}{2\pi i} \oint dz \, z^{m+h-1} \frac{1}{2\pi i} \oint d\bar{z} \, \bar{z}^{n+\bar{h}-1} \phi(z, \bar{z})$$

(6.7)

A straightforward Hermitian conjugation on the real surface yields

$$\phi(z, \bar{z})^{\dagger} = \sum_{m \in \mathbb{Z}} \sum_{n \in \mathbb{Z}} \bar{z}^{-m-h} z^{-n-\bar{h}} \phi_{m,n}^{\dagger}$$

(6.8)

while the definition (6.4) gives instead

$$\phi(z, \bar{z})^{\dagger} = \bar{z}^{-2h} z^{-2\bar{h}} \phi(1/\bar{z}, 1/z)$$

$$= \bar{z}^{-2h} z^{-2\bar{h}} \sum_{m \in \mathbb{Z}} \sum_{n \in \mathbb{Z}} \phi_{m,n} \bar{z}^{m+h} z^{n+\bar{h}}$$

$$= \sum_{m \in \mathbb{Z}} \sum_{n \in \mathbb{Z}} \phi_{-m,-n} \bar{z}^{-m-h} z^{-n-\bar{h}}$$

(6.9)

These two expressions for the Hermitian conjugate of the mode expansion are compatible provided

$$\phi_{m,n}^{\dagger} = \phi_{-m,-n}$$

(6.10)

This is the usual expression for the Hermitian conjugate of modes, and justifies the extra powers of h and $\bar{h}$ occurring in Eq. (6.7). If the "in" and "out" states are to be well-defined, the vacuum must obviously satisfy the condition

$$\phi_{m,n} |0\rangle = 0 \qquad (m > -h, n > -\bar{h})$$

(6.11)

In the following, we shall lighten the notation by dropping the dependence of fields upon the antiholomorphic coordinate. Thus, the mode expansions (6.7) will take the following simplified form:

$$\phi(z) = \sum_{m \in \mathbb{Z}} z^{-m-h} \phi_m$$

$$\phi_m = \frac{1}{2\pi i} \oint dz \, z^{m+h-1} \phi(z)$$

(6.12)

It must be kept in mind, however, that the antiholomorphic dependence is always there. The decoupling between holomorphic and antiholomorphic degrees of freedom that pervades conformal theories makes it a simple task to restore the antiholomorphic dependence when needed.

6.1.2. Radial Ordering and Operator Product Expansion

Within radial quantization, the time ordering that appears in the definition of correlation functions becomes a *radial ordering*, explicitly defined by (cf.

Eq. (2.77))

$$\mathcal{R}\Phi_1(z)\Phi_2(w) = \begin{cases} \Phi_1(z)\Phi_2(w) & \text{if} \quad |z| > |w| \\ \Phi_2(w)\Phi_1(z) & \text{if} \quad |z| < |w| \end{cases} \tag{6.13}$$

If the two fields are fermions, a minus sign is added in front of the second expression. Since all field operators within correlation functions must be radially ordered, so must be the l.h.s. of an OPE if it is to have an operator meaning. In particular, the OPEs written previously have an operator meaning only if $|z| > |w|$. We shall not write the radial ordering symbol $\mathcal{R}$ every time, but radial ordering will be implicit.

We now relate OPEs to commutation relations. Let $a(z)$ and $b(z)$ be two holomorphic fields, and consider the integral

$$\oint_w dz\, a(z)b(w) \tag{6.14}$$

wherein the integration contour circles counterclockwise around w. This expression has an operator meaning within correlation functions as long as it is radially ordered. Accordingly, we split the integration contour into two fixed-time circles (see Fig. 6.2) going in opposite directions. Our integral is now seen to be a commutator:

$$\oint_w dz\, a(z)b(w) = \oint_{C_1} dz\, a(z)b(w) - \oint_{C_2} dz\, b(w)a(z) \tag{6.15}$$
$$= [A, b(w)]$$

where the operator A is the integral over space at fixed time (i.e., a contour integral) of the field $a(z)$:

$$A = \oint a(z)dz \tag{6.16}$$

and where C_1 and C_2 are fixed-time contours (circles centered around the origin) of radii respectively equal to $|w| + \varepsilon$ and $|w| - \varepsilon$, ε being infinitesimal. Naturally, an operator relation cannot be obtained from considering a single correlation function. We must allow an arbitrary number of different fields to lie beside $b(w)$ and $a(z)$ within a generic correlator; the decomposition into two contours is valid as long as $b(w)$ is the only other field having a singular OPE with $a(z)$, which lies between the two circles C_1 and C_2; this is the reason for taking the limit $\varepsilon \to 0$. The commutator obtained is then, in some sense, an *equal time* commutator. We note that if a and b are fermions, the commutator is replaced by an anticommutator. In practice, the integral (6.14) is evaluated by substituting the OPE of $a(z)$ with $b(w)$, of which only the term in $1/(z-w)$ contributes, by the theorem of residues.

The commutator $[A, B]$ of two operators, each the integral of a holomorphic field, is obtained by integrating Eq. (6.15) over w:

$$\boxed{[A, B] = \oint_0 dw \oint_w dz\, a(z)b(w)} \tag{6.17}$$

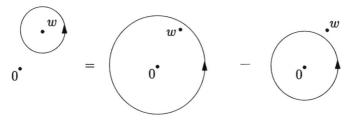

Figure 6.2. Subtraction of contours.

where the integral over z is taken around w, and the integral over w around the origin, and where

$$A = \oint a(z)dz \qquad B = \oint b(z)dz \tag{6.18}$$

Formulas (6.15) and (6.17) are important: They relate OPEs to commutation relations, and allow us to translate into operator language the dynamical or symmetry statements contained in the OPE.

We note that whenever a contour integral is written without a specified contour, it is understood that we integrate at fixed time (i.e., along a circle centered at the origin). Otherwise the relevant points surrounded by the contours are indicated below the integral sign.

§6.2. The Virasoro Algebra

6.2.1. Conformal Generators

We apply Eqs. (6.15) and (6.17) to the conformal identity (5.46). We let $\epsilon(z)$ be the holomorphic component of an infinitesimal conformal change of coordinates. We then define the conformal charge

$$Q_\epsilon = \frac{1}{2\pi i} \oint dz\, \epsilon(z)T(z) \tag{6.19}$$

With the help of Eq. (6.15), the conformal Ward identity translates into

$$\delta_\epsilon \Phi(w) = -[Q_\epsilon, \Phi(w)] \tag{6.20}$$

which means that the operator Q_ϵ is the generator of conformal transformations—in other words, the conformal charge, in the spirit of Eq. (2.163).

We expand the energy-momentum tensor according to (6.7):

$$T(z) = \sum_{n\in\mathbb{Z}} z^{-n-2} L_n \qquad L_n = \frac{1}{2\pi i} \oint dz\, z^{n+1} T(z)$$

$$\bar{T}(\bar{z}) = \sum_{n\in\mathbb{Z}} \bar{z}^{-n-2} \bar{L}_n \qquad \bar{L}_n = \frac{1}{2\pi i} \oint d\bar{z}\, \bar{z}^{n+1} \bar{T}(\bar{z}) \tag{6.21}$$

We also expand the infinitesimal conformal change $\epsilon(z)$ as

$$\epsilon(z) = \sum_{n \in \mathbb{Z}} z^{n+1} \epsilon_n \tag{6.22}$$

Then expression (6.19) for the conformal charge becomes

$$Q_\epsilon = \sum_{n \in \mathbb{Z}} \epsilon_n L_n \tag{6.23}$$

The mode operators L_n and $\bar{L}_n$ of the energy-momentum tensor are the generators of the local conformal transformations on the Hilbert space, exactly like ℓ_n and $\bar{\ell}_n$ of Eq. (5.18) are the generators of conformal mappings on the space of functions. Likewise, the generators of $SL(2, \mathbb{C})$ in the Hilbert space are L_{-1}, L_0, and L_1 (and their antiholomorphic counterparts). In particular, the operator $L_0 + \bar{L}_0$ generates the dilations $(z, \bar{z}) \rightarrow \lambda(z, \bar{z})$, which are nothing but time translations in radial quantization. Thus, $L_0 + \bar{L}_0$ is proportional to the Hamiltonian of the system.

The classical generators of the local conformal transformations obey the algebra (5.19). The quantum generators L_n obey an identical algebra, except for a central term:

$$\boxed{\begin{aligned} [L_n, L_m] &= (n - m)L_{n+m} + \frac{c}{12}n(n^2 - 1)\delta_{n+m,0} \\ [L_n, \bar{L}_m] &= 0 \\ [\bar{L}_n, \bar{L}_m] &= (n - m)\bar{L}_{n+m} + \frac{c}{12}n(n^2 - 1)\delta_{n+m,0} \end{aligned}} \tag{6.24}$$

where c is the central charge of the theory. This is the celebrated *Virasoro algebra*. It may be derived from the mode expansion (6.21), the OPE (5.121) and Eq. (6.17):

$$\begin{aligned} [L_n, L_m] &= \frac{1}{(2\pi i)^2} \oint_0 dw\, w^{m+1} \oint_w dz\, z^{n+1} \left\{ \frac{c/2}{(z-w)^4} \right. \\ &\qquad \left. + \frac{2T(w)}{(z-w)^2} + \frac{\partial T(w)}{(z-w)} + \text{reg.} \right\} \\ &= \frac{1}{2\pi i} \oint_0 dw\, w^{m+1} \left\{ \frac{1}{12}c(n+1)n(n-1)w^{n-2} + \right. \\ &\qquad \left. 2(n+1)w^n T(w) + w^{n+1}\partial T(w) \right\} \\ &= \frac{1}{12}cn(n^2 - 1)\delta_{n+m,0} + 2(n+1)L_{m+n} \\ &\qquad - \frac{1}{2\pi i} \oint_0 dw\,(n + m + 2)w^{n+m+1}T(w) \\ &= \frac{1}{12}cn(n^2 - 1)\delta_{n+m,0} + (n - m)L_{m+n} \end{aligned} \tag{6.25}$$

where, in the third equation, the last term has been integrated by parts. The last equation of (6.24) is demonstrated in exactly the same way, and the second equation of (6.24) follows from the trivial OPE $T(z)\bar{T}(w) \sim 0$.

6.2.2. The Hilbert Space

The Hilbert space of a conformal field theory may have an intricate structure, which will be discussed in Chapter 7. For the moment we shall simply explain some general facts.

The vacuum state $|0\rangle$ must be invariant under global conformal transformations. This means that it must be annihilated by L_{-1}, L_0 and L_1 and their antiholomorphic counterparts (this fixes the ground state energy to zero). This, in turn, can be recovered from the condition that $T(z)|0\rangle$ and $\bar{T}(\bar{z})|0\rangle$ are well-defined as $z, \bar{z} \to 0$, which implies

$$
\begin{aligned}
L_n|0\rangle &= 0 \\
\bar{L}_n|0\rangle &= 0
\end{aligned}
\qquad (n \geq -1)
\tag{6.26}
$$

which includes as a subcondition the invariance of the vacuum $|0\rangle$ with respect to the global conformal group. It also implies the vanishing of the vacuum expectation value of the energy-momentum tensor:

$$
\langle 0|T(z)|0\rangle = \langle 0|\bar{T}(\bar{z})|0\rangle = 0
\tag{6.27}
$$

Primary fields, when acting on the vacuum, create asymptotic states, eigenstates of the Hamiltonian. A simple demonstration follows from the OPE (5.71) between $T(z)$ and a primary field $\phi(z, \bar{z})$ of dimensions $(h, \bar{h})$, translated into operator language:

$$
\begin{aligned}
[L_n, \phi(w, \bar{w})] &= \frac{1}{2\pi i} \oint_w dz\, z^{n+1} T(z)\phi(w, \bar{w}) \\
&= \frac{1}{2\pi i} \oint_w dz\, z^{n+1} \left[\frac{h\phi(w, \bar{w})}{(z-w)^2} + \frac{\partial\phi(w, \bar{w})}{z-w} + \text{reg.} \right] \\
&= h(n+1)w^n \phi(w, \bar{w}) + w^{n+1}\partial\phi(w, \bar{w}) \qquad (n \geq -1)
\end{aligned}
\tag{6.28}
$$

The antiholomorphic counterpart of this relation is

$$
[\bar{L}_n, \phi(w, \bar{w})] = \bar{h}(n+1)\bar{w}^n \phi(w, \bar{w}) + \bar{w}^{n+1}\bar{\partial}\phi(w, \bar{w}) \qquad (n \geq -1)
\tag{6.29}
$$

After applying these relations to the asymptotic state

$$
|h, \bar{h}\rangle \equiv \phi(0, 0)|0\rangle ,
\tag{6.30}
$$

we conclude that

$$
L_0|h, \bar{h}\rangle = h|h, \bar{h}\rangle \qquad \bar{L}_0|h, \bar{h}\rangle = \bar{h}|h, \bar{h}\rangle
\tag{6.31}
$$

Thus $|h, \bar{h}\rangle$ is an eigenstate of the Hamiltonian. Likewise, we have

$$
\begin{aligned}
L_n|h, \bar{h}\rangle &= 0 \\
\bar{L}_n|h, \bar{h}\rangle &= 0
\end{aligned}
\qquad \text{if} \quad n > 0
\tag{6.32}
$$

Excited states above the asymptotic state $|h, \bar{h}\rangle$ may be obtained by applying ladder operators. Explicitly, if we expand the holomorphic field $\phi(w)$ in modes according to (6.12), then we easily find the commutation rule

$$[L_n, \phi_m] = [n(h-1) - m]\phi_{n+m} \tag{6.33}$$

of which a special case is

$$[L_0, \phi_m] = -m\phi_m \tag{6.34}$$

(the antiholomorphic index, if included, would simply be a spectator). This means that the operators ϕ_m act as raising and lowering operators for the eigenstates of L_0: each application of ϕ_{-m} ($m > 0$) increases the conformal dimension of the state by m.

The generators L_{-m} ($m > 0$) also increase the conformal dimension, by virtue of the Virasoro algebra (6.24):

$$[L_0, L_{-m}] = mL_{-m} \tag{6.35}$$

This means that excited states may be obtained by successive applications of these operators on the asymptotic state $|h\rangle$:

$$L_{-k_1} L_{-k_2} \cdots L_{-k_n} |h\rangle \qquad (1 \le k_1 \le \cdots \le k_n) \tag{6.36}$$

By convention the L_{-k_i} appear in increasing order of the k_i; a different ordering can always be brought into a linear combination of the well-ordered states (6.36) by applying the commutation rules (6.24) as necessary. The state (6.36) is an eigenstate of L_0 with eigenvalue

$$h' = h + k_1 + k_2 + \cdots + k_n \equiv h + N \tag{6.37}$$

The states (6.36) are called *descendants* of the asymptotic state $|h\rangle$ and the integer N is called the level of the descendant. The number of distinct, linearly independent states at level N is simply the number $p(N)$ of partitions of the integer N. It is easy to convince oneself, through a Taylor expansion, that the generating function of the partition numbers is (cf. Ex. 6.4)

$$\frac{1}{\varphi(q)} \equiv \prod_{n=1}^{\infty} \frac{1}{1 - q^n} = \sum_{n=0}^{\infty} p(n)q^n \tag{6.38}$$

where $\varphi(q)$ is the Euler function.

The relevance of descendant states lies in their conformal properties: The effect of a conformal transformation on a state is obtained by acting on it with a suitable function of the generators L_m. The subset of the full Hilbert space generated by the asymptotic state $|h\rangle$ and its descendants is closed under the action of the Virasoro generators and thus forms a *representation* (more correctly, a *module*) of the Virasoro algebra. This subspace is called a *Verma module* in the mathematical literature. Chap. 7 will develop these ideas further. We shall come back to descendant states in Sect. 6.6.1.

§6.3. The Free Boson

This section gives a detailed account of the canonical quantization of the free boson on the cylinder. The mode expansions are obtained, after imposing the appropriate boundary conditions. The mapping from the cylinder to the complex plane is used to define the conformal generators and, in particular, the vacuum energies. Free-field theories are of special importance not only because they can be solved explicitly, but also because they are the building blocks of more complicated models, or can be shown to be equivalent to interesting statistical models. This section and the following one will be applied extensively when discussing modular invariance, in Chap. 10. Note that we generally adopt the normalization $g = 1/4\pi$, except when we keep the normalization arbitrary in order to make comparison with other work easier.

6.3.1. Canonical Quantization on the Cylinder

We let $\varphi(x,t)$ be a free Bose field defined on a cylinder of circumference L: $\varphi(x + L, t) \equiv \varphi(x, t)$. This field may be Fourier expanded as follows:

$$\varphi(x,t) = \sum_n e^{2\pi i n x / L} \varphi_n(t)$$

$$\varphi_n(t) = \frac{1}{L} \int dx \, e^{-2\pi i n x / L} \varphi(x,t)$$

(6.39)

In terms of the Fourier coefficients φ_n, the free field Lagrangian

$$\frac{1}{2} g \int dx \, \{(\partial_t \varphi)^2 - (\partial_x \varphi)^2\}$$

(6.40)

becomes

$$\frac{1}{2} g L \sum_n \left\{ \dot\varphi_n \dot\varphi_{-n} - \left(\frac{2\pi n}{L}\right)^2 \varphi_n \varphi_{-n} \right\}$$

(6.41)

The momentum conjugate to φ_n is

$$\pi_n = g L \dot\varphi_{-n} \qquad [\varphi_n, \pi_m] = i\delta_{nm}$$

(6.42)

and the Hamiltonian is

$$H = \frac{1}{2gL} \sum_n \{\pi_n \pi_{-n} + (2\pi n g)^2 \varphi_n \varphi_{-n}\}$$

(6.43)

We notice that $\varphi_n^\dagger = \varphi_{-n}$, and similarly $\pi_n^\dagger = \pi_{-n}$. Of course, this Hamiltonian represents a sum of decoupled harmonic oscillators, of frequencies $\omega_n = 2\pi |n|/L$. The vanishing of one of the frequencies ($n = 0$) is of special importance, since it is a consequence of the absence of a mass term, which, with the boundary conditions chosen, is tantamount to conformal invariance.

The usual procedure is to define creation and annihilation operators $\tilde{a}_n$ and $\tilde{a}_n^\dagger$:

$$\tilde{a}_n = \frac{1}{\sqrt{4\pi g|n|}}\left(2\pi g|n|\varphi_n + i\pi_{-n}\right) \tag{6.44}$$

such that $[\tilde{a}_n, \tilde{a}_m] = 0$ and $[\tilde{a}_n, \tilde{a}_m^\dagger] = \delta_{mn}$; this, of course, does not work for the zero-mode φ_0. Instead of these we shall use the following operators:

$$a_n = \begin{cases} -i\sqrt{n}\,\tilde{a}_n & (n > 0) \\ i\sqrt{-n}\,\tilde{a}_{-n}^\dagger & (n < 0) \end{cases} \qquad \bar{a}_n = \begin{cases} -i\sqrt{n}\,\tilde{a}_{-n} & (n > 0) \\ i\sqrt{-n}\,\tilde{a}_n^\dagger & (n < 0) \end{cases} \tag{6.45}$$

and treat the zero mode φ_0 separately. The associated commutation relations are

$$[a_n, a_m] = n\delta_{n+m} \qquad [a_n, \bar{a}_m] = 0 \qquad [\bar{a}_n, \bar{a}_m] = n\delta_{n+m} \tag{6.46}$$

The Hamiltonian is then expressible as

$$H = \frac{1}{2gL}\pi_0^2 + \frac{2\pi}{L}\sum_{n\neq 0}(a_{-n}a_n + \bar{a}_{-n}\bar{a}_n) \tag{6.47}$$

The commutation relations (6.46) lead to the relation

$$[H, a_{-m}] = \frac{2\pi}{L}ma_{-m} \tag{6.48}$$

which means that a_{-m} $(m > 0)$, when applied to an eigenstate of H of energy E, produces another eigenstate with energy $E + 2m\pi/L$.

Since the Fourier modes are

$$\varphi_n = \frac{i}{n\sqrt{4\pi g}}(a_n - \bar{a}_{-n}) \tag{6.49}$$

the mode expansion at $t = 0$ may be written as

$$\varphi(x) = \varphi_0 + \frac{i}{\sqrt{4\pi g}}\sum_{n\neq 0}\frac{1}{n}(a_n - \bar{a}_{-n})e^{2\pi inx/L} \tag{6.50}$$

The time evolution of the operators $\varphi_0, a_n,$ and $\bar{a}_n$ in the Heisenberg picture follows immediately from the above Hamiltonian:

$$\varphi_0(t) = \varphi_0(0) + \frac{1}{gL}\pi_0 t \qquad \begin{aligned} a_n(t) &= a_n(0)e^{-2\pi int/L} \\ \bar{a}_n(t) &= \bar{a}_n(0)e^{-2\pi int/L} \end{aligned} \tag{6.51}$$

In terms of constant operators, the mode expansion of the field at arbitrary time is then

$$\varphi(x,t) = \varphi_0 + \frac{1}{gL}\pi_0 t + \frac{i}{\sqrt{4\pi g}}\sum_{n\neq 0}\frac{1}{n}\left(a_n e^{2\pi in(x-t)/L} - \bar{a}_{-n}e^{2\pi in(x+t)/L}\right) \tag{6.52}$$

If we go over to Euclidian space-time (i.e., replace t by $-i\tau$) and use the conformal coordinates

$$z = e^{2\pi(\tau-ix)/L} \qquad \bar{z} = e^{2\pi(\tau+ix)/L} \tag{6.53}$$

we finally obtain the expansion

$$\varphi(z,\bar{z}) = \varphi_0 - \frac{i}{4\pi g}\pi_0 \ln(z\bar{z}) + \frac{i}{\sqrt{4\pi g}}\sum_{n\neq 0}\frac{1}{n}\left(a_n z^{-n} + \bar{a}_n \bar{z}^{-n}\right) \tag{6.54}$$

We know that φ is not itself a primary field, but that its derivatives $\partial\varphi$ and $\bar{\partial}\bar{\varphi}$ are. We concentrate on the holomorphic field $\partial\varphi$. From Eq. (6.54) the following expansion follows:

$$i\partial\varphi(z) = \frac{1}{4\pi g}\frac{\pi_0}{z} + \frac{1}{\sqrt{4\pi g}}\sum_{n\neq 0}a_n z^{-n-1} \tag{6.55}$$

(the normalization $g = 1/4\pi$ will usually be used in this work). This expansion coincides with the general conformal mode expansion (6.7). We may introduce two operators a_0 and $\bar{a}_0$:

$$a_0 \equiv \bar{a}_0 \equiv \frac{\pi_0}{\sqrt{4\pi g}} \tag{6.56}$$

which allow us to include the zero-mode term into the sum:

$$i\partial\varphi(z) = \frac{1}{\sqrt{4\pi g}}\sum_n a_n z^{-n-1} \tag{6.57}$$

The periodicity condition on the field φ is the source of the decoupling between holomorphic and antiholomorphic excitations. Thus, the operators a_n create or destroy "right-moving" excitations, whereas the $\bar{a}_n$ are associated with "left-moving" excitations. In string theory applications, these boundary conditions describe a closed string. The zero-mode φ_0 is then the center-of-mass of the string (or, more precisely, one of the components thereof) and π_0 is the string's total momentum.

6.3.2. Vertex Operators

Since the canonical scaling dimension of the boson φ vanishes, it is possible to construct an infinite variety of local fields related to φ without introducing a scale, namely the so-called vertex operators:

$$\mathcal{V}_\alpha(z,\bar{z}) = :e^{i\alpha\varphi(z,\bar{z})}: \tag{6.58}$$

The normal ordering has the following meaning, in terms of the operators appearing in the mode expansion (6.54):

$$\begin{aligned}
\mathcal{V}_\alpha(z,\bar{z}) = {}&\exp\left\{i\alpha\varphi_0 + \frac{\alpha}{\sqrt{4\pi g}}\sum_{n>0}\frac{1}{n}\left(a_{-n}z^n + \bar{a}_{-n}\bar{z}^n\right)\right\} \\
&\times \exp\left\{\frac{\alpha}{4\pi g}\pi_0 - \frac{\alpha}{\sqrt{4\pi g}}\sum_{n>0}\frac{1}{n}\left(a_n z^{-n} + \bar{a}_n \bar{z}^{-n}\right)\right\}
\end{aligned} \tag{6.59}$$

Within each exponential, the different operators commute.

We shall now demonstrate that these fields are primary, with holomorphic and antiholomorphic dimensions

$$h(\alpha) = \bar{h}(\alpha) = \frac{\alpha^2}{8\pi g} \tag{6.60}$$

We first calculate the OPE of $\partial\varphi$ with V_α:

$$\partial\varphi(z)V_\alpha(w,\bar{w}) = \sum_{n=0}^{\infty} \frac{(i\alpha)^n}{n!} \partial\varphi(z) : \varphi(w,\bar{w})^n :$$

$$\sim -\frac{1}{4\pi g} \frac{1}{z-w} \sum_{n=1}^{\infty} \frac{(i\alpha)^n}{(n-1)!} : \varphi(w,\bar{w})^{n-1} : \tag{6.61}$$

$$\sim -\frac{i\alpha}{4\pi g} \frac{V_\alpha(w,\bar{w})}{z-w}$$

Next, we calculate the OPE of V_α with the energy-momentum tensor:

$$T(z)V_\alpha(w,\bar{w}) = -2\pi g \sum_{n=0}^{\infty} \frac{(i\alpha)^n}{n!} : \partial\varphi(z)\partial\varphi(z) :: \varphi(w,\bar{w})^n :$$

$$\sim -\frac{1}{8\pi g} \frac{1}{(z-w)^2} \sum_{n=2}^{\infty} \frac{(i\alpha)^n}{(n-2)!} : \varphi(w,\bar{w})^{n-2} :$$

$$+ \frac{1}{z-w} \sum_{n=1}^{\infty} \frac{(i\alpha)^n}{n!} n : \partial\varphi(z)\varphi(w,\bar{w})^{n-1} :$$

$$\sim \frac{\alpha^2}{8\pi g} \frac{V_\alpha(w,\bar{w})}{(z-w)^2} + \frac{\partial_w V_\alpha(w,\bar{w})}{z-w} \tag{6.62}$$

To the n-th term in the summation we have applied $2n$ single contractions and $n(n-1)$ double contractions. We have replaced $\partial\varphi(z)$ by $\partial\varphi(w)$ in the last equation since the difference between the two leads to a regular term. It is now clear by the form of this OPE that V_α is primary, with the conformal weight given above. The OPE with $\bar{T}$ has exactly the same form.

In order to calculate the OPE of products of vertex operators, we may use the following relation for a single harmonic oscillator:

$$:e^{A_1} ::e^{A_2} : = :e^{A_1+A_2} : e^{\langle A_1 A_2 \rangle} \tag{6.63}$$

where $A_i = \alpha_i a + \beta_i a^\dagger$ is some linear combination of annihilation and creation operators (this relation is demonstrated in App. 6.A). Since a free field is simply an assembly of decoupled harmonic oscillators, the same relation holds if A_1 and A_2 are linear functions of a free field (see also Ex. 6.7). In particular, we may write

$$:e^{a\varphi_1} ::e^{b\varphi_2} : = :e^{a\varphi_1+b\varphi_2} : e^{ab\langle\varphi_1\varphi_2\rangle} \tag{6.64}$$

Applied to vertex operators, this relation yields

$$V_\alpha(z,\bar{z})V_\beta(w,\bar{w}) \sim |z-w|^{2\alpha\beta/4\pi g} V_{\alpha+\beta}(w,\bar{w}) + \cdots \tag{6.65}$$

However, we have seen previously that invariance under the global conformal group forces the fields within a nonzero two-point function to have the same conformal dimension. Furthermore, the requirement that the correlation function $\langle V_\alpha(z,\bar z)V_\beta(w,\bar w)\rangle$ does not grow with distance imposes the constraint $\alpha\beta < 0$, which leaves $\alpha = -\beta$ as the only possibility ($g = 1/4\pi$):

$$V_\alpha(z,\bar z)V_{-\alpha}(w,\bar w) \sim |z-w|^{-2\alpha^2} + \cdots \tag{6.66}$$

In general, the correlator of a string of vertex operators V_{α_i} vanishes unless the sum of the charges vanishes: $\sum_i \alpha_i = 0$; this will be demonstrated in Chap. 9, in which vertex operators will be further studied. From now on, the normal ordering of the vertex operator will not be explicitly written but will always be implicit.

6.3.3. The Fock Space

The independence of the Hamiltonian (6.47) on φ_0 implies that the eigenvalue of π_0 is a "good" quantum number, which may label different sets of eigenstates of H. Since π_0 commutes with all the a_n and $\bar a_n$, these operators cannot change the value of π_0 and the Fock space is built upon a one-parameter family of vacua $|\alpha\rangle$, where α is the continuous eigenvalue of $a_0 = \pi_0/\sqrt{4\pi g}$. As mentioned above, the conformal modes a_n and $\bar a_n$ are annihilation operators for $n > 0$ and creation operators for $n < 0$ (this is also in accordance with the general expansion (6.7) and the definition of the conformal vacuum):

$$a_n|\alpha\rangle = \bar a_n|\alpha\rangle = 0 \quad (n > 0) \quad \text{with} \quad a_0|\alpha\rangle = \bar a_0|\alpha\rangle = \alpha|\alpha\rangle \tag{6.67}$$

As we know, the holomorphic energy-momentum tensor is given by

$$T(z) = -2\pi g : \partial\varphi(z)\partial\varphi(z):$$
$$= \frac{1}{2} \sum_{n,m\in\mathbb{Z}} z^{-n-m-2} :a_n a_m: \tag{6.68}$$

which implies (for g arbitrary)

$$L_n = \frac{1}{2} \sum_{m\in\mathbb{Z}} a_{n-m} a_m \quad (n \neq 0)$$
$$L_0 = \sum_{n>0} a_{-n} a_n + \frac{1}{2}a_0^2 \tag{6.69}$$

and similarly for antiholomorphic modes. The Hamiltonian (6.47) may then be written as

$$H = \frac{2\pi}{L}(L_0 + \bar L_0) \tag{6.70}$$

This confirms the role of $L_0 + \bar L_0$ as a Hamiltonian, modulo some multiplicative factor. The mode operators a_m play a role vis-à-vis L_0 similar to L_m, because of the commutation $[L_0, a_{-m}] = ma_{-m}$. This does not mean that a_m is equivalent to

L_m, but rather that its effect on the conformal dimension (the eigenvalue of L_0) is the same as that of L_m.

From expression (6.69) we see that the vacuum $|\alpha\rangle$ has conformal dimension $\frac{1}{2}\alpha^2$ (we set $g = 1/4\pi$). The elements of the Fock space are, of course, obtained by acting on $|\alpha\rangle$ with the creation operators a_{-n} and $\bar{a}_{-n}$ ($n > 0$):

$$a_{-1}^{n_1} a_{-2}^{n_2} \cdots \bar{a}_{-1}^{m_1} \bar{a}_{-2}^{m_2} \cdots |\alpha\rangle \qquad (n_i, m_j \geq 0) \qquad (6.71)$$

These states are eigenstates of L_0 with conformal dimensions

$$h = \frac{1}{2}\alpha^2 + \sum_j jn_j \qquad \bar{h} = \frac{1}{2}\alpha^2 + \sum_j jm_j \qquad (6.72)$$

Each vacuum $|\alpha\rangle$ may be obtained from the "absolute" vacuum $|0\rangle$ by application of the vertex operator $\mathcal{V}_\alpha(z, \bar{z}) = e^{i\alpha\varphi(z,\bar{z})}$. We now show explicitly that

$$|\alpha\rangle = \mathcal{V}_\alpha(0)|0\rangle \qquad (6.73)$$

We shall proceed by showing that $\mathcal{V}_\alpha(0)|0\rangle$ is an eigenstate of π_0 with eigenvalue α, and that $a_n|\alpha\rangle = 0$ for $n > 0$. For this we need the Hausdorff formula

$$[B, e^A] = e^A[B, A] \qquad (6.74)$$

where $[B, A]$ is assumed to be a constant. If we set $B = \pi_0$ and $A = i\alpha\varphi(z, \bar{z})$, we find

$$[\pi_0, \mathcal{V}_\alpha] = \alpha\mathcal{V}_\alpha \qquad (6.75)$$

This relation, applied at $z = 0$ to the invariant vacuum $|0\rangle$, gives

$$\pi_0\mathcal{V}_\alpha(0)|0\rangle = \alpha\mathcal{V}_\alpha(0)|0\rangle \qquad (6.76)$$

which is one of the desired elements. The other is obtained by setting $B = a_n$; it follows that

$$[a_n, \mathcal{V}_\alpha(z, \bar{z})] = -\alpha z^n \mathcal{V}_\alpha(z, \bar{z}) \qquad (6.77)$$

At $z = 0$, this relation yields $a_n \mathcal{V}_\alpha|0\rangle = 0$ when applied on $|0\rangle$. A similar relation holds for $\bar{a}_n$ ($n > 0$).

6.3.4. Twisted Boundary Conditions

A variant of the free-boson theory may be obtained by assuming antiperiodic boundary conditions on the cylinder: $\varphi(x + L, t) = -\varphi(x, t)$. This is compatible with the Lagrangian (6.40) since the latter is quadratic in φ. This twisted boundary condition will be fully exploited in Chap. 10, when we discuss modular invariance and the orbifold. For the moment, we shall simply be interested in the effect it has on the vacuum energy density. Of course, this boundary condition implies that the field φ is double-valued on the cylinder. Once the cylinder is mapped onto the plane, this amounts to defining the theory on a pair of Riemann sheets.

The mode expansion (6.54) may be retained, except that the zero-mode now disappears, and the summation index n must take half-integral values. This modification naturally incorporates the antiperiodicity of φ, without affecting the commutation relations $[a_n, a_m] = n\delta_{n+m}$. We define the operator G that takes φ into $-\varphi$: $G\varphi G^{-1} = -\varphi$. This operator anticommutes with φ, and with all the mode operators a_n; in fact it brings the system from one Riemann sheet to the other. Since $G^2 = 1$, its eigenvalues are ± 1; since it commutes with the Hamiltonian, every state has a definite value of G, and the two states $|\psi\rangle$ and $G|\psi\rangle$ are degenerate. In particular, the ground state is doubly degenerate, and we must distinguish the vacua $|0_+\rangle$ and $|0_-\rangle$, eigenstates of G with eigenvalues $+1$ and -1 respectively.

We now proceed to calculate the two-point function with the help of the mode expansion. In fact, we also consider the periodic case and verify that the result

$$\langle \partial\varphi(z)\partial\varphi(w)\rangle = -\frac{1}{(z-w)^2} \tag{6.78}$$

obtained by path integral methods, may be recovered by operator methods. From the mode expansion, we find ($|z| > |w|$)

$$\langle \varphi(z)\partial\varphi(w)\rangle = \sum_{m,n\neq 0} \frac{1}{n}\langle a_n a_m\rangle z^{-n}w^{-m-1} \tag{6.79}$$

But $\langle a_n a_m\rangle = n\delta_{n+m}$ if $n > 0$, and 0 otherwise. It follows that

$$\langle \varphi(z)\partial\varphi(w)\rangle = \frac{1}{w}\sum_{n>0}\left(\frac{w}{z}\right)^n \tag{6.80}$$

So far we have not specified the periodicity or antiperiodicity of the field. In the periodic case, the summation index n takes positive integral values, and the correlator becomes

$$\langle \varphi(z)\partial\varphi(w)\rangle = \frac{1}{w}\frac{w/z}{1-w/z} = \frac{1}{z-w} \tag{6.81}$$

If we differentiate with respect to z, we recover the two-point function written above.

In the antiperiodic case, the summation index starts at $n = \frac{1}{2}$ and takes half-integral values thereafter. The vacuum expectation value is taken in one of the two ground states (or a combination thereof) and

$$\langle \varphi(z)\partial\varphi(w)\rangle = \frac{1}{w}\sqrt{\frac{w}{z}}\frac{1}{1-w/z}$$
$$= \sqrt{\frac{z}{w}}\frac{1}{z-w} \tag{6.82}$$

Applying ∂_z yields

$$\langle \partial\varphi(z)\partial\varphi(w)\rangle = -\frac{1}{2}\frac{\sqrt{z/w}+\sqrt{w/z}}{(z-w)^2} \tag{6.83}$$

This expression has branch cuts at $z = 0, \infty$, and $w = 0, \infty$; the antiperiodic boundary condition on φ as z circles around the origin is incorporated in the square roots. The periodic and antiperiodic two-point functions coincide in the limit $z \to w$, meaning that the short distance behavior of the theory is independent of the boundary conditions.

The vacuum energy density may be obtained from the following normal ordering prescription

$$\langle T(z) \rangle = \frac{1}{2} \lim_{\varepsilon \to 0} \left(-\langle \partial\varphi(z+\varepsilon)\partial\varphi(z)\rangle + \frac{1}{\varepsilon^2} \right) \tag{6.84}$$

from which it follows that $\langle T(z) \rangle = 0$ in the periodic case, on the plane. The same calculation applied to Eq. (6.83) gives

$$\langle T(z) \rangle = \frac{1}{16z^2} \tag{6.85}$$

Since L_0 is the coefficient of $1/z^2$ in the mode expansion of the energy-momentum tensor, this nonzero expectation value implies a constant term in the expression for L_0 in terms of modes, in the antiperiodic case:

$$L_0 = \sum_{n>0} a_{-n}a_n + \frac{1}{16} \tag{6.86}$$

On the cylinder, the vacuum expectation value of the energy-momentum tensor must be shifted by a constant, according to Eq. (5.138):

$$\langle T_{\text{cyl.}} \rangle = \begin{cases} -\dfrac{1}{24}\left(\dfrac{2\pi}{L}\right)^2 & \text{(periodic)} \\[4mm] \dfrac{1}{48}\left(\dfrac{2\pi}{L}\right)^2 & \text{(antiperiodic)} \end{cases} \tag{6.87}$$

These vacuum expectation values may be used to fix the constants added to the Hamiltonian when expressed in terms of the mode operators on the cylinder. If we write

$$H = \frac{2\pi}{L}\left((L_0)_{\text{cyl.}} + (\bar{L}_0)_{\text{cyl.}} \right) \tag{6.88}$$

then

$$(L_0)_{\text{cyl.}} = \sum_{n>0} a_{-n}a_n - \frac{1}{24} \quad \text{(periodic)}$$

$$(L_0)_{\text{cyl.}} = \sum_{n>0} a_{-n}a_n + \frac{1}{48} \quad \text{(antiperiodic)} \tag{6.89}$$

This difference between antiperiodic and periodic boundary conditions in the vacuum energies will also appear when considering fermions, although in the opposite manner, as we shall discover.

6.3.5. Compactified Boson

The invariance of the free-boson Lagrangian (6.40) with respect to translations $\varphi \to \varphi+$const. means that it is possible, without modifying too much the dynamics of the field, to restrict the domain of variation of φ to a circle of radius R. In other words, we may identify φ with $\varphi + 2\pi R$, thereby giving φ the character of an angular variable. This brings the following two modifications to our previous analysis: First, the center-of-mass momentum π_0 can no longer take an arbitrary value: it must be an integer multiple of $1/R$, otherwise the vertex operator $\mathcal{V}_\alpha$ is no longer well-defined. Second, we may adopt the more general boundary condition

$$\varphi(x + L, t) \equiv \varphi(x, t) + 2\pi m R \tag{6.90}$$

under which the field φ winds m times as one circles once around the cylinder (m is the *winding number* of the field configuration). These two considerations lead naturally to the following modified mode expansion (cf. Eq. (6.52)):

$$
\begin{aligned}
\varphi(x,t) = \varphi_0 &+ \frac{n}{gRL}t + \frac{2\pi Rm}{L}x \\
&+ \frac{i}{\sqrt{4\pi g}} \sum_{k \neq 0} \frac{1}{k} \left(a_k e^{2\pi i k(x-t)/L} - \bar{a}_{-k} e^{2\pi i k(x+t)/L} \right)
\end{aligned}
\tag{6.91}
$$

If we express this expansion in terms of the complex coordinates z and $\bar{z}$, we find

$$
\begin{aligned}
\varphi(z,\bar{z}) = \varphi_0 &- i\left(n/4\pi gR + \frac{1}{2}mR \right) \ln z + \frac{i}{\sqrt{4\pi g}} \sum_{k \neq 0} \frac{1}{k} a_k z^{-k} \\
&- i\left(n/4\pi gR - \frac{1}{2}mR \right) \ln \bar{z} + \frac{i}{\sqrt{4\pi g}} \sum_{k \neq 0} \frac{1}{k} \bar{a}_k \bar{z}^{-k}
\end{aligned}
\tag{6.92}
$$

The holomorphic derivative $i\partial\varphi$ then has the expansion

$$i\partial\varphi(z) = (n/4\pi gR + \frac{1}{2}mR)\frac{1}{z} + \frac{1}{\sqrt{4\pi g}} \sum_{k \neq 0} a_k z^{-k-1} \tag{6.93}$$

The expression (6.69) for L_0 and that of its antiholomorphic counterpart specialize to

$$
\begin{aligned}
L_0 &= \sum_{n>0} a_{-n}a_n + 2\pi g \left(\frac{n}{4\pi gR} + \frac{1}{2}mR \right)^2 \\
\bar{L}_0 &= \sum_{n>0} \bar{a}_{-n}\bar{a}_n + 2\pi g \left(\frac{n}{4\pi gR} - \frac{1}{2}mR \right)^2
\end{aligned}
\tag{6.94}
$$

Once exploded onto the plane, the winding configurations ($m \neq 0$) are vortices centered at the origin. This is strongly reminiscent of the classical XY spin model, in which similar configurations arise. It is then possible to define an operator creating such a configuration of vorticity m with momentum value n. Such an operator has conformal dimension

$$h_{n,m} = 2\pi g \left(\frac{n}{4\pi gR} + \frac{1}{2} mR \right)^2 \tag{6.95}$$

We shall come back to this matter in Chap. 10. At this point it suffices to say that the vacua (the highest weight states), now labeled $|n, m\rangle$, have conformal weight $h_{n,m}$ and are annihilated by all the $a_{n>0}$.

§6.4. The Free Fermion

In this section we proceed to an analysis similar to what was done in the preceding section, but this time for free fermions.

6.4.1. Canonical Quantization on a Cylinder

The free fermion has the action

$$S = \frac{1}{2} g \int d^2x \, \Psi^\dagger \gamma^0 \gamma^\mu \partial_\mu \Psi \tag{6.96}$$

This system was studied in Sect. 2.1.2 and Sect. 5.3.2. The holomorphic and antiholomorphic fields are the two components of the spinor $\Psi = (\psi, \bar\psi)$. We have found in Sect. 5.3.2 that the OPE between ψ and itself is

$$\psi(z)\psi(w) \sim \frac{1}{z - w} \tag{6.97}$$

wherein the normalization $g = 1/2\pi$ was chosen. This result was, of course, obtained on the plane, with the tacit assumption that the field ψ was single-valued. We also found that the holomorphic energy-momentum tensor is

$$T(z) = -\frac{1}{2} :\psi(z)\partial\psi(z): \tag{6.98}$$

and that the central charge of this system is $c = \frac{1}{2}$, the fermion field ψ having conformal dimension $h = \frac{1}{2}$.

We work on a cylinder of circumference L, and write down the mode expansion of the fermion in terms of creation and annihilation operators, as was done in Sect. 2.1.2. With our choice of normalization, the mode expansion at a fixed time $(t = 0)$ takes the form

$$\psi(x) = \sqrt{\frac{2\pi}{L}} \sum_k b_k \, e^{2\pi i k x/L} \tag{6.99}$$

wherein the operators b_k obey the anticommutation relations

$$\{b_k, b_q\} = \delta_{k+q,0} \tag{6.100}$$

We must distinguish between two types of boundary conditions:

$$\psi(x + 2\pi L) \equiv \quad \psi(x) \qquad \text{Ramond (R)}$$
$$\psi(x + 2\pi L) \equiv -\psi(x) \qquad \text{Neveu-Schwarz (NS)}$$

(6.101)

In the periodic case (R) the mode index k takes integer values, whereas in the antiperiodic case (NS) it takes half-integer values ($k \in \mathbb{Z} + \frac{1}{2}$). Of course, the action is periodic whatever boundary condition we choose (R or NS). We are in the R (resp. NS) *sector* when the boundary conditions are of the Ramond (resp. Neveu-Schwarz) type.

In the limit where the lattice spacing a goes to zero, the Hamiltonian of Sect. 2.1.2 reads

$$H = \sum_{k>0} \omega_k b_{-k} b_k + E_0 \qquad \omega_k = \frac{2\pi|k|}{L}$$

(6.102)

where E_0 is some constant having the meaning of a vacuum energy. There is a similar Hamiltonian for the antiholomorphic component $\bar\psi$, and one must consider the sum of these two Hamiltonians in the complete theory. The time evolution of the mode operators in the Heisenberg picture is

$$b_k(t) = b_k(0)e^{-2\pi i k t/L}$$

(6.103)

The mode expansion of the time-dependent field ψ may then be written as

$$\psi(x, t) = \sqrt{\frac{2\pi}{L}} \sum_k b_k \, e^{-2\pi k w/L}$$

(6.104)

where we have introduced the complex coordinate $w = (\tau - ix)$, τ being the usual Euclidian time.

In the R sector there exists a zero mode b_0 which does not enter the Hamiltonian and leads to a degeneracy of the vacuum: If we define a vacuum $|0\rangle$ annihilated by all the b_k with $k > 0$, then the state $b_0|0\rangle$ is degenerate to $|0\rangle$, and is annihilated by the same b_k. Because of the anticommutation relations (6.100), the zero-mode operator obeys the relation $b_0^2 = \frac{1}{2}$.

6.4.2. Mapping onto the Plane

The cylinder is mapped onto the plane by introducing the coordinate $z = e^{2\pi w/L}$. Since the field ψ has conformal dimension $\frac{1}{2}$, it is affected by this mapping, in contrast with the free boson: according to Eq. (5.22) we have

$$\psi_{\text{cyl.}}(w) \longrightarrow \psi_{\text{cyl.}}(z) = \left(\frac{dz}{dw}\right)^{1/2} \psi_{\text{pl.}}(z)$$

$$= \sqrt{\frac{2\pi z}{L}} \psi_{\text{pl.}}(z)$$

(6.105)

On the plane the field has thus the following mode expansion:

$$\psi(z) = \sum_k b_k \, z^{-k-1/2} \tag{6.106}$$

In the Ramond sector, this coincides with the general mode expansion (6.7). The factor $\sqrt{z}$ picked up in the transformation has interchanged the meanings of the two types of boundary conditions when z is taken around the origin: The NS condition now corresponds to a periodic field ($k \in \mathbb{Z} + \frac{1}{2}$) and the R condition to an antiperiodic field ($k \in \mathbb{Z}$):

$$\begin{aligned}
\psi(e^{2\pi i}z) &= -\psi(z) & \text{Ramond (R)} \\
\psi(e^{2\pi i}z) &= \psi(z) & \text{Neveu-Schwarz (NS)}
\end{aligned} \tag{6.107}$$

The field ψ is double-valued on the plane under Ramond conditions. This has consequences on the two-point function, which will be different from the NS two-point function. We first calculate the two-point function in the NS sector from the mode expansion:

$$\begin{aligned}
\langle\psi(z)\psi(w)\rangle &= \sum_{k,q\in\mathbb{Z}+1/2} z^{-k-1/2}w^{-q-1/2}\langle b_k b_q\rangle \\
&= \sum_{k\in\mathbb{Z}+1/2,\, k>0} z^{-k-1/2}w^{k-1/2} \\
&= \sum_{n=0}^\infty \frac{1}{z}\left(\frac{w}{z}\right)^n \\
&= \frac{1}{z-w}
\end{aligned} \tag{6.108}$$

This agrees with the OPE (6.97) and with the general relation (5.25). However, in the Ramond sector, the result is different:

$$\begin{aligned}
\langle\psi(z)\psi(w)\rangle &= \sum_{k,q\in\mathbb{Z}} z^{-k-1/2}w^{-q-1/2}\langle b_k b_q\rangle \\
&= \frac{1}{2\sqrt{zw}} + \sum_{k=1}^\infty z^{-k-1/2}w^{k-1/2} \\
&= \frac{1}{\sqrt{zw}}\left\{\frac{1}{2} + \sum_{k=1}^\infty\left(\frac{w}{z}\right)^k\right\} \\
&= \frac{1}{2\sqrt{zw}}\frac{z+w}{z-w} \\
&= \frac{1}{2}\frac{\sqrt{z/w}+\sqrt{w/z}}{z-w}
\end{aligned} \tag{6.109}$$

This result coincides with the previous one in the limit $w \to z$. The two-point function picks up a sign when z or w is taken around the origin. Strictly speaking this correlator must be defined using Riemann sheets for the variables z and w.

From the above expression for the two-point function, we may easily show that the energy-momentum tensor has a nonzero vacuum expectation value on the plane, contrary to the NS case. We need to use the same normal-ordering prescription as for the boson:

$$\langle T(z) \rangle = \frac{1}{2} \lim_{\varepsilon \to 0} \left(-\langle \psi(z+\varepsilon) \partial \psi(z) \rangle + \frac{1}{\varepsilon^2} \right) \tag{6.110}$$

which leads to $\langle T(z) \rangle = 0$ in the NS sector, as is trivially verified. In the R sector, the same calculation yields

$$\langle T(z) \rangle = -\frac{1}{4} \lim_{w \to z} \partial_w \left(\frac{\sqrt{z/w} + \sqrt{w/z}}{z - w} \right) + \frac{1}{2(z-w)^2}$$

$$= \frac{1}{16z^2} \tag{6.111}$$

6.4.3. Vacuum Energies

We now find an expression for the conformal generators L_n in terms of the mode operators b_k for the two types of boundary conditions on the plane. The expression (6.98) for the energy-momentum tensor leads to

$$T_{\mathrm{pl.}}(z) = \frac{1}{2} \sum_{k,q} (k + \frac{1}{2}) z^{-q-1/2} z^{-k-3/2} :b_q b_k:$$

$$= \frac{1}{2} \sum_{n,k} (k + \frac{1}{2}) z^{-n-2} :b_{n-k} b_k: \tag{6.112}$$

From this, we extract the conformal generator

$$L_n = \frac{1}{2} \sum_k (k + \frac{1}{2}) :b_{n-k} b_k: \tag{6.113}$$

If we fix the constant to be added to L_0 from the vacuum energy density (like we did for the boson), we find

$$L_0 = \sum_{k>0} k b_{-k} b_k \qquad (\mathrm{NS} : k \in \mathbb{Z} + \frac{1}{2})$$

$$L_0 = \sum_{k>0} k b_{-k} b_k + \frac{1}{16} \qquad (\mathrm{R} : k \in \mathbb{Z}) \tag{6.114}$$

We apply this result to the calculation of the vacuum energies on the cylinder. From Eq. (5.138), we see that the vacuum expectation values of the energy-momentum tensor on the cylinder are

$$\langle T_{\text{cyl.}} \rangle = \begin{cases} -\dfrac{1}{48} \left(\dfrac{2\pi}{L}\right)^2 & \text{NS sector} \\[2ex] \dfrac{1}{24} \left(\dfrac{2\pi}{L}\right)^2 & \text{R sector} \end{cases} \tag{6.115}$$

In general, the Hamiltonian on the cylinder may be written as in Eq. (6.88) or, equivalently, as

$$H = \frac{2\pi}{L} \left(L_0 + \bar{L}_0 - \frac{c}{12} \right) \tag{6.116}$$

We have checked this explicitly for the boson in the last section. The added constant $(c/12)$ ensures that the vacuum energy of the Hamiltonian vanishes in the $L \to \infty$ limit in the NS sector. We could split the Hamiltonian into a contribution H_R from the holomorphic modes plus a contribution H_L from the antiholomorphic modes, with

$$H_R = \frac{2\pi}{L} \left(L_0 - \frac{c}{24} \right) \tag{6.117}$$

From the above considerations, we see that the correct expressions for H_R in terms of modes, in the two sectors, is indeed given by Eq. (6.102), which further confirms Eq. (6.116), with the following vacuum energies:

$$\frac{L}{2\pi} E_0 = \begin{cases} -\dfrac{1}{48} & \text{NS sector} \\[2ex] +\dfrac{1}{24} & \text{R sector} \end{cases} \tag{6.118}$$

The similar result obtained for the boson field had the periodic and antiperiodic values interchanged.

This result could have been obtained in a different way, using ζ-function regularization. We now explain how. The vacuum energy term may be thought of as the result of filling all the states in the Dirac sea (cf. Eq. (2.43)):

$$H_R = \frac{2\pi}{L} \left\{ \frac{1}{2} \sum_k |k| b_{-k} b_k \right\}$$
$$= \frac{2\pi}{L} \left\{ \frac{1}{2} \sum_{k>0} k b_{-k} b_k - \frac{1}{2} \sum_{k>0} k \right\} \tag{6.119}$$

As such, E_0 is formally infinite. However, it may be regularized by means of the generalized Riemann ζ-function:

$$\zeta(s,q) = \sum_{n=0}^{\infty} \frac{1}{(q+n)^s} \tag{6.120}$$

The usual Riemann ζ-function is $\zeta(s) \equiv \zeta(s,1)$. The above series definition is valid provided Re $s > 1$ and q is not a negative integer or zero. However, this function may be analytically continued to other regions of the s plane: its only singular point is $s = 1$. In particular, we have

$$\zeta(-n, q) = -\frac{B_{n+1}(q)}{n+1} \qquad (n \in \mathbb{N}, \, n > 0) \tag{6.121}$$

where $B_n(q)$ is the n-th Bernoulli polynomial, defined by the generating function

$$\frac{te^{xt}}{e^t - 1} \equiv \sum_{n=0}^{\infty} B_n(x) \frac{t^n}{n!}, \tag{6.122}$$

and $B_n(1) = B_n$ is the n-th Bernoulli number. The above expression for the vacuum energy may then be written as

$$\frac{L}{2\pi} E_0 = \begin{cases} -\dfrac{1}{2} \zeta(-1, \dfrac{1}{2}) & \text{NS sector} \\[2ex] -\dfrac{1}{2} \zeta(-1, 1) & \text{R sector} \end{cases} \tag{6.123}$$

Since $B_2(x) = x^2 - x + \frac{1}{6}$, we find $B_2(\frac{1}{2}) = -\frac{1}{12}$ and $B_2(1) = \frac{1}{6}$, and the values (6.118) are recovered.

§6.5. Normal Ordering

Up to now, we have introduced normal-ordered products only for the very special class of *free fields*. The characteristic property of a free field is that its OPE with itself (or various derivatives of this OPE) contains only one singular term, whose coefficient is a constant (cf. Eqs. (5.77), (6.97) and (5.108)). The regularization of a product of two such fields can be done simply by subtracting the corresponding expectation value (cf. (5.80) and (5.98)). In terms of modes, this is equivalent to the usual normal ordering in which the operators annihilating the vacuum are put at the rightmost positions.

However, this is no longer true for fields that are not free in the above sense. For instance, we see what happens when trying to regularize $T(z)T(w)$ by subtracting $\langle T(z)T(w) \rangle$ from the product $T(z)T(w)$ as $z \to w$. This prescription will eliminate the most singular term, proportional to the central charge. However, the two subleading singularities in $T(z)T(w)$ remain: The simple prescription used for free fields does not work in general. It is clear how this prescription should be generalized: Instead of subtracting only the vacuum expectation value, we should subtract all the singular terms of the OPE. To distinguish this generalized definition of normal ordering from that used previously, we shall denote it by parentheses: The normal-ordered version of $A(z)B(z)$ will be written $(AB)(z)$.

More explicitly, if the OPE of A and B is written as

$$A(z)B(w) = \sum_{n=-\infty}^{N} \frac{\{AB\}_n(w)}{(z-w)^n} \tag{6.124}$$

(N is some positive integer), then

$$(AB)(w) = \{AB\}_0(w) \tag{6.125}$$

Our definition of the contraction is generalized to include all the singular terms of the OPE:

$$\overbrace{A(z)B(w)} \equiv \sum_{n=1}^{N} \frac{\{AB\}_n(w)}{(z-w)^n} \tag{6.126}$$

Hence the above expression (6.125) for $(AB)(w)$ may be rewritten as

$$(AB)(w) = \lim_{z \to w} \left[A(z)B(w) - \overbrace{A(z)B(w)} \right] \tag{6.127}$$

and the OPE of $A(z)$ with $B(w)$ is expressed as

$$A(z)B(w) = \overbrace{A(z)B(w)} + (A(z)B(w)) \tag{6.128}$$

where $(A(z)B(w))$ stands for the complete sequence of regular terms whose explicit forms can be extracted from the Taylor expansion of $A(z)$ around w:

$$(A(z)B(w)) = \sum_{k \geq 0} \frac{(z-w)^k}{k!} \left(\partial^k AB \right)(w) \tag{6.129}$$

The method of contour integration provides another useful representation of our newly introduced normal ordering:

$$(AB)(w) = \frac{1}{2\pi i} \oint_w \frac{dz}{z-w} A(z)B(w) \tag{6.130}$$

The equivalence of (6.130) with (6.125) is readily checked by substituting (6.124) into (6.130).

Before translating this expression in terms of modes, a little digression is in order. Until now, all Laurent expansions for fields were made around the point $z = 0$ (cf. Eq. (6.7)). But this point is not special, and it is possible to expand instead around an arbitrary point w as

$$\phi(z) = \sum_{n \in \mathbb{Z}} (z-w)^{-n-h} \phi_n(w) \tag{6.131}$$

In particular, for the energy-momentum tensor, we have

$$T(z) = \sum_{n \in \mathbb{Z}} (z-w)^{-n-2} L_n(w) \tag{6.132}$$

or equivalently

$$L_n(w) = \frac{1}{2\pi i} \oint_w dz \, (z-w)^{n+1} T(z) \tag{6.133}$$

In this way, the OPE of $T(z)$ with an arbitrary field $A(w)$ can be written as

$$T(z)A(w) = \sum_{n \in \mathbb{Z}} (z - w)^{-n-2} (L_n A)(w) \qquad (6.134)$$

This defines the field $(L_n A)$. Comparing this with the expression

$$T(z)A(w) = \cdots + \frac{h_A A(w)}{(z - w)^2} + \frac{\partial A(w)}{(z - w)} + (TA)(w)$$
$$+ (z - w)(\partial TA)(w) + \cdots \qquad (6.135)$$

we see that

$$(L_0 A)(w) = h_A A(w)$$
$$(L_{-1} A)(w) = \partial A(w) \qquad (6.136)$$

as expected, but also

$$(L_{-n-2} A)(w) = \frac{1}{n!} (\partial^n TA)(w) \qquad (6.137)$$

In particular, when A is the identity field $\mathbb{I}$, this reads

$$(L_{-n-2} \mathbb{I})(w) = \frac{1}{n!} \partial^n T(w) \qquad (6.138)$$

We now derive the mode version of (6.130). The contour integration in (6.130) is rearranged along two contours:

$$\oint_w \frac{dz}{z - w} A(z)B(w) = \oint_{|z|>|w|} \frac{dz}{z - w} A(z)B(w) - \oint_{|z|<|w|} \frac{dz}{z - w} B(w)A(z) \qquad (6.139)$$

We consider the first term. Expanding the two fields around an intermediate point x such that $|z| > |x| > |w|$ yields

$$A(z) = \sum_n (z - x)^{-n-h_A} A_n(x)$$
$$B(w) = \sum_p (w - x)^{-p-h_B} B_p(x) \qquad (6.140)$$

Writing $z - w = z - x - (w - x)$, with the expansion

$$\frac{1}{z - w} = \sum_{l \geq 0} \frac{(w - x)^l}{(z - x)^{l+1}} \qquad (6.141)$$

we find

$$\frac{1}{2\pi i} \oint_{|z|>|w|} \frac{dz}{z-w} A(z)B(w)$$

$$= \frac{1}{2\pi i} \oint dz \sum_{\substack{n,p \\ l \geq 0}} (w-x)^{l-p-h_B}(z-x)^{-n-h_A-l-1} A_n(x)B_p(x)$$

$$= \sum_{\substack{p \\ n \leq -h_A}} (w-x)^{-n-p-h_A-h_B} A_n(x)B_p(x) \qquad (6.142)$$

The only singularity inside the contour is at $z = x$, and only the pole contributes; hence $l + n + h_A = 0$. Since $l \geq 0$, it follows that $n \leq -h_A$. For the second term, we proceed in a similar way. With the roles of w and z in (6.141) interchanged, it follows that

$$\frac{1}{2\pi i} \oint_{|w|>|z|} \frac{dz}{z-w} B(w)A(z)$$

$$= \frac{1}{2\pi i} \oint dz \sum_{\substack{n,p \\ l \geq 0}} (w-x)^{-l-1-p-h_B}(z-x)^{l-n-h_A} B_p(x)A_n(x)$$

$$= \sum_{\substack{p \\ n > -h_A}} (w-x)^{-n-p-h_A-h_B} B_p(x)A_n(x). \qquad (6.143)$$

since $l - n - h_A = -1$. Collecting these two results, we find

$$\boxed{(AB)_m = \sum_{n \leq -h_A} A_n B_{m-n} + \sum_{n > -h_A} B_{m-n}A_n} \qquad (6.144)$$

wherein the modes $(AB)_n$ are defined by

$$(AB)(z) = \sum z^{-n-h_A-h_B}(AB)_n. \qquad (6.145)$$

Eq. (6.144) makes manifest the noncommutativity of the normal ordering:

$$(AB)(z) \neq (BA)(z) \qquad (6.146)$$

This generally differs from the usual normal ordering of modes denoted by $:\ :$, in which the operator with larger subindex is placed at the right. A reformulation of Wick's theorem for interacting fields is thus required. This is developed in App. 6.B. The normal order defined above is not associative: $((AB)C) \neq (A(BC))$. Appendix 6.C explains how to go from one form to the other (i.e., how to calculate $((AB)C) - (A(BC)))$.

§6.6. Conformal Families and Operator Algebra

6.6.1. Descendant Fields

Primary fields play a fundamental role in conformal field theory. The asymptotic state $|h\rangle = \phi(0)|0\rangle$ created by a primary field of conformal dimension h is the source of an infinite tower of descendant states of higher conformal dimensions (cf. Sect. 6.2.2). Under a conformal transformation, the state $|h\rangle$ and its descendants transform among themselves.

Each descendant state can be viewed as the result of the application on the vacuum of a *descendant field*. Consider, for instance, the descendant $L_{-n}|h\rangle$:

$$L_{-n}|h\rangle = L_{-n}\phi(0)|0\rangle = \frac{1}{2\pi i}\oint dz\, z^{1-n}T(z)\phi(0)|0\rangle \qquad (6.147)$$

Using the OPE (6.134) this is merely $(L_{-n}\phi)(0)|0\rangle$: descendant states may be obtained by applying on the vacuum the operators appearing in the regular part of the OPE of $T(z)$ with $\phi(0)$ (for a definition of the notation $(L_{-n}\phi)$, see Sect. 6.5). The natural definition of the descendant field associated with the state $L_{-n}|h\rangle$ is

$$\phi^{(-n)}(w) \equiv (L_{-n}\phi)(w) = \frac{1}{2\pi i}\oint_w dz\, \frac{1}{(z-w)^{n-1}}T(z)\phi(w) \qquad (6.148)$$

These are the fields appearing in the OPE (6.134) of $T(z)$ with $\phi(w)$. In particular,

$$\phi^{(0)}(w) = h\phi(w) \quad \text{and} \quad \phi^{(-1)}(w) = \partial\phi(w) \qquad (6.149)$$

The physical properties of these fields (i.e., their correlation functions) may be derived from those of the "ancestor" primary field. Indeed, consider the correlator

$$\langle(L_{-n}\phi)(w)X\rangle \qquad (6.150)$$

where $X = \phi_1(w_1)\cdots\phi_N(w_N)$ is an assembly of primary fields with conformal dimensions h_i. This correlator may be calculated by substituting the definition (6.148) of the descendant, in which the contour circles w only, excluding the positions w_i of the other fields. The residue theorem may be applied by reversing the contour and summing the contributions from the poles at w_i, with the help of the OPE (5.41) of T with primary fields:

$$
\begin{aligned}
\langle\phi^{(-n)}(w)X\rangle &= \frac{1}{2\pi i}\oint_w dz\, (z-w)^{1-n}\langle T(z)\phi(w)X\rangle \\
&= -\frac{1}{2\pi i}\oint_{\{w_i\}} dz\, (z-w)^{1-n}\sum_i\left\{\frac{1}{z-w_i}\partial_{w_i}\langle\phi(w)X\rangle\right. \\
&\qquad\qquad \left.+ \frac{h_i}{(z-w_i)^2}\langle\phi(w)X\rangle\right\} \\
&\equiv \mathcal{L}_{-n}\langle\phi(w)X\rangle \qquad (n \geq 1)
\end{aligned}
\qquad (6.151)
$$

wherein we defined the differential operator

$$\mathcal{L}_{-n} = \sum_i \left\{ \frac{(n-1)h_i}{(w_i - w)^n} - \frac{1}{(w_i - w)^{n-1}} \partial_{w_i} \right\} \tag{6.152}$$

We have thus reduced the evaluation of a correlator containing a descendant field to that of a correlator of primary fields, on which we must apply a differential operator $\mathcal{L}_{-n}$. We note that $\mathcal{L}_{-1}$ is in fact equivalent to $\partial/\partial w$, since the operator

$$\frac{\partial}{\partial w} + \sum_i \frac{\partial}{\partial w_i} \tag{6.153}$$

annihilates any correlator because of translation invariance.

Of course, there are descendant fields more complicated than $\phi^{(-n)}$, corresponding to the more general state (6.36). They may be defined recursively:

$$\begin{aligned}
\phi^{(-k,-n)}(w) &= (L_{-k}L_{-n}\phi)(w) \\
&= \frac{1}{2\pi i} \oint_w dz \, (z-w)^{1-k} T(z)(L_{-n}\phi)(w)
\end{aligned} \tag{6.154}$$

and so on. In particular,

$$\phi^{(0,-n)}(w) = (h+n)\phi^{(-n)}(w) \quad \text{and} \quad \phi^{(-1,-n)}(w) = \partial_w \phi^{(-n)}(w) \tag{6.155}$$

These last two relations follow directly from the roles of L_0 and L_{-1} as generator of dilations and translations, respectively.

It can be shown without difficulty that

$$\langle \phi^{(-k_1,\ldots,-k_n)}(w)X \rangle = \mathcal{L}_{-k_1} \cdots \mathcal{L}_{-k_n} \langle \phi(w)X \rangle \tag{6.156}$$

that is, we simply need to apply the differential operators in succession. We may also consider correlators containing more than one descendant field, but at the end the result is the same: Correlation functions of descendant fields may be reduced to correlation functions of primary fields.

6.6.2. Conformal Families

The set comprising a primary field ϕ and all of its descendants is called a *conformal family*, and is sometimes denoted $[\phi]$. As indicated earlier, the members of a family transform amongst themselves under a conformal transformation. Equivalently, we can say that the OPE of $T(z)$ with any member of the family will be composed solely of other members of the same family.[1]

---

[1] We should keep in mind that conformal fields have an antiholomorphic part as well as a holomorphic part. There will also be descendants of a field through the action of the antiholomorphic generators $\bar{L}_{-n}$.

For instance, we calculate the OPE of $T(z)$ with $\phi^{(-n)}$. Eq. (6.134) implies

$$T(z)\phi^{(-n)}(w) = \sum_{k\geq 0}(z-w)^{k-2}(L_{-k}\phi^{(-n)})(w)$$

$$+ \sum_{k>0}\frac{1}{(z-w)^{k+2}}(L_k\phi^{(-n)})(w) \qquad (6.157)$$

The first sum contains more complex descendant fields, $\phi^{(-k,-n)}$, of the same family. The second sum is made of the most singular terms, and may be calculated by considering the singular part of the OPE of T with itself:

$$T(z)\phi^{(-n)}(w) = \frac{1}{2\pi i}\oint_w dx\,\frac{1}{(x-w)^{n-1}}T(z)T(x)\phi(w)$$

$$\sim \frac{1}{2\pi i}\oint_w dx\,\frac{1}{(x-w)^{n-1}}\left\{\frac{c/2}{(z-x)^4}+\frac{2T(x)}{(z-x)^2}+\frac{\partial T(x)}{z-x}\right\}\phi(w)$$

$$= \frac{cn(n^2-1)/12}{(z-w)^{n+2}}\phi(w)+\oint_w dx\,\frac{1}{(x-w)^{n-1}}\sum_{l=0}^{\infty}\phi^{(-l)}(w)$$

$$\times\left\{\frac{2(x-w)^{l-2}}{(z-x)^2}+\frac{(l-2)(x-w)^{l-3}}{z-x}\right\}$$

$$= \frac{cn(n^2-1)/12}{(z-w)^{n+2}}\phi(w)+\sum_{l=0}^{n+1}\frac{2n-l}{(z-w)^{n+2-l}}\phi^{(-l)}(w)$$

$$\qquad (6.158)$$

where we have used the identity

$$\frac{1}{2\pi i}\oint_w\frac{dx}{(x-w)^n}\frac{F(w)}{(z-x)^m}=\frac{(n+m-2)!}{(n-1)!(m-1)!}\frac{F(w)}{(z-w)^{n+m-1}} \qquad (6.159)$$

Again, the symbol $\sim$ means an equality modulo regular terms. Assembling all the terms and redefining the summation index in the last term, we finally write

$$T(z)\phi^{(-n)}(w) = \frac{cn(n^2-1)/12}{(z-w)^{n+2}}\phi(w)+\sum_{k=1}^{n}\frac{n+k}{(z-w)^{k+2}}\phi^{(k-n)}(w)$$

$$+ \sum_{k\geq 0}(z-w)^{k-2}\phi^{(-k,-n)}(w) \qquad (6.160)$$

For instance, the OPE of $T(z)$ with $\phi^{(-1)}=\partial\phi$ is

$$T(z)\partial\phi(w) \sim \frac{2h\phi(w)}{(z-w)^3}+\frac{(h+1)\partial\phi(w)}{(z-w)^2}+\frac{\partial^2\phi(w)}{z-w} \qquad (6.161)$$

The descendants of a primary field are called *secondary fields*. Under a conformal mapping $z\to f(z)$, a secondary field $A(z)$ transforms like

$$A(z)\to\left(\frac{df}{dz}\right)^{h'}A(f(z))+\quad\text{extra terms} \qquad (6.162)$$

where $h' = h + n$ (n a positive integer) if A is a descendant of a primary field of dimension h. The extra terms translate into pole singularities of degree higher than two in the OPE of $T(z)$ with $A(w)$, as in Eq. (6.160).

6.6.3. The Operator Algebra

The main object of a field theory is the calculation of correlation functions, which are the physically measurable quantities. Conformal invariance helps us in this task: We have seen how the coordinate dependence of two- and three-point functions of primary fields is fixed by global conformal invariance (cf. Eqs. (5.25) and (5.26)). Unfortunately, conformal invariance does not tell us everything, and some dynamical input is necessary to calculate the three-point function coefficient C_{ijk}. Indeed, the information needed in order to write down all correlation functions, and hence solve the theory, is the so-called *operator algebra*: The complete OPE (including all regular terms) of all primary fields with each other. Indeed, applying this OPE within a correlation function allows for a reduction of the number of points, down to two-point functions, which are known. The goal of this section is to spell out the form of this operator algebra and to indicate which of its elements are fixed by conformal invariance, and which are not.

We must first discuss the normalization of fields, that is, the two-point function coefficients C_{12}. We know that the two-point function vanishes if the conformal dimensions of the two fields are different. If the conformal dimensions are the same for a finite set of primary fields ϕ_α, the correlators are

$$\langle \phi_\alpha(w, \bar{w}) \phi_\beta(z, \bar{z}) \rangle = \frac{C_{\alpha\beta}}{(w - z)^{2h}(\bar{w} - \bar{z})^{2\bar{h}}} \tag{6.163}$$

Since the coefficients $C_{\alpha\beta}$ are symmetric, we are free to choose a basis of primary fields such that $C_{\alpha\beta} = \delta_{\alpha\beta}$; it is a simple matter of normalization. We shall adopt this convention in the remainder of this work, unless otherwise indicated. Thus, conformal families associated with different ϕ_α's are orthogonal in the sense of the two-point function. Of course, the same is true of the corresponding Verma modules: By a suitable global conformal transformation, we can always bring the points w and z of a correlator to $w = \infty$ and $z = 0$ respectively. The fields are then asymptotic and the two-point function becomes a bilinear product on the Hilbert space:

$$\lim_{w, \bar{w} \to \infty} w^{2h} \bar{w}^{2\bar{h}} \langle \phi(w, \bar{w}) \phi'(0, 0) \rangle = \langle h | h' \rangle \langle \bar{h} | \bar{h}' \rangle \tag{6.164}$$

The orthogonality of the highest weight states implies the orthogonality of all the descendants of the two fields (i.e., the orthogonality of the Verma modules associated with the two fields).

Invariance under scaling transformations clearly requires the operator algebra to have the following form:

$$\phi_1(z, \bar{z}) \phi_2(0, 0) = \sum_p \sum_{\{k, \bar{k}\}} C_{12}^{p\{k, \bar{k}\}} z^{h_p - h_1 - h_2 + K} \bar{z}^{\bar{h}_p - \bar{h}_1 - \bar{h}_2 + \bar{K}} \phi_p^{\{k, \bar{k}\}}(0, 0) \tag{6.165}$$

where $K = \sum_i k_i$ and $\bar{K} = \sum_i \bar{k}_i$; the expression $\{k\}$ means a collection of indices k_i.

We take the correlator of this relation with a third primary field $\phi_r(w, \bar{w})$ of dimensions $h_r, \bar{h}_r$. Sending $w \to \infty$, we have, on the l.h.s.,

$$\langle \phi_r | \phi_1(z, \bar{z}) | \phi_2 \rangle = \lim_{w, \bar{w} \to \infty} w^{2h_r} \bar{w}^{2\bar{h}_r} \langle \phi_r(w, \bar{w}) \phi_1(z, \bar{z}) \phi_2(0, 0) \rangle$$

$$= \frac{C_{r12}}{z^{h_1 + h_2 - h_r} \bar{z}^{\bar{h}_1 + \bar{h}_2 - \bar{h}_r}} \tag{6.166}$$

The last equality is obtained simply by applying the limit $w \to \infty$ in the general formula (5.26) for the three-point function. On the OPE side, the only contributing term is $p\{k, \bar{k}\} = r\{0, 0\}$, because of the orthogonality of the Verma modules. We conclude that

$$C_{12}^{p\{0,0\}} \equiv C_{12}^p = C_{p12} \tag{6.167}$$

In other words, the most singular term of the operator algebra is the coefficient of the three-point function. The normalization adopted for two-point functions eliminates the distinction between "covariant" and "contravariant" indices. Since the correlations of descendants are built on the correlation of primaries, we expect the coefficients $C_{12}^{p\{k, \bar{k}\}}$ to have the following form:

$$C_{12}^{p\{k, \bar{k}\}} = C_{12}^p \beta_{12}^{p\{k\}} \bar{\beta}_{12}^{p\{\bar{k}\}} \tag{6.168}$$

This simply means that the descendant fields can be correlated to a third field only if the primary itself is correlated, with the holomorphic and antiholomorphic parts factorized. By convention we set $\beta_{ij}^{p\{0\}} = 1$. The other coefficients $\beta_{ij}^{p\{k\}}$ are determined (as functions of the central charge c and of the conformal dimensions) by the requirement that both sides of Eq. (6.165) behave identically upon conformal transformations.

We shall illustrate this statement in the simple case $h_1 = h_2 = h$. When applying Eq. (6.165) on the vacuum, we find

$$\phi_1(z, \bar{z}) | h, \bar{h} \rangle = \sum_p C_{p12} z^{h_p - 2h} \bar{z}^{\bar{h}_p - 2\bar{h}} \varphi(z) \bar{\varphi}(\bar{z}) | h_p, \bar{h}_p \rangle \tag{6.169}$$

wherein we have defined the operator

$$\varphi(z) = \sum_{\{k\}} z^K \beta_{12}^{p\{k\}} L_{-k_1} \cdots L_{-k_N} \tag{6.170}$$

and similarly for $\bar{\varphi}(\bar{z})$. On the holomorphic sector we define the state

$$| z, h_p \rangle \equiv \varphi(z) | h_p \rangle \tag{6.171}$$

which is therefore expressible as a power series:

$$| z, h_p \rangle = \sum_{N=0}^{\infty} z^N | N, h_p \rangle \tag{6.172}$$

The state $|N, h_p\rangle$ is a descendant state at level N in the Verma module $V(h_p)$:

$$L_0|N, h_p\rangle = (h_p + N)|N, h_p\rangle \tag{6.173}$$

(we use the notation $|0, h_p\rangle = |h_p\rangle$). We now apply the operator L_n on both sides of Eq. (6.169). Acting on the l.h.s., L_n yields

$$L_n\phi_1(z, \bar{z})|h, \bar{h}\rangle = [L_n, \phi_1(z, \bar{z})]|h, \bar{h}\rangle$$
$$= \left(z^{n+1}\partial_z + (n+1)h\right)\phi_1(z, \bar{z})|h, \bar{h}\rangle \tag{6.174}$$

Applying this relation on the r.h.s. of Eq. (6.169), we find

$$\sum_p C_{p12}z^{h_p - 2h}\bar{z}^{\bar{h}_p - 2\bar{h}}L_n|z, h_p\rangle|\bar{z}, \bar{h}_p\rangle =$$

$$\sum_p C_{p12}z^{h_p - 2h}\bar{z}^{\bar{h}_p - 2\bar{h}}\left[(h_p + h(n-1))z^n + z^{n+1}\partial_z\right]|z, h_p\rangle|\bar{z}, \bar{h}_p\rangle$$

Substituting the power series (6.172), we finally obtain

$$L_n|N + n, h_p\rangle = (h_p + (n-1)h + N)|N, h_p\rangle \tag{6.175}$$

This relation, together with the Virasoro algebra, allows the recursive calculation of all the $|N, h_p\rangle$, and hence of all the $\beta_{12}^{p\{k\}}$.

We now calculate explicitly the lowest coefficients. First, we know that

$$|1, h_p\rangle = \beta_{12}^{p\{1\}}L_{-1}|h_p\rangle$$

since the r.h.s. is the only state at level 1. Operating with L_1 and applying the relation (6.175), we obtain

$$L_1|1, h_p\rangle = h_p|h_p\rangle = \beta_{12}^{p\{1\}}L_1L_{-1}|h_p\rangle \tag{6.176}$$

Since $L_1L_{-1}|h_p\rangle = [L_1, L_{-1}]|h_p\rangle = 2h_p|h_p\rangle$, we find

$$\beta_{12}^{p\{1\}} = \frac{1}{2} \tag{6.177}$$

At level 2, we have

$$|2, h_p\rangle = \beta_{12}^{p\{1,1\}}L_{-1}^2|h_p\rangle + \beta_{12}^{p\{2\}}L_{-2}|h_p\rangle \tag{6.178}$$

We operate on this equation with L_1 and, separately, with L_2, applying Eq. (6.175). We need the following relations from the Virasoro algebra:

$$L_1L_{-1}^2 = L_{-1}^2L_1 + 4L_{-1}L_0 - 2L_{-2}$$
$$L_1L_{-2} = L_{-2}L_1 + 3L_{-1}$$
$$L_2L_{-1}^2 = L_{-1}^2L_2 + 6L_{-1}L_1 + 6L_0 \tag{6.179}$$
$$L_2L_{-2} = L_{-2}L_2 + 4L_0 + \frac{1}{2}c$$

and we end up with the following matrix equation:

$$\begin{pmatrix} 2(2h_p + 1) & 3 \\ 6h_p & \frac{1}{2}c + 4h_p \end{pmatrix}\begin{pmatrix} \beta_{12}^{p\{1,1\}} \\ \beta_{12}^{p\{2\}} \end{pmatrix} = \begin{pmatrix} \frac{1}{2}(h_p + 1) \\ h_p + h \end{pmatrix} \tag{6.180}$$

whose solution is

$$\beta_{12}^{p\{1,1\}} = \frac{c - 12h - 4h_p + ch_p + 8h_p^2}{4(c - 10h_p + 2ch_p + 16h_p^2)}$$

$$\beta_{12}^{p\{2\}} = \frac{2h - h_p + 4hh_p + h_p^2}{c - 10h_p + 2ch_p + 16h_p^2}$$

(6.181)

At a given level N there are $p(N)$ coefficients $\beta_{12}^{p\{k\}}$ to be found, and accordingly we need $p(N)$ equations for these coefficients. These equations are obtained by considering the $p(N)$ ways to bring $|N, h_p\rangle$ to level 0 with help of the Virasoro operators L_n $(n > 0)$.

In short, we have illustrated how the complete operator algebra of primary fields may be obtained from conformal symmetry, the only necessary ingredients being the central charge c, the conformal dimensions of the primary fields, and the three-point function coefficient C_{pnm}. In principle, any n-point function can be calculated from this operator algebra by successive reduction of the products of primary fields. The correlators of descendant fields thus obtained can be expressed in terms of primary field correlators, and so on. Hence, the theory is then solved, by definition! Of course, the coefficients C_{pnm} must be obtained from another source, for instance through the conformal bootstrap (see below).

6.6.4. Conformal Blocks

In the last subsection we have mentioned that four-point functions can be reduced to three-point functions with the help of the operator algebra (6.165). Here we shall make this point more explicit, and find which part of a four-point function is fixed by conformal invariance and which is not.

We consider the generic four-point function

$$\langle \phi_1(z_1, \bar{z}_1)\phi_2(z_2, \bar{z}_2)\phi_3(z_3, \bar{z}_3)\phi_4(z_4, \bar{z}_4) \rangle$$

(6.182)

We have seen that such a function depends continuously on the anharmonic ratios

$$x = \frac{(z_1 - z_2)(z_3 - z_4)}{(z_1 - z_3)(z_2 - z_4)} \qquad \bar{x} = \frac{(\bar{z}_1 - \bar{z}_2)(\bar{z}_3 - \bar{z}_4)}{(\bar{z}_1 - \bar{z}_3)(\bar{z}_2 - \bar{z}_4)}$$

(6.183)

Since these ratios are invariant under global transformations, we shall perform such a transformation in order to set $z_4 = 0$, $z_1 = \infty$, and $z_2 = 1$; then $z_3 = x$ and the above correlation function may be related to a matrix element between two asymptotic states of a two-field product:

$$\lim_{z_1, \bar{z}_1 \to \infty} z_1^{2h_1} \bar{z}_1^{2\bar{h}_1} \langle \phi_1(z_1, \bar{z}_1)\phi_2(1, 1)\phi_3(x, \bar{x})\phi_4(0, 0) \rangle = G_{34}^{21}(x, \bar{x})$$

wherein we have defined the function

$$G_{34}^{21}(x, \bar{x}) = \langle h_1, \bar{h}_1 | \phi_2(1, 1)\phi_3(x, \bar{x}) | h_4, \bar{h}_4 \rangle$$

(6.184)

(the order in which the indices of G appear is important).

We now use the operator algebra to reduce the products within the four-point function. We write the operator algebra as

$$\phi_3(x,\bar{x})\phi_4(0,0) = \sum_p C_{34}^p x^{h_p-h_3-h_4}\bar{x}^{\bar{h}_p-\bar{h}_3-\bar{h}_4}\Psi_p(x,\bar{x}|0,0) \qquad (6.185)$$

wherein

$$\Psi_p(x,\bar{x}|0,0) = \sum_{\{k,\bar{k}\}} \beta_{34}^{p\{k\}}\bar{\beta}_{34}^{p\{\bar{k}\}} x^K \bar{x}^{\bar{K}} \phi_p^{\{k,\bar{k}\}}(0,0) \qquad \left(K = \sum k_i\right)$$

The function G_{34}^{21} may then be written as

$$G_{34}^{21}(x,\bar{x}) = \sum_p C_{34}^p C_{12}^p A_{34}^{21}(p|x,\bar{x}) \qquad (6.186)$$

where we have introduced the function

$$A_{34}^{21}(p|x,\bar{x}) = (C_{12}^p)^{-1} x^{h_p-h_3-h_4}\bar{x}^{\bar{h}_p-\bar{h}_3-\bar{h}_4} \langle h_1,\bar{h}_1|\phi_2(1,1)\Psi_p(x,\bar{x}|0,0)|0\rangle$$

We have merely rewritten the four-point function as a sum over intermediate conformal families, labeled by the index p. The analogy with the diagrammatic approach to perturbation theory is clear: The intermediate conformal families correspond to the different intermediate states formed during the scattering of the two fields from $(0,x)$ toward $(1,\infty)$. We could therefore represent $A_{34}^{21}(p|x,\bar{x})$ by a tree diagram with four legs (see Fig. 6.3). In the same spirit, we may refer to these functions as *partial waves*.

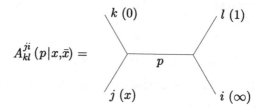

Figure 6.3. Partial wave in diagrammatic language. The same diagram is often used to represent only the holomorphic (or antiholomorphic) part of the partial wave, the conformal block $\mathcal{F}_{34}^{21}(p|x)$.

It is clear from its definition that the partial wave factorizes into a holomorphic and an antiholomorphic part:

$$A_{34}^{21}(p|x,\bar{x}) = \mathcal{F}_{34}^{21}(p|x)\bar{\mathcal{F}}_{34}^{21}(p|\bar{x})$$

where

$$\mathcal{F}_{34}^{21}(p|x) = x^{h_p-h_3-h_4} \sum_{\{k\}} \beta_{34}^{p\{k\}} x^K \frac{\langle h_1|\phi_2(1)L_{-k_1}\cdots L_{-k_N}|h_p\rangle}{\langle h_1|\phi_2(1)|h_p\rangle} \qquad (6.187)$$

The denominator is simply equal to $(C_{21}^p)^{1/2}$. The functions defined in Eq. (6.187) are called *conformal blocks*. They can be calculated simply from the knowledge of the conformal dimensions and the central charge, by commuting the Virasoro generators over the field $\phi_2(1)$ one after the other. The field normalizations and coefficients C_{mn}^p drop out of the conformal block at the end of this process. Going back to the partial wave decomposition (6.186), we see that the conformal blocks represent the element in four-point functions that can be determined from conformal invariance. They depend on the anharmonic ratios through a series expansion. The remaining elements are the three-point function coefficients C_{12}^p and C_{34}^p, which are not fixed by conformal invariance. Therefore, the four-point function (6.184) is expressed as

$$G_{34}^{21}(x,\bar{x}) = \sum_p C_{34}^p C_{12}^p \, \mathcal{F}_{34}^{21}(p|x)\bar{\mathcal{F}}_{34}^{21}(p|\bar{x}) \qquad (6.188)$$

An explicit expression for the conformal blocks is not known in general. Although the formula (6.187) may be applied in principle, its use becomes rapidly tedious. One may write the conformal block as a power series in x:

$$\mathcal{F}_{34}^{21}(p|x) = x^{h_p-h_3-h_4} \sum_{K=0}^{\infty} \mathcal{F}_K x^K \qquad (6.189)$$

where the coefficient $\mathcal{F}_K$ depends on the conformal dimensions h_i ($i = 1,\ldots,4$) and h_p. The normalization fixes $\mathcal{F}_0 = 1$. The next two coefficients may be obtained by blindly applying the definition (6.187):

$$\mathcal{F}_1 = \frac{(h_p+h_2-h_1)(h_p+h_3-h_4)}{2h_p} \qquad (6.190)$$

$$\mathcal{F}_2 = \frac{(h_p+h_2-h_1)(h_p+h_2-h_1+1)(h+h_3-h_4)(h+h_3-h_4+1)}{4h_p(2h_p+1)}$$

$$+2\left(\frac{h_1+h_2}{2}+\frac{h_p(h_p-1)}{2(2h_p+1)}-\frac{3(h_1-h_2)^2}{2(2h_p+1)}\right)^2$$

$$\times\left(\frac{h_3+h_4}{2}+\frac{h_p(h_p-1)}{2(2h_p+1)}-\frac{3(h_3-h_4)^2}{2(2h_p+1)}\right)^2\left(c+\frac{2h_p(8h_p-5)}{2h_p+1}\right)^{-1}$$

$$(6.191)$$

6.6.5. Crossing Symmetry and the Conformal Bootstrap

In defining the function $G_{34}^{21}(x,\bar{x})$, we have chosen a specific order for the four fields ϕ_{1-4} within the correlator. But the ordering of fields within correlators does not matter (except for signs when dealing with fermions); we could have proceeded otherwise, for instance by sending z_2 to 0 and z_4 to 1. Then $z_3 = 1 - x$ and we obtain the identity

$$G_{34}^{21}(x,\bar{x}) = G_{32}^{41}(1 - x, 1 - \bar{x})$$

We could also interchange ϕ_1 and ϕ_4 and obtain

$$G_{34}^{21}(x,\bar{x}) = \frac{1}{x^{2h_3}\bar{x}^{2\bar{h}_3}} G_{31}^{24}(1/x, 1/\bar{x})$$

These conditions on the function G_{34}^{21} are manifestations of *crossing symmetry*.
We write the first of these relations in terms of conformal blocks:

$$\sum_p C_{21}^p C_{34}^p \, \mathcal{F}_{34}^{21}(p|x)\bar{\mathcal{F}}_{34}^{21}(p|\bar{x}) = \sum_q C_{41}^q C_{32}^q \, \mathcal{F}_{32}^{41}(q|1 - x)\bar{\mathcal{F}}_{32}^{41}(q|1 - \bar{x}) \quad (6.192)$$

This relation is represented graphically on Fig. 6.4. Assuming that the conformal blocks are known for arbitrary values of the conformal dimensions, the above expresses a set of constraints that could determine the coefficients C_{mn}^p and the conformal dimensions h_p. Indeed, if we assume the presence of N conformal families in the theory, the above relation yields, through naive counting, N^4 constraints on the $N^3 + N$ parameters C_{mn}^p and h_n. This program of calculating the correlation functions simply by assuming crossing symmetry is known as the *bootstrap approach*. There is no proof that Eq. (6.192) can indeed determine the parameters of the theory in the general case, but there are special cases (the minimal models) in which the bootstrap equations can be solved completely. The bootstrap hypothesis (6.192) is the sole "dynamical input" required to completely solve the theory, once the explicit form of the conformal blocks has been determined from conformal invariance. The crossing symmetry constraint (6.192) is quite natural from the point of view of the operator algebra—rather like the Jacobi identity for Lie algebras or Poisson brackets—and does not constitute a narrow condition excluding interesting theories.

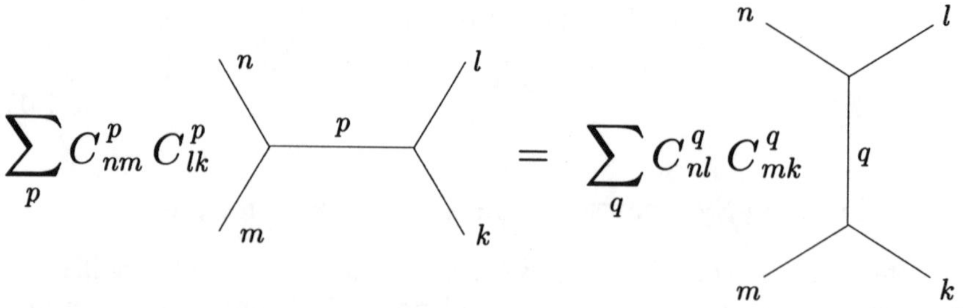

Figure 6.4. Crossing symmetry in diagrammatic language.

Appendix 6.A. Vertex and Coherent States

In this appendix we demonstrate the following formula for the vacuum expectation value of products of n vertex operators involving a single harmonic oscillator:

$$\langle :e^{A_1}::e^{A_2}:\cdots:e^{A_n}:\rangle = \exp\sum_{i<j}^{n}\langle A_i A_j\rangle \qquad (6.193)$$

where $A_i = \alpha_i a + \beta_i a^\dagger$ is a linear combination of creation and annihilation operators.

We first define the harmonic oscillator coherent state

$$|z\rangle \equiv e^{za^\dagger}|0\rangle \qquad (6.194)$$

It is simple to show that $|z\rangle$ is an eigenstate of a:

$$a|z\rangle = z|z\rangle \quad \text{or} \quad f(a)|z\rangle = f(z)|z\rangle \qquad (6.195)$$

Indeed, the Hausdorff relation

$$e^{-A}Be^{A} = B + [B,A] + \frac{1}{2}[[B,A],A] + \cdots \qquad (6.196)$$

applied to $A = za^\dagger$ and $B = a$ yields

$$[a, e^{za^\dagger}] = ze^{za^\dagger} \qquad (6.197)$$

from which Eq. (6.195) follows. If $[B,A]$ is a constant, which is true here, the Hausdorff relation also implies that

$$e^{B}e^{A} = e^{A}e^{B}e^{[B,A]} \qquad (6.198)$$

This, applied to our problem, yields

$$e^{wa}e^{za^\dagger} = e^{za^\dagger}e^{wa}e^{wz} \qquad (6.199)$$

Within a vertex operator A_i, the normal-ordered product reads

$$:e^{A_i}:= e^{\beta_i a^\dagger}e^{\alpha_i a} \qquad (6.200)$$

In calculating the normal-ordered product of a string $:e^{A_1}:\cdots:e^{A_n}:$ of vertex operators, we want to bring all the annihilation operators to the right. For instance, it follows from Eq. (6.199) that

$$e^{\alpha_i a}e^{\beta_{i+1}a^\dagger}\cdots e^{\beta_n a^\dagger} = e^{\beta_{i+1}a^\dagger}\cdots e^{\beta_n a^\dagger}e^{\alpha_i a}e^{\alpha_i(\beta_{i+1}+\beta_{i+2}+\cdots+\beta_n)} \qquad (6.201)$$

Since $[e^{\alpha_i a}, e^{\alpha_j a}] = 0$, this implies

$$e^{\alpha_i a}:e^{A_{i+1}}:\cdots:e^{A_n}:=:e^{A_{i+1}}:\cdots:e^{A_n}:e^{\alpha_i a}e^{\alpha_i(\beta_{i+1}+\beta_{i+2}+\cdots+\beta_n)} \qquad (6.202)$$

Applying this in succession from $i = 1$ to $i = n - 1$, we find

$$:e^{A_1}::e^{A_2}:\cdots:e^{A_n}:= e^{(\beta_1+\cdots+\beta_n)a^\dagger}e^{(\alpha_1+\cdots+\alpha_n)a}\exp\sum_{i<j}^{n}\alpha_i\beta_j \qquad (6.203)$$

Since $\langle A_i A_j \rangle = \alpha_i \beta_j$, one may finally write

$$:e^{A_1}::e^{A_2}:\;\cdots\;:e^{A_n}:=:e^{A_1+\cdots+A_n}:\,\exp\sum_{i<j}^{n}\langle A_i A_j\rangle \qquad (6.204)$$

From the vacuum expectation value of this expression follows the relation (6.193).

Appendix 6.B. The Generalized Wick Theorem

In this appendix we reformulate Wick's theorem for interacting fields, using the generalization of the concept of normal ordering explained in Sect. 6.5. We are not interested in the most general form of Wick's theorem, which gives the relation between the time-ordered (or radial-ordered) product and the normal-ordered product of free fields, illustrated in Eq. (2.109). Such a relation cannot be generalized to interacting fields. Rather, we wish to generalize a specialized form of Wick's theorem for the contraction of a field with a normal-ordered product. For free (commuting) fields, this is

$$\overbrace{\phi_1(x)}^{} :\phi_2\phi_3:(y) = \overbrace{\phi_1(x)\phi_2(y)}^{} :\phi_3(y): + \overbrace{\phi_1(x)\phi_3(y)}^{} :\phi_2(y): \qquad (6.205)$$

The suitable generalization of this relation to interacting fields is

$$\overbrace{A(z)(BC)(w)}^{} = \frac{1}{2\pi i}\oint_w \frac{dx}{x-w}\left\{\overbrace{A(z)B(x)}^{}C(w) + B(x)\overbrace{A(z)C(w)}^{}\right\} \qquad (6.206)$$

In order to demonstrate this relation, one must show that the contractions on the r.h.s. extract all the singular terms of the integral as $z \to w$. But these singular terms can only come from the OPE of $A(z)$ with B and C separately (the integral amounts to a point-splitting procedure). We rewrite this expression as

$$\frac{1}{2\pi i}\oint \frac{dx}{x-w}\left\{\sum_{n>0}\frac{\{AB\}_n(x)C(w)}{(z-x)^n} + \sum_{n>0}\frac{B(x)\{AC\}_n(w)}{(z-w)^n}\right\} \qquad (6.207)$$

From this expression it is manifest that all the inverse powers of $(z-w)$ and $(z-x)$ in the integrand yield inverse powers of $(z-w)$ after integration. Conversely, nonnegative powers of $(z-w)$ and $(z-x)$ in the integrand, if added, would not contribute to inverse powers of $(z-w)$ after integration. Thus the modified Wick rule (6.206) is correct. It is straightforward to check that the rule (6.206), applied to a free boson φ, leads to the same result as the usual Wick theorem. For instance,

$$\overbrace{\partial\varphi(z)(\varphi\varphi)(w)}^{} = \frac{2\varphi(w)}{z-w} \qquad (6.208)$$

The subtlety with formula (6.206) applied to interacting fields is that one is left with full OPEs after one contraction. This is important since the first regular term of the various OPEs always contributes. To see this, we consider the first term on the r.h.s. of Eq. (6.206). Writing the OPE of $\{AB\}_n(x)$ with $C(w)$ as

$$\{AB\}_n(x)C(w) \sim \sum_m (x-w)^{-m} E^{(n,m)}(w) \qquad (6.209)$$

(no restriction on m), the first term on the r.h.s. of Eq. (6.206) becomes

$$\frac{1}{2\pi i} \oint_w dx \sum_{n>0} \sum_m \frac{E^{(n,m)}(w)}{(z-x)^n (x-w)^{m+1}}$$

$$= \sum_{n>0} \sum_{m\geq 0} E^{(n,m)}(w) \frac{(n+m-1)!}{(n-1)!m!} (z-w)^{-n-m} \qquad (6.210)$$

(we have used Eq. (6.159)) and the term $m = 0$ indeed contributes. On the other hand, it is simple to see that only the first regular term contributes to the second term on the r.h.s.of Eq. (6.206). Indeed, since the OPE $B(x)\{AC\}_n(w)$ is expressed in terms of fields evaluated at w, only the pole at $x = w$ contributes.

The main steps of an illustrative application of the Wick rule (6.206) on the energy-momentum tensor follow:

$$\overline{T(z)(T\,T)}(w) = \frac{1}{2\pi i} \oint_w \frac{dx}{x-z} \left\{ \overline{T(z)T}(x)T(w) + T(x)\,\overline{T(z)T}(w) \right\}$$

$$= \frac{1}{2\pi i} \oint_w \frac{dx}{x-w} \left\{ \left[\frac{c/2}{(z-x)^4} + \frac{2T(x)}{(z-x)^2} + \frac{\partial T(x)}{(z-x)} \right] T(w) \right.$$

$$\left. + T(x) \left[\frac{c/2}{(z-w)^4} + \frac{2T(w)}{(z-w)^2} + \frac{\partial T(w)}{(z-w)} \right] \right\} \qquad (6.211)$$

To proceed we need

$$\partial T(x)T(w) = \frac{-2c}{(x-w)^5} - \frac{4T(w)}{(x-w)^3} - \frac{\partial T(w)}{(x-w)^2} + (\partial TT)(w) + \cdots \qquad (6.212)$$

which is obtained by differentiating the OPE $T(x)T(w)$ with respect to x. The OPE $T(x)\partial T(w)$ is obtained in the same way. Substituting in Eq. (6.211) the required OPEs, and using Eq. (6.159), we find that

$$T(z)(TT)(w) \sim \frac{3c}{(z-w)^6} + \frac{(8+c)T(w)}{(z-w)^4} + \frac{3\partial T(w)}{(z-w)^3}$$

$$+ \frac{4(TT)(w)}{(z-w)^2} + \frac{\partial(TT)(w)}{(z-w)} \qquad (6.213)$$

Finally, if we want to calculate $(BC)(z)A(w)$, we should first evaluate $A(z)(BC)(w)$, then interchange $w \leftrightarrow z$, and finally Taylor expand the fields evaluated at z around the point w. For instance, from Eq. (6.213) it is simple to derive

$$(TT)(z)T(w) \sim \frac{3c}{(z-w)^6} + \frac{(8+c)T(w)}{(z-w)^4} + \frac{(5+c)\partial T(w)}{(z-w)^3} + \frac{4(TT)(w)))}{(z-w)^2}$$
$$+ \frac{(1+c/2)\partial^2 T(w)}{(z-w)^2} + \frac{(c-1)\partial^3 T(w)}{6(z-w)} + \frac{3\partial(TT)(w)}{(z-w)} \tag{6.214}$$

Appendix 6.C. A Rearrangement Lemma

We often encounter composite operators involving more than two operators, for instance $(A(BC))(z)$. This notation means that the product of B and C must be first normal ordered and, in a second step, the product of A with the composite (BC) must be normal ordered. This prescription, wherein operators are normal ordered successively from right to left, will be our standard choice. It will be referred to as *right-nested* normal ordering. The necessity of a well-defined prescription is forced by the absence of associativity,

$$(A(BC))(z) \neq ((AB)C)(z) \tag{6.215}$$

which is readily seen from the mode expansions of the two sides of this equation (see also the end of this appendix). Using the contour representation

$$(A(BC))(z) = \frac{1}{(2\pi i)^2} \oint \frac{dy}{y-z} \oint \frac{dx}{x-z} A(y)B(x)C(z) \tag{6.216}$$

we find that

$$(A(BC))(z) = A_{-h_A}B_{-h_B}C_{-h_C}I(z) \tag{6.217}$$

or, equivalently,

$$(A(BC))(0)|0\rangle = A_{-h_A}B_{-h_B}C_{-h_C}|0\rangle \tag{6.218}$$

This correspondence with mode monomials illustrates neatly the naturalness of the chosen prescription.

We now derive some technical results used to compare multi-component composite operators with different ordering of the terms or different normal ordering prescriptions.

The first case to be considered is the relation between $(A(BC))$ and $(B(AC))$. Using the mode monomial representation, we write

$$(A(BC))(z) - (B(AC))(z) = \left[A_{-h_A}, B_{-h_B}\right]C_{-h_C}I(z)$$
$$= (([A,B])C)(z) \tag{6.219}$$

This result can also be verified directly at the level of modes as follows. With the OPE $A(z)B(w)$ given by (6.124), that of $B(z)A(w)$ follows by interchanging

z and w:

$$B(z)A(w) = \sum_n (-1)^n \frac{\{AB\}_n(z)}{(z-w)^n}$$

$$= \sum_n (-1)^n \sum_{m \geq 0} \frac{1}{m!(z-w)^{n-m}} \partial^m \{AB\}_n(w) \tag{6.220}$$

where the second equality is obtained by Taylor expanding $\{AB\}_n(z)$. The normal-ordered product (BA) is the sum of all terms with $n = m$, that is

$$(BA)(w) = \sum_{n \geq 0} \frac{(-1)^n}{n!} \partial^n \{AB\}_n(w) \tag{6.221}$$

This leads to

$$([A,B]) = \sum_{n > 0} \frac{(-1)^{n+1}}{n!} \partial^n \{AB\}_n(w) \tag{6.222}$$

Hence, field-dependent singular terms contribute to the normal-ordered commutator while $\{AB\}_0$ cancels out. In particular, this means that the commutation of two free fields vanishes. For instance, for a free boson ϕ, one has

$$(\partial^n \phi(\phi \partial^m \phi)) = (\phi(\partial^n \phi \partial^m \phi)) = (\partial^m \phi(\phi \partial^n \phi)) \tag{6.223}$$

By use of

$$(A(BC))_n = \sum_{m \leq -h_A} A_m(BC)_{n-m} + \sum_{m > -h_A} (BC)_{n-m} A_m \tag{6.224}$$

in which we substitute back the expression (6.130) for the modes of (BC) in terms of those of B and C, one checks directly that

$$(A(BC))_n - (B(AC))_n = (([A,B]C))_n \tag{6.225}$$

The second case is that of a composite of four operators, normal ordered two by two: We wish to relate $((AB)(CD))$ to $(A(B(CD)))$. One simply treats (CD) as a single operator, say E, and proceeds as follows:

$$((AB)E) = (E(AB)) + ([(AB),E])$$
$$= (A(EB)) + (([E,A])B) + ([(AB),E]) \tag{6.226}$$
$$= (A(BE)) + (A([E,B])) + (([E,A])B] + ([(AB),E]).$$

Replacing E by (CD) gives the desired result. The difference $((AB)E) - (A(BE))$ gives the explicit expression for the violation of associativity:

$$((AB)E) - (A(BE)) = (A([E,B])) + (([E,A])B) + ([(AB),E]) \tag{6.227}$$

Appendix 6.D. Summary of Important Formulas

OPE of the energy-momentum tensor with a primary field ϕ:

$$T(z)\phi(w) \sim \frac{h}{(z-w)^2}\phi(w) + \frac{1}{z-w}\partial\phi(w) \qquad (6.228)$$

OPE of the energy-momentum tensor with itself:

$$T(z)T(w) \sim \frac{c/2}{(z-w)^4} + \frac{2T(w)}{(z-w)^2} + \frac{\partial T(w)}{(z-w)} \qquad (6.229)$$

Normal ordering:

$$(AB)(w) = \frac{1}{2\pi i}\oint \frac{dz}{z-w}A(z)B(w) \qquad (6.230)$$

With this new normal-ordering convention, we rewrite some formulae related to free-field representations for which we make a standard choice of coupling constants.

Free boson ($g = 1/4\pi$, $c = 1$):

$$\varphi(z)\varphi(w) \sim -\ln(z-w) \qquad (6.231)$$

$$T(z) = -\frac{1}{2}(\partial\varphi\partial\varphi)(z) \qquad (6.232)$$

Vertex operators are always assumed to be normal ordered and for these the parentheses are usually omitted. With $\mathcal{V}_\alpha = e^{i\alpha\varphi}$, we have

$$\mathcal{V}_\alpha(z,\bar{z})\mathcal{V}_\beta(w,\bar{w}) \sim |z-w|^{2\alpha\beta}\mathcal{V}_{\alpha+\beta}(w,\bar{w}) + \cdots \qquad (6.233)$$

The conformal dimension of $\mathcal{V}_\alpha$ is $\alpha^2/2$.

Free real fermion ($g = 1/2\pi$, $c = \frac{1}{2}$):

$$\psi(z)\psi(w) \sim \frac{1}{z-w} \qquad (6.234)$$

$$T(z) = -\frac{1}{2}(\psi\partial\psi)(z) \qquad (6.235)$$

Free complex fermion ($c = 1$):

$$\psi^\dagger(z)\psi(w) \sim \frac{1}{z-w} \qquad \psi(z)\psi(w) \sim \psi^\dagger(z)\psi^\dagger(w) \sim 0 \qquad (6.236)$$

$$T(z) = \frac{1}{2}(\partial\psi^\dagger\psi - \psi^\dagger\partial\psi)(z) \qquad (6.237)$$

Ghost system: The two ghosts $\tilde{b}$ and $\tilde{c}$ are either both anticommuting ($\varepsilon = 1$) or both commuting ($\varepsilon = -1$) and have the OPE

$$\tilde{c}(z)\tilde{b}(w) \sim \frac{1}{z-w} \qquad \tilde{b}(z)\tilde{c}(w) \sim \frac{\varepsilon}{z-w} \qquad (6.238)$$

The energy-momentum tensor is

$$T(z) = (1 - \lambda)(\partial \tilde{b} \tilde{c})(z) - \lambda(\tilde{b} \partial \tilde{c})(z) \tag{6.239}$$

with central charge

$$c = -2\varepsilon(6\lambda^2 - 6\lambda + 1). \tag{6.240}$$

The dimensions of $\tilde{b}(z)$ and $\tilde{c}(z)$ are respectively λ and $1 - \lambda$. In Sect. 5.3 we have treated the case $\varepsilon = 1, \lambda = 0$, giving $c = -2$. On the other hand, when $\varepsilon = 1$ and $\lambda = \frac{1}{2}$, we recover the above free complex fermion theory.

Mode expansions:

$$\phi(z) = \sum_{n\in\mathbb{Z}} z^{-n-h}\phi_n \qquad \phi_n = \frac{1}{2\pi i} \oint dz\, z^{n+h-1}\phi(z)$$

Virasoro algebra and mode commutation relations:

$$[L_n, L_m] = (n - m)L_{n+m} + \frac{c}{12}n(n^2 - 1)\delta_{n+m}$$

$$[L_n, \phi_m] = [n(h - 1) - m]\phi_{n+m}$$

Exercises

6.1 Given a primary field $\phi(w)$, demonstrate the following:

$$[L_n(z), \phi(w)] = h(n + 1)(w - z)^n\phi(w) + (w - z)^{n+1}\partial\phi(w)$$

6.2 Find the mode commutation relations for a free real fermion, and for the simple ghost system.

6.3 Demonstrate the identity (6.159).

6.4 *Partition numbers*
Show that the number $p(n)$ of partitions of a nonnegative integer $n \geq 0$ into a sum of nonnegative integers is generated by

$$\sum_{n\geq 0} p(n)q^n = \frac{1}{\prod_{k\geq 1}(1 - q^k)}.$$

Find the generating function for the number $s(n)$ of strictly ordered partitions of a nonnegative integer n into strictly positive integers (we set $s(0) = 1$). Prove that $s(n)$ is equal to the number of partitions of n into positive odd integers.
Hint: Prove and use the identity $\prod_{n\geq 1}(1 - q^{2n-1})(1 + q^n) = 1$.

6.5 *Conformal blocks*
Demonstrate the relation (6.190) for the coefficient $\mathcal{F}_1$ appearing in the power series expansion of the conformal block. If successful, demonstrate the relation (6.191) for the next coefficient ($\mathcal{F}_2$).

6.6 Complete the details of the derivation of Eq. (6.219) in terms of modes.

6.7 *Contraction of two exponentials*

Let A and B be two free fields whose contraction (with themselves and with each other) are c-numbers.

a) Show by recursion that

$$\overbracket{A(z) : B^n}(w) := n\,\overbracket{A(z)B}(w) : B^{n-1}(w) :$$

and therefore

$$\overbracket{A(z) : e^{B(w)}} = \overbracket{A(z)B}(w) : e^{B(w)} :$$

As usual, $: \cdots :$ denotes normal ordering for free fields.

b) By counting correctly multiple contractions, show that

$$e^{\overbracket{A(z)}}e^{B(z)} = \sum_{m,n,k} \frac{k!}{m!n!}\binom{m}{k}\binom{n}{k}[\overbracket{A(z)B}(w)]^k : A^{m-k}(w)B^{n-k}(w) :$$

$$= \exp\left\{\overbracket{A(z)B}(w)\right\} : e^{A(w)}e^{B(w)} :$$

And deduce from this the OPE (6.65) of two vertex operators.

6.8 Calculate $([T,(TT)])$, first using Eq. (6.222) and the OPE $T(z)(TT)(w)$ given in Eq. (6.213), and then directly in terms of modes, from the equality

$$[T,(TT)] = [T_{-2},(TT)_{-4}]$$

with $T_{-2} \equiv L_{-2}$ and

$$(TT)_{-4} = 2\sum_{l\geq 0} L_{-l-3}L_{l-1} + L_{-2}L_{-2}$$

which follows from Eq. (6.213).

6.9 *Rearrangement lemma for free fermions*

a) Rearrange the product of real fermions

$$((\psi_i\psi_j)(\psi_k\psi_l))$$

whose OPE reads

$$\psi_i(z)\psi_j(w) \sim \frac{\delta_{ij}}{(z-w)}$$

in a normal ordering nested toward the right. Before using Eq. (6.226), reconsider the relative signs of the different terms when fermions are present.

b) Same as part (a) for the product of complex free fermions:

$$((\psi_i^\dagger\psi_j)(\psi_k^\dagger\psi_l))$$

with OPE

$$\psi_i(z)\psi_j^\dagger(w) \sim \frac{\delta_{ij}}{z-w} \qquad \psi_i^\dagger(z)\psi_j(w) \sim \frac{\delta_{ij}}{z-w}$$

$$\psi_i(z)\psi_j(w) \sim 0 \qquad \psi_i^\dagger(z)\psi_j^\dagger(w) \sim 0$$

6.10 *The quantum Korteweg–de Vries equation*

Let us introduce an equation of evolution in time for the energy-momentum tensor through the canonical equation of motion

$$\partial_t T = -[H, T] \quad , \quad H = \frac{1}{2\pi i} \oint dw \, (TT)(w)$$

a) Using the OPE (6.213), check that the resulting evolution equation is

$$\partial_t T = \frac{1}{6}(1 - c)\partial^3 T - 3\partial(TT) \tag{6.241}$$

This is called the quantum Korteweg–de Vries (KdV) equation since in the classical limit $c \to -\infty$,[2] the substitution $T = cu(z, t)/6$ and a rescaling of the time variable transforms it into the standard KdV equation:

$$\partial_t u = \partial^3 u + 6u\partial u \tag{6.242}$$

b) The quantum KdV equation (like its classical counterpart) is a completely integrable system in the sense that it has an infinite number of conserved integrals H_n

$$\partial_t H_n = 0$$

(whose densities are polynomial derivatives in T), all commuting with each other. Each of these conserved integrals has a definite spin. The spin of these charges is always odd, and there is one charge for each odd value of the spin. To illustrate this statement, check that there can be no nontrivial conserved integral of spin 2 and 4. A conserved integral is nontrivial if its density is not a total derivative.

c) Show that the first nontrivial conservation law is

$$H_5 = \oint dw \left[(T(TT)) - \frac{(c + 2)}{12} (\partial T \partial T) \right] \tag{6.243}$$

(The subindex indicates the spin of the integral.) To obtain this result, proceed as follows. At first, argue that the above two terms in H_5 are the only possible ones, up to total derivatives. H_5 is thus necessarily of the form

$$H_5 = \oint dw \left[(T(TT)) + a(\partial T \partial T) \right] \tag{6.244}$$

where a is a free parameter to be determined. It is uniquely fixed by requiring $\partial_t H_5 = 0$. Explicitly, in the expression for $\partial_t H_5$ replace $\partial_t T$ by the r.h.s. of the quantum KdV equation, drop total derivatives and cancel the remaining terms by an appropriate choice of a.

d) The conservation of H_5 can also be established independently of the equation of motion, by proving directly the commutativity of H_5 with the defining Hamiltonian H. For this calculation, the following two intermediate results must first be derived:

$$
T(z)(T(TT)(w)) \sim \frac{24c}{(z - w)^8} + \frac{(48 + 9c)\,T(w)}{(z - w)^6} + \frac{15\,\partial T(w)}{(z - w)^5}
$$
$$
+ \frac{(24 + \frac{3}{2}c)\,(TT)(w)}{(z - w)^4} + \frac{\frac{9}{2}\,\partial(TT)(w)}{(z - w)^3} + \frac{\frac{1}{4}\,\partial^3 T(w)}{(z - w)^3} \tag{6.245}
$$
$$
+ \frac{6\,(T(TT))(w)}{(z - w)^2} + \frac{\partial(T(TT))(w)}{(z - w)}
$$

---

[2] The classical limit corresponds to $c \to \pm\infty$, but $-\infty$ can be obtained, by a limit process, from the minimal models to be introduced in the following chapter.

and

$$
\begin{aligned}
T(z)(\partial T \partial T)(w) \sim &\frac{18c}{(z-w)^8} + \frac{28\,T(w)}{(z-w)^6} + \frac{(4c+18)\,\partial T(w)}{(z-w)^5} \\
&+ \frac{5\,\partial^2 T(w)}{(z-w)^5} + \frac{4\,\partial(TT)(w)}{(z-w)^3} + \frac{6\,(\partial T \partial T)(w)}{(z-w)^2} \\
&+ \frac{\partial(\partial T \partial T)(w)}{(z-w)}
\end{aligned}
\tag{6.246}
$$

In these expressions, interchange z and w and then use Eq. (6.206) to calculate the commutator.

6.11 *The quantum Korteweg–de Vries equation at* $c = -2$

a) Verify that for the central charge $c = -2$, T can be represented by the bilinear

$$
T = (\phi\psi)
$$

where ϕ and ψ are both fermions of spin 1 with OPE

$$
\phi(z)\psi(w) = \frac{-1}{(z-w)^2} \quad , \quad \psi(z)\phi(w) = \frac{1}{(z-w)^2}
$$

This is, of course, nothing but a ghost representation (cf. App. 6.D), with $\tilde{c} = \psi$ and $\partial \tilde{b} = \phi$ and $\epsilon = 1$ (i.e., these are anticommuting fields).

b) Using the rearrangement lemma (6.226), show that

$$
(TT)(z) = \frac{1}{2}(\phi''\psi + \phi\psi'')(z)
$$

where a prime stands for a derivative with respect to the complex coordinate.

c) In terms of these variables and the quantum KdV Hamiltonian

$$
H = \frac{1}{2\pi i} \oint dw\,(TT)(w) = \frac{1}{4\pi i} \oint dw\,(\phi''\psi + \phi\psi'')(w)
$$

derive the evolution equations

$$
\partial_t \phi = -[H, \phi] = -\phi''' \quad , \quad \partial_t \psi = -[H, \psi] = -\psi'''
$$

Use these equations to recover the evolution equation of T.

d) Prove that an infinite set of conserved quantities for this system of equations is

$$
H_{k+1} = \oint dz (\phi^{(k)}\psi)(z) \quad \text{with} \quad \partial_t H_{k+1} = 0
$$

where $\phi^{(k)} = \partial_z^k \phi$.

e) Verify the mutual commutativity of these charges.

f) Argue that for k odd these conserved integrals cannot be expressed in terms of T. For k even this can be done as follows:

$$
H_{2n-1} = \frac{2^{n-1}}{n} \oint dz (\overleftarrow{T^n})(z)
$$

where the notation $(\overleftarrow{T^n})$ means a nesting of the normal ordering toward the left:

$$
(\overleftarrow{T^n}) = (\dots(((TT)T)T)\dots T) \quad (n \text{ factors})
$$

The exact expression for $(\overleftarrow{T^n})$ is

$$(\overleftarrow{T^n}) = \frac{n}{2^n}\left(\phi^{(2n-2)}\psi + \phi\psi^{(2n-2)}\right)$$

g) In preparation for establishing the above result for $(\overleftarrow{T^n})$, check the following necessary normal-ordered commutators

$$([(\phi^{(m)}\psi), \psi]) = \frac{(-1)^m}{m+2}\psi^{(m+2)}$$

$$([(\phi^{(m)}\psi), \phi]) = \frac{1}{2}\phi^{(m+2)}$$

$$([(\phi\psi^{(m)}), \phi]) = \frac{(-1)^m}{m+2}\phi^{(m+2)}$$

$$([(\phi\psi^{(m)}), \psi]) = \frac{1}{2}\psi^{(m+2)}$$

h) Prove the above expression for $(\overleftarrow{T^n})$ by an inductive argument. Hint: Assuming the validity of the above expression for $(\overleftarrow{T^n})$, calculate $(\overleftarrow{T^{n+1}}) = ((\overleftarrow{T^n})T)$ in terms of fermions by reordering the terms toward the right using $(A(BC)) = (B(AC)) + (([A,B])C)$ and the commutators calculated in the previous part.

i) Express the charge H_5 obtained in Ex. 6.10 in terms of ϕ and ψ and compare with $\oint dz\,(\overleftarrow{T^3})$.

j) To see that the higher-spin conserved charges cannot be expressed in a simple way with the usual nesting toward the right, compare $(T(T(TT)))$ and $(\overleftarrow{T^4})$.

k) To understand why $c = -2$ is special, consider a general anticommuting ghost system $\{\tilde{b}, \tilde{c}\}$, for which the energy-momentum tensor is given by Eq. (6.238). Show that

$$(TT) = (\frac{4}{3}\lambda(1-\lambda)+1)(\tilde{b}'''\tilde{c}) - 2\lambda(1-\lambda)(\tilde{b}(\tilde{b}'(\tilde{c}\tilde{c}')))$$

Obtain the evolution equation for $\tilde{b}$ and $\tilde{c}$. Observe that unless $\lambda = 0$ or 1, these are coupled equations, for which the integrals of motion cannot be written in a simple bilinear form.

6.12 The classical limit of the Virasoro algebra
The Poisson bracket form of the Virasoro algebra is obtained by replacing the commutator by a Poisson bracket times a factor i, that is,

$$i\{L_n, L_m\} = (n-m)L_{n+m} + \frac{c}{12}n(n^2-1)\delta_{n+m,0}$$

Let $u(x,t)$ be the classical field defined on a cylinder $(u(x+2\pi,t)=u(x,t))$ whose Fourier modes are the L_n's:

$$u(x) = \frac{6}{c}\sum_{n\in\mathbb{Z}}L_n e^{-inx} - \frac{1}{4}$$

(the explicit time-dependence is omitted from now on). This is the classical form of the energy-momentum tensor. Show that its equal-time Poisson bracket is

$$\{u(x), u(y)\} = \frac{6\pi}{c}[-\partial_x^3 + 4u(x)\partial_x + 2(\partial_x u(x))]\delta(x-y) \qquad (6.247)$$

Use:

$$\delta(x - y) = \frac{1}{2\pi} \sum_{n \in \mathbb{Z}} e^{-in(x-y)}$$

Recover the classical Korteweg–de Vries equation

$$\partial_t u = -\partial_x^3 u + 6u\partial_x u$$

from the following canonical formulation:

$$\partial_t u = \{u, H\}, \qquad \text{with} \qquad H = \frac{c}{12\pi} \int dx\, u^2$$

The above Poisson bracket defines the so-called second Hamiltonian structure of the KdV equation. (The relative sign between the two terms on the r.h.s. of the KdV equation is not as in the classical form derived in Ex. 6.10 (cf. Eq. (6.242)); this is explained by a 'space Wick rotation', i.e., the space variables used in the two cases are related by a factor i.)

6.13 *The Feigin-Fuchs transformation and the quantum Korteweg–de Vries equation revisited*

a) Verify that the following deformation of the energy-momentum tensor of a free boson

$$T = -\frac{1}{2}(\partial\varphi\partial\varphi) + i\alpha\partial^2\varphi \qquad (6.248)$$

still satisfies the OPE (5.121), with c related to α by

$$c = 1 - 12\alpha^2$$

This is called the Feigin-Fuchs (or sometimes Feigin-Fuchs-Miura) transformation.

b) Since the relative coefficients of the different terms of the conserved densities of the quantum KdV equation (introduced in Ex. 6.10) depend only upon c, it implies that all these conserved densities, when rewritten in terms of φ via the Feigin-Fuchs transformation, are even functions of φ up to total derivatives. More explicitly, let

$$H_n = \oint dz\, \mathcal{H}_{n+1}[T] \qquad \text{such that} \qquad \partial_t H_n = 0$$

and

$$\mathcal{H}_n[T] = \tilde{\mathcal{H}}_n[\varphi]$$

when T is replaced by Eq. (6.248). Any quantum KdV conserved density satisfies

$$\tilde{\mathcal{H}}_n[\varphi] = \tilde{\mathcal{H}}_n[-\varphi] + \partial(\cdots)$$

Verify that this is indeed so for the defining Hamiltonian $H = \oint(TT)$. It turns out that this criterion characterizes uniquely the quantum KdV conserved densities! Use it to recover H_5 (cf. Ex. 6.10, Eq. (6.243)).

Hint: Start from the ansatz (6.244) of part (c) in Ex. 6.10, replace T by Eq. (6.248), drop total derivatives and terms with an even number of φ factors; fix the relative coefficient a by enforcing the cancellation of the remaining odd terms. Along the way, some normal ordering rearrangements are necessary.

c) Find the canonical evolution equation for the field $B = \partial\varphi$ defined by the Hamiltonian

$$H = \oint dz\,(TT) \qquad \text{with} \qquad T = -\frac{1}{2}(BB) + ia\partial B$$

This is the quantum modified KdV equation.

Remark: The classical version of Feigin-Fuchs transformation is called the Miura transformation:

$$u = \partial_x v + v^2$$

and it is a canonical map from the Poisson bracket

$$\{v(x), v(y)\} = \partial_x \delta(x - y)$$

to the KdV Poisson bracket (6.247) of Ex. 6.12 (up to an irrelevant multiplying factor).

Notes

General references for this chapter are identical to those of the previous chapter.

Radial quantization was introduced by Fubini, Hanson, and Jackiw [145]. It was applied on the complex plane by Friedan [139], who also represented the commutator of two operators by a contour integral.

The Virasoro algebra first appeared in Ref. [344] in the context of dual resonance models. Its general application to conformal field theory was pointed out by Belavin, Polyakov, and Zamolodchikov (BPZ) [36].

The quantization of the free boson is immemorial. Vertex operators were introduced in the context of dual resonance models by Fubini and Veneziano [144]. Fermions were introduced in string theory by Ramond [302] and Neveu and Schwarz [281].

The concepts of a conformal family and conformal block were introduced in BPZ [36]. Analytic properties of the conformal blocks are discussed by Zamolodchikov and Zamolodchikov [367]. The generalized Wick theorem is discussed by Bais, Bouwknegt, Surridge, and Schoutens [20].

Ex. 6.10 is based on Ref. [249], Ex. 6.11 on Ref. [94], Ex. 6.12 on Ref. [175], and Ex. 6.13 on Refs. [316, 121].

CHAPTER 7

Minimal Models I

Chapters 5 and 6 dealt with general properties of two-dimensional conformal field theories. The present chapter is devoted to particularly simple conformal theories called *minimal models*. These theories are characterized by a Hilbert space made of a *finite* number of representations of the Virasoro algebra (Verma modules); in other words, the number of conformal families is finite. Such theories describe discrete statistical models (e.g., Ising, Potts, and so on) at their critical points. Their simplicity in principle allows for a complete solution (i.e., an explicit calculation of all the correlation functions). The discovery of minimal models and their identification with known statistical models at criticality constitutes the greatest application of conformal invariance so far. Since a detailed study of minimal models may rapidly become highly technical, we have split the discussion among two chapters (this one and the next). The present chapter first explains some general features of Verma modules (Sect. 7.1), and in particular the occurrence of states of zero norm, which must be quotiented out. In Sect. 7.2 the question of unitarity is discussed and the Kac determinant is introduced. In Sect. 7.3 a survey of the theory of minimal models is presented. In Sect. 7.4 various examples of the correspondence between minimal theories and statistical systems are described. The next chapter will be devoted to more technical issues and will provide proofs for some assertions of the present chapter.

§7.1. Verma Modules

In a conformal field theory, we expect the energy eigenstates (i.e., the eigenstates of L_0 and $\bar{L}_0$) to fall within representations of the local conformal algebra (the Virasoro algebra) much in the same way as the energy eigenstates of a rotation-invariant system fall into irreducible representations of $su(2)$. In a given theory, the Hilbert space will generically contain several irreducible representations of the Virasoro algebra; this is analogous to the Hilbert space of the hydrogen atom containing an infinite number of $su(2)$ representations.

7.1.1. Highest-Weight Representations

Highest-weight representations are familiar to physicists through the theory of angular momentum. We briefly recall what is done in that context: We assume that the representation space is spanned by the eigenvectors $|m\rangle$ of one of the $su(2)$ generators, which we call J_0 (denoted J_z in most texts on quantum mechanics). An inner product is assumed to exist on the representation space, such that the three generators are Hermitian: $J_a^\dagger = J_a$. The other two generators of $su(2)$ are arranged into *raising* and *lowering* operators $J_\pm = J_x \pm iJ_y$ with the commutation relations

$$[J_0, J_\pm] = \pm J_\pm \qquad [J_+, J_-] = 2J_0 \tag{7.1}$$

We then assume that the eigenvalue m of J_0 has a maximum within the representation space: There is a state $|j\rangle$ such that

$$J_0|j\rangle = j|j\rangle \quad \text{and} \quad J_+|j\rangle = 0$$

The other eigenstates of J_0 are obtained by applying J_- repeatedly on $|j\rangle$; we define the (not normalized) states

$$|m\rangle = (J_-)^{j-m}|j\rangle \tag{7.2}$$

The inner products of these states are easily calculated by using the relations $J_+|j\rangle = 0$ and $\langle j|J_- = 0$:

$$\langle m-1|m-1\rangle = [j(j+1) - m(m-1)]\langle m|m\rangle \tag{7.3}$$

(in order to find this explicit result, we also use the fact that the operator $J^2 = (J_0)^2 + \frac{1}{2}\{J_+, J_-\}$ commutes with all generators, and has therefore a fixed value within the representation, equal to $j(j+1)$). As seen from Eq. (7.3), states of negative norm generically appear when m decreases below $-j$; the representation is then nonunitary. The only exception to this rule occurs when j is an integer or a half-integer: The state $|-j-1\rangle$ then has norm zero, along with all other states obtained by applying J_- on it. We say that these *singular vectors* decouple from the first $(2j+1)$ states for the following reason: Consider any operator A built from the generators J_i; its matrix element $\langle m|A|m'\rangle$ between a positive-norm state $|m\rangle$ and a null state $|m'\rangle$ necessarily vanishes. Indeed, the evaluation of the matrix element proceeds by expressing A in terms of J_0 and $J_\pm$, using the relations (7.3), and it finally reduces to an expression proportional to $\langle m|m'\rangle$, which vanishes. The representation space is thus truncated to the first $(2j+1)$ states of Eq. (7.2), which then form a unitary, finite-dimensional representation of $su(2)$.

We shall proceed in a similar way in order to construct representations of the Virasoro algebra

$$[L_n, L_m] = (n-m)L_{n+m} + \frac{c}{12}n(n^2-1)\delta_{n+m,0} \tag{7.4}$$

Representations of the antiholomorphic counterpart of (7.4) are constructed by the same method. Since the holomorphic and antiholomorphic components of the overall algebra (6.24) decouple, representations of the latter are obtained simply by taking tensor products. Since no pair of generators in (7.4) commute, we choose a single generator (here L_0), which will be diagonal in the representation space, also called a *Verma module*. We denote by $|h\rangle$ the highest-weight state, with eigenvalue h of L_0:

$$L_0|h\rangle = h|h\rangle \qquad (7.5)$$

This state is, of course, the asymptotic state created by applying a primary field operator $\phi(0)$ of dimension h on the vacuum $|0\rangle$ (cf. Sect. 6.2.2). Since $[L_0, L_m] = -mL_m$, L_m $(m > 0)$ is a lowering operator for h, and L_{-m} $(m > 0)$ is a raising operator. We shall adopt the condition

$$L_n|h\rangle = 0 \qquad (n > 0) \qquad (7.6)$$

which is compatible with the regularity condition (6.26). Notice that the above condition follows from the simpler condition $L_1|h\rangle = L_2|h\rangle = 0$, by repeated use of Eq. (7.4).

As discussed in Sect. 6.2.2, a basis for the other states of the representation, the so-called *descendant states*, is obtained by applying the raising operators in all possible ways:

$$L_{-k_1}L_{-k_2}\ldots L_{-k_n}|h\rangle \qquad (1 \le k_1 \le \ldots \le k_n) \qquad (7.7)$$

where, by convention, the L_{-k_i} appear in increasing order of the k_i. Recall that this state is an eigenstate of L_0 with eigenvalue

$$h' = h + k_1 + k_2 + \ldots + k_n = h + N \qquad (7.8)$$

where N is the level of the state. Likewise, the *level* of a string of operators is the level of the state it produces when applied on $|h\rangle$. The first levels are spanned by the states of Table 7.1.

Table 7.1. Lowest states of a Verma module.

l	$p(l)$	
0	1	$|h\rangle$
1	1	$L_{-1}|h\rangle$
2	2	$L_{-1}^2|h\rangle$, $L_{-2}|h\rangle$
3	3	$L_{-1}^3|h\rangle$, $L_{-1}L_{-2}|h\rangle$, $L_{-3}|h\rangle$
4	5	$L_{-1}^4|h\rangle$, $L_{-1}^2L_{-2}|h\rangle$, $L_{-1}L_{-3}|h\rangle$, $L_{-2}^2|h\rangle$, $L_{-4}|h\rangle$

On the Verma module we define an inner product according to our previous definition of the Hermitian conjugate: $L_m^\dagger = L_{-m}$. Thus, the inner product of two states

$$L_{-k_1}\cdots L_{-k_m}|h\rangle \quad \text{and} \quad L_{-l_1}\cdots L_{-l_n}|h\rangle$$

is simply

$$\langle h | L_{k_m} \cdots L_{k_1} L_{-l_1} \cdots L_{-l_n} | h \rangle \tag{7.9}$$

where the dual state $\langle h |$ satisfies

$$\langle h | L_j = 0 \qquad (j < 0) \tag{7.10}$$

The product (7.9) may be evaluated by passing the L_{k_j} over the L_{-l_i} using the Virasoro algebra until they hit $|h\rangle$. Notice that the inner product of two states vanishes unless they belong to the same level. Indeed, two eigenspaces of a Hermitian operator (here L_0) having different eigenvalues are orthogonal. Hermiticity also forces h to be real.

A similar analysis can be done for the Verma modules associated with the anti-holomorphic generators $\bar{L}_n$. Denoting by $V(c,h)$ and $\bar{V}(c,\bar{h})$ the Verma modules generated respectively by the sets $\{L_n\}$ and $\{\bar{L}_n\}$ for a value c of the central charge and with highest weights h and $\bar{h}$, the energy eigenstates belong to the tensor product $V \otimes \bar{V}$. In general, the Hilbert space is a direct sum of such tensor products, over all conformal dimensions of the theory:

$$\sum_{h,\bar{h}} V(c,h) \otimes \bar{V}(c,\bar{h}) \tag{7.11}$$

The number of terms in this sum may be finite or infinite; moreover, there may be several terms with the same conformal dimension.

To conclude this section, we consider the example of the free boson, studied in detail in Sect. 6.3. We recall that the Fock space is constructed by applying the raising operators a_{-n} ($n > 0$) on the vacua $|\alpha\rangle$. The latter are obtained from the "absolute" vacuum $|0\rangle$ by application of the vertex operator $V_\alpha(0)$: $|\alpha\rangle = V_\alpha(0)|0\rangle$. From the expression (6.69) for the Virasoro generators, we immediately see that $L_n|\alpha\rangle = 0$ for $n > 0$, and the vacua $|\alpha\rangle$ form a continuum of highest weight states, with weight $h = \frac{1}{2}\alpha^2$. Thus, to each value of α one associates a Verma module, itself associated with the primary field V_α. The descendant states are obtained by repeated application of the creation operators a_{-n} ($n > 0$), which is equivalent to a repeated application of L_{-n}, since a_{-n} also raises the conformal dimension by n (cf. Sect. 6.3.3).

7.1.2. Virasoro Characters

To a Verma module $V(c,h)$ generated by the Virasoro generators L_{-n} ($n > 0$) acting on the highest-weight state $|h\rangle$, we associate a generating function $\chi_{(c,h)}(\tau)$, called the *character* of the module, defined as

$$\begin{aligned}\chi_{(c,h)}(\tau) &= \mathrm{Tr}\, q^{L_0 - c/24} \qquad (q \equiv e^{2\pi i \tau}) \\ &= \sum_{n=0}^{\infty} \dim(h+n) q^{n+h-c/24}\end{aligned} \tag{7.12}$$

where $\dim(h+n)$ is the number of linearly independent states at level n in the Verma module and τ is a complex variable. The factor $q^{-c/24}$ is introduced in this definition for reasons that will become evident later, when considering modular invariance (Chap. 10). Since $\dim(h+n) \leq p(n)$, the number of (possibly dependent) states at level n, the series (7.12) is uniformly convergent if $|q| < 1$ (i.e., for τ in the upper half-plane), because $|q| < 1$ is the domain of convergence of the series (6.38). The characters are generating functions for the level degeneracy $\dim(h+n)$. In other words, knowing the character amounts to knowing how many states there are at each level. Characters $\bar{\chi}_{c,\bar{h}}(\bar{\tau})$ for the antiholomorphic Verma module are defined in the same manner.

The character of a generic Verma module is easily written. We know that the number of states at level n is $p(n)$, the number of partitions of the integer n. Since the generating function of the partition numbers is (cf. Eq. (6.38) and Ex. 6.4)

$$\frac{1}{\varphi(q)} \equiv \prod_{n=1}^{\infty} \frac{1}{1-q^n} = \sum_{n=0}^{\infty} p(n)q^n \tag{7.13}$$

the generic Virasoro character may be written as

$$\chi_{(c,h)}(\tau) = \frac{q^{h-c/24}}{\varphi(q)} \tag{7.14}$$

In terms of the Dedekind function

$$\eta(\tau) \equiv q^{1/24}\varphi(q) = q^{1/24} \prod_{n=1}^{\infty}(1-q^n) \tag{7.15}$$

the generic Virasoro character becomes

$$\chi_{(c,h)}(\tau) = \frac{q^{h+(1-c)/24}}{\eta(\tau)} \tag{7.16}$$

7.1.3. Singular vectors and Reducible Verma Modules

It may happen that the representation of the Virasoro algebra comprising all the states (7.7) is *reducible*. By this, we mean that there is a subspace (or submodule) that is itself a full-fledged representation of the Virasoro algebra. The states of this submodule transform amongst themselves under any conformal transformation. Such a submodule is also generated from a highest-weight state $|\chi\rangle$, such that $L_n|\chi\rangle = 0$ $(n > 0)$, although this state is also of the form (7.7).

Generally, any state $|\chi\rangle$—other than the highest-weight state—that is annihilated by all L_n $(n > 0)$ is called a *singular vector* (we say also *null vector* or *null state*). Such a state generates its own Verma module V_χ included in the original module $V(c,h)$. Singular vector are orthogonal to the whole Verma module. This follows immediately from the basis states (7.7) and Hermitian conjugation:[1]

$$\langle\chi|L_{-k_1}L_{-k_2}\ldots L_{-k_n}|h\rangle = \langle h|L_{k_n}\ldots L_{k_2}L_{k_1}|\chi\rangle^* = 0 \tag{7.17}$$

---

[1] In the bra-ket notation, non-Hermitian operators act on the ket, by convention.

In particular, $\langle \chi | \chi \rangle = 0$. This observation extends to all the descendants of $|\chi\rangle$: They are also orthogonal to the whole Verma module $V(c, h)$. This last assertion is equivalent to saying that any descendant of $|\chi\rangle$ is orthogonal to all the states of $V(c, h)$ having the same level. Indeed, the relevant inner product has the form

$$\langle h | L_{k_n} \dots L_{k_1} L_{-r_1} \dots L_{-r_m} | \chi \rangle \tag{7.18}$$

where $\sum_i k_i = N + \sum_i r_i$, N being the level of $|\chi\rangle$. By commuting systematically all the L_{k_i} over the L_{-r_j}, one ends up with a sum of products of the form (7.17), since $\sum_i k_i > \sum_i r_i$. Hence the assertion is proven. In particular, all the descendants of $|\chi\rangle$ have zero norm, since the evaluation of their norm leads to an expression proportional to $\langle \chi | \chi \rangle$.

Through the operator-field correspondence, a null state $|\chi\rangle$ is associated with a null *field* $\chi(z)$, which is at the same time primary (meaning that $(L_n \chi)(z) = 0$ if $n > 0$) and secondary, since it is a descendant of a primary field ϕ_h of dimension h (cf. Sect. 6.6.1).

From a Verma module $V(c, h)$ containing one or more singular vectors, one may construct an irreducible representation of the Virasoro algebra by quotienting out of $V(c, h)$ the null submodule or, in other words, by identifying states that differ only by a state of zero norm. These irreducible representations, which will will denote $M(c, h)$ in order not to confuse them with the reducible Verma module, contain relatively "fewer" states than the generic Verma module, and their characters are not given by the simple formula (7.16). Such representations are the building blocks of minimal models.

§7.2. The Kac Determinant

7.2.1. Unitarity and the Kac Determinant

A representation of the Virasoro algebra is said to be *unitary* if it contains no negative-norm states (often called *ghosts* in string theory). Since the explicit value of the inner product (7.9) depends on the highest weight h and the central charge c, the requirement that a representation be unitary imposes some constraints on these parameters. For instance, a simple unitarity bound on (h, c) is obtained by calculating the norm of the state $L_{-n} |h\rangle$:

$$\langle h | L_n L_{-n} | h \rangle = \langle h | \left(L_{-n} L_n + 2nL_0 + \frac{1}{12} cn(n^2 - 1) \right) | h \rangle$$

$$= [2nh + \frac{1}{12} cn(n^2 - 1)] \langle h | h \rangle \tag{7.19}$$

If $c < 0$ the above becomes negative for n sufficiently large. Therefore all representations with negative central charge are nonunitary. Moreover, the case $n = 1$ shows that all representations with negative conformal dimensions are also nonunitary.

The necessary and sufficient conditions for unitarity are found by considering the so-called *Gram matrix* of inner products between all basis states. We denote the basis states (7.7) of the Verma module as $|i\rangle$ and let

$$M_{ij} = \langle i|j\rangle \qquad (M^\dagger = M) \tag{7.20}$$

be the Gram matrix. This matrix is block diagonal, with blocks $M^{(l)}$ corresponding to states of level l. A generic state is a linear combination $|a\rangle = \sum_i a_i|i\rangle$ and its norm is (in matrix notation)

$$\langle a|a\rangle = a^\dagger M a \tag{7.21}$$

Since M is Hermitian it may be diagonalized by a unitary matrix U: $M = U\Lambda U^\dagger$. If $b = Ua$, then

$$\langle a|a\rangle = \sum_i \Lambda_i|b_i|^2 \tag{7.22}$$

Consequently there will be negative-norm states if and only if M has one or more negative eigenvalues. Moreover, there will be singular vectors if one of the eigenvalues Λ_i vanishes and, accordingly, the Verma module will be reducible.

The matrices $M^{(l)}$ associated with the lowest levels of a generic Verma module are easily calculated:

$$M^{(0)} = 1$$
$$M^{(1)} = 2h \tag{7.23}$$
$$M^{(2)} = \begin{pmatrix} 4h(2h+1) & 6h \\ 6h & 4h+c/2 \end{pmatrix}$$

As an illustration of the steps leading to the above expressions, we calculate explicitly a sample matrix element:

$$\begin{aligned} M_{12}^{(2)} &= \langle h|L_1L_1L_{-2}|h\rangle \\ &= \langle h|L_1(L_{-2}L_1 + 3L_{-1})|h\rangle \\ &= 3\langle h|L_1L_{-1}|h\rangle \\ &= 6h\langle h|h\rangle \end{aligned} \tag{7.24}$$

From $M^{(0)}$ we cannot infer any condition for unitarity. From $M^{(1)}$ we recover the condition $h > 0$. The product of the two eigenvalues of $M^{(2)}$ is equal to its determinant:

$$\begin{aligned} \det M^{(2)} &= 32h^3 - 20h^2 + 4h^2c + 2hc \\ &= 32(h - h_{1,1})(h - h_{1,2})(h - h_{2,1}) \end{aligned} \tag{7.25}$$

wherein

$$h_{1,1} = 0$$

$$h_{1,2} = \frac{1}{16}\left(5 - c - \sqrt{(1-c)(25-c)}\right) \tag{7.26}$$

$$h_{2,1} = \frac{1}{16}\left(5 - c + \sqrt{(1-c)(25-c)}\right)$$

The sum of the eigenvalues is equal to the trace:

$$\operatorname{Tr} M^{(2)} = 8h(h+1) + c/2 \tag{7.27}$$

The representation is not unitary whenever $\det M^{(2)}$ or $\operatorname{Tr} M^{(2)}$ is negative. We already know that every doublet (c,h) lying outside the first quadrant leads to a nonunitary representation. We learn here that some regions of the first quadrant also lead to nonunitary representations: As a function of c, the roots $h_{1,2}$ and $h_{2,1}$ describe two curves that join at $c = 1$, as illustrated on the leftmost graph of Fig. 7.1. The determinant $\det M^{(2)}(c,h)$ is negative between these two curves (shaded area) and thus the associated representations are not unitary. We also learn from this exercise that the Verma modules associated with points (c,h) lying on these curves are reducible.

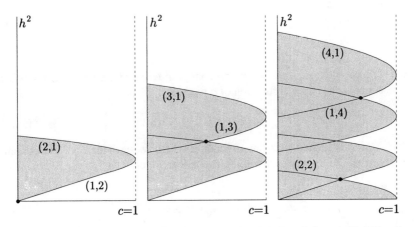

Figure 7.1. The vanishing curves $C_{r,s}$ for levels 2, 3 and 4 (from left to right). The curves at level $l-1$ appear also at level l. The values of r,s are indicated near the curves when they first appear. The black dots are first intersections (defined in the text). The shaded areas correspond to manifestly nonunitary theories.

There exists a general formula, due to Kac, for the determinant of the Gram matrix, the *Kac determinant*:

$$\det M^{(l)} = \alpha_l \prod_{\substack{r,s \geq 1 \\ rs \leq l}} [h - h_{r,s}(c)]^{p(l-rs)} \tag{7.28}$$

where $p(l - rs)$ is the number of partitions of the integer $l - rs$ and α_l is a positive constant independent of h or c:

$$\alpha_l = \prod_{\substack{r,s \geq 1 \\ rs \leq l}} [(2r)^s s!]^{m(r,s)}$$

(7.29)

$$m(r,s) = p(l - rs) - p(l - r(s+1))$$

The functions $h_{r,s}(c)$ may be expressed in various ways. A common expression is the following:

$$h_{r,s}(c) = h_0 + \frac{1}{4}(r\alpha_+ + s\alpha_-)^2$$

$$h_0 = \frac{1}{24}(c-1)$$

$$\alpha_\pm = \frac{\sqrt{1-c} \pm \sqrt{25-c}}{\sqrt{24}}$$

(7.30)

Another convenient way to express the function $h_{r,s}$ is

$$c = 13 - 6\left(t + \frac{1}{t}\right)$$

$$h_{r,s}(t) = \frac{1}{4}(r^2 - 1)t + \frac{1}{4}(s^2 - 1)\frac{1}{t} - \frac{1}{2}(rs - 1)$$

(7.31)

Here we have parametrized the central charge in terms of the (complex) number t. The expression for t as a function of c has two branches:

$$t = 1 + \frac{1}{12}\left\{1 - c \pm \sqrt{(1-c)(25-c)}\right\}$$

(7.32)

Which branch is actually used has no influence on the value of the Kac determinant. If $c < 1$ or $c > 25$, t is real, whereas it lies on the unit circle if $1 < c < 25$. In terms of t,

$$\alpha_+ = \sqrt{t} \qquad\qquad \alpha_- = -\frac{1}{\sqrt{t}}$$

(7.33)

Yet another way of expressing the roots of the Kac determinant is the following:

$$c = 1 - \frac{6}{m(m+1)}$$

$$h_{r,s}(m) = \frac{[(m+1)r - ms]^2 - 1}{4m(m+1)}$$

(7.34)

Again, the expression of m as a function of c has two branches:

$$m = -\frac{1}{2} \pm \frac{1}{2}\sqrt{\frac{25-c}{1-c}}$$

(7.35)

The relation between t and m is not unique:

$$t = \frac{m}{m+1} \quad \text{or} \quad t = \frac{m+1}{m} \tag{7.36}$$

The expressions (7.28) and (7.30) [or (7.31), or (7.34)] are of central importance in the theory of minimal models. The great success of conformal field theory in the study of two-dimensional critical systems is due in great part to the knowledge of the Kac determinant. The expressions (7.28) and (7.31) for the Kac determinant will be demonstrated in the next chapter.

In the (c, h) plane, the Kac determinant vanishes along the curves $h = h_{r,s}(c)$, called *vanishing curves* and denoted $C_{r,s}$. These curves are shown in Fig. 7.1 for all values of r and s allowed at levels 2, 3, and 4. Note that the Kac formula does not provide the eigenvalues of the Gram matrix, but only their product. At each level $l > 1$, the number of roots h_{rs} of the determinant exceeds the number $p(l)$ of eigenvalues. As is clear from the Kac determinant formula (7.30), the first null state in the reducible Verma module $V(c, h_{r,s})$ occurs at level $l = rs$, since $p(l - rs)$ vanishes (by definition) if $l < rs$.

7.2.2. Unitarity of $c \geq 1$ Representations

The explicit expressions (7.28)–(7.34) for the Kac determinant allow us to prove that the representations ($c \geq 1, h \geq 0$) are unitary. The proof is done in three steps: First, we show that the vanishing curves $C_{r,s}$ do not cross the region $R = \{c > 1, h > 0\}$. In a second step we show that $\det M^{(l)} > 0$ throughout this region. Finally we argue that $M^{(l)}$ is positive definite in R. This last statement is in itself equivalent to unitarity, but the first two steps will be useful in proving it.

The first step amounts to showing that the curves $C_{r,s}$ lie below or on the axis $h = 0$ if $c > 1$. An explicit expansion of Eq. (7.30) yields

$$h_{r,s}(c) = \frac{1-c}{96} \left\{ \left[(r+s) + (r-s)\sqrt{\frac{25-c}{1-c}} \right]^2 - 4 \right\} \tag{7.37}$$

If $1 < c < 25$ we see that $h_{r,s}(c)$ is not a real number unless $r = s$, in which case $h_{r,s}(c) \leq 0$. On the other hand, if $c \geq 25$ the choice (7.35) implies that $-1 < m < 0$. Then $m(m+1) < 0$ and

$$[(m+1)r - ms]^2 = [(1-|m|)r + |m|s]^2 \geq 1 \tag{7.38}$$

which implies $h_{r,s}(m) \leq 0$ according to Eq. (7.34). Thus, we have shown that all the curves $C_{r,s}$ are located on or below the $h = 0$ axis if $c > 1$.

When $|h|$ is much larger than $\max\{|h_{r,s}|\}$ for a given level, then $\det M^{(l)} \approx \alpha_l h^r$, for some positive r. Since α_l is a positive constant, the Kac determinant is also positive in this limit. Finally, since none of the roots h_{rs} lies in the region R, the Kac determinant is strictly positive throughout that region. This proves the second point.

In order to prove the last point, we must show that the matrix $M^{(l)}$ is positive definite for at least one point (c, h) in R. Indeed, since the Kac determinant is positive, the number of negative eigenvalues of $M^{(l)}$ must be even. That number can change only across one of the curves $C_{r,s}$, and consequently must stay the same throughout R. It remains to show that this number is 0 at some point in R. To this end, we use a slightly different basis for the level l sector, namely the vectors

$$L_{-k_1} L_{-k_2} \ldots L_{-k_n} |h\rangle \qquad (k_1 \geq k_2 \geq \cdots \geq k_n) \qquad (7.39)$$

and we define the *length* $n(\alpha)$ of a basis vector $|\alpha\rangle$ as the number of operators L_k used to define it. For instance, $L_{-1}^3 |h\rangle$ has length 3 and $L_{-3}|h\rangle$ has length 1. It is then possible to show that the dominant behavior in h of inner products is

$$\langle \alpha | \alpha \rangle = c_\alpha \, h^{n(\alpha)} \left[1 + O(1/h) \right] \qquad (c_\alpha > 0)$$
$$\langle \alpha | \beta \rangle = O(h^{(n(\alpha)+n(\beta))/2 - 1}) + \ldots \qquad (7.40)$$

where $|\alpha\rangle$ and $|\beta\rangle$ are two basis states. We sort our basis in order of decreasing lengths, and consider the submatrices $M_n^{(l)}$ obtained by keeping only vectors of length n. Eq. (7.40) implies that these submatrices are positive definite when h is sufficiently large. In that limit, the eigenvalues of $M^{(l)}$ are those of the submatrices $M_n^{(l)}$, and thus $M^{(l)}$ itself is positive definite.[2]

7.2.3. Unitary $c < 1$ Representations

We mentioned earlier that the points (c, h) located between the two vanishing curves on the leftmost graph of Fig. 7.1 correspond to nonunitary representations. In fact, all the points in the region $R : \{(c, h)|0 < c < 1, h > 0\}$ are associated with nonunitary representations, except the following discrete set:

$$c = 1 - \frac{6}{m(m+1)}$$
$$h_{r,s}(m) = \frac{[(m+1)r - ms]^2 - 1}{4m(m+1)} \qquad (1 \leq r < m, \ 1 \leq s < r) \qquad (7.41)$$

This expression coincides with Eq. (7.34) above, except that m is now an integer greater than or equal to 2, and the integers r and s are bounded as indicated. That the representations defined by Eq. (7.41) are unitary will be proven when discussing cosets in Chap. 18. In the present context, we could prove only that points (c, h) not included in this discrete set correspond to nonunitary representation, that is, Eq. (7.41) is a necessary (not yet sufficient) condition for unitarity. In fact, we shall not give the proof, but simply indicate some of its elements.

It is relatively simple to argue that the points of R that do not lie on a vanishing curve correspond to nonunitary representations. Consider such a point P. Since the Kac determinant does not vanish at P, the associated representation does not

---

[2] It is not sufficient to argue that $M^{(l)}$ is diagonal in the $h \to \infty$ limit, since off-diagonal terms may be of the same (or even larger) degree in h, than some diagonal terms. However, the eigenvalues are equal to the diagonal elements in that limit. Consider, for instance, the simplest example, $M^{(2)}$, as given in Eq. (7.23).

contain zero-norm states, but may contain negative-norm states. In order to show that it does indeed, it is sufficient to demonstrate that the Kac determinant is negative at P, for some level l. This can be done if there is, at some particular level l, a continuous path linking P to the $c > 1, h > 0$ region that crosses a single vanishing curve such that $p(l - rs)$ is odd. For instance, going back to Fig. 7.1, any point left of the curve on the level-2 graph can be linked to the $c > 1, h > 0$ region by a continuous path that crosses either $C_{1,2}$ or $C_{2,1}$, and the factors $(h - h_{1,2}(c))$ and $(h - h_{2,1}(c))$ both appear linearly in the Kac determinant at level 2. Therefore these points are associated with nonunitary representations. This is true of all the points lying left of the vanishing curves represented on that figure. The points not excluded from unitarity at some level by this argument will be excluded at some higher level. Indeed, at $c = 1$, the vanishing curve $C_{r,s}$ ends up at $h_{r,s} = \frac{1}{4}(r - s)^2$. For a given value of $r - s$, the vanishing curve lies closer and closer to the $c = 1$ axis as the product rs increases. Each time rs increases by one step (for a fixed value of $r - s$), a new set of points is excluded from unitarity by this argument at level $l = rs$, since $p(l - rs)$ is then one and no other vanishing curve lies between $C_{r,s}$ and the $c = 1$ axis. This argument excludes from unitarity all the points in the region R, except maybe the points lying on the vanishing curves themselves.

Verma modules associated with points on vanishing curves contain null vectors, but may contain negative-norm states as well. Indeed, the second element of the nonunitarity proof is that points on the vanishing curves also correspond to nonunitary representations, except the so-called *first intersections*. Consider a given vanishing curve, at a given level; the first intersection associated with that curve, if it exists, is the point intersected by another vanishing curve (at the same level) that lies closest to the $c = 1$ axis. At any point on the vanishing curve $C_{r,s}$, the Verma module has a null vector at level rs. The characteristic of first intersections is that this null state is the highest-weight state of a representation that in turn contains a null state. It can be shown that first intersections are indeed located as indicated in Eq. (7.41). On Fig. 7.1, all intersections (indicated by dots) are first intersections, the origin included (it intersects $C_{1,2}$ and $C_{1,1}$).

§7.3. Overview of Minimal Models

This section is a constructive introduction to minimal models. The consequences of the existence of null vectors on correlation functions and the operator algebra are illustrated with the help of simple examples. The general construction of minimal models is presented in a heuristic fashion, formal proofs being reported in the next chapter. Throughout this section we shall work in the holomorphic sector only.

7.3.1. A Simple Example

We study a simple example of reducible Verma module. Consider, in $V(c, h)$, the following state at level 2:

$$|\chi\rangle = \left[L_{-2} + \eta L_{-1}^2\right] |h\rangle \qquad (7.42)$$

We want to tune η and h in such a way that $|\chi\rangle$ is a null state (or singular vector). As mentioned earlier, the conditions $L_1|\chi\rangle = L_2|\chi\rangle = 0$ are sufficient for this, since it then follows from the Virasoro algebra (7.4) that $L_n|\chi\rangle = 0$ for all $n \geq 3$. By acting on this state with L_1 and L_2 and bringing these operators in contact with $|h\rangle$ with the help of the algebra (7.4), we find

$$L_1|\chi\rangle = (3 + 2\eta + 4h\eta)L_{-1}|h\rangle$$
$$L_2|\chi\rangle = (\frac{1}{2}c + 4h + 6h\eta)|h\rangle \tag{7.43}$$

The conditions imposed on η and h for $|\chi\rangle$ to be singular are thus

$$\eta = -\frac{3}{2(2h+1)}$$
$$h = \frac{1}{16}\left\{5 - c \pm \sqrt{(c-1)(c-25)}\right\} \tag{7.44}$$

The latter condition may also be inferred from Eq. (7.30) applied to level-2 states (cf. Eq. (7.26)), since singular vectors exist if and only if the Kac determinant vanishes. In the notation of Eq. (7.30), the above constraint on h is simply $h = h_{1,2}$ or $h = h_{2,1}$ (recall that the first null state in the reducible Verma module $V(c, h_{r,s})$ occurs at level $l = rs$).

As discussed in Sect. 6.6.1, to each state of the Verma module one associates a descendant field, as defined in Eq. (6.148) for the simplest case. In particular, one associates a *null field* $\chi(z)$ with the null state $|\chi\rangle$. This field is a descendant of the primary field $\phi(z)$ of conformal dimension h, but is itself a primary field of dimension $h + 2$. Following the discussion of Sect. 6.6.1, the explicit expression for this null field is

$$\chi(z) = \phi^{(-2)}(z) - \frac{3}{2(2h+1)}\frac{\partial^2}{\partial z^2}\phi(z) \tag{7.45}$$

That the null state is orthogonal to the whole Verma module translates, in the field language, into the vanishing of the correlator $\langle\chi(z)X\rangle$, wherein X is a string of local fields: $X \equiv \phi_1(z_1)\cdots\phi_N(z_N)$. Equivalently, we say that the field χ *decouples* from the other fields. According to Eq. (6.152), this implies the following differential equation for the correlator $\langle\phi(z)X\rangle$:

$$\left\{\mathcal{L}_{-2} - \frac{3}{2(2h+1)}\mathcal{L}_{-1}^2\right\}\langle\phi(z)X\rangle = 0 \tag{7.46}$$

More explicitly, this is

$$\left\{\sum_{i=1}^{N}\left[\frac{1}{z-z_i}\frac{\partial}{\partial z_i} + \frac{h_i}{(z-z_i)^2}\right] - \frac{3}{2(2h+1)}\frac{\partial^2}{\partial z^2}\right\}\langle\phi(z)X\rangle = 0 \tag{7.47}$$

(recall that $\mathcal{L}_{-1}$ is equivalent to ∂_z).

This differential equation should bring nothing new to our knowledge of the two-point function. Indeed, if we let $X = \phi(w)$, Eq. (7.47) becomes simply

$$\left\{\frac{1}{z-w}\partial_w + \frac{h}{(z-w)^2} - \frac{3}{2(2h+1)}\partial_z^2\right\}\langle\phi(z)\phi(w)\rangle = 0 \qquad (7.48)$$

which is trivially satisfied, given the general form (5.25) for the two-point function: $\langle\phi(z)\phi(w)\rangle = (z-w)^{-2h}$.

However, the differential equation (7.47) has a nontrivial effect on the three-point function (5.26). We consider $X = \phi_1(z_1)\phi_2(z_2)$. The three-point function is

$$\langle\phi(z)\phi_1(z_1)\phi_2(z_2)\rangle = \frac{g(h,h_1,h_2)}{(z-z_1)^{h_2-h-h_1}(z_1-z_2)^{h-h_1-h_2}(z-z_2)^{h_1-h-h_2}} \qquad (7.49)$$

where $g(h,h_1,h_2)$ is a constant not fixed by global conformal invariance alone, but by the operator algebra of the theory (cf. Sect. 6.6.3). The differential equation (7.47) imposes constraints on the conformal dimensions h, h_1, and h_2. It turns out, after an explicit calculation, that a single independent constraint remains:

$$2(2h+1)(h+2h_2-h_1) = 3(h-h_1+h_2)(h-h_1+h_2+1) \qquad (7.50)$$

This equation may be solved explicitly for h_2:

$$h_2 = \frac{1}{6} + \frac{1}{3}h + h_1 \pm \frac{2}{3}\sqrt{h^2 + 3hh_1 - \frac{1}{2}h + \frac{3}{2}h_1 + \frac{1}{16}} \qquad (7.51)$$

This solution for h_2 is more elegant if we adopt a notation close to that of Eq. (7.30) and parametrize the conformal dimensions as

$$h(\alpha) \equiv h_0 + \frac{1}{4}\alpha^2 \qquad h_0 = \frac{1}{24}(c-1) \qquad (7.52)$$

If α_1 and α_2 correspond respectively to h_1 and h_2, we then have the following solutions:

$$\begin{aligned}\alpha_2 &= \alpha_1 \pm \alpha_+ \qquad & (h = h_{2,1}) \\ \alpha_2 &= \alpha_1 \pm \alpha_- \qquad & (h = h_{1,2})\end{aligned} \qquad (7.53)$$

Thus, the existence of a null vector at level 2 imposes additional constraints on the three-point functions, which are equivalent to constraints imposed on the operator algebra. If we denote by $\phi_{(\alpha)}$ the primary field of dimension $h(\alpha)$ these constraints on the operator algebra take the following symbolic form:

$$\begin{aligned}\phi_{(2,1)} \times \phi_{(\alpha)} &= \phi_{(\alpha-\alpha_+)} + \phi_{(\alpha+\alpha_+)} \\ \phi_{(1,2)} \times \phi_{(\alpha)} &= \phi_{(\alpha-\alpha_-)} + \phi_{(\alpha+\alpha_-)}\end{aligned} \qquad (7.54)$$

The notation introduced here requires some explanation. By the above, we mean that the operator product expansion of $\phi_{(2,1)}$ with $\phi_{(\alpha)}$ (or of fields belonging to their families) may contain terms belonging only to the conformal families of $\phi_{(\alpha-\alpha_+)}$ and $\phi_{(\alpha+\alpha_+)}$. The symbol $\times$ stands for an operator product expansion, and $\phi_{(\alpha)}$ stands not for the primary field only, but for its entire conformal family. Generally speaking, we call *fusion* the process of taking the short-distance product of two local fields. The conditions under which a given conformal family occurs in the short-distance product of two conformal fields are called the *fusion rules* of the theory. These may be thought of as selection rules for the conformal dimensions of fields appearing in a three-point correlator. We say, for instance, that the fusion of two conformal fields ϕ_1 and ϕ_2 onto a third field ϕ_3 is possible if the three-point function $\langle \phi_1 \phi_2 \phi_3 \rangle$ is not zero. This topic will be examined in more detail in Sect. 8.4. It is implicit that there are coefficients multiplying the families on the r.h.s. of Eq. (7.54): They are the structure constants of the operator algebra. Not only are they not specified here, but they may vanish.[3]

We finally point out the possibility of having a null state at level one. The only state at this level is $L_{-1}|h\rangle$, and its norm vanishes only if $h = h_{1,1} = 0$ (cf. Eq. (7.26)). The corresponding null field is $\partial_z \phi_{(1,1)}(z)$, and the differential equation satisfied by the correlator $\langle \phi_{(1,1)}(z)X \rangle$ is

$$\frac{\partial}{\partial z}\langle \phi_{(1,1)}(z)X \rangle = 0 \tag{7.55}$$

Because the correlator is independent of z, the only conclusion to be drawn is that $\phi_{(1,1)}$ is a constant, since it is, by hypothesis, a purely holomorphic field. We call $\phi_{(1,1)}$ the *identity field* or the *identity operator* (sometimes denoted by $\mathbb{I}$). The obvious consequence of the above differential equation on three-point functions involving $\phi_{(1,1)}$ is the trivial operator algebra:

$$\phi_{(1,1)} \times \phi_{(\alpha)} = \phi_{(\alpha)} \tag{7.56}$$

Incidently, the energy-momentum tensor $T(z)$ is a descendant of the identity field, according to Eq. (6.148): $T(z) = \mathbb{I}^{(-2)}$.

7.3.2. Truncation of the Operator Algebra

The constraint (7.54) on the operator algebra coming from the existence of a null vector at level 2 may be generalized. If $h = h_{r,s}$, then there exists a null vector at level rs, as follows from the Kac determinant formula (7.28). This null vector imposes a similar constraint on the operator algebra:

$$\phi_{(r,s)} \times \phi_{(\alpha)} = \sum_{\substack{k=1-r \\ k+r=1 \text{ mod } 2}}^{k=r-1} \sum_{\substack{l=1-s \\ l+s=1 \text{ mod } 2}}^{l=s-1} \phi_{(\alpha+k\alpha_+ +l\alpha_-)} \tag{7.57}$$

(The summation indices are incremented by 2). In other words, k takes only even values if r is odd and vice versa. We shall not prove this statement here. For the moment, we simply draw its consequences.

---

[3] Of course, when writing the complete fusion rules later on, none of the implicit coefficients will vanish.

The first consequence of Eq. (7.57) is that the conformal families $[\phi_{(r,s)}]$ associated with reducible modules form a closed set under the operator algebra. For instance, we see immediately that

$$\phi_{(1,2)} \times \phi_{(r,s)} = \phi_{(r,s-1)} + \phi_{(r,s+1)}$$
$$\phi_{(2,1)} \times \phi_{(r,s)} = \phi_{(r-1,s)} + \phi_{(r+1,s)} \qquad (7.58)$$

This means that the fields $\phi_{(1,2)}$ and $\phi_{(2,1)}$ act as ladder operators in the operator algebra. That the families $[\phi_{(r,s)}]$ form a closed set under the operator algebra is a profound *dynamical* statement, which holds only for certain values of c and certain highest-weight representations associated with those values. Again, we stress that the coefficients implicit on the r.h.s. of (7.57) may be zero; the above notation simply means that no other conformal family, other that those shown, may appear in the operator product expansion. Indeed, many conformal families can be shown *not* to occur in the OPE, by using the commutativity of the operator algebra. For instance, we write

$$\phi_{(1,2)} \times \phi_{(2,1)} = \phi_{(2,0)} + \phi_{(2,2)}$$
$$\phi_{(2,1)} \times \phi_{(1,2)} = \phi_{(0,2)} + \phi_{(2,2)} \qquad (7.59)$$

Since the two OPEs are equivalent, this shows that $\phi_{(2,0)}$ and $\phi_{(0,2)}$ are excluded from both (their coefficients vanish). Thus, in this example, the operator algebra *truncates* to

$$\phi_{(1,2)} \times \phi_{(2,1)} = \phi_{(2,2)} \qquad (7.60)$$

This *truncation* phenomenon may be generalized, with the following result:

$$\phi_{(r_1,s_1)} \times \phi_{(r_2,s_2)} = \sum_{\substack{k=1+|r_1-r_2| \\ k+r_1+r_2=1 \text{ mod } 2}}^{k=r_1+r_2-1} \sum_{\substack{l=1+|s_1-s_2| \\ l+s_1+s_2=1 \text{ mod } 2}}^{l=s_1+s_2-1} \phi_{(k,l)} \qquad (7.61)$$

Here again, the summation variables k and l are incremented by 2. The truncation is such that only the families $\phi_{(r,s)}$ with positive values of r and s occur on the r.h.s. of (7.61).

7.3.3. Minimal Models

For a generic value of the central charge c, the truncated operator algebra (7.61) implies that an infinite number of conformal families are present in the theory, since families $[\phi_{(r,s)}]$ with r, s arbitrary large are generated by applying repeatedly the fusion rules (7.61). In order to understand the situation graphically, we consider the "diagram of dimensions" of Fig. 7.2. The points (r, s) in the first quadrant label the various conformal dimensions appearing in the Kac formula. The dotted line has a slope $\tan\theta = -\alpha_+/\alpha_-$, fixed by the central charge c. If δ is the Cartesian distance between a point (r, s) and the dotted line, it can easily be shown that (cf.

Ex. 7.6)

$$h_{r,s} = h_0 + \frac{1}{4}\delta^2(\alpha_+^2 + \alpha_-^2)$$ (7.62)

If the slope $\tan\theta$ is irrational, it will never go through any integer point (r,s), although some of these points will be arbitrarily close to it. Thus, given the fusion rules (7.61), there will be an infinite number of distinct primary fields in the theory, and moreover, an infinity of them will have negative conformal dimensions, since $h_0 < 0$ if $c < 1$.

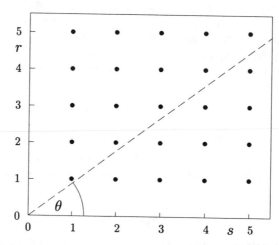

Figure 7.2. The "diagram of dimensions" for a generic value of c. The points on the first quadrant are associated with the conformal dimensions $h_{r,s}$ of the Kac formula. The conformal dimension is related by Eq. (7.62) to the distance between a point (r,s) and the dashed line.

However, if the slope $\tan\theta$ is rational, that is, if there exist two coprime integers p and p' such that

$$p\alpha_- + p'\alpha_+ = 0 ,$$ (7.63)

the dotted line of Fig. (7.2) goes through the point (p',p) and the conformal weights $h_{r,s}$ do not form a dense set. Indeed, we then have the periodicity property:

$$h_{r,s} = h_{r+p',s+p}$$ (7.64)

In terms of these two integers, the central charge and the Kac formula become

$$c = 1 - 6\frac{(p-p')^2}{pp'}$$

$$h_{r,s} = \frac{(pr - p's)^2 - (p-p')^2}{4pp'}$$ (7.65)

If $c \leq 1$, the parameter t of Eq. (7.31) is real and positive, and equal to $t = -\alpha_+/\alpha_- = p/p'$. The two integers p and p' may therefore be taken as positive. Because of the symmetry $t \to 1/t$ of this parametrization, one may also assume that $p > p'$ without loss of generality. Note also the obvious symmetry property

$$h_{r,s} = h_{p'-r,p-s} \tag{7.66}$$

From Eq. (7.65) we easily demonstrate the following identities:

$$h_{r,s} + rs = h_{p'+r,p-s} = h_{p'-r,p+s}$$
$$h_{r,s} + (p'-r)(p-s) = h_{r,2p-s} = h_{2p'-r,s} \tag{7.67}$$

This means that the null vector at level rs contained in the Verma module $V_{r,s}$ is itself the highest weight of a degenerate Verma module, since it fits in the Kac formula! Moreover, the module $V_{r,s}$ also contains a null vector at level $(p'-r)(p-s)$. These two null vectors give rise to submodules that also contain null vectors of the same form, and so on (this is illustrated in Fig. 8.1 of Chap. 8). Thus, there is an infinite number of null vectors within the Verma module $V_{r,s}$ if c is of the form (7.65). Each null vector has its own differential equation acting as a constraint on the correlators and the operator algebra. The net effect is an additional truncation of the operator algebra, yielding a *finite* set of conformal families, which closes under fusion. The corresponding finite set of conformal weights $h_{r,s}$ is delimited by

$$1 \leq r < p' \quad \text{and} \quad 1 \leq s < p \tag{7.68}$$

This rectangle in the (r,s) plane is called the *Kac table*. The symmetry $h_{r,s} = h_{p'-r,p-s}$ makes half of this rectangle redundant:

$$\phi_{(r,s)} = \phi_{(p'-r,p-s)} \tag{7.69}$$

There remain $(p-1)(p'-1)/2$ distinct fields in the theory.

The conformal theories defined by the conditions (7.65) and (7.68) are called *minimal models*, since they contain a finite number of local fields with well-defined scaling behavior. The truncated fusion rules existing between these fields are

$$\phi_{(r,s)} \times \phi_{(m,n)} = \sum_{\substack{k=1+|r-m| \\ k+r+m=1 \bmod 2}}^{k_{max}} \sum_{\substack{l=1+|s-n| \\ k+s+n=1 \bmod 2}}^{l_{max}} \phi_{(k,l)} \tag{7.70}$$

wherein

$$k_{max} = \min(r+m-1, 2p'-1-r-m)$$
$$l_{max} = \min(s+n-1, 2p-1-s-n) \tag{7.71}$$

and k and l are incremented by 2. This expression will be proven in the next chapter.

Of course, the above discussion was restricted to the holomorphic sector. A physical theory is in fact constructed out of tensor products of holomorphic and antiholomorphic modules. A generic Hilbert space has the following form

$$\mathcal{H} = \bigoplus_{h,\bar{h}} M(c,h) \otimes \bar{M}(c,\bar{h}) \tag{7.72}$$

The question of how to combine the components of a minimal model into tensor products will be addressed in detail in Chap. 10, in which conformal field theories on a torus will be studied. However, a particularly simple solution is to associate to each holomorphic module $M(c, h_{r,s})$ the corresponding antiholomorphic module $\bar{M}(c, h_{r,s})$. The Hilbert space of the theory is then

$$\mathcal{H} = \bigoplus_{\substack{1 \le r < p' \\ 1 \le s < p}} M(c, h_{r,s}) \otimes \bar{M}(c, h_{r,s}) \tag{7.73}$$

The resulting theory is termed *diagonal*, since the two factors of each tensor product are identical. We shall symbolically denote a minimal model associated with the pair (p, p') by $\mathcal{M}(p, p')$ and, as mentioned above, will adopt the convention $p > p'$.

7.3.4. Unitary Minimal Models

As seen in Sect. 7.2.1, the constraint of unitarity for a conformal field theory requires that there be no states of negative norm. We have seen that a necessary condition for the unitarity of a representation of the Virasoro algebra with highest weight h is $h \ge 0$. Therefore, a unitary conformal field theory contains only primary fields with nonnegative conformal dimensions. The physical implications of this property are clear: The two-point correlation functions of primary fields (except for the identity operator) have to fall off with distance, instead of exploding at large distances:

$$\langle \phi_{h,\bar{h}}(z, \bar{z}) \phi_{h,\bar{h}}(0, 0) \rangle = \frac{1}{z^{2h} \bar{z}^{2\bar{h}}} \tag{7.74}$$

This is the case for the critical Ising model, to be discussed below: The spin-spin correlator decreases when the separation of the spins increases. That such a behavior is to be expected from any physical spin system with short-range interactions is not quite true in general as we shall see in next section, with the (nonunitary) example of the Yang–Lee edge singularity. It seems that the statistical models of so-called hard objects (i.e., of bulky objects that cannot overlap, subject to simple enough interactions) always admit critical continuum descriptions with nonunitary conformal field theories. Moreover, many other physical systems such as polymers in two dimensions have phases described by nonunitary minimal models. The unitarity condition should therefore not be confused with a physical condition.

We now examine the consequence of the unitarity condition for the $c < 1$ minimal theories discussed above. Recall the form of admissible conformal dimensions (7.65)

$$h_{r,s} = \frac{(pr - p's)^2 - (p - p')^2}{4pp'} \tag{7.75}$$

with $1 \le r \le p' - 1$ and $1 \le s \le p - 1$. The integers p and p' being coprime, Bezout's lemma states that there exists a couple of integers (r_0, s_0) in the above range such that

$$pr_0 - p's_0 = 1 \tag{7.76}$$

Accordingly, the corresponding dimension

$$h_{r_0,s_0} = \frac{1 - (p - p')^2}{4pp'} \tag{7.77}$$

is always negative, except if $|p - p'| = 1$, in which case it vanishes. It turns out that the primary field with smallest dimension (7.77) is always present in the minimal theories. As we shall see in the study of modular invariance (Chap. 10), the primary field with smallest dimension governs the leading anomalous behavior of the free energy of the system through finite size effects. The minimal models can be unitary only if $|p - p'| = 1$. In this case, $h_{r_0,s_0} = h_{1,1} = 0$—that is, $\phi_{(r_0,s_0)}$ is the identity—and the leading finite size effect in the free energy is governed only by the central charge of the theory. That these models are indeed unitary will be proven in Chap. 18 by means of the coset construction; this will provide an explicit unitary realization of each minimal model with $|p - p'| = 1$. With no loss of generality, we label the unitary minimal theories with $c < 1$ by $(p = m + 1, p' = m)$, $m = 2, 3, 4, \dots$. We note that the list of unitary representations given in Eq. (7.41) coincides indeed with the list of highest weights $h_{r,s}$ of unitary minimal models.

§7.4. Examples

7.4.1. The Yang-Lee Singularity

As mentioned in Sect. 3.2.1, the partition function of a lattice theory, such as the Ising model, is an analytic function of the parameters of the model if the number N of sites is finite. Nonanalytic behavior, hence a phase transition, can occur only in the thermodynamic limit ($N \to \infty$). For definiteness, we consider the Ising model at temperature T, in an external field H. The configuration energy is given by Eq. (3.6). As a function of H, the zeros of the partition function cannot lie on the real H-axis for N finite, since Z is then a finite sum of positive terms. These zeros occur at complex values of H and at their complex conjugates. In a generic ferromagnetic spin model, they tend to accumulate on various arcs on the complex plane as $N \to \infty$. In the Ising model, it has been shown that they accumulate on the imaginary axis $H = i\mathfrak{h}$, and the free-energy $F = \ln Z$ may then be expressed in terms of the density $\rho(\mathfrak{h}, T)$ of zeros on the imaginary axis:

$$F(h) = \int_{-\infty}^{\infty} dx \, \rho(x, T) \ln(h - ix) \tag{7.78}$$

The magnetization M is then

$$M = \frac{\partial F}{\partial H} = \int_{-\infty}^{\infty} dx \, \frac{\rho(x, T)}{H - ix} \tag{7.79}$$

Below the critical temperature $(T < T_c)$, the distribution of zeros extends up to the real axis $(\rho(0, T) \neq 0)$ and the magnetization is discontinuous as H crosses the origin along the real axis: There is a first-order phase transition. At $T = T_c$, $\rho(0, T)$ vanishes and this transition becomes continuous.[4] In the paramagnetic phase $(T > T_c)$, the distribution $\rho(\mathfrak{h}, T)$ stops at a critical value $\mathfrak{h}_c(T)$ on either side of the real axis, the so-called *Yang-Lee edge*. We now suppose that, near $\mathfrak{h}_c$, the density of zeros has a power-law behavior:[5]

$$\rho(\mathfrak{h}, T) = (\mathfrak{h} - \mathfrak{h}_c)^\sigma \qquad (7.80)$$

It is then a simple matter to show that the magnetization $M(i\mathfrak{h})$ behaves also like $(\mathfrak{h} - \mathfrak{h}_c)^\sigma$. We may assert, using scaling arguments identical to those explained in Sect. 3.2.2, that the exponent σ is related to the exponent η of the critical correlator by the relation (3.49):

$$\sigma = \frac{1}{\delta} = \frac{d - 2 + \eta}{d + 2 - \eta} = \frac{\eta}{4 - \eta} \qquad (d = 2) \qquad (7.81)$$

Here, however, the correlator with exponent η is not the correlator of the Ising spin at the critical point $(h = 0, T_c)$, but that of another scaling field, yet unspecified, describing the fluctuations of the model in an imaginary field close to $h = i\mathfrak{h}_c$: As $h \to i\mathfrak{h}_c$, the correlation length diverges.

It turns out that the relevant Landau-Ginzburg theory[6] contains a term in $i\Phi^3$:

$$\mathcal{L}_{YL} = \frac{1}{2}(\partial_\mu \Phi)^2 + i(h - h_c)\Phi + i\gamma\Phi^3.$$

This model is, of course, not unitary, because of the imaginary magnetic field, which translates into an imaginary coupling of the Landau-Ginzburg effective field theory.

In trying to identify this critical point with one of the minimal models of conformal field theory, we must keep in mind the following: First, the model is nonunitary. Second, as shown by renormalization-group analyses, the composite field Φ^2 is redundant, which means that the operator product $\Phi\Phi$ does not give rise to any new scaling field. In other words,

$$\Phi \times \Phi = \mathbb{I} + \Phi.$$

The only minimal model with such simple behavior is $\mathcal{M}(5, 2)$, with central charge $c = -22/5$. Its operator content is very simple, with only two primary fields: $\phi_{(1,1)}$ (of dimension 0) and $\phi_{(1,2)} = \phi_{(1,3)}$ (of dimension $-1/5$). These are, of course, the chiral components of the physical operators $\mathbb{I}$ (the identity, of dimensions $(0, 0)$) and Φ (dimensions $(-1/5, -1/5)$). The scaling dimension of the field Φ is thus

---

[4] It is at this critical point $(H = 0, T_c)$ that the Ising model is studied within the formalism of conformal field theory (see the next section).

[5] In this section we adopt the standard notation σ for the critical exponent, not to be confused with the spin operator of the Ising model.

[6] By Landau-Ginzburg theory, we mean a continuous, Lagrangian description of the statistical model. See Sect. 7.4.7.

$\Delta = -2/5$, and the corresponding exponent η is $2\Delta = -4/5$. According to the relation (7.81), the critical exponent σ is then equal to $-1/6$, a result entirely compatible with the outcome of high-temperature series analyses, which yield $\sigma = -0.163 \pm 0.003$.

7.4.2. The Ising Model

The simplest nontrivial unitary minimal model, $\mathcal{M}(4,3)$, describes the critical Ising model. Since Chap. 12 is entirely dedicated to the two-dimensional Ising model, we shall not explain in detail here the precise correspondence between this lattice model and the minimal model $\mathcal{M}(4,3)$. We simply state the results.

In addition to the identity operator, there are two local scaling operators in the critical Ising model: the Ising spin σ (a continuum version of the lattice spin σ_i) and the energy density ε (a continuum version of the interaction energy $\sigma_i\sigma_{i+1}$). The latter is also called the *thermal operator*, since it is coupled to the inverse temperature β in the partition function. The exponents η and ν are defined by the critical behavior of the following correlators ($d = 2$):

$$\langle \sigma_i \sigma_{i+n} \rangle = \frac{1}{|n|^{d-2+\eta}} = \frac{1}{|n|^{\eta}} \qquad \langle \varepsilon_i \varepsilon_{i+n} \rangle = \frac{1}{|n|^{2d-2/\nu}} = \frac{1}{|n|^{4-2/\nu}} \quad (7.82)$$

It is known, from the exact solution, that $\eta = 1/4$ and $\nu = 1$. Therefore, assuming that the scaling fields σ and ε have no spin ($h = \bar{h}$), it follows that their conformal dimensions are

$$(h,\bar{h})_\sigma = (\frac{1}{16}, \frac{1}{16}) \qquad (h,\bar{h})_\varepsilon = (\frac{1}{2}, \frac{1}{2}) \quad (7.83)$$

The three fields making up the holomorphic part of the theory have therefore conformal dimensions 0, $\frac{1}{16}$, and $\frac{1}{2}$. This simple operator content leads to an identification with the minimal model $\mathcal{M}(4,3)$, with central charge $c = \frac{1}{2}$. The operator-field correspondence is

$$\mathbb{I} \Longleftrightarrow \phi_{(1,1)} \quad \text{or} \quad \phi_{(2,3)}$$
$$\sigma \Longleftrightarrow \phi_{(2,2)} \quad \text{or} \quad \phi_{(1,2)} \quad (7.84)$$
$$\varepsilon \Longleftrightarrow \phi_{(2,1)} \quad \text{or} \quad \phi_{(1,3)}$$

The associated diagram of dimensions is illustrated on Fig. 7.3.

The fusion rules following from this identification with $\mathcal{M}(4,3)$ are the following (cf. Eq. (7.70)):

$$\sigma \times \sigma = \mathbb{I} + \varepsilon$$
$$\sigma \times \varepsilon = \sigma \quad (7.85)$$
$$\varepsilon \times \varepsilon = \mathbb{I}$$

Note that these simple fusion rules are compatible with the $\mathbb{Z}_2$ symmetry $\sigma_i \to -\sigma_i$ of the Ising model.

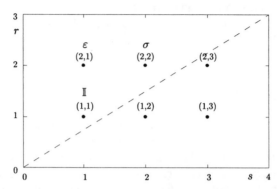

Figure 7.3. The diagram of dimensions for the minimal model $\mathcal{M}_{(4,3)}$, associated with the Ising model. There are six weights in the Kac table, but those below the dashed line are simply a repetition of those above, which correspond to the three scaling fields $\mathbb{I}$, σ and ε.

We know of another unitary conformal field theory with $c = \frac{1}{2}$: the free Majorana fermion ψ (cf. Sect. 5.3.2 and Sect. 6.4). The two theories must be equivalent, and this equivalence is at the origin of Onsager's exact solution of the two-dimensional Ising model. The energy density, as a thermal operator, is readily identified with the fermion mass term $\bar{\psi}\psi$ (recall that $h_\psi = \bar{h}_{\bar{\psi}} = \frac{1}{2}$). However, the expression of the Ising spin σ_i in terms of the fermion field ψ is nonlocal. The questions of locality and *mutual locality* of operators are well illustrated in this model, and will be discussed in more detail in Chap. 12.

7.4.3. The Tricritical Ising Model

Following the Ising model, the next simplest unitary minimal model is $\mathcal{M}(5, 4)$, with central charge $c = \frac{7}{10}$. The associated diagram of dimensions appears in Fig. 7.4. There are six different scaling fields, listed in Table 7.2.

Table 7.2. List of all scaling fields of the minimal model $\mathcal{M}(5, 4)$, which describes the tricritical point of the dilute Ising model.

(r, s)		Dimension	Symbol	Meaning
$(1, 1)$ or	$(3, 4)$	0	$\mathbb{I}$	identity
$(1, 2)$ or	$(3, 3)$	$\frac{1}{10}$	ε	thermal op.
$(1, 3)$ or	$(3, 2)$	$\frac{3}{5}$	ε'	thermal op.
$(1, 4)$ or	$(3, 1)$	$\frac{3}{2}$	ε''	thermal op.
$(2, 2)$ or	$(2, 3)$	$\frac{3}{80}$	σ	spin
$(2, 4)$ or	$(2, 1)$	$\frac{7}{16}$	σ'	spin

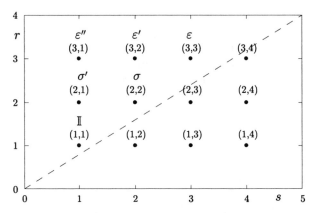

Figure 7.4. The diagram of dimensions for the minimal model $\mathcal{M}_{(5,4)}$, associated with the tricritical Ising model. There are six different weights in the Kac table.

The lattice model associated with this minimal conformal field theory is the *tricritical Ising model*, or more properly said, the dilute Ising model at its tricritical fixed point. This model is defined like an ordinary Ising model, except that vacant sites are allowed and the number of spins on the lattice fluctuates. The configuration energy is

$$E[\sigma_i, t_i] = -\sum_{\langle ij \rangle} t_i t_j (K + \delta_{\sigma_i,\sigma_j}) - \mu \sum_i t_i \tag{7.86}$$

where the variable $t_i = \sigma_i^2$ is 0 if site i is vacant and 1 otherwise. K is the energy of a pair of unlike spins, and $K + 1$ that of a pair of like spins. The chemical potential μ specifies the average number of occupied sites on the lattice. At some value of (β, K, μ), there is a critical point at which three phases meet and coexist critically, hence the epithet *tricritical*. In addition to the identity operator, five scaling operators emerge at this tricritical point: three energy-like operators corresponding to the three terms of the configuration energy and two spin-like operators. The fusion rules of these fields are listed in Table 7.3.

The tricritical Ising model is also one of the few physically relevant theories endowed with supersymmetry. A detailed discussion of supersymmetric conformal field theories does not belong to this chapter, but we nevertheless mention that a supersymmetric generalization of conformal transformations exists (in a superspace formulation) and leads to a supersymmetric generalization of the Virasoro algebra: the so-called *superconformal* or *super-Virasoro algebra*:

$$[L_m, L_n] = (m - n)L_{m+n} + \frac{1}{12}c(m^3 - m)\delta_{m+n}$$

$$\{G_m, G_n\} = 2L_{m+n} + \frac{1}{3}c(m^2 - \frac{1}{4})\delta_{m+n} \tag{7.87}$$

$$[L_m, G_n] = (\frac{1}{2}m - n)G_{m+n}$$

Table 7.3. Nontrivial fusion
rules in the tricritical Ising
model $\mathcal{M}(5, 4)$. It is implicit
here that the symbol used for
the fields stand in fact for the
associated conformal families.

$\varepsilon \times \varepsilon$	$=$	$\mathbb{I} + \varepsilon'$
$\varepsilon \times \varepsilon'$	$=$	$\varepsilon + \varepsilon''$
$\varepsilon \times \varepsilon''$	$=$	ε'
$\varepsilon' \times \varepsilon'$	$=$	$\mathbb{I} + \varepsilon'$
$\varepsilon' \times \varepsilon''$	$=$	ε
$\varepsilon'' \times \varepsilon''$	$=$	$\mathbb{I}$
$\varepsilon \times \sigma$	$=$	$\sigma + \sigma'$
$\varepsilon \times \sigma'$	$=$	σ
$\varepsilon' \times \sigma$	$=$	$\sigma + \sigma'$
$\varepsilon' \times \sigma'$	$=$	σ
$\varepsilon'' \times \sigma$	$=$	σ
$\varepsilon'' \times \sigma'$	$=$	σ'
$\sigma \times \sigma$	$=$	$\mathbb{I} + \varepsilon + \varepsilon' + \varepsilon''$
$\sigma \times \sigma'$	$=$	$\varepsilon + \varepsilon'$
$\sigma' \times \sigma'$	$=$	$\mathbb{I} + \varepsilon''$

Table 7.4. List of all scaling fields of the minimal superconformal model
$m = 3$, associated with the tricritical Ising model. Superpartners are
indicated in the Neveu-Schwarz sector.

(r, s)		Dimension	Symbol	Sector
$(1, 2)$	or $(2, 4)$	0	$[\mathbb{I}, \varepsilon'']$	NS
$(1, 3)$	or $(2, 2)$	$\frac{1}{10}$	$[\varepsilon, \varepsilon']$	NS
$(1, 2)$	or $(2, 3)$	$\frac{3}{80}$	σ	R
$(1, 4)$	or $(2, 1)$	$\frac{7}{16}$	σ'	R

In the above, the modes G_m are the Fourier components of the superpartner $G(z)$ of
the energy-momentum tensor. This anticommuting field has conformal dimension
$\frac{3}{2}$ and corresponds to $\phi_{(1,4)}$ (or $\phi_{(3,1)}$) in Table 7.2.

Depending on the boundary conditions, the index of G_n is either half-integral, in
which case the above algebra is known as the *Neveu-Schwarz algebra*, or integral,
in which case it is known as the *Ramond algebra*. It is possible to identify a discrete
series of unitary, minimal superconformal models, indexed by an integer m, with

the following dimensions:

$$h_{r,s} = \frac{\left[r(m+2) - sm\right]^2 - 4}{8m(m+2)} + \frac{1}{32}\left[1 - (-1)^{r-s}\right]$$

$$(1 \leq r < m, \ 1 \leq s < m+2) \tag{7.88}$$

$$c = \frac{3}{2} - \frac{12}{m(m+2)}$$

Of course, superconformal models are also conformal; however, they possess extra symmetry. A model that is minimal with respect to the superconformal algebra need not be minimal with respect to the plain Virasoro algebra: As is well-known in group theory, when an irreducible representation of some algebra is restricted to a subalgebra, it is generally no longer irreducible. From the relation (7.88), we see that the case $m = 3$ is the only nontrivial model that is both Virasoro and super-Virasoro minimal. This $c = \frac{7}{10}$ model is precisely the tricritical Ising model. The Neveu-Schwarz sector of the theory contains the fields $\mathbb{I}, \varepsilon, \varepsilon'$ and ε'', all even under spin reversal. In terms of superconformal representations, ε'' is a descendant of the identity, exactly like T, and ε' is a descendant of ε. In the Neveu-Schwarz sector, every field has generically a *superpartner* with a conformal dimension differing by $\frac{1}{2}$, and the pair forms what is called a *superfield*. In the case at hand, ε and ε' are superpartners, like T and G. The fusion algebra of these four fields closes onto itself, as may be verified in Table 7.3. The Ramond sector contains the fields σ and σ', which are odd under spin reversal. The field assignments in both sectors according to Eq. (7.88) are indicated on Table 7.4.

7.4.4. The Three-State Potts Model

The next model on the minimal unitary list is $\mathcal{M}(6,5)$, with central charge $c = \frac{4}{5}$ and ten different scaling fields. It turns out that a subset of fields in this model describes the critical point of the three-state Potts model.

The Q-state Potts model is defined in terms of a spin variable σ_i taking Q different values. The configuration energy is

$$E[\sigma_i] = -\sum_{\langle ij \rangle} \delta_{\sigma_i \sigma_j} \tag{7.89}$$

In other words, a nearest-neighbor pair of like spins carries an energy -1 and all other pairs carry no energy. The case $Q = 2$ is equivalent to the Ising model. A related model is the Q-state *clock model*, defined in terms of a spin variable taking its values among the Q-th roots of unity $e^{i\varphi}$, where $Q\varphi \in 2\pi\mathbb{Z}$. Its configuration energy is usually defined as

$$E[\varphi_i] = -\sum_{\langle ij \rangle} \cos(\varphi_i - \varphi_j) \tag{7.90}$$

The clock model has a Z_Q symmetry under $\varphi_j \to e^{2\pi i/Q}\varphi_j$ and a spin-reversal symmetry $\varphi_j \to -\varphi_j$, whereas the Potts model has a permutation symmetry S_Q of the spin labels. Both models are equivalent in the case $Q = 3$, since the clock model Hamiltonian may then be rewritten as follows, modulo additive and multiplicative constants:

$$E[\varphi_i] = -\sum_{\langle ij \rangle} \frac{2}{3} \left[\cos(\varphi_i - \varphi_j) + \frac{1}{2} \right]$$
$$= -\sum_{\langle ij \rangle} \delta_{\varphi_i,\varphi_j} \tag{7.91}$$

The Potts model has a self-duality point at a temperature $1/\beta_c$ given by $e^{\beta_c} = 1 + \sqrt{Q}$. For $Q \leq 4$ this corresponds to a continuous transition, whereas the transition is of first order if $Q > 4$.

From Baxter's exact solution of the three-state Potts model at the critical point, one finds the critical exponents $\nu = 5/6$ and $\eta = 4/15$. It follows that the real field $\cos\varphi \equiv \frac{1}{2}(\sigma + \bar{\sigma})$ must have conformal weights $(h, \bar{h}) = (\frac{1}{15}, \frac{1}{15})$, and the energy density ε has scaling dimension $(\frac{2}{5}, \frac{2}{5})$. These two fields correspond respectively to $\phi_{(3,3)}$ and $\phi_{(2,1)}$ of the minimal model $\mathcal{M}(6,5)$. However, not all scaling fields allowed in this minimal model are actually present in the Potts model. There exists a subset of fields that closes under the fusion rules and forms a minimal, consistent theory. These fields are listed in Table 7.5, and the nontrivial fusion rules appear in Table 7.6. That a subset of the Kac table may in itself form a consistent theory is an unexpected feature; the reasons for this will be discussed in Chap. 10.

Table 7.5. Scaling fields of the minimal model $\mathcal{M}(6,5)$ included in the three-state Potts model.

(r,s)		Dimension	Symbol	Meaning
$(1,1)$ or	$(4,5)$	0	$\mathbb{I}$	identity
$(2,1)$ or	$(3,5)$	$\frac{2}{5}$	ε	thermal op.
$(3,1)$ or	$(2,5)$	$\frac{7}{5}$	X	
$(4,1)$ or	$(1,5)$	3	Y	
$(3,3)$ or	$(2,3)$	$\frac{1}{15}$	σ	spin
$(4,3)$ or	$(1,3)$	$\frac{2}{3}$	Z	

Of course, the physical operators occurring in the Potts model are products of holomorphic and antiholomorphic fields; we denote them by $\Phi_{h,\bar{h}}$, labeling them by their conformal dimensions. The physical operators alluded to in Table 7.5 are in fact the diagonal combinations $\Phi_{h,h}$ ($h = \bar{h}$). In addition to these diagonal (or spinless) operators, the Potts model contains also the following operators with spin:

$$\Phi_{0,3} \qquad \Phi_{3,0} \qquad \Phi_{\frac{2}{5},\frac{7}{5}} \qquad \Phi_{\frac{7}{5},\frac{2}{5}} \tag{7.92}$$

Table 7.6. Nontrivial fusion rules of the
fields for the fields of the minimal model
$\mathcal{M}(6,5)$ included in the three-state Potts
model. It is implicit here that the symbol
used for the fields stand in fact for the
associated conformal families.

$\varepsilon \times \varepsilon$	$=$	$\mathbb{I} + X$
$\varepsilon \times \sigma$	$=$	$\sigma + Z$
$\varepsilon \times X$	$=$	$\varepsilon + Y$
$\varepsilon \times Y$	$=$	X
$\varepsilon \times Z$	$=$	σ
$\sigma \times \sigma$	$=$	$\mathbb{I} + \varepsilon + \sigma + X + Y + Z$
$\sigma \times X$	$=$	$\sigma + Z$
$\sigma \times Y$	$=$	σ
$\sigma \times Z$	$=$	$\varepsilon + \sigma + X$
$X \times X$	$=$	$\mathbb{I} + X$
$X \times Y$	$=$	ε
$X \times Z$	$=$	σ
$Y \times Y$	$=$	$\mathbb{I}$
$Y \times Z$	$=$	Z
$Z \times Z$	$=$	$\mathbb{I} + Y + Z$

The presence in Table 7.5 of a field of conformal dimension 3 indicates the presence of an additional symmetry for which this field is the current, much like the field of dimension $\frac{3}{2}$ in the tricritical Ising model signals the presence of supersymmetry. This additional symmetry is embodied in an infinite-dimensional algebra called the W_3 algebra, which contains the Virasoro algebra as a subset. It is possible to construct a sequence of "minimal models" with representations of this algebra, of which the three-state Potts model is the simplest realization, and the only one that is at the same time a minimal model of the Virasoro algebra. However, we shall not study the W_3 algebra in this volume.

7.4.5. RSOS Models

A correspondence has been suggested, based on known critical exponents, between the unitary minimal models $\mathcal{M}(m + 1, m)$ $(m \geq 3)$ and a sequence of exactly solved statistical models, the RSOS models. A solid-on-solid (SOS) model is defined by associating to each lattice site an integer *height* l_i, constrained by the condition $|l_i - l_j| = 1$ between nearest-neighbor sites. A Boltzmann weight is then associated with each plaquette according to the sequence of heights around the plaquette. In the restricted solid-on-solid (RSOS) model, the heights l_i cannot take all integer values, but only those in the range $1 \leq l_i \leq q - 1$, where q is an integer characterizing the model $(q \geq 4)$. Let l_1, l_2, l_3, and l_4 be the heights associated with the four corners of a plaquette, circled clockwise. Then the Boltzmann weight associated with the plaquette is defined as

$$W(l_1, l_2, l_3, l_4) = w(l_1)w(l_2)w(l_3)w(l_4)y(l_1, l_3)z(l_2, l_4) \qquad (7.93)$$

where the on-site weight $w(l)$ satisfies the relation $w(l) = w(q - l)$ and the next-nearest neighbor interactions y and z are defined as

$$y(l, l') = \begin{cases} 1 & \text{if } l \neq l' \\ y_l = y_{q-l} & \text{if } l' = l \end{cases}$$

$$z(l, l') = \begin{cases} 1 & \text{if } l \neq l' \\ z_l = z_{q-l} & \text{if } l' = l \end{cases} \qquad (7.94)$$

Thus, the number of parameters in this model is $(3q - 8)/2$ (q even) or $(3q - 9)/2$ (q odd).

The constraint $|l_i - l_j| = 1$ naturally divides the lattice into two sublattices, on which the height variables are odd and even, respectively. If q is even, it is possible to define a spin variable $s_i = \frac{1}{4}(q - 2l_i)$, with integer spins on one sublattice and half-integer spins on the other. The parameters z_l and y_l then represent nearest-neighbor interactions between spins on each sublattice. The simplest case ($q = 4$) is then equivalent to the Ising model.

The RSOS model has been solved exactly in a two-dimensional submanifold of the full parameter space, and four different regimes have been identified, denoted I to IV. In regime III, $q - 2$ phases are in coexistence, whereas in regime IV, $q - 3$ phases are in coexistence. Regimes III and IV meet at a multicritical point, which, in the Ising case ($q = 4$) is nothing but the ordinary critical point between the ordered and the disordered phases. A sequence of $q - 3$ order parameters have been constructed for this transition, with exponents

$$\beta_k = \frac{(k + 1)^2 - 1}{8(q - 1)} \qquad (7.95)$$

The heat capacity exponent α has also been calculated:

$$\alpha = 2 - q/2 \qquad (7.96)$$

The scaling laws of Table 3.2 allow us to express the scaling dimension $\Delta = h + \bar{h} = \frac{1}{2}\eta$ in terms of α and β. We thus find a sequence of conformal dimensions (assuming $h = \bar{h}$):

$$h_k = \frac{(k + 1)^2 - 1}{4q(q - 1)} \qquad (1 \leq k \leq q - 3, q \geq 4) \qquad (7.97)$$

These coincide with the dimensions $h_{k+1,k+1}$ of the unitary minimal models $\mathcal{M}(q, q - 1)$, as readily checked from the Kac formula. This is the correspondence between the multicritical points of the RSOS models and the sequence of unitary minimal models.

7.4.6. The $O(n)$ Model

The $O(n)$ model is a generalization of the Ising model in which the spin degree of freedom is a vector S with n components $(S^a)_{a=1,2,...,n}$. For technical reasons, the model is more tractable on a trivalent lattice, and we shall consider the $O(n)$ model on the honeycomb lattice of Fig. 7.5. The configuration energy of the model reads

$$E(S_i) = -J \sum_{\langle ij \rangle} S_i.S_j \qquad (7.98)$$

where $\langle ij \rangle$ denote neighboring sites of the lattice. The partition function is an integral

$$\tilde{Z}_n = \int \prod_i dS_i \, e^{-\beta E[S]} \qquad (7.99)$$

with the following integration rules for the vector components:

$$\int dS^a \, (S^a) = 0$$

$$\int dS^a \, (S^a)^2 = 1 \qquad (7.100)$$

$$\int dS^a \, (S^a)^3 = 0$$

With these rules, we thus have

$$\int dS \, S^2 = n \qquad (7.101)$$

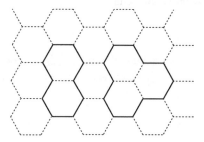

Figure 7.5. A typical loop configuration of the $O(n)$ model on the honeycomb lattice.

The study of the model is greatly simplified if we consider, instead of (7.99), the slightly modified partition function

$$Z_n(K) = \int \prod_i dS_i \prod_{\langle ij \rangle} (1 + KS_i.S_j) \qquad (7.102)$$

Strictly speaking, the two partition functions Z_n and $\tilde{Z}_n$ coincide only in the large $K = \beta J$ limit, but both systems are expected to belong to the same universality class. We shall use the partition function (7.102) in the remainder of this section. The partition function (7.102) of the $O(n)$ model may be perturbatively expanded in powers of K as a sum over loop configurations on the lattice. Indeed, due to the integration rules (7.100), for the integral of a product of spin components to be nonzero, the latter must be taken along a set of closed nonintersecting loops of neighboring sites of the lattice. Moreover, each such loop receives a contribution n from the integration over the spin components, and K per loop bond. For instance, the typical loop configuration of Fig. 7.5 contributes for $n^2 K^{22}$. We may therefore rewrite

$$Z_n(K) = \sum_{\text{loops}} n^{N_L} K^{N_B} \qquad (7.103)$$

where N_L and N_B denote, respectively, the numbers of loops and of bonds in the configuration. The expression (7.103) for the partition function of the $O(n)$ model enables us to analytically continue its definition to any real value of n. The model can be further explored by transforming it into a solid-on-solid (SOS) model. In the latter, the degree of freedom is a height variable l at the center of each hexagon of the lattice, for which the previous loops are domain walls. More precisely, orienting the loops, the height l increases (resp. decreases) by a fixed amount l_0 across a wall pointing to the right (resp. left). In the SOS language, the partition function Z_n is rewritten as a sum over *oriented loops*. The weights in Eq. (7.103) can be reproduced by attaching a weight K per oriented bond of loop, and a weight e^{iv} (resp. e^{-iv}) per right turn (resp. left turn) along the loop at each loop vertex. Summing over the two orientations of each loop gives a net contribution of $2 \cos 6v$ per loop (a loop on the honeycomb lattice always has a difference $n_l - n_r = \pm 6$ between its numbers of left and right turns), which reproduces the factor n provided we take

$$n = 2 \cos 6v \qquad (7.104)$$

This transformation is instrumental in the study of critical properties of the model. It can indeed be shown that the $O(n)$ model undergoes a continuous phase transition at the critical value

$$K = K_c(n) \equiv (2 + \sqrt{2-n})^{-1/2} \quad \text{for } n \in [-2, 2] \qquad (7.105)$$

The continuum limit of the critical model is in turn described for

$$n = -2 \cos \pi(p/p'), \qquad 1 \le p/p' \le 2 \qquad (7.106)$$

by the minimal model (p, p'). More generally, for

$$n = -2 \cos \pi g, \qquad g \in [1, 2] \qquad (7.107)$$

the central charge of the conformal theory describing the continuum limit of the critical $O(n)$ model is

$$c_n = 1 - 6\frac{(g-1)^2}{g} \tag{7.108}$$

For $n = 1$ ($g = 4/3$), we recover the central charge $c_1 = 1/2$ of the Ising model. For $n = 2$ ($g = 1$), the model is called the XY model and is described at criticality (Kosterlitz-Thouless point) by a conformal theory of central charge $c_2 = 1$. When $n = 0$ ($g = 3/2$), the partition function is simply

$$Z_0 = 1 \tag{7.109}$$

Although trivial looking, the model captures the physics of polymers in two dimensions. For instance, nontrivial information such as multipolymer correlations, which exhibit nontrivial scaling behavior, may be obtained by differentiating the critical partition function with respect to n before taking $n \to 0$. The simplest example is the configuration sum of a single polymer, which reads (see Ex. 10.24)

$$\left.\frac{\partial}{\partial n}Z_n\right|_{n=0} \tag{7.110}$$

7.4.7. Effective Landau-Ginzburg Description of Unitary Minimal Models

Most conformal theories have no path-integral formulation based on an action. For a special class of minimal theories, however, there exists a simple effective Lagrangian description, which we now present.

This class is referred to as the $(m+1, m)$ diagonal unitary minimal models with central charge

$$c_m = 1 - \frac{6}{m(m+1)} \qquad m = 2, 3, 4, ... \tag{7.111}$$

and the primary fields have dimensions

$$h_{r,s} = \frac{((m+1)r - ms)^2 - 1}{4m(m+1)} \qquad 1 \leq r \leq m-1,\ 1 \leq s \leq m \tag{7.112}$$

The epithet diagonal means that the primary fields of the theory are built out of identical left and right Virasoro representations, which cover the entirety of the Kac table modulo the equivalence $(r, s) \leftrightarrow (m+1-r, m-s)$. In other words, the primary fields of a diagonal theory are the spinless combinations

$$\Phi_{(r,s)}(z, \bar{z}) = \phi_{(r,s)}(z) \otimes \phi_{(r,s)}(\bar{z}) \tag{7.113}$$

with $h = \bar{h} = h_{r,s}$.

The fusion rules (7.70) for the left part of Virasoro minimal theories extend to the diagonal association of identical left and right Virasoro representations. By setting $p = m + 1$ and $p' = m$, we can write

$$\Phi_{(r,s)} \times \Phi_{(n,k)} = \sum_{\substack{k=|n-r|+1 \\ k-n+r-1 \text{ even}}}^{\min(n+r-1,2m+1-n-r)} \sum_{\substack{l=|k-s|+1 \\ l-k+s-1 \text{ even}}}^{\min(k+s-1,2m-1-k-s)} \Phi_{(k,l)} \qquad (7.114)$$

Eq. (7.114) differs from Eq. (7.70) in that it describes the fusions of the complete left-right association of Virasoro representations, instead of just the left representations of the Virasoro algebra. Nevertheless, the fusions (7.114) are generated by repeated fusions of $X = \Phi_{(2,1)}$ and $Y = \Phi_{(1,2)}$. This provides an effective description of the fusion rules (7.114) by a theory of *two* fields X and Y. A Lagrangian description of the interactions between these two fields would offer an alternative description of the minimal theories, allowing, in particular, to compute correlation functions involving X and Y, directly from the action. Unfortunately, no such description has been found so far. The only known effective description contains *one* self-interacting field Φ, corresponding to $\Phi \equiv \Phi_{(2,2)}$. It is governed by a Lagrangian of the form

$$\mathcal{L} = \int d^2z \left\{ \frac{1}{2}(\partial\Phi)^2 + V(\Phi) \right\} . \qquad (7.115)$$

This Lagrangian is an *effective Landau-Ginzburg Lagrangian*, in which the field Φ stands for the order parameter of some physical system (especially in the continuum formulations of the critical phases of discrete interacting spin systems, such as the archetypical Ising model). The potential term is some general polynomial $V(\Phi)$, whose extrema correspond to the various critical phases of the system. The potential is usually chosen to be invariant under the reflection $\Phi \to -\Phi$. For a polynomial potential $V(\Phi)$ of degree $2(m - 1)$, this ensures the existence of $m - 1$ minima separated by $m - 2$ maxima. Several critical phases of the system can coexist if the corresponding extrema coincide. The most critical potential is therefore just a monomial of the form

$$V_m(\Phi) = \Phi^{2(m-1)} \qquad (7.116)$$

As we shall show, the fusion rules of the $(m+1, m)$ diagonal unitary minimal model are effectively described by the multicritical Landau-Ginzburg theory (7.115), with potential $V = V_m(\Phi)$ as above (we shall denote by $\mathcal{L}_m$ the corresponding Lagrangian). The physical implication of this result is deep: The diagonal unitary minimal models $(m + 1, m)$ can be viewed as the multicritical points of a system described by one scalar field. Clearly, a single scalar field description simplifies substantially the computation of the correlators in the theory, releasing it from all the sophistication (differential equations from singular vectors) encountered so far. Incidentally, Chap. 9 is devoted to another scalar-field representation of minimal conformal field theories—the Coulomb-gas formalism—which allows for an actual computation of correlation functions of conformal fields. However, in that approach, one somehow loses track of the underlying physics which, by contrast, is quite transparent in the present Landau-Ginzburg approach. The advantage is a global treatment of all the relevant operators of the theory, appearing as composite descendants of the order parameter Φ.[7] Moreover, the effective Landau-Ginzburg theory provides an interesting conceptual bridge between the pure field-theoretical problem and the statistical mechanics of related discrete spin systems.

Starting from the field Φ, we can construct *renormalized composite fields* by repeated operator product expansions and subtractions of the most singular terms. For instance, the operator product

$$\Phi(z,\bar{z}) \times \Phi(0,0) = \frac{1}{z^{2d_1}\bar{z}^{2d_1}}\left[\mathbb{I}(0,0) + z^{d_2}\bar{z}^{d_2}\Phi_2(0,0) + \quad \text{less singular} \quad \cdots\right]$$

defines a composite field

$$\Phi_2 \equiv\; :\Phi^2: \tag{7.117}$$

(renormalized square of Φ). The dimensions $2d_1$ and $2d_2$ are the anomalous dimensions of Φ and Φ_2 in the renormalization group sense (and coincide with their respective scaling dimensions $\Delta_1 = 2h_1$ and $\Delta_2 = 2h_2$). We point out that the normal order $: \cdots :$ in (7.117) is not the usual normal ordering, in which *all* the singular terms are subtracted; here only the *most* singular term is subtracted. In particular, note that $d_2 \neq 2d_1$. Composite fields also include renormalized products involving derivatives of Φ. Higher renormalized powers of Φ are obtained by operator expansion and subtraction of only the most singular terms therein, which have already been identified as lower renormalized powers of Φ, namely

$$:\Phi^{k+1}: (0,0) = \lim_{z\to 0} |z|^{d_1+d_k-d_{k+1}}\Big[\Phi(z,\bar{z})\times\; :\Phi^k: (0,0)$$

$$-\sum_{r=1}^{[k+1/2]} C_r|z|^{d_{k+1-2r}-d_1-d_k}\; :\Phi^{k+1-2r}:\Big] \tag{7.118}$$

The even power shifts of $2r$ enforce the $\Phi \to -\Phi$ symmetry, and the constants C_r are completely fixed by the OPE. This construction can be safely repeated, until the equation of motion of the $\mathcal{L}_m$ Landau-Ginzburg theory,

$$:\Phi^{2m-3}: \simeq \partial_z\partial_{\bar{z}}\Phi \tag{7.119}$$

is reached. When compared to the actual operator product expansion of primary fields of the $(m+1,m)$ unitary minimal theory, the definition (7.118) allows for the identification

$$:\Phi^k: \equiv \begin{cases} \Phi_{(k+1,k+1)} & \text{for} \quad k = 0,1,...,m-2 \\ \Phi_{(k+3-m,k+2-m)} & \text{for} \quad k = m-1,m,...,2m-4 \end{cases} \tag{7.120}$$

---

[7] For instance, in the Ising model, the energy operator can be viewed as some composite of the spin operator, as expressed through the fusion rule $\sigma \times \sigma = \mathbb{I} + \varepsilon$. The Coulomb-gas formalism of Chap. 9 does not express this "descendant property", whereas it is crucial in the present Landau-Ginzburg approach.

Indeed, due to the fusion rules

$$\Phi_{(2,2)} \times \Phi_{(r,r)} = \Phi_{(r-1,r-1)} + \Phi_{(r-1,r+1)} + \Phi_{(r+1,r-1)} + \Phi_{(r+1,r+1)} \quad (7.121)$$

the term $\Phi_{(r+1,r+1)}$ is the most singular, after subtraction of $\Phi_{(r-1,r-1)}$. When $r = m - 1$ in Eq. (7.121), the most singular term after subtraction of $\Phi_{(m-2,m-2)}$ is

$$\Phi_{(m-2,m)} = \Phi_{(2,1)} \quad (7.122)$$

($\Phi_{(m,m)}$ and $\Phi_{(m,m-2)}$ lie outside of the Kac table, and do not belong to the theory) which explains the second line of (7.120). Beyond the power $2m - 4$, the identification is more subtle as the equation of motion of the Landau-Ginzburg theory (7.119) introduces a mixing between Φ and its derivatives (descendants of Φ). The order parameters of the theory are the collection of renormalized powers of Φ before the equation of motion is reached; they correspond to the first and second diagonals of the unitary Kac table. To complete the above identification, we check, from the unitary minimal theory point of view, that the Landau-Ginzburg equation of motion (7.119) is satisfied by $\Phi_{(2,2)}$. Due to the unitary minimal fusion rules (7.114), we deduce that

$$\Phi : \Phi^{2m-4} := \Phi_{(2,2)} \times \Phi_{(m-1,m-2)} = \Phi_{(m-2,m-3)} + \Phi_{(m-2,m-1)} \quad (7.123)$$

Defining $: \Phi^{2m-3} :$ by Eq. (7.118), we have to subtract the two most singular contributions allowed by this fusion rule, namely $: \Phi^{2m-5} := \Phi_{(m-3,m-2)}$ and $\Phi = \Phi_{(2,2)} = \Phi_{(m-2,m-1)}$. The most singular contribution after these subtractions comes from the first descendant of the field Φ, $\partial_z \partial_{\bar{z}} \Phi$. This is indeed the operator with lowest dimension among the descendants of Φ and $: \Phi^{2m-5} :$. This establishes, at least formally, the equation of motion (7.119) within the framework of the minimal model.

   In addition to providing a physical picture for the minimal models, the Landau-Ginzburg description sheds some light on the issue of perturbation of conformal theories away from the critical points, and of the renormalization group flows between the various theories. A naive way of interpolating between the $(m+1, m)$ and $(m, m-1)$ unitary minimal theories consists in replacing the potential V_m (7.116) by the linear combination $V_m + \lambda V_{m-1}$. The case $\lambda = 0$ corresponds to the $(m+1, m)$ fixed point, whereas the limit $\lambda \to \infty$ is the fixed point $(m, m-1)$. So a flow between the various unitary theories can be obtained by perturbing the $(m+1, m)$ theory by its most relevant operator (with conformal dimension smaller than 1 but closest to it), namely

$$: \Phi^{2(m-2)} := \Phi_{(m-1,m-2)} = \Phi_{(1,3)} \quad (7.124)$$

   Finally, multiple fusions with $\Phi_{(2,2)}$ *do not* generate the whole unitary minimal fusion algebra, except in the lower cases $m = 2, 3, 4$ (see Ex. 8.20). This is not in conflict with the above results, but points to the subtlety of the actual meaning of the Landau-Ginzburg description. Beyond the first two diagonals of the Kac table, the description of the other primary fields of the theory in terms of Φ is more involved; the equation of motion (7.119) has to be taken into account, and this causes a proliferation of derivatives of Φ.

Exercises

7.1 *Inner product*
Show that the norm of the state $(L_{-1})^n |h\rangle$ is

$$2^n n! \prod_{i=1}^{n} \left(h - (i - 1)/2 \right)$$

7.2 *Gram matrix*
Show that the Gram matrix for level 3 is

$$M^{(3)} = \begin{pmatrix} 24h(1+h)(1+2h) & 12h(1+3h) & 24h \\ 12h(1+3h) & h(8+c+8h) & 10h \\ 24h & 10h & 2c+6h \end{pmatrix}$$

(the states are ordered as in Table 7.1).

7.3 *Gram matrix and vectors of fixed length as $h \to \infty$*
Check explicitly Eq. (7.40) at level 2 with vectors of length 2.

7.4 *Explicit expression of simple null vectors*
Find the explicit expression of the null vectors $\chi_{1,3}$, $\chi_{1,4}$, and $\chi_{2,2}$. Proceed as in the beginning of Sect. 7.3.1, by writing the most general state at level rs and imposing the highest-weight condition

$$L_1|\chi\rangle = L_2|\chi\rangle = 0$$

7.5 *Constraint on the conformal dimensions from the differential equation associated with a null vector*
Show explicitly how the constraint (7.50) follows from applying the differential equation (7.47) to the three-point function (7.49).

7.6 *Diagram of dimensions*
Prove formula (7.62) for the dimensions in the Kac table as a function of the distance δ between the point (r,s) on the plane and the line with slope $-\alpha_+/\alpha_-$ that goes through the origin. To do so, simply subtract from the vector (r,s) its projection on the unit vector $(\cos\theta, \sin\theta)$, where $\tan\theta = -\alpha_+/\alpha_-$, calculate the length squared of the result, and compare with (7.30).

7.7 *Fusion rules in the Ising model*
From the simple ladder operations (7.58), obtain the fusion rules of this Ising model ($\mathcal{M}(4,3)$) by applying repeatedly the truncation procedure leading to Eq. (7.60). Thus, check explicitly the validity of the fusion rules (7.70).

7.8 *Tricritical Potts model*
Write the field content and the fusion rules (given by Eq. (7.70)) for the minimal model $\mathcal{M}(7,6)$. Check that there is a subset of fields that closes under the fusion algebra, like in the minimal model $\mathcal{M}(6,5)$. It turns out that this subset is associated with the tricritical Potts model.

7.9 *Equation of motion for the Yang-Lee model*

Consider the minimal model $\mathcal{M}(5, 2)$ associated with the Yang-Lee edge singularity. The module of the identity operator $\mathbb{I} = \phi_{(1,1)} = \phi_{(1,4)}$ contains a null vector at level four:

$$|\chi\rangle = (L_{-2}^2 - \frac{3}{5}L_{-4})|0\rangle \tag{7.125}$$

a) Show that the field associated with the state $|\chi\rangle$ is

$$T_4(z) = (TT) - \frac{3}{10}T''$$

b) Compute the singular terms in the short-distance product of this field with any primary field Φ of the theory, with dimension h.
Result:

$$T_4(z)\Phi(0) = z^{-4}h(h + \frac{1}{5})\Phi(0)$$

$$+ z^{-3}2(h + \frac{1}{5})\partial\Phi(0)$$

$$+ z^{-2}\left(\frac{5h + 1}{2h + 1}\partial^2\Phi(0) + 2h\Phi^{(2)}(0)\right)$$

$$+ z^{-1}\left(\frac{5h + 1}{(2h + 1)(h + 1)}\partial^3\Phi(0) + \frac{6h}{h + 2}\Phi^{(2)}(0) + 2(h - 1)\Phi^{(3)}(0)\right)$$

where

$$\Phi^{(2)} = (L_{-2} - \frac{3}{2(2h + 1)}L_{-1}^2)\Phi$$

$$\Phi^{(3)} = (L_{-3} - \frac{2}{h + 1}L_{-1}L_{-2} + \frac{1}{(h + 1)(h + 2)}L_{-1}^3)\Phi$$

c) Deduce that the only possible primary fields of the theory have dimensions 0 or $-1/5$. Show, moreover, that when $h = -1/5$, we have $\Phi^{(2)} = \Phi^{(3)} = 0$.

The vanishing of $T_4(z)$ therefore implies most of the structure of the corresponding minimal model: It may be viewed as the equation of motion of the Yang-Lee model. This may be generalized to any minimal model (p, p'). In those cases, the identity has a nontrivial singular descendant at level $(p - 1)(p' - 1)$: It is a composite field $T_{(p-1)(p'-1)}$ of T and its derivatives. Its vanishing forms the equation of motion of the corresponding minimal model and completely determines the spectrum of the theory.

7.10 *Singular vectors of the Ising model*

a) Using the representation of T in terms of the Ising fermion,

$$T = -\frac{1}{2}(\psi\partial\psi)$$

and the rearrangement lemmas of Sect. 6.C, check the following field transcriptions of the $\psi = \phi_{(2,1)} = \phi_{(1,3)}$ singular vector equations:

$$\partial^2\phi_{(2,1)} = \frac{2}{3}(2h_{2,1} + 1)(T\phi_{(2,1)})$$

$$\partial^3\phi_{(1,3)} = (h_{1,3} + 1)[2(T\partial\phi_{(1,3)}) - h_{1,3}(\partial T\phi_{(1,3)})]$$

b) Find the level-6 vacuum singular vector and verify that the corresponding field identity is also satisfied with the above representation of T.

7.11 *Fields dual to each other*
Primary fields that satisfy the condition

$$\left[\oint dz\, \phi_{(r,s)}(z), \oint dw\, \phi_{(r',s')}(w)\right] = 0$$

are said to be dual of each other.

a) Verify that fields whose OPE contains a single family ϕ_h, that is,

$$\phi_{(r,s)}(z)\phi_{(r',s')}(w) \sim (z-w)^{-\Delta}(\phi_h(w) + a\partial\phi_h(w) + \cdots) \qquad (7.126)$$

where a is some constant and

$$\Delta \equiv h_{r,s} + h_{r',s'} - h$$

are dual to each other if $\Delta = 2$. It is crucial here to have a single pole whose residue is a total derivative (and this is the unique possibility when the residue is proportional to the lowest dimensional descendant of a primary field: $L_{-1}\phi_h$ is the unique descendant of ϕ_h at level 1).

b) Find all pairs of primary fields that satisfy Eq. (7.126) with $\Delta = 2$.
Result:

$$\{\phi_{(1,3)}, \phi_{(3,1)}\}, \ \{\phi_{(1,2)}, \phi_{(5,1)}\}, \ \{\phi_{(2,1)}, \phi_{(1,5)}\} \qquad (7.127)$$

c) We will now prove that Eq. (7.126) with $\Delta = 2$ gives all the solutions to the duality condition. Consider first the case where $\Delta = 3$. Argue that the duality requirement can be satisfied only if there exists a relation between $L_{-1}^2\phi_h$ and $L_{-2}\phi_h$, which forces ϕ_h to be either $\phi_{(1,2)}$ or $\phi_{(2,1)}$. But show that this is incompatible with the OPE (7.126) with $\Delta = 3$ and $\phi_{(r,s)}, \phi_{(r',s')}$ being both primary fields. Use a similar argument to rule out $\Delta > 3$.

d) What is the value of the constant a in Eq. (7.126)?

Notes

The representation theory of infinite-dimensional algebras is discussed extensively in the mathematical literature. We note the set of lectures by Kac and Raina [216] and by Saint-Aubin [313].

The formula for the Kac determinant was proposed by Kac [213], and proven by Feigin and Fuchs [127]. The proof is explained in more detail by Kac and Raina [216] and Itzykson and Drouffe [203]. The Kac determinant was also obtained by Thorn [334] in a more physical fashion, in the context of dual resonance models.

The conditions for unitarity of $c < 1$ models were obtained by Friedan, Qiu, and Shenker [140]. A detailed proof of these conditions is provided by the same authors in Ref. [143], where the unitarity of $c > 1, h > 0$ representations is also discussed. Langlands [250] offers a more detailed proof of the unitarity of $c > 1, h > 0$ representations and an alternate proof of the nonunitarity conditions for $c < 1$.

The Yang-Lee edge singularity was studied by Fisher [131], who correctly guessed the relevant Landau-Ginzburg theory. Its relation with the nonunitary minimal model $\mathcal{M}(5,2)$ was pointed out by Cardy [66].

The identification of the Ising model with the minimal model $\mathcal{M}(4,3)$ is due to Belavin, Polyakov, and Zamolodchikov (BPZ) [36].

The tricritical Ising model was related to the minimal model $\mathcal{M}(5, 4)$ and to the simplest model of the superconformal discrete series by Friedan, Qiu, and Shenker [141]. Superconformal models were also discussed by Bershadsky, Knizhnik, and Teitelman [45] and by Eichenherr [122].

The Q-state Potts model was solved by Temperley and Lieb [333] and first studied in the context of conformal field theory by Dotsenko [107] for $Q = 3$. Minimal models based on the W_3 algebra, of which the three-state Potts model is the simplest example, were introduced by Fateev and Zamolodchikov [123]. The three-state Potts model is also a special case of $\mathbb{Z}_N$ parafermionic theories, introduced by Zamolodchikov and Fateev [366].

The RSOS model was introduced and solved for a restricted set of parameters by Andrews, Baxter, and Forrester [14]. Critical exponents for this class of models were obtained by Huse [197] who conjectured the correspondence with unitary minimal models.

The $O(n)$ model was rephrased in Coulomb-gas terms and solved by Nienhuis [283] at criticality. The identification of the precise underlying conformal theories was realized by computing the torus partition function of the model [95, 96]. The physics of polymers $(O(n = 0))$ in solvents was investigated with conformal theory techniques by Duplantier and Saleur [115, 116].

The Landau-Ginzburg description of minimal models was suggested by Zamolodchikov [364]. Exercise 7.11 is based on Ref. [267].

CHAPTER 8

Minimal Models II

This chapter, the second devoted to minimal models, completes the somewhat heuristic point of view adopted in some parts of the previous chapter. We stress at once that the four sections below are to some extent independent. They are intended for an easy reading, the main technical difficulties being left in the large appendix.

In Sect. 8.1, we describe the structure of irreducible Verma modules, as a consequence of the Kac determinant formula (8.1). In particular, we derive the expressions for the characters of the irreducible representations of the Virasoro algebra and give a number of examples to illustrate this point. In Sect. 8.2, we turn to the study of singular vectors of the Virasoro algebra. Instead of proving the Kac determinant formula in an abstract mathematical way, in the spirit of the original proofs (see also the exercises at the end of this chapter), we shall present a more constructive approach, in which we explicitly derive expressions for the singular vectors. These expressions are particularly beautiful for the fields located at the border of the Kac table, namely, of the form $\phi_{(r,1)}$ or $\phi_{(1,s)}$, for which we present a complete construction. The general case $\phi_{(r,s)}$ is presented in the (very large) App. 8.A: after describing all the mathematical implications of the covariance of the operator product expansion of conformal fields (in particular the mechanism of fusion of two Verma modules), we construct the (r,s) singular vectors as a result of the fusion of two particular Verma modules. The proof of the Kac determinant formula is just a by-product of this latter study.

The singular vectors can be used to derive differential equations for the correlation functions of the corresponding fields. The precise mechanism is described in Sect. 8.3. In particular, we derive differential equations of the hypergeometric type for the four-point functions involving $\phi_{(2,1)}$ or $\phi_{(1,2)}$. Section 8.4 is devoted to the complete derivation of the fusion rules for minimal models, hidden in the leading behavior of the differential equations for correlators.

§8.1. Irreducible Modules and Minimal Characters

In this section, we describe the structure of inclusions of Virasoro modules in reducible highest-weight representations of the Virasoro algebra. The result is summarized in the minimal character formula (8.17). This structure is a consequence of the Kac determinant formula[1]

$$\det M^{(l)} = \alpha_l \prod_{\substack{r,s \geq 1 \\ rs \leq l}} [h - h_{r,s}(c)]^{p(l-rs)} \tag{8.1}$$

where

$$h_{r,s}(t) = \frac{1}{4}(r^2 - 1)t + \frac{1}{4}(s^2 - 1)\frac{1}{t} - \frac{1}{2}(rs - 1)$$
$$c = 13 - 6\left(t + \frac{1}{t}\right) \tag{8.2}$$

For the (p', p) minimal model, we have in addition

$$t = p/p' \tag{8.3}$$

hence

$$h_{r,s} = \frac{(pr - p's)^2 - (p - p')^2}{4pp'} \quad 1 \leq r \leq p' - 1, \ 1 \leq s \leq p - 1$$
$$c = 1 - 6\frac{(p - p')^2}{pp'} \tag{8.4}$$

The representation with highest weight h is reducible if and only if h has the form (8.4), for some nonnegative integers $r, s \geq 1$.

8.1.1. The Structure of Reducible Verma Modules for Minimal Models

We consider in detail the structure of the reducible Verma modules for the minimal models specified by Eq. (7.65). Let $V_{r,s}$ denote the Verma module $V(c(p,p'), h_{r,s}(p,p'))$ built on the highest weight $h_{r,s}$ appearing in the Kac table (8.4). According to the Kac determinant formula (8.1), the reducible Verma module with highest weight $h_{r,s}$ has its first singular vector at level $l = rs$. This is the first level at which the determinant vanishes, because of the exponent $p(l-rs)$. We deduce, from the symmetry property (7.66), that it must necessarily have another singular vector at level $(p' - r)(p - s)$. Using the identity

$$h_{r,-s} - h_{r,s} = rs \tag{8.5}$$

---

[1] Note that $p(l - rs)$ denotes here the number of integer partitions of the integer $l - rs$. It should not be confused with the product $p \times (l - rs)$.

and the periodicity property

$$h_{r+p',s+p} = h_{r,s} \tag{8.6}$$

to properly shift the indices, we find that the corresponding dimensions read respectively

$$h_{r,s} + rs = h_{p'+r,p-s} = h_{p'-r,p+s}$$
$$h_{r,s} + (p'-r)(p-s) = h_{r,2p-s} = h_{2p'-r,s} \tag{8.7}$$

The possibilities of labeling the resulting states are exhausted if we insist on having dimensions indexed by pairs of positive integers, which are minimal with respect to translations of (p',p). We may therefore write the following inclusion of submodules

$$V_{p'+r,p-s} \cup V_{r,2p-s} \subset V_{r,s} \tag{8.8}$$

To build an irreducible representation (the irreducible Virasoro module $M_{r,s}$), we have to factor out $V_{r,s}$ by the *direct sum* of these two submodules

$$M_{r,s} = V_{r,s}/[V_{p'+r,p-s} \oplus V_{r,2p-s}] \tag{8.9}$$

Unfortunately, the direct sum $V_{p'+r,p-s} \oplus V_{r,2p-s}$ is a complicated object, as these two Verma modules in turn share *two submodules*. This is readily seen by applying the reducibility condition to each of the two corresponding submodules $V_{p'+r,p-s}$ and $V_{r,2p-s}$. We find two submodules in $V_{p'+r,p-s} \equiv V_{p'-r,p+s}$ at levels $(p'+r)(p-s)$ and $(p'-r)(p+s)$, namely

$$V_{2p'+r,s} \cup V_{r,2p+s} \subset V_{p'+r,p-s} \tag{8.10}$$

Similarly, we find two submodules in $V_{r,2p-s} \equiv V_{2p'-r,s}$ at levels $r(2p-s)$ and $(2p'-r)s$, namely

$$V_{p'-r,3p-s} \cup V_{3p'-r,p-s} \subset V_{r,2p-s} \tag{8.11}$$

Note that the submodules in (8.10) and (8.11) coincide by the symmetry property (7.66): $V_{2p'+r,s} \equiv V_{p'-r,3p-s}$ and $V_{r,2p+s} \equiv V_{3p'-r,p-s}$. Hence the direct sum of the two modules is a quotient

$$V_{p'+r,p-s} \oplus V_{r,2p-s} = V_{p'+r,p-s} \cup V_{r,2p-s}/[V_{2p'+r,s} \oplus V_{r,2p+s}] \tag{8.12}$$

Iterating this, we find the infinite ladder of inclusions of modules, depicted in Fig. 8.1. At each step, the two Verma modules have two common maximal submodules, given by the Kac determinant formula, whose intersection contains in turn two maximal submodules, and so on. The irreducible representation $M_{r,s}$ (8.9) is therefore obtained as the following succession of subtractions and additions of modules

$$M_{r,s} = V_{r,s} - (V_{p'+r,p-s} \cup V_{r,2p-s}) + (V_{2p'+r,s} \cup V_{r,2p+s}) - \cdots \tag{8.13}$$

We note that the first subtraction of $V_{p'+r,p-s} \cup V_{r,2p-s}$ is too large, because we have to subtract the two maximal submodules $V_{2p'+r,s}$ and $V_{r,2p+s}$ from the intersection $V_{p'+r,p-s} \cap V_{r,2p-s}$, and this phenomenon propagates along the ladder of Fig. 8.1.

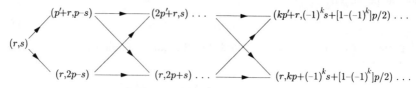

Figure 8.1. The infinite structure of submodules of the Verma module $V_{r,s}(p,p')$. Each module $V_{a,b}$ is represented by a pair of Kac indices (a,b). Each arrow represents an inclusion: $A \to B$ means $B \subset A$, and arrows are transitive. Each module contains two maximal submodules.

8.1.2. Characters

A simple way of summarizing these repeated subtractions is to write the character of the irreducible representation $M_{r,s}$. Each Verma module with highest weight h and central charge c contributes according to the Virasoro character (7.16):

$$\chi_{(c,h)}(q) = \frac{q^{h-c/24}}{\varphi(q)} \tag{8.14}$$

Taking into account all the subtractions of states implied in Eq. (8.13), we find the irreducible character

$$\chi_{(r,s)}(q) = \frac{q^{-c/24}}{\varphi(q)} \left[q^{h_{r,s}} + \sum_{k=1}^{\infty} (-1)^k \left\{ q^{h_{r+kp',(-1)^k s+[1-(-1)^k]p/2}} \right. \right.$$
$$\left. \left. + q^{h_{r,kp+(-1)^k s+[1-(-1)^k]p/2}} \right\} \right] \tag{8.15}$$

where the three terms in the bracket correspond respectively to the Verma module $V_{r,s}$ of the left of the ladder of Fig. 8.1, and the contribution of the modules of the top (resp. bottom) of the ladder weighted by a sign $(-1)^k$ enforcing the successive additions-subtractions along the ladder. The irreducible character (8.15) can be reexpressed in terms of the functions

$$K_{r,s}^{(p,p')}(q) = \frac{q^{-1/24}}{\varphi(q)} \sum_{n \in \mathbb{Z}} q^{(2pp'n + pr - p's)^2/4pp'} \tag{8.16}$$

as

$$\chi_{(r,s)}(q) = K_{r,s}^{(p,p')}(q) - K_{r,-s}^{(p,p')}(q) \tag{8.17}$$

The small q expansions of the characters for the minimal models discussed in Sect. 7.4 are displayed in Table 8.1, up to order 6. The Kac indices (r,s) for the representations have been chosen in such a way that the product rs is minimal. Indeed, comparing these expansions with that of $1/\varphi(q)$

$$\frac{1}{\prod_{n \geq 1}(1 - q^n)} = 1 + q + 2q^2 + 3q^3 + 5q^4 + 7q^5 + 11q^6 + \cdots \tag{8.18}$$

Table 8.1. Expansion of a few minimal characters up to order 6.

(p,p')	$h_{r,s}$	$q^{-h_{rs}+c/24}\chi_{r,s}(q)$
$(5,2)$	$h_{1,1}=0$	$1+q^2+q^3+q^4+q^5+2q^6+\cdots$
Yang-Lee	$h_{1,2}=-2/5$	$1+q+q^2+q^3+2q^4+2q^5+3q^6+\cdots$
$(4,3)$	$h_{1,1}=0$	$1+q^2+q^3+2q^4+2q^5+3q^6\cdots$
Ising	$h_{2,1}=1/16$	$1+q+q^2+q^3+2q^4+2q^5+3q^6+\cdots$
	$h_{1,2}=1/2$	$1+q+q^2+2q^3+2q^4+3q^5+4q^6+\cdots$
$(5,4)$	$h_{1,1}=0$	$1+q^2+q^3+2q^4+2q^5+4q^6+\cdots$
Tricrit.	$h_{2,1}=7/16$	$1+q+q^2+2q^3+3q^4+4q^5+6q^6+\cdots$
Ising	$h_{1,2}=1/10$	$1+q+q^2+2q^3+3q^4+4q^5+6q^6+\cdots$
	$h_{1,3}=3/5$	$1+q+2q^2+2q^3+4q^4+5q^5+7q^6+\cdots$
	$h_{2,2}=3/80$	$1+q+2q^2+3q^3+4q^4+6q^5+8q^6+\cdots$
	$h_{3,1}=3/2$	$1+q+q^2+2q^3+3q^4+4q^5+6q^6+\cdots$
$(6,5)$	$h_{1,1}=0$	$1+q^2+q^3+2q^4+2q^5+4q^6+\cdots$
3-state	$h_{2,1}=2/5$	$1+q+q^2+2q^3+3q^4+4q^5+6q^6+\cdots$
Potts	$h_{3,1}=7/5$	$1+q+2q^2+2q^3+4q^4+5q^5+8q^6+\cdots$
	$h_{1,3}=2/3$	$1+q+2q^2+2q^3+4q^4+5q^5+8q^6+\cdots$
	$h_{4,1}=3$	$1+q+2q^2+3q^3+4q^4+5q^5+8q^6+\cdots$
	$h_{2,3}=1/15$	$1+q+2q^2+3q^3+5q^4+7q^5+10q^6+\cdots$

it is easy to verify that the first singular vector in each representation occurs at level rs, whereas the second one occurs at level $(p-r)(p'-s)$.

§8.2. Explicit Form of Singular Vectors

In this section, we give an explicit construction of the singular vector at level r in the Verma module $V_{r,1}$, with h and c as in Eq. (8.2). The result appears in Eq. (8.26) below, in the form of the determinant $\Delta_{r,1}$ of a matrix operator expressed in a representation of $su(2)$ of spin $(r-1)/2$. More precisely, the singular vector $|h_{r,1}+r\rangle$ is obtained by acting on the highest-weight state $|h_{r,1}\rangle$ with this operator $\Delta_{r,1}$. Such a representation is easily obtained in the *classical limit* $c\to-\infty$ of the Virasoro algebra,[2] where the limits of $(r,1)$ singular vectors are associated with

[2] This limit may be taken by sending $t\to0$ in Eq. (8.2). Note that the dimensions $h_{r,s}(t)$ diverge in this limit, unless $s=1$.

covariant differential operators. In this limit, the highest-weight property translates into the fact that the state behaves as a true differential form, with weight

$$\lim_{c \to -\infty} h_{r,1} = \frac{1-r}{2}$$

The limit of the singular vector expression simply means that a differential form of weight $(1 + r)/2$ (the classical limit of $|h_{r,1} + r\rangle$) is obtained by acting on a differential form of weight $(1 - r)/2$ (the classical limit of $|h_{r,1}\rangle$) with a covariant differential operator of degree r (the classical limit of $\Delta_{r,1}$). The expression of covariant differential operators in the form of a determinant, using $su(2)$ representation matrices, is natural in this classical context, since a covariant differential operator of order r is naturally expressed as the determinant of an $r \times r$ matrix differential operator of first order (see Ex. 8.8 for details).

The idea here is to write the "quantum" singular vector as the result of the action of the formal determinant of a matrix operator, with entries linear in the L_n's, on the highest-weight state of V. This matrix itself lives in a spin $(r - 1)/2$ representation of the Lie algebra $su(2)$. The latter is the r-dimensional irreducible matrix representation of $su(2)$:

$$[J_0]_{i,j} = \frac{1}{2}(r - 2i + 1)\delta_{i,j}$$

$$[J_-]_{i,j} = \begin{cases} \delta_{i,j+1} & (j = 1, 2, .., r - 1) \\ 0 & (j = r) \end{cases}$$

$$[J_+]_{i,j} = \begin{cases} i(r - i)\delta_{i+1,j} & (i = 1, 2, \ldots, r - 1) \\ 0 & (i = r) \end{cases}$$

(8.19)

which satisfy

$$[J_+, J_-] = 2J_0$$
$$[J_0, J_\pm] = \pm J_\pm.$$

(8.20)

For instance, for $r = 4$, these matrices read

$$J_0 = \begin{pmatrix} \frac{3}{2} & 0 & 0 & 0 \\ 0 & \frac{1}{2} & 0 & 0 \\ 0 & 0 & -\frac{1}{2} & 0 \\ 0 & 0 & 0 & -\frac{3}{2} \end{pmatrix}$$

$$J_- = \begin{pmatrix} 0 & 0 & 0 & 0 \\ 1 & 0 & 0 & 0 \\ 0 & 1 & 0 & 0 \\ 0 & 0 & 1 & 0 \end{pmatrix}$$

(8.21)

$$J_+ = \begin{pmatrix} 0 & 3 & 0 & 0 \\ 0 & 0 & 4 & 0 \\ 0 & 0 & 0 & 3 \\ 0 & 0 & 0 & 0 \end{pmatrix}$$

This representation differs from the one generally used in quantum mechanics in that the ladder operators $J_\pm$ are not Hermitian conjugates of each other. Since J_+ is strictly upper triangular and J_- strictly lower triangular, we have $(J_+)^r = (J_-)^r = 0$. We consider the $r \times r$ matrix operator:

$$D_{r,1}(t) = -J_- + \sum_{m=0}^{\infty} (-tJ_+)^m L_{-m-1}. \tag{8.22}$$

whose entries are polynomials in the negative Virasoro modes $L_{-1}, L_{-2}, \ldots$ Only a finite number of terms contributes to the sum, since $J_+^r = 0$. The operator $D_{r,1}$ acts on r-vectors of states of the form $(f_1, f_2, \ldots, f_r)^T$. The formal determinant of this operator, $\Delta_{r,1}(t)$, is defined as follows. The triangular system of linear equations

$$D_{r,1}(t) \begin{pmatrix} f_1 \\ f_2 \\ \vdots \\ f_r \end{pmatrix} = \begin{pmatrix} f_0 \\ 0 \\ \vdots \\ 0 \end{pmatrix} \tag{8.23}$$

can be inverted, and $f_0, f_1, \ldots, f_{r-1}$ become explicit functions of f_r. For instance, we have

$$\begin{aligned}
f_{r-1} &= L_{-1} f_r \\
f_{r-2} &= [L_{-1}^2 - (r-1)tL_{-2}] f_r \\
f_{r-3} &= [L_{-1}^3 - t(r-1)L_{-1}L_{-2} - 2t(r-2)L_{-2}L_{-1} \\
&\quad + 2t^2(r-1)(r-2)L_{-3}] f_r
\end{aligned} \tag{8.24}$$

and so on. The formal determinant operator applies f_r to f_0

$$f_0 = \Delta_{r,1}(t) f_r, \tag{8.25}$$

and by a slight abuse of notation we denote it by

$$\Delta_{r,1}(t) = \det\left[-J_- + \sum_{m=0}^{\infty} (-tJ_+)^m L_{-m-1} \right] \tag{8.26}$$

With this definition, we find

$$\begin{aligned}
\Delta_{1,1}(t) &= L_{-1} \\
\Delta_{2,1}(t) &= L_{-1}^2 - tL_{-2} \\
\Delta_{3,1}(t) &= L_{-1}^3 - 2t(L_{-1}L_{-2} + L_{-2}L_{-1}) + 4t^2 L_{-3}.
\end{aligned} \tag{8.27}$$

The state

$$|\chi_r\rangle = \Delta_{r,1}(t) |h_{r,1}(t)\rangle \tag{8.28}$$

will now be proved to be a singular vector of the Verma module $V_{r,1}$ at level r. To verify this, we need to prove that $|\chi_r\rangle$ satisfies the highest-weight condition

$$L_n|\chi_r\rangle = \delta_{n,0}(h_{r,1} + r)|\chi_r\rangle \qquad\qquad n \geq 0 \tag{8.29}$$

As mentioned earlier, it is actually sufficient to prove it for $n = 0, 1, 2$, since the condition $L_n|\chi_r\rangle = 0$ $(n > 2)$ can be obtained from $L_1|\chi_r\rangle = L_2|\chi_r\rangle = 0$ by commutation of L_1 and L_2. In other words, for $n > 2$,

$$L_n = -\frac{1}{n-2}[L_1, L_{n-1}] = \frac{(-1)^{n-2}}{(n-2)!}\text{ad}(L_1)^{n-2}L_2, \tag{8.30}$$

where the adjoint action of an operator x is defined as

$$\text{ad}(x)y = [x, y] \tag{8.31}$$

We thus proceed in three steps:

(i) $L_0|\chi_r\rangle = (h_{r,1} + r)|\chi_r\rangle$. Using the definition of the formal determinant (8.25), we start from the state $f_r = |h_{r,1}\rangle$, and build f_i, $i = r - 1, r - 2, \ldots, 0, f_0 = |\chi_r\rangle$. It is clear by construction that $L_0 f_j = (h_{r,1} + r - j)f_j$, hence the property follows for $f_0 = |\chi_r\rangle$.

(ii) $L_1|\chi_r\rangle = 0$. The operator L_1 acts on the components f_j as

$$L_1 f_j = \frac{j(r-j)}{2}[(2j + 3 - r)t - 2]f_{j+1}, \tag{8.32}$$

for $j = 0, 1, \ldots, r - 1$. Upon extending the linear space by one dimension, and introducing an extra component f_{r+1}, this holds also for $j = r$. This enables us to set $f_{r+1} \equiv 0$, in which case we simply get the highest-weight condition $L_1 f_r = L_1|h_{r,1}(t)\rangle = 0$. Eq. (8.32) is easily proven by recursion on j. We get the desired result for $j = 0$.

(iii) $L_2|\chi_r\rangle = 0$. As in the previous case, the operator L_2 acts recursively on the components f_j:

$$L_2 f_j = \frac{t}{4}j(r-j)(j+1)(r-j-1)[(4j + 6 - r)t - 7]f_{j+2}, \tag{8.33}$$

for $j = 0, 1, 2, \ldots, r-1$. We extend this to $j = r$ by introducing an extra coordinate f_{r+2}, but we can set $f_{r+2} \equiv 0$, which translates into the second highest-weight condition $L_2 f_r = L_2|h_{r,1}(t)\rangle = 0$. We finally get the desired result for $j = 0$. This completes the proof of Eq. (8.28).

A few comments are in order:

(a) An analogous result holds for the Verma module $V_{1,s}$, if we change simultaneously $r \leftrightarrow s$ and $t \leftrightarrow 1/t$, under which $c(t)$ remains unchanged.

(b) If we perform explicitly the elimination between the components f_j, we get a closed expression for the singular vector:

$$|\chi_r\rangle = \sum_{\substack{p_i \geq 1 \\ p_1 + \cdots + p_k = r}} \frac{[(r-1)!]^2(-t)^{r-k}}{\prod_{i=1}^{k-1}(p_1 + \cdots + p_i)(r - p_1 - \cdots - p_i)}L_{-p_1} \cdots L_{-p_k}|h_{r,1}(t)\rangle \tag{8.34}$$

(c) It is easy to derive the action of L_n for $n > 2$ on the vector $\mathbf{f} = (f_1, \ldots, f_r)^T$ using Eqs. (8.30), (8.32), and (8.33). We find

$$L_n \mathbf{f} = [(J_0 - \frac{3n+1}{2}) + \frac{3n+1}{4t}](-tJ_+)^n \mathbf{f}. \tag{8.35}$$

The expressions for the singular vectors of $V_{r,1}$ are thus relatively simple. They correspond to states located on the boundary of the Kac table. No simple closed expressions are known for states located inside the Kac table. However, we shall develop in App. 8.A an elementary scheme to write them in all generality.

§8.3. Differential Equations for the Correlation Functions

Section 8.2 was dedicated to the study and construction of singular vectors of Verma modules, which carry reducible representations of the Virasoro algebra. As already stated above, the primary fields of a minimal conformal theory are attached to highest-weight vectors of such representations of the Virasoro algebra. But we actually require that the corresponding representation of the Virasoro algebra be *irreducible*. All singular vectors must therefore be set to zero. This results in a highly nontrivial set of constraints. Their consequence on the OPE of conformal fields is described in App. 8.A, Sect. 8.A.1. The subject of this section is to analyze their effect on the correlators of the primary fields, a point briefly addressed in Sect. 7.3.

Before plunging into this analysis, we emphasize a subtlety concerning the field-state equivalence in a conformal theory. Recall that the primary fields $\phi(z, \bar{z})$ are in one-to-one correspondence with *products* of representations of the holomorphic and antiholomorphic (or left and right) Virasoro algebras. In a minimal theory, the primary fields will therefore correspond to a *pair* of Verma modules pertaining respectively to the left and right Virasoro algebras. Minimality requires, as explained before, that both modules have central charge and highest weights of the form (7.65), with r, s in the Kac table: $1 \leq r \leq p', 1 \leq s \leq p$. Moreover, in order to make both representations irreducible, we have to set the singular vectors of both modules to zero and we finally get a decomposition of the Hilbert space of the theory, as in Eq. (7.72). The two sets of constraints obtained this way are factorized, in the sense that they act only on the left (resp. right) part of the primary fields. Actually, as we shall see below, one can solve independently the left and right constraints for any correlator $\langle \phi_0(z_0, \bar{z}_0) \phi_1(z_1, \bar{z}_1) \cdots \rangle$, in the form of several possible left and right conformal blocks $\mathcal{F}(z_0, z_1, \cdots)$ and $\bar{\mathcal{F}}(\bar{z}_0, \bar{z}_1, \cdots)$, so that the full correlator is a sum of products of left×right blocks of this form. The particular association of left×right blocks involves further complications, which will be studied in great detail in Chap. 9. In the following, we mainly concentrate on the consequences of, say, the left singular vector vanishing conditions on the left conformal blocks.

8.3.1. From Singular Vectors to Differential Equations

The basic ingredients in the computation of correlation functions in a field theory are the Ward identities. They summarize the behavior of any correlator under infinitesimal reparametrizations (cf. Sect. 5.2). The Ward identity (5.41) was used in Sect. 6.6.1 to express the correlator of a descendant field in terms of the correlator of the primary, acted on by a string of differential operators. These are Eqs. (6.152) and (6.156), which we repeat here:

$$\langle \phi_0^{(-r_1,...,-r_k)}(z_0)\phi_1(z_1)\cdots\rangle = \mathcal{L}_{-r_1}(z_0)\cdots\mathcal{L}_{-r_k}(z_0)\langle\phi_0(z_0)\phi_1(z_1)\cdots\rangle \quad (8.36)$$

$$\mathcal{L}_{-r}(z) = \sum_{i\geq 1}\left\{\frac{(r-1)h_i}{(z_i-z)^r} - \frac{1}{(z_i-z)^{r-1}}\partial_{z_i}\right\} \quad (8.37)$$

We drop the explicit dependence of the fields on the antiholomorphic variable $\bar{z}$, as the equations involve only the holomorphic dependence.

We suppose that the left Virasoro representation of ϕ_0 is a reducible Verma module $V(c,h_0)$, with a singular vector at level n_0 given by

$$|c,h_0+n_0\rangle = \sum_{Y,|Y|=n_0}\alpha_Y L_{-Y}|c,h_0\rangle$$

where Y stands for[3]

$$
\begin{aligned}
Y &= \{r_1,\ldots,r_k\} \quad \text{with} \quad 1 \leq r_1 \leq r_2 \leq \cdots \leq r_k \\
|Y| &= r_1 + \cdots + r_k \\
L_{-Y} &\equiv L_{-r_1}L_{-r_2}\cdots L_{-r_k}
\end{aligned} \quad (8.38)
$$

Setting to zero this singular vector, we get $\sum\alpha_Y L_{-Y}\phi_0 = 0$, which, inserted into a correlator, leads to

$$\sum_Y \alpha_Y \mathcal{L}_{-Y}(z_0)\langle\phi_0(z_0)\phi_1(z_1)\cdots\rangle = 0 \quad (8.39)$$

We used the Ward identity (8.36) to rewrite the singular vector vanishing condition as a differential equation, with $\mathcal{L}_{-Y} \equiv \mathcal{L}_{-r_1}\cdots\mathcal{L}_{-r_k}$. Let $\Delta_0 = \sum_Y \alpha_Y L_{-Y}$ denote the operator that creates the singular vector in $V(c,h_0)$. The differential equation (8.39) is obtained by acting on the correlator $\langle\phi_0(z_0)\phi_1(z_1)\cdots\rangle$ with the differential operator[4]

$$\gamma_0(z_i,\partial_{z_i}) = \Delta_0\big(L_{-r} \to \mathcal{L}_{-r}(z_0)\big) \quad (8.40)$$

The differential equation (8.39) can be further simplified by using the global conformal invariance of the correlator (see Sect. 5.2.2). The $SL(2,\mathbb{C})$ invariance

---

[3] The symbol Y refers actually to Young tableaux. In this language, r_j denotes the number of boxes in the j line of the tableau, counted from the bottom to the top.

[4] This should be compared with the substitution (8.178) used in App. 8.A, in the discussion of the operator product coefficients: Eq. (8.40) actually coincides with (8.178) in the case of two-point functions, and expresses the *transfer* of the action of Δ_0 from the point z_0 to the other points z_i.

of the correlator can be recast into the three differential equations (5.51), which we reproduce here:

$$\sum_{i=0,1,..} \partial_{z_i} \langle \phi_0(z_0)\phi_1(z_1) \cdots \rangle = 0$$

$$\sum_{i=0,1,..} (z_i\partial_{z_i} + h_i) \langle \phi_0(z_0)\phi_1(z_1) \cdots \rangle = 0 \qquad (8.41)$$

$$\sum_{i=0,1,..} (z_i^2\partial_{z_i} + 2z_ih_i) \langle \phi_0(z_0)\phi_1(z_1) \cdots \rangle = 0$$

They are easily solved as

$$\langle \phi_0(z_0)\phi_1(z_1) \cdots \rangle = \left\{ \prod_{i<j}(z_i - z_j)^{\mu_{ij}} \right\} G(\{z_{ij}^{kl}\}) \qquad (8.42)$$

where μ_{ij} is any solution of

$$\sum_{j\neq i} \mu_{ij} = 2h_i \qquad (8.43)$$

and G is an arbitrary function of the anharmonic ratios

$$z_{ij}^{kl} = \frac{(z_i - z_j)(z_k - z_l)}{(z_i - z_l)(z_k - z_j)} \qquad (8.44)$$

Another way of writing the solution (8.42) is to fix the $SL(2, \mathbb{C})$ gauge, by sending three points of the correlator to fixed values, for instance $z_1 \rightarrow 0, z_2 \rightarrow 1, z_3 \rightarrow \infty$.

We now illustrate Eq. (8.39) in a few cases. For $V(c, h_0) = V_{2,1}$, degenerate at level 2, we have

$$\Delta_0 \equiv \Delta_{2,1}(t) = L_{-1}^2 - tL_{-2} \qquad (8.45)$$

Hence

$$\left\{ \partial_z^2 - t \sum_{i=1,2,..} \left[\frac{h_i}{(z_i - z)^2} - \frac{1}{z_i - z}\partial_{z_i} \right] \right\} \langle \phi_{(2,1)}(z)\phi_1(z_1)\phi_2(z_2) \cdots \rangle = 0 \quad (8.46)$$

This is a second-order partial differential equation, obtained previously in Eq. (7.47). It admits two linearly independent solutions. Singular vector vanishing conditions for the other fields should be implemented as well, further constraining the correlator.

For $V(c, h_0) = V_{r,1}$ (the label 0 is then replaced by the label $(r, 1)$ in γ and Δ), which is degenerate at level r, we have the explicit differential operator (see Sect. 8.2)

$$\gamma_{r,1}(z_i, \partial_{z_i}) =$$

$$\det\left[-J_- + \partial_{z_0} + \sum_{m\geq 1}(-tJ_+)^m \sum_{i\geq 1}\left(\frac{mh_i}{(z_i - z_0)^{m+1}} - \frac{1}{(z_i - z_0)^m}\partial_{z_i} \right) \right]$$

$$\equiv \det[D_{r,1}(z_i, \partial_{z_i})]$$

$$(8.47)$$

expressed as a formal determinant in the manner of Eq. (8.26). It leads to the partial differential equation of order r

$$\gamma_{r,1}(z_i, \partial_{z_i}) \langle \phi_{(r,1)}(z_0)\phi_1(z_1)\phi_2(z_2)\cdots \rangle = 0 \tag{8.48}$$

Using the definition of the formal determinant (8.25)–(8.26), we can translate Eq. (8.48) into a matrix differential system

$$D_{r,1}(z_i, \partial_{z_i})\mathbf{f} = 0 \tag{8.49}$$

for the r-vector $\mathbf{f} = (f_1, f_2, \ldots, f_r)^t$, whose last component is the desired correlator

$$f_r = \langle \phi_0(z_0)\phi_1(z_1)\cdots \rangle$$

Each component f_p is a correlator involving a level $r-p$ descendant of ϕ_0, expressed as the action of a differential operator of order $r - p$ on the correlator f_r.

The differential equation (8.39) is somewhat involved in general. However, in the cases of two-, three- and four-point functions, it can be transformed, using global conformal invariance, into an ordinary differential equation in the variable $z \equiv z_0$.

8.3.2. Differential Equations for Two-Point Functions in Minimal Models

As already noted before (cf. Sect. 4.3.1), the global conformal invariance (8.41) almost fixes the two- and three-point correlators. Actually they are fixed up to some multiplicative constant, which might be zero. The aim of the present section is to exploit the differential equation satisfied by a two-point function of primary fields to get a useful sum rule (Eq. (8.55)) on the coefficients of the corresponding singular vector.

The basic requirement for two-point functions is orthonormality:[5]

$$\langle \phi_{h_0}(z)\phi_{h_1}(0)\rangle = \delta_{h_0, h_1} z^{-2h_0} \tag{8.50}$$

It is instructive to check that this expression is compatible with the differential equation (8.39):

$$\Delta_0 \left(L_{-r} \to \frac{(r-1)h_0}{(w-z)^r} - \frac{1}{(w-z)^{r-1}} \partial_w \right) \langle \phi_0(z)\phi_0(w)\rangle = 0 \tag{8.51}$$

By translational invariance (the first of the three conditions (8.41)), the two-point function is a function of $x = z - w$, subject to

$$\Delta_0 \left(L_{-r} \to \frac{(-1)^r}{x^r} [(r-1)h_0 - x\partial_x] \right) \langle \phi_0(x)\phi_0(0)\rangle = 0 \tag{8.52}$$

---

[5] Here and in the following, we omit the antiholomorphic dependence of the fields. This is harmless, as the differential equations we write are essentially holomorphic.

For any nondecreasing sequence of integers Y, the action of $\mathcal{L}_{-Y}(x)$ on the correlator

$$\langle \phi_0(x)\phi_0(0) \rangle = x^{-2h_0}\bar{x}^{-2\bar{h}_0} \tag{8.53}$$

reads

$$\frac{(-1)^{|Y|}}{x^{|Y|}} \prod_{i=1}^{k}[(r_i - 1)h_0 - x\partial_x]\langle \phi_0(x)\phi_0(0) \rangle$$

$$= \frac{(-1)^{|Y|}}{x^{|Y|}} \prod_{i=1}^{k}[(r_i + 1)h_0 + r_{i+1} + .. + r_k]\langle \phi_0(x)\phi_0(0) \rangle \tag{8.54}$$

Eq. (8.51) is satisfied if and only if the following sum rule for the coefficients of $\Delta_0 = \sum_Y \alpha_Y L_{-Y}$ holds

$$\sum_{\substack{1 \leq r_1 \leq .. \leq r_k \\ \Sigma r_i = n_0}} \alpha_{r_1, r_2, .., r_k} \prod_{i=1}^{k}[(r_i + 1)h_0 + r_{i+1} + .. + r_k] = 0 \tag{8.55}$$

This is indeed the consequence of the following necessary condition for the singular vector

$$L_1^{n_0} \sum_{Y, |Y| = n_0} \alpha_Y L_{-Y}|c, h_0\rangle = 0 \tag{8.56}$$

It is clear that for any sequence $r_1, .., r_k$ (not necessarily ordered) of nonnegative integers with $r_1 + .. + r_k = n_0$, we have

$$L_1^{n_0} L_{-r_1} L_{-r_2} \cdots L_{-r_k}|c, h_0\rangle = P(r_1, \ldots, r_k; h_0)|c, h_0\rangle \tag{8.57}$$

where P is some polynomial of h_0 and the r's. This is due to the highest-weight condition on $|c, h_0\rangle$, ensuring that the result, at level 0, is proportional to $|c, h_0\rangle$. Writing $L_1^{n_0} = L_1^{n_0-1}L_1$, we find a recursion relation for the polynomials P

$$P(r_1, \ldots, r_k; h_0) = \sum_{i=1}^{k}(r_i + 1)P(r_1, .., r_{i-1}, r_i - 1, r_{i+1}, .., r_k; h_0) \tag{8.58}$$

By the definition (8.57), P satisfies

$$P(r_1, .., r_{i-1}, 0, r_{i+1}, .., r_k; h_0) =$$
$$[h_0 + r_{i+1} + .. + r_k]P(r_1, .., r_{i-1}, r_{i+1}, .., r_k; h_0) \tag{8.59}$$

when r_i vanishes. Together with the obvious result for $k = 1, n_0 = r_1 \equiv r$

$$P(r; h_0) = (r + 1)!h_0 \tag{8.60}$$

this determines the P's completely:

$$P(r_1, \ldots, r_k; h_0) = (n_0!) \prod_{i=1}^{k}[(r_i + 1)h_0 + r_{i+1} + .. + r_k] \tag{8.61}$$

With this value of P, the necessary condition (8.56) yields the sum rule (8.55), up to the constant multiplicative factor $(n_0!)$. Note that the fulfillment of the condition (8.39) leads to more sum rules on the coefficients α_Y (see also Ex. 8.22 for a very similar sum rule for OPE coefficients).

Once the normalization of two-point functions is fixed, all the correlators in the theory have fixed normalizations. In particular, the three-point functions are fixed, by SL(2,C) invariance, to be of the form (7.49) ($\times$ its antiholomorphic counterpart). Global conformal invariance does not fix the structure constants $g(h_0, h_1, h_2)$. We have seen in Sect. 7.3 how the differential equations impose constraints on these structure constants. The precise study of these constants, following the lines of App. 8.A, although straightforward in principle, turns out to be tedious. A simpler route consists first in the derivation of the four-point correlators, and then reading off the structure constants at coinciding points.

8.3.3. Differential Equations for Four-Point Functions in Minimal Models

In this section we find the differential equation for the four-point functions of minimal models involving $\phi_{(2,1)}$. It takes the form of the hypergeometric equation (8.71).

The four-point functions have a more complicated structure than the two- and three-point correlators, since global conformal invariance leaves some function of the cross ratio of the points undetermined. More precisely, global conformal invariance forces the correlator to take the form (8.42)

$$\langle \phi_0(z_0)\phi_1(z_1)\phi_2(z_2)\phi_3(z_3)\rangle = \prod_{0 \le i < j \le 3} (z_i - z_j)^{\mu_{ij}} \, G(z) \qquad (8.62)$$

with

$$\mu_{ij} = \frac{1}{3}\left(\sum_{k=1}^{4} h_k\right) - h_i - h_j, \qquad (8.63)$$

We still have to determine the function G of the cross-ratio

$$z = \frac{(z_0 - z_1)(z_2 - z_3)}{(z_0 - z_3)(z_2 - z_1)} \qquad (8.64)$$

With the change of function (8.62), the differential equation (8.39) translates into an ordinary differential equation of order n_0 for $G(z)$.

We illustrate this mechanism in the case $V(c, h_0) = V_{2,1}$, degenerate at level 2. We substitute Eq. (8.62) into Eq. (8.46), and use the action of the derivatives ∂_{z_i} on Eq. (8.62):

$$\partial_{z_0} = \frac{\mu_{01}}{z_0 - z_1} + \frac{\mu_{02}}{z_0 - z_2} + \frac{\mu_{03}}{z_0 - z_3} + \partial_{z_0}(z)\partial_z$$

$$\partial_{z_1} = -\frac{\mu_{01}}{z_0 - z_1} + \frac{\mu_{12}}{z_1 - z_2} + \frac{\mu_{13}}{z_1 - z_3} + \partial_{z_1}(z)\partial_z \qquad (8.65)$$

$$\partial_{z_2} = -\frac{\mu_{02}}{z_0 - z_2} - \frac{\mu_{12}}{z_1 - z_2} + \frac{\mu_{23}}{z_2 - z_3} + \partial_{z_2}(z)\partial_z$$

$$\partial_{z_3} = -\frac{\mu_{03}}{z_0 - z_3} - \frac{\mu_{13}}{z_1 - z_3} - \frac{\mu_{23}}{z_2 - z_3} + \partial_{z_3}(z)\partial_z$$

with

$$\partial_{z_0}(z) = \frac{(z_3 - z_1)(z_2 - z_3)}{(z_2 - z_1)(z_0 - z_3)^2}$$

$$\partial_{z_1}(z) = \frac{(z_0 - z_2)(z_2 - z_3)}{(z_0 - z_3)(z_2 - z_1)^2}$$

$$\partial_{z_2}(z) = -\frac{(z_1 - z_3)(z_0 - z_1)}{(z_0 - z_3)(z_2 - z_1)^2} \tag{8.66}$$

$$\partial_{z_3}(z) = \frac{(z_2 - z_0)(z_0 - z_1)}{(z_2 - z_1)(z_0 - z_3)^2}$$

We also need to rewrite the action of $\partial_{z_0}^2$ in terms of z

$$\partial_{z_0}^2 = \frac{\mu_{01}(\mu_{01} - 1)}{(z_0 - z_1)^2} + \frac{\mu_{02}(\mu_{02} - 1)}{(z_0 - z_2)^2} + \frac{\mu_{03}(\mu_{03} - 1)}{(z_0 - z_3)^2}$$

$$+ 2\left[\frac{\mu_{01}}{z_0 - z_1} + \frac{\mu_{02}}{z_0 - z_2} + \frac{\mu_{03}}{z_0 - z_3}\right]\partial_{z_0}(z)\partial_z \tag{8.67}$$

$$+ \partial_{z_0}^2(z)\partial_z + \left(\partial_{z_0}(z)\right)^2\partial_z^2$$

Upon all these substitutions, the prefactor of $G(z)$ in (8.62) can be factored out of the differential equation. Once this is done, we can take the limits $z_1 \to 0, z_2 \to 1$, and $z_3 \to \infty$, hence $z_0 \to z$, and we are left with an ordinary differential equation for $G(z)$. The latter is then obtained from Eq. (8.46) through the substitutions (8.65), (8.66), and (8.67), which, in the above limits, read

$$\partial_{z_0} = \frac{\mu_{01}}{z} + \frac{\mu_{02}}{z - 1} + \partial_z$$

$$\partial_{z_1} = -\frac{\mu_{01}}{z} - \mu_{12} + (z - 1)\partial_z$$

$$\partial_{z_2} = -\frac{\mu_{02}}{z - 1} + \mu_{12} + z\partial_z$$

$$\partial_{z_3} = 0 \tag{8.68}$$

$$\partial_{z_0}^2 = \frac{\mu_{01}(\mu_{01} - 1)}{z^2} + \frac{\mu_{02}(\mu_{02} - 1)}{(z - 1)^2}$$

$$+ 2\left[\frac{\mu_{01}}{z} + \frac{\mu_{02}}{z - 1}\right]\partial_z + \partial_z^2$$

This leads finally to the equation

$$\left\{\frac{1}{t}\partial_z^2 + \left[2\frac{\mu_{01}}{tz} + 2\frac{\mu_{02}}{t(z - 1)} + \frac{2z - 1}{z(z - 1)}\right]\partial_z + \frac{\mu_{01}(\mu_{01} - 1)}{tz^2}\right.$$

$$\left. + \frac{\mu_{02}(\mu_{02} - 1)}{t(z - 1)^2} + \frac{\mu_{01} - h_1}{z^2} + \frac{\mu_{02} - h_2}{(z - 1)^2} - \frac{\mu_{12}}{z(z - 1)}\right\}G(z) = 0$$

$$\tag{8.69}$$

This equation simplifies a great deal when expressed in terms of the function

$$H(z) = z^{\mu_{01}}(1-z)^{\mu_{02}} G(z) \tag{8.70}$$

for which we have

$$\left\{ \frac{1}{t}\partial_z^2 + \frac{2z-1}{z(z-1)}\partial_z - \frac{h_1}{z^2} - \frac{h_2}{(z-1)^2} + \frac{h_0+h_1+h_2-h_3}{z(z-1)} \right\} H(z) = 0 \tag{8.71}$$

This can be transformed into the so-called hypergeometric equation, and thereby solved in terms of hypergeometric functions (Ex. 8.9, 8.10, 8.11, and 8.12 below, give a detailed illustration of this transformation and show how to obtain the solutions of the differential equation (8.71) in a few explicit cases. A more general study of the solutions to Eq. (8.71) will be performed in Sect. 9.2.3.)

More generally, a correlation function involving the operator $\phi_{(r,s)}$ will satisfy a differential equation of order rs, obtained by transforming the singular vector condition at level rs into a differential operator of order rs. In general, there will be rs independent solutions to this differential equation, referred to as the conformal blocks of the correlation function under study. The full correlator is a sesquilinear combination of these blocks (a sum of holomorphic × antiholomorphic solutions to, respectively, the differential equation and its complex conjugate). Fixing this combination can be done by using the symmetry of the correlation function under the permutation of its fields. This relates different sesquilinear combinations of the same conformal blocks, and completely fixes their relative coefficients. This procedure will be described in Chap. 9.

An important remark is in order: rs may not be the lowest order of the differential equation satisfied by the correlation function (8.42). Indeed, the equivalence $\phi_{(p'-r,p-s)} \equiv \phi_{(r,s)}$ shows that this correlation should also satisfy a differential equation of order $(p'-r)(p-s)$. One can slightly simplify the problem by simple eliminations between the two differential equations. Set $rs = N$, and suppose $(p'-r)(p-s) = N+a$. By differentiating the first equation a times, we can eliminate the highest-order term in the second one by taking a suitable linear combination of the two. This reduces by one the degree of the second equation. Reiterating this process should in principle reduce the degree of the differential equations we started with. Of course, it can happen that at some step the two equations are no longer independent, which means that the lowest possible order has been reached.

Solving the differential equations above should lead to a complete determination of correlators in a minimal theory. However, a more efficient approach to the calculation of conformal correlators is provided by the Coulomb gas formalism described in Chap. 9. The latter is more constructive, in the sense that correlators are directly built from the singular vector conditions. Therefore, the differential equations derived in this section will be automatically satisfied. In the end, this will provide a beautiful and systematic way of solving the equations (8.39) by means of contour integrals of free boson correlators.

§8.4. Fusion Rules

The primary fields of a minimal theory correspond to highest weights in the Kac table (7.65). The object of this section is to derive the fusion rules between all such states, namely, to find which primaries and descendants are created by the short distance product of two given fields. The differential equations of the previous section provide a systematic way of studying fusion rules. Those of the (p,p') minimal theories turn out to be polynomially generated by the fusions of the two fundamental fields $\phi_{(2,1)}$ and $\phi_{(1,2)}$, as stated in Sect. 7.3.

8.4.1. From Differential Equations to Fusion Rules

The differential equations for correlators can be used to derive the *fusion rules* of the minimal conformal theories. In this subsection, we use this path to obtain the fusion rules (8.84) for the degenerate field $\phi_{(r,1)}$.

On the one hand, we have the OPE (8.141):

$$\phi_0(z)\phi_1(w) = (z-w)^{h-h_1-h_0} \sum_h g(h_0,h_1,h)$$
$$\times \sum_Y (z-w)^{|Y|} \beta_Y(h_0,h_1,h) L_{-Y}\phi(w) \qquad (8.72)$$

(we dropped the right, or antiholomorphic contributions for notational simplicity), involving the structure constants $g(h_0,h_1,h)$. Determining the fusion rules amounts to finding the values of h present on the r.h.s. of (8.72) in terms of h_0 and h_1.

On the other hand, we have the differential equation (8.39) for the correlator $\langle \phi_0(z_0)\phi_1(z_1)\cdots \rangle$. Substituting the above OPE in the correlator, we obtain a set of constraints for the coefficients g and β, in the form

$$\sum_h g(h_0,h_1,h) \sum_Y \beta_Y(h_0,h_1,h)\gamma_0(z_i,\partial_{z_i})$$
$$\times (z_0-z_1)^{h-h_1-h_0+|Y|}\langle [L_{-Y}\phi](z_1)\phi_2(z_2)\cdots \rangle = 0 \qquad (8.73)$$

The leading term when $z_0 \to z_1$ corresponds to $|Y| = 0$, $\beta = 1$, h maximal, and also to a maximum number of derivatives with respect to z_0 or z_1, and/or powers of $(z_0-z_1)^{-1}$ taken from γ_0. This leading term yields a nontrivial equation relating h to h_0 and h_1, expressing a fusion rule of the theory.

Take, for instance, the case $V(c,h_0) = V_{2,1}$. In the four-point function (8.62), the leading contribution to Eq. (8.73) is made of three terms: $\partial_{z_0}^2$, $(z_0-z_1)^{-2}$, and $(z_0-z_1)^{-1}\partial_{z_0}$, which give the constraint

$$\frac{1}{t}(h-h_0-h_1)(h-1-h_0-h_1) + (h-h_0-h_1) - h_1 = 0 \qquad (8.74)$$

This constraint is equivalent to Eq. (7.50), written in a different notation. We recall that

$$h_0 = h_{2,1}(t) = \frac{3}{4}t - \frac{1}{2} \tag{8.75}$$

and h_1 is a weight of the minimal Kac table (7.65) of the form $h_1 = h_{r,s}(t)$, with $t = p/p'$. The quadratic equation (8.74) for h is easily solved

$$\left(h - h_0 - h_1 - \frac{1-t}{2}\right)^2 = \left(\frac{1-t}{2}\right)^2 + th_1 \tag{8.76}$$
$$= \left(\frac{tr - s}{2}\right)^2$$

and the two solutions have the simple form

$$h = h_{r+\epsilon,s}(t) \qquad \epsilon = \pm 1 \tag{8.77}$$

This implies the following allowed fusions:

$$\phi_{(2,1)} \times \phi_{(r,s)} = \phi_{(r-1,s)} + \phi_{(r+1,s)} \tag{8.78}$$

As mentioned in Sect. 7.3.1, the above is by no means an equality between fields, it is rather an abusive way of describing the allowed fusions and should be taken as such. Strictly speaking, we derived the fusion rule only for the larger of the two values of h, as we looked only at the leading term in Eq. (8.73). However, we note that

$$h_{r+1,s}(p,p') - h_{r-1,s}(p,p') = r\frac{p}{p'} - s \tag{8.79}$$

which, for (r,s) in the Kac table, never takes an integer value. Therefore, the two leading terms of Eq. (8.73) pertaining to either values of h can never be mixed with descendant contributions, which all have integer-spaced weights with respect to h. Hence both terms are present, except if one of the indices is outside of the Kac table, in which case the corresponding fusion is forbidden (the corresponding state is not in the theory under consideration).

More generally, we take $V(c,h_0) = V_{r,1}$, degenerate at level r. Starting from the differential equation (8.48) for the correlator, with the differential operator $\gamma_{r,1}(z_i, \partial_{z_i})$ defined in Eq. (8.47), we substitute the OPE (8.72) and look for the leading contribution as $z_0 \to z_1$. This procedure is equivalent to retaining only the terms involving powers of $(z_0 - z_1)$, ∂_{z_0}, and ∂_{z_1} in $\gamma_{r,1}$. This amounts to replacing $\gamma_{r,1}$ by $\tilde{\gamma}_{r,1}$, defined by the substitutions

$$L_{-r} \to \frac{(r-1)h_1}{(z_1 - z_0)^r} - \frac{1}{(z_1 - z_0)^{r-1}}\partial_{z_1} \tag{8.80}$$
$$L_{-1} \to \partial_z$$

into the operator $\Delta_{r,1}(t)$ of Eq. (8.26). The operator $\tilde{\gamma}_{r,1}$ acts on the leading term $z^{h-h_0-h_1}$ of the OPE. Comparing this situation with that of Eqs. (8.178)–(8.179), we see that the leading action of the operator $\tilde{\gamma}_{r,1}$ on the leading piece of the correlator is exactly given by the determinant $\theta_{r,1}(\lambda, \mu)$ (see Eqs. (8.195)–(8.196), with $\lambda = -h_1$ and $\mu = h_0 + h_1 - h$), given by

$$(\theta_{r,1})^2 = \prod_{m=1}^{r} \left\{ [h_0 + h_1 - h + (r - m)(1 - tm)] \right.$$

$$\left. \times \left[h_0 + h_1 - h + (m + 1)(1 - t(r + 1 - m)) \right] - 4h_1 t \left(\frac{r+1}{2} - m \right)^2 \right\} \tag{8.81}$$

Therefore the fusion is allowed if and only if

$$\theta_{r,1}^2 = 0 \tag{8.82}$$

Substituting the values

$$h_0 = h_{r,1}(t) = \frac{r^2 - 1}{4}t + \frac{1 - r}{2}$$

$$h_1 = h_{k,l}(t) = \frac{k^2 - 1}{4}t + \frac{l^2 - 1}{4t} + \frac{1 - kl}{2}, \tag{8.83}$$

in the condition (8.82) leads to a very simple result: The allowed fusions for $k \geq r$ and $k + r \leq p'$ are

$$\phi_{(r,1)} \times \phi_{(k,l)} = \sum_{\substack{m=k-r+1 \\ m-k+r-1 \text{ even}}}^{k+r-1} \phi_{(m,l)} \tag{8.84}$$

Working out the details of the derivation of Eq. (8.84) is left as an exercise (Ex. 8.15) at the end of this chapter. If k is larger than r or $p' - r$, Eq. (8.84) becomes more involved. The complete result will be derived in Sect. 8.4.3.

8.4.2. Fusion Algebra

The concept of fusion rules leads to the definition of *fusion numbers* $\mathcal{N}_{ij}{}^k \in \{0, 1\}$[6] as the characteristic functions of the structure constants $g(h_i, h_j, h_k)$ in Eq. (8.72).

$$\mathcal{N}_{ij}{}^k = \begin{cases} 0 & \text{iff} \quad g(h_i, h_j, h_k) = 0 \\ 1 & \text{otherwise} \end{cases} \tag{8.85}$$

We use the indices i, j, k as a shorthand notation for the corresponding conformal dimensions h_i, h_j, h_k. In the particular case of minimal models, the index i stands for Kac indices (r, s), but the concept of a fusion algebra applies to more general situations; hence, one should simply think of the index i as labeling the primary fields (more precisely, its holomorphic, or left part). Here again, we concentrate on the holomorphic dependence of the fields. The same numbers describe allowed

---

[6] We stress that, in full generality, the fusion numbers may be larger than one, but it is not so for the Virasoro minimal models. Ultimately, this reflects the absence of multiplicity greater than one in ordinary tensor products of representations of $su(2)$, as will be made clear in Chap. 18.

fusions of the modules of the right Virasoro algebra. The full fusion numbers factorize as $\mathcal{N}_{\text{left}} \times \mathcal{N}_{\text{right}}$. To these fusion numbers there corresponds an abstract notion of *fusion algebra*, namely a commutative and associative algebra with generators $\phi_j, j = 1, \ldots, r$, an identity element $\phi_1 = I$ (the identity field), and a product $\times$, defined by the multiplication rules

$$\phi_i \times \phi_j = \sum_k \mathcal{N}_{ij}{}^k \, \phi_k \tag{8.86}$$

In particular, the product with the identity ϕ_1 implies that

$$\mathcal{N}_{i1}{}^k = \delta_{i,k} \tag{8.87}$$

and the commutativity of the product simply means that

$$\mathcal{N}_{ij}{}^k = \mathcal{N}_{ji}{}^k \tag{8.88}$$

A direct consequence of the associativity of the OPE of primary fields is the associativity of the fusion algebra. We have

$$\begin{aligned}
\phi_i \times (\phi_j \times \phi_k) &= \phi_i \times \sum_l \mathcal{N}_{jk}{}^l \, \phi_l \\
&= \sum_{l,m} \mathcal{N}_{jk}{}^l \mathcal{N}_{il}{}^m \, \phi_m
\end{aligned} \tag{8.89}$$

and

$$\begin{aligned}
(\phi_i \times \phi_j) \times \phi_k &= \sum_l \mathcal{N}_{ij}{}^l \phi_l \times \phi_k \\
&= \sum_{l,m} \mathcal{N}_{ij}{}^l \mathcal{N}_{lk}{}^m \, \phi_m
\end{aligned} \tag{8.90}$$

Identifying the coefficient of ϕ_m in both expressions yields (using the commutativity (8.88))

$$\sum_l \mathcal{N}_{kj}{}^l \mathcal{N}_{il}{}^m = \sum_l \mathcal{N}_{ij}{}^l \mathcal{N}_{lk}{}^m \tag{8.91}$$

Defining the $r \times r$ matrix operators N_i, with entries

$$(N_i)_{j,k} = \mathcal{N}_{ij}{}^k \tag{8.92}$$

the associativity condition (8.91) can be rephrased in the sense of an ordinary matrix product as

$$N_i \, N_k = N_k \, N_i \tag{8.93}$$

But Eq. (8.91) can also be written in the form

$$N_i \, N_k = \sum_l \mathcal{N}_{ik}{}^l \, N_l \tag{8.94}$$

Hence the $\mathcal{N}$'s form a representation of their own fusion algebra. This has much in common with the notion of adjoint representation for a Lie group. To make the fusion less abstract, it is useful to bear in mind this *adjoint matrix representation* (8.94). Notice that, in this matrix representation, the associativity condition (8.91) takes the form of the commutativity condition (8.93).

8.4.3. Fusion Rules for the Minimal Models

We return to the minimal models $\mathcal{M}(p, p')$ to complete the analysis of their fusion rules. The result (8.131) relies on the assumption that the representations with highest weights $h_{2,1}$ and $h_{1,2}$ are present in the theory.[7] This means that *both* p and p' are larger than, or equal to 3.

For the sake of accuracy, we replace the labels i by the corresponding pairs of Kac indices (r, s) in the fusion numbers (8.85). In particular,

$$\mathcal{N}_{(1,1)(r,s)}{}^{(m,n)} = \delta_{r,m}\delta_{s,n} \tag{8.95}$$

The result (8.78) may be recast as

$$\mathcal{N}_{(2,1)(r,s)}{}^{(m,n)} = \delta_{n,s}\left(\delta_{m,r+1} + \delta_{m,r-1}\right) \tag{8.96}$$

where it is implied that the Kronecker symbols vanish whenever the exterior border of the Kac table ($m = 0$ or $m = p'$) is reached. There is an analogous relation for the $(1, 2)$ fusions, obtained by exchanging all Kac indices within the pairs (which is the same as the $t \to 1/t$ transformation):

$$\mathcal{N}_{(1,2)(r,s)}{}^{(m,n)} = \delta_{m,r}\left(\delta_{n,s+1} + \delta_{n,s-1}\right) \tag{8.97}$$

These are the key relations for the computation of the general fusion numbers $\mathcal{N}_{(r,s)(m,n)}{}^{(k,l)}$, or equivalently, the matrices $N_{(r,s)}$ defined by Eq. (8.92), with (r, s) in the Kac table. Indeed, our main result will be that the fusions of $(2, 1)$ and $(1, 2)$ generate all the others *polynomially*, meaning that the matrix $N_{(r,s)}$ is a polynomial of the matrices $N_{(2,1)}$ and $N_{(1,2)}$. This is intuitively obvious from Fig. 8.2, where the $(2, 1)$ fusion generates the two horizontal moves $r \to r \pm 1$, and the $(1, 2)$ fusion generates the vertical moves $s \to s \pm 1$. By recursion on $m = r + s$, we see that the (r', s') fusions with $r' + s' = m + 1$ are generated by such moves. To make this statement more precise, we set

$$X = N_{(2,1)} \qquad Y = N_{(1,2)} \tag{8.98}$$

Eqs. (8.96) and (8.95) translate respectively into the recursion relation

$$N_{(r+1,1)} = XN_{(r,1)} - N_{(r-1,1)} \tag{8.99}$$

with the initial conditions

$$N_{(1,1)} = \mathbb{I} \quad \text{and} \quad N_{(2,1)} = X \tag{8.100}$$

---

[7] This may not be the case in general (even for $p, p' > 2$), since the conformal weights have to form only a *subset* of the Kac table, with possible repetitions of some weights. The whole issue of determining consistent field contents for minimal models will be studied in Chap. 10.

This is exactly the recursive definition of the Chebyshev polynomials of the second kind, usually defined by

$$U_m(2\cos\theta) = \frac{\sin(m+1)\theta}{\sin\theta} \tag{8.101}$$

with the recursion relation

$$U_m(x) = xU_{m-1}(x) - U_{m-2}(x) \tag{8.102}$$

and initial conditions

$$U_0(x) = 1 \quad \text{and} \quad U_1 = x \tag{8.103}$$

This enables us to identify

$$N_{(r,1)} = U_{r-1}(X) \tag{8.104}$$

Likewise, we find

$$N_{(1,s)} = U_{s-1}(Y) \tag{8.105}$$

The finiteness of the Kac table can be expressed as the vanishing of the states on its exterior border, namely

$$N_{(r,0)} = N_{(r,p)} = N_{(0,s)} = N_{(p',s)} = 0 \tag{8.106}$$

This implies, in particular, the two constraints

$$\boxed{U_{p'-1}(X) = 0 \qquad U_{p-1}(Y) = 0} \tag{8.107}$$

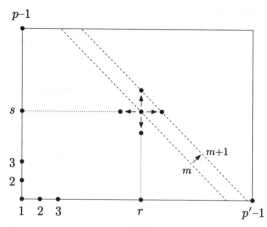

Figure 8.2. The Kac table of minimal (p, p') models. The horizontal arrows describe the effect of fusion by $(2, 1)$, whereas the vertical arrows describe that of fusion by $(1, 2)$. The combination of the two enables us to reach any point of the table by recursion on the value of $m = r + s$.

We now mix the two generators X and Y, and prove by recursion that

$$N_{(r,s)} = U_{r-1}(X) \, U_{s-1}(Y) \tag{8.108}$$

This is true for $N_{(1,1)} = 1$. Suppose Eq. (8.108) is true for any (r,s) such that $r + s \le m$. The fusion by $(2,1)$ then gives

$$X N_{(r,s)} = N_{(r+1,s)} + N_{(r-1,s)} \tag{8.109}$$

Therefore, we have, by the recursion hypothesis,

$$\begin{aligned}
N_{(r+1,s)} &= \big(X U_{r-1}(X) - U_{r-2}(X)\big) U_{s-1}(Y) \\
&= U_r(X) U_{s-1}(Y)
\end{aligned} \tag{8.110}$$

where, in the second step, the Chebyshev recursion has been used. Likewise, we find

$$N_{(r,s+1)} = U_{r-1}(X) \, U_s(Y) \tag{8.111}$$

Hence the recursion hypothesis is proven for $m \rightarrow m + 1$ (see Fig. 8.2), and Eq. (8.108) holds for any (r,s) in the Kac table. The fusions are therefore polynomially generated by X and Y, as announced before.

Of course, X and Y are subject to the constraints (8.107), but this is not all. We still have to implement the symmetry (7.66) of the Kac table

$$N_{(p'-r,p-s)} = N_{(r,s)} \tag{8.112}$$

This is satisfied if and only if

$$U_{p'-r-1}(X) U_{p-s-1}(Y) = U_{r-1}(X) U_{s-1}(Y) \tag{8.113}$$

In particular, if $r = 1$ and $s = p - 1$, we have

$$\boxed{U_{p'-2}(X) = U_{p-2}(Y)} \tag{8.114}$$

We prove that this condition, together with the constraints (8.107), is sufficient to ensure the symmetry (8.113). From Eq. (8.107), it can be shown that

$$U_{p'-2}(X) U_{r-1}(X) = U_{p'-r-1}(X) \tag{8.115}$$

This is trivially true for $r = 1$. Suppose that it is true for r, then

$$\begin{aligned}
U_{p'-2}(X) U_r(X) &= U_{p'-2}(X)\big(X U_{r-1}(X) - U_{r-2}(X)\big) \\
&= X U_{p'-r-1}(X) - U_{p'-r-2}(X) \\
&= U_{p'-r}(X),
\end{aligned} \tag{8.116}$$

which shows that the property (8.115) holds for $r \rightarrow r + 1$, and therefore for any r. A similar argument leads to

$$U_{p-2}(Y) U_{s-1}(Y) = U_{p-s-1}(Y) \tag{8.117}$$

for any s. Now, using Eqs. (8.114), (8.115), and (8.117), we derive

$$
\begin{aligned}
U_{p'-r-1}(X)U_{s-1}(Y) &= U_{p'-2}(X)U_{r-1}(X)U_{s-1}(Y) \\
&= U_{r-1}(X)U_{p-2}(Y)U_{s-1}(Y) \qquad (8.118) \\
&= U_{r-1}(X)U_{p-s-1}(Y)
\end{aligned}
$$

For $s \to p - s$, this is exactly the desired result (8.113).

Summarizing, we found that the fusion algebra $\mathcal{A}_{p,p'}$ of a minimal conformal theory with central charge $c(p,p')$ containing the primary fields $\phi_{(2,1)}$ and $\phi_{(1,2)}$ is polynomially generated by $X = N_{(2,1)}$ and $Y = N_{(1,2)}$ as

$$
N_{(r,s)} = U_{r-1}(X)U_{s-1}(Y) \qquad (8.119)
$$

where U is the Chebyshev polynomial of the second kind, and X and Y are constrained by the three relations

$$
\boxed{U_{p'-1}(X) = U_{p-1}(Y) = U_{p'-2}(X) - U_{p-2}(Y) = 0} \qquad (8.120)
$$

These constraints form an ideal $\mathcal{I}_{p,p'}(X, Y)$ of the ring $\mathbb{C}[X, Y]$ of polynomials of X and Y, and the fusion algebra is endowed with a quotient ring structure[8]

$$
\mathcal{A}_{p,p'} = \mathbb{C}[X, Y]/\mathcal{I}_{p,p'}(X, Y) \qquad (8.121)
$$

This result is in agreement with the direct computation of the $(r, 1)$ fusion rules (8.84). First we note that

$$
N_{(m,n)} = U_{m-1}(X)U_{n-1}(Y) = N_{(m,1)}N_{(1,n)} \qquad (8.122)
$$

for any m, n. Hence

$$
N_{(r,1)}N_{(m,n)} = \big(N_{(r,1)}N_{(m,1)}\big)N_{(1,n)} \qquad (8.123)
$$

We compute the fusion of $(r, 1)$ and $(m, 1)$. First, we extend, for convenience, the definition of the polynomials $U_m(x)$ to negative integers m, by their defining recursion relation. For instance, $U_{-1}(x) = 0$, $U_{-2}(x) = -1$, and so on. It is easy to derive that

$$
U_{-m-2}(X) = -U_m(X) \qquad (8.124)
$$

We can prove by recursion on r that

$$
U_r(X)U_m(X) = \sum_{\substack{k=m-r \\ k-m+r \text{ even}}}^{m+r} U_k(X) \qquad (8.125)
$$

---

[8] See Ex. 8.17 for a summary of the basic definitions of ring, ideal, and quotient.

where the sum may include negative indices. This relation is trivially true for $r = 0$. Suppose Eq. (8.125) is true for r. Then

$$U_{r+1}(X)U_m(X) = X \sum_{\substack{k=m-r \\ k-m+r \text{ even}}}^{m+r} U_k(X) - \sum_{\substack{k=m-r+1 \\ k-m+r-1 \text{ even}}}^{m+r-1} U_k(X)$$

$$= \sum_{\substack{k=m-r \\ k-m+r \text{ even}}}^{m+r} \left(X U_k(X) - U_{k-1}(X) \right) \qquad (8.126)$$

$$= \sum_{\substack{k=m-r-1 \\ k-m+r+1 \text{ even}}}^{m+r+1} U_k(X)$$

The property (8.125) is thus true for $r \to r+1$; hence, it holds for all r. This is in agreement with Eq. (8.84), for $r \leq m$ and $m + r < p'$. We now still have to take care of the possible terms with negative indices in Eq. (8.125). They arise if $r > m$. Thanks to the property (8.124), the net effect of the presence of $U_{m-r}, U_{m-r+2}, \ldots$ with negative indices is to cancel the corresponding terms $U_{r+2-m}, U_{r-m}, \ldots$ with positive indices in the sum (8.125). This results in a modification of the lower bound in Eq. (8.125)

$$U_r(X)U_m(X) = \sum_{\substack{k=|m-r| \\ k-m+r \text{ even}}}^{m+r} U_k(X) \qquad (8.127)$$

Moreover, due to the constraint $U_{p'-1}(X) = 0$, we have still some polynomials whose index goes out of the Kac table whenever $m + r \geq p'$. To take this into account, note that the constraint $U_{p'-1}(X) = 0$ propagates itself through the recursion relations of the Chebyshev polynomials to yield a reflection property

$$U_{p'-1+k}(X) = -U_{p'-1-k}(X) \qquad (8.128)$$

This shows that, in Eq. (8.125), the terms $U_{m+r}, U_{m+r-2}, \ldots$ with indices larger than $p' - 1$ cancel the corresponding terms $U_{2p'-2-m-r}, U_{2p'-m-r}, \ldots$ with indices smaller than $p' - 1$. This results in a modification of the upper bound of the summation in Eq. (8.125). Together with the modification of the lower bound (8.127), this yields

$$U_r(X)U_m(X) = \sum_{\substack{k=|m-r| \\ k-m+r \text{ even}}}^{\min(m+r,2p'-2-m-r)} U_k(X) \qquad (8.129)$$

We finally get the $(r, 1)$ fusion from Eq. (8.123)

$$N_{(r,1)}N_{(m,n)} = \sum_{\substack{k=|m-r|+1 \\ k-m+r-1 \text{ even}}}^{\min(m+r-1,2p'-1-m-r)} N_{(k,n)} \qquad (8.130)$$

and the general fusion rules

$$N_{(r,s)}N_{(m,n)} = \sum_{\substack{k=|m-r|+1 \\ k-m+r-1 \text{ even}}}^{\min(m+r-1,2p'-1-m-r)} \sum_{\substack{l=|n-s|+1 \\ l-n+s-1 \text{ even}}}^{\min(n+s-1,2p-1-n-s)} N_{(k,l)} \qquad (8.131)$$

This is the result announced in Eq. (7.70).

From the above discussion, it is clear that the fusion algebra $\mathcal{A}_{p,p'}$ possesses two subalgebras $\mathcal{X}_{p'}$ and $\mathcal{Y}_p$, generated respectively by $N_{(r,1)}$, $1 \leq r \leq p' - 1$ and $N_{(1,s)}$, $1 \leq s \leq p - 1$, with respective fusion numbers

$$\begin{aligned} \mathcal{X}_{p'} &: \mathcal{N}_{rs}{}^t(q) \equiv \mathcal{N}_{(r,1)(s,1)}{}^{(t,1)} & 1 \leq r,s,t \leq p' - 1 \\ \mathcal{Y}_p &: \mathcal{N}_{rs}{}^t(p) \equiv \mathcal{N}_{(1,r)(1,s)}{}^{(1,t)} & 1 \leq r,s,t \leq p - 1 \end{aligned} \qquad (8.132)$$

where $\mathcal{X}_m$ and $\mathcal{Y}_m$ are isomorphic. Moreover, the relation

$$N_{(r,s)} = N_{(r,1)} N_{(1,s)} \qquad (8.133)$$

expresses that $\mathcal{A}_{p,p'}$ is the product of the two algebras $\mathcal{X}_{p'}$ and $\mathcal{Y}_p$, quotiented by the identification $N_{(r,s)} = N_{(p'-r,p-s)}$, which amounts to the relation (8.114). This tensor product structure for minimal models will also be apparent in the study of modular transformations of the characters of the irreducible representations with weights (7.65), in Chap. 10. The reasons for this are profound, and will be elucidated later in this volume. The subalgebras $\mathcal{X}_{p'}$ and $\mathcal{Y}_p$ are actually the fusion algebras of two Wess-Zumino-Witten models underlying the minimal theory.

A few remarks are in order. We reconsider the Landau-Ginzburg description of diagonal minimal theories presented in Sect. 7.4.7, from the point of view of fusion rules. As the whole structure is generated from powers and descendants of the basic field $\Phi = \Phi_{(2,2)}$, we also expect the fusion rules of the theory to be generated by this field. It turns out that only half of the spectrum of the corresponding minimal theory is generated (see Ex. 8.20). More precisely, the intrinsic $\Phi \rightarrow -\Phi$ symmetry of the Landau-Ginzburg Lagrangian $\mathcal{L}_m$ is easily identified in the $(m+1,m)$ minimal theory. The successive odd powers of Φ are $\mathbb{Z}_2$-odd, and generate a subset of the Kac table. Without loss of generality, we restrict ourselves to the (odd,even) theories $(2k+1,2k)$ and consider the odd powers of the matrix $G = N_{(2,2)}$ in the adjoint representation of the $(2k+1,2k)$ fusion algebra: the matrices G, $G^3,\ldots,$ G^{2N-1} form a linearly independent system for $N = k(k-1)$. This is due to the two following properties of the matrix G:[9] (i) the eigenvalue 0 of G is k times degenerate. (ii) 0 is the only eigenvalue of G that can be degenerate. The dimension of the fusion algebra (size of the matrices $N_{(r,s)}$) is $k(2k-1)$, half the number of points in the Kac table; therefore, the minimal polynomial of G (namely, the polynomial $\Pi(x)$ of lowest degree, such that $\Pi(G) = 0$; cf. Ex. 8.18) is of degree

$$k(2k-1) - (k-1) = 2M+1 \qquad M = k(k-1) \qquad (8.134)$$

---

[9] See Ex. 8.18 for useful basic properties of matrices and Exs. 8.20–8.21 for proofs of (i)–(ii).

which proves the linear independence of $G, G^3, \ldots, G^{2M-1}$. On the other hand, due to the fusion rules (7.114), it is clear that each of these odd powers is itself a linear combination of some $N_{(r,s)}$, with even s. The latter form a subset of M linearly independent matrices; hence, each of them can be expressed as a polynomial of Φ, odd under the transformation $\Phi \to -\Phi$. So the odd powers of G generate the whole $\mathbb{Z}_2$-odd sector of the fusion algebra.

A last remark: as already mentioned, we have assumed in the above derivation that the two primary fields $\phi_{(2,1)}$ and $\phi_{(1,2)}$ belonged to the theory. This might not be the case in general. Actually, in the discussion of modular invariance in Chap. 10, we shall find more solutions than the one discussed here, among which are the so-called D series of minimal models, which contain $\phi_{(2,1)}$, $\phi_{(1,3)}$ but not $\phi_{(1,2)}$ (p is even), and some exceptional theories E_6, E_7, and E_8, with, respectively, $p = 12$, 18, 30, which all contain $\phi_{(2,1)}$ but whose $\mathcal{Y}_p$ subalgebra is generated respectively by $\phi_{(1,4)}$, $\phi_{(1,5)}$, and $\phi_{(1,7)}$. All these theories have different fusion rules. We shall return to this discussion later.

Appendix 8.A. General Singular Vectors from the Covariance of the OPE

This appendix is devoted to the complete derivation of singular vectors of the Virasoro algebra for minimal models, based on the concept of fusion of Verma modules.

We first investigate the properties of the OPE at the level of representations of the Virasoro algebra (Sect. 8.A.1). The principle of the OPE (decomposition of the product of two conformal fields onto conformal fields and their descendants) translates, at the level of Verma modules, into the notion of decomposition of the tensor product of two modules onto other modules. More precisely, the states in the modules over which the decomposition is performed are intimately linked to those in the tensored modules. This link is essentially due to the covariance of the OPE, namely, the fact that both the factor fields and those in the decomposition have compatible transformation properties under conformal mappings. As an application, we show how this covariance constrains the structure coefficients of descendant fields in the OPE.

This leads to the definition of the (covariant) fusion map $\mathcal{F}$ from the tensor product of two Verma modules to a third one, which transfers the action of any operator from the states of the tensored modules to their images in the decomposition (Sect. 8.A.2). In particular, with a suitable choice of modules, the singular vector condition in one of the tensored modules can be transferred to the image modules in the decomposition. Our strategy (Sect. 8.A.3) will be to study the particular fusion of the $V_{r,1}$ module with another one, in such a way that the decomposition includes the module $V_{r,s}$. Then, using the fusion map $\mathcal{F}$, we shall transfer our knowledge of the singular vectors of $V_{r,1}$ onto $V_{r,s}$. This will result in a simple recipe to obtain the singular vectors of the module $V_{r,s}$ (Sects. 8.A.5 and 8.A.6).

In order for the fusion map $\mathcal{F}$ to be efficiently used, we need to evaluate the leading action of the operator $\Delta_{r,1}$ on the OPE of $\phi_{(r,1)}$ with another conformal field when their positions approach each other. This is achieved in Sect. 8.A.4 by proving Eq. (8.81), instrumental in the derivation of the fusion rules (8.84) of $\phi_{(r,1)}$.

8.A.1. Fusion of Irreducible Modules and OPE Coefficients

In this section, we expose a systematic way of computing the OPE coefficients of the algebra of primary fields of a conformal theory. The result takes the form of the recursion formula (8.156). As shown in Sect. 6.6.3, the operator product expansion of two given primary fields gives rise to structure constants (6.168), factorized into left and right coefficients, corresponding to left and right representations of the Virasoro algebra. We shall concentrate on the left part of this structure.

We fix throughout the following discussion the central charge c and attach isomorphic Virasoro algebras to each point x of the complex plane. The corresponding Verma (resp. irreducible) modules acquire an extra dependence on the point x, and are denoted by $V(c, h; x)$ (resp. $M(c, h; x)$), whereas the corresponding highest-weight state reads $|h; x\rangle$. Through the state-operator correspondence, we get holomorphic primary fields $f_h(x)$

$$f_h(x) \leftrightarrow |h; x\rangle, \tag{8.135}$$

subject to the conditions

$$
\begin{aligned}
L_{-1} f_h(x) &= \frac{d}{dx} f_h(x) \\
L_0 f_h(x) &= h f_h(x) \\
L_n f_h(x) &= 0 \qquad (n = 1, 2, \ldots)
\end{aligned}
\tag{8.136}
$$

We denote any finite nondecreasing sequence of positive integers $1 \leq r_1 \leq r_2 \leq \cdots \leq r_k$ by the single letter Y and set $|Y| = r_1 + r_2 + \cdots + r_k$. This provides us with a compact notation for

$$L_{-r_1} L_{-r_2} \cdots L_{-r_k} \equiv L_{-Y} \tag{8.137}$$

and we have the field-state equivalence for descendants of the field f_h

$$L_{-Y} f_h(x) \leftrightarrow L_{-Y} |h; x\rangle \tag{8.138}$$

That states depend upon a complex variable x is somewhat unusual. Indeed, fields are usually considered as operators acting in a fixed vector space with a unique vacuum state invariant under global conformal transformations and a unique dual vacuum linear form (invariant under the same group). The other point of view adopted here is to consider only correlation functions, that is, vacuum expectation values of products of fields at distinct points. The latter are analytic functions of the argument of one of the fields on the complex plane except at the arguments of the other fields. Henceforth identities between fields will be understood as insertions into correlation functions.

The OPE of two conformal fields has been defined in Sect. 6.6.3. By definition, the fusion of two highest-weight modules attached respectively to the fields $f_0(x_0)$ and $f_1(x_1)$ (of conformal dimensions h_0 and h_1) onto a third one relative to the field $f(x)$ (of dimension h) is possible, if the corresponding three-point function

$$\langle f_0(x_0)f_1(x_1)f(x)\rangle = \frac{g(h_0,h_1,h)}{(x_0 - x_1)^{h_0+h_1-h}(x_0 - x)^{h_0+h-h_1}(x_1 - x)^{h_1+h-h_0}} \quad (8.139)$$

is nonzero. This implies selection rules (the fusion rules) on the dimensions h_0, h_1, h, as discussed in Sect. 8.4.

We choose the coordinates as

$$x_0 = x + \frac{z}{2} \qquad x_1 = x - \frac{z}{2} \quad (8.140)$$

This is not restrictive since any triplet of points can be mapped to the three points $(x + z/2, x - z/2, x)$ by a suitable global conformal transformation (we say that global conformal transformations act *transitively* on triplets of points). As $z \to 0$ the product of fields $f_0(x_0)f_1(x_1)$ (inserted in correlations) is equivalent to a sum of expansions of the form[10]

$$f_0(x_0)f_1(x_1) \sim \sum_h \left\{ \frac{g(h_0,h_1,h)}{z^{h_0+h_1-h}} \sum_Y z^{|Y|}\beta_Y(h_0,h_1,h)L_{-Y}f_h(x)\right\} \quad (8.141)$$

As mentioned above, the choice of the midpoint $x = \frac{1}{2}(x_0 + x_1)$ is by no means mandatory and could well be modified using

$$f_h(x + x') = e^{x'L_{-1}}f_h(x) \quad (8.142)$$

without changing the structure of the above expansion (recall that L_{-1} is the translation operator). However, it has the virtue that the coefficients enjoy the symmetry property

$$\beta_Y(h_0,h_1,h) = (-1)^{|Y|}\beta_Y(h_1,h_0,h) \quad (8.143)$$

A global phase, arising from the prefactor $z^{h_0+h_1-h}$, has been conveniently absorbed in the normalization.

Likewise, we may define the fusion process among *irreducible* highest-weight modules as a *covariant* linear map

$$\mathcal{F} : M(c,h_0;x_0) \otimes M(c,h_1;x_1) \to M(c,h;x) \quad (8.144)$$

Since we deal with three isomorphic—but not identical—copies of the Virasoro algebra, it is not surprising that the product of highest-weight states in $M(c,h_0;x_0) \otimes M(c,h_1;x_1)$ *does not* correspond to just the highest-weight state in $M(c,h;x)$, but rather to an infinite linear combination

$$\mathcal{F}\Big(|c,h_0;x_0\rangle \otimes |c,h_1;x_1\rangle\Big) = \sum_Y \beta_Y(h_0,h_1,h|x_0,x_1,x)L_{-Y}|c,h;x\rangle \quad (8.145)$$

---

[10] The coefficients β_Y were introduced in Sect. 6.6.3 (cf. Eq. (6.168)), but the notation used here is slightly different: we do not carry indices for the conformal dimensions (in order to lighten the text) and the index Y appears here as a subscript.

Note that the above equation corresponds to setting $g(h_0, h_1, h) = 1$ (or 0 if it vanishes) in Eq. (8.139), which amounts to a multiplicative redefinition of the fields. We can equivalently absorb $g(h_0, h_1, h)$ in the definition of β_Y.

The covariance of $\mathcal{F}$ means the following. We consider a holomorphic map $x \to \tilde{x} = y(x)$ in a common neighborhood U of x_0, x_1, x, giving a one-to-one map $U \leftrightarrow y(U)$. The various conformal fields transform as

$$f_i(x_i)dx^{h_i} = \tilde{f}_i(\tilde{x}_i)d\tilde{x}_i^{h_i} \tag{8.146}$$

We require that the operator product expansion (8.141) be such that both sides have identical transformation properties, namely that the *same* equation holds for quantities with tildes.

How does this property of fields translate in terms of irreducible modules? We will show that it specifies $\mathcal{F}$ completely; in particular it will enable us to compute all the coefficients β_Y in Eq. (8.145). In other words, local conformal invariance fixes the β_Y's completely. For simplicity, but without loss of generality, we restrict ourselves to the choice of coordinates (8.140). In this case, $\mathcal{F}$ acts on the tensor product of highest-weight states as

$$\mathcal{F}\Big(|c, h_0; x_0\rangle \otimes |c, h_1; x_1\rangle\Big) = \frac{1}{z^{h_0+h_1-h}} \sum_Y z^{|Y|} \beta_Y L_{-Y} |c, h; x\rangle \tag{8.147}$$

The covariance condition translates into

$$\frac{1}{(x_0 - x_1)^{h_0+h_1-h}} \sum_Y (x_0 - x_1)^{|Y|} \beta_Y \left(L_{-Y} f\right)\left(\frac{x_0 + x_1}{2}\right) =$$
$$\left(\frac{d\tilde{x}_0}{dx_0}\right)^{h_0} \left(\frac{d\tilde{x}_1}{dx_1}\right)^{h_1} \frac{1}{(\tilde{x}_0 - \tilde{x}_1)^{h_0+h_1-h}} \times \tag{8.148}$$
$$\sum_Y (\tilde{x}_0 - \tilde{x}_1)^{|Y|} \beta_Y \left(L_{-Y}\tilde{f}\right)\left(\frac{\tilde{x}_0 + \tilde{x}_1}{2}\right)$$

where

$$\left(L_{-Y}\tilde{f}\right)\left(\frac{\tilde{x}_0 + \tilde{x}_1}{2}\right) = L_{-Y} \exp\left[\left(\frac{\tilde{x}_0 + \tilde{x}_1}{2} - \tilde{x}\right)L_{-1}\right]\tilde{f}(\tilde{x}) \tag{8.149}$$

We need the transformation properties of the descendant fields $\left(L_{-Y}f\right)$ with f primary. These are most easily obtained by applying the above formula in infinitesimal form. We let

$$\tilde{x}_i = x_i - \epsilon(x_i) \ , \quad i = 0, 1 \tag{8.150}$$

With $x_0 = x + z/2$ and $x_1 = x - z/2$, Eq. (8.148) reduces to

$$\left\{ h_0\epsilon'(x_0) + \epsilon(x_0)\frac{\partial}{\partial x_0} + h_1\epsilon'(x_1) + \epsilon(x_1)\frac{\partial}{\partial x_1} \right\} \frac{1}{z^{h_0+h_1-h}} \times$$

$$\sum_Y z^{|Y|}\beta_Y L_{-Y}f(x) \tag{8.151}$$

$$= \frac{1}{z^{h_0+h_1-h}} \sum_Y z^{|Y|}\beta_Y \delta_\epsilon \left[L_{-Y}f(x) \right]$$

With the choice

$$\epsilon(y) = \epsilon_k(y-x)^{k+1} \qquad\qquad k \geq -1 \tag{8.152}$$

where ϵ_k is a small constant, we find that

$$\delta_\epsilon L_{-Y}f(x) = \epsilon_k L_k L_{-Y}f(x) \tag{8.153}$$

The covariance condition becomes, with $k \geq -1$,

$$\left[L_k - \left(\frac{z}{2}\right)^k(k+1)(h_0 + (-1)^k h_1) - \left(\frac{z}{2}\right)^{k+1}\left\{ \frac{1-(-1)^k}{2}\frac{\partial}{\partial x} \right.\right.$$
$$\left.\left. + (1+(-1)^k)\frac{\partial}{\partial z} \right\} \right] \times \frac{1}{z^{h_0+h_1-h}} \sum_Y z^{|Y|}\beta_Y L_{-Y}f(x) = 0 \tag{8.154}$$

Since L_1 and L_2 generate by commutators the complete algebra of L_k's ($k \geq 1$), it is sufficient to impose the two relations pertaining to $k = 1$ and $k = 2$. (We note also that Eq. (8.154) is tautological for $k = -1, 0$.). For notational simplicity we define

$$f^{(j)} = \sum_{|Y|=j} \beta_Y L_{-Y}f \qquad\qquad p \geq 0 \tag{8.155}$$

where $f^{(0)}$ is equal to f. The above covariance condition (8.154) translates into

$$\boxed{\begin{aligned} L_1 f^{(j)} &= (h_0 - h_1)f^{(j-1)} + \frac{1}{4}L_{-1}f^{(j-2)} \\ L_2 f^{(j)} &= \frac{h+j+2(h_0+h_1-1)}{4}f^{(j-2)} \end{aligned}} \tag{8.156}$$

It is understood that $f^{(j)} \equiv 0$ when $j < 0$. In particular, this shows that $f^{(0)} = f$ is a primary field (it satisfies the highest-weight conditions). This is also obvious from the $z \to 0$ limit of the $\mathcal{F}$ map

$$\lim_{z \to 0} z^{-h}\mathcal{F}\left(z^{h_0}|c,h_0;x_0\rangle \otimes z^{h_1}|c,h_1;x_1\rangle \right) = |c,h;x\rangle \tag{8.157}$$

which is a map between highest-weight vectors. The two equations (8.156) are recursion (descent) equations, which determine the coefficients β_Y completely.

We illustrate this for the first few values of j. At level 1, we have

$$\begin{aligned} f^{(1)} &= \beta_1 L_{-1}f \\ L_1 f^{(1)} &= 2h\beta_1 f = (h_0 - h_1)f, \end{aligned} \tag{8.158}$$

which implies that

$$\beta_1 = \frac{1}{2h}(h_0 - h_1) \tag{8.159}$$

if the determinant $K_1(c, h) = h$ does not vanish. At level 2,

$$f^{(2)} = (\beta_{1,1}L_{-1}^2 + \beta_2 L_{-2})f$$

$$L_1 f^{(2)} = (2h(2h + 1)\beta_{1,1} + 3h\beta_2)f^{(1)} = (\frac{h}{4} + \frac{(h_0 - h_1)^2}{2})f^{(1)} \tag{8.160}$$

$$L_2 f^{(2)} = (6h\beta_{1,1} + (4h + \frac{c}{2})\beta_2)f = (\frac{h}{4} + \frac{h_0 + h_1}{2})f$$

which is inverted to[11]

$$\beta_{1,1} = \frac{2h^2 + h(c - 12s + 16d^2) + 2cd^2}{8h[16h^2 + 2h(c - 5) + c]},$$

$$\beta_2 = \frac{h^2 + h(2s - 1) + s - 3d^2}{[16h^2 + 2h(c - 5) + c]}, \tag{8.161}$$

where $s = h_0 + h_1$ and $d = h_0 - h_1$, provided the determinant

$$K_2(c, h) = h\left(16h^2 + 2h(c - 5) + c\right) \tag{8.162}$$

is not zero. We recognize here the Kac determinant (8.1).

This method determines recursively $f^{(m)}$, the component at level m, by solving a $p(m) \times p(m)$ linear system ($p(m)$ is the number of partitions of the integer m). The determinant of this linear system is nothing but the Kac determinant at the corresponding level. If the complete determinant does not appear in some of the above expressions, this is due to a cancellation of factors between the numerator and the denominator.

If $M(c, h) \equiv V(c, h)$ (i.e., if $V(c, h)$ is irreducible, meaning that it does not possess singular vectors), then the Kac determinant never vanishes and the coefficients β_Y are determined from (8.156) for arbitrary h_0, h_1. Therefore infinitely many fusions of the type $V(c, h_0) \otimes V(c, h_1) \to V(c, h) \equiv M(c, h)$ are allowed when $V(c, h)$ is irreducible. This is why finite closure of the OPE algebra of a conformal theory is prevented whenever it includes a primary field corresponding to an irreducible Verma module. This is also the reason for which minimal models include only highest-weight states of reducible Verma modules, of the form (8.2).

We now consider the case where $V(c, h)$ is reducible: $M(c, h)$ is then the quotient of $V(c, h)$ by invariant submodules arising from singular vectors. When attempting to solve the system (8.156) at level p or higher, where p is the level of a singular vector in $V(c, h)$, one can at best hope to determine $f^{(p)}$ up to this singular vector or its descendants. This is why we defined fusion among irreducible modules. Hence, in $M(c, h)$, we work modulo singular vectors and their descendants.

---

[11] This result should be compared with Eq. (6.181) of Chap. 6: the difference reflects the difference in the positions at which the OPE is evaluated: $(x + z/2, x - z/2) \to x$ here, whereas $(z, 0) \to 0$ in Chap. 6.

In other words, the singular vectors can be set equal to zero. Still, we have to make sure that the r.h.s. of the linear system (8.156) lies in the range of the linear operator on the left. This imposes conditions on the triplet h_0, h_1, h, which are the fusion rules.

Such selection rules are already apparent at level 1, where the Kac determinant reduces to h. Since $2h\beta_1 = h_0 - h_1$, the vanishing of the Kac determinant, that is $h = 0$, implies $h_0 = h_1$. Thus, two modules can have the vacuum ($h = 0$) sector in their fusion rules only if they have equal weights. We thus recover a well-known property, which entails that the only nonzero two-point functions of primary fields are those involving fields of equal weight.

8.A.2. The Fusion Map $\mathcal{F}$: Transferring the Action of Operators

We recall that, in order for the fusion to be uniquely defined (up to an overall normalization), the target module $M(c, h)$ has to be irreducible. However, we have not yet use the irreducibility of the initial modules $M(c, h_0)$ and $M(c, h_1)$. In the next section we shall examine the consequences of quotienting the initial spaces by descendants of their singular vectors, and this will lead us to a procedure for generating expressions for the singular vectors of the target module. For this purpose, we need to know how the covariant map $\mathcal{F}$ acts on descendant states. This action is calculated below, and the result appears in Eqs. (8.166) and (8.167).

We now compute the action of $\mathcal{F}(L_{-r} \otimes \mathbb{I})$ on $|c, h_0; x_0\rangle \otimes |c, h_1; x_1\rangle$. We use the conformal Ward identities (6.156) of Sect. 6.6.1, which translate the action of the L_m's on a field inside a correlator into the action of differential operators $\mathcal{L}_{-m}$'s on the corresponding correlator:

$$\langle [L_{-r_1} \cdots L_{-r_k} f_0](x_0) f_1(x_1) \cdots \rangle = \mathcal{L}_{-r_1}(x_0) \cdots \mathcal{L}_{-r_k}(x_0) \langle f_0(x_0) f_1(x_1) \cdots \rangle \tag{8.163}$$

The differential operator $\mathcal{L}_{-r}(x_0)$ is defined in Eq. (6.152):

$$\mathcal{L}_{-r}(x_0) = \sum_{i \geq 1} \left\{ \frac{(r-1)h_i}{(x_i - x_0)^r} - \frac{1}{(x_i - x_0)^{r-1}} \frac{\partial}{\partial x_i} \right\} \tag{8.164}$$

Eq. (8.163) translates then the expressions of singular vectors into differential equations for correlators (cf. Sect. 8.3). The effect of $\mathcal{F}$ on $L_{-r} \otimes \mathbb{I}$ is to transfer the action of $\mathcal{L}_{-r}$ from the point x_0 to the point $x = x_0 - \frac{1}{2}z$. A simple way of realizing this is to expand $\mathcal{L}_{-r}(x_0)$ around x as

$$\mathcal{L}_{-r}(x_0) = (-z)^{-r} \left[(r-1)h_1 + \frac{1}{2}z\partial_x - z\partial_z \right] + \sum_{k \geq 0} \left\{ \left(\frac{z}{2}\right)^k \binom{r+k-2}{k} \right.$$

$$\left. \times \sum_{i \geq 2} (-1)^{r+k} \left[\frac{h_i(r+k-1)}{(x-x_i)^{p+k}} + \frac{1}{(x-x_i)^{r+k-1}} \partial_i \right] \right\} \tag{8.165}$$

The last sum is identified with $\mathcal{L}_{-r-k}(x)$ by comparing with the definition (8.164). Therefore, we find

$$
\begin{aligned}
\mathcal{F}(L_{-r} \otimes \mathbb{I}) = \frac{(-1)^r}{z^r} & \left[h_1(r-1) + \frac{z}{2}L_{-1} - z\frac{d}{dz} \right] \\
& + \sum_{k \geq 0} \left(\frac{z}{2}\right)^k \binom{k+r-2}{k} L_{-r-k}
\end{aligned}
\tag{8.166}
$$

where the L_{-m} operators on the r.h.s. act at the point x. Moreover, if we exchange $h_0 \leftrightarrow h_1$ and $z \to -z$ in (8.166), we find the action on the module $M(c, h_1)$

$$
\begin{aligned}
\mathcal{F}(\mathbb{I} \otimes L_{-r}) = \frac{1}{z^r} & \left[h_0(r-1) - \frac{z}{2}L_{-1} - z\frac{d}{dz} \right] \\
& + \sum_{k \geq 0} \left(-\frac{z}{2}\right)^k \binom{k+r-2}{k} L_{-r-k}
\end{aligned}
\tag{8.167}
$$

For instance, for $r = 1$, we have

$$
\begin{aligned}
\mathcal{F}(L_{-1} \otimes \mathbb{I}) &= \frac{d}{dz} + \frac{L_{-1}}{2} \\
\mathcal{F}(\mathbb{I} \otimes L_{-1}) &= -\frac{d}{dz} + \frac{L_{-1}}{2}
\end{aligned}
\tag{8.168}
$$

In order to find the action of $\mathcal{F}$ on any L_{-Y}, we simply have to iterate the transfer process (8.165). This gives a straightforward procedure to write the action of $\mathcal{F}$ on descendant states.

We stress again that nothing is particular to the choice of coordinates (8.140). Taking the points $x_0 = x + z$ and $x_1 = x$ instead, and writing

$$
\mathcal{F}|c, h_0; z+x\rangle \otimes |c, h_1; x\rangle = \frac{1}{z^{h_0+h_1-h}} \sum_Y z^{|Y|} \bar{\beta}_Y L_{-Y} |c, h; x\rangle,
\tag{8.169}
$$

the above procedure would have led us to

$$
\begin{aligned}
\mathcal{F}(L_{-r} \otimes \mathbb{I}) &= \frac{(-1)^r}{z^r} \left[h_1(r-1) + zL_{-1} - z\frac{d}{dz} \right] + \sum_{k \geq 0} z^k \binom{k+r-2}{k} L_{-r-k} \\
\mathcal{F}(\mathbb{I} \otimes L_{-r}) &= \frac{1}{z^r} \left[h_0(r-1) - z\frac{d}{dz} \right] + L_{-r}
\end{aligned}
\tag{8.170}
$$

and

$$
\mathcal{F}(L_{-1} \otimes \mathbb{I}) = \partial_z, \qquad \mathcal{F}(\mathbb{I} \otimes L_{-1}) = -\partial_z + L_{-1}
\tag{8.171}
$$

We note how the action on the right module has been simplified. The new descent equations (8.156) for the determination of the coefficients $\bar{\beta}_Y$ are also simplified.

With

$$\bar{f}^{(j)} = \sum_{|Y|=j} \bar{\beta}_Y L_{-Y} f \tag{8.172}$$

the covariance condition takes the form

$$\left[L_k - \left(h_0(k+1)z^k + z^{k+1}\partial_z \right) \right] \frac{1}{z^{h_0+h_1-h}} \sum_{p \geq 0} z^j \bar{f}^{(j)} = 0 \tag{8.173}$$

and hence

$$L_1 \bar{f}^{(j)} = (j - 1 + h + h_0 - h_1) \bar{f}^{(j-1)}$$
$$L_2 \bar{f}^{(j)} = (j - 2 + h + 2h_0 - h_1) \bar{f}^{(j-2)} \tag{8.174}$$

There is a striking analogy between these last two equations and the action of L_1 and L_2 (Eqs. (8.32) and (8.33)) on the components of the vector $\mathbf{f} = (f_1, \cdots, f_r)^T$. This phenomenon will become clear in the next subsection.

8.A.3. The Singular Vectors $|h_{r,s} + rs\rangle$: General Strategy

In the next few sections, we will derive the singular vector of level rs in the module $V_{r,s}$ by using the results of the two previous sections. The results are summarized in Sect. 8.A.6 below. The main idea is to use the knowledge of the level-r singular vector of $V_{r,1}$, and the following fusion among states of the Verma modules

$$V_{r,1} \otimes V_{1,s} \to V_{r,s}$$

to obtain information about the singular vector at level rs in the target module $V_{r,s}$, by a suitable use of the map $\mathcal{F}$ of Eq. (8.144).

Consider first the fusion of highest-weight states of the Verma modules

$$V(c, h_0; x_0) \otimes V(c, h_1; x_1) \to V(c, h; x),$$

namely the *fusion*

$$f_0(x_0) f_1(x_1) \to \frac{1}{z^{h_0+h_1-h}} \sum_{j \geq 0} z^j f^{(j)}(x) \tag{8.175}$$

where $z = x_0 - x_1$. Suppose that there exists a singular vector $\Delta_0 f_0(x_0)$ at level n_0 in the first Verma module $V(c, h_0; x_0)$. Δ_0 is a polynomial of the L_{-m}'s, of total degree n_0. Then, using the fusion map of the previous section between the highest-weight states of the Verma modules

$$V(c, h_0 + n_0) \otimes V(c, h_1) \to V(c, h') \subset V(c, h)$$

we find

$$(\Delta_0 f_0)(x_0) f_1(x_1) \to \frac{1}{z^{h_0+h_1-h+n_0}} \sum_{j \geq 0} z^j \psi^{(j)}(x) \tag{8.176}$$

where the $\psi^{(j)}$'s are some descendants of f. The leading term $\psi^{(0)}$ on the r.h.s. of Eq. (8.176) is, by definition, a state of $V(c, h)$ at level 0; therefore it is proportional to the highest-weight state f. Two situations may occur. Either $\psi^{(0)} = \text{const.} \times f$, with a nonzero constant; then $h' = h$, from which we do not learn anything about the target module $V(c, h)$. Or $\psi^{(0)} = 0$, in which case the first nonzero $\psi^{(j_0)}$ ($j_0 > 0$) on the r.h.s. of Eq. (8.176) is the highest-weight state of a proper submodule $V(c, h') \subset V(c, h)$, $h' = h + j_0$, and $\psi^{(j_0)}$ is a singular vector of $V(c, h)$.

It is now clear that the knowledge of singular vectors of either module $V(c, h_0)$ or $V(c, h_1)$ gives us information about singular vectors of the target module $V(c, h)$. The only point to clarify is whether the highest-weight state on the r.h.s. of Eq. (8.176) is the highest-weight state f of the target module $V(c, h)$ or one of its descendants. We compute the coefficient of proportionality between $\psi^{(0)}$ and f. We are interested in the *leading* contribution of the action of the operator $\Delta_0(x_0) \otimes \mathbb{I}$ on the tensor product $V(c, h_0) \otimes V(c, h_1)$. Using again the results of the previous section, we can transfer the action of a single L_{-j} at x_0 to the target module at x by using the substitution (8.166) appropriate to the choice of coordinates (8.140), where $x = (x_0 + x_1)/2$. This substitution reads

$$\mathcal{F}(L_{-j} \otimes \mathbb{I}) = \frac{(-1)^j}{z^j}[h_1(j-1) - z\frac{d}{dz}] + O(L_{-1}, L_{-2}, \ldots) \tag{8.177}$$

where we denoted by $O(L_{-1}, L_{-2}, \ldots)$ all the L_{-m}-dependent terms in Eq. (8.166). Since we are interested only in the leading action on the highest-weight state f, we keep only the L_{-m}−independent contributions of (8.177), that is, those preserving the dimension h of f (i.e., the level 0 action of Δ_0). Note that the substitution issued from the relation (8.170) in the case $x_1 = x$ and $z = x_0 - x_1$ leads to exactly the same relation (8.177). The latter is actually independent of the specific choice of coordinates, (cf. Ex. 8.24). Therefore, the leading action of $\Delta_0(x_0) \otimes \mathbb{I}$ on the product (8.175) is that of the operator

$$\gamma_0(z, \frac{d}{dz}) \equiv (-1)^{n_0} z^{h_0+h_1-h+n_0} \Delta_0[L_{-r} \to z^{-r}(h_1(r-1) - z\frac{d}{dz})] \tag{8.178}$$

on the leading term of the r.h.s. of Eq. (8.175), namely

$$\psi^{(0)} = \gamma_0(z, \frac{d}{dz}) \frac{1}{z^{h_0+h_1-h}} f \tag{8.179}$$

The substitution implied in Eq. (8.178) has to be carried out carefully because, like the L_{-r}'s, the substituted operators do not commute and the ordering has to be respected. We examine how each substituted operator acts individually on the components $f^{(p)}$. We define

$$l_{-r}f^{(p)} \equiv z^{h_0+h_1-h}(h_1(r-1) - z\frac{d}{dz}) \frac{1}{z^{h_0+h_1-h}} \sum_{k \geq 0} z^k f^{(k)}\Big|_{z^p}$$

$$= (h_1(r-1) - p + h_0 + h_1 - h)f^{(p+r)} \tag{8.180}$$

The operators l_{-r} ($r > 0$) carry half of a representation of the so-called Witt algebra of the diffeomorphisms of the circle (Virasoro algebra with $c = 0$), with the commutation relations (5.19):

$$[l_m, l_n] = (m - n)l_{m+n} \qquad m, n \in \mathbb{Z} \qquad (8.181)$$

The representations $W(\lambda, \mu)$ of the latter algebra act on an infinite dimensional vector space spanned by $\varphi_p, p \in \mathbb{Z}$, and are labeled by two complex numbers λ and μ. The action of the generators reads

$$l_m(\lambda, \mu)\varphi_p = (p + \mu - \lambda(m - 1))\varphi_{p+m} \qquad (8.182)$$

Interpreting Eq. (8.180) in the language of the representation theory of the Witt algebra, the above substitution (8.178) corresponds to a representation $W(\lambda, \mu)$ with

$$\lambda = -h_1 \qquad \mu = h_0 + h_1 - h \qquad (8.183)$$

8.A.4. The Leading Action of $\Delta_{r,1}$

The computation of the leading action (8.179) is still difficult in general. However, in the case of the operator $\Delta_{r,1}(t)$ of Eqs. (8.22)–(8.25), which creates the singular vector at level r in the Verma module $V_{r,1}$, we can compute it exactly. This is the content of the following result: For the operator $\Delta_{r,1}(t)$ defined in Eqs. (8.22)–(8.25), the Witt algebra substitution (8.178) with $h_0 = h_{r,1}(t)$ and $n_0 = r$, which is

$$\gamma_{r,1}(z, \frac{d}{dz}) \equiv (-1)^r z^{h_0 + h_1 - h + r} \Delta_{r,1}\left[L_{-m} \to z^{-r}(h_1(r - 1) - z\frac{d}{dz})\right] \qquad (8.184)$$

has a leading multiplicative action on the highest-weight state f of the target module $V(c, h)$, which reads

$$\psi^{(0)} = \theta_{r,1}(\lambda = -h_1, \mu = h_0 + h_1 - h)f \qquad (8.185)$$

wherein

$$\boxed{\begin{aligned} (\theta_{r,1})^2 = \prod_{m=1}^{r} & \left\{ [h_0 + h_1 - h + (r - m)(1 - tm)] \right. \\ & \times [h_0 + h_1 - h + (m + 1)(1 - t(r + 1 - m))] \\ & \left. - 4h_1 t\left(\frac{r+1}{2} - m\right)^2 \right\} \end{aligned}} \qquad (8.186)$$

The rest of this section is devoted to a detailed proof of this result. Recall the definition (8.26) of $\Delta_{r,1}(t)$ as the formal determinant of the operator $D_{r,1}(t)$ of Eq. (8.22). The substitution $L_{-r} \to l_{-r}(\lambda, \mu)$ (cf. Eq. (8.182)) leads to the formal determinant of the matrix operator $D_{r,1}(L_{-r} \to l_{-r})$, namely

$$\theta_{r,1}(\lambda, \mu) = \det\left[-J_- + \sum_{k=0}^{r-1}(-t)^k J_+^k \left(\mu + \frac{r-1}{2} + J_0 - \lambda k\right)\right]$$

$$= \det\left[-J_- + \frac{1}{1+tJ_+}\left(\mu + \frac{r-1}{2} + J_0\right) + \lambda \frac{tJ_+}{(1+tJ_+)^2}\right] \quad (8.187)$$

$$= \det\left[tJ_- + \frac{1}{1-J_+}\left(J_0 + \mu + \frac{r-1}{2}\right) + \frac{\lambda tJ_+}{(1-J_+)^2}\right]$$

where we used the automorphism

$$J_\pm \to -t^{\mp 1}J_\pm \qquad J_0 \to J_0 \qquad (8.188)$$

to obtain the last equality. We now proceed as follows. We are left with the computation of the determinant (8.187) of an operator involving $(1-J_+)^{-1}$ and $(1-J_+)^{-2}$. These terms imply a proliferation of powers of J_+, which we wish to eliminate. In a first step, we will "reduce" the operator by performing an appropriate change of basis, in which the term $(1-J_+)^{-2}$ disappears. In a second step, we will dispose of the second term $(1-J_+)^{-1}$ and finally evaluate the determinant. Since the matrix J_+ is nilpotent ($J_+^r = 0$), the matrix

$$U_\gamma = \frac{1}{(1-J_+)^\gamma} = \sum_{k\geq 0}\frac{\gamma(\gamma+1)\cdots(\gamma+k-1)}{k!}(J_+)^k$$

is well defined, as well as its inverse $U_\gamma^{-1} = U_{-\gamma}$. Using the commutation relations

$$[J_0, U_\gamma] = \gamma J_+ U_{\gamma+1}$$
$$[J_-, U_\gamma] = -2\gamma U_{\gamma+1}J_0 - \gamma(\gamma+1)U_{\gamma+2}J_+$$
$$[J_+, U_\gamma] = 0$$

we find that

$$U_\gamma^{-1}\left(tJ_- + \frac{1}{1-J_+}\left(\mu + \frac{r-1}{2} + J_0\right) - \frac{\lambda J_+}{(1-J_+)^2}\right)U_\gamma$$

$$= tJ_- + \frac{1}{1-J_+}\left(\mu + \frac{r-1}{2}\right) + (1-2\gamma t)J_0 + \frac{\gamma - t\gamma(\gamma+1) - \lambda}{(1-J_+)^2}J_+$$

$$(8.189)$$

The formal determinant of an operator D may be evaluated in any new basis preserving f_r. The matrix U_γ is precisely the matrix of such a change of basis: U_γ and its inverse, which are upper triangular with ones on the diagonal, do not modify the highest component f_r, nor

$$f_0 = \det(U_\gamma^{-1}DU_\gamma)f_r = \det(D)f_r \qquad (8.190)$$

To eliminate the $(1-J_+)^{-2}$ term in (8.189), we pick for γ any of the two roots of

$$\gamma - t\gamma(\gamma+1) - \lambda = 0 \qquad (8.191)$$

Then, there follows the simple result:

$$\theta_{r,1}(\lambda, \mu) = \det\left[tJ_- + \frac{1}{1-J_+}(\mu + \frac{r-1}{2} + (1-2\gamma t)J_0\right] \qquad (8.192)$$

Finally, we multiply the above by $1 = \det(1 - J_+)$ and use the action of J_+J_- and J_0 on the components f_j:

$$J_+J_-f_j = [\frac{r^2}{4} - (J_0 - \frac{1}{2})^2]f_j$$

$$J_0f_j = \frac{1}{2}(r - 2j + 1)f_j \qquad (8.193)$$

in order to rewrite

$$\theta_{r,1}(\lambda, \mu) = \det\left[tJ_- t(\frac{r^2-1}{4} + J_0(1-J_0)) + \mu + \frac{r-1}{2} + (1-2\gamma t)J_0\right]$$

$$= \prod_{m=1}^{r}\left[\mu + \frac{r-1}{2} - tm(r-m) + \frac{r+1-2m}{2}(1-2\gamma t)\right] \qquad (8.194)$$

with γ a root of Eq. (8.191). This expression turns out to be independent of which particular solution γ we choose. Grouping the terms with m and $r + 1 - m$ in the product (8.194), we find, for r even,

$$\theta_{r,1}(\lambda, \mu) = \prod_{m=1}^{r/2}\left([\mu + (r-m)(1-tm)]\right.$$

$$\left. \times [\mu + (m+1)(1-t(r+1-m))] + 4\lambda t(\frac{r+1}{2} - m)^2\right) \qquad (8.195)$$

and for r odd,

$$\theta_{r,1}(\lambda, \mu) = \left(\mu + \frac{r-1}{2}(1 - t(r+1)/2)\right)\prod_{m=1}^{\frac{r-1}{2}}\left([\mu + (r-m)(1-tm)]\right.$$

$$\left. \times [\mu + (m+1)(1-t(r+1-m))] + 4\lambda t(\frac{r+1}{2} - m)^2\right) \qquad (8.196)$$

The above two cases (8.195)–(8.196) are summarized in a unique formula for the square of $\theta_{r,1}(\lambda, \mu)$:

$$[\theta_{r,1}(\lambda, \mu)]^2 = \prod_{m=1}^{r}\left([\mu + (r-m)(1-tm)]\right.$$

$$\left. \times [\mu + (m+1)(1-t(r+1-m))] + 4\lambda t(\frac{r+1}{2} - m)^2\right) \qquad (8.197)$$

which, for $\lambda = -h_1$ and $\mu = h_{r,1} + h_1 - h$, yields the desired result (8.186).

8.A.5. Fusion at Work

Knowing the singular vector at level r in $V_{r,1}$ and using it in the fusion $V_{r,1} \otimes M(c, h_1) \to M(c, h)$, we finally arrive at the following result. If $\theta_{r,1} = 0$,

(i) The irreducible module $M(c, h)$ does not occur in the fusion
$M(c, h_{r,1} + r) \otimes M(c, h_1)$.

(ii) The first nonzero term $\psi^{(r_0)}$, $r_0 > 0$ on the r.h.s. of Eq. (8.176) is a singular vector in $V(c, h)$.

(iii) The explicit expression for this singular vector is obtained by transferring the singular vector condition from $V(c, h_0; x_0)$ to the target module $V(c, h; x)$, using the formula (8.166) in the case $x = \frac{1}{2}(x_0 + x_1)$ or its suitable modifications for any other choice of coordinates.

We can use this result in different ways by making various choices for the second module $M(c, h_1)$. In the following, we explore three possibilities, all of them with $h_0 = h_{r,1}(t)$.

(a) $\boxed{V_{2,1} \otimes V_{0,s} \to V_{1,s}}$

First, it is instructive to recover the expression for the singular vector $|\chi_r\rangle$ of Eq. (8.28) in Sect. 8.2. We take $r = 2$, and hence

$$h_0 \equiv h_{2,1}(t) = \frac{3}{4}t - \frac{1}{2}, \qquad (8.198)$$

and we choose

$$h \equiv h_{1,s}(t) = \frac{s^2 - 1}{4t} + \frac{1 - s}{2}. \qquad (8.199)$$

For $\lambda = -h_1$ and $\mu = h_{2,1}(t) + h_1 - h$, the determinant

$$\theta_{2,1}(\lambda, \mu) = \mu(\mu + 1 + t) - \lambda t \qquad (8.200)$$

has two zeros in h_1, namely

$$h_1 = \begin{cases} h_{2,s}(t) = \dfrac{3}{4}t + \dfrac{s^2 - 1}{4t} - s + \dfrac{1}{2} \\[2mm] h_{0,s}(t) = -\dfrac{t}{4} + \dfrac{s^2 - 1}{4t} + \dfrac{1}{2} \end{cases} \qquad (8.201)$$

According to Eq. (8.2), the first value corresponds to a reducible module, with a singular vector at level $2s$, whereas the second one corresponds directly to an irreducible Verma module. We choose the second possibility, $h_1 = h_{0,s}(t)$, which guarantees that no extra information about possible other singular vectors of the second module is overlooked. The above analysis guarantees the existence of a singular vector in the target Verma module $V_{1,s}$. We compute it explicitly. The singular vector of $V_{2,1}$ at level 2 is easily found to be

$$|\chi_2\rangle = \Delta_{2,1}(t)|h_{2,1}(t)\rangle \qquad \Delta_{2,1}(t) = L_{-1}^2 - t L_{-2} \qquad (8.202)$$

We then perform the *transfer* of the action of $\Delta_{2,1}(t) \otimes \mathbb{I}$ on $V_{2,1} \otimes V_{1,s}$ to an action on the target module $V(c,h)$. We fix the coordinates to be $x_1 = x$, $z = x_0 - x_1$. Then, from Eq. (8.170), we have

$$\mathcal{F}(L_{-1} \otimes \mathbb{I}) = \frac{d}{dz}$$

$$\mathcal{F}(L_{-2} \otimes \mathbb{I}) = \frac{h_1}{z^2} - \frac{1}{z}\frac{d}{dz} + \sum_{k=1}^{\infty} z^{k-2} L_{-k}$$

(8.203)

hence the transferred action on $f_0 f_1 \sim f$ is

$$(\Delta_{2,1}(t) f_0)(x_0) f_1(x_1) \rightarrow$$

$$[\frac{d^2}{dz^2} - \frac{t}{z^2}(h_1 - z\frac{d}{dz} + \sum_{k=1}^{\infty} z^k L_{-k}] \frac{1}{z^{h_0+h_1-h}} \sum_{p=0}^{\infty} z^p f^{(p)}(x) = 0$$

(8.204)

whose vanishing is a direct consequence of the vanishing of the singular vector in $M_{2,1}$. Since, in the present case, we have

$$\mu = h_0 + h_1 - h = (r - 1 + t)/2$$

(8.205)

we finally obtain

$$\frac{p(r-p)}{t} f^{(p)} + \sum_{k \geq 1} L_{-k} f^{(p-k)} = 0$$

(8.206)

This yields the descent equations (8.174) for $r \rightarrow s$ and $t \rightarrow 1/t$, which determines the singular vector of level s in $V_{1,s}$.

Now we have another understanding of these descent equations. We can follow step by step the cascade of equations determining the formal determinant of $\Delta_{1,s}(t)$. This requires a slight alteration of the $su(2)$ representation (8.19) used before, by exchanging $tJ_+ \leftrightarrow J_-/t$. Then Eq. (8.206) coincides exactly with the descent equations obtained by writing

$$D_{1,s}(t) \begin{pmatrix} f_1 \\ f_2 \\ \vdots \\ f_s \end{pmatrix} = \begin{pmatrix} f_0 \\ 0 \\ \vdots \\ 0 \end{pmatrix}$$

(8.207)

in components, and identifying $f_j \equiv f^{(r-j)}$.

Eq. (8.206) has yet another interpretation. We define

$$T^{(-)}(z) \equiv \sum_{k \geq 1} z^{k-2} L_{-k}$$

(8.208)

as the negative mode part of the stress tensor $T(z)$. Then we get a second-order differential equation for

$$F(z) = z^{h-h_0-h_1} \sum_{p \geq 0} z^p f^{(p)}$$

(8.209)

namely

$$\left[\frac{d^2}{dz^2} - \frac{t}{z^2}(h_1 - z\frac{d}{dz}) + T^{(-)}(z) \right] F(z) = 0 \tag{8.210}$$

(b) $\boxed{V_{r,1} \otimes V_{1,s} \to V_{r,s}}$

We set

$$h_0 = h_{r,1}(t) = \frac{r^2 - 1}{4}t + \frac{1 - r}{2}$$

$$h_1 = h_{1,s}(t) = \frac{s^2 - 1}{4t} + \frac{1 - s}{2} \tag{8.211}$$

$$h = h_{r,s}(t) = \frac{r^2 - 1}{4}t + \frac{s^2 - 1}{4t} + \frac{1 - rs}{2}$$

for which it is clear that the determinant $\theta_{r,1}(\lambda = -h_1, \mu = h_0 + h_1 - h)$ vanishes (indeed, one easily checks that the factor corresponding to $m = 1$ vanishes in Eq. (8.186)). In this case, we have to use this information about singular vectors of the first and the second module to obtain constraints on the singular vectors of the target module. It is important to notice that this information from the first and second modules is needed to fully characterize the target singular vector, in contrast to the previous case (a) where the singular vector of the first module was sufficient. The best we can hope for here is to obtain a system of coupled equations determining the target singular vector. Due to this intrinsic complication, we prefer to concentrate on the next possibility, in which the second module is directly irreducible, so that all the information is exhausted by implementing the singular vector condition of the first module.

(c) $\boxed{V_{r+1,1} \otimes V_{0,s} \to V_{r,s}}$

We set

$$h_0 = h_{r+1,1}(t) = \frac{(r + 1)^2 - 1}{4}t - \frac{r}{2}$$

$$h_1 = h_{0,s}(t) = -\frac{t}{4} + \frac{s^2 - 1}{4t} + \frac{1}{2} \tag{8.212}$$

$$h = h_{r,s}(t) = \frac{r^2 - 1}{4}t + \frac{s^2 - 1}{4t} + \frac{1 - rs}{2}$$

for which $\theta_{r,1}(\lambda = -h_1, \mu = h_0 + h_1 - h)$ vanishes (as in case (b), the term $m = 1$ vanishes in Eq. (8.186)). As in case (a), the second module is irreducible, so no extra condition has to be implemented except the singular vector condition for $V_{r,1}$. The transfer of these conditions to the target module is readily done by substituting for the L_{-m}'s the equations (8.170)

$$\mathcal{F}(L_{-1} \otimes \mathbb{I}) = \frac{d}{dz}$$

$$\mathcal{F}(L_{-r} \otimes \mathbb{I}) = \frac{(-1)^r}{z^r}[h_1(r-1) + z\frac{L_{-1}-d}{dz}] + \sum_{k \geq 0} z^k \binom{k}{k+r-2} L_{-k-r}$$

$$(8.213)$$

This transfers the action of the operator $\Delta_{r+1,1}(t)$ on the highest-weight state f_0 at x_0 to that of an operator $\gamma_{r+1,1}(t)$ at x, on

$$F(z; x) = z^{h-h_0-h_1} \sum_{p \geq 0} z^p f^{(p)}(x) \qquad (8.214)$$

The target singular vector vanishing condition thus takes the form

$$\gamma_{r+1,1}(t)F(z; x) = 0 \qquad (8.215)$$

This defines in $V(c, h)$ a set of intermediate stages (descent equations) between $f = f^{(0)}$ and $f^{(n)}$, the singular vector of the target module. The p-th stage of these recursions takes the form

$$(-1)^{r-1}\theta_{r,1}(\lambda = -h_1, \mu = h_0 + h_1 - h - p)f^{(p)} = \text{Pol}(L_{-m}; f^{(k<p)}) \quad (8.216)$$

where "Pol" denotes for each stage some polynomial of the L_{-m}'s acting on the higher components $f^{(k)}, k < p$. For $p < rs$, this factor does not vanish, and one can solve recursively for $f^{(p)}$ in terms of the $f^{(k<p)}$. For $p = n = rs$, the determinant factor $\theta_{r+1,1}$ vanishes again: it is responsible for the fact that at that level the $p = rs$ stage expresses directly the vanishing of the singular vector of the target module.

8.A.6. The Singular Vectors $|h_{r,s} + rs\rangle$: Summary

The reducible Verma modules have highest weights parametrized by two non-negative integers (r, s) according to Eq. (8.2). If $r = 1$ or $s = 1$, there exists a singular vector at level s (resp. r), given by $\Delta_{r,1}(t)|h_{r,1}(t)\rangle$, Eq. (8.26). Otherwise, let $f = f^{(0)}$ be the highest-weight state in $V(c, h)$, set $h_0 = h_{r+1,1}(t), h_1 = h_{0,s}(t)$, and $h = h_{r,s}(t)$, and define the operator

$$\gamma_{r+1,1}(t) \equiv \Delta_{r+1,1}\left[L_{-j} \to \frac{(-1)^j}{z^j}[h_1(j-1) + z(L_{-1} - \frac{d}{dz})] + \right.$$
$$\left. + \sum_{k \geq 0} z^k \binom{k+j-2}{k} L_{-k-j}\right]$$

$$(8.217)$$

Then the equation

$$\gamma_{r+1,1}(t)z^{h-h_0-h_1} \sum_{j \geq 0} z^j f^{(j)} = 0 \qquad (8.218)$$

determines recursively the $f^{(j)}$'s, for $0 < j < rs$ in terms of f and yields, at level rs, an equation of the form

$$\Delta_{r,s}(t)f = 0 \qquad (8.219)$$

up to a multiplicative nonzero factor. $\Delta_{r,s}(t)$ is a polynomial of the L_{-m}'s of total degree rs. This equation defines a singular vector of level rs in the module $V_{r,s}$, as $\Delta_{r,s}(t)|h_{r,s}(t)\rangle$. Moreover, the intermediate components $f^{(j)}$, $0 \le j \le rs$ satisfy the descent equations (8.174).

The only fact we did not prove in this section (and which is ensured by the Kac determinant formula (8.1)) is that the Verma modules whose highest weights are *not* of the form (8.2) with (r, s) strictly positive integers are indeed irreducible. Fortunately, we did not need this information to carry out a thorough study of the irreducible representations of the Virasoro algebra for minimal models, as the latter are all based on Verma modules with highest weights of the form (8.2), with $r, s \ge 1$.

We now give an example of the power of this last result to yield explicit expressions for singular vectors. Suppose we want to write the singular vector at level $2s$ in the module $V_{2,s}$. Following the above recipe, we consider the fusion $V_{3,1} \otimes V_{0,s} \to V_{2,s}$. We start from the expression (8.27)

$$\Delta_{3,1}(t) = L_{-1}^3 - 4tL_{-1}L_{-2} + 2t(2t + 1)L_{-3} \qquad (8.220)$$

and perform the substitutions (8.170):

$$L_{-1} \to \frac{d}{dz}$$

$$L_{-2} \to \frac{1}{z^2}[h_1 + z(L_{-1} - \frac{d}{dz})] + \sum_{k \ge 0} z^k L_{-k-2} \qquad (8.221)$$

$$L_{-3} \to -\frac{1}{z^3}[2h_1 + z(L_{-1} - \frac{d}{dz})] + \sum_{k \ge 0}(k + 1)z^k L_{-k-3}$$

With

$$\begin{aligned}
\lambda &= -h_1 = -h_{0,s} \\
\mu &= h_0 + h_1 - h = h_{3,1} + h_{0,s} - h_{2,s} = t - 3s/2
\end{aligned} \qquad (8.222)$$

we find, in components,

$$\theta_{3,1}(\lambda, \mu - p)f^{(p)} = -4t \sum_{k=1}^{p}(p - k - \mu)L_{-k}f^{(k-p)}$$

$$+ 2t(2t + 1)\sum_{k=1}^{p}(k - 2)L_{-k}f^{(k-p)} \qquad (8.223)$$

Using again the slightly modified representation of $su(2)$ obtained from (8.19) by exchanging $tJ_+ \leftrightarrow J_-/t$, we find the operator $\Delta_{2,s}$ which creates the singular vector at level $2s$ in $V_{2,s}$ as the formal determinant

$$\Delta_{2,s}(t) = \det\left[-\frac{1}{4t}J_-(2J_0 + (2t+1)) + \sum_{k\geq 0}(J_+)^k(2J_0 - (2t+1)k)L_{-k-1}\right]$$

$$(8.224)$$

Exercises

8.1 Prove Eq. (8.17).

8.2 *Dyson–Macdonald identity for the* $(3,2)$ *minimal model*
a) We consider the minimal model with $(p = 3, p' = 2)$. Compute its central charge. Check that the Kac table reduces to the identity operator.
b) From the Virasoro algebra commutation relations, show that all the conformal descendants of the identity operator are themselves singular vectors.
c) Deduce the value of the character $\chi_{(1,1)}$ of the $(p = 3, p' = 2)$ minimal model.
Result: $\chi_{(1,1)}(q) = 1$.
d) Use this result to prove the Dyson–Macdonald identity

$$\sum_{n\in\mathbb{Z}} q^{n(6n-1)} - q^{(2n+1)(3n+1)} = \prod_{n\geq 1}(1 - q^n)$$

8.3 *The limit* $c \to 1$ *of the minimal Virasoro representations.*
a) Compute the limit when $m \to \infty$ of the Virasoro minimal characters $\chi_{(r,1)}(q)$ for the minimal models with $(p = m+1, p' = m)$ (with $q < 1$).
b) Assuming that this limit is correct, deduce the structure of the corresponding reducible Verma modules at $c = 1$.
Result: When $h = n^2/4$, n any nonnegative integer, the module $V(c = 1, h)$ is reducible, and contains exactly one submodule $V(c = 1, h')$, with $h' = (n+2)^2/4 = h+n+1$. This result is exact, and it can be shown that these are the only reducible modules at $c = 1$.

8.4 *Characters of the full* $(6,5)$ *minimal model*
Write all the characters of the representations of the minimal model with $p = 6, p' = 5$ (only part of them are given in Table 8.1). Check the singular vector structure on the small q expansion of these characters up to order 6.

8.5 Prove by recursion on j that L_1 and L_2 act on f_j, as claimed in Eqs. (8.32)–(8.33).

8.6 *Benoit–Saint-Aubin formula for* $(r, 1)$ *singular vectors*
Compute the determinant (8.26) by simple elimination. The result should take the form of a sum over monomials $L_{-n_1}L_{-n_2}\cdots L_{-n_k}$ where the indices are not necessarily ordered. Rearrange these terms to recover the Benoit–Saint-Aubin formula (8.34).

8.7 Compute the square of the operator $\Delta_{r,1}(t)$ of Eq. (8.26) modulo $L_{-3}, L_{-4}, \ldots$ as a function of L_{-1} and L_{-2} only ($[L_{-1}, L_{-2}] = L_{-3} = 0$).
Result: $\Delta_{r,1}(t) = \prod_{m=1}^{r}(L_{-1}^2 - t(r+1-2m)^2 L_{-2})$.
Hint: Write $\Delta_{r,1}(t) = \det(L_{-1} + J_2\sqrt{4tL_{-2}})$, where J_2 is defined by

$$J_\pm = J_1 \pm iJ_2 \qquad\qquad (8.225)$$

Conclude by noting that J_2 has the same spectrum as J_0.

8.8 *Covariant differential operators*

Let F_h denote the set of differential forms of weight h in the real variable x. The elements of F_h are those fields transforming, under an infinitesimal change of variables, as

$$\tilde{\phi}(\tilde{x})d\tilde{x}^h = \phi(x)dx^h \qquad \forall \phi \in F_h$$

A differential operator of degree r

$$D_r = d^r + a_2(x)d^{r-2} + \cdots + a_r(x)$$

where $d \equiv d/dx$, is said to be covariant if it is a map from F_h to F_{h+r}.

a) Prove that $h = -(r-1)/2$.

Hint: Let $\phi_1, \ldots, \phi_r \in F_h$ generate the kernel of D_r (the ϕ's form a basis of the set of solutions of the differential equation $D_r f = 0$), then their Wronskian $\det[\phi_i^{(j-1)}(x)]_{1\le i,j\le r}$ is a constant, that is, an element of F_0.

b) Prove that the covariance condition amounts to

$$\tilde{D}_r = \tilde{d}^r + \tilde{a}_2(\tilde{x})\tilde{d}^{r-2} + \cdots + \tilde{a}_r(\tilde{x}) = \varphi^{h+r}D_r\varphi^{-h}$$

where $\varphi = dx/d\tilde{x}$ is the Jacobian of the coordinate transformation and $\tilde{d} = d/d\tilde{x} = \varphi d$.

c) Deduce the following transformation property of the function a_2 under an infinitesimal change of variables

$$\tilde{a}_2(\tilde{x})d\tilde{x}^2 = a_2(x)dx^2 + \frac{r(r^2-1)}{12}\{x,\tilde{x}\}d\tilde{x}^2$$

where the bracket denotes the Schwarzian derivative

$$\{g(x), x\} = \frac{g'''(x)}{g'(x)} - \frac{3}{2}\left(\frac{g''(x)}{g'(x)}\right)^2$$

The prime symbol stands for differentiation with respect to x. The function $a_2(x)$ is the classical analogue of the Virasoro stress tensor. In the following, we set $c_r = r(r^2-1)/12$.

d) Show that for a coordinate $\tilde{x}$ where $a_2(\tilde{x}) = 0$, the function $b(x) = \varphi'(x)/\varphi(x)$ is a solution of the Riccati equation

$$b'(x) - \frac{b^2(x)}{2} = \frac{a_2(x)}{c_n} \equiv 2s(x)$$

e) Prove that the differential operator

$$\Delta_r = (d + \frac{r-1}{2}b(x))(d + \frac{r-3}{2}b(x))\cdots(d - \frac{r-1}{2}b(x)) \tag{8.226}$$

acts covariantly from $F_{-\frac{r-1}{2}}$ to $F_{\frac{r+1}{2}}$.

Hint: It can be written as $\Delta_r = \varphi^{(r+1)/2}(\varphi^{-1}d)^r\varphi^{(r-1)/2}$.

Prove that Δ_r is a function of $s(x) = \frac{1}{2}(b' - b^2/2)$ only.

Hint: In the expression of Δ_r, change $b \to b + \delta b$ while $2s = b' - \frac{b^2}{2}$ remains fixed, i.e., $\delta b' = b\delta b$, then use $(d - (\alpha + 1)b)\delta b = \delta b(d - \alpha b)$ to prove that $\delta\Delta_r = 0$.

f) Prove that Δ_r, defined in (8.226), is the formal determinant of the $r \times r$ matrix differential operator

$$B_r = -J_- + d\mathbb{I} - bJ_0, \tag{8.227}$$

with $J_\pm$ and J_0 defined in (8.19), and the formal determinant defined as in (8.25). Give an alternative proof of the covariance of Δ_r by studying its action on the components $f_1, f_2, \ldots, f_r$ such that

$$B_r \begin{pmatrix} f_1 \\ f_2 \\ \vdots \\ f_r \end{pmatrix} = \begin{pmatrix} f_0 \\ 0 \\ \vdots \\ 0 \end{pmatrix}$$

Hint: Each step $f_j \to f_{j-1}$ is covariant ($d\mathbb{I} - bJ_0$ is called a covariant derivative). Show that the formal determinant is invariant under unitary gauge transformations $B \to B' = U^\dagger B U$, U any $r \times r$ upper triangular unitary matrix. Apply this to $B = B_r$ of Eq. (8.227), and $U = e^{bJ_+/2}$ to show that Δ_r is a function of $s(x) = \frac{1}{2}(b' - \frac{b^2}{2})$ only.
Hint: $B'_r = -J_- + d\mathbb{I} + \frac{1}{2}(b' - \frac{b^2}{2})J_+$.
The operator Δ_r, as a function of $s(x)$, is the classical analogue of the quantum operator $\Delta_{r,1}(t)$, which creates the singular vector of $V(h_{r,1}(t), c(t))$. The correspondence is known as the *classical limit* of the Virasoro algebra, under which $t \to 0^+$ (i.e., $c \to -\infty$), and

$$L_{-1} \to d$$

$$-tL_{-k} \to \frac{s^{(k-2)}(x)}{(k-2)!} \quad k = 2, 3, \ldots,$$

We recover the above B'_r by substituting this in the expression (8.25)–(8.26) of $\Delta_{r,1}(t)$, and taking $t \to 0$.

g) More generally, prove that the matrix differential operator

$$C_r = -J_- + d\mathbb{I} + \sum_{m=2}^{\infty} w_{m+1}(x)J_+^m$$

has a covariant formal determinant, provided that the coefficients $w_m(x) \in F_m$ are differential forms. Deduce from this that the most general covariant differential operator acting from $F_{-(r-1)/2}$ to $F_{(r+1)/2}$ is the formal determinant of

$$E_r = -J_- + d\mathbb{I} + s(x)J_+ + \sum_{m=2}^{\infty} w_{m+1}(x)J_+^m \tag{8.228}$$

where $w_m \in F_m$, and $s(x)$ transforms anomalously under a local change of coordinates as

$$\tilde{s}(\tilde{x})d\tilde{x}^2 = s(x)dx^2 + \frac{1}{2}\{x, \tilde{x}\}d\tilde{x}^2$$

Hint: Go to a coordinate where $\tilde{s}$ vanishes; in this coordinate, the operator has the form C_r, and its formal determinant is covariant. Go back to the initial coordinate using the upper triangular unitary gauge of **(g)**.

8.9 *Hypergeometric differential equation, hypergeometric function, and integral representations*
We look for solutions of the *hypergeometric differential equation*

$$\boxed{z(1-z)\partial_z^2 f(z) + (c - (a+b+1)z)\partial_z f(z) - abf(z) = 0} \tag{8.229}$$

a) Writing a series expansion $f(z) = \sum_{n\geq 0} f_n z^n$ find a solution of (8.229). Express it in terms of $(a)_n$, $(b)_n$ and $(c)_n$, where

$$(x)_n = x(x-1)\cdots(x-n+1) \qquad n = 1, 2, \ldots$$
$$(x)_0 = 1$$

This is the hypergeometric function $F(a, b; c; z)$.
Result:

$$F(a, b; c; z) = \sum_{n\geq 0} \frac{(a)_n (b)_n}{(c)_n} z^n \qquad (8.230)$$

b) Deduce from (8.230) that for $a = -n$, $n = 1, 2, 3, \ldots$, the hypergeometric equation (8.229) admits a polynomial solution of degree n.

c) Show using (8.230) that

$$\frac{(c-a)_n(c-b)_n}{(c)_n}(1-z)^{a+b-c-n}F(a,b;c+n;z) = \frac{d^n}{dz^n}\left[(1-z)^{a+b-c}F(a,b;c;z)\right] \qquad (8.231)$$

d) Show that the result of the action of the differential operator

$$z(1-z)\partial_z^2 + (c - (a+b+1)z)\partial_z - ab$$

on the monomial $t^{b-1}(1-t)^{c-b-1}(1-tz)^{-a}$ is a total derivative with respect to t.

e) Deduce that

$$\int_C dt\, t^{b-1}(1-t)^{c-b-1}(1-tz)^{-a}$$

is a solution of the hypergeometric differential equation (8.229), if the complex integration contour C is closed, or if it originates and terminates at zeros of the monomial $t^b(1-t)^{c-b}(1-tz)^{-a-1}$.

f) Prove *Euler's formula* for the hypergeometric function

$$F(a, b; c; z) = \frac{\Gamma(c)}{\Gamma(b)\Gamma(c-b)} \int_0^1 dt\, t^{b-1}(1-t)^{c-b-1}(1-tz)^{-a} \qquad (8.232)$$

where $\Gamma(x)$ denotes Euler's Gamma function

$$\Gamma(x) = \int_0^\infty dt\, t^{x-1} e^{-t}$$

g) Using the integral representation (8.232), prove that

$$F(a, b; c; z) = (1-z)^{c-a-b} F(c-a, c-b; c; z) \qquad (8.233)$$

h) Using Eqs. (8.231) and (8.233), prove that

$$F(a - 1/2, a; 2a; z) = \left(\frac{1 + \sqrt{1-z}}{2}\right)^{1-2a} \qquad (8.234)$$

8.10 *Transforming Eq. (8.71) into the hypergeometric equation (8.229)*

a) Write the differential equation (8.71) after a change of function $H(z, \bar{z}) = |z|^{2\beta_1} |1 - z|^{2\beta_2} K(z, \bar{z})$.

b) Find the constraint on β_1 and β_2 that allows the differential equation for K to reduce to the hypergeometric equation (8.229).
Result:

$$\beta_i(\beta_i - 1)/t + \beta_i - h_i = 0 , \quad \text{for } i = 1, 2 \qquad (8.235)$$

c) Show that these conditions are equivalent to the fusion rule (7.50), with $h \to h_0, h_1 \to h_i$ $(i = 1, 2)$ and $h_2 \to h$, upon interpreting β_1 (resp. β_2) as the leading power of $z_0 - z_1$ (resp. $z_0 - z_2$) in the OPE of $\phi_0 \times \phi_1$ when z_0 tends to z_1 (resp. $\phi_0 \times \phi_2$ when z_0 tends to z_2) within the four-point correlator (8.42).
Hint: When letting $z_0 \to z_i, i = 1, 2$ in the four-point function (8.42), the corresponding OPE reads $\phi_0(z_0) \times \phi_i(z_i) \sim \sum_h (z_0 - z_i)^{h-h_0-h_i} phi(z_i)$, and hence $\beta_i = h - h_0 - h_i$. The quadratic equations (8.235) for β_i are equivalent to those of the $(2, 1)$ fusion rule (7.50).

8.11 *Yang–Lee four-point correlation as solution of a hypergeometric differential equation*

a) Write the differential equation (8.71) in the case $(p = 2, p' = 5)$, and $h_0 = h_1 = h_2 = h_3 = h_{2,1} = -1/5$.

b) Transform it into a hypergeometric differential equation by following the lines of Ex. 8.10 above.

c) Write the two solutions of the hypergeometric equation corresponding to the two possible small z behaviors ($\sim |z|^{2\beta_1}$) of the function H, in terms of hypergeometric functions. These are the two conformal blocks of the four-point correlation function. The full correlator is a sesquilinear combination of these two functions. The precise determination of this combination is postponed to Chap. 9.
Result: $\beta_1 = \beta_2 = 2/5$ or $1/5$, corresponding respectively to the fusion rules $\phi_{(2,1)} \times \phi_{(2,1)} \to I$ and $\phi_{(2,1)} \times \phi_{(2,1)} \to \phi_{(2,1)}$. The corresponding conformal blocks read respectively

$$F(3/5, 4/5; 6/5; z), \qquad F(3/5, 2/5; 4/5; z) \qquad (8.236)$$

8.12 *Energy and spin four-point correlations in the Ising model, as solutions of hypergeometric differential equations*

a) Write the differential equation (8.71) in the case $(p = 4, p' = 3)$ and $h_0 = h_1 = h_2 = h_3 = h_{2,1} = 1/2$.

b) Transform it into a hypergeometric differential equation by following the lines of Ex. 8.10 above.

c) Write the two solutions of the hypergeometric equation corresponding to the two possible small z behaviors ($\sim |z|^{2\beta_1}$) of the function H, in terms of hypergeometric functions.
Result: $\beta_1 = \beta_2 = -1$ or $2/3$, which would in principle correspond to the fusion rules $\phi_{(2,1)} \times \phi_{(2,1)} \to I$ and $\phi_{(2,1)} \times \phi_{(2,1)} \to \phi_{(3,1)}$. However, for the Ising theory, the $(3, 1)$ representation lies outside of the Kac table, therefore this last fusion is not allowed. This will eventually result in a vanishing coefficient for the second conformal block. The two conformal blocks read respectively

$$F(-2, -1/3; -2/3; z) = 1 - z + z^2, \qquad F(4/3, 3; 8/3; z) \qquad (8.237)$$

d) Repeat the above steps for the four-point function of the spin operator, with $h_0 = h_1 = h_2 = h_3 = h_{1,2} = 1/16$.

Hint: The fields $\phi_{(2,1)}$ and $\phi_{(1,2)}$ are exchanged under the substitution $t \leftrightarrow 1/t$. The two resulting blocks, corresponding respectively to $\beta_1 = \beta_2 = -1/8$ (fusion $\phi_{(1,2)} \times \phi_{(1,2)} \to I$) and $3/8$ (fusion $\phi_{(1,2)} \times \phi_{(1,2)} \to \phi_{(1,3)}$), read

$$F(3/4, 1/4; 1/2; z), \qquad F(1/4, 3/4; 3/2; z)$$

e) Using (8.234), show that the two conformal blocks for the Ising four-spin correlation function read

$$\left(\frac{1 + \sqrt{1-z}}{2} \right)^{\frac{1}{2}}, \qquad \left(2\frac{1 - \sqrt{1-z}}{z} \right)^{\frac{1}{2}}$$

8.13 *Differential equation for $\phi_{(3,1)}$*
Write explicitly the third-order differential equation (8.39) for $V(c, h_0) = V_{3,1}$. In the case of a four-point function, perform the elimination of ∂_{z_0}, ∂_{z_1}, ∂_{z_2} and ∂_{z_3} in terms of the cross-ratio $z = z_{01}z_{23}/z_{02}z_{13}$ to obtain an *ordinary* differential equation of third order for the four-point correlator $H(z, \bar{z}) = |z|^{2\mu_{01}} |1 - z|^{2\mu_{02}} G(z, \bar{z})$.
Result:

$$\left\{ \frac{1}{2t} \partial_z^3 + \frac{2z-1}{z(z-1)} \partial_z^2 + \left[\frac{h - 2h_1}{z^2} + \frac{h - 2h_2}{(z-1)^2} \right] \partial_z \right.$$
$$- 2h \left[\frac{h_1}{z^2} + \frac{h_2}{(z-1)^2} + \frac{2(h+1) + h_1 + h_2 - h_3}{z(z-1)} \right.$$
$$\left. \left. + \frac{(h + h_1 + h_2 - h_3)(2z-1)}{[z(z-1)]^2} \right] \right\} H(z, \bar{z}) = 0$$

8.14 *Sum rule for singular vectors*
Consider a three-point correlator of the form

$$\langle \phi_0(z_0)\phi_1(z_1)\phi_2(z_2) \rangle \tag{8.238}$$

where $V(c, h_0)$ contains a singular vector at level n_0, given by

$$|c, h_0 + n_0\rangle = \sum_Y \alpha_Y L_{-Y} |c, h_0\rangle \tag{8.239}$$

From the explicit form of the correlator and the singular vector vanishing condition for ϕ_0 when $z_0 \to z_1$, deduce a sum rule for the coefficients α_Y.

8.15 *(r,1) fusion rules in minimal models*
Prove the following formula for the determinant $\theta_{r,1}(\lambda, \mu)$ of Eqs. (8.195)–(8.196) with

$$\lambda = -h_1 \equiv -h_{k,l}(t)$$
$$\mu = h_0 + h_1 - h \equiv h_{r,1}(t) + h_{k,l}(t) - h$$

and $h_{k,l}(t)$ as in Eq. (8.2)

$$\theta_{r,1}(\lambda, \mu) = (-1)^r \prod_{m=1}^{r} (h - h_{k-r-1+2m,l}(t))$$

This yields the fusion rules (8.84).

8.16 *Verlinde formula for fusion numbers*

a) We first concentrate on the subalgebra $\mathcal{X}_{p'}$ of the fusion algebra $\mathcal{A}_{p,p'}$ of a minimal model. Show that it is isomorphic to the polynomial ring $\mathbb{C}[x]/U_{p'-1}(x)$, where U are the Chebyshev polynomials of the second kind. Let N_r, $1 \le r \le p' - 1$ be a $(p' - 1) \times (p' - 1)$ matrix representation of $\mathcal{X}_{p'}$, with $N_1 = 1$, $N_2 \equiv N_{(2,1)}$, $N_r \equiv N_{(r,1)}$. Prove that the matrices N_r are simultaneously diagonalizable in an orthonormal basis, and compute the unitary matrix S of the change of basis. Compute the eigenvalues of N_r in terms of the matrix elements $S_{i,j}$, and deduce the following formula result

$$N_{rs}{}^t = \sum_{i=1}^{p'-1} \frac{S_{r,i} S_{s,i} S_{i,t}}{S_{1,i}}$$

known as the *Verlinde formula*. An analogous formula holds for the fusion numbers of $\mathcal{Y}_p$.

Result: $S_{i,j} = \sqrt{\frac{2}{p}} \sin \pi \frac{ij}{p}$.

b) Conclude that the (p, p') minimal fusion rules have the form

$$\mathcal{N}_{(r,s)(m,n)}{}^{(k,l)} = \sum_{i=1}^{p'-1} \sum_{j=1}^{p-1} \frac{S_{(r,s)}^{(i,j)} S_{(m,n)}^{(i,j)} S_{(i,j)}^{(k,l)}}{S_{(1,1)}^{(i,j)}},$$

and compute the matrix elements $S_{(r,s)}^{(i,j)}$.

8.17 *Rings, ideals, and quotients*

A *ring* R is a group (with operation denoted by $+$ and called the addition), endowed with an extra multiplicative law (denoted by $\cdot$ and called the multiplication), which is associative and distributive with respect to the addition. A left (resp. right) ideal $I \subset R$ is a subset of R that is stable by left (resp. right) multiplication by any element of R: $\forall x \in R, y \in I$, $x \cdot y \in I$ (resp. $y \cdot x \in I$). A left *and* right ideal is simply called an ideal. For any ideal I, the *quotient ring* R/I is formed by the equivalence classes of the relation $\simeq$ over R

$$x \simeq y \iff \exists z \in I, \text{ s.t. } x = y + z$$

and endowed with the additive and multiplicative structures inherited from R.

a) Show that the set of polynomials with complex coefficients of N variables, $\mathbb{C}[x_1, \ldots, x_N]$, forms a ring for the usual addition and multiplication of polynomials.

b) Show that for any given polynomial $P(x_1, \ldots, x_N)$, the set of polynomials with P in factor, $I = P(x_1, \ldots, x_N)\mathbb{C}[x_1, \ldots, x_N]$ is an ideal.

c) We now restrict ourselves to $N = 1$. Let $P(x)$ be a polynomial of degree p. Show that the quotient ring $Q = \mathbb{C}[x]/P(x)\mathbb{C}[x]$ (often denoted by $\mathbb{C}[x]/P(x)$) is a finite dimensional vector space. Compute its dimension.

Hint: The equivalence relation $\simeq$ is the identity modulo $P(x)$, namely

$$P_1(x) \simeq P_2(x) \iff \exists P_3(x), \quad \text{such that} \quad P_1(x) = P_2(x) + P_3(x)P(x)$$

By virtue of the polynomial Euclidian division, the representatives of the classes may be taken of degrees $0, 1, \ldots, p - 1$, leading to a p-dimensional vector space.

8.18 *The minimal polynomial of a matrix G*

Let $\chi_G(x) = \det(x\mathbb{I} - G)$ be the *characteristic polynomial* of a given matrix G. G is assumed to be diagonalizable, with r distinct eigenvalues λ_i with multiplicities m_i. In other words, the vector space E over which G acts ($E = \mathbb{R}^p$, p the size of G) is a direct sum of eigenspaces E_i of G for λ_i: $E = \oplus_{i=1}^r E_i$, and $\dim(E_i) = m_i$.

a) Show that

$$\chi_G(x) = \prod_i (x - \lambda_i)^{m_i}$$

b) Let $\Pi_G(x)$ be the *monic* polynomial of smallest degree s (i.e., of the form $\Pi_G(x) = x^s + O(x^{s-1})$), such that $\Pi_G(G) = 0$. Show its existence and uniqueness. $\Pi_G(x)$ is called the *minimal polynomial* of G.

c) Show that the degree of $\Pi_G(x)$ is the dimension of the vector space of matrices generated by the successive powers of G.
Hint: Show that $\mathbb{I}, G, G^2, \ldots, G^{s-1}$ are linearly independent, with $s = \deg(\Pi_G)$.

d) Show that the above vector space is nothing but the quotient ring $\mathbb{C}[x]/\Pi_G(x)$ (see Ex. 8.17 above for a definition).

e) Prove that $s \geq r$.

f) Prove that

$$\Pi_G(x) = \prod_{i=1}^{r} (x - \lambda_i)$$

Hint: Show that $\prod_i (G - \lambda_i \mathbb{I}) = 0$ by restricting its action to the eigenspaces E_j.

8.19 *Fusion algebra attached to a graph*
Given a connected nonoriented graph $\mathcal{G}$ (a set of vertices $v(\mathcal{G})$ with nonoriented links), one defines its adjacency matrix as

$$G_{a,b} = \begin{cases} 1 & \text{if } a \text{ is linked to } b \\ 0 & \text{otherwise} \end{cases}$$

with $a, b \in v(\mathcal{G})$.

a) Show that G is symmetric, therefore diagonalizable in an orthonormal basis. Assuming that all eigenvalues of G are distinct, the unitary matrix S of the change of basis is fixed up to a phase. Let E denote the set of labels of the eigenvalues of G. The matrix elements of S are $S_{a,m}$, with $a \in v(\mathcal{G})$ and $m \in E$. Assume also that G has a *unit* vertex denoted by $1 \in v(\mathcal{G})$, such that $G_{1,a} = \delta_{a,a_0}$: 1 is an endpoint of the graph, linked to only one vertex a_0. Assume also that no matrix element of the form $S_{1,m}$ vanishes. Compute the eigenvalues of G in terms of S.

b) We define graph fusion numbers by the formula

$$\mathcal{N}_{ab}{}^c = \sum_{m \in E} \frac{S_{a,m} S_{b,m} S_{c,m}^*}{S_{1,m}}$$

$a, b, c \in v(\mathcal{G})$. Prove that these numbers define a commutative and associative algebra $\mathcal{A}$, with generators ϕ_a, $a \in v(\mathcal{G})$ and relations

$$\phi_a \phi_b = \sum_c \mathcal{N}_{ab}{}^c \, \phi_c$$

called the graph fusion algebra. Show that the matrices N_a, with entries $[N_a]_{b,c} = \mathcal{N}_{ab}{}^c$, form a representation of $\mathcal{A}$, polynomially generated by $G = N_{a_0}$. Compute the eigenvalues of N_a in terms of S.

c) Let

$$P(x) = \det(x\mathbb{I} - G)$$

be the characteristic polynomial of G. Prove that the graph fusion algebra $\mathcal{A}$ is isomorphic to the quotient ring $\mathbb{C}[x]/P(x)$ of polynomials of x, modulo $P(x)$. In particular, compute the polynomial generators of the ring, $P_a(x)$, $a \in v(\mathcal{G})$, defined by the relation $N_a = P_a(G)$, in terms of the entries of the matrix S.

Hint: P_a is the Lagrange interpolation polynomial between the set of (distinct) eigenvalues of G and that of N_a.

d) Examples. Show that the graph A_{p-1} with adjacency matrix

$$[A_{p-1}]_{i,j} = \delta_{j,i+1} + \delta_{j,i-1} \quad 1 \le i,j \le p-1$$

has a fusion algebra isomorphic to $\mathcal{Y}_p$. Show that the graph E_6 with adjacency matrix

$$G = \begin{pmatrix} 0 & 1 & 0 & 0 & 0 & 0 \\ 1 & 0 & 1 & 0 & 0 & 0 \\ 0 & 1 & 0 & 1 & 0 & 1 \\ 0 & 0 & 1 & 0 & 1 & 0 \\ 0 & 0 & 0 & 1 & 0 & 1 \\ 0 & 0 & 1 & 0 & 1 & 0 \end{pmatrix}$$

admits a fusion algebra, with the first entry of G as unit vertex. Namely, prove that all the eigenvalues of G are distinct. Compute the polynomials $P_a(x)$ and prove that the $N_{ab}{}^c$ are nonnegative integers.

e) Show that the adjacency matrix of the D_4 graph

$$G = \begin{pmatrix} 0 & 1 & 0 & 0 \\ 1 & 0 & 1 & 1 \\ 0 & 1 & 0 & 0 \\ 0 & 1 & 0 & 0 \end{pmatrix}$$

has some degenerate eigenvalue. Strictly speaking, the corresponding graph fusion algebra is ill-defined. Show, however, that there exists one particular choice of unitary change of basis S that diagonalizes G and leads to nonnegative integers $N_{ab}{}^c$ through the relation

$$N_{ab}{}^c = \sum_i \frac{S_{a,i} S_{b,i} S_{c,i}^*}{S_{1,i}},$$

where we choose the unit vertex 1 to be that of the first entry of G (numbered 1, 2, 3, 4). Prove that the corresponding algebra $\mathcal{A}$ is polynomially generated by *two* generators $G = N_2, H = N_3$, and is isomorphic to the quotient ring $\mathbb{C}[x,y]/\mathcal{I}(x,y)$, where the ideal $\mathcal{I}$ is generated by the two polynomials

$$\mathcal{I}(x,y) : \quad x^2 - y^2 - y - 1 \quad \text{and} \quad x(y-1).$$

8.20 $N_{(2,2)}$ does not generate the full fusion algebra of minimal models

We consider the fusion algebra of the $(p = 2l+1, p' = 2k)$ minimal theory, and its *adjoint* matrix representation $N_{(r,s)} = N_{(p'-r,p-s)}$, subject to (8.131). We want to prove that the fusion algebra is generally not polynomially generated by $N_{(2,2)}$. Show that the dimension of the fusion algebra is $N = \frac{1}{2}(p-1)(p'-1)$. The matrices $N_{(r,s)}$ are therefore of size $N \times N$. Show that a diagonalizable $N \times N$ matrix generates a dimension N algebra if and only if all its eigenvalues are distinct. We assume the following form for the eigenvalues of $N_{(2,2)}$ (see Ex. 8.16 for a general proof)

$$\beta_{(2,2)}^{(i,j)} = 4\cos\frac{\pi i}{p'}\cos\frac{\pi j}{p}$$

Prove that the eigenvalue 0 of $N_{(2,2)}$ is degenerate l times, and conclude that for $l > 1$ $N_{(2,2)}$ does not generate the whole fusion algebra. On the other hand, for $p = 3$ ($l = 1$), show that the fusion algebra is generated by $N_{(2,2)}$.

8.21 0 is the only possibly degenerate eigenvalue of $N_{(2,2)}$
The notations are as in the previous exercise ($p = 2l + 1, p' = 2k$). We want to show that the only possibly degenerate eigenvalue of $N_{(2,2)}$ is 0. Due to the form of the eigenvalues of $N_{(2,2)}$ (see previous exercise), we look for solutions to the identity

$$\frac{\cos(\pi r/p)}{\cos(\pi r'/p)} = \frac{\cos(\pi s'/p')}{\cos(\pi s/p')}.$$

a) Prove that the above ratio is necessarily a rational number.
Hint: Let $\xi = e^{i\pi/pp'}$, $\alpha = \xi^{p'}$, $\beta = \xi^p$. Prove that

$$a = \sum a_i \alpha^i = \sum b_i \beta^i, \quad a_i, b_i \in \mathbb{Q}$$

is possible only if a is a rational number, by summing the above identity over conjugates of ξ, ξ^c, which preserve α, but describe all the conjugates of β. Use the fact that a sum over all the conjugates of a root of unity is a rational number.

b) Prove that the polynomial

$$\Pi(x) = \prod_{r,r'=1}^{p} (2x \cos \frac{\pi r'}{p} - 2 \cos \frac{\pi r}{p})$$

has integer coefficients, is monic (the coefficient of the highest-degree term is 1), and reciprocal ($\Pi(x) = x^d \Pi(1/x), d = (p-1)^2$, the degree of Π). Show that the only rational roots of Π are ± 1. Deduce that necessarily $r' = r$ or $r' = p - r$, and consequently $s' = s$ or $s' = p' - s$, which completes the desired proof.
Hint: To prove that Π is monic, show the identity

$$\prod_{r=1}^{p-1} 2 \cos \frac{\pi r}{p} = (-1)^l \quad p = 2l + 1.$$

To prove that a monic reciprocal polynomial with integer coefficients has only rational roots ± 1, suppose a/b is a root. Then it could be written $(a/b)^d = \text{integer}/b^{d-1}$, hence $b = 1$, but by reciprocity, if a is a root, $1/a$ is a root too, hence $a = \pm 1$.

8.22 Sum rules for the coefficients β_Y.
For this exercise, we use the notations of Sect. 8.A.1.

a) Show that in the coordinate (8.140), and for z small enough, the three point function $\langle f_0(x_0)f_1(x_1)f(x)\rangle$ given by Eq. (8.139) can be expanded in a convergent power series of z/x, and compute its coefficients (set $g(h_0, h_1, h) = 1$).

b) Apply Eq. (8.36) to determine $\langle L_{-Y}f(x)f(0)\rangle$ explicitly, where the two-point function f is normalized to $\langle f(x)f(0)\rangle = x^{-2h}$.

c) Deduce a sum rule involving the coefficients β_Y of the OPE

$$f_0(x_0)f_1(x_1) \propto \frac{1}{z^{h_0+h_1-h}} \sum_Y z^{|Y|} \beta_Y L_{-Y}f(x)$$

Hint: Write the three-point function in two ways. On the one hand, the expansion of **(a)**, on the other hand substitute the above OPE expression in the three-point function, and use **(b)** to compute it.

Result:

$$\sum_{\substack{1 \le r_1 \le r_2 \cdots \\ \Sigma r_i = n}} \left[(r_1 + 1)h + \sum_{i \ge 2} r_i \right] \left[(r_2 + 1)h + \sum_{i \ge 3} r_i \right] \cdots \beta_{r_1, r_2, \cdots}$$

$$= \frac{1}{2^n} \sum_{p+q=n} (-1)^p \frac{1}{p! q!} \frac{\Gamma(h_1 - h_0 - h + 1)}{\Gamma(h_1 - h_0 - h + 1 - p)} \frac{\Gamma(h_0 - h_1 - h + 1)}{\Gamma(h_0 - h_1 - h + 1 - q)}$$

8.23 *Fusions at level 2*
We use the notations of App. 8.A. Use the equations (8.161) determining the coefficients β_Y at level $|Y| = 2$ to discuss fusion rules. In particular, find the constraints on h_0 and h_1 for the fusion $M(c, h_0; x_0) \otimes M(c, h_1; x_1) \to M(c, h; x)$ to be allowed in the three cases where h is a zero of the Kac determinant $K_2(c, h)$ (Eq. (8.162)).

8.24 *The dependence of the fusion map $\mathcal{F}$ on the points*
We use the notations of App. 8.A. We consider the fusion at arbitrary points $x_0 = x + vz$, $x_1 = x + (v - 1)z, x$

$$\mathcal{F}|c, h_0; x_0\rangle \otimes |c, h_1; x_1\rangle = \frac{1}{z^{h_0 + h_1 - h}} \sum_Y z^{|Y|} \tilde{\beta}_Y L_{-Y} |c, h; x\rangle$$

a) Compute the corresponding transfer equations for $\mathcal{F}(L_{-r} \otimes \mathbb{I})$ and $\mathcal{F}(\mathbb{I} \otimes L_{-r})$ that generalize (8.166), (8.167), and (8.170) (for $v = \frac{1}{2}$ and 0 respectively).

b) Show that the dominant action of the transferred operators of **(a)** (i.e., their L_{-m}-independent piece) does not depend on v.

c) Find the descent equations determining $\tilde{\beta}_Y$, generalizing Eqs. (8.156) and (8.174) for, respectively, $v = \frac{1}{2}$ and $v = 0$.

Notes

The structure of inclusions of Virasoro modules was found by Kac [213] and proved by Feigin and Fuchs [127], and the corresponding characters have been computed by Rocha-Caridi [308]. The first simple expression (8.34) for the $(r, 1)$ singular vectors of the Virasoro algebra is due to Benoit and Saint-Aubin [37]. Based on analogies with the classical limit of the Virasoro structure (see Ex. 8.8, partly based on Ref. [92]) the matrix determinant formula (8.26) was proposed by Bauer, Di Francesco, Itzykson, and Zuber in [29], where a general procedure was also given to write the (r, s) singular vectors explicitly (see App. 8.A).

The fusion rules for the minimal models first appeared in Ref. [36]. The algebraic proof given in Sect. 8.4.3 is based on Ref. [100] (as well as Exs. 8.20 and 8.21). The differential equations of hypergeometric type for minimal model correlation functions involving (2, 1) and (3, 1) operators appeared in Ref. [36], and were solved for the Ising model ($p = 4, p' = 3$) four-point functions. Analogous solutions involving hypergeometric functions were given for the tricritical Ising model ($p = 5, p' = 4$) (Ref. [141]), the three-state Potts model ($p = 6, p' = 5$) (Ref. [107]), the Yang-Lee edge singularity ($p = 5, p' = 2$) (Ref. [66]), and so on.

CHAPTER 9

The Coulomb-Gas Formalism

This chapter describes a representation of the conformal fields of minimal models in terms of vertex operators built from a free boson with special boundary conditions. This representation bears the name of *Coulomb gas* or *modified Coulomb gas*. This terminology comes from the resemblance of the free boson correlator $\langle \varphi(z, \bar{z})\varphi(w, \bar{w}) \rangle = -\ln |z - w|^2$ with the electric potential energy between two unit charges in two dimensions. In Sect. 9.1, we calculate the correlation function of vertex operators and indicate how the symmetry $\varphi \to \varphi + a$ of the boson theory imposes a constraint (the neutrality condition) on this correlation function. We then modify the free-boson action—or, equivalently, the energy-momentum tensor—and this modifies the central charge and the neutrality condition. This section is supplemented by App. 9.A, where the calculation of the modified energy-momentum tensor is detailed. In Sect. 9.2, we introduce the notion of screening operators and describe how the insertion of such operators in bosonic correlation functions allows for a sort of projection onto minimal-model correlation functions. Examples of correlation functions are calculated. Finally, in Sect. 9.3, we explain the general structure of the minimal-model correlation functions in this formalism. Special attention is devoted to the properties of conformal blocks, and the idea of a conformal field theory defined on a surface of arbitrary genus is introduced. The mathematical setting of the Coulomb-gas representation of minimal models (i.e., BRST cohomology of the bosonic Fock spaces) is described in App. 9.B.

§9.1. Vertex Operators

We have seen that the free boson theory, with action

$$S = \frac{1}{8\pi} \int d^2x \, \partial_\mu \varphi \partial^\mu \varphi \qquad (9.1)$$

is conformal with central charge $c = 1$ and with a holomorphic energy-momentum tensor

$$T(z) = -\frac{1}{2} : \partial\varphi \partial\varphi : \tag{9.2}$$

If we restrict ourselves to the holomorphic sector, the primary fields of this theory are the derivative $\partial\varphi$, with conformal dimension $h = 1$, and the vertex operators[1]

$$\mathcal{V}_\alpha(z, \bar{z}) = e^{i\sqrt{2}\alpha\varphi(z,\bar{z})} \tag{9.3}$$

with dimensions

$$h_\alpha = \bar{h}_\alpha = \alpha^2 \tag{9.4}$$

In contrast with Sect. 6.3, we have included a factor of $\sqrt{2}$ in the definition of the vertex operator. This changes the formula for the conformal dimensions.

The full vertex operator decomposes into a product of left × right chiral vertex operators as follows:

$$\mathcal{V}_\alpha(z, \bar{z}) = V_\alpha(z) \otimes \bar{V}_\alpha(\bar{z}) \tag{9.5}$$

Roughly speaking, $V_\alpha(z)$ contains only the left modes of the free boson, plus the zero-mode (cf. Eq. (6.54)):

$$V_\alpha(z) = :e^{i\sqrt{2}\alpha\phi(z)}:$$

$$\phi(z) = \varphi_0 - ia_0 \ln z + i \sum_{n \neq 0} \frac{1}{n} a_n z^{-n} \tag{9.6}$$

where the various mode operators obey the commutation rules

$$[a_n, a_m] = n\delta_{n+m,0} \qquad [\varphi_0, a_0] = i \tag{9.7}$$

We must keep in mind that $\phi(z)$ is not a purely holomorphic field, because of the zero-mode.[2] It is preferable to regard $V_\alpha(z)$ as containing the holomorphic dependence of the full vertex operator $\mathcal{V}_\alpha(z, \bar{z})$ and to bear in mind that V_α is well-defined only within correlation functions, when matched with its antiholomorphic partner $\bar{V}_\alpha$.

Most of the forthcoming Coulomb-gas construction will be chiral: The chiral vertex operators will be used to represent holomorphic conformal blocks of correlation functions in the minimal models. In the following, we will refer to V_α simply as a *vertex operator*, dropping for simplicity the epithet "chiral."

9.1.1. Correlators of Vertex Operators

Because they are built upon a free boson, correlators of vertex operators are easy to calculate. The only subtlety comes from the zero-frequency mode of the boson.

---

[1] Vertex operators are always implicitly normal-ordered.

[2] Strictly speaking, we cannot write $\varphi(z, \bar{z}) = \phi(z) + \bar{\phi}(\bar{z})$ or $\mathcal{V}_\alpha(z, \bar{z}) = V_\alpha(z)\bar{V}_\alpha(\bar{z})$, because the operator φ_0 would be duplicated in the process.

Here we shall argue that the correlator of a string of vertex operators is given by

$$\langle \mathcal{V}_{\alpha_1}(z_1,\bar{z}_1)\cdots \mathcal{V}_{\alpha_n}(z_n,\bar{z}_n)\rangle = \prod_{i<j}|z_i-z_j|^{4\alpha_i\alpha_j} \tag{9.8}$$

provided the following "neutrality" condition is satisfied (otherwise the correlator vanishes):

$$\alpha_1+\alpha_2+\cdots+\alpha_n=0 \tag{9.9}$$

The correlator (9.8) may equivalently be written as

$$\exp\left\{\sum_{i<j}4\alpha_i\alpha_j\ln|z_i-z_j|\right\} \tag{9.10}$$

The exponent is equal to the electric potential energy between n point charges of strength $2\alpha_i$ in two dimensions, hence the name Coulomb gas associated with correlators of vertex operators. The holomorphic part of Eq. (9.8) is written as

$$\langle V_{\alpha_1}(z_1)V_{\alpha_2}(z_2)\cdots V_{\alpha_n}(z_n)\rangle = \prod_{i<j}(z_i-z_j)^{2\alpha_i\alpha_j} \tag{9.11}$$

In the case of two- and three-point functions, this result may be obtained from global conformal invariance. Indeed, from Eqs. (5.25) and (5.26), it is simple to check that

$$\langle V_{\alpha_1}(z_1)V_{\alpha_2}(z_2)\rangle = (z_1-z_2)^{2\alpha_1\alpha_2} \tag{9.12}$$

and

$$\langle V_{\alpha_1}(z_1)V_{\alpha_2}(z_2)V_{\alpha_3}(z_3)\rangle = (z_1-z_2)^{2\alpha_1\alpha_2}(z_2-z_3)^{2\alpha_2\alpha_3}(z_1-z_3)^{2\alpha_1\alpha_3} \tag{9.13}$$

wherein we have used $h_\alpha = \alpha^2$ for the the conformal dimension as well as the neutrality condition.

The general formula (9.11) is a natural generalization of Eqs. (9.12) and (9.13), but cannot be obtained from global conformal invariance alone. Instead, we shall use the following formula, demonstrated in App. 6.A, for combinations $A_i = \alpha_i a + \beta_i a^\dagger$ of a single specie of creation and annihilation operators:

$$\langle e^{A_1}e^{A_2}\cdots e^{A_n}\rangle = \exp\sum_{i<j}^{n}\langle A_iA_j\rangle \tag{9.14}$$

Taking $A_i = i\sqrt{2}\alpha_i\varphi(z_i,\bar{z}_i)$, we obtain

$$\langle A_iA_j\rangle = -2\alpha_i\alpha_j\langle\varphi(z_i,\bar{z}_i)\varphi(z_j,\bar{z}_j)\rangle$$
$$= 2\alpha_i\alpha_j\ln|z_i-z_j|^2 = \ln|z_i-z_j|^{4\alpha_i\alpha_j} \tag{9.15}$$

from which Eq. (9.8) follows. However, one may question the applicability of the above formula to the zero-mode of the boson. Ex. 9.2 provides a more careful proof of Eq. (9.8), based on the mode expansion and the explicit action of the zero-mode. Also, Ex. 9.1 provides an altogether different proof of Eq. (9.8), based on functional methods.

9.1.2. The Neutrality Condition

The neutrality condition (9.9) does not enter the previous calculation, but follows from considering the zero-mode of φ, which was the only element ignored in the above argument. Instead of considering the zero-mode explicitly, as is done in Ex. 9.2, we shall derive the neutrality condition from symmetry considerations. Indeed, the internal symmetry operation $\varphi \to \varphi + a$ leaves the free boson action invariant because the field is massless. The correlator (9.8) should therefore be invariant when this simple symmetry operation is performed. However, when it is performed on this correlator, a phase $\exp ia\sqrt{2}(\alpha_1 + \cdots + \alpha_n)$ appears. This phase must be unity if the correlator is to be invariant, hence the neutrality condition (9.9) holds since a is arbitrary.

We derive the neutrality condition (9.9) in two other ways, in order to see Ward identities and the operator formalism at work. According to Noether's theorem, the symmetry under $\varphi \to \varphi + a$ implies the classical conservation of the current $j^\mu = -\partial^\mu \varphi / 4\pi$; this is an obvious consequence of the equation of motion $\partial_\mu \partial^\mu \varphi = 0$. We consider the Ward identity associated with the symmetry $\varphi \to \varphi + a$ (we invite the reader to retrace the steps leading to the general Ward identity (2.157)). Since the variation of a vertex under the shift is $\delta V_\alpha = i\sqrt{2}\, a\alpha V_\alpha$, the relation (2.157) becomes here

$$-\frac{1}{4\pi}\partial_\mu \langle \partial^\mu \varphi(x) X \rangle = i\sqrt{2}\sum_{k=1}^{n} \alpha_k \delta(x - x_k)\langle X \rangle \tag{9.16}$$

where X stands for the string of vertex operators appearing in Eq. (9.11). If we integrate this relation over all space, we obtain, according to Eq. (5.35),

$$i\sqrt{2}\langle X \rangle \sum_k \alpha_k = -\frac{1}{4\pi}\oint ds_\mu \langle \partial^\mu \varphi(x) X \rangle$$

$$= \frac{i}{4\pi}\oint dz\, \langle \partial\varphi X \rangle - \frac{i}{4\pi}\oint d\bar{z}\, \langle \bar\partial\varphi X \rangle \tag{9.17}$$

In the first equation Gauss's theorem was applied to the integral, and the surface (or rather contour) integral was expressed in terms of holomorphic components in the second equation. Since the integration contours circle around all space, that is, around the point at infinity, the integrands have no singularity outside the contours (there is no vertex at infinity) and the two contour integrals vanish. The constraint (9.9) follows immediately.

Going back to Eq. (9.16), we mention that the representation (5.33) for the delta function allows us to write that Ward identity in holomorphic form as follows:

$$\langle \partial\varphi X \rangle = -i\sqrt{2}\sum_k \frac{\alpha_k}{z - z_k}\langle X \rangle + \text{reg.} \tag{9.18}$$

where "reg." stands for a term regular in z; a similar equation exists for the antiholomorphic part. This is just but the OPE between $\partial\varphi$ and V_α, as calculated in Sect. 6.3.

Condition (9.9) may also be derived within the operator formalism. To this end we define the holomorphic component of this current as

$$J(z) \equiv i\partial\varphi \qquad (9.19)$$

and the associated holomorphic charge as

$$Q = \frac{1}{2\pi i} \oint dz\, J(z) \qquad (9.20)$$

This charge is conserved since it commutes with the Hamiltonian, as is easily verified:

$$
\begin{aligned}
[L_0, Q] &= \frac{1}{2\pi i} \oint dz\, [L_0, \partial\varphi(z)] \\
&= \frac{1}{2\pi i} \oint dz\, (\partial\varphi(z) + z\partial^2\varphi(z)) \qquad (9.21) \\
&= \frac{1}{2\pi i} \oint dz\, \partial(z\partial\varphi(z)) = 0
\end{aligned}
$$

The last equality holds, of course, because $\partial\varphi$ is a holomorphic field, which is not true of φ (that is why Q is not trivial).

The charge q_A of a field $A(w)$ is then defined by the commutator

$$[Q, A(w)] = q_A A(w) \qquad (9.22)$$

Although the primary field $\partial\varphi$ has charge zero, it is not difficult to show that the vertex V_α has charge $\sqrt{2}\,\alpha$, if we remember the OPE of $\partial\varphi$ with V_α:

$$
\begin{aligned}
[Q, V_\alpha(w)] &= \frac{1}{2\pi i} \oint dz\, i\partial\varphi(z)V_\alpha(w) \\
&= \frac{1}{2\pi i} \oint dz\, (\sqrt{2}\alpha V_\alpha(w)\frac{1}{z-w} + \text{reg.}) \qquad (9.23) \\
&= \sqrt{2}\,\alpha V_\alpha(w)
\end{aligned}
$$

Similarly, the charge of a product of vertices is shown to be the sum of the charges of each vertex. Now, unless the symmetry is spontaneously broken, the vacuum expectation value of any operator of nonzero charge must necessarily vanish. We thus recover condition (9.9), up to an overall factor of $\sqrt{2}$.

In the context of string theory, the vertex operators represent the strings (particles) emitted from an interaction vertex, and the charge conservation law described here is merely the conservation of momentum: In string theory, space-time symmetries are internal symmetries on the world-sheet.

9.1.3. The Background Charge

The basic idea of the Coulomb-gas formalism is to place a background charge in the system, making the $U(1)$ symmetry anomalous. This has the effect of modifying the conformal dimensions of the vertex operators and the central charge. This also spoils unitarity, except for discrete values of the central charge and finite sets of vertex operators corresponding to the minimal models.

This is done by coupling the boson to the scalar curvature R of the manifold on which the theory is defined. In a general coordinate system, the action would have the following form:

$$S = \frac{1}{8\pi} \int d^2x \sqrt{g}(\partial_\mu \varphi \partial^\mu \varphi + 2\gamma\varphi R) \qquad (9.24)$$

where γ is a constant. The above action is no longer invariant upon a translation $\varphi \to \varphi + a$. The variation of the action is

$$\delta S = \frac{\gamma a}{4\pi} \int d^2x \sqrt{g}R \qquad (9.25)$$

But the Gauss-Bonnet theorem states that the above expression is a topological invariant:

$$\int d^2x \sqrt{g}R = 8\pi(1 - h) \qquad (9.26)$$

where h is the number of handles in the manifold. The boundary conditions we normally use on the complex plane give it the topology of the Riemann sphere ($h = 0$). Therefore, the variation of the action upon a shift of φ is $\delta S = 2a\gamma$.

The Ward identity associated with the $U(1)$ symmetry will then be modified. If we take into account the additional variation of the action due to the curvature term, Eq. (9.17) becomes

$$i\sqrt{2}\langle X\rangle \sum_k \alpha_k = \frac{i}{4\pi} \oint dz \, \langle \partial\varphi X\rangle - \frac{i}{4\pi} \oint d\bar{z} \, \langle \bar{\partial}\varphi X\rangle + 2\gamma\langle X\rangle \qquad (9.27)$$

We see that condition (9.9) is modified: it is now $\sum_k \alpha_k = -i\sqrt{2}\gamma$. The coupling to the scalar curvature is then equivalent to putting a charge $i\gamma\sqrt{2}$ at infinity. In order for this model to make sense, we see that γ must be imaginary; otherwise all correlators would vanish. We therefore introduce the notation $\gamma = i\sqrt{2}\alpha_0$ and write the condition

$$\sum_k \alpha_k = 2\alpha_0 \qquad (9.28)$$

(the charge at infinity is $-2\alpha_0$). This, in turn, seems to imply that the theory cannot be unitary, since the action is not real. In fact, the theory is indeed nonunitary for a generic value of α_0. We shall see below how a unitary theory may be extracted for special values of α_0.

Adding the curvature term will, of course, modify the energy-momentum tensor; consequently the central charge and conformal dimensions will be affected. The new energy-momentum tensor can be determined by varying the metric $g_{\mu\nu}$ and evaluating the result in flat space, according to the definition (2.193). This is done in App. 9.A. The result is

$$T_{\mu\nu} = T^{(0)}_{\mu\nu} - \frac{\gamma}{2\pi}\left(\partial_\mu \partial_\nu \varphi - \frac{1}{2}\eta_{\mu\nu}\partial^\sigma \partial_\sigma \varphi\right) \qquad (9.29)$$

where $T_{\mu\nu}^{(0)}$ is the energy-momentum tensor for the free-boson action (9.1). We see that the modified energy-momentum tensor is still traceless, and conserved when the equations of motion are used. The holomorphic component is then

$$T(z) = -2\pi T_{zz} = -\frac{1}{2} :\partial\varphi\partial\varphi: +i\sqrt{2}\alpha_0\partial^2\varphi \tag{9.30}$$

9.1.4. The Anomalous OPEs

We calculate the OPE of the energy-momentum tensor (9.30) with the primary fields of the free boson and with itself. We have to look only at the extra term $i\sqrt{2}\alpha_0\partial^2\varphi$ of T. We easily find that

$$T(z)\partial\varphi(w) \sim \frac{2\sqrt{2}i\alpha_0}{(z-w)^3} + \frac{\partial\varphi(w)}{(z-w)^2} + \frac{\partial^2\varphi(w)}{z-w} \tag{9.31}$$

The first term implies that $\partial\varphi$ is no longer a primary field. However, the vertex operators are still primary: the OPE

$$\partial^2\varphi(z)V_\alpha(w) \sim \frac{i\sqrt{2}\alpha}{(z-w)^2}V_\alpha(w) \tag{9.32}$$

means that the conformal dimension of V_α is now

$$h_\alpha = \alpha^2 - 2\alpha_0\alpha \tag{9.33}$$

Comparing Eq. (9.33) with Eq. (9.4), we see that the dimension is no longer invariant under the transformation $\alpha \to -\alpha$ but instead under $\alpha \to 2\alpha_0 - \alpha$: the vertex operators V_α and $V_{2\alpha_0-\alpha}$ share the same dimension (9.33).

The OPE of T with itself receives the following contributions from the extra term:

$$-\frac{1}{2}i\sqrt{2}\alpha_0 :\partial\varphi(z)\partial\varphi(z): \partial^2\varphi(w)$$

(a)
$$= i\sqrt{2}\alpha_0\partial_w\left\{\frac{\partial\varphi(w)}{(z-w)^2} + \frac{\partial^2\varphi(w)}{z-w} + \text{reg.}\right\}$$

$$= i\sqrt{2}\alpha_0\left\{\frac{2\partial\varphi(w)}{(z-w)^3} + \frac{2\partial^2\varphi(w)}{(z-w)^2} + \frac{\partial^3\varphi(w)}{z-w} + \text{reg.}\right\}$$

(b)
$$-\frac{1}{2}i\sqrt{2}\alpha_0\partial^2\varphi(z) :\partial\varphi(w)\partial\varphi(w): = -i\sqrt{2}\alpha_0\partial_z\left\{-\frac{\partial\varphi(w)}{(z-w)^2}\right\}$$

$$= -2\sqrt{2}i\alpha_0\frac{\partial\varphi(w)}{(z-w)^3}$$

(c)
$$-2\alpha_0^2\partial^2\varphi(z)\partial^2\varphi(w) = \frac{-12\alpha_0^2}{(z-w)^4}$$

Summing these contributions, we recover the usual form for the OPE of T with itself, except that the central charge is now[3]

$$c = 1 - 24\alpha_0^2 \tag{9.34}$$

We know that the theory will be unitary only for those values of α_0 that fit c into the Kac table.

The result (9.11) for the correlator of vertex operators is still applicable, since it depended only on the nonzero frequency modes of the free boson φ. The only difference is the neutrality condition, which is now given by Eq. (9.28). It is worth noticing that the simple correlators (9.12) and (9.13) still follow from global conformal invariance and the new neutrality condition.

§9.2. Screening Operators

9.2.1. Physical and Vertex Operators

If a physical system at criticality is to be treated in the Coulomb gas formalism, the various physical quantities having definite scaling dimensions (such as the energy density, the magnetization, and so on) will be represented by vertex operators. Specifically, a physical operator of conformal dimension $h_\alpha = \alpha^2 - 2\alpha\alpha_0$ will be associated with the vertex operators V_α and $V_{2\alpha_0 - \alpha}$. The same quantity being represented by more than one vertex operator means that the correlator of physical operators may be evaluated in several different but equivalent ways. This equivalence requirement imposes constraints on the vertex operators.

For instance, because the two-point function of a physical operator with itself is nonzero, we would expect the associated vertex correlator $\langle V_\alpha V_\alpha \rangle$ to be nonzero, which is not the case since that correlator violates the neutrality condition (9.28). However, we expect that correlator to be physically equivalent to

$$\langle V_{2\alpha_0 - \alpha}(z) V_\alpha(w) \rangle = \frac{1}{(z-w)^{h_\alpha}} \tag{9.35}$$

The solution to this dilemma is to modify $\langle V_\alpha V_\alpha \rangle$ by changing its charge without affecting its conformal properties. This can be done by inserting in the correlator a *screening operator* with nonzero charge but conformal dimension zero. Such an operator does not exist in local form, but may be obtained by contour integrating a field of conformal dimension 1. Indeed, if ψ is a primary field with $h_\psi = 1$, its integral

$$A = \oint dz \, \psi(z) \tag{9.36}$$

---

[3] See also Ex. 9.11 for an alternative proof using the mode expansions of the fields.

is a nonlocal operator of conformal dimension zero: it is invariant under a conformal mapping $z \rightarrow w$:

$$A \rightarrow \oint dz \, \psi(w) \left(\frac{dw}{dz}\right) = \oint dw \, \psi(w) \tag{9.37}$$

This means that A commutes with all the Virasoro generators, which is confirmed by an explicit calculation:

$$[L_n, A] = \oint dz \, [L_n, \psi(z)]$$

$$= \oint dz \, (n+1)z^n \psi(z) + z^{n+1} \partial\psi(z) \tag{9.38}$$

$$= \oint dz \, \partial(z^{n+1}\psi(z)) = 0$$

There are only two local fields of dimension 1 available for the construction of screening operators: the vertex operators $V_\pm$ defined as

$$V_\pm \equiv V_{\alpha_\pm} \quad \text{where} \quad \alpha_\pm = \alpha_0 \pm \sqrt{\alpha_0^2 + 1} \tag{9.39}$$

We check that the conformal dimension is

$$\alpha_\pm^2 - 2\alpha_\pm\alpha_0 = 1 \tag{9.40}$$

where the charges $\alpha_\pm$ are the same as those introduced in Eq. (7.30) when describing the Kac determinant. We note that

$$\alpha_+ + \alpha_- = 2\alpha_0$$
$$\alpha_+\alpha_- = -1 \tag{9.41}$$

Accordingly, we define the screening operators

$$Q_\pm = \oint dz \, V_\pm(z) = \oint dz \, e^{i\sqrt{2}\alpha_\pm\varphi(z)} \tag{9.42}$$

Inserting Q_+ or Q_- an integer number of times in a correlator will not affect its conformal properties, but will completely screen the charge in some cases, since Q_+ and Q_- carry charges α_+ and α_-, respectively. The modified two-point function

$$\langle V_\alpha(z)V_\alpha(w)Q_-^m Q_+^n \rangle \tag{9.43}$$

is now subjected to the neutrality condition

$$2\alpha + m\alpha_+ + n\alpha_- = 2\alpha_0 = \alpha_+ + \alpha_- \tag{9.44}$$

In other words, the equivalence of V_α and $V_{2\alpha_0-\alpha}$ within two-point functions of physical operators is assured if 2α is an integer combination of α_+ and α_-.

This may also be seen by considering the four-point function of a physical operator with itself. That function should also be nonzero for all operators, contrary to the three-point function. However, it is impossible to write a product of four vertex operators made of V_α and $V_{2\alpha_0-\alpha}$ that satisfies the neutrality condition: we are forced into using screening operators. For instance, the modification

$$\langle V_{2\alpha_0-\alpha}V_\alpha V_\alpha V_\alpha \rangle \longrightarrow \langle V_{2\alpha_0-\alpha}V_\alpha V_\alpha V_\alpha Q_-^n Q_+^m \rangle \qquad (9.45)$$

is needed, on which the neutrality condition yields

$$2\alpha + m\alpha_+ + n\alpha_- = 0 \qquad (9.46)$$

This condition is the same as that discussed previously (with different integers). It is easy to convince oneself that considering different forms of the four-point function (such as $\langle V_\alpha V_\alpha V_\alpha V_\alpha \rangle$) or more complex correlators leads to similar (or less stringent) conditions on the charge α: 2α should be an integer combination of α_+ and α_-. We shall accordingly define the admissible charges as

$$\alpha_{r,s} = \frac{1}{2}(1-r)\alpha_+ + \frac{1}{2}(1-s)\alpha_- \qquad (9.47)$$

and denote

$$V_{r,s} \equiv V_{\alpha_{r,s}} \qquad (9.48)$$

as the corresponding vertex operators. Note that the conjugation operation $\alpha \to 2\alpha_0 - \alpha$ becomes $(r,s) \to (-r,-s)$ in terms of the indices. The conformal dimensions of these fields are then easily seen to be

$$h_{r,s}(c) = \frac{1}{4}(r\alpha_+ + s\alpha_-)^2 - \alpha_0^2 \qquad (9.49)$$

This, of course, is the Kac formula (7.30). However, so far we have not imposed any restriction on the integers r and s, nor on c.

9.2.2. Minimal Models

We will now see how the idea of minimal models can emerge in the context of the Coulomb gas formalism. If a conformal model is to contain a finite number of scaling operators, some condition has to be imposed on c. Indeed, Eq. (9.47) can generate an infinite number of admissible charges for generic values of c. One way of drastically cutting the number of admissible charges is to require α_+/α_- to be rational or, in other words, that there exist two integers p and p' ($p > p'$) such that

$$p'\alpha_+ + p\alpha_- = 0 \qquad (9.50)$$

Then we have the following periodicity relation:

$$\alpha_{r+p',s+p} = \alpha_{r,s} \qquad (9.51)$$

(p and p' may be chosen relatively prime). However, this apparently still leaves an infinite number of admissible charges around, since it imposes a periodicity in only one direction on the (r,s) grid; condition (9.50) allows us to restrict the range of the integer s to $0 \le s < p$, although r remains unrestricted. However, we shall see below that one may apply similar restrictions on r while keeping a closed operator algebra, implying that the truncation thus imposed is legitimate.

Condition (9.50) allows us to write an explicit expression for $\alpha_\pm$:

$$\alpha_+ = \sqrt{p/p'} \quad \text{and} \quad \alpha_- = -\sqrt{p'/p} \tag{9.52}$$

from which it follows that

$$\alpha_{r,s} = \frac{1}{2\sqrt{pp'}}\left\{p(1-r) - p'(1-s)\right\} \qquad \alpha_0 = \frac{p-p'}{2\sqrt{pp'}} \tag{9.53}$$

The relation $c = 1 - 24\alpha_0^2$ then leads to the following expressions for the central charge and the conformal dimensions:

$$c = 1 - \frac{6(p-p')^2}{pp'}$$
$$h_{r,s} = \frac{(rp - sp')^2 - (p-p')^2}{4pp'} \tag{9.54}$$

These are the relations (7.65) for minimal models, except that we have not satisfactorily argued how the restrictions on r and s could be obtained. For this we need to compute explicitly the three-point functions[4]

$$\langle \phi_{(r_1,s_1)}\phi_{(r_2,s_2)}\phi_{(r_3,s_3)} \rangle = \langle V_{r_1,s_1} V_{r_2,s_2} V_{r_3,s_3} Q_+^r Q_-^s \rangle \tag{9.55}$$

wherein r and s are chosen to neutralize the correlator. This three-point function is proportional to the operator algebra coefficient $c_{r_1,s_1;r_2,s_2}^{r_3,s_3}$. In particular, this coefficient vanishes if the fusion rule

$$\phi_{(r_1,s_1)} \times \phi_{(r_2,s_2)} \to \phi_{(r_3,s_3)} \tag{9.56}$$

is not allowed. According to the fusion rules of the minimal models (8.131), the correlator (9.55) vanishes unless

$$|r_1 - r_2| + 1 \leq r_3 < \min(r_1 + r_2 - 1, 2p' - r_1 - r_2 - 1)$$
$$|s_1 - s_2| + 1 \leq s_3 < \min(s_1 + s_2 - 1, 2p - s_1 - s_2 - 1) \tag{9.57}$$

It is then a straightforward matter to see that the following set of indices

$$1 \leq r < p' \qquad\qquad 1 \leq s < p \tag{9.58}$$

closes under the above formula, that is, $c_{r_1,s_1;r_2,s_2}^{r_3,s_3} = 0$ if the three doublets of indices are not taken from the above set. This, therefore, constitutes a legitimate truncation of the set of admissible charges $\alpha_{r,s}$, in the sense that the operator algebra closes within this set. The structure constants for minimal models can be computed according to the scheme presented in the following sections, and independently

---

[4]   Again, this is only the holomorphic part of the actual correlation function. Here and throughout this chapter, we consider only the "diagonal" minimal models, whose operators are all spinless ($h = \bar{h}$). The full operators $\Phi_{(r,s)}(z,\bar{z})$ are actually the left-right symmetric combinations

$$\Phi_{(r,s)}(z,\bar{z}) = \phi_{(r,s)}(z) \otimes \phi_{(r,s)}(\bar{z})$$

For the minimal models under consideration, the full three-point correlator is simply the square modulus of the chiral expression (9.55). Hence the structure constants C of the full OPE are the squares of the numbers $c_{r_1,s_1;r_2,s_2}^{r_3,s_3}$.

shown to vanish unless Eq. (9.57) is satisfied. This confirmation *a posteriori* of the fusion rules of minimal models shows that our argument for the representation of minimal models in terms of vertex operators is complete. An alternative proof of the Coulomb-gas representation of minimal models is described in App. 9.B.

Unfortunately, the calculation of the correlator (9.55) is not straightforward. Although its dependence on the three points is entirely fixed by conformal invariance, the coefficient $c_{r_1,s_1;r_2,s_2}^{r_3,s_3}$ is not. These numbers are usually calculated indirectly, through the computation of the four-point correlations

$$\langle \Phi_{(r_1,s_1)}(z_1,\bar{z}_1)\Phi_{(r_2,s_2)}(z_2,\bar{z}_2)\Phi_{(r_3,s_3)}(z_3,\bar{z}_3)\Phi_{(r_4,s_4)}(z_4,\bar{z}_4)\rangle \qquad (9.59)$$

where the notation $\Phi(z,\bar{z}) = \phi(z)\otimes\phi(\bar{z})$ stands for the full left $\times$ right-symmetric primary fields. The fact that four-point correlation functions turn out to be easier to compute than three-point functions can be explained heuristically as follows. By conformal transformations, the three-point functions are reduced to pure numbers, the structure constants of the theory. Eq. (9.55) gives an integral representation for them. On the other hand, the same transformations can reduce the four-point functions to functions of one complex variable z, the cross-ratio of the four points (see, e.g., Eq. (5.28)). In Chap. 8, we have seen that, as functions of this latter variable, the four-point functions satisfy hypergeometric-type linear differential equations (cf. Eq. (8.71)), whose order is related to the singular vector structure of the various highest-weight modules entering the correlator. The particularly simple properties of the sets of solutions of these differential equations (such as their monodromy properties, discussed below) provide us with a much simpler framework to actually compute them. The structure constants will then be recovered in the limit $z \to 0$. Hence, as is often the case in mathematics, it is useful to first compute an (apparently) more complicated object, and recover the quantity of interest in some suitable limit. Here, the rich structure of the Virasoro symmetry makes four-point functions simpler to calculate than three-point functions.

To recover the structure constants from the data (9.59), we have to take the limits $z_1 \to z_2$ and $z_3 \to z_4$. Then, using the chiral OPE of the primary fields

$$\phi_{(r_1,s_1)}(z_1)\phi_{(r_2,s_2)}(z_2) = \sum_{r,s} \frac{c_{r_1,s_1;r_2,s_2}^{r,s}}{z_{12}^{h_{r_1 s_1}+h_{r_2 s_2}-h_{rs}}}\phi_{(r,s)}(z_2) + \cdots$$

$$\phi_{(r_3,s_3)}(z_3)\phi_{(r_4,s_4)}(z_4) = \sum_{r,s} \frac{c_{r_3,s_3;r_4,s_4}^{r,s}}{z_{34}^{h_{r_3 s_3}+h_{r_4 s_4}-h_{rs}}}\phi_{(r,s)}(z_4) + \cdots \qquad (9.60)$$

and substituting it into Eq. (9.59), we get products of the numbers $C = c^2$ as leading terms in the expansion of the full four-point function

$$\langle \Phi_{(r_1,s_1)}(z_1,\bar{z}_1)\Phi_{(r_2,s_2)}(z_2,\bar{z}_2)\Phi_{(r_3,s_3)}(z_3,\bar{z}_3)\Phi_{(r_4,s_4)}(z_4,\bar{z}_4)\rangle \simeq$$

$$\sum_{r,s} \frac{1}{z_{24}^{h_{rs}}\times\text{c.c.}} \times \frac{C_{r_1,s_1;r_2,s_2}^{r,s}}{z_{12}^{h_{r_1 s_1}+h_{r_2 s_2}-h_{rs}}\times\text{c.c.}} \frac{C_{s_3,r_3;r_4,s_4}^{r,s}}{z_{34}^{h_{r_3 s_3}+h_{r_4 s_4}-h_{rs}}\times\text{c.c.}} \qquad (9.61)$$

$$\times \left(1 + \mathcal{O}(z_{24},\text{c.c.})\right)$$

where $z_{ij}^x \times$ c.c. stands for $z_{ij}^x \times \bar{z}_{ij}^{\bar{x}}$.

9.2.3. Four-Point Functions: Sample Correlators

In this subsection, we present a detailed evaluation of a sample four-point function, using the Coulomb-gas formalism. As explained in Sect. 9.2.1, we consider only holomorphic parts of correlation functions of the form[5]

$$
\begin{aligned}
&\langle \phi_{(r_1,s_1)}(z_1)\phi_{(r_2,s_2)}(z_2)\phi_{(r_3,s_3)}(z_3)\phi_{(r_4,s_4)}(z_4)\rangle \\
&\quad = \langle V_{r_1,s_1}(z_1)V_{r_2,s_2}(z_2)V_{r_3,s_3}(z_3)V_{-r_4,-s_4}(z_4)Q_+^r Q_-^s\rangle
\end{aligned}
\tag{9.62}
$$

The charge neutrality imposes the sum rules

$$
\begin{aligned}
r_4 &= r_1 + r_2 + r_3 - 2r - 2 \\
s_4 &= s_1 + s_2 + s_3 - 2s - 2
\end{aligned}
\tag{9.63}
$$

It appears that the numbers (r,s) of required screening operators are linked to the particular choice of *dual field* (here $\phi_{(r_4,s_4)}$), namely, the one represented by the vertex operator $V_{2\alpha_0-\alpha}$ (here $V_{-r_4,-s_4}$). Another choice $\phi_{(r_i,s_i)}, i = 1, 2,$ or 3, would have exchanged $(r_4, s_4) \leftrightarrow (r_i, s_i)$ in Eq. (9.63), leading to other values of (r, s). The integral representations associated with the different choices must lead to the same answer for the correlation function. The strategy is to look for the choice involving the minimal number of screening operators, leading then to the simplest integral representations. The expression for the conformal block (9.62) can be put into the form

$$
\prod_{1 \leq i < j \leq 4} (z_i - z_j)^{\mu_{ij}} G(z)
\tag{9.64}
$$

with

$$
\mu_{ij} = \frac{1}{3}\left(\sum_{k=1}^{4} h_{r_k,s_k}\right) - h_{r_i,s_i} - h_{r_j,s_j}
\tag{9.65}
$$

and z is the cross-ratio of the four points, which can be recovered by sending $z_1 \to 0, z_2 \to z, z_3 \to 1$ and $z_4 \to \infty$, in which case the function G is identified as

$$
\begin{aligned}
&\langle \phi_{(r_1,s_1)}(0)\phi_{(r_2,s_2)}(z)\phi_{(r_3,s_3)}(1)\phi_{(r_4,s_4)}(\infty)\rangle \\
&\quad = \langle V_{r_1,s_1}(0)V_{r_2,s_2}(z)V_{r_3,s_3}(1)V_{-r_4,-s_4}(\infty)Q_+^r Q_-^s\rangle \\
&\quad = z^{\mu_{12}}(1-z)^{\mu_{23}}G(z)
\end{aligned}
\tag{9.66}
$$

---

[5] We use here an abusive notation $\langle \phi_{(r_1,s_1)}(z_1)\cdots\rangle$ for the (possibly many) holomorphic conformal blocks of the corresponding full correlator (9.61), which will actually appear as a sesquilinear combination of these (cf. Eq. (9.74), where the full correlator is denoted by $G(z,\bar{z})$). A direct nonchiral calculation of the full correlator would have required the computation of two-dimensional integrals involving the corresponding nonchiral vertex operators, in the form of Eq. (9.95).

We are now in a position to illustrate the use of screening operators in an explicit calculation. The four-point function that necessitates the least number of screening operators involves a field $\phi_{(2,1)}$ or $\phi_{(1,2)}$. Consider then

$$\langle V_{r_1,s_1}(0)V_{2,1}(z)V_{r_3,s_3}(1)V_{-r_4,-s_4}(\infty)Q_+\rangle$$

$$= \oint dw\, \langle V_{r_1,s_1}(0)V_{2,1}(z)V_{r_3,s_3}(1)V_{-r_4,-s_4}(\infty)V_+(w)\rangle \quad (9.67)$$

$$= z^{2\alpha_{2,1}\alpha_{r_1,s_1}}(1-z)^{2\alpha_{2,1}\alpha_{r_3,s_3}}G(z)$$

wherein the exact shape of the integration contour is not yet specified. One checks easily that the neutrality condition is satisfied iff

$$\begin{aligned} r_1 + r_2 + r_3 - r_4 &= 4 \\ s_1 + s_2 + s_3 - s_4 &= 2 \end{aligned} \quad (9.68)$$

It is understood that this correlator factorizes into holomorphic and antiholomorphic parts; in what follows we shall pay attention only to the former. Applying Eq. (9.11), the holomorphic part of this correlator becomes

$$\oint dw\, w^a(w-1)^b(w-z)^c \quad (9.69)$$

where

$$a = 2\alpha_+\alpha_{r_1,s_1} \quad b = 2\alpha_+\alpha_{r_3,s_3} \quad c = 2\alpha_+\alpha_{2,1} \quad (9.70)$$

The integrand in Eq. (9.69) has branch cuts at $w = 0, z, 1, \infty$.

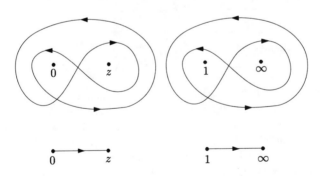

Figure 9.1. A choice of integration contours for Eq. (9.69), leading to two independent solutions. The contours can be shrunk respectively to $[0, z]$ and $[1, \infty[$ as shown.

The contour of integration must cross each branch twice in opposite directions to guarantee its closure. In principle, there is a certain number of different choices for the integration contour. However, only two are independent. We take, for instance, the two contours depicted on Fig. 9.1, which, when the corresponding integrals converge, can be shrunk respectively to

$$\mathcal{C}_1 \to [1, \infty[$$
$$\mathcal{C}_2 \to [0, z] \tag{9.71}$$

This shrinking operation produces an overall phase factor, which is ignored at this point. The precise normalization of each integral will be fixed later. These contours (9.71) lead to the two functions

$$
\begin{aligned}
I_1(a, b, c; z) &= \int_1^\infty dw \, w^a (w - 1)^b (w - z)^c \\
&= \frac{\Gamma(-a - b - c - 1)\Gamma(b + 1)}{\Gamma(-a - c)} F(-c, -a - b - c - 1; -a - c; z), \\
I_2(a, b, c; z) &= \int_0^z dw \, w^a (1 - w)^b (z - w)^c \\
&= z^{1+a+b+c} \int_0^1 dw \, w^a (1 - w)^b (1 - zw)^c \\
&= z^{1+a+b+c} \frac{\Gamma(a + 1)\Gamma(c + 1)}{\Gamma(a + c + 2)} F(-b, a + 1; a + c + 2; z)
\end{aligned}
$$

$$(9.72)$$

Here we denote by $F(\lambda, \mu, \nu; z)$ the hypergeometric function, given for $|z| < 1$ by the series

$$F(\lambda, \mu, \nu; z) = \sum_{k=0}^\infty \frac{(\lambda)_k (\mu)_k}{k! (\nu)_k} z^k \tag{9.73}$$

where

$$(x)_0 = 1$$
$$(x)_k = x(x - 1) \cdots (x - k + 1) \qquad \text{for } k \geq 1$$

These two functions span the space of solutions of the hypergeometric differential equation, derived by using the singular vector structure of the $(r = 2, s = 1)$ highest-weight module associated with $\phi_{(2,1)}$ (see Exs. 8.9 and 8.10 for a proof). Hence, quite remarkably, the screening procedure has somehow taken into account the Virasoro algebra structure, by directly projecting the $(2, 1)$ primary state onto an irreducible representation, and automatically performing the quotient by singular vectors. There lies the real power of the Coulomb-gas formalism.

The physical correlator (with z and $\bar{z}$ dependence) now takes the form

$$
\begin{aligned}
G(z, \bar{z}) &= |z|^{-2\mu_{12}} |1 - z|^{-2\mu_{23}} \\
&\quad \times \left\langle \Phi_{(r_1, s_1)}(0, 0) \Phi_{(2,1)}(z, \bar{z}) \Phi_{(r_3, s_3)}(1, 1) \Phi_{(r_4, s_4)}(\infty, \infty) \right\rangle \\
&= \sum_{j=1,2} X_{ij} \, I_i(z) \overline{I_j(z)}
\end{aligned}
$$

$$(9.74)$$

where X_{ij} is an arbitrary real 2×2 matrix. This is the most general solution to the z- and $\bar{z}$- differential equations obeyed by the full correlator.

To complete the calculation, we need to determine the coefficients X_{ij}. This will be done by enforcing the monodromy invariance of the function G. A monodromy transformation of a function of z consists in letting z circulate around some other point (typically a singular point). Since it represents a physical correlator, the function G must not be affected by an analytical continuation along a contour surrounding any of the branch points 0 and 1 and ∞. It is sufficient to consider the monodromy around the two points 0 and 1 (see Fig. 9.2). Let us denote by c_0 and c_1 the two corresponding transformations. $G(z, \bar{z})$ must then be invariant under the action of c_0 and c_1, namely

$$
\begin{aligned}
c_0 \, G(z, \bar{z}) &= \lim_{t \to 1^-} G(z e^{2i\pi t}, \bar{z} e^{-2i\pi t}) \\
c_1 \, G(z, \bar{z}) &= \lim_{t \to 1^-} G(1 + (z - 1)e^{2i\pi t}, 1 + (\bar{z} - 1)e^{-2i\pi t})
\end{aligned}
\tag{9.75}
$$

These result in nontrivial constraints because the functions I_1 and I_2 are affected by the transformations. The latter are linearly represented in the (I_1, I_2) basis through the monodromy matrices

$$
\begin{aligned}
c_0 \, I_i &= \sum_j (g_0)_{ij} \, I_j \\
c_1 \, I_i &= \sum_j (g_1)_{ij} \, I_j
\end{aligned}
\tag{9.76}
$$

Using the expressions (9.72), we find that the monodromy around 0 is diagonal

$$
g_0 = \begin{pmatrix} 1 & 0 \\ 0 & e^{2i\pi(1+a+c)} \end{pmatrix}
\tag{9.77}
$$

hence for $G(z, \bar{z})$ to be invariant under c_0, the coefficients X_{ij} in Eq. (9.74) must be diagonal, that is $X_{ij} = \delta_{i,j} X_i$, and

$$
G(z, \bar{z}) = \sum_{j=1,2} X_j \, |I_j(z)|^2
\tag{9.78}
$$

The monodromy of $I_j(z)$ around the point 1 is not diagonal. To compute it, it is simpler to reexpress the functions $I_j(z)$ in terms of similar functions $I_i(1 - z)$ for which the monodromy around the point 1 will be diagonal. To get explicit relations, we restrict ourselves to real $z \in]0, 1[$ and contours along the real axis.

Starting from $I_1(z)$ with its contour $[1, +\infty[$, we deform the contour into one going from $-\infty$ to 1, avoiding the two singularities 0 and z. There are two different ways of doing this, as shown on Fig. 9.3 (the multiplicative phase factors are explained below):

(i) *above the real axis*: the half-turn around the point 1 gives a factor $e^{i\pi b}$, the one around z an additional factor of $e^{i\pi c}$, and the one around 0 an additional $e^{i\pi a}$.

(ii) *below the real axis*: all the half-turns have opposite directions (compared to (i)), hence the phase factors picked up are the complex conjugates of those in (i).

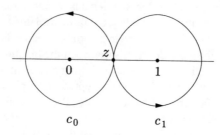

Figure 9.2. Monodromy transformations c_0 and c_1: z circles once around the respective branch points 0 and 1.

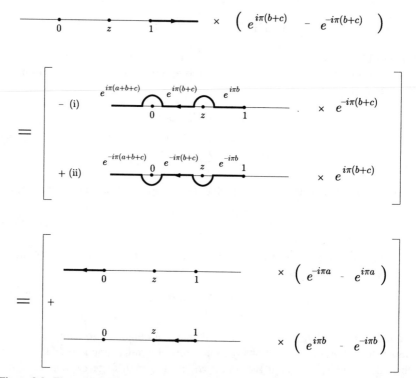

Figure 9.3. The two possible deformations of the contour $[1, \infty[$, (i) and (ii), are combined to yield integrals over the contours $]-\infty, 0]$ and $[z, 1]$.

As indicated on Fig. 9.3, in order to cancel the respective contributions of the $[0, z]$ portion of the contours $\mathcal{C}_{(i)}$ and $\mathcal{C}_{(ii)}$, we must take the following linear combination

$$\sin \pi(b+c) \int_{[1,+\infty[} = \sin \pi a \int_{]-\infty,0]} + 0 \times \int_{[0,z]} - \sin \pi c \int_{[1,z]} \qquad (9.79)$$

which amounts to the relation (for short we denote $s(x) = \sin \pi x$)

$$I_1(a,b,c;z) = \frac{s(a)}{s(b+c)}I_1(b,a,c;1-z) - \frac{s(c)}{s(b+c)}I_2(b,a,c;1-z) \quad (9.80)$$

Analogously, by deforming the contour $[0,z]$ into $]-\infty,0] \cup [z,+\infty[$, we get

$$I_2(a,b,c;z) = -\frac{s(a+b+c)}{s(b+c)}I_1(b,a,c;1-z) - \frac{s(b)}{s(b+c)}I_2(b,a,c;1-z)$$
$$(9.81)$$

Note the interchange $a \leftrightarrow b$, which corresponds to the interchange of the points $z_1 = 0$ and $z_3 = 1$ in the conformal blocks I. With $\tilde{I}_j(z) = I_j(b,a,c;z)$, Eqs. (9.80)–(9.81) take the form

$$I_i(z) = \sum_{j=1,2} f_{ij}\,\tilde{I}_j(1-z) \quad (9.82)$$

where f is a constant 2×2 matrix. Henceforth, we have

$$G(z,\bar{z}) = \sum_{j,k,l=1,2} X_i\, f_{ik}\, f_{il}\, \tilde{I}_k(1-z)\, \overline{\tilde{I}_l(1-z)} \quad (9.83)$$

The monodromy of $\tilde{I}_k(1-z)$ around the point 1 being diagonal, the invariance of G imposes that

$$\tilde{X}_{kl} = \sum_{i=1,2} X_i\, f_{ik}\, f_{il} \quad (9.84)$$

be a diagonal matrix. This forces

$$\sum_{i=1,2} X_i\, f_{ik}\, f_{il} = 0 \quad \forall k \neq l \quad (9.85)$$

from which we read

$$\frac{X_1}{X_2} = -\frac{f_{21}f_{22}}{f_{12}f_{11}} = \frac{s(a+b+c)s(b)}{s(a)s(c)} \quad (9.86)$$

Up to an overall normalization, we finally get

$$G(z,\bar{z}) \sim \left[\frac{s(b)s(a+b+c)}{s(a+c)}|I_1(z)|^2 + \frac{s(a)s(c)}{s(a+c)}|I_2(z)|^2 \right] \quad (9.87)$$

or equivalently

$$\boxed{\begin{array}{l} \langle \phi_{(r_1,s_1)}(0)\phi_{(2,1)}(z)\phi_{(r_3,s_3)}(1)\phi_{(r_4,s_4)}(\infty) \rangle \sim \\ |z|^{4\alpha_{2,1}\alpha_{r_1,s_1}}|1-z|^{4\alpha_{2,1}\alpha_{r_3,s_3}} \times \\ \left[\frac{s(b)s(a+b+c)}{s(a+c)}|I_1(z)|^2 + \frac{s(a)s(c)}{s(a+c)}|I_2(z)|^2 \right] \end{array}} \quad (9.88)$$

We compare this result with Eq. (9.61). In the latter equation, we take the limits $z_1 \to 0$, $z_2 \to z$, $z_3 \to 1$, $z_4 \to \infty$, and let z tend to 0. It appears that only two primary fields occur in the fusion $\phi_{(2,1)} \times \phi_{(r_1,s_1)}$, corresponding to the two leading terms in the $z \to 0$ limit of Eq. (9.87)

$$G(z,\bar{z}) \sim |z|^{4\alpha_{2,1}\alpha_{r_1,s_1}} |1-z|^{4\alpha_{2,1}\alpha_{r_3,s_3}}$$

$$\times \left[\frac{s(b)s(a+b+c)}{s(a+c)} N_1^2 + \frac{s(a)s(c)}{s(a+c)} N_2^2 |z|^{2(1+a+c)} + \cdots \right] \qquad (9.89)$$

The constants N_i are related to the asymptotic behavior of the I's by

$$I_1(z) \simeq N_1 \quad I_2(z) \simeq z^{1+a+c} N_2 \quad \text{when } z \to 0 \qquad (9.90)$$

As $F(\lambda, \mu, \nu; 0) = 1$, they read explicitly

$$N_1 = \frac{\Gamma(-1-a-b-c)\Gamma(b+1)}{\Gamma(-a-c)} \quad N_2 = \frac{\Gamma(a+1)\Gamma(c+1)}{\Gamma(a+c+2)} \qquad (9.91)$$

The leading powers of z and $\bar{z}$ in Eq. (9.89) are easily identified, respectively, with

$$2\alpha_{2,1}\alpha_{r_1,s_1} = \frac{r_1-1}{2}\frac{p}{p'} - \frac{s_1-1}{2}$$

$$= h_{r_1+1,s_1} - h_{r_1,s_1} - h_{2,1}$$

$$2\alpha_{2,1}\alpha_{r_1,s_1} + 1 + a + c = -\frac{1+r_1}{2}\frac{p}{p'} + \frac{1+s_1}{2} \qquad (9.92)$$

$$= h_{r_1-1,s_1} - h_{r_1,s_1} - h_{2,1}$$

Hence the first term in Eq. (9.89) corresponds to $\phi_{(r_1+1,s_1)}$, and the second one to $\phi_{(r_1-1,s_1)}$, leading to the fusion rule

$$\phi_{(2,1)} \times \phi_{(r_1,s_1)} = \phi_{(r_1+1,s_1)} + \phi_{(r_1-1,s_1)} \qquad (9.93)$$

Next, we identify the numerical factors as, respectively, (see Eq. (9.61))

$$X_1 N_1^2 = \frac{s(b)s(a+b+c)}{s(a+c)} N_1^2 \sim C_{r_1,s_1;2,1}^{r_1+1,s_1} C_{r_3,s_3;r_4,s_4}^{r_1+1,s_1}$$

$$X_2 N_2^2 = \frac{s(a)s(c)}{s(a+c)} N_2^2 \sim C_{r_1,s_1;2,1}^{r_1-1,s_1} C_{r_3,s_3;r_4,s_4}^{r_1-1,s_1} \qquad (9.94)$$

up to an overall normalization of G. We get, for instance, all the squares of structure constants of the form $C_{r,s;2,1}^{k,l}$ by taking $r_3 = 1$, $s_3 = 2$, $r_4 = r_1 = r$, $s_4 = s_1 = s$.

We now address the normalization problem for the function G. The correct normalization is required to get the exact expressions for the structure constants. The overall normalization of the function G in Eq. (9.87) can be fixed by directly computing the two-dimensional integral (over $\mathbb{C}$) involving the full left-right vertex operators (9.5)

$$G(z,\bar{z}) = |z|^{-4\alpha_{2,1}\alpha_{r_1,s_1}} |1-z|^{-4\alpha_{2,1}\alpha_{r_3,s_3}} \times$$

$$\int d^2w \, \langle \mathcal{V}_{r_1,s_1}(0,0)\mathcal{V}_{2,1}(z,\bar{z})\mathcal{V}_{r_3,s_3}(1,1)\mathcal{V}_{-r_4,-s_4}(\infty,\infty)\mathcal{V}_+(w,\bar{w}) \rangle \qquad (9.95)$$

but we will not go into this calculation here. Instead we will resort to the crossing symmetry of the four-point correlation to fix this normalization. This amounts to imposing that

$$\langle \Phi_{(r,s)}\Phi_{(2,1)}\Phi_{(2,1)}\Phi_{(r,s)}\rangle = \langle \Phi_{(2,1)}\Phi_{(2,1)}\Phi_{(r,s)}\Phi_{(r,s)}\rangle \qquad (9.96)$$

or equivalently

$$\sum_{j=1,2} X_j \, |I_j(z)|^2 = \sum_{k=1,2} \tilde{X}_k \, |\tilde{I}_k(z)|^2 \qquad (9.97)$$

where the change $z \to 1 - z$ in G accounts for the permutation of the fields in the correlation. Identifying the structure constants (9.61) in the expression of the second correlation function of Eq. (9.96), we get

$$\tilde{X}_1 \tilde{N}_1^2 \sim C_{2,1;2,1}^{3,1} C_{r,s;r,s}^{3,1}$$
$$\tilde{X}_2 \tilde{N}_2^2 \sim C_{2,1;2,1}^{1,1} C_{r,s;r,s}^{1,1} \qquad (9.98)$$

up to *the same* overall normalization constant as implied in Eq. (9.94). Here the normalization constants $\tilde{N}_i$, $i = 1, 2$, are obtained from the $z \to 0$ limit of the $\tilde{I}$'s

$$\tilde{I}_1(z) \simeq \tilde{N}_1, \qquad \tilde{I}_2(z) \simeq \tilde{N}_2 \, z^{1+b+c} \quad \text{when } z \to 0 \qquad (9.99)$$

They are simply related to the original N's. But the second term of Eq. (9.98) involves only the structure constants involving the identity operator $\phi_{(1,1)}$. These two-point function normalization constants have been chosen from the beginning to be 1, that is

$$C_{r,s;r,s}^{1,1} = 1 \qquad (9.100)$$

for any $\Phi_{(r,s)}$ in the theory. This fixes the exact normalization of $\tilde{X}_2$, and henceforth that of all the X's. The correctly normalized structure constants satisfy then

$$\left(C_{r,s;2,1}^{r+1,s}\right)^2 = X_1 N_1^2 / \tilde{X}_2 \tilde{N}_2^2$$
$$\left(C_{r,s;2,1}^{r-1,s}\right)^2 = X_2 N_2^2 / \tilde{X}_2 \tilde{N}_2^2 \qquad (9.101)$$

The general computation of all the structure constants $C_{r_1,s_1;r_2,s_2}^{r_3,s_3}$ relies on a generalization of the above procedure, where $\Phi_{(2,1)}$ is replaced by an arbitrary $\Phi_{(r_2,s_2)}$. The resulting four-point correlation function $G(z,\bar{z})$ of Eq. (9.66) will split again into a block-diagonal sum

$$G(z,\bar{z}) \sim \sum_{j=1}^{N} X_j \, |I_j(z)|^2 \qquad (9.102)$$

over the $N = r_2 \times s_2$ independent solutions of the (r_2, s_2) differential equation of order N obeyed by the correlation function. The solutions I_j take the form of multiple integrals including the necessary screening operators, involving a number of contours of the form $[0, z]$ or $[1, \infty[$. The study of the monodromy of these integrals enables us to fix the values of the coefficients X_j in the sum of Eq. (9.102), and to consequently identify the wanted structure constants. Although the strategy is clear, the calculations quickly become very complicated and will not be pursued here. We leave to the reader the task of computing correlators involving $\Phi_{(1,3)}$ (see Ex. 9.6).

§9.3. Minimal Models: General Structure of Correlation Functions

9.3.1. Conformal Blocks for the Four-Point Functions

The structure of the four-point correlations of minimal models was given in Sect. 9.2.3 above. The result, in the form of Eq. (9.102), strongly suggests the following interpretation. The four-point correlations, as functions of the $SL(2, \mathbb{C})$-invariant cross-ratio z, decompose into a sum of holomorphic × antiholomorphic functions of z and $\bar{z}$, respectively. These functions (denoted I_k in Sect. 9.2.3) are in one-to-one correspondence with the intermediate states $\phi_{(r,s)}$ allowed by the OPE. This leads to the following expression for the correlator

$$\mathcal{G}(z,\bar{z}) = \langle \Phi_{(r_1,s_1)}(0,0)\Phi_{(r_2,s_2)}(z,\bar{z})\Phi_{(r_3,s_3)}(1,1)\Phi_{(r_4,s_4)}(\infty,\infty)\rangle$$
$$= \sum_{r,s;\bar{r},\bar{s}} \mathcal{F}_{r,s}(z)\, \bar{\mathcal{F}}_{\bar{r},\bar{s}}(\bar{z}) \qquad (9.103)$$

where each conformal block $\mathcal{F}_{r,s}$ corresponds to a field $\phi_{(r,s)}$ occurring in the following fusion rules[6]

$$\phi_{(r,s)} \in \phi_{(r_1,s_1)} \times \phi_{(r_2,s_2)}$$
$$\phi_{(r,s)} \in \phi_{(r_3,s_3)} \times \phi_{(r_4,s_4)} \qquad (9.104)$$

This is best seen by inserting a complete set of intermediate states in the correlator (9.103) and taking the limit of coinciding points $z \to 0$ (corresponding to $z_1 \to z_2$ and $z_3 \to z_4$ in the original correlator). In this limit, for each intermediate state, the four-point correlation function $\mathcal{G}(z,\bar{z})$ factorizes into a product of two three-point functions as

$$\mathcal{G}(z,\bar{z}) \sim \langle \Phi_{(r_1,s_1)}\Phi_{(r_2,s_2)}\Phi_{(r,s)}\rangle\langle \Phi_{(r,s)}\Phi_{(r_3,s_3)}\Phi_{(r_4,s_4)}\rangle , \qquad (9.105)$$

which gives the normalization of the corresponding conformal block. We recover the fusion conditions (9.104) for this normalization to be nonzero.

We use the following graphical representation for conformal blocks:

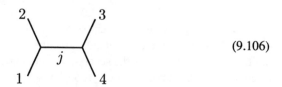

$$(9.106)$$

---

[6] This statement is translated into mathematically rigorous terms in App. 9.B.

where, for simplicity, we trade the Kac indices for a single Latin index, namely $(r_i, s_i) \rightarrow i$ $(i = 1, 2, 3, 4)$ and $(r, s) \rightarrow j$. This graphical representation has the advantage of carrying all the relevant information about the block, namely the fields in the correlator (external legs) and the intermediate state (propagator). Beyond mere nomenclature, the idea of conformal blocks attached to intermediate states provides us with a more physical interpretation of the (contour) integral representations of Sect. 9.2. The intermediate states are not encoded in the integrand (screening operators), but in the contours of integration chosen to represent the blocks. In Sect. 9.2.3, we have seen how monodromy transformations exchanged these contours among themselves. These transformations will have a simple interpretation in terms of conformal blocks.

9.3.2. Conformal Blocks for the N-Point Function on the Plane

By associativity of the OPE, a general correlation function

$$\mathcal{G}_N(z_i, \bar{z}_i) = \langle \Phi_1(z_1, \bar{z}_1)\Phi_2(z_2, \bar{z}_2) \cdots \Phi_N(z_N, \bar{z}_N) \rangle \qquad (9.107)$$

can be inductively decomposed into a sum of holomorphic $\times$ antiholomorphic functions of the z's and $\bar{z}$'s, respectively, which generalize the notion of conformal block already encountered for $N = 4$. This is best seen by inserting complete sets of intermediate states in the correlator (9.107), and decomposing it accordingly into a product of three-point functions, in the limit where all the points coincide, namely

$$\mathcal{G}_N(z_i, \bar{z}_i) \sim \sum_{j_1, \ldots, j_{N-3}} \langle \Phi_1 \Phi_2 \Phi_{j_1} \rangle \langle \Phi_{j_1} \Phi_3 \Phi_{j_2} \rangle \cdots \langle \Phi_{j_{N-3}} \Phi_{N-1} \Phi_N \rangle \qquad (9.108)$$

For the corresponding block to occur, a number of fusion conditions must be satisfied

$$\phi_{j_1} \in \phi_1 \times \phi_2$$
$$\phi_{j_2} \in \phi_{j_1} \times \phi_3$$
$$\cdots \cdots \qquad (9.109)$$
$$\phi_N \in \phi_{j_{N-3}} \times \phi_{N-1}$$

These restrict the possible intermediate states. The graphical representation for the (left) conformal block corresponding to Eq. (9.108) reads

$$(9.110)$$

9.3.3. Monodromy and Exchange Relations for Conformal Blocks

In Sect. 9.2.3, we have computed the monodromy of the conformal blocks of four-point functions involving $\phi_{(2,1)}$. For that purpose, we have used the transformation $z \to 1 - z$ of the conformal blocks. The latter amounts to the exchange of the points $z_1 (= 0)$ and $z_3 (= 1)$ in the original correlation. It is equivalent to the following operation on the conformal blocks, where, again for simplicity, we now denote the various fields by only one Latin index:

$$(9.111)$$

This is called the crossing operation and is easily generalized to any four-point conformal block. The four-point conformal blocks are linearly transformed under crossing as

$$\qquad = \sum_m F \begin{bmatrix} i & j \\ k & l \end{bmatrix}_{n,m} \qquad (9.112)$$

The matrices F are called the *crossing matrices*[7] of the corresponding conformal theory. (In Sect. 9.2.3, F corresponds to the inverse f^{-1} of the matrix f, see, e.g., Eq. (9.82), but with a different normalization.)

Another elementary operation consists in exchanging the two upper external legs, namely $z_2 \leftrightarrow z_3$. In terms of the cross-ratio z, this amounts to $z \to 1/z$. Remarkably, this is also realized linearly on the conformal blocks, as

$$\qquad = \sum_m R \begin{bmatrix} i & j \\ k & l \end{bmatrix}_{n,m} \qquad (9.113)$$

The matrices $R_{r,s}$ are called the *exchange matrices*[8] of the conformal theory.

The crossing and exchange matrices satisfy a number of identities. For instance, rotating the diagram on the l.h.s. of (9.112) by 90 degrees and applying Eq. (9.112), we find the quadratic relation

$$\sum_m F \begin{bmatrix} i & j \\ k & l \end{bmatrix}_{n,m} F \begin{bmatrix} k & i \\ l & j \end{bmatrix}_{m,p} = \delta_{n,p} \qquad (9.114)$$

---

[7] The term *fusion matrices* is also used in the literature.
[8] The term *braiding matrices* is also used.

which allows F to be inverted.

The crossing operation (9.112) can be interpreted as a change of basis of the conformal blocks; recall from Sect. 9.2.3 that crossing amounts to the change of basis $\{I_j\} \to \{\tilde{I}_k\}$. We can use the definition of the matrix F to rewrite the invariance of the full four-point correlation (9.103) under crossing (with the shorthand notation i for (s_i, r_i) and $\bar{i}$ for $(\bar{r}_i, \bar{s}_i)$)

$$
\begin{aligned}
\mathcal{G}(z, \bar{z}) &= \sum_{n, \bar{n}} \mathcal{F}_n(z)\, \mathcal{F}_{\bar{n}}(\bar{z}) \\
&= \mathcal{G}(1 - z, 1 - \bar{z}) \\
&= \sum_{n, \bar{n}} \sum_{m, \bar{m}} F\begin{bmatrix} 1 & 2 \\ 4 & 3 \end{bmatrix}_{n, m} \bar{F}\begin{bmatrix} \bar{1} & \bar{2} \\ \bar{4} & \bar{3} \end{bmatrix}_{\bar{n}, \bar{m}} \tilde{\mathcal{F}}_m(z)\, \tilde{\mathcal{F}}_{\bar{m}}(\bar{z})
\end{aligned}
\tag{9.115}
$$

In the blocks $\tilde{\mathcal{F}}$, the points $z_1 = 0$ and $z_3 = 1$ of $\mathcal{F}$ have been exchanged. The normalization of the conformal blocks is related to the structure constants of the theory through

$$
\begin{aligned}
\mathcal{F}_n(z) &\sim C_{12}{}^n\, C_{34}{}^n\, z^{h_n - h_1 - h_2} \\
\tilde{\mathcal{F}}_n(z) &\sim C_{32}{}^n\, C_{14}{}^n\, z^{h_n - h_3 - h_2}
\end{aligned}
\tag{9.116}
$$

Taking the limit $z \to 0$ in Eq. (9.115), we get a relation between the structure constants of the theory and the matrix F

$$
\begin{aligned}
& C_{12}{}^n\, C_{34}{}^n\, C_{\bar{1}\bar{2}}{}^{\bar{n}}\, C_{\bar{3}\bar{4}}{}^{\bar{n}} \\
&= \sum_{\substack{m, \bar{m} \\ h_m = h_1 + h_3 - h_n}} F\begin{bmatrix} 1 & 2 \\ 4 & 3 \end{bmatrix}_{n, m} \bar{F}\begin{bmatrix} \bar{1} & \bar{2} \\ \bar{4} & \bar{3} \end{bmatrix}_{\bar{n}, \bar{m}} C_{32}{}^m\, C_{14}{}^m\, C_{\bar{3}\bar{2}}{}^{\bar{m}}\, C_{\bar{1}\bar{4}}{}^{\bar{m}}
\end{aligned}
\tag{9.117}
$$

This is an overdetermined system of equations for the C's. Its compatibility is guaranteed by extra relations that are satisfied by F. The matrix F thus contains all the necessary data to compute the structure constants of the theory. It is tempting to consider F as the fundamental data of the theory. A constructive approach to conformal field theory consists of a set of axioms and identities that have to be satisfied by the matrices F and R for the theory to be consistent. One of them is the so-called *Yang-Baxter equation*, which must be satisfied by R, expressing some braiding transformation on the conformal blocks of the five-point function in two inequivalent ways (see Ex. 9.9). Another is the *pentagon identity*, which results from the two distinct but equivalent ways of transforming the conformal blocks of a five-point function (see Ex. 9.10 for the precise statement). But these axioms are incomplete as long as higher topologies are ignored. Enforcing conformal theories to be well-defined on surfaces of arbitrary topology puts more constraints on R and F. This point is briefly addressed in the next subsection.

9.3.4. Conformal Blocks for Correlators on a Surface of Arbitrary Genus

It is possible to define conformal field theories on a two-dimensional closed manifold (surface) with more complicated topology than the plane. In two dimensions, two closed orientable manifolds are topologically equivalent, that is, they can be continuously deformed into each other, iff they have the same genus, that is the same number of handles h. The genus is the only topological invariant of two-dimensional orientable surfaces. We have already considered a free bosonic theory on a genus h surface in Sect. 9.1. More general conformal theories, such as minimal models, can also be formulated on higher genus surfaces, even though a Lagrangian formulation is not available. Indeed the Coulomb-gas representation of minimal theories goes over to surfaces of arbitrary genus, with some extra structure emerging already on the torus, in genus $h = 1$.[9]

A correlation function of conformal fields on a genus h surface will again be decomposed into a sum of left × right conformal blocks, according to the insertion of intermediate states in the original correlator. The influence of the nontrivial topology is readily seen in the intermediate channels, which can be chosen to wrap around the handles, and hence "feel" the topology. Graphically, a conformal block for an N-point function on a genus-h surface is represented as a genus-h ϕ^3 diagram, with N external legs labeled by the fields in the correlator, and h internal loops. The propagators carry intermediate states indices. This is illustrated on Fig. 9.4, in the case of a genus-5 two-point function. Note that there seem to be many inequivalent ways of even drawing the diagram. That these should all be equivalent turns out to be an additional constraint on the theory.

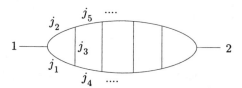

Figure 9.4. A sample conformal block for a two-point correlation on a genus 5 surface.

It is now clear why consistency of the theory at higher genera puts more constraints on R and F: an operation such as the circulation of an argument of a conformal block around a handle of the surface, in addition to capturing the effect of topology, also affects the conformal blocks, thereby relating R and F to the topological structure. The simplest relations of this kind will be studied in Chap. 10, in the case of the torus.

---

[9] The torus topology is studied in detail in Chap. 10, through the physical requirement of modular invariance of the partition function with periodic boundary conditions. Furthermore, in Chap. 12, various correlation functions are calculated for the Ising model defined on a torus.

Appendix 9.A. Calculation of the Energy-Momentum Tensor

In this appendix we show explicitly that the energy-momentum tensor associated with the action (9.24) is $T_{\mu\nu} = T^{(0)}_{\mu\nu} + T^{(1)}_{\mu\nu}$, where $T^{(0)}_{\mu\nu}$ is the energy-momentum tensor for the free boson:

$$T^{(0)}_{\mu\nu} = \frac{1}{4\pi}\left(\partial_\mu\varphi\partial_\nu\varphi - \frac{1}{2}\eta_{\mu\nu}\partial_\sigma\varphi\partial^\sigma\varphi\right) \tag{9.118}$$

whereas $T^{(1)}_{\mu\nu}$ is the energy-momentum tensor associated with the curvature term of Eq. (9.24):

$$T^{(1)}_{\mu\nu} = -\frac{\gamma}{2\pi}\left(\partial_\mu\partial_\nu\varphi - \frac{1}{2}\eta_{\mu\nu}\partial^\sigma\partial_\sigma\varphi\right) \tag{9.119}$$

This calculation will be performed in an arbitrary dimension d.

We use the definition (2.193) of the energy-momentum tensor:

$$\delta S = -\frac{1}{2}\int dx\, T^{\mu\nu}\delta g_{\mu\nu} \tag{9.120}$$

In this definition, the variation $\delta g_{\mu\nu}$ is taken in an arbitrary metric, but evaluated in flat space once $\delta g_{\mu\nu}$ has been isolated. We concentrate on the curvature term, since $T^{(0)}_{\mu\nu}$ has been calculated before (see Chap. 2, following Eq. (2.193), or Sect. 5.3.1). We need to calculate the variation of $\sqrt{g}R\varphi$ under an infinitesimal deformation $\delta g_{\mu\nu}$ of the metric tensor. It is a simple matter to see that

$$\delta\sqrt{g} = \frac{1}{2}\sqrt{g}g^{\mu\nu}\delta g_{\mu\nu}$$
$$\delta g^{\mu\nu} = -g^{\mu\alpha}g^{\nu\beta}\delta g_{\alpha\beta} \tag{9.121}$$

Calculating the variation of the curvature is trickier. We recall the following definitions for the Christoffel symbols $\Gamma^\alpha_{\beta\gamma}$, the Riemann curvature tensor $R^\alpha_{\beta\gamma\lambda}$, the Ricci tensor $R_{\mu\nu}$, and the scalar curvature R:

$$\Gamma^\alpha_{\beta\gamma} = \frac{1}{2}g^{\alpha\delta}\left(\partial_\beta g_{\delta\gamma} + \partial_\gamma g_{\delta\beta} - \partial_\delta g_{\beta\gamma}\right)$$
$$R^\alpha_{\beta\gamma\lambda} = \partial_\gamma\Gamma^\alpha_{\beta\lambda} - \partial_\lambda\Gamma^\alpha_{\beta\gamma} + \Gamma^\alpha_{\delta\gamma}\Gamma^\delta_{\beta\lambda} - \Gamma^\alpha_{\delta\lambda}\Gamma^\delta_{\beta\gamma} \tag{9.122}$$
$$R_{\mu\nu} = R^\alpha_{\mu\alpha\nu}$$
$$R = g^{\mu\nu}R_{\mu\nu}$$

The first step is to express the variation of R as

$$\delta R = \delta g^{\mu\nu}R_{\mu\nu} + g^{\mu\nu}\delta R_{\mu\nu} \tag{9.123}$$

In flat space the first term vanishes; the second term may be evaluated first in a coordinate system that is locally inertial ($\Gamma^\alpha_{\beta\gamma} = 0$ at the point of interest), with the result

$$g^{\mu\nu}\delta R_{\mu\nu} = \partial_\mu\left\{g^{\alpha\beta}\delta\Gamma^\mu_{\alpha\beta} - g^{\alpha\mu}\delta\Gamma^\beta_{\alpha\beta}\right\} \tag{9.124}$$

Since the quantity in braces is a vector (call it w^μ), we may rewrite the above in a general coordinate system as

$$g^{\mu\nu}\delta R_{\mu\nu} = \frac{1}{\sqrt{g}}\partial_\mu\left(\sqrt{g}w^\mu\right) \qquad (9.125)$$

We also use the following properties of the Christoffel symbols, also valid in a general coordinate system:

$$\Gamma^\beta_{\alpha\beta} = \partial_\alpha(\ln\sqrt{g})$$

$$g^{\alpha\beta}\Gamma^\mu_{\alpha\beta} = -\frac{1}{\sqrt{g}}\partial_\alpha\left(g^{\mu\alpha}\sqrt{g}\right) \qquad (9.126)$$

Dropping along the way terms that vanish in flat space once $\delta g_{\mu\nu}$ is isolated, the variation of the curvature term in the action is

$$\begin{aligned}
\delta S_1 &= \frac{\gamma}{4\pi}\int d^2x\,\sqrt{g}\,\varphi g^{\mu\nu}\delta R_{\mu\nu} \\
&= -\frac{\gamma}{4\pi}\int d^2x\,\sqrt{g}\,\partial_\mu\varphi w^\mu \\
&= \frac{\gamma}{4\pi}\int d^2x\,\sqrt{g}\,\partial_\mu\varphi\delta\left\{\frac{1}{\sqrt{g}}\partial_\alpha\left(g^{\mu\alpha}\sqrt{g}\right) + g^{\alpha\mu}\partial_\alpha(\ln\sqrt{g})\right\}
\end{aligned} \qquad (9.127)$$

Integrating by parts once more, we may drop the derivatives acting on the metric (they vanish in the flat limit) and we end up with

$$\delta S_1 = \frac{\gamma}{4\pi}\int d^2x\,\partial_\alpha\partial_\beta\varphi\left(\eta^{\alpha\mu}\eta^{\beta\nu} - \frac{1}{2}\eta^{\alpha\beta}\eta^{\mu\nu}\right)\delta g_{\mu\nu} \qquad (9.128)$$

The energy-momentum tensor (9.119) follows. Note the importance of using the relations (9.126): The variation must be taken in an arbitrary coordinate system. Had we proceeded from Eq. (9.124) directly, the variation δS_1 would have vanished in two-dimensional flat space!

Appendix 9.B. Screened Vertex Operators and BRST Cohomology: A Proof of the Coulomb-Gas Representation of Minimal Models

This appendix presents a sketch of the proof validating the Coulomb-gas approach to minimal models. The proof relies on a detailed study of the action of screening operators on the basic states of a given bosonic Fock space, slightly modified by the addition of a charge. As a result, the irreducible Virasoro modules will appear as the *cohomology* spaces of a particular screening operator, interpreted as a BRST charge for the minimal model. The emergence of a BRST scheme in the context of conformal theory provides us with an interesting parallel with ordinary gauge theory. We recall the origin of the BRST operator in gauge theory: The initial problem is to fix a gauge to regularize the otherwise-divergent path integral of gauge theory. This is done at the cost of introducing a delta-function, bringing an extra (Faddeev-Popov) determinant, which is in turn incorporated into the gauge action upon introducing extra anticommuting (ghost) variables. This modification of the gauge action actually brings an extra symmetry, generated by the BRST operator Q. This symmetry is nilpotent $Q^2 = 0$ and must annihilate the physical states of the theory, which therefore lie in the kernel of Q. The BRST-exact states (lying in the image of Q) have to be further eliminated (they decouple from the theory), hence the physical states belong to the cohomology space of Q[10]

$$H^*(Q) = \operatorname{Ker} Q / \operatorname{Im} Q \qquad (9.129)$$

The BRST operator Q in the present context of minimal theories should be thought of as the result of gauge-fixing of the diffeomorphisms of the circle (reparametrizations) in a would-be free field action functional formulation.

Throughout this appendix, the objects under scrutiny are all *chiral*, namely they pertain only to the left (or right) sector of the corresponding conformal theory. In particular, the vertex operators constructed below are only z-dependent, and will enter in the representation of holomorphic conformal blocks for the correlation functions of the conformal theory.

9.B.1. Charged Bosonic Fock Spaces and Their Virasoro Structure

We start with the notion of charged bosonic Fock space, which is a slight modification of the ordinary bosonic Fock space defined in Sect. 6.3.3. The charged Fock space F_{α,α_0}, with vacuum charge α and background charge α_0, forms a representation of the Heisenberg algebra

$$[a_n, a_m] = n\delta_{n+m,0} \qquad (9.130)$$

The representation is generated by the free action of any product of a_n, $n < 0$, on a highest-weight vector $|\alpha, \alpha_0\rangle$, subject to the conditions

$$a_n|\alpha, \alpha_0\rangle = 0 \quad \forall n > 0$$
$$a_0|\alpha, \alpha_0\rangle = \sqrt{2}\alpha|\alpha, \alpha_0\rangle \qquad (9.131)$$

Moreover, the space F_{α,α_0} is endowed with a structure of Virasoro module, where the Virasoro generators are constructed from the Heisenberg algebra generators as

$$L_n = \frac{1}{2}\sum_{k\in\mathbb{Z}} a_{n-k}a_k - \sqrt{2}\alpha_0(n+1)a_n \quad n \neq 0$$
$$L_0 = \sum_{k=1}^{\infty} a_{-k}a_k + \frac{1}{2}a_0^2 - \sqrt{2}\alpha_0 a_0 \qquad (9.132)$$

---

[10] The concept of cohomology also appears in differential geometry. The BRST operator is analogous to the differential d, acting on differential forms. The cohomology of d is the set of differential forms ω that are closed ($d\omega = 0$) but not exact ($\omega \neq df$).

This is just the mode expansion of the deformed energy-momentum tensor (9.30) of a *chiral* free bosonic theory (that is with the $\bar{z}$-dependence dropped)

$$i\partial\varphi \;=\; \sum_{n\in\mathbb{Z}} a_n z^{-n-1}$$

with a curvature term added to the free action (9.24). (We take the convention $g = 1/4\pi$ in Eq. (6.55).) From Sect. 9.1.4, we know that the modes (9.132) generate a Virasoro algebra with central charge

$$c = 1 - 24\alpha_0^2 \tag{9.133}$$

(See Ex. 9.11 for an alternative proof using the mode expansion.)

The state $|\alpha, \alpha_0\rangle$ is the highest-weight of F_{α,α_0} with conformal dimension $\alpha^2 - 2\alpha\alpha_0$, namely

$$L_0|\alpha, \alpha_0\rangle \;=\; (\alpha^2 - 2\alpha\alpha_0)|\alpha, \alpha_0\rangle \tag{9.134}$$

This state is obtained from the vacuum $|0, \alpha_0\rangle$ by the action of $e^{i\sqrt{2}\alpha\varphi_0}$, where φ_0 is the operator conjugate to a_0: $[\varphi_0, \alpha_0] = i$. From the mode expansion (9.6), this may also be written

$$|\alpha, \alpha_0\rangle = V_\alpha(0)|0, \alpha_0\rangle \tag{9.135}$$

Eq. (9.134) is thus equivalent to Eq. (9.33). The charge Q introduced in Eq. (9.20) is just the zero-mode a_0 of φ, and Eq. (9.23) is equivalent to the second line of Eq. (9.131). Moreover, thanks to the commutation relation

$$[L_0, a_{-n}] \;=\; na_{-n} \quad \forall\, n \geq 0 \tag{9.136}$$

the states in F_{α,α_0} are naturally graded by L_0. By analogy with the ordinary Virasoro case, the eigenvalue of L_0 is called the level. The character of F_{α,α_0} is

$$\chi_{\alpha,\alpha_0}(q) \;=\; \mathrm{Tr}_{F_{\alpha,\alpha_0}}(q^{L_0-c/24}) \;=\; \frac{q^{\alpha^2-2\alpha_0\alpha-c/24}}{\prod_{n\geq 1}(1-q^n)} \tag{9.137}$$

since $p(n)$ states are generated at level n by acting freely on $|\alpha, \alpha_0\rangle$ with arbitrary products of a_k, $k < 0$.

To represent the conformal blocks in correlation functions, we also need to define the dual F^*_{α,α_0} of the charged bosonic Fock space F_{α,α_0}. The latter is built upon a highest-weight (contravariant) vector $\langle \alpha, \alpha_0|$, satisfying

$$\langle \alpha, \alpha_0|\alpha, \alpha_0\rangle \;=\; 1 \tag{9.138}$$

and is given a (dual) Virasoro structure through

$$\langle x|L_{-n}y\rangle = \langle xL_n|y\rangle \quad \forall\, |y\rangle \in F_{\alpha,\alpha_0}, \ \langle x| \in F^*_{\alpha,\alpha_0} \tag{9.139}$$

The transposed A^t of an operator A is defined by

$$\langle x|Ay\rangle \;=\; \langle xA^t|y\rangle \tag{9.140}$$

for any $\langle x| \in F^*_{\alpha,\alpha_0}$, $|y\rangle \in F_{\alpha,\alpha_0}$ (for instance, $L^t_{-n} = L_n$). The dual space F^*_{α,α_0} is thus also a Fock space, obtained by acting freely with the transposed creation operators a^t_{-n} on the highest weight $\langle \alpha, \alpha_0|$. Taking the transpose of the Heisenberg commutation relations (9.130), we may identify $a^t_{-n} \leftrightarrow -a_n$, for $n \neq 0$. Moreover, we must identify $a^t_0 \leftrightarrow 2\sqrt{2}\alpha_0 - a_0$, in order to recover $L^t_{-n} = L_n$. This results in the following identification of Fock spaces:

$$F^*_{\alpha, \alpha_0} \leftrightarrow F_{2\alpha_0 - \alpha, \alpha_0} \tag{9.141}$$

9.B.2. Screened Vertex Operators

Consider the *chiral* vertex operator of Eq. (9.6):

$$V_\alpha(z) = e^{i\sqrt{2}\alpha\varphi_0} z^{\sqrt{2}\alpha a_0} e^{-\sqrt{2}\alpha[\sum_{n\geq 1} a_{-n} z^n/n]} e^{\sqrt{2}\alpha[\sum_{n\geq 1} a_n z^{-n}/n]} \tag{9.142}$$

Vacuum expectation values of such operators are simply the z-dependent part of the full correlator (9.11), namely

$$\langle 0, \alpha_0 | V_{\alpha_1}(z_1) \cdots V_{\alpha_n}(z_n) | 0, \alpha_0 \rangle = \prod_{i<j} (z_i - z_j)^{2\alpha_i \alpha_j} \tag{9.143}$$

Eq. (9.143) is valid only for $|z_1| > |z_2| > \cdots > |z_n|$ (which corresponds to a time-ordering of the successive actions of the vertex operators), and when condition (9.9) is satisfied.

In order to describe the minimal models, we shall restrict ourselves to values of

$$\alpha_0 = \frac{p - p'}{2\sqrt{pp'}} \tag{9.144}$$

with $p > p'$ two coprime integers, and of $\alpha = \alpha_{r,s}$, with

$$\alpha_{r,s} = \frac{1}{2}(1 - r)\alpha_+ + \frac{1}{2}(1 - s)\alpha_- \tag{9.145}$$

and $\alpha_\pm$ as in Eq. (9.52). In the following, the integers r, s are allowed to take arbitrary integer values, not multiples of p', p, respectively. Indeed, although $\alpha_{r+p', s+p} = \alpha_{r,s}$, the Kac formula (9.54) for the conformal dimensions $h_{r,s}(p, p')$ may be applied with arbitrary integer values of r, s (not multiples of p', p resp.) to describe all the null states of the corresponding reducible Verma module $V(h_{r,s}, c)$.

The chiral screened vertex operators $V^{i,j}_{r,s}(z)$, with i positive and j negative screening charges, are defined through the following multiple contour integral

$$V^{i,j}_{r,s}(z) = \oint V_{\alpha_{r,s}}(z) V_{\alpha_+}(u_1) \cdots V_{\alpha_+}(u_i) V_{\alpha_-}(v_1) \cdots V_{\alpha_-}(v_j) \prod du_a dv_b \tag{9.146}$$

where the contours are time-ordered, namely $|z| > |u_1| > \cdots > |u_i| > |v_1| > \cdots > |v_j|$, and all contours pass through the point z. Usually the integrand in Eq. (9.146) has some singularities when arguments approach each other (in a close neighborhood of z), and the integral may be regularized by analytic continuation from a region (with complex values of $\alpha_+, \alpha_- = -1/\alpha_+$) where it converges. This should be equivalent to the subtraction of singularities, for instance by opening each contour at z (point splitting).

By construction, the positive and negative charge screening operators $V_{\alpha_\pm}$ have conformal dimension 1. When integrated on contours as in Eq. (9.146), their conformal dimension is reduced to 0. Therefore, they do not affect the behavior of $V_{\alpha_{rs}}(z)$ under the action of Virasoro generators:

$$[L_k, V_{r,s}^{i,j}(z)] = (z^{k+1}\partial_z + (k+1)h_{rs}z^k)V_{r,s}^{i,j}(z) \qquad (9.147)$$

As an operator acting on a charged bosonic Fock space F_{α,α_0}, $V_{r,s}^{i,j}$ has the effect of modifying the charge $\alpha \to \alpha + \alpha_{rs} + i\alpha_+ + j\alpha_-$. Indeed, taking $\alpha = \alpha_{r_1,s_1}$, the screened vertex operator is a map

$$V_{r,s}^{i,j}(z) : F_{\alpha_{r_1,s_1},\alpha_0} \to F_{\alpha_{r_1+r-2i-1,s_1+s-2j-1},\alpha_0} \qquad (9.148)$$

In the following, we shall denote by $F_{r,s}$ the Fock space F_{α_{rs},α_0}.

9.B.3. The BRST Charge

A screened vertex operator of particular interest is

$$J_s(z) = V_{1,-1}^{0,s-1}(z) \qquad (9.149)$$

The operator $J_s(z)$ is such that when acting on $F_{r,s}$ (or equivalently when multiplied by the operator $V_{\alpha_{rs}}(w)$, the argument w being integrated on a closed contour), it remains a single-valued function of z. This means that no phase is generated when the argument z circles around the origin: $J_s(e^{2i\pi}z) = J_s(z)$. Indeed, when z circles around the origin, all the integrated arguments may be taken to circulate simultaneously, and we get a net phase factor of 1

$$\oint dw J_s(e^{2i\pi}z)V_{\alpha_{rs}}(e^{2i\pi}w) = e^{2i\pi(2s\alpha_-\alpha_{rs}+2\alpha_-^2 s(s-1)/2)}\oint dw J_s(z)V_{\alpha_{rs}}(w)$$

$$= e^{2i\pi(r-1)s}\oint dw J_s(z)V_{\alpha_{rs}}(w)$$

$$= \oint dw J_s(z)V_{\alpha_{rs}}(w)$$

$$(9.150)$$

The phase factor in the first equality comes from the factor $z^{\sqrt{2}\alpha\alpha_0}$ in the expression (9.142) of V_α. Note that the single-valuedness of $J_s(z)$ is true only on $F_{r,s}$; on another Fock space, some spurious phase would be generated in Eq. (9.150). This suggests defining the BRST operator Q_s as the contour integral of $J_s(z)$ over the unit circle, normalized by a prefactor $1/s$:

$$Q_s = \frac{1}{s}\oint_{|v_0|=1>|v_1|>\cdots>|v_{s-1}|} V_{\alpha_-}(v_0)\cdots V_{\alpha_-}(v_{s-1})\prod dv_b \qquad (9.151)$$

The single-valuedness of $J_s(z)$ acting on $F_{r,s}$ together with Eq. (9.147), which implies that $J_s(z)$ has conformal dimension 1, are responsible for the following crucial property of the operator Q_s, when acting on $F_{r,s}$ (or equivalently, multiplied by $V_{\alpha_{r,s}}(w)$, with w integrated over a closed contour):[11]

$$[L_k, Q_s] = 0 \qquad \forall k \in \mathbb{Z} \tag{9.152}$$

This is easily proved by applying Eq. (9.147) to the operator $J_s(z)$ of Eq. (9.149), and then integrating z over a closed contour. With $h_{1,-1} = 1$ ($J_s(z)$ has conformal dimension 1), we get

$$[L_k, Q_s] = \frac{1}{s} \oint dz \partial_z\left(z^{k+1} J_s(z)\right) = 0 \tag{9.153}$$

as the closed contour integration of the total derivative of a single-valued function of z.

9.B.4. BRST Invariance and Cohomology

When acting on the charged Fock space $F_{r,s}$, the BRST charge has a well-defined commutation relation with the screened vertex operator $V^{i,j}_{r',s'}(z)$, which reads

$$Q_{s+s'-2j-1} V^{i,j}_{r',s'}(z) = e^{2i\pi\alpha_{r',s'}\alpha_-(s+s'-2i-1)} V^{i,s-j-1}_{r',s'}(z) Q_s \tag{9.154}$$

To prove this, note that the charge $s\alpha_-$ carried by the BRST operator Q_s can be absorbed by a screened vertex operator, namely that

$$Q_s V^{i,j}_{r',s'}(z) = e^{2i\pi s\alpha_{r',s'}\alpha_-} V^{i,j+s}_{r',s'}(z)$$
$$V^{i,j}_{r',s'}(z) Q_s = V^{i,j+s}_{r',s'}(z) \tag{9.155}$$

The second equality follows from the definition of the screened vertex operator (9.146). In the first one, the phase factor arises from the commutation of the s vertex operators V_{α_-} of Q_s through $V_{\alpha_{r',s'}}$. The various Fock spaces over which the operators act in Eq. (9.154) are represented in the diagram:

$$
\begin{array}{ccc}
& V^{i,j}_{r',s'}(z) & \\
F_{r,s} & \longrightarrow & F_{r+r'-2i-1,s+s'-2j-1} \\
& & \\
Q_s \downarrow & & \downarrow Q_{s+s'-2j-1} \\
& V^{i,s-j-1}_{r',s'}(z) & \\
F_{r,-s} & \longrightarrow & F_{r+r'-2i-1,2p+2j+1-s-s'}
\end{array}
\tag{9.156}
$$

The commutation relation (9.154) is called the BRST invariance of the screened vertex operators. It immediately follows that the screened vertex operators V preserve the Q–vanishing and Q–exactness of the states in the charged Fock spaces, namely

$$Q - \text{vanishing} : Q|x\rangle = 0 \Rightarrow QV|x\rangle = 0$$
$$Q - \text{exactness} : |x\rangle = Q|y\rangle \Rightarrow \exists |z\rangle \text{ s.t. } V|x\rangle = Q|z\rangle \tag{9.157}$$

---

[11] Again, we note that this property is valid only if Q_s acts on $F_{r,s}$. Otherwise, some spurious phases would impair the commutation property with the Virasoro generators.

In the second statement, the state $|z\rangle$ is equal to $V|x\rangle$, up to a phase given by Eq. (9.154). In other words, the screened vertex operator $V^{i,j}_{r',s'}(z)$ maps

$$\text{Ker } Q_s \subset F_{r,s} \longrightarrow \text{Ker } Q_{s+s'-2j-1} \subset F_{r+r'-2i-1,s+s'-2j-1}$$

$$\text{Im } Q_{p-s} \subset F_{r,s} \longrightarrow \text{Im } Q_{p+2j+1-s-s'} \subset F_{r+r'-2i-1,s+s'-2j-1} \tag{9.158}$$

In the second line, the operator Q_{p-s} acts from $F_{r,2p-s} \equiv F_{r-p,p-s}$ to F_{rs} whereas the operator $Q_{p+2j+1-s-s'}$ acts from $F_{r+r'-2i-1,2p+2j+1-s-s'}$ to $F_{r+r'-2i-1,s+s'-2j-1}$. In particular, we may consider the successive action of Q_s and Q_{p-s}

$$F_{r,2p-s} \xrightarrow{\;Q_{p-s}\;} F_{r,s} \xrightarrow{\;Q_s\;} F_{r,-s} \tag{9.159}$$

Then, we have

$$\text{Im } Q_{p-s} \subset \text{Ker } Q_s \tag{9.160}$$

The proof of this fact is a consequence of the general scheme sketched below. This is the so-called BRST property, namely, the square of Q vanishes

$$Q_s \, Q_{p-s} = 0 \tag{9.161}$$

We can therefore define the space of BRST states as

$$B_{r,s} = \text{Ker } Q_s / \text{Im } Q_{p-s} \subset F_{r,s} \tag{9.162}$$

This space is also known as the cohomology space of Q in $F_{r,s}$. According to the two properties (9.158), the screened vertex operator $V^{i,j}_{r',s'}(z)$ is also a map between BRST states

$$V^{ij}_{r',s'}(z): \; B_{r,s} \to B_{r+r'-2i-1,s+s'-2j-1} \tag{9.163}$$

A careful study of the structure of Virasoro singular vectors in the charged Fock spaces leads eventually to the identification of the irreducible Virasoro module $M_{r,s}$ of Eq. (8.13) with the space $B_{r,s}$ of BRST states. We simply sketch the outline of the proof. The BRST charge Q is a tool to generate singular vectors of the Virasoro algebra in charged Fock spaces. In the following infinite chain of actions of Q on Fock spaces

$$\cdots \xrightarrow{\;Q_s\;} F_{r,2p-s} \xrightarrow{\;Q_{p-s}\;} F_{r,s} \xrightarrow{\;Q_s\;} F_{r,-s} \xrightarrow{\;Q_{p-s}\;} \cdots \tag{9.164}$$

the cohomology of Q is trivial, except for the central Fock space $F_{r,s}$. Indeed, $\text{Im } Q_{p-s} = \text{Ker } Q_s$ or $\text{Im } Q_s = \text{Ker } Q_{p-s}$ on all the Fock spaces of the chain except $F_{r,s}$, for which the cohomology space $B_{r,s}$ of Q is nontrivial. The singular vector with conformal dimension $h_{r,2p-s}$ in $F_{r,s}$ is created by the action of Q_{p-s} on the highest-weight vector of $F_{r,2p-s}$. Indeed, as Q commutes with the Virasoro algebra generators, $Q_{p-s}|\alpha_{r,2p-s}, \alpha_0\rangle$ is a singular vector with dimension $h_{r,2p-s}$, and it can be proved that it does not vanish. Moreover, there is no singular vector with dimension $h_{r,-s}$ in $F_{r,s}$. Finally, the highest-weight vector $|\alpha_{r,s}, \alpha_0\rangle$ of $F_{r,s}$ is annihilated by Q_s. Indeed, if it was nonzero, its image by Q_s would be a singular vector in $F_{r,-s}$ with same conformal dimension $h_{r,s}$, but there is no such vector in $F_{r,-s}$. Restricting ourselves to $\mathrm{Ker}\, Q_s$, we are left with a module that contains only the maximal submodule of dimension $h_{r,2p-s}$. As shown in Eq. (8.13), the irreducible Virasoro module $M_{r,s}$ is obtained by factoring the module built on its highest weight by the two sub-modules of conformal dimensions $h_{r,2p-s}$ and $h_{r,-s}$. This is exactly realized by taking the space $B_{r,s} = \mathrm{Ker}\, Q_s/\mathrm{Im}\, Q_{p-s}$ of the BRST states in $F_{r,s}$. Clearly, factoring out by $\mathrm{Im}\, Q_{p-s}$ amounts to factoring out the submodule of dimension $h_{r,2p-s}$, whereas by considering $\mathrm{Ker}\, Q_s$ we have already factored out the submodule of dimension $h_{r,-s}$. We conclude that

$$M_{r,s} = B_{r,s} = \mathrm{Ker}\, Q_s/\mathrm{Im}\, Q_{p-s} \tag{9.165}$$

The BRST states of $F_{r,s}$ form the irreducible Virasoro representation $M_{r,s}$ with conformal dimension $h_{r,s}$.

9.B.5. The Coulomb-Gas Representation

As already mentioned in the previous section, the screened vertex operator $V_{r',s'}^{i,j}(z)$ extends to a map between BRST states (9.163), or equivalently between irreducible Virasoro modules

$$V_{r',s'}^{i,j}(z): \ M_{r,s} \ \to \ M_{r+r'-2i-1,s+s'-2j-1} \tag{9.166}$$

Moreover, it is a primary field of conformal dimension $h_{r',s'}$ (9.147). This map is instrumental in the construction of (left) conformal blocks for correlation functions involving the primary field $\Phi_{(r',s')}(z,\bar{z})$. In the graphical representation (9.106), it corresponds to an intermediate vertex

$$V_{r',s'}^{i,j}(z) = \begin{array}{c} {\scriptstyle (r,s)} \\ \big| \\ {\scriptstyle (r',s')\text{———}(r+r'-2i-1,s+s'-2j-1)} \end{array} \tag{9.167}$$

A general conformal block for the correlation function of primary fields is precisely indexed by the intermediate states (r, s) allowed by the successive OPE of the fields. More precisely, a conformal block for the N-point correlation function

$$\langle \Phi_{(r_1,s_1)}(z_1,\bar{z}_1) \cdots \Phi_{(r_N,s_N)}(z_N,\bar{z}_N) \rangle$$

is indexed by a sequence of allowed intermediate states

$$(\rho_1, \sigma_1), \cdots, (\rho_{N-1}, \sigma_{N-1}) \tag{9.168}$$

such that

$$\phi_{(\rho_k,\sigma_k)} \in \phi_{(\rho_{k-1},\sigma_{k-1})} \times \phi_{(r_k,s_k)} \tag{9.169}$$

with $k = 1,\ldots,N$, and $(\rho_0,\sigma_0) = (\rho_N,\sigma_N) = (1,1)$ (the expectation value is taken over the vacuum state $|h_{11} = 0\rangle$). The condition (9.169) is fulfilled by the action of the screened vertex operator $V_{r_k,s_k}^{i_k,j_k}(z_k)$, provided we take

$$\begin{aligned} \rho_{k-1} &= r_k + \rho_k - 2i_k - 1 \\ \sigma_{k-1} &= s_k + \sigma_k - 2j_k - 1 \end{aligned} \tag{9.170}$$

With this choice, the corresponding conformal block reads

$$\mathcal{F}^{\{(r,s)\}}_{\{\rho\},\{\sigma\}}(z_1,\ldots,z_N) \propto \langle\alpha_{1,1},\alpha_0|\prod_{k=1}^{N}V^{i_k,j_k}_{r_k,s_k}(z_k)|\alpha_{1,1},\alpha_0\rangle \qquad (9.171)$$

up to a multiplicative normalization constant independent of the z's. We note that the chain of identities (9.170) between the indices has to be satisfied in (9.171).

The Coulomb-gas representation (9.171) of conformal blocks is not unique, due to the symmetry $(r,s) \leftrightarrow (p'-r,p-s)$. Taking advantage of this symmetry, it is possible to optimize the calculation of conformal blocks by performing the smallest possible number of integrations. For instance, the outgoing state (ρ_0,σ_0) in (9.171) may be taken to be $(p'-1,p-1)$, in which case the corresponding highest weight reads $\langle\alpha_{p'-1,p-1},\alpha_0| = \langle 2\alpha_0,\alpha_0|$. To recover charge neutrality (i.e., for the chain of identities (9.170) to still be satisfied), the last screened vertex operator also has to be modified. We may take, for instance,

$$\mathcal{F}^{\{(r,s)\}}_{\{\rho\},\{\sigma\}}(z_1,\ldots,z_N) \propto \langle\alpha_{p'-1,p-1},\alpha_0|V^{0,0}_{p'-r_1,p-s_1}(z_1)\prod_{k=2}^{N}V^{i_k,j_k}_{r_k,s_k}(z_k)|\alpha_{1,1},\alpha_0\rangle$$

$$(9.172)$$

The net effect of this manipulation is the reduction of the number of contour integrations by $i_1+j_1 = r_1+s_1-2$. The representation (9.172) is the one used in Sect. 9.2.

Exercises

9.1 *Correlator of vertex operators*

In this exercise we calculate the correlator (9.8) of vertex operators from the general expression (2.107) of the generating functional for the free boson, with a suitably regularized propagator $K(x,y)$. This propagator is

$$K(x,y) = -\ln\left[m^2((x-y)^2+a^2)\right] \qquad (9.173)$$

Here m is the (infinitesimal) mass of the field, which vanishes at the conformal point and serves as a long-distance cutoff, whereas the "lattice spacing" a serves as a short-distance cutoff.

a) Setting the source term of Eq. (2.107) to

$$j(x) = i\sqrt{2}\sum_{n=1}^{N}\alpha_n\varphi(x_n)$$

show that the correlator (9.8) is

$$\langle e^{i\sqrt{2}\alpha_1\varphi(x_1)}\ldots e^{i\sqrt{2}\alpha_N\varphi(x_N)}\rangle = (ma)^{2(\alpha_1+\cdots+\alpha_N)^2}\prod_{n<m}\left(\frac{z_{mn}\bar{z}_{mn}}{a^2}\right)^{2\alpha_n\alpha_m} \qquad (9.174)$$

b) Explain how the neutrality condition (9.9) is recovered at the conformal point. Is the different normalization of the correlator troublesome?

9.2 *Correlator of vertex operators (bis)*
In this exercise we provide yet another way of calculating the correlator (9.8) of vertex operators, based on the mode expansion (6.59). We write the (normal-ordered) vertex operator as

$$\mathcal{V}_\alpha(z,\bar{z}) =: e^{i\sqrt{2}\alpha\tilde{\varphi}(z,\bar{z})}: V'_\alpha(z)\bar{V}'_\alpha(\bar{z}) \tag{9.175}$$

where $\tilde{\varphi}$ is the zero-mode of φ:

$$\tilde{\varphi}(z,\bar{z}) = \varphi_0 - ia_0 \ln(z\bar{z}) \tag{9.176}$$

and where the V'_α contain only the purely holomorphic modes of the expansion (6.54):

$$V'_\alpha(z) = \exp\left\{-\sqrt{2}\alpha\sum_{n>0}\frac{1}{n}a_{-n}z^n\right\}\exp\left\{\sqrt{2}\alpha\sum_{n>0}\frac{1}{n}a_n z^{-n}\right\} \tag{9.177}$$

(likewise for $\bar{V}'_\alpha(\bar{z})$). If $\tilde{\phi}(z)$ stands for the holomorphic part of $\varphi(z,\bar{z})$ (without the zero-mode), then

$$V'_\alpha(z) =: e^{i\sqrt{2}\alpha\tilde{\phi}(z)}: \tag{9.178}$$

a) Using the mode expansion, show that

$$\langle\tilde{\phi}(z)\tilde{\phi}(w)\rangle = -\ln\left(1 - \frac{w}{z}\right) \tag{9.179}$$

b) From Eq. (6.193), which a priori does not hold for the zero-mode, show that

$$\langle V'_{\alpha_1}(z_1)\cdots V'_{\alpha_n}(z_n)\rangle = \prod_{i<j}(z_i - z_j)^{2\alpha_i\alpha_j} z_i^{-2\alpha_i\alpha_j} \tag{9.180}$$

and likewise for the antiholomorphic modes.

c) From the commutation relation $[\varphi_0, a_0] = i$, show that

$$e^{i\sqrt{2}\alpha\varphi_0}|\beta\rangle = |\beta + \alpha\rangle \tag{9.181}$$

where the vacuum $|\beta\rangle$ is an eigenstate of a_0: $a_0|\beta\rangle = \sqrt{2}\beta|\beta\rangle$.

d) Show that

$$\langle:e^{i\sqrt{2}\alpha_1\tilde{\varphi}(z_1,\bar{z}_1)}:\cdots:e^{i\sqrt{2}\alpha_n\tilde{\varphi}(z_n,\bar{z}_n)}:\rangle = \prod_{i<j}|z_i|^{4\alpha_i\alpha_j} \tag{9.182}$$

provided the neutrality condition $\sum_i \alpha_i = 0$ is satisfied (otherwise the result vanishes). Putting everything together, recover the correlator (9.8). Notice that the holomorphic correlator (9.11) may be recovered in the same way if we start with the chiral vertex operator (9.6), which amounts to dropping all the terms involving antiholomorphic coordinates.

9.3 *The Coulomb-gas integrals as solutions of a hypergeometric equation*
Solving Exs. 8.9 and 8.10 first can be of some help here.

a) Rewrite the differential equation (8.71) for the four-point correlation function (9.66) (with $r_2 = 2, s_2 = 1$) as an ordinary (hypergeometric) differential equation for the function $G(z,\bar{z})$.

b) Check that the functions $I_1(a,b,c;z)$ and $I_2(a,b,c;z)$ of Eq. (9.72) generate the two-dimensional linear space of solutions of this differential equation, with a,b,c as in Eq. (9.70).

c) Application: compute the values of a, b, c for the four-point function of $\Phi_{(2,1)}$ in the Yang-Lee model $\mathcal{M}(5,2)$. Check that indeed I_1, I_2 coincide, up to a numerical factor and a well-defined power of z, with the conformal blocks found in Ex. 8.11. Compute the structure constants of the OPE $\phi_{(2,1)} \times \phi_{(2,1)} \to \mathbb{I} + \phi_{(2,1)}$.

d) Repeat the analysis for the Ising model $\mathcal{M}(4,3)$, with the four-point functions of the energy $\varepsilon = \Phi_{(2,1)}$ and spin $\sigma = \Phi_{(1,2)}$ operators. Compute in particular the structure constants of the spin-spin OPE $\Phi_{(1,2)} \times \Phi_{(1,2)} \to \mathbb{I} + \Phi_{(2,1)}$.

9.4 *Fusion rules for the energy operator of the Ising model*

a) Write the two blocks I_1 and I_2 (9.72) for the Ising energy four-point function $\langle \Phi_{(2,1)} \Phi_{(2,1)} \Phi_{(2,1)} \Phi_{(2,1)} \rangle$.
Result: $a = b = c = -4/3$.

b) Show that the monodromy-invariant combination $G(z, \bar{z}) \sim X_1 |I_1|^2 + X_2 |I_2|^2$ reduces to only one term. Deduce that the fusion rule of the Ising energy operator is indeed $\varepsilon \times \varepsilon \to I$. This confirms the fact that the operator $\Phi_{(3,1)}$ does not belong to the Ising theory.

9.5 Check the relation (9.81).

9.6 *Computing the correlation function* $G(z, \bar{z}) = \langle \Phi_{(r,s)} \Phi_{(1,3)} \Phi_{(1,3)} \Phi_{(r,s)} \rangle$

a) Find the number of screening operators needed to represent this correlation function and write the corresponding integral.
Result:

$$I_{(C_1, C_2)}(a, b, c; \rho; z) = z^{2\alpha_1 \alpha_2} (1 - z)^{2\alpha_2 \alpha_3} \int_{C_1} dt_1 \int_{C_2} dt_2$$

$$(t_1 t_2)^a \left((t_1 - 1)(t_2 - 1)\right)^b \left(t_1 - z)(t_2 - z)\right)^c (t_1 - t_2)^\rho \tag{9.183}$$

where $\alpha_1 = \alpha_{r,s} = [(1-r)\alpha_+ + (1-s)\alpha_-]/2$, $\alpha_2 = \alpha_3 = -\alpha_+$, and $a = 2\alpha_1 \alpha_+$, $b = 2\alpha_3 \alpha_+$, $c = 2\alpha_2 \alpha_+$, $\rho = 2\alpha_+^2$.

b) Find a set of independent integration contours leading to a basis $I_j(z)$, $j = 1, 2, \ldots, N$, generalizing (9.72).
Hint: There are $N = 3$ natural couples of contours (C_1, C_2), namely $[0, z] \times [0, z]$, $[0, z] \times [1, \infty[$, and $[1, \infty[\times [1, \infty[$.

c) Show that, with the choice of contours above, the three blocks have the following small z behavior $I_j(z) \sim z^{\rho_j}(1 + O(z))$, where $\rho_1 = 0$, $\rho_2 = 1 + a + c$, and $\rho_3 = 2 + 2a + 2c + \rho$. Deduce that these blocks correspond, respectively, to the fusion rules $\phi_{(1,3)} \times \phi_{(1,3)} \to \mathbb{I}$, $\phi_{(1,3)}$, and $\phi_{(1,5)}$.
Hint: Check that $\rho_j = h_{1,2j-1} - h_{1,3} - h_{r,s}$.

d) Express the monodromy of $I_j(z)$ around the point 0. Show that $G(z, \bar{z})$ has the form $\sum X_j |I_j(z)|^2$.

e) Find a relation $I_j(z) = \sum f_{j,k} \tilde{I}_k(1 - z)$ between I_j and the functions $\tilde{I}_k$ (obtained by interchanging a and b) revealing their monodromy around the point 1.
Hint: Such relations are obtained by moving the contours from $[1, +\infty[$ to $] - \infty, 1]$ and from $[0, z]$ to $] - \infty, 0] \cup [z, +\infty[$. For instance,

$$f_{1,1} = \frac{s(a)s(a + \rho/2)}{s(b + c)s(b + c + \rho/2)}$$

$$f_{2,1} = -\frac{s(a)s(c)}{s(b + c)s(b + c + \rho)} \tag{9.184}$$

$$f_{1,3} = \frac{s(c)s(c + \rho/2)}{s(b + c + \rho/2)s(b + c + \rho)}$$

where $s(x) = \sin \pi x$.

f) Deduce relations among the X's that determine them up to an overall normalization. Write the final result for $G(z, \bar{z})$.
Result:

$$
\begin{aligned}
\frac{X_1}{X_3} &= \frac{f_{3,3}\tilde{f}_{3,1}}{f_{1,3}\tilde{f}_{3,3}} \\
&= \frac{s(a+b+c+\rho)s(a+b+c+\rho/2)s(b)s(b+\rho/2)s(a+c+\rho)}{s(a)s(a+\rho/2)s(c)s(c+\rho/2)s(a+c)} \\
\frac{X_2}{X_3} &= \frac{f_{3,3}\tilde{f}_{3,2}}{f_{2,3}\tilde{f}_{3,3}} \\
&= \frac{s(a+b+c+\rho)s(a+c+\rho/2)s(b)}{2s(c+\rho/2)s(a+\rho/2)s(a+c)c(\rho/2)}
\end{aligned} \tag{9.185}
$$

where $\tilde{f} = f^{-1}$, the inverse matrix of f, $s(x) = \sin \pi x$, and $c(x) = \cos \pi x$.

g) Compute the fusion rule $\phi_{(1,3)} \times \phi_{(r,s)}$, and write relations for the associated structure constants, normalized by crossing symmetry.

h) Check that the four-point correlation function of the energy operator of the Ising model, represented as $\Phi_{(1,3)} (= \Phi_{(2,1)})$, has, according to the above, three possible conformal blocks. Write them as contour integrals. Show that only one of them survives in the monodromy-invariant combination giving the four-point correlator.

i) Application: Compute the conformal blocks for the four-point function of the $\Phi_{(1,3)}$ operator in the Tricritical Ising model $\mathcal{M}(5, 4)$.

9.7 *Crossing and exchange matrices for the conformal blocks of the Ising energy four-point correlation function*
The total conformal block for the energy four-point correlator is

$$
I(z) = \frac{1 - z + z^2}{z(1 - z)} \tag{9.186}
$$

It corresponds to the fusion rule $\Phi_{(2,1)} \times \Phi_{(2,1)} \to \mathbb{I}$. (See, e.g., Ex. 8.12 for a proof.) Compute the matrix elements of the crossing and exchange matrices F and R describing the linear action of the transformations $z \to 1 - z$ and $z \to 1/z$, respectively, on $I(z)$.

9.8 *Crossing and exchange matrices for the conformal blocks of the Ising spin four-point correlation function*
The two total conformal blocks for the Ising four-point spin correlator read (see, e.g., Ex. 8.12 for a proof)

$$
I_1(z) = \frac{(1 + \sqrt{1 - z})^{1/2}}{\sqrt{2}(z(1 - z))^{1/8}} \qquad I_2(z) = \frac{(1 - \sqrt{1 - z})^{1/2}}{\sqrt{2}(z(1 - z))^{1/8}} \tag{9.187}
$$

corresponding, respectively, to the $\phi_{(2,1)} \times \phi_{(2,1)} \to \mathbb{I}$ and $\phi_{(1,2)} \times \phi_{(1,2)} \to \phi_{(2,1)}$ fusion rules.

a) Compute the entries of the crossing matrix F (Eq. (9.112)) for these blocks.

Hint: By identifying the squares of both sides of the equations, show that

$$I_1(1-z) = \frac{1}{\sqrt{2}}\left(I_1(z) + I_2(z)\right)$$

$$I_2(1-z) = \frac{1}{\sqrt{2}}\left(I_1(z) - I_2(z)\right)$$

(9.188)

The overall sign is fixed, for instance, by the $z \rightarrow 0$ limit.

b) Check the inversion formula (9.114).

c) Compute the entries of the exchange matrix R (Eq. (9.113)).
Hint: By identifying the squares of both sides of the equations, show that

$$I_1(1/z) = \frac{1}{\sqrt{2}}\left(\omega I_1(z) + \bar{\omega} I_2(z)\right)$$

$$I_2(1/z) = \frac{1}{\sqrt{2}}\left(\bar{\omega} I_1(z) + \omega I_2(z)\right)$$

(9.189)

where $\omega = \exp(i\pi/4)$; the overall sign is fixed by the $z \rightarrow 0$ limit.

9.9 *Yang–Baxter relation for the exchange matrix R*
Notations are as in Eq. (9.110), with $N = 5$. Show that there are two ways of transforming the corresponding conformal block of a five-point function into one with the fields 2 and 4 exchanged, simply by using the exchange transformation R defined in Eq. (9.113). Deduce a cubic relation between matrix elements of R.
Result: $R(3 \leftrightarrow 4) R(2 \leftrightarrow 4) R(2 \leftrightarrow 3) = R(2 \leftrightarrow 3) R(2 \leftrightarrow 4) R(3 \leftrightarrow 4)$.

9.10 *Pentagon identity*
Notations are as in Eq. (9.110), with $N = 5$. Show that there are two ways of transforming the corresponding conformal block of a five-point function into one with the fields 2 and 4 exchanged as well as 3 and 5. Deduce an identity (quadratic in R and linear in F) between R and F.
Result: $F(2 \leftrightarrow 5) R(2 \leftrightarrow 4) R(3 \leftrightarrow 4) = R(4 \leftrightarrow 2) F(1 \leftrightarrow 3)$.

9.11 *The Virasoro algebra of the charged bosonic Fock space F_{α,α_0}*
Using the commutation relations of the Heisenberg algebra (9.130), show that the generators L_n, $n \in \mathbb{Z}$, defined in Eq. (9.132), satisfy

$$[L_n, L_m] = (n - m)L_{n+m} + \frac{c}{12}n(n^2 - 1)\delta_{n+m,0}$$

with

$$c = 1 - 24\alpha_0^2$$

9.12 *Vertex representation of dual fields*
Dual primary fields satisfy the OPE (cf. Ex. 7.11)

$$\phi_{(r,s)}(z)\phi_{(r',s')}(w) \sim \frac{\phi_h}{(z-w)^2} + \frac{a\partial\phi_h}{(z-w)}$$

where a is some constant. Using the Coulomb-gas representation, show that this reduces to the requirement

$$\alpha_{r',s'} = -1/\alpha_{r,s}$$

Find all those sets of *integers* $\{(r,s)\,(r',s')\}$ that satisfy this condition and use the equivalence between $\phi_{(-r,-s)}$ and $\phi_{(r,s)}$ to reproduce the list of dual primary fields given in Ex. 7.11.

9.13 *The quantum Korteweg–de Vries equation revisited: new characterization of the conservation laws*
As shown in Ex. 6.13, the conserved densities of the quantum Korteweg–de Vries equation $\mathcal{H}_n[T] = \tilde{\mathcal{H}}_n[\varphi]$, where the tilde expression is obtained after the substitution

$$T = -\frac{1}{2}(\partial\varphi\partial\varphi) + i\sqrt{2}\,\alpha_0\partial^2\varphi$$

satisfy

$$\tilde{\mathcal{H}}_n[\varphi] = \tilde{\mathcal{H}}_n[-\varphi] + \partial(\cdots)$$

Given that any differential polynomial $F[T]$ commutes with the screening charges, that is,

$$[F[T], Q_\pm] = [F[T], \oint dz\, e^{i\sqrt{2}\alpha_\pm\varphi}] = 0$$

argue that the conservation laws for the quantum KdV equation satisfy

$$[\oint dz\,\mathcal{H}_n[T], \oint dw\, e^{-i\sqrt{2}\alpha_+\varphi}] = [\oint dz\,\mathcal{H}_n[T], \oint dw\,\phi_{(1,3)}] = 0$$

or equivalently

$$[\oint dz\,\mathcal{H}_n[T], \oint dw\, e^{-i\sqrt{2}\alpha_-\varphi}] = [\oint dz\,\mathcal{H}_n[T], \oint dw\,\phi_{(3,1)}] = 0 \qquad (9.190)$$

Remark: There is also an infinite number of integrals, built out of T, that commutes with the integral of $\phi_{(1,2)}$ or its dual $\phi_{(5,1)}$ (cf. Ex. 7.11) and a distinct infinite family of integrals that commute with $\phi_{(2,1)}$ or $\phi_{(1,5)}$. The spin of these charges is $6n \pm 1$ in both cases. The following exercise substantiates these claims.

9.14 *The quantum Korteweg–de Vries conservation laws and singular vectors*
The starting point of this exercise is the characterization (9.190) (cf. Ex. 9.13) of the conserved integrals of the quantum KdV equation and the duality condition presented in Ex. 7.11, which shows that

$$[\oint dz\,\phi_{(3,1)}, \oint dw\,\phi_{(1,3)}] = 0$$

a) In general $\phi_{(3,1)}$ does not qualify as a conserved density for the quantum KdV equation because it is not a local differential polynomial of T. However, show that for the series $(p,p') = (2k+1,2)$, $\phi_{(3,1)}$ is in fact a singular vector in the vacuum module, and as such it can be expressed locally in terms of T. It is thus necessarily a quantum KdV conserved density. For a fixed value of k, what is the spin of the conserved integral that becomes trivial? Verify that the conserved densities of H_3 and H_5, with

$$H_3 = \oint dz\,(TT) \quad \text{and} \quad H_5 = \oint dz\,[(T(TT)) - \frac{(c+2)}{12}(\partial T\partial T)]$$

do indeed correspond to singular vectors at appropriate values of c.

b) In a similar way, show that for $(p,p') = (3k \pm 1, 3)$, $\phi_{(5,1)}$ lies in the vacuum module. Therefore, it is expressible in terms of T. Since $\phi_{(5,1)}$ is dual to $\phi_{(1,2)}$, this leads to an integral H'_n that commutes with $\oint dz\,\phi_{(1,2)}$. Relate n to k.

Notes

The Coulomb-gas formalism for two-dimensional CFT relies on the so-called Feigin-Fuchs integral representation of conformal blocks for conformal correlation functions (unpublished). It was extensively developed for the minimal models by Dotsenko and Fateev [110, 111].

The deep mathematical structure (BRST cohomology) underlying this construction was unearthed by Felder [128], who extended the Coulomb gas formalism to the case of the torus correlation functions of minimal models as well.

The general structure of minimal model correlation functions is based on the monodromy transformations of the conformal blocks. The first study of the abstract properties of these blocks is due to Rehren and Schroer [305], who introduced the notion of exchange algebra. In parallel, Moore, and Seiberg [272, 273] developed an axiomatic definition of rational conformal field theories (RCFT) based on polynomial relations satisfied by the crossing and exchange matrices. These relations also include additional modular data to ensure that the theory is well defined on any Riemann surface of arbitrary genus.

Ex. 9.14 is based on Refs. [93, 267].

CHAPTER 10

Modular Invariance

We have assumed until now, implicitly or not, that conformal field theories were defined on the whole complex plane. On the infinite plane the holomorphic and antiholomorphic (or left and right) sectors of a conformal theory completely decouple and may be studied separately. In fact, the two sectors may constitute distinct theories on their own since they do not interfere: Correlation functions factorize into holomorphic and antiholomorphic factors with a priori different properties. However, this situation is very unphysical. The decoupling exists only at the fixed point in parameter space (the conformally invariant point) and in the infinite-plane geometry. The physical spectrum of the theory should be continuously deformed as we leave the critical point, and the coupling between right and left sectors away from this point should lead to some constraints on the left and right content of the theory at the fixed point. In operator language, this implies that not every left-right combination of Verma modules is physically sound.

In order to impose physical constraints on the left-right content of a conformal theory without leaving the fixed point, we must couple the left and right sectors through the geometry of the space on which the theory is defined. The infinite plane is topologically equivalent to a sphere, that is, a Riemann surface of genus $h = 0$. In general, one may study conformal field theories defined on a Riemann surface of arbitrary genus h.[1] In the context of critical phenomena, defining Euclidian field theories on arbitrary genus Riemann surfaces may seem unnatural, except in the simplest nonspherical case: that of a torus ($h = 1$), which is equivalent to a plane with periodic boundary conditions in two directions. The goal of this chapter is to study conformal field theories defined on the torus and to extract constraints on the content of the theory coming from the interaction of the holomorphic and antiholomorphic sectors revealed by *modular transformations* (to be defined below).[2]

---

[1] In string theory, this is the basis for calculating multiloop scattering amplitudes.

[2] The requirement that a conformal theory still makes sense on a Riemann surface of arbitrary genus adds in fact many constraints to the theory. Modular invariance is one of them. However, nothing prevents us from considering *projections* of these fully consistent theories—projections that, in general,

In previous chapters, the operator formalism of quantum field theory was applied through radial quantization, namely, the curves of constant time were concentric circles and time was flowing outward from the origin. In this scheme there are two special points, the origin and the point at infinity, at which asymptotic fields are defined and which allow an explicit mapping from the field content of the theory to the abstract representations of the conformal algebra (the Verma modules). This representation is equivalent, via an exponential mapping, to that of a field theory living on a cylinder; the asymptotic points are then really at $\pm\infty$ along the cylinder axis. The operator formalism on the torus is obtained by imposing periodic boundary conditions along this cylinder, that is, by cutting a segment of the cylinder and gluing the ends back together. The Hamiltonian and the momentum operators then propagate states along different directions of the torus, and the spectrum of the theory is embodied in the partition function. Note that in statistical models, the "space" and "time" directions are chosen in order to define a transfer matrix, and their respective orientation is a simple matter of convenience.

This chapter is organized as follows. In Sect. 10.1 the general tools for studying conformal field theories on a torus, the partition function and modular transformations, are introduced. In Sect. 10.2, the partition function of a free boson is calculated. In Sect. 10.3, we calculate the partition functions of a free fermion and show how the various periodic and antiperiodic boundary conditions must be combined to form a consistent theory. In Sect. 10.4 we study variants of the free boson theory with central charge $c = 1$ (the compactified boson and the $\mathbb{Z}_2$ orbifold) and their modular invariant partition functions. In Sect. 10.5 we see how modular invariance forces models with a finite number of fields (minimal models) to have a central charge of the form $c = 1 - 6(p - p')/pp'$, where p and p' are relatively prime nonnegative integers. In Sect. 10.6 the transformation properties of the characters of minimal models under modular transformations are derived. The construction of modular invariant partition functions for minimal models (hence of physically sensible theories) is done in Sect. 10.7. Finally, in Sect. 10.8, we come back to the question of fusion rules in minimal models from the point of view of modular invariance and derive the Verlinde formula, which relates fusion coefficients and modular transformations.

§10.1. Conformal Field Theory on the Torus

A torus may be defined by specifying two linearly independent lattice vectors on the plane and identifying points that differ by an integer combination of these vectors. On the complex plane these lattice vectors may be represented by two complex numbers ω_1 and ω_2, which we call the *periods* of the lattice. Naturally, the properties of conformal field theories defined on a torus do not depend on the

---

do not satisfy these constraints. This is the case for the conformal theories with boundaries studied in Chap. 11.

overall scale of the lattice, nor on the absolute orientation of the lattice vectors. The relevant parameter is the ratio $\tau = \omega_2/\omega_1$, the so-called *modular parameter*.

10.1.1. The Partition Function

A theory defined on a torus may be treated in the path-integral formalism. The essential difference from the infinite plane is the occurrence of local fields obeying periodicity conditions such that the action functional is invariant with respect to translations by the periods $\omega_{1,2}$. This does not necessarily mean that the conformal fields themselves are simply periodic. For instance, a real fermion living on a torus may pick up a factor of -1 when translated by a period: this constitutes the Neveu-Schwarz (NS) condition, whereas the Ramond (R) condition demands that the fermion be periodic. Since there are two periods, a fermion field may be defined according to four types of boundary conditions: (R,R), (R,NS), (NS,R), and (NS,NS). In any case, the path-integral formulation of a field theory on a torus is well defined provided the boundary conditions chosen for the dynamical fields leave the action invariant.

However, we shall work mainly in the operator formalism. The relevant quantity in this scheme is the partition function Z (or the vacuum functional, in Minkowski space-time) and its dependence on the modular parameter τ. We find an expression for the partition function of the theory in terms of the Virasoro generators L_0 and $\bar{L}_0$. We need to define space and time directions, which we shall take to run along the real and imaginary axes, respectively; it is the orientation of the periods relative to these space and time axes that matters. If H and P denote, respectively, the Hamiltonian and the total momentum of the theory (generating translations along the time and space directions), then the operator that translates the system parallel to the period ω_2 over a distance a in Euclidian space-time is

$$\exp -\frac{a}{|\omega_2|}\left\{H \operatorname{Im} \omega_2 - iP \operatorname{Re} \omega_2\right\} \tag{10.1}$$

If we regard a as a lattice spacing, the above translation takes us from one row of a lattice to the next, but parallel to the period ω_2. If the complete period contains m lattice spacings ($|\omega_2| = ma$) then the partition function is obtained by taking the trace of the above translation operator to the m-th power:

$$Z(\omega_1, \omega_2) = \operatorname{Tr} \exp -\left\{H \operatorname{Im} \omega_2 - iP \operatorname{Re} \omega_2\right\} \tag{10.2}$$

We need to express the operators H and P in terms of the Virasoro generators L_0 and $\bar{L}_0$. This can be done by regarding the torus as a cylinder of finite length whose ends have been glued back together. We know that on a cylinder of circumference L the Hamiltonian operator is $H = (2\pi/L)(L_0 + \bar{L}_0 - c/12)$, wherein the Virasoro generators are defined on the whole complex plane after an exponential map; the constant term has been added to make the vacuum energy density vanish in the $L \to \infty$ limit. Likewise, the momentum operator, which generates translations along the circumference of the cylinder, is $P = (2\pi i/L)(L_0 - \bar{L}_0)$. Since we have chosen ω_1 to be real (and equal to L), we may finally write the partition function as[3]

$$
\begin{aligned}
Z(\tau) &= \text{Tr} \exp \pi i \{ (\tau - \bar{\tau})(L_0 + \bar{L}_0 - c/12) + (\tau + \bar{\tau})(L_0 - \bar{L}_0) \} \\
&= \text{Tr} \exp 2\pi i \{ \tau(L_0 - c/24) - \bar{\tau}(\bar{L}_0 - c/24) \}
\end{aligned}
\tag{10.3}
$$

We define the parameters

$$
q = \exp 2\pi i \tau \qquad \bar{q} = \exp -2\pi i \bar{\tau}
\tag{10.4}
$$

We may then express the partition function as

$$
\boxed{Z(\tau) = \text{Tr} \left(q^{L_0 - c/24} \, \bar{q}^{\bar{L}_0 - c/24} \right)}
\tag{10.5}
$$

Note that the partition function depends on the periods $\omega_{1,2}$ only through their ratio τ. This expression for the partition function involves the characters defined in Eq. (7.12). We thus expect that the partition function will be expressible as a bilinear combination of characters of the Verma modules forming the Hilbert space of the theory.

10.1.2. Modular Invariance

The main advantage of studying conformal field theories on a torus is the imposition of constraints on the operator content of the theory from the requirement that the partition function be independent of the choice of periods $\omega_{1,2}$ for a given torus.

We let $\omega'_{1,2}$ be two periods describing the same lattice as $\omega_{1,2}$. Since the points ω'_1 and ω'_2 belong to the lattice, they must be expressible as integer combinations of ω_1 and ω_2:

$$
\begin{pmatrix} \omega'_1 \\ \omega'_2 \end{pmatrix} = \begin{pmatrix} a & b \\ c & d \end{pmatrix} \begin{pmatrix} \omega_1 \\ \omega_2 \end{pmatrix} \qquad \begin{array}{l} a, b, c, d \in \mathbb{Z} \\ ad - bc = 1 \end{array}
\tag{10.6}
$$

Of course, the same may be said of $\omega_{1,2}$ in terms of $\omega'_{1,2}$, which implies that the above matrix should have an inverse with integer components. Since the unit cell of the lattice should have the same area whatever the periods we use, the determinant of that matrix should be unity. We are therefore led to consider the group of integer, invertible matrices with unit determinant, or $SL(2, \mathbb{Z})$. Such matrices evidently form a group, since the unit determinant guarantees that the matrix has an integer inverse.

Under the change of period (10.6), the modular parameter transforms as

$$
\tau \to \frac{a\tau + b}{c\tau + d} \qquad ad - bc = 1
\tag{10.7}
$$

Insofar as τ is concerned, the sign of all the parameters a, b, c, d may be simultaneously changed without affecting the transformation. The symmetry of interest here

---

[3] Here and in the following, we drop the $\bar{\tau}$ dependence of the partition function, since τ and $\bar{\tau}$ are not independent, the latter being the complex conjugate of the former. In particular, a modular transformation will act on τ and $\bar{\tau}$ *simultaneously*.

is therefore the *modular group* $SL(2, \mathbb{Z})/\mathbb{Z}_2$, or $PSL(2, \mathbb{Z})$. We are thus interested in finding partition functions $Z(\tau)$ that are invariant under modular transformations of the torus modular parameter τ. Note that the modular group will keep τ on the upper half-plane.

10.1.3. Generators and the Fundamental Domain

We consider the particular modular transformations

$$
\begin{aligned}
T : \tau \to \tau + 1 \quad &\text{or} \quad T = \begin{pmatrix} 1 & 0 \\ 1 & 1 \end{pmatrix} \\
S : \tau \to -\frac{1}{\tau} \quad &\text{or} \quad S = \begin{pmatrix} 0 & 1 \\ -1 & 0 \end{pmatrix}
\end{aligned}
\tag{10.8}
$$

It can be shown that these two transformations, satisfying

$$
(ST)^3 = S^2 = 1
\tag{10.9}
$$

generate the whole modular group, namely, each modular transformation may be reduced to successive applications of S and T (see Ex. 10.2 for a detailed proof).

This result is easier to understand geometrically by considering the so-called *Dehn twists*. Consider a torus specified by the periods ω_1 and ω_2. The modular transformation $T : \tau \to \tau + 1$ amounts to changing the second period as follows: $\omega_2 \to \omega_2 + \omega_1$. On the torus, this is equivalent to cutting the torus at a fixed time, turning one of the ends by 2π and gluing the two ends back together. Likewise, the modular transformation $\mathcal{U} : \tau \to \tau/(\tau + 1)$ is equivalent to a similar operation, but after cutting along a fixed space coordinate. These two operations on the torus are called Dehn twists. They are in fact finite diffeomorphisms of the torus that cannot be obtained continuously from the identity. It is intuitively clear that any redefinition of the periods $\omega_{1,2}$ may be obtained by a succession of operations of the above type. It is easy to verify that the modular transformation $\mathcal{U}$ may be written as $\mathcal{U} = TST$. Therefore S and T are indeed generators of the modular group. The mappings T and $\mathcal{U}$ are illustrated on Fig. 10.1.

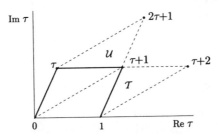

Figure 10.1. The modular parameter τ and the unit cell of the lattice. The unit cells obtained under the modular transformations T and $\mathcal{U} = TST$ are illustrated by dashed lines.

The action of the modular group Γ on the upper half of the τ-plane is rather complicated. A *fundamental domain* of Γ is a domain of the upper half-plane such that no pair of points within it can be reached through a modular transformation, and any point outside it can be reached from a unique point inside, by some modular transformation. Of course, the action of any modular transformation on a fundamental domain as a whole yields another fundamental domain. The usual convention is to pick the fundamental domain denoted F_0 defined as follows:

$$z \in F_0 \quad \text{if} \quad \begin{cases} \text{Im } z > 0, \quad -\dfrac{1}{2} \le \text{Re } z \le 0 \quad \text{and} \quad |z| \ge 1 \\ \text{or} \\ \text{Im } z > 0, \ 0 < \text{Re } z < \dfrac{1}{2} \quad \text{and} \quad |z| > 1 \end{cases} \tag{10.10}$$

We note the use of strict inequalities where appropriate. This domain is illustrated on Fig. 10.2, as well as some other domains obtained by applying simple modular transformations on F_0.

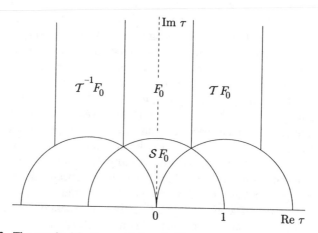

Figure 10.2. The standard fundamental domain F_0 of the modular group, and some other domains obtained by applying modular transformations on F_0.

§10.2. The Free Boson on the Torus

In this section we shall calculate the partition function of a free boson. Some care is needed because of the zero-mode, which should be discarded since it contributes an infinite amount to Z. Discarding the zero-mode should be equivalent to evaluating the trace (10.5) over the Fock space associated with the identity operator (i.e., the vertex operator of charge zero). From Eqs. (10.5) and (7.16), we therefore expect the partition to be of the following form:

$$Z_{\text{bos}} \propto \frac{1}{|\eta(\tau)|^2} \qquad (10.11)$$

The proportionality constant is important, since the above expression is not modular invariant! Indeed, as is shown in App. 10.A, Dedekind's η function transforms as follows under the modular transformations $\mathcal{T}$ and $\mathcal{S}$:

$$\begin{aligned} \eta(\tau + 1) &= e^{i\pi/12}\, \eta(\tau) \\ \eta(-1/\tau) &= \sqrt{-i\tau}\, \eta(\tau) \end{aligned} \qquad (10.12)$$

Although $|\eta(\tau)|$ is not modular invariant, it is a simple matter to check that the product $(\text{Im }\tau)^{1/4}\,|\eta(\tau)|$ is. Therefore, with a suitable proportionality constant ensuring modular invariance, the free-boson partition function (without zero-mode) is

$$\boxed{Z_{\text{bos}}(\tau) = \frac{1}{(\text{Im }\tau)^{1/2}|\eta(\tau)|^2}} \qquad (10.13)$$

In the remainder of this section we shall see how this result may be obtained directly from the path-integral formalism. Readers willing to skip the rather technical discussion that follows may proceed to the next section.

In the path-integral approach, the free-boson partition function without the zero-mode may be written as follows:

$$Z_{\text{bos}}(\tau) = \int [d\varphi]\sqrt{A}\, \delta\left(\int d^2x\, \varphi\varphi_0\right) \exp\left(-\frac{1}{2}\int d^2x\, (\nabla\varphi)^2\right) \qquad (10.14)$$

where the coordinate integrals are carried over the torus. Here A denotes the area of the torus, equal to Im $(\omega_2\omega_1^*)$, and $\varphi_0 = A^{-1/2}$ is the normalized eigenfunction of the zero-mode. The argument of the delta function is the coefficient of the zero-mode in an arbitrary field configuration; this delta function therefore ensures that the zero-mode is not integrated. The $\sqrt{A}$ in front was put there to make the whole expression dimensionless (in fact, the delta function with its properly normalized argument was introduced mainly in order to justify this factor). We have chosen the normalization $g = 1$ for the free-boson action, instead of our standard $g = 1/4\pi$. However, changing this normalization would only result in a constant multiplicative factor, which we are ignoring anyway (only τ-dependent multiplicative factors matter).

We expand the field φ along the normalized eigenfunctions φ_n of the Laplacian operator ∇^2, with eigenvalues $-\lambda_n$:

$$\varphi(x) = \sum_n c_n\varphi_n(x) \qquad (10.15)$$

The functional integral over the nonzero modes ($\lambda_n \neq 0$) is then

$$Z_{\text{bos}}(\tau) = \sqrt{A} \int \prod_i dc_i \, \exp -\frac{1}{2} \sum_n \lambda_n c_n^2$$

$$= \sqrt{A} \prod_n \left(\frac{2\pi}{\lambda_n}\right)^{1/2} \tag{10.16}$$

In general, this product diverges and must be regularized. We shall use the so-called ζ-function regularization technique, which is based upon the definition of the following function:

$$G(s) \equiv \sum_n' \frac{1}{\lambda_n^s} \tag{10.17}$$

where the primed sum means that the eigenvalue $\lambda = 0$ is excluded. The function $G(s)$ is analytic for sufficiently large values of s, and may be analytically continued to lower values of s, in particular $s = 0$, for which the above series definition is no longer valid. The partition function is then formally equal to

$$Z_{\text{bos}}(\tau) = \sqrt{A} \, \exp \frac{1}{2} G'(0) \tag{10.18}$$

(we have discarded the irrelevant numeric factor $(2\pi)^{1/2}$ coming from each mode).

In the case under consideration, the eigenvalues of the Laplacian are labeled by two integers m and n:

$$\lambda_{n,m} = (2\pi)^2 |nk_1 + mk_2|^2 \tag{10.19}$$

where $k_{1,2}$ are the basis vectors of the lattice dual to the one defined by the periods $\omega_{1,2}$:

$$k_1 = -i\omega_2/A \qquad k_2 = i\omega_1/A \tag{10.20}$$

It follows that

$$\left|\frac{2\pi\omega_1}{A}\right|^{2s} G(s) = \sum_{m,n}' \frac{1}{|m + n\tau|^{2s}}$$

$$= 2\zeta(2s) + \sum_n' \left(\sum_m \frac{1}{|m + n\tau|^{2s}}\right) \tag{10.21}$$

where $\zeta(z)$ is the Riemann ζ-function. The second term of the last expression is a periodic function of $n\tau$ with unit period, since all values of m are summed upon; it may therefore be Fourier expanded (we write $\tau = \tau_1 + i\tau_2$):

$$\sum_m \frac{1}{|m + n\tau|^{2s}}$$

$$= \sum_p e^{2i\pi p n\tau_1} \int_0^1 dy \, e^{-2\pi i p y} \sum_m \frac{1}{[(m+y)^2 + n^2\tau_2^2]^s}$$

$$= \sum_p e^{2i\pi pn\tau_1} \sum_m \int_m^{m+1} dy\, e^{-2\pi i p y} \frac{1}{[y^2 + n^2\tau_2^2]^s}$$

$$= \sum_p \int_{-\infty}^{\infty} dy\, e^{2\pi i p(n\tau_1 - y)} \frac{1}{[y^2 + n^2\tau_2^2]^s}$$

$$= \frac{1}{\Gamma(s)} \sum_p \int_{-\infty}^{\infty} dy\, e^{2\pi i p(n\tau_1 - y)} \int_0^{\infty} dt\, t^{s-1} e^{-t(y^2 + n^2\tau_2^2)}$$

$$= \frac{\sqrt{\pi}}{\Gamma(s)} \sum_p \int_0^{\infty} dt\, t^{s-3/2} e^{-[tn^2\tau_2^2 + \pi^2 p^2/t - 2\pi i p n\tau_1]} \qquad (10.22)$$

where, in the fourth line, we have used the integral representation of the Euler Γ function

$$\Gamma(s) = \int_0^{+\infty} dt\, t^{s-1} e^{-t}$$

to rewrite

$$\frac{1}{z^s} = \frac{1}{\Gamma(s)} \int_0^{+\infty} dt\, t^{s-1} e^{-zt}$$

with $z = y^2 + n^2\tau_2^2$. We separate the contribution of $p = 0$ from the rest. At $p = 0$ the above reduces to $\Gamma(s - \frac{1}{2})|n\tau_2|^{1-2s}$. Summing this term over $n \neq 0$, we find

$$\sqrt{\pi} \frac{\Gamma(s - \frac{1}{2})}{\Gamma(s)} \sum_n' |n\tau_2|^{1-2s} = 2\sqrt{\pi} \frac{\Gamma(s - \frac{1}{2})}{\Gamma(s)} |\tau_2|^{1-2s} \zeta(2s - 1) \qquad (10.23)$$

We use, on the above expression, the following functional relation for the ζ-function:

$$\pi^{-s/2}\Gamma(s/2)\zeta(s) = \pi^{(1-s)/2}\Gamma((1 - s)/2)\zeta(1 - s) \qquad (10.24)$$

We may therefore write the following relation, after changing the integration variable from t to $t\pi p/n\tau_2$:

$$\Gamma(s) \left(\frac{\tau_2}{\pi}\right)^{s-1/2} \left|\frac{2\pi\omega_1}{A}\right|^{2s} G(s)$$

$$= 2\Gamma(s)\zeta(2s) \left(\frac{\tau_2}{\pi}\right)^{s-1/2} + 2\Gamma(1 - s)\zeta(2 - 2s) \left(\frac{\tau_2}{\pi}\right)^{1/2-s} \qquad (10.25)$$

$$+ \sqrt{\pi} \sum_p' \sum_n' e^{2\pi i p n\tau_1} \int_0^{\infty} \frac{dt}{t} t^{s-1/2} \left|\frac{p}{n}\right|^{s-1/2} e^{-\pi|np|\tau_2(t+1/t)}$$

This expression for $G(s)$ has the merit of being explicitly even in $s - \frac{1}{2}$, namely, it is symmetric under $s \to 1 - s$. Since it is well-defined for $s > 1$ and coincides then with the original series expansion, we shall use this expression to extract the value of $G(0)$ and $G'(0)$. We need only expand $G(s)$ to first order around $s = 0$. Since $\Gamma(s) \sim 1/s$, the integral is needed only at $s = 0$:

$$\int_0^\infty \frac{dt}{t} t^{-1/2} e^{-\pi|np|\tau_2(t+1/t)} = \frac{1}{|pn|^{1/2}} e^{-2\pi|pn|\tau_2} \tag{10.26}$$

Using the special value $\zeta(2) = \pi^2/6$, we write

$$G(s) = -1 - 2s \ln|A/\omega_1| + \frac{1}{3} s\pi\tau_2$$
$$+ s \sum_p {}' \sum_n {}' \frac{1}{|p|} e^{2\pi i p n \tau_1 - 2\pi|pn|\tau_2} + O(s^2) \tag{10.27}$$

We still need to work on the double sum. It may be written as

$$\sum_{n,p>0} \frac{2}{p} \left(e^{2\pi i p n \tau_1 - 2\pi p n \tau_2} + e^{-2\pi i p n \tau_1 - 2\pi p n \tau_2} \right)$$

$$= \sum_{n,p>0} \frac{2}{p} (q^n + \bar{q}^n)$$

$$= -2 \sum_{n>0} \left(\ln(1 - q^n) + \ln(1 - \bar{q}^n) \right) \tag{10.28}$$

$$= -2 \ln|\eta(q)|^2 - \frac{1}{3}\pi\tau_2$$

Since $\sqrt{A}/|\omega_1| = \sqrt{\tau_2}$, one may finally write

$$G'(0) = -2 \ln\left(\sqrt{A}\tau_2 |\eta(\tau)|^2 \right) \tag{10.29}$$

According to Eq. (10.18) the free-boson partition function is then

$$Z_{\text{bos}}(\tau) = \frac{1}{\sqrt{\text{Im }\tau}|\eta(\tau)|^2} \tag{10.30}$$

which is the desired result, with the correct multiplicative factor ensuring modular invariance.

§10.3. Free Fermions on the Torus

The path-integral calculation of the free-fermion partition function could in principle be obtained with the same method as for the free boson. Indeed, the free-fermion action may be written as (cf. Sect. 5.3.2)

$$S = \frac{1}{2\pi} \int d^2x \left(\bar{\psi}\partial\bar{\psi} + \psi\bar{\partial}\psi \right) \tag{10.31}$$

Since the two fields ψ and $\bar{\psi}$ are decoupled, the partition function is simply the product of the Pfaffians (defined in App. 2.B) of the differential operators ∂ and $\bar{\partial}$:

$$Z = \mathrm{Pf}(\partial)\mathrm{Pf}(\bar{\partial}) \tag{10.32}$$

Since the Pfaffian is the square-root of the determinant, and since the product $\partial\bar{\partial}$ is the Laplacian, we find

$$Z = (\det \nabla^2)^{1/2} \tag{10.33}$$

We have not yet specified the periodicity conditions to be imposed on the fermions. These conditions affect the admissible eigenvalues of the Laplacian and the associated determinant. We shall assume that the fermions pick up a phase when translated by a period:

$$\psi(z + \omega_1) = e^{2i\pi v}\psi(z) \qquad \psi(z + \omega_2) = e^{2i\pi u}\psi(z) \tag{10.34}$$

We suppose that the same periodicity conditions are satisfied by the antiholomorphic component $\bar{\psi}$. Since the action must be periodic when the torus coordinate z is shifted by a period, we are restricted to the following four possibilities:

$$
\begin{aligned}
(v, u) &= (0, 0) & \text{or} \quad & (R, R) \\
(v, u) &= (0, \tfrac{1}{2}) & \text{or} \quad & (R, NS) \\
(v, u) &= (\tfrac{1}{2}, 0) & \text{or} \quad & (NS, R) \\
(v, u) &= (\tfrac{1}{2}, \tfrac{1}{2}) & \text{or} \quad & (NS, NS)
\end{aligned}
\tag{10.35}
$$

Again, we associate the names of Ramond (R) to the periodic boundary condition and Neveu-Schwarz (NS) to the antiperiodic one. We shall denote by $Z_{v,u}$ the partition function associated with the periodicity condition (v, u). A set (v, u) of periodicity conditions is called a *spin structure* for the fermion. Because of the decoupling between ψ and $\bar{\psi}$, we may consider the partition function obtained by integrating the holomorphic field only, which we call $d_{v,u}$. It follows that

$$Z_{v,u} = |d_{v,u}|^2 \tag{10.36}$$

If an eigenfunction of the Laplacian satisfies the periodicity conditions (v, u),

$$\varphi(z + k\omega_1 + l\omega_2) = e^{2\pi i(kv + lu)}\varphi(z) \tag{10.37}$$

then the associated eigenvalue has the form

$$\frac{i}{A}\Big((m + u)\omega_1 + (n + v)\omega_2\Big) \tag{10.38}$$

(A is the area of the torus). We could proceed as in the previous section, and consider the following function:

$$G_{v,u}(s) = \left(\frac{A}{2\pi\omega_1}\right) \sum_{m,n} \frac{1}{|m + n\tau + (u + v\tau)|^{2s}} \qquad (10.39)$$

The partition function, after this ζ-function regularization, is equal to

$$Z_{v,u} = \exp{-\frac{1}{2}G'_{v,u}(0)} \qquad (10.40)$$

If $(v, u) \neq (0, 0)$, there is no Laplacian zero-mode, and the subtlety associated with it disappears. That zero-mode exists only for the (R, R) sector, in which case the partition function vanishes: $Z_{0,0} = 0$. Accordingly, we shall not use the path-integral method to calculate the partition functions, but rather the operator method, following Eq. (10.5).

We need to implement the periodicity conditions in the time direction within the operator formalism. These conditions are rather unusual in the context of field theory, but may be expressed as conditions on correlation functions on the torus. Consider the generic correlation function of fermions

$$\langle \psi(z)X \rangle \qquad (10.41)$$

where X stands for the product of an *odd* number of fermion fields at various positions, so that the correlator is nonzero (the correlator of an odd number of fermions is zero, since they are Grassmann numbers). We take this fermion from its position z to $z + \omega_2$ via some continuous path. Within the operator formalism, this means that the $\psi(z)$ will go through all the possible instants of time (modulo the periodicity) and will have to be passed over all the other fermions in X in succession, because of the time-ordering. Since a minus sign is generated each time, there will be an overall factor of -1 generated by this translation, and therefore the usual correspondence between the path-integral and the Hamiltonian approach leads naturally to the antiperiodic condition ($u = \frac{1}{2}$) when the theory is defined on a torus. To implement the periodic condition ($u = 0$) we need to modify the usual correspondence by inserting in all the correlators an operator that anticommutes with $\psi(z)$, whatever the value of z. Such an operator is $(-1)^F$, where F is the fermion number

$$F = \sum_{k \geq 0} F_k \qquad F_k = b_{-k}b_k \ (k > 0) \qquad (10.42)$$

and where F_0 is an operator defined in the space-periodic case, equal to 0 when acting on $|0\rangle$ and to 1 when acting on $b_0|0\rangle$. A fermion number $\bar{F}$ is defined in the same way for the antiholomorphic component $\bar{\psi}$. This amounts to multiplying the time-evolution operator over a time L by a factor $\exp{-i\pi F} = (-1)^F$. To make sure that this feature is built into the partition function, we simply insert $(-1)^F$ in the definition of the partition function, within the trace, in the time-periodic case.

This prescription implies the following expressions for the holomorphic partition functions $d_{\nu,\mu}$ associated with each periodicity condition:

$$
\begin{aligned}
d_{0,0} &= \frac{1}{\sqrt{2}} \operatorname{Tr}(-1)^F q^{L_0 - 1/48} = \frac{1}{\sqrt{2}} \operatorname{Tr}(-1)^F q^{\sum_k k b_{-k} b_k + 1/24} \\
d_{0,\frac{1}{2}} &= \frac{1}{\sqrt{2}} \operatorname{Tr} q^{L_0 - 1/48} \qquad = \frac{1}{\sqrt{2}} \operatorname{Tr} q^{\sum_k k b_{-k} b_k + 1/24} \\
d_{\frac{1}{2},0} &= \operatorname{Tr}(-1)^F q^{L_0 - 1/48} \qquad = \operatorname{Tr}(-1)^F q^{\sum_k k b_{-k} b_k - 1/48} \\
d_{\frac{1}{2},\frac{1}{2}} &= \operatorname{Tr} q^{L_0 - 1/48} \qquad\quad = \operatorname{Tr} q^{\sum_k k b_{-k} b_k - 1/48}
\end{aligned}
$$

(10.43)

The expressions (6.114) for L_0 were used. The factors of $\sqrt{2}$ in the first two lines are conventional and are introduced in order to simplify the modular properties later on.

These partition functions may easily be calculated, since q^{L_0} factorizes into an infinite product of operators, one for each fermion mode (the same is true of $(-1)^F$). For instance,

$$
\begin{aligned}
d_{\frac{1}{2},0} &= q^{-1/48} \operatorname{Tr} \prod_{k>0} q^{k b_{-k} b_k} (-1)^{F_k} \\
&= q^{-1/48} \prod_{k>0} \left(\operatorname{Tr} q^{k b_{-k} b_k} (-1)^{F_k} \right)
\end{aligned}
$$

(10.44)

wherein we have used the fact the the trace $\operatorname{Tr}(AB)$ of a product of two operators acting on different factors of a tensor product is simply the product $(\operatorname{Tr} A)(\operatorname{Tr} B)$, the latter traces being taken only over the restricted spaces on which A and B specifically act. For a given fermion mode, there are only two states and the traces are trivially calculated:

$$
\begin{aligned}
\operatorname{Tr} q^{k b_{-k} b_k} &= 1 + q^k \\
\operatorname{Tr} q^{k b_{-k} b_k} (-1)^{F_k} &= 1 - q^k
\end{aligned}
$$

(10.45)

We may therefore write the following infinite products for the partition functions, and relate them to the theta functions defined in App. (10.A):

$$
\begin{aligned}
d_{0,0} &= \frac{1}{\sqrt{2}} q^{1/24} \prod_{n=0}^{\infty} (1 - q^n) = 0 \\[2mm]
d_{0,\frac{1}{2}} &= \frac{1}{\sqrt{2}} q^{1/24} \prod_{n=0}^{\infty} (1 + q^n) = \sqrt{\frac{\theta_2(\tau)}{\eta(\tau)}} \\[2mm]
d_{\frac{1}{2},0} &= q^{-1/48} \prod_{r=1/2}^{\infty} (1 - q^r) = \sqrt{\frac{\theta_4(\tau)}{\eta(\tau)}} \\[2mm]
d_{\frac{1}{2},\frac{1}{2}} &= q^{-1/48} \prod_{r=1/2}^{\infty} (1 + q^r) = \sqrt{\frac{\theta_3(\tau)}{\eta(\tau)}}
\end{aligned}
$$

(10.46)

How does this relate to the Virasoro characters? These characters are not defined with the torus in mind, and so they do not take into account the periodicity in the time direction: only the boundary condition on the cylinder matters; we therefore distinguish the R and NS sectors, and for each we define the characters according to Eq. (7.12) with the help of the expressions (6.114) for L_0.

Table 10.1. Lowest energy states in the NS sector.

L_0	State(s)
0	$|0\rangle$
$\frac{1}{2}$	$b_{-1/2}|0\rangle$
$\frac{3}{2}$	$b_{-3/2}|0\rangle$
2	$b_{-3/2}b_{-1/2}|0\rangle$
$\frac{5}{2}$	$b_{-5/2}|0\rangle$
3	$b_{-5/2}b_{-1/2}|0\rangle$
$\frac{7}{2}$	$b_{-7/2}|0\rangle$
4	$b_{-5/2}b_{-3/2}|0\rangle$, $b_{-7/2}b_{-1/2}|0\rangle$

We first consider the NS sector. The lowest eigenstates of L_0 are listed in Table 10.1. There are states with integral values of L_0, others with half-integral values. This means that the trace of q^{L_0} is not a pure character in this case, but the sum of two (or more) simple Virasoro characters. Since this system has $c = c(4,3) = \frac{1}{2}$, we know exactly what the allowed Verma modules are: they have conformal weights 0, $\frac{1}{2}$ and $\frac{1}{16}$, according to the Kac table. We therefore have a sum of the Virasoro characters $\chi_{1,1}$ and $\chi_{2,1}$, occurring each with multiplicity one, as may be seen from the lowest two states. The states contributing to $\chi_{1,1}$ have an even fermion number and vice versa; this allows us to write the Virasoro characters as follows:

$$\chi_{1,1} = q^{-1/48}\frac{1}{2}\operatorname{Tr}(1+(-1)^F)q^{L_0}$$
$$\chi_{2,1} = q^{-1/48}\frac{1}{2}\operatorname{Tr}(1-(-1)^F)q^{L_0}$$
(10.47)

Comparing with the partition functions calculated above, we have the relations

$$\chi_{1,1} = \frac{1}{2}\left(d_{\frac{1}{2},\frac{1}{2}}+d_{\frac{1}{2},0}\right) \qquad \chi_{2,1} = \frac{1}{2}\left(d_{\frac{1}{2},\frac{1}{2}}-d_{\frac{1}{2},0}\right)$$
(10.48)

We then consider the Ramond sector. Here there are two degenerate ground states, differing by the fermion number F. Moreover, the eigenvalues of L_0 in this case are integer offsets of $\frac{1}{16}$, according to Eq. (6.114). The character $\chi_{1,2}$ is obtained by choosing one of the ground states, and obviously $\chi_{1,2} = d_{0,\frac{1}{2}}/\sqrt{2}$.

From the expression (10.46) for the partition functions, it is a simple matter, with the help of App. 10.A, to determine their modular transformations: under $\tau \to -1/\tau$, we have

$$d_{0,\frac{1}{2}}(-1/\tau) = d_{\frac{1}{2},0}(\tau)$$
$$d_{\frac{1}{2},0}(-1/\tau) = d_{0,\frac{1}{2}}(\tau) \qquad\qquad (10.49)$$
$$d_{\frac{1}{2},\frac{1}{2}}(-1/\tau) = d_{\frac{1}{2},\frac{1}{2}}(\tau)$$

Under the transformation $\tau \to \tau + 1$, we have instead

$$d_{0,\frac{1}{2}}(\tau + 1) = e^{i\pi/8} d_{0,\frac{1}{2}}(\tau)$$
$$d_{\frac{1}{2},0}(\tau + 1) = e^{-i\pi/24} d_{\frac{1}{2},\frac{1}{2}}(\tau) \qquad\qquad (10.50)$$
$$d_{\frac{1}{2},\frac{1}{2}}(\tau + 1) = e^{-i\pi/24} d_{\frac{1}{2},0}(\tau)$$

Since the full partition functions are simply $Z_{v,u} = |d_{v,u}|^2$, they transform exactly like the $d_{v,u}$'s, but without the phase factors.

It is now evident that the only ways to obtain a modular-invariant partition function are (1) to impose periodic boundary conditions on the fermion (R,R), in which case the partition function vanishes because of the zero-mode, and (2) to include in the theory the three possibilities (NS,R), (R,NS) and (NS,NS), leading to the modular-invariant combination

$$Z = Z_{\frac{1}{2},\frac{1}{2}} + Z_{0,\frac{1}{2}} + Z_{\frac{1}{2},0}$$
$$= \left|\frac{\theta_2}{\eta}\right| + \left|\frac{\theta_3}{\eta}\right| + \left|\frac{\theta_4}{\eta}\right| \qquad\qquad (10.51)$$
$$= 2 \left(|\chi_{1,1}|^2 + |\chi_{2,1}|^2 + |\chi_{1,2}|^2\right)$$

Thus, modular invariance requires that all three conformal fields associated with $c = \frac{1}{2}$ actually be present in the theory. Eq. (10.51) is merely twice the partition function of the Ising model on a torus.

§10.4. Models with $c = 1$

10.4.1. Compactified Boson

We have seen in Sect. 6.3.5 how the restriction of the domain of variation of a free boson to a circle of radius R restricts the allowed values of the charge α of the vertex operators, and how it allows new configurations with nonzero winding number. On the torus, such windings can occur when going from a point z to the equivalent points $z + \omega_1$ and $z + \omega_2$. There are thus two types of winding, and we must generally consider configurations with the following boundary conditions:

$$\varphi(z + k\omega_1 + k'\omega_2) = \varphi(z) + 2\pi R(km + k'm') \qquad k, k' \in \mathbb{Z} \qquad (10.52)$$

A doublet of integers (m, m') then specifies a topological class of configurations obeying the above periodicity conditions, and a partition function $Z_{m,m'}$ is defined by integrating over the configurations of such a class. The integration may be done by decomposing φ into a special configuration, which is also a classical solution to the equation of motion, $\varphi_{m,m'}^{cl}$ (with vanishing Laplacian, hence we take it to be the imaginary part of a holomorphic function), and a periodic field $\tilde{\varphi}$ (the "free part" of φ). This reads

$$\varphi = \varphi_{m,m'}^{cl} + \tilde{\varphi}$$

$$\varphi_{m,m'}^{cl} = 2\pi R \left\{ \frac{z}{\omega_1} \frac{m\bar{\tau} - m'}{\bar{\tau} - \tau} - \frac{\bar{z}}{\omega_1^*} \frac{m\tau - m'}{\bar{\tau} - \tau} \right\} \tag{10.53}$$

We check that the above configuration has indeed the right periodicity conditions and is real. The action $S(\varphi)$ is then the sum of $S[\tilde{\varphi}]$ (the action of the periodic field) plus the action $S[\varphi_{m,m'}^{cl}]$ of the classical linear configuration. Indeed, since $\Delta\varphi_{m,m'}^{cl} = 0$, the crossed terms in the action $S[\varphi]$ are proportional to

$$\int d^2x \, \nabla\varphi_{m,m'}^{cl} \nabla\tilde{\varphi} = -\int d^2x \, \tilde{\varphi} \, \Delta\varphi_{m,m'}^{cl} = 0 \tag{10.54}$$

where we have performed an integration by parts. $S[\varphi_{m,m'}^{cl}]$ is easily calculated as:

$$\begin{aligned}
S[\varphi_{m,m'}^{cl}] &= \frac{1}{8\pi} \int d^2x \, (\nabla\varphi_{m,m'}^{cl})^2 \\
&= \frac{1}{2\pi} \int dz d\bar{z} \, \partial\varphi_{m,m'}^{cl} \bar{\partial}\varphi_{m,m'}^{cl} \\
&= 2\pi R^2 A \frac{1}{|\omega_1|^2} \left| \frac{m\tau - m'}{\tau - \bar{\tau}} \right|^2 \\
&= \pi R^2 \frac{|m\tau - m'|^2}{2\,\mathrm{Im}\,\tau}
\end{aligned} \tag{10.55}$$

wherein $A = \mathrm{Im}\,(\omega_2\omega_1^*)$ is the area of the torus. The functional integration over the periodic field $\tilde{\varphi}$ gives a prefactor Z_{bos} (cf. Eq. (10.13)), leading to the following partition function:

$$Z_{m,m'}(\tau) = Z_{\text{bos}}(\tau) \exp -\frac{\pi R^2 |m\tau - m'|^2}{2\,\mathrm{Im}\,\tau} \tag{10.56}$$

It is then a simple matter to determine the modular properties of this partition function, since Z_{bos} is invariant under modular transformations. Under a general $SL(2, \mathbb{Z})$ mapping $\tau \to (a\tau + b)/(c\tau + d)$, the τ-dependent part of the exponent becomes

$$\begin{aligned}
\frac{|m\tau - m'|^2}{\mathrm{Im}\,\tau} &\longrightarrow \frac{|(ma\tau + bm)/(c\tau + d) - m'|^2 \, |c\tau + d|^2}{\mathrm{Im}\,[(a\tau + b)(c\bar{\tau} + d)]} \\
&= \frac{|ma\tau + bm - m'c\tau - m'd|^2}{\mathrm{Im}\,\tau}
\end{aligned} \tag{10.57}$$

wherein we have used

$$\text{Im}\,[(a\tau + b)(c\bar{\tau} + d)] = \text{Im}\,(ad\tau - bc\tau) = \text{Im}\,\tau \qquad (ad - bc = 1) \quad (10.58)$$

Under modular transformations, the doublet (m, m') transforms like

$$\begin{pmatrix} m \\ m' \end{pmatrix} \longrightarrow \begin{pmatrix} a & -c \\ -b & d \end{pmatrix} \begin{pmatrix} m \\ m' \end{pmatrix} \quad (10.59)$$

where the matrix is the inverse of the original $SL(2, \mathbb{Z})$ matrix. The doublet (m, m') thus transforms like the periods (k_1, k_2) of the reciprocal lattice. That the set of modular transformations forms a group implies that a sum of the partition functions over all the doublets (m, m') with equal weights is a modular invariant. Indeed, the $\mathcal{T}$ and $\mathcal{S}$ transformations on $Z_{m,m'}(\tau)$ read

$$\begin{aligned} Z_{m,m'}(\tau + 1) &= Z_{m,m'-m} \\ Z_{m,m'}(-1/\tau) &= Z_{-m',m} \end{aligned} \quad (10.60)$$

hence the sum over all $(m, m') \in \mathbb{Z}^2$ forms a modular-invariant partition function

$$Z(R) = \frac{R}{\sqrt{2}} Z_{\text{bos}}(\tau) \sum_{m,m'} \exp - \frac{\pi R^2 |m\tau - m'|^2}{2\,\text{Im}\,\tau} \quad (10.61)$$

The factor of $R/\sqrt{2}$ in front can actually be derived from a careful zero-mode integration of φ. It also gives the correct normalization 1 of the Virasoro character of the identity at $c = 1$ in the transformed expression (10.62). Poisson's resummation formula (cf. App. 10.A) may be used to reexpress this partition function in a different form. Setting

$$a = R^2/2\tau_2, \qquad b = \pi m R^2 \tau_1/\tau_2 \quad (\tau = \tau_1 + i\tau_2)$$

in Poisson's formula (10.264) leads to

$$Z(R) = \frac{1}{|\eta(\tau)|^2} \sum_{e,m \in \mathbb{Z}} q^{(e/R + mR/2)^2/2}\, \bar{q}^{(e/R - mR/2)^2/2} \quad (10.62)$$

In this form the partition function is manifestly compatible with the expressions (6.94) for L_0 and $\bar{L}_0$. It is simply the sum over all possible (electric) charges of vertex operators and all possible winding numbers (magnetic charges) of the $c = 1$ Virasoro characters squared, with conformal dimensions

$$h_{e,m} = \frac{1}{2}(e/R + mR/2)^2 \qquad \bar{h}_{e,m} = \frac{1}{2}(e/R - mR/2)^2 \quad (10.63)$$

These dimensions give the spectrum of primary fields in this model. The $m \neq 0$ fields represent vortex configurations of the field φ, namely lines of defect along which φ has a discontinuity of $2\pi m R$. The $e \neq 0$ fields correspond to electrically charged vertex operators $\exp ie\varphi/R$. A general field with $e, m \neq 0$ is a superposition of these two. (The computation of some correlation functions of these fields on the plane and on the torus will be presented in Chap. 12, Sect. 12.6.2.). The factor $(\text{Im } \tau)^{-1/2}$ has disappeared, since it resulted from the exclusion of the zero-mode, whereas the sum over charge and winding number sectors means that the zero-mode has been naturally incorporated. From the conformal dimensions we see that the scaling dimensions $\Delta_{e,m}$ and the spins $s_{e,m}$ of the primary fields are (cf. Eq. (5.21))

$$\Delta_{e,m} = e^2/R^2 + m^2 R^2/4 \qquad s_{e,m} = em \qquad (10.64)$$

The spins thus take integral values, as they should for a boson. The scaling dimensions are all positive (or zero) and vary continuously with R. As it stands, the model exhibits a remarkable electric-magnetic ($e \leftrightarrow m$) duality, which results in the invariance of the partition function and the spectrum of states under the interchange $R \leftrightarrow 2/R$

$$\boxed{Z(2/R) = Z(R)} \qquad (10.65)$$

10.4.2. Multi-Component Chiral Boson

In this section we shall indicate how to form modular-invariant partition functions out of an assembly of compactified free bosons. We first need to introduce the notion of a multidimensional lattice.

A lattice Γ of dimension n is a set of points in $\mathbb{R}^n$ with the property that its elements may be expressed as an integer linear combination of a set of n basis vectors ϵ_i:

$$\Gamma = \left\{ x = \sum_i x_i \epsilon_i \;\middle|\; x_i \in \mathbb{Z} \right\} \qquad (10.66)$$

The lattice is said to be Lorentzian with signature $(s, \bar{s})$ if it possesses (through $\mathbb{R}^n$) an indefinite inner product with signature $(+ \cdots + | - \cdots -)$, with s $(+)$ signs and $\bar{s}$ $(-)$ signs. If s or $\bar{s}$ is zero the lattice is, of course, Euclidian; we shall denote by $x \cdot y$ the inner product between two elements x and y of $\mathbb{R}^n$. The volume $\text{vol}(\Gamma)$ of the unit cell is the determinant of the matrix formed by the components of the basis vectors: $\text{vol}(\Gamma) = \det[\epsilon_i \cdot \epsilon_j]$. The lattice Γ^* dual to Γ is the set of points p such that $x \cdot p \in \mathbb{Z}$. Of course, Γ^* is also a lattice in the above sense, and may be generated by the dual basis $\{\epsilon_i^*\}$ satisfying the relation $\epsilon_i \cdot \epsilon_j^* = \delta_{ij}$; the volume of its unit cell is $\text{vol}(\Gamma^*) = 1/\text{vol}(\Gamma)$. A lattice is said to be self-dual if $\Gamma = \Gamma^*$; then, of course, $\text{vol}(\Gamma) = 1$. An integer lattice is defined to satisfy the property $x \cdot y \in \mathbb{Z}$ for all its elements x, y; it follows in that case that $\Gamma \in \Gamma^*$. An even-integer lattice is such that all its elements have even norm: $x^2 \in 2\mathbb{Z}$.

Now we go back to the partition function (10.62), which may be written as

$$Z(R) = \frac{1}{|\eta(\tau)|^2} \sum_{p, \bar{p}} e^{i\pi\tau p^2 - i\pi\bar{\tau}\bar{p}^2} \qquad (10.67)$$

wherein we have defined

$$p = e/R + mR/2 \qquad \bar{p} = e/R - mR/2 \qquad (10.68)$$

and the sum is taken over all integer values of e and m. The doublet $(p, \bar{p})$ may be expressed as $(p, \bar{p}) = ee_1 + me_2$, with

$$e_1 = (1/R, 1/R) \qquad e_2 = (R/2, -R/2) \qquad (10.69)$$

After defining the Lorentzian product

$$(x, y) \cdot (x', y') = xx' - yy' \qquad (10.70)$$

we see that the set of points $(p, \bar{p})$ forms an *even, self-dual, Lorentzian integer lattice*, since $e_1 \cdot e_1 = e_2 \cdot e_2 = 0$ and $e_1 \cdot e_2 = -1$. As we shall now demonstrate, this fact is closely related to the modular invariance of the partition function.

We consider a set of n bosons of which we keep only the holomorphic modes, and an a priori distinct set of $\bar{n}$ bosons of which we keep only the antiholomorphic modes. The theory is in fact defined by the following expression for the Virasoro generator:

$$
\begin{aligned}
L_0 &= \frac{1}{2} p^2 + \sum_{i=1}^{n} \sum_{k>0} a_{-k}^{(i)} a_k^{(i)} \\
\bar{L}_0 &= \frac{1}{2} \bar{p}^2 + \sum_{i=1}^{\bar{n}} \sum_{k>0} \bar{a}_{-k}^{(i)} \bar{a}_k^{(i)}
\end{aligned}
\qquad (10.71)
$$

where p belongs to some lattice Γ, and $\bar{p}$ to a lattice $\bar{\Gamma}$. The partition function of such a system would then be

$$Z_\Gamma(\tau) = \frac{1}{\eta(\tau)^n \bar{\eta}(\tau)^{\bar{n}}} \sum_{p \in \Gamma, \bar{p} \in \bar{\Gamma}} e^{i\pi \tau p^2 - i\pi \bar{\tau} \bar{p}^2} \qquad (10.72)$$

We are interested in knowing under what conditions this partition function is modular invariant. The effect of the modular transformation $\tau \to \tau + 1$ is easily seen to be

$$Z_\Gamma(\tau + 1) = Z_\Gamma(\tau) \exp \frac{2\pi i (n - \bar{n})}{24} \qquad (10.73)$$

provided that $p^2 - \bar{p}^2$ be always an even integer. Thus the Lorentzian lattice $\Gamma \oplus \bar{\Gamma}$ must be an even-integer lattice. In order to investigate the transformation $\tau \to -1/\tau$, we need to use a generalization of Poisson's resummation formula:

$$\sum_{q \in \Gamma} \exp\left(-\pi a q^2 + q \cdot b\right) = \frac{1}{\text{vol}(\Gamma)} \frac{1}{a^{n/2}} \sum_{p \in \Gamma^*} \exp -\frac{\pi}{a}\left(p + \frac{b}{2\pi i}\right) \qquad (10.74)$$

wherein a is some constant with positive real part, and b is some constant n-component vector. This formula may be easily demonstrated by using the closure relation[4]

$$\sum_{q \in \Gamma} \delta(x - q) = \frac{1}{\text{vol}(\Gamma)} \sum_{p \in \Gamma^*} e^{2\pi i x \cdot p} \tag{10.75}$$

and by integrating it over $\mathbb{R}^n$ against the function $\exp(-\pi a x^2 + b \cdot x)$. Applying Poisson's formula to the partition function (10.72), we easily find that it is invariant under the mapping $\tau \rightarrow -1/\tau$, provided $\Gamma = \Gamma^*$, that is, provided the lattice is self-dual. We conclude that models built from n holomorphic and $\bar{n}$ antiholomorphic bosons are modular invariant provided the *charge lattice* (i.e., the lattice of the charge (or momentum) vectors $(p, \bar{p})$) is an even-integer self-dual lattice, with $n - \bar{n} = 0$ mod 24.

This issue of modular invariance of a multicomponent boson system arises in the compactification of the bosonic string. Indeed, compactifying the extra dimensions in a consistent way is a task that has drawn a lot of attention in string theory. In that context the target space of the boson (the space in which the boson takes its values) is assumed to be physically very compact (of the order of the Planck length). The momenta of the string then take discrete values, and nonzero winding numbers must be considered.

10.4.3. $\mathbb{Z}_2$ Orbifold

A variation of the compactified free boson theory is obtained by assuming that the field φ does not take its values on the full circle, but on the object defined by identifying the angle φ with $-\varphi$, namely, by performing a quotient by the natural action of $\mathbb{Z}_2$. Such an object is called a $\mathbb{Z}_2$ *orbifold*. When taken across a period ω_1 or ω_2, the field φ may then be "twisted", resulting in the more general boundary condition

$$\varphi(z + k\omega_1 + l\omega_2) = e^{2\pi i (kv + lu)} \varphi(z) \tag{10.76}$$

already encountered when dealing with fermions, with u, v being equal either to 0 or to $\frac{1}{2}$. In the case of fermions, these boundary conditions were allowed by the fermionic nature of the fields, whereas here they follow from the topology of the space on which φ resides. Since the action for the free boson is symmetric under the interchange $\varphi \rightarrow -\varphi$, we may proceed as if the field were defined on the circle, except that we must integrate over half the range of φ in the path integral. Partition functions may then be calculated within the path-integral formalism, as before.

However, we shall again work within the operator formalism and calculate traces in order to obtain explicit expressions for the partition functions. For $(v, u) \neq (0, 0)$, we shall denote the traces over the holomorphic modes by $f_{v,u}$; the partition functions $Z_{v,u}$ are then equal to $|f_{v,u}|^2$. The partition function $Z_{0,0}$ in the untwisted sector is $Z(R)$. In Sect. 6.3.4, we defined an operator G that

---

[4] See App.14.C for a detailed derivation.

takes φ into $-\varphi$. This operator anticommutes with φ and may play a role similar to that played by $(-1)^F$ in the case of fermions, except that it must be inserted in the trace in the time-antiperiodic case ($u = \frac{1}{2}$). It also anticommutes with the mode operators a_n and $\bar{a}_n$ and has the following action on the vacua: $G|m, n\rangle = |-m, -n\rangle$. The Fock spaces built upon $|m, n\rangle$ and $|-m, -n\rangle$ must be combined into sectors $\mathcal{F}_\pm$, respectively symmetric and antisymmetric under the twist $\varphi \to -\varphi$. Explicitly, $\mathcal{F}_+$ is obtained by acting with an even number of creation operators a_n or $\bar{a}_n$ on the symmetric combination $|m, n\rangle + |-m, -n\rangle$, or with an odd number of creation operators on the antisymmetric combination $|m, n\rangle - |-m, -n\rangle$. $\mathcal{F}_-$ is built likewise, with the opposite combinations. The case $m = n = 0$ is special since the vacuum $|0, 0\rangle$ is doubly degenerate: $G|0, 0\rangle_\pm = \pm|0, 0\rangle_\pm$.

The holomorphic partition functions $f_{\nu, u}$ may then be calculated like their fermionic counterparts:

$$f_{0, \frac{1}{2}} = \operatorname{Tr} Gq^{L_0 - 1/24} = \operatorname{Tr} Gq^{\sum_n a_{-n} a_n - 1/24}$$

$$f_{\frac{1}{2}, 0} = \operatorname{Tr} q^{L_0 - 1/48} = \operatorname{Tr} q^{\sum_n a_{-n} a_n + 1/48} \tag{10.77}$$

$$f_{\frac{1}{2}, \frac{1}{2}} = \operatorname{Tr} Gq^{L_0 - 1/48} = \operatorname{Tr} Gq^{\sum_n a_{-n} a_n + 1/48}$$

Of course, the trace must also include a sum over the different vacua $|m, n\rangle$, including the two vacua $|0, 0\rangle_\pm$ in the space-antiperiodic case ($\nu = \frac{1}{2}$). Regarding $f_{0, 1/2}$, the insertion of G within the trace implies that only the sector $m = n = 0$ will contribute; indeed, each state obtained by acting on $|m, n\rangle + |-m, -n\rangle$ with creation operators has a counterpart with the same L_0 eigenvalue obtained by acting on $|m, n\rangle - |-m, -n\rangle$ with the same creation operators; however, these two states have opposite G values, and their contributions cancel in the trace. Thus, only the states obtained from the vacuum $|0, 0\rangle$ contribute, and the sign of their contribution is -1 if they are obtained from an odd number of creation operators ($G = -1$) and $+1$ otherwise. It follows that

$$f_{0, \frac{1}{2}} = q^{-1/24} \prod_{n=1}^{\infty} \frac{1}{(1 + q^n)} = 2\sqrt{\frac{\eta(\tau)}{\theta_2(\tau)}} \tag{10.78}$$

We now consider the space-antiperiodic case ($\nu = \frac{1}{2}$). Here we need consider only the two vacua $|0, 0\rangle_\pm$, each giving identical results, resulting in a factor of 2. The difference here lies in the vacuum energy, and in the fact that the mode indices take half-integer values. We therefore have

$$f_{\frac{1}{2}, 0} = 2q^{1/48} \prod_{r \in \mathbb{N} + 1/2} \frac{1}{(1 - q^r)} = 2\sqrt{\frac{\eta(\tau)}{\theta_4(\tau)}}$$

$$\tag{10.79}$$

$$f_{\frac{1}{2}, \frac{1}{2}} = 2q^{1/48} \prod_{r \in \mathbb{N} + 1/2} \frac{1}{(1 + q^r)} = 2\sqrt{\frac{\eta(\tau)}{\theta_3(\tau)}}$$

As in the case of fermions, the modular properties of these quantities are easily obtained:

$$f_{0,\frac{1}{2}}(-1/\tau) = f_{\frac{1}{2},0}(\tau)$$
$$f_{\frac{1}{2},0}(-1/\tau) = f_{0,\frac{1}{2}}(\tau) \qquad (10.80)$$
$$f_{\frac{1}{2},\frac{1}{2}}(-1/\tau) = f_{\frac{1}{2},\frac{1}{2}}(\tau)$$

and

$$f_{0,\frac{1}{2}}(\tau+1) = e^{-i\pi/8} f_{0,\frac{1}{2}}(\tau)$$
$$f_{\frac{1}{2},0}(\tau+1) = e^{i\pi/24} f_{\frac{1}{2},\frac{1}{2}}(\tau) \qquad (10.81)$$
$$f_{\frac{1}{2},\frac{1}{2}}(\tau+1) = e^{i\pi/24} f_{\frac{1}{2},0}(\tau)$$

The only modular-invariant combinations are thus $Z_{0,0} = Z(R)$ and

$$|f_{0,\frac{1}{2}}|^2 + |f_{\frac{1}{2},0}|^2 + |f_{\frac{1}{2},\frac{1}{2}}|^2 \qquad (10.82)$$

What we call the orbifold partition function $Z_{\text{orb}}(R)$ is obtained by summing over all types of boundary conditions and projecting on G-invariant states. This amounts to the calculation

$$Z_{\text{orb}}(R) = |q|^{-1/12} \frac{1}{2} \operatorname{Tr}_+ (1+G) q^{L_0} \bar{q}^{\bar{L}_0} + |q|^{-1/12} \frac{1}{2} \operatorname{Tr}_- (1+G) q^{L_0} \bar{q}^{\bar{L}_0}$$

$$= \frac{1}{2}(Z_{0,0} + |f_{0,\frac{1}{2}}|^2 + |f_{\frac{1}{2},0}|^2 + |f_{\frac{1}{2},\frac{1}{2}}|^2)$$

$$= \frac{1}{2}\left(Z(R) + 4\frac{|\eta|}{|\theta_2|} + 4\frac{|\eta|}{|\theta_3|} + 4\frac{|\eta|}{|\theta_4|} \right)$$

$$(10.83)$$

In the first line, $\operatorname{Tr}_{\pm}$ means a trace in the space-periodic and space-antiperiodic sectors, respectively. By using the identity $\theta_2\theta_3\theta_4 = 2\eta^3$ proven in App. 10.A, Eq. (10.260), the result can be finally written in the form

$$\boxed{Z_{\text{orb}}(R) = \frac{1}{2}\left(Z(R) + \frac{|\theta_2\theta_3|}{|\eta|^2} + \frac{|\theta_2\theta_4|}{|\eta|^2} + \frac{|\theta_3\theta_4|}{|\eta|^2} \right)} \qquad (10.84)$$

§10.5. Minimal Models: Modular Invariance and Operator Content

After having treated various free-field examples, we now turn to the study of modular invariance in the context of minimal models. In this section, we show that, for a theory to have only a finite number of primary fields, the modular invariance of the partition function forces its central charge to be strictly less than one. Conversely, we will prove that if c is not of the form $1 - 6(p - p')^2/pp'$, for relatively prime nonnegative integers p and p', the model cannot be minimal, that is, it contains an infinite number of Virasoro primary fields.

We recall that the Hilbert space of a minimal model with central charge c is a finite collection of irreducible left-right Virasoro modules

$$\mathcal{H} = \bigoplus_{h,\bar{h}} M(c,h) \otimes M(c,\bar{h}) \tag{10.85}$$

The modular invariance of the partition function of the theory on a torus turns out to be a very strong constraint on the operator content of the theory itself. The torus partition function reads, following Eq. (10.5),

$$Z(\tau) = \sum_{h,\bar{h}} \mathcal{M}_{h,\bar{h}} \, \chi_h(\tau) \, \bar{\chi}_{\bar{h}}(\bar{\tau}) \tag{10.86}$$

where $\mathcal{M}_{h,\bar{h}}$ denotes the multiplicity of occurrence of $M(c,h) \otimes M(c,\bar{h})$ in $\mathcal{H}$, and we identified the left Virasoro characters

$$\chi_h(\tau) = \text{Tr}_{M(c,h)}(q^{L_0 - c/24}) = q^{h-c/24} \sum_{n \geq 0} d(n) q^n \tag{10.87}$$

and their right counterparts. To make contact with Sect. 9.3.4, the characters can be viewed as the conformal blocks of the (nonnormalized) zero-point correlation function on the torus, namely the torus partition function (10.86), where the left-right decomposition is manifest.

For the following discussion, we take τ to be purely imaginary (corresponding to a rectangular torus), namely $\tau = i\theta$, in order to make q real. Due to the presence of singular vectors, the number $d(n)$ of independent vectors at level n in $M(c,h)$ is bounded by $p(n)$, the number of partitions of n. This results in the following upper bound on χ_h:

$$\chi_h(i\theta) \leq q^{h-c/24} \sum_{n \geq 0} p(n) q^n = \frac{q^{h-(c-1)/24}}{\eta(i\theta)} \tag{10.88}$$

In the limit $\theta \to 0^+$ (hence $q \to 1^-$), and since $\eta(i\theta) = \sqrt{\theta} \, \eta(i/\theta)$ (cf. Eq. (10.12)), we have

$$\chi_h(i\theta) \leq \frac{\theta^{\frac{1}{2}}}{\eta(i/\theta)} \simeq \theta^{\frac{1}{2}} e^{\pi/12\theta} \tag{10.89}$$

In the last step, we keep only the leading term of η. Consequently, the modular-invariant partition function satisfies the bound

$$Z(i\theta) = Z(i/\theta) \leq \theta e^{\pi/6\theta} \sum_{h,\bar{h}} \mathcal{M}_{h,\bar{h}} \tag{10.90}$$

The last sum is the total number $\mathcal{M}$ of primary fields in the theory

$$\mathcal{M} = \sum_{h,\bar{h}} \mathcal{M}_{h,\bar{h}} \tag{10.91}$$

On the other hand, the leading behavior of $Z(i/\theta)$ when $\theta \to 0^+$ is given by the contribution of the smallest dimension operators. Defining

$$
\begin{aligned}
h_{\min} &= \frac{1}{2} \min \{h + \bar{h} | \mathcal{M}_{h,\bar{h}} \neq 0\} \\
&= \frac{1}{2}(h_0 + \bar{h}_0)
\end{aligned}
\tag{10.92}
$$

we have

$$
Z(i/\theta) \simeq \mathcal{M}_{h_0, \bar{h}_0} e^{-4\pi/\theta(h_{\min} - c/24)}
\tag{10.93}
$$

With the above upper bound (10.90), this implies

$$
\mathcal{M}_{h_0, \bar{h}_0} e^{-\frac{4\pi}{\theta}(h_{\min} - (c-1)/24)} \leq \theta \mathcal{M}
\tag{10.94}
$$

If the theory is minimal, the number $\mathcal{M}$ is finite, and the r.h.s. of (10.94) goes to 0 in the limit $\theta \to 0^+$. The bound then forces the strict inequality

$$
h_{\min} > \frac{(c-1)}{24}
\tag{10.95}
$$

or equivalently

$$
c < 1 + 24 h_{\min}
\tag{10.96}
$$

Since the identity operator with $h = \bar{h} = 0$ always belongs to the theory, $h_{\min} \leq 0$ (one would have $h_{\min} = 0$ in a unitary theory). As a consequence, we find that all minimal theories must have

$$
c < 1
\tag{10.97}
$$

We now refine the analysis to find which values of $c < 1$ can lead to minimal theories. For this, we need a lower bound for the torus partition function of the theory. We consider a theory with central charge c *not of the form* $1 - 6(p-p')^2/pp'$, and write again its modular-invariant partition function Z, in the form (10.86). Then two situations may occur for the Verma module $V(c, h)$:[5]
(i) The module is irreducible, and its character reads

$$
\chi_h(\tau) = \frac{q^{h-(c-1)/24}}{\eta(\tau)}
\tag{10.98}
$$

(ii) There is a unique singular vector at some level N in $V(c, h)$ (see Ex. 8.3)); therefore, the character of the associated irreducible module $M(c, h)$ reads

$$
\chi_h(\tau) = \frac{q^{h-(c-1)/24}}{\eta(\tau)} (1 - q^N)
\tag{10.99}
$$

In both cases, since $N \geq 1$, we have the following lower bound for the characters (with $\tau = i\theta$):

$$
\chi_h(i\theta) \geq \frac{q^{h-(c-1)/24}}{\eta(i\theta)} (1 - q)
\tag{10.100}
$$

---

[5] This is a theorem of the representation theory of the Virasoro algebra (7.4).

When $\theta \to 0^+$, we find

$$\chi_h(i\theta) \geq (1 - e^{-2\pi\theta})\theta^{\frac{1}{2}}e^{\pi/12\theta}$$
$$\geq 2\pi\theta^{\frac{3}{2}}e^{\pi/12\theta} \qquad (10.101)$$

This yields a lower bound on any modular invariant partition function built from these characters, when $\theta \to 0^+$:

$$Z(i\theta) = Z(i/\theta) \geq 4\pi^2\theta^3 e^{\pi/6\theta}\mathcal{M} \qquad (10.102)$$

Again, the leading behavior of $Z(i/\theta)$ when $\theta \to 0^+$ gives

$$\mathcal{M}_{h_0,\bar{h}_0}e^{-(4\pi/\theta)(h_{min}-c/24)} \geq 4\pi^2\theta^3 e^{\pi/6\theta}\mathcal{M} \qquad (10.103)$$

Therefore, if the theory is assumed to be minimal, with c not of the form $1 - 6(p - p')^2/pp'$, the r.h.s. of the relation (10.103) goes to $+\infty$, which imposes the strict inequality

$$h_{min} < \frac{(c-1)}{24} \qquad (10.104)$$

in contradiction with inequality (10.95).

We have thus proven that all the minimal theories have a central charge of the form $c = 1 - 6(p - p')^2/pp'$, where p and p' are two relatively prime integers.

§10.6. Minimal Models: Modular Transformations of the Characters

We recall the expression of the characters of the minimal models with central charge

$$c(p,p') = 1 - 6\frac{(p-p')^2}{pp'} \qquad (10.105)$$

pertaining to the irreducible representation with Kac indices (r, s) in the range

$$1 \leq r \leq p' - 1$$
$$1 \leq s \leq p - 1 \qquad (10.106)$$
$$p's < pr$$

From now on, we denote by $E_{p,p'}$ the set of pairs (r, s) in the range (10.106). The characters can be written in the form (8.17)

$$\chi_{r,s}(\tau) \equiv \chi_{\lambda_{r,s}}(\tau) = K_{\lambda_{r,s}}(\tau) - K_{\lambda_{r,-s}}(\tau) \qquad (10.107)$$

where

$$\lambda_{r,s} = pr - p's \qquad \lambda_{r,-s} = pr + p's \qquad (10.108)$$

and

$$K_\lambda(\tau) = \frac{1}{\eta(\tau)} \sum_{n\in\mathbb{Z}} q^{(Nn+\lambda)^2/2N} \qquad (10.109)$$

with

$$N = 2pp' \qquad (10.110)$$

The transformation $T : \tau \to \tau+1$ can be read directly from the expression of $K_\lambda(\tau)$. Using the T transformation of η given by Eq. (10.12) and the fact that

$$\frac{(Nn+\lambda)^2}{2N} = \frac{\lambda^2}{2N} \mod 1 \qquad (10.111)$$

we readily obtain

$$K_\lambda(\tau+1) = e^{2i\pi[(\lambda^2/2N)-1/24]} K_\lambda(\tau) \qquad (10.112)$$

The relation

$$\frac{\lambda_{r,-s}^2}{2N} - \frac{\lambda_{r,s}^2}{2N} = rs = 0 \mod 1 \qquad (10.113)$$

allows us to write

$$K_{\lambda_{r,-s}}(\tau+1) = e^{2i\pi[\lambda_{r,s}^2/2N-1/24]} K_{\lambda_{r,-s}}(\tau) \qquad (10.114)$$

Hence both $K_{\lambda_{r,s}}$ and $K_{\lambda_{r,-s}}$ transform in the same way. The action of T on the minimal characters reads then

$$\chi_{r,s}(\tau+1) = e^{2i\pi[\lambda_{r,s}^2/2N-1/24]} \chi_{r,s}(\tau) \qquad (10.115)$$

Writing

$$\chi_{r,s}(\tau+1) = \sum_{(\rho,\sigma)\in E_{p,p'}} T_{rs,\rho\sigma} \chi_{\rho,\sigma}(\tau) \qquad (10.116)$$

we obtain the matrix element of T in the basis of minimal characters

$$T_{rs;\rho\sigma} = \delta_{r,\rho}\delta_{s,\sigma} e^{2i\pi(h_{r,s}-c/24)} \qquad (10.117)$$

with the conformal dimension $h_{r,s}$ given by the Kac formula (7.65). Note that the original definition of characters (10.87) immediately yields (10.115). The use of functions K is, however, instrumental in the computation of the S transformation.

In order to compute the action of $S : \tau \to -1/\tau$, we first need a close look at the change of indices from (r,s) to $\lambda_{r,s} = pr - p's$. For two relatively prime integers p and p', there exists a unique pair (r_0, s_0) in the range (10.106), such that[6]

$$pr_0 - p's_0 = 1 \qquad (10.118)$$

---

[6] This result is known in France as the Bezout lemma. See Ex. 10.1 for a detailed proof.

We define

$$\omega_0 = pr_0 + p's_0 \mod N \tag{10.119}$$

for which

$$\omega_0^2 = 1 \mod 2N \tag{10.120}$$

The integer ω_0 has been designed to generate the transformation $s \to -s$ in $\lambda_{r,s}$, namely

$$\lambda_{r,-s} = \omega_0 \lambda_{r,s} \mod N \tag{10.121}$$

The minimal characters can then be reexpressed in the form

$$\chi_\lambda(\tau) = K_\lambda(\tau) - K_{\omega_0\lambda}(\tau) \tag{10.122}$$

From the obvious symmetries

$$K_{\lambda+N} = K_\lambda = K_{-\lambda} \tag{10.123}$$

we see that K_λ defines a set of $\frac{1}{2}N + 1$ independent functions. These relations immediately imply that

$$\chi_\lambda = \chi_{\lambda+N} = \chi_{-\lambda} = -\chi_{\omega_0\lambda} \tag{10.124}$$

Therefore, χ_λ takes $(p-1)(p'-1)/2$ independent values, which can be taken in the fundamental domain $\{\lambda_{r,s}|(r,s) \in E_{p,p'}\}$.

The modular transformation S will now be shown to act linearly on K_λ. For this we apply the Poisson resummation formula (10.264) to $K_\lambda(-1/\tau)$, $\lambda = 0, 1, \ldots, N - 1$:

$$
\begin{aligned}
K_\lambda(-1/\tau) &= \frac{1}{\sqrt{-i\tau}\eta(\tau)} \sum_{n\in\mathbb{Z}} \exp\left[-\frac{2i\pi}{\tau}\frac{(Nn+\lambda)^2}{2N}\right] \\
&= \frac{1}{\sqrt{-i\tau}\eta(\tau)} \int_{\mathbb{R}} dx \sum_{k\in\mathbb{Z}} \exp 2i\pi\left[kx - \frac{N}{2\tau}(x+\frac{\lambda}{N})^2\right] \\
&= \frac{1}{\sqrt{-i\tau}\eta(\tau)} \int_{\mathbb{R}} dx \sum_{k\in\mathbb{Z}} \exp 2i\pi\left[\frac{\tau}{2N}k^2 - \frac{k\lambda}{N} - \frac{N}{2\tau}(x+\frac{\lambda}{N}-\frac{k\tau}{N})^2\right] \\
&= \frac{1}{\sqrt{2N}\eta(\tau)} \sum_{k\in\mathbb{Z}} \exp\left[-2i\pi\frac{k\lambda}{N}\right] q^{k^2/2N}
\end{aligned}
\tag{10.125}
$$

Writing $k = \mu + Nm$, $m \in \mathbb{Z}$, $\mu \in [0, N-1]$, we get

$$K_\lambda(-1/\tau) = \sum_{\mu=0}^{N-1} \frac{1}{\sqrt{N}} e^{2i\pi\lambda\mu/N} K_\mu(\tau) \tag{10.126}$$

We have changed the sign of the phase in the exponential factor, which does not affect the summation as $K_\mu = K_{\mu+N} = K_{-\mu}$. Likewise, we get

$$K_{\omega_0\lambda}(-1/\tau) = \sum_{\mu=0}^{N-1} \frac{1}{\sqrt{N}} e^{-2i\pi\omega_0\lambda\mu/N} K_\mu(\tau)$$

$$= \sum_{\nu=0}^{N-1} \frac{1}{\sqrt{N}} e^{2i\pi\lambda\nu/N} K_{\omega_0\nu}(\tau)$$

(10.127)

where we have performed the change of summation index $\nu = \omega_0\mu$, and changed the sign in the phase factor, using again $K_{-\mu} = K_\mu$. The minimal characters are therefore transformed as

$$\chi_\lambda(-1/\tau) = \sum_{\mu=0}^{N-1} \frac{1}{\sqrt{N}} e^{2i\pi\lambda\mu/N} \chi_\mu(\tau)$$

(10.128)

We are not quite finished since the range of summation is still not the desired one: we must restrict the sum over μ on the r.h.s. of Eq. (10.128) to the fundamental domain $E_{p,p'}$ associated with (10.106). In the interval over which μ is summed, there are points at which χ_λ vanishes, namely when

$$\omega_0\mu = \pm\mu \mod N$$

(10.129)

a consequence of the (anti)symmetry relations of Eq. (10.124). This corresponds to the case when μ is a multiple of p or p'. The set of μ's for which $\omega_0\lambda \neq \pm\lambda$ mod N can be decomposed into four sets of an equal number of elements: (i) a fundamental domain for the action of ω_0, namely $\{\lambda_{r,s}|(r,s) \in E_{p,p'}\}$, (ii) its image under multiplication by ω_0 modulo N, and (iii) and (iv) their respective images under $\mu \to N - \mu$. This enables to reorganize the r.h.s. of Eq. (10.128) into

$$\chi_\lambda(-1/\tau) =$$

$$\sum_{\substack{\mu=\mu_{\rho,\sigma} \\ (\rho,\sigma)\in E_{p,p'}}} \chi_\mu(\tau)\frac{1}{\sqrt{N}} \left[e^{2i\pi\lambda\mu/N} - e^{2i\pi\lambda\omega_0\mu/N} + e^{-2i\pi\lambda\mu/N} - e^{-2i\pi\lambda\omega_0\mu/N} \right]$$

(10.130)

Writing $\lambda = pr - p's$ and $\mu = p\rho - p'\sigma$, we get the sum of exponentials

$$2\cos(2\pi\lambda(p\rho - p'\sigma)/N) - 2\cos(2\pi\lambda(p\rho + p'\sigma)/N)$$

$$= 4(-1)^{1+s\rho+r\sigma} \sin(\pi\frac{p}{p'}r\rho) \sin(\pi\frac{p'}{p}s\sigma)$$

(10.131)

This leads to the modular transformation

$$\chi_{r,s}(-1/\tau) = 2\sqrt{\frac{2}{pp'}} \sum_{(\rho,\sigma)\in E_{p,p'}} (-1)^{1+s\rho+r\sigma} \sin(\pi\frac{p}{p'}r\rho) \sin(\pi\frac{p'}{p}s\sigma) \chi_{\rho,\sigma}(\tau)$$

(10.132)

This is usually written in the form

$$\chi_{r,s}(-1/\tau) = \sum_{(\rho,\sigma)\in E_{p,p'}} S_{rs,\rho\sigma} \, \chi_{\rho,\sigma}(\tau) \tag{10.133}$$

with

$$S_{rs;\rho\sigma} = 2\sqrt{\frac{2}{pp'}}(-1)^{1+s\rho+r\sigma} \sin(\pi\frac{p}{p'}r\rho) \sin(\pi\frac{p'}{p}s\sigma) \tag{10.134}$$

The matrix elements of the transformation S on the basis of minimal characters are clearly symmetric and real. In addition, the transformation S is unitary, which implies that

$$S^2 = 1 \tag{10.135}$$

This can be checked directly on the expression (10.134) by simple trigonometric manipulations (see Ex. 10.4). Notice also that (see Ex. 10.5)

$$S_{11;\rho\sigma} \neq 0 \quad \text{for all} \quad (\rho,\sigma) \in E_{p,p'} \tag{10.136}$$

We conclude this section by giving the explicit form of the modular matrix S for the simplest minimal models.

(i) The Yang-Lee model $\mathcal{M}(5,2)$. As mentioned in Sect. (7.4.1), this nonunitary minimal model is built out of two primary fields: the identity $\mathbb{I}$ and $\phi_{(1,2)}$, in this order. The modular matrix (10.134) is

$$S = \frac{2}{\sqrt{5}}\begin{pmatrix} -\sin(2\pi/5) & \sin(4\pi/5) \\ \sin(4\pi/5) & \sin(2\pi/5) \end{pmatrix} \tag{10.137}$$

(ii) The Ising model $\mathcal{M}(4,3)$. The three primary fields are, in this order, the identity $\mathbb{I}$, the energy field ε, and the spin field σ. The modular matrix is

$$S = \frac{1}{2}\begin{pmatrix} 1 & 1 & \sqrt{2} \\ 1 & 1 & -\sqrt{2} \\ \sqrt{2} & -\sqrt{2} & 0 \end{pmatrix} \tag{10.138}$$

(iii) The tricritical Ising model $\mathcal{M}(5,4)$. The six primary fields are listed in Table 7.2, in that order. The corresponding modular matrix is

$$S = \begin{pmatrix} s_2 & s_1 & s_1 & s_2 & \sqrt{2}\,s_1 & \sqrt{2}\,s_2 \\ s_1 & -s_2 & -s_2 & s_1 & \sqrt{2}\,s_2 & -\sqrt{2}\,s_1 \\ s_1 & -s_2 & -s_2 & s_1 & -\sqrt{2}\,s_2 & \sqrt{2}\,s_1 \\ s_2 & s_1 & s_1 & s_2 & -\sqrt{2}\,s_1 & -\sqrt{2}\,s_2 \\ \sqrt{2}\,s_1 & \sqrt{2}\,s_2 & -\sqrt{2}\,s_2 & -\sqrt{2}\,s_1 & 0 & 0 \\ \sqrt{2}\,s_2 & -\sqrt{2}\,s_1 & \sqrt{2}\,s_1 & -\sqrt{2}\,s_2 & 0 & 0 \end{pmatrix} \tag{10.139}$$

where

$$s_1 \equiv \sin(2\pi/5) \qquad s_2 \equiv \sin(4\pi/5) \tag{10.140}$$

§10.7. Minimal Models: Modular Invariant Partition Functions

In this section, we exhibit the modular-invariant partition functions of the minimal models (p, p'). They turn out to be in one-to-one correspondence with pairs (G, H) of simply laced Lie algebras[7] $(A_n, D_n, E_6, E_7, E_8)$ with respective dual Coxeter numbers p' and p. We do not intend to develop the Lie-algebraic interpretation at this point but merely wish to justify the notation $Z_{G,H}$ adopted in the following discussion.

The expression (10.86) for the partition function of a minimal theory reads

$$Z(\tau) = \sum_{(r,s),(t,u)\in E_{p,p'}} \mathcal{M}_{rs;tu} \, \chi_{r,s}(\tau) \, \bar{\chi}_{t,u}(\bar{\tau}) \tag{10.141}$$

The multiplicities $\mathcal{M}_{r,s;t,u}$ of occurrence of the corresponding left-right representation modules $V_{rs} \otimes V_{tu}$ are nonnegative integers, and the identity is nondegenerate, that is, $\mathcal{M}_{1,1;1,1} = 1$.

Constructing a modular-invariant partition function amounts to finding a set of multiplicities $\mathcal{M}_{r,s;t,u}$, such that

$$\begin{aligned} \mathcal{M}_{1,1;1,1} &= 1 \\ \mathcal{M}\mathcal{T} &= \mathcal{T}\mathcal{M} \\ \mathcal{M}\mathcal{S} &= \mathcal{S}\mathcal{M} \end{aligned} \tag{10.142}$$

The last two conditions express in matrix form the invariance of the partition function (10.141) under, respectively

$$\mathcal{T} : Z(\tau + 1) = Z(\tau) \quad \text{and} \quad \mathcal{S} : Z(-1/\tau) = Z(\tau) \tag{10.143}$$

(The unitarity of the matrices $\mathcal{T}$ and $\mathcal{S}$ has been used.)

In this section, the results are presented in a rather sketchy way since the detailed mechanism behind the construction of these invariants will be exposed in full length in Part C, where classification issues are also discussed. One of the main features of the classification of minimal models is that, except for p or $p' = 2, 4$, there always exist *more than one* modular-invariant theory at a given value of the central charge $c(p, p') = 1 - 6(p - p')^2/pp'$. This means that one can find different operator algebras, closed under OPE, and built out of the same set of primary fields. The conformal theories discussed so far correspond only to one of these invariants, namely the diagonal invariant, which we shall denote by $Z_{A_{p'-1}, A_{p-1}}$. In particular, the fusion rules discussed in Chap. 8 apply only to these theories, and we expect different fusion rules for the other theories. Fusion rules will be addressed in all generality in Sect. 10.8 below.

---

[7] Simply-laced Lie algebras are fully discussed in relation to conformal field theory in Part C.

10.7.1. Diagonal Modular Invariants

The weakest condition, the T invariance, restricts the possible left-right association of modules $(h, \bar{h})$ by the condition that

$$h - \bar{h} = 0 \mod 1 \qquad (10.144)$$

An obvious solution consists of only left-right symmetric states with $h = \bar{h}$. The corresponding partition function reads[8]

$$Z_{A_{p'-1}, A_{p-1}} = \sum_{(r,s) \in E_{p,p'}} |\chi_{r,s}|^2 \qquad (10.145)$$

which is indeed a modular invariant, thanks to the unitarity of the matrix S, Eq. (10.134). This modular invariant is said to be *diagonal*. It is the partition function of the minimal $\mathcal{M}(p, p')$ model on a torus. The operator content of such a theory is read off directly from the invariant (10.145): each field of the Kac table $E(p, p')$ appears exactly once in the spinless left-right combination $\Phi_{(r,s)} = \phi_{(r,s)} \otimes \bar{\phi}_{(r,s)}$.

10.7.2. Nondiagonal Modular Invariants: Example of the Three-state Potts Model

The modular invariants of the form (10.145) are the torus (doubly periodic) partition functions of the diagonal minimal conformal theories, referred to as the $\mathcal{M}(p, p')$ models. However, there exist minimal theories in which *not all* the fields of the Kac table are present. This is the case for the three-state Potts model, already mentioned in Sect. 7.4.4, whose (left or right) field content is a subset of the $\mathcal{M}(6, 5)$ minimal model. It has been observed that only the fields $\phi_{(r,s)}$, with $s = 1, 3, 5$, are present in the (left or right) theory. In looking for a modular invariant corresponding to this theory, the simplest thing to do is to group the fields into *blocks* having nice modular transformation properties. In particular, for a block I of representations of the form $\bigoplus_{j \in I} V_j$ to be invariant, up to a phase, under T, it is necessary and sufficient that the corresponding dimensions h_j be integer-spaced, i.e., $h_j - h_k \in \mathbb{Z}$, for any $i, j \in I$. Such blocks are easily found for the three-state Potts model, by noticing that

$$h_{r,5} - h_{r,1} = 5 - 2r \qquad (10.146)$$

The corresponding block-characters

$$\begin{aligned} C_{r,1}(\tau) &= \chi_{r,1}(\tau) + \chi_{r,5}(\tau) \\ C_{r,3}(\tau) &= \chi_{r,3}(\tau) \end{aligned} \qquad (10.147)$$

---

[8] The subscripts $A_{p'-1}$ and A_{p-1} refer to the Lie-algebraic interpretation mentioned before. Here they should be viewed only as labeling indices.

defined for $r = 1, 2$, are invariant up to a phase under the action of $\mathcal{T}$:

$$
\begin{aligned}
C_{r,1}(\tau + 1) &= e^{2i\pi(h_{r,1}-c/24)} \, C_{r,1}(\tau) \\
C_{r,3}(\tau + 1) &= e^{2i\pi(h_{r,3}-c/24)} \, C_{r,3}(\tau)
\end{aligned}
\tag{10.148}
$$

Under the action of S, with the matrix elements (10.134), $p = 6$, $p' = 5$, they transform as

$$
C_{1,1}(-1/\tau) = \frac{2}{\sqrt{15}}\Big[s_1 C_{1,1}(\tau) + s_2 C_{2,1}(\tau) + 2\big(s_1 C_{1,3}(\tau) + s_2 C_{2,3}(\tau)\big)\Big]
$$

$$
C_{2,1}(-1/\tau) = \frac{2}{\sqrt{15}}\Big[s_2 C_{1,1}(\tau) - s_1 C_{2,1}(\tau) + 2\big(s_2 C_{1,3}(\tau) - s_1 C_{2,3}(\tau)\big)\Big]
$$

$$
C_{1,3}(-1/\tau) = \frac{2}{\sqrt{15}}\Big[s_1 C_{1,1}(\tau) + s_2 C_{2,1}(\tau) - \big(s_1 C_{1,3}(\tau) + s_2 C_{2,3}(\tau)\big)\Big]
$$

$$
C_{2,3}(-1/\tau) = \frac{2}{\sqrt{15}}\Big[s_2 C_{1,1}(\tau) - s_1 C_{2,1}(\tau) - \big(s_2 C_{1,3}(\tau) - s_1 C_{2,3}(\tau)\big)\Big]
\tag{10.149}
$$

with the notation $s_1 = \sin \pi/5$, $s_2 = \sin 2\pi/5$. In view of (10.149), and noting that $s_1^2 + s_2^2 = 5/4$, the sesquilinear combination

$$
\begin{aligned}
Z_{\text{Potts 3}}(\tau) &= \sum_{r=1,2} \Big\{ |C_{r,1}(\tau)|^2 + 2|C_{r,3}(\tau)|^2 \Big\} \\
&= \sum_{r=1,2} \Big\{ |\chi_{r,1} + \chi_{r,5}|^2 + 2|\chi_{r,3}|^2 \Big\}
\end{aligned}
\tag{10.150}
$$

is seen to be a modular invariant. It is indeed the modular-invariant partition function of the three-state Potts model on a torus. This partition function is different from that of the $\mathcal{M}(6,5)$ minimal model, of the form (10.145). It exhibits an operator content different from that of the $\mathcal{M}(6,5)$ model. From the partition function (10.150), we read that only the operators $\phi_{(r,s)}$, with $s = 1, 5$ and $r = 1, 2$, are present in the (left or right) theory, together with *two copies* of the operators $\phi_{(r,3)}$, $r = 1, 2$. This last fact is crucial for the modular invariance of Eq. (10.150). This is the first occurrence of a multiplicity 2 in a modular-invariant combination of the form (10.86). This multiplicity shows that the three-state Potts model is not just a subtheory of the minimal $\mathcal{M}(6,5)$ model, as it contains more copies of some of its fields. This is reflected in the nontrivial structure of the three-state Potts fusion rules, which are not just a subset of the $\mathcal{M}(6,5)$ fusion rules, as the naive analysis of Sect. 7.4.4 first suggested. Moreover, two sets of nonsymmetric left-right combinations of fields occur, namely $\phi_{(r,1)} \otimes \bar{\phi}_{(r,5)}$ and their complex conjugates, which have a nonvanishing spin $\pm(2r - 5)$.

To study the fusion rules of the three-state Potts model, we use the same notations as in Sect. 7.4.4. We must take into account the multiplicity 2 for the two operators denoted σ and Z in Table 7.5. We denote by σ_1, σ_2 and Z_1, Z_2 the corresponding copies. This multiplicity can in fact be understood in relation to the $\mathbb{Z}_3$ symmetry of the three-state Potts model, under which the fields are transformed as

$$
\begin{aligned}
\mathbb{I} &\rightarrow \mathbb{I} \\
\varepsilon &\rightarrow \varepsilon \\
X &\rightarrow X \\
Y &\rightarrow Y \\
\sigma_1 &\rightarrow e^{2i\pi/3}\,\sigma_1 \\
\sigma_2 &\rightarrow e^{-2i\pi/3}\,\sigma_2 \\
Z_1 &\rightarrow e^{2i\pi/3}\,Z_1 \\
Z_2 &\rightarrow e^{-2i\pi/3}\,Z_2
\end{aligned}
\tag{10.151}
$$

The fusion rules of Table 7.6 do not preserve this symmetry. To account for Eq. (10.151), it is a simple matter to verify that they have to be changed into those appearing on Table 10.2.

Table 10.2. $\mathbb{Z}_3$-invariant fusion rules of the three-state Potts model.

$\varepsilon \times \varepsilon$	$= \mathbb{I} + X$	
$\varepsilon \times \sigma_i$	$= \sigma_i + Z_i$	$(i = 1, 2)$
$\varepsilon \times X$	$= \varepsilon + Y$	
$\varepsilon \times Y$	$= X$	
$\varepsilon \times Z_i$	$= \sigma_i$	$(i = 1, 2)$
$\sigma_i \times \sigma_{2-i}$	$= \mathbb{I} + \varepsilon + X + Y$	
$\sigma_i \times \sigma_i$	$= \sigma_{2-i} + Z_{2-i}$	$(i = 1, 2)$
$\sigma_i \times X$	$= \sigma_i + Z_i$	$(i = 1, 2)$
$\sigma_i \times Y$	$= \sigma_i$	$(i = 1, 2)$
$\sigma_i \times Z_{2-i}$	$= \varepsilon + X$	$(i = 1, 2)$
$\sigma_i \times Z_i$	$= \sigma_{2-i}$	$(i = 1, 2)$
$X \times X$	$= \mathbb{I} + X$	
$X \times Y$	$= \varepsilon$	
$X \times Z_i$	$= \sigma_i$	$(i = 1, 2)$
$Y \times Y$	$= \mathbb{I}$	
$Y \times Z_i$	$= Z_i$	$(i = 1, 2)$
$Z_i \times Z_{2-i}$	$= \mathbb{I} + Y$	$(i = 1, 2)$
$Z_i \times Z_i$	$= Z_{2-i}$	$(i = 1, 2)$

10.7.3. Block-Diagonal Modular Invariants

In the framework of a general minimal theory $\mathcal{M}(p, p')$, the Potts example suggests looking for sets (blocks) of representations containing fields with dimensions differing by integers, which are linearly transformed among each other under the action of $\mathcal{S}$, as in Eq. (10.149). If we let

$$C_\lambda(\tau) = \sum_{(r,s) \in I_\lambda} \chi_{r,s}(\tau) \qquad (10.152)$$

be the corresponding block-characters, then

$$C_\lambda(-1/\tau) = \sum_\mu S_{\lambda,\mu} \, C_\mu(\tau) \qquad (10.153)$$

If the restriction of $\mathcal{S}$ to this space is still unitary, we get a modular invariant by taking the sum of the moduli square of these blocks C_λ. This is indeed the case whenever p or p' is of the form $4m+2$, for some integer m. Using the Kac formula for the dimensions, one readily sees that, in either case, the field $\phi_{(p'-1,1)} = \phi_{(1,p-1)}$ has an integer dimension. In all these cases, the block-character of the identity operator can indeed be taken to be $C_{1,1} = \chi_{1,1} + \chi_{p'-1,1}$, and consequently all the other blocks are also made of two representations of the Virasoro algebra. However, when p or $p' = 12$ or 30, a *third* modular invariant can be constructed with a different block of the identity $C_{1,1}$. The explicit form of all these block-diagonal modular invariants appears in Table 10.3.

The partition function (10.150) of the three-state Potts model corresponds to Z_{A_4, D_4} here. The next simplest nondiagonal modular invariant describes the tricritical three-state Potts model, and it reads

$$Z_{D_4, A_6} = \sum_{s=1,2,3} |\chi_{1,s} + \chi_{5,s}|^2 + 2|\chi_{3,s}|^2 \qquad (10.154)$$

We note the appearance of many fields with multiplicity 2, namely those of the central column (or line) of the Kac table, that is, with either $r = p/2$, $s = 1, 2, \ldots, (p'-1)/2$ or $s = p'/2$, $r = 1, 2, \ldots, (p-1)/2$. The case-by-case proof of $\mathcal{T}$ and $\mathcal{S}$ invariance of the partition functions of Table 10.3 is performed in Exs 10.6, 10.7 and 10.8, for, respectively, the $(D_{p'/2+1}, A_{p-1})$, (E_6, A_{p-1}), and (E_8, A_{p-1}) theories. The modular invariants of Table 10.3 are said to be *block-diagonal*.

The fusion rules for the $(D_{p'/2+1}, A_{p-1})$ and $(A_{p'-1}, D_{p/2+1})$ models can be inferred from the restrictions of the corresponding $\mathcal{M}(p, p')$ diagonal models, with the extra constraint that they should be invariant under a generic $\mathbb{Z}_2$ symmetry, which distinguishes between the two copies of each doubly degenerate field, while the other fields are left unchanged (see Ex. 10.12 for a complete derivation). This generic $\mathbb{Z}_2$ symmetry is hidden in the Potts example, as an artifact of the $\mathbb{Z}_3$ symmetry: in that case, it corresponds to the charge conjugation, which indeed exchanges $\sigma_1 \leftrightarrow \sigma_2$ and $Z_1 \leftrightarrow Z_2$.

Table 10.3. Explicit form of block-diagonal modular invariants.

Label	Modular Invariant
$p' = 2(2m+1)$	$Z_{D_{p'/2+1},A_{p-1}} = \dfrac{1}{2} \displaystyle\sum_{\substack{(r,s)\in E_{p,p'} \\ r \text{ odd}}} \left\lvert \chi_{r,s} + \chi_{p'-r,s} \right\rvert^2$
$p = 2(2m+1)$	$Z_{A_{p'-1},D_{p/2+1}} = \dfrac{1}{2} \displaystyle\sum_{\substack{(r,s)\in E_{p,p'} \\ s \text{ odd}}} \left\lvert \chi_{r,s} + \chi_{r,p-s} \right\rvert^2$
$p' = 12$	$Z_{E_6,A_{p-1}} = \dfrac{1}{2} \displaystyle\sum_{s=1}^{p-1} \left[\left\lvert \chi_{1,s} + \chi_{7,s} \right\rvert^2 + \left\lvert \chi_{4,s} + \chi_{8,s} \right\rvert^2 + \left\lvert \chi_{5,s} + \chi_{11,s} \right\rvert^2 \right]$
$p = 12$	$Z_{A_{p'-1},E_6} = \dfrac{1}{2} \displaystyle\sum_{r=1}^{p'-1} \left[\left\lvert \chi_{r,1} + \chi_{r,7} \right\rvert^2 + \left\lvert \chi_{r,4} + \chi_{r,8} \right\rvert^2 + \left\lvert \chi_{r,5} + \chi_{r,11} \right\rvert^2 \right]$
$p' = 30$	$Z_{E_8,A_{p-1}} = \dfrac{1}{2} \displaystyle\sum_{s=1}^{p-1} \left[\left\lvert \chi_{1,s} + \chi_{11,s} + \chi_{19,s} + \chi_{29,s} \right\rvert^2 + \left\lvert \chi_{7,s} + \chi_{13,s} + \chi_{17,s} + \chi_{23,s} \right\rvert^2 \right]$
$p = 30$	$Z_{A_{p'-1},E_8} = \dfrac{1}{2} \displaystyle\sum_{r=1}^{p'-1} \left[\left\lvert \chi_{r,1} + \chi_{r,11} + \chi_{r,19} + \chi_{r,29} \right\rvert^2 + \left\lvert \chi_{r,7} + \chi_{r,13} + \chi_{r,17} + \chi_{r,23} \right\rvert^2 \right]$

When p or p' is a multiple of 4, we shall find another way of deriving a nondiagonal modular invariant, using symmetries of the matrix S. This is the subject of the next section.

10.7.4. Nondiagonal Modular Invariants Related to an Automorphism

Starting from a block-diagonal modular invariant of the form

$$\sum_\lambda |C_\lambda|^2 \tag{10.155}$$

where the C's are either minimal characters or linear combinations thereof as in (10.152), it is sometimes possible to build a *permutation* modular invariant

$$Z_\Pi = \sum_\lambda C_\lambda \, \bar{C}_{\Pi(\lambda)} \tag{10.156}$$

for some special permutation (or automorphism) Π. In order to get $\mathcal{M}_{1,1;1,1} = 1$, Π must satisfy $\Pi(1,1) = (1,1)$. If the matrix elements $S_{\lambda,\mu}$ of the transformation $S : \tau \to -1/\tau$ of the C's,

$$C_\lambda(-1/\tau) = \sum_\mu S_{\lambda,\mu} \, C_\mu(\tau) \tag{10.157}$$

satisfy the property

$$S_{\Pi(\lambda),\Pi(\mu)} = S_{\lambda,\mu} \tag{10.158}$$

for some automorphism Π acting on the blocks, then the partition function Z_Π of Eq. (10.156) is invariant under the transformation S. The T invariance will be granted if in addition

$$h_{\Pi(\lambda)} - h_\lambda = 0 \quad \text{mod } 1 \tag{10.159}$$

for all λ's, which amounts to

$$T_{\Pi(\lambda),\Pi(\mu)} = T_{\lambda,\mu} \tag{10.160}$$

The modular invariants Z_Π are usually called *permutation or automorphism invariants*.

The modular invariants appearing in Table 10.4 can all be described in this way. They are the respective conjugates of the modular invariants $Z_{A_{p'-1},A_{p-1}}$ when $p' = 4m$ or $p = 4m$, of $Z_{D_{10},A_{p-1}}$ when $p' = 18$, and of $Z_{A_{p'-1},D_{10}}$ when $p = 18$. The automorphism of $Z_{A_{p'-1},A_{p-1}}$ leading to $Z_{D_{p'/2+1},A_{p-1}}$ is studied in detail in Ex. 10.9 below.

10.7.5. D Series from $\mathbb{Z}_2$ Orbifolds

The $\mathbb{Z}_2$ orbifold method, applied previously to the $c = 1$ model described by a compactified boson (see Sect. 10.4.3), can equally be applied to the $(A_{p'-1}, A_{p-1})$ minimal theory. The $\mathbb{Z}_2$ symmetry of the diagonal minimal theories is identified as

Table 10.4. Explicit form of nondiagonal permutation modular invariants.

Label	Modular Invariant
$p' = 4m$	$Z_{D_{p'/2+1},A_{p-1}} = \dfrac{1}{2}\displaystyle\sum_{s=1}^{p-1}\left[\left(\sum_{r\ \text{odd}}|\chi_{r,s}|^2\right)+|\chi_{p'/2,s}|^2 + \sum_{r\ \text{even}}\chi_{r,s}\bar\chi_{p'-r,s}\right]$
$p = 4m$	$Z_{A_{p'-1},D_{p/2+1}} = \dfrac{1}{2}\displaystyle\sum_{r=1}^{p'-1}\left[\left(\sum_{s\ \text{odd}}|\chi_{r,s}|^2\right)+|\chi_{r,p/2}|^2 + \sum_{s\ \text{even}}\chi_{r,s}\bar\chi_{r,p-s}\right]$
$p' = 18$	$Z_{E_7,A_{p-1}} = \dfrac{1}{2}\displaystyle\sum_{s=1}^{p-1}\left[|\chi_{1,s}+\chi_{17,s}|^2 + |\chi_{5,s}+\chi_{13,s}|^2 + |\chi_{7,s}+\chi_{11,s}|^2 + |\chi_{9,s}|^2\right.$ $\left. + \chi_{9,s}(\bar\chi_{3,s}+\bar\chi_{15,s})+(\chi_{3,s}+\chi_{15,s})\bar\chi_{9,s}\right]$
$p = 18$	$Z_{A_{p'-1},E_7} = \dfrac{1}{2}\displaystyle\sum_{r=1}^{p'-1}\left[|\chi_{r,1}+\chi_{r,17}|^2 + |\chi_{r,5}+\chi_{r,13}|^2 + |\chi_{r,7}+\chi_{r,11}|^2 + |\chi_{r,9}|^2\right.$ $\left. + \chi_{r,9}(\bar\chi_{r,3}+\bar\chi_{r,15})+(\chi_{r,3}+\chi_{r,15})\bar\chi_{r,9}\right]$

follows. Instead of taking the usual fundamental domain $E_{p,p'}$ for the Kac indices (r,s) of the irreducible representations that form the theory, we instead take the following equivalent set $E'_{p,p'}$:

$$1 \leq r \leq p' - 1$$
$$1 \leq s \leq p - 1 \qquad (10.161)$$
$$r + s = 0 \mod 2$$

The equivalence to $E_{p,p'}$ is a consequence of the symmetry $(r,s) \to (p'-r,p-s)$ of the Kac table.

For $(r,s) \in E'_{p,p'}$, the $\mathbb{Z}_2$ symmetry acts as follows on the (chiral) primary fields $\phi_{(r,s)}$:

$$\phi_{(r,s)} \to (-1)^{r+1} \phi_{(r,s)} \qquad (10.162)$$

This leaves the minimal fusion rules (7.61) invariant, since the only nonzero structure constants are of the type C_{+++} and C_{--+} (the symbol $\pm$ stands for the $\mathbb{Z}_2$ charge of the respective fields). In addition to the *untwisted* partition function[9]

$$Z_{++} = Z_{A_{p'-1},A_{p-1}} \qquad (10.163)$$

we may construct a *twisted* one

$$Z_{+-} = \sum_{(r,s)\in F_{p,p'}} (-1)^{r+1} |\chi_{r,s}|^2 \qquad (10.164)$$

As in the $c = 1$ case of Sect. 10.4.3, the modular-invariant orbifold partition function is obtained by including also two other twisted sectors, images of (10.164) by modular transformations, namely

$$Z_{-+} = S Z_{+-}$$
$$Z_{--} = T Z_{-+} = T S Z_{+-} \qquad (10.165)$$

The modular-invariant $\mathbb{Z}_2$ orbifold partition function finally reads

$$Z_{\text{orb}} = \frac{1}{2} \left(Z_{++} + Z_{+-} + Z_{-+} + Z_{--} \right) \qquad (10.166)$$

Comparing the partition functions obtained by this procedure with those listed in Tables 10.3 and 10.4, we find that

$$Z_{\text{orb}} = Z_{D_{p'/2+1},A_{p-1}} \quad \text{when} \quad p' = 0 \mod 2$$
$$= Z_{A_{p'-1},D_{p/2+1}} \quad \text{when} \quad p = 0 \mod 2 \qquad (10.167)$$

10.7.6. The Classification of Minimal Models

The mathematical classification of all modular invariants for minimal models can be carried out explicitly. The result is the list of block-diagonal and nondiagonal

---

[9] We denote here by $+$ and $-$ the boundary conditions on the partition function according to the $\mathbb{Z}_2$ charge. They correspond, respectively, to the labels 0 and $\frac{1}{2}$ in the $c = 1$ bosonic case (10.84).

invariants given in Sect. 10.7.1 and Sect. 10.7.4 above; it is usually known as the ADE classification, for reasons to become clear later. This list exhausts the modular-invariant minimal conformal theories. The case $c = 1$ will be treated in Chap. 17 (see App. 17.B), and will also display a remarkable relation to the ADE classification. However, the classification of $c > 1$ theories, as well as nonminimal $c < 1$ theories, is still an open question.[10]

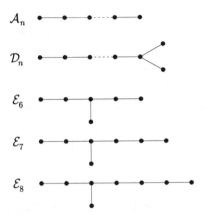

Figure 10.3. The $\mathcal{A}_n$, $\mathcal{D}_n$, $\mathcal{E}_6$, $\mathcal{E}_7$, and $\mathcal{E}_8$ diagrams, with, respectively, n, n, 6, 7, and 8 nodes.

To understand at least how the labeling of these modular invariants goes, we illustrate on Fig. 10.3 a set of diagrams[11] indexed by $\mathcal{A}_n$, $\mathcal{D}_n$, $\mathcal{E}_6$, $\mathcal{E}_7$ and $\mathcal{E}_8$. To each diagram we associate an *adjacency matrix* G_{ab}, with entries indexed by the nodes a, b of the diagram and such that

$$G_{ab} = \text{\# of links between } a \text{ and } b \qquad (10.168)$$

For instance, the adjacency matrix for $\mathcal{A}_3$ reads

$$A_3 = \begin{pmatrix} 0 & 1 & 0 \\ 1 & 0 & 1 \\ 0 & 1 & 0 \end{pmatrix} \qquad (10.169)$$

The matrices corresponding to the diagrams of Fig. 10.3 are exactly *all the adjacency matrices of connected diagrams, whose eigenvalues are strictly less than* 2.[12]

---

[10] More precise statements on the classification issues are given in Chap. 17.

[11] These diagrams will appear naturally in the discussion of the classical simply laced Lie algebras in Chap. 13. Notice that the diagrams are denoted by calligraphic letters.

[12] This is one of the many occurrences of ADE in general classification problems. A few other occurrences of ADE will appear in Part C: classification of modular invariants of conformal theories with Lie-algebra symmetry, classification of $c = 1$ compactified bosonic theories, classification of the finite subgroups of $SU(2)$, and so on. A proof of the present ADE graph classification is sketched in Ex. 10.10 below.

The eigenvalues of the A, D, and E matrices are of the form

$$2 \cos \pi \frac{m}{g} \tag{10.170}$$

where the values of g and m are listed in Table 10.5.[13]

Table 10.5. Values of g and m for the $\mathcal{A}$, $\mathcal{D}$, $\mathcal{E}$ diagrams.

Diagram	g	Values of m
$\mathcal{A}_n$	$n+1$	$1, 2, 3, \cdots, n$
$\mathcal{D}_n$	$2n-2$	$1, 3, 5, \cdots, 2n-3$, and $n-1$
$\mathcal{E}_6$	12	$1, 4, 5, 7, 8, 11$
$\mathcal{E}_7$	18	$1, 5, 7, 9, 11, 13, 17$
$\mathcal{E}_8$	30	$1, 7, 11, 13, 17, 19, 23, 29$

This table provides the rationale for the previous labeling of the modular invariants: the Kac indices (r, s) of the spinless $(h = \bar{h})$ operators in the modular-invariant theory (G, H) are exactly the values of m labeling the eigenvalues of the associated adjacency matrices G (values of r) and H (values of s). The corresponding values of p' and p match exactly the corresponding values of g for G and H, respectively.

§10.8. Fusion Rules and Modular Invariance

The fusion rules of the minimal theories have been studied in detail in Sect. 8.4, and revisited in the light of correlation functions in Sect. 9.2. Here we propose yet another approach, based on modular transformations. The whole idea might seem paradoxical at first sight. Modular invariance really states how the left and right representations of the Virasoro algebra should be paired. In contrast, the fusion rules are essentially chiral, namely they pertain to only the left (or right) part of the theory. This apparent incompatibility is resolved by the highly nontrivial and very constraining fact that the characters of the left (or right) Virasoro representations form a unitary linear representation of the modular group. This is actually the main reason for the relation between the modular transformation properties of the characters of these representations and the fusion rules. In this section, only the chiral properties of the theories will be used.

---

[13] For reference, we indicate their Lie-algebraic meaning: g is the dual Coxeter number and m runs over the Coxeter exponents of the algebra.

In the following discussion, we find an explicit relation between the modular transformation S of the characters and the fusion numbers $\mathcal{N}$. This proves to be a very general fact, and extends to the block-diagonal theories of Sect. 10.7.1. This leads naturally to the concept of *rational conformal field theory* (RCFT), namely, theories that are not necessarily minimal but whose (possibly infinite) collection of primary fields can be reorganized into a *finite* number of blocks corresponding to an *extended symmetry algebra*.

10.8.1. Verlinde's Formula for Minimal Theories

The relation between the fusion numbers $\mathcal{N}_{rs,mn}{}^{kl}$ of the minimal theories Eq. (8.131) and the S matrix elements of Sect. 10.6 reads

$$\mathcal{N}_{rs,mn}{}^{kl} = \sum_{(i,j)\in E_{p,p'}} \frac{S_{rs,ij}S_{mn,ij}S_{ij,kl}}{S_{11,ij}} \qquad (10.171)$$

(the division by $S_{11;ij}$ is allowed, thanks to the positivity property (10.136)). Eq. (10.171) is known as the *Verlinde formula* for minimal models.[14] Its particularly simple form is rooted in commutative algebra theory, as illustrated in Ex. 10.18 in the case of the algebras of representations and classes of a finite group. This formula can be proven directly, by using the expression of S, Eq. (10.134), and some trigonometric sum rules. Another way consists in interpreting the Verlinde formula (10.171) as expressing the *simultaneous* diagonalization of the commuting matrices[15] $N_{(r,s)}$, with entries

$$[N_{(r,s)}]_{mn,kl} = \mathcal{N}_{rs,mn}{}^{kl} \qquad (10.172)$$

Indeed, Eq. (10.171) amounts to the eigenvector equation

$$\sum_{(k,l)\in E_{p,p'}} [N_{(r,s)}]_{mn,kl}\, S_{kl,\rho\sigma} = \left(\frac{S_{rs,\rho\sigma}}{S_{11,\rho\sigma}}\right) S_{mn,\rho\sigma} \qquad (10.173)$$

In order to prove this relation, we use the fact that the matrices $N_{(r,s)}$ are entirely determined (polynomially) by the matrices $X = N_{(2,1)}$ and $Y = N_{(1,2)}$, acting as

$$\begin{aligned} N_{(2,1)}\, N_{(r,s)} &= N_{(r+1,s)} + N_{(r-1,s)} \quad \text{with} \quad N_{(0,s)} = N_{(p',s)} = 0 \\ N_{(1,2)}\, N_{(r,s)} &= N_{(r,s+1)} + N_{(r,s-1)} \quad \text{with} \quad N_{(r,0)} = N_{(r,p)} = 0 \end{aligned} \qquad (10.174)$$

---

[14] This is actually a specialized version of the Verlinde formula. The general form is presented in the next section.

[15] We recall that these matrices form a *regular* representation of the fusion algebra, as commutation of these matrices amounts to associativity of the fusion rules (see Sect. 8.4.2 for details).

(cf. Eqs. (8.96)–(8.97)). Thus, using Eq. (10.134), we have

$$
\begin{aligned}
\sum_{m,n} X_{rs;mn} \mathcal{S}_{mn,\rho\sigma} \\
&= \mathcal{S}_{r+1,s;\rho,\sigma} + \mathcal{S}_{r-1,s;\rho,\sigma} \\
&= \frac{4}{\sqrt{2pp'}} (-1)^{s\rho+1+(r+1)\sigma} \, \sin\left[\pi \frac{p'}{p} s\sigma\right] \\
&\quad \times \left(\sin\left[\pi \frac{p}{p'}(r+1)\rho\right] + \sin\left[\pi \frac{p}{p'}(r-1)\rho\right] \right) \\
&= 2\,(-1)^{\sigma} \, \cos\left[\pi \frac{p}{p'}\rho\right] \mathcal{S}_{rs,\rho\sigma}
\end{aligned}
\tag{10.175}
$$

Likewise, Y acts on $\mathcal{S}$ as

$$
\sum_{m,n} Y_{rs,mn} \mathcal{S}_{mn,\rho\sigma} = 2(-1)^{\rho} \, \cos\left[\pi \frac{p'}{p}\sigma\right] \mathcal{S}_{rs,\rho\sigma}
$$

Hence, the vectors $v^{(\rho,\sigma)}$ with components

$$
[v^{(\rho,\sigma)}]_{r,s} = \mathcal{S}_{rs,\rho\sigma}
\tag{10.176}
$$

are simultaneously eigenvectors of X and Y, and therefore of all the matrices $N_{(r,s)}$, since these are polynomials of X and Y. We finally compute the eigenvalues of $N_{(r,s)}$ in this basis, denoted by $\gamma_{r,s}^{(\rho,\sigma)}$. They satisfy the following condition (expressing that $N_{(1,1)} = \mathbb{I}$, Eqs. (8.87)–(8.95))

$$
\mathcal{N}_{11,rs}{}^{mn} = \delta_{r,m}\,\delta_{s,n} = \sum_{(k,l)\in E_{p,p'}} \mathcal{S}_{11,kl}\, \gamma_{rs}^{(kl)}\, \mathcal{S}_{kl,mn}
\tag{10.177}
$$

Using the unitarity of the matrix $\mathcal{S}$ (Eq. (10.135)), multiplying both sides of Eq. (10.177) by $\mathcal{S}_{mn,\rho\sigma}$, and summing over m, n, we find

$$
\gamma_{r,s}^{(\rho,\sigma)} = \frac{\mathcal{S}_{rs,\rho\sigma}}{\mathcal{S}_{11,\rho\sigma}}
\tag{10.178}
$$

This completes the proof of Eq. (10.173). However, a more conceptual proof is certainly desirable. This will be the object of Sect. 10.8.3 below, which relies on some special transformations of the conformal blocks of the theory. In preparation for this, we first recall a few facts on conformal blocks.

10.8.2. Counting Conformal Blocks

In Chaps. 8 and 9, we have seen how the correlation functions of a minimal conformal theory factorized into holomorphic × antiholomorphic conformal blocks corresponding to the various intermediate state projections allowed by OPE. More precisely, to a correlation function of primary fields $\langle \phi_1 \phi_2 \cdots \phi_N \rangle$ on the plane, there correspond all the conformal blocks

$$(10.179)$$

allowed by OPE, namely such that

$$\phi_{j_1} \in \phi_1 \times \phi_2$$
$$\phi_{j_2} \in \phi_{j_1} \times \phi_3$$
$$\cdots \quad \cdots \qquad\qquad\qquad (10.180)$$
$$\phi_N \in \phi_{j_{N-3}} \times \phi_{N-1}$$

In all generality, a basis of conformal blocks is associated with a ϕ^3 diagram (i.e., a graph with only trivalent vertices), with external legs carrying the indices of the fields of the correlation function, and propagators carrying the indices of intermediate states allowed by OPE. The number of such allowed states $\phi_k \in \phi_i \times \phi_j$ is simply the fusion number $\mathcal{N}_{ij}{}^k$, counting the number of independent couplings (ijk). So the conformal blocks can be counted by associating a factor $\mathcal{N}_{ij}{}^k$ to each vertex with legs carrying the indices (ijk), and summing over internal indices (intermediate states on the internal propagators). Of course, the number of independent conformal blocks should be the same in the various bases corresponding to the various ϕ^3 diagrams.

For a four-point function $\langle \phi_i \phi_j \phi_k \phi_l \rangle$ on the plane, the number of conformal blocks is

$$\mathcal{N} = \sum_m \mathcal{N}_{ij}{}^m \mathcal{N}_{mk}{}^l \qquad\qquad (10.181)$$

Equivalently, the correlator $\langle \phi_i \phi_l \phi_k \phi_j \rangle$ has

$$\mathcal{N} = \sum_m \mathcal{N}_{il}{}^m \mathcal{N}_{mk}{}^j \qquad\qquad (10.182)$$

conformal blocks in a different basis. The identity between these two numbers (10.181) and (10.182) expresses simply the associativity of the fusion algebra

$$\phi_i \times \phi_j = \sum_k \mathcal{N}_{ij}{}^k \phi_k \qquad\qquad (10.183)$$

This is sufficient to ensure that any choice of basis for conformal blocks (hence any choice of ϕ^3 diagram) leads to the same number of independent conformal blocks (see Ex. 10.19 for a simple proof).

For the N-point function of Eq. (10.179), we find the following number of conformal blocks

$$\mathcal{N} = \sum_{j_1, j_2, \dots, j_{N-3}} \mathcal{N}_{12}{}^{j_1} \mathcal{N}_{j_1 3}{}^{j_2} \cdots \mathcal{N}_{j_{N-3} N-1}{}^N \qquad (10.184)$$

Remarkably, this recipe goes over to correlations on a Riemann surface of arbitrary genus (the genus is then also that of the ϕ^3 graph). Take, for instance, the one-point function of the field ϕ_i on the torus. The corresponding diagram yields immediately

$$\rightarrow \mathcal{N} = \sum_j \mathcal{N}_{ij}^{\;\;j} \qquad (10.185)$$

10.8.3. A General Proof of Verlinde's Formula

We are now ready to prove Eq. (10.171) in all generality. Let a and b denote the two basic homotopy cycles of the torus, depicted on Fig. 10.4. They are exchanged under the action of S. For any cycle c on the torus and any primary field ϕ_i, let $\phi_i(c)$ denote an operator acting on the character χ_j of the representation associated with ϕ_j according to the following steps:

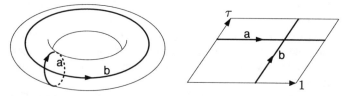

Figure 10.4. The homotopy cycles a and b on the torus. They are homotopically inequivalent (they cannot be continuously deformed into each other), and are exchanged under the modular transformation $S : \tau \rightarrow -1/\tau$.

(i) As mentioned above, the character χ_j is itself a conformal block for the zero-point correlation on the torus. As such, it is also equal to the corresponding conformal block of the one-point function of the identity operator[16] $\phi_0 = \mathbb{I}$ on the torus, namely

$$\chi_j = \bigcirc_j = Tr_j(\mathbb{I}\, q^{L_0 - c/24}) = \bigcirc_j^{\;|0} \qquad (10.186)$$

(In the above graphical representation, the circle corresponds to a b cycle.) We now write the identity operator $\phi_0 = \mathbb{I}$ as the result of the fusion of an operator ϕ_i with its conjugate[17] $\mathbb{I} = \phi_i \times \phi_{i^*}$. This amounts to replacing the character χ_j by

---

[16] Throughout this section, the identity operator is indexed by the label 0. For minimal theories, this would correspond to the double index $0 \equiv (1, 1)$.

[17] This is slightly more general than the minimal case we are used to, in which all representations are selfconjugate $i = i^*$. The conjugation is defined by the normalization of the nonzero two-point functions on the plane: $\langle \phi_i \phi_j \rangle \sim \delta_{j,i^*}$.

the conformal block

$$\mathcal{F}_j^{i,i^*}(z-w) = \qquad (10.187)$$

for the two-point correlation $\langle\phi_i(z)\phi_{i^*}(w)\rangle$ on the torus (by translational invariance, this is a function of $z-w$ only). The character χ_j is recovered from the conformal block, in the limit when the two points coincide

$$\chi_j = \lim_{z\to w} (z-w)^{2h_i}\mathcal{F}_j^{i,i^*}(z-w)$$

The prefactor ensures that the limit is finite (h_i is the conformal dimension of ϕ_i). (ii) In the conformal block (10.187), we move the operator ϕ_i around the torus, along the cycle c. This amounts to letting z circulate along the closed contour c in the conformal block (10.187), namely, to compute the monodromy of the block $\mathcal{F}_j^{i,i^*}$ along c.
(iii) We take again the limit of coinciding points. This yields

$$\phi_i(c)\,\chi_j \qquad (10.188)$$

We shall study this operator for the special choices $c = a$ or b. The interplay between the precise definition of $\phi_i(c)$ in terms of conformal blocks and the action of the modular group, through S, which exchanges a and b, will eventually give a relation between S and the fusion numbers $\mathcal{N}$.

For $c = a$, when ϕ_i is moved along the space (horizontal) direction of the torus, step (ii) above amounts to the following operation:

$$\mathcal{F}_j^{i,i^*}(z+1) = \qquad$$

The representation j is not affected, and we simply get a proportionality factor:

$$\boxed{\phi_i(a)\,\chi_j = \gamma_i^{(j)}\,\chi_j} \qquad (10.189)$$

For the operator $\phi_i(b)$, step (ii) consists in taking the following path:

$$\mathcal{F}_j^{i,i^*}(z+\tau) = \qquad (10.190)$$

The operator $\phi_i(b)$ acts on χ_0, the character of the identity, in a simple way. It replaces the identity representation by i. This enables us to fix the normalization of the operator $\phi_i(b)$ as

$$\phi_i(b)\,\chi_0 \;=\; \chi_i \tag{10.191}$$

We will now show that

$$\phi_i(b)\,\chi_j \;=\; \sum_k \mathcal{N}_{ij}{}^k\,\chi_k \ . \tag{10.192}$$

The operator $\phi_i(b)$ is, up to a normalization factor μ fixed by (10.191), equal to the composition of the two elementary operators A and B

$$\tag{10.193}$$

Here we write only the k-th component of this action. The full action of $\phi_i(b)$ is obtained by summing over all the possible intermediate states k. The operators A and B, as well as the normalization constant μ, act on conformal blocks of four-point functions as

$$\tag{10.194}$$

The Greek indices α, β label the different couplings[18] of the three fields (ijk), hence $\alpha, \beta = 1, 2 \ldots, \mathcal{N}_{ij}{}^k$. The action of the operators A, B, μ is to pick one particular component of the action of the crossing matrix or its inverse (see Sect. 9.3.3). Since

---

[18] Although for the minimal models, $\mathcal{N}_{ij}{}^k = 0$ or 1, in more complicated theories there could be more than one distinct coupling among the three fields. That would correspond to a situation where there are distinct conformal blocks with the same asymptotic behavior. Many examples of theories with some $\mathcal{N}$'s larger than 1 are considered in Part C.

a projection is implied, the transformations A and B are not invertible, whereas μ is just a scalar. In A, only a sum over the index α is implied: this accounts for the fact that only ϕ_i is moved along the b cycle (ϕ_{i^*} remains fixed). We finally show that

$$\mu = \sum_{\beta=1}^{\mathcal{N}_{ij}{}^k} A_{\alpha,\beta}\, B_{\alpha,\beta} \,. \tag{10.195}$$

This is readily seen to be a consequence of the equality of the two sequences of transformations depicted on Fig. 10.5.

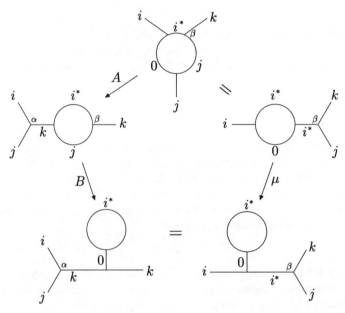

Figure 10.5. Graphical proof of Eq. (10.195). The equality between the two bottom diagrams is a consequence of the associativity of the OPE (crossing symmetry of the conformal blocks of the four-point function). Note that the coupling β is fixed, whereas α requires a summation.

As the l.h.s. of (10.195) is independent of β, we can sum over it from 1 to $\mathcal{N}_{ij}{}^k$, which gives

$$\mu\, \mathcal{N}_{ij}{}^k = \sum_{\alpha,\beta} A_{\alpha,\beta}\, B_{\alpha,\beta} \tag{10.196}$$

Hence, by summing Eq. (10.193) over k, and taking the correct finite limit $z \to w$ of the last conformal block

$$\lim_{z \to w} (z - w)^{2h_k}\, \mathcal{F}_k^{i,i^*}(z - w) = \chi_k \tag{10.197}$$

we get the desired result (10.192).

The fact that S exchanges $\mathbf{a}$ and $\mathbf{b}$, means that $\phi_i(\mathbf{a})$ (Eq. (10.189)) and $\phi_i(\mathbf{b})$ (Eq. (10.192)) are conjugate under the action of S

$$\phi_i(\mathbf{b}) \;=\; S\,\phi_i(\mathbf{a})\,S^{-1} \tag{10.198}$$

By comparing Eqs. (10.189) and (10.192), we see that the fusion numbers are diagonalized by S in the form[19]

$$\mathcal{N}_{ij}{}^k \;=\; \sum_m S_{im}\,\gamma_j^{(m)}\,\bar{S}_{mk} \tag{10.199}$$

where we used the unitarity of S. Setting $i = 0$ (the identity representation) in Eq. (10.199) and using the relation $\mathcal{N}_{0i}{}^k = \delta_{i,k}$ (Eq. (8.87)), we finally get

$$S_{0m}\,\gamma_j^{(m)} \;=\; S_{jm} \tag{10.200}$$

If the matrix element S_{0m} vanished, we would have $S_{jm} = 0$ for all j, which contradicts the fact that S is invertible. Hence, we can divide by S_{0m} in the above relation. Substituting this value in Eq. (10.199) yields the Verlinde formula

$$\boxed{\mathcal{N}_{ij}{}^k \;=\; \sum_m \frac{S_{im}\,S_{jm}\,\bar{S}_{mk}}{S_{0m}}} \tag{10.201}$$

In this proof, only very general facts about conformal blocks have been used. In fact, formula (10.201) extends beyond the minimal theories. In the latter case, the S matrix elements are real, hence (10.201) reduces to (10.171).

As a consequence of Eq. (10.201), we can directly recover the unitarity of S, and, in addition, prove that it is symmetric. Note first that one can lower the index k in $\mathcal{N}_{ij}{}^k$ by conjugation of the representation:

$$\mathcal{N}_{ij}{}^k \;=\; \mathcal{N}_{ijk^*} \tag{10.202}$$

meaning that the fusion $\phi_k \in \phi_i \times \phi_j$ is allowed iff $\langle \phi_i \phi_j \phi_{k^*} \rangle \neq 0$. More precisely, $\mathcal{N}_{ijk^*}$ is the number of copies of the identity $\mathbb{I}$ occurring in the fusion $\phi_i \times \phi_j \times \phi_{k^*}$. The numbers $\mathcal{N}_{ijk}$ being symmetric, we get from (10.201) that

$$S_{k,m} \;=\; \bar{S}_{m,k^*} \tag{10.203}$$

Using the conjugation matrix

$$C_{i,j} \;=\; \delta_{j,i^*} \tag{10.204}$$

such that $C^2 = 1$, we can rewrite this as

$$C\,S \;=\; \bar{S}^t = S^\dagger \tag{10.205}$$

On the other hand, we reexamine the action of S on the cycles $\mathbf{a}$ and $\mathbf{b}$ of the torus. The exact transformation is indeed given by Eq. (10.8)

$$S : (\mathbf{a}, \mathbf{b}) \;\rightarrow\; (-\mathbf{b}, \mathbf{a})$$

---

[19] Here we do not assume that S is real symmetric, but just that it is unitary. This is responsible for the complex conjugation of the matrix element of S on the right.

In addition to the interchange of a and b, the direction of the cycle b has been reversed. This means that S^2 inverts the space and time directions (a $\rightarrow$ $-$a and b $\rightarrow$ $-$b) on the torus. Therefore, by CPT invariance,[20] it transforms a character χ_i into its conjugate χ_{i^*}, pertaining to the conjugate representation. Hence we have

$$\boxed{S^2 = C} \tag{10.206}$$

This equation should be compared with the relation (10.9) satisfied by the representation S of the modular group. In general we have only $S^4 = 1$ but not $S^2 = 1$ when acting on characters, but there is no contradiction with Eq. (10.9): this simply means that the characters form an unfaithful representation of the modular group (in which *stricto sensu* one should have $S^2 = 1$), but rather a representation of a double covering of the modular group, for which only (10.206) holds.

Using (10.205) and (10.206) we finally obtain the unitarity condition

$$\boxed{S\,S^\dagger = \mathbb{I}} \tag{10.207}$$

The symmetry of S is readily seen from the relation

$$\bar{\mathcal{N}}_{ijk} = \mathcal{N}_{i^*j^*k^*} = \mathcal{N}^{ijk} = \mathcal{N}_{ijk} \tag{10.208}$$

where, in the last step, we have used the fact that the numbers $\mathcal{N}$ are integral, and therefore real. Hence

$$\boxed{S = S^t} \tag{10.209}$$

The Verlinde formula (10.201) shows that the matrices N_i, with entries $[N_i]_{j,k} = N_{ij}{}^k$ are simultaneously diagonalizable. Since they are integral matrices, they satisfy the Perron-Frobenius theorem. This means that their common eigenvector $S_{j,\max}$, whose eigenvalues $\gamma_i^{(\max)}$ are maximal for all the N_i's (and this property uniquely characterizes $S_{j,\max}$), has only positive entries. This common maximal eigenvector is called the Perron-Frobenius eigenvector of the N_i. (A sketch of the proof of the Perron-Frobenius theorem for any symmetric integral matrix G is proposed in Ex. 10.10.). We will prove that, in a unitary theory, the field label max must correspond to the identity, i.e., max $= 0$. The starting point is the equality

$$\chi_i(-1/\tau) = \sum_m S_{i,m}\,\chi_m(\tau) \tag{10.210}$$

When evaluating in the limit $\tau \rightarrow i\infty$ ($q \rightarrow 0^+$), we can keep only the leading contribution of each character:

$$\chi_i(-1/\tau) \sim \sum_m S_{i,m}\,q^{h_m - c/24} \tag{10.211}$$

---

[20] A reasonable field theory should always be invariant under simultaneous *charge, parity,* and *time reversal*. The three corresponding nilpotent operators $C, P,$ and T satisfy thus $CPT = 1$. Here the time and parity have been reversed: the conjugation must act too.

Moreover, the leading contribution of this sum comes from the field with lowest dimension; in a unitary theory, this is the identity with $h = 0$ (the conformal dimension of all the other fields being strictly positive). Hence, we have

$$\chi_i(-1/\tau) \sim q^{-c/24} \, S_{i0} \tag{10.212}$$

When $\tau \to i\infty$, $q \to 1^-$ and the l.h.s. of (10.212) being an infinite sum of positive integers, diverges to $+\infty$. We deduce that $S_{i0} > 0$ for all i. This is simply the above mentioned Perron-Frobenius property, which enables us to identify $\max = 0$. We have thus shown that in any unitary theory,

$$\boxed{S_{i0} = S_{0i} > 0} \tag{10.213}$$

In a nonunitary theory, the same argument leads to $\max = \min$, where $\min$ labels the representation with lowest conformal dimension $h_{\min} < 0$. Indeed, the r.h.s. in the second line of (10.211) is dominated by the term with $m = \min$, when $q \to 0^+$; hence, we find that $S_{i,\min} > 0$ for all i. This characterizes the Perron-Frobenius eigenvector completely, and therefore proves that $\max = \min$. For a nonunitary theory with a unique field of lowest dimension, the positivity condition (10.213) therefore generalizes to

$$\boxed{S_{i,\min} = S_{\min,i} > 0} \tag{10.214}$$

This property is checked for the (p, p') minimal theories in Ex. 10.5. If there is more than one field with lowest dimension, we have instead

$$\sum_{m \in \min} S_{i,m} > 0 \tag{10.215}$$

Finally, we note that the relation between S and T becomes

$$\boxed{(ST)^3 = C} \tag{10.216}$$

These general relations can be used to infer some constraints on conformal field theories, as illustrated in Ex. 10.16.

10.8.4. Extended Symmetries and Fusion Rules

As already mentioned, the Verlinde formula (10.201) applies to minimal diagonal theories, with modular invariants listed in Sect. 10.7.1 in the form (10.171). We now see to what extent we can describe the fusion rules of nondiagonal theories. The chiral fusion rules of the nondiagonal theories related to an automorphism (Sect. 10.7.4) are not distinguishable from those of their ancestor, which is always block-diagonal (Sect. 10.7.3). Indeed, the same chiral fields are present in both theories, the difference consisting only in their left-right association. Hence the present discussion is relevant only for the block-diagonal theories of Sect. 10.7.3.

The minimal block-diagonal theories (G, H) $(G \neq A_{p'-1}$ or $H \neq A_{p-1})$, whose modular invariants are listed in Sect. 10.7.1, have an operator content different from that of the $(G, H) = (A_{p'-1}, A_{p-1})$ theories. The operators present in the block-diagonal theories are indicated in the modular invariant by the nonzero multiplicities $\mathcal{M}_{rs,tu}$ of Eq. (10.141). How can we find their fusion rules? In principle, to answer this question, one should reexamine those cases, compute the new correlation functions, and extract the fusion coefficients. But here the corresponding fusion rules must take into account the (nonsymmetric) left-right pairing of Virasoro primary fields, and will not be chiral in general. On the other hand, motivated by the example of the three-state Potts model of Sect. 10.7.2, we shall proceed in a much simpler way, by describing the *extended chiral fusion rules* of the theory. The price to be paid for this simplification is that we get only a sort of *average* description of the fusion rules of the Virasoro primary fields, considered as blocks rather than as individual entities.

The common feature to all the block-diagonal models is that one can define *extended* characters C_λ, through Eq. (10.152), to rewrite the modular invariant as $Z = \sum |C_\lambda|^2$. The functions C_λ are characters of reducible representations of the Virasoro algebra, themselves direct sums of irreducible ones. They are believed to be the irreducible characters of some extended symmetry algebra, enhancing the Virasoro symmetry. As such, they correspond to some extended operators ϕ_λ, for which the proof of the Verlinde formula (10.201) still applies: the conformal blocks are replaced by extended conformal blocks, sums of the former, and the proof essentially goes through, since the extended theory is diagonal. This means that the extended operators ϕ_λ satisfy extended fusion rules

$$\phi_\lambda \times \phi_\mu = \sum_\nu \mathcal{N}_{\lambda\mu}^{(\text{ext})\,\nu} \, \phi_\nu \qquad (10.217)$$

On the other hand, the extended characters transform modularly as

$$C_\lambda(-1/\tau) = \sum_\mu S_{\lambda,\mu}^{(\text{ext})} \, C_\mu(\tau) \qquad (10.218)$$

with $S^{(\text{ext})}$ as in (10.153). The extension of the Verlinde formula is simply

$$\mathcal{N}_{\lambda\mu}^{(\text{ext})\,\nu} = \sum_\rho \frac{S_{\lambda,\rho}^{(\text{ext})} S_{\mu,\rho}^{(\text{ext})}}{S_{0,\rho}^{(\text{ext})}} \bar{S}_{\rho,\nu}^{(\text{ext})} \qquad (10.219)$$

Here the index 0 stands for the *extended* identity block.

As explained in the beginning of Sect. 10.8, the Verlinde formula, as well as the fusion numbers, are essentially chiral and based on the chiral operator content of the theory. This means that the fusion rules associated with the nondiagonal modular-invariant theories of Sect. 10.7.4 should be the same as those of the associated block-diagonal ones. For instance, the fusion rules of the (E_7, A_{p-1}) theories are the same as those of the (D_{10}, A_{p-1}) theories. We recall that the link between the two is an automorphism Π acting on S and T as

$$S_{\Pi(\lambda),\Pi(\mu)} = S_{\lambda,\mu} \qquad T_{\Pi(\lambda),\Pi(\mu)} = T_{\lambda,\mu} \qquad \Pi(0) = 0 \qquad (10.220)$$

The substitution of the first relation into the Verlinde formula (10.201) shows that Π is also an automorphism of the fusion rules, namely

$$\mathcal{N}_{\Pi(\lambda)\Pi(\mu)}{}^{\Pi(\nu)} = \mathcal{N}_{\lambda\mu}{}^{\nu} \tag{10.221}$$

Therefore nondiagonal modular invariants are generally built using automorphisms of the fusion rules of a block-diagonal theory. However, it should be stressed that, although automorphisms are most easily identified as automorphisms of the fusion rules, Eq. (10.221) does not imply Eq. (10.220) (and there are cases where (10.221) is satisfied but Eq. (10.220) is not.[21]) The construction is therefore valid only for those Π's that satisfy Eq. (10.220).

10.8.5. Fusion Rules of the Extended Theory of the Three-State Potts Model

In Sect. 10.7.2, we have already defined the extended characters of the three-state Potts model. We denote the extended fields corresponding to the extended characters $C_{i,j}$ of Eq. (10.147) as follows:

$$\begin{aligned} \mathbb{I} &\leftrightarrow C_{1,1} & \varepsilon &\leftrightarrow C_{2,1} \\ \mathbb{Z}_i &\leftrightarrow C_{1,3}^{(i)} & \sigma_i &\leftrightarrow C_{2,3}^{(i)} & i=1,2 \end{aligned} \tag{10.222}$$

The modular transformations (10.149) do not suffice to completely determine the fusion of these fields, as the two doubly-degenerated blocks $C^{(i)}$ appear only through their sums $C^{(1)} + C^{(2)}$, and not individually. We first have to split them into two distinct characters, in such a way that the symmetric 6×6 S matrix reads (in the basis $C_{1,1}, C_{2,1}, C_{1,3}^{(1)}, C_{1,3}^{(2)}, C_{2,3}^{(1)}, C_{2,3}^{(2)}$):

$$S^{(\text{ext})} = \frac{2}{\sqrt{15}} \begin{pmatrix} s_1 & s_2 & s_1 & s_1 & s_2 & s_2 \\ s_2 & -s_1 & s_2 & s_2 & -s_1 & -s_1 \\ s_1 & s_2 & \omega s_1 & \bar\omega s_1 & \omega s_2 & \bar\omega s_2 \\ s_1 & s_2 & \bar\omega s_1 & \omega s_1 & \bar\omega s_2 & \omega s_2 \\ s_2 & -s_1 & \omega s_2 & \bar\omega s_2 & -\omega s_1 & -\bar\omega s_1 \\ s_2 & -s_1 & \bar\omega s_2 & \omega s_2 & -\bar\omega s_1 & -\omega s_1 \end{pmatrix} \tag{10.223}$$

As before, we have $s_i = \sin(\pi i/5)$, $i = 1, 2$, and $\omega = e^{2i\pi/3}$. This is the simplest way of splitting the S matrix entries (10.149), restoring the symmetry (10.209) and the unitarity (10.207) of the matrix S, and preserving the $\mathbb{Z}_3$ symmetry of the model. Indeed, the conjugation matrix $C = S^2$ of Eq. (10.206) reads

$$C^{(\text{ext})} = \begin{pmatrix} 1 & 0 & 0 & 0 & 0 & 0 \\ 0 & 1 & 0 & 0 & 0 & 0 \\ 0 & 0 & 0 & 1 & 0 & 0 \\ 0 & 0 & 1 & 0 & 0 & 0 \\ 0 & 0 & 0 & 0 & 0 & 1 \\ 0 & 0 & 0 & 0 & 1 & 0 \end{pmatrix} \tag{10.224}$$

---

[21] The Galois permutation invariants described in Chap. 17 provide many examples of such a situation.

In other words, the only nonvanishing two-point correlation functions of extended fields are the $\mathbb{Z}_3$-neutral combinations

$$\langle \varepsilon\varepsilon \rangle , \qquad \langle \mathbf{Z}_1\mathbf{Z}_2 \rangle , \qquad \langle \sigma_1\sigma_2 \rangle \qquad (10.225)$$

The extended fusion rules of the three-state Potts model, displayed in Table 10.6, follow from the formula (10.219), with the extended S matrix (10.223).

Table 10.6. Extended fusion rules of the three-state Potts model.

$\varepsilon \times \varepsilon$	$=$	$\mathbb{I} + \varepsilon$	
$\varepsilon \times \mathbf{Z}_i$	$=$	σ_i	$(i = 1, 2)$
$\varepsilon \times \sigma_i$	$=$	$\sigma_i + \mathbf{Z}_i$	$(i = 1, 2)$
$\mathbf{Z}_i \times \mathbf{Z}_i$	$=$	$\mathbf{Z}_{2-i}$	$(i = 1, 2)$
$\mathbf{Z}_i \times \mathbf{Z}_{2-i}$	$=$	$\mathbb{I}$	$(i = 1, 2)$
$\mathbf{Z}_i \times \sigma_i$	$=$	σ_{2-i}	$(i = 1, 2)$
$\mathbf{Z}_i \times \sigma_{2-i}$	$=$	$\mathbb{I} + \varepsilon$	$(i = 1, 2)$
$\sigma_i \times \sigma_i$	$=$	$\mathbf{Z}_{2-i} + \sigma_{2-i}$	$(i = 1, 2)$
$\sigma_i \times \sigma_{2-i}$	$=$	$\mathbb{I} + \varepsilon$	$(i = 1, 2)$

A practical way of applying (10.219) is to write down its one-dimensional representations, indexed by the label α

$$\rho_\alpha^{(\text{ext})}(\lambda) = \frac{S_{\lambda,\alpha}^{(\text{ext})}}{S_{0,\alpha}^{(\text{ext})}} \qquad (10.226)$$

which satisfy

$$\rho_\alpha^{(\text{ext})}(\lambda)\, \rho_\alpha^{(\text{ext})}(\mu) = \sum_\nu \mathcal{N}_{\lambda\mu}^{(\text{ext})\ \nu}\, \rho_\alpha^{(\text{ext})}(\nu) \qquad (10.227)$$

In many cases, the fusion rules can be read off directly from these one-dimensional representations (see Exs. 10.13, 10.14, and 10.15 for the cases (A_6, D_4), (E_6, A_{p-1}), and (E_8, A_{p-1}), respectively). For instance, in the three-state Potts case, they read

$$
\begin{array}{llllllllll}
\rho_{(1,1)}^{(\text{ext})} & = & \rho_{\mathbb{I}} & = & (& 1 & 1 & 1 & 1 & 1 & 1 &) \\
\rho_{(2,1)}^{(\text{ext})} & = & \rho_\varepsilon & = & (& a_+ & a_- & a_+ & a_+ & a_- & a_- &) \\
\rho_{(1,3)^{(1)}}^{(\text{ext})} & = & \rho_{\mathbf{Z}_1} & = & (& 1 & 1 & \omega & \bar\omega & \omega & \bar\omega &) \\
\rho_{(1,3)^{(2)}}^{(\text{ext})} & = & \rho_{\mathbf{Z}_2} & = & (& 1 & 1 & \bar\omega & \omega & \bar\omega & \omega &) \\
\rho_{(2,3)^{(1)}}^{(\text{ext})} & = & \rho_{\sigma_1} & = & (& a_+ & a_- & \omega a_+ & \bar\omega a_+ & \omega a_- & \bar\omega a_- &) \\
\rho_{(2,3)^{(2)}}^{(\text{ext})} & = & \rho_{\sigma_2} & = & (& a_+ & a_- & \bar\omega a_+ & \omega a_+ & \bar\omega a_- & \omega a_- &) \\
\end{array}
$$
$$(10.228)$$

where $a_\pm = (1 \pm \sqrt{5})/2$. Note that the conjugation matrix $\mathcal{C}$ acts on these one-dimensional representations as the complex conjugation, namely

$$\mathcal{C} \, \rho^{(\text{ext})} = \left(\rho^{(\text{ext})}\right)^* \tag{10.229}$$

10.8.6. A Simple Example of Nonminimal Extended Theory: The Free Boson at the Self-Dual Radius

The notion of extended symmetry applies also to nonminimal theories. Take the simplest example of the $c = 1$ bosonic theory compactified on a circle of radius $R = \sqrt{2}$, invariant under the duality transformation $R \rightarrow 2/R$ mentioned in Sect. 10.4.1 (Eq. (10.65)). This theory is certainly nonminimal,[22] and the partition function on a torus reads

$$Z(\sqrt{2}) = \frac{1}{|\eta(\tau)|^2} \sum_{n,m \in \mathbb{Z}} q^{\frac{1}{4}(n+m)^2} \, \bar{q}^{\frac{1}{4}(n-m)^2} \tag{10.230}$$

However, changing the summation variables to:

$$\lambda = n + m$$
$$\mu = n - m \tag{10.231}$$
$$\lambda - \mu = 0 \qquad \text{mod } 2 \,,$$

we can reexpress the partition function as

$$Z(\sqrt{2}) = \frac{1}{|\eta(\tau)|^2} \sum_{\substack{\lambda,\mu \in \mathbb{Z} \\ \lambda-\mu=0 \text{ mod } 2}} q^{\lambda^2/4} \, \bar{q}^{\mu^2/4} \tag{10.232}$$

We define the extended characters

$$C_0(\tau) = \frac{1}{\eta} \sum_{\lambda \text{ even}} q^{\lambda^2/4} = \frac{1}{\eta} \sum_{m \in \mathbb{Z}} q^{m^2} = \frac{\theta_3(2\tau)}{\eta(\tau)}$$
$$C_1(\tau) = \frac{1}{\eta} \sum_{\lambda \text{ odd}} q^{\lambda^2/4} = \frac{1}{\eta} \sum_{m \in \mathbb{Z}} q^{(m+\frac{1}{2})^2} = \frac{\theta_2(2\tau)}{\eta(\tau)} \tag{10.233}$$

where we have identified the Jacobi theta functions defined in App. 10.A. (Note that here the argument is 2τ.) We can write the partition function as

$$Z(\sqrt{2}) = |C_0|^2 + |C_1|^2 \tag{10.234}$$

This has a finite block-diagonal form, although the theory is not minimal. The extended characters C_0, C_1 transform under $\mathcal{T}$ as

$$C_0(\tau + 1) = e^{-i\pi/12} \, C_0(\tau)$$
$$C_1(\tau + 1) = e^{5i\pi/12} \, C_1(\tau) \,. \tag{10.235}$$

---

[22] We showed in Sect. 10.5 that the minimal theories can have only central charges $c = 1 - 6(p - p')^2/pp' < 1$.

The S transformation is a special case of Eq. (10.126), with $N = 2$, upon the identification:

$$C_0(\tau) = K_0(\tau) \qquad C_1(\tau) = K_1(\tau) \tag{10.236}$$

so that

$$C_0(-1/\tau) = \frac{C_0(\tau) + C_1(\tau)}{\sqrt{2}}$$

$$C_1(-1/\tau) = \frac{C_0(\tau) - C_1(\tau)}{\sqrt{2}} \tag{10.237}$$

Hence, the extended S matrix reads

$$S^{(ext)} = \frac{1}{\sqrt{2}} \begin{pmatrix} 1 & 1 \\ 1 & -1 \end{pmatrix} \tag{10.238}$$

and the associated extended fusion rules are again given by the obvious generalization (10.219) of Eq. (10.201), with $\phi_0 = \mathbb{I}$, the extended identity. A simple calculation yields[23]

$$\phi_1 \times \phi_1 = \phi_0 \tag{10.239}$$

We can repeat this construction whenever the *square* of the radius R of the bosonic theory is a rational number (see Exs. 10.21 and 10.23 below).

10.8.7. Rational Conformal Field Theory: A Definition

A conformal field theory is said to be *rational* if its (possibly infinite) irreducible Virasoro representations can be reorganized into a *finite* number of extended blocks, linearly transformed into each other under the modular group. More precisely, we let

$$C_\lambda = \sum_{i \in I_\lambda} \chi_i \qquad \lambda = 1, 2, \dots, N \tag{10.240}$$

denote the corresponding finite set of extended characters, where i denotes the irreducible Virasoro representations, and I_λ some (possibly infinite) sets. Diagonal RCFTs have modular-invariant partition functions of the form

$$\sum_{\lambda=1}^{N} |C_\lambda|^2 \tag{10.241}$$

whereas the nondiagonal RCFTs have partition functions of the form

$$\sum_{\lambda=1}^{N} C_\lambda C^*_{\Pi(\lambda)} \tag{10.242}$$

---

[23] In part C, we shall identify the extended symmetry of the $c = 1$ bosonic theory on a circle of radius $\sqrt{2}$ as the affine Lie algebra $\widehat{su}(2)_1$, whereas C_0 and C_1 will be identified as, respectively, the characters of the identity and spin $\frac{1}{2}$ representations.

for some automorphism Π of the extended fusion rules. The latter are obtained through Eq. (10.219).

The classification of all RCFTs is a formidable task, and it will probably remain an open problem for a while. A possible attack would be to first start by classifying all the possible fusion rules, and to use the information provided by the Verlinde formula (10.219) to get some clues concerning the operator content of the theory. Only partial results have been obtained so far; more details can be found in Chap. 17.

Appendix 10.A. Theta Functions

This appendix describes some of the properties of theta functions. We begin by explaining Jacobi's triple product formula; then we define the theta functions in terms of series and infinite products. We also express the Dedekind η function in terms of the theta functions. We finally derive the conformal properties of theta functions and of the Dedekind η function.

10.A.1. The Jacobi Triple Product

In order to prepare ourselves for some theta function manipulations, we consider Jacobi's triple product identity:

$$\prod_{n=1}^{\infty}(1-q^n)(1+q^{n-1/2}t)(1+q^{n-1/2}/t) = \sum_{n\in\mathbb{Z}}q^{n^2/2}t^n \qquad (10.243)$$

This identity is valid for $|q| < 1$ and $t \neq 0$, and can be demonstrated by combinatorial methods or in the context of Lie algebras (cf. Ex. 14.7). We shall argue that this identity is correct by analogy with a fermion-antifermion system. Consider a set of fermion oscillators b_n and their antifermion counterparts $\bar{b}_n$, with the Hamiltonian

$$H = E_0 \sum_{r\in\mathbb{N}+1/2} r(b_r^\dagger b_r + \bar{b}_r^\dagger \bar{b}_r) \qquad (10.244)$$

The fermion number operator is

$$N = \sum_{r\in\mathbb{N}+1/2} (b_r^\dagger b_r - \bar{b}_r^\dagger \bar{b}_r) \qquad (10.245)$$

Now we consider the grand partition function

$$Z(q,t) = \sum_{\text{states}} e^{-\beta(E-\mu N)} \qquad q = e^{-\beta E_0}, \ t = e^{\beta\mu} \qquad (10.246)$$

We shall evaluate this quantity in two different ways, leading to the two sides of the following equation, which is manifestly equivalent to Eq. (10.243):

$$\prod_{r\in\mathbb{N}+1/2}^{\infty}(1+q^r t)(1+q^r/t) = \prod_{n=1}^{\infty}\frac{1}{1-q^n}\sum_{n\in\mathbb{Z}}q^{n^2/2}t^n \qquad (10.247)$$

First, the grand partition function factorizes into a product of grand partition functions, each associated with a single fermion oscillator. This, of course, follows from the fact that H and N decouple into sums over different fermion modes. Since the grand partition functions for a fermion and antifermion modes labeled r are, respectively, $(1+q^r t)$ and $(1+q^r/t)$ (there are two occupation states), the complete grand partition function coincides with the l.h.s. of Eq. (10.247).

Second, the grand partition function may be written as

$$Z(q,t) = \sum_{n \in \mathbb{Z}} t^n Z_n(q) \tag{10.248}$$

where $Z_n(q)$ is the ordinary partition function for a fixed fermion number n. We consider first Z_0, the partition function with no net fermion number. The lowest energy states are given in Table 10.7.

Table 10.7. Lowest energy states of the fermion system with $N = 0$.

Energy	Degeneracy	States
0	1	$\lvert 0 \rangle$
1	1	$b_{1/2}^\dagger \bar{b}_{1/2}^\dagger \lvert 0 \rangle$
2	2	$b_{3/2}^\dagger \bar{b}_{1/2}^\dagger \lvert 0 \rangle$, $b_{1/2}^\dagger \bar{b}_{3/2}^\dagger \lvert 0 \rangle$
3	3	$b_{5/2}^\dagger \bar{b}_{1/2}^\dagger \lvert 0 \rangle$, $b_{3/2}^\dagger \bar{b}_{3/2}^\dagger \lvert 0 \rangle$, $b_{1/2}^\dagger \bar{b}_{5/2}^\dagger \lvert 0 \rangle$
4	5	$b_{7/2}^\dagger \bar{b}_{1/2}^\dagger \lvert 0 \rangle$, $b_{5/2}^\dagger \bar{b}_{3/2}^\dagger \lvert 0 \rangle$, $b_{3/2}^\dagger \bar{b}_{5/2}^\dagger \lvert 0 \rangle$, $b_{1/2}^\dagger \bar{b}_{7/2}^\dagger \lvert 0 \rangle$, $b_{3/2}^\dagger \bar{b}_{3/2}^\dagger b_{1/2}^\dagger \bar{b}_{1/2}^\dagger \lvert 0 \rangle$

The number of creation operators in these states is always even, and the sum of their indices is equal to the normalized energy level $m = E/E_0$. We notice so far that the degeneracy at level m is equal to the partition number $p(m)$. This may be shown in general. Therefore Z_0 is equal to

$$Z_0 = \sum_{m=0}^{\infty} p(m) q^m = \prod_{n=1}^{\infty} \frac{1}{1-q^n} \tag{10.249}$$

This confirms Eq. (10.247) as far as the t^0 term is concerned. We now consider Z_n. The lowest energy state with fermion number n is obtained by exciting the lowest n oscillators:

$$b_{1/2}^\dagger b_{3/2}^\dagger \cdots b_{n-1/2}^\dagger \lvert 0 \rangle \qquad E/E_0 = \sum_{r=1}^{n}\left(n - \frac{1}{2}\right) = \frac{1}{2}n^2 \tag{10.250}$$

It turns out that the excitations on top of this ground state have exactly the same structure as the excitations of the $n = 0$ sector. Therefore

$$Z_n = q^{n^2/2} \prod_{m=1}^{\infty} \frac{1}{1 - q^m} \qquad (10.251)$$

and Eq. (10.247) follows.

10.A.2. Theta Functions

Jacobi's theta functions are defined as follows:

$$\theta_1(z|\tau) = -i \sum_{r\in\mathbb{Z}+1/2} (-1)^{r-1/2} y^r q^{r^2/2}$$

$$\theta_2(z|\tau) = \sum_{r\in\mathbb{Z}+1/2} y^r q^{r^2/2}$$

$$\theta_3(z|\tau) = \sum_{n\in\mathbb{Z}} y^n q^{n^2/2} \qquad (10.252)$$

$$\theta_4(z|\tau) = \sum_{n\in\mathbb{Z}} (-1)^n y^n q^{n^2/2}$$

where z is a complex variable and τ a complex parameter living on the upper half-plane. We have defined $q = \exp 2\pi i \tau$ and $y = \exp 2\pi i z$.

Jacobi's triple product allows us to rewrite these functions in the form of infinite products:

$$\theta_1(z|\tau) = -iy^{1/2} q^{1/8} \prod_{n=1}^{\infty}(1 - q^n) \prod_{n=0}^{\infty}(1 - yq^{n+1})(1 - y^{-1}q^n)$$

$$\theta_2(z|\tau) = y^{1/2} q^{1/8} \prod_{n=1}^{\infty}(1 - q^n) \prod_{n=0}^{\infty}(1 + yq^{n+1})(1 + y^{-1}q^n)$$

$$\theta_3(z|\tau) = \prod_{n=1}^{\infty}(1 - q^n) \prod_{r\in\mathbb{N}+1/2}(1 + yq^r)(1 + y^{-1}q^r) \qquad (10.253)$$

$$\theta_4(z|\tau) = \prod_{n=1}^{\infty}(1 - q^n) \prod_{r\in\mathbb{N}+1/2}(1 - yq^r)(1 - y^{-1}q^r)$$

For instance, the equivalence of the two expressions for θ_1 is obtained by setting $t = yq^{1/2}$ in Eq. (10.243).

By shifting their arguments, theta functions may all be related to each other; from their definitions it is a simple matter to check that

$$\theta_4(z|\tau) = \theta_3(z + \tfrac{1}{2}|\tau)$$

$$\theta_1(z|\tau) = -ie^{i\pi z} q^{1/8} \theta_4(z + \tfrac{1}{2}\tau|\tau) \qquad (10.254)$$

$$\theta_2(z|\tau) = \theta_1(z + \frac{1}{2}|\tau)$$

Theta functions are used to define doubly periodic functions on the complex plane. One sees that they are not periodic under $z \to z+1$ or $z \to z+\tau$, but obey the simple relations:

$$
\begin{array}{ll}
\theta_1(z + 1|\tau) = -\theta_1(z|\tau) & \theta_1(z + \tau|\tau) = -\dfrac{1}{yq}\theta_1(z|\tau) \\[2mm]
\theta_2(z + 1|\tau) = -\theta_2(z|\tau) & \theta_2(z + \tau|\tau) = \dfrac{1}{yq}\theta_2(z|\tau) \\[2mm]
\theta_3(z + 1|\tau) = \theta_3(z|\tau) & \theta_3(z + \tau|\tau) = \dfrac{1}{yq^{1/2}}\theta_3(z|\tau) \\[2mm]
\theta_4(z + 1|\tau) = \theta_4(z|\tau) & \theta_4(z + \tau|\tau) = -\dfrac{1}{yq^{1/2}}\theta_4(z|\tau)
\end{array}
\tag{10.255}
$$

It follows that doubly periodic functions may be easily constructed out of ratios or logarithmic derivatives of theta functions. The best-known example is the Weierstrass function:

$$\wp(z|\tau) = -\frac{\partial^2}{\partial z^2} \ln \theta_1(z|\tau) - 2\eta_1 \tag{10.256}$$

where the constant η_1 depends only on τ

$$\eta_1 = -\frac{1}{6}\frac{\partial_z^3\theta_1(0|\tau)}{\partial_z\theta_1(0|\tau)} \tag{10.257}$$

We shall also use the theta functions at $z = 0$:

$$\theta_i(\tau) \equiv \theta_i(0|\tau)$$

for $i = 2, 3, 4$ (one easily checks that $\theta_1(0|\tau) = 0$). Their explicit expressions, in terms of sums and products, are

$$
\begin{array}{lll}
\theta_2(\tau) = \displaystyle\sum_{n\in\mathbb{Z}} q^{(n+1/2)^2/2} & = 2q^{1/8}\displaystyle\prod_{n=1}^{\infty}(1 - q^n)(1 + q^n)^2 \\[4mm]
\theta_3(\tau) = \displaystyle\sum_{n\in\mathbb{Z}} q^{n^2/2} & = \displaystyle\prod_{n=1}^{\infty}(1 - q^n)(1 + q^{n-1/2})^2 \\[4mm]
\theta_4(\tau) = \displaystyle\sum_{n\in\mathbb{Z}} (-1)^n q^{n^2/2} & = \displaystyle\prod_{n=1}^{\infty}(1 - q^n)(1 - q^{n-1/2})^2
\end{array}
\tag{10.258}
$$

10.A.3. Dedekind's η Function

Dedekind's η function is defined as

$$\eta(\tau) = q^{1/24}\varphi(q) = q^{1/24}\prod_{n=1}^{\infty}(1 - q^n) \tag{10.259}$$

where $\varphi(q)$ is the Euler function. This function is related to theta functions as follows:

$$\eta^3(\tau) = \frac{1}{2}\theta_2(\tau)\theta_3(\tau)\theta_4(\tau) \tag{10.260}$$

This identity is an immediate consequence of the infinite product expressions for the theta functions at $z = 0$; we simply need to show that the function

$$f(q) = \prod_{n=1}^{\infty}(1 + q^n)(1 + q^{n-1/2})(1 - q^{n-1/2}) \tag{10.261}$$

is equal to unity. But we may write

$$f(q) = \prod_{n=1}^{\infty}(1 + q^n)(1 - q^{2n-1}) \tag{10.262}$$

The first factor may be written in the product as $(1 + q^{2n})(1 + q^{2n-1})$. Combining the second factor of this last expression with $(1 - q^{2n-1})$, one finds

$$f(q) = \prod_{n=1}^{\infty}(1 + q^{2n})(1 - q^{4n-2}) = f(q^2) \tag{10.263}$$

Since $f(0) = 1$, it follows that $f(q) = 1$ if $|q| < 1$.

10.A.4. Modular Transformations of Theta Functions

We are now interested in the behavior of theta functions $\theta_i(\tau)$ under the modular transformation $\tau \to -1/\tau$. For this we need the following formula, called the Poisson *resummation formula*:

$$\sum_{n\in\mathbb{Z}}\exp(-\pi a n^2 + bn) = \frac{1}{\sqrt{a}}\sum_{k\in\mathbb{Z}}\exp-\frac{\pi}{a}\left(k + b/2\pi i\right)^2 \tag{10.264}$$

This formula is easily demonstrated by using the identity[24]

$$\sum_{n\in\mathbb{Z}}\delta(x - n) = \sum_{k\in\mathbb{Z}}e^{2\pi i k x} \tag{10.265}$$

and by integrating it over $\exp(-\pi a x^2 + bx)$.

We consider now the infinite series expression for $\theta_3(\tau)$. Applying the formula (10.264) with $a = -i\tau$ and $b = 0$, we immediately find

$$\theta_3(-1/\tau) = \sqrt{-i\tau}\,\theta_3(\tau) \tag{10.266}$$

---

[24] A detailed derivation of a generalization of this identity is given in App. 14.B.

If we set $a = -i\tau$ and $b = -i\pi$, we obtain the modular transformation of θ_2:

$$\theta_2(-1/\tau) = \sqrt{-i\tau}\theta_4(\tau) \tag{10.267}$$

Applying the modular transformation a second time, we find

$$\theta_4(-1/\tau) = \sqrt{-i\tau}\theta_2(\tau) \tag{10.268}$$

These simple transformation properties, as well as the relation (10.260) for the η function, give us directly the modular transformation of that function:

$$\eta(-1/\tau) = \sqrt{-i\tau}\,\eta(\tau) \tag{10.269}$$

The modular properties under the shift $\tau \to \tau + 1$ are easily derived from Eq. (10.258). The infinite product expression for θ_2 implies that $\theta_2(\tau + 1) = e^{i\pi/4}\theta_2(\tau)$. On the other hand, the infinite series expressions for θ_3 and θ_4 yield:

$$\theta_3(\tau + 1) = \sum_{n\in\mathbb{Z}} q^{n^2/2}e^{i\pi n^2}$$
$$= \sum_{n\in\mathbb{Z}} q^{n^2/2}e^{i\pi n^2}(-1)^n \tag{10.270}$$
$$= \theta_4(\tau)$$

Likewise, we find that $\theta_4(\tau + 1) = \theta_3(\tau)$.

We can group these results, as well as the transformation of the Dedekind η function, as follows:

$$\begin{array}{|ll|}
\hline
\theta_2(\tau + 1) = e^{i\pi/4}\theta_2(\tau) & \theta_2(-1/\tau) = \sqrt{-i\tau}\theta_4(\tau) \\
\theta_3(\tau + 1) = \theta_4(\tau) & \theta_3(-1/\tau) = \sqrt{-i\tau}\theta_3(\tau) \\
\theta_4(\tau + 1) = \theta_3(\tau) & \theta_4(-1/\tau) = \sqrt{-i\tau}\theta_2(\tau) \\
\eta(\tau + 1) = e^{i\pi/12}\eta(\tau) & \eta(-1/\tau) = \sqrt{-i\tau}\,\eta(\tau) \\
\hline
\end{array} \tag{10.271}$$

10.A.5. Doubling Identities

The Jacobi theta functions satisfy the following doubling identities

$$\theta_2(2\tau) = \sqrt{\frac{\theta_3(\tau)^2 - \theta_4(\tau)^2}{2}}$$
$$\theta_3(2\tau) = \sqrt{\frac{\theta_3(\tau)^2 + \theta_4(\tau)^2}{2}} \tag{10.272}$$
$$\theta_4(2\tau) = \sqrt{\theta_3(\tau)\theta_4(\tau)}$$

whereas θ_1 satisfies

$$\partial_z\theta_1(2\tau) = \frac{1}{2}\frac{\theta_2(\tau)\partial_z\theta_1(\tau)}{\sqrt{\theta_3(\tau)\theta_4(\tau)}} \tag{10.273}$$

Exercises

10.1 *Euclidian division and the Bezout lemma*
For any two positive integers a, c, we denote by $\gcd(a,c)$ the greatest common divisor of a and c.

a) *Euclidian division*: Show that for two given integers $a > c > 0$, there is a unique couple a_1, c_1 of integers, such that

$$a = a_1 c + c_1$$
$$0 \leq c_1 < c$$

b) Show that $\gcd(a,c) = \gcd(c,c_1)$.

c) *Bezout lemma*: Show that there exist two integers a_0 and c_0 such that

$$c_0 a - a_0 c = \gcd(a,c)$$

Hint: Repeat the Euclidian division, namely write $c = a_2 c_1 + c_2$, and so on. The sequence $c > c_1 > c_2 > \cdots \geq 0$ is strictly decreasing, hence there exists a finite k, such that $c_k = \gcd(a,c)$ and $c_{k+1} = 0$.

d) Deduce that the two integers a, c are coprime iff there exist two integers a_0 and c_0 such that $c_0 a - a_0 c = 1$. (Two integers are said to be coprime iff their only common divisor is 1.)

10.2 *The modular group PSL(2, $\mathbb{Z}$)*
The modular group is defined as

$$PSL(2,\mathbb{Z}) = \left\{ \begin{pmatrix} a & b \\ c & d \end{pmatrix} \ a,b,c,d \in \mathbb{Z} \mid ad - bc = 1 \right\}$$

The elements of $PSL(2,\mathbb{Z})$ are also often labeled by the fractions $(a\tau + b)/(c\tau + d)$. The aim of this exercise is to show that the transformations S and T of Eq. 10.8 generate the modular group.

a) Prove that any product of S and T is an element of $PSL(2,\mathbb{Z})$.

b) In the following, we consider a generic element $x = (a\tau + b)/(c\tau + d)$ of $PSL(2,\mathbb{Z})$. Show that a and c are coprime.
Hint: Use the Bezout lemma of Ex. 10.1 above.

c) If $a > c$, show that there exists an integer ρ_0, such that

$$\frac{a\tau + b}{c\tau + d} = \rho_0 + \frac{a_1\tau + b_1}{c_1\tau + d_1}$$

with $c_1 = c, d_1 = d$, and $0 \leq a_1 < c$.
Hint: Perform the Euclidian division of a by c, to get $a = \rho_0 c + a_1$, and therefore $b_1 = b - \rho_0 d$.

d) If $a_1 = 0$, show that one can take $-b_1 = c_1 = 1$, and write x as a composition of S and T actions.
Result: $x = T^{\rho_0} S T^{\rho_1}$, with $\rho_1 = d_1$.

e) If $a_1 > 0$, write

$$\frac{a\tau + b}{c\tau + d} = \rho_0 - 1 \Bigg/ \left(\frac{-c_1\tau - d_1}{a_1\tau + b_1} \right)$$

and repeat the above division procedure to rewrite

$$\frac{a\tau + b}{c\tau + d} = \rho_0 - 1 \left/ \left(\rho_1 + \frac{a_2\tau + b_2}{c_2\tau + d_2} \right) \right.$$

where $c_2 = a_1, d_2 = b_1$, and $0 \le a_2 < a_1$.

f) Repeating this division procedure (leading to five sequences $\rho_i, a_i, b_i, c_i, d_i, i = 1, 2, \ldots$ of integers), show that there exists a finite integer k, such that $a_k = 0$, $a_{k-1} \ne 0$. Show that one can take $-b_k = c_k = 1$, and conclude that the element x may be written as a composition of S and T actions.
Result: $x = T^{\rho_0} S T^{\rho_1} S \cdots T^{\rho_{k-1}} S T^{\rho_k}$, where $\rho_k = d_k$. This completes the proof that $PSL(2, \mathbb{Z})$ is generated by S and T actions.

10.3 *Smallest dimension in a minimal theory*

a) Find the smallest dimension in the Kac table of the minimal (p, p') theory.
Hint: Use the Bezout lemma of Ex. 10.1 c.

b) Check the strict inequality (10.95).

10.4 $S^2 = 1$ *for minimal models*
We wish to compute the matrix C

$$C_{rs,\rho\sigma} = \sum_{(n,m)\in E_{p,p'}} S_{rs,nm} \, S_{nm,\rho\sigma}$$

for minimal models, with the S matrix elements given by (10.134).

a) Write the transformations of S under $(r, s) \rightarrow (p' - r, p - s)$, $(p' + r, p - s)$, and $(p' - r, p + s)$, which correspond, respectively, to the transformations of $\lambda = \lambda_{r,s} = pr - p's \rightarrow -\lambda$, $\omega_0\lambda$ and $-\omega_0\lambda$.

b) Rewrite $S_{rs,nm}$ as a function of $\lambda_{n,m} = pn - p'm$ and r, s only. Call this function $S_{rs}(\lambda)$.

c) Deduce that for $N = 2pp'$

$$C_{rs,\rho\sigma} = \sum_{\mu=0}^{N-1} S_{rs}(\mu) \, S_{\rho\sigma}(\mu)$$

d) Show that

$$\sum_{\mu=0}^{N-1} S_{rs}(\mu) \, S_{\rho\sigma}(\mu) = \sum_{\epsilon_i=\pm 1} \delta^{(N)} \Big(p(\epsilon_1 r + \epsilon_2 \rho) + p'(\epsilon_3 s + \epsilon_4 \sigma) \Big)$$

where the delta function modulo N reads

$$\delta^{(N)}(x) = \frac{1}{N} \sum_{\mu=0}^{N-1} e^{2i\pi x/N}$$

e) Conclude that

$$C_{rs,\rho\sigma} = \delta_{r,\rho} \delta_{s,\sigma}$$

by restricting C back to the fundamental domain $E_{p,p'}$.

10.5 *Positivity and nonvanishing of basic S matrix elements for minimal models*

a) Use the expression (10.134) to directly prove Eq. (10.136), for any minimal model.

b) For unitary theories (i.e., with $|p - p'| - 1$), show further that

$$S_{\rho\sigma;11} > 0 \qquad \text{for all } (\rho, \sigma) \in E_{p,p'}$$

c) For nonunitary theories (i.e., with $|p - p'| > 1$), prove that the matrix elements

$$S_{\rho\sigma;r_0s_0} > 0 \qquad \text{for all } (\rho, \sigma) \in E_{p,p'}$$

where (r_0, s_0) are the Kac labels of the smallest dimension of the theory (see Ex. 10.3).

10.6 *Modular invariance of* $Z_{D_{p'/2+1},A_{p-1}}$

a) For $p' = 4m + 2$ and p an odd integer, compute the quantities

$$h_{p'-2r-1,s} - h_{2r+1,s} \quad \text{for} \quad r = 0, 1, \ldots, m-1 \quad \text{and} \quad s = 1, 2, \ldots, (p-1)/2$$

Deduce the T invariance of the partition function $Z_{D_{p'/2+1},A_{p-1}}$.

b) Show that the extended characters

$$C_{2r+1,s} = \chi_{2r+1,s} + \chi_{p'-2r-1,s} \quad (r = 0, 1, \ldots, m-1), \ (s = 1, 2, \ldots, (p-1)/2)$$

and

$$C_{2m+1,s} = \chi_{2m+1,s} \quad \text{for} \quad s = 1, 2, \ldots, (p-1)/2$$

form an $(m+1)(p-1)/2$–dimensional space invariant under the linear action of S. Write the matrix elements of the restriction of S to this basis, and check that this restriction is unitary. Deduce the modular invariance of $Z_{D_{p'/2+1},A_{p-1}}$.

10.7 *Modular invariance of* $Z_{E_6,A_{p-1}}$

a) For $p' = 12$ and p an arbitrary odd integer, compute

$$h_{7,s} - h_{1,s}, \qquad h_{8,s} - h_{4,s}, \qquad h_{11,s} - h_{5,s}$$

Deduce the T invariance of the $Z_{E_6,A_{p-1}}$ partition function of Table 10.3.

b) Show that

$$C_{1,s} = \chi_{1,s} + \chi_{7,s}, \quad C_{4,s} = \chi_{4,s} + \chi_{8,s}, \quad C_{5,s} = \chi_{5,s} + \chi_{11,s}$$

for $1 \leq s \leq (p-1)/2$ form a basis of a $3(p-1)/2$–dimensional space invariant under the action of S. Write the matrix elements of the restriction of S to this basis, and check that this restriction is unitary. Deduce the modular invariance of $Z_{E_6,A_{p-1}}$.

10.8 *Modular invariance of* $Z_{E_8,A_{p-1}}$

a) For $p' = 30$ and p an arbitrary odd integer, compute

$$h_{11,s} - h_{1,s}, \quad h_{19,s} - h_{1,s}, \quad h_{29,s} - h_{1,s},$$
$$h_{13,s} - h_{7,s}, \quad h_{17,s} - h_{7,s}, \quad h_{23,s} - h_{7,s}$$

for $s = 1, \ldots, p-1$. Deduce the T invariance of the $Z_{E_8,A_{p-1}}$ partition function of Table 10.3.

b) Show that

$$C_{1,s} = \chi_{1,s} + \chi_{11,s} + \chi_{9,s} + \chi_{29,s},$$
$$C_{7,s} = \chi_{7,s} + \chi_{13,s} + \chi_{17,s} + \chi_{23,s}$$

for $1 \leq s \leq (p-1)/2$ form a basis of a $(p-1)$–dimensional space invariant under the action of S. Write the matrix elements of the restriction of S to this basis, and check that this restriction is unitary. Deduce the modular invariance of $Z_{E_6,A_{p-1}}$.

10.9 *Modular invariance of* $Z_{D_{p'/2+1},A_{p-1}}$ *for* $p' = 4m$

a) From the expression of the matrix elements of S on minimal characters Eq. (10.134), show that

$$S_{\Upsilon(r,s);\rho,\sigma} = (-1)^{1+\rho} S_{rs,\rho\sigma} = (-1)^{r+\rho} S_{r,s;\Upsilon(\rho,\sigma)} ,$$

where Υ is the automorphism

$$\Upsilon(r,s) = (p' - r, s)$$

b) Find an automorphism Π leading from $Z_{A_{p'-1},A_{p-1}}$ to $Z_{D_{p'/2+1},A_{p-1}}$ and deduce the modular invariance of the latter.
Result: $\Pi = \Upsilon$ for even r, $\Pi = \mathbb{I}$ for odd r.

c) Why does the construction fail in the case $p' = 2(2m + 1)$?

10.10 *ADE classification of integer matrices with eigenvalues < 2*
Let G denote the adjacency matrix of a connected graph $\mathcal{G}$ (cf. Sect. 10.7.6). It is therefore a nondecomposable symmetric matrix with nonnegative integer entries $G_{a,b} \in \mathbb{N}$. We assume that the largest eigenvalue of G, denoted by $\lambda_{\max}$, is strictly less than 2. We denote by $v_{\max}$ the corresponding eigenvector.

a) Show that the maximum eigenvalue $\lambda_{\max}$ of G is positive.
Hint: $\lambda_{\max} = \max_x (x \cdot Gx)/(x \cdot x)$, where x is any nonzero vector.

b) Prove that if a component of $v_{\max}$ is strictly negative, say $[v_{\max}]_{a_0} < 0$, then there exists $a_1 \neq a_0$ such that $[v_{\max}]_{a_1} < 0$, and $G_{a_0,a_1} \neq 0$. (We assume that $\mathcal{G}$ is not made of a single point.) Show further that if $\mathcal{G}$ has at least 3 nodes, then there exists $a_2 \neq a_0, a_1$ such that $[v_{\max}]_{a_2} < 0$.
Hint: Prove and use the fact that $\lambda_{\max} > G_{a_0,a_1}$.

c) Deduce from this that the eigenvector $v_{\max}$ for $\lambda_{\max}$ can be chosen with all components positive. It is called the Perron–Frobenius eigenvector of G, and is fully characterized, among the eigenvectors of G, by this positivity condition.

d) Show that if a nonzero entry of G is reduced by a small quantity, namely $G \to G(\epsilon)$, where $G_{a,b}(\epsilon) = G_{a,b} - \epsilon$ for some particular pair of vertices (a,b) such that $G_{a,b} \geq 1$ and $G_{c,d}(\epsilon) = G_{c,d}$ for all other matrix elements of $G(\epsilon)$, then $\lambda_{\max}$ is also *reduced* by a quantity of the order of ϵ, namely $\lambda_{\max}(\epsilon) < \lambda_{\max}$ for small $\epsilon > 0$. We denote by $v_{\max}(\epsilon)$ the Perron–Frobenius eigenvector of $G(\epsilon)$. (The reasoning for **(b)** holds for any matrix G with nonnegative real entries.)
Hint: Use the first-order perturbation theory of quantum mechanics.

e) Using the result of **(d)**, show that the removal of a link from the graph $\mathcal{G}$ has the effect of lowering the maximal eigenvalue of G.
Hint: Suppose that $\lambda_{\max}$ has increased to $\lambda_{\max}(1) > \lambda_{\max}$. Since it started by decreasing with ϵ, there must exist a *finite* positive value ϵ_0 of ϵ such that $\lambda_{\max}(\epsilon_0) = \lambda_{\max}$. If we denote by $v_0 = v_{\max}(\epsilon_0)$, this amounts to

$$0 = v_{\max} \cdot (G - \lambda_{\max}) v_{\max}$$
$$0 = v_0 \cdot (G(\varepsilon_0) - \lambda_{\max}) v_0$$
$$\Rightarrow 0 = v_0 \cdot (G - \lambda_{\max}) v_0 - 2\epsilon_0 [v_0]_a [v_0]_b$$

where we use the explicit form of $G(\varepsilon_0)$

$$G(\varepsilon_0)_{i,j} = G_{i,j} - \varepsilon_0(\delta_{i,a}\delta_{j,b} + \delta_{i,b}\delta_{j,a})$$

We get a contradiction to the fact that $x \cdot (G - \lambda_{max})x \le 0$ for all vectors x (see hint of (a)).

f) Show that the following graphs $\hat{A}, \hat{D}, \hat{E}$ have 2 as eigenvalue:

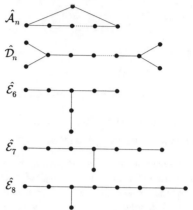

g) From these graphs, deduce that: (i) G has no internal cycle, i.e., it is a tree; and (ii) except for at most one trivalent node, it only has one- or two-valent nodes. Conclude, by inspection of the last three cases, that G is necessarily of the form A, D or E of Fig. 10.3.

10.11 *Fusion rules of several minimal models*
Use the Verlinde formula (10.171) to compute the fusion rules of the following models. (The reader can write a simple computer program to generate the fusion numbers.)

a) The Yang–Lee model $\mathcal{M}(5,2)$.

b) The Ising model $\mathcal{M}(4,3)$.

c) The tricritical Ising model $\mathcal{M}(5,4)$.

10.12 *Fusion rules for the models* $(A_{p'-1}, D_{p/2+1})$, $p = 4m + 2$ *and* $(D_{p'/2+1}, A_{p-1})$, $p' = 4m + 2$

a) Let $p = 4m+2$. Write the fusion rules for the subset of fields $\phi_{(r,2s+1)}, r = 1, 2, \ldots, (p'-1)/2, s = 0, 1, \ldots, p/2 - 1$ of the $\mathcal{M}(p,p')$ model.

b) Taking into account the multiplicity 2 of the fields $\phi_{(r,p/2-1)}, r = 1, 2, \ldots, (p'-1)/2$, obtain fusion rules that are invariant under the $\mathbb{Z}_2$ symmetry, which leaves all the fields invariant except one of the two copies of each degenerate field, which is changed into its opposite.

c) Repeat the above analysis with the $(D_{p'/2+1}, A_{p-1})$ models, for $p' = 4m + 2$.

10.13 *Extended fusion rules of the tricritical three-state Potts model from the Verlinde formula*
The tricritical three-state Potts model is the (A_6, D_4) theory, with $p' = 7, p = 6$.

a) Write the modular transformations of the extended characters of the theory.

b) Following the three-state Potts example treated in Sect. 10.8.5, split the doubly degenerate characters in such a way that the symmetry and unitarity of the S matrix are restored.

c) Compute the conjugation matrix $C = S^2$. Check that the conjugation interchanges the two copies of each degenerate field and leaves the other extended fields invariant.

d) Write the one-dimensional representations $\rho_\alpha(\lambda)$ of the extended fusion rules of the model.

e) Compute the extended fusion rules of the tricritical three-state Potts model.

10.14 *Extended fusion rules of the* (E_6, A_{p-1}) *model*

a) Compute the matrix elements of the modular transformation S of the extended characters

$$C_{1,s} = \chi_{1,s} + \chi_{7,s}, \quad C_{4,s} = \chi_{4,s} + \chi_{8,s}, \quad C_{5,s} = \chi_{5,s} + \chi_{11,s}$$

for $1 \leq s \leq (p-1)/2$.

b) Write the corresponding one-dimensional representations of the extended fusion algebra of the model (cf. Sect. 10.8.5).

c) Deduce the extended fusion rules of the (E_6, A_{p-1}) model.

10.15 *Extended fusion rules of the* (E_8, A_{p-1}) *model*

a) Compute the matrix elements of the modular transformation S of the extended characters

$$C_{1,s} = \chi_{1,s} + \chi_{11,s} + \chi_{19,s} + \chi_{29,s},$$
$$C_{7,s} = \chi_{7,s} + \chi_{13,s} + \chi_{17,s} + \chi_{23,s}$$

for $1 \leq s \leq (p-1)/2$.

b) Write the corresponding one-dimensional representations of the extended fusion algebra of the model (cf. Sect. 10.8.5).

c) Deduce the extended fusion rules of the (E_8, A_{p-1}) model.

10.16 *Constraints on an RCFT from the data of its fusion rules*

We start with an RCFT having two (extended) primary fields $\mathbb{I}$ and ϕ, satisfying the fusion rule

$$\phi \times \phi = \mathbb{I} + \phi$$

a) Compute the matrix S using the Verlinde formula (10.201).
Hint: The Verlinde formula should be used in the reverse order: The matrix S is the matrix of the change of basis that diagonalizes the above fusion rules.

b) Use the condition (10.216) to find constraints on the central charge c and on the conformal dimension h of ϕ.
Result: $12h - c = 2 \mod 8$ and $h = \frac{m}{5} \mod 1$, with $m = 1, 2, 3$, or 4.

c) Check these constraints for the minimal model $\mathcal{M}(2, 5)$.
Result: $h_\phi = h_{2,1} = -1/5$ and $12h_\phi - c = 2$.

d) Check these constraints on the extended theory of the (E_8, A_2) model (cf. Ex. 10.15, with $p = 3$): First prove that the extended fusion rules of this theory are indeed $\phi \times \phi = \mathbb{I} + \phi$, where $\mathbb{I}$ and ϕ have the respective extended characters

$$C_{\mathbb{I}} \equiv C_{1,1} = \chi_{1,1} + \chi_{11,1} + \chi_{19,1} + \chi_{29,1}$$
$$C_\phi \equiv C_{7,1} = \chi_{7,1} + \chi_{13,1} + \chi_{17,1} + \chi_{23,1}$$

Result: $h_\phi = h_{7,1} = -9/5$ and $12h_\phi - c = 26$.

10.17 *Constraints on a general RCFT*

We start from an arbitrary RCFT, with extended modular transformation matrices S and $\mathcal{T}$. Let h_i, $i = 1, \ldots, N$ denote the conformal dimensions of the corresponding blocks (h_i is the smallest conformal dimension of the fields forming the i-th block).

a) Using the identity $(ST)^3 = C$, show that the matrix T satisfies

$$\det\left[(T)^6\right] = 1$$

b) Using the result of **(a)**, derive the general sum rule

$$Nc/4 = 6\sum_{i=1}^{N} h_i \bmod 1$$

c) Check this result for the (diagonal) minimal models $\mathcal{M}(p,p')$, and for the three-state Potts model.

d) Check the above sum rule for the extended theories of (D_4, A_4) (three-state Potts model), (E_6, A_{p-1}), and (E_8, A_{p-1}) (cf. Sect. 10.8.5, and Exs. 10.14, 10.15).

e) Check the above sum rule for the extended (nonminimal) theory of the free boson compactified on a circle of self-dual radius (cf. Sect. 10.8.6).
Result: $2c/4 = 1/2 = 6(0 + 1/4) \bmod 1$.

10.18 *Verlinde formula for a finite group*
Let G be a finite group, with unit e and multiplication law $\circ$. The conjugacy class of an element g of G is defined as

$$C(g) = \{x \in G \mid \exists y \in G, \ x = ygy^{-1}\}$$

The irreducible linear unitary representations of G are denoted by ρ_j, $j = 1, 2, \ldots, d_G$. One can think of them as unitary matrices $\rho_j(x)$ attached to each element x of G (with size $\dim_j \times \dim_j$, where $\dim_j$ is the dimension of the representation) such that

$$\rho_j(x)\, \rho_j(y) = \rho_j(x \circ y)$$

for any two group elements x and y. In other words, a representation translates the group multiplication $\circ$ into matrix multiplication. For a given representation ρ_j, we define the corresponding *character* by the function

$$g \in G \to \chi_j(g) = \operatorname{Tr} \rho_j(g) = \sum_{i=1}^{\dim_j} [\rho_j(g)]_{ii}$$

where the trace yields a function independent of the choice of basis of the representation.

a) Show that the distinct conjugacy classes of the elements of G (called simply the classes of G from now on) form a partition of G. We choose a representative $\alpha \in G$ in each class and denote by C_α the corresponding class. We take for granted that there are d_G distinct classes, where d_G is also the number of irreducible representations of G. As an illustrative example, enumerate the classes of the permutation group of three objects, S_3, and compute the corresponding value of d_{S_3}.

b) Show that the characters χ_j are constant functions on each class C_α. We now denote by $\chi_j(\alpha)$ the corresponding functions. Compute $\chi_j(e)$ and $\chi_0(\alpha)$, where 0 denotes the identity representation.
Result: $\chi_j(e) = \dim_j$, the size of the corresponding matrix, and $\chi_0(\alpha) = 1$ for all classes C_α.

c) Assume the following orthogonality relations for characters

$$\sum_{j=1}^{d_G} \chi_j(\alpha)\,\bar{\chi}_j(\beta) = \frac{|G|}{|C_\alpha|}\,\delta_{\alpha,\beta}$$

(10.274)

$$\sum_\alpha |C_\alpha|\,\chi_j(\alpha)\,\bar{\chi}_k(\alpha) = |G|\,\delta_{j,k}$$

where $|G|$ and $|C_\alpha|$ denote, respectively, the orders of G and of the conjugacy class C_α. The last identity is actually a particular case of the following

$$\sum_\alpha |C_\alpha|\,\chi_j(\alpha)\,\bar{\chi}_k(\alpha\beta) = |G|\frac{\chi_j(\beta)}{\dim_j}\,\delta_{j,k}$$

(10.275)

Prove the following relations

$$\sum_{j=1}^{d_G} (\dim_j)^2 = |G|$$

$$\sum_\alpha |C_\alpha| = |G|$$

and deduce the dimensions $\dim_j$ of the irreducible representations of S_3. Use the orthogonality relations (10.274) to compute all the characters of S_3.

d) The tensor product of two representations $\rho_i(\alpha)$ and $\rho_j(\alpha)$ is a reducible representation of G of dimension $\dim_i + \dim_j$. It can be decomposed onto irreducible ones, as

$$\rho_i \otimes \rho_j = \bigoplus_k \mathcal{N}_{ij}^{\ k}\,\rho_k$$

where the $\mathcal{N}_{ij}^{\ k}$'s are nonnegative integer multiplicities, independent of the class (this is why we dropped the class index α in the decomposition formula). From this relation, deduce a product decomposition formula for characters. Check it in the case of S_3, and compute the numbers $N_{ij}^{\ k}$.
Hint: Take the trace of the tensor product decomposition formula. This yields

$$\chi_i(\alpha)\,\chi_j(\alpha) = \sum_k \mathcal{N}_{ij}^{\ k}\,\chi_k(\alpha)$$

for any class C_α.

e) From the orthogonality relations (10.274) between characters, deduce an expression for $\mathcal{N}_{ij}^{\ k}$ in terms of characters. This is the group *Verlinde formula* for tensor products of irreducible representations.
Result:

$$\mathcal{N}_{ij}^{\ k} = \frac{1}{|G|}\sum_\alpha |C_\alpha|\chi_i(\alpha)\chi_j(\alpha)\bar{\chi}_k(\alpha)$$

(10.276)

f) We define the *group S matrix* as

$$S_j(\alpha) = \left(\frac{|C_\alpha|}{|G|}\right)^{\frac{1}{2}} \chi_j(\alpha)$$

(10.277)

Show that S is unitary. Rewrite the formula (10.276) in terms of S. Is the matrix S symmetric in the case of S_3?

Result:

$$\mathcal{N}_{ij}{}^k = \sum_{\alpha} S_i(\alpha) \frac{S_j(\alpha)}{S_0(\alpha)} \bar{S}_k(\alpha)$$

g) Multiplying two classes C_α and C_β just means performing the group product

$$[\sum_{x\in C_\alpha} x] \circ [\sum_{y\in C_\beta} y]$$

Together with the usual addition of classes, this endows the group with an algebra structure, called the *group algebra*. We denote by $C_\alpha * C_\beta$ the corresponding product. Decompose this product in G and reorganize the result into sums of elements of G over classes to get the class algebra

$$C_\alpha * C_\beta = \sum_{\gamma} \mathcal{N}_{\alpha\beta}{}^\gamma C_\gamma$$

where $\mathcal{N}_{\alpha\beta}{}^\gamma$ are integer multiplicities. Find the numbers $\mathcal{N}_{\alpha\beta}{}^\gamma$ for S_3.

h) Using the class algebra of **(g)**, find another decomposition formula for characters. Hint: Take first the representation ρ_j of the class algebra relation, then compute its trace in terms of characters. The result reads

$$\chi_j(\alpha\beta) = \sum_{\gamma} \mathcal{N}_{\alpha\beta}{}^\gamma \chi_j(\gamma)$$

i) Using the orthogonality relations for characters (10.275) and (10.274), deduce an expression of the numbers $\mathcal{N}_{\alpha\beta}{}^\gamma$ in terms of characters. This is the *group Verlinde formula* for products of classes.
Result:

$$\boxed{\mathcal{N}_{\alpha\beta}{}^\gamma = \frac{|C_\alpha||C_\beta|}{|G|} \sum_j \chi_j(\alpha) \frac{\chi_j(\beta)}{\dim_j} \bar{\chi}_j(\gamma)}$$

j) Rewrite this in terms of the *group S matrix* (10.277).
Result:

$$\mathcal{N}_{\alpha\beta}{}^\gamma = \left(\frac{|C_\alpha||C_\beta|}{|C_\gamma|}\right)^{\frac{1}{2}} \sum_j S_j(\alpha) \frac{S_j(\beta)}{S_j(e)} \bar{S}_j(\gamma)$$

k) Conclude that, in general, the numbers

$$\mathcal{M}_{\alpha\beta}{}^\gamma = \sum_j S_j(\alpha) \frac{S_j(\beta)}{S_j(e)} \bar{S}_j(\gamma)$$

and the numbers $\mathcal{N}_{\alpha\beta}{}^\gamma$ cannot be *simultaneously* integers. Exemplify this with S_3. Therefore in general the tensor product algebra for irreducible representations of a group G is a bad candidate for the fusion algebra of a conformal theory. Show however that when G is Abelian, then

$$\mathcal{N}_{\alpha\beta}{}^\gamma = \mathcal{M}_{\alpha\beta}{}^\gamma$$

In that case, representations and classes are isomorphic, and the matrix S is symmetric. Therefore only Abelian groups provide good candidates for conformal fusion rules. However, many conformal fusion rules have no Abelian group interpretation. In this respect, one can think of the structure of the fusion rules in conformal theory as generalizing that of an Abelian group.

10.19 *Conformal blocks and ϕ^3 diagrams*

a) We study the *crossing* transformation acting on a given ϕ^3 diagram of genus h. This is a local transformation, which acts on pairs of neighboring trivalent vertices, linked by a propagator, as

Argue heuristically that this connects all the possible genus h ϕ^3 diagrams.

b) Prove that the number of conformal blocks in a basis for some correlation function on a surface of genus h is independent on the (genus h)-ϕ^3 diagram encoding the basis elements.

10.20 *Modular invariance of the $c = 1$ theory at the self-dual radius $R = \sqrt{2}$*

a) Express the extended characters C_0 and C_1 of Eq. (10.233) in terms of Jacobi theta functions at the value τ (instead of 2τ) by using the doubling formulae (10.272).

b) Using the fact that

$$\theta_2(\tau)^4 = \theta_3(\tau)^4 - \theta_4(\tau)^4 \qquad (10.278)$$

deduce the modular invariance of the partition function, and check the modular transformations of the characters C_0 and C_1.

10.21 *Examples of RCFTs: the boson on a circle of rational square radius*

a) Prove that the $c = 1$ bosonic theory on a circle of radius $R = \sqrt{2n}$ is rational.
Hint: The extended characters are the functions K_λ, $\lambda = 0, 1, \dots, N - 1$ defined in Eq. (10.109), with $N = 2n$, and the partition function reads

$$Z(\sqrt{2n}) = \sum_{\lambda=0}^{N-1} |K_\lambda(\tau)|^2$$

b) Check the following sum rule for the dimensions of the extended operators (cf. Ex. 10.17)

$$(1 + N/2)c/4 = 6\sum_\lambda h_\lambda \bmod 1 .$$

c) Compute the conjugation matrix $\mathcal{C} = \mathcal{S}^2$.
Result: $\mathcal{C}_{\lambda,\mu} = \delta_{\mu,N-\lambda}$ ($\lambda = 0$ is self-conjugate).

d) Compute the extended fusion rules of this RCFT.
Result: $\mathcal{N}_{\lambda\mu}{}^\nu = \delta_{\nu,\lambda+\mu \bmod N}$, for any $0 \le \lambda, \mu, \nu \le N - 1$.

e) We now consider the bosonic $c = 1$ theory on a circle of radius $R = \sqrt{2p'/p}$. Let $N = 2pp'$ and K_λ as in (10.109). Show that the partition function on the torus reads

$$Z(\sqrt{\frac{2p'}{p}}) = \sum_{\lambda=0}^{N-1} K_\lambda(\tau) K_{\omega_0\lambda}(\bar{\tau})$$

with ω_0 defined in Eqs. (10.118)–(10.119). Conclude that for p' and $p \ne 1$, the corresponding RCFT is nondiagonal.

10.22 *Bosonic representation of minimal theories on a torus*
Using the expression (10.122) for the minimal characters, prove that the partition function
of the $(A_{p'-1}, A_{p-1})$ minimal theory can be reexpressed as the half difference of two $c = 1$
bosonic theories on circles of respective radii $\sqrt{2pp'}$ and $\sqrt{2p'/p}$.
Hint: Use the results of the previous exercise. This representation may be generalized to all
the modular invariants of the ADE classification. The results read

$$
\begin{aligned}
Z_{A_{p'-1},A_{p-1}} &= \frac{1}{2}(Z(\sqrt{2pp'}) - Z(\sqrt{2p'/p})) \\
Z_{A_{p-1},D_{p/2+1}} &= \frac{1}{2}(Z(\sqrt{8p'/p}) - Z(\sqrt{2p'/p}) - Z(\sqrt{pp'/2}) + Z(\sqrt{2pp'})) \\
Z_{A_{p-1},E_6} &= \frac{1}{2}(Z(2\sqrt{6p}) - Z(\sqrt{6p}) - Z(2\sqrt{2p/3}) \\
&\quad + Z(\sqrt{2p/3}) + Z(2\sqrt{6p/9}) - Z(\sqrt{6p/9})) \\
Z_{A_{p-1},E_7} &= \frac{1}{2}(Z(6\sqrt{p}) - Z(3\sqrt{p}) - Z(2\sqrt{p}) \\
&\quad + Z(\sqrt{p}) - Z(2\sqrt{p}/3) + Z(\sqrt{p}/3)) \\
Z_{A_{p-1},E_8} &= \frac{1}{2}(Z(2\sqrt{15p}) - Z(\sqrt{15p}) - Z(2\sqrt{5p/3}) - Z(2\sqrt{3p/5}) \\
&\quad + Z(\sqrt{5p/3}) + Z(\sqrt{3p/5}) + Z(2\sqrt{p/15}) - Z(\sqrt{p/15}))
\end{aligned}
$$

10.23 *Example of bosonic orbifold RCFT: orbifold at radius $R = 1$*
a) Prove the identity

$$ Z_I^2 = Z_{\text{orb}}(1) $$

using the expression (10.51) for the partition function Z_I of the free fermion (Ising model)
on a torus, and the identity (10.260) on theta functions.

b) Deduce that the orbifold bosonic theory on a circle of radius $R = 1$ is a RCFT. Compute
the corresponding extended characters and fusion rules.

10.24 *The $O(n)$ model on a torus, for $-2 \le n \le 2$*
It is possible to show that the partition function of the $O(n)$ model on a rectangle of size
$L \times T$ with periodic boundary conditions in both (time and space) directions has a well-
defined thermodynamic limit when $L, T \to \infty$, while the (purely imaginary) parameter
$\tau = iT/L$ is kept fixed. The result reads

$$ Z_n(\tau) = \frac{1}{|\eta(\tau)|^2} \sum_{m,m' \in \mathbb{Z}} \cos\left(\pi e_0 \gcd(|m|, |m'|)\right) Z_{m,m'}(R; \tau) $$

where $\gcd(|m|, |m'|)$ stands for the greatest common divisor of the two integers $|m|$ and
$|m'|$,

$$ Z_{m,m'}(R; \tau) = \frac{R}{\sqrt{2}|\eta|^2 \text{Im } \tau} \exp -\frac{\pi R^2 |m\tau - m'|^2}{2\text{Im } \tau} $$

and

$$ \cos(\pi e_0) = n \qquad -2\cos(\pi R^2/2) = n $$

with $0 \le e_0 \le 1$ and $\sqrt{2} \le R \le 2$.

a) Check the modular invariance of this partition function. Hint: Show that $\gcd(|m|, |m'|)$ is invariant under the action of the modular group.

b) Compute the central charge of the system. Compare the result with Eq. (7.108). Hint: Use the Poisson resummation formula (10.264) to extract the small-$q = \exp(2i\pi\tau)$ behavior of the partition function.

c) Compute the partition function for $n = 2, 1, 0$.

d) When $R = \sqrt{2p'/p}$, with $p' > p$ two coprime integers, rewrite the partition function of the $O(n)$ model as

$$ Z_n(\tau) = \frac{1}{2}(Z(\sqrt{2pp'}) - Z(\sqrt{2p'/p})) $$

Compare this with the results of Ex. 10.22. Conclude that for $n = -2\cos(\pi p'/p)$, the continuum limit of the $O(n)$ model is described by the (diagonal) minimal model $(A_{p'-1}, A_{p-1})$.

e) Compute the one-polymer configuration sum on the torus, namely

$$ \left.\frac{\partial Z_n(\tau)}{\partial n}\right|_{n=0} $$

Notes

The concept of modular invariance in conformal theory was first stressed as a fundamental requirement by Cardy [73], who studied the minimal models in a finite geometry (strip, torus), and derived constraints on the possible operator content. The hunt for modular invariant partition functions started with Refs. [207, 172], and reached a climax with the conjecture of Cappelli, Itzykson, and Zuber on the ADE classification of modular invariants for minimal models [63], subsequently proved in Refs. [64] and [231]. A parallel construction of nondiagonal statistical RSOS models, indexed by the Dynkin ADE diagrams, was performed by Pasquier [289]. It was argued that the continuum limit of the latter are described by the ADE minimal models.

   The Coulomb-gas models have been studied (Ref. [224]) in the description of the critical lines of the Ashkin-Teller statistical model (made of two interacting Ising models), and of various integrable lattice models, such as Baxter's eight-vertex model [283]. Modular invariant partition functions for these $c = 1$ theories were built in Refs. [95, 96, 314, 362]. The list of $c = 1$ modular invariant partition functions was further completed in Ref. [288], by the construction of RSOS lattice models based on extended Dynkin diagrams, and in Ref. [176] by using the orbifold procedure to build additional $c = 1$ theories that do not lie on the critical lines of the Ashkin-Teller model, and correspond to the continuum limit of the RSOS models based on the exceptional extended Dynkin diagrams. An extension of these results to arbitrary Riemann surfaces was also performed in Ref. [103].

   Making the connection between the modular properties of the minimal characters and their fusion rules, E. Verlinde proposed the celebrated *Verlinde formula* [340] (Eq. 10.171), which expresses the fusion numbers as a function of the modular S matrix of the theory. A general proof was derived in the extended context of rational conformal theory in Refs. [102, 272] and in Ref. [273] (see Sect. 10.8.3). This formula attracted much attention in the mathematical literature.

The natural notion of extended symmetry first arose in the context of modular invariance, in relation to block-diagonal modular invariants, and was further developed by explicitly constructing enhanced symmetry algebras (called W-algebras) extending the Virasoro algebra, and governing the corresponding theories. The axiomatic definition of rational conformal theories is due to Moore and Seiberg [272].

Many identities on theta functions can be found in Refs. [27, 358]. The ADE classification of integer matrices with largest eigenvalue < 2 is due to Cartan, in the context of Lie algebras (see Ref. [185] for a graph-theoretic proof). Exs. 10.16 and 10.17 are based on Ref. [338]. Exs. 10.21, 10.22 and 10.23 are based on Refs. [95, 96].

CHAPTER 11

Finite-Size Scaling and Boundaries

Until now, with the notable exception of Chap. 10, we have concentrated our attention on conformal field theories defined on the infinite plane, which is equivalent to a sphere. In this chapter we shall study the consequences of conformal invariance on models defined on portions of the plane delimited by one or more boundaries, with various types of boundary conditions. We shall proceed mainly by applying local conformal mappings from the infinite plane or the upper half-plane to these restricted geometries. This will prove to be a particularly useful application of *local* conformal invariance, as these mappings do not belong to the global conformal group.

The relevance of studying models over a finite-size region is manifold. For instance, a lot of information on two-dimensional statistical models or one-dimensional quantum models is derived from computer simulations, which are necessarily limited to systems of finite size L. The properties of the model in the thermodynamic limit ($L \to \infty$) are inferred from the finite-size properties. Conformal invariance can in many cases provide the L-dependence of these properties, thus allowing a more precise inference of the thermodynamic limit. This comparison with numerical work may also provide an otherwise unknown correspondence between a model at criticality and a conformal field theory. In quantum systems (e.g., spin chains), the finite size may be in the (imaginary) time direction, which corresponds to finite temperature (cf. Sect. 3.1.2). Conformal invariance is then useful in studying the finite-temperature behavior of a 1D quantum model, which is critical at $T = 0$.

Local conformal transformations may also provide the behavior of a critical system near a boundary, when free or fixed boundary conditions are used. The problem is then to find out the effect of the boundary on the decay of correlation functions. The prototype of a manifold with a boundary is the upper half-plane. From there, other geometries with boundaries may be obtained via conformal transformations. Often, such as in the study of percolation across a rectangle, the boundary is part of the definition of the model itself, and not a limitation brought by the finite means of the investigator. The study of conformal field theories defined on an infinite strip has, of course, direct implications in open string theory, which will not be discussed in this work.

This chapter is organized as follows. In Sect. 11.1, we come back to the issue of conformal theories defined on an infinite cylinder. Our aim is to illustrate the effect of the boundary on the two-point function of conformal fields, in particular how it introduces a correlation length along the direction of the cylinder. Section 11.2 discusses the general issue of a conformal theory defined on the upper half-plane— or on a manifold with boundary that can be obtained from the upper half-plane via conformal mapping—with conformally invariant boundary conditions. We describe such a theory with the "method of images", by which the holomorphic sector of a theory defined on the whole complex plane replaces the coupled holomorphic and antiholomorphic sectors of the theory defined on the upper half-plane, and where each field insertion in the physical region is compensated by the insertion of an "image" in the unphysical region. We apply this method to the Ising model and to the behavior of the spin field two-point function as one approaches the boundary. In Sect. 11.3 we introduce the notion of a boundary operator which, when inserted at a point on the boundary, changes the boundary condition from this point onward. These operators must belong to the same set as the bulk operators, a restriction imposed by the condition of conformal invariance on the boundary conditions. The significance of boundary operators is established by a close analogy with the Verlinde formula. Finally, in Sect. 11.4, we apply these ideas to the study of critical percolation and obtain the aspect-ratio dependence of the crossing probability, an analytic result fully confirmed by numerical simulations and giving spectacular support to the hypothesis of local conformal invariance in two-dimensional critical systems.

§11.1. Conformal Invariance on a Cylinder

Before embarking on a study of critical systems with boundaries, we consider a field theory or statistical model at criticality defined on an infinite cylinder of circumference L. This geometry is useful in providing a physical motivation for the procedure of radial quantization (cf. Sect. 6.1.1) and for the quantization of free fields (cf. Sects. 6.3 and 6.4). Although the infinite cylinder has no boundary, it is the source of finite-size effects analogous to those observed on other manifolds with boundaries (e.g., the infinite strip). These effects have important implications in the practical study of critical quantum systems at finite temperature or of finite length.

The mapping from the infinite plane (with holomorphic coordinate z) to the cylinder (with coordinate w) is

$$w = \frac{L}{2\pi} \ln z \qquad \text{or} \qquad z = e^{2\pi w/L} \tag{11.1}$$

We have already performed a finite-size scaling analysis in this geometry when we obtained the expression (5.143) in Sect. 5.4.2 for the free energy per unit length F_L as a function of L:

$$F_L = f_0 L - \frac{\pi c}{6L} \tag{11.2}$$

Here f_0 is the free energy per unit area in the $L \to \infty$ limit.

Another quantity of interest is the two-point function of a primary field ϕ of conformal dimension h. Its form on the plane (Eq. (5.25)) was fixed by invariance under the global conformal transformations (5.12). Those were defined as one-to-one mappings from the infinite plane onto itself. In order to write the two-point function on the cylinder we need to use the covariance relation (5.24) for primary fields, with the mapping (11.1). Here we write the holomorphic part only:

$$
\begin{aligned}
\langle \phi(w_1)\phi(w_2)\rangle &= \left(\frac{dw}{dz}\right)^{-h}_{w=w_1} \left(\frac{dw}{dz}\right)^{-h}_{w=w_2} \langle \phi(z_1)\phi(z_2)\rangle \\
&= \left(\frac{2\pi}{L}\right)^{2h} \frac{e^{2\pi h(w_1+w_2)/L}}{(z_1-z_2)^{2h}} \\
&= \left(\frac{2\pi}{L}\right)^{2h} \left(2\sinh[\pi(w_1-w_2)/L]\right)^{-2h}
\end{aligned} \tag{11.3}
$$

The full correlator is the product of the above with its antiholomorphic counterpart:

$$\langle \phi(w_1,\bar{w}_1)\phi(w_2,\bar{w}_2)\rangle =$$
$$\left(\frac{2\pi}{L}\right)^{2h+2\bar{h}} \left(2\sinh[\pi(w_1-w_2)/L]\right)^{-2h} \left(2\sinh[\pi(\bar{w}_1-\bar{w}_2)/L]\right)^{-2\bar{h}} \tag{11.4}$$

For simplicity, we assume that the field ϕ has no spin: $h = \bar{h} = \Delta/2$. Then the above reduces to

$$\left(\frac{2\pi}{L}\right)^{2\Delta} \left[4\sinh\frac{\pi w}{L}\sinh\frac{\pi\bar{w}}{L}\right]^{-\Delta} \tag{11.5}$$

where $w \equiv w_1 - w_2$ and $\bar{w} \equiv \bar{w}_1 - \bar{w}_2$ are the relative coordinates. We express this result in terms of real coordinates u and v, respectively, along and across the cylinder: $w = u + iv$ and $\bar{w} = u - iv$. After using standard identities for hyperbolic functions, we end up with

$$\langle \phi(u_1,v_1)\phi(u_2,v_2)\rangle = \left(\frac{2\pi}{L}\right)^{2\Delta} \left[2\cosh\frac{2\pi u}{L} - 2\cos\frac{2\pi v}{L}\right]^{-\Delta} \tag{11.6}$$

As expected, the effect of the finite size L disappears if the distance $|u+iv|$ is much smaller than L. Then $\sinh(\pi w/L) \sim \pi w/L$ and we recover the infinite plane result (5.25). On the other hand, when $u \gg L$, then $2\cosh(2\pi u/L) \sim e^{2\pi u/L}$ and the correlator becomes

$$\langle \phi(u_1,v_1)\phi(u_2,v_2)\rangle \sim \left(\frac{2\pi}{L}\right)^{2\Delta} \exp-\frac{2\pi u\Delta}{L} \qquad (u \gg L) \tag{11.7}$$

Thus, correlations along the cylinder decay exponentially, with a correlation length $\xi = L/2\pi\Delta$, proportional to the size L. The appearance of a correlation length in a critical system is here entirely due to the existence of a macroscopic scale L.

When dealing with a quantum chain, the infinite cylinder geometry may correspond either to a finite chain at zero temperature with periodic boundary conditions, or to an infinite chain at a finite temperature $T = 1/L$ (cf. Sect. 3.1.2). In the first case, the correlation length $\xi = L/2\pi\Delta$ in the time direction is the signature of an energy gap between the ground state and the first excited state, a gap induced by the system's finite size, as routinely observed in numerical simulations. To make this point more explicit, we consider the two-point function $\langle\phi(x,0)\phi(x,\tau)\rangle$, where x and τ are, respectively, the space and imaginary time coordinates. In the operator formalism, this two-point function may be expressed as a ground-state expectation value:

$$
\begin{aligned}
\langle\phi(x,0)\phi(x,\tau)\rangle &= \langle 0|\phi(x,0)e^{-H\tau}\phi(x,0)e^{H\tau}|0\rangle \\
&= \sum_n \langle 0|\phi(x,0)e^{-H\tau}|n\rangle\langle n|\phi(x,0)e^{H\tau}|0\rangle \\
&= \sum_n e^{-(E_n-E_0)\tau}|\langle 0|\phi(x,0)|n\rangle|^2
\end{aligned}
\tag{11.8}
$$

Here H is the Hamiltonian, the states $|n\rangle$ are the energy eigenstates (in increasing order of energy), and E_n is the eigenvalue of H associated with $|n\rangle$. In the first line we have performed a time translation to make the two fields simultaneous. In the second line we have inserted a completeness relation for the basis of eigenstates of H. In the absence of spontaneous symmetry breaking—which is generally true at the critical point—the expectation value $\langle 0|\phi(x,0)|0\rangle$ vanishes. Otherwise, the above equation may be rewritten as follows for the connected two-point function:

$$
\langle\phi(x,0)\phi(x,\tau)\rangle - \langle\phi(x,0)\rangle\langle\phi(x,\tau)\rangle = \sum_{n>0} e^{-(E_n-E_0)\tau}|\langle 0|\phi(x,0)|n\rangle|^2 \tag{11.9}
$$

The term that dominates the above sum when τ is large is associated with the first excited state $|1\rangle$, with an energy $\delta E = E_1 - E_0$ above the ground state:

$$
\langle\phi(x,0)\phi(x,\tau)\rangle_c \propto e^{-\delta E\tau} \qquad (\tau \to \infty) \tag{11.10}
$$

Comparing Eq. (11.10) with Eq. (11.7), we conclude that

$$
\delta E = \frac{2\pi\Delta}{L} \tag{11.11}
$$

Of course, this relation holds in "natural units", in which Planck's constant and the characteristic velocity v of the system—the equivalent of the speed of light in a Lorentz invariant theory—are set to unity. In order to restore the correct dimensions, one must multiply the r.h.s. by $\hbar v$.

If the finite extent of the system is in the imaginary time direction, the size L is equal to the inverse temperature $1/T$ and the correlation length ξ becomes

$$
\xi = \frac{1}{2\pi T\Delta} \tag{11.12}
$$

This length has the physical interpretation of a coherence length, giving the spatial extent over which quantum coherence is not destroyed by thermal fluctuations. It is the thermal de Broglie wavelength characterizing the system at a temperature T. The interpretation of Eq. (11.2) for the free energy is then different: In a one-dimensional quantum system, this formula gives the vacuum functional W per unit length, which is related to the free energy f by $W = fL$, L being the extent of the time direction. Thus, the free energy per unit length is

$$f = f_0 - \frac{1}{6}\pi c T^2 \qquad \text{(quantum chain)} \qquad (11.13)$$

From this we infer the specific heat C per unit length:

$$C \equiv -T\frac{\partial^2 f}{\partial T^2} = \frac{1}{3}\pi c T \qquad (11.14)$$

Of course, we must divide this result by $\hbar v$ to restore the correct units.

§11.2. Surface Critical Behavior

In this section we apply conformal invariance to a two-dimensional system with a boundary, of which the prototype is the upper half-plane. The goal is to determine the behavior of correlation functions near the boundary when the bulk is critical: we call this *surface critical behavior*—even though the surface (or boundary) is here a one-dimensional object. A given statistical or quantum model is characterized by a set of boundary conditions at the surface. If the model is to have some form of conformal symmetry at criticality, conformal transformations must map the boundary onto itself and preserve the boundary conditions. This restricts the overall symmetry of the model: holomorphic and antiholomorphic fields no longer decouple and only half of the conformal generators remain.

11.2.1. Conformal Field Theory on the Upper Half-Plane

The simplest two-dimensional manifold with a boundary on which to apply the formalism of conformal field theory is the upper half-plane. A model defined on the upper half-plane may have conformal invariance only if the conformal transformations keep the boundary (the real axis) and the boundary conditions invariant. Among the conformal transformations (5.12), those that map the real axis onto itself are obtained by keeping the parameters a, b, and c real. Thus, the global conformal group is half as large as it is for the entire plane. Likewise, infinitesimal local conformal transformations of the form $z \rightarrow z + \epsilon(z)$ will map the real axis onto itself if and only if $\epsilon(\bar{z}) = \bar{\epsilon}(z)$ (i.e., ϵ is real on the real axis). This is a strong constraint that eliminates half of the conformal generators: the holomorphic and antiholomorphic sectors of the theory are no longer independent.

As for the boundary conditions on a scaling field ϕ, invariance under conformal transformations requires them to be homogeneous, for instance as follows:

$$\phi\big|_{\mathbb{R}} = 0 \quad , \quad \phi\big|_{\mathbb{R}} = \infty \quad , \quad \frac{\partial\phi}{dn}\big|_{\mathbb{R}} = 0 \qquad (11.15)$$

Indeed, the transformation law (5.22) for primary fields is multiplicative and therefore leaves the above boundary conditions unchanged. The Dirichlet boundary condition $\phi\big|_{\mathbb{R}} = 0$ is also referred to as the "free" boundary condition, since the vanishing of the order parameter at the boundary generally follows from the absence of constraints on the microscopic degrees of freedom. A critical system obeying such a boundary condition is said to undergo an *ordinary* transition. On the other hand, it may happen in some systems that the surface orders before the bulk, for instance because of stronger interactions at the boundary. In that case the order parameter is infinite[1] at the boundary ($\phi\big|_{\mathbb{R}} = \infty$) and the system is said to undergo an *extraordinary* transition.

The conformal Ward identity (5.46) embodies the effects of local conformal invariance on correlation functions:

$$\delta_{\epsilon,\bar{\epsilon}}\langle X\rangle = -\frac{1}{2\pi i}\oint_C dz\,\epsilon(z)\langle T(z)X\rangle + \frac{1}{2\pi i}\oint_C d\bar{z}\,\bar{\epsilon}(\bar{z})\langle \bar{T}(\bar{z})X\rangle \qquad (11.16)$$

where, as usual, X stands for a product of local fields. Without loss of generality, we may assume that it is a product of primary fields:

$$X = \phi_{h_1,\bar{h}_1}(z_1,\bar{z}_1)\cdots\phi_{h_n,\bar{h}_n}(z_n,\bar{z}_n) \qquad (11.17)$$

(the indices on each field are the holomorphic and antiholomorphic conformal dimensions). On the infinite plane, the infinitesimal coordinate variations $\epsilon(z)$ and $\bar{\epsilon}(\bar{z})$ are independent, and therefore this identity is in fact a pair of identities giving the independent variations $\delta_\epsilon\langle X\rangle$ and $\delta_{\bar{\epsilon}}\langle X\rangle$ of a correlation function under an infinitesimal conformal transformation. On the upper half-plane the conformal Ward identity is still applicable, except that the integration contour C must lie entirely in the upper half-plane and the coordinate variation $\bar{\epsilon}$ is the complex conjugate of ϵ: we no longer have a decoupling into holomorphic and antiholomorphic identities.

In order to apply the machinery developed in the previous chapters to a theory defined on the upper half-plane, we shall regard the dependence of the correlators on antiholomorphic coordinates $\bar{z}_i$ on the upper half-plane as a dependence on holomorphic coordinates $z_i^* = \bar{z}_i$ on the lower half-plane. We thus introduce a mirror image of the system on the lower half-plane, via a parity transformation. In going from the upper to the lower half-plane, vector and tensor fields change their holomorphic indices into antiholomorphic indices and vice versa. Thus $T(z^*) = \bar{T}(z), \bar{T}(z^*) = T(z)$, and so on. Of course, such an extension is compatible with the boundary conditions only if $\bar{T} = T$ on the real axis (and likewise for all vector or tensor fields). The boundary condition $\bar{T} = T$ becomes $T_{xy} = 0$ when expressed in terms of Cartesian coordinates, in which its meaning becomes clear: no energy or

---

[1] In fact, the divergence of the order parameter near the boundary is cut off at a microscopic distance from the surface, of the order of the lattice spacing.

momentum flows across the real axis. This general condition is obviously satisfied in a physical system with a boundary, and in particular it is compatible with the homogeneous boundary conditions cited above.

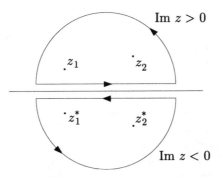

Figure 11.1. Contour used in the conformal Ward identity for two fields at points z_1 and z_2 on the upper half-plane, with the mirror contour and points on the lower half-plane.

It is then possible to rewrite the conformal identity (11.16) on the upper half-plane as a purely holomorphic expression on the infinite plane. The second term of Eq. (11.16) becomes an integration along a mirror image of the contour C, as indicated in Fig. 11.1. That figure shows an example contour to be used with this identity, with singularities of the integrand at the locations z_1 and z_2 of two local fields. The "mirror images" of this contour and points on the lower-half plane are also shown. The direction of the mirror contour is reversed, because of the relative sign of the two integrals appearing in the conformal Ward identity (11.16). Since, by hypothesis, $\bar{T} = T$ on the real axis, the two disjoint contours may be fused into one, their horizontal parts canceling each other, and we end up with a single contour circling around twice the number of points. Thus the original conformal Ward identity now takes the simpler form

$$\delta_\epsilon \langle X \rangle = -\frac{1}{2\pi i} \oint_C dz \, \epsilon(z) \langle T(z) X' \rangle \tag{11.18}$$

where X' stands for

$$X' \equiv \phi_{h_1}(z_1) \bar{\phi}_{\bar{h}_1}(z_1^*) \cdots \phi_{h_n}(z_n) \bar{\phi}_{\bar{h}_n}(z_n^*) \tag{11.19}$$

Here $\phi_h(z)$ stands for the holomorphic part of the field $\phi_{h,\bar{h}}(z,\bar{z})$ and $\bar{\phi}_{\bar{h}}(z^*)$ stands for its antiholomorphic part, after a parity transformation on the lower half-plane making it a holomorphic field with *holomorphic* dimension $\bar{h}$. For instance, the parity transformation has the following effect on the free boson and the free fermion:

$$\bar{\partial}\varphi(\bar{z}) \longrightarrow \pm\partial\varphi(z^*)$$
$$\bar{\psi}(\bar{z}) \longrightarrow \pm\psi(z^*) \tag{11.20}$$

In the free fermion case, the parity transformation interchanges the two components of the spinor $\Psi = (\psi, \bar{\psi})$. There is a certain freedom in the definition of the parity transformation, which translates into different boundary conditions on the real axis (cf. Ex. 11.6 and 11.8).

In other words, the correlator $\langle X \rangle$ on the upper half-plane, as a function of the $2n$ variables $z_1, \bar{z}_1, \cdots, z_n, \bar{z}_n$, satisfies the same differential equation (coming from local conformal invariance) as the correlator $\langle X' \rangle$ on the entire plane, regarded as a function of the $2n$ holomorphic variables $z_1, \cdots, z_{2n}$ where $z_{n+i} = z_i^*$. We have effectively replaced the antiholomorphic degrees of freedom on the upper half-plane by holomorphic degrees of freedom on the lower half-plane.[2] An n-point function on the upper half-plane—the object of interest—is replaced here by the holomorphic part of a $2n$-point function on the infinite plane.[3] The interaction of the local fields with the boundary (in the form of the boundary conditions) is simulated by the interaction between mirror images of the same holomorphic field. Considering Fig. 11.1 for the two-point function, we expect to feel the effect of the boundary when the separation $|z_1 - z_2|$ is larger than the distance from the real axis, while the bulk result is recovered in the other limit. Notice that, even for minimal models, the four-point function and higher correlators are not uniquely determined by conformal invariance and singular vectors: we need to specify some boundary or asymptotic conditions. Here, it is the role of the particular boundary condition on the real axis to determine which linear combinations of the conformal blocks of the $2n$-point function is chosen.

All this is reminiscent of the method of images used in electrostatics, in which fictitious electric charges are placed in an unphysical region of space in order to produce, in the physical region, a contribution to the electric potential that fulfills the boundary conditions, without affecting the differential equation obeyed by the potential in the presence of real charges (Poisson's equation). Accordingly, we may call the procedure described above the "method of images."

The simplest application of the method of images is the determination of the order parameter profile near the boundary. By this we mean the dependence of the expectation value $\langle \phi(z) \rangle$ on the distance from the boundary. It is assumed here that the local fields fluctuate about zero, that is, $\langle \phi(z) \rangle = 0$ in the bulk (no symmetry breaking at criticality). However, in an "extraordinary transition", the boundary condition is that $\phi \to \infty$ on the real axis. According to the above analysis, the one-point function $\langle \phi(z, \bar{z}) \rangle$ on the upper half-plane is given by the two-point function $\langle \phi(z) \phi(\bar{z}) \rangle$ on the infinite plane. The latter is known to be equal to $(z - \bar{z})^{-2h}$. Thus,

---

[2] Conformal invariance does not fix the overall normalization of correlation functions. The doubling of fields in going from X to X' certainly affects this normalization: as an operation, the renormalization $\phi \to$ const. $\times \phi$ does not commute with it. Thus, the method explained here may fix the coordinate dependence of correlators on the upper half-plane from those on the entire plane, but not their overall normalization.

[3] If a primary field is purely holomorphic ($\bar{h} = 0$), its antiholomorphic part is the identity and has no effect on the correlator. Thus, depending on the number of purely holomorphic (or antiholomorphic) fields, the effective number of points on the entire plane varies between n and $2n$.

if y is the distance from the real axis and if $h = \bar{h}$, the order parameter profile is

$$\langle \phi(y) \rangle \sim \frac{1}{y^\Delta} \qquad (11.21)$$

where $\Delta = h + \bar{h}$ is the scaling dimension of the field ϕ.

11.2.2. The Ising Model on the Upper Half-Plane

An interesting, yet simple application of the method of images is the calculation of the spin-spin correlation function of the Ising model on the upper half-plane (UHP). This function may be written as

$$G_s(y_1, y_2, \rho) \equiv \langle \sigma(z_1, \bar{z}_1)\sigma(z_2, \bar{z}_2) \rangle_{\text{UHP}}$$
$$= \langle \sigma(z_1)\sigma(z_2)\sigma(z_1^*)\sigma(z_2^*) \rangle \qquad (11.22)$$

Here y_1 and y_2 are the distances of the two points from the real axis and $\rho \equiv x_2 - x_1$ is the horizontal distance between the two points (cf. Fig. 11.2). The r.h.s. of the second line is the holomorphic part of the four-spin correlator on the infinite plane.

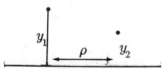

Figure 11.2. Real coordinates y_1, y_2, and ρ for the two-point function near the boundary.

The Ising model is one of the minimal models discussed at length in Chaps. 7 and 8. Its correlation functions satisfy special linear differential equations, which allow us, in principle, to write them down explicitly.[4] For the sake of computing the correlator (11.22), it is preferable to apply the differential equation rather than to borrow directly the result (12.61), because different boundary conditions are needed (cf. also Ex. 8.12). The differential equation obeyed by the four-spin correlation function is particularly simple: it is a special case of Eq. (7.47), in which $X = \sigma(z_1)\sigma(z_2)\sigma(z_3)$ and $\phi = \sigma$:

$$\left\{ \sum_{i=1}^{3} \left[\frac{1}{z - z_i} \frac{\partial}{\partial z_i} + \frac{1/16}{(z - z_i)^2} \right] - \frac{4}{3}\frac{\partial^2}{\partial z^2} \right\} \langle \sigma(z_1)\sigma(z_2)\sigma(z_3)\sigma(z) \rangle = 0 \quad (11.23)$$

Indeed, the primary field σ has conformal dimension $h_{1,2} = \frac{1}{16}$ and is precisely the null field studied in Sect. 7.3.1.

---

[4] This is not the way correlation functions are found when studying the Ising model in detail in Chap. 12, where other methods—in particular bosonization—are used.

We know from Chap. 5 (Eq. (5.31)) that the holomorphic part of the four-point function may be expressed as follows:

$$\langle \sigma(z_1)\sigma(z_2)\sigma(z_3)\sigma(z_4)\rangle = \left(\frac{z_{13}z_{24}}{z_{12}z_{23}z_{14}z_{34}}\right)^{\frac{1}{8}} F(x) \qquad (11.24)$$

where F is some function of the anharmonic ratio $x \equiv z_{12}z_{34}/z_{13}z_{24}$ and where $z_{ij} \equiv z_i - z_j$ (here $z_4 \equiv z$). If we substitute this form into Eq. (11.23), we end up with an ordinary differential equation in the variable x:

$$\left[x(1-x)\frac{d^2}{dx^2} + (\frac{1}{2} - x)\frac{d}{dx} + \frac{1}{16}\right] F(x) = 0 \qquad (11.25)$$

This is a special case of the hypergeometric equation, which may be solved by a simple change of variables: $x = \sin^2 \theta$; this substitution yields

$$\left[\frac{d^2}{d\theta^2} + \frac{1}{4}\right] F(\theta) = 0 \qquad (11.26)$$

The two linearly independent solutions are $\cos \frac{1}{2}\theta$ and $\sin \frac{1}{2}\theta$ or, equivalently, $\sqrt{1 \pm \cos \theta} = \sqrt{1 \pm \sqrt{1-x}}$. Appropriate linear combinations of these two solutions[5] must be taken in order to satisfy the boundary conditions. Alternately, if one borrows directly the infinite-plane correlation function obtained by different means (e.g., bosonization), the two solutions correspond to two different definitions of the parity transformation on the Ising model (cf. Ex. 11.6).

These boundary conditions are fixed by the asymptotic behavior of the spin-spin correlator (11.22) near the real axis. In a so-called "ordinary transition", the surface is disordered, which means that $G_s(y_1, y_2, \rho) \to 0$ as $\rho \to \infty$ for fixed values of y_1 and y_2, which corresponds to $x \to -\infty$. On the other hand, in an "extraordinary transition", the surface orders before the bulk, which means that, in the same limit,

$$G_s(y_1, y_2, \rho) \sim \langle \sigma(z_1, \bar{z}_1)\rangle_{\text{UHP}} \langle \sigma(z_2, \bar{z}_2)\rangle_{\text{UHP}}$$
$$\propto \frac{1}{(y_1 y_2)^{\frac{1}{8}}} \qquad (11.27)$$

It follows that the correct linear combinations are

$$F(x) = \sqrt{\sqrt{1-x}+1} \mp \sqrt{\sqrt{1-x}-1} \qquad (11.28)$$

where the upper (resp. lower) sign corresponds to the ordinary (resp. extraordinary) transition. If we express these four-point functions in terms of y_1, y_2, and ρ, we find

$$G_s(y_1, y_2, \rho) \propto \frac{1}{(y_1 y_2)^{\frac{1}{8}}} \sqrt{\tau^{1/4} \mp \tau^{-1/4}} \qquad (11.29)$$

---

[5] These are the two conformal blocks of the Ising spin four-point function (see Ex. 12.7).

where

$$\tau \equiv \frac{\rho^2 + (y_1 + y_2)^2}{\rho^2 + (y_1 - y_2)^2} \tag{11.30}$$

The asymptotic behavior of the correlator as $\rho \to \infty$ (y_1 and y_2 fixed) is characterized by an exponent $\eta_\parallel$ defined as

$$G_s(y_1, y_2, \rho) \sim \frac{1}{\rho^{\eta_\parallel}} \qquad (\rho \gg y_1, y_2) \tag{11.31}$$

It follows from Eq. (11.29) that

$$\eta_\parallel = \begin{cases} 1 & \text{(ordinary)} \\ 4 & \text{(extraordinary)} \end{cases} \tag{11.32}$$

11.2.3. The Infinite Strip

We now consider the infinite strip of width L. It is understood that the strip does not support periodic or antiperiodic boundary conditions across its width, otherwise it would effectively be a cylinder. This manifold may be obtained from the upper half-plane by the following conformal map:

$$w = \frac{L}{\pi} \ln z \tag{11.33}$$

where w and z are the holomorphic coordinates on the strip and the upper half-plane, respectively. Notice the difference from the map (11.1), going from the infinite plane to the cylinder. Here the positive real axis is mapped onto the lower edge of the strip and the negative real axis onto the upper edge. Therefore the two edges must support the same boundary conditions (e.g., both fixed to the same value, or both free) if the results obtained on the upper half-plane are to be imported here.

We first determine the order parameter profile near the boundary, in the case of an extraordinary transition. This is obtained by transforming the one-point function

$$\langle \phi(z, \bar{z}) \rangle_{\text{UHP}} = \langle \phi(z) \phi(\bar{z}) \rangle$$

$$= \frac{1}{(z - \bar{z})^{2h}} \tag{11.34}$$

onto the strip, with the help of Eq. (5.24). The result is

$$\langle \phi(w, \bar{w}) \rangle_{\text{strip}} = \left(\frac{\pi}{L} \right)^{2h} \frac{e^{\pi h(w + \bar{w})/L}}{\left[e^{\pi w/L} - e^{\pi \bar{w}/L} \right]^{2h}}$$

$$= \left(\frac{2iL}{\pi} \right)^{-\Delta} \frac{1}{\left[\sin(\pi v/L) \right]^{\Delta}} \tag{11.35}$$

where we have used real coordinates u and v (respectively, longitudinal and transverse) defined by $w = u + iv$. This profile is symmetric about the middle of the strip, where it reaches its minimum. In the limit $v \ll L$, we may write

$$\langle \phi(v) \rangle_{\text{strip}} \propto \frac{1}{v^\Delta} \left[1 + \frac{1}{6}\pi^2 \Delta(v/L)^2 + \cdots \right] \tag{11.36}$$

This is compatible with a more general result of Fisher and de Gennes, obtained through a scaling analysis in dimension d:

$$\langle \phi(v) \rangle \sim \frac{1}{v^\Delta} \left[1 + \text{const.}(v/L)^d \right] \qquad (v \ll L) \tag{11.37}$$

It is also interesting to look at the two-point function of a primary field on the strip. We shall limit ourselves to the spin-spin correlation function in the Ising model, in the limit of large separation u along the strip. We let $w_1 = u_1 + iv_1$ and $w_2 = u_2 + iv_2$ be the locations of the two points on the strip and $u = u_2 - u_1$. According to the covariance relation (5.24), the spin-spin correlation function is

$$\langle \sigma(w_1, \bar{w}_1)\sigma(w_2, \bar{w}_2) \rangle_{\text{strip}} =$$

$$\left(\frac{\pi}{L} \right)^{\frac{1}{4}} \left[e^{2\pi u_1/L} e^{\pi u/L} \right]^{\frac{1}{8}} \langle \sigma(z_1)\sigma(z_2)\sigma(\bar{z}_1)\sigma(\bar{z}_2) \rangle \tag{11.38}$$

The last factor is given by

$$\langle \sigma\sigma\sigma\sigma \rangle = \left(\frac{z_{13}z_{24}}{z_{12}z_{23}z_{14}z_{34}} \right)^{\frac{1}{8}} F(x)$$

$$= \frac{1}{(z_{13}z_{24})^{\frac{1}{8}}} \left(\frac{x^3}{1-x} \right)^{-\frac{1}{8}} F(x) \tag{11.39}$$

In terms of the strip coordinates, the anharmonic ratio x is

$$x = \frac{z_{12}z_{34}}{z_{13}z_{24}} = -\frac{1 + e^{2\pi u/L} - 2e^{\pi u/L}\cos[\pi(v_2 - v_1)/L]}{Ae^{\pi u/L}} \tag{11.40}$$

where

$$A \equiv 4\sin\frac{\pi v_1}{L} \sin\frac{\pi v_2}{L} \tag{11.41}$$

Likewise,

$$z_{13}z_{24} = -4Ae^{2\pi u_1/L}e^{\pi u/L} \tag{11.42}$$

In the limit $u \gg L$, x is proportional to $e^{\pi u/L}$. Combining all the factors, we find, in this limit and for an ordinary transition,

$$\langle \sigma(w_1, \bar{w}_1)\sigma(w_2, \bar{w}_2) \rangle_{\text{strip}} \propto e^{-\pi u/2L} \qquad (u \gg L) \tag{11.43}$$

The two-point function decays exponentially in the longitudinal direction, with a correlation length $\xi = 2L/\pi$. This may be argued to be a special case of the more general relation

$$\xi = \frac{2L}{\pi\eta_\parallel} \tag{11.44}$$

This is to be compared with the relation $\xi = L/2\pi\Delta = L/\pi\eta$ on the cylinder geometry. It is the surface exponent $\eta_\parallel$ that now determines the correlation length.

Before leaving the strip geometry, we mention the finite-size correction to the free energy. The reasoning leading to Eq. (5.143) is still applicable here, except that the mapping is now slightly different (L is replaced by $2L$). The net result is

$$F_L = f_0 L - \frac{\pi c}{24L} \tag{11.45}$$

§11.3. Boundary Operators

11.3.1. Introduction

In this section we shall see how the methods of conformal field theory may be applied to a system limited by two (or more) boundaries with possibly different boundary conditions. We shall find an explicit formula, in the operator formalism, for all the boundary conditions compatible with conformal symmetry. The key concept in the treatment of boundary conditions is that of a *boundary operator*, or *boundary field*. This will be applied to critical percolation later in this chapter.

The existence of scaling fields living on the boundary appears naturally within the method of images. We consider a bulk scaling field $\phi(z)$ on the upper half-plane, which we bring closer and closer to the boundary (the real axis). As it approaches the real axis, this field interacts with its mirror image $\phi(z^*)$ (i.e., with the boundary itself) and can be replaced by its OPE with its image:

$$\phi(z)\phi(z^*) \approx \sum_i (z - z^*)^{(h_i - 2h)} \phi_B^{(i)}(x) \tag{11.46}$$

where we have dropped the higher terms and where $x \equiv (z + z^*)/2$. The fields $\phi_B^{(i)}$ live on the boundary, but belong to the same operator algebra as the bulk fields. As we shall see, these boundary fields, when inserted at a point on the boundary, change the boundary condition thereafter. In fact, the goal of this section is to justify this interpretation.

For the moment, we accept this interpretation of boundary operators and see how it applies in practice. We consider again the infinite strip of width L. We let t and σ be the coordinates, respectively, along and across the strip, so that the complex coordinate is $w = t + i\sigma$. We denote the boundary conditions at $\sigma = 0$ and $\sigma = L$, respectively, by the symbols α and β. In a concrete system such as the Ising model, α and β could stand for fixed boundary conditions—under which the boundary spins are $+1$ or -1—or free boundary conditions. If we choose the time direction to be along the strip, then the Hamiltonian depends on the boundary conditions: we denote it $(\pi/L)H_{\alpha\beta}$ (the prefactor is inserted so that $H_{\alpha\beta}$ has the same normalization as L_0 on the plane). We assume that the system has local conformal invariance under those transformations that preserve the boundary conditions.

If we map the strip back to the upper half-plane using the transformation (11.33), the boundary condition on the real axis changes from α to β at the origin. According to the above interpretation of boundary operators, this change may be obtained from a uniform boundary condition by the insertion at the origin of a boundary operator $\phi_{\alpha\beta}(0)$. In this notation, $\phi_{\alpha\beta}(x)$ is a scaling field of dimension $h_{\alpha\beta}$ living on the boundary and which, when inserted at a point x on the real axis, changes the boundary condition from α to β. In the context of radial quantization, this means that the vacuum is no longer invariant under translations (i.e., is no longer annihilated by L_{-1}), but is obtained from the $SL(2,\mathbb{Z})$-invariant vacuum $|0\rangle$ by the application of $\phi_{\alpha\beta}(0)$. For an infinite strip, it is clear that a boundary operator $\phi_{\beta\alpha}$ is also inserted at infinity. In fact, the fields $\phi_{\alpha\beta}$ and $\phi_{\beta\alpha}$ are conjugate and the two-point function $\langle\phi_{\alpha\beta}(x_1)\phi_{\beta\alpha}(x_2)\rangle$ is nonzero.

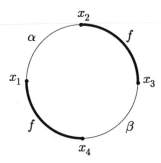

Figure 11.3. Bounded region with changing boundary conditions.

The introduction of boundary operators allows us to relate the partition function of a system with changing boundary conditions to a correlator of boundary operators on the upper half-plane. We consider a general bounded geometry, such as a rectangle or a circle. The interior of such a region may be mapped onto the upper half-plane, while its boundary is mapped onto the real axis. We suppose that the boundary condition is α on a segment $[x_1, x_2]$ of the boundary, β on a segment $[x_3, x_4]$, and free (f) everywhere else (cf. Fig. 11.3). The partition function $Z_{\alpha\beta}$ of this system will be expressed as

$$Z_{\alpha\beta} = Z_f \langle \phi_{f\alpha}(x_1)\phi_{\alpha f}(x_2)\phi_{f\beta}(x_3)\phi_{\beta f}(x_4)\rangle \qquad (11.47)$$

where Z_f is the partition function for free boundary conditions throughout.

11.3.2. Boundary States and the Verlinde Formula

In this subsection we justify the interpretation of boundary operators described above. The basic idea is to describe a conformal field theory defined on a finite cylinder within two equivalent quantization schemes, one in which time flows *around* the cylinder, another one in which it flows *along* the cylinder. In the first scheme, the Hamiltonian $H_{\alpha\beta}$ depends on the boundary conditions on the edges of the cylinder. In the second scheme, the boundary conditions are embodied in initial and final states $|a\rangle$ and $|\beta\rangle$, while the Hamiltonian is obtained directly from the whole complex plane.

If we go back to the strip and impose periodic boundary conditions in the time direction along the strip, after a period T, we have transformed the strip into a finite cylinder of circumference T and length L. The boundary conditions α and β are still imposed on the two edges of the cylinder. Because of the finite extent of the system it is now convenient to introduce a partition function

$$Z_{\alpha\beta}(q) = \operatorname{Tr} \exp -(\pi T/L)H_{\alpha\beta} \qquad q \equiv e^{2\pi i\tau} \ , \quad \tau \equiv iT/2L \qquad (11.48)$$
$$= \operatorname{Tr} q^{H_{\alpha\beta}}$$

where we have borrowed the notation of Chap. 10. Local conformal invariance implies that the spectrum of $H_{\alpha\beta}$ falls into irreducible representations of the Virasoro algebra (Verma modules). If we call $n^i_{\alpha\beta}$ the number of copies of the representation labeled i occurring in the spectrum, then the partition function may be written as

$$Z_{\alpha\beta}(q) = \sum_i n^i_{\alpha\beta}\chi_i(q) \qquad (11.49)$$

where χ_i is the Virasoro character of the representation i:

$$\chi_i(q) = q^{-c/24} \operatorname{Tr}_i q^{L_0} \qquad (11.50)$$

Since the full theory resides on the holomorphic sector only, the partition function is a linear, not bilinear, combination of characters.

In Chap. 10 it was pointed out that there are minimal conformal field theories, termed *rational*, which are made up of a finite number of Verma modules and for which, under a modular transformation $\tau \to -1/\tau$, the holomorphic characters transform as follows:

$$\chi_i(q) = \sum_j S_{ij}\chi_j(\tilde{q}) \qquad \tilde{q} \equiv e^{-2\pi i/\tau} \qquad (11.51)$$

The partition function $Z_{\alpha\beta}(q)$ may therefore be expressed as

$$Z_{\alpha\beta}(q) = \sum_{ij} n^i_{\alpha\beta} S_{ij}\chi_j(\tilde{q}) \qquad (11.52)$$

In the present context, such a modular transformation interchanges the roles of L and T. It is therefore possible to switch axes and to regard the partition function as a trace of a Hamiltonian generating translations along σ. To this end, we map the cylinder onto the plane via the coordinate transformation

$$\zeta = \exp\left\{-2\pi i(t + i\sigma)/T\right\} \qquad \text{or} \qquad w = i\frac{T}{2\pi} \ln \zeta \qquad (11.53)$$

The ζ-plane is, of course, distinct from the z-plane defined by the mapping (11.33). We let L_n^ζ and $\bar{L}_n^\zeta$ be the Virasoro generators on the ζ-plane. The Hamiltonian $\tilde{H}$ needed to perform the translations in the σ-direction is then

$$\tilde{H} = \frac{2\pi}{T}\left(L_0^\zeta + \bar{L}_0^\zeta - \frac{c}{12}\right) \tag{11.54}$$

On the ζ-plane the boundaries are concentric circles centered at the origin. In radial quantization, the boundary conditions are imposed by propagating states from an initial state $|\alpha\rangle$ residing on the inner boundary, toward a final state $|\beta\rangle$ on the outer boundary. The precise form of these states depends on the specific boundary conditions used. The partition function is then expressed as

$$Z_{\alpha\beta}(q) = \langle\alpha|e^{L\tilde{H}}|\beta\rangle$$
$$= \langle\alpha|(\tilde{q}^{1/2})^{L_0^\zeta + \bar{L}_0^\zeta - c/12}|\beta\rangle \tag{11.55}$$

The advantage of such a formulation is that we are familiar with the Hilbert space on the ζ-plane, where the holomorphic and antiholomorphic sectors propagate separately.

For all boundary conditions, it is imperative that there be no flow of energy across the edges of the finite cylinder, a condition that translates into

$$T_{\text{cyl.}}(0,t) = \bar{T}_{\text{cyl.}}(0,t) \quad \text{and} \quad T_{\text{cyl.}}(L,t) = \bar{T}_{\text{cyl.}}(L,t) \tag{11.56}$$

Here $T_{\text{cyl.}}$ and $\bar{T}_{\text{cyl.}}$ are the holomorphic and antiholomorphic components of the energy-momentum tensor on the cylinder. If we map this condition onto the ζ-plane, it takes the form

$$T_{\text{pl.}}(\zeta)\zeta^2 = \bar{T}_{\text{pl.}}(\bar{\zeta})\bar{\zeta}^2 \qquad \zeta = e^{-2\pi it/T} \tag{11.57}$$

on the boundary. In terms of the Virasoro generators acting on the boundary state $|\alpha\rangle$, this condition becomes

$$(L_n^\zeta - \bar{L}_{-n}^\zeta)|\alpha\rangle = 0 \tag{11.58}$$

A similar condition holds on the final state $|\beta\rangle$. We note that the condition (11.56) also enforces the invariance of the boundary condition (or boundary state) under conformal transformations that leave the boundary unchanged.

It turns out that the constraint (11.58) is quite rigid and that very few states satisfy it. We will give the general solution here, without proving its uniqueness. We let $|j; N\rangle$ be a holomorphic state belonging to the Verma module j (N labels the different states within that module) and $|\bar{j}; N\rangle$ be the corresponding antiholomorphic state. We introduce an antiunitary operator U such that

$$U|\bar{j}; 0\rangle = |\bar{j}; 0\rangle^* \qquad U\bar{L}_n^\zeta = \bar{L}_n^\zeta U \tag{11.59}$$

Then the solution to (11.58) is

$$|j\rangle \equiv \sum_N |j; N\rangle \otimes U|\bar{j}; N\rangle \tag{11.60}$$

In order to show that this state is indeed a solution to the constraint (11.58), it is enough to project the constraint onto each basis state of the Hilbert space (cf. Ex. 11.10). We thus have a complete list of boundary states compatible with local conformal invariance.

The boundary states $|\alpha\rangle$ and $|\beta\rangle$ will then be linear combinations of the states $|j\rangle$ associated with different Verma modules. Assuming the states $|j\rangle$ have been normalized in some way, we may then write the partition function as

$$
\begin{aligned}
Z_{\alpha\beta}(q) &= \sum_{i,j} \langle\alpha|i\rangle\langle i|(\tilde{q}^{1/2})^{L_0^c+\bar{L}_0^c-c/12}|j\rangle\langle j|\beta\rangle \\
&= \sum_j \langle\alpha|j\rangle\langle j|\beta\rangle\chi_j(\tilde{q})
\end{aligned}
\tag{11.61}
$$

In the second line we have restricted ourselves to diagonal theories, that is, theories whose partition function on the torus is a diagonal combination of characters: $Z = \sum_i \chi_i(\tau)\chi_i(\bar{\tau})$. Because of this, it is $\tilde{q}$ that appears in the last line of the above equation, not $\tilde{q}^{1/2}$. Comparing the above result with Eq. (11.51) leads to the following relation:

$$
\sum_i S_{ij}n_{\alpha\beta}^i = \langle\alpha|j\rangle\langle j|\beta\rangle
\tag{11.62}
$$

To proceed, we first identify a boundary state $|\tilde{0}\rangle$ such that the only representation occurring in the Hamiltonian $H_{\tilde{0}\tilde{0}}$ is the identity: $n_{\tilde{0}\tilde{0}}^i = \delta_0^i$. From Eq. (11.62), such a state satisfies the relation $|\langle\tilde{0}|j\rangle|^2 = S_{0j}$. In a *unitary* model, S_{0j} can be shown to be positive (cf Ex. 10.5) and therefore this state indeed exists and can be taken as

$$
|\tilde{0}\rangle = \sum_j \sqrt{S_{0j}}|j\rangle
\tag{11.63}
$$

Likewise, we define a state

$$
|\tilde{l}\rangle = \sum_j \frac{S_{lj}}{\sqrt{S_{0j}}}|j\rangle
\tag{11.64}
$$

From Eq. (11.62), this state is such that $n_{\tilde{0}\tilde{l}}^i = \delta_l^i$: only the representation l propagates in $H_{\tilde{0}\tilde{l}}$. We may then apply Eq. (11.62) one last time and find the following relation:

$$
\begin{aligned}
\sum_i S_{ij}n_{\tilde{k}\tilde{l}}^i &= \langle\tilde{k}|j\rangle\langle j|\tilde{l}\rangle \\
&= \frac{S_{ki}S_{lj}}{S_{0j}}
\end{aligned}
\tag{11.65}
$$

Here the matrix S is real: the Virasoro representations are self-conjugate.[6] This relation is identical to the Verlinde formula, which relates fusion coefficients and

---

[6] If this is not true, for instance if there are extended symmetries present, then the argument is only slightly modified, by replacing the representation k by its conjugate.

the modular matrix:

$$\sum_i S_{ij} \mathcal{N}^i{}_{kl} = \frac{S_{kj} S_{lj}}{S_{0j}} \tag{11.66}$$

We conclude from this exercise that

$$n^i_{\tilde{k}\tilde{l}} = \mathcal{N}^i{}_{kl} \tag{11.67}$$

that is, the number of times representation i occurs in the Hamiltonian $H_{\tilde{k}\tilde{l}}$ is precisely the fusion coefficient $\mathcal{N}^i{}_{kl}$.

This result warrants the interpretation that boundary conditions may be changed by inserting a local operator on the boundary. Consider Fig. 11.4. Initially the Hamiltonian is $H_{\tilde{l}\tilde{0}}$ and only the states belonging to representation l propagate. At time t_0 there is a change in boundary conditions to $(\tilde{l}, \tilde{k})$ and there will be $\mathcal{N}^i{}_{lk}$ copies of representation i that will propagate. Viewed differently, a boundary operator $\phi_{\tilde{0}\tilde{k}}$ has been applied at time t_0 on the states of representation l; since $\phi_{\tilde{0}\tilde{k}}$ transforms in the representation k of the Virasoro algebra, the resulting states will fall into a variety of representations, of which representation i occurs $\mathcal{N}^i{}_{lk}$ times, according to the usual fusion rules.

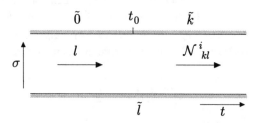

Figure 11.4. Insertion of the boundary operator $\phi_{\tilde{0}\tilde{k}}$ at an instant t_0 on the strip and consequence on the propagating modes.

EXAMPLE: THE ISING MODEL

In order to illustrate the above results, we apply them to the Ising model. According to Eq. (10.134) of Chap. 10, the modular matrix S is in this case

$$S = \begin{pmatrix} \frac{1}{2} & \frac{1}{2} & \sqrt{\frac{1}{2}} \\ \frac{1}{2} & \frac{1}{2} & -\sqrt{\frac{1}{2}} \\ \sqrt{\frac{1}{2}} & -\sqrt{\frac{1}{2}} & 0 \end{pmatrix} \tag{11.68}$$

where the three rows correspond, respectively, to the representations with highest weights $h = 0$ (0), $h = \frac{1}{2}$ (ε), and $h = \frac{1}{16}$ (σ); we indicated within parentheses the symbols used for the corresponding bulk operators. The number of possible conformally invariant boundary conditions is equal to the number of admissible boundary states defined in Eq. (11.64):

$$|\tilde{0}\rangle = \frac{1}{\sqrt{2}}|0\rangle + \frac{1}{\sqrt{2}}|\varepsilon\rangle + \frac{1}{\sqrt[4]{2}}|\sigma\rangle$$

$$|\widetilde{\frac{1}{2}}\rangle = \frac{1}{\sqrt{2}}|0\rangle + \frac{1}{\sqrt{2}}|\varepsilon\rangle - \frac{1}{\sqrt[4]{2}}|\sigma\rangle \qquad (11.69)$$

$$|\widetilde{\frac{1}{16}}\rangle = |0\rangle - |\varepsilon\rangle$$

Here we have designated by $|0\rangle$, $|\varepsilon\rangle$, and $|\sigma\rangle$ the three states defined in Eq. (11.60) for the three possible values of j.

Each of the three states defined in Eq. (11.69) is the realization, in radial quantization on the ζ-plane, of a particular type of conformally invariant boundary condition. In the Ising model, the three possible boundary conditions are to fix the boundary spins at $+$, $-$, or to let them free. Since the first two states of (11.69) differ only by the sign of the state associated with the odd operator σ, we infer that these two boundary states correspond to the two types of fixed boundary conditions, whereas the third state represents free boundary conditions. Which of the first two states of (11.69) represents the $+$ boundary condition is really a matter of choice.

Identifying the boundary operators $\phi_{\alpha\beta}$ taking us from one boundary condition to the other is not difficult. The operator ϕ_{+-} producing a transition from the $(+)$ boundary condition to the $(-)$ boundary condition could be written $\phi_{0\frac{1}{2}}$ in the notation of this subsection. Thus, it transforms under the representation of weight $\frac{1}{2}$ of the Virasoro algebra. In other words, it is the scaling field $\phi_{(2,1)} = \phi_{(1,3)}$. Likewise, the boundary operator ϕ_{+f} is identified with $\phi_{(1,2)} = \phi_{(2,2)}$.

§11.4. Critical Percolation

11.4.1. Statement of the Problem

We explain briefly the problem of *bond percolation*.[7] We consider a finite lattice (for definiteness, a rectangular lattice) and call G the set of bonds (or links) between nearest-neighbor sites. We suppose now that each bond has a probability p of being "activated"—graphically, one may represent activated bonds by thick lines, and "inert" bonds by thin lines (cf. Fig. 11.5). In a given configuration, activated bonds will fall into clusters. The greater the probability p, the bigger will be the average cluster. The central question of percolation theory is the following: for a given value of p, what is the probability $\pi_h(p)$ that there is a cluster spanning the whole lattice, from left to right? In other words, what is the probability that one can cross the lattice by walking continuously on activated bonds? Here the index h stands

---

[7] A detailed introduction to the general problem of percolation lies outside the scope of this work.

for *horizontal*; one also defines the probability $\pi_v(p)$ for a vertical crossing of the lattice. In fact, more general probabilities may be defined for crossings from a definite portion of the boundary to another. These probabilities depend, of course, on the size and aspect ratio (i.e., width over height) of the lattice. The central result of percolation theory is that in the limit of infinite lattice size (the thermodynamic limit) there exists a critical value p_c of the activation probability such that the crossing probability $\pi_h(p)$ vanishes if $p < p_c$ and is unity if $p > p_c$. At $p = p_c$, the crossing probability depends on the shape (aspect ratio) of the lattice.

Figure 11.5. Typical configuration of bonds on a finite rectangular lattice. In this specific case, there is a horizontal crossing but no vertical crossing.

By critical percolation, we mean the study of percolation at the critical value $p = p_c$. For a square lattice, it is known exactly that $p_c = \frac{1}{2}$.[8] Let r be the aspect ratio of the rectangular lattice. One of the main questions in the theory of critical percolation is then the calculation of the crossing probability $\pi_h(r)$ as a function of aspect ratio, in the thermodynamic limit. This function has been "measured" quite accurately by computer simulations. The goal of this section is to explain how to calculate it with the methods of conformal invariance. As we shall show, the agreement between the theoretical and the measured values is striking and provides a remarkable validation of the assumptions behind conformal field theory, in particular in the treatment of boundaries.

---

[8] We can also define percolation *by site*, in which sites, not bonds, are activated. Crossing is then possible only by hopping between nearest-neighbor activated sites. The critical probabilities are different from those of the bond-percolation problem. For a square lattice, one finds $p_c \approx 0.5927460 \pm 0.0000005$ with the help of computer simulations.

11.4.2. Bond Percolation and the Q-state Potts Model

The bond percolation problem fits naturally into the family of Potts models. Recall that the Q-state Potts model is defined as follows: on each site of the lattice lives a discrete variable σ_i—call it *spin*—taking one of Q possible values, and the energy of a spin configuration is

$$E = J \sum_{\langle ij \rangle} \delta_{\sigma_i \sigma_j} \tag{11.70}$$

In other words, each bond linking two like-spins has an energy J, other bonds have no energy. The total energy being a sum over bonds, the partition function may be expressed as follows:

$$Z = \sum_{\{\sigma\}} \prod_{\langle ij \rangle} \left(1 + x\delta_{\sigma_i \sigma_j}\right) \tag{11.71}$$

where $x = \exp -\beta J$. In a generic configuration of spins, the bonds are arranged in clusters containing like-spins. The expression (11.71) for the partition function allows a different interpretation of the Q-state Potts model, closer to the percolation problem: each bond has a probability p of being activated and $1-p$ of being inert, with the ratio $x = p/(1-p)$. However, each activated bond has a "color" taking Q possible values, so that a given cluster of bonds is "colored" according to the value of σ it supports. The partition function (11.71) may in fact be reformulated as follows, up to a multiplicative constant: If we let B be the total number of bonds on the lattice, R the subset of bonds that are activated, $B(R)$ the number of activated bonds, and $N_c(R)$ the number of disjoint clusters in the subset R, then, up to a multiplicative constant, the partition function is

$$Z = \sum_R p^{B(R)}(1-p)^{B-B(R)} Q^{N_c(R)} \tag{11.72}$$

where the sum is taken over all possible sets of activated bonds. A given configuration of activated bonds has a probability $p^{B(R)}(1-p)^{B-B(R)}$ of being realized and, once it exists, there are $Q^{N_c(R)}$ ways of distributing the Q colors among the $N_c(R)$ clusters. In order to have a perfect correspondence with the Q-state Potts model, we must count clusters of size zero (i.e., isolated spins of any color). Thus, the usual bond percolation problem appears as a special case ($Q = 1$) of the Q-state Potts model.[9] We can easily check that the partition function (11.72) is normalized (i.e., $Z = 1$) if $Q = 1$.

This correspondence between the Q-state Potts model and the bond percolation problem allows us to formulate the problem of the crossing probability $\pi_h(r)$ in terms of partition functions of the Q-state Potts model with different boundary conditions. Specifically, we let $Z_{\alpha\beta}$ be the partition function on a rectangular lattice with fixed boundary conditions—spins in state α on the left edge and in state β

---

[9] The self-dual point x_c of the Q-state Potts model is $x_c = 1 + \sqrt{n}$. However, this represents a critical point only for $n = 2, 3, 4$. The simple bond percolation problem has a critical point at $x_c = 1$ (or $p_c = \frac{1}{2}$), not $x = 2$.

on the right edge—while the boundary conditions are free on the top and bottom edges. We assert that the crossing probability $\pi_h(r)$ is given by

$$\pi_h(r) = \lim_{n \to 1} (Z_{\alpha\alpha} - Z_{\alpha\beta}) \tag{11.73}$$

where $\alpha \neq \beta$. Indeed, the first term $(Z_{\alpha\alpha})$ is a sum over colored-bond configurations containing clusters of color α that cross from left to right, whereas the second term $(Z_{\alpha\beta})$ excludes those very configurations (which fixed boundary conditions are chosen for α and β does not matter, provided they are different). The difference $Z_{\alpha\alpha} - Z_{\alpha\beta}$ contains only configurations with crossings of color α, as expressed below with the primed sum:

$$Z_{\alpha\alpha} - Z_{\alpha\beta} = \sum_{R}' p^{B(R)}(1-p)^{B-B(R)} Q^{N_c(R)} \tag{11.74}$$

Because of the normalization of Eq. (11.72) when $Q = 1$, this is indeed the crossing probability. Of course, this expression makes no sense if Q is set to one from the start ($Z_{\alpha\beta}$ has then no meaning), but it yields the correct answer if the limit is taken after having expressed the partition functions in terms of correlators of boundary operators.

11.4.3. Boundary Operators and Crossing Probabilities

We have seen earlier in this chapter that partition functions with specific boundary conditions can be expressed as correlators of boundary operators inserted at the points on the boundary where the boundary conditions change from one type to the other. If $x_1 \ldots x_4$ are the coordinates of the four corners of the rectangular lattice in the thermodynamic limit—and, of course, at the critical point p_c—the above partition functions have the following representation:

$$\begin{aligned} Z_{\alpha\alpha} &= Z_f \langle \phi_{f\alpha}(x_1)\phi_{\alpha f}(x_2)\phi_{f\alpha}(x_3)\phi_{\alpha f}(x_4) \rangle \\ Z_{\alpha\beta} &= Z_f \langle \phi_{f\alpha}(x_1)\phi_{\alpha f}(x_2)\phi_{f\beta}(x_3)\phi_{\beta f}(x_4) \rangle \end{aligned} \tag{11.75}$$

where Z_f is the partition function for free boundary conditions. The problem is now to identify correctly the boundary operators.

We have shown in Chap. 7 how the critical Q-state Potts model is related to the unitary minimal model $\mathcal{M}(m+1, m)$ with $m = 3$ for $Q = 2$ (the Ising model) and $m = 5$ for $Q = 3$. It turns out that the four-state Potts model is related to the $c = 1$ model obtained in the limit $m \to \infty$. The precise correspondence between Q and m is embodied in the relation $Q = 4\cos^2(\pi/(m+1))$, the three cases $m = 3, 5$ and ∞, corresponding, respectively, to $Q = 2, 3$ and 4. The bond percolation case $(Q = 1)$ is thus associated with the minimal theory $\mathcal{M}(3, 2)$, with central charge $c = 0$.

In the absence of any rigorous argument identifying the boundary operator $\phi_{f\alpha}$ with a specific field $\phi_{(r,s)}$ of the minimal theory, one must proceed in a heuristic way. We suppose that the points x and x' on the boundary where the boundary conditions change from α to f, and then from f to β, are brought together. This procedure must be equivalent to the following schematic operator product expansion:

$$\phi_{\alpha f}\phi_{f\beta} \sim \delta_{\alpha\beta} + \phi_{\alpha\beta} + \cdots \tag{11.76}$$

We notice that the operators $\phi_{\alpha f}$, $\phi_{\beta f}$, $\phi_{f\alpha}$ and $\phi_{f\beta}$ are all equivalent in the limit $n \to 1$. We have established earlier that the operator $\phi_{\alpha\beta}$ associated with a change of fixed boundary conditions in the Ising model is $\phi_{(1,3)}$. This turns out to be the case also in the three-state Potts model. The above OPE then leaves no choice but to take $\phi_{\alpha f} = \phi_{(1,2)}$. The conformal dimension of $\phi_{(1,2)}$ vanishes for $c = 0$, and this supports our choice. Indeed, the crossing probability should be invariant under uniform scalings of the lattice at $p = p_c$, and the only way the four-point functions $\langle \phi\phi\phi\phi \rangle$ can be invariant under such rescaling is if $h = 0$.

Since $h_{1,2} = 0$, the four-point functions (11.75) are truly invariant under local conformal transformations. We now map the rectangular boundary of the lattice onto the real axis. This can be done in many ways, in particular using a Schwarz-Christoffel transformation. We let $z_1 \ldots z_4$ be the four points on the real axis corresponding to the four corners $x_1 \ldots x_4$ of the rectangle. After transformation, the four-point function becomes simply

$$\langle \phi(z_1)\phi(z_2)\phi(z_3)\phi(z_4) \rangle \tag{11.77}$$

where $\phi = \phi_{(1,2)}$. This function has been studied before, in particular in the context of the Ising model on the upper half-plane (cf. Sect. 11.2.2). In the present context, $h = 0$, and Eq. (7.47) becomes

$$\left\{ \mathcal{L}_{-2} - \frac{3}{2}\mathcal{L}_{-1}^2 \right\} \langle \phi(z_1)\phi(z_2)\phi(z_3)\phi(z_4) \rangle = 0 \tag{11.78}$$

where the operator $\mathcal{L}_{-1}$ stands for $\partial/\partial z_4$ and

$$\mathcal{L}_{-2} = \sum_{1=1}^{3} \frac{1}{z_4 - z_i} \frac{\partial}{\partial z_i} \tag{11.79}$$

The differential equation obeyed by the correlator (11.77) is thus

$$\partial_4^2 + \frac{2}{3}\left[\frac{1}{z_{14}}\partial_1 + \frac{1}{z_{24}}\partial_2 + \frac{1}{z_{34}}\partial_3 \right] \langle \phi(z_1)\phi(z_2)\phi(z_3)\phi(z_4) \rangle = 0 \tag{11.80}$$

where, as usual, $z_{ij} \equiv z_i - z_j$. Since $h = 0$, the correlator on the infinite plane is simply a function $g(x)$ of the anharmonic ratio $x = (z_{12}z_{34})/(z_{13}z_{24})$. After some algebra, the above differential equation reduces to

$$x(1 - x)g'' + \frac{2}{3}(1 - 2x)g' = 0 \tag{11.81}$$

The two independent solutions to this linear equation are $g(x) = 1$ and $g(x) = x^{1/3}F(\frac{1}{3}, \frac{2}{3}, \frac{4}{3}; x)$, where F is the hypergeometric function. It remains to determine which linear combination of these two solutions is equal to $Z_{\alpha\alpha} - Z_{\alpha\beta}$ (the crossing probability) in the $n \to 1$ limit.

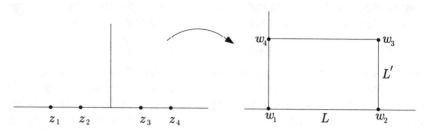

Figure 11.6. The Schwarz-Christoffel transformation mapping the upper half-plane to the interior of a rectangle.

Before going further, we need the precise correspondence between the aspect ratio r of the rectangle and the anharmonic ratio x. A possible mapping from the upper half-plane to the interior of the rectangle is the following Schwarz-Christoffel transformation:

$$w = A \int^z \frac{dt}{\sqrt{(t - z_1)(t - z_2)(t - z_3)(t - z_4)}} \tag{11.82}$$

where the four points z_i are the images of the four corners w_i of the rectangle (cf. Fig. (11.6)) and A is a constant proportional to L. This transformation is singular at the four points z_i and is not conformal precisely at these points, since it does not preserve angles there. We let these four points be respectively $z_1 = -k^{-1}$, $z_2 = -1, z_3 = 1$ and $z_4 = k^{-1}$, where $0 < k < 1$. It follows that the height of the rectangle is

$$L' = \frac{w_3 - w_2}{i}$$

$$= 2Ak \int_0^1 \frac{dt}{\sqrt{(1 - t^2)(1 - k^2t^2)}} \tag{11.83}$$

$$= 2AkK(k^2)$$

Likewise, the width is

$$L = w_3 - w_4$$

$$= Ak \int_1^{1/k} \frac{dt}{\sqrt{(1 - k^2t^2)(t^2 - 1)}} \tag{11.84}$$

$$= AkK(1 - k^2)$$

Here we have used the definitions of the complete elliptic integrals of the first kind K and K':

$$K(k^2) \equiv \int_0^1 \frac{dt}{\sqrt{(1 - t^2)(1 - k^2t^2)}} = \frac{1}{2}\pi F(\frac{1}{2}, \frac{1}{2}, 1; k^2)$$

$$\tag{11.85}$$

$$K'(k^2) \equiv \int_1^{1/k} \frac{dt}{\sqrt{(1 - k^2t^2)(t^2 - 1)}} = K(1 - k^2)$$

Therefore, the aspect ratio $r = L'/L$ is

$$r = \frac{K(1 - k^2)}{2K(k^2)} \tag{11.86}$$

The anharmonic ratio x has a simpler expression:

$$x = \frac{z_{12}z_{34}}{z_{13}z_{24}} = \frac{(1 - k)^2}{(1 + k)^2} \tag{11.87}$$

When $r \to 0$ (infinitely narrow lattice), $k \to 1$ and $x \to 0$. On the other hand, when $r \to \infty$ (infinitely wide lattice), $k \to 0$ and $x \to 1$. Obviously, the crossing probability $\pi_h(r)$ should be 1 if $r = 0$ ($x = 0$) and zero if $r = \infty$ ($x = 1$). However, the hypergeometric function satisfies the following identity:

$$\frac{3\Gamma(\frac{2}{3})}{\Gamma(\frac{1}{3})^2} x^{1/3} F(\frac{1}{3}, \frac{2}{3}, \frac{4}{3}; x) = 1 - (1 - x)^{1/3} F(\frac{1}{3}, \frac{2}{3}, \frac{4}{3}; 1 - x) \tag{11.88}$$

This identity makes it clear, from the the values at $x = 0$ and $x = 1$, that the appropriate combination describing $\pi_h(r)$ is

$$\pi_h(r) = \frac{3\Gamma(\frac{2}{3})}{\Gamma(\frac{1}{3})^2} x^{1/3} F(\frac{1}{3}, \frac{2}{3}, \frac{4}{3}; x) \tag{11.89}$$

The comparison between this expression and the crossing probabilities obtained in numerical simulations[10] is illustrated on Fig. 11.7. The striking agreement between simulation and theory is one of the most convincing confirmations to date of the validity of the hypothesis of local conformal invariance in two-dimensional critical systems.

Exercises

11.1 *Conformal theory on an infinite cylinder*

a) Show that any correlation function on the torus has a well-defined "infinite cylinder" limit by taking $\tau = iT/L \to i\infty$ (namely $T \to \infty$ while L, the transverse size of the cylinder, remains fixed).

b) Derive the infinite cylinder limit of the torus Ward identity (12.79). In particular, compute the expectation value of the energy-momentum tensor on the infinite cylinder in terms of the central charge c.

c) Use Eqs. (12.93) and (12.108) to compute the energy and spin two-point functions of the Ising model on an infinite cylinder.

d) Check that the result satisfies the infinite cylinder Ward identity derived in part (b).

---

[10] The simulations were done on percolation by sites, not by bonds. However, even if the critical value p_c of the "activation" probability is different, the critical behavior—and thus the crossing probability—are expected to be identical, because of universality.

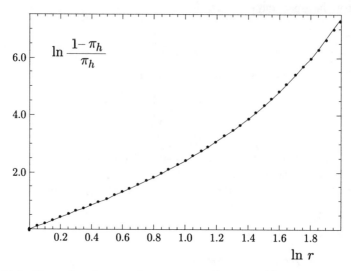

Figure 11.7. Comparison between the crossing probability $\pi_h(r)$ measured in computer simulations (dots) and the prediction of Eq. (11.89). The errors on the simulation data are smaller than the point size used. The last point on the right deviates slightly from the theoretical curve, although this is barely visible on this graph. This deviation may be attributed to the finite number of sites used in the simulation.

11.2 *Compactified boson on an infinite cylinder*
We consider the free bosonic theory compactified on a circle of radius R, with partition function on the torus

$$Z(R) = \frac{1}{\operatorname{Im}\tau \, |\eta(\tau)|^2} \sum_{m,m'\in\mathbb{Z}} \exp -\frac{\pi R^2 |m' - m\tau|^2}{2\operatorname{Im}\tau}$$

corresponding to the possible winding numbers (m,m') of the boson around the a and b-cycles of the torus (see Sect. 10.4.1 for details).

a) In the infinite cylinder limit $\tau = iT/L, T \to \infty, L$ fixed, show that the leading contribution to any correlation function comes from the doubly periodic sector $(m,m') = (0,0)$.

b) Use Eq. (12.148) to compute the electromagnetic operator two-point function on the infinite cylinder.

c) Compare this result with the cylinder limit of the same two-point function in the $\mathbb{Z}_2$-orbifold theory (Eq. (12.152)).

d) Use Eq. (12.150) to compute the n-point function of electromagnetic operators on the cylinder.

11.3 Verify all the steps leading to Eq. (11.29) and check the asymptotic result (11.32).

11.4 *Conformal invariance on the unit disk*

a) Show that the mapping $w = (z - i)/(z + i)$ maps the upper half-plane (z) to the interior of the unit circle (w) centered at the origin.

b) Show that the order parameter profile on the unit disk is the following:

$$\langle\phi(w,\bar{w})\rangle_{\text{disk}} = \frac{1}{(1-r^2)^{2h}} \tag{11.90}$$

where r is the distance from the origin.

c) We denote a generic two-point function on the unit disk by $G(r_1,\theta_1;r_2,\theta_2)$ where (r_i,θ_i) $(i=1,2)$ are the polar coordinates of the two points considered. Show that conformal invariance implies the following universal ratio:

$$\frac{G(a,0;a,\pi)}{G(0,0;2a/(1+a^2),0)} = (1+a^2)^{2\Delta} \tag{11.91}$$

(Δ is the scaling dimension of the local field under study).

11.5 *Ising energy correlator on the upper half-plane*

a) Show that the energy correlation function of the Ising model on the upper half-plane is

$$\langle\varepsilon(z,\bar{z})\varepsilon(w,\bar{w})\rangle_{\text{UHP}} = \frac{1}{4y_1y_2} + \frac{1}{|z-w|^2} - \frac{1}{|z-w^*|^2} \tag{11.92}$$

up to a multiplicative factor. The energy operator is defined as $\varepsilon(z,\bar{z}) = i\psi(z)\bar{\psi}(\bar{z})$.

b) Show that the corresponding surface exponent $\eta_\parallel$ is equal to 4.

c) Show that, once mapped onto the infinite strip, the energy-energy correlator decays with a correlation length $\xi = 2L/\pi\eta_\parallel$.

11.6 *Parity transformation of the Ising model*
The effect of a parity transformation $z \to \bar{z}$ on the operators of the Ising model is not unique: it is possible to define two such parity transformations, which correspond to different boundary conditions on the real axis and distinguish between ordinary and extraordinary transitions.

a) Show that the parity transformations

$$\begin{aligned} \psi(z) &\to \bar{\psi}(\bar{z}) & \sigma(z,\bar{z}) &\to \mu(\bar{z},z) \\ \bar{\psi}(z) &\to \psi(\bar{z}) & \mu(z,\bar{z}) &\to \sigma(\bar{z},z) \end{aligned} \tag{11.93}$$

leave unchanged the OPEs of $\psi,\bar{\psi},\sigma$, and μ. These OPEs are given in Chap. 12 (Eq. (12.68) and the preceding one).[11] Note that some of those OPEs are defined up to sign, because of a branch cut. This parity transformation amounts to a duality transformation of the Ising model.

b) Show that the parity transformation

$$\begin{aligned} \psi(z) &\to -\bar{\psi}(\bar{z}) & \sigma(z,\bar{z}) &\to \sigma(\bar{z},z) \\ \bar{\psi}(z) &\to \psi(\bar{z}) & \mu(z,\bar{z}) &\to \mu(\bar{z},z) \end{aligned} \tag{11.94}$$

also leaves unchanged the same OPEs, provided it is antiunitary, that is, provided the coefficients of the OPE are complex-conjugated.

---

[11] Note that it is not necessary to work through Chap. 12 in detail in order to do the exercises of the present chapter pertaining to the Ising model. The necessary identities of Chap. 12 may be borrowed without reference to their derivation.

c) Let $F_\pm(x) = \sqrt{1 \pm \sqrt{1-x}}$. Show that the results (12.63) and (12.66) may be, respectively, written as

$$\langle \sigma(z_1, \bar{z}_1), \sigma(z_2, \bar{z}_2), \sigma(z_3, \bar{z}_3), \sigma(z_4, \bar{z}_4) \rangle = \frac{1}{2} \left| \frac{z_{13}z_{24}}{z_{12}z_{14}z_{23}z_{34}} \right|^{\frac{1}{4}} (F_+ \bar{F}_+ + F_- \bar{F}_-)$$

$$\langle \sigma(z_1, \bar{z}_1), \mu(z_2, \bar{z}_2), \sigma(z_3, \bar{z}_3), \mu(z_4, \bar{z}_4) \rangle = \frac{i}{2} \left| \frac{z_{13}z_{24}}{z_{12}z_{14}z_{23}z_{34}} \right|^{\frac{1}{4}} (F_+ \bar{F}_- - F_- \bar{F}_+)$$

d) Referring to Sect. 11.2.2, argue that the holomorphic part of the above correlators correspond, respectively, to the extraordinary and ordinary transitions when applied to the spin-spin correlator on the upper half-plane, and are obtained, respectively, by applying the parity transformations (11.94) and (11.93). Of course, the infinite-plane correlator does not factorize into holomorphic × antiholomorphic factors, but rather like a sum thereof. By "holomorphic part", we mean what is obtained by setting $\bar{F}_\pm(\bar{x})$ to a constant (e.g., unity).

11.7 Spin-energy correlator on the upper half-plane

a) On the infinite plane, the spin-energy function $\langle \varepsilon(z, \bar{z}) \sigma(w, \bar{w}) \rangle$ vanishes. In the case of an ordinary transition, show that this is again the case on the upper half-plane (you must use the parity transformation (11.93)).

b) In the case of an extraordinary transition (with the parity transformation (11.94)), show that, on the upper half-plane, the spin-energy function is

$$\langle \varepsilon(z, \bar{z}) \sigma(w, \bar{w}) \rangle_{\text{UHP}} \propto \frac{1}{(\text{Im } z)(\text{Im } w)^{1/4}} \left\{ \left| \frac{z-w}{z-w^*} \right| + \left| \frac{z-w^*}{z-w} \right| \right\}$$

Hint: Use Eq. (12.30).
Extract the corresponding surface exponent $\eta_\parallel$.
Result: $\eta_\parallel = 4$.

c) The result of part (b) is incompatible with spin-reversal symmetry $\sigma \to -\sigma$. Why is this acceptable here, for an extraordinary transition?

11.8 Parity transformation of the free boson

a) Show that, under parity, the free boson transforms as

$$\varphi(z, \bar{z}) \to \eta \varphi(\bar{z}, z) \qquad \eta = \pm 1$$

and that the choice $\eta = +1$ corresponds to the boundary condition $\partial \varphi = 0$ on the real axis, while the choice $\eta = -1$ corresponds to the boundary condition $\varphi = 0$ on the real axis.

b) With the choice $\eta = +1$, show that the two-point function of vertex operators on the upper half-plane is

$$\langle e^{i\alpha\varphi(z,\bar{z})} e^{i\beta\varphi(w,\bar{w})} \rangle_{\text{UHP}} = \begin{cases} \left(\dfrac{\text{Im } z \, \text{Im } w}{|z-w|^2 |z-w^*|^2} \right)^{\alpha^2} & \text{if } \alpha = -\beta \\ 0 & \text{otherwise} \end{cases} \qquad (11.95)$$

and extract the surface exponent $\eta_\parallel$. Does this correspond to an ordinary or extraordinary transition?

c) With the choice $\eta = -1$, show that

$$\langle e^{i\alpha\varphi(z,\bar{z})} e^{i\beta\varphi(w,\bar{w})} \rangle_{\text{UHP}} \propto \frac{1}{(\text{Im } z)^{\alpha^2} (\text{Im } w)^{\beta^2}} \left| \frac{z-w}{z-w^*} \right|^{2\alpha\beta} \qquad (11.96)$$

and extract the surface exponent $\eta_\|$. Does this correspond to an ordinary or extraordinary transition? Why is the neutrality condition $\alpha + \beta = 0$ not necessary here?

11.9 *Free boson on a cylinder with fixed boundary conditions*
The aim of this exercise is to compute the partition function of the free boson on a cylinder of size $L \times T$, that is, subject to the periodicity condition in the space direction $\varphi(x+L, y) = \varphi(x, y)$, and with fixed boundary conditions (a, b) in the time direction, namely

$$\varphi(x, 0) = 2\pi Ra \qquad \varphi(x, T) = 2\pi Rb \qquad \forall x \in \mathbb{R}$$

(φ is compactified on a circle of radius R, and a, b are two integers). The corresponding partition function $Z_{(a,b)}(L, T; R)$ will be computed using the zeta regularization scheme presented in Sect. 10.2.

a) Write the eigenvalues of the Laplace operator Δ on the cylinder, with the boundary conditions $(a, b) = (0, 0)$. What is the main difference with the doubly periodic case of Sect. 10.2?

b) Follow the lines of Sect. 10.2 to derive the partition function

$$Z_{(0,0)}(L, T; R) = \frac{1}{\eta(iT/L)}$$

where η is the Dedekind eta function (see App. 10.A).

c) Compute the partition function $Z_{(a,b)}(L, T; R)$ with nonzero fixed boundary conditions (a, b). Show that

$$Z_{(a,b)}(L, T) = \frac{q^{R^2(b-a)^2/2}}{\eta(iT/L)}$$

where $q = \exp(-2\pi T/l)$.
Hint: Use the path integral formulation, with action $S(\varphi) = (1/8\pi) \int (\nabla\varphi)^2 d^2x$, and write $\varphi = \tilde{\varphi} + \varphi^{\rm cl}$, with $\tilde{\varphi}$ subject to the $(a, b) = (0, 0)$ boundary condition, and $\varphi^{\rm cl}$ a classical solution of the equation of motion ($\Delta\varphi^{\rm cl} = 0$), with the (a, b) boundary condition. When $2R^2$ is not the square of an integer, $Z_{(a,b)}(L, T; R)$ is just the irreducible character of the $c = 1$ representation with highest weight $h = R^2(b - a)^2/2$. This exhibits a collection of $c = 1$ characters as free-boson cylindric partition functions with fixed boundary conditions.

11.10 *Solution to the reparametrization constraint*
Show that the state $|j\rangle$ defined in Eq. (11.60) is indeed a solution to the constraint (11.58). Project the constraint on the generic basis state $\langle k; N_1| \otimes U\langle l; N_2|$ and show that the result vanishes.

11.11 *Normalization of the percolation partition function*
Show that the partition function (11.72) is normalized ($Z = 1$) in the simple bond percolation problem ($n = 1$). This is a simple combinatoric problem, based on the expansion of $(p+q)^N$, where N is the number of bonds, p is the activation probability of a bond, and $q = 1 - p$.

Notes

Finite-size corrections to the free energy and other thermodynamic quantities are discussed by Blöte, Cardy and Nightingale [49] and by Affleck [1]. The name of John Cardy is associated with most applications of conformal invariance to systems with boundaries. Surface critical behavior was discussed in [65] (see also the review article [68]). The restriction on the operator content of a theory imposed by the boundary conditions (the coefficients $n^i_{\alpha\beta}$ of Sect. 11.3.2) were discussed in [67]. The relation with the Verlinde formula, described in Sect. 11.3.2, is described in [70]. The application of this formalism to critical percolation is found in [71]. Monte Carlo simulations of critical percolation were performed by Langlands and collaborators [251,252]. The data of Fig. 11.7 are borrowed from [252].

Exercise 11.9 is in part based on [315].

CHAPTER 12

The Two-Dimensional Ising Model

The two-dimensional Ising model is probably one of the most famous statistical models, and it has been extensively studied in the literature. Our aim in this chapter is to present a detailed study of its continuum limit, in the framework of conformally invariant (free fermionic or bosonic) field theories. After reviewing basic facts on the statistical-mechanical model, we concentrate on its continuum fermionic representation. This framework is particularly suitable for the computation of correlation functions of the energy operator on the plane. For correlations involving the spin operator, it is more convenient to consider a bosonic field theory, made of two independent Ising models. In this bosonic formulation, the spin operator has a simple realization in terms of the free field. To complete the study of correlators, we also present the solution of the continuum Ising model on the torus, and use it as an illustrative example of the general theory of conformal blocks covered in Chaps. 9 and 10.

§12.1. The Statistical Model

The two-dimensional Ising model is defined as follows (cf. Chap. 3). Spin variables $\sigma_i \in \{-1, 1\}$ sit at the nodes of a square lattice of size $N \times M$, and interact through a nearest-neighbor energy[1] per link $\langle ij \rangle$

$$E_{\langle ij \rangle} = -J\sigma_i\sigma_j \tag{12.1}$$

leading to the partition function

$$Z = \sum_{\{\sigma\}} e^{-\beta \sum_{\langle ij \rangle} E_{\langle ij \rangle}} \tag{12.2}$$

---

[1] More generally, the system can be submitted to an external magnetic field B, resulting in the energy $B\sigma_i$ per node. However, here and in the following, we concentrate on the zero magnetic field case $B = 0$.

where $\beta = 1/(k_B T)$. This system undergoes a second-order phase transition at a critical value K_c of the coupling $K = \beta J$. We are interested mainly in the continuum limit formulation of the model at this critical point. The latter separates a low temperature ordered phase $(K > K_c)$ from a high-temperature disordered phase $(K < K_c)$. The partition function can be expanded in power series of $1/K$ and K, respectively, in these two phases.

In the high-temperature phase (small K), we write

$$Z = \sum_{\{\sigma\}} \prod_{\langle ij \rangle} \cosh(K)(1 + \sigma_i \sigma_j \tanh(K)) \tag{12.3}$$

and expand the product on the r.h.s. into monomials. When summing over all $\sigma_i \in \{-1, 1\}$, the only monomials with a nonvanishing contribution are products of σ_i^2 only (a term σ_i sums to $1 - 1 = 0$). These monomials come thus from spins forming closed chains of neighbors. Hence, the sum over all spins can be replaced by a sum over all closed (possibly disconnected) loops on the square lattice, namely

$$Z_{\text{high}} = [2\cosh(K)]^{NM} \sum_{\text{loops}} [\tanh(K)]^{\text{length}} \tag{12.4}$$

This is the so-called high-temperature expansion of the Ising model. A typical term in this expansion is illustrated on Fig. 12.1(a).

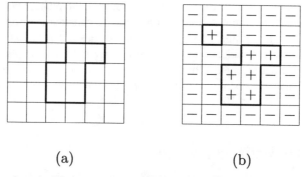

(a)   (b)

Figure 12.1. A typical term in the high- and low-temperature expansions of the Ising model. We display in (a) a loop configuration on the square lattice, with contribution $[2\cosh K]^{NM}[\tanh K]^{16}$ in the high-temperature expansion, and in (b) a spin configuration corresponding to the *same* loop configuration in the low-temperature expansion, with contribution $e^{KNM}e^{-32K}$.

In the low-temperature phase (large K), a given spin configuration is characterized by the borders of, say, all the spin $+1$ areas in a spin -1 background. Since the borders form loops, the sum (12.2) can be replaced by

$$Z_{\text{low}} = 2e^{NMK} \sum_{\text{loops}} e^{-2K(\text{length})} \tag{12.5}$$

where the contribution of all spins down has been factored out of the sum. The factor 2 accounts for the degeneracy under the reversal of all spins. This is the low-temperature expansion of the Ising model. A typical term in this expansion is illustrated on Fig. 12.1(b).

From expressions (12.4) and (12.5), we see that the two phases are mapped into each other through the identification

$$e^{-2K'} = \tanh K \tag{12.6}$$

which leads to

$$Z_{\text{low}}(K') = 2\,(\sinh 2K)^{-NM/2} Z_{\text{high}}(K) \tag{12.7}$$

At the phase transition point, where singular behavior is expected in the thermodynamic limit $(N, M \to \infty)$, we see that the two couplings K and K' of Eq. (12.6) should be identical, since the r.h.s. and l.h.s. of Eq. (12.7) must become *simultaneously* singular. This defines the critical coupling

$$K_c = -\frac{1}{2}\ln(\sqrt{2} - 1) \simeq 0.440686... \tag{12.8}$$

as the self-dual point of the duality relation (12.6) between the high- and low-temperature phases of the Ising model.

The duality transformation (12.6) relates the ordered and disordered phases of the Ising model. Actually it enables us to define an operator dual to the spin operator, called the disorder operator and denoted by μ, as follows.

The correlation of two spin operators sitting at positions r_1 and r_2 on the lattice, reads

$$\langle \sigma(r_1)\sigma(r_2) \rangle = \frac{1}{Z} \sum_{\{\sigma\}} \sigma(r_1)\sigma(r_2) \exp \left\{ K \sum_{\langle ij \rangle} \sigma_i \sigma_j \right\} \tag{12.9}$$

This is equivalent to picking an arbitrary path of n steps from r_1 to r_2 along the lattice bonds and changing the coupling $K \to K + i\pi/2$ for each bond of the path. Indeed, this introduces a factor $e^{i\pi\sigma_i\sigma_j/2} = i\,\sigma_i\sigma_j$ per bond $\langle ij \rangle$ of the path, leaving us with $i^n\,\sigma(r_1)\sigma(r_2)$, as each intermediary spin σ_i appears twice and $\sigma_i^2 = 1$. The result is actually independent, up to a sign, of the path chosen (see Ex. 12.4 for a proof).

In the spirit of the low-temperature expansion (12.5) we can also consider a correlation function of *disorder* operators $\langle \mu(r_1)\mu(r_2) \rangle$ defined as follows. We pick any path from r_1 to r_2 and change $K \to -K$ on all the bonds along the path; then we compute the sum over spin configurations as in Eq. (12.2), and normalize the result by dividing it by the partition function Z. This operation yields a result independent of the path (see Ex. 12.5). Actually, it is easy to see that the duality transformation (12.6) maps $\langle \sigma(r_1)\sigma(r_2) \rangle$ into $\langle \mu(r_1)\mu(r_2) \rangle$. This is a direct consequence of the transformations

$$(-K)^* = K^* + i\frac{\pi}{2} \qquad (K + i\frac{\pi}{2})^* = -(K^*) \tag{12.10}$$

where the duality transformation $*$ is defined by Eq. (12.6):

$$e^{-2K^*} = \tanh K \tag{12.11}$$

Therefore the high-low temperature duality of the Ising model exchanges the spin and disorder operators $\sigma \leftrightarrow \mu$.

The actual study of the transition requires much more work. It is far beyond the scope of this chapter to describe the original solution of the model by Onsager, or its modern formulation as a particular case of the eight-vertex model due to Baxter.

As our main task will be the study of the continuum limit of the model, we now briefly exhibit its fermionic character. The spin operator is actually not sufficient to describe the continuum limit of the model. The precise solution of the model involves some nonlocal observables, which are built out of the spin variables *along a whole line*, which crosses the lattice all the way to its border. This is the fermion operator, built out of the spin operator through the Jordan-Wigner transformation.[2] Whereas the spin operator is local (only defined at a point), the fermion operator is nonlocal because it depends on the values of the spin operator along a whole line starting from the boundary and ending at a given point. Moreover, the correlation function of two fermion operators must depend on their *order* of insertion, namely on the relative positions of the two lines defining them. It can be shown that the exchange of their positions results in an overall change of sign of the fermion correlator: this justifies the identification of the basic field in the continuum limit of the critical Ising model with a fermionic field.

Moreover, the exact Jordan-Wigner transformation of the partition function results, in the continuum limit, in a real free-fermion action functional. This action will be the subject of our study for the remainder of this chapter. We stress that, in this description, the spin operator is nonlocal with respect to the fermion operator. Although it does not appear explicitly in the free-fermion action, it has an expression (inverse Jordan-Wigner transformation) in terms of the values of the fermion operators along a line ending at the argument of the operator. The effects of nonlocality appear in the OPE of the spin and fermion operators.

§12.2. The Underlying Fermionic Theory

As mentioned before, the continuum critical Ising model is described by a free massless real fermion, governed by the action

$$S = \frac{1}{2\pi} \int d^2z \, (\psi\bar{\partial}\psi + \bar{\psi}\partial\bar{\psi}) \tag{12.12}$$

---

[2] This is best seen by performing the following sequence of transformations of the Ising model. In a first step, the two-dimensional Ising model is related to the *one-dimensional Ising quantum spin chain* (see Ex. 12.1). In a second step, the Ising spin chain is shown to be equivalent to the *XZ* spin chain (see Ex. 12.1 for a proof). Finally, the Jordan-Wigner transformation is performed on the equivalent $XY \equiv XZ$ model (cf. Ex. 12.2).

The above action is conformally invariant and was studied in Sect. 5.3.2. Actually, the vicinity of the critical point is also described by a free fermionic action, but with the addition of a mass term $m\psi\bar{\psi} \propto (K-K_c)\psi\bar{\psi}$, which spoils the conformal invariance. Here we concentrate on the massless case only.

The conformally invariant action (12.12) leads to a theory with central charge

$$c_{\text{Ising}} = c(4,3) = \frac{1}{2} \tag{12.13}$$

and the various operators are identified as

$$
\begin{aligned}
\text{Fermions}: \quad & \psi(z) && \propto \phi_{(2,1)}(z) \otimes \phi_{(1,1)}(\bar{z}) \\
& \bar{\psi}(\bar{z}) && \propto \phi_{(1,1)}(z) \otimes \phi_{(2,1)}(\bar{z}) \\
\text{Spin}: \quad & \sigma(z,\bar{z}) && \propto \phi_{(1,2)}(z) \otimes \phi_{(1,2)}(\bar{z})
\end{aligned}
\tag{12.14}
$$

12.2.1. Fermion: Energy and Energy-Momentum Tensor

As shown in Sect. 5.3.2, the free fermion action (12.12) leads to the following propagators

$$
\begin{aligned}
\langle \psi(z)\psi(w) \rangle &= \frac{1}{z-w} \\
\langle \bar{\psi}(\bar{z})\bar{\psi}(\bar{w}) \rangle &= \frac{1}{\bar{z}-\bar{w}}
\end{aligned}
\tag{12.15}
$$

Both functions are antisymmetric under the exchange of arguments $z \leftrightarrow w$ (resp. $\bar{z} \leftrightarrow \bar{w}$), and exhibit the conformal dimensions of ψ ($h = \frac{1}{2}, \bar{h} = 0$) and $\bar{\psi}$ ($h = 0$, $\bar{h} = \frac{1}{2}$). The energy operator is just a composite of the two fermionic fields (with $h = \bar{h} = \frac{1}{2}$), namely[3]

$$\varepsilon(z,\bar{z}) = i : \psi\bar{\psi}: \propto \phi_{(2,1)}(z) \otimes \phi_{(2,1)}(\bar{z}) \tag{12.16}$$

with the usual convention for the normal-ordered product. The normal ordering is, however, purely formal here, as the OPE of ψ and $\bar{\psi}$ is regular.

The correlation functions of the energy operator on the plane are easily derived by means of the fermionic version of Wick's theorem. Since the latter involves only pairings of the fermion operators, only correlators of an *even number* of energy operators survive. This is a manifestation of an underlying $\mathbb{Z}_2$ symmetry under which the sign of the energy operator is reversed $\varepsilon \to -\varepsilon$. This symmetry indeed reflects the high-low temperature duality discussed above: slightly away from criticality, the Ising action (12.12) acquires a mass term

$$im \int \psi(z)\bar{\psi}(\bar{z}) \propto (K-K_c) \int \varepsilon(z,\bar{z}) \tag{12.17}$$

---

[3] We recall that only left-right symmetric primary fields appear in the diagonal minimal theory $\mathcal{M}(4,3)$. The fermion can be seen as a nondiagonal projection of the energy field.

This can be viewed as a perturbation of the free theory by the energy operator ε. For $K \simeq K_c$, the high-low temperature duality (12.6) just amounts to $K^* - K_c = K_c - K$, that is, a change of sign in the perturbation, which can be absorbed into a change of sign of the energy operator. We thus wish to compute

$$
\begin{aligned}
E_{2n} &= \langle \varepsilon(z_1, \bar{z}_1) \cdots \varepsilon(z_{2n}, \bar{z}_{2n}) \rangle \\
&= (-1)^n \langle \psi(z_1) \bar{\psi}(\bar{z}_1) \cdots \psi(z_{2n}) \bar{\psi}(\bar{z}_{2n}) \rangle \\
&= \langle \psi(z_1) \cdots \psi(z_{2n}) \rangle \langle \bar{\psi}(\bar{z}_1) \cdots \bar{\psi}(\bar{z}_{2n}) \rangle
\end{aligned}
\tag{12.18}
$$

where we used the decoupling and anticommutation of ψ and $\bar{\psi}$ to factorize the correlator into holomorphic $\times$ antiholomorphic parts (the $(-1)^n$ disappeared in the third line after grouping the holomorphic fields together). According to Wick's theorem, we have to sum over all the possible pairings of fermionic operators ψ (resp. $\bar{\psi}$), and weigh the contribution of each pairing with the signature of the corresponding permutation. We end up with

$$
E_{2n} = \mathrm{Pf}\big[\langle \psi(z_i)\psi(z_j) \rangle \big]_{1 \le i,j \le 2n} \times \mathrm{Pf}\big[\langle \bar{\psi}(\bar{z}_i)\bar{\psi}(\bar{z}_j) \rangle \big]_{1 \le i,j \le 2n}
\tag{12.19}
$$

where we used the notation $\mathrm{Pf}(A)$ for the Pfaffian of a $(2n) \times (2n)$ antisymmetric matrix $A_{ij} = -A_{ji}$, defined as[4]

$$
\mathrm{Pf}(A) = \frac{1}{n!2^n} \sum_{\pi \in S_{2n}} \mathrm{sgn}\,(\pi) \prod_{i=1}^{n} A_{\pi(2i-1),\pi(2i)}
\tag{12.20}
$$

where the sum extends over the permutation group of the $2n$ indices, S_{2n}, and $\mathrm{sgn}\,(\pi)$ denotes the signature of the permutation π. The prefactor avoids over-counting pairs. In Eq. (12.19), it is understood that the matrix $A_{ij} = \langle \psi(z_i)\psi(z_j) \rangle$ has vanishing diagonal elements. By using the propagators (12.15), we finally obtain

$$
\langle \varepsilon(z_1, \bar{z}_1) \cdots \varepsilon(z_{2n}, \bar{z}_{2n}) \rangle = \left| \mathrm{Pf}\left[\frac{1}{z_i - z_j} \right]_{1 \le i,j \le 2n} \right|^2
\tag{12.21}
$$

The energy-momentum tensor for the real free fermion reads

$$
\begin{aligned}
T(z) &= -\frac{1}{2} : \psi(z)\partial_z \psi(z) : \\
&= -\frac{1}{2} \lim_{w \to z} \left[\frac{1}{2}(\psi(z)\partial_w \psi(w) - \partial_z \psi(z)\psi(w)) - \frac{1}{(z-w)^2} \right]
\end{aligned}
\tag{12.22}
$$

and a similar expression for $\bar{T}$ in terms of $\bar{\psi}$. Here the normal ordering prescription amounts to subtracting the divergence when the two points z and w coincide. It is easy to recover the Ward identities (5.41) of Chap. 5 expressing the insertion of the energy-momentum operator in an energy correlator by direct use of Wick's theorem (see Ex. 12.6 below).

---

[4] Note that this definition is equivalent to Eq. 2.225, upon regrouping permutations in the r.h.s. of (12.20).

12.2.2. Spin

The spin operator $\sigma(z, \bar{z})$ is in many respects more subtle to deal with. From the knowledge of its conformal dimensions $h = \bar{h} = 1/16$, we immediately write the two-point correlator

$$\langle \sigma(z_1, \bar{z}_1) \sigma(z_2, \bar{z}_2) \rangle = \frac{1}{|z_1 - z_2|^{\frac{1}{4}}} \quad (12.23)$$

The $\mathbb{Z}_2$ symmetry of the Ising model (under reversal of all spins) implies that the correlators should all be invariant under the change $\sigma \to -\sigma$. Hence only the correlators involving an even number of spin operators will survive. To compute higher-order correlators, we need the OPE of the various fields. The fusion rules predicted by conformal theory, namely

$$\sigma\sigma \to \mathbb{I} + \varepsilon \qquad \varepsilon\varepsilon \to \mathbb{I} \quad (12.24)$$

are expressed as

$$\varepsilon(z, \bar{z})\varepsilon(w, \bar{w}) = \frac{1}{|z - w|^2} + \cdots$$

$$\sigma(z, \bar{z})\sigma(w, \bar{w}) = \frac{1}{|z - w|^{\frac{1}{4}}} + C_{\sigma\sigma\varepsilon}|z - w|^{\frac{3}{4}}\varepsilon(w, \bar{w}) + \cdots \quad (12.25)$$

The structure constant $C_{\sigma\sigma\varepsilon}$ will be computed later as a limit of the four-spin correlator. The OPE with fermion operators expected from the statistical-model analysis read (up to some multiplicative factors, which will be derived later)

$$\psi(z)\sigma(w, \bar{w}) \sim \frac{1}{(z - w)^{\frac{1}{2}}} \mu(w, \bar{w})$$

$$\bar{\psi}(\bar{z})\mu(w, \bar{w}) \sim \frac{1}{(\bar{z} - \bar{w})^{\frac{1}{2}}} \sigma(w, \bar{w}) \quad (12.26)$$

where μ denotes the disorder operator dual to the spin operator. μ has the same OPE and conformal dimensions as σ, except for a sign: $C_{\mu\mu\varepsilon} = -C_{\sigma\sigma\varepsilon}$, since the sign of the thermal operator ε must change in the duality transformation. The relative nonlocality of the fermion and spin operators translates into the noninteger power ($\frac{1}{2}$) of the singular term. Inside a correlator, circulating the argument of the fermion around that of the spin, will result in a phase $e^{2i\pi/2} = -1$, hence a global change of sign.

We now illustrate the use of the OPE (12.26) in the computation of the mixed correlator

$$G(z, w|z_1, \bar{z}_1, z_2, \bar{z}_2) = \frac{\langle \psi(z)\psi(w)\sigma(z_1, \bar{z}_1)\sigma(z_2, \bar{z}_2) \rangle}{\langle \sigma(z_1, \bar{z}_1)\sigma(z_2, \bar{z}_2) \rangle} \quad (12.27)$$

When $z \to w$, we should get the limit

$$G \to \frac{1}{z - w} \quad (12.28)$$

When $z \to z_1$, we have

$$G \to \frac{1}{(z-z_1)^{\frac{1}{2}}|z_1-z_2|^{\frac{1}{4}}} \langle \psi(w)\mu(z_1,\bar{z}_1)\sigma(z_2,\bar{z}_2) \rangle \tag{12.29}$$

and a similar expression when $z \to z_2$. Moreover, G must be antisymmetric under the exchange of $z \leftrightarrow w$. These properties fix the function G. By global conformal invariance, $(z-w)G$ must be a function of the cross ratio of the four points. The precise dependence on the cross-ratio is completely determined by the above limits and the fact that G is antisymmetric under the exchange $z \leftrightarrow w$. We find

$$G = \frac{1}{2(z-w)}\left[\sqrt{\frac{(z-z_1)(w-z_2)}{(z-z_2)(w-z_1)}} + \sqrt{\frac{(z-z_2)(w-z_1)}{(z-z_1)(w-z_2)}}\right] \tag{12.30}$$

As a nontrivial check of the coherence of the theory, we recover the dimension $h_\sigma = 1/16$ of the spin operator by computing

$$\frac{\langle T(z)\sigma(z_1,\bar{z}_1)\sigma(z_2,\bar{z}_2)\rangle}{\langle \sigma(z_1,\bar{z}_1)\sigma(z_2,\bar{z}_2)\rangle} = -\frac{1}{4}\lim_{z\to w}\left(\partial_w G - \partial_z G - \frac{1}{(z-w)^2}\right) \tag{12.31}$$

and taking the $z \to z_1$ limit, with the leading term identified as

$$\frac{h_\sigma}{(z-z_1)} \tag{12.32}$$

A direct way of computing more general correlators is by solving the differential equations they satisfy. We recall that these differential equations are consequences of the singular vector structure of the Verma modules associated with the primary fields ε, σ. Actually, both the energy and the spin fields are degenerate at level 2, so that the associated highest weight vectors $|h\rangle$ have to satisfy the null vector condition

$$\left[L_{-2} - \frac{3}{2(2h+1)}L_{-1}^2\right]|h\rangle = 0 \tag{12.33}$$

with

$$h_\varepsilon = h_{2,1} = \frac{1}{2} \qquad h_\sigma = h_{1,2} = \frac{1}{16} \tag{12.34}$$

Combined with the Ward identity (5.41), this yields a second-order differential equation of the form (8.71) for any correlator involving ε or σ (see Ex. 12.8 for an illustrative example). But, instead of pursuing this rather technical program, we present below a simpler alternative for computing Ising correlators on the plane.

§12.3. Correlation Functions on the Plane by Bosonization

12.3.1. The Bosonization Rules

The superposition of two critical continuum Ising models on the *same* square lattice must have central charge $c = \frac{1}{2} + \frac{1}{2} = 1$. Indeed, since the two theories do not interact with each other, the total energy-momentum tensor is the sum of the energy-momentum tensors of each theory, and the central charges simply add up. This is the essence of the bosonization of the Ising model: to take two copies of the Ising model and to find a description of all the operators in terms of the free bosonic field at $c = 1$.

One way of explicitly realizing this is to consider the theory of a free *Dirac* (complex) fermion,

$$\mathcal{D}(z, \bar{z}) = \begin{pmatrix} D(z) \\ \bar{D}(\bar{z}) \end{pmatrix} = \frac{1}{\sqrt{2}} \begin{pmatrix} \psi_1 + i\psi_2 \\ \bar{\psi}_1 + i\bar{\psi}_2 \end{pmatrix} \tag{12.35}$$

the components of which are expressed in terms of two *real* fermions (indexed by 1 and 2). This theory is conformally invariant, with central charge $c = 1$: its energy-momentum tensor T is the sum of the energy-momentum tensors associated with the real fermions ψ_1 and ψ_2 (cf. Ex. 5.5):

$$T(z) = \frac{1}{2}(\partial D^\dagger D - D^\dagger \partial D) = -\frac{1}{2}\psi_1 \partial \psi_1 - \frac{1}{2}\psi_2 \partial \psi_2 \tag{12.36}$$

where $D^\dagger(z) = (\psi_1 - i\psi_2)/\sqrt{2}$.

Since the Dirac fermion is a $c = 1$ theory, a relation with the free boson may seem reasonable. Such a representation of the free complex fermion in terms of a free boson is the object of *bosonization*. We write

$$D(z) = e^{i\phi(z)} \qquad \bar{D}(\bar{z}) = e^{i\bar{\phi}(\bar{z})} \tag{12.37}$$

where $\phi(z)$ is a chiral (holomorphic) bosonic field with propagator

$$\langle \phi(z)\phi(0) \rangle = -\ln z \tag{12.38}$$

The fields ϕ and $\bar{\phi}$ are the chiral components of the free boson of Sect. 6.3:[5]

$$\varphi(z, \bar{z}) = \phi(z) - \bar{\phi}(\bar{z}) \qquad \langle \varphi(z, \bar{z})\varphi(w, \bar{w}) \rangle = -\ln|z - w|^2 \tag{12.39}$$

The properties of the chiral vertex operator $e^{i\alpha\phi(z)}$ are those found in Sect. 6.3, except that they pertain to the holomorphic sector only: its conformal dimension is $\frac{1}{2}\alpha^2$ (with the normalization chosen above for the propagator) and its OPE is

$$e^{i\alpha\phi(z)}e^{i\beta\phi(w)} \sim e^{i(\alpha\phi(z)+\beta\phi(w))}(z - w)^{\alpha\beta} \tag{12.40}$$

---

[5] We ignore for the moment subtle issues related to the zero-mode of φ.

The vertex operators of Eq. (12.37) have conformal dimensions $(\frac{1}{2}, 0)$ and $(0, \frac{1}{2})$ respectively and their OPEs are indeed compatible with those of complex fermions:

$$e^{i\phi(z)}e^{i\phi(w)} \sim e^{i(\phi(z)+\phi(w))}(z-w)$$

$$e^{i\phi(z)}e^{-i\phi(w)} \sim e^{i(\phi(z)-\phi(w))}\frac{1}{z-w} \tag{12.41}$$

In the limit $z \to w$, the first equation corresponds to $D(z)^2 = 0$, whereas the second gives

$$D(z)D^\dagger(w) \sim \frac{1}{z-w} + i\partial\phi(w) \tag{12.42}$$

The relation between the Dirac fermion and the boson φ may also be expressed in terms of the Dirac current J^μ:

$$J^\mu = \bar{D}\gamma^\mu D = i\varepsilon^{\mu\nu}\partial_\nu\varphi \tag{12.43}$$

where $\bar{D} = D^\dagger\gamma^0$. Indeed, if we adopt the convention of Sect. 5.3.2 for Dirac matrices, then

$$J^0 = i\psi_1\psi_2 + i\bar{\psi}_1\bar{\psi}_2$$
$$J^1 = -\psi_1\psi_2 + \bar{\psi}_1\bar{\psi}_2 \tag{12.44}$$

But

$$\psi_1(z)\psi_2(z) = -\frac{1}{2}i(D^\dagger(z)D(z) - D(z)D^\dagger(z))$$
$$= -\frac{1}{2}i \lim_{w \to z}\left\{e^{-i\phi(z)}e^{i\phi(w)} - e^{i\phi(z)}e^{-i\phi(w)}\right\} \tag{12.45}$$
$$= i\partial\phi(z)$$

and, likewise, $\bar{\psi}_1\bar{\psi}_2 = i\bar{\partial}\bar{\phi}$. Therefore, the holomorphic components of the current are (cf. Eqs. (5.7) and (5.8))

$$J_z = \frac{1}{2}(J^0 - iJ^1) = -\partial\phi = -\partial\varphi$$
$$J_{\bar{z}} = \frac{1}{2}(J^0 + iJ^1) = -\bar{\partial}\bar{\phi} = +\bar{\partial}\varphi \tag{12.46}$$

This confirms Eq. (12.43).

12.3.2. Energy Correlators

Two different mass terms may be considered for the Dirac fermion D. The usual Dirac mass term is

$$\bar{D}D = D\gamma^0 D = D^\dagger\bar{D} + \bar{D}^\dagger D = i(\psi_1\bar{\psi}_2 + \bar{\psi}_1\psi_2) \tag{12.47}$$

On the other hand, another (pseudoscalar) mass term exists:

$$\bar{D}\gamma^5 D = D\gamma^0\gamma^5 D = D^\dagger\bar{D} - \bar{D}^\dagger D = -(\psi_1\bar{\psi}_1 + \psi_2\bar{\psi}_2) \tag{12.48}$$

where $\gamma^5 \equiv i\gamma^0\gamma^1 = -\sigma_3$. This mass term is proportional to the total energy operator of the two copies of the Ising model:

$$\varepsilon_1 + \varepsilon_2 = i(\psi_1\bar{\psi}_1 + \psi_2\bar{\psi}_2) = -i\bar{D}\gamma^5 D \tag{12.49}$$

On the other hand,

$$\begin{aligned}-i\bar{D}\gamma^5 D &= -i(D^\dagger\bar{D} - \bar{D}^\dagger D) \\ &= -i(D^\dagger\bar{D} + D\bar{D}^\dagger) \\ &= -i\left(e^{-i\phi}e^{i\bar{\phi}} + e^{i\phi}e^{-i\bar{\phi}}\right) \\ &= -2i\cos\varphi(z,\bar{z})\end{aligned} \tag{12.50}$$

Hence we may represent the correlation functions of the energy operator using the bosonic field φ in the form

$$\langle(\varepsilon_1 + \varepsilon_2)(z_1,\bar{z}_1)\cdots(\varepsilon_1 + \varepsilon_2)(z_{2n},\bar{z}_{2n})\rangle = M_n\langle\prod_{i=1}^{2n}\cos\varphi(z_i,\bar{z}_i)\rangle \tag{12.51}$$

Of course, the mixed correlators of ε_1 and ε_2 factorize into a product of correlators pertaining to each Ising model. Hence, the l.h.s. of Eq. (12.51) decomposes into a sum of products of energy correlators of each theory. The normalization factor M_n is fixed by the short-distance limits (see Ex. 12.9).

There is a more direct relation between the energy correlators of the Ising model and those of the free field φ, which uses the result (12.19). Guided by the idea of duplication of the Ising model in order to bosonize it, we compute the square of the energy correlator

$$\langle\varepsilon(z_1,\bar{z}_1)\cdots\varepsilon(z_{2n},\bar{z}_{2n})\rangle^2 = \left|\text{Pf}\left[\frac{1}{z_i - z_j}\right]\right|^4 \tag{12.52}$$

The square of the Pfaffian of an antisymmetric matrix A is equal to its determinant. Actually we can write

$$\det\left[\frac{1}{z_i - z_j}\right] = \text{Pf}^2\left[\frac{1}{z_i - z_j}\right] = \text{Hf}\left[\frac{1}{(z_i - z_j)^2}\right] \tag{12.53}$$

where $\text{Hf}(B)$ denotes the Haffnian[6] of a *symmetric* matrix B

$$\text{Hf}(B) = \frac{1}{2^n n!}\sum_{\sigma\in S_{2n}}\prod_{i=1}^{n}B_{\sigma(2i-1)\sigma(2i)} \tag{12.54}$$

The Haffnian expression for Eq. (12.52) enables us to rewrite the square of Eq. (12.19) as the free-field correlator

$$\langle\varepsilon(z_1,\bar{z}_1)\cdots\varepsilon(z_{2n},\bar{z}_{2n})\rangle^2 = P_n\left\langle\prod_{i=1}^{2n}(\nabla\varphi/2)^2(z_i,\bar{z}_i)\right\rangle \tag{12.55}$$

---

[6] See Ex. 12.12 for a proof of this relation and more identities involving the fermion propagator.

where

$$(\nabla\varphi/2)^2 = \partial\varphi\bar\partial\varphi \tag{12.56}$$

Again, the normalization factor P_n is fixed by the short-distance behavior. This suggests that the composite energy operator $\varepsilon_1 \times \varepsilon_2$ should be identified with the operator $(\nabla\varphi)^2$ in the bosonized Dirac theory. Indeed, a correlator of composite energies factorizes into the product of the corresponding correlators in the two Ising theories

$$\langle\varepsilon_1\varepsilon_2(1)\cdots\varepsilon_1\varepsilon_2(2n)\rangle = \langle\varepsilon_1(1)\cdots\varepsilon_1(2n)\rangle \times \langle\varepsilon_2(1)\cdots\varepsilon_2(2n)\rangle$$
$$= \langle\varepsilon(1)\cdots\varepsilon(2n)\rangle^2 \tag{12.57}$$

The identification of $\varepsilon_1\varepsilon_2$ with $\partial\varphi\bar\partial\varphi$ may also be obtained directly from the bosonization procedure. Since $\psi_1\psi_2 = i\partial\phi$, we find

$$\varepsilon_1\varepsilon_2 = (i\psi_1\bar\psi_1)(i\psi_2\bar\psi_2) = \psi_1\psi_2\bar\psi_1\bar\psi_2$$
$$= -\partial\phi\bar\partial\bar\phi = \partial\varphi\bar\partial\varphi \tag{12.58}$$

12.3.3. Spin and General Correlators

A careful study of the Jordan-Wigner transformation (see Ex. 12.2) enables us to rewrite the correlation function of spin operators directly in terms of the bosonic field φ. We do not work out the detail of this calculation here, but simply give the result. The spin-spin correlation function actually appears only squared, because of the duplication of the model, just like in the energy case: the correlation functions of the composite spin operator $\sigma_1 \times \sigma_2$ factorize into a product of the correlators for each Ising theory. The result reads

$$\langle\sigma(z,\bar z)\sigma(w,\bar w)\rangle^2 = N_1\langle\cos\frac{\varphi}{2}(z,\bar z)\cos\frac{\varphi}{2}(w,\bar w)\rangle \tag{12.59}$$

By using the free-field propagator (12.39) and the spin-spin OPE, we can fix the normalization constant $N_1 = 2$ and recover

$$\langle\sigma(z,\bar z)\sigma(w,\bar w)\rangle^2 = \frac{1}{|z-w|^{\frac{1}{2}}} \tag{12.60}$$

This generalizes to

$$\langle\sigma(z_1,\bar z_1)\cdots\sigma(z_{2n},\bar z_{2n})\rangle^2 = N_n\left\langle\prod_{j=1}^{2n}\cos\frac{\varphi}{2}(z_j,\bar z_j)\right\rangle$$
$$= \frac{N_n}{2^{2n}}\left\langle\prod_{j=1}^{2n}(e^{i\varphi(j)/2}+e^{-i\varphi(j)/2})\right\rangle \tag{12.61}$$
$$= \frac{N_n}{2^{2n}}\sum_{\substack{\varepsilon_i=\pm1\ i=1,\dots,2n\\ \Sigma\varepsilon_i=0}}\prod_{i<j}|z_i-z_j|^{\varepsilon_i\varepsilon_j/2}$$

where we used the notation $\varphi(j) \equiv \varphi(z_j, \bar{z}_j)$, and the last equation is simply the sum over all the charge-neutral products of vertex operators, computed by Wick's theorem. The overall normalization is fixed to

$$N_n = 2^n \tag{12.62}$$

by the OPE. Note that we did not use all the information contained in the OPE (12.25). Actually we can compute the structure constant $C_{\sigma\sigma\varepsilon}$ from the above result (12.61). We take the four-spin correlator ($n = 2$) and let $z_1 \to z_2$ and $z_3 \to z_4$, then[7]

$$\langle \sigma(z_1, \bar{z}_1)\sigma(z_2, \bar{z}_2)\sigma(z_3, \bar{z}_3)\sigma(z_4, \bar{z}_4) \rangle^2$$

$$= \frac{1}{2} \frac{|z_{13}z_{24}|^{\frac{1}{2}}}{|z_{14}z_{23}z_{12}z_{34}|^{\frac{1}{2}}} \left[1 + \frac{|z_{12}z_{34}|}{|z_{13}z_{24}|} + \frac{|z_{14}z_{23}|}{|z_{13}z_{24}|} \right]$$

$$\simeq \frac{1}{|z_{12}z_{34}|^{\frac{1}{2}}} \left[1 + \frac{1}{2} \frac{|z_{12}z_{34}|}{|z_{24}|^2} \right] \tag{12.63}$$

$$= \frac{1}{|z_{12}z_{34}|^{\frac{1}{2}}} \left[C_{\sigma\sigma\mathbb{I}}^2 + 2|z_{12}z_{34}|C_{\sigma\sigma\varepsilon}^2 \langle \varepsilon(z_2, \bar{z}_2)\varepsilon(z_4, \bar{z}_4) \rangle \right]$$

This shows that $C_{\sigma\sigma\mathbb{I}}^2 = 1$ and $C_{\sigma\sigma\varepsilon}^2 = 1/4$. Hence, up to a multiplicative redefinition of the operator ε, we find

$$C_{\sigma\sigma\mathbb{I}} = 1 \quad \text{and} \quad C_{\sigma\sigma\varepsilon} = \frac{1}{2} \tag{12.64}$$

The high-low temperature duality of the Ising model reverses the sign of the energy operator, whereas it exchanges spin and disorder operators. It is therefore easily identified as $\varphi \to \pi - \varphi$ in the bosonized Dirac fermion theory. Consequently, replacing σ by μ in the square of a correlator just amounts to replacing $\cos \varphi/2 \to \sin \varphi/2$ in the corresponding free-field correlator. This results in

$$\langle \sigma(z_1, \bar{z}_1) \cdots \sigma(z_{2n}, \bar{z}_{2n})\mu(w_1, \bar{w}_1) \cdots \mu(w_{2m}, \bar{w}_{2m}) \rangle^2$$

$$= 2^{n+m} \left\langle \prod_{i=1}^{2n} \cos \frac{\varphi}{2}(z_i, \bar{z}_i) \prod_{j=1}^{2m} \sin \frac{\varphi}{2}(w_j, \bar{w}_j) \right\rangle$$

$$= \frac{(-1)^m}{2^{n+m}} \sum_{\substack{\varepsilon_i, \eta_k = \pm 1 \\ \Sigma\varepsilon_i + \Sigma\eta_k = 0}} \prod_k \eta_k \prod_{i<j} |z_i - z_j|^{\varepsilon_i \varepsilon_j/2} \tag{12.65}$$

$$\times \prod_{k<l} |w_k - w_l|^{\eta_k \eta_l/2} \prod_{r,s} |z_r - w_s|^{\varepsilon_r \eta_s/2}$$

---

[7] Notice that the first equality in Eq. (12.63) exhibits the dependence (5.31) on the cross-ratio of the four points.

For example, in the case $m = n = 1$, we find (see also Eq. (5.31))

$$\langle \sigma(z_1, \bar{z}_1)\mu(z_2, \bar{z}_2)\sigma(z_3, \bar{z}_3)\mu(z_4, \bar{z}_4)\rangle^2$$

$$= \frac{1}{2} \frac{|z_{13}z_{24}|^{\frac{1}{2}}}{|z_{14}z_{23}z_{12}z_{34}|^{\frac{1}{2}}} \left[-1 + \frac{|z_{12}z_{34}|}{|z_{13}z_{24}|} + \frac{|z_{14}z_{23}|}{|z_{13}z_{24}|} \right] \quad (12.66)$$

We note the change of sign for the first term when compared to the four-spin result (12.63). This again teaches us something about the OPE of σ and μ. Considering for instance the above expression in the limit $z_1 \to z_2$ and $z_3 \to z_4$, we find

$$\sigma(z, \bar{z})\mu(w, \bar{w}) = \frac{e^{i\pi/4}(z - w)^{\frac{1}{2}}\psi(w) + e^{-i\pi/4}(\bar{z} - \bar{w})^{\frac{1}{2}}\bar{\psi}(\bar{w})}{\sqrt{2}|z - w|^{\frac{1}{4}}} + \cdots \quad (12.67)$$

Using the associativity of the OPE, we deduce the exact multiplicative normalization factors in the OPE (12.26), which read

$$\psi(z)\sigma(w, \bar{w}) = \frac{e^{i\pi/4}}{\sqrt{2}(z - w)^{\frac{1}{2}}}\mu(w, \bar{w})$$

$$\psi(z)\mu(w, \bar{w}) = \frac{e^{-i\pi/4}}{\sqrt{2}(z - w)^{\frac{1}{2}}}\sigma(w, \bar{w})$$

$$\bar{\psi}(\bar{z})\sigma(w, \bar{w}) = \frac{e^{-i\pi/4}}{\sqrt{2}(\bar{z} - \bar{w})^{\frac{1}{2}}}\mu(w, \bar{w}) \quad (12.68)$$

$$\bar{\psi}(\bar{z})\mu(w, \bar{w}) = \frac{e^{i\pi/4}}{\sqrt{2}(\bar{z} - \bar{w})^{\frac{1}{2}}}\sigma(w, \bar{w})$$

Note that the phase factors agree with the definition of the energy operator as $\varepsilon = i : \psi\bar{\psi} :$, and guarantee that the operators ε, σ, and μ are real. For instance, we recover

$$\varepsilon(z, \bar{z})\sigma(w, \bar{w}) = i\psi(z)[\bar{\psi}(\bar{z})\sigma(w, \bar{w})]$$

$$= i\psi(z)\frac{e^{-i\pi/4}}{(\bar{z} - \bar{w})^{\frac{1}{2}}}\mu(w, \bar{w})$$

$$= i\frac{e^{-i\pi/4}}{\sqrt{2}(\bar{z} - \bar{w})^{\frac{1}{2}}}[\psi(z)\mu(w, \bar{w})]$$

$$= i\frac{e^{-i\pi/4}}{\sqrt{2}(\bar{z} - \bar{w})^{\frac{1}{2}}}\frac{e^{-i\pi/4}}{\sqrt{2}(z - w)^{\frac{1}{2}}}\sigma(w, \bar{w})$$

$$= \frac{1}{2|z - w|}\sigma(w, \bar{w}) \quad (12.69)$$

hence $C_{\varepsilon\sigma}{}^{\sigma} = \frac{1}{2}$.

The bosonization formulae for the Ising correlation functions may be summarized in a unique equation

$$
\begin{aligned}
&\langle \sigma(1) \cdots \sigma(2m) \mu(2m+1) \cdots \mu(2n) \varepsilon(2n+1) \cdots \varepsilon(2n+p) \rangle^2 \\
&= N_{n,p} \left\langle \prod_{i=1}^{2m} \cos \frac{\varphi}{2}(i) \prod_{j=2m+1}^{2n} \sin \frac{\varphi}{2}(j) \prod_{k=2n+1}^{2n+p} (\nabla \varphi / 2)^2(k) \right\rangle
\end{aligned}
$$

(12.70)

where we used again the notation $\sigma(j) \equiv \sigma(z_j, \bar{z}_j)$ and similarly for μ and ε. The normalization factor is again fixed to be

$$
N_{n,p} = (-1)^p 2^n \tag{12.71}
$$

by the short-distance limit.

§12.4. The Ising Model on the Torus

So far we have dealt with the Ising theory only on the complex plane. However, with this completely solvable case, it is interesting to study the effect of finite geometry. Actually, numerical calculations for spin models are always carried out, using the transfer matrix method, on strips of finite width (related to the size of the transfer matrix). If no transfer matrix is available, the calculations must be performed on some finite rectangle of size, say, $N \times M$. In each case, various boundary conditions can be imposed.

In fact, we can define the square Ising model on any closed Riemann surface by replacing the surface by a "quadrangulation", namely, a tessellation with possibly slightly deformed squares, which wraps around the handles of the surface, incorporating thus the effect of topology. To include also the effects of curvature, a given vertex of the tessellation may be common to an arbitrary number of deformed squares (not necessarily 4; for instance, a vertex common to only 3 squares in the tessellation looks like the vertex of a cube and therefore indicates some positive curvature, whereas a vertex common to 5 or more squares indicates negative curvature). Such spin models are instrumental in investigations of quantum gravity, more precisely of the coupling of a matter theory (here the Ising model) to fluctuations of space-time geometry (topology and curvature). These are actually toy models for quantum string theory. In this sense, conformal theories on the plane are often considered as *string vacua*, namely the nonfluctuating flat space version of some string theories.

The simplest example of a surface with nontrivial topology (one handle) is the torus or, equivalently, a parallelogram with doubly periodic boundary conditions (cf. Chap. 10). One example is the rectangle of $N \times M$ bonds with the identification of the N horizontal bonds at the top and bottom of the rectangle and the identification of the M vertical ones at the left and right of the rectangle. Such a torus is parametrized by a complex parameter, $\tau = iM/N$. When taking a suitable thermodynamic limit ($M, N \rightarrow \infty$, M/N finite), we end up with a continuum model with doubly periodic boundary conditions *on the fields*. The most general torus is characterized by a complex parameter τ, with Im $\tau > 0$ (a nonvanishing real part of τ means that the two lines, along which periodic boundary conditions are to be taken, form an angle $\alpha \neq \pi/2$).

12.4.1. The Partition Function

The partition function of the free real fermion on the torus was calculated in Chap. 10. We recall some relevant results. Along the two directions of the torus (1 and τ), the fermions may have periodic (R) or antiperiodic (NS) boundary conditions. This leads to four sectors denoted by (v, u), $u, v \in \{0, \frac{1}{2}\}$, according to the boundary conditions

$$\psi(z + 1) = e^{2i\pi v} \psi(z) \qquad \psi(z + \tau) = e^{2i\pi u} \psi(z) \qquad (12.72)$$

and hence four contributions $Z_{v,u}$ to the partition function. According to Eq. (10.46), $Z_{0,0}$ vanishes identically, and we are left with

$$
\begin{aligned}
Z_{\text{Ising}} &= Z_{0,\frac{1}{2}} + Z_{\frac{1}{2},\frac{1}{2}} + Z_{\frac{1}{2},0} \\
&= \frac{1}{2}\left[\left|\frac{\theta_2(0|\tau)}{\eta(\tau)}\right| + \left|\frac{\theta_3(0|\tau)}{\eta(\tau)}\right| + \left|\frac{\theta_4(0|\tau)}{\eta(\tau)}\right|\right] \qquad (12.73) \\
&= |\chi_{1,1}(\tau)|^2 + |\chi_{2,1}(\tau)|^2 + |\chi_{1,2}(\tau)|^2
\end{aligned}
$$

The notations for the θ functions are defined in App. 10.A. The conformal characters of the identity, energy, and spin operators are respectively identified as

$$
\begin{aligned}
\chi_{1,1}(\tau) &= \frac{1}{2\sqrt{\eta(\tau)}}\left[\sqrt{\theta_3(0|\tau)} + \sqrt{\theta_4(0|\tau)}\right] \\
\chi_{2,1}(\tau) &= \frac{1}{2\sqrt{\eta(\tau)}}\left[\sqrt{\theta_3(0|\tau)} - \sqrt{\theta_4(0|\tau)}\right] \qquad (12.74) \\
\chi_{1,2}(\tau) &= \frac{1}{\sqrt{2\eta(\tau)}}\sqrt{\theta_2(0|\tau)}
\end{aligned}
$$

These are also the three conformal blocks of any correlation of the identity operator on the torus.

The vanishing of the partition function $Z_{0,0}$ in the periodic-periodic sector is simply due to the zero-mode of the Laplacian in this sector, namely, the contribution of the Grassmannian integral $\int d\psi_0 = 0$. However, some correlation functions will receive nonvanishing contributions from this sector whenever some field insertion compensates the zero-mode, simply because $\int \psi_0 d\psi_0 = 1$. This is the case for the expectation value of the energy operator in the periodic-periodic sector

$$Z_{0,0}\langle\varepsilon(z,\bar{z})\rangle_{0,0} = \int d\psi d\bar{\psi}\,\psi\bar{\psi}\,e^{-S} \qquad (12.75)$$

The insertion of the fermions has the effect of canceling exactly the zero-mode contribution $n = 0$ responsible for the vanishing of $d_{0,0}$ (cf. Eq. (10.46)), and we are left with

$$|d'_{0,0}|^2 = \left| q^{1/24} \prod_{n=1}^{\infty} (1 - q^n) \right|^2 = |\eta(\tau)|^2 \qquad (12.76)$$

Conversely, the insertion of ε introduces zero-modes in the other sectors $(v, u) \neq (0, 0)$, causing their contributions to vanish. As a result, the expectation value of the energy operator receives contributions only from the $(0, 0)$ sector, so that

$$Z_{\text{Ising}} \langle \varepsilon(z, \bar{z}) \rangle \propto |\eta(\tau)|^2 \qquad (12.77)$$

We note that the expectation value (12.77) is identified with the modulus square of the only conformal block of the one-point function of the energy on the torus, namely

$$\eta(\tau) = \quad \begin{array}{c} \varepsilon \\ \mid \\ \bigcirc \end{array} {}_{\sigma} \qquad (12.78)$$

also expressing the fusion rule $\sigma \times \sigma \to \varepsilon$. In general, all four sectors are expected to contribute to correlation functions.

12.4.2. General Ward Identities on the Torus

The conformal Ward identity (5.41) of Chap. 5 admits an extension on any Riemann surface. It expresses the transformation of correlators under a change of coordinates. On the torus, two new ingredients must be incorporated: (i) since the correlation functions are now elliptic (doubly periodic) functions, short-distance singularities give rise to an infinite number of poles on a doubly periodic lattice; (ii) reparametrizations of the torus (i.e., changes of τ) have to be incorporated. A direct consequence of the presence of this new parameter τ is that the expectation value of the energy-momentum tensor does not vanish on the torus. In fact, the energy-momentum tensor generates the reparametrizations of the torus itself and its expectation value is related to the variation of the partition function with respect to τ. The torus Ward identity for the insertion of the energy-momentum tensor, in a correlator of primary fields ϕ_i with conformal dimensions h_i, reads

$$
\begin{aligned}
&\langle T(z)\phi_1(z_1, \bar{z}_1) \cdots \phi_n(z_n, \bar{z}_n) \rangle - \langle T \rangle \langle \phi_1(z_1, \bar{z}_1) \cdots \phi_n(z_n, \bar{z}_n) \rangle \\
&= \left\{ \sum_{i=1}^{n} \left[h_i (\wp(z - z_i) + 2\eta_1) + (\zeta(z - z_i) + 2\eta_1 z_i) \partial_{z_i} \right] \right. \\
&\qquad \left. + 2i\pi \partial_\tau \right\} \langle \phi_1(z_1, \bar{z}_1) \cdots \phi_n(z_n, \bar{z}_n) \rangle
\end{aligned}
\qquad (12.79)
$$

with

$$\langle T \rangle = 2i\pi \partial_\tau \ln Z \tag{12.80}$$

where Z is the torus partition function of the theory. The zeta function ζ and the Weierstrass function $\wp$ are the elliptic generalizations of $1/z$ and $1/z^2$, respectively, namely

$$\zeta(z) = \frac{\partial_z \theta_1(z|\tau)}{\theta_1(z|\tau)} + 2\eta_1 z$$

$$\wp(z) = -\partial_z \zeta(z) \tag{12.81}$$

with

$$\eta_1 = \zeta(\tfrac{1}{2}) = (2\pi)^2 \left(\frac{1}{24} - \sum_{n\geq 1} \frac{nq^n}{(1-q^n)} \right) = -\frac{1}{6}\frac{\partial_z^3 \theta_1(0|\tau)}{\partial_z \theta_1(0|\tau)} \tag{12.82}$$

We will not prove the Ward identities (12.79) here. However, it will be an instructive exercise for the reader to directly check that they are indeed satisfied by the Ising torus correlators calculated below (see, e.g., Ex. 12.20). In the Ising case, we have, sector by sector,

$$\langle T \rangle_\nu = 2i\pi \partial_\tau \ln Z_\nu = i\pi \partial_\tau \ln \frac{\theta_\nu(0|\tau)}{\eta(\tau)} \tag{12.83}$$

for $\nu = 2,3,4$ (corresponding resp. to $(\nu,u) = (0,\tfrac{1}{2}), (\tfrac{1}{2},\tfrac{1}{2}), (\tfrac{1}{2},0)$). Summing over all sectors yields

$$\langle T \rangle = 2i\pi \partial_\tau \ln Z_{\text{Ising}} \tag{12.84}$$

We see that the periodic-periodic sector $\nu = 1$ does not contribute: $Z_1 \langle T \rangle_1 = 0$.

When combined with the singular-vector conditions for primary fields, the Ward identity (12.79) leads to elliptic differential equations for correlators on the torus. In the case of a field ϕ with conformal dimension $h = h_{1,2}$ or $h_{2,1}$, where the highest weight vector $|h\rangle$ is degenerate at level 2, this leads to

$$\left[\frac{3}{2(2h+1)}\partial_z^2 - 2\eta_1(h+z\partial_z) - 2i\pi\partial_\tau - \sum_{i=1}^{n}(\zeta(z-z_i) + 2\eta_1 z_i)\partial_{z_i} \right.$$
$$\left. - \sum_{i=1}^{n}(\wp(z-z_i)+2\eta_1) \right]\left\{ Z\langle\phi(z,\bar z)\phi_1(z_1,\bar z_1)\cdots\phi_n(z_n,\bar z_n)\rangle \right\} = 0 \tag{12.85}$$

where the differential operator acts on $Z \times \langle \cdots \rangle$. Instead of trying to solve Eq. (12.85) directly, we will resort to other methods for computing the correlators. Checking that these equations are indeed satisfied will result from/in a variety of theta function identities.

§12.5. Correlation Functions on the Torus

12.5.1. Fermion and Energy Correlators

The fermion propagator

$$G_v(z - w) = \langle \psi(z)\psi(w) \rangle_v \tag{12.86}$$

on the torus in a given sector $v \equiv (v, u)$ has the following properties:
(i) it has a single pole at $z \to w$, with residue 1 (at short distances, we must recover the propagator on the plane).
(ii) it is a meromorphic function with periodicity conditions

$$G_v(z + 1) = e^{2i\pi v}G_v(z) \qquad G_v(z + \tau) = e^{2i\pi u}G_v(z) \tag{12.87}$$

The function

$$\wp_v(z) = \frac{\theta_v(z|\tau)\partial_z\theta_1(0|\tau)}{\theta_v(0|\tau)\theta_1(z|\tau)} \qquad (v = 2, 3, 4) \tag{12.88}$$

satisfies (i) and (ii), due to the theta function transformation properties under $z \to z + 1, z + \tau$ (cf. App. 10.A). Therefore, the ratio

$$r_v = G_v/\wp_v \tag{12.89}$$

is an elliptic (doubly periodic) function, whose possible poles could come only from the zeros of $\wp_v$. But $\wp_v$ has only one zero on the torus, as a consequence of Eq. (12.190), hence the ratio r_v has *at most one pole*. By the standard theory of elliptic functions,[8] this implies that the ratio r_v is a constant, fixed by the residue condition (i) and hence $r_v = 1$. Therefore, in the three sectors $v = 2, 3, 4$, the fermion propagator reads

$$\langle \psi(z)\psi(w) \rangle_v = \wp_v(z - w) = \frac{\theta_v(z - w|\tau)\partial_z\theta_1(0|\tau)}{\theta_v(0|\tau)\theta_1(z - w|\tau)} \qquad (v = 2, 3, 4) \quad (12.90)$$

The fermion propagator receives no contribution from the doubly periodic sector $v = 1$, because of the fermion zero-modes. Hence, the total propagator on the torus reads

$$\langle \psi(z)\psi(w) \rangle = \frac{1}{Z_{\text{Ising}}} \sum_{v=2}^{4} Z_v \wp_v(z - w) \tag{12.91}$$

The propagator of $\bar{\psi}$ is simply the complex conjugate of that of ψ. It is instructive to check that, sector by sector, the propagator (12.90) satisfies the differential equation (12.85) for $h = \frac{1}{2}$ (cf. Ex. 12.13).

---

[8] An elliptic function must have the same number of zeros and poles, and if it is not a constant, this number is at least 2. See App. 12.A and particularly Sect. 12.A.1 for a detailed proof of this statement.

From the expression (12.90), we derive the expectation value of the energy-momentum tensor in each sector $\nu = 2, 3, 4$, by applying the definition (12.22)

$$
\begin{aligned}
\langle T \rangle_\nu &= -\frac{1}{2} \left\langle \lim_{z \to w} \left[\frac{1}{2} (\psi(z) \partial_w \psi(w) - \partial_z \psi(z) \psi(w)) - \frac{1}{(z-w)^2} \right] \right\rangle_\nu \\
&= \frac{1}{2} \lim_{z \to 0} \left[\partial_z G_\nu(z) - \frac{1}{z^2} \right] \\
&= \frac{1}{4} \frac{\partial_z^2 \theta_\nu(0|\tau)}{\theta_\nu(0|\tau)} - \frac{1}{12} \frac{\partial_z^3 \theta_1(0|\tau)}{\partial_z \theta_1(0|\tau)}
\end{aligned}
\tag{12.92}
$$

By using the results of App. 10.A, we can readily check Eq. (12.92) to be in agreement with Eq. (12.83).

The torus correlators of an even number of energy operators follow directly from Wick's theorem, using the propagators (12.90). Only the sectors $\nu = 2, 3, 4$ contribute. For the two-point function, this gives

$$
\langle \varepsilon(z, \bar{z}) \varepsilon(0, 0) \rangle_\nu = |\wp_\nu(z)|^2
$$

$$
\begin{aligned}
\langle \varepsilon(z, \bar{z}) \varepsilon(0, 0) \rangle &= \frac{1}{Z_{\text{Ising}}} \sum_{\nu=2}^{4} Z_\nu |\wp_\nu(z)|^2 \\
&= \frac{1}{Z_{\text{Ising}}} \left| \frac{\partial_z \theta_1(0|\tau)}{\theta_1(z|\tau)} \right|^2 \sum_{\nu=2}^{4} \left| \frac{\theta_\nu(z|\tau)^2}{2\eta(\tau)\theta_\nu(0|\tau)} \right|
\end{aligned}
\tag{12.93}
$$

We easily identify this equation as the sum of the moduli squared of the three conformal blocks of the energy two-point function on the torus, which read respectively

$$
\text{(diagram)} \quad = \quad \frac{1}{2\sqrt{\eta(\tau)}} \frac{\partial_z \theta_1(0|\tau)}{\theta_1(z|\tau)} \left(\frac{\theta_3(z|\tau)}{\sqrt{\theta_3(0|\tau)}} + \frac{\theta_4(z|\tau)}{\sqrt{\theta_4(0|\tau)}} \right)
$$

$$
\text{(diagram)} \quad = \quad \frac{1}{2\sqrt{\eta(\tau)}} \frac{\partial_z \theta_1(0|\tau)}{\theta_1(z|\tau)} \left(\frac{\theta_3(z|\tau)}{\sqrt{\theta_3(0|\tau)}} - \frac{\theta_4(z|\tau)}{\sqrt{\theta_4(0|\tau)}} \right) \tag{12.94}
$$

$$
\text{(diagram)} \quad = \quad \frac{1}{\sqrt{2\eta(\tau)}} \frac{\partial_z \theta_1(0|\tau)}{\theta_1(z|\tau)} \frac{\theta_2(z|\tau)}{\sqrt{\theta_2(0|\tau)}}
$$

in agreement with the Ising fusion rules

$$
\begin{aligned}
\varepsilon \times \mathbb{I} &= \varepsilon \\
\varepsilon \times \varepsilon &= \mathbb{I} \\
\varepsilon \times \sigma &= \sigma
\end{aligned}
\tag{12.95}
$$

We note that the blocks (12.94) are normalized in such a way that the corresponding Ising characters (12.74) are recovered in the $z \to 0$ limit. We also note that the monodromy properties of these blocks along the homology cycles a and b of the torus agree with the general analysis of Sect. 10.8.3. More precisely, using the transformations of ratios of θ functions given in Eq. (12.188), we find that the monodromy is diagonal along the cycle a (i.e., the transformation $z \to z+1$ leaves the blocks unchanged up to a multiplicative phase factor), whereas the monodromy along the cycle b ($z \to z + \tau$) exchanges the two first blocks in Eq. (12.94) and leaves the third one unchanged up to a multiplicative phase. The latter agrees with the interpretation of the circle as the time (cycle b) direction in the pictorial representation of the conformal blocks. Letting the variable z circulate along this cycle indeed results in the exchange of the upper and lower intermediate states.

The torus generalization of Eq. (12.21) for the $2n$-point function in each sector $\nu = 2, 3, 4$, is

$$\langle \varepsilon(z_1, \bar{z}_1) \cdots \varepsilon(z_{2n}, \bar{z}_{2n}) \rangle_\nu = \left| \mathrm{Pf}\left[\wp_\nu(z_i - z_j) \right] \right|^2 \qquad (12.96)$$

where the matrix of propagators is understood to have vanishing diagonal entries, and the total energy correlator is then

$$\langle \varepsilon(z_1, \bar{z}_1) \cdots \varepsilon(z_{2n}, \bar{z}_{2n}) \rangle = \frac{1}{Z_{\mathrm{Ising}}} \sum_{\nu=2}^{4} Z_\nu \langle \varepsilon(z_1, \bar{z}_1) \cdots \varepsilon(z_{2n}, \bar{z}_{2n}) \rangle_\nu \qquad (12.97)$$

As mentioned above, the one-point function of the energy on the torus receives a contribution only from the $\nu = 1$ doubly periodic sector. Actually, the computation of the correlator of an *odd* number of energy operators requires more work. This will be treated in the next section by means of bosonization techniques.

12.5.2. Spin and Disorder-Field Correlators

As pointed out previously, the spin correlations are more involved, due to the non-locality of the spin operator with respect to the fermion operator (cf. Eq. (12.26)). In this section, we use a trick, known as the energy-momentum-tensor technique, to compute directly the two-point function of the spin operator on the torus. It is some straightforward elliptic generalization of the plane calculation performed in Sect. 12.2.2. It is simpler to first compute the insertion of the energy-momentum tensor in the spin-spin correlator, and then deduce the spin-spin correlator by short-distance limits.

The first step is the computation of the ratio of correlators

$$G_\nu(z, w, z_1, \bar{z}_1, z_2, \bar{z}_2) = \frac{\langle \psi(z)\psi(w)\sigma(z_1, \bar{z}_1)\sigma(z_2, \bar{z}_2) \rangle_\nu}{\langle \sigma(z_1, \bar{z}_1)\sigma(z_2, \bar{z}_2) \rangle_\nu} \qquad (12.98)$$

sector by sector, for $\nu = 1, 2, 3, 4$. This function G_ν satisfies the following properties as a function of the complex variable z:
(i) It is an analytic function, except for a single pole at $z = w$ with residue 1, and some inverse square root branch cuts at $z = z_1, z_2$ (due to the OPE (12.26)).

(ii) Under $z \to z+1, z+\tau$, it transforms with the (anti-)periodic boundary conditions pertaining to the sector v, and is antisymmetric under the exchange $z \leftrightarrow w$.

These properties fix the value of G_v. It is not too difficult to figure out how to modify the plane solution (12.30) in order to satisfy these conditions. We find the candidate

$$H_v(z, w, z_1, \bar{z}_1, z_2, \bar{z}_2) = \frac{1}{2} \frac{\partial_z \theta(0|\tau)}{\theta_1(z-w|\tau)}$$

$$\times \left[\frac{\theta_v(z-w+(z_{12}/2)|\tau)}{\theta_v(z_{12}/2|\tau)} \left(\frac{\theta_1(z-z_1|\tau)\theta_1(w-z_2|\tau)}{\theta_1(z-z_2|\tau)\theta_1(w-z_1|\tau)} \right)^{\frac{1}{2}} + (z \leftrightarrow w) \right]$$

$$(12.99)$$

where $z_{12} = z_1 - z_2$. The reader will easily check the properties (i)–(ii) using Apps. 10.A and 12.A. We introduce the auxiliary function

$$\alpha_v(z) = \frac{\theta_v(z - \frac{1}{2}(z_1+z_2)|\tau)}{\left[\theta_1(z-z_1|\tau)\theta_1(z-z_2|\tau) \right]^{\frac{1}{2}}} \tag{12.100}$$

and consider the normalized difference

$$\frac{G_v - H_v}{\alpha_v(z)\alpha_v(w)} \tag{12.101}$$

This is an elliptic function of z, with *at most* one pole, located at $z = w$. By standard elliptic function theory, this must be a constant. Due to the requirement of antisymmetry under the exchange $z \leftrightarrow w$ (property (ii)), this constant must vanish. We have thus proven that $G_v = H_v$.

In a second step, we use the fermionic expression of the energy-momentum tensor (12.22) to derive

$$\frac{\langle T(z)\sigma(z_1, \bar{z}_1)\sigma(z_2, \bar{z}_2)\rangle_v}{\langle \sigma(z_1, \bar{z}_1)\sigma(z_2, \bar{z}_2)\rangle_v} = \frac{1}{2} \lim_{z \to w} \left[\partial_z G_v(z, w, z_1, \bar{z}_1, z_2, \bar{z}_2) + \frac{1}{(z-w)^2} \right]$$

$$(12.102)$$

The expansion of the l.h.s. of (12.102) around $z \to z_1$ and the usual expression for the OPE of $T(z)$ with the primary field σ lead to

$$\frac{\langle T(z)\sigma(z_1, \bar{z}_1)\sigma(z_2, \bar{z}_2)\rangle_v}{\langle \sigma(z_1, \bar{z}_1)\sigma(z_2, \bar{z}_2)\rangle_v} = \frac{h_\sigma}{(z-z_1)^2}$$

$$+ \frac{1}{z-z_1} \partial_{z_1} \ln\langle \sigma(z_1, \bar{z}_1)\sigma(z_2, \bar{z}_2)\rangle_v + \text{reg.}$$

$$(12.103)$$

Expanding the r.h.s. of (12.102) using the exact expression (12.99), we identify $h_\sigma = \frac{1}{16}$ (we recover the planar case) and the z_1, z_2-dependence of the spin-spin correlator:

$$\langle \sigma(z_1, \bar{z}_1)\sigma(z_2, \bar{z}_2)\rangle_v \propto C_v \frac{\theta_v(z_{12}/2|\tau)^{\frac{1}{2}}}{\theta_1(z_{12}|\tau)^{\frac{1}{8}}} \tag{12.104}$$

where C_ν is a proportionality constant (possibly a function of τ), not fixed by the above limiting procedure. A similar argument yields the antiholomorphic part of the correlator. For $\nu = 2, 3, 4$, the normalization of the two-point function is fixed by the planar limit (12.23) when $z_1 \to z_2$, which fixes

$$|C_\nu|^2 = \frac{|\partial_z \theta_1(0|\tau)|^{\frac{1}{4}}}{|\theta_\nu(0|\tau)|} \tag{12.105}$$

We end up with

$$\langle \sigma(z_1, \bar{z}_1) \sigma(z_2, \bar{z}_2) \rangle_\nu = \left| \frac{\theta_\nu(z_{12}/2|\tau)}{\theta_\nu(0|\tau)} \right| \left| \frac{\partial_z \theta_1(0|\tau)}{\theta_1(z_{12}|\tau)} \right|^{\frac{1}{4}} \tag{12.106}$$

This expression would lead naively to an infinite contribution of the $\nu = 1$ sector. However, in the full correlator, this should be weighed by $Z_1 = 0$. Indeed, the $\nu = 1$ sector must contribute, because the OPE (12.25) of two spin operators contains the energy operator, and the latter has a nonvanishing expectation value (Eq. (12.77)). Therefore, the limit $z_1 \to z_2$ of the total spin-spin correlator must lead to the energy expectation value (12.77), which receives contributions from the $\nu = 1$ sector only. To fix the contribution of the $\nu = 1$ sector to the spin-spin correlator, we first notice that the functions $\langle \sigma\sigma \rangle_\nu(z_{12}, \bar{z}_{12})$ are not periodic under $z_1 \to z_1 + 1, z_1 + \tau$ (cf. Ex. 12.15). This is again a manifestation of the nonlocality of the spin operator with respect to the fermion. Letting $z_1 \to z_1 + 1, z_1 + \tau$ creates a "frustration line" winding around the torus, which reverses the sign of the boundary condition for the fermion. As a result, this exchanges the various sectors ν. This involves the $\nu = 1$ sector as well and fixes its normalization (see Ex. 12.15 for details). We finally get the general answer for $\nu = 1, 2, 3, 4$:

$$Z_\nu \langle \sigma(z_1, \bar{z}_1) \sigma(z_2, \bar{z}_2) \rangle_\nu = \frac{1}{2} (2\pi)^{\frac{1}{3}} |\partial_z \theta_1(0|\tau)|^{-\frac{1}{12}} \frac{|\theta_\nu(z_{12}/2|\tau)|}{|\theta_1(z_{12}|\tau)|^{\frac{1}{4}}} \tag{12.107}$$

This function is indeed a solution of the differential equation (12.85), sector by sector (see Ex. 12.16 for a proof). Furthermore, in the $\nu = 1$ sector, it gives the precise normalization of the energy expectation value (12.77) (see Ex. 12.17).

Finally the complete spin-spin correlator on the torus reads

$$\langle \sigma(z, \bar{z}) \sigma(0, 0) \rangle = \left| \frac{\partial_z \theta_1(0|\tau)}{\theta_1(z|\tau)} \right|^{\frac{1}{4}} \frac{\sum_{\nu=1}^{4} |\theta_\nu(z/2|\tau)|}{\sum_{\nu=2}^{4} |\theta_\nu(0|\tau)|} \tag{12.108}$$

We identify the four conformal blocks of the spin-spin correlator on the torus as

$$= \frac{1}{2\sqrt{\eta(\tau)}} \left(\frac{\partial_z \theta_1(0|\tau)}{\theta_1(z|\tau)} \right)^{\frac{1}{8}} \left[\sqrt{\theta_3(z/2|\tau)} + \sqrt{\theta_4(z/2|\tau)} \right]$$

$$\frac{1}{2\sqrt{\eta(\tau)}}\left(\frac{\partial_z\theta_1(0|\tau)}{\theta_1(z|\tau)}\right)^{\frac{1}{8}}\left[\sqrt{\theta_3(z/2|\tau)}-\sqrt{\theta_4(z/2|\tau)}\right]$$

$$\frac{1}{2\sqrt{\eta(\tau)}}\left(\frac{\partial_z\theta_1(0|\tau)}{\theta_1(z|\tau)}\right)^{\frac{1}{8}}\left[\sqrt{\theta_2(z/2|\tau)}+e^{i\pi/4}\sqrt{\theta_1(z/2|\tau)}\right]$$

$$\frac{1}{2\sqrt{\eta(\tau)}}\left(\frac{\partial_z\theta_1(0|\tau)}{\theta_1(z|\tau)}\right)^{\frac{1}{8}}\left[\sqrt{\theta_2(z/2|\tau)}-e^{i\pi/4}\sqrt{\theta_1(z/2|\tau)}\right]$$

$$(12.109)$$

Again the blocks are normalized so as to recover the corresponding characters in the limit $z \to 0$ (an extra factor $1/\sqrt{2}$ is added in the two last blocks, each of which contributes to the spin character in this limit). The precise relative normalization of the terms $\sqrt{\theta_1}$ in the two last blocks is fixed by the monodromy properties of the blocks, when $z \to z+1$ and $z \to z+\tau$. We recall that, from the general analysis of Sect. 10.8.3, we expect the monodromy to be diagonal around the cycle a (i.e., under $z \to z+1$), and to exchange the upper and lower intermediate states around the cycle b (i.e., under $z \to z+\tau$). To compute the monodromy of the conformal blocks of Eq. (12.109), we need the transformation properties of the θ functions under shifts of z by half periods listed in Eq. (12.189). Under the monodromy transformation $z \to z+1$, we have

$$\begin{aligned}
\sqrt{\theta_3(z/2|\tau)} &\to \sqrt{\theta_4(z/2|\tau)} \\
\sqrt{\theta_4(z/2|\tau)} &\to \sqrt{\theta_3(z/2|\tau)} \\
\sqrt{\theta_2(z/2|\tau)} &\to i\sqrt{\theta_1(z/2|\tau)} \\
\sqrt{\theta_1(z/2|\tau)} &\to \sqrt{\theta_2(z/2|\tau)}
\end{aligned} \qquad (12.110)$$

This transformation is diagonalized by taking the combinations

$$\begin{aligned}
\sqrt{\theta_3(z/2|\tau)} &\pm \sqrt{\theta_4(z/2|\tau)} \\
\sqrt{\theta_2(z/2|\tau)} &\pm e^{i\pi/4}\sqrt{\theta_1(z/2|\tau)}
\end{aligned} \qquad (12.111)$$

which fix, up to a global normalization, the form of the four conformal blocks in Eq. (12.109). With these combinations, the monodromy transformation $z \to z+\tau$ is readily seen to exchange, up to global multiplicative phases, the first and second blocks in Eq. (12.109), as well as the third and the fourth. The above blocks agree with the conformal fusion rule $\sigma \times \sigma = \mathbb{I} + \varepsilon$.

The two-point correlation function for the disorder operator on the torus is easily obtained from the function G_ν (12.98). From the OPE (cf. Eq. (12.68))

$$\psi(z)\sigma(w,\bar{w}) = \frac{e^{i\pi/4}}{\sqrt{2}} \frac{\mu(w,\bar{w})}{(z-w)^{\frac{1}{2}}} + \cdots \tag{12.112}$$

we see that the ratio of the disorder-disorder correlator to the spin-spin correlator can be extracted by taking the limits $z \to z_2$ and $w \to z_1$ in $G_\nu(z,w,z_1,\bar{z}_1,z_2,\bar{z}_2)$. The result is

$$\frac{\langle \mu(z_1,\bar{z}_1)\mu(z_2,\bar{z}_2)\rangle_\nu}{\langle \sigma(z_1,\bar{z}_1)\sigma(z_2,\bar{z}_2)\rangle_\nu} = \pm\epsilon_\nu \tag{12.113}$$

where ϵ_ν is the parity of the function θ_ν as a function of z, namely $\epsilon_1 = -1$ and $\epsilon_\nu = 1$ for $\nu = 2,3,4$. The overall sign is fixed to be (+) by the OPE

$$\mu(z_1,\bar{z}_1)\mu(z_2,\bar{z}_2) = \frac{1}{|z_1-z_2|^{\frac{1}{4}}} - \frac{1}{2}|z_1-z_2|^{\frac{3}{4}}\varepsilon(z_2,\bar{z}_2) + \cdots \tag{12.114}$$

The OPE (12.114) is obtained from the spin-spin one (12.25) by the duality transformation $\sigma \leftrightarrow \mu$ and $\varepsilon \to -\varepsilon$. This leads to the total disorder-disorder correlator on the torus:

$$\langle \mu(z_1,\bar{z}_1)\mu(z_2,\bar{z}_2)\rangle = \left|\frac{\partial_z\theta_1(0|\tau)}{\theta_1(z|\tau)}\right|^{\frac{1}{4}} \frac{\sum_{\nu=1}^{4}\epsilon_\nu|\theta_\nu(z/2|\tau)|}{\sum_{\nu=2}^{4}|\theta_\nu(0|\tau)|} \tag{12.115}$$

This function, however, has a new peculiarity: it is *not* doubly periodic on the torus. Actually, the transformations $z_1 \to z_1 + 1, z_1 + \tau$ and $z_1 + 1 + \tau$ generate three other correlators. As pointed out in Sect. 12.1, the disorder-disorder correlator can be viewed as the ratio of a "frustrated" Ising partition function to the Ising partition function. Indeed, the insertion of the disorder operators creates a frustration line joining z_1 and z_2. In Eq. (12.115), this line is shrunk to a point when $z_1 \to z_2$.

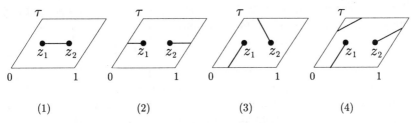

$$(1) \qquad\qquad (2) \qquad\qquad (3) \qquad\qquad (4)$$

Figure 12.2. The four possible frustration lines for the disorder-disorder correlator on the torus. In (1), the cycle is contractible. In (2), (3), and (4), we get, respectively, the noncontractible cycles a, b, and a + b, corresponding to the translations of z_1 by 1, τ, and $1 + \tau$, respectively.

However, after any of the above three translations, this line winds around the torus and can no longer be shrunk to a point. Actually, the three noncontractible frustration lines correspond respectively to the cycles a, b, and a + b of Fig. 12.2.

The resulting frustrated partition functions are therefore different, since the boundary conditions are affected by the presence of the frustration line. More precisely, letting $z_1 \to z_2$ after each translation, we get the three limits (labeled by an index $\alpha = 2, 3, 4$):

$$\langle \mu(z_1, \bar{z}_1) \mu(z_2, \bar{z}_2) \rangle^{(\alpha)} = \frac{1}{|z_1 - z_2|^{\frac{1}{4}}} \frac{1}{Z_{\text{Ising}}} \sum_{\nu=2}^{4} \epsilon_\nu^{(\alpha)} Z_\nu \qquad (12.116)$$

where $\epsilon_\nu^{(\alpha)} = -1$ if $\alpha = \nu$ and $\epsilon_\nu^{(\alpha)} = 1$ otherwise. The numerator in Eq. (12.116) is identified as the corresponding frustrated partition function.

§12.6. Bosonization on the Torus

Except for the correlation functions of an even number of energy operators, for which Wick's theorem can be applied directly, the techniques sketched so far are not powerful enough to yield the most general torus correlators. In this last section, we show how the bosonization techniques introduced in Sect. 12.3 go over to the torus. Given the relation between a theory made of two copies of the Ising model and a free bosonic field, the main question boils down to the calculation of correlations involving the bosonic field on a torus.

12.6.1. The Two Bosonizations of the Ising Model: Partition Functions and Operators

The Dirac field (12.35) is a linear combination of two real fermions ψ_1, ψ_2 and their antiholomorphic counterparts. However, due to the precise relation (12.35), the two real fermions turn out to be coupled in a subtle way. The partition function of the free Dirac fermion on a torus receives contributions from various (anti-)periodicity sectors, again labeled $\nu = 1, 2, 3, 4$. In each of these sectors, the partition function Z_ν^{Dirac} is the determinant of the Laplace operator *squared*. It is thus expressed very simply in terms of the Ising partition functions Z_ν as

$$Z_\nu^{\text{Dirac}} \propto Z_\nu^2 \qquad \nu = 1, 2, 3, 4 \qquad (12.117)$$

we note that, if $\nu = 1$, the zero-modes of the Dirac fermion cause Z_1^{Dirac} to vanish. The result (12.117) shows how the two underlying copies of the Ising model turn out to be coupled in this scheme: the two real fermions ψ_1 and ψ_2 must have *the same* boundary conditions on the torus. This phenomenon could not be observed on the plane, where the two fermions appeared totally decoupled. The total partition function Z_{Dirac} exhibits the central charge $c = 1$ in its small-q behavior:

$$Z_{\text{Dirac}} = 2 \sum_{\nu=2}^{4} Z_\nu^2 \to (q\bar{q})^{-1/24}(1 + O(q, \bar{q})) \qquad (12.118)$$

The prefactor 2 normalizes the leading term to 1 (the identity operator is nondegenerate in the corresponding conformal theory).

We now show that the partition function (12.118) is equivalent to that of a free boson compactified on a circle of radius $R = 1$ (cf. Sect. 10.4.1). The partition function of the latter reads

$$Z(R = 1) \equiv Z(1) = \frac{1}{|\eta(\tau)|^2} \sum_{e,m \, \in \, \mathbb{Z}} q^{(e+m/2)^2/2} \bar{q}^{(e-m/2)^2/2} \qquad (12.119)$$

We introduce the new summation variables

$$\begin{aligned}
r = e + m/2 \qquad & s = e - m/2 \qquad && (m \text{ even}) \\
r + \frac{1}{2} = e + m/2 \quad & s - 1/2 = e - m/2 \quad && (m \text{ odd})
\end{aligned} \qquad (12.120)$$

The only constraint left on r and s is that they must have the same parity; hence we rewrite

$$Z(1) = \frac{1}{|\eta(\tau)|^2} \sum_{\substack{r,s \\ r-s=0 \bmod 2}} \left[q^{r^2/2} \bar{q}^{s^2/2} + q^{(r+\frac{1}{2})^2/2} \bar{q}^{(s-\frac{1}{2})^2/2} \right] \qquad (12.121)$$

On the other hand, using the expressions $Z_\nu = |\theta_\nu(0|\tau)/2\eta(\tau)|$ and the series expansions for the Jacobi theta function of App. 10.A, we have

$$\begin{aligned}
Z_{\text{Dirac}} &= \frac{1}{2|\eta(\tau)|^2} \sum_{\nu=2}^{4} |\theta_\nu(0|\tau)|^2 \\
&= \frac{1}{2|\eta(\tau)|^2} \left[\left| \sum_n q^{(n+1/2)^2/2} \right|^2 + \left| \sum_n q^{n^2/2} \right|^2 + \left| \sum_n (-1)^n q^{n^2/2} \right|^2 \right] \\
&= \frac{1}{2|\eta(\tau)|^2} \left[\left| \sum_n q^{(n+1/2)^2/2} \right|^2 + 2 \sum_{\substack{n,m \\ n-m=0 \bmod 2}} q^{n^2/2} \bar{q}^{m^2/2} \right] \\
&= Z(1)
\end{aligned} \qquad (12.122)$$

by direct comparison with Eq. (12.121).

This $c = 1$ conformal theory is somewhat different from the ordinary square of the Ising model, with partition function

$$Z_{\text{Ising}}^2 = \left(\sum_{\nu=2}^{4} Z_\nu \right)^2 \qquad (12.123)$$

since in this case the two real fermions are completely decoupled. The decoupling leads to 16 possible boundary conditions (among which only 9 contribute to the torus partition function) instead of 4 for the Dirac case (among which only 3 contribute to the torus partition function). The partition function (12.123) is actually the $\mathbb{Z}_2$ *orbifold* partition function of the free boson compactified on a circle of radius $R = 1$ introduced in Sect. 10.4.3. Indeed, from the expression (10.84),

$$Z_{\text{orb}}(1) = \frac{1}{2}\left[Z(1) + 4(Z_2 Z_3 + Z_2 Z_4 + Z_3 Z_4)\right]$$
$$= Z_{\text{Ising}}^2 \tag{12.124}$$

where we used $Z(1) = 2(Z_2^2 + Z_3^2 + Z_4^2)$. In the following, we will use the notation

$$\tilde{Z}_\nu = 4 \prod_{\substack{\nu' \in \{2,3,4\} \\ \nu' \neq \nu}} Z_{\nu'} \tag{12.125}$$

for the partition function of the twisted $\nu \equiv (v, u)$ sector of the boson.

Therefore, we have at our disposal two different ways of bosonizing the Ising model, which, of course, must lead to the same correlators. The operator correspondence with the Ising observables within the two bosonization schemes is summarized in Table 12.1. The main difference between the two schemes is that for the Dirac bosonization, the correlators will be obtained in the form of squares *sector by sector* ($\nu = 1, 2, 3, 4$), whereas for the Ising squared they will appear as a whole, and squared.

Table 12.1. Ising model observables in the two bosonization schemes.

Dirac:	Operator	Ising 1 and 2	boson at $R = 1$
	energy	$\varepsilon_1 \times \varepsilon_2$	$(\nabla\varphi/2)^2$
	spin	$\sigma_1 \times \sigma_2$	$\cos\varphi/2$
	disorder	$\mu_1 \times \mu_2$	$\sin\varphi/2$
Ising^2:	Operator	Ising 1 and 2	orb. boson at $R = 1$
	tot. energy	$\varepsilon_1 + \varepsilon_2$	$\cos\varphi$
	energy	$\varepsilon_1 \times \varepsilon_2$	$(\nabla\varphi/2)^2$
	spin	$\sigma_1 \times \sigma_2$	$\cos\varphi/2$
	disorder	$\mu_1 \times \mu_2$	$\sin\varphi/2$

12.6.2. Compactified Boson Correlations on the Plane and on the Torus

Before applying the bosonization scheme, we must first explain how to compute correlators in a $c = 1$ bosonic theory on the torus. In this subsection, we derive the torus two-point functions of electromagnetic operators for the bosonic theory compactified on a circle of radius R, as well as for the $\mathbb{Z}_2$ orbifold of this theory.

We start from the free boson compactified on a circle of radius R, with torus partition function (10.62)

$$Z(R) = \frac{1}{|\eta(\tau)|^2} \sum_{e,m} q^{h_{e,m}} \bar{q}^{\bar{h}_{e,m}} \qquad (12.126)$$

which exhibits the "electromagnetic" operator content, namely an infinite collection of operators $\mathcal{O}_{e,m}$, e, m arbitrary relative integers, with conformal dimensions

$$h_{e,m} = \frac{1}{2}(e/R + mR/2)^2 \qquad \bar{h}_{e,m} = \frac{1}{2}(e/R - mR/2)^2 \qquad (12.127)$$

In Sect. 10.4.1, we have identified the integers e, m as the electric and magnetic charges in the following sense. Returning to the plane for a while, we see that the purely electric operator of charge $e \in \mathbb{Z}$ is identified as

$$\mathcal{O}_{e,0} = e^{ie\varphi/R} \qquad (12.128)$$

This is indeed a single-valued function on the range $\varphi \in \mathbb{R}/2\pi\mathbb{Z}$ for any integer e. On the other hand, the purely magnetic operator $\mathcal{O}_{0,m}(z,\bar{z})$ creates a semi-infinite line of defect starting at the point z, along which the bosonic field φ has a jump discontinuity of $2\pi Rm$. The general operator $\mathcal{O}_{e,m}$ is a combination of these two basic operators. From this definition, we get very simple fusion rules, namely the charges are additive under the short-distance product

$$\mathcal{O}_{e,m} \times \mathcal{O}_{e',m'} \to \mathcal{O}_{e+e',m+m'} \qquad (12.129)$$

The conformal dimensions (12.127) can be read off the two-point function on the plane

$$\langle \mathcal{O}_{e,m}(z_1,\bar{z}_1)\mathcal{O}_{-e,-m}(z_2,\bar{z}_2)\rangle = \frac{1}{z_{12}^{2h_{e,m}}\bar{z}_{12}^{2\bar{h}_{e,m}}} \qquad (12.130)$$

In this correlator, the magnetic charges create a discontinuity of $2\pi m$ along a segment joining z_1 to z_2. Electric and magnetic neutrality force the total charges to be zero.

We now prove Eq. (12.130). The correlator (12.130) has the following path-integral expression

$$\int d\varphi\, e^{-S(\varphi)}\, \mathcal{O}_{e,m}(z_1,\bar{z}_1)\mathcal{O}_{-e,-m}(z_2,\bar{z}_2) \qquad (12.131)$$

with the usual free bosonic action

$$S(\varphi) = (1/8\pi)\int (\nabla\varphi)^2 d^2x \qquad (12.132)$$

If $m = 0$ (Eq. (12.131)), we get the usual two-point function of vertex operators

$$\langle e^{ie\varphi(z_1,\bar{z}_1)/R}e^{-ie\varphi(z_2,\bar{z}_2)/R}\rangle = |z_1 - z_2|^{-2e^2/R^2} \qquad (12.133)$$

The difficulty arises when $m \neq 0$, as the boson φ is no longer free but constrained because it must have a discontinuity of $2\pi mR$ along the segment joining z_1 to z_2.

To incorporate this constraint in the path-integral calculation, we first decompose the bosonic field into[9]

$$\varphi = \varphi_m^{cl} + \tilde{\varphi} \tag{12.134}$$

The classical part φ_m^{cl} has a vanishing Laplacian (we take it to be the imaginary part of an holomorphic function), and it incorporates the discontinuity condition. This classical solution of the equation of motion reads

$$\varphi_m^{cl}(z, \bar{z}) = mR \text{ Im } \ln \left(\frac{z - z_1}{z - z_2} \right) \tag{12.135}$$

It has a branch cut joining z_1 to z_2, across which it satisfies the desired discontinuity property. $\tilde{\varphi}$ is the "free part" of the boson (i.e., since the field φ_m^{cl} incorporates all the boundary conditions, $\tilde{\varphi}$ is periodic), with propagator

$$\langle \tilde{\varphi}(z, \bar{z}) \tilde{\varphi}(0, 0) \rangle = -\ln|z|^2 \tag{12.136}$$

We have the following decomposition of the action

$$\begin{aligned} S(\varphi) &= S(\varphi_m^{cl}) + S(\tilde{\varphi}) - (1/4\pi) \int \tilde{\varphi} \Delta \varphi_m^{cl} \\ &= S(\varphi_m^{cl}) + S(\tilde{\varphi}) \end{aligned} \tag{12.137}$$

due to the vanishing of the Laplacian of φ_m^{cl}. It is then straightforward to reproduce Eq. (12.130).

For the purpose of this chapter, we will need to compute only correlations of purely electric operators of electric charges 1 and 2, at radius $R = 1$ (see Table 12.1), and correlations including square gradient terms $(\nabla\varphi/2)^2$. It is, however, instructive to compute the generalization of Eq. (12.130) on the torus. It can be expressed as

$$\begin{aligned} &\langle \mathcal{O}_{e,m}(z_1, \bar{z}_1) \mathcal{O}_{-e,-m}(z_2, \bar{z}_2) \rangle \\ &\qquad = \frac{1}{Z(R)} \sum_{n,n'} Z_{n,n'} \langle \mathcal{O}_{e,m}(z_1, \bar{z}_1) \mathcal{O}_{-e,-m}(z_2, \bar{z}_2) \rangle_{n,n'} \end{aligned} \tag{12.138}$$

expressed as a sum over the periodicity sectors (n, n') of the compactified boson

$$(n, n') : \quad \varphi(z + 1) = \varphi(z) + 2\pi R n \quad \varphi(z + \tau) = \varphi(z) + 2\pi R n' \tag{12.139}$$

The partition functions $Z_{n,n'}$ were obtained in Sect. 10.4.1. To compute the corresponding path integral in the sector (n, n'), we again decompose φ into

$$\varphi = \varphi_{n,n'}^{cl} + \varphi_m^{cl} + \tilde{\varphi} \tag{12.140}$$

---

[9] Here we use the same method as in Sect. 10.4.1, where we have incorporated the two winding conditions (m, m') in a classical solution to the equations of motion (10.53), leaving us with a path integral over the free part of the boson.

The φ^{cl}'s are the classical solutions of the Laplace equation $\Delta\varphi^{cl} = 0$ on the torus, which incorporate respectively the (n, n') boundary conditions, and the defect line condition m:

$$\varphi^{cl}_{n,n'}(z, \bar{z}) = 2\pi R \operatorname{Im}\left(z\frac{n' - n\bar{\tau}}{\operatorname{Im}\tau}\right)$$

$$\varphi^{cl}_m(z, \bar{z}) = mR \operatorname{Im}\left[\ln\frac{\theta_1(z - z_1|\tau)}{\theta_1(z - z_2|\tau)} - \frac{2\pi}{\operatorname{Im}\tau}z\operatorname{Re} z_{12}\right]$$

(12.141)

$\tilde{\varphi}$ is the doubly periodic free part of the boson, with propagator

$$\langle\tilde{\varphi}(z, \bar{z})\tilde{\varphi}(0, 0)\rangle = -\ln\left|\frac{\theta_1(z|\tau)}{\partial_z\theta_1(0|\tau)}e^{-\pi\frac{(\operatorname{Im} z)^2}{\operatorname{Im}\tau}}\right|^2$$

$$\equiv \ln|\mathcal{E}(z|\tau)|^2$$

(12.142)

This last expression is the doubly periodic elementary solution of the Laplacian on the torus, often called the *prime form*, namely

$$-\Delta\langle\tilde{\varphi}(z, \bar{z})\tilde{\varphi}(0, 0)\rangle = 2\pi\delta^{(2)}(z) - \frac{4\pi}{\operatorname{Im}\tau}$$

(12.143)

where

$$\delta^{(2)}(z) = \sum_{n,n' \in \mathbb{Z}} \delta(z - n - n'\tau)$$

(12.144)

is the doubly periodic delta function on the torus, and the subtraction takes care of the zero modes of the Laplacian on the torus. This is just a more physical reformulation of the computation carried out in Sect. 10.4.1. We now compute the path integral (12.131) on the torus, with the total action

$$S(\varphi) = S(\varphi^{cl}_{n,n'}) + S(\varphi^{cl}_m) + S(\tilde{\varphi})$$

(12.145)

We obtain

$$\langle\mathcal{O}_{e,m}(z_1, \bar{z}_1)\mathcal{O}_{-e,-m}(z_2, \bar{z}_2)\rangle_{n,n'} = \langle\mathcal{O}_{e,m}(z_1, \bar{z}_1)\mathcal{O}_{-e,-m}(z_2, \bar{z}_2)\rangle_{0,0}$$

$$\times \exp\left[2i\pi e\frac{\operatorname{Im}(z_{12}(n' - n\bar{\tau}))}{\operatorname{Im}\tau} + \pi mR\frac{\operatorname{Re}(z_{12}(n' - n\bar{\tau}))}{\operatorname{Im}\tau}\right]$$

(12.146)

and the $(0, 0)$ sector contribution reads

$$\langle\mathcal{O}_{e,m}(z_1, \bar{z}_1)\mathcal{O}_{-e,-m}(z_2, \bar{z}_2)\rangle_{0,0} = \left(\frac{\partial_z\theta_1(0|\tau)}{\theta_1(z_{12}|\tau)}\right)^{2h_{e,m}} \times \text{c.c.}$$

$$\times \exp\left[\frac{8\pi}{R^2\operatorname{Im}\tau}\left(e\operatorname{Im} z_{12} - im\frac{R^2}{8}\operatorname{Re} z_{12}\right)^2\right]$$

(12.147)

where by "c.c." we mean the corresponding antiholomorphic counterpart, with barred quantities. Applying the Poisson resummation formula (10.264) of App. 10.A in the final result, we can trade the sum over the winding numbers n' for a sum over electric charges e' (we also rename the winding numbers $n \to m'$).

This enables us to write the final result for the correlator (12.138) in a symmetric form

$$Z(R)\langle \mathcal{O}_{e,m}(z_1,\bar{z}_1)\mathcal{O}_{-e,-m}(z_2,\bar{z}_2)\rangle = \left| \frac{\partial_z \theta_1(0|\tau)}{\theta_1(z_{12}|\tau)} \right|^{4h_{e,m}}$$

$$\times \frac{1}{|\eta(\tau)|^2} \sum_{e',m'} q^{h_{e',m'}} \bar{q}^{\bar{h}_{e',m'}} e^{4i\pi[\alpha_{e',m'}\alpha_{e,m}z_{12}-\bar{\alpha}_{e',m'}\bar{\alpha}_{e,m}\bar{z}_{12}]} \tag{12.148}$$

where

$$\alpha_{e,m} = \frac{1}{\sqrt{2}}(e/R + Rm/2) \qquad \bar{\alpha}_{e,m} = \frac{1}{\sqrt{2}}(e/R - Rm/2) \tag{12.149}$$

The formula (12.148) can be generalized to any number of electromagnetic operators. We leave its detailed derivation as an exercise to the courageous reader. The result reads

$$Z(R)\langle \mathcal{O}_1(z_1,\bar{z}_1)\cdots\mathcal{O}_n(z_n,\bar{z}_n)\rangle = \prod_{i<j} \left| \frac{\partial_z \theta_1(0|\tau)}{\theta_1(z_{ij}|\tau)} \right|^{4\alpha_i\alpha_j}$$

$$\times \frac{1}{|\eta(\tau)|^2} \sum_{e,m} q^{h_{e,m}} \bar{q}^{\bar{h}_{e,m}} e^{4i\pi[\Sigma_k \alpha_{e,m}\alpha_k z_k - \bar{\alpha}_{e,m}\bar{\alpha}_k \bar{z}_k]} \tag{12.150}$$

with the obvious notations $\mathcal{O}_k \equiv \mathcal{O}_{e_k,m_k}$ and $\alpha_k \equiv \alpha_{e_k,m_k}$ and the condition of global electric and magnetic neutrality

$$\sum_{k=1}^{n} e_k = \sum_{k=1}^{n} m_k = 0 \tag{12.151}$$

Since the square of the Ising theory is described by a $\mathbb{Z}_2$ orbifold theory at $c = 1$, we must also explain how to calculate correlators in the twisted sectors $\nu = 2, 3, 4$. More precisely, the total orbifold correlator decomposes into

$$Z_{\text{orb}}(R)\langle \mathcal{O}_{e,m}(z_1,\bar{z}_1)\mathcal{O}_{-e,-m}(z_2,\bar{z}_2)\rangle_{\text{orb}}$$

$$= \frac{1}{2}\left[Z(R)\langle \mathcal{O}_{e,m}(z_1,\bar{z}_1)\mathcal{O}_{-e,-m}(z_2,\bar{z}_2)\rangle \right.$$

$$\left. + \sum_{\nu=2}^{4} \tilde{Z}_\nu \langle \mathcal{O}_{e,m}(z_1,\bar{z}_1)\mathcal{O}_{-e,-m}(z_2,\bar{z}_2)\rangle_\nu \right] \tag{12.152}$$

The three sectors $\nu = 2, 3, 4$ correspond to (anti-)periodic boundary conditions on the boson, namely

$$\nu = 2 , (v,u) = (0,\tfrac{1}{2}): \quad \varphi(z+1) = -\varphi(z) \qquad \varphi(z+\tau) = \varphi(z)$$

$$\nu = 3 , (v,u) = (\tfrac{1}{2},\tfrac{1}{2}): \quad \varphi(z+1) = -\varphi(z) \qquad \varphi(z+\tau) = -\varphi(z) \tag{12.153}$$

$$\nu = 4 , (v,u) = (\tfrac{1}{2},0): \quad \varphi(z+1) = \varphi(z) \qquad \varphi(z+\tau) = -\varphi(z)$$

The corresponding partition functions (12.125) read

$$\tilde{Z}_v = 4 \prod_{\substack{v' \in \{2,3,4\} \\ v' \neq v}} Z_{v'} = 4 \left| \frac{\eta(\tau)}{\theta_v(0|\tau)} \right| \tag{12.154}$$

In order to compute the correlator of two electromagnetic operators in the sector v, we must perform the corresponding path integral, with twisted boundary conditions on φ. As usual, we decompose φ into

$$\varphi = \varphi^{\mathrm{cl}}_{m,v} + \tilde{\varphi}_v \tag{12.155}$$

where the classical part is a solution of the Laplace equation on the torus, with the (anti-)periodicity conditions pertaining to the sector v, that is

$$\varphi^{\mathrm{cl}}_{m,v}(z,\bar{z}) = mR \operatorname{Im} \ln \left(\frac{\theta_1((z-z_1)/2|\tau/2)\theta_v((z-z_2)/2|\tau/2)}{\theta_1((z-z_2)/2|\tau/2)\theta_v((z-z_1)/2|\tau/2)} \right) \tag{12.156}$$

and the free part of the boson in the sector $v = (v,u)$ has the propagator

$$\langle \tilde{\varphi}(z,\bar{z})\tilde{\varphi}(0,0) \rangle_v = -\ln \left| \frac{\mathcal{E}(z/2|\tau/2)}{\mathcal{E}(z/2+v+u\tau|\tau/2)} \right|^2 \tag{12.157}$$

The prime form $\mathcal{E}(z|\tau)$ is defined in Eq. (12.142). We thus finally get

$$\langle \mathcal{O}_{e,m}(z_1,\bar{z}_1)\mathcal{O}_{-e,-m}(z_2,\bar{z}_2) \rangle_v$$
$$= \left(\frac{\theta_v(z_{12}/2|\tau/2)\partial_z\theta_1(0|\tau)}{2\theta_v(0|\tau/2)\theta_1(z_{12}/2|\tau)} \right)^{2h_{e,m}} \times \text{c.c.} \tag{12.158}$$

This has a straightforward generalization to the n-point correlator

$$\langle \mathcal{O}_1(z_1,\bar{z}_1) \cdots \mathcal{O}_n(z_n,\bar{z}_n) \rangle_v$$
$$= \prod_{1 \leq i < j \leq n} \left(\frac{\theta_v(z_{ij}/2|\tau/2)\partial_z\theta_1(0|\tau)}{2\theta_v(0|\tau/2)\theta_1(z_{ij}/2|\tau)} \right)^{2\alpha_i\alpha_j} \times \text{c.c.} \tag{12.159}$$

with the same notations as in Eq. (12.150) and under the condition (12.151).

12.6.3. Ising Correlators from the Bosonization of the Dirac Fermion

In principle, the correspondence between operators in the Ising models and in the bosonized Dirac fermion theory (Table 12.1) enables us to compute, sector by sector, the square of any Ising correlator. The only problem is to relate a given sector v for the fermion to some particular subset of the boson winding modes. But this can be done easily by examining the partition function (12.118), sector by sector. We have

$$Z_v = \left| \frac{\theta_v(0|\tau)}{2\eta(\tau)} \right| = \left| \frac{1}{2\eta(\tau)} \sum_{n \in \mathbb{Z}} q^{\frac{1}{2}(n+v+\frac{1}{2})^2} e^{2i\pi(n+v+\frac{1}{2})(u+\frac{1}{2})} \right| \tag{12.160}$$

for all $\nu \equiv (v, u)$. This readily specifies the set of winding sectors of the boson that reproduces the sector ν for the Dirac fermion and builds the corresponding piece of the boson partition function:

$$Z_\nu^{bos} = 2Z_\nu^2 \tag{12.161}$$

with

$$\sum_{\nu=2}^{4} Z_\nu^{bos} = 2\sum_{\nu=2}^{4} Z_\nu^2 = Z(1) \tag{12.162}$$

$Z(1)$ is the compactified boson partition function at radius $R = 1$. In particular, a total correlator of observables in the bosonic theory can be expressed as the sum over the four sectors $\nu = 1, 2, 3, 4$ of the squares of the corresponding Ising model correlators (through Table 12.1) as

$$Z(1)\langle \cdots \rangle = \sum_{\nu=1}^{4} 2Z_\nu^2 \langle \cdots \rangle_\nu \tag{12.163}$$

Instead of computing the *total* correlator in the bosonic theory, it is desirable to get it *sector by sector*. This is done by the so-called *chiral bosonization* procedure. Before we proceed any further, we should recall the duality $R \leftrightarrow 2/R$ of the bosonic theory on a circle of radius R, under which electric and magnetic charges are exchanged. Since this leaves the partition function invariant, we have

$$Z(1) = Z(2) \tag{12.164}$$

To represent the Ising correlation functions, according to Table 12.1, we will need only the electric operators

$$e^{\pm i\varphi/2} \quad \text{and} \quad e^{\pm i\varphi} \tag{12.165}$$

Comparing Eq. (12.165) with Eq. (12.128), these operators correspond to electric charges of ± 1 and ± 2, respectively, *provided* we work in the framework of the $R = 2$ theory. For this reason, we shall use the $R = 2$ bosonic theory in the following discussion.[10]

Guided by the expression (12.160), we get a set of rules for computing the bosonic correlator in each sector ν. We start from the bosonization formula (12.37). To compute any correlator in the sector $\nu \equiv (v, u)$, we expand the chiral (holomorphic) component $\phi(z)$ of the boson $\varphi(z, \bar{z}) = \phi(z) - \bar{\phi}(\bar{z})$ into

$$\phi(z) = 2\pi z P + \hat{\phi}(z) \tag{12.166}$$

The operator P has eigenvalues $n + v + \frac{1}{2}, n \in \mathbb{Z}$, and $\hat{\phi}$ is the "free part" of the boson, with propagator

$$\langle \hat{\phi}(z)\hat{\phi}(0) \rangle = -\ln \frac{\theta_1(z|\tau)}{\partial_z \theta_1(0|\tau)} \tag{12.167}$$

---

[10] If we insisted on using the $R = 1$ theory, these operators would be represented as purely magnetic operators of respective magnetic charges ± 1 and ± 2.

which approaches the plane expression (12.38) when $z \to 0$. We must weigh each eigenvalue $n + v + \frac{1}{2}$ by the factor

$$\frac{1}{2\eta(\tau)} q^{\frac{1}{2}(n+v+\frac{1}{2})^2} e^{2i\pi(n+v+\frac{1}{2})(u+\frac{1}{2})} \tag{12.168}$$

The operation is repeated for the antiholomorphic part of the boson $\bar{\phi}$, and the total answer is the product of the two. This set of rules enables us to compute the bosonic correlators in each sector v. They have been defined in order to be compatible with the result of the direct calculation of the *total* correlators (12.163), using Sect. 12.6.2.

For instance, for the purely electric correlator with $e_1 + \cdots + e_n = 0, R = 2$, we find

$$Z_v^{bos} \langle \mathcal{O}_{e_1,0}(z_1, \bar{z}_1) \cdots \mathcal{O}_{e_n,0}(z_n, \bar{z}_n) \rangle_v = \left[\frac{1}{2\eta(\tau)} \sum_n q^{\frac{1}{2}(n+v+\frac{1}{2})^2} \right.$$

$$\left. \times e^{2i\pi(n+v+\frac{1}{2})(u+\frac{1}{2})} e^{i\pi(n+v+\frac{1}{2})(\Sigma e_i z_i)} \prod_{i<j} \left(\frac{\partial_z \theta_1(0|\tau)}{\theta_1(z_{ij}|\tau)} \right)^{e_i e_j/4} \right] \times c.c. \tag{12.169}$$

We have decomposed the contributions according to the above rules into the weighing factor (12.168), the contribution of the P operator, and the free part of the correlator obtained by using Wick's theorem with the propagator (12.167). By $\times$c.c., we mean the product by the corresponding antiholomorphic part with $\phi \to \bar{\phi}$. The result (12.169) agrees with the total correlator (12.150) for $m_i = 0$ and $R = 2$. We note in particular that when $n = 2$ and $e_1 = -e_2 = e$, Eq. (12.169) takes the simple form

$$Z_v^{bos} \langle \mathcal{O}_{e,0}(z_1, \bar{z}_1) \mathcal{O}_{-e,0}(z_2, \bar{z}_2) \rangle_v$$

$$= \left[\frac{1}{2\eta(\tau)} \theta_v(ez_{12}/2|\tau) \left(\frac{\partial_z \theta_1(0|\tau)}{\theta_1(z_{12}|\tau)} \right)^{e^2/4} \right] \times c.c. \tag{12.170}$$

As a first application of the above rules, we rederive the expression (12.107) for the spin-spin correlator in the sector v. In the bosonized version, it reads

$$\left[Z_v \langle \sigma(z_1, \bar{z}_1) \sigma(z_2, \bar{z}_2) \rangle_v \right]^2 = 2Z_v^2 \langle \cos \frac{\varphi}{2}(z_1, \bar{z}_1) \cos \frac{\varphi}{2}(z_2, \bar{z}_2) \rangle_v$$

$$= Z_v^{bos} \langle \mathcal{O}_{1,0}(z_1, \bar{z}_1) \mathcal{O}_{-1,0}(z_2, \bar{z}_2) \rangle_v \tag{12.171}$$

$$= \left| \frac{1}{2\eta(\tau)} \theta_v(z_{12}/2|\tau) \right|^2 \left| \frac{\partial_z \theta_1(0|\tau)}{\theta_1(z_{12}|\tau)} \right|^{\frac{1}{2}}$$

where the first equality is simply Eq. (12.59), valid in each sector v with $N_1 = 2$. The second equality uses Eq. (12.161) and the definition of the purely electric operators (12.128). The last equality is a direct application of Eq. (12.170) for $e = 1$. This result agrees with the previous expression (12.107), due to the identity

$$\eta(\tau) = \left[\frac{\partial_z \theta_1(0|\tau)}{2\pi} \right]^{1/3} \tag{12.172}$$

We are now in position to compute more general correlation functions. The square of the correlator of an arbitrary even number of spin operators reads, for $\nu = 1, 2, 3, 4$,

$$
\begin{aligned}
\left[Z_\nu \langle \sigma(z_1, \bar{z}_1) \cdots \sigma(z_{2n}, \bar{z}_{2n}) \rangle_\nu \right]^2 \\
= 2^n \, Z_\nu^2 \langle \cos \frac{\varphi}{2}(z_1, \bar{z}_1) \cdots \cos \frac{\varphi}{2}(z_{2n}, \bar{z}_{2n}) \rangle_\nu \\
= \frac{1}{|\eta(\tau)|^2} \sum_{\substack{\epsilon_1, \ldots, \epsilon_{2n} = \pm 1 \\ \Sigma_i \epsilon_i = 0}} |\theta_\nu((\Sigma_i \epsilon_i z_i)/2|\tau)|^2 \prod_{i<j} \left| \frac{\theta_1(z_{ij}|\tau)}{\partial_z \theta_1(0|\tau)} \right|^{\epsilon_i \epsilon_j / 2}
\end{aligned}
$$
(12.173)

Similarly, the square of the correlator of an even number of energy operators is easily obtained by direct application of the chiral bosonization rules and of Wick's theorem:

$$
\begin{aligned}
\left[Z_\nu \langle \varepsilon(z_1, \bar{z}_1) \cdots \varepsilon(z_{2n}, \bar{z}_{2n}) \rangle_\nu \right]^2 \\
= Z_\nu^2 \langle (\nabla \varphi(z_1, \bar{z}_1)/2)^2 \cdots (\nabla \varphi(z_{2n}, \bar{z}_{2n})/2)^2 \\
= \left| \sum_{k=0}^n \frac{\partial_z^{2n-2k} \theta_\nu(0|\tau)}{2\eta(\tau)} \sum_{\sigma \in S_{2n}} \prod_{i=1}^n [\wp(z_{\sigma(2i-1)} - z_{\sigma(2i)}) + 2\eta_1] \right|^2
\end{aligned}
$$
(12.174)

The last sum extends over the permutation group S_{2n} of the $2n$ indices. The identity between this result and the previous expression (12.96) follows from the torus version of the Cauchy determinant formula (see Ex. 12.19 for a detailed proof). The advantage of the chiral bosonization approach is that it gives the correlator of an *odd* number of energy operators as well. The latter receives contributions from the $\nu = 1$ sector only and reads

$$
\begin{aligned}
\left[Z_1 \langle \varepsilon(z_1, \bar{z}_1) \cdots \varepsilon(z_{2n+1}, \bar{z}_{2n+1}) \rangle_1 \right]^2 \\
= -Z_1^2 \langle (\nabla \varphi(z_1, \bar{z}_1)/2)^2 \cdots (\nabla \varphi(z_{2n+1}, \bar{z}_{2n+1})/2)^2 \\
= \left| \sum_{k=0}^n \frac{\partial_z^{2n-2k+1} \theta_\nu(0|\tau)}{2\eta(\tau)} \sum_{\sigma \in S_{2n+1}} \prod_{i=1}^n [\wp(z_{\sigma(2i-1)} - z_{\sigma(2i)}) + 2\eta_1] \right|^2
\end{aligned}
$$
(12.175)

so that the total energy correlator reads

$$
\begin{aligned}
\langle \varepsilon(z_1, \bar{z}_1) \cdots \varepsilon(z_{2n+1}, \bar{z}_{2n+1}) \rangle \\
= \frac{1}{Z_{\text{Ising}}} \left| \sum_{k=0}^n \frac{\partial_z^{2n-2k+1} \theta_\nu(0|\tau)}{2\eta(\tau)} \sum_{\sigma \in S_{2n+1}} \prod_{i=1}^n [\wp(z_{\sigma(2i-1)} - z_{\sigma(2i)}) + 2\eta_1] \right|
\end{aligned}
$$
(12.176)

More generally, the square of the mixed spin and energy correlator reads

$$
\begin{aligned}
\left[Z_\nu \langle \sigma(1) \cdots \sigma(2n) \varepsilon(2n+1) \cdots \varepsilon(2n+p) \rangle_\nu \right]^2 \\
= 2^n (-1)^p Z_\nu^2 \Big\langle \cos \frac{\varphi}{2}(1) \cdots \cos \frac{\varphi}{2}(2n) \\
\times (\nabla \varphi(2n+1)/2)^2 \cdots (\nabla \varphi(2n+p)/2)^2 \Big\rangle_\nu
\end{aligned}
$$
(12.177)

and it is readily evaluated using the chiral bosonization prescriptions.

12.6.4. Ising Correlators from the Bosonization of Two Real Fermions

In this section, we use the direct bosonization scheme for the product of two decoupled Ising models. As already shown in Sect. 12.6.1, this is precisely the $c = 1$ orbifold model of a boson compactified on a circle of radius $R = 1$, quotiented by the extra $\mathbb{Z}_2$ symmetry $\varphi \to -\varphi$. As in the previous section, we shall work within the dual theory at radius $R = 2$, for which $Z_{\text{orb}}(2) = Z_{\text{orb}}(1)$. A first quantity of interest is the *total* energy operator which is, according to Table 12.1, $\varepsilon_{\text{tot}} = \varepsilon_1 + \varepsilon_2 = \cos\varphi$. Using the bosonic correlators (12.148) in the untwisted sector and (12.158) in the twisted ones for $\nu = 2, 3, 4$, and setting $e = 1, m = 0, R = 2$, we obtain the respective contributions to the total two-point correlator of total energy:

$$Z_{\text{orb}}(2)\langle\varepsilon_{\text{tot}}(z_1,\bar{z}_1)\varepsilon_{\text{tot}}(z_2,\bar{z}_2)\rangle_{\text{orb}}$$

$$= \frac{1}{2}\left[Z(2)\langle\varepsilon_{\text{tot}}(z_1,\bar{z}_1)\varepsilon_{\text{tot}}(z_2,\bar{z}_2)\rangle + \sum_{\nu=2}^{4}\tilde{Z}_\nu\langle\varepsilon_{\text{tot}}(z_1,\bar{z}_1)\varepsilon_{\text{tot}}(z_2,\bar{z}_2)\rangle_\nu\right]$$

$$(12.178)$$

The contributing parts are

$$Z(2)\langle\varepsilon_{\text{tot}}(z_1,\bar{z}_1)\varepsilon_{\text{tot}}(z_2,\bar{z}_2)\rangle$$

$$= 2Z(2)\langle\cos\varphi(z_1,\bar{z}_1)\cos\varphi(z_2,\bar{z}_2)\rangle \qquad (12.179)$$

$$= \frac{1}{4|\eta(\tau)|^2}\left|\frac{\partial_z\theta_1(0|\tau)}{\theta_1(z_{12}|\tau)}\right|^2\sum_{\nu=2}^{4}|\theta_\nu(z_{12}|\tau)|^2$$

and

$$\tilde{Z}_\nu\langle\varepsilon_{\text{tot}}(z_1,\bar{z}_1)\varepsilon_{\text{tot}}(z_2,\bar{z}_2)\rangle_\nu$$

$$= \tilde{Z}_\nu\left[\left|\frac{\theta_\nu(z_{12}/2|\tau/2)\partial_z\theta_1(0|\tau)}{\theta_\nu(0|\tau)\theta_1(z_{12}|\tau)}\right|^2 + \left|\frac{\theta_\nu(0|\tau)\theta_1(z_{12}|\tau)}{\theta_\nu(z_{12}/2|\tau/2)\partial_z\theta_1(0|\tau)}\right|^2\right] \qquad (12.180)$$

In the twisted sector, we have included both contributions $\exp-4\langle\varphi\varphi\rangle_\nu$ and $\exp 4\langle\varphi\varphi\rangle_\nu$ since neither of the two is periodic by itself, whereas the result must be periodic. This amounts to including electrically nonneutral contributions to the correlator, which are now allowed as we work in a twisted sector (with antiperiodic boundary conditions on the field φ, which do not affect $\varepsilon_{\text{tot}} \propto \cos\varphi$).

We can check that this agrees with the interpretation of ε as the total energy operator, namely, that

$$Z_{\text{orb}}(2)\langle\varepsilon_{\text{tot}}(z_1,\bar{z}_1)\varepsilon_{\text{tot}}(z_2,\bar{z}_2)\rangle_{\text{orb}}$$

$$= 2Z_{\text{Ising}}^2\left[\langle\varepsilon(z_1,\bar{z}_1)\varepsilon(z_2,\bar{z}_2)\rangle_{\text{Ising}} + \langle\varepsilon\rangle_{\text{Ising}}^2\right] \qquad (12.181)$$

The untwisted sector contribution of Eq. (12.180) is easily identified as

$$
Z(2)\langle\varepsilon_{\mathrm{tot}}(z_1,\bar{z}_1)\varepsilon_{\mathrm{tot}}(z_2,\bar{z}_2)\rangle
$$
$$
= 2\sum_{\nu=2}^{4} Z_\nu^2\langle\varepsilon(z_1,\bar{z}_1)\varepsilon(z_2,\bar{z}_2)\rangle_\nu + 2\big[Z_1\langle\varepsilon\rangle_1\big]^2
\tag{12.182}
$$

On the other hand, we can show that the twisted sectors contribute for

$$
\tilde{Z}_\nu\langle\varepsilon_{\mathrm{tot}}(z_1,\bar{z}_1)\varepsilon_{\mathrm{tot}}(z_2,\bar{z}_2)\rangle_\nu
$$
$$
= \frac{1}{2}\tilde{Z}_\nu \sum_{\substack{\nu'\in\{2,3,4\} \\ \nu'\neq\nu}} \langle\varepsilon(z_1,\bar{z}_1)\varepsilon(z_2,\bar{z}_2)\rangle_{\nu'}
\tag{12.183}
$$

This is easily checked by means of the doubling identities of App. 12.A. For instance, in the sector $\nu=4$, by substituting the values of $\theta_2(z|\tau)$ and $\theta_3(z|\tau)$ in terms of theta functions evaluated at $z/2,\tau$ (cf. Eq. (12.191)), we can rewrite

$$
\langle\varepsilon(1)\varepsilon(2)\rangle_2 + \langle\varepsilon(1)\varepsilon(2)\rangle_3
$$
$$
= \left|\frac{\partial_z\theta_1(0|\tau)}{\theta_1(z_{12}|\tau)}\right|^2 \left[\left|\frac{\theta_2(z_{12}|\tau)}{\theta_2(0|\tau)}\right|^2 + \left|\frac{\theta_3(z_{12}|\tau)}{\theta_3(0|\tau)}\right|^2\right]
$$
$$
= 2\pi^2 \frac{|\theta_4(0|\tau)|^2}{|\theta_2(0|\tau)\theta_3(0|\tau)|^2}
\tag{12.184}
$$
$$
\times \frac{|\theta_2(z_{12}/2|\tau)\theta_3(z_{12}/2|\tau)|^4 + |\theta_1(z_{12}/2|\tau)\theta_4(z_{12}/2|\tau)|^4}{|\theta_1(z_{12}|\tau)|^2}
$$

By using the expressions of the theta functions of period τ in terms of those of half period $\tau/2$ (Eq. (12.192)), we get the desired result (12.183) for $\nu=4$. The other sectors are obtained by modular transformations of the variable τ: $\tau \rightarrow \tau+1$ gives the $\nu=3$ contribution, whereas $\tau \rightarrow -1/\tau$ gives the $\nu=2$ one, up to an overall factor $|\tau|^2$. This completes the general proof of Eq. (12.183). Together with the untwisted sector correspondence (12.182), this entails the compatibility (12.181) between the direct computation of one- and two-point functions of the Ising energy operator (12.77)–(12.93) and the $c=1$ orbifold bosonization result.

The comparison of the expressions obtained from the two bosonization methods leads to very interesting and nontrivial identities between theta functions. Clearly, higher correlators become more and more involved, although in principle the results (12.150) and (12.159) can be used directly to write the most general Ising correlators, through the correspondence of Table 12.1. However the analysis of the resulting identities goes beyond the scope of this work.

Appendix 12.A. Elliptic and Theta Function Identities

12.A.1. Generalities on Elliptic Functions

A meromorphic[11] function $f(z|\tau)$ of z is said to be elliptic if it is doubly periodic on the torus, namely

$$f(z + 1|\tau) = f(z|\tau) \qquad f(z + \tau|\tau) = f(z|\tau) \tag{12.185}$$

The *finite* zero and pole structure of such a function on the torus is infinitely duplicated, by periodicity, in the whole complex plane. However, the Cauchy residue theorem holds, namely

$$\oint_C dz\, f(z|\tau) = 2i\pi \sum_j \mathrm{Res}_j(f) \tag{12.186}$$

where the sum extends over all the residues of the poles encompassed by the closed (counterclockwise) contour C. What makes the use of this theorem particularly interesting on the torus is the possibility of nontrivial closed contours C.

For instance, take C to be the closed boundary of the torus, made of the segments $[0, 1]$, $[1, 1 + \tau]$, $[1 + \tau, \tau]$ and $[\tau, 1]$. Computing the above integral (12.186), we find

$$\oint_C dz\, f(z|\tau) = \left[\int_{[0,1]} + \int_{[1,1+\tau]} + \int_{[1+\tau,\tau]} + \int_{[\tau,1]} \right] dz\, f(z|\tau)$$
$$= 0 \tag{12.187}$$
$$= 2i\pi \sum_j \mathrm{Res}_j(f)$$

The second equality follows by grouping the first and third integrals and the second and fourth respectively: by periodicity, the value of the function is the same along any two opposed edges of C; but since the direction of circulation is opposite, the contributions cancel each other. This leads to a first theorem on elliptic functions: *The sum of residues of the poles of an elliptic function on the torus vanishes.*

We now relate the numbers of zeros and poles of f. By applying the above result (12.187) to the elliptic function $\partial_z f/f$, we find

$$\oint_C dz\, \frac{\partial_z f(z|\tau)}{f(z|\tau)} = 0$$

The pole structure of $\partial_z f/f$ is entirely determined by the zeros and poles of f. More precisely, near a zero z_0 of f, with order n, we have

$$\frac{\partial_z f(z|\tau)}{f(z|\tau)} = \frac{n}{z - z_0} + \mathrm{reg}.$$

---

[11] A meromorphic function is a function whose sole singularities in the complex plane are poles.

Near a pole z_1 of order m, we have

$$\frac{\partial_z f(z|\tau)}{f(z|\tau)} = -\frac{m}{z - z_1} + \text{reg.}$$

Hence, applying the Cauchy formula (12.186), we find the second theorem on elliptic functions:

An elliptic function has same number of zeros and poles, counted with their multiplicities.

Combining the two above theorems, we see that a nonconstant elliptic function must have at least two poles. Indeed, if it has no pole, it is an analytic function, which is bounded on the compact fundamental domain of the torus. By periodicity, it is bounded on the whole complex plane; it is therefore a constant by Liouville's theorem. An elliptic function cannot have a single pole, because the first theorem implies that its residue vanishes. So we get the third theorem:

An elliptic function with at most one pole must be a constant.

12.A.2. Periodicity and Zeros of the Jacobi Theta Functions

The Jacobi theta functions $\theta_\nu(z|\tau)$ have been defined in App. 10.A. Here we list a number of their properties, as functions of the complex variable z, which are useful when dealing with torus correlation functions.

The periodicity relations (10.255) induce the following behavior of the ratios $r_\nu(z|\tau) = \theta_\nu(z|\tau)/\theta_1(z|\tau)$ under translations of z by 1 and τ

$$\begin{aligned}
r_2(z+1|\tau) &= r_2(z|\tau) & r_2(z+\tau|\tau) &= -r_2(z|\tau) \\
r_3(z+1|\tau) &= -r_3(z|\tau) & r_3(z+\tau|\tau) &= -r_3(z|\tau) \\
r_4(z+1|\tau) &= -r_4(z|\tau) & r_4(z+\tau|\tau) &= r_4(z|\tau)
\end{aligned} \quad (12.188)$$

The θ functions also have simple transformations under translations of half periods (see Eq. (10.254)). These transformations read

$$\begin{aligned}
\theta_1(z+\tfrac{1}{2}|\tau) &= \theta_2(z|\tau) & \theta_1(z+\tau/2|\tau) &= ie^{-i\pi z}q^{-1/8}\theta_4(z|\tau) \\
\theta_2(z+\tfrac{1}{2}|\tau) &= -\theta_1(z|\tau) & \theta_2(z+\tau/2|\tau) &= e^{-i\pi z}q^{-1/8}\theta_3(z|\tau) \\
\theta_3(z+\tfrac{1}{2}|\tau) &= \theta_4(z|\tau) & \theta_3(z+\tau/2|\tau) &= e^{-i\pi z}q^{-1/8}\theta_2(z|\tau) \\
\theta_4(z+\tfrac{1}{2}|\tau) &= \theta_3(z|\tau) & \theta_4(z+\tau/2|\tau) &= ie^{-i\pi z}q^{-1/8}\theta_1(z|\tau)
\end{aligned} \quad (12.189)$$

The expression of $\theta_1(z|\tau)$ as an infinite product (10.253) shows explicitly that its zeros are all simple and lie on the lattice $m + n\tau$, for m, n arbitrary relative integers. Hence, when restricted to the torus, the function $\theta_1(z|\tau)$ has only one single zero at $z = 0$. Together with the half-period translation identities (12.189), this shows immediately that the other $\theta_\nu(z|\tau)$ have one single zero on the torus lying, respectively, at

$$\theta_2 \ : \ z = \frac{1}{2}$$

$$\theta_3 \ : \ z = \frac{1}{2}(1 + \tau) \qquad\qquad (12.190)$$

$$\theta_4 \ : \ z = \frac{1}{2}\tau$$

12.A.3. Doubling Identities

The direct application of the theorems of Sect. 12.A.1 leads to a number of identities between Jacobi theta functions. Of particular interest are the so-called doubling identities. Their proof is left as an exercise to the reader. It goes generically as follows: we first identify the zero and pole structure of the two expressions we want to equate. It is usually a straightforward calculation to get the exact residues and check that they coincide for the two expressions. In a second step, we study the periodicity of the functions, to finally conclude that the ratio of the two is elliptic and has no pole, hence it is a constant, fixed to be 1 by the residue identity. In this way, we get the following doubling identities:

$$\frac{\theta_2(z|\tau)}{\theta_2(0|\tau)} = \frac{[\theta_2(z/2|\tau)\theta_3(z/2|\tau)]^2 - [\theta_1(z/2|\tau)\theta_4(z/2|\tau)]^2}{[\theta_2(0|\tau)\theta_3(0|\tau)]^2}$$

$$\frac{\theta_3(z|\tau)}{\theta_3(0|\tau)} = \frac{[\theta_2(z/2|\tau)\theta_3(z/2|\tau)]^2 + [\theta_1(z/2|\tau)\theta_4(z/2|\tau)]^2}{[\theta_2(0|\tau)\theta_3(0|\tau)]^2}$$

$$= \frac{[\theta_3(z/2|\tau)\theta_4(z/2|\tau)]^2 - [\theta_1(z/2|\tau)\theta_2(z/2|\tau)]^2}{[\theta_3(0|\tau)\theta_4(0|\tau)]^2} \qquad (12.191)$$

$$\frac{\theta_4(z|\tau)}{\theta_4(0|\tau)} = \frac{[\theta_3(z/2|\tau)\theta_4(z/2|\tau)]^2 + [\theta_1(z/2|\tau)\theta_2(z/2|\tau)]^2}{[\theta_3(0|\tau)\theta_4(0|\tau)]^2}$$

and

$$\theta_2(z/2|\tau)\theta_3(z/2|\tau) = \left(\frac{\theta_2(0|\tau)\theta_3(0|\tau)}{2}\right)^{\frac{1}{2}} \theta_2(z/2|\tau/2)$$

$$\qquad\qquad (12.192)$$

$$\theta_1(z/2|\tau)\theta_4(z/2|\tau) = \left(\frac{\theta_2(0|\tau)\theta_3(0|\tau)}{2}\right)^{\frac{1}{2}} \theta_1(z/2|\tau/2)$$

The latter can also be derived directly by using the infinite product expressions (10.253) of App. 10.A.

Exercises

12.1 *The correspondence between the quantum Ising spin chain and the XZ spin chain*
The transfer matrix of the two-dimensional Ising model is related to the *one-dimensional quantum Ising spin-chain* Hamiltonian in a transverse magnetic field. (Through Eq. (3.114),

the logarithm of the two-dimensional Ising transfer matrix is actually proportional to the Ising spin chain Hamiltonian in a scaling region around T_c.) The latter model is defined as follows. On each site $i = 1, ..., N$ of a finite one-dimensional lattice, we consider a two-component spin-$\frac{1}{2}$ variable $S_i^a, a = x, z$, where

$$S_i^x = \begin{pmatrix} 0 & 1 \\ 1 & 0 \end{pmatrix} \qquad S_i^z = \begin{pmatrix} 1 & 0 \\ 0 & -1 \end{pmatrix}$$

The Ising quantum spin-chain Hamiltonian reads

$$H_{\text{Ising}} = \sum_{i=1}^{N} K S_i^z S_{i+1}^z + \gamma S_i^x$$

($S_{N+1}^a \equiv 0$). The aim of this exercise is to prove the equivalence of this model with the XZ spin chain, defined with the same spin variables by the Hamiltonian

$$H_{XZ} = \sum_{i=1}^{N} K_x S_i^x S_{i+1}^x + K_z S_i^z S_{i+1}^z$$

In order to prove the equivalence, we define new spin variables $T_j^a, a = x, z$, as

$$T_j^x = S_j^z S_{j+1}^z, \quad j = 1, 2, ..., N-1, \qquad T_N^x = S_N^z$$

and

$$T_j^z = \prod_{k=1}^{j} S_k^x$$

a) Show that these new variables have the same commutation relations as the S's.

b) Reexpress the Hamiltonian H_{XZ} in terms of these new variables. Show that, up to unimportant boundary terms, it reads

$$H_{XZ}' = \sum_{j=1}^{N} K_x T_{j-1}^z T_{j+1}^z + K_z S_j^x$$

c) Show that the Hamiltonian H_{XZ}' obtained above splits into the sum of two noninteracting Hamiltonians of the form H_{Ising} for the sets of spin variables T_{2j}^a and T_{2j-1}^a, respectively.

12.2 *Jordan-Wigner transformation in the Heisenberg model*
The Heisenberg XY model is a one-dimensional *quantum* two-component spin chain. More precisely, on each site $i = 1, ..., N$ of a finite one-dimensional lattice, we consider a two-component spin $\frac{1}{2}$ variable $S_i^a, a = x, y$, represented by Pauli matrices as:

$$S_i^x = \frac{1}{2} \begin{pmatrix} 0 & 1 \\ 1 & 0 \end{pmatrix} \qquad S_i^y = \frac{1}{2} \begin{pmatrix} 0 & -i \\ i & 0 \end{pmatrix}$$

The XY model Hamiltonian contains only nearest-neighbor interactions. It reads

$$H = \sum_{i=1}^{N-1} (1 + \gamma) S_i^x S_{i+1}^x + (1 - \gamma) S_i^y S_{i+1}^y$$

where γ is a real parameter.

a) Show that

$$\{a_j, a_j^\dagger\} = 1, \qquad\qquad (a_j)^2 = (a_j^\dagger)^2 = 0$$

$$[a_i^\dagger, a_j] = [a_i^\dagger, a_j^\dagger] = [a_i, a_j] = 0 \qquad i \neq j$$

The a's and $a^\dagger$'s have both fermionic (anti-commuting) and bosonic (commuting) characters.

b) The *Jordan-Wigner transformation* consists in rewriting the Hamiltonian H in terms of new variables $c_j, c_j^\dagger$, defined as

$$c_j = \left(e^{i\pi \sum_{k=1}^{j-1} a_k^\dagger a_k}\right) a_j$$

$$c_j^\dagger = a_j^\dagger \left(e^{i\pi \sum_{k=1}^{j-1} a_k^\dagger a_k}\right)$$

(This is a *nonlocal* transformation; these fermionic operators go over to a free-fermion field in the continuum limit.) Show that

$$c_j c_j^\dagger = a_j a_j^\dagger$$

c) Show further that c_j and $c_j^\dagger$ are true fermionic operators, namely that

$$\{c_i, c_j^\dagger\} = \delta_{i,j} \qquad \{c_i, c_j\} = \{c_i^\dagger, c_j^\dagger\} = 0$$

d) Show that

$$a_i^\dagger a_{i+1} = c_i^\dagger c_{i+1}$$
$$a_i^\dagger a_{i+1}^\dagger = c_i^\dagger c_{i+1}^\dagger$$

e) Deduce that

$$H = \frac{1}{2} \sum_{i=1}^{N-1} (c_i^\dagger c_{i+1} + \gamma c_i^\dagger c_{i+1}^\dagger + \text{h.c.})$$

The Hamiltonian is therefore expressed, through the Jordan-Wigner transformation, as a bilinear form of the fermion operators c_i, $c_i^\dagger$. This transformation is instrumental in the diagonalization of H.

12.3 Write explicitly the high-temperature expansion of the Ising model up to degree 8 in $\tanh K$. Show that the corresponding infinite series expansion of the partition function is an even function of K.

12.4 We consider the correlation function of two Ising spins $\langle \sigma(r_1)\sigma(r_2) \rangle$ (12.9) on the lattice, obtained by changing $K \to K + i\pi/2$ for each bond of a path joining r_1 to r_2. Show that the result is independent, up to a sign, of the choice of path joining r_1 to r_2.
Hint: A path of length n contributes for $i^n \sigma(r_1)\sigma(r_2)$, $i^2 = -1$. Any two paths have lengths differing by a multiple of 2.

12.5 Prove that the correlation function $\langle \mu(r_1)\mu(r_2) \rangle$ of disorder operators is actually independent of the choice of path from r_1 to r_2 along which the couplings K are reversed.
Hint: Consider two paths from r_1 to r_2; the corresponding correlations are exchanged by reversing the sign of all spins along *both* paths, but this operation leaves the result invariant.

12.6 Derive the Ward identity for the insertion of the energy-momentum tensor in a correlator of $2n$ energy operators.

12.7 *Ising energy and spin four-point correlation functions on the plane and conformal blocks*

a) Write the four-point energy correlator on the plane (12.21) in terms of the cross-ratio $z = z_{12}z_{34}/z_{13}z_{24}$ of the four points. Deduce the form of the corresponding unique conformal block. Compare this result with that of Ex. 8.12.

Result:

$$\langle \varepsilon(1)\varepsilon(2)\varepsilon(3)\varepsilon(4)\rangle = \frac{1}{|z(1-z)|^2}|H(z)|^2$$

with the conformal block $H(z) = 1 - z + z^2$, corresponding to the fusion rule $\varepsilon \times \varepsilon \rightarrow \mathbb{I}$.

b) Write the four-point spin correlator on the plane (12.63) in terms of the cross-ratio z. Deduce the form of the corresponding two conformal blocks. Compare this result with that of Ex. 8.12.

Result:

$$\langle \sigma(1)\sigma(2)\sigma(3)\sigma(4)\rangle = \frac{1}{|z(1-z)|^{\frac{1}{2}}}\left(|H_1(z)|^2 + \frac{1}{4}|z||H_2(z)|^2\right)$$

with the conformal blocks

$$H_1(z) = \left(\frac{1+\sqrt{1-z}}{2}\right)^{\frac{1}{2}} \qquad H_2(z) = \left(2\frac{1-\sqrt{1-z}}{z}\right)^{\frac{1}{2}}$$

12.8 Write a differential equation for the correlator $\langle \psi\psi\sigma\sigma\rangle$, expressing the presence of a singular vector of level 2 in the Verma module $V_{2,1}$ of ψ. Check that the function G of Eq. (12.30) is indeed a solution of this equation.

12.9 Express the l.h.s. of Eq. (12.51) as a sum of products of energy correlators of one Ising model only. Using the expression (12.19), derive the most singular contribution of this sum when all arguments approach each other. Compute the free-field correlator on the r.h.s. of Eq. (12.51). Compare the two sides of the equation to deduce the value of the normalization factor $M_n = 2^{n-1}$.

12.10 Along the lines of Ex. 12.9, check that the normalization factor P_n of Eq. (12.55) is $(-1)^n$.

12.11 *Ward identity for the plane 2n-spin correlator*

a) Express the Ising energy-momentum tensor in terms of the real fermion.

b) Express the Dirac energy-momentum tensor in terms of the Dirac fermion.
Result: We set $D = \frac{1}{\sqrt{2}}(\psi_1 + i\psi_2)$ and $D^\dagger = \frac{1}{\sqrt{2}}(\psi_1 - i\psi_2)$. The energy-momentum tensor reads

$$T_{\mathrm{Dir}}(z) = -\lim_{z \to w}\left[\frac{1}{2}\left(D^\dagger(z)\partial_w D(w) - \partial_z D^\dagger(z)D(w)\right) - \frac{1}{(z-w)^2}\right] \qquad (12.193)$$

c) Use the bosonization formulae (12.37) to reexpress Eq. (12.193) in terms of the free field φ.

Result:

$$T_{\mathrm{Dir}}(z) = -\frac{1}{2}\lim_{z \to w}\left[\partial_z\varphi(z,\bar{z})\partial_w\varphi(w,\bar{w}) + \frac{1}{(z-w)^2}\right] \qquad (12.194)$$

d) Express the correlator of the (bosonized) Dirac theory with insertions of the Dirac fermion

$$G(z,w) = \langle D^\dagger(z)D(w)\cos\frac{\varphi}{2}(1)\cdots\cos\frac{\varphi}{2}(2n)\rangle$$

in terms of Ising spin correlators with real fermion insertions only.
Result: $G(z,w) = \frac{1}{2^n}\langle\psi(z)\psi(w)\sigma(1)\cdots\sigma(2n)\rangle\langle\sigma(1)\cdots\sigma(2n)\rangle$.

e) Relate the insertion of the Ising energy-momentum tensor $T_{\text{Ising}}(z)$ in a correlator of $2n$ spins for the Ising model to that of the $c = 1$ energy-momentum tensor $T_{\text{Dir}}(z)$ of the Dirac theory in the corresponding correlator of $2n \cos(\varphi)$ operators.

Result: $\langle T_{\text{Dir}}(z) \cos\frac{\varphi}{2}(1) \cdots \cos\frac{\varphi}{2}(2n) \rangle = 2^{n-1} \langle T_{\text{Ising}}(z)\sigma(1) \cdots \sigma(2n) \rangle$.

f) Deduce that the conformal Ward identity (5.41) of Chap. 5 is satisfied for the insertion of the Ising energy-momentum tensor in the Ising spin correlator iff it is satisfied for the insertion of the Dirac energy-momentum tensor in the $\cos \varphi$ correlator. Check the latter.

12.12 *Cauchy determinant formula, Pfaffian, and Haffnian.*

a) Prove the Cauchy determinant formula

$$\det\left[\frac{1}{z_i - w_j}\right]_{1\le i,j\le n} = (-1)^{n(n-1)/2} \frac{\prod_{i<j}(z_i - z_j)(w_i - w_j)}{\prod_{i,j}(z_i - w_j)} \quad (12.195)$$

Hint: Consider the above determinant as a function of the complex variable $z = z_1$ and analyze its pole and zero structures. Conclude by using Liouville's theorem (a bounded analytic function is necessarily constant).

b) Setting $w_j = z_j + \epsilon_j$ in Eq. (12.195), compute the limit when all $\epsilon_i \to 0$. Deduce the relation between determinant, Pfaffian, and Haffnian expressed in Eq. (12.53).

12.13 *Differential equation for the fermion propagator*

a) Write the differential equation satisfied by the nonnormalized fermion propagator $Z_\nu\langle\psi(z)\psi(0)\rangle_\nu$ in any sector $\nu \in \{2,3,4\}$.

b) From the definition of the Jacobi theta functions (App. 10.A) show that

$$i\pi\partial_\tau\theta_\nu = \partial_z^2\theta_\nu$$

c) Using

$$\zeta(z) - 2\eta_1 z = \partial_z\ln\theta_1, \qquad \wp(z) + 2\eta_1 = -\partial_z^2\ln\theta_1$$

rewrite the differential equation of **(a)** in terms of theta functions only.

d) Use the results for the partition function Z_ν and the expression for the fermion propagator $\langle\psi(z)\psi(0)\rangle_\nu$ to rewrite the differential equation as a differential relation between θ_ν and θ_1. Hint: For the differential equation to be satisfied, the following relation must hold

$$2i\pi\partial_\tau\ln\left[\frac{\partial_z\theta_1(0|\tau)}{\theta_\nu(0|\tau)^{\frac12}}\right] = \left[\frac{\partial_z^2\theta_1\theta_\nu + \theta_1\partial_z^2\theta_\nu - 2\partial_z\theta_1\partial_z\theta_\nu - \eta_1\theta_1\theta_\nu}{\theta_1^2}\right](z|\tau) \quad (12.196)$$

e) Prove this last relation using the standard property of elliptic functions, namely that an elliptic function with at most one pole must be a constant.

Hint: Show that the r.h.s. of Eq. (12.196) is doubly periodic in z and that the residue of the double pole at $z \to 0$ vanishes. Conclude by the standard ellipticity argument.

12.14 *Torus conformal blocks and the proof of the Verlinde formula for the Ising model* Notations are as in Chap. 10, Sect. 10.8.3.

a) Using the conformal blocks (12.94) of the energy-energy correlation on the torus, compute the monodromy transformations $\phi_\varepsilon(a)$ and $\phi_\varepsilon(b)$ of Eq. (10.188) on any Ising character. Check Eqs. (10.189)–(10.192).

b) Repeat the calculation with the conformal blocks (12.109) of the spin-spin correlation function. Recover the Ising fusion rules and the Verlinde formula.

484   12. The Two-Dimensional Ising Model

Hint: The monodromy transformations $\phi_i(a)$ and $\phi_i(b)$ are generated by letting $z \to z+1$ and $z \to z+\tau$, respectively, in the conformal blocks of the torus two-point correlator $\langle ii \rangle$, $i = \varepsilon, \sigma$. The action on characters is obtained by letting $z \to 0$ in the end.

12.15 *Modular covariance of the spin-spin correlator*

a) Using App. 10.A, derive the transformations of the three functions

$$Z_\nu \langle \sigma(z,\bar z)\sigma(0,0)\rangle_\nu \qquad (\nu = 2,3,4)$$

of Eq. (12.106) under $z \to z+1$ and $z \to z+\tau$.

b) Show that (a) fixes $Z_1\langle\sigma\sigma\rangle_1$ uniquely. Compare with Eq. (12.107).

c) For $\nu = 1,2,3,4$, relate the functions $Z_\nu\langle\sigma\sigma\rangle_\nu(z,\bar z,\tau)$ to their "modular transforms" $Z_\nu\langle\sigma\sigma\rangle_\nu(z/\tau; -1/\tau)$. Deduce the "modular covariance" of the spin-spin correlator

$$\langle\sigma(z/\tau,\bar z/\bar\tau)\sigma(0,0)\rangle_{-1/\tau} = |z|^{\frac14}\langle\sigma(z,\bar z)\sigma(0,0)\rangle$$

12.16 *Differential equation for the spin-spin correlator*

a) Write the differential equation for the torus spin-spin correlator $Z_\nu\langle\sigma\sigma\rangle_\nu$ in each sector $\nu = 1,2,3,4$.

b) Prove that this equation is indeed satisfied by $Z_\nu\langle\sigma\sigma\rangle_\nu$, given by Eq. (12.107).
Hint: The differential equation reduces to the identity

$$\left[\frac{\partial_z^2\theta_\nu(z/2|\tau)}{\theta_\nu(z/2|\tau)} + \left(\frac{\partial_z\theta_\nu(z/2|\tau)}{\theta_\nu(z/2|\tau)}\right)^2 - \frac{\partial_z\theta_1(z|\tau)}{\theta_1(z|\tau)}\frac{\partial_z\theta_\nu(z/2|\tau)}{\theta_\nu(z/2|\tau)}\right] + \frac{\partial_z^2\theta_1(z|\tau)}{\theta_1(z|\tau)} + 6\eta_1 = 0$$

This identity is proven by standard elliptic function techniques: We show that the l.h.s. is doubly periodic, with at most one pole, hence a constant, which is readily evaluated.

12.17 *Energy expectation value from the spin-spin correlator*

a) In the sector $\nu = 1$, show that the short-distance limit

$$\lim_{z_1\to z_2}|z_{12}|^{-\frac34}Z_1\langle\sigma(z_1,\bar z_1)\sigma(z_2,\bar z_2)\rangle_1$$

is proportional to the energy one-point function $Z_1\langle\varepsilon\rangle_1$.

b) Deduce the energy expectation value $\langle\varepsilon\rangle$ on the torus.
Result: $\langle\varepsilon\rangle = \pi|\eta(\tau)|^2/Z_{\text{Ising}}$.

12.18 *Correlators for rational $c = 1$ theories on the torus*
This exercise is a sequel of Ex. 10.21. Consider here the $c = 1$ theory of a boson compactified on a circle of radius $R = \sqrt{2p'/p}$. Recall that this theory is a rational conformal theory, namely it can be reorganized into a *finite* number of sectors indexed by $2pp'$ integers $\lambda \equiv \lambda_{e,m} = pe - p'm \bmod 2pp'$. The torus conformal blocks have been derived in Ex. 10.21, and read

$$K_\lambda(\tau) = \frac{1}{\eta(\tau)}\sum_n q^{(2pp'n+\lambda)^2/4pp'}$$

for $\lambda = 0,1,...,2pp'-1$. The torus partition function can be reexpressed as a finite sum

$$Z\left(\sqrt{\frac{2p'}{p}}\right) = \sum_{\lambda=0}^{2pp'-1} K_\lambda(\tau)\bar K_{\omega_0\lambda}(\bar\tau)$$

where ω_0 is defined by

$$pr_0 - p's_0 = 1 \quad \text{and} \quad \omega_0 = pr_0 + p's_0 \mod 2pp'$$

a) Rewrite the correlator (12.148) as a finite sum over the conformal blocks of the corresponding rational conformal theory.

Result: The conformal blocks of the torus two-point function read

$$\equiv \mathcal{F}_\lambda^{e,m}(z) = K_\lambda(4\alpha_{e,m}z_{12}|\tau)\left(\frac{\partial_z\theta_1(0|\tau)}{\theta_1(z|\tau)}\right)^{2h_{e,m}}$$

with $\alpha_{e,m}$ defined in Eq. (12.149), with $R = \sqrt{2p'/p}$, and

$$K_\lambda(z|\tau) = \frac{1}{\eta(\tau)}\sum_n q^{(2pp'n+\lambda)^2/4pp'}e^{i\pi(2pp'n+\lambda)z/2}\sqrt{pp'}$$

The correlator reads

$$Z\left(\sqrt{\frac{2p'}{p}}\right)\langle \mathcal{O}_{e,m}(z,\bar{z})\mathcal{O}_{-e,-m}(0,0)\rangle = \sum_{\lambda=0}^{2pp'-1} \mathcal{F}_\lambda^{e,m}(z) \times \bar{\mathcal{F}}_{\omega_0\lambda}^{-e,-m}(\bar{z})$$

b) Check that the monodromy of the conformal block $\mathcal{F}_\lambda^{e,m}(z)$ along the a cycle is diagonal.

c) Check that the monodromy transformation of $\mathcal{F}_\lambda^{e,m}(z)$ along the cycle b exchanges the upper and lower intermediate states $\lambda \leftrightarrow \lambda + \lambda_{e,m}$.

d) Compute the torus conformal blocks of the n-point function (12.150).

12.19 *Energy correlators and Cauchy determinant formula on the torus.*

a) Prove the Cauchy determinant formula on the torus

$$\det[\wp_\nu(z_i - w_j)]_{1\le i,j\le n} = (-1)^{n(n-1)/2}(\partial_z\theta_1(0|\tau))^n \frac{\theta_\nu(\Sigma z_i - w_i|\tau)}{\theta_\nu(0|\tau)}$$

$$\times \frac{\prod_{i<j}\theta_1(z_{ij}|\tau)\theta_1(w_{ij}|\tau)}{\prod_{i,j}\theta_1(z_i - w_j|\tau)} \qquad (12.197)$$

Hint: Examine the zero and pole structure of the l.h.s. of Eq. (12.197) using the short-distance behavior, given by Eq. (12.195). Conclude by the standard elliptic function argument.

b) Prove the identity between Eqs. (12.96) and (12.174).

Hint: Take the multiple limit $\epsilon_1, ..., \epsilon_n \to 0$ after the substitution $w_i = z_i + \epsilon_i$ in the Cauchy determinant formula (12.197).

12.20 *Ward identity for the torus 2n-spin correlator*

This is the torus version of Ex. 12.11, whose results are assumed here.

a) First take $n = 1$. Using chiral bosonization, compute the insertion of the Dirac energy-momentum tensor in the correlation of two $\cos\varphi$ operators in any sector ν of the bosonized Dirac-fermion theory.

Result:

$$
\begin{aligned}
Z_v^2 \langle T_{\text{Dir}}(z) \cos \frac{\varphi}{2}(z_1, \bar{z}_1) \cos \frac{\varphi}{2}(z_2, \bar{z}_2) \rangle_v &= 2 \left[\eta_1 + \frac{\partial_{z_1}^2 \theta_v(z_{12}/2|\tau)}{\theta_v(z_{12}/2|\tau)} \right. \\
&+ \frac{1}{2} \frac{\partial_{z_1} \theta_v(z_{12}/2|\tau)}{\theta_v(z_{12}/2|\tau)} \left(\frac{\partial_z \theta_1(z - z_1|\tau)}{\theta_1(z - z_1|\tau)} - \frac{\partial_z \theta_1(z - z_2|\tau)}{\theta_1(z - z_2|\tau)} \right) \\
&+ \left. \frac{1}{16} \left(\frac{\partial_z \theta_1(z - z_1|\tau)}{\theta_1(z - z_1|\tau)} - \frac{\partial_z \theta_1(z - z_2|\tau)}{\theta_1(z - z_2|\tau)} \right)^2 \right] Z_v^2 \langle \cos \frac{\varphi}{2}(z_1, \bar{z}_1) \cos \frac{\varphi}{2}(z_2, \bar{z}_2) \rangle_v
\end{aligned}
\tag{12.198}
$$

b) Prove the Ward identity (12.79) on the torus, in the case of the $\cos \varphi$ two-point correlator.

c) Deduce the analogous result for the Ising spin-spin correlator on the torus.

d) Repeat the above calculation for arbitrary n.

Notes

The two-dimensional Ising model on a square lattice was solved by Onsager in 1944 [286] by diagonalization of the transfer matrix. Several combinatorial improvements, among which the rewriting of the partition function as a determinant, have simplified the solution and prepared the route for the contact with a fermionic theory in the continuum limit [211, 229]. In the late 1970s, Baxter and Enting [32] gave yet another solution, based on some recursion relations for commuting transfer matrices, granting the integrability of the model. This treatment could actually be generalized [31] to a large class of models, including the six- and eight-vertex models. In this respect, the Ising model is the simplest example of an integrable lattice model.

The fermionic continuum formulation arises from the use of the Jordan-Wigner transformation of the original spin variables, to build a set of *anticommuting* variables [325, 262]. These were identified in the continuum as the two components of a free real (Majorana) fermion field. In this presentation, the spin operator is a nonlocal observable, as well as its dual operator under high/low temperature duality, the disorder operator [225]. Whereas the correlation functions for the fermionic operators, or composites such as the energy operator, are very easily calculated, the correlations of spin or disorder fields are more difficult to obtain.

The conformal invariance of the free fermionic theory provides, however, a powerful machinery to unify all these results [36]. In this framework, the energy-momentum technique was introduced in Ref. [105]. The concept of bosonization was first applied to the Ising model in Ref. [206].

The continuum Ising model on a torus was extensively studied in Ref. [97], where most correlation functions were computed either using the chiral bosonization techniques of Refs. [119, 341] or some direct bosonization scheme in the framework of $c = 1$ Coulomb gas on the torus. In Ref. [97], use is also made of the torus version of the energy-momentum technique [15]. An interesting issue was the elliptic differential equations satisfied by the torus correlators, due to a combination of the torus Ward identities [120], and the singular vector structure of the energy and spin Virasoro modules. In fact, just like in the plane case, the general solutions of these differential equations can be written à la Feigin-Fuchs, as complex contour integrals, with screening operator insertions [128].

CONFORMAL FIELD THEORIES WITH LIE-GROUP SYMMETRY

CHAPTER 13

Simple Lie Algebras

This chapter presents a survey of the theory of Lie algebras. This might appear somewhat remote from our main subject of interest: affine Lie algebras and their applications to conformal field theory. However, it turns out that in many respects the theory of affine Lie algebras is a natural extension of the theory of simple Lie algebras, and as such cannot be studied efficiently in isolation. This is an immediate motivation for devoting a complete chapter to Lie algebras. But as subsequent developments will show, conformal field theories with nonaffine additional symmetries, such as W algebras, parafermions, and son on, as well as related exactly solvable statistical models, also have a deep Lie-algebraic underlying structure, which can only be appreciated with a minimal background on simple Lie algebras.

No previous knowledge on Lie algebras is assumed except for a first encounter with $su(2)$ and the theory of angular momentum in quantum mechanics. Admittedly, for those readers unfamiliar with the subject, this chapter will appear to be somewhat dense. Nevertheless the presentation is conceptually self-contained. This is not so at the technical level, since some statements and constructions are given without proofs. Furthermore, the choice of material is not completely standard, being dictated by our subsequent applications.

Section 13.1 covers the basic elements of the theory of simple Lie algebras: roots, weights, Cartan matrices, Dynkin diagrams, and the Weyl group. The subsequent section is devoted to the study of highest-weight representations. This is followed by an explicit description of states in $su(N)$ highest-weight representations, in terms of tableaux and patterns. Characters of irreducible representations are introduced in Sect. 13.4. From the results of Part B, it should already be clear that characters play a central role in some aspects of conformal field theory, such as modular invariance.

One of the central problems in conformal field theory is the calculation of fusion rules. Given that most conformal field theories have a Lie-algebraic core, the fusion rules are, to a large extent, determined by the tensor-product coefficients of this Lie algebra. It is thus mandatory to review in detail some methods for calculating tensor products. This is the subject of two sections, Sects 13.5 and 13.6. In the first we present efficient techniques for tensor-product calculations, and in the second one we reconsider the problem from a conformal-field-theoretical angle.

Quotienting two affine Lie algebras will prove to be one the key tools in constructing conformal field theories. At the heart of this construction, there is a finite Lie algebra embedding, to which Sect. 13.7 is dedicated.

The basic properties of simple Lie algebras are displayed in App. 13.A, in a form that should facilitate later consultation. Finally, all symbols used in this chapter are collected in App. 13.B.

Readers familiar with Lie algebras may skip this chapter and use it only as a reference—except for a glance at App. 13.B in order to fix the notations. To those readers, we indicate that only Sects. 13.5.4 and 13.6 do not present standard material. On the other hand, those who wish to learn the basics of Lie algebra in this chapter should not necessarily read it linearly. Sections 13.1, 13.2, and 13.4.1 are essential and must be read sequentially. But the rest can be consulted when needed. Furthermore, it is not essential to master all the techniques for calculating tensor products in order to proceed. The description in the main text of tools particular to $su(N)$ (tableaux, the Littlewood-Richardson rule, Berenstein-Zelevinsky triangles) has the main purpose of lightening the presentation.

§13.1. The Structure of Simple Lie Algebras

13.1.1. The Cartan-Weyl Basis

A Lie algebra g is a vector space equipped with an antisymmetric binary operation [,], called a commutator, mapping g × g into g, and further constrained to satisfy the Jacobi identity

$$[X, [Y, Z]] + [Z, [X, Y]] + [Y, [Z, X]] = 0 \qquad \text{for} \quad X, Y, Z \in g \qquad (13.1)$$

Roughly speaking, the exponential of g is the Lie group G (more precisely, its connected component containing the unit element): to $X \in g$, there corresponds the group elements e^{iaX} where a is some parameter and the exponential is defined from its power expansion. Hence, the algebra describes the group in the vicinity of the identity.

A *representation* refers to the association of every element of g to a linear operator acting on some vector space V, which respects the commutation relations of the algebra. The maximal number of linearly independent states that generate V is the *dimension of the representation*. Relative to a given basis, each element of g can thus be represented in terms of a square matrix and the basis vectors are represented by column matrices. (In the representation, the commutator corresponds to the usual matrix commutation.) A representation is said to be *irreducible* if the matrices representing the elements of g cannot all be brought in a block-diagonal form by a change of basis.

These elementary notions are sufficient to start analyzing the structure of Lie algebras. A Lie algebra can be specified by a set of generators $\{J^a\}$ and their commutation relations

$$[J^a, J^b] = \sum_c if^{ab}{}_c J^c \tag{13.2}$$

The number of generators is the *dimension of the algebra*. The constants $f^{ab}{}_c$ are the *structure constants*, real parameters when $(J^a)^\dagger = J^a$.[1] We are concerned with *simple* Lie algebras, that is, Lie algebras that contain no proper ideal (meaning no proper subset of generators $\{L^a\}$ such that $[L^a, J^b] \in \{L^a\}$ for any J^b). A direct sum of simple algebras is said to be *semisimple*.

In the standard *Cartan-Weyl basis*, the generators are constructed as follows. We first find the maximal set of commuting Hermitian generators H^i, $i = 1, \cdots, r$ (r is the *rank* of the algebra):

$$[H^i, H^j] = 0 \tag{13.3}$$

This set of generators form the *Cartan subalgebra* h. The generators of the Cartan subalgebra can all be diagonalized simultaneously. The remaining generators are chosen to be those particular combinations of the J^a's that satisfy the following eigenvalue equation:

$$[H^i, E^\alpha] = \alpha^i E^\alpha \tag{13.4}$$

The vector $\alpha = (\alpha^1, \cdots, \alpha^r)$ is called a *root* and E^α is the corresponding *ladder operator*. Because h is the maximal Abelian subalgebra of g, the roots are non-degenerate. The root α naturally maps an element $H^i \in$ h to the number α^i by $\alpha(H^i) = \alpha^i$. Hence, the roots are elements of the dual of the Cartan subalgebra: $\alpha \in$ h*.

Equation (13.4), through its Hermitian conjugate, shows that $-\alpha$ is necessarily a root whenever α is, with

$$E^{-\alpha} = (E^\alpha)^\dagger \tag{13.5}$$

In the following, Δ will denote the set of all roots.

Root components can be regarded as the nonzero eigenvalues of the H^i in the particular representation, called the *adjoint*, for which the Lie algebra itself serves as the vector space on which the generators act. In this representation, we have an identification

$$\begin{aligned} E^\alpha &\longmapsto &|E^\alpha\rangle \equiv |\alpha\rangle \\ H^i &\longmapsto &|H^i\rangle \end{aligned} \tag{13.6}$$

between the generators and the states of the representation. It follows from Eq. (13.4) that in the adjoint representation the action of a generator X is represented by ad(X), defined as

$$\text{ad}(X)Y = [X, Y] \tag{13.7}$$

---

[1] In physics, it is usual to work with Hermitian operators; these are the observables, whose eigenvalues are real.

so that

$$\text{ad}(H^i)E^\alpha = \alpha^i E^\alpha \qquad \longmapsto \qquad H^i|\alpha\rangle = \alpha^i|\alpha\rangle \qquad (13.8)$$

The one-to-one correspondence between the states $|\alpha\rangle$ and the ladder operators E^α reflects the nondegenerate character of roots. In this representation, the zero eigenvalue has degeneracy r (associated with the different states $|H^i\rangle$). By construction, the dimension of the adjoint is equal to the dimension of the algebra, itself equal to the total number of roots plus r.

In view of specifying the remaining commutators, we first observe that the Jacobi identity implies

$$[H^i, [E^\alpha, E^\beta]] = (\alpha^i + \beta^i)[E^\alpha, E^\beta] \qquad (13.9)$$

If $\alpha + \beta \in \Delta$, the commutator $[E^\alpha, E^\beta]$ must be proportional to $E^{\alpha+\beta}$, and it must vanish if $\alpha + \beta \notin \Delta$. When $\alpha = -\beta$, $[E^\alpha, E^{-\alpha}]$ commutes with all H^i, which is possible only if it is a linear combination of the generators of the Cartan subalgebra. The normalization of the ladder operators is fixed by setting this commutator equal to $2\alpha \cdot H/|\alpha|^2$, where

$$\alpha \cdot H = \sum_{i=1}^{r} \alpha^i H^i \qquad |\alpha|^2 = \sum_{i=1}^{r} \alpha^i \alpha^i \qquad (13.10)$$

Summarizing, the full set of commutation relations in the Cartan-Weyl basis is

$$
\begin{aligned}
[H^i, H^j] &= 0 \\
[H^i, E^\alpha] &= \alpha^i E^\alpha \\
[E^\alpha, E^\beta] &= N_{\alpha,\beta} E^{\alpha+\beta} \quad &&\text{if} \quad \alpha + \beta \in \Delta \\
&= \frac{2}{|\alpha|^2}\alpha \cdot H \quad &&\text{if} \quad \alpha = -\beta \\
&= 0 \quad &&\text{otherwise}
\end{aligned}
\qquad (13.11)
$$

where $N_{\alpha,\beta}$ is a constant.

13.1.2. The Killing Form

The normalization used to fix the commutators is usually introduced by means of the *Killing form*

$$\tilde{K}(X, Y) = \text{Tr}(\text{ad}X \text{ad}Y) \qquad (13.12)$$

which gives a sort of scalar product for the Lie algebra. To calculate this trace in some basis of generators $\{T^a\}$, we first evaluate $[X, [Y, T^b]]$ in terms of the elements of this basis; the coefficient of T^b in the result gives the contribution of this term to the trace. For semisimple Lie algebras, the Killing form is nondegenerate: $\tilde{K}(X, Y) = 0$ for all Y implies that $X = 0$. This is in fact an alternate way of defining semisimplicity.

In the following, we will mainly use a renormalized version of the Killing form defined as

$$K(X,Y) \equiv \frac{1}{2g}\mathrm{Tr}(\mathrm{ad}X\,\mathrm{ad}Y) \qquad (13.13)$$

where g is a constant that will be defined later (it is the dual Coxeter number of the algebra g). The standard basis $\{J^a\}$ is understood to be orthonormal with respect to K:

$$K(J^a,J^b) = \delta^{a,b} \qquad (13.14)$$

The same normalization holds for the generators of the Cartan subalgebra

$$K(H^i,H^j) = \delta^{i,j} \qquad (13.15)$$

Since the Killing form defines a scalar product, it can be used to lower or raise the indices, e.g.,

$$f_{abc} = \sum_d f^{ad}{}_c \, [K(J^d,J^b)]^{-1} \qquad (13.16)$$

We note that f_{abc} is antisymmetric in all three indices. In the $\{J^a\}$ orthonormal basis, the position of the indices (up or down) is thus irrelevant.

The cyclic property of the trace yields the identity

$$K([Z,X],Y) + K(X,[Z,Y]) = 0 \qquad (13.17)$$

The Killing form is actually uniquely characterized by this property. With appropriate choices for $X, Y, Z \in$ g, it follows that

$$[E^\alpha, E^{-\alpha}] = K(E^\alpha, E^{-\alpha})\,\alpha \cdot H \qquad (13.18)$$

(all other pairs involving a ladder operator have zero Killing form). Hence, the previously introduced normalization corresponds to

$$K(E^\alpha, E^{-\alpha}) = \frac{2}{|\alpha|^2} \qquad (13.19)$$

However, the fundamental role of the Killing form is to establish an isomorphism between the Cartan subalgebra h and its dual h*: the form $K(H^i, \cdot)$ (i fixed) maps every element of the Cartan subalgebra onto a number. Hence, to every element $\gamma \in$ h*, there corresponds a $H^\gamma \in$ h through

$$\gamma(H^i) = K(H^i, H^\gamma) \qquad (13.20)$$

(in particular for a root α, $H^\alpha = \alpha \cdot H = \sum_i \alpha^i H^i$). With this isomorphism, the Killing form can be transferred into a positive definite scalar product in the dual space

$$(\gamma, \beta) = K(H^\beta, H^\gamma) \qquad (13.21)$$

Since roots are elements of h*, this defines a scalar product in the root space. From now on, the scalar product between roots will be denoted as above, with the understanding that $|\alpha|^2 = (\alpha, \alpha)$.

13.1.3. Weights

Up to this point, we have analyzed the structure of the algebra from the point of view of a particular representation (the adjoint), that for which the algebra itself plays the role of the vector space.[2] In this representation, the eigenvalues of the Cartan generators are called the roots and the scalar product between roots is induced by the Killing form. Since the essential structure of the algebra is coded in this representation, it needs to be studied in more detail. For this, it is useful to first recast the problem in the general context of a finite-dimensional representation.

For an arbitrary representation, a basis $\{|\lambda\rangle\}$ can always be found such that

$$H^i|\lambda\rangle = \lambda^i|\lambda\rangle \tag{13.22}$$

The eigenvalues λ^i build the vector $\lambda = (\lambda^1, \cdots, \lambda^r)$, called a *weight*. Weights live in the space h^*: $\lambda(H^i) = \lambda^i$. Hence, the scalar product between weights is also fixed by the Killing form. In the adjoint representation, the weights deserve the special name of roots. The commutator (13.4) shows that E^α changes the eigenvalue of a state by α:

$$H^i E^\alpha|\lambda\rangle = [H^i, E^\alpha]|\lambda\rangle + E^\alpha H^i|\lambda\rangle = (\lambda^i + \alpha^i)E^\alpha|\lambda\rangle \tag{13.23}$$

so that $E^\alpha|\lambda\rangle$, if nonzero, must be proportional to a state $|\lambda + \alpha\rangle$. This justifies the name ladder (or step) operator for E^α.

Representations of interest are the finite-dimensional ones. For these, we will derive an important relation, to be used shortly for the adjoint representation. For any state $|\lambda\rangle$ in a finite-dimensional representation, there are necessarily two positive integers p and q, such that

$$\begin{aligned}
(E^\alpha)^{p+1}|\lambda\rangle &\sim E^\alpha|\lambda + p\alpha\rangle = 0 \\
(E^{-\alpha})^{q+1}|\lambda\rangle &\sim E^\alpha|\lambda - q\alpha\rangle = 0
\end{aligned} \tag{13.24}$$

for any root α. Indeed, notice that the triplet of generators E^α, $E^{-\alpha}$, and $\alpha \cdot H/|\alpha|^2$ forms an $su(2)$ subalgebra analogue to the set $\{J^+, J^-, J^3\}$, with commutation relations

$$[J^+, J^-] = 2J^3, \qquad [J^3, J^\pm] = \pm J^\pm \tag{13.25}$$

Therefore, if $|\lambda\rangle$ belongs to a finite-dimensional representation, its projection onto the $su(2)$ subalgebra associated with the root α must also be finite dimensional. Let the dimension of the latter be $2j + 1$; then from the state $|\lambda\rangle$, the state with highest $J^3 = \alpha \cdot H/|\alpha|^2$ projection ($m = j$) can be reached by a finite number, say p, applications of $J^+ = E^\alpha$, whereas, say, q applications of $J^- = E^{-\alpha}$ lead to the

---

[2] A matrix realization of the adjoint representation in the basis $\{J^a\}$ is given by

$$(J^a)_{bc} = -if_{abc}$$

state with $m = -j$:

$$j = \frac{(\alpha, \lambda)}{|\alpha|^2} + p, \qquad -j = \frac{(\alpha, \lambda)}{|\alpha|^2} - q \qquad (13.26)$$

Eliminating j from the above two equations yields

$$2\frac{(\alpha, \lambda)}{|\alpha|^2} = -(p - q) \qquad (13.27)$$

This is the relation we were looking for: any weight λ in a finite-dimensional representation is such that $(\alpha, \lambda)/|\alpha|^2$ is an integer. This is true in particular for $\lambda = \beta$, where β is any root of the algebra. We now return to the analysis of the root properties.

13.1.4. Simple Roots and the Cartan Matrix

As already mentioned, the number of roots is equal to the dimension of the algebra minus its rank, and this number is in general much larger than the rank itself. This means that the roots are linearly dependent. We then fix a basis $\{\beta_1, \beta_2, \cdots, \beta_r\}$ in the space h*, so that any root can be expanded as

$$\alpha = \sum_{i=1}^{r} n_i \beta_i \qquad (13.28)$$

In this basis, an ordering can be defined as follows: α is said to be positive if the first nonzero number in the sequence $(n_1, n_2, \cdots, n_r)$ is positive. Denote by Δ_+ the set of positive roots. The set of negative roots Δ_- is defined in the obvious way. We have already observed that whenever α is a root, $-\alpha$ is also a root; hence $\Delta_- = -\Delta_+$.

A *simple root* α_i is defined to be a root that cannot be written as the sum of two positive roots. There are necessarily r simple roots, and their set $\{\alpha_1, \cdots, \alpha_r\}$ provides the most convenient basis for the r-dimensional space of roots. Notice that the subindex is a labeling index: it does not refer to a root component. Two immediate consequences of the definition of simple roots are : (i) $\alpha_i - \alpha_j \notin \Delta$ (otherwise, if $\alpha_i - \alpha_j > 0$, say, we would conclude that $\alpha_i = \alpha_j + (\alpha_i - \alpha_j)$, a contradiction); (ii) any positive root is a sum of positive roots (indeed, if a positive root is not simple, it can be written as a sum of two positive roots, which, if not simple, can also be written as the sum of two positive roots, and so on).[3]

The scalar products of simple roots define the *Cartan matrix*

$$A_{ij} = \frac{2(\alpha_i, \alpha_j)}{\alpha_j^2} \qquad (13.29)$$

---

[3] It should be clear from the construction that the choice of a set of simple roots is not unique since it depends upon the initial choice of basis $\{\beta_i\}$ and the ordering used. The precise relation between different sets of simple roots will be clarified later.

In view of Eq. (13.27), the entries of this matrix are necessarily integers. Its diagonal elements are all equal to 2 and it is not symmetric in general. The Schwarz inequality implies that $A_{ij}A_{ji} < 4$ for $i \neq j$. Since $\alpha_i - \alpha_j$ is not a root, $E^{-\alpha_i}|\alpha_i\rangle = 0$, and $q = 0$ in Eq. (13.24) for $\lambda = \alpha_i$ and $\alpha = \alpha_j$. Hence, from Eq. (13.27) it follows that

$$(\alpha_i, \alpha_j) \leq 0, \quad i \neq j \tag{13.30}$$

Thus for $i \neq j$, A_{ij} is a nonpositive integer, and in view of the above inequality, it can only be $0, -1, -2$, or -3. If $A_{ij} \neq 0$, the inequality forces at least one of A_{ij} or A_{ji} to be -1.

It can be shown that in the set of roots of a simple Lie algebra, at most two different lengths (long and short) are possible. The ratio of the length of the long roots over the short roots is bound to be 2 or 3, if different from 1. When all the roots have the same length, the algebra is said to be *simply laced*.

It is convenient for us to introduce a special notation for the quantity $2\alpha_i/|\alpha_i|^2$:

$$\alpha_i^\vee = \frac{2\alpha_i}{|\alpha_i|^2} \tag{13.31}$$

$\alpha_i^\vee$ is called the *coroot* associated with the root α_i. The scalar product between roots and coroots is thus always an integer. The Cartan matrix now takes the compact form

$$\boxed{A_{ij} = (\alpha_i, \alpha_j^\vee)} \tag{13.32}$$

A distinguished element of Δ is the *highest root* θ. It is the unique root for which, in the expansion $\sum m_i \alpha_i$, the sum $\sum m_i$ is maximized. All elements of Δ can be obtained by repeated subtraction of simple roots from θ. The coefficients of the decomposition of θ in the bases $\{\alpha_i\}$ and $\{\alpha_i^\vee\}$ bear special names, being called, respectively, the *marks* (a_i) and the *comarks* $(a_i^\vee)$:[4]

$$\boxed{\theta = \sum_{i=1}^r a_i \alpha_i = \sum_{i=1}^r a_i^\vee \alpha_i^\vee, \quad a_i, a_i^\vee \in \mathbb{N}} \tag{13.33}$$

Marks and comarks are related by

$$a_i = a_i^\vee \frac{2}{|\alpha_i|^2} \tag{13.34}$$

The *dual Coxeter number* is defined as

$$\boxed{g = \sum_{i=1}^r a_i^\vee + 1} \tag{13.35}$$

---

[4] These are also called Kac labels and dual Kac labels, respectively.

(The Coxeter number can be defined similarly, but it will not be used here. The superscript $\vee$, which would naturally appear in the notation for the dual Coxeter number, is thus omitted.)

13.1.5. The Chevalley Basis

As will be shown below, the full set of roots can be reconstructed from the set of simple roots, and the latter can be extracted from the Cartan matrix in a very simple way. Moreover, the Cartan matrix fixes completely the commutation relations of the algebra. This point is made fully manifest in the *Chevalley basis* where to each simple root α_i there corresponds the three generators

$$e^i = E^{\alpha_i} \qquad f^i = E^{-\alpha_i} \qquad h^i = \frac{2\alpha_i \cdot H}{|\alpha_i|^2} \tag{13.36}$$

whose commutation relations are

$$\boxed{\begin{aligned} &[h^i, h^j] = 0 \\ &[h^i, e^j] = A_{ji} e^j \\ &[h^i, f^j] = -A_{ji} f^j \\ &[e^i, f^j] = \delta_{ij} h^j \end{aligned}} \tag{13.37}$$

The remaining step operators are obtained by repeated commutations of these basic generators, subject to the *Serre relations*

$$[\mathrm{ad}(e^i)]^{1-A_{ji}} e^j = 0 \\ [\mathrm{ad}(f^i)]^{1-A_{ji}} f^j = 0 \tag{13.38}$$

For instance, $[\mathrm{ad}(e^i)]^2 e^j = [e^i, [e^i, e^j]]$. These constraints—the analogues of relations (13.24) for the adjoint representation—encode the rules for reconstructing the full root system from the simple roots. (For this specific problem, still another approach will be presented later.) The Serre relations do not mix the e^i's and the f^i's and this reflects the separation of the roots into two disjoint sets $\Delta_\pm$. That the Serre relations and the basic commutation relations can be expressed in terms of the Cartan matrix shows that A contains all the information on the structure of g. Actually, the abstract formulation of Lie algebras in terms of Cartan matrices is the most efficient starting point for generalizations.

The Killing form of the generators of the Cartan subalgebra is easily transcribed from the Cartan-Weyl to the Chevalley basis:

$$K(h^i, h^j) = (\alpha_i^\vee, \alpha_j^\vee) \tag{13.39}$$

13.1.6. Dynkin Diagrams

All the information contained in the Cartan matrix can be encapsulated in a simple planar diagram: the *Dynkin diagram*. To every simple root α_i, we associate a node (white for a long root and black for a short one) and join the nodes i and j with $A_{ij}A_{ji}$ lines. Hence orthogonal simple roots are disconnected, and those sustaining an angle of 120, 135, or 150 degrees are linked by one, two, or three lines, respectively.

The classification of simple Lie algebras boils down to a classification of Dynkin diagrams. The complete list contains four infinite families, the algebras A_r, B_r, C_r and D_r (the classical algebras, whose compact real forms are respectively $su(r+1), so(2r+1), sp(2r)$, and $so(2r))$, and five exceptional cases: E_6, E_7, E_8, F_4, and G_2.[5] The subscript gives the rank of the algebra. The Dynkin diagrams as well as basic properties of these Lie algebras are displayed in App. 13.A. Note that the A, D, E algebras are simply laced. (The classification of simply-laced algebras has already been considered in Ex. 10.10.)

13.1.7. Fundamental Weights

As already pointed out, weights and roots live in the same r-dimensional vector space. The weights can thus be expanded in the basis of simple roots. However, this expansion is not very useful since for irreducible finite-dimensional representations—the representations of interest—its coefficients are not integers. The convenient basis for weights is in fact the one dual to the simple coroot basis. It is denoted by $\{\omega_i\}$ and defined by

$$(\omega_i, \alpha_j^\vee) = \delta_{ij} \tag{13.40}$$

The ω_i are called the *fundamental weights*.

The expansion coefficients λ_i of a weight λ in the fundamental weight basis are called *Dynkin labels*. Hence,

$$\boxed{\lambda = \sum_{i=1}^{r} \lambda_i \omega_i \quad \Longleftrightarrow \quad \lambda_i = (\lambda, \alpha_i^\vee)} \tag{13.41}$$

The Dynkin labels of weights in finite-dimensional irreducible representations are always integers (this follows from Eq. (13.27) and it will be made explicit in the next section); such weights are said to be *integral*. From now on, whenever a weight is written in component form

$$\lambda = (\lambda_1, \cdots, \lambda_r) \tag{13.42}$$

(with entries separated by commas) it is understood that these components are the Dynkin labels. Note that the elements of the Cartan matrix are the Dynkin labels

---

[5] We have not been careful about the specification of the field over which g is a vector space. The result of this classification holds for the field $\mathbb{C}$, which is algebraically closed. Hence, A_r stands for $sl_{r+1}(\mathbb{C})$, the complex $(r+1) \times (r+1)$ traceless matrices. Among its real forms, only $su_{r+1}(\mathbb{R}) \equiv su(r+1)$ is compact.

of the simple roots

$$\alpha_i = \sum_j A_{ij}\omega_j \qquad (13.43)$$

that is, the i-th row of A is the set of Dynkin labels for the simple root α_i.

The Dynkin labels are the eigenvalues of the Chevalley generators of the Cartan subalgebra:

$$h^i|\lambda\rangle = \lambda(h^i)|\lambda\rangle = (\lambda, \alpha_i^\vee)|\lambda\rangle \qquad (13.44)$$

that is

$$\boxed{h^i|\lambda\rangle = \lambda_i|\lambda\rangle} \qquad (13.45)$$

The position of the index has the following meaning: λ_i refers to an eigenvalue of h^i (a Dynkin label), whereas λ^i is an eigenvalue of H^i.)

A weight of special importance, thus deserving a special notation, is the one for which all Dynkin labels are unity:

$$\rho = \sum_i \omega_i = (1, 1, \cdots, 1) \qquad (13.46)$$

This is called the *Weyl vector* (or principal vector) and has the following alternate definition (to be proved later):

$$\rho = \frac{1}{2} \sum_{\alpha \in \Delta_+} \alpha \qquad (13.47)$$

The scalar product of weights can be expressed in terms of a symmetric *quadratic form matrix F_{ij}*

$$(\omega_i, \omega_j) = F_{ij} \qquad (13.48)$$

The definition implies that F_{ij} is the transformation matrix relating the two bases $\{\omega_i\}$ and $\{\alpha_i^\vee\}$

$$\omega_i = \sum_j F_{ij}\alpha_j^\vee \qquad (13.49)$$

Indeed, the product of this equation with $\alpha_j^\vee$ reproduces (13.48). Hence F_{ij} is the inverse of the matrix whose rows are the Dynkin labels of the simple coroots, and these can be read off the following rescaled version of (13.43)

$$\alpha_i^\vee = \sum_j \frac{2}{|\alpha_i|^2} A_{ij}\omega_j \qquad (13.50)$$

This leads to an explicit relation between the quadratic form and the Cartan matrix:

$$F_{ij} = (A^{-1})_{ij}\frac{\alpha_j^2}{2} \qquad (13.51)$$

The scalar product of the two weights $\lambda = \sum \lambda_i \omega_i$ and $\mu = \sum \mu_i \omega_i$ reads

$$(\lambda, \mu) = \sum_{i,j} \lambda_i \mu_j (\omega_i, \omega_j) = \sum_{i,j} \lambda_i \mu_j F_{ij} \qquad (13.52)$$

The quadratic form matrices of all the simple Lie algebras are tabulated in App. 13.A, with the normalization convention defined in Sect. 13.1.10.

13.1.8. The Weyl Group

We return for a moment to the projection of the adjoint representation onto the $su(2)$ subalgebra associated with the root α. Let m be the eigenvalue of the J^3 operator $\alpha \cdot H/|\alpha|^2$ on the state $|\beta\rangle$; that is,

$$2m = (\alpha^\vee, \beta) \qquad (13.53)$$

If $m \neq 0$, this state must be paired with another one with J^3 eigenvalue $-m$. Therefore, there must exist another state in the multiplet, say $|\beta + \ell\alpha\rangle$, whose projection on the J^3 axis is equal to

$$(\alpha^\vee, \beta + \ell\alpha) = (\alpha^\vee, \beta) + 2\ell = -(\alpha^\vee, \beta) \qquad (13.54)$$

This shows that if β is a root, $\beta - (\alpha^\vee, \beta)\alpha$ is also a root.

The operation s_α defined by

$$s_\alpha \beta = \beta - (\alpha^\vee, \beta)\,\alpha \qquad (13.55)$$

is a reflection with respect to the hyperplane perpendicular to α. The set of all such reflections with respect to roots forms a group, called the *Weyl group* of the algebra, denoted W. It is generated by the r elements s_i, the *simple Weyl reflections*,

$$s_i \equiv s_{\alpha_i} \qquad (13.56)$$

in the sense that every element $w \in W$ can be decomposed as

$$w = s_i s_j \cdots s_k \qquad (13.57)$$

For the simple Weyl reflections, the following relations are easily checked

$$s_i^2 = 1, \qquad s_i s_j = s_j s_i \quad \text{if} \quad A_{ij} = 0 \qquad (13.58)$$

These generalize to[6]

$$(s_i s_j)^{m_{ij}} = 1 \qquad \text{where} \quad m_{ij} = \begin{cases} 2 & \text{if} \quad i = j \\[2mm] \dfrac{\pi}{\pi - \theta_{ij}} & \text{if} \quad i \neq j \end{cases} \qquad (13.59)$$

with θ_{ij} the angle between the simple root α_i and α_j.[7] Eq. (13.59) can be regarded as the defining relation of the Weyl group. We note again that it is expressed in

---

[6] Any group having such a representation is called a *Coxeter group* .

[7] Thus $m_{ij} = 2, 3, 4$, or 6, corresponding to the number of lines joining the i-th and the j-th node being 0, 1, 2, or 3.

terms of data directly related to the Cartan matrix. On the simple roots, the action of s_i takes the simple form

$$\boxed{s_i \alpha_j = \alpha_j - A_{ji} \alpha_i} \tag{13.60}$$

It has just been shown that W maps Δ into itself. In fact, it provides a simple way to generate the complete set Δ from the simple roots by acting with all the elements of W on the set $\{\alpha_i\}$:

$$\Delta = \{w\alpha_1, \cdots, w\alpha_r | w \in W\} \tag{13.61}$$

From this construction, it is clear that any set $\{w'\alpha_i\}$ with w' fixed, could serve as a basis of simple roots. (This gives the announced relation between the different bases of simple roots.)

As a short digression, we now prove, using the Weyl group, the equivalence between (13.46) and (13.47). From (13.46) it follows that $(\rho, \alpha_i^\vee) = 1$ for all i. We want to show that the same result follows from the second definition. We set $\sigma = \sum_{\alpha>0} \alpha/2$ and consider $s_i\sigma$. Since s_i permutes all the positive roots—that is, $A_{ij} \leq 0$ if $i \neq j$ (except α_i which is mapped to $-\alpha_i$), we can write

$$s_i \sigma = \frac{1}{2} \sum_{\substack{\alpha>0 \\ \alpha \neq \alpha_i}} \alpha - \frac{1}{2}\alpha_i = \frac{1}{2} \sum_{\alpha>0} \alpha - \alpha_i \tag{13.62}$$

implying that

$$(s_i\sigma, \alpha_i^\vee) = (\sigma - \alpha_i, \alpha_i^\vee) = (\sigma, \alpha_i^\vee) - 2 \tag{13.63}$$

On the other hand, from the invariance of the scalar product with respect to Weyl transformations, the same product can be written as

$$(s_i\sigma, \alpha_i^\vee) = (\sigma, s_i\alpha_i^\vee) = -(\sigma, \alpha_i^\vee) \tag{13.64}$$

The compatibility of these two equations gives the desired result, namely $(\sigma, \alpha_i^\vee) = 1$ and thus $\sigma = \rho$.

The action of the Weyl group, defined so far only for roots, extends naturally to weights:

$$\boxed{s_\alpha \lambda = \lambda - (\alpha^\vee, \lambda)\, \alpha} \tag{13.65}$$

It is straightforward to verify from the above relation that the Weyl group leaves the scalar product invariant

$$(s_\alpha\lambda, s_\alpha\mu) = (\lambda, \mu) \tag{13.66}$$

or more generally

$$(w\lambda, \mu) = (\lambda, w^{-1}\mu) \tag{13.67}$$

The Weyl group induces a natural splitting of the r-dimensional weight vector space into *chambers*, whose number is equal to the order of W. These are simplicial cones defined as

$$C_w = \{\lambda|\ (w\lambda, \alpha_i) \geq 0,\ i = 1, \cdots, r\}, \quad w \in W \qquad (13.68)$$

These chambers intersect only at their boundaries $(w\lambda, \alpha_i) = 0$, the reflecting hyperplanes of the s_i's. The chamber corresponding to the identity element of the Weyl group is called the *fundamental chamber*, and it will be denoted by C_0. An obvious but fundamental consequence of this splitting is that for any weight $\lambda \notin C_0$, there exists a $w \in W$ such that $w\lambda \in C_0$. More precisely, the W orbit of every weight has exactly one point in the fundamental chamber. The W orbit of λ is the set of all weights $\{w\lambda|\ w \in W\}$. A weight in the fundamental chamber and whose Dynkin labels are all integers, $\lambda_i \in \mathbb{Z}_+$, is said to be *dominant*. (A dominant weight is thus understood to be integral.) θ is an example of a dominant weight.

To conclude this section, we present some notation that will be used extensively in the sequel. The modified Weyl reflection

$$w \cdot \lambda \equiv w(\lambda + \rho) - \rho \qquad (13.69)$$

denoted by a dot, will be referred to as a *shifted Weyl reflection*. Here ρ is the Weyl vector. It is simple to verify that

$$w \cdot (w' \cdot \lambda) = (ww') \cdot \lambda \qquad (13.70)$$

The *length* of w, denoted $\ell(w)$, is the minimum number of s_i among all possible decompositions of $w = \prod_i s_i$. The *signature* of w is defined as

$$\epsilon(w) = (-1)^{\ell(w)} \qquad (13.71)$$

In the linear representation of w, this is simply $\det(w)$ (cf. Ex. 13.3). Finally, the longest element of the Weyl group will be denoted by w_0. It is the unique element of W that maps Δ_+ to Δ_-.

13.1.9. Lattices

In terms of a basis $(\epsilon_1, \cdots, \epsilon_d)$ of the d-dimensional Euclidian space $\mathbb{R}^d$, a *lattice* is the set of all points whose expansion coefficients, in terms of the specified basis, are all integers:

$$\mathbb{Z}\epsilon_1 + \mathbb{Z}\epsilon_2 + \cdots + \mathbb{Z}\epsilon_d \qquad (13.72)$$

In other words, it is the $\mathbb{Z}$ span of $\{\epsilon_i\}$. Three r-dimensional lattices are important for Lie algebras. These are the weight lattice

$$P = \mathbb{Z}\omega_1 + \cdots + \mathbb{Z}\omega_r \qquad (13.73)$$

the root lattice

$$Q = \mathbb{Z}\alpha_1 + \cdots + \mathbb{Z}\alpha_r \qquad (13.74)$$

and the coroot lattice

$$Q^\vee = \mathbb{Z}\alpha_1^\vee + \cdots + \mathbb{Z}\alpha_r^\vee \qquad (13.75)$$

The relevance of the weight lattice lies in that the weights in finite-dimensional representations have integer Dynkin labels (cf. Eq. (13.27)), hence they belong to P. The connection between P and the generators of g is twofold. First, the integers specifying the position of a weight in P are the eigenvalues of the Chevalley generators h^i. Second, the effect of the other generators is to shift the eigenvalues by an element of the root lattice Q. Since roots are weights in a particular finite-dimensional representation, $Q \subseteq P$. Hence, upon the action of E^α, a point of P is translated to another point of P. In the following, we denote by P_+ the set of dominant weights

$$P_+ = \mathbb{Z}_+\omega_1 + \cdots + \mathbb{Z}_+\omega_r \qquad (13.76)$$

For the algebras G_2, F_4, and E_8, it turns out that $Q = P$. In all other cases, Q is a proper subset of P, and the ratio P/Q is a finite group. Its order, $|P/Q|$, is equal to the determinant of the Cartan matrix. Actually, it is isomorphic to the center of the group of the algebra under consideration (whose structure will be studied in more detail later). The distinct elements of the coset P/Q define the so-called *congruence classes* (often called conjugacy classes). A weight λ lies in exactly one congruence class. For instance, for $su(2)$ there are two congruence classes given by λ_1 mod 2 (integer or half-integer spins). For $su(3)$, there are three classes, defined by the triality: $\lambda_1 + 2\lambda_2$ mod 3. The $su(N)$ generalization is

$$\lambda_1 + 2\lambda_2 + \cdots + (N-1)\lambda_{N-1} \text{ mod } N \qquad (13.77)$$

For any algebra g, the congruence classes take the form

$$\lambda \cdot v = \sum_{i=1}^{r} \lambda_i v_i \quad \text{mod } |P/Q| \quad (\text{mod } \mathbb{Z}_2 \quad \text{for} \quad g = D_{2\ell}) \qquad (13.78)$$

where the vector $(v_1, \cdots, v_r)$, equal to $(1, 2, \cdots, N-1)$ for $su(N)$, is called the *congruence vector*. The congruence classes are tabulated in App. 13.A for all simple Lie algebras.

On the other hand, since the bases $\{\omega_i\}$ and $\{\alpha_i^\vee\}$ are dual, P and $Q^\vee$ are dual lattices. A lattice is said to be self-dual if it is equal to its dual. For simple Lie algebras, the weight lattice is self-dual only for E_8.

13.1.10. Normalization Convention

Up to now, all the normalizations have been fixed with respect to the root square lengths. In order to fully fix the normalization, it is necessary to give a specific value to these lengths. We follow the standard convention in which the square length of the long roots is set equal to two. Given that θ is necessarily a long root, we thus fix our normalization by setting

$$|\theta|^2 = 2 \qquad (13.79)$$

With $|\alpha_i|^2 \le 2$, it follows from Eq. (13.34) that

$$a_i \ge a_i^\vee \quad \Rightarrow \quad a_i^\vee = 1 \quad \text{if} \quad a_i = 1 \tag{13.80}$$

and similarly

$$\alpha_i^\vee = \alpha_i \frac{a_i}{a_i^\vee} \quad \Rightarrow \quad \alpha_i^\vee \ge \alpha_i \tag{13.81}$$

13.1.11. Examples

EXAMPLE 1: $su(2)$

This is the only simple Lie algebra of rank 1. Its Cartan matrix is $A = (2)$, meaning that the simple root α_1 is related to the fundamental weight ω_1 by

$$\alpha_1 = 2\omega_1 \tag{13.82}$$

Since $|\alpha_1|^2 = 2$, it follows that

$$(\omega_1, \omega_1) = \frac{1}{2} \tag{13.83}$$

The Weyl group is generated by the simple reflection s_1, whose action on a weight $\lambda = \lambda_1 \omega_1$ is

$$s_1(\lambda_1\omega_1) = \lambda_1\omega_1 - \lambda_1\alpha_1 = -\lambda_1\omega_1 \tag{13.84}$$

Because $s_1^2 = 1$, W contains only the two elements $\{1, s_1\}$. The full system of roots is then seen to be given by $\Delta = \{\alpha_1, -\alpha_1\}$.

The weight and the root lattices are displayed in Fig. 13.1. The weight lattice is composed of all the nodes, whereas the root lattice contains only those with a cross. The fundamental Weyl chamber is the positive part of the weight lattice (here one-dimensional).

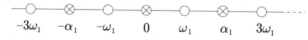

$$-3\omega_1 \quad -\alpha_1 \quad -\omega_1 \quad 0 \quad \omega_1 \quad \alpha_1 \quad 3\omega_1$$

Figure 13.1. Weight and root lattices for $su(2)$.

For subsequent reference, we give the explicit form of the commutation relations in different bases. In the Chevalley basis, it reads (dropping the superscript 1):

$$[e, f] = h \quad , \qquad [h, e] = 2e \quad , \qquad [h, f] = -2f \tag{13.85}$$

On a state $|\lambda\rangle$ of weight λ, the action of h is:

$$h|\lambda\rangle = \lambda_1|\lambda\rangle \tag{13.86}$$

In the Cartan-Weyl basis, the generators are (cf. Eq. (13.36) with $\alpha_1 = \sqrt{2}$):

$$H = h/\sqrt{2}, \quad E^+ = e, \quad E^- = f \tag{13.87}$$

with $E^\pm \equiv E^{\pm\alpha_1}$. The commutation relations are thus

$$[E^+, E^-] = \sqrt{2}H \quad , \qquad [H, E^\pm] = \pm\sqrt{2}E^\pm \tag{13.88}$$

and

$$H|\lambda\rangle = \lambda^1|\lambda\rangle = (\lambda_1/\sqrt{2})|\lambda\rangle \tag{13.89}$$

Another frequently used basis in the case of $su(2)$, which we call the *spin basis*, is defined by

$$J^0 = H/\sqrt{2}, \quad J^\pm = E^\pm \tag{13.90}$$

This yields

$$[J^+, J^-] = 2J^0 \quad , \qquad [J^0, J^\pm] = \pm J^\pm \tag{13.91}$$

and on the state $|\lambda\rangle = |j, m\rangle$, the action of the generators is

$$J^0 |j, m\rangle = m |j, m\rangle$$
$$J^\pm |j, m\rangle = \sqrt{(j(j+1) - m(m \pm 1)} \, |j, m \pm 1\rangle \tag{13.92}$$

EXAMPLE 2: $su(3)$

The Cartan matrix for this rank-2 algebra is

$$A = \begin{pmatrix} 2 & -1 \\ -1 & 2 \end{pmatrix} \tag{13.93}$$

The simple roots α_1 and α_2 have the same length (the algebra is simply laced) and they are related to the fundamental weights by

$$\alpha_1 = \alpha_1^\vee = 2\omega_1 - \omega_2 = (2, -1)$$
$$\alpha_2 = \alpha_2^\vee = -\omega_1 + 2\omega_2 = (-1, 2) \tag{13.94}$$

The scalar products between fundamental weights are

$$(\omega_1, \omega_1) = (\omega_2, \omega_2) = \frac{2}{3}, \qquad (\omega_1, \omega_2) = \frac{1}{3} \tag{13.95}$$

The full Weyl group is given by

$$W = \{1, s_1, s_2, s_1 s_2, s_2 s_1, s_1 s_2 s_1\} \tag{13.96}$$

This follows from the relation

$$(s_1 s_2)^3 = 1 \quad \Longrightarrow \quad s_1 s_2 s_1 = s_2 s_1 s_2 \tag{13.97}$$

a consequence of Eq. (13.59), which implies that there are no strings of s_i with more than three elements. This identity can also be checked directly by acting on an arbitrary weight:

$$
\begin{aligned}
s_1(\lambda_1, \lambda_2) &= (\lambda_1, \lambda_2) - \lambda_1 \alpha_1 = (-\lambda_1, \lambda_1 + \lambda_2) \\
s_2(\lambda_1, \lambda_2) &= (\lambda_1, \lambda_2) - \lambda_2 \alpha_2 = (\lambda_1 + \lambda_2, -\lambda_2) \\
s_1 s_2(\lambda_1, \lambda_2) &= (-\lambda_1 - \lambda_2, \lambda_1) \\
s_2 s_1(\lambda_1, \lambda_2) &= (\lambda_2, -\lambda_1 - \lambda_2) \\
s_1 s_2 s_1(\lambda_1, \lambda_2) &= s_2 s_1 s_2(\lambda_1, \lambda_2) = (-\lambda_2, -\lambda_1)
\end{aligned}
\tag{13.98}
$$

The action of the different elements of the Weyl group on the two simple roots gives all possible roots. For instance, $-\alpha_1$ and $\alpha_1 + \alpha_2$ are roots because

$$
s_1 \alpha_1 = -\alpha_1, \qquad s_1 \alpha_2 = \alpha_1 + \alpha_2
\tag{13.99}
$$

In this way, Δ is found to be

$$
\Delta = \{\alpha_1, \alpha_2, \alpha_1 + \alpha_2, -\alpha_1, -\alpha_2, -\alpha_1 - \alpha_2\}
\tag{13.100}
$$

The highest root is

$$
\theta = \alpha_1 + \alpha_2 \quad \Longrightarrow \quad a_i = a_i^\vee = 1, \ i = 1, 2
\tag{13.101}
$$

The root system and the Weyl chambers are presented in Fig. 13.2. The Weyl chambers are the regions separated by the dashed lines and they are specified here in terms of the elements of the Weyl group.

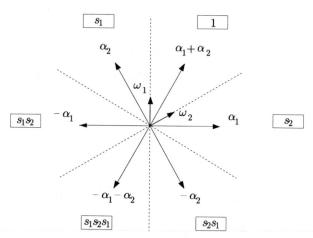

Figure 13.2. Root system and Weyl chambers $su(3)$.

EXAMPLE 3: $sp(4)$

This is again a rank-2 algebra, but it is not simply laced. The Cartan matrix is

$$
A = \begin{pmatrix} 2 & -1 \\ -2 & 2 \end{pmatrix}
\tag{13.102}
$$

so that

$$\alpha_1 = \frac{1}{2}\alpha_1^\vee = 2\omega_1 - \omega_2 = (2, -1)$$
$$\alpha_2 = \alpha_2^\vee = -2\omega_1 + 2\omega_2 = (-2, 2)$$
(13.103)

Because the long root is α_2,

$$|\alpha_2|^2 = 2 \quad \Longrightarrow \quad |\alpha_1|^2 = 1 \qquad (13.104)$$

The components of the quadratic form matrix are

$$(\omega_1, \omega_1) = (\omega_1, \omega_2) = \frac{1}{2}, \qquad (\omega_2, \omega_2) = 1 \qquad (13.105)$$

On the other hand, the complete structure of the Weyl group is easily recovered from the equality

$$(s_1 s_2)^4 = 1 \qquad (13.106)$$

meaning that the longest element is $s_1 s_2 s_1 s_2$; hence,

$$W = \{1, s_1, s_2, s_1 s_2, s_2 s_1, s_1 s_2 s_1, s_2 s_1 s_2, s_1 s_2 s_1 s_2\} \qquad (13.107)$$

Having determined the Weyl group, the set Δ can be constructed

$$\Delta = \{\alpha_1, \alpha_2, \alpha_1 + \alpha_2, 2\alpha_1 + \alpha_2, -\alpha_1, -\alpha_2, -\alpha_1 - \alpha_2, -2\alpha_1 - \alpha_2\} \quad (13.108)$$

The highest root is thus

$$\theta = 2\alpha_1 + \alpha_2 = 2\alpha_1^\vee + \alpha_2^\vee \quad \Longrightarrow \quad a_1 = 2, \; a_2 = a_1^\vee = a_2^\vee = 1 \quad (13.109)$$

In this case, the root vectors separate the Weyl chambers, as can be seen in Fig. 13.3.

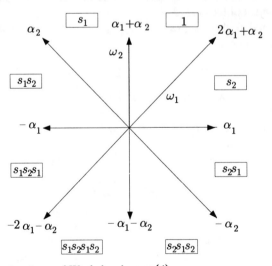

Figure 13.3. Root system and Weyl chambers $sp(4)$.

§13.2. Highest-Weight Representations

Any finite-dimensional irreducible representation has a unique *highest-weight state* $|\lambda\rangle$. Being nondegenerate, $|\lambda\rangle$ is completely specified by its eigenvalues (Dynkin labels) $\lambda(h^i) = \lambda_i$. Among all the weights in the representation, the highest weight is the one for which the sum of the coefficient expansions in the basis of simple roots is maximal. As a result, for any $\alpha > 0$, $\lambda + \alpha$ cannot be a weight in the representation, so that

$$E^\alpha|\lambda\rangle = 0, \qquad \forall\, \alpha > 0 \tag{13.110}$$

From Eq. (13.27), it is clear that the highest weight of a finite-dimensional representation is necessarily dominant (i.e., with positive-integer Dynkin labels). Moreover, to each dominant weight λ there corresponds a unique irreducible finite-dimensional representation L_λ whose highest weight is λ. By abuse of notation, we will often specify a representation by its highest weight. The highest weight for the adjoint representation is θ.

13.2.1. Weights and Their Multiplicities

Starting from the highest-weight state $|\lambda\rangle$, all the states in the representation space (or irreducible module) L_λ can be obtained by the action of the lowering operators of g as

$$E^{-\beta}E^{-\gamma}\cdots E^{-\eta}|\lambda\rangle \quad \text{for} \quad \beta, \gamma, \eta \in \Delta_+ \tag{13.111}$$

The set of eigenvalues of all the states in L_λ is the weight system, written Ω_λ. Any weight λ' in the set Ω_λ is such that $\lambda - \lambda' \in \Delta_+$. An immediate consequence is that all the weights of a given representation lie in exactly one congruence class, that is, one element of the coset P/Q.

In order to find all the weights $\lambda' \in \Omega_\lambda$, the key relation is again Eq. (13.27), which can be rewritten as

$$(\lambda', \alpha_i^\vee) = \lambda_i' = -(p_i - q_i), \qquad p_i, q_i \in \mathbb{Z}_+ \tag{13.112}$$

As already mentioned, λ' is necessarily of the form $\lambda - \sum n_i\alpha_i$, with $n_i \in \mathbb{Z}_+$. If we call $\sum n_i$ the level of the weight λ' in the representation λ, proceeding level by level, we know at each step the value of p_i. Clearly, $\lambda' - \alpha_i$ is also a weight if q_i is nonzero, that is, if $\lambda_i' - p_i > 0$.

With this criterion, the systematic construction of all the weights in the representation can be done by means of the following algorithm. We start with the highest weight $\lambda = (\lambda_1, \cdots, \lambda_r)$. For each positive Dynkin label $\lambda_i > 0$, we construct the sequence of weights $\lambda - \alpha_i, \lambda - 2\alpha_i, \cdots, \lambda - \lambda_i\alpha_i$, which all belong to Ω_λ. The process is then repeated with λ replaced by each of the weights just obtained, and iterated until no more weights with positive Dynkin labels are produced. Simple examples will clarify the method. Consider the adjoint representation of $su(3)$, whose highest weight is $(1, 1)$. The weights obtained at each step can be read from

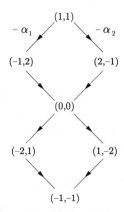

Figure 13.4. Weights in the adjoint representation of $su(3)$.

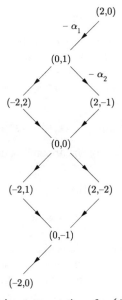

Figure 13.5. Weights in the adjoint representation of $sp(4)$.

Fig. 13.4. Similarly, Fig. 13.5 displays the weights in the adjoint representation of $sp(4)$.

However, this procedure does not keep track of multiplicities. For this, one can use the *Freudenthal recursion formula*, whose origin will be indicated in Sect. 13.2.3, and which gives the multiplicity of λ' in the representation λ in terms of the multiplicity of all the weights above it:

$$[|\lambda + \rho|^2 - |\lambda' + \rho|^2] \, \mathrm{mult}_\lambda(\lambda') = 2 \sum_{\alpha>0} \sum_{k=1}^\infty (\lambda' + k\alpha, \alpha) \, \mathrm{mult}_\lambda(\lambda' + k\alpha)$$

(13.113)

To illustrate the formula, we calculate the multiplicity of the weight $(0,0)$ in the adjoint representation of $su(3)$. Having proceeded recursively, we know that k can only be 1 and the three weights above $(0,0)$ have multiplicity 1. Furthermore, $(\lambda' + \alpha, \alpha) = 2$ for the three positive roots. Then, using $\lambda = \theta = \rho = \alpha_1 + \alpha_2$, we easily find that

$$(8 - 2)\mathrm{mult}_\theta(0,0) = 2(2 + 2 + 2) \quad \Longrightarrow \quad \mathrm{mult}_\theta(0,0) = 2 \qquad (13.114)$$

Indeed, the zero eigenvalue in the adjoint representation always has multiplicity r, being associated with the generators of the Cartan subalgebra (whereas the nonzero weights (roots) are nondegenerate). Another multiplicity formula is presented in Ex. 13.17.

We note that all the weights in a given W orbit have the same multiplicity:

$$\mathrm{mult}_\lambda(w\lambda') = \mathrm{mult}_\lambda(\lambda') \quad \text{for all} \quad w \in W \qquad (13.115)$$

This ultimately reflects the arbitrariness of the basis of simple roots, that is, that any set $\{w\alpha_i\}$ with w fixed, could serve as a basis.

Finally, we mention that a finite-dimensional irreducible module L_λ is always *unitary*. This means that, with $(H^i)^\dagger = H^i$ and $(E^\alpha)^\dagger = E^{-\alpha}$, the norm of any state $|\lambda'\rangle$ in L_λ is positive definite:

$$|\lambda'\rangle = E^{-\beta} \cdots E^{-\gamma}|\lambda\rangle \Longrightarrow \langle\lambda'|\lambda'\rangle = \langle\lambda|E^\gamma \cdots E^\beta E^{-\beta} \cdots E^{-\gamma}|\lambda\rangle > 0 \quad (13.116)$$

with $\beta, \gamma \in \Delta_+$. This also holds for linear combinations of such states.

13.2.2. Conjugate Representations

In an irreducible finite-dimensional representation, there is obviously a lowest state, also unique. It lies in the W orbit of the highest state, in the chamber exactly opposite to the fundamental one. This chamber is specified by the longest element of the Weyl group w_0. In terms of the highest state λ, the lowest state is thus given by $w_0\lambda$. Turning a representation "upside down" produces the *conjugate representation*, indicated by λ^*. Its highest-weight state is the negative of the lowest state of the original representation

$$\lambda^* = -(w_0\lambda) = (-w_0) \cdot \lambda \qquad (13.117)$$

since ρ is the highest weight of a self-conjugate representation: $\rho = -w_0\rho$. More generally, all the weights in Ω_{λ^*} are the negatives of those in Ω_λ. For $su(N)$, w_0 is given by

$$w_0 = s_1 s_2 \cdots s_{N-1} s_1 s_2 \cdots s_{N-2} \cdots s_1 s_2 s_1 \qquad (13.118)$$

With $N = 3$, it yields

$$(-w_0) \cdot (\lambda_1, \lambda_2) = -s_1 s_2 s_1 (\lambda_1 + 1, \lambda_2 + 1) - (1, 1) = (\lambda_2, \lambda_1) \qquad (13.119)$$

The conjugation is related to the reflection symmetry of the Dynkin diagram. This readily shows that for $su(N)$, the conjugation amounts to reversing the order of the finite Dynkin labels. Because the Dynkin diagram of $so(2r + 1), sp(2r), so(4r), G_2, F_4, E_7$, and E_8 have no symmetry, all representations of these algebras are self-conjugate. For the other algebras, self-conjugate representations are those with highest weight satisfying:

$$
\begin{aligned}
su(r + 1): && \lambda_i &= \lambda_{r-i} \\
so(4r + 2): && \lambda_r &= \lambda_{r-1} && (13.120) \\
E_6: && \lambda_1 &= \lambda_5, \quad \lambda_2 = \lambda_4
\end{aligned}
$$

13.2.3. Quadratic Casimir Operator

A generalization of the $su(2)$ quadratic *Casimir operator* Q can be constructed for any semisimple Lie algebra. Up to a scale factor, it is uniquely characterized by its commutativity with all the generators of the algebra. In a generic basis $\{\mathcal{L}^a\}$, it can be checked to be given by

$$Q = \sum_{a,b} [K(\mathcal{L}^a, \mathcal{L}^b)]^{-1} \mathcal{L}^a \mathcal{L}^b \qquad (13.121)$$

where K is the Killing form (which, as already mentioned, is nondegenerate for semisimple Lie algebras). In the orthonormal $\{J^a\}$ basis, it is thus

$$Q = \sum_a J^a J^a \qquad (13.122)$$

On the other hand, in the Cartan-Weyl basis, it reads

$$Q = \sum_i H^i H^i + \sum_{\alpha>0} \frac{|\alpha|^2}{2} (E^\alpha E^{-\alpha} + E^{-\alpha} E^\alpha) \qquad (13.123)$$

We note that Q is not an element of g itself; it lies in its universal enveloping algebra, which is the set of all formal power series in elements of g.

Since Q commutes with all the generators of the algebra, its eigenvalue is the same on all the states of an irreducible representation. It is most easily evaluated on the highest-weight state, using the Cartan-Weyl basis. First, we have

$$\sum_i H^i H^i \, |\lambda\rangle = \sum_i \lambda^i \lambda^i \, |\lambda\rangle = (\lambda, \lambda) |\lambda\rangle \qquad (13.124)$$

Because $E^\alpha |\lambda\rangle = 0$ for $\alpha > 0$, the term $E^{-\alpha} E^\alpha$ does not contribute . For the remaining term, we move E^α to the right of $E^{-\alpha}$ using

$$[E^\alpha, E^{-\alpha}] \, |\lambda\rangle = \frac{2}{|\alpha|^2} \alpha \cdot H \, |\lambda\rangle = \frac{2}{|\alpha|^2} (\alpha, \lambda) |\lambda\rangle \qquad (13.125)$$

The result is

$$\mathcal{Q}|\lambda\rangle = [(\lambda, \lambda) + \sum_{\alpha > 0} (\alpha, \lambda)] \, |\lambda\rangle \qquad (13.126)$$

By using the definition (13.47) of the Weyl vector, we can write

$$\boxed{\mathcal{Q}|\lambda\rangle = (\lambda, \lambda + 2\rho) \, |\lambda\rangle} \qquad (13.127)$$

In the adjoint representation, the eigenvalue of the Casimir operator is

$$(\theta, \theta + 2\rho) = 2 + 2(\theta, \rho) = 2 + 2 \sum_{i,j} a_i^\vee (\alpha_i^\vee, \omega_j)$$
$$= 2 + 2 \sum_i \alpha_i^\vee = 2 + 2(g - 1) = 2g \qquad (13.128)$$

The quadratic Casimir operator does not distinguish a representation from its conjugate

$$\mathcal{Q}|\lambda^*\rangle = \mathcal{Q}|\lambda\rangle \qquad (13.129)$$

This follows from the equality

$$|\lambda^* + \rho|^2 = |\lambda + \rho|^2 \qquad (13.130)$$

which is itself a simple consequence of Eq. (13.117):

$$\lambda^* + \rho = (-w_0) \cdot \lambda + \rho = -w_0(\lambda + \rho) \qquad (13.131)$$

and of the invariance of the scalar product with respect to the Weyl group: $|w\mu|^2 = |\mu|^2$.

The Freudenthal formula (13.113) is obtained by evaluating the trace of $\mathcal{Q}$ in the subspace associated with the weight λ', first using the eigenvalue just obtained and then using the explicit form of $\mathcal{Q}$ in the Cartan-Weyl basis.

For $su(2)$, the quadratic Casimir operator is the unique operator that commutes with all the generators. However, we mention that for higher-rank algebras there exist Casimir operators of higher degree. Their degrees minus one are called the *exponents* of the algebras (tabulated in App. 13.A).[8]

13.2.4. Index of a Representation

The quadratic Casimir operator enters in the definition of an important quantity, the *index of a representation*, which gives the relative normalization of invariant bilinear products taken in different representations.

As already stressed, once a normalization is fixed for the length of the long roots, every product is uniquely determined. In particular, the normalization of the invariant bilinear form $\mathrm{Tr}_\lambda(\mathcal{R}(J^a)\mathcal{R}(J^b))$, for $\mathcal{R}(J^a)$ standing for a matrix

---

[8] For simply laced Lie algebras, the exponents can be defined more simply in terms of the eigenvalues of the adjacency matrix $G_{ij} = 2 - A_{ij}$: the eigenvalues are of the form $2\cos(n_i \pi / g)$ where the n_i's are the exponents.

representation of the generator J^a, must be fixed. Here the trace is evaluated in L_λ. The relative normalization of this product with respect to $|\theta|^2$ defines the Dynkin index x_λ of the representation λ

$$\text{Tr}_\lambda(\mathcal{R}(J^a)\mathcal{R}(J^b)) = |\theta|^2 x_\lambda \delta_{ab} = 2 x_\lambda \delta_{ab} \tag{13.132}$$

An explicit expression for x_λ can be easily obtained by setting $a = b$ and summing over all values of a. The l.h.s. becomes equal to the trace of the quadratic Casimir, so that

$$\boxed{x_\lambda = \frac{\dim|\lambda|\,(\lambda, \lambda + 2\rho)}{2 \dim g}} \tag{13.133}$$

We note that the Dynkin index of the adjoint representation, $\lambda = \theta$, is simply the dual Coxeter number

$$x_\theta = g \tag{13.134}$$

since $\dim |\theta| = \dim g$ and $(\theta, \rho) = g - 1$ (cf. Eq (13.128)).

§13.3. Tableaux and Patterns $(su(N))$

In this section, we introduce a useful diagrammatic representation of highest weights, which will also turn out to be a powerful combinatorial tool, particularly efficient in tensor-product calculations. A refinement of this diagrammatic representation leads to the simple construction of a complete basis of states in a finite-dimensional representation. This will be shown to be equivalent to a description of states in terms of triangular arrays of numbers, the so-called Gelfand-Tsetlin patterns. For simplicity, we restrict the whole discussion to $su(N)$.

13.3.1. Young Tableaux

A $su(N)$ integrable highest weight λ, with Dynkin labels

$$\lambda = (\lambda_1, \cdots, \lambda_{N-1}) \tag{13.135}$$

can equally well be specified in terms of its *partition*

$$\lambda = \{\ell_1; \ell_2; \cdots; \ell_{N-1}\} \tag{13.136}$$

where

$$\ell_i = \lambda_i + \lambda_{i+1} + \cdots + \lambda_{N-1} \tag{13.137}$$

To a partition, we associate a *Young tableau*, which is a box array of rows lined up on the left, such that the length of the i-th row is equal to ℓ_i. For example, to the $su(5)$ weight

$$\lambda = (2, 0, 2, 0) = \{4; 2; 2\} \tag{13.138}$$

(zero entries in partitions being generally omitted) corresponds the Young tableau

$$(13.139)$$

Dynkin labels provide a dual description of the tableau: λ_i gives the number of columns of i boxes. The fundamental representation ω_ℓ is described by a single column of ℓ boxes. To the scalar representation corresponds a void tableau or, equivalently, a single column of N boxes. Allowing for columns of N boxes, partitions are fixed by N integers. But clearly, when $\ell_N \neq 0$, we can always subtract $\{\ell_N; \ell_N; \cdots; \ell_N\}$ (N entries) from the partition, which just amounts to eliminating columns of N boxes; for instance,

$$\{5; 3; 3; 1; 1\} = \{4; 2; 2\} \tag{13.140}$$

Tableaux with $\ell_N = 0$ will be referred to as reduced tableaux, and likewise for partitions.

The *transpose* of a Young tableau is obtained by interchanging rows and columns. We denote by λ^t the corresponding weight. For instance, the transpose of $\lambda = \{4; 2; 2\}$ is

$$\leftrightarrow \lambda^t = \{3; 3; 1; 1\} \tag{13.141}$$

With

$$\lambda^t = \{\tilde{\ell}_1; \tilde{\ell}_2; \cdots\} \tag{13.142}$$

it is not difficult to see that

$$\tilde{\ell}_i = \quad \text{number of } \ell_j \text{ such that } \ell_j \geq i \tag{13.143}$$

13.3.2. Partitions and Orthonormal Bases

The entries of the partition (13.136) are the expansion coefficients of a dominant weight in a certain basis, which we now describe. We also indicate how partitions can be associated with nondominant integral weights, providing a rationale for the construction of the next section.

Elements of $su(N)$ can be represented by $N \times N$ traceless matrices. In this representation, the Cartan subalgebra is spanned by the set of all diagonal traceless matrices. We let e_{ij} stand for the matrix with 0 everywhere, except for a single 1 at position (i, j) (i-th row, j-th column). With this notation, the elements of the Cartan subalgebra are of the form $\sum_{i=1}^{N} \epsilon_i e_{ii}$ with $\sum_{i=1}^{N} \epsilon_i = 0$. The ladder operators are represented by the matrices $e_{ij}, i \neq j$. The roots are then given by $\epsilon_i - \epsilon_j, i \neq j$, and a basis of simple roots is

$$\alpha_i = \epsilon_i - \epsilon_{i+1}, \qquad i = 1, \cdots, N - 1 \tag{13.144}$$

Generalizing this point of view, we can consider the ϵ_i as orthonormal vectors in an $(r+1)$-dimensional space, and in terms of these vectors, the root lattice is simply

$$Q = \sum_{i=1}^{N} n_i \epsilon_i \qquad \text{with} \quad n_i \in \mathbb{Z} \quad \text{and} \quad \sum_{i=1}^{N} n_i = 0 \qquad (13.145)$$

With $\epsilon_i^2 = 1$ and $\epsilon_i \cdot \epsilon_j = 0$ for $i \neq j$, we see that $|\alpha|^2 = 2$ for any root. The fundamental weights are related to the simple roots by the quadratic form matrix (since here roots are the same as coroots), which leads to

$$\omega_i = \epsilon_1 + \epsilon_2 + \cdots + \epsilon_i - \frac{i}{N} \sum_{i=1}^{N} \epsilon_i \qquad (13.146)$$

Hence, the expansion coefficients of a highest weight in the $\{\epsilon_i\}$ basis are exactly the entries of the partition

$$\lambda = \sum_{i=1}^{N-1} \lambda_i \omega_i = \sum_{i=1}^{N} (\ell_i - \kappa)\epsilon_i \qquad (13.147)$$

where the ℓ_i are related to the Dynkin labels by Eq. (13.137) and κ is

$$\kappa = \frac{1}{N} \sum_{j=1}^{N-1} j\lambda_j \qquad (13.148)$$

A well-defined partition is thus associated with the highest weight λ of each representation. The other weights in the representation are obtained by subtracting from λ the positive roots $\epsilon_i - \epsilon_j$, $i < j$. This construction gives directly their expansion coefficients in the $\{\epsilon_i\}$ basis. A weight λ' can thus be described by a partition $\{\ell_1'; \ell_2'; \cdots; \ell_N'\}$. We stress that the partition of a weight that is not a highest weight is not related to the shape of a Young tableau. In particular, such a partition is no longer bound to satisfy $\ell_i' \geq \ell_{i+1}'$.

13.3.3. Semistandard Tableaux

We now indicate how tableau techniques can be used to explicitly describe all the states in a representation. This involves filling the boxes of a Young tableau with positive integers, generating the so-called *semistandard tableaux*. They are defined as follows. We let $c_{i,j}$ be the integer appearing in the box on the i-th row (from top) and the j-th column (from left), and satisfying

$$1 \leq c_{i,j} \leq N \quad , \quad c_{i,j} \leq c_{i,j+1} \quad , \quad c_{i,j} < c_{i+1,j} \qquad (13.149)$$

In other words, the numbers are nondecreasing from left to right and strictly increasing from top to bottom.

Semistandard tableaux of shape λ are in one-to-one correspondence with the states in the module L_λ. The numbering in the semistandard tableaux encodes the partition of the corresponding weight. We can think of a box with number i as representing ϵ_i. The number of i's in the semistandard tableau of weight $\lambda' = \{\ell'_1; \cdots; \ell'_N\}$ is given by ℓ'_i. In the semistandard tableau representing the highest weight λ, all boxes of the i-th row have number i. The weight of a semistandard tableau is clearly obtained by adding the weights of all its boxes. The weight of a box marked with a i is

$$\epsilon_i = \omega_i - \omega_{i-1}, \qquad i = 1, \cdots, N \tag{13.150}$$

(modulo $\sum \epsilon_i / N$) with $\omega_0 = \omega_N = 0$.

The number of semistandard tableaux of a fixed shape that can be constructed with a given partition gives the multiplicity of the corresponding weight in the representation. In other words, the rules for constructing semistandard tableaux provide a combinatorial realization of the Freudenthal multiplicity formula (13.113).

We consider for example $su(3)$. The semistandard tableaux of the three states in the representation ω_1 are

$$\boxed{1} \leftrightarrow (1,0) \quad , \quad \boxed{2} \leftrightarrow (-1,1) \quad , \quad \boxed{3} \leftrightarrow (0,-1) \tag{13.151}$$

whereas those in the representation ω_2 are

$$\boxed{\begin{array}{c}1\\2\end{array}} \leftrightarrow (0,1) \quad , \quad \boxed{\begin{array}{c}1\\3\end{array}} \leftrightarrow (1,-1) \quad , \quad \boxed{\begin{array}{c}2\\3\end{array}} \leftrightarrow (-1,0) \tag{13.152}$$

The adjoint representation $(1,1)$ contains the 8 semistandard tableaux:

$$
\begin{array}{cccccccc}
\boxed{\begin{array}{cc}1&1\\2&\end{array}} & \boxed{\begin{array}{cc}1&2\\2&\end{array}} & \boxed{\begin{array}{cc}1&3\\2&\end{array}} & \boxed{\begin{array}{cc}1&1\\3&\end{array}} & \boxed{\begin{array}{cc}1&2\\3&\end{array}} & \boxed{\begin{array}{cc}1&3\\3&\end{array}} & \boxed{\begin{array}{cc}2&2\\3&\end{array}} & \boxed{\begin{array}{cc}2&3\\3&\end{array}} \\
(1,1) & (-1,2) & (0,0) & (2,-1) & (0,0) & (1,-2) & (-2,1) & (-1,-1)
\end{array}
$$
$$\tag{13.153}$$

Two distinct semistandard tableaux, that is, two distinct states, correspond to the doubly degenerate weight $(0,0)$. Similarly, to the weight $(0,0,0)$ of the $su(4)$ representation $(2,0,2)$ (with partition $\{2;2;2;2\}$) correspond 6 semistandard tableaux:

$$
\boxed{\begin{array}{cccc}1&1&4&4\\2&2&&\\3&3&&\end{array}} \quad
\boxed{\begin{array}{cccc}1&1&3&4\\2&2&&\\3&4&&\end{array}} \quad
\boxed{\begin{array}{cccc}1&1&3&3\\2&2&&\\4&4&&\end{array}} \quad
\boxed{\begin{array}{cccc}1&1&2&4\\2&3&&\\3&4&&\end{array}} \quad
\boxed{\begin{array}{cccc}1&1&2&3\\2&3&&\\4&4&&\end{array}} \quad
\boxed{\begin{array}{cccc}1&1&2&2\\3&3&&\\4&4&&\end{array}}
$$
$$\tag{13.154}$$

13.3.4. Gelfand-Tsetlin Patterns

An equivalent representation of the basis of semistandard tableaux is given by the *Gelfand-Tsetlin patterns*. To a given semistandard tableau we associate the following triangular array of numbers:

$$
\begin{array}{ccccc}
\beta_1^{(N)} & \beta_2^{(N)} & \cdots\cdots & \beta_N^{(N)} & \\
& \beta_1^{(N-1)} & \cdots & \beta_{N-1}^{(N-1)} & \\
& & \cdots\cdots & & \\
& \beta_1^{(2)} & \beta_2^{(2)} & & \\
& & \beta_1^{(1)} & &
\end{array}
\tag{13.155}
$$

such that $\beta_i^{(j)}$ is the number of boxes containing numbers less or equal to j in the i-th row (from top) of the semistandard tableau.[9] For instance, the following semistandard tableau and Gelfand-Tsetlin pattern corresponding to the weight $(-2, 1, 0)$ in the representation $(1, 2, 1)$ of $su(4)$ are equivalent:

$$
\begin{array}{|c|c|c|c|}
\hline
1 & 2 & 2 & 4 \\
\hline
\end{array}
\quad\leftrightarrow\quad
\begin{array}{c}
4\ 3\ 1\ 0 \\
3\ 2\ 1 \\
3\ 1 \\
1
\end{array}
\tag{13.156}
$$

The first line in the Gelfand-Tsetlin pattern is common to all patterns in the representation, being simply the partition of the tableau

$$
\beta_i^{(N)} = \ell_i
\tag{13.157}
$$

All the states in a representation are then generated by filling a triangular array of N lines with integers $\beta_i^{(j)}$ satisfying

$$
\beta_i^{(j)} \geq \beta_{i+1}^{(j)} \qquad \beta_i^{(j)} \geq \beta_{i+1}^{(j+1)}
\tag{13.158}
$$

with the first line fixed by the partition. In this way, the 8 patterns of the $(1, 1)$ representation of $su(3)$ are found to be

$$
\begin{array}{cccccccc}
2\ 1\ 0 & 2\ 1\ 0 & 2\ 1\ 0 & 2\ 1\ 0 & 2\ 1\ 0 & 2\ 1\ 0 & 2\ 1\ 0 & 2\ 1\ 0 \\
2\ 1 & 2\ 1 & 1\ 1 & 2\ 0 & 2\ 0 & 1\ 0 & 2\ 0 & 1\ 0 \\
2 & 1 & 1 & 2 & 1 & 1 & 0 & 0
\end{array}
\tag{13.159}
$$

Their ordering corresponds to the semistandard tableaux (13.153). We note that the Gelfand-Tsetlin pattern of the highest-weight state in the representation is completely fixed by the partition, being

$$
\begin{array}{c}
\ell_1\ \ell_2\ \cdots\cdots\ \ell_N \\
\ell_1\ \cdots\ \ell_{N-1} \\
\cdots\cdots \\
\ell_1\ \ell_2 \\
\ell_1
\end{array}
\tag{13.160}
$$

§13.4. Characters

13.4.1. Weyl's Character Formula

A character is a useful functional way of coding the whole content of a representation. The character of the representation of highest weight λ is formally defined

---

[9] Stated more simply, the numbers in the first row of the triangular array give the lengths of the rows of the semistandard tableau (from top to bottom); the numbers $\beta_i^{(N-1)}$ in the second row give the lengths of the rows of the semistandard tableau obtained by deleting all the boxes marked by N; the third row of the pattern is obtained similarly by deleting all the boxes marked by N and $N-1$, and so on; the last number $\beta_1^{(1)}$ is thus the number of 1's in the semistandard tableau.

as

$$\chi_\lambda = \sum_{\lambda' \in \Omega_\lambda} \text{mult}_\lambda(\lambda') e^{\lambda'} \tag{13.161}$$

where the sum is over all the weights of the representation. e^λ denotes a formal exponential satisfying

$$e^\lambda e^\mu = e^{\lambda+\mu}$$
$$e^\lambda(\xi) = e^{(\lambda,\xi)} \tag{13.162}$$

On the r.h.s. of the last expression, e is a genuine exponential function, and ξ is an arbitrary element of the dual Cartan subalgebra (i.e., an arbitrary weight).

This formal character is related to the familiar character in the representation theory of groups as follows. Let G be the Lie group of g and H an element of the Cartan subgroup of G. The character of H in some representation is simply its trace evaluated in the corresponding module V:

$$\text{Tr}_V H = \sum_\gamma \text{mult}(\gamma)[\gamma(H)] \tag{13.163}$$

where $\gamma(H)$ denotes the eigenvalues of H. Complete information about the representation is obtained by considering the group character as restricted to the full Cartan subgroup. Since H is associated with an element h of the Cartan subalgebra of g by $H = \exp(h)$, spanning the full Cartan subgroup amounts to replacing the single element h by the vector $\vec{h} = (h^1, h^2, \cdots, h^r)$. Thus $\gamma(H)$ is replaced by $e^{\lambda'(\vec{h})} = e^{(\lambda'_1, \cdots, \lambda'_r)}$ where λ' is a weight. For a vector exponent, e must be regarded as a formal exponential.

The expression (13.161) for the character can be brought into a more manageable form in two steps (we omit the details). At first, the auxiliary quantity

$$D_\rho = \prod_{\alpha>0} (e^{\alpha/2} - e^{-\alpha/2}) \tag{13.164}$$

is introduced, and shown to be expressible as a sum over the elements of the Weyl group

$$D_\rho = \sum_{w \in W} \epsilon(w) e^{w\rho} \tag{13.165}$$

The second step (more involved) consists of showing, using the Freudenthal multiplicity formula (13.113), that

$$D_\rho \chi_\lambda = D_{\lambda+\rho} \tag{13.166}$$

where $D_{\lambda+\rho}$ is defined from Eq. (13.165) with ρ replaced by $\lambda + \rho$. This last result is the famous *Weyl character formula*

$$\chi_\lambda = \frac{D_{\lambda+\rho}}{D_\rho} = \frac{\sum_{w \in W} \epsilon(w) e^{w(\lambda+\rho)}}{\sum_{w \in W} \epsilon(w) e^{w\rho}} \tag{13.167}$$

For $su(2)$, with $t = e^{\omega_1}$, this becomes

$$\chi_\lambda = \frac{t^{\lambda_1+1} - t^{-\lambda_1-1}}{t - t^{-1}} = t^{\lambda_1} + t^{\lambda_1-2} + \cdots + t^{-\lambda_1} \qquad (13.168)$$

For some manipulations, it is more convenient to work with the character evaluated at a special but arbitrary value ξ.

$$\chi_\lambda(\xi) = \frac{\sum_{w \in W} \epsilon(w) \, e^{(w(\lambda+\rho),\xi)}}{\sum_{w \in W} \epsilon(w) \, e^{(w\rho,\xi)}} \qquad (13.169)$$

13.4.2. The Dimension and the Strange Formulae

As an immediate application, we derive a formula for the dimension of a representation. From Eq. (13.161), it is clear that this amounts to evaluating the character at the special point $\xi = 0$. But setting $\xi = 0$ in Eq. (13.169) leads to an indeterminate expression since $\sum \epsilon(w) = 0$ (W has the same number of even and odd elements). Rather, a limiting process must be used. For this we set $\xi = t\rho$ and consider the limit $t \to 0$. For ξ proportional to ρ, the character takes the simple form

$$\chi_\lambda(t\rho) = \frac{D_{\lambda+\rho}(t\rho)}{D_\rho(t\rho)} = \frac{D_\rho(t(\lambda+\rho))}{D_\rho(t\rho)} = \prod_{\alpha>0} \frac{\sinh(\alpha,(\lambda+\rho)t/2)}{\sinh(\alpha,\rho t/2)} \qquad (13.170)$$

which yields

$$\boxed{\dim |\lambda| = \lim_{t\to 0} \chi_\lambda(t\rho) = \prod_{\alpha>0} \frac{(\lambda+\rho,\alpha)}{(\rho,\alpha)}} \qquad (13.171)$$

For instance, the application of this formula to $su(2), su(3)$, and $sp(4)$ gives

$su(2):\quad \dim |\lambda| = \lambda_1 + 1$

$su(3):\quad \dim |\lambda| = \frac{1}{2}(\lambda_1 + 1)(\lambda_2 + 1)(\lambda_1 + \lambda_2 + 2)$ \qquad (13.172)

$sp(4):\quad \dim |\lambda| = \frac{1}{6}(\lambda_1 + 1)(\lambda_2 + 1)(\lambda_1 + 2\lambda_2 + 3)(\lambda_1 + \lambda_2 + 2)$

Keeping track of the subleading term in Eq. (13.170) leads to another interesting formula. At first, we have

$$\chi_\lambda(t\rho) = \prod_{\alpha>0} \frac{(\lambda+\rho,\alpha)}{(\rho,\alpha)} \left\{ 1 + \frac{t^2}{24}[(\alpha,\lambda+\rho)^2 - (\alpha,\rho)^2] \right\}$$

$$= \dim |\lambda| \left\{ 1 + \frac{t^2}{24} \sum_{\alpha>0}[(\alpha,\lambda+\rho)^2 - (\alpha,\rho)^2] \right\} \qquad (13.173)$$

Now, as demonstrated in the following paragraph, we can always write

$$(\lambda,\mu) = \frac{1}{y} \sum_{\alpha \in \Delta}(\lambda,\alpha)(\alpha,\mu) = \frac{2}{y} \sum_{\alpha \in \Delta_+}(\lambda,\alpha)(\alpha,\mu) \qquad (13.174)$$

where the constant y, evaluated below, depends upon the algebra. Thus we have

$$\chi_\lambda(t\rho) = \dim |\lambda| \left\{ 1 + \frac{t^2 y}{48} [|\lambda + \rho|^2 - |\rho|^2] \right\} \qquad (13.175)$$

The comparison of this expression with the t expansion of

$$\chi_\lambda(t\rho) = \sum_{\lambda' \in \Omega_\lambda} \text{mult}_\lambda(\lambda') e^{(\lambda',\rho)t} \qquad (13.176)$$

yields

$$\frac{1}{2} \sum_{\lambda' \in \Omega_\lambda} \text{mult}_\lambda(\lambda') (\lambda',\rho)^2 = \frac{y}{48} \dim |\lambda| \, [\, |\lambda + \rho|^2 - |\rho|^2] \qquad (13.177)$$

For the adjoint representation, $\lambda = \theta$, and the different nonzero weights λ' are the roots, which all have multiplicity 1; the l.h.s. then becomes

$$\frac{1}{2} \sum_{\lambda' \in \Omega_\theta} \text{mult}_\theta(\lambda') (\lambda',\rho)^2 = \frac{1}{2} \sum_{\alpha \in \Delta} (\rho,\alpha)(\alpha,\rho) = \frac{1}{2} y |\rho|^2$$

where in the last step we used Eq. (13.174). The r.h.s. is

$$\frac{y}{48} \dim |\theta| \, (\theta, \theta + 2\rho) = \frac{yg}{24} \dim g \qquad (13.178)$$

(cf. Eq. (13.128)). This yields the *Freudenthal–de Vries strange formula*:

$$\boxed{|\rho|^2 = \frac{g}{12} \dim g} \qquad (13.179)$$

We now return to Eq. (13.174). The product $\sum_{\alpha \in \Delta} \alpha \alpha^t$ (where α^t stands for the transpose of α) is necessarily proportional to the $r \times r$ identity matrix I_r:

$$\sum_{\alpha \in \Delta} \alpha \alpha^t = y I_r \qquad (13.180)$$

Indeed, the l.h.s. commutes with any element of the Weyl group since the action of the latter simply amounts to a permutation of the roots. Since the action of the Weyl group on $\sum_{\alpha \in \Delta} \alpha \alpha^t$ is irreducible, this latter quantity must then be proportional to the identity. The proportionality constant y is evaluated by taking the trace of this equation. With $\text{Tr} \, \alpha \alpha^t = |\alpha|^2$, this yields

$$\sum_{\alpha \in \Delta} |\alpha|^2 = yr \qquad (13.181)$$

The l.h.s. can be evaluated from Eq. (13.132) restricted to the generators of the Cartan subalgebra:

$$\text{Tr}_\theta H^i H^j = 2g \delta^{ij} \qquad (13.182)$$

Setting $j = i$ and summing over i yields

$$\sum_i \sum_{\alpha \in \Delta} \alpha^i \alpha^i = \sum_{\alpha \in \Delta} |\alpha|^2 = 2gr \qquad (13.183)$$

This fixes the value of y:

$$y = 2g \tag{13.184}$$

13.4.3. Schur Functions

In the $su(N)$ orthogonal basis $\{\epsilon_i\}$ introduced in Sect. 13.3.2, the characters are called *Schur functions*. In this basis, there is a simple combinatorial formula for the dimension of a representation.

In the orthonormal basis, the Weyl group acts as the permutation group S_N of the N basis vectors. For instance, the action of s_α with $\alpha = \epsilon_i - \epsilon_j$ simply amounts to interchanging ϵ_i and ϵ_j. This observation allows us to rewrite the character as a ratio of matrix determinants, that is, as a Schur function. To this end we introduce the variables

$$q_i = e^{\epsilon_i} \tag{13.185}$$

subject to the constraint

$$\prod_{i=1}^{N} q_i = 1 \tag{13.186}$$

The formal exponential can thus be written as

$$e^\lambda = q_1^{\ell_1} q_2^{\ell_2} \cdots q_N^{\ell_N} \tag{13.187}$$

and we have

$$D_\lambda = \sum_{w \in W} e^{w\lambda} = \sum_{\sigma \in S_N} \epsilon(\sigma) \prod_{i=1}^{N} q^{\ell_{\sigma(i)}} = \det q_j^{\ell_i} \tag{13.188}$$

Since the i-th entry of the partition of ρ is $N - i$, the character can be written as

$$\chi_\lambda = S_\lambda(q_1, \cdots, q_N) = \frac{\det q_j^{\ell_i + N - i}}{\det q_j^{N-i}} \tag{13.189}$$

where S_λ stands for the Schur function. The above denominator is the ubiquitous Vandermonde determinant:

$$\det q_j^{N-i} = \det q_j^{i-1} = \det \begin{pmatrix} 1 & 1 & \cdots & 1 \\ q_1 & q_2 & \cdots & q_N \\ \vdots & & & \\ q_1^{N-1} & q_2^{N-1} & \cdots & q_N^{N-1} \end{pmatrix} \tag{13.190}$$

which can also be written under the more familiar form

$$\det q_j^{N-i} = \prod_{1 \le i < j \le N} (q_i - q_j) \tag{13.191}$$

The dimension of the representation is calculated by letting all q_i approach 1 in the expression for the character, with the result

$$\dim |\lambda| = \prod_{1 \le i < j \le N} \frac{(\ell_i - \ell_j + j - i)}{(j - i)} \qquad (13.192)$$

For $su(2)$ and $su(3)$ reduced tableaux, this is easily seen to reproduce Eq. (13.172). (See also Ex. 13.13 for another dimension formula.)

§13.5. Tensor Products: Computational Tools

In principle, the problem of calculating tensor products is straightforward. In order to calculate the product $L_\lambda \otimes L_\mu$, usually written as $\lambda \otimes \mu$, we simply add together all pairs of weights λ', μ' (belonging respectively to the weight systems Ω_λ and Ω_μ), taking care of their multiplicities, and reorganize the full set of dim $|\lambda| \times$ dim $|\mu|$ resulting weights in irreducible representations. We write the result under the form

$$\lambda \otimes \mu = \bigoplus_{\nu \in P_+} \mathcal{N}_{\lambda\mu}{}^\nu \, \nu \qquad (13.193)$$

where the sum is taken over all dominant weights, and $\mathcal{N}_{\lambda\mu}{}^\nu$, called a *tensor-product coefficient*, gives the multiplicity of the representation ν in the decomposition of the tensor product $\lambda \otimes \mu$.

In practice, this method is obviously too cumbersome, and more efficient techniques are required. Such methods are described in the next subsections. The first one follows directly from manipulations of the Weyl character formula and it is completely general. Although theoretically important, as a computational tool it is not very powerful. For this reason we introduce two other methods, which, however, are presented more like recipes. These are the famous Littlewood-Richardson rule and the more novel Berenstein-Zelevinsky method of triangles. For these, however, the discussion is again restricted to $su(N)$. Another motivation for introducing these last two methods is that they allow us to determine precisely those very states that contribute to the tensor product, a point on which we will expand in due time. Furthermore, in their framework, a particular tensor-product coefficient can be studied in isolation, that is without necessarily having to compute the full tensor-product decomposition.

But before turning to techniques, we list some general properties of tensor-product coefficients. It is clear that

$$\mathcal{N}_{\lambda 0}{}^\nu = \delta_\lambda^\nu \qquad (13.194)$$

where 0 stands for the scalar representation (i.e., the representation whose highest weight has all Dynkin labels equal to zero and which thus contains a single state). On the other hand, the tensor product of a representation with its conjugate always contains the scalar representation. It is obtained from the pairing of all the states of λ with their negatives, which necessarily lie in λ^*. In other words,

$$\mathcal{N}_{\lambda\lambda^*}{}^0 = 1 \qquad (13.195)$$

These two relations show that lower and upper indices in $\mathcal{N}$ can be interchanged by means of the conjugate operation:

$$\mathcal{N}_{\lambda\mu}{}^{\nu} = \mathcal{N}_{\lambda\nu*}{}^{\mu*} \tag{13.196}$$

Let the coefficient $\mathcal{N}_{\lambda\mu\sigma}$, with three lower indices, correspond to the multiplicity of the scalar representation in the triple product $\lambda \otimes \mu \otimes \sigma$. We thus have

$$\mathcal{N}_{\lambda\mu}{}^{\nu} = \mathcal{N}_{\lambda\mu\nu*} \tag{13.197}$$

13.5.1. The Character Method

The first method that will be described is based on the specification of a representation by its character. In consequence, Eq. (13.193) must also hold in character form (since the trace of a tensor product is the product of the trace)

$$\chi_\lambda \chi_\mu = \sum_{\nu \in P_+} \mathcal{N}_{\lambda\mu}{}^{\nu} \chi_\nu \tag{13.198}$$

Using this character equation, we can derive a simple relation between $\mathcal{N}_{\lambda\mu}{}^{\nu}$ and the multiplicities of the weights μ' in the representation μ, which will lead to an efficient way of calculating tensor-product coefficients. We rewrite Eq. (13.198) under the form

$$\sum_{w \in W} \epsilon(w) e^{w(\lambda+\rho)} \sum_{\mu' \in \Omega_\mu} \mathrm{mult}_\mu(\mu') e^{\mu'} = \sum_{\nu \in P_+} \mathcal{N}_{\lambda\mu}{}^{\nu} \sum_{w \in W} \epsilon(w) e^{w(\nu+\rho)} \tag{13.199}$$

(using Eq. (13.167) for χ_λ and χ_ν and Eq. (13.161) for χ_μ) and compare the contributions of both sides restricted to the fundamental chamber. Since $\nu \in P_+$, only the identity element of the Weyl group contributes on the r.h.s. If $\lambda + \mu' \in P_+$ on the l.h.s., then again only $w = 1$ contributes. Otherwise, we first rewrite the second sum on the l.h.s. as

$$\sum_{\mu' \in \Omega_\mu} \mathrm{mult}_\mu(\mu') e^{\mu'} = \sum_{\mu' \in \Omega_\mu} \mathrm{mult}_\mu(\mu') e^{w_{\mu'}\mu'} \tag{13.200}$$

for any $w_{\mu'} \in W$ (the multiplicity being constant along a W orbit). The contributing element is the particular element of the Weyl group $w_{\mu'}$ that reflects the weight $\lambda + \mu'$ in the fundamental chamber; it contributes with the sign $\epsilon(w_{\mu'})$. This proves the relation

$$\mathcal{N}_{\lambda\mu}{}^{\nu} = \sum_{\substack{\mu' \in \Omega_\mu \\ w \in W \ w \cdot (\lambda+\mu')=\nu \in P_+}} \epsilon(w) \, \mathrm{mult}_\mu(\mu') \tag{13.201}$$

where we dropped the index μ' from w for simplicity. There are two summations here: a sum over all the weights in the representation μ and a sum over those elements of the Weyl group that satisfy the condition $w \cdot (\lambda + \mu') = \nu \in P_+$. The result can be rewritten more simply, with a single summation, as

$$\boxed{\mathcal{N}_{\lambda\mu}{}^{\nu} = \sum_{w \in W} \epsilon(w) \, \mathrm{mult}_\mu(w \cdot \nu - \lambda)} \tag{13.202}$$

This method will be referred to as the *character method*. Its theoretical interest lies in its generality and in that it has a direct extension for affine fusion rules.

13.5.2. Algorithm for the Calculation of Tensor Products

Formula (13.202) can be translated into the following algorithm. In order to calculate the product $\lambda \otimes \mu$, we first write down all the weights μ' in the representation μ and add each of them to $\lambda + \rho$. Degenerate weights are treated separately. The resulting weights $\lambda + \rho + \mu'$ are of two types:

(i) those that can be reflected into dominant weights by an element $w \in W$ of the finite Weyl group;

(ii) those in the W orbit of a weight with some vanishing Dynkin labels.

Weights of type (i) contribute $\epsilon(w)$ to the tensor-product coefficient $\mathcal{N}_{\lambda\mu}{}^\nu$, where ν is the resulting dominant weight. $\mathcal{N}_{\lambda\mu}{}^\nu$ is obtained from the sum of all these contributions.

By definition, a weight ξ of type (ii) is such that there is a $w \in W$ for which $w\xi$ has at least one vanishing Dynkin label. If, for instance, $(w\xi)_i = 0$, then $s_i(w\xi) = 0$. Such weights can be ignored since they could be counted with both $\epsilon(w)$ and $\epsilon(s_i w) = -\epsilon(w)$. They are located at one boundary, or a Weyl reflection thereof, of the fundamental chamber.

It should be stressed that reflecting a weight in the fundamental chamber is a finite process: at most $|\Delta_+|$ (the number of positive roots) reflections are needed.

Reformulated in terms of the shifted action of the Weyl group, the procedure is as follows: If $\lambda + \mu'$ can be reflected into a dominant weight by the shifted action of the Weyl group—that is, if there exists a $w \in W$ such that $w \cdot (\lambda + \mu') \in P_+$—it contributes $\epsilon(w)$ to $\mathcal{N}_{\lambda\mu}{}^\nu$; if it cannot, it is ignored.

su(2) EXAMPLE

As a simple illustration of this procedure, consider the $su(2)$ tensor product $(2) \otimes (7)$. We display on the $su(2)$ weight lattice all the weights of the representation (7), $(-7\omega_1, -5\omega_1, \cdots, 7\omega_1)$, augmented by $2\omega_1$. A shifted Weyl reflection here is a reflection with respect to the weight $-\omega_1$ (as $\rho = \omega_1$). The weight $-\omega_1$ is of type (ii) and it is thus ignored. By reflection, the nondominant weights $-5\omega_1$ and $-3\omega_1$ are sent respectively onto $3\omega_1$ and ω_1, and contribute with a minus sign, which cancels the contribution of the representations (1) and (3). This is illustrated in Fig. 13.6, from which the result of the tensor-product decomposition is directly read off:

$$(2) \otimes (7) = (5) \oplus (7) \oplus (9) \tag{13.203}$$

This agrees with the familiar rules of angular-momentum addition.

su(3) EXAMPLE

Consider the $su(3)$ tensor product $(1,0) \otimes (2,0)$. The six weights in the representation $(2,0)$ are $\{(2,0), (0,1), (1,-1), (-2,2), (-1,0), (0,-2)\}$. Adding $(1,0)$

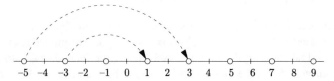

Figure 13.6. The $su(2)$ tensor product $(2) \otimes (7)$. The weights of the representation (7) are centered around $2\omega_1$ and the nondominant weights are Weyl reflected back into the dominant sector.

to each of them yields:

$$(3,0), \ (1,1), \ (2,-1), \ (-1,2), \ (0,0), \ (1,-2) \tag{13.204}$$

The third and fourth of the weights (13.204) are ignored since they are respectively invariant under the shifted action of s_2 and s_1 (and are therefore of type (ii)). Acting on the sixth one with s_2· yields

$$s_2 \cdot (1,-2) = s_2(2,-1) - (1,1) = (2,-1) + (-1,2) - (1,1) = (0,0)$$

Hence the reflection of the sixth weight into the fundamental chamber contributes to $\epsilon(s_2)(0,0) = -(0,0)$, and consequently cancels the contribution of the fifth weight in (13.204). The final result is

$$(1,0) \otimes (2,0) = (3,0) \oplus (1,1) \tag{13.205}$$

as illustrated on Fig. 13.7.

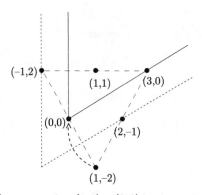

Figure 13.7. The $su(3)$ tensor product $(1,0) \otimes (2,0)$ by the method of Weyl reflections.

In these two examples, it would have been wiser to interchange the roles of the two representations. For instance, adding $(2,0)$ to the three weights of the representation $(1,0)$ gives directly $(3,0)$, $(1,1)$, $(2,-1)$, and the last one is ignored. Choosing for μ the highest weight of the smallest of the two representations simplifies the calculation in two respects: fewer states need to be considered and most of the weights in the representation μ, when added to λ, are dominant.

13.5.3. The Littlewood-Richardson Rule

The *Littlewood-Richardson rule* is a simple and powerful algorithm, formulated in terms of the product of Young tableaux. This algorithm proceeds as follows: In the second tableau, we fill the first row with 1's, the second row with 2's, and so on. Then we add all the boxes with a 1 to the first tableau and keep only the resulting tableaux that satisfy the following two conditions:

(i) They must be regular: the number of boxes in a given row must be smaller or equal to the number of boxes in the row just above.

(ii) They must not contain two boxes marked by 1 in the same column.

   Tableaux that do not satisfy these conditions are ignored. To the resulting tableaux, we then add all the boxes marked by a 2 and again we keep only the tableaux that satisfy (i) and (ii), where in (ii), 1 is replaced by 2. We continue until all the boxes of the second tableau in the original product have been used. In this process an additional rule must be respected:

(iii) In counting from right to left and top to bottom, the number of 1's must always be greater or equal to the number of 2's, the number of 2's must always be greater or equal to the number of 3's, and so on.

   The resulting Littlewood-Richardson tableaux are the Young tableaux of the irreducible representations occurring in the decomposition.

   A warning: In this process, we do not construct semistandard tableaux! However, in Littlewood-Richardson tableaux it is clear that the numbers are strictly increasing in each column and they are nondecreasing in rows.

   For example, consider the $su(3)$ tensor product $(2, 0) \otimes (1, 1)$:

The tableaux obtained after the first step are

Adding now the box marked by a 2 yields[10]

from which we read off

$$(2, 0) \otimes (1, 1) = (3, 1) \oplus (1, 2) \oplus (2, 0) \oplus (0, 1) \qquad (13.206)$$

(for $su(3)$, columns of three boxes are ignored).

   The multiplicity of a given representation ν in the tensor product $\lambda \otimes \mu$ can be evaluated directly, without necessarily having to calculate the full decomposition. For this we simply add to the Young tableau representing λ all boxes of the tableau μ such that the resulting tableau has weight ν. The added boxes are then filled

---

[10] Rule (iii) prevents us from adding a box marked by a 2 at the end of the first row.

with the following set of numbers: 1 ($\mu_1 + \cdots + \mu_{N-1}$ times), 2 ($\mu_2 + \cdots + \mu_{N-1}$ times), up to $N-1$ (μ_{N-1} times), in a way that respects the Littlewood-Richardson rule. $\mathcal{N}_{\lambda\mu}{}^\nu$ is the number of distinct Littlewood-Richardson tableaux that can be produced in this way.

For instance, to the $su(4)$ tensor product $(1,2,1) \otimes (1,2,1) \supset (1,2,1)$, there correspond 5 Littlewood-Richardson tableaux:

$$
\begin{array}{ccc}
\begin{array}{c}
\begin{matrix} & & 1 & 1 \\ & & 1 & 2 \\ 2 & 2 \\ 1 & 3 \end{matrix}
\end{array}
&
\begin{array}{c}
\begin{matrix} & & 1 & 1 \\ & & 1 & 2 \\ 1 & 2 \\ 2 & 3 \end{matrix}
\end{array}
&
\begin{array}{c}
\begin{matrix} & & 1 & 1 \\ & & 1 & 2 \\ 1 & 3 \\ 2 & 2 \end{matrix}
\end{array}
\end{array}
$$

$$
\begin{array}{cc}
\begin{array}{c}
\begin{matrix} & & 1 & 1 \\ & 2 & 2 \\ 1 & 1 \\ 2 & 3 \end{matrix}
\end{array}
&
\begin{array}{c}
\begin{matrix} & & 1 & 1 \\ & 2 & 2 \\ 1 & 3 \\ 1 & 2 \end{matrix}
\end{array}
\end{array}
$$

which means that the tensor-product coefficient $\mathcal{N}_{(121)(121)}{}^{(121)}$ is 5.

In some applications, it is necessary to know which states contribute to the tensor product. It turns out that this information is coded in the Littlewood-Richardson tableaux. More precisely, there is a one-to-one correspondence between a Littlewood-Richardson tableau associated with the product $\lambda \otimes \mu \supset \nu$ and a Gelfand-Tsetlin pattern $\{\beta_j^{(i)}\}$ of weight $\mu' = \nu - \lambda$ in the representation μ. The entries $\beta_j^{(i)}$ of the Gelfand-Tsetlin pattern can be read off the Littlewood-Richardson tableau as follows:

$$
\begin{aligned}
\beta_j^{(i)} = \quad & \text{number of } j\text{'s in the first } i \text{ rows of} \\
& \text{the Littlewood-Richardson tableau}
\end{aligned}
\tag{13.207}
$$

The states associated with each Littlewood-Richardson tableau in the previous example are

$$
\begin{array}{c}
\boxed{\begin{array}{cccc} & & 1 & 1 \\ & 2 & 2 & \\ & 1 & 3 & \\ 1 & 2 & & \end{array}} \quad \rightarrow \quad
\begin{array}{l} 4\ 3\ 1\ 0 \\ \ 3\ 2\ 1 \\ \ \ 2\ 2 \\ \ \ \ 2 \end{array} \quad \leftrightarrow \quad
\boxed{\begin{array}{cccc} 1 & 1 & 3 & 4 \\ 2 & 2 & 4 & \\ 3 & & & \end{array}}
\end{array}
\qquad (13.208)
$$

The weight $\mu' = \nu - \lambda = (0,0,0)$ in the representation $(1,2,1)$ has multiplicity 7. The two states that do not contribute to the tensor product are

$$
\begin{array}{l} 4\ 3\ 1\ 0 \\ \ 3\ 2\ 1 \\ \ \ 3\ 1 \\ \ \ \ 2 \end{array} \quad \leftrightarrow \quad
\boxed{\begin{array}{cccc} 1 & 1 & 2 & 4 \\ 2 & 3 & 4 & \\ 3 & & & \end{array}}
\qquad
\begin{array}{l} 4\ 3\ 1\ 0 \\ \ 4\ 2\ 0 \\ \ \ 4\ 0 \\ \ \ \ 2 \end{array} \quad \leftrightarrow \quad
\boxed{\begin{array}{cccc} 1 & 1 & 2 & 2 \\ 3 & 3 & 4 & \\ 4 & & & \end{array}}
$$

For completeness, we mention that this relationship between states and Littlewood-Richardson tableaux can be used to obtain an algebraic description of the tensor-product coefficients:

$$
\mathcal{N}_{\lambda\mu}{}^{\nu} = \text{number of Gelfand-Tsetlin patterns } \{\beta_j^{(i)}\} \qquad (13.209)
$$
$$
\text{of weight } \mu' = \nu - \lambda \text{ in the representation}
$$
$$
\mu \text{ that satisfy the conditions } d_j^{(i)} \le \lambda_i \text{ for}
$$
$$
\text{all values of } j,\ 1 \le j \le i \le N - 1
$$

where

$$
d_j^{(i)} = \sum_{1 \le n < j} \left(\beta_n^{(i+1)} - 2\beta_n^{(i)} + \beta_n^{(i-1)} \right) + \left(\beta_j^{(i+1)} - \beta_j^{(i)} \right) \qquad (13.210)
$$

For the first noncontributing pattern of the previous example: $d_2^{(3)} = 2 > \lambda_3 = 1$, and for the other one: $d_1^{(1)} = 2 > \lambda_1 = 1$.

13.5.4. Berenstein-Zelevinsky Triangles

Berenstein-Zelevinsky triangles (BZ) provide a powerful way to calculate the multiplicity of a triple product, that is, the multiplicity of the scalar representation in $\lambda \otimes \mu \otimes \nu$. (We point out the slight change in the notation for the third weight: we take it to be ν instead of ν^*.) They also contain information on the states contributing to the product. We first describe the construction for $su(3)$.

We consider the set of three $su(3)$ highest weights (λ_1, λ_2), (μ_1, μ_2), and (ν_1, ν_2). We construct triangles according to the following rules:

$$
\begin{array}{ccccc}
 & & m_{13} & & \\
 & n_{12} & & l_{23} & \\
 & & m_{23} \qquad\qquad m_{12} & & \\
 n_{13} & l_{12} & & n_{23} & l_{13}
\end{array}
\qquad (13.211)
$$

where the nine nonnegative integers l_{ij}, m_{ij}, n_{ij} are related to the Dynkin labels of the three integrable weights by

$$
\begin{array}{lll}
m_{13} + n_{12} = \lambda_1 & n_{13} + l_{12} = \mu_1 & l_{13} + m_{12} = \nu_1 \\
m_{23} + n_{13} = \lambda_2 & n_{23} + l_{13} = \mu_2 & l_{23} + m_{13} = \nu_2
\end{array}
\qquad (13.212)
$$

They must further satisfy the so-called hexagon conditions

$$
\begin{aligned}
n_{12} + m_{23} &= n_{23} + m_{12} \\
l_{12} + m_{23} &= l_{23} + m_{12} \\
l_{12} + n_{23} &= l_{23} + n_{12}
\end{aligned}
\tag{13.213}
$$

This means that the length of opposite sides in the hexagon formed by $n_{12}, l_{23}, m_{12}, n_{23}, l_{12}$ and m_{23} in (13.211) are equal, the length of a segment being defined as the sum of its two vertices.

The number of such triangles gives the value of $\mathcal{N}_{\lambda\mu\nu}$. If it is not possible to construct such a triangle, it means that ν^* does not occur in the tensor product $\lambda \otimes \mu$.

The integers in the BZ triangles have the following origin. Each pair of indices ij, $i < j$, on the labels of the triangle is related to a positive root of $su(3)$. We recall that the positive roots of $su(N)$ can be written as $\epsilon_i - \epsilon_j$, $1 \le i < j \le N$ in terms of orthonormal vectors ϵ_i in $\mathbb{R}^N$. The triangle encodes three sums of positive roots:

$$
\begin{aligned}
\mu + \nu - \lambda^* &= \sum_{i<j} l_{ij}(\epsilon_i - \epsilon_j) \\
\nu + \lambda - \mu^* &= \sum_{i<j} m_{ij}(\epsilon_i - \epsilon_j) \\
\lambda + \mu - \nu^* &= \sum_{i<j} n_{ij}(\epsilon_i - \epsilon_j)
\end{aligned}
\tag{13.214}
$$

The hexagon relations (13.213) can be seen as consistency conditions for these three expansions.

The four triangles for the example (13.206) are

$$
\begin{array}{cccc}
\begin{matrix}
2 \\
0 \ \ 1 \\
0 \quad\quad 0 \\
0 \ \ 1 \ \ 0 \ \ 1
\end{matrix}
&
\begin{matrix}
1 \\
1 \ \ 0 \\
0 \quad\quad 1 \\
0 \ \ 1 \ \ 0 \ \ 1
\end{matrix}
&
\begin{matrix}
1 \\
1 \ \ 1 \\
0 \quad\quad 0 \\
0 \ \ 1 \ \ 1 \ \ 0
\end{matrix}
&
\begin{matrix}
0 \\
2 \ \ 0 \\
0 \quad\quad 1 \\
0 \ \ 1 \ \ 1 \ \ 0
\end{matrix}
\end{array}
\tag{13.215}
$$

On the other hand, corresponding to the coupling $(2,2) \otimes (2,2) \otimes (2,2)$, three triangles can be constructed:

$$
\begin{array}{ccc}
\begin{matrix}
0 \\
2 \ \ 2 \\
2 \quad\quad 2 \\
0 \ \ 2 \ \ 2 \ \ 0
\end{matrix}
&
\begin{matrix}
1 \\
1 \ \ 1 \\
1 \quad\quad 1 \\
1 \ \ 1 \ \ 1 \ \ 1
\end{matrix}
&
\begin{matrix}
2 \\
0 \ \ 0 \\
0 \quad\quad 0 \\
2 \ \ 0 \ \ 0 \ \ 2
\end{matrix}
\end{array}
\tag{13.216}
$$

and accordingly the multiplicity of the scalar representation in this triple product is 3.

The states involved in a specific coupling can be read off a triangle as follows. Consider the product $\lambda \otimes \mu \supset \nu^*$ associated with the triangle (13.211). The state of weight $\mu' = \nu^* - \lambda$ in this coupling is described by the Gelfand-Tsetlin pattern

$$\mu_1 + \mu_2 \qquad \mu_2 \qquad 0$$

$$\mu_1 + \mu_2 - n_{13} \qquad \mu_2 - n_{23} \qquad (13.217)$$

$$\mu_1 + \mu_2 - n_{13} - n_{12}$$

For example, the Gelfand-Tsetlin patterns and corresponding semistandard tableaux in the representation $\mu = (2,2)$ associated with the three triangles of the last example (13.216) are (in the same order)

$$
\begin{array}{l} 4\ 2\ 0 \\ \ \ 4\ 0 \\ \ \ \ \ 2 \end{array} \leftrightarrow
\begin{array}{|c|c|c|c|} \hline 1 & 1 & 2 & 2 \\ \hline 3 & 3 \\ \cline{1-2} \end{array}
\qquad
\begin{array}{l} 4\ 2\ 0 \\ \ \ 3\ 1 \\ \ \ \ \ 2 \end{array} \leftrightarrow
\begin{array}{|c|c|c|c|} \hline 1 & 1 & 2 & 3 \\ \hline 2 & 3 \\ \cline{1-2} \end{array}
\qquad
\begin{array}{l} 4\ 2\ 0 \\ \ \ 2\ 2 \\ \ \ \ \ 2 \end{array} \leftrightarrow
\begin{array}{|c|c|c|c|} \hline 1 & 1 & 3 & 3 \\ \hline 2 & 2 \\ \cline{1-2} \end{array}
$$

For $su(4)$, the BZ triangles are defined in a similar way, in terms of eighteen nonnegative integers:

$$
\begin{array}{ccccccc}
 & & & m_{14} & & & \\
 & & n_{12} & & l_{34} & & \\
 & m_{24} & & & & m_{13} & \\
 & n_{13} & l_{23} & & n_{23} & l_{24} & \\
m_{34} & & & m_{23} & & & m_{12} \\
n_{14} & l_{12} & & n_{24} & l_{13} & n_{34} & l_{14}
\end{array}
\qquad (13.218)
$$

related to the Dynkin labels by

$$
\begin{array}{lll}
m_{14} + n_{12} = \lambda_1 & n_{14} + l_{12} = \mu_1 & l_{14} + m_{12} = \nu_1 \\
m_{24} + n_{13} = \lambda_2 & n_{24} + l_{13} = \mu_2 & l_{24} + m_{13} = \nu_2 \\
m_{34} + n_{14} = \lambda_3 & n_{34} + l_{14} = \mu_3 & l_{34} + m_{14} = \nu_3
\end{array}
\qquad (13.219)
$$

Furthermore, a $su(4)$ BZ triangle has 3 hexagons:

$$
\begin{array}{lll}
n_{12}+m_{24} = m_{13}+n_{23} & n_{13}+l_{23} = l_{12}+n_{24} & l_{24}+n_{23} = l_{13}+n_{34} \\
n_{12}+l_{34} = l_{23}+n_{23} & n_{13}+m_{34} = n_{24}+m_{23} & n_{23}+m_{23} = m_{12}+n_{34} \\
m_{24}+l_{23} = l_{34}+m_{13} & m_{34}+l_{12} = l_{23}+m_{23} & l_{13}+m_{23} = l_{24}+m_{12}
\end{array}
$$
$$(13.220)$$

The $su(N)$ generalization is straightforward; the triangles are built out of $(N-1)(N-2)/2$ hexagons and three corner points. On the other hand, for $su(2)$ there are no hexagons: the tensor products are described by the simple triangles

$$
\begin{array}{ccc}
 & m_{12} & \\
n_{12} & & l_{12}
\end{array}
\qquad (13.221)
$$

written in terms of three nonnegative integers constrained by

$$m_{12} + n_{12} = \lambda_1$$

$$n_{12} + l_{12} = \mu_1 \qquad (13.222)$$

$$l_{12} + m_{12} = \nu_1$$

With λ_1 and μ_1 fixed, ν_1 satisfies

$$\nu_1 = \lambda_1 + \mu_1 - 2n_{12} \tag{13.223}$$

which reproduces the rule for $su(2)$ tensor products in a very simple way.

With the last two methods described, it is possible to study a particular triple product in isolation, that is, without necessarily computing the full product $\lambda \otimes \mu$. This is a clear advantage when reasonably large representations are involved. The BZ triangles have the further advantage of preserving most of the symmetries of the tensor-product coefficients. In fact, the only symmetry that is not manifest is $\mathcal{N}_{\lambda\mu\nu} = \mathcal{N}_{\mu\lambda\nu}$.

We note finally that, in contradistinction with the Littlewood-Richardson rule, the generalization of the BZ triangles to $so(N)$ and $sp(N)$ is unknown at this time.

§13.6. Tensor Products: A Fusion-Rule Point of View

In this section we discuss tensor products from a point of view close in spirit to the approach used in fusion-rule calculations. At first, we indicate how generic tensor-product coefficients are fixed by associativity in terms of tensor-product coefficients involving the fundamental representations. We recall that for minimal models, the fusion ring was found to be generated by $\phi_{(1,2)}$ and $\phi_{(2,1)}$. In that context, Chebyshev polynomials appeared naturally. These polynomials and their generalizations resurface here. In a second step, we derive the Lie algebra version of the Verlinde formula (10.201).

The associativity of tensor products translates into the following condition

$$\sum_{\sigma} \mathcal{N}_{\lambda\mu}{}^{\sigma} \mathcal{N}_{\sigma\nu\xi} = \sum_{\zeta} \mathcal{N}_{\mu\nu}{}^{\zeta} \mathcal{N}_{\zeta\lambda\xi} \tag{13.224}$$

It is clear that if λ and μ are fundamental representations, any general coefficient $\mathcal{N}_{\sigma\nu\xi}$ can be deduced from this condition whenever all tensor-product coefficients involving at least one fundamental representation are known. Again, by introducing a matrix N_λ with entries

$$(N_\lambda)_\mu{}^{\sigma} = \mathcal{N}_{\lambda\mu}{}^{\sigma} \tag{13.225}$$

we see that the associativity requirement boils down to the commutativity of the matrices N:

$$(N_\lambda N_\nu)_{\mu\xi} = (N_\nu N_\lambda)_{\mu\xi} \tag{13.226}$$

These matrices provide a representation of the tensor-product algebra:

$$N_\lambda N_\nu = \sum_{\sigma \in P_+} \mathcal{N}_{\lambda\nu}{}^{\sigma} N_\sigma \tag{13.227}$$

Here we did nothing but rewrite (13.224) in matrix form. We note that these matrices are infinite.

We will now see how Chebyshev-like polynomials arise in this picture. We consider first the $su(2)$ case. From the Littlewood-Richardson rule (or the angular-momentum addition theory), we easily see that

$$(1) \otimes (n) = (n+1) \oplus (n-1) \tag{13.228}$$

where it is understood that if the Dynkin label $n-1$ is negative, the second representation on the r.h.s. is omitted. The comparison of this product rule with Eq. (13.227) shows that the matrix N_1 is simply:

$$(N_1)_j^k = \delta_{j,k+1} + \delta_{j,k-1} \tag{13.229}$$

Mutually commuting matrices associated with other representations can be constructed as follows. One first observes that Eq. (13.228) translates into the following relation

$$N_1 N_n = N_{n+1} + N_{n-1} \tag{13.230}$$

which can be regarded as a recurrence relation to be solved for N_n in terms of N_1. This becomes clearer if we replace N_1 by x, N_n by $U_n(x)$ and rewrite the above equation in the form

$$x U_n = U_{n+1} + U_{n-1} \tag{13.231}$$

With $U_0 = 1$, $U_1 = x$, this is the defining relation for Chebyshev polynomials of the second kind, which already arose in the context of minimal-model fusion rules (cf. Eq. (8.101)). The desired expression for N_n is thus

$$N_n = U_n(N_1) \tag{13.232}$$

For instance, given the matrix N_1—which fixes all tensor products with the fundamental representation—the matrix N_2 describing the products with the adjoint representation is

$$N_2 = N_1^2 - 1 \tag{13.233}$$

That is,

$$N_1 = \begin{pmatrix} 0 & 1 & 0 & 0 & 0 & \cdots \\ 1 & 0 & 1 & 0 & 0 & \cdots \\ 0 & 1 & 0 & 1 & 0 & \cdots \\ 0 & 0 & 1 & 0 & 1 & \cdots \\ \cdots & \cdots & \cdots & & & \end{pmatrix} \Longrightarrow N_2 = \begin{pmatrix} 0 & 0 & 1 & 0 & 0 & \cdots \\ 0 & 1 & 0 & 1 & 0 & \cdots \\ 1 & 0 & 1 & 0 & 1 & \cdots \\ 0 & 1 & 0 & 1 & 0 & \cdots \\ \cdots & \cdots & \cdots & & & \end{pmatrix}$$

$$\tag{13.234}$$

from which we read off directly that

$$(2) \otimes (0) = (2)$$
$$(2) \otimes (1) = (1) \oplus (3) \tag{13.235}$$
$$(2) \otimes (2) = (0) \oplus (2) \oplus (4)$$

and so on. Because they are all constructed out of polynomials in N_1, the matrices N_n necessarily commute among themselves.

It is interesting to construct the generating function of the Chebyshev polynomials. This is done in the standard way: one multiplies Eq. (13.231) by t^n and sums the result from $n = 0$ to $n = \infty$; by simple manipulations, each term can be reexpressed in terms of

$$F(x; t) = \sum_{n=0}^{\infty} U_n t^n \tag{13.236}$$

with the result

$$xF = (F - 1)/t + tF \tag{13.237}$$

that is,

$$F(x; t) = \frac{1}{1 - xt + t^2} \tag{13.238}$$

A similar analysis can be done for any Lie algebra. For instance, for $su(3)$, the Littlewood-Richardson rule immediately tells us that

$$(1, 0) \otimes (\lambda_1, \lambda_2) = (\lambda_1 + 1, \lambda_2) \oplus (\lambda_1, \lambda_2 - 1) \oplus (\lambda_1 - 1, \lambda_2 + 1)$$
$$(0, 1) \otimes (\lambda_1, \lambda_2) = (\lambda_1, \lambda_2 + 1) \oplus (\lambda_1 - 1, \lambda_2) \oplus (\lambda_1 + 1, \lambda_2 - 1) \tag{13.239}$$

Again, we can replace the representations in these expressions by their corresponding tensor-product matrices $N_{(\lambda_1, \lambda_2)}$. These matrices turn out be expressible in terms of some generalized Chebyshev polynomials $U_{(\lambda_1, \lambda_2)}$, a function of two variables x_1, x_2 associated respectively with $N_{(1,0)}$ and $N_{(0,1)}$, as follows:

$$N_{(\lambda_1, \lambda_2)} = U_{(\lambda_1, \lambda_2)}(N_{(1,0)}, N_{(0,1)}) \equiv U_{(\lambda_1, \lambda_2)}(x_1, x_2) \tag{13.240}$$

These polynomials are defined in terms of the generating function

$$F(x_1, x_2; t, s) = \sum_{\lambda_1, \lambda_2 = 0}^{\infty} U_{(\lambda_1, \lambda_2)}(x_1, x_2) \, t^{\lambda_1} s^{\lambda_2}$$
$$= \frac{1 - ts}{(1 - tx_1 + t^2 x_2 - t^3)(1 - sx_2 + s^2 x_1 - s^3)} \tag{13.241}$$

The details of this analysis are left to the reader (cf. Ex. 13.20).

We now turn to a Lie algebra version of the Verlinde formula (10.201). The starting point is the character product form (13.198), in which all the characters are supposed to be evaluated at the particular point

$$X = -2\pi i \sum_{i=1}^{r} t_i \alpha_i^{\vee} \tag{13.242}$$

where the t_i's are real numbers valued in the range $[0, 1]$. With $\chi_\lambda = D_{\lambda+\rho}/D_\rho$, this becomes

$$\frac{D_{\lambda+\rho}(X) D_{\mu+\rho}(X)}{D_\rho(X)} = \sum_{\nu} \mathcal{N}_{\lambda\mu}{}^{\nu} D_{\nu+\rho}(X) \tag{13.243}$$

The $D_{\lambda+\rho}(X)$'s satisfy the following orthogonality relation

$$\int_0^1 \left(\prod_i dt_i\right) D_{\lambda+\rho}(X) D_{\mu+\rho}(X) = |W| \, \delta_{\mu,\lambda^*} \qquad (13.244)$$

where $|W|$ is the order of the Weyl group. This follows directly from the definition (13.165) and the expression (13.117) for the conjugate of a representation, which implies that $D_{\lambda^*+\rho}(X)$ is the complex conjugate of $D_{\lambda+\rho}(X)$. To proceed, we multiply Eq. (13.243) by $D_{\sigma^*+\rho}(X)$ and integrate the result over X (i.e., integrate over all t_i's from 0 to 1). This gives the desired Verlinde-type formula

$$\mathcal{N}_{\lambda\mu}{}^\sigma = \int_0^1 (\prod_i dt_i) \frac{S_\lambda(X) S_\mu(X) \bar{S}_\sigma(X)}{S_0(X)} \qquad (13.245)$$

where

$$S_\lambda(X) = \frac{1}{\sqrt{|W|}} D_{\lambda+\rho}(X) \qquad (13.246)$$

Such an S matrix is analogous to the one obtained for a finite group in Ex. 10.18: it is indexed by r discrete numbers, the Dynkin labels λ_i, and r continuous ones, the t_i. This immediately tells us that such an S matrix cannot be a transformation matrix of characters into themselves, like the modular transformation matrix. For $su(2)$, it takes the simple form

$$S_{\lambda_1}(t_1) = -i\sqrt{2} \, \sin[2\pi t_1(\lambda_1 + 1)] \qquad (13.247)$$

§13.7. Algebra Embeddings and Branching Rules

As mentioned in the introduction, we will often encounter "affine" generalizations of simple Lie algebra embeddings. This fact motivates the general remarks of this section.

13.7.1. Embedding Index

We first present different ways of characterizing an embedding $p \subset g$, deferring classification issues to the next subsection.

i) Branching rules:

Viewed from the standpoint of the smaller algebra p, an irreducible representation of g usually breaks down into many irreducible representations of p. Such decompositions are called *branching rules* and are noted as

$$L_\lambda \mapsto \bigoplus_{\mu \in P_+} b_{\lambda\mu} L_\mu \qquad (13.248)$$

or simply as

$$\lambda \mapsto \bigoplus_{\mu \in P_+} b_{\lambda\mu} \, \mu \qquad (13.249)$$

The branching coefficient $b_{\lambda\mu}$ gives the multiplicity of the irreducible representation μ of p in the decomposition of the irreducible representation λ of g. The decomposition of the lowest-dimensional nontrivial representation is sufficient to characterize an embedding. To each of its inequivalent branching rules corresponds a distinct embedding.

ii) Projection matrix:
A *projection matrix* $\mathcal{P}$ gives the explicit projection of every weight of g onto a weight of p. Hence, to calculate the branching rules one first projects all the weights of a given irreducible representation of g into p-weights and reorganizes them into irreducible representations. Projection matrices are not unique: a Weyl reflection of the root diagram modifies them without affecting the embedding.

iii) Embedding index:
The *embedding index* x_e is defined as the ratio of the square length of the projection of θ, the highest root of g, to the square length of the highest root of p, which is denoted by ϑ:

$$x_e = \frac{|\mathcal{P}\theta|^2}{|\vartheta|^2} \qquad (13.250)$$

Given a branching rule, the embedding index can also be calculated from

$$x_e = \sum_{\mu \in P_+} b_{\lambda\mu} \frac{x_\mu}{x_\lambda} \qquad (13.251)$$

where x_λ is the index of the representation λ of g defined in Eq. (13.133). The proof of this relation is left as an exercise (cf. Ex. 13.22).

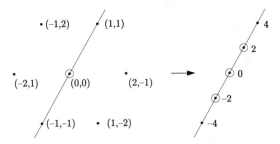

Figure 13.8. Projection of the $su(3)$ adjoint representation onto $su(2)$.

As an example, we show how $su(2)$ can be embedded into $su(3)$. Fig. 13.8 shows how the $su(3)$ root system is projected along the highest root vector and gives a possible assignment of the $su(2)$ weights. The representation $(1, 1)$ of $su(3)$ decomposes into the $su(2)$ representations $(2) \oplus (4)$ (of respective dimension 3 and 5); it is thus characterized by the branching rule:

$$(1, 1) \mapsto (4) \oplus (2) \qquad (13.252)$$

The embedding index is easily found by noticing that the highest weight of $su(3)$, $\alpha_1 + \alpha_2$, is projected onto $2\alpha_1$. The ratio of highest roots is thus 4, and $x_e = 4$. This can also be seen from Eq. (13.251) using the branching rule (13.252). The required representation indices are:

$$su(3) : x_{(1,1)} = 3$$
$$su(2) : x_{(4)} = 10, \quad x_{(2)} = 2 \tag{13.253}$$

Their substitution in Eq. (13.251) reproduces the value $x_e = 4$. The projection matrix for this embedding can be chosen as

$$\mathcal{P}_{(4)} = (2,2) \tag{13.254}$$

if the $su(3)$ weight is written in a column matrix whose entries are its Dynkin labels. Hence, the $su(3)$ weight (λ_1, λ_2) is projected into the $su(2)$ weight of Dynkin label $2\lambda_1 + 2\lambda_2$ (which is thus always even):

$$(2,2) \begin{pmatrix} \lambda_1 \\ \lambda_2 \end{pmatrix} = (2\lambda_1 + 2\lambda_2) \tag{13.255}$$

Using this matrix, the branching rules $(1,0) \mapsto (2)$, $(0,1) \mapsto (2)$ are easily derived.[11]

Dividing all $su(2)$ Dynkin labels by 2 in Fig 13.8 leads to another possible assignment for the $su(2)$ weights. The branching rule specifying this embedding is

$$(1,1) \mapsto (2) \oplus 2(1) \oplus (0) \tag{13.256}$$

Because $\alpha_1 + \alpha_2$ is projected onto α_1, the embedding index is equal to 1. A candidate projection matrix for this embedding is

$$\mathcal{P}_{(1)} = (1,1) \tag{13.257}$$

The basic branching rule is $(1,0) \mapsto (1) + (0)$.

These two embeddings are most conveniently described by means of the following generating functions:

$$F_{(1)} = \frac{1}{(1 - L_1 M)(1 - L_2 M)(1 - L_1)(1 - L_2)}$$
$$F_{(4)} = \frac{(1 + L_1 L_2 M^2)}{(1 - L_1 M^2)(1 - L_2 M^2)(1 - L_1^2)(1 - L_2^2)} \tag{13.258}$$

where the subscript indicates the embedding index. To obtain the decomposition of the $su(3)$ weight (λ_1, λ_2), we expand F and collect all the terms multiplying $L_1^{\lambda_1} L_2^{\lambda_2}$; its coefficient, of the form $a M^m + b M^n + \cdots$, codes the decomposition of (λ_1, λ_2):

$$(\lambda_1, \lambda_2) \mapsto a(m) \oplus b(n) \oplus \cdots \tag{13.259}$$

---

[11] This shows the simple origin of this embedding: The three-dimensional representation of $su(2)$, which is the fundamental representation of $so(3)$, can be realized in terms of 3×3 $su(3)$ matrices.

We take for instance the embedding with $x_e = 4$. In the power expansion of $F_{(4)}$, the term $L_1^3 L_2^0$ is multiplied by $M^6 + M^2$, so that

$$(3,0) \mapsto (6) \oplus (2) \tag{13.260}$$

A few remarks complete this discussion. An obvious necessary condition for the branching coefficient $b_{\lambda\mu}$ to be nonzero is

$$\mathcal{P}\lambda - \mu \in \mathcal{P}Q \tag{13.261}$$

where Q is the root lattice of g. This simply means that the integrable weight μ must lie somewhere in the integrable representation λ, after projection; since any weight in Ω_λ can be obtained from λ by subtracting an appropriate number of positive roots, the condition follows. In the examples above, the root lattice projects as follows. For the first embedding,

$$x_e = 4: \quad \mathcal{P}Q_{su(3)} = Q_{su(2)} \tag{13.262}$$

since the $su(3)$ roots α_1 and α_2 are projected onto the $su(2)$ weight $2\omega_1$, that is, onto the $su(2)$ simple root α_1. The condition (13.261) forces then

$$(\mathcal{P}\lambda)_1 = \mu_1 \bmod 2 \tag{13.263}$$

where $(\mathcal{P}\lambda)_1$ is the Dynkin label of the projected $su(3)$ weight. For the other embedding, we find that both α_1 and α_2 are mapped onto ω_1 of $su(2)$, so that

$$x_e = 1: \quad \mathcal{P}Q_{su(3)} = P_{su(2)} \tag{13.264}$$

where, as usual, (noncalligraphic) P stands for the weight lattice. As a result Eq. (13.261) gives no constraint: both λ and μ are integrable weights.

We note finally that a useful tool for the computation of branching rules uses tensor products. If

$$\lambda \mapsto \bigoplus_\mu b_{\lambda\mu}\, \mu \qquad \text{and} \qquad \xi \mapsto \bigoplus_\nu b_{\xi\nu}\, \nu \tag{13.265}$$

then

$$\lambda \otimes \xi \mapsto \bigoplus_{\mu,\nu} b_{\lambda\mu} b_{\xi\nu}\, \mu \otimes \nu \tag{13.266}$$

For instance, given the branching rules $(1,0) \mapsto (1) \oplus (0)$ and $(0,1) \mapsto (1) \oplus (0)$ for $su(2) \subset su(3)$ with $x_e = 1$, we find the branching rule for $(2,0)$ from

$$(1,0) \otimes (1,0) = (2,0) \oplus (0,1) \mapsto [(1) \oplus (0)] \otimes [(1) \oplus (0)] = (2) \oplus 2(1) \oplus 2(0) \tag{13.267}$$

that is, $(2,0) \mapsto (2) \oplus (1) \oplus (0)$.

13.7.2. Classification of Embeddings

We now briefly address the question of classifying the possible embeddings. The following discussion is restricted to maximal embeddings; these are embeddings p $\subset$ g for which there is no p' such that p $\subset$ p' $\subset$ g. All nonmaximal embeddings can be obtained from a chain of maximal ones. We also suppose that g is semisimple; that also makes p semisimple up to a possible $u(1)$ factor.

The simplest embeddings are those for which there exists a basis of g in which a subset of generators form the generators of p. In other words, if the p generators are denoted by a tilde, we have $\{\tilde{E}^\alpha\} \subset \{E^\alpha\}$ and $\{\tilde{H}^i\} \subset \{H^i\}$. These are called the *regular subalgebras*. The maximal regular subalgebras have the same rank as the algebra g and they are easily described in terms of the root system of g.

We first construct the *extended* Dynkin diagram of g by adding an extra node, associated with $-\theta$. The extended Dynkin diagrams are displayed in Fig. 14.1 of Chap. 14. Promoting $-\theta$ to a "simple root" preserves the characteristic property that the difference between two simple roots is not a root (i.e., $\alpha_i + \theta$ cannot be a root since θ is the highest root). However in order to restore the linear independence of the simple roots, at least one α_i has to be removed from this augmented set of simple roots. All semisimple maximal regular subalgebras are obtained by removing from the extended Dynkin diagram of g any node whose mark is a prime number.[12] Maximal regular algebras that are not semisimple are constructed from the removal of two nodes with mark $a_i = 1$ and the addition of a $u(1)$ factor. The embedding $su(2) \subset su(3)$ with $x_e = 1$ is a regular embedding because it can be obtained from $su(3)$ by dropping one simple root (hence two simple roots with unit mark from the extended $su(3)$ diagram). However, it is not a maximal embedding, being associated with the regular chain $su(2) \subset su(2) \oplus u(1) \subset su(3)$. As another example, consider the extended Dynkin diagram of E_8 out of which one of the simple roots $\{\alpha_1, \alpha_2, \alpha_4, \alpha_7, \alpha_8\}$ is removed. The resulting algebras in each case are, respectively, $su(2) \oplus E_7$, $su(3) \oplus E_6$, $su(4) \oplus so(11)$, $so(16)$ and $su(9)$. Since E_8 has no simple root with unit mark, its maximal regular subalgebras are all semisimple.

The calculation of branching rules in a regular embedding proceeds as follows. We first add to all the weights in the representation L_λ an extra Dynkin label, associated with the extra simple root $-\theta$. Since the decomposition of θ in terms of the simple coroots is known (the expansion coefficients being the comarks), this extra Dynkin label is simply

$$\lambda_{-\theta} = -\sum_i a_i^\vee \lambda_i \tag{13.268}$$

---

[12] With the exceptional algebras, deleting a simple root from the extended diagram may give an algebra contained in one obtained from the removal of another simple root from the same extended diagram. The following list is exhaustive:

$$\widehat{F}_4/\alpha_3 \subset \widehat{F}_4/\alpha_4, \qquad \widehat{E}_7/\alpha_3 \subset \widehat{E}_7/\alpha_1, \qquad \widehat{E}_8/\alpha_6 \subset \widehat{E}_8/\alpha_1$$
$$\widehat{E}_8/\alpha_5 \subset \widehat{E}_8/\alpha_2, \qquad \widehat{E}_8/\alpha_3 \subset \widehat{E}_8/\alpha_7$$

Here $\hat{g}/\alpha_i$ stands for the algebra associated with the extended diagram with the simple root α_i deleted. For each case in which the removed root does not lead to a maximal algebra, we see that the corresponding mark is not a prime number.

If the regular subalgebra p is obtained by deleting the simple root α_j, we simply delete the Dynkin label λ_j from all the weights. The resulting weights are exactly the projected weights, and they can be reorganized into irreducible representations of p. This is illustrated in Ex. 13.25. The same procedure works for the semisimple algebra obtained from the removal of two nodes.

Table 13.1. Maximal semisimple special algebras of exceptional Lie algebras. The upper index gives the value of the embedding index.

Exceptional g	Maximal special p
G_2	$A_1^{(28)}$
F_4	$A_1^{(156)}$, $G_2^{(1)} \oplus A_1^{(8)}$
E_6	$A_1^{(9)}, G_2^{(3)}, C_4^{(1)}$, $G_2^{(1)} \oplus A_2^{(2)}, F_4^{(1)}$
E_7	$A_1^{(399)}, A_1^{(231)}, A_2^{(21)}$, $G_2^{(1)} \oplus C_3^{(1)}$, $F_4^{(1)} \oplus A_2^{(3)}$, $G_2^{(2)} \oplus A_1^{(7)}$, $A_1^{(24)} \oplus A_1^{(15)}$
E_8	$A_1^{(1240)}, A_1^{(760)}, A_1^{(520)}$, $G_2^{(1)} \oplus F_4^{(1)}$, $A_2^{(6)} \oplus A_1^{(16)}, B_2^{(12)}$

Nonregular subalgebras are called *special subalgebras*. There is still a general method for obtaining the special embeddings of the classical algebras, but the exceptional ones require a case-by-case analysis, whose result is given in Table 13.1. For the classical algebras, we use the realization of the corresponding compact groups as a group of matrices. This makes the following embeddings almost immediate:

$$su(p) \oplus su(q) \subset su(pq)$$
$$so(p) \oplus so(q) \subset so(pq)$$
$$sp(2p) \oplus sp(2q) \subset so(4pq) \qquad (13.269)$$
$$sp(2p) \oplus so(q) \subset sp(2pq)$$
$$so(p) \oplus so(q) \subset so(p+q) \qquad \text{for } p \text{ and } q \text{ odd}$$

On the other hand, if the algebra p has an N-dimensional representation with an invariant bilinear form, p can be embedded in $so(N)$ (resp. $sp(N)$) if this bilinear form is symmetric (resp. antisymmetric). If the representation has no invariant bilinear form, it realizes an embedding into $su(N)$.[13] A necessary condition for L_λ to have an invariant bilinear form is that $-\lambda \in \Omega_\lambda$, which means that L_λ must be self-conjugate. These representations have already been identified in Sect. 13.2.2. The symmetry of the bilinear form is determined by the height of the representation, defined in terms of a *height vector u*, tabulated in App. 13.A. The form is symmetric

---

[13] The subalgebra obtained in this way is generally maximal, but there are a few exceptions.

(resp. antisymmetric) if $\lambda \cdot u = \sum_i \lambda_i u_i = 0$ (resp. 1) mod 2. For instance, all representations of $su(2)$ are self-conjugate ($u_1 = 1$), so the representation L_λ is symmetric when λ_1 is even, in which case it leads to the special embeddings $su(2) \subset so(\lambda_1 + 1)$. An interesting generic example is $p \subset so(\dim p)$. Indeed, the adjoint representation is always self-conjugate (i.e., the highest weight is θ and $-\theta$ is also a root) and it has a symmetric bilinear form (the Killing form).

In the following chapters we often encounter a special embedding, which, although not maximal, deserves a particular mention. This is the diagonal embedding $g \subset g \oplus g$, in which the two weights (λ, μ) of $g \oplus g$ are projected onto the weight $\lambda + \mu$. Because the highest root is of the same length for the algebra and its subalgebra, the embedding index of a diagonal embedding is always equal to 1.

Appendix 13.A. Properties of Simple Lie Algebras

The following summaries present the essential information needed for all simple Lie algebras. The Cartan notation is used, and, for the classical algebras, the compact real form is also given in parentheses. For each algebra, we present the Dynkin diagram, a short list of basic properties, the Cartan matrix and the quadratic form matrix. Black nodes in the Dynkin diagrams refer to short roots. The numbers appearing beside the nodes of the Dynkin diagrams give (in this order) the numbering of the corresponding simple root, its mark, and its comark. For simply laced algebras, the third entry is omitted (marks and comarks are identical). The numbering of the simple roots also gives the numbering of the fundamental weights; this is the numbering used when a weight is specified in terms of a sequence of Dynkin labels as $\lambda = (\lambda_1, \cdots, \lambda_r)$. Marks and comarks are defined in Eq. (13.33). The list of properties includes the dimension of the algebra, the dual Coxeter number g, the order of the Weyl group $|W|$, the highest root θ (in Dynkin label notation), the finite group that corresponds to the ratio of the weight lattice P to the root lattice Q, the associated congruence vector (v, defined in Sect. 13.1.9), the height vector (u, defined in Sect. 13.7.2) and the exponents (defined at the end of Sect. 13.2.3). For some algebras, the entry "P/Q" and "congruence vector" do not appear; for those cases, $P/Q = I$.

$A_{r \geq 2}$ $(su(r + 1))$

(1;1)
(2;1)
(3;1)

(r;1)

$\dim g = r^2 + 2r$
$g = r + 1$
$|W| = (r + 1)!$
$\theta = (1, 0, \cdots, 1)$
$P/Q = Z_{r+1}$
$v = (1, 2, \cdots, r)$
$u = (r, 2(r - 1), \cdots, r)$
exponents $= 1, 2, \cdots, r$

Cartan matrix:

$$\begin{pmatrix} 2 & -1 & 0 & \cdots & 0 & 0 \\ -1 & 2 & -1 & \cdots & 0 & 0 \\ 0 & -1 & 2 & \cdots & 0 & 0 \\ \cdot & \cdot & \cdot & \cdots & \cdot & \cdot \\ 0 & 0 & 0 & \cdots & 2 & -1 \\ 0 & 0 & 0 & \cdots & -1 & 2 \end{pmatrix}$$

Quadratic form matrix:

$$\frac{1}{r+1} \begin{pmatrix} r & r-1 & r-2 & \cdots & 2 & 1 \\ r-1 & 2(r-1) & 2(r-2) & \cdots & 4 & 2 \\ r-2 & 2(r-2) & 3(r-2) & \cdots & 6 & 3 \\ \cdot & \cdot & \cdot & \cdots & \cdot & \cdot \\ 2 & 4 & 2(r-1) & \cdots & 2(r-1) & r-1 \\ 1 & 2 & r-1 & \cdots & r-1 & r \end{pmatrix}$$

$\mathbf{B_{r \geq 3}} \ (so(2r+1))$

$\dim g = 2r^2 + r$

$g = 2r - 1$

$|W| = 2^r r!$

$\theta = (0, 1, \cdots, 0)$

$P/Q = \mathbb{Z}_2$

$v = (0, \cdots, 0, 1)$

$u = (2r, 2(2r-1), \cdots, (r-1)(r+2), \frac{1}{2}r(r+1))$

$\text{exponents} = 1, 3, \cdots, 2r-1$

Cartan matrix:

$$\begin{pmatrix} 2 & -1 & 0 & \cdots & 0 & 0 \\ -1 & 2 & -1 & \cdots & 0 & 0 \\ 0 & -1 & 2 & \cdots & 0 & 0 \\ \cdot & \cdot & \cdot & \cdots & \cdot & \cdot \\ 0 & 0 & 0 & \cdots & 2 & -2 \\ 0 & 0 & 0 & \cdots & -1 & 2 \end{pmatrix}$$

Quadratic form matrix:

$$\frac{1}{2} \begin{pmatrix} 2 & 2 & 2 & \cdots & 2 & 1 \\ 2 & 4 & 4 & \cdots & 4 & 2 \\ 2 & 4 & 6 & \cdots & 6 & 3 \\ \cdot & \cdot & \cdot & \cdots & \cdot & \cdot \\ 2 & 4 & 6 & \cdots & 2(r-1) & r-1 \\ 1 & 2 & 3 & \cdots & r-1 & r/2 \end{pmatrix}$$

$\mathbf{C_{r \geq 2}} \ (sp(2r))$

● $(1;2;1)$

● $(2;2;1)$

● $(r\text{-}1;2;1)$

○ $(r;1;1)$

$\dim g = 2r^2 + r$

$g = r + 1$

$|W| = 2^r r!$

$\theta = (2, 0, \cdots, 0)$

$P/Q = \mathbb{Z}_2$

$v = (1, 2, \cdots, r-1, r)$

$u = ((2r-2), 2(2r-2), \cdots, (r-1)(r+1), r^2)$

exponents $= 1, 3, \cdots, 2r-1$

Cartan matrix:

$$
\begin{pmatrix}
2 & -1 & 0 & \cdots & 0 & 0 \\
-1 & 2 & -1 & \cdots & 0 & 0 \\
0 & -1 & 2 & \cdots & 0 & 0 \\
\cdot & \cdot & \cdot & \cdots & \cdot & \cdot \\
0 & 0 & 0 & \cdots & 2 & -1 \\
0 & 0 & 0 & \cdots & -2 & 2
\end{pmatrix}
$$

Quadratic form matrix:

$$
\frac{1}{2}
\begin{pmatrix}
1 & 1 & 1 & \cdots & 1 & 1 \\
1 & 2 & 2 & \cdots & 2 & 2 \\
1 & 2 & 3 & \cdots & 3 & 3 \\
\cdot & \cdot & \cdot & \cdots & \cdot & \cdot \\
1 & 2 & 3 & \cdots & r-1 & r-1 \\
1 & 2 & 3 & \cdots & r-1 & r
\end{pmatrix}
$$

$D_{r \geq 4}$ ($so(2r)$)

○ $(1;1)$

○ $(2;2)$

○ $(r\text{-}2;2)$

○ $(r\text{-}1;1)$

○ $(r;1)$

$\dim g = 2r^2 - r$

$g = 2r - 2$

$|W| = 2^{r-1} r!$

$\theta = (0, 1, \cdots, 0)$

$P/Q = \mathbb{Z}_4$ $(r \text{ odd})$

     $= \mathbb{Z}_2 \times \mathbb{Z}_2$ $(r \text{ even})$

$v = (2, 4, \cdots, 2r-4, r-2, r)$ $(r \text{ odd})$

   $= (0, \cdots, 0, 1, 1)$ $(r \text{ even})$

$u = ((2r-2), 2(2r-3), \cdots,$

    $(r-2)(r+1), \frac{1}{2}r(r-1), \frac{1}{2}r(r-1))$

exponents $= 1, 3, \cdots, 2r-3, r-1$

Cartan matrix:

$$
\begin{pmatrix}
2 & -1 & 0 & \cdots & 0 & 0 & 0 \\
-1 & 2 & -1 & \cdots & 0 & 0 & 0 \\
0 & -1 & 2 & \cdots & 0 & 0 & 0 \\
\cdot & \cdot & \cdot & \cdots & \cdot & \cdot & \cdot \\
0 & 0 & 0 & \cdots & 2 & -1 & -1 \\
0 & 0 & 0 & \cdots & -1 & 2 & 0 \\
0 & 0 & 0 & \cdots & -1 & 0 & 2
\end{pmatrix}
$$

Quadratic form matrix:

$$\frac{1}{2}\begin{pmatrix}
2 & 2 & 2 & \cdots & 2 & 1 & 1 \\
2 & 4 & 4 & \cdots & 4 & 2 & 2 \\
2 & 4 & 6 & \cdots & 6 & 3 & 3 \\
\cdot & \cdot & \cdot & \cdots & \cdot & \cdot & \cdot \\
2 & 4 & 6 & \cdots & 2(r-2) & r-2 & r-2 \\
1 & 2 & 3 & \cdots & r-2 & r/2 & (r-2)/2 \\
1 & 2 & 3 & \cdots & r-2 & (r-2)/2 & r/2
\end{pmatrix}$$

E₈

dim g = 248
g = 30
|W| = 696729600
$\theta = (1, 0, \cdots, 0)$
$u = (58, 114, 168, 220, 270, 182, 92, 136)$
exponents = 1, 7, 11, 13, 17, 19, 23, 29

Cartan matrix:

$$\begin{pmatrix}
2 & -1 & 0 & 0 & 0 & 0 & 0 & 0 \\
-1 & 2 & -1 & 0 & 0 & 0 & 0 & 0 \\
0 & -1 & 2 & -1 & 0 & 0 & 0 & 0 \\
0 & 0 & -1 & 2 & -1 & 0 & 0 & 0 \\
0 & 0 & 0 & -1 & 2 & -1 & 0 & -1 \\
0 & 0 & 0 & 0 & -1 & 2 & -1 & 0 \\
0 & 0 & 0 & 0 & 0 & -1 & 2 & 0 \\
0 & 0 & 0 & 0 & -1 & 0 & 0 & 2
\end{pmatrix}$$

Quadratic form matrix:

$$\begin{pmatrix}
2 & 3 & 4 & 5 & 6 & 4 & 2 & 3 \\
3 & 6 & 8 & 10 & 12 & 8 & 4 & 6 \\
4 & 8 & 12 & 15 & 18 & 12 & 6 & 9 \\
5 & 10 & 15 & 20 & 24 & 16 & 8 & 12 \\
6 & 12 & 18 & 24 & 30 & 20 & 10 & 15 \\
4 & 8 & 12 & 16 & 20 & 14 & 7 & 10 \\
2 & 4 & 6 & 8 & 10 & 7 & 4 & 5 \\
3 & 6 & 9 & 12 & 15 & 10 & 5 & 8
\end{pmatrix}$$

E₇

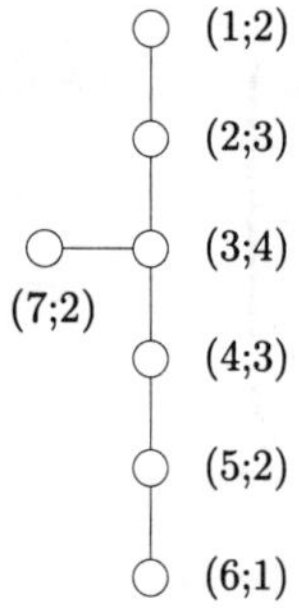

Dynkin diagram labels:
(1;2)
(2;3)
(3;4)
(7;2)
(4;3)
(5;2)
(6;1)

$\dim g = 133$
$g = 18$
$|W| = 2903040$
$\theta = (1, 0, \cdots, 0)$
$P/Q = \mathbb{Z}_2$
$v = (0, 0, 0, 1, 0, 1, 1)$
$u = (34, 66, 96, 75, 52, 27, 49)$
exponents $= 1,5,7,9,11,13,17$

Cartan matrix:

$$\begin{pmatrix} 2 & -1 & 0 & 0 & 0 & 0 & 0 \\ -1 & 2 & -1 & 0 & 0 & 0 & 0 \\ 0 & -1 & 2 & -1 & 0 & 0 & -1 \\ 0 & 0 & -1 & 2 & -1 & 0 & 0 \\ 0 & 0 & 0 & -1 & 2 & -1 & 0 \\ 0 & 0 & 0 & 0 & -1 & 2 & 0 \\ 0 & 0 & -1 & 0 & 0 & 0 & 2 \end{pmatrix}$$

Quadratic form matrix:

$$\frac{1}{2}\begin{pmatrix} 4 & 6 & 8 & 6 & 4 & 2 & 4 \\ 6 & 12 & 16 & 12 & 8 & 4 & 8 \\ 8 & 16 & 24 & 18 & 12 & 6 & 12 \\ 6 & 12 & 18 & 15 & 10 & 5 & 9 \\ 4 & 8 & 12 & 10 & 8 & 4 & 6 \\ 2 & 4 & 6 & 5 & 4 & 3 & 3 \\ 4 & 8 & 12 & 9 & 6 & 3 & 7 \end{pmatrix}$$

E₆

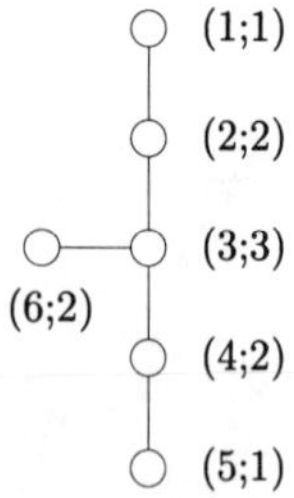

Dynkin diagram labels:
(1;1)
(2;2)
(3;3)
(6;2)
(4;2)
(5;1)

$\dim g = 78$
$g = 12$
$|W| = 51840$
$\theta = (0, 0, \cdots, 1)$
$P/Q = \mathbb{Z}_3$
$v = (1, 2, 0, 1, 2, 0)$
$u = (16, 30, 42, 30, 16, 22)$
exponents $= 1, 4, 5, 7, 8, 11$

Cartan matrix:

$$\begin{pmatrix} 2 & -1 & 0 & 0 & 0 & 0 \\ -1 & 2 & -1 & 0 & 0 & 0 \\ 0 & -1 & 2 & -1 & 0 & -1 \\ 0 & 0 & -1 & 2 & -1 & 0 \\ 0 & 0 & 0 & -1 & 2 & 0 \\ 0 & 0 & -1 & 0 & 0 & 2 \end{pmatrix}$$

Quadratic form matrix:

$$\frac{1}{3}\begin{pmatrix} 4 & 5 & 6 & 4 & 2 & 3 \\ 5 & 10 & 12 & 8 & 4 & 6 \\ 6 & 12 & 18 & 12 & 6 & 9 \\ 4 & 8 & 12 & 10 & 5 & 6 \\ 2 & 4 & 6 & 5 & 4 & 3 \\ 3 & 6 & 9 & 6 & 3 & 6 \end{pmatrix}$$

F₄

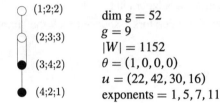

(1;2;2)
(2;3;3)
(3;4;2)
(4;2;1)

dim g = 52
g = 9
|W| = 1152
$\theta = (1,0,0,0)$
$u = (22, 42, 30, 16)$
exponents $= 1, 5, 7, 11$

Cartan matrix:

$$\begin{pmatrix} 2 & -1 & 0 & 0 \\ -1 & 2 & -2 & 0 \\ 0 & -1 & 2 & -1 \\ 0 & 0 & -1 & 2 \end{pmatrix}$$

Quadratic form matrix:

$$\begin{pmatrix} 2 & 3 & 2 & 1 \\ 3 & 6 & 4 & 2 \\ 2 & 4 & 3 & \frac{3}{2} \\ 1 & 2 & \frac{3}{2} & 1 \end{pmatrix}$$

G₂

dim g = 14
g = 4
|W| = 12
$\theta = (1,0)$
$u = (10, 6)$
exponents $= 1, 5$

(1;2;2)
(2;3;1)

Cartan matrix:

$$\begin{pmatrix} 2 & -3 \\ -1 & 2 \end{pmatrix}$$

Quadratic form matrix:

$$\frac{1}{3}\begin{pmatrix} 6 & 3 \\ 3 & 2 \end{pmatrix}$$

Appendix 13.B. Notation for Simple Lie Algebras

g, h : finite Lie algebras

G, H : corresponding Lie groups

dim g : dimension of the algebra g

r : rank

J^a $(a = 1, \cdots, \dim$ g$)$: generators of g

H^i $(i = 1, \cdots, r)$: generators of the Cartan subalgebra in the Cartan-Weyl basis

E^α : ladder generators in the Cartan-Weyl basis

h^i $(i = 1, \cdots, r)$: generators of the Cartan subalgebra in the Chevalley basis

e^i, f^i $(i = 1, \cdots, r)$: raising and lowering operators associated with the simple roots in the Chevalley basis

α, β : roots

α^i : i-th component of α in the Cartan-Weyl basis

α_i : simple roots

$\alpha_i^\vee$: simple coroots $= 2\alpha_i/\alpha_i^2$

A_{ij} : Cartan matrix element $= 2(\alpha_i, \alpha_j^\vee)$

Δ : set of roots

Δ_+, Δ_- : set of positive, negative roots

$|\Delta|, |\Delta_+|$: number of roots, number of positive roots

θ : highest root

a_i, $a_i^\vee$: marks and comarks; $\theta = \sum_{i=1}^r a_i\alpha_i = \sum_{i=1}^r a_i^\vee \alpha_i^\vee$

g : dual Coxeter number $= \sum_{i=1}^r a_i^\vee + 1$

λ, μ, ν : finite weights (usually highest weights)

dim$|\lambda|$: dimension of the representation of highest weight λ

Ω_λ : weight system of the representation of highest weight λ

L_λ : irreducible module of highest weight λ

λ', λ'' : particular weights in Ω_λ

$|\lambda'\rangle$: particular state, of weight λ', in the module L_λ

mult$_\lambda(\lambda')$: multiplicity of λ' in the highest-weight representation λ

ρ : Weyl vector (half-sum of positive roots)

ω_i : fundamental weights

F_{ij} : quadratic form matrix $= (\omega_i, \omega_j)$

λ_i : Dynkin labels: $\lambda = \sum_{i=1}^r \lambda_i\omega_i = (\lambda_1, \ldots, \lambda_r)$; h^i eigenvalues of $|\lambda\rangle$

λ^i : H^i eigenvalues of $|\lambda\rangle$

ϵ_i : orthonormal vectors

$\{\ell_1; \ell_2; \ldots; \ell_r\}$: partition of the Young tableau associated with the $su(N)$ weight $\lambda = \sum_i \ell_i\epsilon_i = \sum_i \lambda_i\omega_i$ so that $\ell_i = \lambda_i + \lambda_{i+1} + \ldots + \lambda_r = $ length of the i-th row (from top)

$\{\tilde{\ell}_1; \tilde{\ell}_2; \ldots; \tilde{\ell}_s\}$: transposed partition (change rows and columns)

λ^t : transposed weight

λ^* : conjugate of λ

χ_λ : character of the representation λ

W : Weyl group

$|W|$: order of the Weyl group

s_α : reflection with respect to the root α

s_i : reflection with respect to the simple root α_i (a simple Weyl reflection)

w : element of the Weyl group (a Weyl reflection)

w_0 : longest element of the Weyl group

$\epsilon(w)$: signature of w

$\ell(w)$: length of w

$w\cdot$: shifted Weyl reflection: $w \cdot \lambda = w(\lambda + \rho) - \rho$

C_w : Weyl chamber associated with the element w

Q : root lattice

$Q^\vee$: coroot lattice

P : weight lattice

P_+ : set of dominant weights (= set of highest weights for irreducible representations)

$\mathcal{N}_{\lambda\mu\nu} = \mathcal{N}_{\lambda\mu}{}^{\nu^*}$: tensor-product coefficients

$\mathcal{Q}$: quadratic Casimir operator

ad : adjoint operator; $\mathrm{ad}(X)Y = [X, Y]$

$K(\ ,\ )$: (normalized) Killing form; $K(X, Y) = \mathrm{Tr}(\mathrm{ad}X, \mathrm{ad}Y)/2g$

$\mathcal{K}$: Kostant partition function

x_λ : Dynkin index of the representation λ

x_e : embedding index

$b_{\lambda\mu}$: branching coefficient (multiplicity of L_μ in L_λ)

v : congruence vector

u : height vector

$B(G)$: center of the group G[14]

Exercises

13.1 *The Killing form*

a) Verify Eq. (13.18) and check that the only nonzero Killing norms are $K(H^i, H^i)$ and $K(E^\alpha, E^{-\alpha})$.

---

[14] Although the center of a group has not yet been defined, it has been included in order to make this list of symbols referring to Lie algebras complete.

b) Calculate the $su(2)$ Killing form $\tilde{K}$ in the Chevalley basis (13.85).
Result: With the ordering e, h, f, it reads

$$\tilde{K} = \begin{pmatrix} 0 & 0 & 4 \\ 0 & 8 & 0 \\ 4 & 0 & 0 \end{pmatrix}$$

A rescaling by a factor $\frac{1}{4} = 1/(2g)$ yields the standard normalization:

$$K(e,f) = K(f,e) = \frac{1}{2}K(h,h) = 1$$

13.2 *Weyl group for G_2 and $su(4)$*
Starting from the corresponding Cartan matrix given in App. 13.A, find the Weyl group and the set of all roots for:

a) G_2

b) $su(4)$

13.3 *Linear representation of the Weyl group*
The linear representation of the simple Weyl reflection s_i is the $r \times r$ matrix that maps the column vector with components λ_i to that with components $(s_i\lambda)_i$.

a) Show that $\det s_i = -1$. Deduce that for a general Weyl reflection w,

$$\det w = (-1)^\ell$$

where ℓ is the number of simple reflections in the decomposition of w.

b) Find the matrix representation of the simple reflections of G_2 and verify the relations (13.59).

c) Same as **(b)** for the algebra F_4.

13.4 *Order of the Weyl group*
Verify the following formula for the order of the Weyl group of a simple Lie algebra of rank r with marks $\{a_i\}$:

$$|W| = |P/Q|\, r! \prod_{i=1}^{r} a_i$$

Proceed case by case, using the data of App. 13.A.

13.5 *Weight systems*
Write all weights in the representation of highest weight:

a) $(1,0)$ of G_2,

b) $(0,0,1)$ of $so(7)$,

c) $(0,0,0,0,1)$ of $so(10)$.

13.6 *Weight multiplicities*
Find the multiplicity of the $su(4)$ weight $(-2,3,0)$ in the representation $(3,1,1)$ using:

a) the Freudenthal formula (13.113);

b) semistandard tableaux (cf. Sect. 13.3.3).

Hint: The calculation in **(a)** is greatly simplified if the weight is first transformed into a dominant one.

13.7 $su(3)$ *Gelfand-Tsetlin patterns*

a) Write all the Gelfand-Tsetlin patterns for the $su(3)$ representation of highest weight (2,2).

b) For a $su(3)$ weight $\lambda' \in \Omega_\lambda$, there corresponds $\text{mult}_\lambda(\lambda')$ Gelfand-Tsetlin patterns of the form

$$\begin{matrix} \lambda_1 + \lambda_2 & \lambda_2 & 0 \\ & a & b \\ & & c \end{matrix}$$

Relate the parameters a, b, c to the two Dynkin labels λ'_1, λ'_2. Find inequalities satisfied by the free parameter of the Gelfand-Tsetlin pattern, and deduce a simple formula for $\text{mult}_\lambda(\lambda')$. Compare with the example of part (**a**).

13.8 *The Demazure character formula*
An expression equivalent to Eq. (13.167) is given by

$$\chi_\lambda = M_{w_0}(e^\lambda)$$

where w_0 is the longest element of the Weyl group, and for $w_0 = s_i \cdots s_j$, $M_{w_0}(e^\lambda)$ is defined by

$$M_{w_0} = M_i \cdots M_j$$

with

$$M_i(e^\lambda) = \frac{e^\lambda - e^{s_i \cdot \lambda}}{1 - e^{-\alpha_i}}$$

(notice that the Weyl reflection is shifted), where, as usual, α_i stands for a simple root and

$$M_i M_j(e^\lambda) \equiv M_i(M_j(e^\lambda))$$

This is called the *Demazure character formula*.

a) Verify the following properties of M_i:

$$\begin{aligned} M_i(e^\lambda) &= e^\lambda + e^{\lambda+1} + \cdots + e^{\lambda-\lambda_i\alpha_i} & &\text{if} \quad \lambda_i \geq 0 \\ &= 0 & &\text{if} \quad \lambda_i = -1 \\ &= -e^{\lambda+\alpha_i} - e^{\lambda+\alpha_i+1} - \cdots - e^{\lambda-(\lambda_i+1)\alpha_i} & &\text{if} \quad \lambda_i \leq -2 \end{aligned}$$

and

$$(M_i)^2 = M_i$$

b) For $su(2)$, show that the Demazure formula is equivalent to the Weyl character formula.

c) Check the formula for the $su(3)$ representation $(1,2)$ (compare the result with Eq. (13.161)). For this representation, verify also that

$$M_{s_1 s_2 s_1}(e^\lambda) = M_{s_2 s_1 s_2}(e^\lambda)$$

d) Another version of the Demazure formula is

$$\chi_\lambda = \sum_{w \in W} N_w(e^\lambda)$$

where, in terms of a (minimal) decomposition of w in simple Weyl reflections, e.g., if $w = s_l \cdots s_k$, N_w is given by

$$N_w(e^\lambda) = N_l \cdots N_k(e^\lambda)$$

and

$$N_i(e^\lambda) = \frac{e^{s_i\lambda} - e^\lambda}{1 - e^{\alpha_i}}$$

Express N_i as a sum, as done in part (a) for M_i.

e) Evaluate the different $N_w(e^\lambda)$'s for the $su(3)$ highest weight $\lambda = (1, 2)$. Observe that each $N_w(e^\lambda)$ is a *positive sum*.

f) Prove the relation:

$$(1 + N_i)(e^\lambda) = M_i(e^\lambda)$$

13.9 *Dimension of G_2 representations*
Derive the dimension formula for the irreducible representations of G_2 and check that $L_{(0,1)}$ and $L_{(1,0)}$ have respective dimensions 7 and 14.

13.10 *Another expression for the dual Coxeter number*
Equations (13.181) and (13.184) lead to the following expression for the dual Coxeter number:

$$g = (2n_L + n_S)/2r \qquad \text{for} \quad g \neq G_2$$
$$= (3n_L + n_S)/3r \qquad \text{for} \quad G_2$$

where $n_{L,S}$ denotes the number of long and short roots, respectively. Verify this result for $sp(4)$ and G_2.
Remark: For simply laced algebras, this reduces to the relation: $|\Delta| = gr$.

13.11 *Cauchy determinant and Schur functions*
a) Show that

$$\phi(\{x\}, \{y\}) = \frac{\Delta(y)}{\prod_{1 \leq i,j \leq N}(1 - x_i y_j)}$$

where $\Delta(x) = \prod_{1 \leq i < j \leq N}(x_i - x_j)$, is a generating function for the Schur functions (13.189), namely

$$\phi(\{x\}, \{y\}) = \sum_{m_1, m_2, \ldots, m_N \geq 0} y_1^{m_1} \ldots y_N^{m_N} S_\lambda(x_1, \ldots, x_N)$$

where $\lambda = \{\ell_i\}$, and $\ell_i = m_i + i - N$.

b) By means of the Cauchy determinant formula (see Ex. 12.12 for a proof; take $z_i = 1/x_i$ and $w_j = y_j$ in the formula (12.195))

$$\det\left[\frac{1}{1 - x_i y_j}\right]_{1 \leq i,j \leq N} = \frac{\Delta(x)\Delta(y)}{\prod_{1 \leq i,j \leq N}(1 - x_i y_j)}$$

rewrite the generating function $\phi(\{x\}, \{y\})$ as the single determinant

$$\phi(\{x\}, \{y\}) = \frac{\Delta(y)}{\prod_{1 \leq i,j \leq N}(1 - x_i y_j)}$$

$$= \det\left[\frac{y_i^{N-j}}{\prod_{k=1}^{N}(1 - y_i x_k)}\right]_{1 \leq i,j \leq N}$$

Hint: Represent the quantity $\Delta(y)$ as a determinant (13.191).

c) The Schur polynomials of the variables $t_1, t_2, \ldots$ are defined through the generating function

$$F(y) = \sum_{m \geq 0} y^m P_m(t.) = e^{\sum_{k=1}^{\infty} y^k \frac{t_k}{k}}$$

This definition is supplemented by the convention that $P_m(t.) = 0$ for $m \leq -1$. Show that

$$F(y) = \prod_{k=1}^{N} \frac{1}{(1 - yx_k)}$$

iff the t_k are expressed as

$$t_k = \sum_{i=1}^{N} x_i^k$$

for some integer N.

d) Prove the following properties of the Schur polynomials

$$\frac{\partial}{\partial t_k} P_m(t.) = P_{m-k}(t.)$$

$$P_m(1) = \frac{1}{m!}$$

where 1 stands for $t_k = 1$ for all $k \geq 1$.

e) Express the generating function $\phi(\{x\}, \{y\})$ in terms of Schur polynomials. Deduce the following formula expressing the Schur functions as determinants of Schur polynomials of the variable $t_k = \sum_{i=1}^{N} x_i^k$.

$$S_\lambda(x_1, \ldots, x_N) = \det\left[P_{\ell_i + j - i}(t.) \right]_{1 \leq i, j \leq N}$$

13.12 Partitions and Schur functions

a) Work out the details of the derivation of Eqs. (13.189) and (13.192).

b) Prove directly the equivalence of Eqs. (13.192) and (13.172) by evaluating the scalar products in Eq. (13.172) in the orthogonal basis.

c) Find the action of the s_i's on the partitions.

13.13 Dimension of su(N) representations and hooks

The dimension of a representation can be read off a Young tableau in a rather simple way using *hooks*. The hook associated with the box at position (i, j) (i-th row, j-th column) is composed of two lines joined at right angle in the box (i, j) and leaving the tableau downward and toward the right. Its length, denoted by $h_{i,j}$, is the number of boxes it crosses. The following tableau is filled with the numbers $h_{i,j}$

$$h_{i,j}: \quad \begin{array}{|c|c|c|c|} \hline 6 & 5 & 2 & 1 \\ \hline 3 & 2 \\ \cline{1-2} 2 & 1 \\ \cline{1-2} \end{array}$$

In terms of hooks, the dimension of a $su(N)$ representation reads

$$\dim |\lambda| = \prod_{i,j} \frac{(N - i + j)}{h_{i,j}}$$

where the product is taken over all the boxes of the tableau.

a) Verify the equivalence of this formula with Eq. (13.192) for the above $su(4)$ tableau.

b) Using this expression, reproduce the $su(2)$ and $su(3)$ dimension formulae (13.172).

13.14 *sp(4) tensor product: character method*
Calculate the $sp(4)$ tensor product $(1, 1) \otimes (2, 0)$ using the character method and check the result by calculating the total dimension of each sides.

13.15 *Weyl-group folding in the character method*
Extending the validity of Eq. (13.171) to nondominant weights, prove that

$$\dim |w \cdot \lambda| = \epsilon(w) \dim |\lambda|$$

In the character method for tensor-product calculations, this shows that weights that are ignored have zero dimension, and two weights cancel each other if their dimensions add up to zero. Check this explicitly for the $su(3)$ example $(3, 2) \otimes (2, 4)$, to be worked out graphically using the algorithm underlying the character method.

13.16 *Littlewood-Richardson and Berenstein-Zelevinsky methods*

a) Using the Littlewood-Richardson method once and then the BZ triangles, calculate the following tensor products:

$$su(3) : (3, 2) \otimes (0, 3)$$
$$su(4) : (1, 0, 1) \otimes (1, 0, 1)$$

b) Using Littlewood-Richardson tableaux once and then the BZ triangles, find the multiplicity of the scalar representation in the following triple tensor products

$$su(3) : (4, 4) \otimes (4, 4) \otimes (4, 4)$$
$$su(4) : (2, 1, 1) \otimes (1, 2, 1) \otimes (1, 1, 2)$$

c) Observe that all the $su(3)$ triangles in **(b)** are related to each other by addition or subtraction of the "basic" triangle

$$\Omega = \begin{array}{ccccccc} & & & 1 & & & \\ & & -1 & & -1 & & \\ & -1 & & & & -1 & \\ 1 & & -1 & & -1 & & 1 \end{array}$$

Hence, once a triangle is found, all the others are readily generated. Relate this to a one-parameter indeterminacy in (13.212). Find the analogous result for $su(4)$ and compare with the example worked out in **(b)**.

d) Prove, using either Littlewood-Richardson tableaux or BZ triangles, that the $su(3)$ tensor-product coefficient $\mathcal{N}_{\lambda\mu\nu}$ is at most 1 if one of the three weights has at least one vanishing Dynkin label.

13.17 *Kostant's multiplicity formula*
The Weyl character formula leads directly to a new expression for weight multiplicities, Kostant's formula. For this, we introduce the partition function $\mathcal{K}(\mu)$ defined to be the number of distinct decompositions of μ in terms of positive roots. In other words, $\mathcal{K}(\mu)$ is

the number of solutions $\{k_\alpha\}$, $\alpha \in \Delta_+$ of the equation $\sum_{\alpha>0} k_\alpha \alpha = \mu$, with all $k_\alpha \geq 0$. Of course, if there is no such decomposition, $\mathcal{K}(\mu) = 0$. Setting $\mathcal{K}(0) = 1$, we have

$$\prod_{\alpha>0} \frac{1}{1 - e^\alpha} = \sum_\mu \mathcal{K}(\mu) e^\mu$$

In terms of this partition function, show that the multiplicity of the weight λ' in the representation λ is given by

$$\text{mult}_\lambda(\lambda') = \sum_{w \in W} \epsilon(w) \mathcal{K}(w(\lambda + \rho) - (\lambda' + \rho))$$

Hint: Use the product form of D_ρ^{-1} to relate it to the partition function $\mathcal{K}$.
The advantage of Kostant's formula over Freudenthal's is that a given weight can be treated in isolation. The price that has to be paid is a sum over the whole Weyl group. Nevertheless, in favorable circumstances only a few terms contribute. Illustrate this by calculating the multiplicity of the weight $(0,0)$ in the adjoint representation of $su(3)$.

13.18 *Steinberg formula for tensor products*
Use the Kostant multiplicity formula to obtain the Steinberg formula for tensor-product coefficients:

$$\mathcal{N}_{\lambda\mu}{}^\nu = \sum_{w,w' \in W} \epsilon(ww') \mathcal{K}(w \cdot \lambda + w' \cdot \mu - \nu)$$

13.19 *Associativity in tensor products*
Tensor product coefficients can be calculated from the fusion coefficients involving fundamental weights, that is, $\{\mathcal{N}_{\lambda\mu}{}^{\omega_i}\}$ for $i = 1, \cdots, r$ and any λ, μ, and the associativity condition (13.224). Illustrate this by calculating, from these data, the $su(3)$ coefficient $\mathcal{N}_{(1,1)(1,1)}{}^{(1,1)}$.

13.20 *Generalized Chebyshev polynomials and tensor products*

a) Verify the relations (13.239), regarded as the defining recursion relations for the generalized Chebyshev polynomials $U_{(\lambda_1,\lambda_2)}$, associated with the tensor-product matrix $N_{(\lambda_1,\lambda_2)}$. Check further that

$$U_{(\lambda_1,\lambda_2)} = U_{(\lambda_1,0)}U_{(0,\lambda_2)} - U_{(\lambda_1-1,0)}U_{(0,\lambda_2-1)}$$

for $\lambda_1, \lambda_2 > 1$. Argue that the matrices $N_{(1,0)}$ and $N_{(0,1)}$ must commute. Use these relations to obtain the generating function (13.241).

b) Derive analogous results for $sp(4)$. With $N_{(1,0)} = x_1$ and $N_{(0,1)} = x_2$, the generating function $F(x_1,x_2;t,s)$ is

$$\frac{1 + s(t^2 + 1) + s^2t^2 - tsx_1}{(1 + t^2 + t^4 - x_1(t^3 + t) + t^2x_2)(1 + s + s^3 + s^4 - x_2(s + 2s^2 + s^3) + s^2x_1^2)}$$

13.21 *Verlinde formula for a Lie algebra*
Check carefully the derivation of the orthogonality relation (13.244). Use the Verlinde formula (13.245) to recover the $su(2)$ tensor-product matrices N_1 and N_2.

13.22 *Embedding index*
a) Prove the relation (13.251).

b) For the embedding $E_8 \supset su(2) \oplus su(3)$, calculate the embedding index, using the branching rule:

$$(1,0,0,0,0,0,0,0) \mapsto \{(6) \otimes (1,1)\} \oplus \{(4) \otimes (3,0)\}$$

c) For the embedding $so(7) \supset su(4)$, calculate the embedding index, using the projection matrix:

$$\mathcal{P} = \begin{pmatrix} 0 & 1 & 1 \\ 1 & 0 & 0 \\ 0 & 1 & 0 \end{pmatrix}$$

13.23 *Embeddings of su(2)*

a) Describe all possible embeddings of $su(2)$ in $sp(4)$. In each case, find the branching rule for (1,0), the projection matrix and the embedding index.

b) Same as (**a**) for the embeddings $su(2) \subset G_2$, using the representation (0,1).

13.24 *Regular maximal subalgebras*

Find all regular maximal subalgebras of F_4, E_6, and E_7.

13.25 *Branching rules in regular embeddings*

a) Consider the regular embedding $su(3) \subset G_2$. Draw the extended Dynkin diagram of G_2 (i.e., calculate the number of links between the new root $-\theta$ and α_1, α_2). Identify the node that must be deleted to recover the $su(3)$ Dynkin diagram. Write all the weights in the (0, 1) representation of G_2 and their extended Dynkin labels $[\lambda_{-\theta}, \lambda_1, \lambda_2]$, where

$$\lambda_{-\theta} = -2\lambda_1 - \lambda_2$$

(cf. Eq. (13.268)). Delete the Dynkin label appropriate for the $su(3)$ embedding and reorganize the resulting $su(3)$ weights in irreducible representations. This gives the branching of the (0, 1) G_2 representation into $su(3)$ ones.

b) By proceeding similarly for the regular embedding $su(4) \subset so(7)$, find the branching of the $so(7)$ representation $(1,0,0)$.

Notes

Except for some aspects of tensor-product calculations and tableaux techniques, the content of this chapter is rather standard. It is covered, for instance, in Cahn [61], Wybourne [361], Fulton and Harris [155], Jacobson [209], Humphreys [196], Bourbaki [56], and Zelobenko [368]. The book of Cahn provides a clear and concise first introduction to the subject, and that of Fulton and Harris is a particularly readable mathematical textbook; tableaux techniques are well covered there. A sharp focus on the material presented in Sects. 13.1 and 13.2 can be found in those sections of Kass et al. [228] related to finite Lie algebras. The theory of semisimple Lie algebras is also well summarized in the first chapter of Fuchs [148]. The proof of the strange formula follows Freudenthal and de Vries [138]. The relation between semistandard tableaux and Gelfand-Tsetlin patterns can be found in Ref. [193].

The character method for tensor products is presented in Racah [301], Speiser [329], and Klimyk [239]. The relation between Littlewood-Richardson tableaux and Gelfand-Tsetlin patterns can be found in Gelfand and Zelevinsky [164]. It is equivalent to the method for calculating tensor-product coefficients by means of semistandard tableaux, which is presented in [257, 354, 278]. Berenstein-Zelevinsky triangles were introduced in Ref. [38] and further developed in Refs. [74, 39].

The basics of algebra embeddings are explained in Cahn [61]. For a more detailed discussion, the reader is referred to the original articles of Dynkin [117, 118]. The generating functions for the embeddings of $su(2)$ into $su(3)$ (and many others) can be found in Patera and Sharp [291].

The Demazure formula of Ex. 13.8 is proved in Ref. [90] (see also Ref. [163]).

Our conventions and most of our notations follow mainly that of Patera and collaborators [268, 59], which makes easier the consultation of these extensive and very useful tables of weight multiplicities, dimensions of representations, branching rules, and so forth.

CHAPTER 14

Affine Lie Algebras

This chapter is a basic introduction to affine Lie algebras, preparing the stage for their application to conformal field theory. In Sect. 14.1.1, after having introduced the affine Lie algebras per se, we show how the fundamental concepts of roots, weights, Cartan matrices, and Weyl groups are extended to the affine case. Section 14.2 introduces the outer automorphism group of affine Lie algebras, which is generated by the new symmetry transformations of the extended Dynkin diagram. The following section describes highest-weight representations, focusing on those whose highest weight is dominant. Characters for these representations are introduced in Sect. 14.4. Their modular properties are presented in the following sections, where various properties of their modular S matrices are also reported. The affine extension of finite Lie algebra embeddings is presented in Sect. 14.7. Four appendices complete the chapter. The first one contains the proof of a technical identity related to outer automorphism groups. The second appendix displays an explicit basis (in terms of semi-infinite paths) for the states in integrable representations of affine $su(N)$. In the third one, the modular transformation properties of the affine characters are derived. The final appendix lists all the symbols pertaining to affine Lie algebras.

The minimal background required for proceeding to Chap. 15, which initiates the analysis of affine Lie algebras in the context of conformal field theory, is contained in Sects. 14.1.1, 14.3.1, 14.4.1, and 14.5. The remaining sections could be consulted when needed.

The next few sentences give a flavor of the relevant aspects of the theory of affine Lie algebras. To every (finite) Lie algebra g, we associate an affine extension $\hat{g}$ by adding to the Dynkin diagram of g an extra node, related to the highest root θ. The introduction of this particular simple root has the immediate effect of making the root system (and thereby the Weyl group) of $\hat{g}$ infinite. As a result, highest-weight representations are infinite dimensional. However, as a simplifying feature, these representations are organized in terms of a new parameter, called the level, which plays a role analogous to that of the central charge in the Virasoro algebra. The level of a weight, described now by $r + 1$ Dynkin labels, is the sum of all its Dynkin labels, each multiplied by its corresponding comark. For affine algebras, comarks are thus data of prime importance. Integrable highest-weight representations occur for positive integer values of the level. Moreover, the corresponding highest weights have nonnegative integer Dynkin labels. For a fixed level, there is thus a finite number of integrable representations. Quite remarkably, their characters transform into each other under modular transformations.

§14.1. The Structure of Affine Lie Algebras

14.1.1. From Simple Lie Algebras to Affine Lie Algebras

We consider the generalization of g in which the elements of the algebra are also Laurent polynomials in some variable t. The set of such polynomials is denoted by $\mathbb{C}[t, t^{-1}]$. This generalization is called the *loop* algebra $\tilde{g}$:[1]

$$\tilde{g} = g \otimes \mathbb{C}[t, t^{-1}] \tag{14.1}$$

with generators $J^a \otimes t^n$. The algebra multiplication rule extends naturally from g to $\tilde{g}$ as

$$[J^a \otimes t^n, J^b \otimes t^m] = \sum_c if^{ab}{}_c J^c \otimes t^{n+m} \tag{14.2}$$

A central extension is obtained by adjoining to $\tilde{g}$ a *central element*

$$[J^a \otimes t^n, J^b \otimes t^m] = \sum_c if^{ab}{}_c J^c \otimes t^{n+m} + \hat{k} n K(J^a, J^b)\delta_{n+m,0} \tag{14.3}$$

where $\hat{k}$ commutes with all J^a's, and K is the Killing form of g. Assuming as usual that the generators J^a are orthonormal with respect to the Killing form, and using the notation

$$J_n^a \equiv J^a \otimes t^n \tag{14.4}$$

we can rewrite the above commutation relation in the form

$$\boxed{[J_n^a, J_m^b] = \sum_c if^{ab}{}_c J_{n+m}^c + \hat{k} n \delta_{ab}\delta_{n+m,0}} \tag{14.5}$$

This must be supplemented by

$$[J_n^a, \hat{k}] = 0 \tag{14.6}$$

The above introduction of the central extension may appear to be somewhat ad hoc. The following considerations demonstrate its uniqueness. We start with the generic commutator

$$[J_n^a, J_m^b] = \sum_c if^{ab}{}_c J_{n+m}^c + \sum_{i=1}^{\ell} \hat{k}^i (d_i^{ab})_{nm} \tag{14.7}$$

---

[1] With $t = e^{i\gamma}$ and γ real, this yields a map from the circle S^1 to g, hence the name "loop."

containing ℓ central terms. With the representation (14.4), it is clear that the central terms can occur only for $n + m = 0$. (Otherwise they could be eliminated by a redefinition of the generators, exactly as in the finite case in which central extensions are trivial.) This shows that

$$[J_0^a, J_n^b] = \sum_c i f^{ab}{}_c J_n^c \tag{14.8}$$

meaning that the generators $\{J_n^a\}$ transform in the adjoint representation of g (i.e., under the action of $\mathrm{ad}(J_0^a)$, where $\mathrm{ad}(X)Y = [X, Y]$, J_n^b transforms exactly like J_0^b). That the central extensions commute with all the generators J_n^a means that they are invariant tensors of the adjoint representation. But up to normalization, there is only one such tensor, the Killing form itself.[2] Hence, only one central element can possibly be added to the loop extension of a simple Lie algebra. In a basis in which the generators are orthonormal with respect to the Killing form, it is simple to check that the only central extension compatible with the antisymmetry of the commutators and the Jacobi identities is the one given in Eq. (14.5).

To analyze this new algebra, it is useful to rewrite the commutation relations (14.5) in the affine Cartan-Weyl basis. With the nonzero Killing norms being

$$K(H^i, H^j) = \delta^{i,j}, \quad K(E^\alpha, E^{-\alpha}) = \frac{2}{|\alpha|^2} \tag{14.9}$$

the commutation relations read

$$
\begin{aligned}
&[H_n^i, H_m^j] = \hat{k} n \delta^{ij} \delta_{n+m,0} \\
&[H_n^i, E_m^\alpha] = \alpha^i E_{n+m}^\alpha \\
&[E_n^\alpha, E_m^\beta] = \frac{2}{\alpha^2}\left(\alpha \cdot H_{n+m} + \hat{k} n \delta_{n+m,0}\right) \quad \text{if} \quad \alpha = -\beta \\
&\qquad\quad = \mathcal{N}_{\alpha,\beta} E_{n+m}^{\alpha+\beta} \quad\qquad\qquad \text{if} \quad \alpha + \beta \in \Delta \\
&\qquad\quad = 0 \qquad\qquad\qquad\qquad\qquad \text{otherwise}
\end{aligned}
\tag{14.10}
$$

with the generators H_n^i and E_n^α defined as in Eq. (14.4) (Δ is the set of roots of g).

The set of generators $\{H_0^1, \cdots, H_0^r, \hat{k}\}$ is manifestly Abelian. In the adjoint representation, in which the action of a generator X is represented by $\mathrm{ad}(X)$, the eigenvalues of $\mathrm{ad}(H_0^i)$ and $\mathrm{ad}(\hat{k})$ on the generator E_n^α are respectively α^i and 0. Being independent of n, the eigenvector $(\alpha^1, \cdots, \alpha^r, 0)$ is thus infinitely degenerate (i.e., it is the same for all the E_m^α's). Hence, $\{H_0^1, \cdots, H_0^r, \hat{k}\}$ is not a maximal Abelian subalgebra. It must be augmented by the addition of a new grading operator L_0, whose eigenvalues in the adjoint representation depend upon n; it is defined as follows:[3]

$$L_0 = -t\frac{d}{dt} \tag{14.11}$$

[2] This invariance property is essentially Eq. (13.17), which characterizes $K(\cdot, \cdot)$ up to a rescaling.
[3] $-L_0$ is usually denoted by d in the mathematical literature.

Its action on the generators is

$$\text{ad}(L_0)J^a \otimes t^n = [L_0, J^a \otimes t^n] = -nJ^a \otimes t^n \quad \Longrightarrow \quad [L_0, J_n^a] = -nJ_n^a \quad (14.12)$$

The maximal Cartan subalgebra is generated by $\{H_0^1, \cdots, H_0^r, \hat{k}, L_0\}$. The other generators, E_n^α for any n and H_n^i for $n \neq 0$, play the role of ladder operators.

With the addition of the operator L_0, the resulting algebra is denoted by $\hat{g}$

$$\hat{g} = \tilde{g} \oplus \mathbb{C}\hat{k} \oplus \mathbb{C}L_0 \quad (14.13)$$

It will be referred to as an *affine* Lie algebra.[4] It is clearly an infinite dimensional algebra, given that it has an infinite number of generators $\{J_n^a\}, n \in \mathbb{Z}$. From the perspective of the affine algebra, g will be referred to as the corresponding *finite* algebra. Its generators are the zero modes $\{J_0^a\}$.

An already familiar infinite-dimensional algebra is the one generated by the modes of a free boson:

$$[a_n, a_m] = n\delta_{n+m,o} \quad (14.14)$$

It is usually referred to as the *Heisenberg algebra*, and is simply the affine extension of the $u(1)$ algebra generated by the element a_0. Comparison of the above commutation relation with Eq. (14.5) seems to indicate that the level is equal to one. However, the central term can be changed at will by a rescaling of the modes: this shows that the level has no meaning in the $\widehat{u}(1)$ case.

14.1.2. The Killing Form

To parallel the development of the theory of Lie algebra, we must first equip $\hat{g}$ with a scalar product. This amounts to extending the definition of the Killing form from g to $\hat{g}$. Again the key relation is the extension of (13.17) to $\hat{g}$, which expresses the $\hat{g}$ invariance of this bilinear form—with now $X, Y, Z \in \hat{g}$. With $X, Y \in \{J_n^a\}$ and $Z = L_0$, we have

$$K(J_n^a, J_m^b) = 0 \quad \text{unless} \quad n + m = 0 \quad (14.15)$$

The identification (14.4) shows that when $n + m = 0$ the t factors disappear; we are thus left with the g Killing form, implying that

$$K(J_n^a, J_m^b) = \delta^{ab}\delta_{n+m,0} \qu(14.16)$$

We note that the affine Killing form is still orthonormal with respect to the finite algebra indices; from now on, we will no longer care about the position of these indices.[5] The choice $X, Z \in \{J_n^a\}$ and $Y = \hat{k}$ yields

$$K(J_n^a, \hat{k}) = 0 \qquad \text{and} \qquad K(\hat{k}, \hat{k}) = 0 \quad (14.17)$$

---

[4] In the mathematical literature, it is called a nontwisted (or direct) affine Lie algebra. (Twisted algebras will not be considered here.) In the physics literature, they are often called Kac-Moody algebras. However, the name Kac-Moody is usually attached to a more general construction.

[5] In particular, we will identify f_{abc} and $f^{ab}{}_c$ as in the finite case.

whereas $Y = L_0$ leads to

$$K(J_n^a, L_0) = 0 \qquad \text{and} \qquad K(L_0, \hat{k}) = -1 \qquad (14.18)$$

The only unspecified norm is $K(L_0, L_0)$, which, by convention, is chosen to be zero

$$K(L_0, L_0) = 0 \qquad (14.19)$$

The arbitrariness of this norm is related to the possibility of redefining L_0 as

$$L_0 \to L_0' = L_0 + a\hat{k} \qquad (14.20)$$

where a is some constant, without affecting the algebra. It changes its Killing norm by only $-2a$.

As in the finite case, the Killing form leads to an isomorphism between the elements of the Cartan subalgebra and those of its dual, and defines for the latter a scalar product. Let the components of the vector $\hat{\lambda}$ be the eigenvalues of a state that is a simultaneous eigenvector of all the generators of the Cartan subalgebra:

$$\hat{\lambda} = (\hat{\lambda}(H_0^1), \hat{\lambda}(H_0^2), \cdots, \hat{\lambda}(H_0^r); \hat{\lambda}(\hat{k}); \hat{\lambda}(-L_0)) \qquad (14.21)$$

The first r components characterize the finite part λ of the weight $\hat{\lambda}$ [6]

$$\hat{\lambda} = (\lambda; k_\lambda; n_\lambda) \qquad (14.22)$$

(We note that the last entry refers to $-L_0$). The scalar product induced by the extended Killing form is

$$(\hat{\lambda}, \hat{\mu}) = (\lambda, \mu) + k_\lambda n_\mu + k_\mu n_\lambda \qquad (14.23)$$

$\hat{\lambda}$ is called an *affine weight*.

As for Lie algebras, weights in the adjoint representation are called roots. Since $\hat{k}$ commutes with all the generators of $\hat{g}$, its eigenvalue on the states of the adjoint representation is equal to zero. Hence, affine roots are of the form

$$\hat{\beta} = (\beta; 0; n) \qquad (14.24)$$

Their scalar product is thus exactly the same as in the finite case

$$(\hat{\beta}, \hat{\alpha}) = (\beta, \alpha) \qquad (14.25)$$

The affine root associated with the generator E_n^α is

$$\hat{\alpha} = (\alpha; 0; n) \qquad n \in \mathbb{Z}, \quad \alpha \in \Delta \qquad (14.26)$$

If we let

$$\delta = (0; 0; 1) \qquad (14.27)$$

---

[6] The weight λ is given here in terms of its components in the Cartan-Weyl basis of g. It could equally well be expressed in terms of its components in the Chevalley basis, in which case the first r components of $\hat{\lambda}$ would be the Dynkin labels of λ.

then, $n\delta$ is the root associated with H^i_n. In the following we write

$$\alpha \equiv (\alpha; 0; 0) \tag{14.28}$$

so that the roots (14.26) can be reexpressed as

$$\boxed{\hat{\alpha} = \alpha + n\delta} \tag{14.29}$$

The full set of roots is

$$\hat{\Delta} = \{\alpha + n\delta| \; n \in \mathbb{Z}, \alpha \in \Delta\} \cup \{n\delta| \; n \in \mathbb{Z}, n \neq 0\} \tag{14.30}$$

The root δ is rather unusual since it has zero length

$$(\delta, \delta) = 0 \tag{14.31}$$

For this reason, it is often called an *imaginary* root. Likewise all the roots in the set $\{n\delta\}$ are imaginary and $(n\delta, m\delta) = 0$ for all n, m. All these imaginary roots have multiplicity r. The other roots are then said to be *real*, and they have multiplicity 1.

14.1.3. Simple Roots, the Cartan Matrix and Dynkin Diagrams

The next step is the identification of a basis of simple roots for the affine algebra. In such a basis, the expansion coefficients of any root are either all positive or all negative. This basis must contain $r + 1$ elements, r of which are necessarily the finite simple roots α_i, whereas the remaining simple root must be a linear combination involving δ. The proper choice for this extra simple root is

$$\boxed{\alpha_0 \equiv (-\theta; 0; 1) = -\theta + \delta} \tag{14.32}$$

where θ is the highest root of g. The correct basis of simple roots is thus $\{\alpha_i\}, i = 0, \cdots, r$. The set of positive roots is

$$\hat{\Delta}_+ = \{\alpha + n\delta| \; n > 0, \alpha \in \Delta\} \cup \{\alpha| \alpha \in \Delta_+\} \tag{14.33}$$

Indeed, for $n > 0$ and $\alpha \in \Delta$,

$$\alpha + n\delta = \alpha + n\alpha_0 + n\theta = n\alpha_0 + (n - 1)\theta + (\theta + \alpha) \tag{14.34}$$

and the expansion coefficients of the last two factors in terms of finite simple roots are necessarily nonnegative. Notice that in the affine case there is no highest root (i.e., the adjoint representation is not a highest-weight representation).

Given a set of affine simple roots and a scalar product, we can define the *extended Cartan matrix* as

$$\boxed{\widehat{A}_{ij} = (\alpha_i, \alpha_j^\vee) \qquad 0 \leq i, j \leq r} \tag{14.35}$$

where affine coroots are given by

$$\hat{\alpha}^\vee = \frac{2}{|\hat{\alpha}|^2}(\alpha; 0; n) = \frac{2}{|\alpha|^2}(\alpha; 0; n) = (\alpha^\vee; 0; \frac{2}{|\alpha|^2}n) \tag{14.36}$$

As for simple roots, the hat is omitted over the simple coroots, e.g.,

$$\alpha_0^\vee = \alpha_0 \qquad \alpha_i^\vee \equiv (\alpha_i^\vee; 0; 0) \quad i \neq 0 \tag{14.37}$$

Compared to the finite Cartan matrix, $\widehat{A}_{ij}$ contains an extra row and column. These additional entries are easily calculated in terms of the marks defined in Eq. (13.33) since $(\alpha_0, \alpha_0^\vee) = |\theta|^2 = 2$ and

$$(\alpha_0, \alpha_j^\vee) = -(\theta, \alpha_j^\vee) = -\sum_{i=1}^{r} a_i (\alpha_i, \alpha_j^\vee) \tag{14.38}$$

Again, all the information contained in extended Cartan matrices can be encoded in *extended Dynkin diagrams*. The Dynkin diagram of $\hat{g}$ is obtained from that of g by the addition of an extra node representing α_0. This extra node is linked to the α_i-nodes by $\widehat{A}_{0i}\widehat{A}_{i0}$ lines. Since the finite part of α_0 is not linearly independent of the finite simple roots, the product $\widehat{A}_{0i}\widehat{A}_{i0}$ can now take the value 4 (although this occurs only for $\widehat{su}(2)$). The affine extension of the simple Lie Dynkin diagrams are displayed in Fig. 14.1. The numbers next to each node are respectively the numbering of the simple roots, the marks, and the comarks. For simply-laced algebras, for which marks and comarks are identical, the third entry is omitted. Extended Dynkin diagrams obviously have more symmetry than their finite version, a point we will discuss in some detail later.

For future reference, we mention that the zeroth mark a_0 is defined to be 1. Since the finite part of α_0 is a long root, so that $|\alpha_0|^2 = 2$, the zeroth comark is also 1:

$$a_0^\vee = a_0 \frac{|\alpha_0|^2}{2} = 1 \tag{14.39}$$

By construction the extended Cartan matrix satisfies

$$\sum_{i=0}^{r} a_i \widehat{A}_{ij} = \sum_{i=0}^{r} \widehat{A}_{ij} a_j^\vee = 0 \tag{14.40}$$

The linear dependence between the rows of the extended Cartan matrix means that it has one zero eigenvalue, a reflection of the semipositive character of the affine scalar product.[7] The imaginary root can now be written in the form

$$\delta = \sum_{i=0}^{r} a_i \alpha_i = \sum_{i=0}^{r} a_i^\vee \alpha_i^\vee \tag{14.41}$$

Similarly, the dual Coxeter number reads

$$g = \sum_{i=0}^{r} a_i^\vee \tag{14.42}$$

---

[7] From the point of view of the Cartan matrix A, the generalization from simple to affine Lie algebras can be described as follows. Let D be the diagonal matrix with entries $2/|\alpha_i|^2$; the product DA is thus symmetric. For simple Lie algebras, DA is positive definite; this constraint is relaxed to positive semidefiniteness in the affine case (meaning then that DA has one zero eigenvalue). Affine Lie algebras are special cases of Kac-Moody algebras, which allow for even more general Cartan matrices.

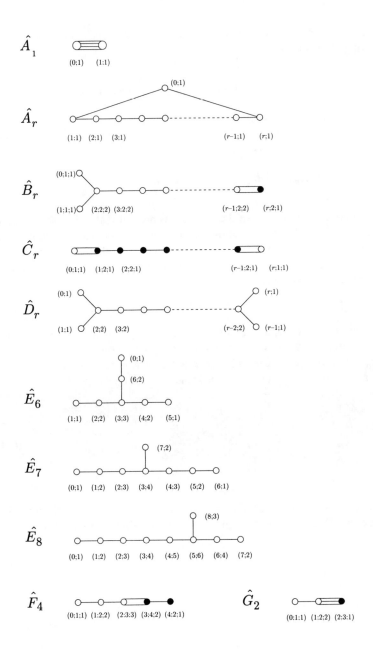

Figure 14.1. Affine Dynkin diagrams. The numbers give respectively the ordering of the simple roots, their mark, and comark (written only for the nonsimply-laced algebras). Black dots refer to short roots.

14.1.4. The Chevalley Basis

The commutation relations of the generators in the Chevalley basis have the following affine extension

$$
\begin{aligned}
[h_n^i, h_m^j] &= (\alpha_i^\vee, \alpha_j^\vee) k n \delta_{ij} \delta_{n+m,0} = \frac{4}{\alpha_i^2} k n \delta_{ij} \delta_{n+m,0} \\
[h_n^i, e_m^j] &= A_{ji} e_{n+m}^j \\
[h_n^i, f_m^j] &= -A_{ji} f_{n+m}^j \\
[e_n^i, f_m^j] &= \delta_{ij} h_{n+m}^i + \frac{2}{\alpha_i^2} k n \delta_{ij} \delta_{n+m,0}
\end{aligned}
\tag{14.43}
$$

with $i, j = 1, \cdots, r$. However, these relations are not really the affine analogues of the finite Chevalley commutation relations in the sense that they do not involve only the generators of the $r + 1$ simple roots of $\hat{g}$ and they are not expressed in terms of the $\hat{g}$ Cartan matrix. In order to construct a genuine affine Chevalley basis, we need only to add the generators

$$
e^0 = E_1^{-\theta}, \qquad f^0 = E_{-1}^{\theta}, \qquad h^0 = \hat{k} - \theta \cdot H_0 \tag{14.44}
$$

to the set of finite generators e^i and f^i (i.e., e^0 and f^0 are respectively the raising and lowering operators for α_0). From now on, we will omit the mode index of e_0^i and f_0^i with $i \neq 0$ (the g Chevalley generators). The commutation relation for the generators associated with the simple roots of $\hat{g}$ can be written as

$$
\begin{aligned}
[h^i, h^j] &= 0 \\
[h^i, e^j] &= \widehat{A}_{ji} e^j \\
[h^i, f^j] &= -\widehat{A}_{ji} f^j \\
[e^i, f^j] &= \delta_{ij} h^j
\end{aligned}
\tag{14.45}
$$

where now $i, j = 0, 1, \cdots, r$. For instance, $[e^0, f^i] = 0$ if $i \neq 0$ because $-\theta - \alpha_i$ is not a root. These are to be supplemented by the affine Serre relations

$$
\begin{aligned}
[\mathrm{ad}(e^i)]^{1-\widehat{A}_{ji}} e^j &= 0 \\
[\mathrm{ad}(f^i)]^{1-\widehat{A}_{ji}} f^j &= 0
\end{aligned}
\tag{14.46}
$$

with $i \neq j$. This form makes manifest that $\widehat{A}$ encodes the whole structure of $\hat{g}$. However, it does not make apparent the infinite-dimensional nature of $\hat{g}$.

14.1.5. Fundamental Weights

As in the finite case, the fundamental weights $\{\hat{\omega}_i\}, 0 \le i \le r$ are defined to be the elements of the basis dual to the simple coroots. The fundamental weights are assumed to be eigenstates of L_0 with zero eigenvalue. For $i \neq 0$, these are

$$\boxed{\hat{\omega}_i = (\omega_i; a_i^\vee; 0) \qquad (i \neq 0)} \qquad (14.47)$$

Their finite part makes them dual to the finite simple roots, while the $\hat{k}$ eigenvalue is fixed by the condition

$$(\hat{\omega}_i, \alpha_0^\vee) = 0 \qquad (i \neq 0) \qquad (14.48)$$

The zeroth fundamental weight, on the other hand, must have zero scalar product with all finite α_i's and satisfy $(\hat{\omega}_0, \alpha_0^\vee) = 1$. Hence, it must be

$$\boxed{\hat{\omega}_0 = (0; 1; 0)} \qquad (14.49)$$

It is called the *basic fundamental weight*. With

$$\omega_i \equiv (\omega_i; 0; 0) \qquad (14.50)$$

it follows that

$$\hat{\omega}_i = a_i^\vee \hat{\omega}_0 + \omega_i \qquad (14.51)$$

The scalar product between the fundamental weights is

$$\begin{aligned} (\hat{\omega}_i, \hat{\omega}_j) &= (\omega_i, \omega_j) = F_{ij} \qquad (i, j \neq 0) \\ (\hat{\omega}_0, \hat{\omega}_i) &= (\hat{\omega}_0, \hat{\omega}_0) = 0 \qquad (i \neq 0) \end{aligned} \qquad (14.52)$$

where F_{ij} is the quadratic form matrix of g.

Affine weights can thus be expanded in terms of the affine fundamental weights and δ as

$$\hat{\lambda} = \sum_{i=0}^r \lambda_i \hat{\omega}_i + \ell \delta \qquad \ell \in \mathbb{R} \qquad (14.53)$$

Since each fundamental weight contributes to the $\hat{k}$ eigenvalue by a factor $a_i^\vee$, we have

$$\boxed{k \equiv \hat{\lambda}(\hat{k}) = \sum_{i=0}^r a_i^\vee \lambda_i} \qquad (14.54)$$

k is called the *level*. This relation could also have been derived directly as follows:

$$(\hat{\lambda}, \delta) = \hat{\lambda}(\hat{k}) = \sum_{i=0}^r a_i^\vee (\hat{\lambda}, \alpha_i^\vee) = \sum_{i=0}^r a_i^\vee \lambda_i \qquad (14.55)$$

The first equality is obtained from $\delta = (0; 0; 1)$, $\hat{\lambda}$ defined by Eq. (14.21) and the scalar product (14.23), whereas the second one uses $\delta = \sum_{i=0}^r a_i^\vee \alpha_i^\vee$ and the expansion of $\hat{\lambda}$ in terms of fundamental weights. It implies that the zeroth Dynkin label λ_0 is related to the finite Dynkin labels $\{\lambda_i\}$, $i = 1, \cdots, r$ and the level by

$$\lambda_0 = \hat{\lambda}(\hat{k}) - \sum_{i=1}^{r} a_i^{\vee} \lambda_i \tag{14.56}$$

(because $a_0^{\vee} = 1$), that is,

$$\boxed{\lambda_0 = k - (\lambda, \theta)} \tag{14.57}$$

Modulo a possible δ factor, the relation between $\hat{\lambda}$ and its finite counterpart is simply

$$\hat{\lambda} = k\hat{\omega}_0 + \lambda \tag{14.58}$$

We note that roots are weights at level zero.

Affine weights will generally be given in terms of Dynkin labels under the form

$$\hat{\lambda} = [\lambda_0, \lambda_1, \cdots, \lambda_r] \tag{14.59}$$

(However, we stress that this notation does not keep track of the eigenvalue of L_0.) For instance,

$$\hat{\omega}_0 = [1, 0, \cdots, 0], \quad \hat{\omega}_1 = [0, 1, \cdots, 0], \quad \hat{\omega}_r = [0, 0, \cdots, 1] \tag{14.60}$$

The Dynkin labels of simple roots are given by the rows of the affine Cartan matrix

$$\alpha_i = [\widehat{A}_{i0}, \widehat{A}_{i1}, \cdots, \widehat{A}_{ir}] \tag{14.61}$$

Finally, the *affine Weyl vector* is defined as

$$\hat{\rho} = \sum_{i=0}^{r} \hat{\omega}_i = [1, 1, \cdots, 1], \qquad \hat{\rho}(\hat{k}) = g \tag{14.62}$$

We note that it cannot be written as the half sum of the positive affine roots.

As in the finite case, affine weights whose Dynkin labels are all nonnegative integers will play a special role (cf. Sect. 14.3). These weights are called *dominant*. Since the zeroth Dynkin label is fixed by k and the finite Dynkin labels through Eq.(14.57), this characteristic is clearly level-dependent. The set of all dominant weights at level k is denoted P_+^k. Clearly, the finite part of an affine dominant weight is itself a dominant weight: $\hat{\lambda} \in P_+^k$ implies that $\lambda \in P_+$ (but not vice versa).

14.1.6. The Affine Weyl Group

The Weyl reflection with respect to the real affine root $\hat{\alpha}$ is defined exactly as in the finite case:

$$\boxed{s_{\hat{\alpha}}\hat{\lambda} = \hat{\lambda} - (\hat{\lambda}, \hat{\alpha}^{\vee})\hat{\alpha}} \tag{14.63}$$

and the set of all such reflections generates the *affine Weyl group* $\hat{W}$. With $\hat{\lambda} = (\lambda; k; n)$ and $\hat{\alpha} = (\alpha; 0; m)$, a direct calculation yields

$$s_{\hat{\alpha}}\hat{\lambda} = (\lambda - [(\lambda, \alpha) + km]\alpha^\vee; k; n - [(\lambda, \alpha) + km]\frac{2m}{|\alpha|^2})$$

$$= (s_\alpha(\lambda + km\alpha^\vee); k; n - [(\lambda, \alpha) + km]\frac{2m}{|\alpha|^2}) \tag{14.64}$$

As a consistency check, we see that for $\hat{\lambda} = \hat{\alpha}$,

$$s_{\hat{\alpha}}\hat{\alpha} = (s_\alpha\alpha; 0; m - (\alpha, \alpha^\vee)m) = (-\alpha; 0; -m) = -\hat{\alpha} \tag{14.65}$$

On the other hand, since $(\delta, \hat{\alpha}) = 0$, imaginary roots are unaffected by affine Weyl reflections

$$s_{\hat{\alpha}}\delta = \delta \tag{14.66}$$

To analyze the structure of $\hat{W}$, we rewrite Eq. (14.64) under the form

$$s_{\hat{\alpha}}\hat{\lambda} = s_\alpha(t_{\alpha^\vee})^m\,\hat{\lambda} \tag{14.67}$$

with $t_{\alpha^\vee}$ defined as

$$t_{\alpha^\vee} = s_{-\alpha+\delta}s_\alpha = s_\alpha s_{\alpha+\delta} \tag{14.68}$$

That is,

$$t_{\alpha^\vee}\hat{\lambda} = (\lambda + k\alpha^\vee; k; n + [|\lambda|^2 - |\lambda + k\alpha^\vee|^2]/2k) \tag{14.69}$$

The action of $t_{\alpha^\vee}$ on the finite part λ of $\hat{\lambda}$ corresponds to a translation by the coroot $\alpha^\vee$. Since

$$(t_{\alpha^\vee})(t_{\beta^\vee}) = t_{\alpha^\vee+\beta^\vee} \tag{14.70}$$

(and in particular $(t_{\alpha^\vee})^m = t_{m\alpha^\vee}$) the set of all $t_{\alpha^\vee}$'s generates the coroot lattice $Q^\vee$. An affine Weyl reflection is thus a product of a finite Weyl reflection times a translation by an appropriate coroot. The group of such translations being infinite, the affine Weyl group is infinite dimensional. Actually, the affine Weyl group has a semidirect product structure since $Q^\vee$ and W have only the identity in common and $Q^\vee$ is an invariant subgroup of $\hat{W}$:

$$w(t_{\alpha^\vee})w^{-1} = t_{w\alpha^\vee} \qquad \forall w \in \hat{W} \tag{14.71}$$

a relation easily verified. We note its following implication:

$$w'(t_{\alpha^\vee})w(t_{\beta^\vee}) = w'w(t_{w^{-1}\alpha^\vee})(t_{\beta^\vee}) \tag{14.72}$$

The generators for the group $\hat{W}$ are the reflections s_i with respect to the simple roots. For $i \neq 0$, the definition of s_i does not differ from the finite case, whereas for s_0, Eq. (14.64) gives

$$s_0\hat{\lambda} = (\lambda + k\theta - (\lambda, \theta)\theta; k; n - k + (\lambda, \theta)) = s_\theta t_{-\theta}(\hat{\lambda}) \tag{14.73}$$

(Clearly $s_{-\theta} = s_\theta$.) With $s_\theta\theta = -\theta$, the finite part of $s_0\hat{\lambda}$ is $s_\theta\lambda + k\theta$.

The affine Weyl group divides the vector space of affine weights in an infinite number of *affine Weyl chambers* defined as

$$\hat{C}_w = \{\hat{\lambda} \mid (w\hat{\lambda}, \alpha_i) \geq 0, \ i = 0, 1, \cdots, r\}, \quad w \in \hat{W} \tag{14.74}$$

The fundamental chamber corresponds to the element $w = 1$. Weights in the fundamental chamber are then of the form

$$\hat{\lambda} = \sum_{i=0}^{r} \lambda_i \hat{\omega}_i + \ell\delta, \quad \text{with} \quad \lambda_i \geq 0, \quad \ell \in \mathbb{R} \tag{14.75}$$

Once the δ part of the weights is projected out, affine Weyl chambers have finite area, in contrast to the finite case where the chambers are simplicial cones extending to infinity.

By definition, the affine Weyl group preserves the scalar product (14.23), e.g., using Eq. (14.64)

$$(s_{\hat{\alpha}}\hat{\lambda}, s_{\hat{\alpha}}\hat{\lambda}) = (s_\alpha(\lambda + km\alpha^\vee), s_\alpha(\lambda + km\alpha^\vee)) + 2k(n - [(\lambda, \alpha) + km]\frac{2m}{|\alpha|^2})$$
$$= (\lambda, \lambda) + 2kn$$
$$= (\hat{\lambda}, \hat{\lambda})$$

$$\tag{14.76}$$

Thus, all the weights in a given Weyl orbit have the same length. A $\hat{W}$ orbit contains an infinite number of weights and it has a unique weight in the fundamental chamber.

We note finally that shifted Weyl reflections are defined as in the finite case, but now in terms of the affine Weyl vector:

$$w \cdot \hat{\lambda} = w(\hat{\lambda} + \hat{\rho}) - \hat{\rho} \tag{14.77}$$

14.1.7. Examples

EXAMPLE 1: $\widehat{su}(2)$

Here $\theta = \alpha_1$, the only positive root of $su(2)$. Since

$$(\alpha_0, \alpha_1^\vee) = (\alpha_1, \alpha_0^\vee) = (\alpha_1, \alpha_0) = -\alpha_1^2 = -2 \tag{14.78}$$

the extended Cartan matrix reads

$$\hat{A} = \begin{pmatrix} 2 & -2 \\ -2 & 2 \end{pmatrix} \tag{14.79}$$

The Dynkin labels of the simple roots are then

$$\alpha_0 = [2, -2], \quad \alpha_1 = [-2, 2] \tag{14.80}$$

For $\widehat{su}(N)$, all marks and comarks are one. The level is thus obtained from the sum of all Dynkin labels. For the $\widehat{su}(2)$ simple roots, these add up to zero as they should. The complete set of roots is

$$\hat{\Delta} = \{\pm\alpha_1, \pm\alpha_1 + n\delta, n\delta | \; n \in \mathbb{Z}, \; n \neq 0\} \tag{14.81}$$

With $\delta = \alpha_0 + \alpha_1$, this is the same as

$$\hat{\Delta} = \{n\alpha_0 + m\alpha_1 | \; |n - m| \leq 1, \; n, m \in \mathbb{Z}\} \tag{14.82}$$

This structure is also encoded in the Serre relations (14.46). For instance, since $\alpha_1 + 3\alpha_0$, associated with the commutator $[e^0, [e^0, [e^0, e^1]]]$, is not a root, means that

$$[e^0, [e^0, [e^0, e^1]]] = [\mathrm{ad}(e^0)]^3 e^1 = 0 \tag{14.83}$$

But this is a consequence of the Serre relations since $1 - \widehat{A}_{10} = 3$.

The affine Weyl group is generated by the reflections s_0, s_1, whose actions on a weight $\hat{\lambda} = [\lambda_0, \lambda_1]$ read

$$\begin{aligned} s_0\hat{\lambda} &= \hat{\lambda} - \lambda_0\alpha_0 = [\lambda_0, \lambda_1] - \lambda_0[2, -2] = [-\lambda_0, \lambda_1 + 2\lambda_0] \\ s_1\hat{\lambda} &= \hat{\lambda} - \lambda_1\alpha_1 = [\lambda_0, \lambda_1] - \lambda_1[-2, 2] = [\lambda_0 + 2\lambda_1, -\lambda_1] \end{aligned} \tag{14.84}$$

The action of s_0 amounts to subtracting λ_0 copies of α_0 from $\hat{\lambda}$. Since the L_0 eigenvalue of α_0 is -1, $s_0\hat{\lambda}$ increases the L_0 eigenvalue of $\hat{\lambda}$ by λ_0. Let the level of $\hat{\lambda}$ be k. The zeroth Dynkin label is thus

$$\lambda_0 = k - \lambda_1 \tag{14.85}$$

and the simple affine Weyl reflections can be written as

$$s_0\hat{\lambda} = [-k + \lambda_1, 2k - \lambda_1], \qquad s_1\hat{\lambda} = [k + \lambda_1, -\lambda_1] \tag{14.86}$$

so that

$$s_0s_1\hat{\lambda} = [-k - \lambda_1, 2k + \lambda_1] \tag{14.87}$$

This shows that s_0s_1 translates the finite part of $\hat{\lambda}$ by $2k\omega_1$, that is by $k\alpha_1 = k\alpha_1^\vee$. Therefore, it is the basic translation operator $t_{\alpha_1^\vee}$

$$t_{\alpha_1^\vee} = s_0s_1 \tag{14.88}$$

The structure of the affine Weyl group is thus

$$\hat{W} = \{(s_0s_1)^n, s_1(s_0s_1)^n | \; n \in \mathbb{Z}\} \tag{14.89}$$

In the particular case of $\widehat{su}(2)$, Eq. (13.59) implies that the Weyl group is infinite: the angle between the finite parts of the simple roots is π, and therefore (s_0s_1) has no finite order.

A few affine chambers at various integer levels are displayed in Fig. 14.2 (modulo $\mathbb{R}\delta$). The affine Weyl chambers at level $k \in \mathbb{Z}_+$ are the segments of the weight axis at level k that are separated by the dashed lines. The size of the chambers obviously increases with the level. In the limit $k \to \infty$, if we can restrict ourselves

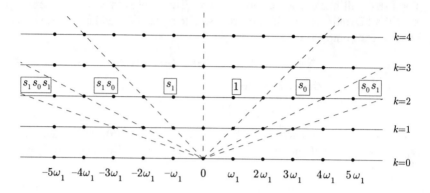

Figure 14.2. Affine Weyl chambers for $\widehat{su}(2)$.

to "small weights", there is effectively no difference between the affine and the finite fundamental chambers.

EXAMPLE 2: $\widehat{su}(3)$

With $\theta = \alpha_1 + \alpha_2$, the extended Cartan matrix is

$$\widehat{A} = \begin{pmatrix} 2 & -1 & -1 \\ -1 & 2 & -1 \\ -1 & -1 & 2 \end{pmatrix} \tag{14.90}$$

so that

$$\alpha_0 = [2, -1, -1], \qquad \alpha_1 = [-1, 2, -1], \qquad \alpha_2 = [-1, -1, 2] \tag{14.91}$$

The full set of roots can be described by an infinite pile of hexagons, each layer representing the roots of $su(3)$, with two adjacent hexagons being separated by δ.

The reflections of $\widehat{\lambda} = [\lambda_0, \lambda_1, \lambda_2]$ with respect to the simple roots are

$$\begin{aligned} s_0\widehat{\lambda} &= [-\lambda_0, \lambda_0 + \lambda_1, \lambda_0 + \lambda_2] \\ s_1\widehat{\lambda} &= [\lambda_0 + \lambda_1, -\lambda_1, \lambda_1 + \lambda_2] \\ s_2\widehat{\lambda} &= [\lambda_0 + \lambda_2, \lambda_1 + \lambda_2, -\lambda_2] \end{aligned} \tag{14.92}$$

With $\lambda_0 = k - \lambda_1 - \lambda_2$, the basic translation operators are found to be

$$\begin{aligned} t_{\alpha_1^\vee} &= s_2 s_0 s_2 s_1 \\ t_{\alpha_2^\vee} &= s_1 s_0 s_1 s_2 \end{aligned} \tag{14.93}$$

The relations

$$(s_0 s_1 s_0)^3 = (s_0 s_2 s_0)^3 = (s_1 s_2 s_1)^3 = 1 \tag{14.94}$$

are easily checked (cf. Eq. (13.59)). However they do not make manifest the infinite order of the Weyl group.

§14.2. Outer Automorphisms

In this section, we introduce outer automorphisms from the symmetry properties of Dynkin diagrams. Although this is a somewhat technical issue, it will turn out to be an important tool in many applications.

14.2.1. Symmetry of the Extended Diagram and Group of Outer Automorphisms

We let $D(g)$ and $D(\hat{g})$ stand, respectively, for the symmetry group of the g and $\hat{g}$ Dynkin diagrams. These are the sets of symmetry transformations of the simple roots that preserve the scalar products, hence the Cartan matrices. In the affine case, we need to consider only the finite projection of the system of simple roots since scalar products of roots depend only on their finite parts. This also implies that a simple root is mapped into another simple root having the same mark and comark. By inspection, we see that $D(g) = 1$ except for $A_{r>1}$, $D_{r>4}$, E_6, for which it is respectively $\mathbb{Z}_2$, D_4, and S_3, the permutation group of three objects.

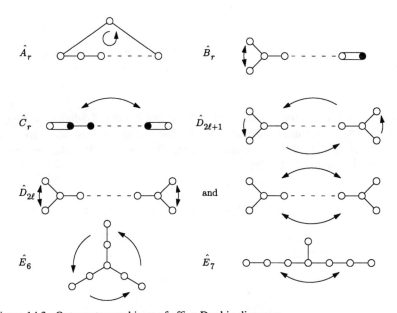

Figure 14.3. Outer automorphisms of affine Dynkin diagrams.

Define the group of *outer automorphisms* of $\hat{g}$, $\mathcal{O}(\hat{g})$, as

$$\mathcal{O}(\hat{g}) = D(\hat{g})/D(g) \qquad (14.95)$$

This quotient is sensible because $D(g)$ is the set of elements of $D(\hat{g})$ that leaves the zeroth node of the extended Dynkin diagram fixed, that is, $D(g)$ is a subgroup of $D(\hat{g})$. $\mathcal{O}(\hat{g})$ is thus the set of symmetry transformations of the Dynkin diagram of

$\hat{g}$ that are not symmetry transformations of the Dynkin diagram of g. For instance, for $\widehat{su}(N)$, we do not consider reflections leaving the zeroth root fixed; the only remaining symmetry transformations are the cyclic permutations. The symmetry transformations of the extended Dynkin diagrams are displayed in Fig. 14.3, when $\mathcal{O}(\hat{g})$ is not simply the identity.

An explicit description of outer-automorphism groups of affine Lie algebras in terms of their generating element on an arbitrary weight $\hat{\lambda} = [\lambda_0, \cdots, \lambda_r]$ is presented in Table 14.1.

Table 14.1. Outer automorphisms of affine Lie algebras

g	$\mathcal{O}(\hat{g})$	Action of the $\mathcal{O}(\hat{g})$ generators
A_r	$\mathbb{Z}_{r+1}$	$a[\lambda_0, \lambda_1, \cdots, \lambda_{r-1}, \lambda_r] = [\lambda_r, \lambda_0, \cdots, \lambda_{r-2}, \lambda_{r-1}]$
B_r	$\mathbb{Z}_2$	$a[\lambda_0, \lambda_1, \cdots, \lambda_{r-1}, \lambda_r] = [\lambda_1, \lambda_0, \cdots, \lambda_{r-1}, \lambda_r]$
C_r	$\mathbb{Z}_2$	$a[\lambda_0, \lambda_1, \cdots, \lambda_{r-1}, \lambda_r] = [\lambda_r, \lambda_{r-1}, \cdots, \lambda_1, \lambda_0]$
$D_{r=2\ell}$	$\mathbb{Z}_2 \times \mathbb{Z}_2$	$a[\lambda_0, \lambda_1, \lambda_2, \cdots, \lambda_{r-1}, \lambda_r] = [\lambda_1, \lambda_0, \lambda_2, \cdots, \lambda_r, \lambda_{r-1}]$
		$\tilde{a}[\lambda_0, \lambda_1, \lambda_2, \cdots, \lambda_{r-1}, \lambda_r] = [\lambda_r, \lambda_{r-1}, \lambda_{r-2}, \cdots, \lambda_1, \lambda_0]$
$D_{r=2\ell+1}$	$\mathbb{Z}_4$	$a[\lambda_0, \lambda_1, \lambda_2, \cdots, \lambda_{r-1}, \lambda_r] = [\lambda_{r-1}, \lambda_r, \lambda_{r-2}, \cdots, \lambda_1, \lambda_0]$
E_6	$\mathbb{Z}_3$	$a[\lambda_0, \lambda_1, \cdots, \lambda_6] = [\lambda_1, \lambda_5, \lambda_4, \lambda_3, \lambda_6, \lambda_0, \lambda_2]$
E_7	$\mathbb{Z}_2$	$a[\lambda_0, \lambda_1, \cdots, \lambda_7] = [\lambda_6, \lambda_5, \lambda_4, \lambda_3, \lambda_2, \lambda_1, \lambda_0, \lambda_7]$

Since every fundamental weight is mapped into another fundamental weight having the same comark, the action of A does not change the level. Moreover, it is clear that $\mathcal{O}(\hat{g})$ maps the set of dominant weights P_+^k into itself. It thus preserves the affine fundamental chamber.

14.2.2. Action of Outer Automorphisms on Weights

We let $A\hat{\omega}_0$ denote the fundamental weight to which $\hat{\omega}_0$ is mapped by the action of A, a generic element of $\mathcal{O}(\hat{g})$. Its action on an affine weight is written as

$$A\hat{\lambda} = kA\hat{\omega}_0 + \sum_{i=1}^{r} \lambda_i A(\hat{\omega}_i - a_i^\vee \hat{\omega}_0) \tag{14.96}$$

where k is the level of $\hat{\lambda}$. This result follows directly from the definition of λ_0 as

$$\lambda_0 = k - \sum_{i=1}^{r} a_i^\vee \lambda_i \tag{14.97}$$

The second term in the r.h.s. of Eq. (14.96) acts, on the finite part of $\hat{\lambda}$, like an automorphism of the finite weight lattice that leaves its origin fixed. It is actually an element of the finite Weyl group.[8] We denote this element by w_A (it is described below). The sum in Eq. (14.96) can then be written as $w_A\lambda$. More precisely, it is the affine extension of $w_A\lambda$ at level zero, which is equivalent to $w_A\hat{\lambda} - k\hat{\omega}_0$. This yields

$$A\hat{\lambda} = k(A-1)\hat{\omega}_0 + w_A\hat{\lambda} \tag{14.98}$$

This important relation will have many applications.

It is usually easy to find the element w_A associated with a given A by a simple calculation. But there is a general way of characterizing w_A. Define w_i to be the longest element of $W_{(i)}$, the subgroup of the finite Weyl group generated by all s_j ($j \neq i$). Then

$$w_A = w_i w_0 \quad \text{for } i \text{ such that} \quad A\hat{\omega}_0 = \hat{\omega}_i \tag{14.99}$$

(as before, w_0 stands for the longest element of W). In future calculations, we will often need to know the sign of the element w_A. In App. 14.A, it is shown to be

$$\epsilon(w_A) = e^{2\pi i(A\hat{\omega}_0, \rho)} = e^{-\pi i g|A\hat{\omega}_0|^2} \tag{14.100}$$

Consider some examples. For $\widehat{su}(2)$, the only nontrivial outer automorphism is $a : \hat{\omega}_0 \leftrightarrow \hat{\omega}_1$. Since $W = \{1, s_1\}$, w_a is simply s_1. The comparison of

$$a[\lambda_0, \lambda_1] = [\lambda_1, \lambda_0] = [\lambda_1, k - \lambda_1] \tag{14.101}$$

with

$$
\begin{aligned}
a[\lambda_0, \lambda_1] &= k(a-1)\hat{\omega}_0 + s_1[\lambda_0, \lambda_1] \\
&= k(\hat{\omega}_1 - \hat{\omega}_0) + [\lambda_0 + 2\lambda_1, -\lambda_1] \\
&= [\lambda_1, k - \lambda_1]
\end{aligned}
\tag{14.102}
$$

shows that this is indeed correct. We turn to $\widehat{su}(3)$, in which the basic element a maps $\hat{\omega}_0 \to \hat{\omega}_1 \to \hat{\omega}_2 \to \hat{\omega}_0$. Here $i = 1$ and the longest element of the group $W_{(1)}$ is s_2. Recalling that $w_0 = s_1 s_2 s_1 = s_2 s_1 s_2$, the above construction yields

$$w_a = s_2 s_2 s_1 s_2 = s_1 s_2 \tag{14.103}$$

This is again easily confirmed by a direct calculation:

$$
\begin{aligned}
a[\lambda_0, \lambda_1, \lambda_2] &= k(a-1)\hat{\omega}_0 + s_1 s_2[\lambda_0, \lambda_1, \lambda_2] \\
&= k(\hat{\omega}_1 - \hat{\omega}_0) + [\lambda_0 + 2\lambda_2 + \lambda_1, -\lambda_1 - \lambda_2, \lambda_1] \\
&= [\lambda_2, k - \lambda_1 - \lambda_2, \lambda_1] \\
&= [\lambda_2, \lambda_0, \lambda_1]
\end{aligned}
\tag{14.104}
$$

It is also simple to verify that a^2 corresponds to the element $s_2 s_1$:

$$w_{a^2} = (w_a)^2 = (s_1 s_2)^2 = s_1 s_2 s_1 s_2 = s_1 s_1 s_2 s_1 = s_2 s_1 = (w_a)^{-1} = w_{a^{-1}} \tag{14.105}$$

---

[8] A general automorphism of the finite weight lattice that leaves its origin fixed is a product of a Weyl reflection by a conjugation.

The general result for $\widehat{su}(N)$ is

$$w_a = s_1 s_2 \cdots s_{N-1} \tag{14.106}$$

where the action of a is defined in Table 14.1.

We finally stress that outer automorphisms must preserve the commutation relations of the algebra. We illustrate this feature by rederiving for $\widehat{su}(2)$ the action of a on weights by enforcing the invariance of the commutator under the interchange $\alpha_0 \leftrightarrow \alpha_1$. We use the spin basis, in which the commutation relations read:

$$
\begin{aligned}
[J_m^0, J_n^0] &= 2km\delta_{m+n,0} \\
[J_m^0, J_n^\pm] &= \pm J_{n+m}^\pm \\
[J_m^+, J_n^-] &= 2J_{n+m}^0 + km\delta_{m+n,0}
\end{aligned}
\tag{14.107}
$$

with $2J^0|\lambda\rangle = \lambda_1|\lambda\rangle$. Acting on a weight, J_0^+ adds α_1 whereas J_1^- adds α_0 (the superscript $-$ because the finite part of α_0 is $-\alpha_1$ and the subscript 1 because α_0 has grade 1). Hence, the interchange $\alpha_0 \leftrightarrow \alpha_1$ translates into $J_0^+ \leftrightarrow J_1^-$, or more generally $J_m^+ \leftrightarrow J_{m+1}^-$. The commutator $[J_m^+, J_n^-]$ is thus transformed into

$$
\begin{aligned}
[J_m^+, J_n^-] &\to [J_{m+1}^-, J_{n-1}^+] \\
&= -2J_{n+m}^0 + k(m+1)\delta_{m+n,0}
\end{aligned}
\tag{14.108}
$$

In order to preserve the third commutator, J_m^0 must transform as

$$J_m^0 \to k\delta_{m,0} - 2J_m^0 \tag{14.109}$$

which means that $\lambda_1 \to k - \lambda_1$ under the action of a. (The other two commutators are also invariant under this transformation.)

14.2.3. Relation with the Center of the Group

It is easily checked, case by case, that $\mathcal{O}(\hat{g})$ is isomorphic to the *center of the group* of g, denoted $B(G)$

$$\boxed{\mathcal{O}(\hat{g}) \simeq B(G)} \tag{14.110}$$

The center $B(G)$ is composed of those particular elements of G that commute with all the elements of the group. This set is easily obtained in terms of a matrix representation of the group. Consider, for instance, the group $SU(N)$ represented by unitary matrices of determinant one. The elements of the center are the unit matrices multiplied by any N-th root of unity, so that $B(SU(N)) = \mathbb{Z}_N$. The centers of the other groups can be read from Table 14.1.

The isomorphism between $B(G)$ and $\mathcal{O}(\hat{g})$ can be realized as follows. To every element $A \in \mathcal{O}(\hat{g})$, there corresponds an element $b \in B(G)$, given by

$$b = e^{-2\pi i A\hat{\omega}_0 \cdot H} \tag{14.111}$$

This manifestly commutes with the generators of the Cartan subalgebra. Moreover, since

$$b \, E^\beta = e^{-2\pi i (A\hat{\omega}_0, \beta)} \, E^\beta \, b \tag{14.112}$$

commutativity with a ladder operator requires

$$(A\hat{\omega}_0, \beta) \in \mathbb{Z} \qquad \text{for any} \quad \beta \in Q \tag{14.113}$$

But this is certainly true, being a direct consequence of $A\hat{\omega}_0$ having unit mark: the simple coroot to which it is dual is equal to its corresponding root. Thus, $A\hat{\omega}_0$ is dual to the root lattice, from which the result follows. (We note here a special case of this relation, which will be used frequently in the following discussion:

$$(A\hat{\omega}_0, \xi - w\xi) \in \mathbb{Z} \tag{14.114}$$

valid for any integral weight ξ and $w \in W$.) The action of b on any state $|\lambda'\rangle$ in the highest-weight module L_λ of g is given by

$$b|\lambda'\rangle = e^{-2\pi i (A\hat{\omega}_0, \lambda')} |\lambda'\rangle \tag{14.115}$$

or equivalently

$$b\lambda' = \lambda' e^{-2\pi i (A\hat{\omega}_0, \lambda')} = \lambda' e^{-2\pi i (A\hat{\omega}_0, \lambda)} \tag{14.116}$$

Indeed, all the states in the representation have the same eigenvalue because the generators of the algebra are unaffected by the action of the center.

Since the center element b commutes also with all the affine generators, the action (14.116) extends uniquely to the affine case, with the same eigenvalue:

$$b\hat{\lambda}' = \hat{\lambda}' e^{-2\pi i (A\hat{\omega}_0, \lambda)} \tag{14.117}$$

If b corresponds to A, b^q will correspond to A^q. The comparison of Eq. (14.117), with b replaced by b^q, with the same equation applied q times, yields the relation

$$(A^q \hat{\omega}_0, \lambda) = q(A\hat{\omega}_0, \lambda) \mod 1 \tag{14.118}$$

If $\mathcal{O}(\hat{g})$ is a cyclic group of order N, $A^N = 1$ so that $(A^N \hat{\omega}_0, \lambda) = 0$. Together with the above equation, this implies that

$$N(A\hat{\omega}_0, \lambda) \in \mathbb{Z} \tag{14.119}$$

This verifies that the eigenvalue of b is a N-th root of unity.

§14.3. Highest-Weight Representations

Highest-weight representations are characterized by a unique highest state $|\hat{\lambda}\rangle$ annihilated by the action of all ladder operators for positive roots

$$E_0^\alpha |\hat{\lambda}\rangle = E_n^{\pm\alpha} |\hat{\lambda}\rangle = H_n^i |\hat{\lambda}\rangle = 0, \quad \text{for} \quad n > 0, \ \alpha > 0 \tag{14.120}$$

The eigenvalue of this state, $\hat{\lambda}$, is the highest weight of the representation

$$H_0^i |\hat{\lambda}\rangle = \lambda^i |\hat{\lambda}\rangle \ (i \neq 0), \qquad \hat{k} |\hat{\lambda}\rangle = k |\hat{\lambda}\rangle, \qquad L_0 |\hat{\lambda}\rangle = 0 \tag{14.121}$$

Setting the L_0 eigenvalue to zero is merely a matter of convention; a redefinition of L_0 would yield any desired value. In the Chevalley basis, the eigenvalues are the Dynkin labels:

$$h^i|\hat{\lambda}\rangle = \lambda_i|\hat{\lambda}\rangle \quad i = 0, 1, \cdots, r \tag{14.122}$$

All the states in the module are generated by the action of the lowering operators on $|\hat{\lambda}\rangle$. They obviously have the same $\hat{k}$-eigenvalue, given that $\hat{k}$ commutes with all the generators. From now on, $\hat{k}$ will be identified with its eigenvalue k, the level. In most applications of interest, k is fixed from the outset.

14.3.1. Integrable Highest-Weight Representations

The analogues of the irreducible finite-dimensional representations of g are those representations whose projections onto the $su(2)$ algebra associated with any real root are finite. It is clearly sufficient to concentrate on simple roots. An analysis in all points similar to the one that led to Eq. (13.27) shows that any weight $\hat{\lambda}'$ in the weight system $\Omega_{\hat{\lambda}}$ (the set of all weights in the representation of highest-weight state $|\hat{\lambda}\rangle$) satisfies

$$(\hat{\lambda}', \alpha_i^\vee) = -(p_i - q_i) \qquad i = 0, 1, \cdots, r \tag{14.123}$$

for some positive integers p_i, q_i, which thereby implies that

$$\lambda_i' \in \mathbb{Z}, \qquad i = 0, 1, \cdots, r \tag{14.124}$$

For the highest weight $\hat{\lambda}$, all p_i's are zero, and therefore

$$\lambda_i \in \mathbb{Z}_+, \qquad i = 0, 1, \cdots, r \tag{14.125}$$

This requires in particular that (cf. Eq. (14.57))

$$\lambda_0 = k - (\lambda, \theta) \in \mathbb{Z}_+ \tag{14.126}$$

With $(\lambda, \theta) \in \mathbb{Z}_+$, this immediately shows that k must be a positive integer, bounded from below by (λ, θ):

$$\boxed{k \in \mathbb{Z}_+, \qquad k \geq (\lambda, \theta)} \tag{14.127}$$

As we already mentioned, an affine weight for which all Dynkin labels are non-negative integers is said to be dominant, and the set of all dominant weights will be denoted P_+^k.

A far-reaching consequence of the constraints (14.127) is that for a fixed value of k, there can be only a finite number of dominant highest-weight representations. For instance, for $k = 1$, the only such representations are those with highest weight $\hat{\omega}_i$ such that the corresponding simple root α_i has unit comark. Since $a_0^\vee = 1$ for all $\hat{g}$, $\hat{\omega}_0$ is always dominant. The level-1 representation of highest weight $\hat{\omega}_0$ is called the *basic representation*. For $\widehat{su}(N)$, all comarks are one; there are thus N possible dominant highest-weight representations at level 1 whose highest weights are the

$\hat{\omega}_i$'s, $i = 0, \cdots, r$. As another example, the set of all dominant highest-weight representations of $\widehat{su}(3)$ at level 2 is:

$$[2,0,0], \quad [0,2,0], \quad [0,0,2], \quad [1,1,0], \quad [1,0,1], \quad [0,1,1] \qquad (14.128)$$

On the other hand, the possibilities for $\widehat{G}_2$ at level 2 are

$$[2,0,0], \quad [0,1,0], \quad [0,0,2], \quad [1,0,1] \qquad (14.129)$$

since $a_1^\vee = 2a_0^\vee = 2a_2^\vee = 2$. In the following, the algebra $\hat{g}$ at level k will be denoted $\hat{g}_k$.

Representations that decompose into finite irreducible representations of $su(2)$ and can further be written as a direct sum of finite-dimensional weight spaces are said to be *integrable*. The adjoint representation, although not a highest-weight representation, is integrable. The first condition is obviously satisfied and the direct-sum decomposition in the second requirement is simply the root-space decomposition (i.e., the decomposition of the root space into a sum of finite roots and imaginary roots). Dominant highest-weight representations are also integrable. We will show shortly that the second condition is verified. Moreover, if

$$(J_n^a)^\dagger = J_{-n}^a, \qquad \text{or} \qquad (H_n^i)^\dagger = H_{-n}^i \qquad (E_n^\alpha)^\dagger = E_{-n}^{-\alpha} \qquad (14.130)$$

dominant highest-weight representations are easily checked to be unitary. For instance,

$$|E_{-n}^\alpha|\hat{\lambda}\rangle|^2 = \langle\hat{\lambda}|E_n^{-\alpha}E_{-n}^\alpha|\hat{\lambda}\rangle$$
$$= \frac{2}{|\alpha|^2}[nk - (\alpha,\lambda)]\langle\hat{\lambda}|\hat{\lambda}\rangle \geq 0 \qquad (14.131)$$

since for $n > 0$, any α, and $\hat{\lambda}$ dominant,

$$nk - (\alpha,\lambda) = [k - (\theta,\lambda)] + (n-1)k + (\theta - \alpha,\lambda) \geq 0 \qquad (14.132)$$

For dominant highest weights, the conditions (13.27) are equivalent to the existence of the following singular vectors in the Verma module $V_{\hat{\lambda}}$ of highest-weight state $|\hat{\lambda}\rangle$:

$$E_0^{\alpha_i}|\hat{\lambda}\rangle = E_1^{-\theta}|\hat{\lambda}\rangle = 0 \qquad (14.133)$$

and

$$(E_0^{-\alpha_i})^{\lambda_i+1}|\hat{\lambda}\rangle = (E_{-1}^\theta)^{k-(\lambda,\theta)+1}|\hat{\lambda}\rangle = 0 \qquad (14.134)$$

with $i \neq 0$. In the Chevalley basis, these vectors read

$$e^i|\hat{\lambda}\rangle = (f^i)^{\lambda_i+1}|\hat{\lambda}\rangle = 0, \qquad i = 0,1,\cdots,r \qquad (14.135)$$

In sharp contrast with simple Lie algebras, when these singular vectors are quotiented out from the dominant highest-weight Verma module $V_{\hat{\lambda}}$ (modulo their intersections), the resulting irreducible module, to be denoted $L_{\hat{\lambda}}$, is not finite-dimensional. The imaginary root can be subtracted from any weight without leaving the representation, that is, if $\hat{\lambda}' \in \Omega_{\hat{\lambda}}$, then $\hat{\lambda}' - n\delta \in \Omega_{\hat{\lambda}}$ for any $n > 0$.

The source of infinity clearly lies in the absence of a singular vector similar to Eq.(14.134) related to the imaginary root δ, that is, a singular vector that would involve H_n^i for $n < 0$.

We now show how the various weights in $\Omega_{\hat{\lambda}}$ can be obtained. The algorithm that gives the list of weights in irreducible highest-weight representations of g also works for $\hat{g}$. We simply have to keep track of an additional Dynkin label. However, in the affine case this algorithm never terminates.

We define the *grade* to be the L_0 eigenvalue, shifted so that $L_0|\hat{\lambda}\rangle = 0$ for the highest state $|\hat{\lambda}\rangle$. At grade zero, the various states are obtained from $|\hat{\lambda}\rangle$ by applications of the finite Lie algebra generators (the only generators of $\hat{g}$ that do not change the L_0 eigenvalue). Hence, the finite projection of weights at grade zero are all the weights in the g irreducible finite-dimensional representation of highest weight λ. Weights at grade one are obtained from those weights at grade zero that have positive zeroth Dynkin labels, by the subtraction of α_0 (which augments the L_0 eigenvalue by one), followed again by all possible subtractions of the finite simple roots. The analysis of the higher grades proceeds along the same lines. An important point is that the finite projections of the affine weights at a fixed value of the grade are organized into a direct sum of irreducible finite-dimensional representations of g.

Since at each grade there is a finite number of weights, the weight space decomposes into a direct sum of finite dimensional weight spaces, which shows that dominant highest-weight representations are integrable.

To complete the description of the representation, we must give the multiplicity of each weight. The multiplicity of weights, when the L_0 eigenvalue is taken into account, is clearly finite. In fact, it can be calculated from the following modification of the Freudenthal recursion formula, which keeps track of the root multiplicities,

$$[|\hat{\lambda} + \hat{\rho}|^2 - |\hat{\lambda}' + \hat{\rho}|^2] \, \text{mult}_{\hat{\lambda}}(\hat{\lambda}') = 2 \sum_{\hat{\alpha}>0} \text{mult}(\hat{\alpha})$$

$$\times \sum_{p=1}^{\infty} (\hat{\lambda}' + p\hat{\alpha}, \hat{\alpha}) \, \text{mult}_{\hat{\lambda}}(\hat{\lambda}' + p\hat{\alpha}) \tag{14.136}$$

We recall that real roots have multiplicity one whereas imaginary ones have multiplicity r. (We stress that the scalar product in this formula is the one defined in Eq. (14.23).) With our convention for the L_0 eigenvalue of highest-weight states, the scalar product of two affine highest weights does not differ from its finite form:

$$(\hat{\lambda}, \hat{\mu}) = (\lambda, \mu), \qquad \text{for} \quad \hat{\lambda}(L_0) = \hat{\mu}(L_0) = 0 \tag{14.137}$$

Thus, with $\hat{\lambda} = (\lambda; k; 0)$ and $\hat{\rho} = (\rho; g; 0)$,

$$|\hat{\lambda} + \hat{\rho}|^2 = |\lambda + \rho|^2 \tag{14.138}$$

However, for a weight at grade m, $\hat{\lambda}' = (\lambda'; k; -m)$ and

$$|\hat{\lambda}' + \hat{\rho}|^2 = |\lambda' + \rho|^2 - 2m(k + g) \tag{14.139}$$

14.3.2. The Basic Representation of $\widehat{su}(2)_1$

We now consider a detailed example, the basic representation of $\widehat{su}(2)_1$, of highest weight $[1, 0]$. Using the algorithm mentioned above, it is simple to write down the different weights at the first few grades. These are displayed in Fig. 14.4, with their multiplicities given by subscripts. The three weights at grade one can be reached in only one way. Their multiplicity is thus bounded to be one. The same applies for the weights at grade two, except for $[1, 0]$ which can be obtained in two distinct ways. This means that its multiplicity can be either one or two. We can calculate it with the Freudenthal formula (14.136). With $\hat{\lambda} = (0; 1; 0)$ and $\hat{\lambda}' = (0; 1; -2)$, we have

$$|\hat{\lambda} + \hat{\rho}|^2 = |\rho|^2 = \frac{1}{2}$$

$$|\hat{\lambda}' + \hat{\rho}|^2 = |\rho|^2 - 4(1+2) = \frac{1}{2} - 12$$

$$(14.140)$$

Hence, the l.h.s. of Eq. (14.136) reads

$$[|\hat{\lambda} + \hat{\rho}|^2 - |\hat{\lambda}' + \hat{\rho}|^2] \, \text{mult}_{\hat{\lambda}}(\hat{\lambda}') = 12 \, \text{mult}_{\hat{\lambda}}(0; 1; -2) \qquad (14.141)$$

To calculate the r.h.s., we need to consider all the weights $\hat{\lambda}' + p\hat{\alpha}$ for $p, \hat{\alpha} > 0$, up to grade zero. We recall that the positive roots of $\widehat{su}(2)$ are α_1, $\pm\alpha_1 + n\delta$, $n\delta$ with $n > 0$, and they all have multiplicity one. The list of all possible contributions is given in Table 14.2, with their multiplicities and the required scalar products. From it, the r.h.s. of Eq. (14.136) is seen to be

$$\sum_{\hat{\alpha}>0} \sum_{p=1}^{\infty} (\hat{\lambda}' + p\hat{\alpha}, \hat{\alpha}) \, \text{mult}_{\hat{\lambda}}(\hat{\lambda}' + p\hat{\alpha}) = 24 \qquad (14.142)$$

We thus find that $\text{mult}_{\hat{\lambda}}(0; 1; -2) = 2$.

14.3.3. String Functions

Needless to say, multiplicity calculations are rather involved. However, the constancy of the weight multiplicities along $\hat{W}$ orbits greatly simplifies the analysis. For instance, in the above example, all the weights in the representation can be obtained from the different weights $(0; 1; -m)$ by Weyl reflections. This is easily seen, at least at the first few grades. Therefore, the information on the multiplicities of all the weights in the representation is coded in $\text{mult}(0; 1; -m)$ for all $m \geq 0$. The generating function for such multiplicities is called a *string function*. More precisely, let $\hat{\mu}$ be a weight in $\Omega_{\hat{\lambda}}$ such that $\hat{\mu} + \delta \notin \Omega_{\hat{\lambda}}$, and denote the set of such weights as $\Omega_{\hat{\lambda}}^{\text{max}}$. The multiplicity of the various weights in the string $\hat{\mu}, \hat{\mu} - \delta, \hat{\mu} - 2\delta, \cdots$ is given by the string function

$$\sigma_{\hat{\mu}}^{(\hat{\lambda})}(q) = \sum_{n=0}^{\infty} \text{mult}_{\hat{\lambda}}(\hat{\mu} - n\delta) \, q^n \qquad (14.143)$$

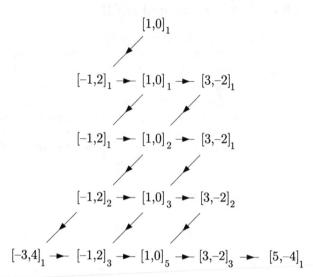

Figure 14.4. Weights at the first few grades of the basic representation of $\widehat{su}(2)_1$. The subscript gives the multiplicity.

Table 14.2. Weights above $(0; 1; -2)$ in the basic module of $\widehat{su}(2)_1$

p	$\hat{\alpha}$	$\hat{\lambda}' + p\hat{\alpha}$	$(\hat{\lambda}' + p\hat{\alpha}, \hat{\alpha})$	$\text{mult}_{\hat{\lambda}}(\hat{\lambda}' + p\hat{\alpha})$
1	$(\alpha_1; 0; 0)$	$(\alpha_1; 1; -2)$	2	1
1	$(\alpha_1; 0; 1)$	$(\alpha_1; 1; -1)$	3	1
1	$(\alpha_1; 0; 2)$	$(\alpha_1; 1; 0)$	4	0
1	$(-\alpha_1; 0; 1)$	$(-\alpha_1; 1; -1)$	3	1
1	$(-\alpha_1; 0; 2)$	$(-\alpha_1; 1; 0)$	4	0
1	$(0; 0; 1)$	$(0; 1; -1)$	1	1
1	$(0; 0; 2)$	$(0; 1; 0)$	2	1
2	$(\alpha_1; 0; 0)$	$(2\alpha_1; 1; -2)$	4	0
2	$(\alpha_1; 0; 1)$	$(2\alpha_1; 1; 0)$	5	0
2	$(-\alpha_1; 0; 1)$	$(-2\alpha_1; 1; 0)$	5	0
2	$(0; 0; 1)$	$(0; 1; 0)$	2	1

The string function for the weight $[1, 0]$ in our example turns out to be the inverse of the Euler function (this will be demonstrated later):

$$\sigma_{[1,0]}^{([1,0])}(q) = \varphi(q)^{-1} = \prod_{n=1}^{\infty}(1 - q^n)^{-1} = \sum_{n=0}^{\infty} p(n)q^n \tag{14.144}$$

where $p(n)$ is the number of inequivalent decompositions of n into positive integers. The first few coefficients are $1, 1, 2, 3, 5, 7, 11, 15, \cdots$.

For more complicated representations, more than one string function will be required. We now see how many of them are actually needed. The complete information about the multiplicity of all the weights in the representation is contained in the set of string functions $\sigma_{\hat{\mu}}^{(\hat{\lambda})}(q)$ for all $\hat{\mu} \in \Omega_{\hat{\lambda}}^{max}$. However, since weight multiplicities are constant along Weyl orbits, that is,

$$\sigma_{w\hat{\mu}}^{(\hat{\lambda})}(q) = \sigma_{\hat{\mu}}^{(\hat{\lambda})}(q) \tag{14.145}$$

it is sufficient to know the string functions for those weights in $\Omega_{\hat{\lambda}}^{max}$ that are also dominant. (Recall that a Weyl orbit contains exactly one element in the fundamental chamber.) We note further that all the weights in $\Omega_{\hat{\lambda}}$ must also be in the same congruence class as $\hat{\lambda}$ (or λ). The number of independent string functions required to fully specify the representation of highest weight $\hat{\lambda}$ is thus equal to the number of integrable weights at level k that are in the same congruence class as $\hat{\lambda}$. Take for instance $\widehat{su}(2)_2$; there are three integrable weights: $[2, 0], [0, 2], [1, 1]$. The first two belong to the same class. Thus, two string functions are required for the module $L_{[2,0]}$.

The consideration of string functions brings us naturally to the characters of the integrable representations.

§14.4. Characters

14.4.1. Weyl-Kac Character Formula

The character of an integrable highest-weight representation is defined as

$$\mathrm{ch}_{\hat{\lambda}} = \sum_{\hat{\lambda}' \in \Omega_{\hat{\lambda}}} \mathrm{mult}_{\hat{\lambda}}(\hat{\lambda}') e^{\hat{\lambda}'} \tag{14.146}$$

In terms of string functions, this is just

$$\mathrm{ch}_{\hat{\lambda}} = \sum_{\hat{\mu} \in \Omega_{\hat{\lambda}}^{max}} \sigma_{\hat{\mu}}^{(\hat{\lambda})}(e^{-\delta}) e^{\hat{\mu}} \tag{14.147}$$

This expression can be rewritten as

$$\boxed{\mathrm{ch}_{\hat{\lambda}} = \frac{\sum_{w \in \hat{W}} \epsilon(w) e^{w(\hat{\lambda}+\hat{\rho})}}{\sum_{w \in \hat{W}} \epsilon(w) e^{w\hat{\rho}}}} \tag{14.148}$$

a formula known as the *Weyl-Kac character formula*. An alternate expression of the character is

$$\mathrm{ch}_{\hat{\lambda}} = \frac{\sum_{w\in\hat{W}} \epsilon(w)e^{w(\hat{\lambda}+\hat{\rho})}}{e^{\hat{\rho}} \prod_{\hat{\alpha}>0}(1 - e^{-\hat{\alpha}})^{\mathrm{mult}(\hat{\alpha})}} \tag{14.149}$$

Since the character for the representation of highest weight $\hat{\lambda} = 0$ is 1, there follows the famous *Macdonald-Weyl denominator identity*

$$\boxed{\sum_{w\in\hat{W}} \epsilon(w)e^{w\hat{\rho}} = e^{\hat{\rho}} \prod_{\hat{\alpha}>0}(1 - e^{-\hat{\alpha}})^{\mathrm{mult}(\hat{\alpha})}} \tag{14.150}$$

which is the root of many combinatorial formulae. For instance, applying it to $\widehat{su}(2)$,

$$\prod_{n=1}^{\infty}(1 - x^n y^n)(1 - x^n y^{n-1})(1 - x^{n-1}y^n) = \sum_{n\in\mathbb{Z}}(-1)^n x^{n(n+1)/2}y^{n(n-1)/2} \tag{14.151}$$

with $x = e^{-\alpha_0}$ and $y = e^{-\alpha_1}$. Upon specialization, it reduces to various classical partition identities, including the Jacobi triple-product identity (cf. Exs. 14.7 and 14.8).

In terms of summations over the full affine Weyl group, the character formula (14.148) is not very useful. A more convenient expression is obtained by taking advantage of the semidirect product structure of the affine Weyl group, which allows factorization of a summation over the finite Weyl group:

$$\sum_{w\in\hat{W}} \epsilon(w)e^{w\hat{\lambda}} = \sum_{w\in W} \epsilon(w) \sum_{\alpha^\vee\in Q^\vee} e^{w(t_{\alpha^\vee})\hat{\lambda}} = \sum_{w\in W} \epsilon(w) \sum_{\alpha^\vee\in Q^\vee} e^{(t_{\alpha^\vee})w\hat{\lambda}} \tag{14.152}$$

In the last equality, we have used the identity

$$w(t_{\hat{\alpha}^\vee}) = (t_{w\hat{\alpha}^\vee})w \tag{14.153}$$

and the invariance of the coroot lattice under the action of the finite Weyl group. Next, we introduce the *generalized theta function*

$$\boxed{\Theta_{\hat{\lambda}} = e^{-\frac{1}{2k}|\hat{\lambda}|^2\delta} \sum_{\alpha^\vee\in Q^\vee} e^{(t_{\alpha^\vee})\hat{\lambda}}} \tag{14.154}$$

With $\hat{\lambda} = (\lambda; k; 0)$ and the explicit expression (14.69) for $t_{\hat{\alpha}^\vee}$, the above expression can be manipulated as follows:

$$\Theta_{\hat{\lambda}} = \sum_{\alpha^\vee\in Q^\vee} e^{(\lambda+k\alpha^\vee;k;-\frac{1}{2k}|\lambda+k\alpha^\vee|^2)}$$

$$= e^{k\hat{\omega}_0} \sum_{\alpha^\vee\in Q^\vee} e^{k(\alpha^\vee+\lambda/k;0;-\frac{1}{2}|\alpha^\vee+\lambda/k|^2)} \tag{14.155}$$

$$= e^{k\hat{\omega}_0} \sum_{\alpha^\vee\in Q^\vee+\lambda/k} e^{k[\alpha^\vee-\frac{1}{2}|\alpha^\vee|^2\delta]}$$

In the second and third equality we used $\hat{\omega}_0 = (0; 1; 0)$ and $\alpha^\vee = (\alpha^\vee; 0; 0)$, $\alpha^\vee$ being a finite coroot. This result has been obtained for a highest weight such that $\hat{\lambda}(L_0) = 0$. But since the action of an element of the finite Weyl group cannot change the L_0 eigenvalue, we write

$$\sum_{w \in \hat{W}} \epsilon(w) e^{w\hat{\lambda}} = \sum_{w \in W} \epsilon(w) e^{\frac{1}{2k}|\lambda|^2 \delta} \Theta_{w\hat{\lambda}} \tag{14.156}$$

If the level of $\hat{\lambda}$ is k, that of $\hat{\lambda} + \hat{\rho}$ is $k + g$, and the character takes the compact form

$$\text{ch}_{\hat{\lambda}} = e^{m_{\hat{\lambda}} \delta} \frac{\sum_{w \in W} \epsilon(w) \Theta_{w(\hat{\lambda}+\hat{\rho})}}{\sum_{w \in W} \epsilon(w) \Theta_{w\hat{\rho}}} \tag{14.157}$$

where the quantity $m_{\hat{\lambda}}$, the *modular anomaly*, is defined as

$$m_{\hat{\lambda}} = \frac{|\hat{\lambda} + \hat{\rho}|^2}{2(k+g)} - \frac{|\hat{\rho}|^2}{2g} = \frac{|\lambda + \rho|^2}{2(k+g)} - \frac{|\rho|^2}{2g} \tag{14.158}$$

As in the finite case, it is often useful to evaluate the characters at an arbitrary point, denoted

$$\hat{\xi} = -2\pi i(\zeta; \tau; t) \tag{14.159}$$

We then use the notation

$$\text{ch}_{\hat{\lambda}}(\hat{\xi}) \equiv \text{ch}_{\hat{\lambda}}(\zeta; \tau; t) \tag{14.160}$$

The explicit expression of the theta function at this point is

$$\Theta_{\hat{\lambda}}(\hat{\xi}) = e^{-2\pi i k t} \sum_{\alpha^\vee \in Q^\vee} e^{-\pi i [2k(\alpha^\vee, \zeta) + 2(\lambda, \zeta) - \tau k |\alpha^\vee + \lambda/k|^2]} \tag{14.161}$$

Another widely used form for the character evaluated at a special point follows from expanding ζ as

$$\zeta = \sum_{i=1}^{r} z_i \alpha_i^\vee \tag{14.162}$$

so that

$$(\lambda, \zeta) = \lambda \cdot z = \sum_{i=1}^{r} z_i \lambda_i \tag{14.163}$$

Since little information is contained in the t dependence of the characters, they are usually presented as functions of the parameters τ and $z = (z_1, z_2, \cdots, z_r)$:

$$\text{ch}_{\hat{\lambda}}(z; \tau) \equiv \text{ch}_{\hat{\lambda}}\left(\sum_{i=1}^{r} z_i \lambda_i; \tau; 0\right) \tag{14.164}$$

This expression is the same as

$$
\mathrm{ch}_{\hat{\lambda}}(z; \tau) = \sum_{\hat{\lambda}'} \mathrm{mult}_{\hat{\lambda}}(\hat{\lambda}')\, e^{-2\pi i \tau(\hat{\lambda}',\hat{\omega}_0)} e^{-2\pi i z_i \lambda_i'}
$$
$$
= \sum_{n} \sum_{\lambda'} \mathrm{mult}_{\hat{\lambda}}(\lambda')|_n e^{2\pi i \tau n} e^{-2\pi i z_i \lambda_i'} \tag{14.165}
$$

where, in the second equation, we used $\hat{\lambda}' = (\lambda'; k; -n)$, and denoted by $\mathrm{mult}_{\hat{\lambda}}(\lambda')|_n$ the multiplicity of λ' at grade n. More compactly, it reads

$$
\boxed{\mathrm{ch}_{\hat{\lambda}}(z; \tau) = \mathrm{Tr}_{\hat{\lambda}}\, e^{2\pi i \tau L_0} e^{-2\pi i \sum_j z_j h^j}} \tag{14.166}
$$

where h^j is a Chevalley generator, whose action on $|\hat{\lambda}'\rangle$ yields λ_j'. For $\widehat{su}(2)$, it is more standard to write $2z_1 = z$, since in the spin basis $J^0 = h/2$:

$$
\widehat{su}(2): \qquad \mathrm{ch}_{\hat{\lambda}}(z; \tau) = \mathrm{Tr}_{\hat{\lambda}}\, e^{2\pi i \tau L_0} e^{-2\pi i z J^0} \tag{14.167}
$$

When evaluated at $\zeta = t = 0$, that is for $\hat{\xi} = -2\pi i \tau \hat{\omega}_0$, the character is said to be *specialized*:

$$
\mathrm{ch}_{\hat{\lambda}}(\tau) \equiv \mathrm{ch}_{\hat{\lambda}}(0; \tau; 0) = \sum_{n \geq 0} d(n) q^n \qquad (q = e^{2\pi i \tau}) \tag{14.168}
$$

where $d(n)$ gives the total number of states at grade n. Another useful specialization is

$$
\hat{\xi} = -2\pi i \hat{\rho} x \tag{14.169}
$$

where x is some constant. This is called the *principal specialization*. By means of the denominator identity, the principally specialized character can be expressed in a product form, hence its usefulness.

Expression (14.157) makes natural the introduction of the *normalized* characters defined as

$$
\boxed{\chi_{\hat{\lambda}} = e^{-m_{\hat{\lambda}} \delta} \mathrm{ch}_{\hat{\lambda}} = \frac{\sum_{w \in W} \epsilon(w) \Theta_{w(\hat{\lambda} + \hat{\rho})}}{\sum_{w \in W} \epsilon(w) \Theta_{w\hat{\rho}}}} \tag{14.170}
$$

The specialized form of $\chi_{\hat{\lambda}}$ is

$$
\chi_{\hat{\lambda}}(q) = q^{m_{\hat{\lambda}}}\, \mathrm{Tr}_{\hat{\lambda}} q^{L_0} \tag{14.171}
$$

From now on we will mostly use the normalized characters, and the epithet *normalized* will frequently be omitted.

Extending the validity of the formula (14.170) to arbitrary weights (i.e., not necessarily dominant ones), and using the relation

$$
w \cdot \hat{\lambda} + \hat{\rho} = w(\hat{\lambda} + \hat{\rho}) \tag{14.172}
$$

leads directly to

$$
\chi_{w \cdot \hat{\lambda}} = \epsilon(w) \chi_{\hat{\lambda}} \tag{14.173}
$$

This simple formula will be used extensively. It has the following remarkable implication: if a weight $\hat{\lambda}$ is fixed under the action of an odd element of the Weyl group, which means that there is a $w \in \hat{W}$ with $\epsilon(w) = -1$ such that $\hat{\lambda} = w \cdot \hat{\lambda}$, then $\chi_{\hat{\lambda}} = 0$.

14.4.2. The $\widehat{su}(2)_k$ Characters

The character of the $\widehat{su}(2)_k$ integrable module $L_{\hat{\lambda}}$ is

$$\chi_{\lambda_1}^{(k)} = \frac{\Theta_{\lambda_1+1}^{(k+2)} - \Theta_{-\lambda_1-1}^{(k+2)}}{\Theta_1^{(2)} - \Theta_{-1}^{(2)}} \tag{14.174}$$

with the notation

$$\chi_{\lambda_1}^{(k)} \equiv \chi_{\hat{\lambda}} \qquad \text{and} \qquad \Theta_{\lambda_1}^{(k)} \equiv \Theta_{\hat{\lambda}} \qquad (\hat{\lambda} = [k - \lambda_1, \lambda_1]) \tag{14.175}$$

Although it is not manifest, the specialized form of this character is indeterminate. Indeed, consider the generic expression (14.161) for the $\widehat{su}(2)_k$ theta function evaluated at $\hat{\xi}$, where $\alpha^\vee = n\alpha_1$ and $\alpha_1^2 = 2$, with $\zeta = z\omega_1$:

$$\begin{aligned}
\Theta_{\lambda_1}^{(k)}(\zeta; \tau; t) &= e^{-2\pi i k t} \sum_{n \in \mathbb{Z} + \lambda_1/2k} e^{2\pi i k \tau n^2} e^{-2\pi i k n z} \\
&= e^{-2\pi i k t} \sum_{n \in \mathbb{Z}} e^{-2\pi i [knz + \frac{1}{2}\lambda_1 z - kn^2\tau - n\lambda_1\tau - \lambda_1^2\tau/4k]}
\end{aligned} \tag{14.176}$$

For $z = 0$ and $t = 0$, changing the sign of λ_1 affects only the sign of the exponent of $e^{2\pi i k n \lambda_1 \tau}$; however, this is irrelevant given that n is summed over all integers. Therefore, both the numerator and the denominator of the specialized character vanish. The specialized form of Eq. (14.174) must be obtained in a limit process, with $z \to 0$. We find

$$\chi_{\lambda_1}^{(k)}(\tau) = \frac{\sum_{n \in \mathbb{Z}} [\lambda_1 + 1 + 2n(k+2)] e^{2\pi i \tau [\lambda_1+1+2(k+2)n]^2/4(k+2)}}{\sum_{n \in \mathbb{Z}} [1 + 4n] e^{2\pi i \tau [1+4n]^2/8}} \tag{14.177}$$

In terms of q, it reads:

$$\chi_{\lambda_1}^{(k)}(q) = q^{(\lambda_1+1)^2/4(k+2)-\frac{1}{8}} \frac{\sum_{n \in \mathbb{Z}} [\lambda_1 + 1 + 2n(k+2)] q^{n[\lambda_1+1+(k+2)n]}}{\sum_{n \in \mathbb{Z}} [1 + 4n] q^{n[1+2n]}} \tag{14.178}$$

where the prefactor is recognized as being $q^{m_{\hat{\lambda}}}$.

The first few terms of the simple case $k = \lambda_1 = 1$ are:

$$\begin{aligned}
\chi_1^{(1)}(q) &= q^{\frac{5}{24}} \frac{(2 - 4q + 8q^5 - 10q^8 + \cdots)}{(1 - 3q + 5q^3 + 9q^{10} + \cdots)} \\
&= q^{\frac{5}{24}} (2 - 4q + 8q^5 - 10q^8 + \cdots)(1 + 3q + 9q^2 + 22q^3 + \cdots) \\
&= q^{\frac{5}{24}} (2 + 2q + 6q^2 + 8q^3 + \cdots)
\end{aligned} \tag{14.179}$$

The $su(2)$ irreducible content of the module $L_{[0,1]}$ for the first four grades is: $(1), (1), (3) \oplus (1), (3) \oplus 2(1)$, so that the number of states at these different grades is indeed 2, 2, 6, and 8.

We note finally that by means of the Jacobi identity[9]

$$[\varphi(q)]^3 = \sum_{n \in \mathbb{Z}_+} (-1)^n (2n+1) q^{n(n+1)/2} \tag{14.180}$$

the $\widehat{su}(2)_k$ characters can be written somewhat more compactly. First, we express the above sum with even and odd terms separated:

$$[\varphi(q)]^3 = \sum_{n \in \mathbb{Z}} (1 + 4n) q^{2n(n+1)} \tag{14.181}$$

Then, by using the relation between the Dedekind and the Euler functions

$$\eta(q) = q^{\frac{1}{24}} \varphi(q) \tag{14.182}$$

we find

$$\chi_{\lambda_1}^{(k)}(q) = \frac{q^{(\lambda_1+1)^2/4(k+2)}}{[\eta(q)]^3} \sum_{n \in \mathbb{Z}} [\lambda_1 + 1 + 2n(k+2)] q^{n[\lambda_1+1+(k+2)n]} \tag{14.183}$$

14.4.3. Characters of Heisenberg Algebra Modules

To finalize the presentation of the affine algebra characters, we consider the $\widehat{u}(1)$ case, also called the Heisenberg algebra. The module of the Heisenberg algebra is simply the usual Fock space of a free boson (cf. Sect. 6.3.3). The highest-weight state of the Fock space is $|\ell; \{0\}\rangle$ for any fixed eigenvalue ℓ of the operator a_0. It is annihilated by all raising operators $a_{n>0}$. Such modules are always irreducible, and states are of the form

$$a_{-1}^{n_1} a_{-2}^{n_2} \cdots |\ell; \{0\}\rangle \propto |\ell; n_1, n_2, \cdots\rangle \tag{14.184}$$

As for affine simple Lie algebras, the differentiation operator L_0 is included in the Cartan subalgebra of the affine extension of $u(1)$. The L_0 eigenvalue of a state in the Fock space is simply the sum of all occupation numbers n_i.[10] The number of states at a fixed grade m is thus given by the number of partitions of m (written $p(m)$) into positive integers, the occupation numbers n_i. The specialized character of the Heisenberg module is thus equal to the inverse of the Euler function:

$$\text{ch}_{\widehat{u}(1)}(q) = \sum_{n=0}^{\infty} p(n) q^n = \varphi(q)^{-1} \tag{14.185}$$

---

[9] A field-theoretical proof of the Jacobi identity is presented in App. 10.A, and the Lie-algebraic proof is sketched in Ex. 14.7.

[10] This is due to the convention that sets the L_0 eigenvalue equal to zero for the highest state. In a field-theory context, this eigenvalue will be shifted by a factor $\ell^2/2$.

We note that Eq. (14.185) is equal to the string function $\sigma_{[1,0]}^{([1,0])}(q)$; this equivalence will be explained in Sect. 15.6.2.

With r copies of Eq. (14.14), we construct an r-fold Heisenberg algebra, with the defining commutation relations

$$[a_n^i, a_m^j] = n\delta_{n+m,0}\delta_{i,j} \qquad i,j = 1,.,r \tag{14.186}$$

The restricted character for the module of a r-fold Heisenberg algebra is

$$\mathrm{ch}_{\hat{u}(1)^r}(q) = [\mathrm{ch}_{\hat{u}(1)}(q)]^r = \varphi(q)^{-r} = \sum_{n=0}^{\infty} p_r(n)q^n \tag{14.187}$$

where $p_r(m)$ is the number of ways of separating m into positive integers of "r colors." This character is the same as the string function for the weight $\hat{\omega}_0$ in the basic module of all simply-laced algebras of rank r.

14.4.4. The $\hat{u}(1)$ Characters Associated with the Free Boson on a Circle of Rational Square Radius

A complete discussion of $\hat{u}(1)$ characters requires a brief field-theoretical digression. A $\hat{u}(1)$ current algebra is simply the commutation relations for the modes of a free boson. However, if the boson is compactified on a circle of radius R, possible windings must be taken into account (cf. Sect. 10.4.1). The corresponding partition function has already been evaluated in Eq. (10.62). Writing it in the form

$$Z = \sum_{n,m} \chi_{n,m}(q)\bar{\chi}_{n,m}(\bar{q}) \tag{14.188}$$

we can extract the chiral character $\chi_{n,m}(q)$ as

$$\chi_{n,m}(q) = \frac{1}{\eta(q)}q^{\frac{1}{2}(n/R+mR/2)^2} \tag{14.189}$$

This is a simple $\hat{u}(1)$ character, related to that of the previous section by

$$\chi_{n,m}(q) = q^{h_{n,m}-1/24}\mathrm{ch}(q) \tag{14.190}$$

with

$$h_{n,m} = \frac{1}{2}(n/R + mR/2)^2 \tag{14.191}$$

The extra power of q has to be equal to the modular anomaly for $\chi_{n,m}$ to be a genuine affine normalized character. This is indeed the case, as indicated at the end of this section.

It turns out that when R^2 is rational, say[11]

$$R = \sqrt{\frac{2p'}{p}} \tag{14.192}$$

---

[11] Due to the $R \leftrightarrow 2/R$ duality, this form is not restrictive: The case $R = \sqrt{p'/q'}$ with p' odd is transformed to $2/R = \sqrt{4q'/p'}$.

this infinite number of $\widehat{u}(1)$ characters can be reorganized into a finite number of generalized characters (cf. also Ex. 10.21). With Eq. (14.192), the character reads:

$$\chi_{n,m}(q) = \frac{1}{\eta(q)} q^{pp'(n/2p'+m/2p)^2} \tag{14.193}$$

Introducing then the new integers

$$
\begin{aligned}
n &= 2p'n' + r \quad 0 \le r \le 2p' - 1 \quad n' \in \mathbb{Z} \\
m &= 2pm' + s \quad 0 \le s \le 2p - 1 \quad m' \in \mathbb{Z}
\end{aligned}
\tag{14.194}
$$

and setting

$$
\begin{aligned}
u &= n' + m' \in \mathbb{Z} \\
\ell &= pr + p's \in \mathbb{Z}
\end{aligned}
\tag{14.195}
$$

so that $0 \le \ell \le 4pp' - p - p'$, we find

$$\chi_{\ell,u}(q) = \frac{1}{\eta(q)} q^{pp'(u+\ell/2pp')^2} \tag{14.196}$$

The announced reorganization is obtained by summing over u:[12]

$$\boxed{\chi_\ell^{(pp')}(q) = \frac{1}{\eta(q)} \sum_{u \in \mathbb{Z}} q^{pp'(u+\ell/2pp')^2}} \tag{14.197}$$

The range of ℓ can be further restricted to $0 \le \ell < 2pp'$, or equivalently

$$-pp' + 1 \le \ell \le pp' \tag{14.198}$$

since

$$\chi_{\ell+2pp'}(q) = \chi_\ell(q) \tag{14.199}$$

These generalized characters are no longer simple $\widehat{u}(1)$ characters. The latter have been rearranged in terms of an extended algebra. This extended algebra is generated by the $\widehat{u}(1)$ current $i\partial\varphi$, whose modes yield the Heisenberg algebra, together with some extra currents. These currents are necessarily vertex operators of the form $e^{i\alpha\varphi}$, since these are the only available operators in a free-boson theory. Simple conditions fix the value of α. At first, these operators must be well defined when φ is replaced by $\varphi + 2\pi R$; this forces $\alpha R \in \mathbb{Z}$. Moreover, the currents generating an extended algebra are required to have integer conformal dimension: $|\alpha|^2/2 \in \mathbb{Z}$. Finally, the dimension of these currents must be symmetric in p, p' in order to respect the duality $R \leftrightarrow 2/R$, which amounts to the interchange $p \leftrightarrow p'$. Looking for the lowest nonzero dimensional solution of these requirements, we find

$$\alpha = \pm\sqrt{2pp'} \tag{14.200}$$

----

[12] The resulting characters are exactly the functions K_λ introduced in Eq. (10.109), with λ replaced by ℓ.

Hence, the generators of the extended algebras are

$$i\partial\varphi \qquad \text{and} \qquad \Gamma^\pm = e^{\pm i\sqrt{2pp'}\varphi} \tag{14.201}$$

(For $p = p' = 1$, this extended algebra is simply $\widehat{su}(2)_1$, as shown in Sect. 15.6). The primary fields of the extended theory are those vertex operators $e^{i\gamma\varphi}$ whose operator product expansions with the generators are local. In other words, they must have trivial monodromy with the currents. This fixes γ to be

$$\gamma = \frac{\ell}{\sqrt{2pp'}} \qquad \ell \in \mathbb{Z} \tag{14.202}$$

Their conformal dimension is

$$h_\ell = \frac{\ell^2}{4pp'} \tag{14.203}$$

For primary fields, the range of ℓ must be restricted to Eq. (14.198) since a shift of ℓ by $2pp'$ in $e^{i\ell\varphi/\sqrt{2pp'}}$ amounts to an insertion of the ladder operator Γ^+, which thereby produces a descendant field.

From the point of view of the extended conformal field theory, the characters are easily derived. A factor $q^{h_\ell - 1/24}/\eta(q)$—since the central charge is unity—takes care of the action of the free boson generators. To account for the effect of the distinct multiple applications of the generators $\Gamma^\pm$, which yield shifts of the momentum ℓ by integer multiples of $2pp'$, we must replace ℓ by $\ell + m2pp'$ and sum over m. The net result is

$$\begin{aligned} \chi_\ell^{(pp')}(q) &= \frac{1}{\eta(q)} \sum_{m\in\mathbb{Z}} q^{\frac{1}{4pp'}(\ell+m2pp')^2} \\ &= \frac{1}{\eta(q)} \sum_{m\in\mathbb{Z}} q^{pp'(m+\ell/2pp')^2} \end{aligned} \tag{14.204}$$

This is exactly Eq. (14.197).

In the following, the extended $\widehat{u}(1)$ theory defined on a rational square radius $R = \sqrt{2p'/p}$ will be denoted by $\widehat{u}(1)_{pp'}$. For an ordinary $\widehat{u}(1)$ theory, there is no notion of level.

We now reinterpret the extended chiral structure in a Lie-algebraic language. The action of Γ^+ on $e^{i\gamma\varphi}$ amounts to adding a "simple root" $\alpha = \sqrt{2pp'}$ to the weight γ. There is thus a natural root lattice:

$$Q = \sqrt{2pp'}\,\mathbb{Z} \tag{14.205}$$

whose dual is the weight lattice:

$$P = \mathbb{Z}/\sqrt{2pp'} \tag{14.206}$$

The ratio $P/Q = \mathbb{Z}/2pp'$ shows that there are $2pp'$ congruence classes, which can be labeled by an integer ℓ restricted by Eq. (14.198). To ℓ, we can associate the state $|\ell; \{0\}\rangle$, which is a highest-weight state with respect to the Heisenberg algebra.

The full representation is obtained by adding all positive and negative roots and taking into account all distinct actions of the $a_{n<0}$'s.

There is still another Lie-algebraic approach to the construction of the $\widehat{u}(1)_{pp'}$ characters, this time rooted in a $\widehat{su}(2)_k$ framework. When positive, ℓ can be viewed as the finite Dynkin label of an integrable $\widehat{su}(2)_{pp'}$ weight. This suggests evaluating the character $\mathrm{ch}_{\ell}^{(pp')}$ from the $\widehat{su}(2)_{pp'}$ character $\mathrm{ch}_{\hat{\lambda}}$, by ignoring the finite roots and their corresponding Weyl reflections. Equation (14.156) with $\rho = 0$ and $w = 1$ as the only finite Weyl contribution gives

$$\sum_{w \in \hat{W}} \epsilon(w) e^{w\hat{\lambda}} = e^{\frac{1}{2pp'}|\lambda|^2 \delta} \Theta_{\hat{\lambda}} \tag{14.207}$$

For the denominator, we use the product form (14.150), with $\hat{\rho} = 0$ and $\hat{\alpha} = n\delta$:

$$e^{\hat{\rho}} \prod_{\hat{\alpha}>0} (1 - e^{-\hat{\alpha}})^{\mathrm{mult}(\hat{\alpha})} = (1 - e^{-n\delta}) \tag{14.208}$$

When evaluated at $\xi = -2\pi i(z\omega_1; \tau; 0)$, this becomes $\varphi(q)$, and the full character is

$$\mathrm{ch}_{\ell}^{(pp')}(z; q) = \frac{q^{-\ell^2/4pp'}}{\varphi(q)} \Theta_{\ell}^{(pp')}(z; \tau) \tag{14.209}$$

where $\Theta_{\ell}^{(pp')}(z; \tau)$ is $\Theta_{\lambda_1}^{(k)}(\zeta; \tau; t)$ defined in Eq. (14.176), with $k = pp'$, $\lambda_1 = \ell$, $\zeta = z\omega_1$, and $t = 0$.

In the present context, the evaluation of the modular anomaly $m_{\hat{\lambda}} \equiv m_\ell$ looks ambiguous. By using a relation to be derived later (Eq. (15.122), with pp' playing the role of k), we can show m_ℓ to be

$$m_\ell = \frac{\ell^2}{4pp'} - \frac{1}{24} \tag{14.210}$$

The normalized character thus reads

$$\chi_{\ell}^{(pp')}(z; q) = \frac{1}{\eta(q)} \Theta_{\ell}^{(pp')}(z; \tau) \tag{14.211}$$

This expression holds for ℓ negative as well. At $z = 0$, it takes the simple form

$$\chi_{\ell}^{(pp')}(q) = \frac{1}{\eta(q)} \sum_{m \in \mathbb{Z} + \ell/2pp'} q^{pp'm^2} \tag{14.212}$$

This is again the same as Eq. (14.197).

This equivalence relies on the fact that the modular anomaly is simply $h - c/24$. As already mentioned, this links the nonnormalized $\widehat{u}(1)$ characters derived in the previous section to the normalized $\widehat{u}(1)$ characters written at the beginning of the present section. However, to understand this relation, we must be able to associate a conformal dimension to the highest weight of an integrable representation of an affine algebra. This will be done in the following chapter.

§14.5. Modular Transformations

The action of the modular group

$$\tau \to \frac{a\tau + b}{c\tau + d} \tag{14.213}$$

on the weight $(\zeta; \tau; t)$ is defined as

$$(\zeta; \tau; t) \to \left(\frac{\zeta}{c\tau + d}; \frac{a\tau + b}{c\tau + d}; t + \frac{c|\zeta|^2}{2(c\tau + d)} \right) \tag{14.214}$$

The modular transformation properties of the characters $\chi_{\hat{\lambda}}$ are derived in App. 14.B. They take the form

$$\chi_{\hat{\lambda}}(\zeta; \tau + 1; t) = \sum_{\hat{\mu} \in P_+^k} T_{\hat{\lambda}\hat{\mu}} \, \chi_{\hat{\mu}}(\zeta; \tau; t)$$

$$\chi_{\hat{\lambda}}(\zeta/\tau; -1/\tau; t + |\zeta|^2/2\tau) = \sum_{\hat{\mu} \in P_+^k} S_{\hat{\lambda}\hat{\mu}} \, \chi_{\hat{\mu}}(\zeta; \tau; t) \tag{14.215}$$

Even before looking at the explicit expressions for the matrices T and S, the crucial observation we should make from these transformation properties is that the characters of dominant highest-weight representations at some fixed level k transform into each other under the action of the modular group, that is, the summation on the r.h.s. is restricted to weights in P_+^k. This fact turns out to be at the core of modular covariance in (unitary) rational conformal field theories.

The matrix T is given by

$$T_{\hat{\lambda}\hat{\mu}} = \delta_{\hat{\lambda}\hat{\mu}} e^{2\pi i m_{\hat{\lambda}}} \tag{14.216}$$

showing that the transformation $\tau \to \tau + 1$ induces only a phase change. The matrix S reads

$$S_{\hat{\lambda}\hat{\mu}} = i^{|\Delta_+|} |P/Q^\vee|^{-\frac{1}{2}} (k + g)^{-r/2} \sum_{w \in W} \epsilon(w) e^{-2\pi i (w(\lambda + \rho), \mu + \rho)/(k + g)} \tag{14.217}$$

where $P/Q^\vee$ stands for the set of lattice points of P lying in an elementary cell of $Q^\vee$; $|P/Q^\vee|$ is the number of points in this set (e.g., for $su(2)$, it is 2 since $\alpha_1^\vee = \alpha_1 = 2\omega_1$). This number is easily calculated from the determinant of the matrix whose rows are the Dynkin labels of the coroots (cf. Eqs. (14.296)–(14.319) in App. 14.B):

$$|P/Q^\vee| = \det(\alpha_i^\vee, \alpha_j^\vee) = \det[(\alpha_i^\vee)_j] \tag{14.218}$$

For simply-laced algebras, this is the determinant of the Cartan matrix:

$$|P/Q^\vee| = |P/Q| = \det A_{ij} \qquad \text{(simply-laced)} \tag{14.219}$$

As before, $|\Delta_+|$ is the number of positive roots in the finite Lie algebra g.

Both matrices are unitary:

$$\boxed{T^{-1} = T^\dagger, \qquad S^{-1} = S^\dagger}$$

(14.220)

For the T matrix, this is obvious. For S, it follows from the unitarity of the $\tilde{S}$ matrices of theta functions, a verification that is left as an exercise (cf. Ex. 14.13).

Some simple examples of modular S matrices are presented next. For $\widehat{su}(2)_k$, since $|\Delta_+| = 1$, $|P/Q^\vee| = 2$, $g = 2$, and $|\omega_1|^2 = \frac{1}{2}$, the S matrix becomes

$$\boxed{S_{\hat{\lambda}\hat{\mu}} = \left[\frac{2}{k+2}\right]^{\frac{1}{2}} \sin\left[\frac{\pi(\lambda_1+1)(\mu_1+1)}{(k+2)}\right] \qquad 0 \le \lambda_1, \mu_1 \le k}$$

(14.221)

On the other hand, the S matrix for $\widehat{su}(3)_1$ is

$$S = \frac{1}{\sqrt{3}}\begin{pmatrix} 1 & 1 & 1 \\ 1 & \kappa & \kappa^2 \\ 1 & \kappa^2 & \kappa \end{pmatrix} \qquad \kappa = e^{2\pi i/3}$$

(14.222)

where the fields are ordered as $[1, 0, 0]$, $[0, 1, 0]$, $[0, 0, 1]$.

It should be stressed that these are the *normalized characters* that transform covariantly with respect to modular transformations. The string functions can be made modular covariant by a simple rescaling. These *normalized string functions* are defined as follows:

$$\boxed{c_{\hat{\mu}}^{(\hat{\lambda})}(q) \equiv q^{m_{\hat{\lambda}}(\hat{\mu})} \sigma_{\hat{\mu}}^{(\hat{\lambda})}(q)}$$

(14.223)

where $m_{\hat{\lambda}}(\hat{\mu})$ is the *relative modular anomaly*,

$$m_{\hat{\lambda}}(\hat{\mu}) = m_{\hat{\lambda}} - \frac{|\mu|^2}{2k}$$

(14.224)

with $m_{\hat{\lambda}}$ defined in Eq. (14.158). The various $c_{\hat{\mu}}^{(\hat{\lambda})}(q)$ with $\hat{\lambda}, \hat{\mu} \in P_+^k$ transform into themselves under modular transformations.

§14.6. Properties of the Modular S Matrix

14.6.1. The S Matrix and the Charge Conjugation Matrix

From the explicit form of the modular transformation S, we see that $S^2 \ne 1$. Indeed, under two successive transformations $\tau \to -1/\tau$,

$$(\zeta; \tau; t) \to (\frac{\zeta}{\tau}; \frac{-1}{\tau}; t + \frac{|\zeta|^2}{2\tau})$$

$$\to (-\zeta; \tau; t + \frac{|\zeta|^2}{2\tau} + \frac{|\zeta/\tau|^2}{2(-1/\tau)})$$

(14.225)

$$= (-\zeta; \tau; t)$$

Evaluating $\chi_{\hat{\lambda}}$ at $-2\pi i(-\zeta; \tau; t)$ yields the same result as evaluating $\chi_{\hat{\lambda}^*}$ at $-2\pi i(\zeta; \tau; t)$:

$$\chi_{\hat{\lambda}}(-\zeta; \tau; t) = \chi_{\hat{\lambda}^*}(\zeta; \tau; t) \tag{14.226}$$

This can be seen from the expression (14.170): the minus sign in front of ζ can be absorbed in a redefinition of the summation variable $w \to -\tilde{w}w_0$, so that

$$w \cdot \hat{\lambda} = (-\tilde{w}w_0) \cdot \hat{\lambda} = \tilde{w} \cdot (-w_0) \cdot \hat{\lambda} = \tilde{w} \cdot \hat{\lambda}^* \tag{14.227}$$

This produces an extra phase $\epsilon(w_0)$ in both the numerator and the denominator, which cancels out. This shows that S^2 is indeed the charge conjugation matrix (cf. Eq. (10.206))

$$\boxed{S^2 = C} \tag{14.228}$$

with

$$C\chi_{\hat{\lambda}} = \chi_{\hat{\lambda}^*} \tag{14.229}$$

This is simply illustrated with the $\widehat{su}(3)_1$ S matrix (14.222):

$$S^2 = \begin{pmatrix} 1 & 0 & 0 \\ 0 & 0 & 1 \\ 0 & 1 & 0 \end{pmatrix} \tag{14.230}$$

On S, the action of C is simply the usual complex conjugation:

$$\bar{S} = CS = SC \tag{14.231}$$

which is equivalent to

$$\bar{S}_{\hat{\lambda}\hat{\mu}} = S_{\hat{\lambda}^*,\hat{\mu}} = S_{\hat{\lambda},\hat{\mu}^*} \tag{14.232}$$

An alternative direct check of the above relation is obtained by comparing

$$\bar{S}_{\hat{\lambda}\hat{\mu}} = (-i)^{|\Delta_+|} C \sum_{w\in W} \epsilon(w) e^{2\pi i (w(\lambda+\rho),\mu+\rho)/(k+g)} \tag{14.233}$$

with

$$S_{\hat{\lambda}^*\hat{\mu}} = i^{|\Delta_+|} C \sum_{w\in W} \epsilon(w) e^{-2\pi i (w(\lambda^*+\rho),\mu+\rho)/(k+g)} \tag{14.234}$$

where C is some real constant. With $\lambda^* = -w_0 \cdot \lambda$, the change of the summation variable from w to ww_0 yields a residual factor $\epsilon(w_0)$. This factor is simply equal to $(-1)^{|\Delta_+|}$ (see App. 14.A), exactly what is needed to change $(-i)^{|\Delta_+|}$ into $i^{|\Delta_+|}$.

14.6.2. The S Matrix and the Asymptotic Form of Characters

For restricted characters, the second part of Eq. (14.215) reduces to:

$$\chi_{\hat{\lambda}}(-1/\tau) = \sum_{\hat{\mu}\in P_+^k} S_{\hat{\lambda}\hat{\mu}} \, \chi_{\hat{\mu}}(\tau) \tag{14.235}$$

which can also be written as

$$\chi_{\hat{\lambda}}(\tau) = \sum_{\hat{\mu} \in P_+^k} S_{\hat{\lambda}\hat{\mu}} \, \chi_{\hat{\mu}}(-1/\tau) \tag{14.236}$$

In this latter form, consider the limit $\tau \to i0^+$:

$$\lim_{\tau \to i0^+} \chi_{\hat{\mu}}(-1/\tau) = e^{-(2\pi i m_{\hat{\mu}}/\tau)}[\dim |\mu| + O(e^{-2\pi i/\tau})] \tag{14.237}$$

The dominant term in the sum (14.236) comes from the contribution of the representation with lowest modular anomaly, namely $\hat{\mu} = k\hat{\omega}_0$, whose representation contains only one term at grade zero. This yields

$$\boxed{\lim_{\tau \to i0^+} \chi_{\hat{\lambda}}(\tau) = S_{\hat{\lambda}0}e^{\pi i c/12\tau}} \tag{14.238}$$

where the parameter c is given by:

$$c = -24 \, m_{k\hat{\omega}_0} = \frac{k \dim g}{(k+g)} \tag{14.239}$$

(for the second equality, we used the strange formula (13.179)).

Equation (14.238) shows that the matrix element $S_{\hat{\lambda}0}$ is a real positive number (cf. the argument leading to Eq. (10.213)). That $S_{\hat{\lambda}0}$ is real is actually a simple consequence of Eq. (14.232) and the fact that the vacuum representation is self-conjugate. It can also be seen as follows. From Eq. (14.217), $S_{\hat{\lambda}0}$ is seen to be proportional to the sum over the finite Weyl group of $\epsilon(w)e^{w\rho}$, evaluated at the point ξ_σ:

$$\xi_\sigma = \frac{-2\pi i}{k+g}(\sigma + \rho) \tag{14.240}$$

This sum can be rewritten in product form by means of the Weyl denominator formula (13.165) (where D_ρ is defined in Eq. (13.164)):

$$S_{\hat{\lambda}0} = |P/Q^\vee|^{-\frac{1}{2}}(k+g)^{-r/2} \prod_{\alpha \in \Delta_+} 2 \sin\left(\frac{\pi(\alpha, \lambda + \rho)}{k+g}\right) \tag{14.241}$$

All factors of i have disappeared, as expected. Furthermore, for any positive root α, and $\hat{\lambda}$ an integrable weight, we have

$$0 \le (\alpha, \lambda + \rho) \le k + g \tag{14.242}$$

which shows that

$$S_{\hat{\lambda}0} > 0 \tag{14.243}$$

Actually, the following, stronger result is also a direct consequence of Eq. (14.241):

$$\boxed{S_{\hat{\lambda}0} \ge S_{00} > 0} \tag{14.244}$$

We note that the 0-th row of S is the only positive one.

14.6.3. The S Matrix and Finite Characters

By comparing the ratio

$$\gamma_{\hat{\lambda}}^{(\hat{\sigma})} \equiv \frac{S_{\hat{\sigma}\hat{\lambda}}}{S_{\hat{\sigma}0}} = \frac{\sum_{w\in W}\epsilon(w)e^{-2\pi i(\sigma+\rho,w(\lambda+\rho))/(k+g)}}{\sum_{w\in W}\epsilon(w)e^{-2\pi i(\sigma+\rho,w\rho)/(k+g)}} \tag{14.245}$$

with the finite character formula for the g representation of highest weight ν evaluated at ζ

$$\chi_\lambda(\zeta) = \frac{\sum_{w\in W}\epsilon(w)e^{(w(\lambda+\rho),\zeta)}}{\sum_{w\in W}\epsilon(w)e^{(w\rho,\zeta)}} \tag{14.246}$$

we conclude that

$$\boxed{\gamma_{\hat{\lambda}}^{(\hat{\sigma})} = \chi_\lambda(\xi_\sigma)} \tag{14.247}$$

where the character is evaluated at the special point ξ_σ defined in Eq. (14.240). In due course, we will extract important consequences from this remarkable relationship.

14.6.4. Outer Automorphisms and the Modular S Matrix

We now evaluate the action of the outer-automorphism group on the S matrix. This will bring new light on the isomorphism between the center of the group $B(G)$ and the outer-automorphism group $\mathcal{O}(\hat{g})$.

We consider then the action of $A \in \mathcal{O}(\hat{g})$ on the modular matrix S

$$AS_{\hat{\lambda}\hat{\mu}} = S_{A(\hat{\lambda})\hat{\mu}} \propto \sum_{w\in W}\epsilon(w)e^{-2\pi i(wA(\hat{\lambda}+\hat{\rho}),\mu+\rho)/(k+g)}$$

$$\propto \sum_{w\in W}\epsilon(w)e^{-2\pi i((k+g)wA\hat{\omega}_0+ww_A(\lambda+\rho),\mu+\rho)/(k+g)} \tag{14.248}$$

In the first step, we use the invariance of $\hat{\rho}$ under the action of A, so that

$$A\hat{\lambda} + \hat{\rho} = A(\hat{\lambda} + \hat{\rho}) \tag{14.249}$$

We then apply the formula (14.98), with $\hat{\lambda}$ replaced by $\hat{\lambda} + \hat{\rho}$ (whose level is $k+g$):

$$A(\hat{\lambda} + \hat{\rho}) = (k+g)(A-1)\hat{\omega}_0 + w_A(\hat{\lambda} + \hat{\rho}) \tag{14.250}$$

Since, from Eq. (14.114),

$$(wA\hat{\omega}_0, \mu+\rho) = (A\hat{\omega}_0, \mu+\rho) \mod \mathbb{Z} \tag{14.251}$$

the first exponential factor can be put outside the sum. Then, we let $\tilde{w} = ww_A$, with

$$\epsilon(\tilde{w}) = \epsilon(ww_A) = \epsilon(w)\epsilon(w_A) \tag{14.252}$$

and sum over $\tilde{w}$, to obtain

$$AS_{\hat{\lambda}\hat{\mu}} = \epsilon(w_A)S_{\hat{\lambda}\hat{\mu}}e^{-2\pi i(A\hat{\omega}_0,\mu+\rho)} \tag{14.253}$$

Recall that the signature of w_A can be written explicitly as

$$\epsilon(w_A) = e^{2\pi i(A\hat{\omega}_0, \rho)} \tag{14.254}$$

(cf. Eq. (14.100)), so that

$$\boxed{A\mathcal{S}_{\hat{\lambda}\hat{\mu}} = \mathcal{S}_{\hat{\lambda}\hat{\mu}} e^{-2\pi i(A\hat{\omega}_0, \mu)}} \tag{14.255}$$

The remaining phase factor is exactly the eigenvalue b, the element of the center associated with A (cf. Sect. 14.2.3 and in particular Eq. (14.117)) on $\hat{\mu}$, the second label of the modular $\mathcal{S}$ matrix. In a matrix notation, we can thus write

$$A\mathcal{S} = \mathcal{S}b \tag{14.256}$$

or, equivalently,

$$\boxed{\mathcal{S}^\dagger A\mathcal{S} = b} \tag{14.257}$$

This shows that the outer-automorphism group $\mathcal{O}(\hat{g})$ and the center $B(G)$ are not only isomorphic, they are also exactly $\mathcal{S}$ duals of each other! This key relation will be the cornerstone of a general method for the construction of modular-invariant partition functions in conformal field theories with Lie group symmetry.

§14.7. Affine Embeddings

14.7.1. Level of the Embedded Algebra

We consider an embedding of semisimple Lie algebras $p \subset g$. Since the generators of the Lie algebra p are linear combinations of those of g, this finite Lie algebra embedding has an affine extension of the form $\hat{p}_{\tilde{k}} \subset \hat{g}_k$. What has to be determined is the exact relationship between the levels $\tilde{k}$ and k, which amounts to a comparison of normalizations in two different algebras. Hence, the first step is to reinsert the appropriate factors of $|\theta|^2$ in the commutation relation (14.5). Recall that the level is obtained from the scalar product of $\hat{\lambda}$ with δ (see Eq. (14.55)), and that all scalar products have been normalized in terms of $|\theta|^2 = 2$. If we avoid fixing the normalization, we must write

$$k = k' \frac{|\theta|^2}{2} \tag{14.258}$$

where now k' is called the level. It is still an integer given by

$$k' = \sum_{i=0}^{r} a_i^\vee \lambda_i \in \mathbb{Z}_+ \tag{14.259}$$

With this modification, the commutator (14.5) takes the form

$$[J_n^a, J_m^b] = i f_{abc} J_{n+m}^c + k' n \frac{|\theta|^2}{2} \delta_{a,b} \delta_{n+m,0} \tag{14.260}$$

We denote the p generators by $\tilde{J}^{a'}$ (in an orthonormal basis), with $a' = 1, \cdots , \dim$ p. They are generically related to the g generators through a linear combination of the form

$$\tilde{J}^{a'} = \sum_a m_{a'a} J^a \tag{14.261}$$

Orthonormality with respect to the Killing form forces

$$\sum_a m_{a'a} m_{b'a} = \delta_{a'b'} \tag{14.262}$$

These expressions together with Eq. (14.260) give the $\hat{p}$ commutation relations. The coefficient of the central term remains unaffected (now proportional to $\delta_{a'b'}\delta_{n+m,0}$). This coefficient can be expressed in terms of ϑ, the highest root of p:

$$k' \frac{|\theta|^2}{2} = k' \frac{|\vartheta|^2}{2} \frac{|\theta|^2}{|\vartheta|^2} \equiv \tilde{k}' \frac{|\vartheta|^2}{2} \tag{14.263}$$

In other words, the level $\tilde{k}$ of the embedded affine algebra $\hat{p}$ is related to that of $\hat{g}$ by the simple relation

$$\tilde{k} = k \frac{|\theta|^2}{|\vartheta|^2} \tag{14.264}$$

Of course, in order to compare the length of the two roots, we should project the g root onto the p algebra. The ratio $|\mathcal{P}\theta|^2/|\vartheta|^2$ is simply the embedding index x_e defined in Eq. (13.250):

$$\boxed{\tilde{k} = k x_e} \tag{14.265}$$

Clearly, $x_e \geq 1$, which implies that $\tilde{k} \geq k$. (Note also that x_e is always an integer, which is made manifest by Eq. (13.251).)

14.7.2. Affine Branching Rules

The next point of interest is the calculation of affine branching rules, that is, the coefficients $b_{\hat{\lambda},\hat{\mu}}$ in the decomposition

$$L_{\hat{\lambda}} \mapsto \bigoplus_{\hat{\mu}} b_{\hat{\lambda},\hat{\mu}} L_{\hat{\mu}} \tag{14.266}$$

In principle, this calculation is straightforward but tedious. We decompose grade by grade the module $L_{\hat{\lambda}}$ into irreducible representations of p and then reorganize the results into a direct sum of affine $\hat{p}$ modules $L_{\hat{\mu}}$. A few simple examples will clarify the procedure.

Consider first the decomposition of the $\widehat{su}(3)_1$ module $L_{[1,0,0]}$ into the $\widehat{su}(2)_1$ modules $L_{[1,0]}$ and $L_{[0,1]}$. (This is the affine extension of the regular embedding $su(2) \subset su(3)$ with $x_e = 1$.) The first step is the construction of the modules under consideration. The result is presented in Tables 14.3 and 14.4, for the first five grades. The third column of Table 14.3 gives the appropriate grade-by-grade $su(2)$ decomposition of the irreducible $su(3)$ representations of the module $L_{[1,0,0]}$. These states should then be reorganized in terms of the two integrable $\widehat{su}(2)_1$ modules $L_{[1,0]}$ and $L_{[0,1]}$. To proceed, it is convenient to express a module decomposition into irreducible representations of the corresponding finite Lie algebra:

$$L_{\hat{\lambda}} = \sum_n q^n \bigoplus_i L_{\lambda^{(i,n)}} \qquad (14.267)$$

where q is a parameter keeping track of the grade and the $L_{\lambda^{(i,n)}}$'s denote the irreducible representations of g at grade n. For instance,

$$[1,0,0] = (0,0) + q\,(1,1) + q^2\,[2(1,1) \oplus (0,0)] + \cdots \qquad (14.268)$$

Hence, for the embedding under consideration,

$$[1,0,0] \mapsto (0) + q\,[(2) \oplus 2(1) \oplus (0)] + q^2\,[2(2) \oplus 4(1) \oplus 3(0)] + \cdots \quad (14.269)$$

This has to be reexpressed in terms of the $\widehat{su}(2)_1$ representations $[1,0]$ and $[0,1]$. Using the decomposition presented in Table 14.4, we find

$$\begin{aligned}[1,0,0] \mapsto &(1 + q + 2q^2 + 5q^3 + 6q^4 + \cdots)\,[1,0] \\ &\oplus (2q + 2q^2 + 4q^3 + 6q^4 + \cdots)\,[0,1]\end{aligned} \qquad (14.270)$$

The branching functions are thus:

$$\begin{aligned}b_{[1,0,0],[1,0]} &= (1 + q + 2q^2 + 5q^3 + 6q^4 + \cdots) \\ b_{[1,0,0],[0,1]} &= (2q + 2q^2 + 4q^3 + 6q^4 + \cdots)\end{aligned} \qquad (14.271)$$

These are infinite series in q, meaning that there is an infinite number of terms in the decomposition of $L_{[1,0,0]}$. This is a generic feature. However, there are exceptions, as the next example demonstrates.

Table 14.3. The $su(2)$ decomposition of each grade of the $\widehat{su}(3)_1$ module $L_{[1,0,0]}$ for the two embeddings $su(2) \subset su(3)$.

	$L_{[1,0,0]}$	module $su(2)$ decomposition	
Grade	$su(3)$ content	$x_e = 1$	$x_e = 4$
0	$(0,0)$	(0)	(0)
1	$(1,1)$	$(2) \oplus 2(1) \oplus (0)$	$(4) \oplus (2)$
2	$2(1,1) \oplus (0,0)$	$2(2) \oplus 4(1) \oplus 3(0)$	$2(4) \oplus 2(2) \oplus (0)$
3	$(3,0) \oplus (0,3)$	$2(3) \oplus 5(2)$	$2(6) \oplus 3(4)$
	$\oplus 3(1,1) \oplus 2(0,0)$	$\oplus 8(1) \oplus 7(0)$	$\oplus 5(2) \oplus 2(0)$
4	$(2,2) \oplus (3,0) \oplus (0,3)$	$(4) \oplus 4(3) \oplus 11(2)$	$(8) \oplus 3(6) \oplus 8(4)$
	$\oplus 6(1,1) \oplus 3(0,0)$	$\oplus 16(1) \oplus 12(0)$	$\oplus 8(2) \oplus 4(0)$

Table 14.4. The $su(2)$ irreducible representations of the first five grades in the two integrable $\widehat{su}(2)_1$ modules.

Grade	$L_{[1,0]}$	$L_{[0,1]}$
0	(0)	(1)
1	(2)	(1)
2	$(2) \oplus (0)$	$(3) \oplus (1)$
3	$2(2) \oplus (0)$	$(3) \oplus 2(1)$
4	$(4) \oplus 2(2) \oplus 2(0)$	$2(3) \oplus 3(1)$

Consider now the affine extension of the special embedding $su(2) \subset su(3)$ (for which $x_e = 4$) at level 1, that is, the affine embedding $\widehat{su}(2)_4 \subset \widehat{su}(3)_1$. We again concentrate our attention on the module $L_{[1,0,0]}$. The appropriate $su(2)$ decomposition at the first five grades is given by the fourth column of Table 14.3. Since the corresponding finite embedding is really an embedding of $so(3)$ into $su(3)$, all $su(3)$ weights are projected onto $su(2)$ weights with even finite Dynkin labels. Hence, only the modules $L_{[4,0]}$, $L_{[0,4]}$, and $L_{[2,2]}$ are relevant in the decomposition of $L_{[1,0,0]}$. Their $su(2)$ decomposition is presented in Table (14.5). Quite remarkably, we find that

$$[1,0,0] \mapsto [4,0] \oplus q[0,4] \tag{14.272}$$

so that the two branching functions contain a single term. A similar analysis with the other integrable modules $\widehat{su}(3)_1$ $L_{[0,1,0]}$ and $L_{[0,0,1]}$ yields

$$[0,1,0] \mapsto [2,2], \qquad [0,0,1] \mapsto [2,2] \tag{14.273}$$

This finite reducibility is particular to the level-1 decomposition of the $\widehat{su}(3)$ modules: for all $k > 1$, the branching functions for $\widehat{su}(2)_{4k} \subset \widehat{su}(3)_k$ contain an infinite number of terms. The precise conditions under which finite reducibility is possible will be presented in Sect. 17.5.

14.7.3. Branching of Outer Automorphism Groups

The construction of conformal field theories in terms of affine Lie algebra embeddings will raise a tricky issue concerning the precise determination of the physical spectra. To address this technicality, we need to know how group centers branch.

To an element $b \in B(G)$, there corresponds an element $\tilde{b}$ in the center of the group of p if, for any $\hat{\lambda} \in \hat{g}$, the eigenvalue of b on $\hat{\lambda}$ equals the eigenvalue of $\tilde{b}$ on the projection of $\hat{\lambda}$ onto a p weight. This relation is noted as follows:

$$b\hat{\lambda} \mapsto \tilde{b}\mathcal{P}\hat{\lambda}, \qquad \forall \hat{\lambda} \in \hat{g} \tag{14.274}$$

Table 14.5. $su(2)$ irreducible representations of each grade for the three integrable $\widehat{su}(2)_4$ modules with even finite Dynkin labels.

Grade	$L_{[4,0]}$	$L_{[0,4]}$	$L_{[2,2]}$
0	(0)	(4)	(2)
1	(2)	$(4) \oplus (2)$	$(4) \oplus (2) \oplus (0)$
2	$(4) \oplus (2)$	$(6) \oplus 2(4)$	$(6) \oplus 2(4)$
	$\oplus (0)$	$\oplus 2(2) \oplus (0)$	$\oplus 3(2) \oplus (0)$
3	$(6) \oplus (4)$	$2(6) \oplus 4(4)$	$2(6) \oplus 5(4)$
	$\oplus 3(2) \oplus (0)$	$\oplus 4(2) \oplus (0)$	$\oplus 5(2) \oplus 3(0)$
4	$(8) \oplus (6) \oplus 4(4)$	$(8) \oplus 4(6) \oplus 8(4)$	$(8) \oplus 5(6) \oplus 9(4)$
	$\oplus 4(2) \oplus 3(0)$	$\oplus 7(2) \oplus 3(0)$	$\oplus 11(2) \oplus 4(0)$

This implies

$$\mathcal{P}b\hat{\lambda} = \tilde{b}\mathcal{P}\hat{\lambda}, \qquad \forall\,\hat{\lambda} \in \hat{\mathfrak{g}} \tag{14.275}$$

or, more compactly,

$$\mathcal{P}b = \tilde{b}\mathcal{P} \tag{14.276}$$

Let A (resp. $\tilde{A}$) be the element of $\mathcal{O}(\hat{\mathfrak{g}})$ (resp. $\mathcal{O}(\hat{\mathfrak{p}})$) related to b (resp. $\tilde{b}$) through the eigenvalue equation

$$b\hat{\lambda} = \hat{\lambda}e^{2\pi i(A\hat{\omega}_0, \lambda)}$$
$$\tilde{b}\mathcal{P}\hat{\lambda} = \mathcal{P}\hat{\lambda}e^{2\pi i(\tilde{A}\hat{\omega}_0, \mathcal{P}\lambda)} \tag{14.277}$$

These eigenvalues are equal if and only if

$$(A\hat{\omega}_0, \lambda) = (\tilde{A}\hat{\omega}_0, \mathcal{P}\lambda) \qquad \text{mod } 1 \qquad \forall\hat{\lambda} \in \hat{\mathfrak{g}} \tag{14.278}$$

This correspondence between elements of $\mathcal{O}(\hat{\mathfrak{g}})$ and $\mathcal{O}(\hat{\mathfrak{p}})$ will be written under either the form

$$A \mapsto \tilde{A} \qquad \text{or} \qquad \mathcal{P}A = \tilde{A}\mathcal{P} \tag{14.279}$$

We note that such relations are level independent. From the center point of view, they characterize the finite embeddings as well as their affine extensions.

We look again at our familiar $su(2) \subset su(3)$ embeddings. For $su(3)$, A can be either $1, a$, or a^2, and

$$(\hat{\omega}_0, \lambda) = 0 \qquad (a\hat{\omega}_0, \lambda) = \frac{1}{3}(\lambda_1 + 2\lambda_2) \qquad (a^2\hat{\omega}_0, \lambda) = \frac{1}{3}(2\lambda_1 + \lambda_2) \tag{14.280}$$

In the regular embedding ($x_e = 1$), $\mathcal{P}_{(1)}\lambda = (\lambda_1 + \lambda_2)\omega_1$, and

$$(\hat{\omega}_0, \mathcal{P}_{(1)}\lambda) = 0 \qquad (\tilde{a}\hat{\omega}_0, \mathcal{P}_{(1)}\lambda) = \frac{1}{2}(\lambda_1 + \lambda_2) \tag{14.281}$$

since for $su(2)$ $\tilde{A}$ can be 1 or $\tilde{a}$. The comparison of these results shows the nonexistence of relations between $\mathcal{O}(\widehat{su}(2))$ and $\mathcal{O}(\widehat{su}(3))$ for the regular embedding, besides $1 \mapsto 1$. However, for the special embedding, $\mathcal{P}_{(4)}\lambda = 2(\lambda_1 + \lambda_2)\omega_1$ so that

$$(\tilde{a}\hat{\omega}_0, \mathcal{P}_{(4)}\lambda) = \lambda_1 + \lambda_2 = 0 \text{ mod } 1 \tag{14.282}$$

which reveals a nontrivial branching $1 \mapsto \tilde{a}$.[13]

There may exist more than one nontrivial relation, as the following diagonal embedding example shows. A diagonal embedding is of the form

$$\hat{g}_{k+k'} \subset \hat{g}_k \oplus \hat{g}_{k'} \tag{14.283}$$

Here, the pair of weights $\hat{\lambda}$, $\hat{\mu}$ of $\hat{g}_k \oplus \hat{g}_{k'}$ is projected onto the weight $\hat{\lambda} + \hat{\mu}$ of $\hat{g}_{k+k'}$. It is then easily checked that to any element A of $\mathcal{O}(\hat{g})$ corresponds the element $A \otimes A$ of $\mathcal{O}(\hat{g} \oplus \hat{g})$, because in that case Eq. (14.278) becomes

$$(A\hat{\omega}_0, \lambda) + (A\hat{\omega}_0, \mu) = (A\hat{\omega}_0, \lambda + \mu) \tag{14.284}$$

If $\mathcal{O}(\hat{g})$ has order N, this produces $N - 1$ nontrivial relations.

Appendix 14.A. A Technical Identity

In this appendix, we show that the signature of w_A, the element of the Weyl group associated with the outer automorphism A through Eq. (14.98), is

$$\epsilon(w_A) = e^{2\pi i(A\hat{\omega}_0, \rho)} = e^{-\pi i g |A\hat{\omega}_0|^2} \tag{14.285}$$

We recall that $w_A = w_i w_0$, where w_0 is the longest element of W, w_i is the longest element of $W_{(i)}$, and i is such that $A\hat{\omega}_0 = \hat{\omega}_i$. The signature of w_A is thus

$$\epsilon(w_A) = (-1)^{\ell(w_0)}(-1)^{\ell(w_i)} \tag{14.286}$$

where $\ell(w)$ stands for the length of w. It is clear that

$$\ell(w_0) = |\Delta_+| \tag{14.287}$$

where $|\Delta_+|$ is the number of positive roots. Indeed, the weight space is divided into $2|\Delta_+|$ chambers. The chamber that is the farthest from the fundamental one is associated with w_0, which is thus an element of length $|\Delta_+|$. Similarly, $\ell(w_i) = |\Delta_+^{(i)}|$, where $\Delta_+^{(i)}$ is the set of positive roots generated by all simple roots except α_i:

$$\Delta_+^{(i)} = \{\alpha \in \Delta_+ |\ (\alpha, \omega_i) = 0\} \tag{14.288}$$

---

[13] Since the full symmetry group of $\widehat{su}(3)$ is the dihedral group D_3 (generated by charge conjugation and the outer automorphism group), we found that $D_3 \mapsto \mathbb{Z}_2$. The $\mathbb{Z}_2$ part of D_3 that survives the projection must be the charge conjugation, so that $* \mapsto \tilde{a}$ (where the star indicates the charge conjugation operation); in the present case, this cannot be distinguished from $1 \mapsto \tilde{a}$.

The roots of Δ_+ not belonging to $\Delta_+^{(i)}$ necessarily satisfy $(\alpha, \omega_i) = 1$ since $a_i = 1$.[14] We can thus write

$$
\begin{aligned}
|\Delta_+^{(i)}| &= |\Delta_+| - \sum_{\substack{\alpha \in \Delta_+ \\ (\alpha, \omega_i) = 1}} (\alpha, \omega_i) \\
&= |\Delta_+| - \sum_{\alpha \in \Delta_+} (\alpha, \omega_i) \\
&= |\Delta_+| - 2(\rho, \omega_i)
\end{aligned}
\tag{14.289}
$$

In the second equality we removed out the constraint $(\alpha, \omega_i) = 1$ since it does not affect the sum, and in the last step we used the definition of the Weyl vector as the half sum of positive roots. With $\hat{\omega}_i = A\hat{\omega}_0$, this verifies the first part of Eq. (14.285). Consider now Eq. (14.98) with $\hat{\lambda} = \hat{\rho} = \sum_{i=0}^{r} \hat{\omega}_i$. Using the invariance of $\hat{\rho}$ under A, we find

$$
\hat{\rho} - w_A \hat{\rho} = g(A - 1)\hat{\omega}_0
\tag{14.290}
$$

Taking the scalar product of both sides with $A\hat{\omega}_0$ yields

$$
(\rho - w_A \rho, A\hat{\omega}_0) = g|A\hat{\omega}_0|^2
\tag{14.291}
$$

(We recall that in the scalar product of an affine weight with a finite one, only the finite part of the former weight contributes.) Another property of ρ is that it is self-conjugate:

$$
\rho^* = -w_0 \rho = \rho
\tag{14.292}
$$

so that $\rho - w_A \rho = \rho + w_i \rho$. But since w_i involves reflections with respect to roots orthogonal to ω_i, $(w_i \rho, \omega_i) = (\rho, \omega_i)$, and finally

$$
2(\rho, A\hat{\omega}_0) = g|A\hat{\omega}_0|^2
\tag{14.293}
$$

completing the proof of Eq. (14.285).

Appendix 14.B. Modular Transformation Properties of Affine Characters

We recall that a d-dimensional Euclidian lattice Γ is defined in terms of some basis $\{\epsilon_i\}$ of $\mathbb{R}^d$ as

$$
\Gamma = \left\{ \sum_{i=1}^{d} n_i \epsilon_i \mid n_i \in \mathbb{Z} \right\}
\tag{14.294}
$$

---

[14] The mark associated with the i-th simple root is necessarily 1 because α_i is related to the zeroth root by an outer automorphism. $a_i = 1$ prevents $2\alpha_i + something$ from being a root (otherwise θ, whose expansion coefficient on α_i is $a_i = 1$, would not be the highest root), which justifies the above claim.

Its unit cell Γ_0 is

$$\Gamma_0 = \left\{ \sum_{i=1}^{d} x_i \epsilon_i | 0 \leq x_i < 1 \right\} \tag{14.295}$$

so that the intersection $\Gamma_0 \cap \Gamma$ contains a single point, the origin. The volume of the cell is

$$\text{vol}(\Gamma) = \sqrt{\det g_{ij}} \tag{14.296}$$

(it is standard to not write the subindex zero in the volume) where the metric g_{ij} is naturally given by

$$g_{ij} = \epsilon_i \cdot \epsilon_j \tag{14.297}$$

The dual lattice Γ^* is similarly defined, in terms of the dual basis $\{\epsilon_i\}$:

$$\epsilon_i \cdot \epsilon_j^* = \delta_{ij} \tag{14.298}$$

The volume of the unit cell of the dual lattice is related to that of Γ by

$$\text{vol}(\Gamma^*) = [\text{vol}(\Gamma)]^{-1} \tag{14.299}$$

The main tool in the analysis of modular properties of functions defined by a summation over a lattice is the *Poisson resummation formula*:

$$\sum_{x \in \Gamma} f(x) = \frac{1}{\text{vol}(\Gamma)} \sum_{p \in \Gamma^*} \tilde{f}(p) \tag{14.300}$$

where $\tilde{f}(p)$ is the Fourier transform of $f(x)$:

$$\tilde{f}(p) = \int_{\mathbb{R}^d} dx \, e^{2\pi i x \cdot p} f(x) \tag{14.301}$$

To prove this relation, we introduce the auxiliary function

$$F(z) = \sum_{x \in \Gamma} f(x + z) \tag{14.302}$$

This is manifestly a periodic function of z, which can thus be Fourier expanded as

$$F(z) = \sum_{p \in \Gamma^*} e^{-2\pi i z \cdot p} \tilde{F}(p) \tag{14.303}$$

with

$$\tilde{F}(p) = \frac{1}{\text{vol}(\Gamma)} \int_{\Gamma_0} dy \, e^{2\pi i y \cdot p} F(y) \tag{14.304}$$

We note that the range of integration is restricted to Γ_0, the fundamental period of $F(z)$. Substituting Eq. (14.302) into Eq. (14.304) and feeding back the result in Eq. (14.303) yields

$$F(z) = \sum_{p \in \Gamma^*} \frac{e^{-2\pi i z \cdot p}}{\text{vol}(\Gamma)} \int_{\Gamma_0} dy \sum_{x \in \Gamma} e^{2\pi i y \cdot p} f(x + y) \tag{14.305}$$

The integration over Γ_0 combined with the summation over all lattice points produces an integration over the whole space $\mathbb{R}^d$:

$$\int_{\Gamma_0} dy \sum_{x \in \Gamma} e^{2\pi i y \cdot p} f(x+y) = \int_{\mathbb{R}^d} dx' e^{2\pi i x' \cdot p} f(x') = \tilde{f}(p) \tag{14.306}$$

Here we used the fact that $x \cdot p \in \mathbb{Z}$ to replace $y \cdot p$ by $(x+y) \cdot p$ in the exponential factor. Thus,

$$\sum_{x \in \Gamma} f(x+z) = \frac{1}{\text{vol}(\Gamma)} \sum_{p \in \Gamma^*} e^{-2\pi i z \cdot p} \tilde{f}(p) \tag{14.307}$$

and the desired result follows by setting $z = 0$. Applied to a Gaussian-like function of the form

$$f(x) = e^{-i\pi x^2/\tau + 2\pi i x \cdot a} \tag{14.308}$$

Eq. (14.307) yields

$$\sum_{x \in \Gamma} e^{-i\pi x^2/\tau + 2\pi i x \cdot a} = \frac{1}{\text{vol}(\Gamma)} (-i\tau)^{d/2} \sum_{p \in \Gamma^*} e^{i\pi\tau(p+a)^2} \tag{14.309}$$

We recall that x is a d-dimensional vector; the factor $(-i\tau)^{d/2}$ comes from the d-independent Gaussian integrals of the Fourier transform:

$$\int_{\mathbb{R}^d} dx \, e^{-ix^2\pi/\tau} = \int_{-\infty}^{+\infty} (\prod_{i=1}^{d} dx^i) e^{-i\pi \sum_{j=1}^d x^j x^j/\tau} = (-i\tau)^{d/2} \tag{14.310}$$

The use of Gaussian integrals is justified since the imaginary part of τ is assumed to be positive.

We are now in a position to study the modular properties of the theta function

$$\Theta_{\hat{\lambda}} \equiv \Theta_\lambda^{(k)} \tag{14.311}$$

evaluated at $\hat{\xi} \equiv -2\pi i(\zeta; \tau; t)$:

$$\Theta_\lambda^{(k)}(\zeta; \tau; t) = e^{-2\pi i k t} \sum_{\alpha^\vee \in Q^\vee} e^{-2\pi i k(\alpha^\vee + \lambda/k, \zeta)} e^{i\pi k\tau|\alpha^\vee + \lambda/k|^2} \tag{14.312}$$

We consider first the S transformation:

$$\Theta_\lambda^{(k)}(\frac{\zeta}{\tau}; -\frac{1}{\tau}; t + \frac{|\zeta|^2}{2\tau}) = e^{-2\pi i t} \sum_{\alpha^\vee \in Q^\vee} e^{-i\pi k|\alpha^\vee + \lambda/k + \zeta|^2/\tau} \tag{14.313}$$

The Fourier transform of the function $f(\alpha^\vee)$ that follows the summation symbol is

$$\tilde{f}(p) = \int_{\mathbb{R}^r} d\alpha^\vee e^{2\pi i(\alpha^\vee, p)} e^{-i\pi k|\alpha^\vee + \lambda/k + \zeta|^2/\tau}$$
$$= (-\frac{i\tau}{k})^{r/2} e^{-2\pi i(p, \zeta + \lambda/k)} e^{i\pi\tau|p|^2/k} \tag{14.314}$$

(The dimension of the Euclidian lattice is now r, the rank of g). A direct application of Eq. (14.309) yields

$$\Theta_\lambda^{(k)}(\frac{\zeta}{\tau}; -\frac{1}{\tau}, t + \frac{|\zeta|^2}{2\tau}) = (-\frac{i\tau}{k})^{r/2} \frac{e^{-2\pi ikt}}{\text{vol}(Q^\vee)} \sum_{p \in P} e^{-2\pi i(p, \zeta + \lambda/k) + i\pi|p|^2 \tau/k} \quad (14.315)$$

the weight lattice P being the dual of $Q^\vee$. Note that $Q^\vee \subset P$ and *a fortiori* $kQ^\vee \subset P$ for any positive integer k. This allows a separation of the summation over P into two parts:

$$\sum_{p \in P} \tilde{f}(p) = \sum_{\beta^\vee \in kQ^\vee} \sum_{\mu \in P/kQ^\vee} \tilde{f}(\beta^\vee + \mu) = \sum_{\alpha^\vee \in Q^\vee} \sum_{\mu \in P/kQ^\vee} \tilde{f}(k\alpha^\vee + \mu) \quad (14.316)$$

In this way, the S transform of $\Theta_\lambda^{(k)}$ can be reexpressed as

$$\Theta_\lambda^{(k)}(\frac{\zeta}{\tau}; -\frac{1}{\tau}, t + \frac{|\zeta|^2}{2\tau}) = (-\frac{i\tau}{k})^{r/2} \frac{e^{-2\pi ikt}}{\text{vol}(Q^\vee)} \sum_{\mu \in P/kQ^\vee} e^{-2\pi i(\mu, \lambda)/k}$$
$$\times \sum_{\alpha^\vee \in Q^\vee} e^{-2\pi i(\alpha^\vee, \lambda)} e^{-2\pi ik|\alpha^\vee + \mu/k, \zeta|} e^{i\pi k\tau|\alpha^\vee + \mu/k|^2} \quad (14.317)$$

Since λ is an integrable weight, $(\alpha^\vee, \lambda) \in \mathbb{Z}$ and $e^{-2\pi i(\alpha^\vee, \lambda)} = 1$; the function that is summed over $Q^\vee$ is then just $\Theta_\mu^{(k)}$ (up to a multiplicative factor) evaluated at $\hat{\xi}$:

$$\Theta_\lambda^{(k)}(\frac{\zeta}{\tau}; -\frac{1}{\tau}; t + \frac{|\zeta|^2}{2\tau}) = (-\frac{i\tau}{k})^{r/2} \frac{1}{\text{vol}(Q^\vee)} \sum_{\mu \in P/kQ^\vee} e^{-2\pi i(\mu, \lambda)/k} \Theta_\mu^{(k)}(\zeta; \tau; t)$$
$$(14.318)$$

The factor $[\text{vol}(Q^\vee)]^{-1}$ can be written as

$$\frac{1}{\text{vol}(Q^\vee)} = \left[\frac{\text{vol}(P)}{\text{vol}(Q^\vee)}\right]^{\frac{1}{2}} \equiv |P/Q^\vee|^{-\frac{1}{2}} \quad (14.319)$$

This is a direct consequence of Eq. (14.299). $|P/Q^\vee|$ is the number of points of P in a unit cell of $Q^\vee$.

It turns out that the S matrix defined by

$$\Theta_\lambda^{(k)}(\frac{\zeta}{\tau}; -\frac{1}{\tau}; t + \frac{|\zeta|^2}{2\tau}) = \sum_{\mu \in P/kQ^\vee} \tilde{S}_{\lambda\mu}^{(k)} \Theta_\mu^{(k)}(\zeta; \tau; t) \quad (14.320)$$

is unitary. This can be justified as follows (dropping the superscript k):

$$(\tilde{S}\tilde{S}^\dagger)_{\lambda\mu} = \sum_{\sigma \in P/kQ^\vee} \tilde{S}_{\lambda\sigma} \tilde{S}_{\sigma\mu} = |P/Q^\vee|^{-1} k^{-r} \sum_{\sigma \in P/kQ^\vee} e^{-2\pi i(\lambda - \mu, \sigma)/k} \quad (14.321)$$

The sum over σ can be decomposed into different sums over roots of unity, which all vanish when $\lambda - \mu = 0$.[15] In that case, the contribution of the sum is simply

---

[15] More precisely, the sum $\sum \exp(2\pi i(\mu - \lambda, \sigma)/k)$ with $\sigma \in P/kQ^\vee$ splits into $|P/kQ^\vee|/n$ identical sums over the n n-th roots of unity (when $\lambda - \mu \neq 0$). Clearly, for $w = e^{2\pi i/n}$, $\sum_{j=1}^n w^j = 0$. The number n is defined as the smallest integer such that $n(\mu - \lambda, \sigma)/k \in \mathbb{Z}$ for all $\sigma \in P/kQ^\vee$. Here is

the number of terms it contains, namely

$$\sum_{\sigma\in(P/kQ^\vee)} 1 = \frac{\text{vol}(kQ^\vee)}{\text{vol}(P)} = k^r\frac{\text{vol}(Q^\vee)}{\text{vol}(P)} \equiv k^r\left|P/Q^\vee\right| \tag{14.322}$$

from which it follows that

$$\tilde{S}\tilde{S}^\dagger = I \tag{14.323}$$

We are actually interested in the modular transformation properties of the sum:

$$F_{\hat\lambda+\hat\rho} \equiv \sum_{w\in W} \epsilon(w)\Theta^{(k+g)}_{w(\lambda+\rho)} \tag{14.324}$$

From Eq. (14.318) it follows that

$$F_{\hat\lambda+\hat\rho}\left(\frac{\zeta}{\tau}; -\frac{1}{\tau}; t + \frac{|\zeta|^2}{2\tau}\right) = \left(-\frac{i\tau}{k+g}\right)^{r/2} |P/Q^\vee|^{-\frac12} \sum_{w\in W}\epsilon(w)$$

$$\times \sum_{\mu\in P/(k+g)Q^\vee} e^{-2\pi i(w(\lambda+\rho),\mu)/(k+g)}\Theta^{(k+g)}_{\mu}(\zeta;\tau;t) \tag{14.325}$$

Since $\mu \in P$, it is an integral weight. Further, it lives within a unit cell of $(k+g)Q^\vee$, which, for convenience, is supposed to be centered on the origin of P. All μ in $P/(k+g)Q^\vee$ satisfy $(\theta,\mu) \le (k+g)$. Notice that $\mu = 0$ does not contribute in Eq. (14.325) since in that case the summation over the finite Weyl group reduces to $\sum_{w\in W}\epsilon(w) = 0$. The sum over all $\mu \ne 0$ within $P/(k+g)Q^\vee$ can then be split into a sum over all μ in that part of the fundamental (finite) chamber delimited by $(k+g)Q^\vee$, plus a sum over the finite Weyl group

$$\sum_{\mu\in P/(k+g)Q^\vee} f(\mu) = \sum_{\substack{v\in P_+\\(v,\theta)\le k}} \sum_{w'\in W} f(w'(v+\rho)) \tag{14.326}$$

The weight v extends to integrable affine weight at level k. This yields

$$F_{\hat\lambda+\hat\rho}\left(\frac{\zeta}{\tau}; -\frac{1}{\tau}; t + \frac{|\zeta|^2}{2\tau}\right)$$

$$= K \sum_{w',w\in W}\epsilon(w) \sum_{\substack{v\in P_+\\(v,\theta)\le k}} e^{\frac{-2\pi i}{(k+g)}(w(\lambda+\rho),w'(v+\rho))}\Theta^{(k+g)}_{w'(v+\rho)}(\zeta;\tau;t) \tag{14.327}$$

$$= K \sum_{\substack{v\in P_+\\(v,\theta)\le k}} \sum_{w''\in W}\epsilon(w'')e^{-2\pi i(w''(\lambda+\rho),v+\rho)/(k+g)} F_{\hat v+\hat\rho}(\zeta;\tau;t)$$

where

$$K = \left(-\frac{i\tau}{k+g}\right)^{r/2}|P/Q^\vee|^{-\frac12} \tag{14.328}$$

---

another argument: for the present purpose, $\mu - \lambda$ can be restricted to $P/kQ^\vee$ (because of the scalar product with an integrable weight); the result is thus equivalent to the orthogonality of the characters of the Abelian group of $|P/kQ^\vee|$, interpreted as a finite group under addition.

To obtain the second equality, we set $w'' = (w')^{-1}w$ so that $\epsilon(w) = \epsilon(w')\epsilon(w'')$.

This takes care of the modular transformation property of the numerator of the normalized character. For the denominator, $\lambda = 0$, so that in Eq. (14.327) only $\nu = 0$ contributes:

$$F_{\hat{\rho}}(\frac{\zeta}{\tau}; -\frac{1}{\tau}; t + \frac{|\zeta|^2}{2\tau}) =$$
$$(-\frac{i\tau}{g})^{r/2} |P/Q^\vee|^{-\frac{1}{2}} \left(\sum_{w \in W} \epsilon(w) e^{-2\pi i(w\rho,\rho)/g} \right) F_{\hat{\rho}}(\zeta; \tau; t) \tag{14.329}$$

Performing another S transformation yields $F_{\hat{\rho}}(-\zeta; \tau; t)$, whose absolute value equals that of $F_{\hat{\rho}}(\zeta; \tau; t)$. This forces

$$\left| \sum_{w \in W} \epsilon(w) e^{-2\pi i(w\rho,\rho)/g} \right| \equiv |D_\rho(-2\pi i\rho/g)| = g^{r/2} |P/Q^\vee|^{\frac{1}{2}} \tag{14.330}$$

where we used the notation (13.165). The phase of $D_\rho(-2\pi i\rho/g)$ can be extracted as follows. From Eq. (13.165),

$$D_\rho(-2\pi i\rho/g) = \prod_{\alpha>0}(-2i)\,\sin(\pi(\alpha, \rho)/g) \tag{14.331}$$

For any positive root α, $(\alpha, \rho) > 0$. Furthermore, since

$$(\alpha, \rho) \le (\theta, \rho) = g - 1 \tag{14.332}$$

it follows that

$$0 < \sin(\pi(\alpha, \rho)/g) < 1 \qquad \text{for} \quad \alpha > 0 \tag{14.333}$$

This implies

$$D_\rho(-2\pi i\rho/g) = (-i)^{|\Delta_+|} |\prod_{\alpha>0} 2\,\sin(\pi(\alpha, \rho)/g)|$$
$$= (-i)^{|\Delta_+|} |D_\rho(-2\pi i\rho/g)| \tag{14.334}$$

and leads us to

$$F_{\hat{\rho}}(\frac{\zeta}{\tau}; -\frac{1}{\tau}; t + \frac{|\zeta|^2}{2\tau}) = (-i\tau)^{r/2}(-i)^{|\Delta_+|} F_{\hat{\rho}}(\zeta; \tau; t) \tag{14.335}$$

The final result is obtained from the ratio of Eqs. (14.327) and (14.335):

$$\chi_{\hat{\lambda}}(\frac{\zeta}{\tau}; -\frac{1}{\tau}; t + \frac{|\zeta|^2}{2\tau}) = i^{|\Delta_+|} |P/Q^\vee|^{-\frac{1}{2}} (k + g)^{-r/2}$$
$$\times \sum_{\hat{\nu} \in P_+^k} \sum_{w \in W} \epsilon(w) e^{-2\pi i(w(\lambda+\rho),\nu+\rho)/(k+g)} \chi_{\hat{\nu}}(\zeta; \tau; t) \tag{14.336}$$

The unitarity of this S matrix follows directly from that of $\tilde{S}$ (cf. Ex. 14.13).

For the T transformation, the analysis is much simpler. The replacement $\tau \to \tau + 1$ in Eq. (14.312) produces the extra factor $e^{i\pi k|\alpha^\vee + \lambda/k|^2}$, which reduces to

$e^{i\pi|\lambda|^2/k}$ because $|\alpha^\vee|^2 \in 2\mathbb{Z}$ and $(\alpha^\vee, \lambda) \in \mathbb{Z}$. Since this factor is unaffected by the replacement $\lambda \to w\lambda$, we have

$$F_{\hat{\lambda}+\hat{\rho}}(\zeta; \tau + 1; t) = e^{i\pi|\lambda+\rho|^2/(k+g)} F_{\hat{\lambda}+\hat{\rho}}(\zeta; \tau; t) \qquad (14.337)$$

and therefore

$$\chi_{\hat{\lambda}}(\zeta; \tau + 1; t) = e^{2\pi i m_{\hat{\lambda}}} \chi_{\hat{\lambda}}(\zeta; \tau; t) \qquad (14.338)$$

where $m_{\hat{\lambda}}$ is given by Eq. (14.158).

Appendix 14.C. Paths as a Basis of States

14.C.1. Basis for the Integrable Representations of $\widehat{su}(2)_1$

We now present an explicit basis for the $\widehat{su}(2)_1$ representation of highest weight [1, 0]. The states are associated with the various integer-spin excitations of a semi-infinite spin-$\frac{1}{2}$ quantum Ising chain in the antiferromagnetic regime. One of the two possible ground states is

$$|0\rangle = - + - + - + - + \cdots \qquad (14.339)$$

where $+$ or $-$ refer to spin up or down, respectively. We define the action of some generators $\tilde{f}^i$ and $\tilde{e}^i$ on the semiinfinite spin chain as follows:
$\tilde{f}^0$: freezes all pairs $(+ -)$ (in this order) and changes the rightmost (unfrozen) $-$ into a $+$.[16]
$\tilde{f}^1$: freezes all pairs $(- +)$ (in this order) and changes the rightmost (unfrozen) $+$ into a $-$.
In the first case, if there are no $-$'s left, the state is annihilated by $\tilde{f}^0$. A similar statement applies for the action of $\tilde{f}^1$. The action of the generators $\tilde{e}^0$ and $\tilde{e}^1$ is identical except that rightmost is replaced by leftmost and $+$ by $-$ (but the frozen pairs are the same). The freezing operation starts with adjacent pairs, but nonadjacent ones must also be considered. For instance, in order to calculate the action of $\tilde{f}^0$ on

$$+ + - + - + + - - + - \cdots$$

the pairs $(+ -)$ must be frozen as follows

$$+ (+ -) (+ -)(+ (+ -) -) (+ -) \cdots$$

so that only the initial $+$ is left.

Applied on the highest state, the $\tilde{f}^i$'s generate all the states in the representation. The highest state of the representation is the ground state $|0\rangle$ of the semiinfinite

---

[16] Parentheses can be fitted into each other, as illustrated below.

spin chain. Indeed, at grade zero this is the only state since $|0\rangle$ is annihilated by the action of $\tilde{f}^1$:

$$\tilde{f}^1 |0\rangle = \tilde{f}^1 - + - + - + - + \cdots$$
$$= \tilde{f}^1 (-+)(-+)(-+)(-+)\cdots = 0 \tag{14.340}$$

The different states at grade one are

$$\tilde{f}^0 |0\rangle = \tilde{f}^0 - (+-)(+-)(+-)\cdots = + + - + - + - \cdots$$
$$\tilde{f}^1 \tilde{f}^0 |0\rangle = \tilde{f}^1 + +(-+)(-+)(-+)\cdots = + - - + - + - \cdots$$
$$(\tilde{f}^1)^2 \tilde{f}^0 |0\rangle = \tilde{f}^1 + -(-+)(-+)(-+)\cdots = - - - + - + - \cdots \tag{14.341}$$

The action of $\tilde{f}^0$ on the second state generates a string of three states at grade two

$$\tilde{f}^0 \tilde{f}^1 \tilde{f}^0 |0\rangle = + - + + - + \cdots$$
$$\tilde{f}^1 \tilde{f}^0 \tilde{f}^1 \tilde{f}^0 |0\rangle = + - + - - + \cdots \tag{14.342}$$
$$(\tilde{f}^1)^2 \tilde{f}^0 \tilde{f}^1 \tilde{f}^0 |0\rangle = - - + - - + \cdots$$

The last state is annihilated by a further action of $\tilde{f}^1$. However, there is another state at grade two, obtained from the action of $\tilde{f}^0$ on $(\tilde{f}^1)^2 \tilde{f}^0 |0\rangle$,

$$\tilde{f}^0 (\tilde{f}^1)^2 \tilde{f}^0 |0\rangle = - - + + - + \cdots \tag{14.343}$$

We thus find two distinct states at grade two corresponding to the weight $[1,0]$, namely

$$+ - + - - + \cdots , \qquad - - + + - + \cdots \tag{14.344}$$

(The finite weight is given by twice the number of $\tilde{f}^0$'s minus the number of $\tilde{f}^1$'s.) Similarly, at grade three this weight is associated with the three different states:

$$+ - + - + - - + \cdots$$
$$- + + - - + - + \cdots \tag{14.345}$$
$$- - + - + + - + \cdots$$

The states of the other integrable representations of $\widehat{su}(2)_1$ are obtained by application of the operators $\tilde{f}^i$ on the vacuum

$$|\tilde{0}\rangle = + - + - + - + - \cdots \tag{14.346}$$

14.C.2. $\widehat{su}(N)_1$ Paths

The semiinfinite spin chains provide examples of $\widehat{su}(N)_1$ *paths*. Such a path is a sequence of positive integers taking values in the set $\{1, \cdots, N\}$ (for $su(2)$ replace $-$ and $+$ by 2 and 1, respectively) such that at large distances the path is a periodic repetition, in appropriate position, of the basic sequence $N, N-1, \cdots, 1$. These paths provide a basis for the integrable modules of $\widehat{su}(N)_1$.

Given a path representation for the highest state, the explicit path for the other states in the module can be obtained from the action of the operators $\tilde{f}^i$ defined as follows:

$\tilde{f}^i$: freezes all pairs $(i + 1, i)$ (in this order) and changes the rightmost (unfrozen) i into a $i + 1$.

Here i is defined modulo N, so that $\tilde{f}^0 = \tilde{f}^N$.

Stripping off the "vacuum tail" of a path, (i.e., the repeated sequences of $N, N - 1, \cdots, 1$), leaves a finite sequence of numbers, called a *word*, which characterizes a particular state in an irreducible representation of $su(N)$. Indeed, there is a one-to-one relation between a word and a semistandard tableau: the word of a semistandard tableau is the sequence of numbers appearing in the tableau, read column after column from bottom to top and from left to right. For instance, to the semistandard tableau

$$\begin{array}{|c|c|c|c|c|} \hline 1 & 1 & 2 & 3 & 5 \\ \hline 3 & 4 & 4 \\ \cline{1-3} 5 \\ \cline{1-1} \end{array}$$

we associate the word [531414235]. On the other hand, given a word, we can reconstruct the corresponding semistandard tableau by the "bumping" method. We place the leftmost number in the upper left box of the tableau being constructed. For the next number we proceed as follows: If it is greater than, or equal to, the number already placed in the first box, it is put in the second box of the first row; otherwise it takes the place of the first number in the first box, the latter being bumped in the first box of the second row. Proceeding in this way for all numbers of the word and allowing bumping in every row, we easily reconstruct the corresponding semistandard tableau. This relation between semistandard tableaux and words is known as the Robinson-Schensted correspondence. The step-by-step reconstruction of the tableau associated with the word [3121] is thus:

$$\boxed{3} \leftarrow [121] = \begin{array}{|c|} \hline 1 \\ \hline 3 \\ \hline \end{array} \leftarrow [21] = \begin{array}{|c|c|} \hline 1 & 2 \\ \hline 3 \\ \cline{1-1} \end{array} \leftarrow [1] = \begin{array}{|c|c|} \hline 1 & 1 \\ \hline 2 \\ \cline{1-1} 3 \\ \cline{1-1} \end{array}$$

In this way, the relation path $\to$ word $\to$ semistandard tableau for the two spin-zero states at grade two in the $\widehat{su}(2)_1$ module $L_{[1,0]}$ is:

$$1\,2\,1\,2\,2\,1\,2\,1\cdots \quad \to \quad 1\,2\,1\,2 \quad \to \quad \begin{array}{|c|c|c|} \hline 1 & 1 & 2 \\ \hline 2 \\ \cline{1-1} \end{array} = \begin{array}{|c|c|} \hline 1 & 2 \\ \hline \end{array} \quad \in (2)$$

$$2\,2\,1\,1\,2\,1\,2\,1\cdots \quad \to \quad 2\,2\,1\,1 \quad \to \quad \begin{array}{|c|c|} \hline 1 & 1 \\ \hline 2 & 2 \\ \hline \end{array} \quad \in (0)$$

$$(14.347)$$

This procedure provides a very nice way of recovering the irreducible $su(N)$ content of the representation at a fixed grade.

The path construction has a generalization for all positive levels but this will not be considered here.

Appendix 14.D. Notation for Affine Lie Algebras

$\tilde{g}$: loop extension of g

$\hat{k}$: central extension

k : level ($\hat{k}$ eigenvalue)

$\hat{g}_k$, $\hat{h}_k$: affine Lie algebras at fixed level k

J_n^a : generators of the affine Lie algebra

E_n^α, H_n^i ($i = 1, \cdots, r$) : generators in the Cartan-Weyl basis

e^i, f^i, h^i ($i = 0, \cdots, r$) : generators in the affine Chevalley basis

$\hat{\lambda}$, $\hat{\mu}$, $\hat{\nu}$: affine weights; $\hat{\lambda} = (\lambda; \hat{\lambda}(\hat{k}); \hat{\lambda}(-L_0))$ where λ is a finite weight

$\hat{\omega}_i \equiv (\omega_i; a_i^\vee; 0)$, $\hat{\omega}_0 = (0; 1; 0)$: fundamental weights

$\hat{\lambda} = \sum_{i=0}^r \lambda_i \hat{\omega}_i = [\lambda_0, \lambda_1, \ldots, \lambda_r]$: affine Dynkin labels

$\Omega_{\hat{\lambda}}$: weight system for the representation of highest weight $\hat{\lambda}$

$L_{\hat{\lambda}}$: irreducible module of highest weight $\hat{\lambda}$

P_+^k : set of integrable affine weights at level k

$\hat{\alpha} = (\alpha; 0; n)$ (for $\alpha \in \Delta$): real root

$\delta = (0; 0; 1)$: imaginary root

$\alpha_i \equiv (\alpha_i; 0; 0)$, $\alpha_0 = (-\theta; 0; 1)$: simple roots

$\hat{\Delta}$: set of affine roots

$\hat{\Delta}_+$: set of positive affine roots

$\hat{A}_{ij}$: affine Cartan matrix

$\hat{W}$: affine Weyl group

$\text{ch}_{\hat{\lambda}}$, $\chi_{\hat{\lambda}}$: character and normalized character of the module $L_{\hat{\lambda}}$

$\mathcal{S}$, $\mathcal{T}$: modular matrices

$\mathcal{C}$: charge conjugation matrix: $\mathcal{C} = \mathcal{S}^2$

$\mathcal{O}(\hat{g})$: group of outer automorphisms

a : generator of the outer-automorphism group (except for $D_{2\ell}$).

A : generic element of the outer-automorphism group

w_A : element of the finite Weyl group associated with A

Exercises

14.1 *Central extensions*

a) Prove that the central extension of a finite Lie algebra is necessarily trivial.

b) Verify the uniqueness of the central term in Eq. (14.5).

14.2 *Translation generators*

For $\widehat{sp}(4)$ and $\widehat{G}_2$, express the translation generators $t_{\alpha_1^\vee}$ and $t_{\alpha_2^\vee}$ in terms of the simple affine Weyl reflections s_0, s_1, s_2.

14. Affine Lie Algebras

14.3 *Basic representation of $\widehat{su}(2)_1$*
Extend Fig. 14.4 up to $L_0 = 6$ and relate together weights that are on the same Weyl orbit.

14.4 *Grade decompositions*
Find the finite Lie algebra irreducible content of the following representations at grade 1:
a) $(\widehat{G_2})_2$: $[0, 1, 0]$
b) $(\hat{B}_3)_1$: $[0, 1, 0, 0]$.

14.5 *Fixed points of the outer automorphisms*
A fixed point of an outer automorphism A is a weight $\hat{\lambda}$ such that $A\hat{\lambda} = \hat{\lambda}$. For instance, all $\widehat{su}(2)$ weights of the form $[n, n]$ are fixed points of the outer-automorphism group. For each simple affine algebra, find the general form of all integrable weights that are fixed under the generating element a whose action is described in Table 14.1.

14.6 *Identity involving outer automorphisms*
Prove the identity (14.293) by a case-by-case analysis.

14.7 *Euler-Jacobi identities*
a) Check the expression (14.151), using Eqs. (14.152) and (14.69).

b) By appropriate specialization of this identity, obtain the Jacobi triple-product identity, as well as

$$\varphi(q) = \sum_{n\in\mathbb{Z}}(-1)^n q^{n(3n-1)/2}$$

$$[\varphi(q)]^3 = \sum_{n\in\mathbb{Z}_+}(-1)^n(2n+1)q^{n(n+1)/2}$$

(respectively due to Euler and Jacobi).

14.8 *Rogers-Ramanujan-Gordon identities*
a) Evaluate Eq. (14.151) at the special point $-2\pi i\tau(\hat{\lambda} + \hat{\rho})$. Setting $\hat{\lambda} = k\hat{\omega}_0$ and $q = e^{2\pi i\tau}$ yields identities related to the Gordon identities ($k > 3$) and the Rogers-Ramanujan identity ($k = 3$). (Generalizations follow by considering a $\hat{\lambda} \in P_+^k$ different from $k\hat{\omega}_0$.)

b) Show that the expression obtained in **(a)** is invariant under the action of the outer-automorphism group (i.e., under the substitution $\lambda_1 \to k - \lambda_1$).

c) The character evaluated at the particular point $-2\pi i\tau\hat{\rho}$ is expressible as the ratio of the sum $\sum_{w\in\hat{W}} \epsilon(w)e^{w\hat{\rho}}$ evaluated once at the point $-2\pi i\tau(\hat{\lambda} + \hat{\rho})$ and then at $-2\pi i\tau\hat{\rho}$. Using the denominator identity, rewrite the characters for the modules $L_{[3,0]}$ and $L_{[2,1]}$ in a product form. (According to **(b)**, these are the two independent irreducible characters for $\widehat{su}(2)_3$ in this particular specialization.) Compare the results with the characters of two irreducible representations of the Yang-Lee singularity (i.e., the Virasoro minimal model of central charge $c = -\frac{22}{5}$).

14.9 *Macdonald identities for $\widehat{su}(2)_k$*
Given that the character of the representation with highest weight $\lambda = 0$ at $k = 0$ is one:

$$\chi_{[0,0]}(q) = 1 = \eta^{-3}(q)\sum_{n\in\mathbb{Z}}(1 + 4n)q^{(1+4n)^2/8}$$

derive the following identity:

$$\sum_{j=0}^{p-1}\chi_{[2p^2-1-(4j+1)p,(4j+1)p-1]}(q) = p$$

(Hint: Set $n = pm + j$ for some fixed positive integer p, $m \in \mathbb{Z}$ and $j = 0, 1, 2, \cdots, p-1$.)
If $(4j+1)p - 1 > k = 2p^2 - 2$, we can perform a shifted Weyl reflection. Check then that
the explicit form of this identity for the cases $p = 2, 3$ reads:

$$\chi_{[5,1]}(q) - \chi_{[1,5]}(q) = 2$$

$$\chi_{[14,2]}(q) + \chi_{[2,14]}(q) - \chi_{[8,8]}(q) = 3$$

Similar identities can be obtained for higher-rank algebras.

14.10 *Affine Kostant multiplicity formula*
The affine Kostant partition function is defined similarly as in the finite case:

$$\prod_{\hat{\alpha} \in \hat{\Delta}_+} (1 - e^{\hat{\alpha}})^{-\text{mult}(\hat{\alpha})} = \sum_{\hat{\mu}} \hat{\mathcal{K}}(\hat{\mu}) e^{\hat{\mu}}$$

a) Show that for $\widehat{su}(2)$ it reads

$$\hat{\mathcal{K}}(n_0 \alpha_0 + n_1 \alpha_1) = \sum_{k \geq 0} (-1)^k p_3((k+1)n_0 - kn_1 - \tfrac{1}{2}k(k+1))$$

where $p_3(n)$ is defined by $\sum_{n \geq 0} p_3(n) q^n = [\varphi(q)]^{-3}$.

b) Derive the affine form of the Kostant multiplicity formula and use it to calculate
$\text{mult}(0; 1; -2)$ in the module $L_{[1,0]}$.

14.11 *String functions*
The three independent $\widehat{su}(2)_2$ normalized string functions are

$$c_1 \equiv c_{[2,0]}^{([2,0])}(q) = q^{\frac{-1}{16}}(1 + q + 3q^2 + 5q^3 + 10q^4 + \cdots)$$

$$c_2 \equiv c_{[0,2]}^{([2,0])}(q) = q^{\frac{-9}{16}}(q + 2q^2 + 4q^3 + 7q^4 + 13q^5 + \cdots)$$

$$c_3 \equiv c_{[1,1]}^{([1,1])}(q) = (1 + q + 2q^2 + 3q^3 + 5q^4 + \cdots)$$

a) Check the correctness of the prefactors and the coefficient of q^2 in c_3 using the Freudenthal
multiplicity formula.

b) The exact form of these string functions is

$$c_+ \equiv c_1 + c_2 = \frac{\eta(q)}{\eta(\sqrt{q})\eta(q^2)}, \qquad c_- \equiv c_1 - c_2 = \frac{\eta(\sqrt{q})}{[\eta(q)]^2}, \qquad c_3 = \frac{\eta(q^2)}{[\eta(q)]^2}$$

where η is the Dedekind function $\eta(q) = q^{\frac{1}{24}} \prod_1^\infty (1 - q^n)$. Use the modular transformation
properties of $\eta(q)$ (cf. Eq. (10.12)) to derive those of c_+, c_-, and c_3.

14.12 *Generalized and Jacobi theta functions*
Find the relation between the $su(2)$ version (14.176) of the generalized theta function and
the Jacobi theta function $\theta_1(z, \tau)$ defined in Eq. (10.252).

14.13 *Modular transformation matrices*
a) Using the unitarity of the modular transformation matrix $\tilde{S}$ for the theta functions (given
in Eq. (14.318)), prove the unitarity of the S matrix (14.217) for the affine characters.

b) From the modular transformation matrices for the characters, derive those of the
normalized string functions.

14.14 $\widehat{su}(3)_1$ *modular matrix*
For $\widehat{su}(3)_1$, verify that the S matrix is indeed given by Eq. (14.222) and that $(ST)^3 = C$.

14.15 *Action of outer automorphisms on the modular T matrix*
Find the relation between $T_{A(\hat{\lambda})\hat{\mu}}$ and $T_{\hat{\lambda}\hat{\mu}}$, where A is an outer automorphism and T is the modular matrix (14.216).

14.16 *Branching of outer automorphisms*
For the following embeddings, find relations between elements of the corresponding outer-automorphism groups:

a) $su(2) \subset sp(4)$, $\mathcal{P} = (3, 4)$

b) $su(4) \subset so(7)$, $\mathcal{P} = \begin{pmatrix} 0 & 1 & 1 \\ 1 & 0 & 0 \\ 0 & 1 & 0 \end{pmatrix}$

c) $su(2) \oplus su(3) \subset su(6)$, $\mathcal{P} = \begin{pmatrix} 1 & 0 & 1 & 0 & 1 \\ 1 & 2 & 1 & 0 & 0 \\ 0 & 0 & 1 & 2 & 1 \end{pmatrix}$

14.17 *A regular embedding in E_6*
Consider the regular embedding:

$$su(2) \oplus su(6) \subset E_6$$

obtained by deleting from the extended Dynkin diagram of E_6 the simple root α_6. The projection of an E_6 weight λ onto $su(2) \oplus su(6)$ is then easily described: the $su(2)$ Dynkin label is the zeroth label of the affine extension of the E_6 weight at level zero, namely $-\sum_{i=1}^{6} a_i^\vee \lambda_i$, and the first 5 Dynkin labels of λ are the labels of the $su(6)$ weight. For instance, it is easily checked that

$$(100000) \mapsto (1) - (10000) \oplus (0) - (00010)$$

a) Calculate the embedding index, directly from Eq. (13.250).

b) Show that the outer automorphism a of $\widehat{E}_6$, which maps $\hat{\omega}_0$ to $\hat{\omega}_1$, branches to

$$a \mapsto 1 \otimes \tilde{a}^4$$

and moreover that

$$1 \mapsto a' \otimes \tilde{a}^3$$

where a' and $\tilde{a}$ stand, respectively, for the $\widehat{su}(2)$ and $\widehat{su}(6)$ basic outer automorphisms.

14.18 *The special embedding $\widehat{su}(2)_8 \subset \widehat{su}(3)_2$*
Find the irreducible $su(3)$ content of the $\widehat{su}(3)_2$ modules $L_{[2,0,0]}$ and $L_{[1,1,0]}$ and the irreducible $su(2)$ content of the $\widehat{su}(2)_8$ modules $L_{[8,0]}$, $L_{[2,6]}$, $L_{[4,4]}$, $L_{[6,2]}$, and $L_{[0,8]}$ at grades 0, 1, 2 and obtain the first few terms in the branching rule of $[2, 0, 0]$ and $[1, 1, 0]$.

14.19 *Branching functions*

a) Show that Eq. (14.266) implies the character identity:

$$\chi_{\mathcal{P}\hat{\lambda}}(\zeta, \tau, t) = \sum_{\hat{\mu} \in P_+^k} \chi_{\{\hat{\lambda};\hat{\mu}\}}(\tau) \chi_{\hat{\mu}}(\zeta, \tau, t)$$

with ζ an element of the $\mathbb{C}$ span of the p weight lattice and where we have introduced the new notation

$$\chi_{\{\hat{\lambda};\hat{\mu}\}}(\tau) = e^{2\pi i \tau(m_{\hat{\lambda}} - m_{\hat{\mu}})} b_{\hat{\lambda}\hat{\mu}}(\tau)$$

with $m_{\hat{\lambda}}$ defined in Eq. (14.158). Argue that the normalized branching function $\chi_{\{\hat{\lambda};\hat{\mu}\}}(\tau)$ must be independent of the parameters ζ and t. With $q = e^{2\pi i \tau}$, find

$$\chi_{\{[1,0,0];[4,0]\}} \quad \text{and} \quad \chi_{\{[1,0,0];[0,4]\}}$$

for the embedding $\widehat{su}(2)_4 \subset \widehat{su}(3)_1$, using the branching functions already calculated in Sect. 14.7.2.

b) By using the simple modular transformation properties of the normalized affine characters, find the relation between $\chi_{\{\hat{\lambda};\hat{\mu}\}}(\tau + 1)$ and $\chi_{\{\hat{\lambda};\hat{\mu}\}}(\tau)$. From this result, obtain a simple necessary condition for having a constant normalized branching function. (This condition illustrates neatly the fact that finite reducibility depends crucially upon the level.)

14.20 $\widehat{su}(3)_1$ paths
For the three integrable $\widehat{su}(3)_1$ modules, write the path associated with each highest weight. (Hint: The highest weight $\hat{\omega}_i$ is annihilated by all f^i with $j \neq i$.) Using the action of the f^i's, obtain all the states at grade 0 and 1. Write the finite word and the semistandard tableau corresponding to each of these paths, and from this recover the irreducible $su(3)$ content of three representations at grade 1.

14.21 *Paths and bicolored tableaux*
There exists a one-to-one correspondence between semiinfinite spin chains of the modules $L_{[1,0]}$ and $L_{[0,1]}$, defined by the sequence $(\varepsilon_i), i \geq 1$, with $\varepsilon_i \in \{+, -\}$, and bicolored tableaux. These are tableaux with boxes marked by 0 or 1, such that:
  (i)   the first box (i.e., the box in the upper left corner) has 0 for tableaux in the module $L_{[1,0]}$ and 1 for those in $L_{[0,1]}$;
  (ii) the box numbers must alternate in rows and in columns;
  (iii) the row lengths must be strictly decreasing.
(Warning: These are not Young tableaux: even though we are dealing with $su(2)$, columns of arbitrary length are allowed. Furthermore, rows of equal length are forbidden. The $su(2)$ character of the construction lies in the number of colors.) The correspondence is simply:

$$\varepsilon_i = (-1)^{i+g_i}$$

where g_i is the length of the i-th column.

a) Write down all bicolored tableaux associated with the states at the first few grades ($L_0 \leq 3$) of the modules $L_{[1,0]}$ and $L_{[0,1]}$.

b) What is the weight and the grade of a bicolored tableau in terms of the number of boxes of each color?

c) The bicolored tableau representation of states leads to a simple combinatorial description of the weight multiplicities. Let n_i stand for the number of boxes marked with an i in a given tableau. Then, the number of partitions of $n_0 + n_1$ into distinct positive integers and compatible with the coloring appropriate to the module under consideration gives the multiplicity of the weight characterized by the pair (n_0, n_1) in that module. In this way, check that in $L_{[1,0]}$, the multiplicity of the weight $[1, 0]$ at grade 6 is 7.

Notes

Affine Lie algebras were discovered independently in mathematics (by V. Kac [212] and R.V. Moody [270] in 1967) and in physics (by Bardacki and Halpern [25]) in the context of dual models, the string ancestors. Actually, the occurrence of affine Lie algebras in physics (in disguised form) can be traced back to the early 1960s, with the introduction by Gell-Mann of charge algebras, later extended to current algebras. They were used to derive interaction-independent results, to circumvent the absence of suitable dynamical models for the description of nonelectromagnetic interactions. Curiously, the central extension was introduced even earlier (1959) by Schwinger.

The standard mathematical reference on affine Lie algebras is the book of Kac [214], which also contains an extensive list of references to the mathematical literature. The recent book of Moody and Pianzola [271] provides detailed coverage of the more general Kac-Moody algebras. A good introduction to affine Lie algebras is the book of Fuchs [148]. Another very readable text is Kass et al. [228], in which the first part is a "physicist's run"; the second volume provides extensive tables, in particular of string functions and branching rules (part of the latter can be interpreted as the decomposition of integrable affine representations into a sum of finite ones, grade by grade). A succinct survey of affine Lie algebras, which has been very influential in the physics literature, is the article of Goddard and Olive [183].

Modular transformation properties of characters of integrable representations were derived by Kac and Peterson in [215] (see also Ref. [214]). Related results on lattices are discussed thoroughly in the review paper of Lerche, Schellekens, and Warner [254]. The group of outer automorphisms for extended Dynkin diagrams is described in Olive and Turok [285] and Bernard [41] (from which expressions for $\epsilon(w_A)$ have first been obtained); the standard mathematical reference is Bourbaki [56]. Our presentation of the isomorphism between the group of outer automorphisms and the center of the group follows that of Ahn and Walton [5]. Further technical details can be found in the work of Felder, Gawedzki, and Kupiainen [129] (from which the result of App. 14.A has been extracted). The path basis for the integrable $\widehat{su}(N)$ modules has been discovered by Date et al. [87] (Ex 14.21 on bicolored tableaux is adapted from this reference). However, the action of the tilde generators is the one appropriate for a crystal base—see, for instance, Ref. [132]. The Robinson-Schensted correspondence is described in Schensted [323] and Knuth [241].

CHAPTER 15

WZW Models

This chapter initiates the analysis of conformal field theories with Lie-algebraic symmetry, for which an affine Lie algebra $\hat{g}$ arises as the spectrum-generating algebra. Such models are somewhat peculiar among conformal field theories in that they can be formulated directly in terms of an action. We will thus introduce them by means of this action and show how to extract from it their algebraic structure, which provides them with an alternative algebraic definition. Special emphasis is placed on the formulation of the concept of primary field, and its relation with the integrable representations of the affine algebra $\hat{g}$. A key construction along this program is that of the Sugawara energy-momentum tensor, which is presented in great detail. It leads directly to a differential equation for the correlation functions, the Knizhnik-Zamolodchikov equation, of which simple solutions are presented. In the second part of the chapter, we present various free-field representations. In addition to being extremely useful computational tools, they provide an illustration of the different concepts introduced in the first part.

§15.1. Introducing WZW Models

15.1.1. Nonlinear Sigma Models

In searching for an explicit conformal field theory with additional conserved currents generating an affine Lie algebra, it is natural to first consider the *nonlinear sigma model*

$$S_0 = \frac{1}{4a^2} \int d^2x \, \mathrm{Tr}'(\partial^\mu g^{-1} \partial_\mu g) \tag{15.1}$$

where a^2 is a positive, dimensionless coupling constant. This action describes a matrix bosonic field $g(x)$ living on the (semisimple) group manifold G associated with the Lie algebra g.

For the action to be real, $g(x)$ must be valued in a unitary representation. The trace is thus taken in such a representation; the prime indicates a representation-independent normalization, namely[1]

$$\text{Tr}'(t^a t^b) = 2\delta_{a,b} \qquad \text{where} \qquad \left[t^a, t^b\right] = \sum_c i f_{abc} t^c \tag{15.2}$$

and the t^a's stand for any matrix representation of the Lie algebra generators. This trace is related to the usual trace by

$$\text{Tr}' = \frac{1}{x_{\text{rep}}} \text{Tr} \tag{15.3}$$

where x_{rep} is the Dynkin index of the representation, which has been defined in Eq. (13.133).

With g unitary, $g^{-1}\partial_\mu g$ is antihermitian:

$$(g^{-1}\partial_\mu g)^\dagger = \partial_\mu g^{-1} g = -g^{-1}\partial_\mu g \tag{15.4}$$

since $\partial_\mu g^{-1} = -g^{-1}\partial_\mu g g^{-1}$, a direct consequence of $\partial_\mu(gg^{-1}) = 0$. This in turn ensures the positivity of the action, since the Lagrangian density is proportional to

$$\text{Tr}'(\partial^\mu g^{-1}\partial_\mu g) = \text{Tr}'((\partial^\mu g)^\dagger \partial_\mu g) \geq 0 \tag{15.5}$$

Although conformally invariant classically, it has been known for a while that this theory is effectively massive: the dimensionless coupling constant acquires a scale dependence at the quantum level. Therefore, the β function is nonzero and the quantum theory is not scale invariant. (In fact, it is asymptotically free.)

That this is not the kind of theory we are looking for can also be seen classically, at the level of conserved currents. Due to the holomorphic factorization property of a conformal field theory, these conserved currents must similarly factorize into a holomorphic and an antiholomorphic part. From the equation of motion, it is simple to check that there are not two conserved currents, one independent of z and the other independent of $\bar{z}$, where as usual

$$z = x^0 + ix^1 \qquad \bar{z} = x^0 - ix^1 \tag{15.6}$$

Indeed, under the substitution $g \to g + \delta g$, the variation of the action is

$$\delta S_0 = \frac{1}{2a^2} \int d^2x \, \text{Tr}' \left(g^{-1}\delta g \partial^\mu (g^{-1}\partial_\mu g)\right) \tag{15.7}$$

which results in the following equation of motion

$$\partial^\mu (g^{-1}\partial_\mu g) = 0 \tag{15.8}$$

Some technical aspects of the above derivation deserve clarification. First, to calculate $\frac{\delta}{\delta g}\text{Tr}'[Ag^{-1}B]$ for A, B independent of g, we should proceed as follows:

$$\delta \text{Tr}'[Ag^{-1}B] = -\text{Tr}'[Ag^{-1}\delta g g^{-1}B] = -\text{Tr}'[\delta g g^{-1}BAg^{-1}] \tag{15.9}$$

---

[1] We recall that once the Killing form has been defined, whether the indices are up or down is irrelevant in the orthonormal basis.

(since $\delta(gg^{-1}) = 0$) so that

$$\frac{\delta}{\delta g}\text{Tr}'[Ag^{-1}B] = -\frac{1}{x_{\text{rep}}}g^{-1}BAg^{-1} \tag{15.10}$$

Also, we stress that we use a matrix notation; when we differentiate with respect to g, we mean differentiation with respect to a particular matrix element of g.

The equation of motion implies the conservation of the currents

$$J_\mu = g^{-1}\partial_\mu g \tag{15.11}$$

Thus, if we write $\tilde{J}_z = g^{-1}\partial_z g$ and $\tilde{J}_{\bar{z}} = g^{-1}\partial_{\bar{z}}g$, we find

$$\partial_z \tilde{J}_{\bar{z}} + \partial_{\bar{z}}\tilde{J}_z = 0 \tag{15.12}$$

For the holomorphic and antiholomorphic currents to be separately conserved, the two terms of the above equation need to vanish separately. For this, the dual current $\epsilon^{\mu\nu}J_\nu$ must also be conserved ($\epsilon^{\mu\nu}$ is the rank-two antisymmetric tensor). However, the definition of J_μ implies that

$$[J_\mu, J_\nu] + \partial_\mu J_\nu - \partial_\nu J_\mu = 0 \tag{15.13}$$

which yields

$$\partial_\mu(\epsilon^{\mu\nu}J_\nu) = -\epsilon^{\mu\nu}J_\mu J_\nu \neq 0 \tag{15.14}$$

Actually, the separate conservation of the two currents $\tilde{J}_z$ and $\tilde{J}_{\bar{z}}$ would be inconsistent in that $\partial_z(g^{-1}\partial_{\bar{z}}g) = 0$ would force:

$$\partial_z \partial_{\bar{z}} g = \partial_{\bar{z}}gg^{-1}\partial_z g \tag{15.15}$$

Since the l.h.s. is symmetric with respect to the interchange of z and $\bar{z}$, it implies that

$$\partial_{\bar{z}}gg^{-1}\partial_z g = \partial_z gg^{-1}\partial_{\bar{z}}g \tag{15.16}$$

This is an equality of the form $abc = cba$ for some group elements, which is not expected to hold in general for a non-Abelian algebra.

The correct choice for these conserved currents turns out to be:

$$J_z = \partial_z gg^{-1} \quad J_{\bar{z}} = g^{-1}\partial_{\bar{z}}g \tag{15.17}$$

(or the dual of these expressions, obtained by interchanging z and $\bar{z}$.) The conservation of either J_z or $J_{\bar{z}}$ implies that of the other:

$$\partial_z(g^{-1}\partial_{\bar{z}}g) = g^{-1}\partial_{\bar{z}}(\partial_z gg^{-1})g \tag{15.18}$$

However, neither of these is conserved separately in the sigma model (15.1).

15.1.2. Wess-Zumino-Witten Models

A more complicated action must be considered in order to enhance the symmetry and recover the conserved currents (15.17). The solution is not entirely obvious:

we should add to the above action a *Wess-Zumino term*,

$$\Gamma = \frac{-i}{24\pi} \int_B d^3y \; \epsilon_{\alpha\beta\gamma} \text{Tr}' \left(\tilde{g}^{-1} \partial^\alpha \tilde{g} \tilde{g}^{-1} \partial^\beta \tilde{g} \tilde{g}^{-1} \partial^\gamma \tilde{g} \right) \tag{15.19}$$

This is defined on a three-dimensional manifold B, whose boundary is the compact-ification of our original two-dimensional space. We have denoted by $\tilde{g}$ the extension of the field g to this three-dimensional manifold. However, this extension is not unique, hence there is a potential ambiguity in the definition of Γ. Indeed, in a compactified three-dimensional space, a compact two-dimensional space delimits two distinct three-manifolds. The difference between these two choices quantifies the ambiguity. Taking the orientation into account, this difference $\Delta\Gamma$ is given by the r.h.s. of Eq. (15.19) but with the integration range extended over the whole compact three-dimensional space. Since the latter is topologically equivalent to the three-sphere, we can write

$$\Delta\Gamma = \frac{-i}{24\pi} \int_{S^3} d^3y \; \epsilon_{\alpha\beta\gamma} \text{Tr}' \left(\tilde{g}^{-1} \partial^\alpha \tilde{g} \tilde{g}^{-1} \partial^\beta \tilde{g} \tilde{g}^{-1} \partial^\gamma \tilde{g} \right) \tag{15.20}$$

It is shown in App. 15.A that our choice of normalization for Tr' ensures that $\Delta\Gamma$ is defined modulo $2\pi i$ (πi for $SO(3)$). Therefore the Euclidian functional integral, with weight $\exp(-\Gamma)$, is perfectly well defined. Clearly, any coupling constant multiplying this term must be "quantized": it has to be an integer (or an even integer for $SO(3)$).

We then consider the action

$$S = S_0 + k\Gamma \tag{15.21}$$

where k is an integer. Although the Wess-Zumino term is expressed as a three-dimensional integral, its variation under $g \to g + \delta g$ is a two-dimensional func-tional, because the variation of its density can be written as a total derivative and

$$\int_B d^3y \; \epsilon_{\alpha\beta\gamma} \partial^\gamma(\cdots) = \int d^2x \; \epsilon_{\alpha\beta}(\cdots) \tag{15.22}$$

The details of this derivation are left as an exercise (see Ex. 15.1). The final result is

$$\delta\Gamma = \frac{i}{8\pi} \int d^2x \; \epsilon_{\mu\nu} \text{Tr}' \left(g^{-1} \delta g \partial^\mu (g^{-1} \partial^\nu g) \right) \tag{15.23}$$

The equation of motion for the full action (15.21) is then

$$\partial^\mu (g^{-1}\partial_\mu g) + \frac{a^2 ik}{4\pi} \epsilon_{\mu\nu} \partial^\mu (g^{-1}\partial^\nu g) = 0 \tag{15.24}$$

In terms of the complex variables $z, \bar{z}$ (with $\partial^z = 2\partial_{\bar{z}}$, $\epsilon_{z\bar{z}} = i/2$, etc.), the equation of motion becomes

$$\left(1 + \frac{a^2 k}{4\pi}\right) \partial_z(g^{-1}\partial_{\bar{z}}g) + \left(1 - \frac{a^2 k}{4\pi}\right) \partial_{\bar{z}}(g^{-1}\partial_z g) = 0 \tag{15.25}$$

Thus, for

$$a^2 = 4\pi/k \qquad (15.26)$$

we find the desired conservation law

$$\partial_z(g^{-1}\partial_{\bar{z}}g) = 0 \qquad (15.27)$$

Since a^2 is positive, k must be a positive integer. The other solution $a^2 = -4\pi/k$, which requires $k < 0$, implies the conservation of the dual currents.

The solution of the classical field equation is simply

$$g(z, \bar{z}) = f(z)\bar{f}(\bar{z}) \qquad (15.28)$$

for arbitrary functions $f(z)$ and $\bar{f}(\bar{z})$. This factorization indicates that this model is very much like a free-field theory. Actually, it is reminiscent of a "bilinear" fermion free-field theory, a point that will be pursued below.

The separate conservation of the currents J_z and $J_{\bar{z}}$ implies the invariance of the action under

$$g(z, \bar{z}) \to \Omega(z)g(z, \bar{z})\bar{\Omega}^{-1}(\bar{z}) \qquad (15.29)$$

where Ω and $\bar{\Omega}$ are two arbitrary matrices valued in G. Indeed, under the infinitesimal transformation

$$\Omega(z) = 1 + \omega(z) \qquad \bar{\Omega}(\bar{z}) = 1 + \bar{\omega}(\bar{z}) \qquad (15.30)$$

g transforms as follows

$$\delta_\omega g = \omega g \qquad \delta_{\bar{\omega}} g = -g\bar{\omega} \qquad (15.31)$$

With $a^2 = 4\pi/k$, the variation of the action for $g \to g + \delta_\omega g + \delta_{\bar{\omega}} g$ is

$$
\begin{aligned}
\delta S &= \frac{k}{2\pi} \int d^2x \, \mathrm{Tr}'(g^{-1}\delta g[\partial_z(g^{-1}\partial_{\bar{z}}g)]) \\
&= \frac{k}{2\pi} \int d^2x \, \mathrm{Tr}'[\omega(z)\partial_{\bar{z}}(\partial_z g g^{-1}) - \bar{\omega}(\bar{z})\partial_z(g^{-1}\partial_{\bar{z}}g)]
\end{aligned}
\qquad (15.32)
$$

which clearly vanishes after an integration by parts. The global $G \times G$ invariance of the sigma model has thus been extended to a local $G(z) \times G(\bar{z})$ invariance.

The holomorphic factorization of the conserved currents is the first signal that Eq. (15.21), together with Eq. (15.26), is indeed the desired model. From now on this model will be referred to as the *Wess-Zumino-Witten (WZW) model*, or more precisely, the $\hat{g}_k$ WZW model.[2] For later reference, we rewrite its action

$$\boxed{\, S^{\mathrm{WZW}} = \frac{k}{16\pi} \int d^2x \, \mathrm{Tr}'(\partial^\mu g^{-1}\partial_\mu g) + k\Gamma \,} \qquad (15.33)$$

We now carry the analysis to the quantum level.

---

[2] In the literature, it is also called the Wess-Zumino-Novikov-Witten model.

15.1.3. Ward Identity and Affine Lie Algebras

By rescaling the conserved currents as

$$J(z) \equiv -kJ_z(z) = -k\partial_z g g^{-1}$$
$$\bar{J}(\bar{z}) \equiv kJ_{\bar{z}}(\bar{z}) = kg^{-1}\partial_{\bar{z}}g \tag{15.34}$$

we can rewrite Eq. (15.32) in the form

$$\delta S = -\frac{1}{2\pi} \int d^2x \, \{\partial_{\bar{z}}(\mathrm{Tr}'[\omega(z)J(z)]) + \partial_z(\mathrm{Tr}'[\bar{\omega}(\bar{z})\bar{J}(\bar{z})])\} \tag{15.35}$$

We replace d^2x by $(-i/2)dzd\bar{z}$, and, after integration by parts, take the holomorphic contour to be counterclockwise and the antiholomorphic contour to be clockwise (compare with Eq. (5.35)). This leads to

$$\delta_{\omega,\bar{\omega}}S = \frac{i}{4\pi} \oint dz \, \mathrm{Tr}'[\omega(z)J(z)] - \frac{i}{4\pi} \oint d\bar{z} \, \mathrm{Tr}'[\bar{\omega}(\bar{z})\bar{J}(\bar{z})] \tag{15.36}$$

where both contours are understood to be counterclockwise, which explains the relative sign between the two terms. With

$$J = \sum_a J^a t^a, \qquad \omega = \sum_a \omega^a t^a \tag{15.37}$$

and the normalization (15.2) for Tr', this yields

$$\delta_{\omega,\bar{\omega}}S = -\frac{1}{2\pi i} \oint dz \sum_a \omega^a J^a + \frac{1}{2\pi i} \oint d\bar{z} \sum_a \bar{\omega}^a \bar{J}^a \tag{15.38}$$

Finally, following the method of Sect. 5.2.2, we have

$$\delta\langle X \rangle = \langle (\delta s)X \rangle \tag{15.39}$$

where δs means the density of δS and as usual, $\langle X \rangle$ stands for the correlation function of a number of fields, which leads to the Ward identity

$$\delta_{\omega,\bar{\omega}}\langle X \rangle = -\frac{1}{2\pi i} \oint dz \sum_a \omega^a \langle J^a X \rangle + \frac{1}{2\pi i} \oint d\bar{z} \sum_a \bar{\omega}^a \langle \bar{J}^a X \rangle \tag{15.40}$$

The transformation law for the current follows from Eqs. (15.34) and (15.31):

$$\delta_\omega J = -k\left[\partial_z(\delta_\omega g)g^{-1} - \partial_z g g^{-1}\delta_\omega g g^{-1}\right]$$
$$= -k(\partial_z \omega g + \omega\partial_z g)g^{-1} + k\partial_z g g^{-1}\omega \tag{15.41}$$
$$= [\omega, J] - k\partial_z\omega$$

It can be rewritten as

$$\delta_\omega J^a = \sum_{b,c} if_{abc}\omega^b J^c - k\partial_z\omega^a \tag{15.42}$$

The substitution of this transformation into Eq. (15.40) leads to the OPE

$$J^a(z)J^b(w) \sim \frac{k\delta_{ab}}{(z-w)^2} + \sum_c if_{abc}\frac{J^c(w)}{(z-w)} \tag{15.43}$$

This will be called a *current algebra*. Introducing the modes J_n^a from the Laurent expansion

$$J^a(z) = \sum_{n\in\mathbb{Z}} z^{-n-1}J_n^a \tag{15.44}$$

we can easily check (proceeding as in the analysis of the Virasoro case, cf. Eq. (6.25)) the equivalence between Eq. (15.43) and the commutation relations of the $\hat{g}$ affine Lie algebra at level k

$$\left[J_n^a, J_m^b\right] = \sum_c if_{abc}J_{n+m}^c + kn\delta_{ab}\delta_{n+m,0} \tag{15.45}$$

The transformation property of $\bar{J}$ is

$$\delta_{\bar{\omega}}\bar{J} = [\bar{\omega}, \bar{J}] - k\partial_z\bar{\omega} \tag{15.46}$$

This yields another copy of the affine algebra (15.45) for the modes $\bar{J}_m^b$. Since $\bar{\omega}(\bar{z})$ is independent of z,

$$\delta_{\bar{\omega}}J = 0 \tag{15.47}$$

Eq. (15.47) implies that the OPE of $J^a(z)$ with $\bar{J}^b(\bar{w})$ contains only regular terms, so that the corresponding modes commute:

$$\left[J_n^a, \bar{J}_m^b\right] = 0 \tag{15.48}$$

The two algebras are thus independent.

The occurrence of two independent conserved currents generating independent affine Lie algebras is the fundamental property of the model (15.33). As will be seen in the next section, this leads directly to conformal invariance.

We conclude this section with a clarifying remark concerning the field $g(z, \bar{z})$. We mentioned earlier that $g(z, \bar{z})$ transforms in some unitary representation of G. In fact, we just said that it transforms independently with respect to left and right G transformations. However, we need not specify these representations since the WZW action is formulated in a representation-independent way. The full spectrum of the quantum theory is uniquely fixed by the group G (which need not be simply connected). This spectrum can in principle be obtained by canonical quantization, and global considerations determine which combinations of left and right representations can appear in the physical spectrum. But this approach will not be followed here. Instead, we will turn to an algebraic formulation of WZW models that parallels the development of minimal models. Once primary fields are identified, physical spectra will be obtained from modular invariance.

Nevertheless, it is useful to bear in mind some sort of effective description of WZW models (although not adequate for all cases) in which $g(z, \bar{z})$ is a basic field and the different fields of the theory are obtained by appropriate multiple products of $g(z, \bar{z})$ with itself. In this view, it is natural to regard $g(z, \bar{z})$ as transforming in the lowest-dimensional fundamental representation of the algebra g. This representation will be called the *minimal representation*. In the following, $g(z, \bar{z})$ will be understood in this sense.

§15.2. The Sugawara Construction

In this section, we initiate the analysis of the conformal aspects of the WZW models. The starting point is the construction of the energy-momentum tensor, with classical form $(1/2k) \sum_a J^a J^a$ (see Ex. 15.1). We look for a normal-ordered version thereof, namely

$$T(z) = \gamma \sum_a (J^a J^a)(z) \qquad (15.49)$$

where, as usual, $(\cdots)$ stands for the normal ordering defined in Eq. (6.130). However, the constant γ cannot be fixed from the classical theory since it is renormalized by quantum effects (because the currents are not free fields). It can be fixed uniquely by requiring the OPE of the above T with itself to be of the form

$$T(z)T(w) = \frac{c/2}{(z-w)^4} + \frac{2T(w)}{(z-w)^2} + \frac{\partial T(w)}{(z-w)} \qquad (15.50)$$

for some c, or, equivalently, by requiring J^a to be a primary field of dimension 1. Since this is both important and an interesting example of OPE techniques, we present this calculation in some detail. First, we need to calculate the contraction

$$\overline{J^a(z)(J^b J^b)}(w) = \frac{1}{2\pi i} \oint_w \frac{dx}{x-w} \left[\overline{J^a(z)J^b(x)}J^b(w) + J^b(x)\overline{J^a(z)J^b(w)} \right]$$

$$= \frac{1}{2\pi i} \oint_w \frac{dx}{x-w} \left\{ \left[\frac{k\delta_{ab}}{(z-x)^2} + \sum_c if_{abc} \frac{J^c(x)}{(z-x)} \right] J^b(w) \right.$$

$$\left. + J^b(x) \left[\frac{k\delta_{ab}}{(z-w)^2} + \sum_c if_{abc} \frac{J^c(w)}{(z-w)} \right] \right\}$$

$$(15.51)$$

We recall that the remaining products are not just contractions but full OPEs, for which the first regular term contributes (cf. the discussion in App. 6.B). We recall also the simplification that arises for the OPE $J^b(x)J^c(w)$. It produces field-dependent terms evaluated at w; hence nothing depends upon x except for inverse powers of $(x-w)$. As a result, only the single pole gives a nonzero contribution, and since one factor of $(x-w)^{-1}$ is already present, only the first regular term in $J^b(x)J^c(w)$ survives—i.e., $(J^b J^c)(w)$. Accordingly, after developing the OPEs, we obtain:

$$
J^a(z)(J^b J^b)(w) = \frac{1}{2\pi i} \oint_w \frac{dx}{x-w} \left\{ \frac{k\delta_{ab}J^b(w)}{(z-x)^2} \right.
$$

$$
+ \sum_c \frac{if_{abc}}{(z-x)} \left[if_{cbd} \frac{J^d(w)}{(x-w)} + \frac{k\delta_{cb}}{(x-w)^2} + (J^c J^b)(w) \right]
$$

$$
\left. + \frac{k\delta_{ab}J^b(w)}{(z-w)^2} + \sum_c if_{abc} \frac{(J^b J^c)(w)}{(z-w)} \right\}
$$

$$
\tag{15.52}
$$

Due to the antisymmetry of the structure constants f_{abc}, the term $f_{abc}\delta_{bc}$ vanishes. We now sum the result over b and use

$$
-\sum_{b,c} f_{abc}f_{cbd} = \sum_{b,c} f_{abc}f_{dbc} = 2g\delta_{ad} \tag{15.53}
$$

where g is the dual Coxeter number[3] (not to be confused with the field $g(z,\bar z)$). Moreover, we also have

$$
\sum_{b,c} f_{abc}[(J^b J^c) + (J^c J^b)] = 0 \tag{15.55}
$$

We thus end up with

$$
J^a(z) \sum_b (J^b J^b)(w) = 2(k+g)\frac{J^a(w)}{(z-w)^2} \tag{15.56}
$$

By inverting the order of the two contracted fields (with $z \leftrightarrow w$) and multiplying the result by γ, we find that

$$
T(z)J^a(w) = 2\gamma(k+g)\frac{J^a(z)}{(z-w)^2}
$$

$$
= 2\gamma(k+g) \left\{ \frac{J^a(w)}{(z-w)^2} + \frac{\partial J^a(w)}{(z-w)} \right\} \tag{15.57}
$$

For T to be a genuine energy-momentum tensor, the coefficient of the second term in the last expression must be 1, which fixes γ to be

$$
\gamma = \frac{1}{2(k+g)} \tag{15.58}
$$

---

[3] We recall that the matrices $(t^a)_{bc} = -if_{abc}$ generate the adjoint representation, whose Casimir

$$
\sum_a (t^a t^a)_{b,d} = \sum_{a,c} (t^a)_{bc}(t^a)_{cd} = -\sum_{a,c} f_{abc}f_{acd} = \sum_{a,c} f_{bca}f_{dca} \tag{15.54}
$$

is twice the dual Coxeter number when the squared length of the highest root is normalized to 2 (cf. Eq. (13.128)).

that is,

$$T(z) = \frac{1}{2(k+g)} \sum_a (J^a J^a)(z) \qquad (15.59)$$

This also implies that J^a has conformal dimension 1, as expected, and that it is a Virasoro primary field.

It is clear from Eq. (15.51) that the factor g in Eq. (15.58) arises from a double contraction, which means that it is a quantum effect.[4] Hence, in the quantum case, the product $J^a J^a$ becomes normal ordered and the multiplicative factor is renormalized by $k \to k + g$.

Having calculated $T(z)J^a(w)$, we are in position to compute the singular terms in the product $T(z)T(w)$:

$$\overbrace{T(z)T}(w) = \frac{1}{2(k+g)} \frac{1}{2\pi i} \oint_w \frac{dx}{x-w} \sum_a \left\{ \overbrace{T(z)J^a}(x)J^a(w) \right.$$

$$\left. + J^a(x)\,\overbrace{T(z)J^a}(w) \right\} \qquad (15.60)$$

$$= \frac{c/2}{(z-w)^4} + \frac{2T(w)}{(z-w)^2} + \frac{\partial T(w)}{(z-w)}$$

with T given by Eq. (15.59) and

$$c \equiv c(\hat{g}_k) = \frac{k \dim g}{k+g} \qquad (15.61)$$

($\dim g = \delta_{aa}$). To obtain this last OPE, we used

$$\partial J^a(z)J^b(w) \sim \frac{-2k\delta_{ab}}{(z-w)^3} - \sum_c if_{abc} \frac{J^c(w)}{(z-w)^2} \qquad (15.62)$$

and its analogue with the operators exchanged.

Of course, the OPE of $\bar{T}$ with itself has the same form, with exactly the same value of the central charge. The above calculation establishes indirectly the scale invariance of WZW models. Actually, the expression (15.59) for the energy-momentum tensor can be regarded as an alternative definition of the $\hat{g}_k$ WZW model.

It should be clear that the result we just proved goes much beyond WZW models. Stated in full generality, we have shown that the Virasoro algebra belongs to the enveloping algebra of the affine Lie algebra $\hat{g}_k$, a result known in the physics literature as the *Sugawara construction*.

---

[4] As a rule, single contraction terms simply reproduce the classical contributions.

For later use, we will express the Sugawara energy-momentum tensor in terms of modes. Using Eq. (6.144), we obtain

$$L_n = \frac{1}{2(k+g)} \sum_a \left\{ \sum_{m \leq -1} J_m^a J_{n-m}^a + \sum_{m \geq 0} J_{n-m}^a J_m^a \right\} \tag{15.63}$$

For $n \neq 0$, J_m^a and J_{n-m}^a commute, and so the order of the terms is irrelevant. However, for $n = 0$ the above expression shows that the term with larger subindex must be placed at the rightmost position. But this is just the definition of the usual normal ordering for modes. We can thus write

$$L_n = \frac{1}{2(k+g)} \sum_a \sum_m \, : J_m^a J_{n-m}^a : \tag{15.64}$$

In terms of modes, the complete affine Lie and Virasoro algebra is

$$
\boxed{
\begin{aligned}
[L_n, L_m] &= (n-m)L_{n+m} + \frac{c}{12}(n^3 - n)\delta_{n+m,0} \\
[L_n, J_m^a] &= -m J_{n+m}^a \\
[J_n^a, J_m^b] &= \sum_c i f_{abc} J_{n+m}^c + k n \delta_{ab} \delta_{n+m,0}
\end{aligned}
} \tag{15.65}
$$

The commutativity of the zero modes of the affine algebra with the Virasoro generators (and in particular with L_0) reflects the built-in g invariance. However, the full affine Lie algebra is not a symmetry algebra since its generators do not all commute with L_0. It will turn out to be the *spectrum-generating algebra* of the theory.

The Sugawara construction has been presented in terms of the particular currents $J^a(z)$ whose modes are orthonormal with respect to the Killing form, that is

$$K(J_n^a, J_m^b) = K(J_0^a, J_0^b)\delta_{n+m,0} = \delta^{a,b}\delta_{n+m,0} \tag{15.66}$$

In a generic basis $\mathcal{J}^a$, the affine Lie commutator is changed to

$$[\mathcal{J}_n^a, \mathcal{J}_m^b] = \sum_c i \tilde{f}^{ab}{}_c \mathcal{J}_{n+m}^c + k n K(\mathcal{J}_0^a, \mathcal{J}_0^b)\delta_{n+m,0} \tag{15.67}$$

where $\tilde{f}^{ab}{}_c$ are the corresponding structure constants. The energy-momentum tensor reads then

$$T(z) = \frac{1}{2(k+g)} \sum_{a,b} [K(\mathcal{J}_0^a, \mathcal{J}_0^b)]^{-1}(\mathcal{J}^a \mathcal{J}^b)(z) \tag{15.68}$$

Indeed, L_0 contains the (normal-ordered) quadratic Casimir operator of g, itself defined in terms of the inverse of the Killing form; this directly implies the above generalization. In the Cartan-Weyl basis, T becomes

$$T(z) = \frac{1}{2(k+g)} \left\{ \sum_i (H^i H^i)(z) + \sum_{\alpha > 0} \frac{|\alpha|^2}{2} \left[(E^\alpha E^{-\alpha})(z) + (E^{-\alpha} E^\alpha)(z) \right] \right\} \tag{15.69}$$

For further illustration, the $\widehat{su}(2)_k$ energy-momentum tensor in the Chevalley basis (13.85) is (cf. Eqs. (13.36)–(13.39))

$$T(z) = \frac{1}{2(k+2)} \left\{ \frac{1}{2}(hh)(z) + (ef)(z) + (fe)(z) \right\} \tag{15.70}$$

This could also have been written directly from the relation between the generators in the Chevalley and the Cartan-Weyl bases:

$$h = \sqrt{2}H, \quad e = E^+, \quad f = E^- \tag{15.71}$$

with $E^\pm \equiv E^{\pm\alpha_1}$, and $\alpha_1 = \sqrt{2}$. In the spin basis, defined by

$$J^0 = H/\sqrt{2}, \quad J^\pm = E^\pm \tag{15.72}$$

the OPEs take the form

$$J^0(z)J^0(w) \sim \frac{k/2}{(z-w)^2}$$

$$J^0(z)J^\pm(w) \sim \frac{\pm J^\pm(w)}{(z-w)} \tag{15.73}$$

$$J^+(z)J^-(w) \sim \frac{k}{(z-w)^2} + \frac{2J^0(w)}{(z-w)}$$

and T reads

$$T(z) = \frac{1}{2(k+2)} \left\{ 2(J^0 J^0)(z) + (J^+ J^-)(z) + (J^- J^+)(z) \right\} \tag{15.74}$$

The Sugawara construction generalizes directly to the case of a semisimple Lie algebra $g = \oplus_i g_i$. Then, the energy-momentum tensor is the sum of the Sugawara energy-momentum tensors associated with each simple Lie algebra in the direct sum. Since these distinct energy-momentum tensors all commute among themselves, the total central charge is the sum of the central charges of the contributing pieces.

We note that the central charge (15.61) is bounded by

$$r \le c \le \dim g \tag{15.75}$$

The lower bound is saturated only for the simply-laced algebras at $k = 1$. The upper bound is obtained in the limit $k \to \infty$.

§15.3. WZW Primary Fields

15.3.1. Primary Fields as Covariant Fields

By analogy with the purely conformal case, where a primary field transforms covariantly with respect to a scale transformation, a WZW primary field is defined as a field that transforms covariantly with respect to a $G(z) \times G(\bar{z})$ transformation, exactly as $g(z, \bar{z})$ in Eq. (15.29). From Eqs. (15.31) and (15.40), we reformulate this property for the field $g(z, \bar{z})$ in terms of the OPE

$$
\begin{aligned}
J^a(z)\, g(w, \bar{w}) &\sim \frac{-t^a\, g(w, \bar{w})}{z - w} \\
\bar{J}^a(z)\, g(w, \bar{w}) &\sim \frac{g(w, \bar{w})\, t^a}{z - w}
\end{aligned}
\tag{15.76}
$$

The field $g(z, \bar{z})$ transforms in the minimal representation of g, to which t^a refers, for both the z and $\bar{z}$ sectors. More generally, any field $\phi_{\lambda,\mu}$ transforming covariantly with respect to some representation specified by λ in the holomorphic sector and by μ in the antiholomorphic sector (i.e., λ and μ are the highest weights of the representations), will be a WZW primary field. It can be characterized by the OPE

$$
\begin{aligned}
J^a(z)\, \phi_{\lambda,\mu}(w, \bar{w}) &\sim \frac{-t_\lambda^a\, \phi_{\lambda,\mu}(w, \bar{w})}{z - w} \\
\bar{J}^a(\bar{z})\, \phi_{\lambda,\mu}(w, \bar{w}) &\sim \frac{\phi_{\lambda,\mu}(w, \bar{w})\, t_\mu^a}{\bar{z} - \bar{w}}
\end{aligned}
\tag{15.77}
$$

where t_λ^a is the matrix t^a in the λ representation and similarly for t_μ^a. In principle, λ and μ can be different, which is certainly the case for nonscalar fields (i.e., fields with nonzero spin). It should be stressed that here we use a compact matrix notation. With all indices reinserted, Eq. (15.77) would read

$$
J^a(z)(\phi_{\lambda,\mu})_{ru}(w, \bar{w}) \sim -\sum_s \frac{(t_\lambda^a)_{rs}\, (\phi_{\lambda,\mu})_{su}(w, \bar{w})}{z - w}
\tag{15.78}
$$

with $r, s = 1, \cdots, \dim |\lambda|$ and $u = 1, \cdots, \dim |\mu|$.

We have implicitly assumed that primary fields are associated with finite-dimensional representations of g. This is certainly natural in our effective description: $g(z, \bar{z})$ is associated with a finite-dimensional matrix, and since the other fields are some sort of composites of $g(z, \bar{z})$, they also transform in finite-dimensional representations. More generally, we will show below that fields transforming in Lie algebra representations that are not finite dimensional simply decouple in correlation functions.

By expanding the currents in terms of the modes evaluated at w,

$$
J^a(z) = \sum_n (z - w)^{-n-1} J_n^a(w)
\tag{15.79}
$$

we write their OPE with an arbitrary field A as (cf. Eq. (6.134) in the Virasoro case):

$$
J^a(z)\, A(w) = \sum_n (z - w)^{-n-1} (J_n^a\, A)(w)
\tag{15.80}
$$

Thus, for the primary field ϕ_λ (from now on we concentrate on the holomorphic sector) we have

$$
\begin{aligned}
(J_0^a\, \phi_\lambda) &= -t_\lambda^a\, \phi_\lambda \\
(J_n^a\, \phi_\lambda) &= 0 \qquad \text{for} \quad n > 0
\end{aligned}
\tag{15.81}
$$

Associating the state $|\phi_\lambda\rangle$ to the field ϕ_λ via

$$\phi_\lambda(0)|0\rangle = |\phi_\lambda\rangle \tag{15.82}$$

the conditions for a WZW primary field translate into[5]

$$
\boxed{
\begin{aligned}
J_0^a\,|\phi_\lambda\rangle &= -t_\lambda^a\,|\phi_\lambda\rangle \\
J_n^a\,|\phi_\lambda\rangle &= 0 \qquad \text{for} \quad n > 0
\end{aligned}
}
\tag{15.83}
$$

A remarkable aspect of WZW models is that WZW primary fields are also Virasoro primary fields. Indeed, from Eq. (15.64) we see that in the expression for L_n with $n > 0$, the rightmost factor J_m^a has $m > 0$, which implies that

$$L_n\,|\phi_\lambda\rangle = 0 \quad \text{for} \quad n > 0 \tag{15.84}$$

On the other hand, the action of L_0 on $|\phi_\lambda\rangle$ becomes very simple since only the zero modes of the current contribute:

$$L_0\,|\phi_\lambda\rangle = \frac{1}{2(k+g)} \sum_a J_0^a J_0^a\,|\phi_\lambda\rangle \tag{15.85}$$

Thus, L_0 acting on $|\phi_\lambda\rangle$ is proportional to the quadratic Casimir operator of the finite Lie algebra. Associating a conformal weight to $|\phi_\lambda\rangle$ according to

$$L_0\,|\phi_\lambda\rangle = h_\lambda\,|\phi_\lambda\rangle \tag{15.86}$$

we find

$$h_\lambda = \frac{\sum_a t_\lambda^a t_\lambda^a}{2(k+g)} = \frac{(\lambda, \lambda + 2\rho)}{2(k+g)} \tag{15.87}$$

In the last equality we used the explicit form of the quadratic Casimir eigenvalue obtained in Eq. (13.127). Equations (15.84) and (15.86) show that ϕ_λ is indeed a Virasoro primary field. However, the inverse is not true: a Virasoro primary field can be a WZW descendant. The field J^a is such an example:

$$J^a(z) = (J_{-1}^a \mathbb{1})(z) \tag{15.88}$$

The other states in the theory are then of the form

$$J_{-n_1}^a J_{-n_2}^b \cdots |\phi_\lambda\rangle \tag{15.89}$$

---

[5] The minus sign may seem odd, but the following calculation testifies its naturalness:

$$
\begin{aligned}
[J_0^a, J_0^b]|\phi_\lambda\rangle &= (-J_0^a t_\lambda^b + J_0^b t_\lambda^a)|\phi_\lambda\rangle = (-t_\lambda^b J_0^a + t_\lambda^a J_0^b)|\phi_\lambda\rangle = -[t_\lambda^a, t_\lambda^b]|\phi_\lambda\rangle \\
&= -\sum_c if_{abc} t_\lambda^c |\phi_\lambda\rangle = \sum_c if_{abc} J_0^c |\phi_\lambda\rangle
\end{aligned}
$$

In the antiholomorphic sector, we have:

$$
\begin{aligned}
[\bar{J}_0^a, \bar{J}_0^b]|\bar{\phi}_\lambda\rangle &= \bar{J}_0^a |\phi_\lambda\rangle t_\lambda^b - \bar{J}_0^b |\bar{\phi}_\lambda\rangle t_\lambda^a = |\bar{\phi}_\lambda\rangle [t_\lambda^a, t_\lambda^b] \\
&= \sum_c if_{abc} |\bar{\phi}_\lambda\rangle t_\lambda^c = \sum_c if_{abc} \bar{J}_0^c |\bar{\phi}_\lambda\rangle
\end{aligned}
$$

with $n_1, n_2, \cdots$ positive integers. These states are associated with descendant fields. The insertion of negative Virasoro modes is unnecessary because the energy-momentum tensor is constructed out of the currents.

15.3.2. The Knizhnik-Zamolodchikov Equation

As usual, the Virasoro primary nature of the fields ϕ_i (here i is used to denote the representation) leads directly to the Ward identities (see Eq. 5.51)

$$\sum_{i=1}^{n} \{z_i^m(z_i\partial_{z_i} + (m+1)h_i)\}\langle\phi_1(z_1)\cdots\phi_n(z_n)\rangle = 0 \quad \text{for} \quad m = 0, \pm 1 \quad (15.90)$$

expressing the global $SL(2,\mathbb{C})$ invariance of the theory. On the other hand, the global G invariance requires that

$$\delta_\omega\langle\phi_1(z_1)\cdots\phi_n(z_n)\rangle = 0 \qquad (15.91)$$

for a constant ω. From the Ward identity (15.40), one can rewrite this condition in the form

$$\oint dz \sum_a \omega^a \langle J^a(z)\phi_1(z_1)\cdots.\phi_n(z_n)\rangle$$
$$= -\sum_{i=1}^{n} \frac{1}{2\pi i} \oint \frac{dz}{z - z_i} \sum_a \omega^a t_i^a \langle\phi_1(z_1)\cdots\phi_n(z_n)\rangle = 0 \qquad (15.92)$$

Since ω is independent of z and otherwise arbitrary, it becomes

$$\boxed{\sum_{i=1}^{n} t_i^a \langle\phi_1(z_1)\cdots\phi_n(z_n)\rangle = 0} \qquad (15.93)$$

Here t_i^a is a matrix acting on the vector field ϕ_i, in the representation labeled by i. These Ward identities fix the structure of the two- and three-point functions.

We note that to avoid a contributing pole at infinity in Eq. (15.92), the following asymptotic behavior

$$J^a(z) \sim \frac{1}{z^2} \quad \text{as} \quad z \to \infty \qquad (15.94)$$

needs to be assumed.

Further constraints follow from the null fields in the primary field representation, that is, the affine singular vectors.[6] These will be considered later.

Also, additional constraints are rooted in the very definition of the Sugawara energy-momentum tensor. In particular, acting on a state $|\phi_i\rangle$ satisfying Eq.(15.83),

---

[6] There can be no purely Virasoro singular vectors since these occur only for the Virasoro minimal models, that is, for $c = 1 - 6(p - p')^2/pp'$, which is clearly outside of the range (15.75).

Eq. (15.64) yields, for $n = -1$

$$L_{-1} |\phi_i\rangle = \frac{1}{k+g} \sum_a (J^a_{-1} J^a_0) |\phi_i\rangle = \frac{-1}{k+g} \sum_a (J^a_{-1} t^a_i) |\phi_i\rangle \qquad (15.95)$$

We consider the insertion of the zero vector

$$|\chi\rangle = \left[L_{-1} + \frac{1}{k+g} \sum_a (J^a_{-1} t^a_i) \right] |\phi_i\rangle = 0 \qquad (15.96)$$

inside the correlation function of a set of primary fields. We note that the insertion of the operator J^a_{-1} in the correlator can be expressed as

$$\langle \phi_1(z_1) \cdots (J^a_{-1} \phi_i)(z_i) \cdots \phi_n(z_n) \rangle$$

$$= \frac{1}{2\pi i} \oint_{z_i} \frac{dz}{z - z_i} \langle J^a(z) \phi_1(z_1) \cdots \phi_n(z_n) \rangle$$

$$= \frac{1}{2\pi i} \oint_{z_j, j \neq i} \frac{dz}{z - z_i} \sum_{j \neq i} \frac{t^a_j}{z - z_j} \langle \phi_1(z_1) \cdots \phi_n(z_n) \rangle$$

$$= \sum_{j \neq i} \frac{t^a_j}{z_i - z_j} \langle \phi_1(z_1) \cdots \phi_n(z_n) \rangle$$

$$(15.97)$$

For the second equality, the integration contour has been reversed in order to circle around all $z_j \neq z_i$. This produces a minus sign. Another minus sign comes from the use of the OPE (15.77). Therefore[7]

$$\langle \phi_1(z_1) \cdots \chi(z_i) \cdots \phi_n(z_n) \rangle$$

$$= \left[\partial_{z_i} + \frac{1}{k+g} \sum_{j \neq i} \frac{\sum_a t^a_i \otimes t^a_j}{z_i - z_j} \right] \langle \phi_1(z_1) \cdots \phi_n(z_n) \rangle$$

$$(15.98)$$

and by construction this must vanish:

$$\boxed{\left[\partial_{z_i} + \frac{1}{k+g} \sum_{j \neq i} \frac{\sum_a t^a_i \otimes t^a_j}{z_i - z_j} \right] \langle \phi_1(z_1) \cdots \phi_n(z_n) \rangle = 0} \qquad (15.99)$$

This is the *Knizhnik-Zamolodchikov equation*. The solutions to this equation are the correlation functions of primary fields. However, in practice they are rather difficult to solve directly, except for the four-point functions, in which case the partial differential equation can be reduced to an ordinary differential equation by means of the global $SL(2, \mathbb{C})$ invariance. An example is worked out in the next

---

[7] The necessity of the tensor product should be clear: we recall that t^a_i acts on the field ϕ_i at the point z_i; thus, on the doublet (ϕ_i, ϕ_j), $t^a_i \otimes t^a_j$ acts as $t^a_i \otimes t^a_j (\phi_i, \phi_j) = (t^a_i \phi_i, t^a_j \phi_j)$. Furthermore, we stress that there is an identity matrix with appropriate dimension multiplying the z_i derivative.

section. As in the purely Virasoro case, the correlation functions involving the descendant fields (15.89) can be obtained directly from those of the primary fields.

15.3.3. Primary Fields as Highest-Weight States

Before analyzing the affine singular vectors, the concept of a WZW primary field has to be sharpened. Up to now we have characterized a primary field by two properties: (i) it is annihilated by all the raising operators for positive roots of nonzero grade, that is, by the operators $H_n^i, E_n^\alpha, (n > 0)$;[8] and (ii) it is valued in some Lie algebra representation. However, all the states in a highest-weight representation can be obtained from the highest-weight state by application of the various (zero-grade) lowering operators. (We note that the application of zero-grade operators on a state does not change its conformal dimension since such operators commute with L_0.) Hence the genuine primary fields, that is the fundamental fields from which all other fields can be obtained by application of the Virasoro or affine Lie generators, are those associated with the highest-weight states of g representations. Denote by $|\lambda\rangle$ such a highest-weight state, and by $|\hat{\lambda}\rangle$ its extension to $\hat{g}_k$. Since $|\hat{\lambda}\rangle$ can be used to characterize a primary field, we denote the corresponding field by the same symbol $\hat{\lambda}(z)$:

$$\hat{\lambda}(z)|0\rangle = |\hat{\lambda}\rangle \qquad (15.100)$$

This state satisfies the following conditions,

$$E_n^{\pm\alpha}|\hat{\lambda}\rangle = H_n^i|\hat{\lambda}\rangle = E_0^\alpha|\hat{\lambda}\rangle = 0 \quad \text{for} \quad n, \alpha > 0 \qquad (15.101)$$

which means that it is annihilated by all raising operators for positive roots, together with

$$H_0^i|\hat{\lambda}\rangle = \lambda^i|\hat{\lambda}\rangle \qquad (15.102)$$

or, in the Chevalley basis,

$$h_0^i|\hat{\lambda}\rangle = \lambda_i|\hat{\lambda}\rangle \qquad (15.103)$$

(we recall that λ^i is not a Dynkin label, but λ_i is.)

From now on, by a WZW primary state we will mean a state that satisfies Eqs. (15.101) and (15.102), and we write (cf. Eq. (15.87))

$$\boxed{h_{\hat{\lambda}} \equiv h_\lambda = \frac{(\lambda, \lambda + 2\rho)}{2(k+g)}} \qquad (15.104)$$

With this clarification concerning primary fields, we see that the basic n-point functions of the theory, from which those incorporating positive-grade descendants can be derived, are of the form

$$\langle \hat{\mu}'(z_1) \cdots \hat{\nu}'(z_n) \rangle \qquad (15.105)$$

---

[8] We recall that $\alpha + n\delta$ with $n > 0$ is a positive root, irrespective of the positivity of α (cf. Eq. (14.33)).

where here the prime indicates a zero-grade descendant of the corresponding (unprimed) primary field, that is

$$|\hat{\mu}'\rangle = E_0^{-\alpha} \cdots E_0^{-\beta} |\hat{\mu}\rangle \qquad (15.106)$$

for $\alpha, \cdots, \beta > 0$. In other words, μ' is some finite weight in the highest-weight representation μ of g: $\mu' \in \Omega_\mu$. That basic correlation functions contain zero-grade descendants and not just primary fields is forced by the requirement of global G invariance (cf. Eq. (15.93)). This implies the vanishing of the sum of the finite weights associated with the fields in the correlator:

$$\mu' + \cdots + \nu' = 0 \qquad (15.107)$$

which is the non-Abelian analogue of the charge neutrality condition in the Coulomb-gas formalism (cf. Eq. (9.9)). This condition can be rederived directly from a global G transformation for which the only nonzero transformation parameters ω^a are those associated with the generators of the Cartan subalgebra. The condition (15.107) can obviously not be satisfied when the Dynkin labels of all the fields are positive.

15.3.4. Affine Lie Algebra Singular Vectors

A special class of primary fields is formed by the highest-weight states of *integrable* representations $\hat{\lambda} \in P_+^k$. Those states generate finite representations with respect to any $su(2)$ subalgebra of $\hat{g}_k$, which implies the conditions

$$(E_0^{-\alpha})^{(\alpha^\vee,\lambda)+1}|\hat{\lambda}\rangle = 0, \qquad \alpha > 0 \qquad (15.108)$$

$$(E_{-n}^{\alpha})^{2kn/|\alpha|^2-(\alpha^\vee,\lambda)+1}|\hat{\lambda}\rangle = 0, \qquad n > 0 \qquad (15.109)$$

Equation (15.109) defines purely affine singular vectors in the Verma module of highest-weight state $|\hat{\lambda}\rangle$, for $\hat{\lambda} \in P_+^k$. An example of such a state is provided by the vacuum $|0\rangle = |k\hat{\omega}_0\rangle$.

Since all $su(2)$ subalgebras of nonzero grade (i.e., with generators E_{-n}^{α}, $E_n^{-\alpha}$, and $\alpha \cdot H_0/|\alpha|^2$) are equivalent, we can consider only the case $n = 1$ in Eq. (15.109), which gives the most stringent constraint (smallest number of applications of the lowering operator). In the same respect, the condition is obviously optimized for $\alpha = \theta$. Actually, all the conditions (15.109) are properly taken into account by

$$\boxed{(E_{-1}^{\theta})^{k-(\theta,\lambda)+1}|\hat{\lambda}\rangle = 0} \qquad (15.110)$$

together with the set (15.108). In other words, for singular vectors involving nonzero modes, it is sufficient to consider the generator associated with the simple root α_0.

Consider the constraints on the correlation functions that follow from the singular vector (15.110). Inserting Eq. (15.110) into Eq. (15.105), we find

$$\langle((E_{-1}^{\theta})^p \hat{\lambda})(z)\hat{\mu}'(z_1)\cdots.\hat{\nu}'(z_n)\rangle = 0 \quad \text{with} \quad p \geq k - (\lambda,\theta) + 1 \qquad (15.111)$$

Together with

$$E^\theta_{-1}(z) = \frac{1}{2\pi i} \oint_z \frac{d\zeta}{\zeta - z} E^\theta(\zeta) \qquad (15.112)$$

Eq. (15.111) can be rewritten as

$$\oint_z \frac{d\zeta_1}{\zeta_1 - z} \cdots \oint_z \frac{d\zeta_p}{\zeta_p - z} \langle E^\theta(\zeta_1) \cdots E^\theta(\zeta_p) \hat\lambda(z) \hat\mu'(z_1) \cdots \hat\nu'(z_n) \rangle = 0 \quad (15.113)$$

We now deform the integration contour of every variable ζ_i such that it circles around the singular points $z_1, \cdots, z_n$. Hence, only the singular terms in the OPE of the various $E^\theta(\zeta_i)$ with $\mu'(z_1) \cdots \nu'(z_n)$ will contribute. By expanding the fields $E^\theta(\zeta_i)$ in terms of their modes and performing the integrations (in which only the mode $m = 0$ contributes), we obtain

$$\sum_{\substack{l_i = 0 \\ l_1 + \cdots + l_n = p}}^{p} \frac{p!}{l_1! \cdots l_n!} \frac{1}{(z - z_1)^{l_1}} \cdots \frac{1}{(z - z_n)^{l_n}}$$

$$\times \langle \hat\lambda(z) [(E^\theta_0)^{l_1} \hat\mu'](z_1) \cdots [(E^\theta_0)^{l_n} \hat\nu'](z_n) \rangle = 0 \qquad (15.114)$$

Let $\hat\lambda(z)$ be the identity field: $\hat\lambda = \mathbb{I}$. By multiplying the above result by $(z - z_n)^{p-1}$ and integrating with respect to z, we pick up the term with all $l_i = 0$, except for $l_n = p$, which finally yields

$$\langle \hat\mu'(z_1) \cdots \hat\nu''(z_n) \rangle = 0 \qquad (15.115)$$

where the state $|\hat\nu''\rangle$ is defined as

$$|\nu''\rangle = (E^\theta_0)^p |\hat\nu'\rangle \qquad (15.116)$$

This state $|\hat\nu''\rangle$, if nonzero, can be acted on with $(E^{-\theta}_0)^p$, with an integer p as large as desired, without leaving the module $L_{\hat\nu}$. That means that $|\hat\nu''\rangle$ does not belong to an integrable representation. Eq. (15.115) shows that correlation functions involving states in nonintegrable representations vanish.

Starting with a correlation function of which all the fields belong to highest-weight representations, we thus obtain the very remarkable result that all the states in nonintegrable representations decouple from the theory; that is, their correlations with arbitrary fields vanish.

We stress that the above derivation is quite general in that it requires only the existence in the theory of at least one field in an integrable representation. As already pointed out, the identity field is such an example. Therefore, the basic assumption boils down to the existence of a vacuum state.

One more comment is warranted before leaving the analysis of general constraints on correlation functions. By means of OPEs, an n-point function can be reduced to a sum of lower-point functions, and ultimately to a sum of one-point functions. From the global G invariance, we know that the only contributing one-point function is that associated with the scalar representation. Therefore, global G invariance entails a constraint slightly stronger than Eq. (15.107), namely that the tensor product of all the fields involved in the correlator must contain at least one copy of the scalar representation,

$$\mu \otimes \cdots \otimes \nu = 0 \oplus \cdots \qquad (15.117)$$

Condition (15.107) ensures that the zero weight will occur in the product of all the representations of the correlator, but Eq. (15.117) is stronger in that it forces at least one copy of the zero weight to be the highest weight of the scalar representation.

We now look at the consequences of this for two-point functions. We consider $\langle \hat{\lambda}'(z)\phi(w) \rangle$, where λ' is in the integrable representation λ, and look for the field ϕ, which yields a nonzero correlator. Global $SL(2, \mathbb{C})$ invariance requires $h_\phi = h_{\hat{\lambda}}$, which implies that ϕ belongs to the highest-weight representation λ or its conjugate (because $h_{\hat{\lambda}} = h_{\hat{\lambda}^*}$). Condition (15.107) shows that ϕ must correspond to the finite weight $-\lambda'$. But generically, if λ' is in the representation λ, $-\lambda'$ is in the conjugate representation λ^*. This thus fixes ϕ, up to weight degeneracies. We note that the tensor product of a representation with its conjugate always contains the scalar representation, and Eqs. (15.107) and (15.117) are identical in this case.

We have not exhausted all the information contained in Eq.(15.114). It leads to very interesting constraints on the OPE of primary fields, which, in favorable circumstances, allow us to determine completely the fusion rules of the theory. These additional implications of Eq. (15.114) will be analyzed in Chap. 16.

15.3.5. WZW Models as Rational Conformal Field Theories

We now come back to the description of the WZW primary fields. Having established the decoupling of all the states in nonintegrable representations, we conclude that the only physically relevant fields are those in integrable representations. Therefore, for the WZW model with underlying $\hat{g}_k$ affine structure, *the primary fields are in correspondence with the weights* $\hat{\lambda} \in P_+^k$, the highest weights of integrable representations. Since there is a finite number of such weights for a fixed positive integer k, we end up with the important conclusion that there is a finite number of primary fields in the $\hat{g}_k$ WZW model.

We stress that, even though there is a finite number of WZW primary fields, that is, primary fields with respect to the affine Lie algebra, there is an infinite number of Virasoro primary fields. Indeed, as shown in Sect. 10.5, only for very special values of the central charge, all strictly less than 1, is a finite number of Virasoro primary fields possible. For WZW models, c is necessarily greater than or equal to 1 (cf. Eq. (15.75)).

But where are all these Virasoro primary fields? Take, for instance, the $\widehat{su}(2)_1$ model, for which the first few states in the vacuum module are displayed in Fig. 14.4. In this representation, we can already see an infinite number of Virasoro primary states: these are the top states in each vertical strip of the module.[9] Indeed, each vertical string of states (i.e., the set of all the states with the same finite weight

---

[9] A more direct argument is presented in Ex. 15.17, based on the character expressions derived in Sect. 15.6.

at different grades) forms an irreducible Virasoro representation at $c = 1$. For instance, the state $[-1, 2]$ at grade 1, that is, $J_{-1}^+|[1, 0]\rangle$, is a Virasoro primary state:

$$L_n J_{-1}^+|[1, 0]\rangle = J_{-1}^+ L_n|[1, 0]\rangle + J_{n-1}^+|[1, 0]\rangle = 0 \quad \text{for} \quad n \geq 1 \quad (15.118)$$

where we used the commutation relation (15.65). However, it is easily checked that the action of L_1 on the the state $[-1, 2]$ at grade 2, which can be written as $J_{-1}^+ J_0^- J_{-1}^+|[1, 0]\rangle$, is not zero, showing that it is not a Virasoro highest-weight state.

That each vertical string of states corresponds to exactly one Virasoro module is particular to the $\widehat{su}(2)_1$ case. Generally, there is more than one Virasoro module in each string. But the key point is that all those states at the tip of vertical strings are necessarily Virasoro highest-weight states. These are the states whose weights $\hat{\mu}$ in $\Omega_{\hat{\lambda}}$ are such that $\hat{\mu} + \delta \notin \Omega_{\hat{\lambda}}$. And this set, called $\Omega_{\hat{\lambda}}^{\max}$ previously, is infinite: Fix a $\hat{\mu} \in \Omega_{\hat{\lambda}}^{\max}$; acting with the affine Weyl group (whose order is infinite) on $\hat{\mu}$ still produces an element of $\Omega_{\hat{\lambda}}^{\max}$.

In WZW models, the *infinite* number of Virasoro primary fields are thus reorganized into a *finite* number of affine Lie algebra representations. The rational character of a conformal theory with an additional symmetry entails a rearrangement of the fields with respect to the full extended algebra. In the present context, because the Virasoro algebra belongs to the universal enveloping algebra of the affine algebra, it suffices to classify the fields in terms of the irreducible representations of the latter.

Having identified the primary fields of the theory, we should inquire about modular covariance. In analogy with the Virasoro case, we introduce the character of the integrable representation of $|\hat{\lambda}\rangle$ as

$$\boxed{\chi_{\hat{\lambda}}(\tau) = \text{Tr}_{\hat{\lambda}} e^{[2\pi i \tau(L_0 - c/24)]}} \quad (15.119)$$

with L_0 given by Eq. (15.64) and c by Eq. (15.61). Since the states at level n in the module of $|\hat{\lambda}\rangle$ have dimension $h_{\hat{\lambda}} + n$, where $h_{\hat{\lambda}}$ is given by Eq. (15.104), we can rewrite the expression for the character as

$$\chi_{\hat{\lambda}}(\tau) = e^{[2\pi i \tau(h_{\hat{\lambda}} - c/24)]} \sum_n d(n) e^{2\pi i n \tau} \quad (15.120)$$

where $d(n)$ is the number of states at level n. Using the Freudenthal–de Vries strange formula

$$12|\rho|^2 = g \dim g \quad (15.121)$$

it is easy to check that

$$\boxed{h_{\hat{\lambda}} - c/24 = m_{\hat{\lambda}}} \quad (15.122)$$

Thus, Eq. (15.120) is simply the expression for the specialized (and normalized) characters of the irreducible highest-weight representations of the affine Lie algebra $\hat{g}_k$. We recall that in our study of affine Lie algebras, we found the L_0 eigenvalue of the highest-weight states to be arbitrary, and conventionally chosen to be zero.

This is obviously no longer the case here since L_0 also appears in the Virasoro algebra, and its eigenvalue is unambiguously fixed to be $h_{\hat{\lambda}}$. To distinguish conjugate representations, we introduce, as usual, a ζ dependence; with $\zeta = \sum_{i=1}^{r} z_i \alpha_i^{\vee}$, this gives

$$\chi_{\hat{\lambda}}(z; \tau) = \text{Tr}_{\hat{\lambda}} \, e^{2\pi i \tau (L_0 - c/24)} e^{-2\pi i \sum_j z_j h_j} \tag{15.123}$$

where here h_j is a Chevalley generator.

The identification of the WZW characters with those of the integrable representations of $\hat{g}_k$ readily tells us that the characters of WZW primary fields transform into each other under modular transformations, with the explicit form of the modular matrices given by Eqs. (14.215), (14.216), and (14.217). Hence, the modular covariance of the $\hat{g}_k$ WZW model in each sector is established automatically. From the Verlinde formula, this also implies that the set of holomorphic and antiholomorphic primary fields are closed independently under OPE. (The detailed analysis of fusion rules is reported in the next chapter.)

A WZW model is thus very similar to a minimal Virasoro model. Actually, it is also a unitary theory. This follows from the unitarity of the integrable representations of an affine Lie algebra at integer level. This is also manifest from the positivity of the conformal dimensions of the WZW primary fields, Eq. (15.104): to each primary field there corresponds a module with $h \geq 0$ and $c > 1$, and these are always unitary (cf. Sect. 7.3.4).

To obtain physical spectra, we have to determine how left and right representations are tied together. This amounts to constructing a modular-invariant partition function. This subject will be studied in full in Chap. 17. However, from our past experience with modular invariants, we can guess that a diagonal theory, in which all the primary fields transform with respect to the same representation in the holomorphic and antiholomorphic sectors, with each integrable representation appearing exactly once, is always possible.

§15.4. Four-Point Functions and the Knizhnik-Zamolodchikov Equation

In this section we present the detailed calculation of the four-point function

$$\mathcal{G}(z_i, \bar{z}_i) = \langle g(z_1, \bar{z}_1) \, g^{-1}(z_2, \bar{z}_2) \, g^{-1}(z_3, \bar{z}_3) \, g(z_4, \bar{z}_4) \rangle \tag{15.124}$$

for $\widehat{su}(N)_k$. We note that in order for correlations of $g(z, \bar{z})$ fields to be $SU(N)$ invariants, $g(z, \bar{z})$ and $g^{-1}(z, \bar{z})$ must appear in equal numbers.[10]

---

[10] For instance, under the right action of $SU(N)$, $g(z, \bar{z})$ transform as $g\Omega$ with $\Omega \in SU(N)$, and the product $gg^{-1}g^{-1}g$ is manifestly invariant. The invariance with respect to the left action of $SU(N)$ follows from the invariance of the correlator with respect to the reordering of its fields.

15.4.1. Introductory Comments

In preparation for this calculation, we present a series of preliminary remarks. First, we recall that $g(z, \bar{z})$ transforms in the fundamental representation ω_1 in both sectors:

$$g(z, \bar{z}): \quad (\omega_1; \omega_1) \tag{15.125}$$

Since $g(z, \bar{z})$ is unitary, $g^{-1} = g^\dagger$, $g^{-1}(z, \bar{z})$ transforms in the representation conjugate to ω_1:

$$g^{-1}(z, \bar{z}): \quad (\omega_1^*; \omega_1^*) = (\omega_{N-1}; \omega_{N-1}) \tag{15.126}$$

Moreover, in the Knizhnik-Zamolodchikov equation, a factor t^a associated with a $g(z, \bar{z})$ factor acts from the left, whereas if it is associated with a g^{-1} factor, it acts from the right (and it is the same t^a in both cases because $(t^a)^\dagger = t^a$); that is,

$$t_1^a \otimes t_2^a \langle g(z_1, \bar{z}_1)\, g^{-1}(z_2, \bar{z}_2) \cdots \rangle = \langle [t^a g(z_1, \bar{z}_1)]\, [g^{-1}(z_2, \bar{z}_2) t^a] \cdots \rangle \tag{15.127}$$

The matrix $g(z, \bar{z})$ is actually a tensor product of a column vector made from the fields associated with all the states in the fundamental representation ω_1 times a row vector made out of the same states:

$$g(z, \bar{z}) = \begin{pmatrix} \omega_1(z) \\ (\omega_1 - \alpha_1)(z) \\ \cdots \\ \cdots \\ (\omega_1 - \theta)(z) \end{pmatrix} \otimes \left(\omega_1(\bar{z}), (\omega_1 - \alpha_1)(\bar{z}), \cdots, (\omega_1 - \theta)(\bar{z}) \right) \tag{15.128}$$

Here fields and weights are denoted by the same symbol since all weights have multiplicity 1. Take, for instance, $\widehat{su}(2)_1$: in terms of the fields

$$(\phi_{j=\frac{1}{2}, m}(z); \bar{\phi}_{j=\frac{1}{2}, \bar{m}}(\bar{z})) \equiv (m; \bar{m}) \tag{15.129}$$

$g(z, \bar{z})$ reads

$$g(z, \bar{z}) = \begin{pmatrix} (\frac{1}{2}; \frac{1}{2}) & (\frac{1}{2}; -\frac{1}{2}) \\ (-\frac{1}{2}; \frac{1}{2}) & (-\frac{1}{2}; -\frac{1}{2}) \end{pmatrix} \tag{15.130}$$

From now on, the matrix entries of $g(z_i, \bar{z}_i)$ will be written as $g_{m_i, \bar{m}_i}(z_i, \bar{z}_i)$ with m_i (resp. $\bar{m}_i$) referring to the holomorphic (resp. antiholomorphic) sector.

The group theoretical content of the correlator (15.124) is coded in the tensor product

$$\omega_1 \otimes \omega_{N-1} \otimes \omega_{N-1} \otimes \omega_1 \tag{15.131}$$

Using the Littlewood-Richardson rule described in Sect. 13.5.3, it is easy to verify that:

$$\omega_1 \otimes \omega_{N-1} = \theta \oplus \mathbb{I} \tag{15.132}$$

where θ refers to the adjoint representation and $\mathbb{I}$ to the scalar representation of zero weight. The above four-term tensor product reduces then to

$$(\theta \oplus \mathbb{I}) \otimes (\theta \oplus \mathbb{I}) = \mathbb{I} \oplus 2\theta \oplus (\theta \otimes \theta)$$
$$= 2\mathbb{I} \oplus 2\theta \oplus \cdots \qquad (15.133)$$

since, in the product of two adjoint representations, the scalar representation appears only once. Thus, in the product (15.131), the identity occurs twice. Equivalently, the two identities could be characterized as follows: one comes from the identity in each of the two products

$$g(z_1, \bar{z}_1) \, g^{-1}(z_2, \bar{z}_2) \quad \text{and} \quad g^{-1}(z_3, \bar{z}_3) \, g(z_4, \bar{z}_4) \qquad (15.134)$$

and the other from similar terms with the indices 2 and 3 interchanged. Therefore, each conformal block of the four-point function, in addition to being decomposable into holomorphic and antiholomorphic sectors as

$$\mathcal{F}(z_i, \bar{z}_i) = \mathcal{F}(z_i)\bar{\mathcal{F}}(\bar{z}_i) \qquad (15.135)$$

also decomposes into the $SU(N)$ invariant factors

$$I_1 \equiv \delta_{m_1,m_2}\delta_{m_3,m_4}, \qquad I_2 \equiv \delta_{m_1,m_3}\delta_{m_2,m_4} \qquad (15.136)$$

(where, by an abuse of notation, we identify the matrix with its components) as

$$\mathcal{F}(z_i) = I_1\mathcal{F}_1(z_i) + I_2\mathcal{F}_2(z_i) \qquad (15.137)$$

and likewise for $\bar{\mathcal{F}}(\bar{z}_i)$.

The product (15.131) yields two copies of the identity, which means that the correlation function (15.124) will decompose into a linear combination of at most two conformal blocks. To obtain the precise number of contributing blocks, we need to evaluate the level k fusion rules

$$\hat{\lambda} \times \hat{\lambda}^* \times \hat{\lambda}^* \times \hat{\lambda} \qquad (15.138)$$

with

$$\hat{\lambda} = [k-1, 1, 0, \cdots, 0] = (k-1)\hat{\omega}_0 + \hat{\omega}_1 \qquad (15.139)$$

Techniques for calculating fusion rules will be presented in the next chapter. However, it is not difficult to convince oneself that for $k > 1$,

$$\hat{\lambda} \times \hat{\lambda}^* = \mathbb{I} + \hat{\theta} \qquad (k > 1) \qquad (15.140)$$

with

$$\hat{\theta} = k\hat{\omega}_0 + \theta \qquad (15.141)$$

This means that there is no truncation of the tensor product. The complete four-term fusion rule then produces two copies of the identity. However, for $k = 1$, the field transforming in the adjoint representation decouples since it is no longer primary (i.e., $\hat{\theta}$ at level 1 is $[-1, 1, 0, \cdots, 0, 1]$, and it has one negative Dynkin label); we thus have

$$\hat{\lambda} \times \hat{\lambda}^* = \mathbb{I} \qquad (k = 1) \qquad (15.142)$$

or, equivalently,

$$\hat{\omega}_1 \times \hat{\omega}_{N-1} = \mathbb{I} \tag{15.143}$$

so that for $k = 1$ there is only one channel, that is, only one contributing conformal block.

15.4.2. The Four-Point $\widehat{su}(N)_k$ Knizhnik-Zamolodchikov Equation

As usual, the projective Ward identities allow us to write

$$\mathcal{G}(z_i, \bar{z}_i) = [(z_1 - z_4)(z_2 - z_3)(\bar{z}_1 - \bar{z}_4)(\bar{z}_2 - \bar{z}_3)]^{-2h} \mathcal{G}(x, \bar{x}) \tag{15.144}$$

where h is the dimension of the field $g(z, \bar{z})$:[11]

$$h = \bar{h} \equiv h_g = \frac{(\omega_1, \omega_1 + 2\rho)}{2(k + N)} = \frac{N^2 - 1}{2N(k + N)} \tag{15.145}$$

and

$$x = \frac{(z_1 - z_2)(z_3 - z_4)}{(z_1 - z_4)(z_3 - z_2)} \qquad \bar{x} = \frac{(\bar{z}_1 - \bar{z}_2)(\bar{z}_3 - \bar{z}_4)}{(\bar{z}_1 - \bar{z}_4)(\bar{z}_3 - \bar{z}_3)} \tag{15.146}$$

In the following, we set $z_{ij} = z_i - z_j$.

In terms of the variables $F_j(x)$ defined by

$$\mathcal{F}_j(z_i) = [z_{14}z_{23}]^{-2h} F_j(x) \qquad j = 1, 2 \tag{15.147}$$

the holomorphic part of the Knizhnik-Zamolodchikov equation (15.99) becomes

$$\left(\partial_{z_i} + \frac{1}{k + N} \sum_{j \neq i} \frac{\sum_a t_i^a \otimes t_j^a}{z_i - z_j} \right) [z_{14}z_{23}]^{-4h} (I_1 F_1(x) + I_2 F_2(x)) = 0 \tag{15.148}$$

The next step is to transform these partial differential equations into an ordinary differential equation in the variable x. Consider the case $i = 1$; since

$$\partial_{z_1} = \left(\frac{x}{z_{12}} - \frac{x}{z_{14}} \right) \partial_x \tag{15.149}$$

the equation takes the form

$$\left\{ \frac{-2h}{z_{14}} + \left(\frac{x}{z_{12}} - \frac{x}{z_{14}} \right) \partial_x + \frac{1}{k + N} \sum_{j=2,3,4} \frac{\sum_a t_1^a \otimes t_j^a}{z_1 - z_j} \right\} (I_1 F_1(x) + I_2 F_2(x)) = 0 \tag{15.150}$$

Once the z_1 derivative has been evaluated explicitly, and after taking out the common factor $[z_{14}z_{23}]^{-2h}$, we can fix three values of the z_i's. A convenient choice is

$$z_1 = x, \qquad z_2 = 0, \qquad z_3 = 1, \qquad z_2 = \infty \tag{15.151}$$

---

[11] We recall that $(\omega_i, \omega_j) = i(N-j)/N$ for $i \leq j$; we then have $(\omega_1, \rho) = \sum_{j=1}^N (N-j)/N = (N-1)/2$.

This reduces the equation to

$$\left\{\partial_x + \frac{1}{k+N} \frac{\sum_a t_1^a \otimes t_2^a}{x} + \frac{1}{k+N} \frac{\sum_a t_1^a \otimes t_3^a}{x-1}\right\} (I_1 F_1(x) + I_2 F_2(x)) = 0$$

$$(15.152)$$

The next step is to evaluate the various terms $\sum_a t_i^a \otimes t_j^a$ on the I_i's. For this, we must recall that when t^a is associated with a g (resp. g^{-1}) factor, it acts on the left (resp. right). Thus, for instance,

$$t_1^a \otimes t_4^a I_1 = \sum_{m_1', m_4'} t_{m_1, m_1'}^a t_{m_4, m_4'}^a \delta_{m_1', m_2} \delta_{m_4', m_3} = t_{m_1, m_2}^a t_{m_4, m_3}^a \qquad (15.153)$$

On the other hand, if say $i = 1\, j = 3$, we should write:

$$t_1^a \otimes t_3^a I_1 = \sum_{m_1', m_3'} t_{m_1, m_1'}^a t_{m_3', m_3}^a \delta_{m_1', m_2} \delta_{m_4, m_3'} = t_{m_1, m_2}^a t_{m_4, m_3}^a \qquad (15.154)$$

(We note the position of the repeated indices in each case.) To decompose the product $t^a t^a$ in terms of I_1 and I_2, we need the formula:

$$\sum_a t_{ij}^a t_{kl}^a = \delta_{il} \delta_{jk} - \frac{1}{N} \delta_{ij} \delta_{kl} \qquad (15.155)$$

which holds for $su(N)$ matrices in the fundamental representation. The quadratic Casimir operator is

$$\sum_a (t^a t^a)_{ij} = (\omega_1, \omega_1 + 2\rho)\delta_{ij} = \frac{N^2 - 1}{N} \delta_{ij} \qquad (15.156)$$

We are now in position to evaluate the different terms occurring in Eq. (15.152):

$$\sum_a t_1^a \otimes t_2^a I_1 = \sum_a (t^a t^a)_{m_1, m_2} \delta_{m_3, m_4}$$

$$= \frac{N^2 - 1}{N} \delta_{m_1, m_2} \delta_{m_4, m_3} = \frac{N^2 - 1}{N} I_1$$

$$\sum_a t_1^a \otimes t_2^a I_2 = \sum_a t_{m_1, m_3}^a t_{m_4, m_2}^a$$

$$= \delta_{m_1, m_2} \delta_{m_3, m_4} - \frac{1}{N} \delta_{m_1, m_3} \delta_{m_2, m_4} = I_1 - \frac{1}{N} I_2$$

$$\sum_a t_1^a \otimes t_3^a I_1 = \sum_a t_{m_1, m_2}^a t_{m_4, m_2}^a \qquad (15.157)$$

$$= \delta_{m_1, m_3} \delta_{m_2, m_4} - \frac{1}{N} \delta_{m_1, m_2} \delta_{m_3, m_4} = I_2 - \frac{1}{N} I_1$$

$$\sum_a t_1^a \otimes t_3^a I_2 = \sum_a (t^a t^a)_{m_1, m_3} \delta_{m_2, m_4}$$

$$= \frac{N^2 - 1}{N} \delta_{m_1, m_3} \delta_{m_2, m_4} = \frac{N^2 - 1}{N} I_2$$

After substituting these results into Eq. (15.152), since the terms I_1 and I_2 are independent—so that their multiplying factors must vanish separately—, we obtain the two equations:

$$\partial_x F_1 = \frac{-1}{k+N}\left\{\frac{(N^2-1)F_1}{N}+\frac{F_2}{x}-\frac{1}{N}\frac{F_1}{x-1}\right\}$$

$$\partial_x F_2 = \frac{-1}{k+N}\left\{\frac{(N^2-1)}{N}\frac{F_2}{x-1}+\frac{F_1}{x-1}-\frac{1}{N}\frac{F_2}{x}\right\}$$

(15.158)

To proceed, we use the first of these equations to express F_2 in terms of F_1 and substitute the result into the second equation. We then set

$$F_1 = x^r(1-x)^s f_1 \tag{15.159}$$

to obtain the following second-order differential equation for f_1:

$$\frac{x(1-x)}{N^2}\{N^2\kappa^2\partial_x^2 + A(x)\partial_x + B(x)\}f_1 = 0 \tag{15.160}$$

where $\kappa = k + N$,

$$A(x) = \left(\frac{2r\kappa N + N^2 - 2 + \kappa N}{x} - \frac{2s\kappa N + N^2 - 2}{1-x}\right)N\kappa \tag{15.161}$$

and

$$\begin{aligned}
B(x) =\ & \left(\frac{\kappa^2 N^2 r(r-1) + r\kappa N(N^2-2) + r\kappa^2 N^2 - (N^2-1)}{x^2}\right)\\
& + \left(\frac{\kappa^2 N^2 s(s-1) + s\kappa N(N^2-2) + \kappa N - (N^2-1)}{(1-x)^2}\right)\\
& + \left(\frac{-2rs\kappa^2 N^2 - (s+r)\kappa N(N^2-1) - s\kappa^2 N^2}{x(1-x)}\right)\\
& + \left(\frac{\kappa N - 1 - (N^2-1)^2}{x(1-x)}\right) + \frac{N^2}{x(1-x)}
\end{aligned} \tag{15.162}$$

Any solution of a linear second-order differential equation can be expressed in terms of two independent ones. Simple solutions can be obtained by making appropriate choices of the parameters r and s. In this way, the above equation can be transformed into the hypergeometric equation (cf. Ex. 8.9):

$$x(1-x)\partial_x^2 f + [c - (a+b+1)]\partial_x f - abf = 0 \tag{15.163}$$

For this, we must eliminate the coefficients of the x^{-2} and $(1-x)^{-2}$ terms; this fixes the possible values of r and s to be

$$r = r_+ = \frac{1}{\kappa N} \qquad \text{or} \quad r = r_- = -\frac{N^2-1}{\kappa N} \tag{15.164}$$

and

$$s = s_+ = \frac{1}{\kappa N} \qquad \text{or} \quad s = s_- = 1 - \frac{N^2-1}{\kappa N} \tag{15.165}$$

Out of these four possible choices, two are independent. We then fix the value of s to be s_+. For the two choices of r, the solutions are:

$$r = r_- : \quad f_1^{(-)} = F\left(\frac{1}{\kappa}, -\frac{1}{\kappa}; 1 - \frac{N}{\kappa}; x\right)$$

$$r = r_+ : \quad f_1^{(+)} = F\left(\frac{N-1}{\kappa}, \frac{N+1}{\kappa}; 1 + \frac{N}{\kappa}; x\right) \tag{15.166}$$

where $F(a, b; c; x)$ is the hypergeometric function (defined in Ex. 8.9). Notice that $r_- = 2h$, where h is the dimension of the field $g(z, \bar{z})$ given in Eq. (15.145). In the same vein, we can express r_+ in terms of the dimension of the adjoint field (with weight $\hat{\theta}$) and that of $g(z, \bar{z})$: Since

$$h_{\hat{\theta}} = \frac{(\theta, \theta + 2\rho)}{2(N+k)} = \frac{N}{N+k} \tag{15.167}$$

where we used Eq. (13.128) with $g = N$, we can write

$$r_+ = s_+ = h_{\hat{\theta}} - 2h \tag{15.168}$$

The two solutions for F_1 are thus:

$$F_1^{(-)} = x^{-2h}(1-x)^{h_{\hat{\theta}}-2h} F\left(\frac{1}{\kappa}, -\frac{1}{\kappa}; 1 - \frac{N}{\kappa}; x\right)$$

$$F_1^{(+)} = x^{h_{\hat{\theta}}-2h}(1-x)^{h_{\hat{\theta}}-2h} F\left(\frac{N-1}{\kappa}, \frac{N+1}{\kappa}; 1 + \frac{N}{\kappa}; x\right) \tag{15.169}$$

Solutions for F_2 are obtained in exactly the same way. A suitable choice of solutions (incorporating a convenient normalization) is

$$F_2^{(-)} = \frac{1}{k} x^{1-2h}(1-x)^{h_{\hat{\theta}}-2h} F\left(1+\frac{1}{\kappa}, 1 - \frac{1}{\kappa}; 1 - \frac{N}{\kappa}; x\right)$$

$$F_2^{(+)} = -N x^{h_{\hat{\theta}}-2h}(1-x)^{h_{\hat{\theta}}-2h} F\left(\frac{N-1}{\kappa}, \frac{N+1}{\kappa}; \frac{N}{\kappa}; x\right) \tag{15.170}$$

From the leading x power, we can identify the two conformal blocks, which have two components each, as follows:

$$\mathcal{F}_{\mathbb{I}} \equiv \mathcal{F}^{(-)} = \begin{pmatrix} F_1^{(-)} \\ F_2^{(-)} \end{pmatrix} \tag{15.171}$$

is the identity conformal block, while

$$\mathcal{F}_{\hat{\theta}} \equiv \mathcal{F}^{(+)} = \begin{pmatrix} F_1^{(+)} \\ F_2^{(+)} \end{pmatrix} \tag{15.172}$$

is the adjoint-field conformal block.

15.4.3. The Crossing-Symmetry Constraint

The final step of the calculation consists in fixing the appropriate linear combination of our solutions making the correlator well-defined. As illustrated in Sect. 9.2.3, this amounts to enforcing the monodromy invariance of the correlator. Equivalently, the emphasis can be placed on crossing symmetry. More precisely, we require the correlator to be single valued and invariant under the crossing symmetry:

$$\mathcal{G}(x,\bar{x}) = \mathcal{G}(1-x,1-\bar{x}) \tag{15.173}$$

where $\mathcal{G}(x,\bar{x})$ has been defined in Eq. (15.144). This is the point of view adopted in the present calculation.

Let

$$\mathcal{G}(x,\bar{x}) = \sum_{ij=1,2} I_i \bar{I}_j \mathcal{G}_{i,j}(x,\bar{x}) \tag{15.174}$$

and

$$\mathcal{G}_{i,j}(x,\bar{x}) = \sum_{n,m=-,+} X_{nm} F_i^{(n)}(x) F_j^{(m)}(\bar{x}) \tag{15.175}$$

A first condition on the constants X_{nm} follows from the requirement of single-valuedness. Consider the sector where x is very small; for a single valued solution, a fractional power of x is permitted only if it appears in absolute value. This forces

$$X_{-+} = X_{+-} = 0 \tag{15.176}$$

Crossing symmetry requires:

$$\mathcal{G}_{i,j}(x,\bar{x}) = \mathcal{G}_{3-i,3-j}(1-x,1-\bar{x}) \tag{15.177}$$

This equality simply translates the fact $x \leftrightarrow 1-x$ interchanges the channels I_1 and I_2. We now look for the particular linear combinations of our solutions which satisfy Eq. (15.177). For this, we need the following transformation property of the hypergeometric functions:

$$F(a,b;c;x) = A_1\, F(a,b;a+b-c+1;1-x)$$
$$+A_2\,(1-x)^{c-a-b}\, F(c-a,c-b;c-a-b+1;1-x) \tag{15.178}$$

with

$$A_1 = \frac{\Gamma(c)\Gamma(c-a-b)}{\Gamma(c-a)\Gamma(c-b)} \qquad A_2 = \frac{\Gamma(c)\Gamma(a+b-c)}{\Gamma(a)\Gamma(b)} \tag{15.179}$$

To characterize the transformation properties of the conformal block, we introduce the coefficients c_{nm} defined by:

$$F_i^{(n)}(x) = \sum_m c_{nm} F_{3-i}^{(m)}(1-x) \tag{15.180}$$

They are determined from the explicit expressions of the conformal-block components and by Eq. (15.178). Consistency requirements on these coefficients follow from a double use of Eq. (15.178):

$$c_{--} = -c_{++} \qquad c_{-+}c_{+-} + c_{++}c_{++} = 1 \tag{15.181}$$

Hence, only two coefficients need to be evaluated. A simple calculation gives

$$c_{--} = N\frac{\Gamma(N/\kappa)\Gamma(-N/\kappa)}{\Gamma(1/\kappa)\Gamma(-1/\kappa)} \tag{15.182}$$

and

$$c_{+-} = -N\frac{\Gamma^2(N/\kappa)}{\Gamma((N+1)/\kappa)\Gamma((N-1)/\kappa)} \tag{15.183}$$

The crossing-symmetry requirement takes the following form. To avoid terms mixing the two conformal blocks we must have:

$$X_{--}c_{--}c_{-+} + X_{++}c_{+-}c_{++} = 0 \tag{15.184}$$

and the equality of the nonmixed terms forces

$$\begin{aligned} X_{--}c_{--}^2 + X_{++}c_{+-}^2 &= X_{--} \\ X_{--}c_{-+}^2 + X_{++}c_{++}^2 &= X_{++} \end{aligned} \tag{15.185}$$

In view of Eq. (15.181), all these requirements are compatible. We then set

$$X_{++} = \frac{c_{--}^2 - 1}{c_{+-}^2}X_{--} \tag{15.186}$$

With our choice of normalization for the solutions of the differential equations, the correlation function is properly normalized when

$$X_{--} = 1 \tag{15.187}$$

The final result is then

$$\mathcal{G}_{i,j}(x,\bar{x}) = F_i^{(-)}(x)\,F_j^{(-)}(\bar{x}) + \frac{c_{--}^2 - 1}{c_{+-}^2}F_i^{(+)}(x)\,F_j^{(+)}(\bar{x}) \tag{15.188}$$

We have argued previously that for $k = 1$, the contribution of the second conformal block should decouple. It is indeed easily verified that for $k = 1, c_{--} = 1$.

§15.5. Free-Fermion Representations

15.5.1. Free-Field Representations and Quantum Equivalence

In the study of the Virasoro algebra, we found various representations in terms of free fields, based on bosons, fermions, and ghosts. In particular, the free-boson representation, including vertex operators, proved to be very useful for practical calculations, especially for the evaluation of correlation functions. In this section and the next two, we show that similar representations exist for WZW models. The simplest ones are the free-fermion representations of the $\widehat{so}(N)_1$ and $\widehat{su}(N)_1 \times \widehat{u}(1) \equiv \widehat{u}(N)_1$ WZW models, famous examples of non-Abelian bosonization.

For models with simply-laced algebras at level 1, there is also a representation in terms of r (=rank) free bosons and vertex operators, the so-called vertex representation. Its extension to nonsimply-laced algebras requires the introduction of a suitable set of fermionic fields. This type of representation also exists for $k \neq 1$, but it is much more complicated and will not be discussed here.

A generic representation for all $\hat{g}_k$ WZW models, in terms of r free bosons and $|\Delta_+|$ pairs of ghosts, will be described in Sect. 15.7.

It is certainly a remarkable fact that two theories with completely different classical actions can be equivalent at the quantum level. We stress that the quantum equivalence of two theories means the equivalence of their correlation functions. This implies directly that the two theories have the same symmetry algebra (which means the same Ward identities associated with the symmetry) and the same primary-field spectrum (which is encoded in the two-point correlation functions).

Although these last two characteristics are in general easily compared, this is not so for the full set of correlation functions. Even though it is sufficient to restrict ourselves to four-point correlation functions of primary fields (out of which the various operator coefficients can be extracted), the evaluation of these functions remains a tedious task. In practice, it appears that establishing the equivalence of the two (quantum) energy-momentum tensors is generally sufficient[12] and from now on we will understand quantum equivalence in this weak sense. An easily verified necessary condition for the equivalence of the energy-momentum tensors of two theories is the equality of their central charges.

15.5.2. The $\widehat{so}(N)_1$ Current Algebra From Real Free Fermions

Consider N real independent free fermions, with OPEs

$$\psi_i(z)\psi_j(w) \sim \frac{\delta_{ij}}{(z-w)} \tag{15.189}$$

transforming in the vector representation of $so(N)$ (i.e., with highest weight ω_1), with transformation matrices t_{ij}^a. In this case, the index a actually stands for a pair of integers (r,s) such that $1 \leq r < s \leq N$ and an explicit realization of these matrices is given by:

$$t_{ij}^a \equiv t_{ij}^{(rs)} = i(\delta_i^r \delta_j^s - \delta_j^r \delta_i^s)$$

$$\mathrm{Tr}(t^a t^b) = 2\delta_{ab} \tag{15.190}$$

$$\sum_a t_{ij}^a t_{kl}^a = -\delta_{ik}\delta_{jl} + \delta_{il}\delta_{jk}$$

They satisfy $[t^a, t^b] = \sum_c i f_{abc} t^c$ with

$$f_{abc} \equiv f_{(rs)(pq)(mn)} = \delta_{mr}(\delta_{nq}\delta_{sp} - \delta_{np}\delta_{sq}) + \delta_{ms}(\delta_{rq}\delta_{np} - \delta_{nq}\delta_{rp}) \tag{15.191}$$

$$\sum_{a,b} f_{abc} f_{abd} = 2(N-2)\delta_{cd} \tag{15.192}$$

---

[12] There are simple counterexamples, however: free-boson theories defined on circles of different radii have the same energy-momentum tensor but different spectra.

This is indeed consistent with our normalization $|\theta|^2 = 2$ (see, for instance, Eq. (15.53)), since for $so(N)$ g is equal to $N - 2$.

Out of these fermions, a bosonic spin-1 field is easily constructed as

$$J^a(z) = \beta \sum_{i,j} (\psi_i t^a_{ij} \psi_j)(z) \tag{15.193}$$

for some constant β. We now check that this provides a realization of the OPE (15.43), thereby fixing β and the level. Since we have free fields, we can use Wick's theorem to calculate contractions:

$$\overbrace{J^a(z)J^b}(w) = \beta^2 \sum_{i,j,k,l} t^a_{ij} t^b_{kl} \left[\frac{1}{(z-w)} \{ -\delta_{ik}(\psi_j(z)\psi_l(w)) + \delta_{il}(\psi_j(z)\psi_k(w)) \right.$$

$$+ \delta_{jk}(\psi_i(z)\psi_l(w)) - \delta_{jl}(\psi_i(z)\psi_k(w)) \}$$

$$\left. + \frac{1}{(z-w)^2} \{ -\delta_{ik}\delta_{jl} + \delta_{il}\delta_{jk} \} \right] \tag{15.194}$$

The two brackets result from the single and double contractions, respectively. (We recall that in order to contract ψ_k with ψ_i, we must pass ψ_k over ψ_j, and this induces a minus sign.) By Taylor-expanding the fields evaluated at z around the point w, we obtain

$$\overbrace{J^a(z)J^b}(w) = 2\beta \sum_c \frac{if_{abc}J^c(w)}{(z-w)} + 2\beta^2 \frac{\mathrm{Tr}(t^a t^b)}{(z-w)^2} \tag{15.195}$$

From the first term, we see that β must be equal to $\frac{1}{2}$, and since by Eq. (15.190) the trace is 2 when $a = b$, we conclude from the second term that $k = 1$. The corresponding Sugawara central charge is given by (cf. Eq. (15.61))

$$c(\widehat{so}(N)_1) = \frac{\frac{1}{2}N(N-1)}{(N-2)+1} = \frac{N}{2} \tag{15.196}$$

This certainly looks natural at first sight since a real free fermion is known to contribute $\frac{1}{2}$ to the Virasoro central charge. But this result has been derived previously from an energy-momentum tensor of the form

$$T(z) = -\frac{1}{2} \sum_i (\psi_i \partial \psi_i)(z) \tag{15.197}$$

while here the energy-momentum tensor is given by the Sugawara construction, which is quartic in the fermion fields:

$$T(z) = \frac{1}{8(N-1)} \sum_a \sum_{i,j,k,l} ((\psi_i t^a_{ij} \psi_j)(\psi_k t^a_{kl} \psi_l))(z)$$

$$= \frac{1}{8(N-1)} \sum_{i,j,k,l} [-\delta_{ik}\delta_{jl} + \delta_{il}\delta_{jk}] ((\psi_i \psi_j)(\psi_k \psi_l))(z) \tag{15.198}$$

In fact, these expressions are only superficially different. The exact equivalence of Eqs. (15.197) and (15.198) can be established by means of the rearrangement lemma of App. 6.C, which we reproduce for convenience:

$$((AB)(CD)) = (A(B(CD))) + (A([(CD), B]))$$
$$+ (([[(CD), A])B) + ([(AB), (CD)])$$
(15.199)

Note however that this result was derived for bosonic fields. Since it is to be applied to fermionic fields, the signs of the different terms must be reconsidered. But in the present context (CD) is bilinear in the fermionic fields, thus bosonic, and because the order of the factors A and B is nowhere changed, the formula is not modified. We then rewrite Eq. (15.199) as

$$((\psi_i \psi_j)(\psi_k \psi_l)) = (\psi_i(\psi_j(\psi_k \psi_l))) + ([(\psi_i \psi_j), (\psi_k \psi_l)])$$
$$+ (([(\psi_k \psi_l), \psi_i)\psi_j) + (\psi_i([(\psi_k \psi_l), \psi_j]))$$
(15.200)

From Eqs. (6.222) and (15.194), we find

$$([(\psi_i \psi_j), (\psi_k \psi_l)]) = \partial \left[\delta_{jk}(\psi_i \psi_l) - \delta_{jl}(\psi_i \psi_k) + \delta_{il}(\psi_j \psi_k) - \delta_{ik}(\psi_j \psi_l) \right]$$
(15.201)

Similarly, with

$$(\psi_k \psi_l)(z)\psi_i(w) \sim \frac{1}{z - w}(\delta_{li}\psi_k(w) - \delta_{ki}\psi_l(w))$$
(15.202)

we have

$$([(\psi_k \psi_l), \psi_i]) = \partial[\delta_{li}\psi_k - \delta_{ik}\psi_l]$$
(15.203)

Collecting all these results, a simple calculation shows that

$$\sum_{i,j,k,l}[-\delta_{ik}\delta_{jl} + \delta_{il}\delta_{jk}]((\psi_i \psi_j)(\psi_k \psi_l)) = \sum_i -4(N - 1)(\psi_i \partial \psi_i)$$
(15.204)

whose substitution into Eq. (15.198) reproduces exactly Eq. (15.197).

15.5.3. Description of the $\widehat{so}(N)_1$ Primary Fields

As a first application of a current algebra free-field representation, we present a description of the WZW primary fields for the $\widehat{so}(N)_1$ model in terms of the fields in the free-fermion theory. It turns out that not all the primary fields can be expressed directly in terms of the free fermions. But a free-fermion theory is in fact equivalent to the Ising model. It is then in terms of the primary fields of N decoupled Ising models that all the WZW primaries for the $\widehat{so}(N)_1$ model can be written. The dimension spectrum of the $\widehat{so}(N)_1$ model is easily calculated, using the quadratic form matrix given in App. 13.A, to be

$$N = 2r + 1: \quad h_{\hat{\omega}_1} = \frac{1}{2} \quad h_{\hat{\omega}_r} = \frac{2r + 1}{16}$$
$$N = 2r: \quad h_{\hat{\omega}_1} = \frac{1}{2} \quad h_{\hat{\omega}_r} = h_{\hat{\omega}_{r-1}} = \frac{r}{8}$$
(15.205)

together with the vacuum state, related to the weight $\hat{\omega}_0$. (We recall that the only integrable representations of $\widehat{so}(N)_1$ are $\hat{\omega}_0, \hat{\omega}_1$, and $\hat{\omega}_r$ for B_r, and $\hat{\omega}_0, \hat{\omega}_1, \hat{\omega}_{r-1}$, and $\hat{\omega}_r$, for D_r. The corresponding nodes in the Dynkin diagram have comark equal to one.) The primary field transforming in the vector representation must be $g(z, \bar{z})$ itself. Since it has holomorphic and antiholomorphic dimension $\frac{1}{2}$, it should be[13]

$$g_{kl}(z, \bar{z}) = \psi_k(z)\bar{\psi}_l(\bar{z}) \tag{15.206}$$

This must reproduce the OPE

$$g_{kl}(z, \bar{z})J^a(w) \sim \sum_n \frac{t^a_{kn} g_{nl}(w, \bar{z})}{z - w} \tag{15.207}$$

In terms of free fermions, this reads

$$\psi_k(z)\frac{1}{2}\sum_{i,j}(\psi_i t^a_{ij} \psi_j)(w) \sim \sum_l \frac{t^a_{kl}\psi_l(w)}{z - w} \tag{15.208}$$

and is indeed easily checked to be true. The remaining fields (transforming in the spinor representations of $so(N)$) all have conformal dimension $\frac{1}{16}N$. Hence, they must be made of N copies of the Ising spin field $\sigma(z, \bar{z})$ or its dual $\mu(z, \bar{z})$, which both have conformal dimension $\frac{1}{16}$. The number of possible fields that can be constructed in this way from N copies of the Ising model is 2^N. This number matches exactly the dimension of the associated representations:

$$B_r: \quad \dim|\hat{\omega}_r| = 2^{2r+1} \quad D_r: \quad \dim|\hat{\omega}_{r-1}| = \dim|\hat{\omega}_r| = 2^{2r-1} \tag{15.209}$$

15.5.4. $\widehat{so}(N)_1$ Characters

A free-field representation provides a simple way of finding an expression for the characters of the theory. Since $\widehat{so}(N)_1$ can be represented in terms of N free fermions with the same boundary conditions (ensuring the periodicity of the different components $g_{kl}(z, \bar{z})$ as given by Eq. (15.206)), its partition function must be of the form

$$Z \propto \sum_{\nu=2,3,4} Z_\nu^N \tag{15.210}$$

where

$$Z_\nu = \frac{1}{2}\left|\frac{\theta_\nu}{\eta}\right| \tag{15.211}$$

---

[13] Such a fermionic representation of a WZW model makes natural the effective description presented at the end of Sect. 15.1.1. Here, the comparison of the $\widehat{so}(N)_1$-WZW action with a free-fermion theory leads to a relation between the field $g(z, \bar{z})$ and the fermions that makes $g(z, \bar{z})$ transform in the vectorial representation (the minimal representation of $so(N)$) in both sectors.

(a notation used frequently in Chap. 12). The characters are obtained by rewriting this equality in the form

$$Z = \sum_{\hat{\lambda} \in P_+^!} |\chi_{\hat{\lambda}}|^2 \tag{15.212}$$

by requiring the $\chi_{\hat{\lambda}}$'s to transform covariantly among themselves. For $\widehat{so}(2r)_1$, the primary-field characters are thus found to be

$$\chi_{\hat{\omega}_0} = \frac{1}{2} \left(\frac{\theta_3^r + \theta_4^r}{\eta^r} \right)$$

$$\chi_{\hat{\omega}_1} = \frac{1}{2} \left(\frac{\theta_3^r - \theta_4^r}{\eta^r} \right) \tag{15.213}$$

$$\chi_{\hat{\omega}_{r-1}} = \chi_{\hat{\omega}_r} = \frac{1}{2} \frac{\theta_2^r}{\eta^r}$$

All functions are understood to be evaluated at $q = e^{2\pi i \tau}$ and $\theta_\nu(q) \equiv \theta_\nu(0|\tau)$, $\nu = 2, 3, 4$ stand for the Jacobi theta functions defined in App. 10.A. The relative normalizations of the characters are fixed by modular invariance, and the absolute normalization of $\chi_{\hat{\omega}_0}$ is fixed from the first two terms of its expansion:

$$\chi_{\hat{\omega}_0} = 1 + [\dim so(N)]\, q + \cdots \tag{15.214}$$

For $\widehat{so}(2r+1)_1$ the characters are

$$\chi_{\hat{\omega}_0} = \frac{1}{2} \left(\frac{\theta_3^{r+\frac{1}{2}} + \theta_4^{r+\frac{1}{2}}}{\eta^{r+\frac{1}{2}}} \right)$$

$$\chi_{\hat{\omega}_1} = \frac{1}{2} \left(\frac{\theta_3^{r+\frac{1}{2}} - \theta_4^{r+\frac{1}{2}}}{\eta^{r+\frac{1}{2}}} \right) \tag{15.215}$$

$$\chi_{\hat{\omega}_r} = \frac{1}{\sqrt{2}} \left(\frac{\theta_2}{\eta} \right)^{r+\frac{1}{2}}$$

For $N = 1$, we recover the Ising characters, as we should.

15.5.5. $\widehat{so}(N)$ Representations at Higher Levels

With N free fermions transforming in the vector representation of $so(N)$, we have constructed a representation of the $\widehat{so}(N)_1$ affine algebra. Is it possible to obtain higher levels, by considering other representations than the vectorial one? Clearly, the derivation of Eq. (15.195) is representation independent, so that the whole dependence upon a specific representation is encapsulated in the value of the level read off the equality: $\mathrm{Tr}\, t^a t^b = 2k\delta_{ab}$. We know that this trace, evaluated in the highest-weight representation λ, is related to the Dynkin index x_λ defined by

$$\mathrm{Tr}\, t_\lambda^a t_\lambda^b = 2x_\lambda \delta_{ab} \tag{15.216}$$

and for which we already found the explicit expression (13.133). (For $so(N)$, x_λ is always an integer.) Therefore Eq. (15.193) with $\beta = \frac{1}{2}$ and $t^a = t^a_\lambda$ yields a representation of $\widehat{so}(N)_{x_\lambda}$.

A particularly interesting case is $\widehat{so}(3)$. It is equivalent to $\widehat{su}(2)$ with its representations restricted to those whose highest weights have even finite Dynkin label. For the adjoint representation, the index is found to be 2. Hence, $\widehat{su}(2)_2$ has a representation in terms of three real fermions.

15.5.6. Complex Free-Fermion Representations: $\widehat{u}(N)_k$

Consider now N complex free fermions ψ_i, $\psi_i^\dagger$, with OPEs

$$\psi_i(z)\psi_j^\dagger(w) \sim \frac{\delta_{ij}}{z-w} \qquad \psi_i^\dagger(z)\psi_j(w) \sim \frac{\delta_{ij}}{z-w}$$
$$\psi_i(z)\psi_j(w) \sim 0 \qquad \psi_i^\dagger(z)\psi_j^\dagger(w) \sim 0 \tag{15.217}$$

They provide a natural realization of $\widehat{u}(1)$ by

$$J^0(z) = \beta \sum_i (\psi_i^\dagger \psi_i)(z) \tag{15.218}$$

where β is some constant. The level is found to be $\beta^2 N$. It is clear that β is not fixed, which makes the value of the level arbitrary. In spite of this, the contribution of such a $\widehat{u}(1)$ factor to the Virasoro central charge is unambiguous and equal to 1. This can be seen from Eq. (15.61), with $g = 0$ and dim $g = 1$. The corresponding Sugawara energy-momentum tensor is

$$T_{\widehat{u}(1)}(z) = \frac{1}{2\beta^2 N}(J^0 J^0)(z) = \frac{1}{2N} \sum_{i,j} ((\psi_i^\dagger \psi_i)(\psi_j^\dagger \psi_j))(z) \tag{15.219}$$

and the arbitrary factor β has disappeared.

If these fields transform in the defining representation of $su(N)$ (i.e., with highest weight ω_1), with transformation matrices t^a_{ij} such that

$$\text{Tr } t^a t^b = \delta_{ab}$$

$$\sum_a t^a_{ij} t^a_{kl} = \delta_{il}\delta_{jk} - \frac{1}{N}\delta_{ij}\delta_{kl} \tag{15.220}$$

$$\sum_{a,b} f_{abc} f_{abd} = 2N\delta_{cd}$$

we can construct a representation of $\widehat{su}(N)$ with

$$J^a(z) = \sum_{i,j} (\psi_i^\dagger t^a_{ij} \psi_j)(z) \tag{15.221}$$

This is readily checked, and the level is found to be 1. It is also simple to verify that

$$J^0(z)J^a(w) \sim 0 \tag{15.222}$$

so that from N complex free fermions, we actually get a representation of $\widehat{u}(N)_1$.

The relation with a standard theory of complex free fermions can be obtained as in the $so(N)$ case. The starting point is again the rearrangement lemma (6.226), which yields

$$((\psi_i^\dagger \psi_j)(\psi_k^\dagger \psi_l)) = (\psi_i^\dagger(\psi_j(\psi_k^\dagger \psi_l))) + \partial[\delta_{jk}(\psi_i^\dagger \psi_l) - \delta_{il}(\psi_k^\dagger \psi_j)]$$
$$+ (\partial\psi_k^\dagger \psi_j)\delta_{il} - (\psi_i^\dagger \partial\psi_l)\delta_{kj} \tag{15.223}$$

Notice that for $N > 1$, the quartic term is nonzero. Using this relation, the $\widehat{u}(1)$ Sugawara energy-momentum tensor (15.219) and that for $\widehat{su}(N)_1$, namely

$$T_{\widehat{su}(N)_1}(z) = \frac{1}{2(N+1)} \sum_a \sum_{i,j,k,l} ((\psi_i^\dagger t_{ij}^a \psi_j)(\psi_k^\dagger t_{kl}^a \psi_l))(z) \tag{15.224}$$

can be reexpressed respectively as

$$T_{\widehat{u}(1)} = \frac{1}{2N} \sum_{i,j}(\psi_i^\dagger(\psi_i(\psi_j^\dagger \psi_j))) + \frac{1}{2N} \sum_i [(\partial\psi_i^\dagger \psi_i) - (\psi_i^\dagger \partial\psi_i)] \tag{15.225}$$

$$T_{\widehat{su}(N)_1} = -\frac{1}{2N} \sum_{i,j}(\psi_i^\dagger(\psi_i(\psi_j^\dagger \psi_j))) + \frac{(N-1)}{2N} \sum_i [(\partial\psi_i^\dagger \psi_i) - (\psi_i^\dagger \partial\psi_i)] \tag{15.226}$$

Adding these, we find that the quartic piece cancels out, with the result

$$T_{\widehat{u}(1)} + T_{\widehat{su}(N)_1} = \frac{1}{2} \sum_i [(\partial\psi_i^\dagger \psi_i) - (\psi_i^\dagger \partial\psi_i)] \tag{15.227}$$

This is exactly the energy-momentum tensor for N free complex fermions. Since a free complex fermion (which is equivalent to two real fermions) contributes to $c = 1$, the central charge for Eq. (15.227) is equal to N. This agrees with the sum of the Sugawara central charges

$$c(\widehat{u}(1)) + c(\widehat{su}(N)_1) = 1 + \frac{N^2 - 1}{N + 1} = N \tag{15.228}$$

Here again, higher levels can be obtained from $su(N)$ representations larger than the defining representation.

§15.6. Vertex Representations

15.6.1. The $\widehat{su}(2)_1$ Case

Consider a theory with a single free boson $\varphi(z)$,[14] where as usual

$$\varphi(z)\varphi(w) \sim -\ln(z - w) \tag{15.229}$$

---

[14] In this chapter and the following ones, when considering the free-boson theory, we will always refer to its chiral (holomorphic) part.

Clearly $i\partial\varphi(z)$ defines a $\hat{u}(1)$ current. But as we know, this is not the only spin-1 operator that can be constructed from this free boson: the vertex operators $e^{\pm i\sqrt{2}\varphi}$ also have conformal dimension 1 (cf. Sect. 6.3.2). It turns out that these three spin-1 operators

$$H = i\partial\varphi \qquad E^{\pm} = e^{\pm i\sqrt{2}\varphi} \tag{15.230}$$

satisfy the $\hat{su}(2)$ algebra at level 1. Using, for instance (cf. App. 6.A),

$$E^+(z)E^-(w) = \exp\{2\,\overline{\varphi(z)\varphi(w)}\}\exp\{i\sqrt{2}(\varphi(z)-\varphi(w))\}$$

$$= \frac{1}{(z-w)^2}\exp\left\{i\sqrt{2}\sum_{n\geq 1}\frac{(z-w)^n}{n!}\partial^n\varphi(w)\right\} \tag{15.231}$$

the full set of OPE is readily found to be

$$E^+(z)E^-(w) \sim \frac{1}{(z-w)^2} + \frac{\sqrt{2}H}{z-w}$$

$$H(z)E^{\pm}(w) \sim \frac{\pm\sqrt{2}E^{\pm}(w)}{z-w} \tag{15.232}$$

$$H(z)H(w) \sim \frac{1}{(z-w)^2}$$

These are the expected OPE in the Cartan-Weyl basis (with $E^{\pm} \equiv E^{\pm\alpha_1}$), since α_1 is $\sqrt{2}$. (Unpleasant factors of $\sqrt{2}$ can be avoided by moving to either the Chevalley or the spin basis.)

The Sugawara central charge is equal to 1 (cf. Eq. (15.61) with $k = 1$ and $g = 2$), as expected for a theory with a single boson. Indeed, a direct calculation shows that the Sugawara energy-momentum tensor reduces to that of a free boson:

$$T(z) = \frac{1}{6}\left[(HH) + (E^+E^-) + (E^-E^+)\right]$$

$$= \frac{1}{6}\left[-(\partial\varphi\partial\varphi) + \frac{i}{\sqrt{2}}\partial^2\varphi - (\partial\varphi\partial\varphi) - \frac{i}{\sqrt{2}}\partial^2\varphi - (\partial\varphi\partial\varphi)\right] \tag{15.233}$$

$$= -\frac{1}{2}(\partial\varphi\partial\varphi)$$

where (E^+E^-) can be read off the first regular term in the OPE (15.231), and similarly for (E^-E^+).

For this WZW model, there is only one field beside the identity, and it is bound to be $g(z,\bar{z})$ itself, with conformal dimension $h = \bar{h} = \frac{1}{4}$. It has four components, two for each sector. Those in the holomorphic sector are $e^{\pm i\varphi(z)/\sqrt{2}}$, corresponding, respectively, to the two states $j = \frac{1}{2}, m = \pm\frac{1}{2}$ (where $j = \lambda_1/2$):

$$E^+(z)e^{i\varphi(w)/\sqrt{2}} = (z-w)e^{i(\varphi(z)+\frac{1}{2}\varphi(w))/\sqrt{2}} \sim 0$$

$$E^-(z)e^{i\varphi(w)/\sqrt{2}} \sim \frac{e^{-i\varphi(w)/\sqrt{2}}}{z-w} \tag{15.234}$$

In the first OPE, the absence of a pole and of a nonzero regular term as $z \to w$ implies that

$$(E_0^+ e^{i\varphi/\sqrt{2}}) = 0 \qquad (E_{-1}^+ e^{i\varphi/\sqrt{2}}) = 0 \qquad (15.235)$$

They correspond respectively to the last condition in Eq. (15.101) (the other equalities of condition (15.101) are trivially checked) and the singular vector (15.110) (since here $k = (\lambda, \theta) = 1$).

15.6.2. Fock Construction of the $\widehat{su}(2)_1$ Integrable Modules

We now show that the above construction provides a representation of the integrable $\widehat{su}(2)_1$ modules. We start with a description of the identity module $L_{[1,0]}$ (i.e., $\lambda_1 = 0$).

We consider then the Fock space generated by the states

$$|p; \{n_i\}\rangle = N_{\{n_i\}} a_{-1}^{n_1} a_{-2}^{n_2} \cdots |p; \{0\}\rangle \qquad (15.236)$$

where $N_{\{n_i\}}$ is some constant, and $|p; \{0\}\rangle$ is obtained from the vacuum state $|0; \{0\}\rangle$ by

$$|p; \{0\}\rangle = (E^+(0))^m \, |0; \{0\}\rangle \qquad \text{if} \quad p = \sqrt{2}m > 0$$
$$|p; \{0\}\rangle = (E^-(0))^m \, |0; \{0\}\rangle \qquad \text{if} \quad p = -\sqrt{2}m < 0 \qquad (15.237)$$

for some positive integer m. The L_0 eigenvalue of these states is (cf. Eq. (6.72)):

$$L_0|p; \{n_i\}\rangle = \left[\frac{1}{2}p^2 + \sum_{k=1}^{\infty} k n_k \right] |p; \{n_i\}\rangle$$
$$= \left[m^2 + \sum_{k=1}^{\infty} k n_k \right] |\sqrt{2}m; \{n_i\}\rangle \qquad (15.238)$$

now with $m \in \mathbb{Z}$. For the first few values of m, the number of distinct states at each grade (i.e., at each eigenvalue of L_0) is given in Table 15.1. For instance, the five states with $m = 0$ and $L_0 = 4$ are

$$|0; 4, 0, \cdots\rangle \qquad |0; 2, 1, 0, \cdots\rangle \qquad |0; 1, 0, 1, 0, \cdots\rangle$$
$$|0; 0, 2, 0, \cdots\rangle \qquad |0; 0, 0, 0, 1, 0, \cdots\rangle \qquad (15.239)$$

The states $|p; \{n_i\}\rangle$ are seen to build up the module $L_{[1,0]}$. They are associated with weights whose finite part is $2m\omega_1$. (Note that since the $su(2)$ root α_1 can be taken to be $\sqrt{2}$, we can reinterpret $p = \sqrt{2}m$ as $p = \alpha_1 m = 2m\omega_1$.) At each grade, the finite weights are easily reorganized into irreducible representations.

This field theoretical construction of the module $L_{[1,0]}$ shows clearly that the string function for the weight $[1, 0]$ is given by the inverse of the Euler function: the states with $m = 0$ and $L_0 = N$ are in one-to-one correspondence with the various partitions of the integer N. This demonstrates the identity (14.144).

Table 15.1. States in the lowest grades of the $\widehat{su}(2)_1$ module $L_{[1,0]}$.

L_0			m			$su(2)$
	-2	-1	0	1	2	decomposition
0			1			(0)
1		1	1	1		(2)
2		1	2	1		(2)+(0)
3		2	3	2		2(2)+(0)
4	1	3	5	3	1	(4)+2(2)+2(0)
5	1	5	7	5	1	(4)+4(2)+2(0)
6	2	7	11	7	2	2(4)+5(2)+4(0)

For the construction of the module $L_{[0,1]}$, with highest weight state $e^{i\varphi(0)/\sqrt{2}}|0\rangle$ (i.e., $\lambda_1 = 1$), the procedure is similar. We consider again states of the form (15.236) with now

$$p \to p' = p + \frac{1}{\sqrt{2}} \tag{15.240}$$

and p still living in the root lattice (i.e., the eigenvalue p has been shifted by ω_1). The action of L_0 is then

$$L_0|p'; \{n_i\}\rangle = [\frac{1}{4} + m^2 - m + \sum_{k=1}^{\infty} kn_k]\,|\sqrt{2}m + \frac{1}{\sqrt{2}}; \{n_i\}\rangle \tag{15.241}$$

There are two states of lowest energy, with L_0 eigenvalue $\frac{1}{4}$; they both have all $n_i = 0$ and they are distinguished by their value of m, which can be 0 or 1. The number of states and their $su(2)$ decomposition at the first few values of L_0 are given in Table 15.2.

For these two representations, the vertex construction leads to a simple expression for the characters. Using

$$\chi_{\hat{\lambda}}(z; \tau) = \text{Tr}_{\hat{\lambda}}\, e^{2\pi i \tau(L_0 - c/24)} e^{-2\pi i z J^0} \tag{15.242}$$

with

$$q = e^{2\pi i \tau} \qquad x = e^{-2\pi i z} \tag{15.243}$$

the characters read

$$\chi_{[1-\lambda_1, \lambda_1]}(z; \tau) = \eta(q)^{-1} \sum_{m \in \mathbb{Z}+\lambda_1/2} q^{m^2} x^m \tag{15.244}$$

Their specialized forms, obtained by setting $z = 0$, are exactly the extended characters of a free-boson theory compactified on a circle of radius $R = \sqrt{2}$ (compare with Eq. (10.233)). This is actually the only value of R that leaves both the generators and the primary fields invariant under $\varphi \to \varphi + 2\pi R$.

Table 15.2. States in the lowest grades of the $\widehat{su}(2)_1$ module $L_{[0,1]}$.

L_0	m						$su(2)$ decomposition
	-2	-1	0	1	2	3	
$\frac{1}{4}$			1	1			(1)
$\frac{5}{4}$			1	1			(1)
$\frac{9}{4}$		1	2	2	1		(3)+(1)
$\frac{13}{4}$		1	3	3	1		(3)+2(1)
$\frac{17}{4}$		2	5	5	2		2(3)+3(1)
$\frac{21}{4}$		3	7	7	3		3(3)+4(1)
$\frac{25}{4}$	1	5	11	11	5	1	(5)+4(3)+6(1)

15.6.3. Generalization: Vertex Representations of Simply-Laced Algebras at Level 1

The above construction has a natural generalization for all simply-laced algebras, where to every simple root we associate an independent free boson φ^i, with

$$\varphi^i(z)\varphi^j(w) \sim -\delta_{ij}\ln(z-w) \qquad (15.245)$$

The Cartan currents are

$$H^j(z) = i\partial\varphi^j(z) \qquad (15.246)$$

For the other generators, the natural guess is

$$\tilde{E}^\alpha(z) = e^{i\alpha\cdot\varphi(z)} \qquad (15.247)$$

with[15]

$$\alpha \cdot \varphi = \sum_{i=1}^{r} \alpha^i \varphi^i \qquad (15.248)$$

(This almost works, but since this is not the final story, we have put a tilde over E^α.) For a simply-laced algebra, $|\alpha|^2 = 2$ for all roots α; the dimension of all the $\tilde{E}^\alpha$'s is thus 1. This vertex has the right OPE with H^i:

$$H^i(z)\tilde{E}^\alpha(w) \sim \frac{\alpha^i\tilde{E}^\alpha(w)}{z-w} \qquad (15.249)$$

---

[15] We recall that upper indices refer to eigenvalues in the Cartan-Weyl basis.

For the product $\tilde{E}^\alpha$ and $\tilde{E}^\beta$, proceeding as for $E^+(z)E^-(w)$ in Eq. (15.232), we find

$$\tilde{E}^\alpha(z)\tilde{E}^\beta(w) \sim (z-w)^{(\alpha,\beta)}\tilde{E}^{\alpha+\beta}(w) + (z-w)^{(\alpha,\beta)-1}\alpha \cdot i\partial\varphi\,\tilde{E}^{\alpha+\beta}(w) + \cdots$$
(15.250)

Singular terms can appear only if $(\alpha,\beta) = -1$, in which case $\alpha + \beta$ is a root (i.e., $\alpha + \beta \neq 0$), or if $(\alpha,\beta) = -2$, which means that $\beta = -\alpha$. When $(\alpha,\beta) = -2$, Eq. (15.250) yields the expected result

$$\tilde{E}^\alpha(z)\tilde{E}^{-\alpha}(w) \sim \frac{1}{(z-w)^2} + \frac{\alpha \cdot H}{z-w}$$
(15.251)

and when $(\alpha,\beta) = -1$, it gives

$$\tilde{E}^\alpha(z)\tilde{E}^\beta(w) \sim \frac{\tilde{E}^{\alpha+\beta}}{z-w}$$
(15.252)

This may seem right at first sight. However, we note that the interchange of $\tilde{E}^\alpha(z)$ and $\tilde{E}^\beta(w)$, with $z \leftrightarrow w$, should leave the result invariant, but that is not the case here since the simple pole picks up a minus sign. The cure is to introduce a correction factor c_α to compensate for the extra sign:

$$E^\alpha(z) = c_\alpha \tilde{E}^\alpha(z)$$
(15.253)

c_α is an operator acting on the Fock space and it depends only upon the momentum part of the free-boson zero modes. Explicitly, if the mode expansion of φ^j is written in the form (cf. Eq. (6.54))

$$\varphi^j(z) = \hat{q}^j - i\hat{p}^j\,\ln z + i\sum_{n\neq 0}\frac{a_n^j}{n}z^{-n}$$
(15.254)

where in particular

$$[\hat{p}^j, \hat{q}^k] = -i\delta_{jk}$$
(15.255)

then $c_\alpha = c_\alpha(\hat{p})$. As a result, when $c_\alpha(\hat{p})$ passes over a vertex, its argument is shifted:

$$e^{i\alpha\cdot\varphi}c_\beta(\hat{p}) = c_\beta(\hat{p} - \alpha)e^{i\alpha\cdot\varphi}$$
(15.256)

To have all signs right in the OPE, we thus need

$$c_\alpha(\hat{p})c_\beta(\hat{p} - \alpha) = (-1)^{(\alpha,\beta)}c_\beta(\hat{p})c_\alpha(\hat{p} - \beta)$$
(15.257)

Furthermore, in order to obtain a closed algebra, we also require that

$$c_\alpha(\hat{p})c_\beta(\hat{p} - \alpha) = \epsilon(\alpha,\beta)c_{\alpha+\beta}(\hat{p})$$
(15.258)

with $\epsilon(\alpha,\beta) = \pm 1$.[16] We now present a simple solution to these conditions. For $\alpha,\beta \in Q$,

$$\alpha = \sum n_i\alpha_i, \qquad \beta = \sum m_i\alpha_i, \qquad n_i, m_i \in \mathbb{Z}$$
(15.259)

---

[16] The following result is used here: for simply-laced algebras, the constants $N_{\alpha,\beta}$ that appear in Eq. (13.11) are equal to ± 1.

we introduce the product

$$\alpha * \beta = \sum_{i>j} n_i m_j (\alpha_i, \alpha_j) \tag{15.260}$$

Now we denote by $p = (p^1, \cdots, p^r)$ the eigenvalues of $(\hat{p}^1, \cdots, \hat{p}^r)$. First we consider the basic representation (i.e., with highest weight $\hat{\omega}_0$). In that case, consistency forces p to lie in the root lattice Q. This makes the following definition sensible

$$c_\alpha(\hat{p}) = (-1)^{\hat{p} * \alpha} \tag{15.261}$$

Since $\alpha * \beta \in \mathbb{Z}$ and $|\alpha_i|^2 = 2$ for all i, it is easy to check that Eq. (15.257) is satisfied, and that further

$$\epsilon(\alpha, \beta) = (-1)^{\alpha * \beta} \tag{15.262}$$

As indicated in Ex. 15.19, $\epsilon(\alpha, \beta)$ is a $\mathbb{Z}_2$-valued two-cocycle.

For the other level-1 representations, the highest weight is a fundamental weight $\tilde{\omega}_i$ with unit comark. For these representations, the $\hat{p}$ eigenvalue must lie in $Q + \tilde{\omega}_i$. This requires a slight modification of the correction factor, namely

$$c_\alpha(\hat{p}) = (-1)^{(\hat{p} - \tilde{\omega}_i) * \alpha} \tag{15.263}$$

The quantum equivalence between the theory of r free bosons and the (simply-laced) $\hat{g}_1$ WZW model can again be established directly at the level of the energy-momentum tensors. For the WZW model, it reads

$$T(z) = \frac{1}{2(1+g)} \left\{ \sum_i (H^i H^i)(z) + \sum_{\alpha>0} \left[(E^\alpha E^{-\alpha})(z) + (E^{-\alpha} E^\alpha)(z) \right] \right\} \tag{15.264}$$

The evaluation the first regular term in the OPE $E^\alpha(z) E^{-\alpha}(w)$ shows that the second member in the r.h.s. of the above equation is equal to

$$\sum_{\alpha \in \Delta_+} (\alpha \cdot H)^2 = \frac{1}{2} H^i (\sum_{\alpha \in \Delta} \alpha^i \alpha^j) H^j = g H^i H^j \delta_{ij} = g \, H \cdot H \tag{15.265}$$

For the last equality we have used Eqs. (13.180) and (13.184). This readily implies that

$$T(z) = \frac{1}{2} (H \cdot H)(z) = -\frac{1}{2} (\partial \varphi^i \partial \varphi^i)(z) \tag{15.266}$$

as expected. The equality of the corresponding central charges is a consequence of the identity

$$\dim g = r(g + 1) \tag{15.267}$$

valid for simply-laced algebras.

For nonsimply-laced algebras, the situation is more complicated due to the presence of short roots. For these, the operators $e^{\pm i\alpha \cdot \varphi}$ have conformal dimension $\frac{1}{2}$. To produce an operator of dimension 1, we must multiply it by a free fermion.

For instance, a vertex representation of the B_r algebra requires r bosons and one fermion.

For $k > 1$, vertex representations still exist. The level-k Cartan generators can be represented by $H^i = i\sqrt{k}\partial\varphi_i$. We then expect that $E^{\pm\alpha}$ will contain the piece $e^{\pm i\alpha\cdot\varphi/\sqrt{k}}$, which indeed reproduces the OPE $H^i(z)E^{\pm\alpha}(w)$. However, this vertex operator has dimension $|\alpha|^2/2k$. In order to build a spin-1 current, it has to be combined with a field of spin $1 - |\alpha|^2/2k$. This representation requires thus the introduction of fields that are neither fermions nor bosons (the so-called parafermions). We will not pursue this matter, and consider instead a third type of free-field representation, whose field content is level independent.

§15.7. The Wakimoto Free-Field Representation

We now turn to the generic free-field representation of affine Lie algebras. This is the starting point for a systematic analysis of WZW models, which parallels the Coulomb-gas description of the Virasoro minimal models.

15.7.1. From the $su(2)$ Monomial Representation to the Affine Case

The free-field representation to be introduced presently can be regarded as an affine extension of the monomial representation of finite Lie algebras. For $su(2)$, the spin-j irreducible representation is spanned by the $2j + 1$ monomials $\{1, x, x^2, \cdots, x^{2j}\}$ in some complex variable x. In this basis, the Chevalley generators are represented as

$$e_0 = \frac{\partial}{\partial x}$$
$$h_0 = 2j - 2x\frac{\partial}{\partial x} \qquad (15.268)$$
$$f_0 = 2jx - x^2\frac{\partial}{\partial x}$$

and their action on the monomial x^{j-m}, representing the state $|j, m\rangle$, is

$$e_0 x^{j-m} = (j - m)x^{j-m-1}$$
$$h_0 x^{j-m} = 2m x^{j-m} \qquad (15.269)$$
$$f_0 x^{j-m} = (j + m)x^{j-m+1}$$

In view of constructing a Fock space, the operators x and $\partial/\partial x$ can be interpreted as creation and annihilation operators denoted respectively by $-\gamma_0$ and β_0, with $[\gamma_0, \beta_0] = 1$. In addition, we introduce a pair of canonical variables $\hat{p}, \hat{q}$ with commutation relation $[\hat{p}, \hat{q}] = -i$. A Fock space can be constructed by the repeated application of the creation operator γ_0 on the vacuum state $|p; 0\rangle$. This vacuum state is characterized by its momentum eigenvalue p, which takes its values in the weight lattice (i.e., $p = 2j\omega_1 \leftrightarrow \sqrt{2}j$), and its property of being annihilated by β_0. Henceforth, with $(x, \partial/\partial x, 2j)$ replaced by $(-\gamma_0, \beta_0, \sqrt{2}p)$, the above relation for the generators reads

$$e_0 = \beta_0$$
$$h_0 = \sqrt{2}\,\hat{p} + 2\gamma_0\beta_0 \qquad\qquad (15.270)$$
$$f_0 = -\sqrt{2}\,\hat{p}\,\gamma_0 - \gamma_0^2\beta_0$$

When supplemented with the singular vector

$$f_0^{2j+1}|p;0\rangle = 0 \qquad\qquad (15.271)$$

the Fock space is equivalent to the representation space of the spin-j module of $su(2)$.

We consider the affine extension of this representation. In the same way as the finite Lie algebra generators correspond to the zero modes of the affine generators, we interpret the set of operators $\{\gamma_0, \beta_0, \hat{p}, \hat{q}\}$ as the zero modes of appropriate free fields and replace these zero modes by the corresponding fields. In this way, the operators γ_0 and β_0 are transformed into bosonic ghosts $\beta(z)$ and $\gamma(z)$ (of respective spins 1 and 0) (cf. Sect. 5.3.3)

$$\beta(z) = \sum_n \beta_n z^{-n-1} \qquad \gamma(z) = \sum_n \gamma_n z^{-n} \qquad\qquad (15.272)$$

with

$$\gamma(z)\beta(w) \sim \frac{1}{z-w} \qquad\qquad (15.273)$$

Similarly, the conjugate variables $\hat{q}$ and $\hat{p}$ are identified with the zero modes of a free-bosonic field $\varphi(z)$, whose mode expansion takes the form (cf. Eq. (6.54))

$$\varphi(z) = \hat{q} - i\hat{p}\,\ln z + i\sum_{n\neq 0} \frac{a_n}{n} z^{-n} \qquad\qquad (15.274)$$

The commutator $[\hat{q}, \hat{p}]$ fixes the normalization of the OPE to be

$$\varphi(z)\varphi(w) \sim -\ln(z-w) \qquad\qquad (15.275)$$

The affine extension consists in replacing $\hat{p}$ by $i\partial\varphi/\alpha_+$, where α_+ is some multiplicative factor to be fixed below. This process yields the following candidate for the $\widehat{su}(2)_k$ currents in the Chevalley basis

$$\tilde{e}(z) = \beta(z)$$
$$\tilde{h}(z) = \frac{i\sqrt{2}}{\alpha_+}\partial\varphi(z) + 2(\gamma\beta)(z) \qquad\qquad (15.276)$$
$$\tilde{f}(z) = \frac{-i\sqrt{2}}{\alpha_+}(\partial\varphi\gamma)(z) - (\beta(\gamma\gamma))(z)$$

where the tilde reminds us of their tentative character. Of course, as for the zero modes, the products are normal ordered. The OPE of $\tilde{h}(z)$ with itself is

$$\tilde{h}(z)\tilde{h}(w) \sim \frac{2(\alpha_+^{-2} - 2)}{(z-w)^2} \tag{15.277}$$

In order to reproduce the desired factor $2k$ on the numerator (the 2 because we are in the Chevalley basis), we must set

$$\alpha_+^2 = \frac{1}{k+2} \tag{15.278}$$

Next, calculating the OPE of $\tilde{e}(z)$ with $\tilde{f}(w)$, we find the correct first singular term, $h(w)/(z-w)$, but the central term is missing. To cure this, we add an extra term in $f(z)$, proportional to $\partial \gamma$ (the proportionality factor is easily found to be $-k$). Such a term could not have been predicted from the finite representation since its zero mode contribution vanishes. The correct form of the currents is thus

$$
\begin{aligned}
e(z) &= \beta(z) \\
h(z) &= \frac{i\sqrt{2}}{\alpha_+}\partial\varphi(z) + 2(\gamma\beta)(z) \\
f(z) &= \frac{-i\sqrt{2}}{\alpha_+}(\partial\varphi\gamma)(z) - k\partial\gamma - (\beta(\gamma\gamma))(z)
\end{aligned}
\tag{15.279}
$$

and the full set of OPEs reads

$$
\begin{aligned}
h(z)h(w) &\sim \frac{2k}{(z-w)^2} \\
h(z)e(w) &\sim \frac{2e(w)}{z-w} \\
h(z)f(w) &\sim \frac{-2f(w)}{z-w} \\
e(z)f(w) &\sim \frac{k}{(z-w)^2} + \frac{h(w)}{z-w}
\end{aligned}
\tag{15.280}
$$

This shows that the currents in Eq. (15.279) are indeed $\widehat{su}(2)_k$ generators. The representation (15.279) is usually referred to as a Wakimoto representation.

Here again, it is possible to check, at the level of the energy-momentum tensor, the equivalence with a theory of free fields. With the (by now familiar) rearrangements

$$
\begin{aligned}
((\gamma\beta)\partial\varphi) &= (\partial\varphi(\gamma\beta)) = (\partial\varphi(\beta\gamma)) = (\beta(\partial\varphi\gamma)) \\
((\gamma\partial\varphi)\beta) &= (\beta(\gamma\partial\varphi)) + \partial^2\varphi \\
((\gamma\beta)(\gamma\beta)) &= (\beta(\beta(\gamma\gamma))) + (\partial\beta\gamma) - (\beta\partial\gamma) \\
((\beta(\gamma\gamma))\beta) &= (\beta(\beta(\gamma\gamma))) + 2(\partial\beta\gamma) + 2(\beta\partial\gamma)
\end{aligned}
\tag{15.281}
$$

the Sugawara energy-momentum tensor becomes

$$T = \frac{1}{2(k+2)} \left[\frac{1}{2}(hh) + (ef) + (fe) \right]$$

$$= -(\beta \partial \gamma) - \frac{1}{2}(\partial \varphi \partial \varphi) - \frac{i\alpha_+}{\sqrt{2}} \partial^2 \varphi$$

$$(15.282)$$

If we express the free-boson part of the energy-momentum tensor as in Chap. 9 (cf. Eq. (9.30)):

$$T_\varphi = -\frac{1}{2}(\partial \varphi \partial \varphi) + i\sqrt{2} \, \tilde{\alpha}_0 \partial^2 \varphi \qquad (15.283)$$

(the tilde over α_0 is designed to avoid confusion with a simple root), we see that for the bosonic field, there is a nonzero background charge $-\alpha_+/2$. The contribution to the central charge of each term in the total energy-momentum tensor is

$$c = 2 + 1 - 24\tilde{\alpha}_0^2 = \frac{3k}{k+2} \qquad (15.284)$$

whose sum reproduces the Sugawara central charge.

15.7.2. $\widehat{su}(2)_k$ Primary Fields

We now turn to a description of the primary fields for the $\widehat{su}(2)_k$ WZW model. From the $su(2)$ monomial basis, we expect the spin-j irreducible vacuum representation to be spanned by the $2j + 1$ monomials $(-\gamma)^{j-m}$. However, in order to have a nonzero conformal dimension (and actually, the correct $su(2)$ spin as probed by the affine generators), it must be accompanied by a vertex operator. We recall that for a free boson with a background charge $\tilde{\alpha}_0$, the dimension of the vertex $e^{ia\varphi}$ is $a^2/2 - \sqrt{2}a\tilde{\alpha}_0$. To reproduce the conformal dimension (15.104), we require

$$\frac{a^2}{2} + \frac{a\alpha_+}{\sqrt{2}} = \frac{j(j+1)}{k+2} \qquad (15.285)$$

which fixes a to be

$$a = \sqrt{2}j\alpha_+ \qquad (15.286)$$

The primary field of finite weight $2j\omega_1$ at level k is then $e^{i\sqrt{2}j\alpha_+\varphi}$ and successive OPEs with the field $f(z)$ generate the fields

$$\phi_{j,m} \equiv (-\gamma)^{j-m} e^{i\sqrt{2}j\alpha_+\varphi} \qquad (15.287)$$

The action of the currents on $\phi_{j,m}$ is

$$f(z)\phi_{j,m}(w) \sim \frac{(j+m)\phi_{j,m-1}(w)}{z-w}$$

$$h(z)\phi_{j,m}(w) \sim \frac{2m\phi_{j,m}(w)}{z-w} \qquad (15.288)$$

$$e(z)\phi_{j,m}(w) \sim \frac{(j-m)\phi_{j,m+1}(w)}{z-w}$$

15.7.3. Calculation of Correlation Functions

Since the Wakimoto representation provides the simplest way of evaluating correlation functions, we outline the calculation procedure for the $\widehat{su}(2)_k$ case.

We recall that in the Coulomb-gas formalism, for calculating correlation functions, it proves convenient to introduce a dual description of the fields: a vertex operator $V_a = e^{i\sqrt{2}a\,\varphi}$ is equally well described by the dual vertex $\tilde{V}_a \equiv V_{2\tilde{\alpha}_0 - a}$. Hence the two-point correlation function $\langle V_a \tilde{V}_a \rangle = \langle V_a V_{2\tilde{\alpha}_0 - a}\rangle$ satisfies the charge neutrality condition.

The first step in the $\widehat{su}(2)_k$ WZW model is to find the analogue of this dual description for the primary fields $\phi_{j,m}$. A simple calculation shows that

$$\tilde{\phi}_{j,j} = \beta^{s+2j} e^{-i\sqrt{2}\alpha_+\,(s+j)\,\varphi} \tag{15.289}$$

with

$$s = -k - 1 \tag{15.290}$$

has dimension $j(j+1)/(k+2)$. Moreover, the OPE

$$e(z)\tilde{\phi}_{j,j}(w) \sim 0$$
$$h(z)\tilde{\phi}_{j,j}(w) \sim \frac{2j\tilde{\phi}_{j,j}(w)}{z-w} \tag{15.291}$$

indicates that it also transforms as the highest weight of a spin-j representation. $\tilde{\phi}_{j,j}$ is uniquely determined by these requirements (cf. Ex. 15.22). The two-point correlation function is reproduced provided we project out (i.e., ignore) the extra β^s factors:

$$\langle \phi_{j,-j}(z)\tilde{\phi}_{j,j}(w)\rangle \sim \langle \gamma^{2j}(z)\beta^{2j}(w)\rangle\langle V_{\alpha_+ j}(z)V_{\alpha_+(-s-j)}(w)\rangle$$
$$\sim \frac{C}{(z-w)^{2j(j+1)/(k+2)}} \tag{15.292}$$

where C is a constant. The different $\tilde{\phi}_{j,m}$'s are obtained by successive applications of $f(z)$ on $\tilde{\phi}_{j,j}$.

The dual of the vertex operator part of $\phi_{j,j}$ is found to be

$$\tilde{V}_{\alpha_+ j} = V_{\alpha_+(-s-j)} = V_{2\tilde{\alpha}_0 s - \alpha_+ j} \tag{15.293}$$

From the example of the two-point function, we see that the Coulomb-gas charge-neutrality condition (i.e., $\sum_i \alpha_i = 2\tilde{\alpha}_0$) is now replaced by the requirement

$$\sum_i \alpha_i = 2\tilde{\alpha}_0 s \tag{15.294}$$

where s is the difference between the number of β and γ factors:

$$s = \#\beta - \#\gamma \tag{15.295}$$

It is left as an exercise (see Ex. 15.23) to verify that the above prescription reproduces the $\widehat{su}(2)_k$ three-point functions (in which one of the fields is represented by its dual). For four-point functions, as in the Coulomb-gas formalism, a screening charge is required. This is the integral of a dimension-1 field that commutes, up to a total derivative, with the three current generators. Such a field is easily found to be:

$$V_+ = \beta e^{-i\sqrt{2}\alpha_+\varphi} \tag{15.296}$$

since it satisfies

$$e(z)V_+(w) \sim 0$$
$$h(z)V_+(w) \sim 0$$
$$f(z)V_+(w) \sim (k+2)\partial_w \left(\frac{e^{-i\sqrt{2}\alpha_+\varphi(w)}}{z-w} \right) \tag{15.297}$$

The method for calculating four-point functions with proper insertions of screening charges is then essentially the same as for Virasoro primary fields and it will not be detailed.

15.7.4. Wakimoto Representation for $\widehat{su}(3)_k$

We now consider the generalization of the Wakimoto free-field representation to the $\widehat{su}(3)$ case.

The finite-dimensional description of the $su(3)$ states requires three variables $x_1, x_2,$ and x_3, one for each positive root (with $x_{1,2}$ associated with $\alpha_{1,2}$, and x_3 with $\alpha_1 + \alpha_2$). We can start by looking at the simplest possible form for the raising operator e_0^i satisfying the commutation relations

$$[e_0^1, e_0^2] = e_0^3, \qquad [e_0^1, e_0^3] = [e_0^2, e_0^3] = 0 \tag{15.298}$$

with $e_0^i = \partial/\partial x_i + \cdots$. A simple solution is

$$e_0^1 = \frac{\partial}{\partial x_1} \qquad e_0^2 = \frac{\partial}{\partial x_2} + x_1 \frac{\partial}{\partial x_3} \qquad e_0^3 = \frac{\partial}{\partial x_3} \tag{15.299}$$

By searching for the Cartan generators h_0^i in the form $c_{ij}x_j\partial/\partial x_j + \alpha_i \cdot p$, for some constants c_{ij} and $p = (p_1, p_2)$, we readily find

$$h_0^1 = \alpha_1 \cdot p - 2x_1 \frac{\partial}{\partial x_1} + x_2 \frac{\partial}{\partial x_2} - x_3 \frac{\partial}{\partial x_3}$$

$$h_0^2 = \alpha_2 \cdot p + x_1 \frac{\partial}{\partial x_1} - 2x_2 \frac{\partial}{\partial x_2} - x_3 \frac{\partial}{\partial x_3} \tag{15.300}$$

The lowering operators are now completely determined by the remaining commutators:

$$f_0^1 = \alpha_1 \cdot p \, x_1 - x_3 \frac{\partial}{\partial x_2} - x_1^2 \frac{\partial}{\partial x_1} - x_1 x_3 \frac{\partial}{\partial x_3} + x_1 x_2 \frac{\partial}{\partial x_2}$$

$$f_0^2 = \alpha_2 \cdot p \, x_2 + x_3 \frac{\partial}{\partial x_1} - x_2^2 \frac{\partial}{\partial x_2}$$

$$f_0^3 = (\alpha_1 + \alpha_2) \cdot p \, x_3 - \alpha_2 \cdot p \, x_1 x_2 - x_1 x_3 \frac{\partial}{\partial x_1}$$

$$- x_2 x_3 \frac{\partial}{\partial x_2} - x_3^2 \frac{\partial}{\partial x_3} + x_1 x_2^2 \frac{\partial}{\partial x_2}$$

(15.301)

With this choice of basis, the states are generated by monomials of the form $x_1^r x_2^s x_3^t$ with $r + t \le \alpha_1 \cdot p$ and $s \le \alpha_2 \cdot p$. It is easily verified that when the bounds are saturated, the action of f_i on such monomials vanishes.

With the relabeling $x_i \to -\gamma_0^i$, $\partial/\partial x_i \to \beta_0^i$, and the replacement

$$(\gamma_0^i, \beta_0^i, p_j) \to (\gamma^i(z), \beta^i(z), \frac{i}{\alpha_+} \partial \varphi^i(z))$$

(15.302)

with

$$\partial \varphi^i(z) \partial \varphi^j(w) \sim -\frac{\delta_{ij}}{(z-w)^2} \qquad i,j = 1,2$$

$$\gamma^i(z) \beta^j(w) \sim \frac{\delta_{ij}}{z-w} \qquad i,j = 1,2,3$$

(15.303)

we lift this representation to a representation of the full affine Lie algebra, up to a finite number of terms of the form $\gamma^n \partial \gamma$ (for some power n) in f_i, which are uniquely determined by forcing the OPEs to reproduce the $\widehat{su}(3)_k$ algebra. The final result is (cf. Ex.15.26):

$$e^1 = \beta^1$$

$$e^2 = \beta^2 - \gamma^1 \beta^3$$

$$e^3 = \beta^3$$

$$h^1 = \frac{i}{\alpha_+} \alpha_1 \cdot \partial \varphi + 2\gamma^1 \beta^1 - \gamma^2 \beta^2 + \gamma^3 \beta^3$$

$$h^2 = \frac{i}{\alpha_+} \alpha_2 \cdot \partial \varphi - \gamma^1 \beta^1 + 2\gamma^2 \beta^2 + \gamma^3 \beta^3$$

(15.304)

$$f^1 = -\frac{i}{\alpha_+} \alpha_1 \cdot \partial \varphi \, \gamma^1 - k \partial \gamma^1 + \gamma^3 \beta^2 - \gamma^1 \gamma^1 \beta^1 + \gamma^1 \gamma^2 \beta^2 - \gamma^1 \gamma^3 \beta^3$$

$$f^2 = -\frac{i}{\alpha_+} \alpha_2 \cdot \partial \varphi \, \gamma^2 - (k+1) \partial \gamma^2 - \gamma^3 \beta^1 - \gamma^2 \gamma^2 \beta^2$$

$$f^3 = -\frac{i}{\alpha_+} (\alpha_1 + \alpha_2) \cdot \partial \varphi \, \gamma^3 - \frac{i}{\alpha_+} \alpha_2 \cdot \partial \varphi \, \gamma^1 \gamma^2 - k \partial \gamma^3 - (k+1) \gamma^1 \partial \gamma^2$$

$$-\gamma^1\gamma^3\beta^1 - \gamma^2\gamma^3\beta^2 - \gamma^3\gamma^3\beta^3 - \gamma^1\gamma^2\gamma^2\beta^2$$

where now $\alpha_+^2 = 1/(k+3)$. Here, in order to lighten the notation, we have omitted the parentheses indicating the normal ordering, but these are understood, with the usual nesting toward the right. We note further that since the bosonic fields are orthogonal, the scalar products are naturally evaluated in the orthonormal basis, where

$$\alpha_1 = \frac{1}{\sqrt{2}}(1,\sqrt{3}), \qquad \alpha_2 = \frac{1}{\sqrt{2}}(1,-\sqrt{3}) \tag{15.305}$$

15.7.5. Generalization

Representations similar to those described for $su(2)$ and $su(3)$ exist for any finite Lie algebra: it requires $|\Delta_+|$ variables x_α and r pairs of conjugate variables $\hat{p}_i, \hat{q}_i$. For the Fock space construction, we also need vacua that are eigenstates of $\hat{p}_i$ and annihilated by the modes $\beta_{n\geq 0}$ and $\gamma_{n>0}$. The passage from the finite Lie algebra representation to the free-field representation is done in the general case exactly as in the above two examples. Although this will not be proven in full generality, the Sugawara form of the energy-momentum tensor can be reduced to that of the corresponding free fields, as

$$T = -\frac{1}{2}(\partial\varphi \cdot \partial\varphi) - i\alpha_+\rho \cdot \partial^2\varphi + \sum_{\alpha\in\Delta_+}(\gamma_\alpha\beta_\alpha) \tag{15.306}$$

where the generic expression of α_+ is

$$\alpha_+^2 = \frac{1}{k+g} \tag{15.307}$$

and

$$\partial\varphi^i(z)\partial\varphi^j(w) \sim -\frac{\delta_{ij}}{(z-w)^2} \qquad \gamma_\alpha(z)\beta_{\alpha'}(w) \sim \frac{\delta_{\alpha,\alpha'}}{z-w} \tag{15.308}$$

We will be content here with a simple "numerical" check, at the level of the central charge. The respective contribution of each term in the above expression of T is

$$c = r - 12\alpha_+^2|\rho|^2 + 2|\Delta_+| \tag{15.309}$$

By using the Freudenthal–de Vries strange formula (15.121) and the obvious equality

$$\dim g = 2|\Delta_+| + r \tag{15.310}$$

we see that these terms add up to the expected value of c, namely $k \dim g/(k+g)$.

Appendix 15.A. Normalization of the Wess-Zumino Term

In this appendix, we demonstrate that the ambiguity in the Wess-Zumino term

$$\Delta\Gamma = \frac{-i}{24\pi} \int_{S^3} d^3y \; \epsilon_{\alpha\beta\gamma} \mathrm{Tr}' \left(\tilde{g}^{-1} \partial^\alpha \tilde{g} \tilde{g}^{-1} \partial^\beta \tilde{g} \tilde{g}^{-1} \partial^\gamma \tilde{g} \right) \tag{15.311}$$

is an integer multiple of $2\pi i$. It is sufficient to focus on a $SU(2)$ subgroup of the symmetry group G.[17] We take a point $y \in S^3$ (assumed to be of radius 1) to be parametrized by the variables $y^{0,1,2,3}$ satisfying $(y^0)^2 + (y^i)^2 = 1$ and introduce the uniform map

$$\tilde{g}(y) = y^0 - iy^k \sigma_k \tag{15.312}$$

appropriate to the $su(2)$ lowest-dimensional representation: here the σ's are the usual Pauli matrices

$$\sigma_1 = \begin{pmatrix} 0 & 1 \\ 1 & 0 \end{pmatrix} \qquad \sigma_2 = \begin{pmatrix} 0 & -i \\ i & 0 \end{pmatrix} \qquad \sigma_3 = \begin{pmatrix} 1 & 0 \\ 0 & -1 \end{pmatrix} \tag{15.313}$$

The index of the $su(2)$ fundamental representation (cf. Eq. (13.133)) is:

$$x_{\omega_1} = \frac{\dim |\omega_1| (\omega_1, 3\omega_1)}{2 \dim [su(2)]} = \frac{1}{2} \tag{15.314}$$

so that

$$\mathrm{Tr}' \sigma_i \sigma_j = 2\mathrm{Tr}\, \sigma_i \sigma_j = 4\delta_{ij} \tag{15.315}$$

The uniformity of the map allows us to evaluate the integrand at one particular point, say $y^0 = 1$ and all $y^i = 0$, where it is

$$\sum_{i,j,k} \frac{1}{24\pi} \epsilon_{ijk} \mathrm{Tr}'(\sigma_i \sigma_j \sigma_k) = \frac{i}{\pi} \tag{15.316}$$

Since

$$\int_{S^3} d^3y = 2\pi^2 \tag{15.317}$$

the anticipated result follows:

$$\Delta\Gamma = 2\pi i \tag{15.318}$$

Here we have considered a situation in which the physical space S^3 is mapped once onto the group $SU(2)$, itself topologically equivalent to S^3. The n-th power of this application would have described n coverings of S^3 by S^3, affecting $\Delta\Gamma$ by an extra multiplicative factor n.

The argument fails when G has no $SU(2)$ subgroup, namely for $SO(3)$. That case can be dealt with easily by noticing that $SO(3)$ is like $SU(2)$, but with a

---

[17] This relies on a theorem of Bott, which states that any continuous mapping of S^3 into a general simple Lie group G can be continuously deformed into a mapping of S^3 into an $SU(2)$ subgroup of G.

short root, so that $|\theta|^2 = 1$. Referring to Eq. (15.216), we see that the index of the representation is proportional to $1/|\theta|^2$. Hence for $SO(3)$, $\Delta\Gamma = \pi i$.

Exercises

15.1 *Classical aspects of WZW models*

a) Derive explicitly the equations of motion for the $\hat{g}_k$ WZW model, obtaining first Eq. (15.23).

b) Derive the classical energy-momentum tensor in terms of the currents (15.34).

15.2 *Polyakov-Wiegman identity*
Writing the action (15.33) as $S(g)$, show that

$$S(gh^{-1}) = S(g) + S(h^{-1}) + \frac{k}{2\pi} \int d^2x \, \mathrm{Tr}' \left(g^{-1}\partial_{\bar{z}}gh^{-1}\partial_z h \right)$$

which makes manifest the invariance property (15.29).

15.3 *Sugawara construction in terms of modes*
Verify the Sugawara construction directly at the level of modes, that is, check that the L_n's given by Eq. (15.64) satisfy the Virasoro algebra with central charge (15.61), where the commutation relation for the current modes is given by Eq. (15.45).

15.4 *Normalization of the energy-momentum tensor in the Sugawara construction*

a) Given that J^a is a primary field of conformal dimension 1, which translates into

$$\left[L_n, J_m^a\right] = -mJ_{n+m}^a$$

obtain the value of γ in

$$L_n = \gamma \sum_a \sum_m : J_m^a J_{n-m}^a :$$

simply from the application of $J_1^b L_{-1}$ on a state $|\phi_\lambda\rangle$ satisfying Eq. (15.83).

b) Similarly, obtain the central charge by evaluating the norm of $L_{-2}|0\rangle$, first using the Virasoro commutation relations and then using the expression of L_{-2} in terms of the affine modes.

15.5 *Action of the $\hat{g}_k$ outer automorphism on the energy-momentum tensor*
Find the action of the $\widehat{su}(2)_k$ outer automorphism a on the Sugawara Virasoro modes. Note that on the affine generators, this action is (see the end of Sect. 14.2.2):

$$J_m^+ \leftrightarrow J_{m+1}^-, \quad 2J_m^0 \to k\delta_{m,0} - 2J_m^0 \tag{15.319}$$

15.6 *Generalized Sugawara construction*

a) Show that the following deformed Sugawara construction

$$T = \sum_a \left\{ \frac{1}{2(k+g)}(J^a J^a) + p^a \partial J^a \right\}$$

where the p^a are some constants, still defines an energy-momentum tensor. Calculate the corresponding central charge.

b) In its full generality, this construction is not physically interesting. Why? For the relevant specialization, the constants p^a are nonzero only for the generators of the Cartan subalgebra:

$$T = T^{\text{Sug}} + p^i \partial H^i$$

Evaluate the change in the conformal dimension of primary fields induced by this modification.

15.7 *Current four-point function*
Calculate the current four-point function $\langle J^a(z_1)J^b(z_2)J^c(z_3)J^d(z_4)\rangle$.

15.8 *Ward identities and the Knizhnik-Zamolodchikov equation*
Show that the Knizhnik-Zamolodchikov equation (15.99), with Eq. (15.91) expressing the global G invariance of the correlation functions, implies the projective Ward identities (15.90).

15.9 *Integrability of the Knizhnik-Zamolodchikov equation*
Check the integrability condition of the Knizhnik-Zamolodchikov equation, i.e., that

$$[\partial_{z_i}, \partial_{z_j}] = 0$$

when evaluated on a correlation of primary fields. This integrability condition is equivalent to the infinitesimal braiding relations:

$$[A_{ij}, A_{kl}] = 0 \qquad [A_{ij}, A_{ik} + A_{jk}] = 0$$

for i, j, k, l all distinct, where here $A_{ij} = t_i \otimes t_j$.

15.10 *The $\widehat{su}(N)_k$ Knizhnik-Zamolodchikov equation*
a) Derive the $i = 2, 3, 4$ analogue of Eq. (15.152) and show that the resulting equations are not independent.

b) Work out the details of the derivation of the F_1 and F_2 solutions.

c) Verify that the normalization used for the F_2 solutions makes the correlation function (15.188) compatible with

$$\langle g(z, \bar{z})g^{-1}(0,0)\rangle = (z\bar{z})^{-2h}$$

d) Determine the leading behavior of the correlation function (15.188) in the limit $x \to \infty$. Identify the contribution of the conformal blocks of fields with finite weights $2\omega_1$ and ω_2. (The corresponding conformal dimensions will then have to be determined.)

15.11 *$\widehat{su}(N)_k$ correlators and complex free fermions*
a) Write down explicitly the correlation function (15.188) for $k = 1$. Redefine each field $g(z, \bar{z})$ by a free-boson vertex, that is

$$f(z, \bar{z}) = e^{ip\phi(z,\bar{z})}g(z, \bar{z})$$

and find the expression for the correlation function $\langle ff^{-1}f^{-1}f\rangle$.

b) Find the representation of the field f in terms of the complex free fermions described in Sect. 15.5.6, and fix the value of p. (To fix the fermionic dependence of $g(z, \bar{z})$, and thereby f, force the OPE of the currents with $g(z, \bar{z})$ to be of the form (15.76). The parameter p is fixed by the dimension.)

c) Recalculate the correlator of four f-fields in terms of complex free fermions and compare with the result of part **(a)**.

15.12 *The $\widehat{su}(2)_1$ four-point correlation and free boson*
Compare the correlation function (15.188) for the $\widehat{su}(2)_1$ case with that obtained from the vertex representation of Sect. 15.6.

15.13 *The $\widehat{so}(N)_1$ Knizhnik-Zamolodchikov equation and real fermions*

a) Calculate the correlation functions for four fields in the defining (ω_1) representation of $\widehat{so}(N)_1$, by solving the Knizhnik-Zamolodchikov equation.

b) Compare the result with the correlation function calculated in terms of the free-fermion representation of Sect. 15.5.2.

15.14 *Free-fermion representation of $\widehat{so}(N)_g$*
Verify that real free fermions transforming in the adjoint representation of $so(N)$ realize the OPE (15.43) for $k = g$. Consider

$$J^a(z) = \beta \sum_{b,c} f_{abc} (\psi_b \psi_c)(z)$$

for some constant β, and check the equivalence with a theory of $\frac{1}{2}N(N-1)$ free fermions.

15.15 *Free-fermion representation of $\widehat{su}(2)_2$*

a) In terms of three real fermions, construct explicitly the operators E^+, E^-, and H satisfying the level-2 version of Eq. (15.232). (Hint: Introduce $\psi_\pm = \psi_1 \pm i\psi_2$).

b) Write down all possible states that can be obtained from three sets of Neveu-Schwarz fermionic modes, up to grade $L_0 = 2$, and decompose them into irreducible representations of $su(2)$. Display their relationship through the action of the modes of the current. Verify that the states at integer grades build up the module $L_{[2,0]}$ whereas those at half-integer grades build up $L_{[0,2]}$.

c) Find another representation of the $\widehat{su}(2)_2$ current algebra, in terms of a free fermion and a free boson.

15.16 *$\widehat{su}(2)_2$ characters from the free-fermion representation*
Given that $\widehat{su}(2)_2$ can be represented by three fermions (cf. Ex. 15.15), obtain the characters of the irreducible representations. Hint: Proceed as in the $\widehat{so}(N)_1$ case, and start from the partition function

$$Z \propto \sum_{v=2,3,4} Z_v^3 \qquad (15.320)$$

with Z_v defined in Eq. (15.211).

15.17 *Decomposition of the irreducible $\widehat{su}(2)_1$ representations into an infinite number of Virasoro irreducible modules*
In terms of the Virasoro irreducible characters at $c = 1$, which are (cf. Ex. 8.3):

$$\chi_h(q) = \frac{q^h}{\eta(q)} \qquad \text{for} \quad h \neq n^2/4$$

$$\chi'_h(q) = \frac{q^{n^2/4} - q^{(n+2)^2/4}}{\eta(q)} \qquad \text{for} \quad h = n^2/4$$

show that the specialized $\widehat{su}(2)_1$ characters (15.244) can be expressed as

$$\chi_{[1-\lambda_1,\lambda_1]}(\tau) = \sum_{\substack{m\geq 0 \\ m\in\mathbb{Z}+\lambda_1/2}} (2m+1)\chi'_{m^2}(q)$$

This illustrates neatly the reducibility of affine Lie algebra representations in terms of an infinite number of Virasoro irreducible representations, or, equivalently, that there is an infinite number of Virasoro primary fields in $c \geq 1$ theories.

15.18 $\widehat{su}(3)_1$ *modules from the vertex representations*
In the framework of the vertex representation, construct all the states

$$|p; \{n_i^{(1)}\}, \{n_j^{(2)}\}\rangle$$

with $L_0 < 3$ in the $\widehat{su}(3)_1$ modules $L_{[1,0,0]}$ and $L_{[0,1,0]}$; at each grade, organize the states into $su(3)$ irreducible representations.

15.19 *Cocycles*
Using Eq. (15.258), evaluate the product

$$c_\alpha(p - \beta - \gamma)c_\beta(p - \gamma)c_\gamma(p)$$

in different orders and show that

$$\epsilon(\alpha, \beta)\epsilon(\alpha + \beta, \gamma) = \epsilon(\beta, \gamma)\epsilon(\alpha, \beta + \gamma)$$

This is the characterizing property of a two-cocycle.

15.20 *Sugawara form of* L_{-1}
In the $su(2)$ spin basis, the expression for L_{-1} is

$$L_{-1} = \frac{1}{k+2}(2J_{-1}^0 J_0^0 + J_{-1}^+ J_0^- + J_{-1}^- J_0^+)$$

a) Using the free-boson representation at level 1, verify that the action of the r.h.s. on $e^{\pm i\varphi(z)/\sqrt{2}}$ is indeed the same as ∂_z.

b) Generalize the computation to an arbitrary level k, by using the Wakimoto representation and by considering the action on a primary field $\phi_{j,m}$.

15.21 *Bosonic version of the Wakimoto representation*
a) Verify the following ghost bosonization:

$$\beta = i\partial v\, e^{-u+iv} \qquad \gamma = e^{u-iv}$$

where u and v are standard free bosons

$$u(z)u(w) \sim -\ln(z - w) \qquad v(z)v(w) \sim -\ln(z - w)$$

b) In terms of the three bosons u, v, and φ, show that the currents and the energy-momentum tensor become

$$e(z) = i\partial v\, e^{-u+iv}$$

$$h(z) = \frac{i\sqrt{2}}{\alpha_+}\partial\varphi + 2\partial u$$

$$f(z) = \left[\frac{-i\sqrt{2}}{\alpha_+}\partial\varphi - (k+2)\partial u + (k+1)\partial v\right]e^{u-iv}$$

$$T(z) = -\frac{1}{2}(\partial\varphi\partial\varphi) - \frac{-i\alpha_+}{\sqrt{2}}\partial^2\varphi - \frac{1}{2}(\partial u\partial u) - \frac{1}{2}\partial^2 u - \frac{1}{2}(\partial v\partial v) + \frac{i}{2}\partial^2 v$$

(Warning: The normal-ordered product $(\beta\gamma)$ must be understood as follows:

$$\beta\gamma = ((i\partial v e^{-u+iv})e^{u-iv}) = (i\partial v\, (e^{-u+iv}e^{u-iv})) + \cdots = i\partial v + \cdots$$

where the dots stand for terms resulting from the necessary rearrangements.)

c) Consider the vertex operator

$$V_{(a,b,b')} = e^{ia\varphi - bu + ib'v}$$

What is its conformal dimension? Calculate the most singular term of its product with the currents and find the values of the constants a, b, and b' for which such vertex operators generate an irreducible spin-j representation.

15.22 *Dual* $\widehat{su}(2)_k$ *fields*

a) Argue that Eq. (15.289) is the only possible choice for the dual of the field $\phi_{j,j}$.

b) Find the explicit expression of $\tilde{\phi}_{1/2,-1/2}$.

15.23 $\widehat{su}(2)_k$ *three-point correlation functions*
Show that the three-point function of $\widehat{su}(2)_k$ fields are appropriately reproduced by the Wakimoto free-field representation, when the correlation is evaluated under the form $\langle \phi\phi\tilde{\phi} \rangle$ by projecting out extra factors β^s (cf. Sect. 15.7.3).

15.24 *A second* $\widehat{su}(2)_k$ *screening field*
For the $\widehat{su}(2)_k$ WZW model in the Wakimoto representation, test the following candidate screening field

$$V_- = \beta^{-k-2} e^{i\sqrt{2}\alpha_+(k+2)\varphi}$$

Prove that V_+ and V_- exhaust the possibilities for screening fields.

15.25 *Screening operators*
Show that the operator $e^{-i\sqrt{2}\hat{q}}$, for the representation (15.270) of the $su(2)$ generators, is the finite analogue of the screening field V_+ defined in Eq. (15.296). Find the two analogous operators for $su(3)$ and their affine extension.

15.26 $\widehat{su}(3)_k$ *Wakimoto representation*

a) Verify that Eq. (15.304) yields the correct $\widehat{su}(3)_k$ current commutation relations.

b) Check that for this $\widehat{su}(3)_k$ representation, the energy-momentum tensor can be brought into the form (15.306).

15.27 *Generators of the Cartan subalgebra in the Wakimoto free-field representation*   For a general algebra $\hat{g}$, the Wakimoto free-field expression of the Chevalley generators of the Cartan subalgebra is

$$h^i = \sum_{\alpha \in \Delta_+} (\alpha_i^\vee, \alpha)(\gamma^\alpha \beta^\alpha) + \frac{i}{\alpha_+}\alpha_i^\vee \cdot \partial\varphi$$

with $\alpha_+^2 = 1/(k+g)$. Check that these are genuine primary fields of dimension 1 with respect to the energy-momentum tensor (15.306), and calculate the OPE $h^i(z) h^j(w)$.

Notes

The sigma model (15.1) was solved exactly by Polyakov and Wiegman [299]. Its property of asymptotic freedom, akin to QCD, was a partial motivation for its study. The Wess-Zumino term was first introduced in a four-dimensional context by Wess and Zumino in 1971 [352]. Ten years later, Novikov [284] pointed out its multivaluedness, which forces the coupling constant of such a topological term to be an integer for the quantum consistency of the theory. The topological theorem of Bott mentioned in App. 15.A can be found in Ref. [57].

The consideration of a Wess-Zumino term in the framework of the two-dimensional sigma model is due to Witten [359], who was also the first to notice the necessary asymmetry of the conserved currents J_z and $J_{\bar{z}}$ (cf. Eq. (15.17)). In the same work, he derived the current algebra commutation relations canonically at the level of Poisson brackets, and presented the first evidence for the conformal invariance of WZW models by displaying, through a one-loop calculation, the vanishing of the β function. The model was further studied by Polyakov and Wiegman [300] who derived in particular the relation given in Ex. 15.2.

The Sugawara construction goes back to the late 1960s, discovered independently by Sugawara [332] and Sommerfield [328]. However its correct quantum version, that is, the correct value of γ in Eq. (15.49), was found somewhat later: by Dashen and Frishman [86] for $\widehat{su}(N)$ at level 1, and by Goddard and Olive [182], Knizhnik and Zamolodchikov [240], and Todorov [336] for the general case. (The method used in Ref. [240] is presented in Ex. 15.4.) This construction also appeared in mathematics (Ref. [326]).

The study of the conformal aspects of WZW models was initiated by Knizhnik and Zamolodchikov [240]. They introduced key concepts such as WZW primary fields, the basic equation which bears their name and they calculated the $\widehat{su}(N)_1$ correlation functions. Section 15.4 follows rather closely (but with much more detail) their presentation. (Actually, the Knizhnik-Zamolodchikov equation first appeared in Ref. [86].) This important work was extended by Gepner and Witten [172]. They elucidated the structure of the WZW primary fields and derived the constraints that follow from affine singular vectors. The works of Tsuchiya and Kanie [337] and Felder, Gawedzki, and Kupiainen [129] presented significant developments.

The solutions of the Knizhnik-Zamolodchikov equation for the $\widehat{su}(2)_k$ WZW model were obtained by Zamolodchikov and Fateev [365] and Christe and Flume [75].

The first free-fermion representation of a current algebra was found by Bardacki and Halpern in 1971 [25]. The quantum equivalence between the $\widehat{so}(N)_1$ WZW model and a theory of N real free fermions was shown in Ref. [359], while the analogous statement for $\widehat{su}(N)_1$ was proved in Ref. [240]. General conditions for the equivalence of the $\hat{g}_k$ WZW model and a theory of free fermions were derived by Goddard, Nahm, and Olive [180]. Many aspects of Sects. 15.2–15.5 are covered in the book of Fuchs [148].

Vertex operators originated from string theory (see, for instance, the book of Green, Schwarz and Witten [187]). Their use in the representation of current algebras goes back to the work of Halpern [190] and Banks, Horn, and Neuberger [23]. This construction was generalized by Frenkel and Kac [137] and Segal [326]. Vertex representations for nonsimply-laced algebras were found in Ref. [181, 43]. The cocycle construction presented here is taken from [187] and the construction of the $\widehat{su}(2)_1$ representations follows [228]. Free-fermion and vertex representations are reviewed in Ref. [183].

The $\widehat{su}(2)$ Wakimoto free-field representation spaces appeared in Wakimoto [345], and the Feigen-Fuchs-type construction is due to A.B. Zamolodchikov (unpublished). Further analysis of the $\widehat{su}(2)_k$ model can be found in Bernard and Felder [42] (in particular, the analysis of the BRST cohomology). The generalization to higher-rank algebras was found by various groups (Refs. [125, 126, 46, 174, 53, 55]). Our presentation follows Refs. [174, 55] for the most part. (Exercise 15.21 is taken from the first of these references.) The calculation of the $\widehat{su}(2)_k$ correlation functions was done by Dotsenko [108, 109] (the second screening operator presented in Ex. 15.24 is mentioned in the second of these references and in Ref. [46]).

CHAPTER 16

Fusion Rules in WZW Models

This chapter is mainly concerned with the calculation of fusion rules in WZW models. As in any rational conformal field theory, fusion rules can be calculated from the Verlinde formula, in terms of the modular transformation matrices S. For WZW models, all these matrices are known explicitly. However, working out a few examples is convincing enough that the Verlinde formula is not very convenient for explicit calculations. Indeed, the dimension of the matrix S grows very quickly with the level and the rank of the algebra. Nevertheless, the formula itself allows us to derive useful identities (which is done in Sect. 16.1), in addition to being the starting point for a more efficient approach: the affine extension of the character method used for the calculation of tensor products in finite Lie algebra. This method is described in Sect. 16.2. It leads to a very nice relation between tensor-product coefficients and fusion coefficients. The concept of quantum dimension, naturally related to the character method, is introduced in Sect. 16.3.

Another method of calculating WZW fusion rules—historically the first technique to be found—amounts to implementing the vanishing of three-point functions containing affine singular vectors. It leads to the so-called depth rule (Sect. 16.4). Although not very practical, this method leads to the useful concept of a threshold level $k_0^{(i)}$, below which the coupling indexed by i is absent and above which it is always present. The knowledge of threshold levels and ordinary tensor-product coefficients suffices to fully determine the fusion coefficients.

The depth rule boils down to a formula for threshold levels, but in terms of quantities difficult to evaluate. On the other hand, there is a very explicit approach to the calculation of fusion rules based on the idea that any coupling can be decomposed into a finite number of basic or elementary couplings, such that the threshold level of a given coupling is simply the sum of the threshold levels of the elementary couplings occurring in its decomposition. This method is described in App. 16.A. However, it is limited to the lower-rank algebras, namely those algebras for which the basis of elementary couplings can be described explicitly.

In Sect. 16.5, the interesting concept of fusion potential is introduced. It provides a framework in which fusion rule calculations are reduced to polynomial multiplications, subject to some constraints. These constraints are captured in the simple conditions that the derivatives of some function (the fusion potential) with respect to each variable (one for each fundamental weight) should vanish.

In the final section, a remarkable quantum symmetry (with no classical, i.e., infinite-level, analogue) is exhibited. It relates theories whose level and rank are interchanged. The most spectacular manifestation of this duality is an exact equivalence between the fusion coefficients of the two models.

At this stage, the importance of fusion rules should be clear. Not only do they give the structure of the operator product algebra, but, in theories with boundaries, they are closely related to partition functions. Subsequent developments, in particular the construction of modular-invariant partition functions presented in Chap. 17, will provide further applications.

Finally, we stress that only diagonal theories will be considered in this chapter. We recall that in diagonal theories each primary field transforms in the same representation of g in both holomorphic and antiholomorphic sectors, and each integrable representation has multiplicity one. The naturalness of such a spectrum was already indicated in Sect. 15.3.5 and its modular invariance will be proved in Chap. 17. The necessity of this assumption reflects the intrinsically chiral nature of fusion rules. In a chiral sector, the unitarity of the S matrix requires each representation to be present. The whole presentation is further restricted to the holomorphic sector.

Readers who are not interested in all the facets of fusion rule calculations may restrict themselves to the first two sections. The remaining sections are all independent of each other.

§16.1. Symmetries of Fusion Coefficients

We first recall that, in a rational conformal field theory, the fusion coefficient $\mathcal{N}_{\phi_i, \phi_j, \phi_k}$ counts the number of independent couplings between the three primary fields ϕ_i, ϕ_j, and ϕ_k. This is equivalent to the multiplicity of the conjugate field ϕ_k^* in the operator product expansion of $\phi_i(z)$ with $\phi_j(w)$. Indeed, the substitution of this OPE in the three-point function transforms it into a sum of two-point functions involving ϕ_k, which are nonzero only when the other field is ϕ_k^*. The fusion coefficients can thus be described as the decomposition coefficients of the formal product $\times$ defined as

$$\phi_i \times \phi_j = \sum_{\phi_k^*} \mathcal{N}_{\phi_i \phi_j}{}^{\phi_k^*} \phi_k^* \tag{16.1}$$

Such a decomposition is known as a fusion rule.

For the WZW model with $\hat{g}_k$ spectrum-generating algebra, we know that the primary fields correspond to the integrable representations $\hat{\lambda} \in P_+^k$ of $\hat{g}_k$. The fusion rules thus take the form:

$$\hat{\lambda} \times \hat{\mu} = \bigoplus_{\hat{\nu} \in P_+^k} \mathcal{N}_{\hat{\lambda}\hat{\mu}}^{(k)\,\hat{\nu}} \hat{\nu} \tag{16.2}$$

and the fusion coefficients can be calculated from the Verlinde formula:

$$\boxed{\mathcal{N}_{\hat{\lambda}\hat{\mu}}^{(k)\,\hat{\nu}} = \sum_{\hat{\sigma} \in P_+^k} \frac{S_{\hat{\lambda}\hat{\sigma}} S_{\hat{\mu}\hat{\sigma}} \bar{S}_{\hat{\nu}\hat{\sigma}}}{S_{0\hat{\sigma}}}} \tag{16.3}$$

where, as usual, 0 stands for $k\hat{\omega}_0$, the weight associated with the identity field.

In this section, we derive simple properties of fusion coefficients, derived mostly from the Verlinde formula. In order to simplify the notation in the following, we omit the index k.

We recall that the unitarity of S implies directly that

$$\mathcal{N}_{0\hat{\mu}}^{\hat{\nu}} = \delta_{\hat{\mu}}^{\hat{\nu}} \tag{16.4}$$

which means that the identity field is the neutral element of fusions.

In Eq. (16.3), the lower indices of $\mathcal{N}_{\hat{\lambda}\hat{\mu}}^{\hat{\nu}}$ are associated with S whereas the upper index is associated with $\bar{S}$. Since

$$\bar{S}_{\hat{\lambda}\hat{\mu}} = S_{\hat{\lambda}^*,\hat{\mu}} \tag{16.5}$$

(cf. Eq. (14.232)), this implies

$$\mathcal{N}_{\hat{\lambda}\hat{\mu}}^{\hat{\nu}} = \mathcal{N}_{\hat{\lambda}\hat{\mu}\hat{\nu}^*} = \mathcal{N}_{\hat{\lambda}\hat{\nu}^*}^{\hat{\mu}^*} \tag{16.6}$$

where, as before, $\hat{\nu}^*$ denotes the affine weight whose finite part ν^* is the weight conjugate to ν. Hence, in WZW models as in any general conformal field theory, indices are raised and lowered by the (charge) conjugation matrix $C = S^2$.

We consider now the action of the outer automorphism group on fusion rules. For $A \in \mathcal{O}(\hat{g})$, we have shown (cf. Eq. (14.255)) that

$$A S_{\hat{\lambda}\hat{\sigma}} = S_{A(\hat{\lambda})\hat{\sigma}} = S_{\hat{\lambda}\hat{\sigma}} e^{-2\pi i (A\hat{\omega}_0, \sigma)} \tag{16.7}$$

Consequently,

$$\mathcal{N}_{A(\hat{\lambda})\hat{\mu}}^{A(\hat{\nu})} = \mathcal{N}_{\hat{\lambda}\hat{\mu}}^{\hat{\nu}} \qquad \text{and} \qquad \mathcal{N}_{\hat{\lambda}A'(\hat{\mu})}^{A'(\hat{\nu})} = \mathcal{N}_{\hat{\lambda}\hat{\mu}}^{\hat{\nu}} \tag{16.8}$$

implying that

$$\boxed{\mathcal{N}_{A(\hat{\lambda})A'(\hat{\mu})}^{AA'(\hat{\nu})} = \mathcal{N}_{\hat{\lambda}\hat{\mu}}^{\hat{\nu}}} \tag{16.9}$$

A similar argument leads to

$$\mathcal{N}_{A(\hat{\lambda})\hat{\mu}}^{\hat{\nu}} = \mathcal{N}_{\hat{\lambda}A(\hat{\mu})}^{\hat{\nu}} \tag{16.10}$$

of which a special case is

$$\mathcal{N}_{A(0)\hat{\mu}}^{\hat{\nu}} = \mathcal{N}_{0A(\hat{\mu})}^{\hat{\nu}} = \delta_{A(\hat{\mu})}^{\hat{\nu}} \tag{16.11}$$

Thus, the fusion rule of the field $A(0) = kA\hat{\omega}_0$ with any field $\hat{\mu}$ contains a single term, which is simply the field obtained by the action of A on $\hat{\mu}$. In other words, the field $A(0)$ acts in fusion rules in exactly the same way as A does on the weights.

On the other hand, if we replace $\hat{\sigma}$ by $A\hat{\sigma}$ in the summation (16.3)—which is perfectly justified, since whenever $\hat{\sigma} \in P_+^k$, it automatically follows that $A\hat{\sigma} \in P_+^k$—we find

$$\mathcal{N}_{\hat{\lambda}\hat{\mu}}^{\hat{\nu}} = \mathcal{N}_{\hat{\lambda}\hat{\mu}}^{\hat{\nu}} e^{-2\pi i(A\hat{\omega}_0, \lambda+\mu-\nu)} \tag{16.12}$$

For the fusion coefficient to be nonzero, the phase factor must necessarily be one, that is,

$$(A\hat{\omega}_0, \lambda + \mu - \nu) \in \mathbb{Z} \tag{16.13}$$

This translates into the condition

$$\lambda + \mu - \nu \in Q \tag{16.14}$$

where Q is the root lattice of g. We recall (cf. the analysis following Eq. (14.114)) that we can write Q instead of $Q^\vee$ because the mark associated with the dual of $A\hat{\omega}_0$ is 1, meaning that the corresponding root equals its coroot. It is certainly very natural to recover this condition, since it is also necessary for the nonvanishing of the corresponding tensor-product coefficient $\mathcal{N}_{\lambda\mu}^\nu$.

We now work out a simple application of the above results. We first calculate the fusion rules in the $\widehat{so}(2N+1)_1$ model, for which the modular S matrix is

$$S = \frac{1}{2} \begin{pmatrix} 1 & 1 & \sqrt{2} \\ 1 & 1 & -\sqrt{2} \\ \sqrt{2} & -\sqrt{2} & 0 \end{pmatrix} \tag{16.15}$$

with the ordering $\hat{\omega}_0, \hat{\omega}_1, \hat{\omega}_N$ (which are the only integrable representations at level one). A direct calculation using the Verlinde formula gives

$$\hat{\omega}_0 \times \hat{\omega}_0 = \hat{\omega}_0$$
$$\hat{\omega}_0 \times \hat{\omega}_1 = \hat{\omega}_1$$
$$\hat{\omega}_0 \times \hat{\omega}_N = \hat{\omega}_N$$
$$\hat{\omega}_1 \times \hat{\omega}_1 = \hat{\omega}_0 \tag{16.16}$$
$$\hat{\omega}_1 \times \hat{\omega}_N = \hat{\omega}_N$$
$$\hat{\omega}_N \times \hat{\omega}_N = \hat{\omega}_0 + \hat{\omega}_1$$

A sample calculation follows:

$$\mathcal{N}_{\hat{\omega}_N \hat{\omega}_N}^{\hat{\omega}_1} = \sum_{i=\hat{\omega}_0, \hat{\omega}_1, \hat{\omega}_N} \frac{S_{\hat{\omega}_N, i} S_{\hat{\omega}_N, i} \bar{S}_{\hat{\omega}_1, i}}{S_{\hat{\omega}_0, i}} = \frac{1}{2} + \frac{1}{2} + 0 = 1 \tag{16.17}$$

The first three relations could have been written directly from Eq. (16.4). All others are consequences of the first three and of the relations (16.6) and (16.9). Indeed, for the B_r algebra, the only nontrivial outer automorphism a exchanges $\hat{\omega}_0$ and $\hat{\omega}_1$, without affecting $\hat{\omega}_N$. Therefore, from $\hat{\omega}_0 \times \hat{\omega}_0 = \hat{\omega}_0$ and $a^2 = 1$, we find

$$a\hat{\omega}_0 \times a\hat{\omega}_0 = \hat{\omega}_1 \times \hat{\omega}_1 = \hat{\omega}_0 \tag{16.18}$$

Similarly, we have

$$a\hat{\omega}_0 \times \hat{\omega}_N = \hat{\omega}_1 \times \hat{\omega}_N = a\hat{\omega}_N = \hat{\omega}_N \tag{16.19}$$

Finally, since $C = I$,

$$\mathcal{N}_{\hat{\omega}_N \hat{\omega}_N}{}^{\hat{\sigma}} = \mathcal{N}_{\hat{\omega}_N \hat{\sigma}}{}^{\hat{\omega}_N} \tag{16.20}$$

We already know that the rightmost fusion coefficient in the above formula is unity if $\hat{\sigma}$ is $\hat{\omega}_0$ or $\hat{\omega}_1$, and this gives the last fusion rule in Eq. (16.16).

We conclude this preliminary section with a general comment. Fields that act as permutations in fusion rules are usually called *simple currents*. $A(0)$ is thus an example of simple current. For WZW models based on simple Lie algebras, it turns out that all simple currents are related to outer automorphisms, with a single exception at level 2 in the $\hat{E}_8$ model. Simple currents provide a way of defining the center of a rational conformal field theory as the set of elements that are dual to simple currents. The center of a WZW model is simply the center of the corresponding symmetry group.

§16.2. Fusion Rules Using the Affine Weyl Group

16.2.1. The Kac-Walton Formula

We now derive a precise relation between tensor-product and fusion coefficients. It essentially follows from a straightforward affine extension of the character method for tensor products. But there is a shortcut: the simple connection that exists between fusion and tensor-product coefficients is actually rooted in the remarkable relation between ratios of elements of the modular S matrix and characters of integrable representations of the finite Lie algebra, evaluated at special points. This observation has already been recorded in Sect. 14.6.3. Its symbolic transcription is:

$$\gamma_{\hat{\mu}}^{(\hat{\sigma})} \equiv \frac{S_{\hat{\sigma}\hat{\mu}}}{S_{\hat{\sigma}0}} = \chi_\mu(\xi_\sigma) \tag{16.21}$$

where the character is evaluated at the special point

$$\xi_\sigma = \frac{-2\pi i}{k+g}(\sigma + \rho) \tag{16.22}$$

There is a direct relation between the fusion and tensor-product coefficients since the $\gamma_{\hat{\mu}}^{(\hat{\sigma})}$ satisfy the fusion algebra (cf. Sect. 10.8.3):

$$\gamma_{\hat{\lambda}}^{(\hat{\sigma})} \gamma_{\hat{\mu}}^{(\hat{\sigma})} = \sum_{\hat{\nu} \in P_+^k} \mathcal{N}_{\hat{\lambda}\hat{\mu}}^{(k)\ \hat{\nu}} \gamma_{\hat{\nu}}^{(\hat{\sigma})} \tag{16.23}$$

This identity is easily verified by substituting the expression (16.3) for $\mathcal{N}_{\hat{\lambda}\hat{\mu}}^{(k)\ \hat{\nu}}$ and using the unitarity of $\mathcal{S}$. The $\chi_\mu(\xi_\sigma)$ satisfy the tensor-product algebra

$$\chi_\lambda(\zeta)\chi_\mu(\zeta) = \sum_{\varphi \in P_+} \mathcal{N}_{\lambda\mu}{}^\varphi \chi_\varphi(\zeta) \tag{16.24}$$

When evaluated at $\zeta = \xi_\sigma$, the last expression must be equivalent to Eq. (16.23). We conclude that

$$\sum_{\hat{\nu} \in P_+^k} \mathcal{N}_{\hat{\lambda}\hat{\mu}}^{(k)\ \hat{\nu}} \chi_\nu(\xi_\sigma) = \sum_{\varphi \in P_+} \mathcal{N}_{\lambda\mu}{}^\varphi \chi_\varphi(\xi_\sigma) \tag{16.25}$$

A relation between $\mathcal{N}_{\hat{\lambda}\hat{\mu}}^{(k)\ \hat{\nu}}$ and $\mathcal{N}_{\lambda\mu}{}^\nu$ follows once the two summations are made comparable. The clue for this lies in the following observation. Any weight φ in the fundamental chamber (defined by $\varphi \in P_+$) can be mapped into a weight in the affine fundamental chamber (defined by $\nu \in P_+$ and $(\nu, \theta) \leq k$, i.e., $\hat{\nu} \in P_+^k$) by an appropriate affine Weyl reflection. (This, of course, is not true for weights that lie on the boundary of P_+^k. But in view of Eq. (14.173), these weights can be ignored: they have zero character.) Hence, for any $\varphi \in P_+$ not on the boundary of P_+^k (or a Weyl reflection thereof), there is a $w \in \hat{W}$ and a $\nu \in P_+$ satisfying $(\nu, \theta) \leq k$, related by

$$\hat{\varphi} = w \cdot \hat{\nu} \equiv w(\hat{\nu} + \hat{\rho}) - \hat{\rho} \tag{16.26}$$

where as usual $\hat{\nu}$ and $\hat{\varphi}$ are the affine extension of ν and φ at level k.

Denoting the finite part of $\hat{\varphi} = w \cdot \hat{\nu}$ as $\varphi = w \cdot \nu$, we can rewrite the right-hand side of Eq. (16.25) as

$$\sum_{\varphi \in P_+} \mathcal{N}_{\lambda\mu}{}^\varphi \chi_\varphi(\xi_\sigma) = \sum_{\substack{\hat{\nu} \in P_+^k \\ w \cdot \nu \in P_+}} \sum_{w \in \hat{W}} \mathcal{N}_{\lambda\mu}{}^{w \cdot \nu} \chi_{w \cdot \nu}(\xi_\sigma) \tag{16.27}$$

The replacement of μ by $w \cdot \mu$ in the Weyl character formula (14.246):

$$\chi_\mu(\xi_\sigma) = \frac{\sum_{w \in W} \epsilon(w) e^{(w(\mu+\rho),\xi_\sigma)}}{\sum_{w \in W} \epsilon(w) e^{(w\rho,\xi_\sigma)}} \tag{16.28}$$

with $w \cdot \mu + \rho = w(\mu + \rho)$ and a redefinition of the summation variable, yield

$$\chi_{w \cdot \mu}(\xi_\sigma) = \epsilon(w) \chi_\mu(\xi_\sigma) \tag{16.29}$$

This allows us to rewrite Eq. (16.25) in the form

$$\sum_{\hat{\nu} \in P_+^k} \mathcal{N}_{\hat{\lambda}\hat{\mu}}^{(k)\ \hat{\nu}} \chi_\nu(\xi_\sigma) = \sum_{\substack{\hat{\nu} \in P_+^k \\ w \cdot \nu \in P_+}} \sum_{w \in \hat{W}} \mathcal{N}_{\lambda\mu}{}^{w \cdot \nu} \epsilon(w) \chi_\nu(\xi_\sigma) \tag{16.30}$$

from which we obtain

$$\boxed{\mathcal{N}_{\hat{\lambda}\hat{\mu}}^{(k)\ \hat{\nu}} = \sum_{\substack{w \in \hat{W} \\ w \cdot \nu \in P_+}} \mathcal{N}_{\lambda\mu}{}^{w \cdot \nu} \epsilon(w)} \tag{16.31}$$

This is the *Kac-Walton formula*.[1]

16.2.2. Algorithm for Fusion Rules

The Kac-Walton formula leads to a simple algorithm for the calculation of fusion coefficients, which is a direct extension of the character method for tensor products (described in Sect. 13.5.1), in which the finite Weyl group is replaced by its affine extension.

The procedure for evaluating the fusion rules $\hat{\lambda} \times \hat{\mu}$ for $\hat{\lambda}, \hat{\mu} \in P_+^k$ may be formulated as follows. We first add each weight μ' of the representation μ to λ, and sets $\varphi = \lambda + \mu'$. Next, we construct the affine extension $\hat{\varphi}$ of φ at level k. Then there are two possibilities:

(i) $\hat{\varphi}$ can be reflected into the affine fundamental chamber, in which all Dynkin labels (including the zeroth one) are nonnegative integers, by the shifted action of the affine Weyl group. Hence, there is a $w \in \hat{W}$ such that $w \cdot \hat{\varphi} = \hat{\nu} \in P_+^k$. Then $\hat{\nu}$ gives a contribution $\epsilon(w)$ to the fusion coefficient $\mathcal{N}_{\hat{\lambda}\hat{\mu}}^{(k)\ \hat{\nu}}$.

(ii) $\hat{\varphi}$ is in the $\hat{W}$ orbit of some weight fixed under the shifted action of a simple reflection. This means that $(s_i w) \cdot \hat{\varphi} = w \cdot \hat{\varphi}$ for some $w \in \hat{W}$ and some $s_i, i = 0, 1, \cdots, r$. Such weights are ignored: their characters simply vanish.

In practical calculations, the algorithm is broken into two parts, the contribution of the finite Weyl group being first considered. In other words, in the first step we simply calculate tensor-product coefficients. Next, for a fusion at some fixed level k, we write the affine extension of every weight ν occurring in the tensor product $\lambda \otimes \mu$. Those with a negative zeroth Dynkin label are reflected into the affine fundamental chamber by means of the shifted action of an element w of the affine Weyl group, whose rightmost part is now necessarily s_0. The result gives a contribution $\epsilon(w)$ to the fusion coefficient. Furthermore, weights with $\nu_0 = -1$ are ignored since

$$s_0 \cdot [-1, \nu] = s_0[0, \nu + \rho] - [1, \rho] = [-1, \nu] \tag{16.32}$$

so that the same weight would contribute with signatures $\epsilon(1) = 1$ and $\epsilon(s_0) = -1$. The same holds for all weights that can be reflected into a weight having -1 as one of its Dynkin labels.

Consider for instance the $su(3)$ fusion

$$[k - 3, 2, 1] \times [k - 3, 1, 2] \tag{16.33}$$

---

[1] If tensor-product coefficients are defined for arbitrary weights by requiring that

$$\mathcal{N}_{\lambda\mu}{}^{\nu} = 0 \quad \text{if any of} \quad \lambda, \nu, \mu \notin P_+$$

then the formula (16.31) can be simplified somewhat in that the restriction $w \cdot \nu \in P_+$ is no longer necessary.

The corresponding tensor product is easily evaluated by means of the Littlewood-Richardson rule

$$(2,1) \otimes (1,2) = (0,0) \oplus (0,3) \oplus 2(1,1) \oplus (1,4) \oplus 2(2,2) \oplus (3,0) \oplus (3,3) \oplus (4,1) \tag{16.34}$$

We let $k = 3$, the lowest level for which the affine extension of the weights on the l.h.s. are integrable. The affine extensions of those on the r.h.s. are respectively

$$\begin{array}{c} [3,0,0], \quad [0,0,3], \quad 2[1,1,1], \quad [-2,1,4], \\ 2[-1,2,2], \quad [0,3,0], \quad [-3,3,3], \quad [-2,4,1] \end{array} \tag{16.35}$$

The two weights $[-1,2,2]$ are ignored. For the other integrable weights, we use

$$s_0 \cdot \hat{\lambda} = \hat{\lambda} + \hat{\rho} - (\lambda_0 + 1)\alpha_0 - \hat{\rho} \tag{16.36}$$

with

$$\alpha_0 = [2,-1,-1] \quad \text{and} \quad \hat{\rho} = [1,1,1] \tag{16.37}$$

to find

$$\begin{aligned} s_0 \cdot [-2,1,4] &= [-1,2,5] + [2,-1,-1] - [1,1,1] = [0,0,3] \\ s_0 \cdot [-2,4,1] &= [0,3,0] \\ s_0 \cdot [-3,3,3] &= [1,1,1] \end{aligned} \tag{16.38}$$

The complete fusion rule is then

$$[0,2,1] \times [0,1,2] = [3,0,0] + [1,1,1] \tag{16.39}$$

This result is illustrated in Fig. 16.1. Similarly, at $k = 4$ we find

$$[1,2,1] \times [1,1,2] = [4,0,0] + [1,0,3] + 2[2,1,1] + [0,2,2] + [1,3,0] \tag{16.40}$$

At $k = 5$, only the affine extension of $(3,3)$ is absent, whereas for $k = 6$, all weights appearing in the tensor product are also present in the fusion rule.

In general, the sole action of s_0 is not sufficient to transform nonintegrable weights with $\nu_0 < -1$ into integrable ones. Take, for instance, the $\widehat{su}(3)$ weight $[-3,6,0]$:

$$s_0 \cdot [-3,6,0] = [1,4,-2] \tag{16.41}$$

while

$$s_2 s_0 \cdot [-3,6,0] = [0,3,0] \tag{16.42}$$

An example of weight $\hat{\nu}$ that would be ignored in the present context, even though $\nu_0 < -1$, is $[-2,0,6]$, since

$$s_0 \cdot [-2,0,6] = [0,-1,5] \tag{16.43}$$

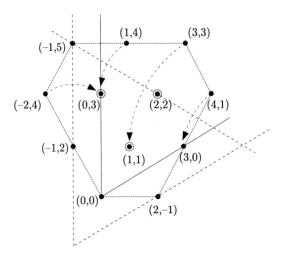

Figure 16.1. The $\widehat{su}(3)_3$ fusion rule $[0, 2, 1] \times [0, 1, 2]$. Weights are reflected with respect to the lines $[-1, \cdot, \cdot], [\cdot, -1, \cdot], [\cdot, \cdot, -1]$; weights sitting on those lines are ignored (circled dots correspond to weights with multiplicity 2).

The Kac-Walton formula makes transparent a certain number of fundamental properties of fusion coefficients. First, we see that fusion coefficients reduce to tensor-product coefficients in the limit $k \to \infty$:

$$\lim_{k \to \infty} \mathcal{N}_{\hat{\lambda}\hat{\mu}}^{(k)\ \hat{\nu}} = \mathcal{N}_{\lambda\mu}{}^{\nu} \tag{16.44}$$

In that limit the affine fundamental chamber and the finite Lie algebra fundamental chamber become essentially equivalent. (To be more precise, here we consider the limit where k is large with respect to (λ, θ) and (ν, θ), such that only a small part of the affine fundamental Weyl chamber is probed.) In that case, there is no need for affine Weyl reflection. Second, that the fusion coefficients are some sort of truncated tensor-product coefficients, where the degree of truncation is a sole function of the level, is very neatly illustrated by Eq. (16.31).

However, this approach has an obvious drawback: in order to calculate a specific fusion coefficient $\mathcal{N}_{\hat{\lambda}\hat{\mu}}^{(k)\ \hat{\nu}}$, we have to first evaluate the full product $\lambda \otimes \mu$, and reflect all the terms in the decomposition whose affine extension at a fixed level k are not in P_+^k. This is a rather long process when reasonably large representations are involved. Furthermore, the formula does not make manifest the nonnegativity of fusion coefficients.

This is a good place to stress that a fusion rule differs from a standard tensor product. Indeed, in order to calculate the tensor product of two affine weights $\hat{\lambda}$ and $\hat{\mu}$, both at level k, we can proceed as in the finite case, by adding all the weights $\hat{\mu}'$ of the representation $L_{\hat{\mu}}$ to $\hat{\lambda}$ and decomposing the result into irreducible representations $L_{\hat{\nu}}$. However, the mere addition of two affine weights shows that in this process the levels add up, that is, the decomposed representations are at level $2k$. But this is clearly not suited to the physical problem at hand, for which we want to describe the general pattern of operator products in a theory with a fixed value of the central charge, hence a fixed value of the level k.

16.2.3. The $\widehat{su}(2)_k$ Fusion Coefficients

It is very easy to obtain a general formula for the $\widehat{su}(2)_k$ fusion coefficients by the above method. The starting point is the well-known formula for angular momentum tensor-product coefficients:

$$\lambda \otimes \mu = \bigoplus_{\substack{\nu_1 = |\lambda_1 - \mu_1| \\ \lambda_1 + \mu_1 + \nu_1 = 0 \bmod 2}}^{\lambda_1 + \mu_1} \nu \qquad (16.45)$$

We fix the level k so that:

$$k \geq \max\{\lambda_1, \mu_1\} \qquad (16.46)$$

ensuring the integrability of both $\hat{\lambda}$ and $\hat{\mu}$. If $\lambda_1 + \mu_1 \leq k$, the affine extension of all the weights on the r.h.s. is integrable and the above expression extends without modification to the affine case. If, however, $\lambda_1 + \mu_1 > k$, it is necessary to perform a shifted Weyl reflection of some weights on the r.h.s. to recover integrable weights. The boundary weight $\nu_1 = k + 1$, if allowed by the parity constraint, is ignored as usual. For the other ones, it is clearly sufficient to act with $(s_0 \cdot)$. Indeed

$$s_0 \cdot [k - \nu_1, \nu_1] = [-k - 2 + \nu_1, 2k + 2 - \nu_1] \qquad (16.47)$$

Even though $\lambda_1 + \mu_1 > k + 1$, it is nevertheless true that $2k \geq \lambda_1 + \mu_1$ (which follows from Eq. (16.46)), so that $2k + 2 \geq \nu_1$ for all ν_1 appearing in Eq. (16.45). Hence, all weights with $\nu_1 > k + 1$ are reflected back into the dominant sector and contribute with a negative sign. They thus cancel all terms in the range $2k + 2 - \lambda_1 - \mu_1 \leq \nu_1 \leq k$. The final result for the fusion coefficient is

$$\mathcal{N}_{\hat{\lambda}\hat{\mu}}^{(k)\,\hat{\nu}} = \begin{cases} 1 & \text{if} \quad |\lambda_1 - \mu_1| \leq \nu_1 \leq \min\{\lambda_1 + \mu_1, 2k - \lambda_1 - \mu_1\} \\ & \text{and} \quad \lambda_1 + \mu_1 + \nu_1 = 0 \mod 2 \\ 0 & \text{otherwise} \end{cases}$$

$$(16.48)$$

This expression illustrates neatly the relation (16.44).

16.2.4. $\widehat{su}(N)_k$ Fusion Rules: Combinatorial Description

The algorithm described in Sect. 16.2.2 has a simple combinatorial interpretation in terms of Young tableaux. For simplicity, we only describe the $\widehat{su}(N)_k$ case. The tensor product of the fusion rules under consideration is first calculated by the Littlewood-Richardson rule. Those tableaux whose first row has length $\ell_1 \leq k$

correspond to integrable $\widehat{su}(N)_k$ affine weights. Otherwise, the weights either lie on the boundary of an affine Weyl chamber (and are ignored) or can be reflected back into integrable weights by appropriate shifted Weyl reflections. Such a transformation is described combinatorially as follows. We remove a boundary strip of length

$$t = \ell_1 - k - 1 \geq 0 \qquad (16.49)$$

starting from the end of the first row and moving downward and to the left; then, we add a boundary strip of the same length at the bottom of the first column, starting from the N-th row and moving upward and to the right. We repeat these modifications until either $t \leq 0$ or an irregular tableau is produced. ($t = 0$ or the occurrence of an irregular tableau signals a boundary weight.) The sign of the corresponding Weyl reflection is given by $(-1)^{r_- + r_+ + 1}$, where r_- is the number of columns crossed by the removed strip and r_+ is the number of columns crossed by the added strip. An example will clarify the rule: we consider the $su(4)$ weight $(3, 6, 4)$ at level 5, whose tableau has rows of length $\ell_1 = 3 + 6 + 4 = 13$, $\ell_2 = 3 + 6 = 9$, $\ell_3 = 3$. This is not integrable since $\ell_1 > k$; the strip to be removed has length $t = 13 - 5 - 1 = 7$ and it is indicated by boxes marked by 0's:

$$(16.50)$$

We now add a strip of length 7 at the bottom of the tableau (again indicated with boxes $\boxed{0}$):

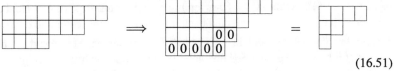

$$(16.51)$$

where in the last step, columns of 4 boxes have been deleted. The sign of the Weyl reflection is -1: 6 columns have been crossed in each step. This result is confirmed by a direct calculation:

$$(s_2 s_1 s_0) \cdot [-8, 3, 6, 4] = [1, 2, 1, 1] \qquad (16.52)$$

Boundary weights are easily identified by this procedure: for instance, $6\omega_4$ is a boundary affine weight for $\widehat{su}(N > 5)_2$, since $t = 6 - 2 - 1 = 3 < 4 = \tilde{\ell}_1$, so that the strip removal produces an irregular tableau:

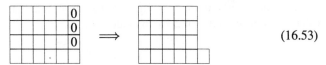

$$(16.53)$$

686   16. Fusion Rules in WZW Models

Returning to the $\widehat{su}(3)_3$ example (16.34), the three relevant "nonintegrable tableaux" and their modifications are

$$(16.54)$$

and they all contribute with a minus sign. In all the above examples, the $(N-1)$-th row is always present. A final example deals with a different case: we consider the $\widehat{su}(3)_3$ weight $[-3, 6, 0]$ and its modification:

$$(16.55)$$

We observe that the added strip must start at the third row $(N = 3)$ of the first column. The sign of the corresponding Weyl reflection is positive. This agrees with a previous calculation: $s_2 s_0 \cdot [-3, 6, 0] = [0, 3, 0]$.

§16.3. Quantum Dimensions

Verifying the equality

$$(\dim |\lambda|) \times (\dim |\mu|) = \sum_{\nu \in P_+} \mathcal{N}_{\lambda\mu}{}^{\nu} \dim |\nu| \tag{16.56}$$

is a simple consistency check in tensor-product calculations. At first sight, it might seem that it has no analogue for fusion rules, given that the highest-weight representations are infinite dimensional. However, we can define a dimension, relative to the reference vacuum representation, as

$$\mathcal{D}_{\hat{\lambda}} = \frac{\mathrm{Tr}_{\hat{\lambda}}(1)}{\mathrm{Tr}_0(1)} \tag{16.57}$$

This can be reexpressed in terms of specialized characters evaluated in the limit $q \to 1^-$ or $\tau \to i0^+$

$$\mathcal{D}_{\hat{\lambda}} = \lim_{q \to 1^-} \frac{\chi_{\hat{\lambda}}(q)}{\chi_0(q)} = \lim_{\tau \to i0^+} \frac{\chi_{\hat{\lambda}}(\tau)}{\chi_0(\tau)} \tag{16.58}$$

By using the asymptotic behavior of the characters (14.238), the relative dimension can be written as a ratio of S matrix elements

$$\boxed{\mathcal{D}_{\hat{\lambda}} = \frac{S_{\hat{\lambda}0}}{S_{00}}} \tag{16.59}$$

This last expression shows that $\mathcal{D}_{\hat{\lambda}}$ is merely the $\gamma_{\hat{\lambda}}^{(0)}$ defined in Eq. (16.21). The desired dimension formula for fusion coefficients is thus simply Eq. (16.23) for $\sigma = 0$

$$\mathcal{D}_{\hat{\lambda}}\mathcal{D}_{\hat{\mu}} = \sum_{\hat{\nu} \in P_+^k} N_{\hat{\lambda}\hat{\mu}}^{(k)\;\hat{\nu}}\mathcal{D}_{\hat{\nu}} \qquad (16.60)$$

(which is simply the one-dimensional representation of fusion rules). For reasons to become clear later, $\mathcal{D}_{\hat{\lambda}}$ is referred to as a *quantum dimension*.

Intuitively, we expect $\mathcal{D}_{\hat{\lambda}}$ to be a real number, greater or equal to one. We showed the reality of $\mathcal{D}_{\hat{\lambda}}$ in Sect. 14.6.2; in the same section, we proved the inequality $S_{\hat{\lambda}0} \geq S_{00} \geq 0$, which ensures that $\mathcal{D}_{\hat{\lambda}} \geq 1$. Being the reference representation, the vacuum has quantum dimension one. The same is true for all fields obtained from the vacuum by the action of the elements of the outer-automorphism group. This is a direct outcome of Eq. (14.255):

$$AS_{\hat{\lambda}0} = S_{A\hat{\lambda},0} = S_{\hat{\lambda}0}e^{-2\pi i(A\hat{\omega}_0,0)} = S_{\hat{\lambda}0} \qquad (16.61)$$

which implies that

$$\mathcal{D}_{A\hat{\lambda}} = \mathcal{D}_{\hat{\lambda}} \qquad (16.62)$$

That $S_{\hat{\lambda}0}$ is real also has the consequence that conjugate fields have the same quantum dimension

$$\mathcal{D}_{\hat{\lambda}^*} = \mathcal{D}_{\hat{\lambda}} \qquad (16.63)$$

Whenever there is a small number of fields, the sum rule (16.60) may be sufficient to fix the fusion coefficients. For instance, we can consider $\widehat{G}_2$ at level 1, for which the S matrix is

$$S = \sqrt{\frac{4}{5}}\begin{pmatrix} \sin\frac{\pi}{5} & \sin\frac{3\pi}{5} \\ \sin\frac{3\pi}{5} & -\sin\frac{\pi}{5} \end{pmatrix} \qquad (16.64)$$

The two fields are $\hat{\omega}_0$ and $\hat{\omega}_2$, with respective quantum dimensions 1 and $(1+\sqrt{5})/2$. Equation (16.60) implies directly the "Yang-Lee" type fusion rules

$$\hat{\omega}_2 \times \hat{\omega}_2 = \hat{\omega}_0 + \hat{\omega}_2 \qquad (16.65)$$

An apparent flaw of the above example is that we need to calculate the S matrix to extract the quantum dimensions before applying the sum rule. But when the theory is simple, it is just as simple to use the Verlinde formula directly. However, we stress that only one column of the S matrix is actually required in order to calculate the quantum dimensions. Furthermore, there is a more direct way to calculate the quantum dimensions, which ultimately does not rely on the modular S matrix.

A simple expression for the quantum dimension comes from Eqs. (16.59) and (14.241):

$$\mathcal{D}_{\hat{\lambda}} = \prod_{\alpha>0} \frac{\sin\left(\frac{\pi(\lambda+\rho,\alpha)}{k+g}\right)}{\sin\left(\frac{\pi(\rho,\alpha)}{k+g}\right)} \qquad (16.66)$$

To proceed further, we introduce the so-called q-*numbers* defined by

$$[x] = \frac{q^{x/2} - q^{-x/2}}{q^{1/2} - q^{-1/2}}, \quad \text{where} \quad q = e^{2\pi i/(k+g)} \qquad (16.67)$$

Some obvious properties of these numbers are:[2]

$$[x] = [k+g-x], \quad [-x] = -[x], \quad [k+g] = 0 \qquad (16.68)$$

In terms of q-numbers, the quantum dimension takes the compact form

$$\mathcal{D}_{\hat{\lambda}} = \prod_{\alpha>0} \frac{[(\lambda+\rho,\alpha)]}{[(\rho,\alpha)]} \qquad (16.69)$$

In the limit $q \to 1$ $(k \to \infty)$, this reduces to the usual Weyl dimension formula, since

$$\lim_{q \to 1} [x] = x \qquad (16.70)$$

For $su(N)$, the Weyl formula has a simple transcription in terms of partitions (cf. Eq. (13.192)), which is simply obtained by evaluating the scalar products in the orthonormal basis. Its "quantum" or affine counterpart is

$$\mathcal{D}_{\hat{\lambda}} = \prod_{1 \leq i < j \leq N} \frac{[\ell_i - \ell_j + j - i]}{[j-i]} \qquad (16.71)$$

Apart from $q \neq 1$, the affine aspect of this formula lies in the integrability requirement on $\hat{\lambda}$, which translates into the constraint $\ell_1 - \ell_N \leq k$.

The quantum dimensions of nonintegrable weights need not be positive. Actually, from the explicit expression of the S matrix or directly from Eq. (16.29), we find

$$\mathcal{D}_{w\cdot\hat{\lambda}} = \epsilon(w)\mathcal{D}_{\hat{\lambda}} \qquad (16.72)$$

for any element w of the affine Weyl group. Hence, all weights in the Weyl orbit of a weight fixed under the action of the affine Weyl group necessarily have zero quantum dimension. In the approach described in the previous section, where fusion coefficients are obtained from tensor products, we see that those weights

---

[2] For integer values of $x = n + 1$, the q-number $[x]$ reduces to the n-th Chebyshev polynomial of the second kind, introduced in Eq. (8.101):

$$[n+1] = U_n\left(\sqrt{q} + \frac{1}{\sqrt{q}}\right)$$

that are ignored are exactly those with quantum dimension equal to zero. From Eqs. (16.71) and (16.68), it is manifest that representations with $\ell_1 - \ell_N = k+1$ have zero quantum dimension since the above product contains a factor $[k+N] = 0$ in that case. Furthermore, the cancellation resulting from shifted affine Weyl reflections is a cancellation between weights in the same orbit and with opposite quantum dimensions.

An example will illustrate the cancellation of representations with opposite quantum dimensions in fusion rules. We consider the affine extension of the tensor product

$$(2) \otimes (2) = (0) \oplus (2) \oplus (4) \tag{16.73}$$

at $k = 2$. Since the $\widehat{su}(2)_k$ representation with finite Dynkin label λ_1 has quantum dimension $[\lambda_1 + 1]$, the quantum dimensions of the various terms on the r.h.s. are:

$$\mathcal{D}_{[2,0]} = [1] = 1$$
$$\mathcal{D}_{[0,2]} = [3] = [4 - 3] = [1] = 1 \tag{16.74}$$
$$\mathcal{D}_{[-2,4]} = [5] = [4 - 5] = [-1] = -1$$

The representations $[0, 2]$ and $[-2, 4]$ are on the same s_0 orbit

$$s_0 \cdot [-2, 4] = s_0[-1, 5] - [1, 1] = [0, 2] \tag{16.75}$$

so that they cancel each other; that is, their quantum dimensions add up to zero.

§16.4. The Depth Rule and Threshold Levels

16.4.1. The Depth Rule

We now complete the analysis, initiated in Sect. 15.3.4, of the constraints imposed by the affine Lie singular vectors on the correlation functions. Concentrating on the three-point functions, we saw that

$$\langle ((E^\theta_{-1})^p \hat{\lambda})(z) \hat{\mu}'(z_1) \hat{\nu}'(z_2) \rangle = 0 \quad \text{if} \quad p \geq k - (\lambda, \theta) + 1 \tag{16.76}$$

since it incorporates a singular vector. We recall that the primed field is obtained from the (unprimed) primary field associated with the highest-weight state by application of some finite Lie algebra lowering operators. We have seen that Eq. (16.76) implies that

$$\sum_{\substack{l_1,l_2=0 \\ l_1+l_2=p}}^{p} \frac{p!}{l_1! l_2!} \frac{1}{(z - z_1)^{l_1}} \frac{1}{(z - z_2)^{l_2}} \langle \hat{\lambda}(z)[(E^\theta_0)^{l_1} \hat{\mu}'](z_1)[(E^\theta_0)^{l_2} \hat{\nu}'](z_2) \rangle = 0 \tag{16.77}$$

The three-point functions are particularly simple in that all the terms in the sum are independent. Indeed, a three-point function contains a single term whose dependence upon the variables z, z_1, and z_2 is given by powers of their differences, powers that are sole functions of the conformal dimensions of the three fields.

Since the action of finite Lie generators does not modify the conformal dimension, these powers are thus independent of the l_i's. Therefore, the coefficients of the different terms $(z - z_i)^{-l_i}$ in the above equation must vanish separately:

$$\langle \hat{\lambda}(z)[(E_0^\theta)^{l_1}\hat{\mu}'](z_1)[(E_0^\theta)^{l_2}\hat{v}'](z_2)\rangle = 0$$

$$\text{if} \quad l_1 + l_2 = p > k - (\lambda, \theta) \tag{16.78}$$

Therefore, whenever

$$(E_0^\theta)^{l_1}|\hat{\mu}'\rangle \equiv |\hat{\mu}''\rangle \neq 0$$
$$(E_0^\theta)^{l_2}|\hat{v}'\rangle \equiv |\hat{v}''\rangle \neq 0 \tag{16.79}$$

for integers l_1, l_2 satisfying Eq. (16.78), the nontrivial three-point function

$$\langle \hat{\lambda}(z)\hat{\mu}''(z_1)\hat{v}''(z_2)\rangle \tag{16.80}$$

must vanish. The relevant characterizing property of the state $|\hat{\mu}''\rangle$ is that $(E_0^{-\theta})^{l_1}|\hat{\mu}''\rangle$, for some value of l_1, is still in the representation $\hat{\mu}$ (i.e., it is a nonzero state, $|\hat{\mu}'\rangle$)—and similarly for $|\hat{v}''\rangle$.

The condition we just obtained can be formulated in a more intrinsic way by introducing the *depth* $d_{\mu''}$ of the state $|\hat{\mu}''\rangle$, as the maximal value of l_1 for which Eq. (16.79) is true for $|\hat{\mu}''\rangle$ fixed. However, this definition needs to be made more precise since the depth is a property of a state, and not that of the associated weight; in general, the finite weights in a highest-weight representation of g can be degenerate. Therefore, we have to introduce a subindex i distinguishing the different fields that could be associated with the same finite weight. The proper definition of the depth is thus

$$d_{\mu''_{(i)}} = \max(l) \quad \text{such that} \quad (E_0^{-\theta})^l|\hat{\mu}''_{(i)}\rangle \neq 0 \tag{16.81}$$

The different states sharing the same finite weight generically have different depths. For instance, in the $su(3)$ representation of highest weight $(1, 1)$, the weight $(0, 0)$ has multiplicity two, corresponding to the two distinct states $E^{-\alpha_1}E^{-\alpha_2}|(1, 1)\rangle$ and $E^{-\alpha_2}E^{-\alpha_1}|(1, 1)\rangle$. From a linear combination of these states, we can form a $su(2)_\theta$-singlet and a $su(2)_\theta$-triplet, with respective depths of zero and one.

Therefore, a coupling should really be denoted as $\langle \hat{\lambda}\hat{\mu}''_{(j)}\hat{v}''_{(\ell)}\rangle$, and the doublet (j, ℓ) can take $\mathcal{N}_{\lambda\mu v}$ values. Since

$$l_1 + l_2 \le d_{\mu''_{(j)}} + d_{v''_{(\ell)}} \tag{16.82}$$

we are forced to conclude that

$$\langle \hat{\lambda}(z)\hat{\mu}''_{(j)}(z_1)\hat{v}''_{(\ell)}(z_2)\rangle = 0 \quad \text{if} \quad k < d_{\mu''_{(j)}} + d_{v''_{(\ell)}} + (\lambda, \theta) \tag{16.83}$$

This is the explicit form of the constraint we were looking for. It is called the *depth rule*.

The depth constraint can be simplified by using the g-invariance of the three-point function to fix one of the two primed states. Indeed, $|\hat{v}''_{(j)}\rangle$ can always be expressed in terms of the action of some raising operators on the lowest state of the representation, $| - \hat{v}^* \rangle$; that is,

$$|\hat{v}''_{(j)}\rangle = E^{\alpha_{\ell_1}} E^{\alpha_{\ell_2}} \cdots |-\hat{v}^* \rangle \tag{16.84}$$

with α_{ℓ_i} some positive roots. Then, by moving the action of these finite Lie algebra generators on the other two states in the three-point function, and recalling that their action on $\hat{\lambda}$ vanishes because λ is a highest weight, we can rewrite $\langle \hat{\lambda} \hat{\mu}''_{(j)} \hat{v}''_{(\ell)} \rangle$ in the form $\langle \hat{\lambda} \hat{\mu}'_{(i)} (-\hat{v}^*) \rangle$ for some state $|\hat{\mu}'_{(i)}\rangle$ (from now on we drop one prime to lighten the notation). Hence, there is now only one state that is neither a highest nor a lowest state. The different states $|\hat{\mu}'_{(i)}\rangle$ characterize the $\mathcal{N}_{\lambda\mu\nu}$ couplings. Since the depth of the lowest state in a representation is necessarily zero, the constraint (16.83) can be reexpressed as

$$\langle \hat{\lambda}(z) \hat{\mu}'_{(i)}(z_1)(-\hat{v}^*)(z_2) \rangle = 0 \quad \text{if} \quad k < k_0^{(i)} \tag{16.85}$$

where

$$k_0^{(i)} \equiv d_{\mu'_{(i)}} + (\lambda, \theta) \tag{16.86}$$

Clearly, the different couplings associated with a given triple product are characterized by (a priori) different constraints, that is, different values of $k_0^{(i)}$.

We now look at what happens when $k \geq k_0^{(i)}$. In that case, there are no more affine constraints, no possibility of hitting an affine singular vector. The only constraints are those associated with the finite Lie algebra singular vectors. These, in turn, fix completely the tensor-product coefficient $\mathcal{N}_{\lambda\mu\nu}$. Thus, for $k \geq k_0^{(i)}$,

$$\langle \hat{\lambda}(z) \hat{\mu}'_{(i)}(z_1)(-\hat{v}^*)(z_2) \rangle \neq 0 \quad \text{if} \quad \mathcal{N}_{\lambda\mu\nu} \neq 0 \tag{16.87}$$

which means that $k_0^{(i)}$ is the *threshold level* for this coupling. Of course, irrespective of the value of k, it is always true that

$$\langle \hat{\lambda}(z) \hat{\mu}'_{(i)}(z_1)(-v^*)(z_2) \rangle = 0 \quad \text{if} \quad \mathcal{N}_{\lambda\mu\nu} = 0 \tag{16.88}$$

Finally, how do we read off fusion coefficients in this approach? For fixed k, the number of allowed couplings is simply determined by the number of values of $k_0^{(i)}$ that lie below k. Suppose that the set $\{k_0^{(i)}\}$ is ordered such that $k_0^{(i)} \leq k_0^{(i+1)}$. Then

$$\mathcal{N}_{\hat{\lambda}\hat{\mu}}^{(k)\,\hat{v}} = \begin{cases} \max(i) \text{ such that } k \geq k_0^{(i)} \text{ and } \mathcal{N}_{\lambda\mu\nu} \neq 0 \\ 0 \text{ if } k < k_0^{(1)} \text{ or } \mathcal{N}_{\lambda\mu\nu} = 0 \end{cases} \tag{16.89}$$

with $k_0^{(i)}$ defined by Eq. (16.86), in terms of the depth (16.81).

This is a manifestly positive expression for the fusion coefficients. Furthermore, it implies the following remarkable inequality:

$$N_{\hat\lambda\hat\mu}^{(k)\ \hat\nu} \leq N_{\hat\lambda\hat\mu}^{(k+1)\ \hat\nu} \tag{16.90}$$

which, from the point of view of the Kac-Walton formula, was totally unexpected.

For practical calculations, the rule (16.89) is not very useful because in general the depth is difficult to calculate. Furthermore, the determination of a proper basis, appropriate for the calculation of both the tensor-product coefficients and the depth constraint, is still an open problem. This difficulty is obviously absent for $su(2)$, where finite weights always have multiplicity 1. In that case, the depth is also easily calculated. [3]

We rederive the $\widehat{su}(2)_k$ fusion rules from the depth rule. We consider the fusion $\hat\lambda \times \hat\mu \times \hat\nu$. For the coupling to be nonvanishing, the following equality must be satisfied

$$\lambda_1 + \mu_1' + \nu_1' = 0 \tag{16.91}$$

The fusion coefficient $N_{\hat\lambda\hat\mu\hat\nu}^{(k)}$ will be nonzero (and then the only value it can take is one) if

$$|\lambda_1 - \mu_1| \leq \nu_1 \leq \lambda_1 + \mu_1 \quad \text{and} \quad \lambda_1 + \mu_1 + \nu_1 = 0 \ \mathrm{mod}\ 2 \tag{16.92}$$

and

$$k \geq (\lambda, \theta) + d_{\mu'} + d_{\nu'} \tag{16.93}$$

The depth of μ' is the number of times we can subtract $\theta = \alpha_1$ from μ' without leaving the representation, whose limit is $-\mu$:

$$\mu_1' - 2d_{\mu'} = -\mu_1 \tag{16.94}$$

so that

$$d_{\mu'} = \frac{\mu_1}{2} + \frac{\mu_1'}{2} \tag{16.95}$$

Therefore, using Eq. (16.91), we find

$$k \geq \frac{1}{2}(\lambda_1 + \mu_1 + \nu_1) \equiv k_0 \tag{16.96}$$

The expression for the $\widehat{su}(2)$ threshold level is thus very simple. Adding the constraint

$$\nu_1 \leq 2k - \lambda_1 - \mu_1 \tag{16.97}$$

to Eq. (16.92), we recover the $\widehat{su}(2)_k$ fusion rules (16.48).

There is another circumstance in which the depth is easily calculated, namely for weights in fundamental representations. Consider for instance the fundamental representation of highest weight ω_1 of $su(N)$. The lowest state is obtained by subtracting θ from ω_1. This means that the depth of ω_1 is one. For all other weights

---

[3] At present, we can calculate directly *all* the fusion coefficients from the depth rule *only* for $\widehat{su}(2)_k$.

in the representation the depth vanishes. Other fundamental representations are treated similarly.

In terms of semistandard tableaux, the different weights ω'_ℓ in the representation ω_ℓ of $su(N)$ are obtained by filling, in all possible ways, the ℓ boxes with numbers $i_1, \cdots, i_\ell$ such that

$$1 \leq i_1 < i_2 < \cdots < i_\ell \leq N \qquad (16.98)$$

In terms of this numbering, it is not difficult to check that the depth is simply

$$d_{\omega'_\ell} = \frac{1}{2}[\,|1's - N's| + (\omega'_\ell, \theta)] \qquad (16.99)$$

where $|1's - N's|$ means the absolute value of the number of 1's minus the number of N's.

16.4.2. Threshold Levels and $\widehat{su}(3)_k$ Fusion Coefficients

The depth rule, through Eq. (16.89), shows clearly that the fusion coefficients are fully determined by the data $\mathcal{N}_{\lambda\mu\nu}$ and $\{k_0^{(i)}\}$. Hence, fusion coefficients can be encoded in expressions for tensor products by simply introducing a subindex indicating the threshold level of the various possible couplings. For instance, the fusion rules for Eq. (16.33) at different levels are easily recovered from

$$(2,1) \otimes (1,2) = (0,0)_3 \oplus (0,3)_4 \oplus 2(1,1)_{3,4} \oplus (1,4)_5$$
$$\oplus\, 2(2,2)_{4,5} \oplus (3,0)_4 \oplus (3,3)_6 \oplus (4,1)_5 \qquad (16.100)$$

The notation $2(2,2)_{4,5}$ means that one copy of $(2,2)$ appears at level 4 and the other one appears at level 5 (so that for $k \geq 5$, the two copies are always present). Fusion coefficients at a fixed value of the level k are obtained by considering all terms with subindex $\leq k$.

We note that the very existence of a threshold level for each coupling is in fact a consequence of Eq. (16.90). This inequality in turn can be established directly as follows. The truncation on tensor-product coefficients is solely governed by the extra singular vector (15.109). The strength of this constraint decreases as k increases, which implies the inequality (16.90).

Unfortunately, as already pointed out, we do not know how to calculate the set of values $k_0^{(i)}$ from the depth rule. Nevertheless, another approach is possible, based on the concept of *elementary couplings*. The idea is to decompose a coupling into a set of elementary couplings in such a way that the threshold level for the coupling under consideration is simply the sum of the threshold levels of the elementary couplings appearing in the decomposition. The latter are always easily calculated. Explicit expressions for the set $\{k_0^{(i)}\}$ can be obtained in this way for low-rank algebras.

For $su(N)$, this idea is most neatly formulated in terms of the Berenstein-Zelevinsky triangles. In App. 16.A, the complete analysis for the $su(3)$ case is presented. It culminates in an explicit formula for both $\mathcal{N}_{\lambda\mu\nu}$ and $\{k_0^{(i)}\}$, which we report here.

The number of scalars contained in the $su(3)$ triple product $\lambda \otimes \mu \otimes \nu$ is

$$\mathcal{N}_{\lambda\mu\nu} = \delta_{k_0} (k_0^{\max} - k_0^{\min} + 1) \tag{16.101}$$

where

$$
\begin{aligned}
k_0^{\min} &= \max\big(\lambda_1 + \lambda_2, \mu_1 + \mu_2, \nu_1 + \nu_2, L_1 - \min(\lambda_1, \mu_1, \nu_1), \\
&\qquad\quad L_2 - \min(\lambda_2, \mu_2, \nu_2)\big) \\
k_0^{\max} &= \min(L_1, L_2)
\end{aligned}
\tag{16.102}
$$

and

$$
\delta_{k_0} =
\begin{cases}
1 & \text{if} \quad k_0^{\max} \geq k_0^{\min} \quad \text{and} \quad L_1, L_2 \in \mathbb{Z}_+ \\
0 & \text{otherwise}
\end{cases}
\tag{16.103}
$$

The quantities L_i are defined as

$$L_i = (\lambda + \mu + \nu, \omega_i) \tag{16.104}$$

that is,

$$
\begin{aligned}
L_1 &= \frac{1}{3}[2(\lambda_1 + \mu_1 + \nu_1) + \lambda_2 + \mu_2 + \nu_2] \\
L_2 &= \frac{1}{3}[\lambda_1 + \mu_1 + \nu_1 + 2(\lambda_2 + \mu_2 + \nu_2)]
\end{aligned}
\tag{16.105}
$$

The corresponding fusion coefficients are given by Eq. (16.89) with the above $\mathcal{N}_{\lambda\mu\nu}$ and the following values of $k_0^{(i)}$:

$$\{k_0^{(i)}\}(\lambda \otimes \mu \otimes \nu) = \{k_0^{\min}, k_0^{\min} + 1, k_0^{\min} + 2, \cdots, k_0^{\max}\} \tag{16.106}$$

where $k_0^{\min}$ and $k_0^{\max}$ are given in Eq. (16.102). This yields

$$
\mathcal{N}_{\hat{\lambda}\hat{\mu}\hat{\nu}}^{(k)} = \delta_{k_0}
\begin{cases}
0 & \text{if} \quad k < k_0^{\min} \\
\mathcal{N}_{\lambda\mu\nu} - (k_0^{\max} - k) & \text{if} \quad k_0^{\min} \leq k \leq k_0^{\max} \\
\mathcal{N}_{\lambda\mu\nu} & \text{if} \quad k > k_0^{\max}
\end{cases}
\tag{16.107}
$$

In a sense, the degeneracy of the tensor product is completely lifted in the $\widehat{su}(3)$ fusion rules in that all $\mathcal{N}_{\lambda\mu\nu}$ values $k_0^{(i)}$ are distinct. In fact, $k_0^{(i+1)} - k_0^{(i)}$ is always equal to one, a special property of $\widehat{su}(3)$.

Notice that the inequality

$$k_0^{\min} \geq \max(\lambda_1 + \lambda_2, \mu_1 + \mu_2, \nu_1 + \nu_2) \tag{16.108}$$

simply means that the three affine weights $\hat{\lambda}, \hat{\mu}, \hat{\nu}$ must be integrable.

To illustrate the power of this formula, consider the product of $\lambda = \mu = \nu = (8, 8)$. In this case, $L_i = 24$, so that $k_0^{\min} = 16$ and $k_0^{\max} = 24$. The multiplicity of the product is then 9, and the fusion coefficient at level 22, say, is equal to 7.

§16.5. Fusion Potentials ($\widehat{su}(N)$)

An established result in group theory is that tensor products can be reduced to products of polynomials in r variables, one for each fundamental weight. This is based on two formulae: the Pieri formula, which gives the decomposition of the tensor product of any representation with a fundamental weight, and the Giambelli (or Jacobi-Trudy) formula, which gives the polynomial representation of any irreducible representation. Extended to affine Lie algebras, this leads to a conceptually very interesting approach to fusion rules. The level-dependent truncation of tensor-product coefficients induces polynomial constraints among the r variables. Quite remarkably, these constraints can be integrated to a potential. To simplify the presentation of these results, we restrict the whole discussion to $\widehat{su}(N)$.

16.5.1. Tensor-Product Coefficients Revisited

We start by reviewing the classical results mentioned above. These are most naturally formulated in terms of Young tableaux. We recall that the Young tableau for the highest weight $\lambda = (\lambda_1, \cdots, \lambda_{N-1})$ is specified by the partition[4]

$$\lambda = \{\ell_1; \ell_2; \ldots; \ell_{N-1}\} = \{\lambda_1 + \lambda_2 + \ldots \lambda_{N-1}; \ldots; \lambda_{N-1}\} \qquad (16.109)$$

where ℓ_i is the length of the i-th row. We recall that $\lambda^t = \{\tilde{\ell}_1; \tilde{\ell}_2; \ldots; \tilde{\ell}_{N-1}\}$ stands for the partition of its transpose, obtained by interchanging rows and columns. A special notation is introduced for the fundamental weights

$$x_j = \{1; 1; \ldots; 1\} \quad (j \text{ entries}) \qquad (16.110)$$

described by a column of j boxes, and their transpose,

$$y_j = \{j\} \qquad (16.111)$$

a single row of j boxes.

The *Pieri formula* reads

$$x_j \otimes \{\ell_1; \ldots; \ell_n\} = \bigoplus_{\substack{\ell_i \leq p_i \leq \ell_i + 1 \leq p_{i-1} \\ \sum p_i = \sum \ell_i + j}} \{p_1; \ldots; p_N\} \qquad (16.112)$$

where the missing entries $\ell_{n+1}, \ldots, \ell_{N-1}$ are just zero. This is simply a specialization of the Littlewood-Richardson rule. Here j boxes are added to the tableau with partition $\{\ell_1; \cdots; \ell_{N-1}\}$, with at most one box per row. A simple $su(3)$ example

---

[4] To such a tableau, we can add an arbitrary number of columns of N boxes, which makes $\ell_N \neq 0$. A tableau is said to be *reduced* if these irrelevant columns of N boxes are removed. The vanishing entries of a partition are usually omitted.

is

$$x_2 \otimes \{2; 1\} = \bigoplus_{\substack{2 \leq p_1 \leq 3 \\ 1 \leq p_2 \leq 2 \\ 0 \leq p_3 \leq 1 \\ p_1 + p_2 + p_3 = 5}} \{p_1; p_2; p_3\} = \{3; 2\} \oplus \{3; 1; 1\} \oplus \{2; 2; 1\}$$

(16.113)

$$= \{3; 2\} \oplus \{2\} \oplus \{1; 1\}$$

where the last equality is obtained by dropping columns of 3 boxes, whose partition is $\{1; 1; 1\}$.

On the other hand, the *Giambelli formula* gives the polynomial decomposition of a partition $\{\ell_1; \ldots; \ell_n\}$ as a matrix determinant whose entries are specified in terms of the transposed partition: $\{\tilde{\ell}_1; \ldots; \tilde{\ell}_s\}$

$$\{\ell_1; \ldots; \ell_n\} = \det x_{\tilde{\ell}_i + j - i} = \det \begin{pmatrix} x_{\tilde{\ell}_1} & x_{\tilde{\ell}_1+1} & \cdots & x_{\tilde{\ell}_1+s-1} \\ x_{\tilde{\ell}_2-1} & x_{\tilde{\ell}_2} & \cdots & \\ \vdots & & & \\ x_{\tilde{\ell}_s-s+1} & & \cdots & x_{\tilde{\ell}_s} \end{pmatrix}$$

(16.114)

Here the convention is that $x_0 = x_N = 1$ and $x_i = 0$ for $i < 0$ and $i > N$.[5] For example, with $N = 5$, the decomposition of the partition $\{3; 3; 1\}$, whose transpose is $\{3; 2; 2\}$, is

$$\{3; 3; 1\} = \det \begin{pmatrix} x_3 & x_4 & 1 \\ x_1 & x_2 & x_3 \\ 1 & x_1 & x_2 \end{pmatrix}$$

(16.115)

A simple inductive argument shows that this formula is a direct consequence of Eq. (16.112). We start by expanding the determinants of the $s \times s$ matrix in terms of $(s - 1) \times (s - 1)$ determinants, multiplying elements of the first column. For these lower-order determinants, the validity of the formula is assumed and, as a consequence, they can be rewritten as partitions. The final step amounts to multiplying the resulting partitions with the appropriate $x_{\tilde{\ell}_i - i + 1}$ using the Pieri formula.

To calculate a given tensor product $\lambda \otimes \mu$, the formulae are used as follows. The weight λ is first expressed as a polynomial in the x_j's via Eq. (16.114) and then the product of the x_j's with the partition corresponding to μ is calculated by means of Eq. (16.112). Consider the simple $su(3)$ example:

$$(2, 0) \otimes (0, 2) = \{2\} \otimes \{2; 2\}$$

Since the transpose of the partition $\{2\}$ is $\{1; 1\}$,

$$\{2\} = \det \begin{pmatrix} x_1 & x_2 \\ 1 & x_1 \end{pmatrix} = x_1{}^2 - x_2$$

(16.116)

---

[5] We stress that in the decomposition of the determinant, x_i^2 is to be understood as $x_i \otimes x_i$.

This result should be obvious to the reader:

$$\square \otimes \square = \square\square \oplus \begin{array}{c}\square\\\square\end{array} \quad \Rightarrow \quad \square\square = \square \otimes \square \ominus \begin{array}{c}\square\\\square\end{array} \tag{16.117}$$

Hence, from Eq. (16.112)

$$
\begin{aligned}
x_1 \otimes \{2; 2\} &= \{3; 2\} \oplus \{1; 1\}\\
x_1{}^2 \otimes \{2; 2\} &= \{4; 2\} \oplus \{3; 3\} \oplus 2\,\{2; 1\} \oplus \{0\} \\
x_2 \otimes \{2; 2\} &= \{3; 3\} \oplus \{2; 1\}
\end{aligned} \tag{16.118}
$$

(the results being given in terms of reduced partitions, i.e., after subtracting $\{1; 1; 1\}$ from partitions whose third entry is one), so that

$$\{2\} \otimes \{2; 2\} = (x_1{}^2 - x_2) \otimes \{2; 2\} = \{0\} \oplus \{2; 1\} \oplus \{4; 2\} \tag{16.119}$$

which, in terms of tableaux, reads

$$\begin{array}{c}\square\square\\\square\square\end{array} \otimes \square\square = 1 \oplus \begin{array}{c}\square\square\\\square\end{array} \oplus \begin{array}{c}\square\square\square\square\\\square\square\end{array}$$

This way of calculating tensor products is sometimes referred to as the *Weyl determinant method*.

16.5.2. Level Truncation in the Determinant Method

We now show how this extends to fusion rules. Clearly, the only required modification is at the level of the Pieri formula. An integrable representation of $\widehat{su}(N)$ at level k is associated with a reduced tableau with at most k boxes in the first row. Its product with a fundamental weight gives tableaux with at most $k + 1$ boxes in the first row. The affine weight at level k associated with a reduced tableau with $k + 1$ boxes in the first row has its zeroth Dynkin label equal to

$$\lambda_0 = k - (\lambda_1 + \ldots + \lambda_{N-1}) = k - \ell_1 = -1 \tag{16.120}$$

According to the Kac-Walton formula, such weights are ignored (these are representations with vanishing quantum dimension or, equivalently, those representations whose characters vanish). Hence, the level-k truncation of the Pieri formula (16.112) is obtained simply by imposing the extra condition on the resulting partition $\{p_1; \ldots; p_n\}$:

$$p_1 - p_N \le k \tag{16.121}$$

Fusion rule calculations can thus be done exactly as in the Weyl determinant method described above, by taking into account, at each step, this very simple constraint. For instance, in the above example, it is easily seen that only the scalar representation survives the $k = 2$ truncation.

Formulated differently, the truncation amounts to setting equal to zero all reduced tableaux with $\ell_1 = k + 1$. This turns out to be equivalent to the condition

$$y_{k+1} = y_{k+2} = \ldots = y_{k+N-1} = 0 \tag{16.122}$$

This is a straightforward consequence of another determinant formula, essentially the transposed version of the above Giambelli formula:

$$\{\ell_1; \ldots; \ell_{N-1}\} = \det y_{\ell_i + j - i} = \det \begin{pmatrix} y_{\ell_1} & y_{\ell_1 + 1} & \cdots & y_{\ell_1 + N - 2} \\ y_{\ell_2 - 1} & y_{\ell_2} & & \cdots \\ \vdots & & & \\ y_{\ell_{N-1} - N + 2} & \cdots & & y_{\ell_{N-1}} \end{pmatrix}$$

$$(16.123)$$

It is then clear that all partitions with a fixed value of ℓ_1 vanish whenever $y_{\ell_1} = \cdots = y_{\ell_1 + N - 2} = 0$.

With $N = 3$, $k = 2$, the fusion constraints are

$$y_3 = \det \begin{pmatrix} x_1 & x_2 & 1 \\ 1 & x_1 & x_2 \\ 0 & 1 & x_1 \end{pmatrix} = x_1{}^3 - 2x_1 x_2 + 1 = 0$$

$$(16.124)$$

$$y_4 = \det \begin{pmatrix} x_1 & x_2 & 1 & 0 \\ 1 & x_1 & x_2 & 1 \\ 0 & 1 & x_1 & x_2 \\ 0 & 0 & 1 & x_1 \end{pmatrix} = x_1{}^4 - 3x_1{}^2 x_2 + x_2{}^2 + 2x_1 = 0$$

These imply in particular that

$$x_2^2 = x_1^2 x_2 - x_1 \qquad\qquad (16.125)$$

There are thus only 6 independent monomials:

$$1, \ x_1, \ x_1^2, \ x_2, \ x_1 x_2, \ x_1^2 x_2 \qquad\qquad (16.126)$$

corresponding to the 6 integrable representations of $\widehat{su}(3)_2$:

$$[2,0,0] = 1, \qquad\qquad [1,1,0] = x_1, \qquad\qquad [1,0,1] = x_2$$
$$[0,1,1] = x_1 x_2 - 1, \quad [0,2,0] = x_1^2 - x_2, \quad [0,0,2] = x_2^2 - x_1 \qquad (16.127)$$

Returning once more to the $\widehat{su}(3)_2$ product $\{2\} \times \{2; 2\}$, taking into account the above constraints, we find:

$$\{2\} \times \{2; 2\} = (x_1{}^2 - x_2)(x_2{}^2 - x_1)$$
$$= x_1^2 x_2^2 - x_1^3 - x_2^3 + x_1 x_2 \qquad\qquad (16.128)$$
$$= 1 + x_2(x_1^2 x_2 - x_1 - x_2^2) = 1$$

As another example, we consider how the $\widehat{su}(2)_k$ fusion rules fit into this framework. The irreducible representation with Dynkin label $\lambda_1 = n$ is associated with a single row of n boxes, that is, with y_n. In this case, the Pieri formula reduces to the recursion relation $(x = x_1)$

$$x y_n = y_{n+1} + y_{n-1} \qquad\qquad (16.129)$$

Given that there is only one variable, this is must also be the exact content of the Giambelli formula. The recurrence is fixed by the conditions

$$y_0 = 1 \qquad y_1 = x \qquad\qquad (16.130)$$

These relations define the Chebyshev polynomials of the second kind (cf. Eq. (8.101)):

$$y_n(2\cos\theta) = \frac{\sin(n+1)\theta}{\sin\theta} \qquad (16.131)$$

The fusion constraint is simply

$$y_{k+1}(x) = 0 \qquad (16.132)$$

The $\widehat{su}(2)_k$ truncated Pieri formula has a simple representation in terms of an A-type Dynkin diagram with $k+1$ nodes, as illustrated in Fig. 16.2.

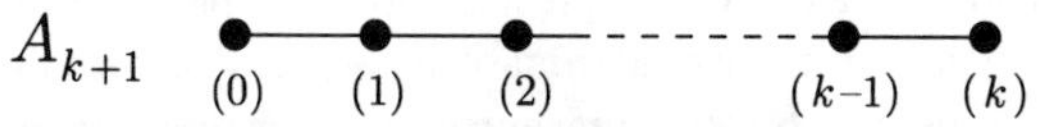

Figure 16.2. The A_{k+1} Dynkin diagram coding the $\widehat{su}(2)_k$ fusion rule by the fundamental representation (1). Vertices are associated with integrable representations of $\widehat{su}(2)_k$. The fusion $(1) \times (j)$ is given by the sum of those representation vertices that are directly linked to the vertex j.

The diagram encodes fusion by the fundamental representation (1). Its adjacency matrix is $G_{ij} = 2 - A_{ij} = \mathcal{N}_{1i}{}^j$. It encodes as well higher fusions by (j), $j > 1$ as follows. The corresponding matrix N_j with elements $[N_j]_{kl} = \mathcal{N}_{jk}{}^l$ (with $N_1 = G$) reads

$$N_j = y_j(N_1) \qquad (16.133)$$

16.5.3. The Constraint-Generating Function

As a computational tool, the reformulation of the simple constraint (16.121) in the form (16.122) may seem to be a step backward. However, the real theoretical significance of this second point of view lies in that the various constraints $y_j = 0$, $k+1 \le j \le N-1$, can be integrated to a potential V_{k+N}, the *fusion potential*, defined as

$$\boxed{y_{k+N-i} = (-1)^{i+1}\frac{\partial V_{k+N}}{\partial x_i}} \qquad (16.134)$$

Indeed, in our $\widehat{su}(3)_2$ example, the two constraints (16.124) correspond to the extrema of the potential:

$$V_5 = \frac{1}{5}x_1^5 - x_1^3 x_2 + x_1 x_2^2 + x_1^2 - x_2 \qquad (16.135)$$

In order to derive Eq. (16.134), we first perform a change of variable. Introduce the variables q_i, $i = 1, \ldots, N$ satisfying

$$\prod_{i=1}^{N} q_i = 1 \qquad (16.136)$$

and defined in terms of the x_ℓ by

$$x_\ell = \sum_{1 \le i_1 < i_2 \ldots < i_\ell \le N} q_{i_1} \ldots q_{i_\ell} \tag{16.137}$$

The variables q_i are simply the character variables introduced previously in Eq. (13.185) to define the Schur functions. We recall that q_i is related to the orthonormal vector ϵ_i by $q_i = e^{\epsilon_i}$. In terms of semistandard tableaux, where a box marked by i is associated with a factor ϵ_i, and hence a factor q_i, the left-hand side has the following interpretation. Filling a column of ℓ boxes with the increasing sequence of numbers $i_1, \ldots, i_\ell$, such that $1 \le i_1 < i_2 \ldots < i_\ell \le N$, in all possible ways, generates all the states in the representation x_ℓ. The constraint (16.136) simply means that the unique semistandard tableau associated with the Young tableau with a single column of N boxes represents the scalar representation. In this way, it is clear that the expression of the variables y_ℓ in terms of the q_i is

$$y_\ell = \sum_{1 \le i_1 \le \ldots \le i_\ell \le N} q_{i_1} \ldots q_{i_\ell} \tag{16.138}$$

corresponding to all semistandard tableaux obtained by filling a row of ℓ boxes with a nondecreasing sequence of integers ranging from 1 to N. The generating functions for these expansions are easily found to be

$$\sum_{i=0}^{N} x_i t^i = \prod_{i=1}^{N} (1 + q_i t) \tag{16.139}$$

$$\sum_{i=0}^{\infty} y_i t^i = \prod_{i=1}^{N} (1 - q_i t)^{-1} \tag{16.140}$$

We now define the potential V_m by

$$V_m = \frac{1}{m} \sum_{i=1}^{N} q_i^m \tag{16.141}$$

(corresponding to the sum of all possible semistandard tableaux of m boxes, each filled with the same number), whose generating function is

$$V(t) = \sum_{m=1}^{\infty} (-1)^{m-1} V_m t^m = \sum_{m=1}^{\infty} \sum_{i=1}^{N} \frac{1}{m} (-1)^{m-1} q_i^m t^m \tag{16.142}$$

Interchanging the summations and summing over m gives

$$V(t) = \sum_{i=1}^{N} \ln(1 + q_i t) = \ln \prod_{i=1}^{N} (1 + q_i t) = \ln \sum_{i=0}^{N} x_i t^i \tag{16.143}$$

The derivative of $V(t)$ with respect to q_i computed from Eq. (16.142) is

$$\frac{\partial V(t)}{\partial q_i} = \sum_{m=1}^{\infty} (-1)^{m-1} t^m q_i^{m-1} \tag{16.144}$$

whereas Eq. (16.143) leads to

$$\frac{\partial V(t)}{\partial q_i} = \left(\prod_{n=1}^{N}(1+q_n t)^{-1}\right)\frac{\partial}{\partial q_i}\left(\prod_{j=1}^{N}(1+q_j t)\right)$$

$$= \sum_{\ell=0}^{\infty}(-1)^\ell y_\ell t^\ell \sum_{j=0}^{N} t^j \frac{\partial x_j}{\partial q_i}$$

(16.145)

The compatibility of these two expressions forces the equality

$$(-1)^{m-1}q_i^{m-1} = \sum_{\ell+j=m}(-1)^\ell y_\ell \frac{\partial x_j}{\partial q_i}$$

(16.146)

Multiplying the result by $\partial q_i/\partial x_n$ and summing over i yields

$$\sum_i (-1)^{m-1}q_i^{m-1}\frac{\partial q_i}{\partial x_n} = \sum_{\ell+j=m}(-1)^\ell y_\ell \sum_i \frac{\partial q_i}{\partial x_n}\frac{\partial x_j}{\partial q_i}$$

$$= \sum_{\ell+j=m}(-1)^\ell y_\ell\, \delta_{n,j}$$

(16.147)

$$= (-1)^{m-n}y_{m-n}$$

Finally, the definition of the potential leads to Eq. (16.134).

We note that the equivalence between Eq. (16.142) and Eq. (16.143) provides an explicit expression of V_m as a function of the variables x_i:

$$V_m = \frac{(-1)^{m-1}}{m!}\frac{d^m}{dt^m}\ln\left(\sum_{i=0}^{N}x_i t^i\right)\Big|_{t=0}$$

(16.148)

For $su(2)$ and $su(3)$, Eq. (16.148) reads respectively

$$V_{k+2} = \frac{(-1)^{k+1}}{(k+2)!}\frac{d^{k+2}}{dt^{k+2}}\ln(1+xt+t^2)|_{t=0}$$

$$V_{k+3} = \frac{(-1)^k}{(k+3)!}\frac{d^{k+3}}{dt^{k+3}}\ln(1+x_1 t+x_2 t^2+t^3)|_{t=0}$$

(16.149)

We detail the two simplest $\widehat{su}(2)_k$ examples. For $k=1$, with $[0,1]=x$, the only nontrivial fusion rule is $x^2=1$; the constraint is thus

$$x^2 - 1 = 0$$

(16.150)

which is the extremum of the potential

$$V_3 = \frac{1}{3}x^3 - x$$

(16.151)

For $k=2$, with $[1,1]=x$ and $[0,2]=z$, the fusion rules take the form

$$x^2 = 1+z, \qquad xz = x, \qquad z^2 = 1$$

(16.152)

Hence, $z = x^2 - 1$, and the fusion constraint is

$$x^3 = 2x \qquad (16.153)$$

(needed to reproduce $zx = x$), so that the associated potential is

$$V_4 = \frac{1}{4}x^4 - x^2 \qquad (16.154)$$

which is in agreement with Eq. (16.148), up to an irrelevant additive constant.

   In summary, we have found that the precise level of truncation on tensor-product coefficients is captured in a set of r constraints, which can be expressed as the vanishing of the potential (16.141) with respect to its r independent variables.

   It should be stressed that the fusion potential is not unique. To see this, let us return to the $\widehat{su}(3)_2$ example. Eliminating the variable x_2 from the two constraints (16.124) leaves the single condition

$$x_1^6 - 4x_1^3 - 1 = 0 \qquad (16.155)$$

associated with the potential

$$V = \frac{1}{7}x_1^7 - x_1^4 - x_1 \qquad (16.156)$$

Hence, the $\widehat{su}(3)_2$ fusion constraints can be described by two different fusion potentials. This situation is not exceptional: in fact all $\widehat{su}(3)$ fusion rules can be reformulated in terms of a single-variable potential (cf. Ex. 16.11). A generic sufficient condition for the existence of a one-variable description is the nondegeneracy of all the eigenvalues of the fusion matrix of at least one primary field. To finish with our example, notice that with the condition (16.155), $x_2 = (x_1^3 + 1)/2x_1$ can be rewritten as

$$x_2 = \frac{1}{2}(x_1^5 - 3x_1^2) \qquad (16.157)$$

The expression for the different integrable representations of $\widehat{su}(3)_2$, in terms of the single variable x_1, is then

$$[2,0,0] = 1, \qquad [1,1,0] = x_1, \qquad [1,0,1] = \frac{1}{2}(x_1^5 - 3x_1^2)$$

$$[0,1,1] = \frac{1}{2}(x_1^3 - 1), \quad [0,2,0] = \frac{1}{2}(5x_1^2 - x_1^5), \quad [0,0,2] = \frac{1}{2}(x_1^4 - 3x_1)$$

$$\qquad (16.158)$$

§16.6. Level-Rank Duality

In this section, we introduce a remarkable level-rank duality relation between WZW models associated with classical Lie algebras. For simplicity, the discussion is restricted to $su(N)$ algebras.

   A first hint for the existence of a relation between the $\widehat{su}(N)_k$ and $\widehat{su}(k)_N$ models is provided by the following observation. An integrable representation $\hat{\lambda}$

of $\widehat{su}(N)_k$ is associated with a Young tableau whose first row contains at most k boxes. (We recall that the Dynkin label λ_i gives the number of columns of i boxes in the tableau. Therefore the sum of all finite Dynkin labels gives the number of boxes in the first row. For an integrable representation, $\sum_{i=1}^{N-1} \lambda_i \leq k$. Here the tableau is assumed to be reduced, i.e., there are no columns of N boxes.) The transposed tableau, obtained by interchanging rows and columns, has at most $N-1$ boxes in the first row and no columns with more than k boxes. It can thus be associated with an integrable weight of $\widehat{su}(k)_{N-1}$, and therefore of $\widehat{su}(k)_N$. A simple counting argument shows that the number of integrable representations in $\widehat{su}(N)_k$ and $\widehat{su}(k)_N$ are

$$\widehat{su}(N)_k : \quad \frac{(k+N-1)!}{k!(N-1)!} \,, \qquad \widehat{su}(k)_N : \quad \frac{(k+N-1)!}{N!(k-1)!} \qquad (16.159)$$

Because these numbers are distinct, the relation between integrable weights of the two models cannot be one-to-one. However, if the above numbers are divided, respectively, by N and k, they become equal. This division amounts to factorizing the algebras by the action of their respective centers, which is equivalent to considering only equivalence classes of the outer-automorphism groups: $\hat{\lambda} \sim \hat{\mu}$ if $\hat{\mu} = a^j \hat{\lambda}$ for some j, with a standing for the basic outer automorphism. The numerical coincidence just found suggests that the objects that can be made in one-to-one correspondence are the orbits of the outer-automorphism groups. The following direct argument shows that this is indeed so.

We construct a circle divided into $N+k$ equal parts by means of dashed lines. We consider an integrable weight $\hat{\lambda} \in \widehat{su}(N)_k$. Starting anywhere on the circle, fill the dashed lines separating $\lambda_0+1, \lambda_1+1, \cdots, \lambda_{N-1}+1$ pieces, in a clockwise ordering. That the initial position on the circle is irrelevant shows that this splitting of the circle describes a full $\mathbb{Z}_N$ orbit rather that a single weight. Now, the dual splitting, namely the sequence of slices separated by dashed lines and read counterclockwise, represents the affine Dynkin labels of a weight $\hat{\mu} + \hat{\rho}$, with $\hat{\mu} \in \widehat{su}(k)_N$ lying in the orbit of the $\widehat{su}(k)_N$ affine extension of λ^t.

As is an illustrative example, let $N = 5, k = 3$ and

$$\lambda = (1,0,1,0) \longleftrightarrow \quad \boxed{} \quad \Rightarrow \quad \hat{\lambda} + \hat{\rho} = [2,2,1,2,1] \qquad (16.160)$$

The corresponding splitting of the circle is displayed in Fig. 16.3; the dual splitting is seen to be $[3,2,3]$:

$$\hat{\mu} + \hat{\rho} = [3,2,3] \quad \Rightarrow \quad \hat{\mu} = [2,1,2] \qquad (16.161)$$

The transpose of λ is

$$\lambda^t = \boxed{} \longleftrightarrow (2,1) \qquad (16.162)$$

whose affine $\widehat{su}(3)_5$ extension $[2,2,1]$ belongs to the orbits

$$\{[2,2,1], [1,2,2], [2,1,2]\} \qquad (16.163)$$

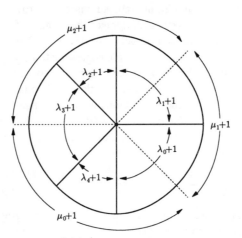

Figure 16.3. The shifted affine $\widehat{su}(5)_3$ weights are the number of slices enclosed within the full lines. Reading the diagram from its lowest point in a clockwise direction, we obtain $[2, 2, 1, 2, 1]$. Read in the opposite direction, the diagram gives the shifted affine $\widehat{su}(3)_5$ weights as the number of slices enclosed within the dashed lines; starting from the leftmost one, we have $[3, 2, 3]$.

which, indeed, also contains $[2, 1, 2]$.

As a side remark, we point out that the action of the outer automorphism has a direct transcription in terms of Young tableaux. The action of a—the basic outer automorphism of $\widehat{su}(N)$—on a reduced tableau associated with an affine weight at level k, amounts to adding on the top of the tableau a row of k boxes. Indeed, the Dynkin labels λ_i are related to the partition's entries by

$$\lambda_i = \ell_i - \ell_{i+1} \tag{16.164}$$

where ℓ_i gives the length of the i-th row:

$$\ell_i = \lambda_1 + \cdots + \lambda_{N-1} \tag{16.165}$$

We denote by ℓ_i^a the partition entries of the finite part of the affine weight $a\hat{\lambda}$,

$$a\hat{\lambda} = a[k - \sum_{i=1}^{N-1} \lambda_i, \lambda_1, \cdots, \lambda_{N-1}] = [\lambda_{N-1}, k - \sum_{i=1}^{N-1} \lambda_i, \lambda_1, \cdots, \lambda_{N-2}] \tag{16.166}$$

Starting with a reduced tableau, with $\ell_N = 0$, we obtain

$$\ell_1^a = k, \qquad \ell_{i+1}^a = \ell_i, \quad \text{for} \quad i \geq 1 \tag{16.167}$$

For repeated applications of a, the tableau must be reduced after each step. In this way, $\mathcal{O}(\widehat{su}(N))$ orbits of tableaux are easily constructed.

That outer-automorphism orbits are one-to-one in theories related by level-rank duality is interesting. But more interesting is that fusion coefficients can be made equal. This relation is rooted in the $\mathcal{S}$ matrix duality

$$\mathcal{S}_{\hat{\lambda}\hat{\mu}} = \sqrt{\frac{k}{N}}\, e^{\{2\pi i |\lambda||\mu|/Nk\}}\, \bar{\mathcal{S}}_{\hat{\lambda}'\hat{\mu}'} \tag{16.168}$$

for

$$\hat{\lambda}, \hat{\mu} \in \widehat{su}(N)_k, \quad \hat{\lambda}', \hat{\mu}' \in \widehat{su}(k)_N \tag{16.169}$$

and $|\lambda|$ stands for the total number of boxes in the tableau representing λ. The proof of this relation, being somewhat technical, is omitted.

In the expression for the fusion rules of the $\widehat{su}(N)_k$ theory, the sum over all in-tegrable representations can be decomposed into orbits of the outer-automorphism groups as follows:

$$\mathcal{N}_{\hat{\lambda}\hat{\mu}}^{(k)\,\hat{\nu}} = \sum_{\hat{\sigma}\in P_+^k} \frac{\mathcal{S}_{\hat{\lambda}\hat{\sigma}}\mathcal{S}_{\hat{\mu}\hat{\sigma}}\bar{\mathcal{S}}_{\hat{\nu}\hat{\sigma}}}{\mathcal{S}_{0\hat{\sigma}}}$$

$$= \sum_{\hat{\sigma}\in P_+^k/\mathbb{Z}_N} \frac{1}{m_N(\hat{\sigma})} \sum_{\ell=0}^{N-1} \frac{\mathcal{S}_{\hat{\lambda},a^\ell(\hat{\sigma})}\mathcal{S}_{\hat{\mu},a^\ell(\hat{\sigma})}\bar{\mathcal{S}}_{\hat{\nu},a^\ell(\hat{\sigma})}}{\mathcal{S}_{0,a^\ell(\hat{\sigma})}} \tag{16.170}$$

Here $m_N(\hat{\sigma})$ denotes the multiplicity of $\hat{\sigma}$ in the orbit $\{\hat{\sigma}, a\hat{\sigma}, \cdots, a^{N-1}\hat{\sigma}\}$ (necessarily a divisor of N). Observe that

$$\mathcal{S}_{a^\ell(\hat{\sigma}),\hat{\mu}} = \mathcal{S}_{\hat{\sigma}\hat{\mu}}e^{-2\pi i\ell(a\hat{\omega}_0,\mu)} = \mathcal{S}_{\hat{\sigma}\hat{\mu}}e^{2\pi i\ell|\mu|/N} \tag{16.171}$$

Indeed

$$(a\hat{\omega}_0,\mu) = (\omega_1,\mu) = \sum_{j=1}^{N-1} \mu_j \frac{(N-j)}{N} \tag{16.172}$$

and with $\mu_j = m_j - m_{j+1}$ (i.e., $\mu = \{m_1; \cdots; m_{N-1}\}$), this becomes

$$(\omega_1,\mu) = \sum_{j=1}^{N-1} \mu_j - \frac{1}{N}\sum_{j=1}^{N-1} m_j = -\frac{1}{N}|\mu| \bmod 1 \tag{16.173}$$

The expression for fusion coefficients then reduces to

$$\mathcal{N}_{\hat{\lambda}\hat{\mu}}^{(k)\,\hat{\nu}} = \sum_{\hat{\sigma}\in P_+^k/\mathbb{Z}_N} \frac{1}{m_N(\hat{\sigma})} \frac{\mathcal{S}_{\hat{\lambda}\hat{\sigma}}\mathcal{S}_{\hat{\mu},\hat{\sigma}}\bar{\mathcal{S}}_{\hat{\nu},\hat{\sigma}}}{\mathcal{S}_{0,\hat{\sigma}}} \sum_{\ell=0}^{N-1} e^{(2\pi i\ell/N)(|\lambda|+|\mu|-|\nu|)} \tag{16.174}$$

If the corresponding tensor-product coefficient is nonzero, an immediate consequence of the Littlewood-Richardson rule is that

$$|\lambda| + |\mu| - |\nu| = 0 \quad \bmod N \tag{16.175}$$

so the last sum contributes to a factor N. Since there is a one-to-one correspondence between orbits of $\widehat{su}(N)_k$ and $\widehat{su}(k)_N$, we can write

$$\sum_{\hat{\sigma}\in P_+^k/\mathbb{Z}_N} = \sum_{\hat{\sigma}'\in P_+^N/\mathbb{Z}_k} \tag{16.176}$$

The substitution of Eq. (16.168) into the expression for fusion coefficients leads to

$$
\begin{aligned}
\mathcal{N}_{\hat{\lambda}\hat{\mu}}^{(k)\,\hat{\nu}} &= \sum_{\hat{\sigma}' \in P_+^N / \mathbb{Z}_k} \frac{k}{m_N(\hat{\sigma})}\, e^{(2\pi i/Nk)|\sigma|(|\lambda|+|\mu|-|\nu|)} \frac{\bar{S}_{\hat{\lambda}'\hat{\sigma}'}\bar{S}_{\hat{\mu}'\hat{\sigma}'}S_{\hat{\nu}'\hat{\sigma}'}}{\bar{S}_{0\hat{\sigma}'}} \\[2mm]
&= \sum_{\hat{\sigma}' \in P_+^N / \mathbb{Z}_k} \frac{1}{m_k(\hat{\sigma}')}\, e^{(2\pi i/Nk)|\sigma|(|\lambda|+|\mu|-|\nu|)} \\[2mm]
&\qquad \sum_{\ell=0}^{k-1} \frac{\bar{S}_{\hat{\lambda}',\tilde{a}^\ell(\hat{\sigma}')}\bar{S}_{\hat{\mu}',\tilde{a}^\ell(\hat{\sigma}')}S_{\hat{\nu}',\tilde{a}^\ell(\hat{\sigma}')}}{\bar{S}_{0,\tilde{a}^\ell(\hat{\sigma}')}} \\[2mm]
&= \sum_{\hat{\sigma}' \in P_+^N} \frac{\bar{S}_{\hat{\lambda}'\hat{\sigma}'}\bar{S}_{\hat{\mu}'\hat{\sigma}'}S_{\hat{\nu}'\hat{\sigma}'}}{\bar{S}_{0\hat{\sigma}'}} e^{(2\pi i/Nk)|\sigma|(|\lambda|+|\mu|-|\nu|)}
\end{aligned}
\tag{16.177}
$$

where $\tilde{a}$ denotes the basic outer automorphism of $\widehat{su}(k)$. In the second equality, we used the identity $m_N(\hat{\sigma}) = m_k(\hat{\sigma}')$ (see Ex. 16.14) and a reorganization of the result into orbits of the $\mathcal{O}(\widehat{su}(k))$ outer-automorphism group. The extra phase factor can be absorbed as follows

$$
S_{\hat{\nu}'\hat{\sigma}'}\, e^{(2\pi i/Nk)|\sigma|(|\lambda|+|\mu|-|\nu|)} = S_{\tilde{a}^p(\hat{\nu}')\hat{\sigma}'}
\tag{16.178}
$$

with

$$
p = \frac{1}{N}(|\lambda| + |\mu| - |\nu|)
\tag{16.179}
$$

Finally, the reality of fusion coefficients allows us to write

$$
\boxed{\mathcal{N}_{\hat{\lambda}\hat{\mu}}^{(k)\,\hat{\nu}} = \mathcal{N}_{\hat{\lambda}'\hat{\mu}'}^{(N)\,\tilde{a}^p\hat{\nu}'}}
\tag{16.180}
$$

As a simple illustration of this formula, consider the case $N = 3, k = 5$, and the fusion

$$
[1,3,1] \times [2,1,2] = [4,1,0]+[3,0,2]+[1,4,0]+2\,[2,2,1]+[1,1,3]+[0,3,2]
\tag{16.181}
$$

whose tableau representation reads:

$$
\tag{16.182}
$$

The transposed product, with appropriate action of $\tilde{a}$ on the right-hand side terms (the required powers of $\tilde{a}$ being respectively $3, 2, 2, 2, 1, 1$), is

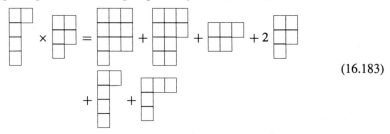

$$(16.183)$$

or equivalently

$$[1, 1, 0, 0, 1] \times [1, 0, 1, 1, 0] = [0, 0, 0, 2, 1] + [0, 0, 1, 0, 2] + [0, 1, 2, 0, 0]$$
$$+ 2\,[1, 0, 1, 1, 0] + [1, 1, 0, 0, 1] + [0, 2, 0, 1, 0]$$
$$(16.184)$$

The relation (16.180) has the following interesting consequences:

(i) The fusion coefficients for $\widehat{su}(N)_2$, for any value of N, can be only 0 or 1, as for $\widehat{su}(2)$.

(ii) The threshold levels for all $\widehat{su}(N)_3$ fusion rules are always nondegenerate.

Appendix 16.A. Fusion Elementary Couplings in $\widehat{su}(N)$

In order to introduce the idea of fusion elementary couplings, we consider first $su(2)$ tensor products. In that case, the elementary couplings are

$$E_1 = (1)(1)(0), \quad E_2 = (1)(0)(1), \quad E_3 = (0)(1)(1) \qquad (16.185)$$

where the notation $(1)(1)(0)$ stands for $(1) \otimes (1) \otimes (0) \supset (0)$. Notice that elementary couplings that differ by a permutation of the weights are considered as distinct. The $su(2)$ elementary couplings are just the various permutations of the product of the fundamental representation with the scalar one. In products of the E_i's, the corresponding Dynkin labels are added. The decomposition of a generic coupling into elementary ones is then

$$\lambda \otimes \mu \otimes \nu = E_1^{\frac{1}{2}(\lambda_1 + \mu_1 - \nu_1)}\, E_2^{\frac{1}{2}(\nu_1 + \lambda_1 - \mu_1)}\, E_3^{\frac{1}{2}(\mu_1 + \nu_1 - \lambda_1)} \qquad (16.186)$$

Of course, for the coupling to be allowed, each exponent must be a positive integer. With λ_1 and μ_1 fixed, this sole condition determines the possible values of ν_1.

To extend this construction to fusion rules, we need the following assumption: the threshold level for the coupling is the sum of the threshold levels in the decomposition into elementary couplings.[6] Translated into equations, this means that the

---

[6] This must still be regarded as a conjecture; it has been proved only for $\widehat{su}(2)$ and $\widehat{su}(3)$. The conjecture is expected to apply to affine Lie algebra. Although there is no analogue of the Berenstein-Zelevinsky triangles (used below) for algebras other than $su(N)$, elementary couplings can be defined for any algebra.

coupling

$$\{\lambda \otimes \mu \otimes \nu\}^{(i)} = \prod_\ell E_\ell^{e_\ell} \tag{16.187}$$

(where (i) is a degeneracy label) has threshold level

$$k_0^{(i)}(\lambda \otimes \mu \otimes \nu) = \sum_\ell e_\ell k_0(E_\ell) \tag{16.188}$$

where $k_0(E_\ell)$ is the threshold level for the elementary coupling E_ℓ.

Applying this to $su(2)$, we must first calculate the value of $k_0(E_\ell)$, using for instance the Kac-Walton formula. This gives

$$k_0(E_\ell) = 1 \quad \text{for} \quad \ell = 1, 2, 3 \tag{16.189}$$

From Eqs. (16.188) and (16.186), it then follows that

$$k_0(\lambda \otimes \mu \otimes \nu) = \frac{1}{2}(\lambda_1 + \mu_1 + \nu_1) \tag{16.190}$$

which, of course, reproduces Eq. (16.48).

The Berenstein-Zelevinsky triangles corresponding to elementary couplings, for which the counterclockwise ordering λ, μ, ν is assumed, are respectively

$$
\begin{array}{ccccccc}
& 0 & & & 1 & & & 0 & \\
1 & & 0 & & 0 & & 0 & & 0 & & 1
\end{array}
\tag{16.191}
$$

A decomposition into a product of elementary couplings is equivalent to a decomposition into a sum of the corresponding basic triangles. In terms of the triangle data (13.221), the formula (16.190) reads

$$k_0(\lambda \otimes \mu \otimes \nu) = m_{12} + n_{12} + l_{12} \tag{16.192}$$

We consider now the $su(3)$ case, for which there are eight elementary couplings:

$E_1 = (1,0)(0,1)(0,0)$ $\qquad$ $E_3 = (0,1)(0,0)(1,0)$ $\qquad$ $E_5 = (0,0)(1,0)(0,1)$

$$
\begin{array}{cccc}
& 0 & & \\
& 1 & 0 & \\
0 & & & 0 \\
0 & 0 & 1 & 0
\end{array}
\qquad
\begin{array}{cccc}
& 0 & & \\
& 0 & 0 & \\
1 & & & 1 \\
0 & 0 & 0 & 0
\end{array}
\qquad
\begin{array}{cccc}
& 0 & & \\
& 0 & 1 & \\
0 & & & 0 \\
0 & 1 & 0 & 0
\end{array}
$$

$E_2 = (1,0)(0,0)(0,1)$ $\qquad$ $E_4 = (0,1)(1,0)(0,0)$ $\qquad$ $E_6 = (0,0)(0,1)(1,0)$

$$
\begin{array}{cccc}
& 1 & & \\
& 0 & 0 & \\
0 & & & 0 \\
0 & 0 & 0 & 0
\end{array}
\qquad
\begin{array}{cccc}
& 0 & & \\
& 0 & 0 & \\
0 & & & 0 \\
1 & 0 & 0 & 0
\end{array}
\qquad
\begin{array}{cccc}
& 0 & & \\
& 0 & 0 & \\
0 & & & 0 \\
0 & 0 & 0 & 1
\end{array}
$$

$$E_7 = (1,0)(1,0)(1,0) \qquad E_8 = (0,1)(0,1)(0,1)$$

$$
\begin{array}{cccc}
 & 0 & & \\
 & 1 & 0 & \\
0 & & 1 & \\
0 & 1 & 0 & 0
\end{array}
\qquad
\begin{array}{cccc}
 & 0 & & \\
 & 0 & 1 & \\
 1 & & 0 & \\
0 & 0 & 1 & 0
\end{array}
$$

The direct extension of the $su(2)$ approach faces an immediate problem in that the decomposition of a generic coupling into elementary ones is no longer unique. Indeed, the two products $E_1E_3E_5$ and E_7E_8 are equivalent since they yield exactly the same triangle:

$$E_1E_3E_5 = E_7E_8$$

$$
\begin{array}{cccc}
 & 0 & & \\
 1 & & 1 & \\
 1 & & 1 & \\
0 & 1 & 1 & 0
\end{array}
\tag{16.193}
$$

For $su(3)$ this is the only redundancy. In order to proceed, it must be eliminated by forbidding either $E_1E_3E_5$ or E_7E_8. For tensor products, whichever is eliminated is immaterial. However, for fusion rules, the situation is quite different. First, the threshold level of all the elementary couplings is easily found to be equal to one, exactly as in the $\widehat{su}(2)$ case. Then, according to the above assumption, the two products $E_1E_3E_5$ and E_7E_8 have different threshold levels

$$k_0(E_7E_8) = 2, \qquad k_0(E_1E_3E_5) = 3 \tag{16.194}$$

We must then determine which one must be eliminated in order for the decomposition into elementary couplings to reproduce the correct threshold level. These two products give two equivalent descriptions of a single coupling describing the tensor product $(1, 1)(1, 1)(1, 1)$. Using the algorithm described in Sect. 16.2.2, it is simple to check that

$$(1,1) \otimes (1,1) = (0,0)_2 \oplus 2(1,1)_{2,3} \oplus (3,0)_3 \oplus (0,3)_3 \oplus (2,2)_4 \tag{16.195}$$

Hence, the triple product $(1,1)(1,1)(1,1)$ has multiplicity 2, with threshold levels 2 and 3. The representation of these two couplings in terms of Berenstein-Zelevinsky triangles is

$$
\begin{array}{cccc}
 & 1 & & \\
 0 & & 0 & \\
 0 & & 0 & \\
 1 & 0 & 0 & 1
\end{array}
\qquad
\begin{array}{cccc}
 & 0 & & \\
 1 & & 1 & \\
 1 & & 1 & \\
 0 & 1 & 1 & 0
\end{array}
\tag{16.196}
$$

The first triangle is unambiguously associated with the product $E_2E_4E_6$, and consequently its threshold level is $k_0 = 3$. The second triangle must then necessarily be decomposable into a product of two elementary couplings in order to have $k_0 = 2$. Hence, it must correspond to E_7E_8, which means that the product $E_1E_3E_5$ must be forbidden. This eliminates all possible redundancies, and as such the decomposition of any coupling into elementary ones is unique. The decomposition is either of the form

$$E_2^a E_3^b E_4^c E_5^d E_6^e E_7^f E_8^g \tag{16.197}$$

where $a, \cdots, g$ are nonnegative integers, or one of its two rotated versions, obtained by the replacements $(E_3, E_5) \to (E_5, E_1) \to (E_1, E_3)$. Whichever form is realized is uniquely fixed by the values of the weights under consideration. The threshold level of the above decomposition is simply the sum of all the exponents

$$k_0 = a + b + c + d + e + f + g \tag{16.198}$$

We let this decomposition describe a particular coupling of the triple product $(\lambda_1, \lambda_2) \otimes (\mu_1, \mu_2) \otimes (\nu_1, \nu_2)$. From the expressions for the elementary couplings in terms of the Dynkin labels, it follows that

$$\begin{array}{lll} a + f = \lambda_1 & b + c + g = \lambda_2 & c + d + f = \mu_1 \\ e + g = \mu_2 & b + e + f = \nu_1 & a + d + g = \nu_2 \end{array} \tag{16.199}$$

Having six relations for seven parameters, we can express everything in terms of one parameter, say c. In particular, k_0 can be rewritten as

$$k_0 = c + \nu_1 + \nu_2 \tag{16.200}$$

However, c is not a completely free parameter. It is constrained by a set of inequalities, consequences of Eq. (16.199) and the positivity requirement imposed on all the parameters. Let the result be of the form

$$c_{min} \le c \le c_{max} \tag{16.201}$$

Clearly, any integer satisfying these bounds leads to an allowed decomposition. This immediately implies that the multiplicity of the tensor product is $c_{max} - c_{min} + 1$. With c fixed, the decomposition and the associated triangle are uniquely specified. Denote the corresponding triangle as Δ_c. All triangles associated with the triple product under consideration are given by

$$\Delta_{c_{min}} + n\Omega, \qquad 0 \le n \le c_{max} - c_{min} \tag{16.202}$$

where

$$\Omega = \begin{array}{ccccc} & & 1 & & \\ & -1 & & -1 & \\ & & -1 & & -1 \\ 1 & & -1 & & -1 & 1 \end{array} \tag{16.203}$$

Since k_0 is a linear function of c, the threshold level of the different couplings reads

$$\{k_0\}^{(i)} = (c_{min} + \nu_1 + \nu_2, \; c_{min} + \nu_1 + \nu_2 + 1, \cdots, \; c_{max} + \nu_1 + \nu_2) \tag{16.204}$$

Evaluating explicitly the range of values of c, and considering the two other possible decompositions, leads to formulæ (16.101)–(16.106). Filling the gaps in the argument is left as an exercise.

Formulated in general terms, the assumption underlying this construction is the existence of a set of forbidden couplings ensuring that the threshold levels in the decomposition into elementary couplings can simply be added. This assumption has been proved for $\widehat{su}(3)$, which places the above results on a firm basis.

Finally, it should be pointed out that the threshold level of a particular coupling is not an observable. The only observable is the full set of values of $k_0^{(i)}$, or equivalently, the fusion coefficients. Here we used a particular basis for the couplings, namely the Berenstein-Zelevinsky triangles. Any other basis will lead to the same set of values of $k_0^{(i)}$. However, two different descriptions of the same coupling, that is, two representations in different bases, may yield a different value for a specific threshold level $k_0^{(i)}$ (i.e., the results in a given basis can be permuted in another one).

Exercises

16.1 *Applications of the Verlinde formula*
Use the Verlinde formula to calculate the fusion rules for:
a) $\widehat{so}(2N)_1$;
b) $(\widehat{F}_4)_1$.

16.2 *A simple derivation of the $\widehat{su}(2)_k$ fusion rules using the outer automorphism*
Show that the $\widehat{su}(2)_k$ fusion rules can be obtained from the intersection of the tensor product of $\lambda_1 \omega_1 \otimes \mu_1 \omega_1$ and $(k - \lambda_1)\omega_1 \otimes (k - \mu_1)\omega_1$.[7]

16.3 *Aspects of $\widehat{su}(3)_k$ fusion rules*
Prove that the affine extension of any finite weight ν occurring in a tensor product of two weights whose extensions at some level k are integrable is itself always integrable at level $2k$. Use this to identify the $\widehat{su}(3)$ affine chambers, in terms of elements of the affine Weyl group, which can contribute to fusion coefficients, when the latter are expressed in terms of tensor-product coefficients.

16.4 *Multiple fusions and Fibonacci numbers*
For $\widehat{su}(2)_3$, show that the coefficients of the decomposition of the multiple product

$$[1, 2]^{\times \ell} = [1, 2] \times [1, 2] \times \cdots \times [1, 2]$$

in terms of the integrable $\widehat{su}(2)_3$ representations, are given by the Fibonacci numbers, defined by the recursion relation

$$a_{\ell+1} = a_{\ell-1} + a_\ell \qquad a_0 = a_1 = 1$$

---

[7] It is not generically true that intersecting the results obtained by all possible actions of the outer-automorphism group provides the correct fusion coefficients. But this is so for $\widehat{su}(2)$ and $\widehat{su}(3)$. A counterexample, for $\widehat{su}(4)_{16}$, is

$$[4, 4, 4, 4] \times [4, 4, 4, 4] \supset 45 \, [4, 4, 4, 4]$$

This is clearly invariant under the action of $\mathcal{O}(\hat{g})$, whereas

$$(4, 4, 4) \otimes (4, 4, 4) \supset 50 \, (4, 4, 4)$$

Find the analogous decomposition of [2, 2] at level 4.

16.5 $\widehat{su}(4)$ *fusion rules by combinatorial methods*
Use the Littlewood-Richardson rule and strip alterations (cf. Sect.16.2.4) to calculate the
fusion rules at levels 3 and 4 associated with the $su(4)$ tensor product $(1, 1, 1) \otimes (1, 1, 1)$.
Result:

$$k = 3 : \quad (0,0,0) \oplus (0,1,2) \oplus (0,2,0) \oplus 2\,(1,0,1) \oplus (2,1,0)$$

$$k = 4 : \quad (0,0,0) \oplus (0,0,4) \oplus 3\,(0,1,2) \oplus 2\,(0,2,0) \oplus (0,4,0) \oplus 3\,(1,0,1)$$
$$\oplus\, 3\,(1,2,1) \oplus 2\,(2,0,2) \oplus 3\,(2,1,0) \oplus (4,0,0)$$

16.6 *Threshold level calculations*

a) With the algorithm underlying the Kac-Walton formula, calculate the threshold level of
the various terms occurring in the $su(3)$ tensor products

$$(3, 1) \otimes (3, 1)$$

$$(2, 2) \otimes (2, 2)$$

and check the results by means of the formulæ(16.101)–(16.106).

b) Find the fusion coefficients (or equivalently, the threshold levels) associated with the
$sp(4)$ tensor product

$$(1, 1) \otimes (1, 1) = (0,0) \oplus (0,1) \oplus (0,2) \oplus (0,3) \oplus 2(2,0) \oplus 2(2,1) \oplus (2,2) \oplus (4,0)$$

c) Same as (**b**) for the G_2 tensor product

$$(0, 1) \otimes (0, 1) = (0,0) \oplus (1,0) \oplus (0,1) \oplus (0,2)$$

16.7 *Fusions and quantum dimensions*

a) Show that simple currents have quantum dimension one, directly from their definition as
fields that act as permutations in fusion rules.

b) From the dimension sum rule for fusions (cf. Eq. (16.60)), argue that the smallest possible
quantum dimension, apart from one, is $\sqrt{2}$.

c) Calculate the quantum dimension of the primary fields in the $\widehat{E}_8$ model at level two, and
show how they fix the fusion rules. Identify the simple current, which is not of the outer
automorphism type. Compare the fusion rules with those of the Ising model (cf. Sect. 7.4).

16.8 *Handle operators*
The *handle operator* is defined in terms of the fusion matrices by

$$R = \sum_{\hat{\nu} \in P^k_+} N^{\hat{\nu}} \, \mathrm{Tr}(N_{\hat{\nu}})$$

The component $R^{\hat{\mu}}_{\hat{\lambda}}$ has the following graphical representation:

The inverse of R turns out to have a simple Lie-algebraic expression:

$$R^{-1} = \frac{(-1)^{|\Delta_+|}}{|P/Q^\vee|(k+g)^r} \, D^2_\rho \Big|_{\chi_\lambda \to N^{(k)}_\lambda} \qquad (16.205)$$

where as before (cf. Eq. (13.165)) D_ρ is defined by

$$D_\rho = \sum_{w \in W} e^{w\rho}$$

and the meaning of $\chi_\lambda \to N_\lambda^{(k)}$ is the following: D_ρ^2 is first reorganized in terms of the irreducible formal characters and each χ_λ is then replaced by the corresponding fusion matrix at level k. (This general result is established in the last part of the exercise.) For instance, for $su(2)$,

$$D_\rho^2 = (e^{(1)} - e^{(-1)})^2 = (e^{(2)} - 2e^{(0)} + e^{(-2)})$$
$$= \chi_{(2)} - 3\chi_{(0)}$$

so that R^{-1} reads

$$R_{\widehat{su}(2)_k}^{-1} = \frac{1}{2(k+2)}\{3N_{(0)}^{(k)} - N_{(2)}^{(k)}\}$$

For a fixed value of k, if the affine extension of the weight λ is not integrable, $N_\lambda^{(k)}$ is replaced by $\epsilon(w)N_{w\cdot\lambda}^{(k)}$ for the particular element of the affine Weyl group w that makes $w \cdot \hat{\lambda}$ integrable, and it is ignored if this is not possible. For instance, at $k = 1$, $N_{(2)}^{(k=1)} \equiv N_{[-1,2]}$ must be dropped, so that

$$R_{\widehat{su}(2)_1}^{-1} = \frac{1}{2}N_{(0)}$$

The result is easily verified: the $\widehat{su}(2)_1$ fusion matrices are

$$N_{(0)} = \begin{pmatrix} 1 & 0 \\ 0 & 1 \end{pmatrix}, \quad N_{(1)} = \begin{pmatrix} 0 & 1 \\ 1 & 0 \end{pmatrix}$$

so that $R = 2N_{(0)}$, hence $R^{-1} = N_{(0)}/2_,$.

a) Prove the general form of the $\widehat{su}(2)_k$ expression (16.205) for R^{-1}. Hint: First, observe that $\text{Tr}\, N_{(n)}$ vanishes if n is odd. For n even, show that $\text{Tr}\, N_{(n)} = k + 1 - n$, so that

$$R = \sum_{\substack{0 \le n \le k \\ n \text{ even}}} (k + 1 - n)N_{(n)}$$

Then, using the $\widehat{su}(2)_k$ fusion rules, show directly that

$$\frac{1}{2(k+2)}\{3N_{(0)} - N_{(2)}\} \sum_{\substack{0 \le n \le k \\ n \text{ even}}} (k + 1 - n)N_{(n)} = 1$$

b) Check that for $\widehat{su}(3)_k$, the expression (16.205) for R^{-1} reduces to

$$R_{\widehat{su}(3)_k}^{-1} = \frac{1}{3(k+3)^2}\{15N_{(0,0)} - 6N_{(1,1)} + 3N_{(0,3)} + 3N_{(3,0)} - N_{(2,2)}\}$$

To facilitate the reorganization of D_ρ^2 in terms of characters, we display the dominant-weight multiplicities for the relevant representations:

$$\chi_{(2,2)}^{\text{dom}} = 3e^{(0,0)} + 2e^{(1,1)} + e^{(3,0)} + e^{(0,3)} + e^{(2,2)}$$
$$\chi_{(0,3)}^{\text{dom}} = e^{(0,0)} + e^{(1,1)} + e^{(0,3)}$$
$$\chi_{(3,0)}^{\text{dom}} = e^{(0,0)} + e^{(1,1)} + e^{(3,0)}$$
$$\chi_{(1,1)}^{\text{dom}} = 2e^{(0,0)} + e^{(1,1)}$$
$$\chi_{(0,0)}^{\text{dom}} = e^{(1,1)}$$

c) Verify explicitly the $\widehat{su}(3)_k$ result by calculating R for $k = 1, 2$ and inverting the result.

d) Using the $sp(4)$ multiplicities:

$$\chi^{\text{dom}}_{(2,2)} = 5e^{(0,0)} + 4e^{(0,1)} + 4e^{(2,0)} + 3e^{(0,2)} + 2e^{(2,1)} + e^{(0,3)} + e^{(4,0)} + e^{(2,2)}$$

$$\chi^{\text{dom}}_{(4,0)} = 3e^{(0,0)} + 2e^{(0,1)} + 2e^{(2,0)} + e^{(0,2)} + e^{(2,1)} + e^{(4,0)}$$

$$\chi^{\text{dom}}_{(0,3)} = 2e^{(0,0)} + 2e^{(0,1)} + e^{(2,0)} + e^{(0,2)} + e^{(2,1)} + e^{(0,3)}$$

$$\chi^{\text{dom}}_{(2,1)} = 3e^{(0,0)} + 3e^{(0,1)} + 2e^{(2,0)} + e^{(0,2)} + e^{(2,1)}$$

$$\chi^{\text{dom}}_{(0,2)} = 2e^{(0,0)} + e^{(0,1)} + e^{(2,0)} + e^{(0,2)}$$

$$\chi^{\text{dom}}_{(2,0)} = 2e^{(0,0)} + e^{(0,1)} + e^{(2,0)}$$

$$\chi^{\text{dom}}_{(0,1)} = e^{(0,0)} + e^{(0,1)}$$

$$\chi^{\text{dom}}_{(0,0)} = e^{(0,0)}$$

show that for $\widehat{sp}(4)_k$, R^{-1} takes the form

$$R^{-1}_{\widehat{sp}(4)_k} = \frac{1}{4(k+3)^2} \{21N_{(0,0)} - 3N_{(0,1)} - 6N_{(0,2)} - 3N_{(0,3)}$$

$$+ 6N_{(2,1)} - 3N_{(4,0)} - 3N_{(0,3)} + N_{(2,2)}\}$$

e) Check this result for $k = 1, 2$.

f) The aim of the last part of the exercise is to demonstrate the general expression of R^{-1}. Use the Verlinde formula to obtain:

$$R_{\hat\lambda\,\hat\mu} = \sum_{\hat\nu,\hat\sigma} \mathcal{N}_{\hat\lambda\,\hat\mu}{}^{\hat\nu}\, \mathcal{N}_{\hat\nu\,\hat\sigma}{}^{\hat\sigma} = \sum_{\hat\zeta} \frac{S_{\hat\lambda\,\hat\zeta}\, S_{\hat\mu\,\hat\zeta}}{S^2_{0\,\hat\zeta}}$$

Deduce the following form for $(R^{-1})^{\hat\mu}_{\hat\lambda}$:

$$(R^{-1})^{\hat\mu}_{\hat\lambda} = \sum_{\hat\zeta} S_{\hat\lambda\,\hat\zeta}\, \bar S_{\hat\mu\,\hat\zeta}\, S^2_{0\,\hat\zeta}$$

Finally, observe that D_ρ evaluated at the special point

$$\xi_\zeta = -\frac{2\pi i}{k+g}(\zeta, \rho)$$

is related to $S^2_{0\,\hat\zeta}$:

$$\frac{(-1)^{|\Delta_+|}}{|P/Q^\vee|(k+g)^r} \, [D_\rho(\xi_\zeta)]^2 = S^2_{0\,\hat\zeta}$$

$S^2_{0\,\hat\zeta}$ can thus be expanded in terms of the finite characters. Setting

$$S^2_{0\,\hat\zeta} = \sum_{\gamma \in P_+} d^\gamma\, \chi_\gamma(\xi_\zeta)$$

for some numbers d^γ, obtain

$$(R^{-1})^{\hat\mu}_{\hat\lambda} x = \sum_{\gamma \in P_+} d^\gamma\, \mathcal{N}_{\hat\lambda\,\gamma}{}^{\hat\mu}$$

which, in matrix notation, reads

$$R^{-1} = \sum_{\gamma \in P_+} d^\gamma\, N^{(k)}_\gamma$$

16.9 *su(N) depths*
Prove that the depth of the weights in the $su(N)$ fundamental representations is given by Eq. (16.99).

16.10 *Pieri and Giambelli formulae*
a) Complete the detail of the proof of the Giambelli formula (16.114) from Eq. (16.112).
b) From the Littlewood-Richardson rule, derive the Pieri formula for the product

$$y_j \otimes \{\ell_1; \cdots; \ell_n\}$$

c) Using this new Pieri formula, prove Eq. (16.123).

16.11 *One-variable description of the fusion ring*
We recall that the associativity of fusion rules implies that the fusion matrices $N_{\hat{\lambda}}^{(k)}$, with entries $N_{\hat{\lambda}\hat{\mu}}^{(k)\,\hat{\nu}}$, form a representation of the fusion algebra

$$N_{\hat{\lambda}}^{(k)} N_{\hat{\mu}}^{(k)} = \sum_{\hat{\nu}\in P_+^k} N_{\hat{\lambda}\hat{\mu}}^{(k)\,\hat{\nu}} N_{\hat{\nu}}^{(k)}$$

and that their eigenvalues are the numbers $\gamma_{\hat{\lambda}}^{(\hat{\sigma})}$.
a) Show that a sufficient condition for the existence of a one-variable polynomial ring is the nondegeneracy of all the eigenvalues of at least one fusion matrix.
Hint: The minimal polynomial of this matrix (see Ex. 8.18) is equal to its characteristic polynomial, hence the dimension of the ring it generates is equal to the degree of this polynomial, and therefore to that of the full fusion ring.

b) We wish now to use this criterion to prove that $\widehat{su}(3)_k$ fusion rules can always be described in terms of a single variable. Show that the eigenvalues $\gamma_{\hat{\lambda}}^{(\hat{\sigma})}$, $\hat{\sigma} \in P_+^k$ and $\hat{\lambda} = k\hat{\omega}_0 + \omega_1$, of the fundamental fusion matrix may be expressed as the sum of three complex numbers of modulus 1, whose product is 1.
c) Show that if

$$e^{ia} + e^{ib} + e^{-i(a+b)} = e^{ic} + e^{id} + e^{-i(c+d)} \tag{16.206}$$

for some real numbers a,b,c,d, then

$$\sin(c' - d') \sin(a' + b') = \sin(b' - a') \sin(c' + d')$$
$$\sin a' \sin b' \sin(c' - d') = \sin c' \sin d' \sin(b' - a')$$

with

$$a' = \frac{1}{2}(a+b+c) \qquad b' = \frac{1}{2}(c+d+a)$$
$$c' = \frac{1}{2}(a+b+d) \qquad d' = \frac{1}{2}(c+d+b)$$

Hint: Multiply both sides of Eq. (16.206) by $e^{i(a+b+c+d)/2}$ and take the real and imaginary parts.

d) Conclude that necessarily $(c,d) = (a,b) \bmod \pi$ (or any of the five other permutations of $(b,a),(a,-a-b),(-a-b,a),(b,-a-b),(-a-b,b)$) and deduce the property **(a)** for $\widehat{su}(3)_k$ fusions.

e) Use Eq. (14.255) to argue that the condition of **(a)** always fails for $su(N)$ when $N > 2$ and k are both even, due to the presence of pairs of fixed points associated with the zero eigenvalue (again for a fundamental fusion matrix, with $\lambda = \omega_i$ for some i). Generalize this statement to all values of N for which the center of $su(N)$ has a nontrivial subgroup.

16.12 *Fusion coefficients by level-rank duality*
Using the $\widehat{su}(3)$ fusion rules given in Eq. (16.195) at level 4, obtain the fusion coefficients for

$$[1, 2, 1] \times [2, 1, 1]$$

by an appropriate action of the outer automorphism and derive, by level-rank duality, the $\widehat{su}(4)_3$ coefficients for

$$[1, 1, 0, 1] \times [1, 1, 1, 0]$$

16.13 *Level-rank duality of quantum dimensions*
Show that

$$\mathcal{D}_{\hat{\lambda}} = \mathcal{D}_{\hat{\lambda}'}$$

with $\hat{\lambda} \in \widehat{su}(N)_k$ and $\hat{\lambda}' \in \widehat{su}(k)_N$. Use this relation to compute the quantum dimension of an arbitrary $\widehat{su}(N)_2$ integrable weight.

16.14 *Level-rank duality for multiplicities on outer-automorphism orbits*
For $\hat{\lambda} \in \widehat{su}(N)_k$ and $\hat{\lambda}' \in \widehat{su}(k)_N$, show that

$$m_N(\hat{\lambda}) = m_k(\hat{\lambda}')$$

where $m_N(\hat{\lambda})$ denotes the multiplicity of $\hat{\lambda}$ in the orbit $\{\hat{\lambda}, a\hat{\lambda}, \cdots, a^{N-1}\hat{\lambda}\}$ and $m_k(\hat{\lambda}')$ is the multiplicity of $\hat{\lambda}'$ in the orbit $\{\hat{\lambda}', \tilde{a}\hat{\lambda}', \cdots, \tilde{a}^{k-1}\hat{\lambda}'\}$.

16.15 $\widehat{su}(3)$ *fusion threshold level*
Complete the derivation of the formulae (16.101)–(16.106) sketched in App. 16.A. In particular, show that when the product of elementary couplings $E_1 E_3 E_5$ is forbidden, whichever of E_1, E_3, or E_5 does not appear in the decomposition can be determined in terms of the Dynkin labels of the weights of the triple product. Obtain the explicit values of $c_{\min}$ and $c_{\max}$, derive similar bounds for a and e associated with the other possible decompositions, and recover Eq. (16.106) with $k_0^{\min}$ and $k_0^{\max}$ defined in Eq. (16.102). Finally, show that the threshold level associated with the coupling described by the $su(3)$ Berenstein-Zelevinsky triangle (13.211) can be written explicitly in terms of the entries of the triangle as

$$k_0 = \max(l_{13} + \lambda_1 + \lambda_2, m_{13} + \mu_1 + \mu_2, n_{13} + \nu_1 + \nu_2)$$

16.16 *Generating function for fusion rules*
The generating function for tensor-product coefficients is given by a sum over all possible couplings

$$G(L, M, N) = \sum_{\lambda, \mu, \nu \in P_+} \mathcal{N}_{\lambda\mu\nu} \, L^\lambda M^\mu N^\nu$$

in terms of some dummy variables L_i, M_i, N_i, where

$$L^\lambda = \prod_{i=1}^{r} L_i^{\lambda_i}$$

and similarly for M^μ and N^ν. The $su(2)$ generating function reads

$$G = [(1 - L_1 M_1)(1 - L_1 N_1)(1 - M_1 N_1)]^{-1} = \prod_{i=1}^{3} (1 - E_i)^{-1}$$

in terms of the elementary couplings (16.185). (More precisely, the elementary couplings are $E_1 = L_1 M_1$, $E_2 = L_1 N_1$, $E_3 = M_1 N_1$. This shows why when taking products of the E_i's, we add the corresponding Dynkin labels.) Fusion rule generating functions can be defined in a similar way as

$$\hat{G}(L, M, N; d) = \sum_{k=0}^{\infty} d^k \sum_{\lambda, \mu, \nu \in P_+^k} N_{\hat{\lambda}\hat{\mu}\hat{\nu}}^{(k)} L^\lambda M^\mu N^\nu$$

where the extra dummy variable d keeps track of the level.

a) Prove that for $\widehat{su}(2)_k$

$$\hat{G}(L, M, N; d) = (1 - d)^{-1} \prod_{i=1}^{3} (1 - dE_i)^{-1}$$

by showing that this reproduces the $\widehat{su}(2)_k$ fusion rules.

b) What is the origin of the prefactor $(1 - d)^{-1}$? Argue that it must be present for any affine Lie algebra, and that it automatically implies the inequality (16.90).

c) Obtain the fusion rule generating function for $\widehat{su}(3)$.
Hint: The generating function for tensor products can be written as

$$G = (\prod_{i=1}^{8} (1 - E_i)^{-1}(1 - F)$$

where F denotes the forbidden coupling.

Notes

The simple way outer automorphisms act on fusion coefficients was found by Gepner and Fuchs [169]. The Kac-Walton formula was discovered in Refs. [348, 214] and proved in Ref. [349, 153]. The same result has been obtained independently in Ref. [156, 184] from the point of view of quantum groups. The construction was also anticipated in Ref. [330]. (Exercise 16.2 is in the spirit of a conjecture of Ref. [156], which turned out to be wrong: a counterexample is given in the footnote to Ex. 16.2, which is taken from Ref. [246].) The concept of quantum dimension in conformal field theory is due to Verlinde [340] (see also Refs. [11, 290] and the review [147]). It was used in fusion rules by Fuchs and van Driel [154] (Ex. 16.7 is adapted from this reference). The depth rule is due to Gepner and Witten [172] and its formulation was sharpened by Kirillov et al. [237]. The idea of a threshold level first appeared in the work of Cummins, Mathieu, and Walton [84], where fusion-generating functions are also introduced (cf. Ex. 16.16).

The explicit formula for the $\widehat{su}(2)_k$ fusion coefficients was obtained by Gepner and Witten [172] and by Zamolodchikov and Fateev [365]. (Exercise 16.4 is taken from Ref. [236].) The $\widehat{su}(3)_k$ formula was derived by Bégin, Mathieu, and Walton [34]. (The related Ex. 16.15 is adapted from this reference and Ref. [237]). Berenstein-Zelevinsky triangles were applied to fusion-rule calculations in Ref. [237] and this was further developed in Ref. [33]. The combinatorial description of fusion rules presented in Sect. 16.2.4 is taken from Refs. [184, 83]. Handle operators are studied along the lines of Ex. 16.8 in Refs. [81, 82, 200].

The concept of simple currents is due to Intriligator [199] and Schellekens and Yankielowicz [319, 320]. Simple currents can be used to define the center of a rational conformal field theory by S duality. The complete classification of simple currents in WZW models is due to Fuchs [146].

Fusion potentials were introduced by Gepner [168]. Some of the underlying ideas can also be found in the mathematical literature (Ref. [184]). Our presentation follows closely the spirit of Ref. [168], but the notation and the discussion of the classical results are in the line of Fulton and Harris [155]. Further developments on fusion potentials (as far as WZW models are directly concerned) can be found in Refs. [100, 51, 81, 171, 200, 3, 28] (Ex. 16.11 is taken from the first reference).

The theoretical importance of fusion potentials has not been addressed in the main text. It points in two main directions. One is a close connection with the chiral algebra in $N = 2$ supersymmetric conformal field theories and the $\hat{g}_k$ fusion algebra. The other related object is more mathematical: the cohomology of Grassmannian manifolds (the set of all L-dimensional planes in $\mathbb{C}^{L+M}$; see for instance Ref. [188]). Briefly stated, the determinant method for tensor-product calculations is simply the multiplication structure giving a ring structure to the cohomology group of these manifolds. The multiplication in this context is dual to the intersection of the basic homology cycles (the so-called Schubert cycles). Hence, to the k truncation, we can associate a new geometric object, the quantum cohomology of Grassmannians (Refs. [200, 339, 360]).

Level-rank duality was first discovered in mathematics, in the context of affine branching rules (see the notes for the following chapter). A level-rank duality was later observed for the Boltzmann weights of a generic class of two-dimensional statistical models (RSOS) based on $\widehat{su}(N)_k$ by Kuniba and Nakanishi [247], and it was shown that this equivalence implies the equality of the fusion coefficients of the dual theories. The duality between $\widehat{su}(N)_k$ and $\widehat{su}(k)_N$ WZW models was discovered independently by Naculich and Schnitzer [276] from a different point of view, namely, as a duality between holomorphic blocks of four-point functions. Generalizations are presented in Mlawer et al. [269]. The proof of Eq. (16.168) can be found in Altschuler, Bauer, and Itzykson [9]. A method for calculating $\widehat{su}(N)_k$ and $\widehat{sp}(N)_k$ fusion rules by exploiting the level-rank duality is presented in Ref. [83].

CHAPTER 17

Modular Invariants in WZW Models

The characters of the integrable representations of an affine Lie algebra at fixed level have been shown to be covariant under modular transformations. These characters have subsequently been identified with those of the WZW primary fields. More precisely, a WZW primary field is associated with both a left and a right integrable highest weight, and its descendants are associated with the different states in the tensor product of the two modules. To a large extent, the holomorphic and the antiholomorphic sectors can be studied independently. However, they must ultimately be combined to form a modular-invariant partition function. The construction of modular invariant partition functions is the subject of the present chapter.

There is a natural way of combining the holomorphic and antiholomorphic sectors in the WZW model with spectrum-generating algebra $\hat{g}_k$ for which each primary field has the same transformation property (i.e., it transforms in the same representation of g) in both sectors and each integrable representation occurs once and only once. As previously stressed, in order to see whether this can produce a physically sensible theory, we must verify the modular invariance of the corresponding partition function. We show in Sect. 17.1.2 that it is indeed so. This should not be too surprising: it is a feature that has already been encountered with minimal models. But this reference also suggests that other spectra could be compatible with modular invariance. Indeed, we recall that for some values of the central charge, it has been found that a subset of the fields in the Kac table can be closed under fusion and have a modular-invariant partition function. The physical spectrum of the three-state Potts model provides the simplest example of this phenomenon. It also illustrates the physical importance of nondiagonal spectra. The essence of this chapter is the elaboration of techniques for generating analogous invariants in the class of WZW models.[1]

---

[1] To avoid any confusion on terminology, we stress that models with diagonal or nondiagonal spectra are both referred to as WZW models.

In a first stage, two general ways of producing physical nondiagonal modular invariants are presented in detail: one is based on outer-automorphism groups (Sect. 17.3) and the other relies on algebra embeddings (Sect. 17.5). Almost all modular invariants can be produced by one of these two methods. The complete list of modular invariants in the $\hat{g}_k$-WZW model is known only for $\widehat{su}(2)$ and $\widehat{su}(3)$. In the $\widehat{su}(2)$ case, a remarkable relation with ADE algebras emerges. The result of these classifications is given in Sect. 17.7 (but their completeness is not proven).

The classification of modular-invariant WZW partition functions is an important problem. Anticipating the basic role of conformal-model building blocks played by WZW models, we can glimpse that the classification of rational conformal field theories with affine Lie spectrum-generating algebra is a key step in the classification of all physical modular invariants. In particular, once the $\widehat{su}(2)$ structure underlying the minimal models has been unraveled (cf. Chap. 18), the results presented here on the classification of the $\widehat{su}(2)_k$ modular invariants will be directly reinterpreted as a classification of minimal models modular invariants. On the other hand, despite limited success on the general classification issue, the results for $\widehat{su}(2)$ and $\widehat{su}(3)$ hint at the existence of a deep and rich mathematical structure underlying these problems.

On this subject of classification, a powerful result states that any modular invariant is either obtained from a diagonal invariant by a permutation of the fields in one sector (necessarily an automorphism of the fusion rules), or it is a diagonal invariant (more generally a permutation thereof) in a hidden form, namely in a theory with a larger symmetry algebra. This result is discussed in Sect. 17.8. Although remarkably simple and general, it marks out neatly the limit of our understanding of the fine structure of the theories underlying nondiagonal modular invariants: most of these extended algebras are not affine Lie algebras, and almost nothing is known about their structure.

A straightforward way of generating permutation invariants is to use the Galois symmetries of the WZW models. Galois symmetries are introduced in Sect. 17.9. In addition to being a constructive tool for obtaining new invariants, these symmetries yield efficient constraints in modular-invariant hunting.

In a final section, we describe a striking relationship between graphs and modular invariants. In particular, we show that the ADE structure underlying the $\widehat{su}(2)_k$ modular invariants can be traced back to the basic properties of the adjacency matrix of the corresponding Dynkin diagrams. This observation allows us to associate a generalized Dynkin diagram with each $\widehat{su}(3)_k$ invariant. For those $\widehat{su}(2)_k$ and $\widehat{su}(3)_k$ invariants that do not originate from an automorphism of the fusion rules, it is also found that the graph algebra coded in the adjacency matrix (which is viewed as the fusion matrix of the fundamental representation) has a graph subalgebra that is exactly the same as the fusion algebra of the extended theory.

A method for calculating the branching rules in semisimple conformal embeddings (defined in Sect. 17.5) is revealed in App. 17.A and illustrated for $su(N)$ algebras. In App. 17.B, we present the classification of the $c = 1$ theories, by using orbifold techniques; it reveals another remarkable ADE classification scheme. The operator content of these different theories is also displayed.

§17.1. Modular Invariance in WZW Models

17.1.1. The Construction of Modular-Invariant Partition Functions

In a theory with affine spectrum-generating algebra $\hat{g}_k$, the full Hilbert space $\mathcal{H}$ decomposes as:

$$\mathcal{H} = \bigoplus_{\hat{\lambda}, \hat{\xi} \in P_+^{(k)}} \mathcal{M}_{\hat{\lambda}\hat{\xi}} \, L_{\hat{\lambda}} \otimes L_{\hat{\xi}} \tag{17.1}$$

where, as usual, the tensor product reflects the separation into holomorphic and antiholomorphic sectors and $\mathcal{M}_{\hat{\lambda}\hat{\xi}}$ gives the multiplicity of the combined modules $L_{\hat{\lambda}} \otimes L_{\hat{\xi}}$ in the Hilbert space. The partition function of a conformal field theory takes the form

$$Z(q) = \mathrm{Tr}_{\mathcal{H}} \, q^{L_0 - c/24} \, \bar{q}^{\bar{L}_0 - c/24} \tag{17.2}$$

In the present case, the trace over $\mathcal{H}$ can be broken into a sum of traces over irreducible affine modules. The partition function can thus be expressed in terms of the affine characters. With $q = e^{2\pi i \tau}$ and $\bar{q} = e^{-2\pi i \bar{\tau}}$, it reads:

$$Z(\tau) = \sum_{\hat{\lambda}, \hat{\xi} \in P_+^k} \chi_{\hat{\lambda}}(\tau) \, \mathcal{M}_{\hat{\lambda}\hat{\xi}} \, \bar{\chi}_{\hat{\xi}}(\bar{\tau}) \tag{17.3}$$

where $\mathcal{M}_{\hat{\lambda}\hat{\xi}}$ can be reinterpreted as the multiplicity of the primary field which, under $G(z) \times G(\bar{z})$, transforms with respect to the λ and ξ representations of g, respectively. Since it specifies the physical spectrum of the model, $\mathcal{M}$ is usually called the *mass matrix*. This form of the partition function is not quite satisfying for WZW models because the extra Lie algebra symmetry is not fully taken into account. In other words, some parameters required for a full characterization of the spectrum are missing. These are simply the parameters ζ and t appearing in the expression of the specialized characters (cf. Sect. 14.4). Although t is not necessary, the ζ dependence is needed to distinguish charge conjugate characters. However, in order to lighten the notation, we omit these extra parameters in most of the following expressions.[2]

We recall the modular transformation matrices for the characters of the integrable representations of $\hat{g}_k$:

$$\chi_{\hat{\lambda}}(\tau + 1) = \sum_{\hat{\mu} \in P_+^k} T_{\hat{\lambda}\hat{\mu}} \chi_{\hat{\mu}}(\tau)$$

$$\chi_{\hat{\lambda}}(-1/\tau) = \sum_{\hat{\mu} \in P_+^k} S_{\hat{\lambda}\hat{\mu}} \chi_{\hat{\mu}}(\tau) \tag{17.4}$$

---

[2] This complete partition function is often written under the form

$$Z(q, x_i) = \mathrm{Tr}_{\mathcal{H}} \, q^{L_0 - c/24} x_i^{h_0^i} \, \bar{q}^{\bar{L}_0 - c/24} \bar{x}_i^{\bar{h}_0^i}$$

where $x_i = e^{-2\pi i z_i}$ with z_i defined by $\zeta = \sum_i z_i \alpha_i^\vee$.

where

$$T_{\hat{\lambda}\hat{\mu}} = \delta_{\hat{\lambda}\hat{\mu}} e^{2\pi i (h_{\hat{\lambda}} - c/24)}$$

$$S_{\hat{\lambda}\hat{\mu}} = K \sum_{w \in W} \epsilon(w) \exp\left\{ -\frac{2\pi i}{k+g} (w(\lambda + \rho), \mu + \rho) \right\} \tag{17.5}$$

The multiplicative constant K is fixed by unitarity and the requirement $S_{00} > 0$.

We recall also, from the general discussion of Chap. 10, that the modular invariance of the partition function translates into the following conditions on $\mathcal{M}$:

$$T^\dagger \mathcal{M} T = S^\dagger \mathcal{M} S = \mathcal{M} \tag{17.6}$$

or equivalently

$$[\mathcal{M}, S] = [\mathcal{M}, T] = 0 \tag{17.7}$$

Hence, $\mathcal{M}$ must lie in the *commutant* of S and T.

In addition to being modular invariant the partition functions must satisfy the following obvious physical conditions:
(a) $\mathcal{M}_{\hat{\lambda}\hat{\xi}}$ must be a nonnegative integer.
(b) $\mathcal{M}_{00} = 1$ for the vacuum state to be unique.
(As usual, 0 stands for $k\hat{\omega}_0$.) An invariant satisfying conditions (a) and (b), is said to be a *physical invariant*.[3] The next subsection deals with the simplest physical invariant.

17.1.2. Diagonal Modular Invariants

The natural spectrum alluded to in the introduction corresponds to a mass matrix with entries

$$\mathcal{M}_{\hat{\lambda}\hat{\xi}} = \delta_{\hat{\lambda}\hat{\xi}} \tag{17.8}$$

that is, $\mathcal{M} = I$. The partition function is thus simply

$$Z(\tau) = \sum_{\hat{\lambda} \in P_+^k} \chi_{\hat{\lambda}}(\tau) \bar{\chi}_{\hat{\lambda}}(\bar{\tau}) \tag{17.9}$$

Given the unitarity of both T and S, this partition function is obviously modular invariant, i.e., T and S obviously commute with I. Such a theory, in which all integrable representations appear exactly once and all the fields have equal holomorphic and antiholomorphic conformal dimensions, is said to be *diagonal*.

$\mathcal{M} = I$ gives thus an example of a mass matrix associated with a physical invariant. Modular invariants with $\mathcal{M}$ different from I will be called *nondiagonal*.

---

[3] It appears that conditions (a) and (b) are not quite sufficient to fully specify well-defined theories. For example, a physical invariant could lead to a theory with noninteger fusion coefficients, which is clearly not acceptable. At this time, a complete set of conditions that must be satisfied by a physical invariant to qualify it as a genuine rational conformal field theory is not known.

17.1.3. The Search for New Modular Invariants

A direct and constructive attack to the problem of finding all modular invariants would consist in first obtaining all the mass matrices that lie in the commutant of the modular group and then selecting among the solutions those for which the above two physical conditions are satisfied. Although tractable for $\widehat{su}(2)_k$, this approach is not useful in general because the dimension of the commutant grows very rapidly with the rank and the level. For instance, for $\widehat{su}(N)_k$, it behaves like $(k + N)^{2N}$ for large k and N. In sharp contrast, the number of physical solutions at each level is very small (typically 1 or 2). This clearly indicates the inefficiency of this approach.[4]

We will instead approach this problem from a more modest (and more physical) point of view, by developing methods by which new physical invariants can be constructed out of the diagonal ones just found. Three such methods are known: one involves a sort of twisting (the method of outer automorphisms, which is an Abelian orbifold construction), another one amounts to an immersion into a larger theory (conformal embeddings), and the third one corresponds to a modular-invariant permutation of the fields associated with an automorphism of the fusion rules (Galois permutations). These methods always produce sensible physical theories. Unfortunately, none of these prove to be complete; that is, all known physical invariants cannot be generated by only one of these techniques. There are other methods for generating modular invariants (for example, block-diagonal invariants can be obtained by Galois symmetries), but these are not guaranteed to be physical.

§17.2. A Simple Nondiagonal Modular Invariant

The first method for constructing nondiagonal modular invariants will be introduced in a very pedestrian way, by means of a simple $\widehat{su}(2)$ example.

Consider the $\widehat{su}(2)_2$ modular S matrix

$$S = \frac{1}{2} \begin{pmatrix} 1 & \sqrt{2} & 1 \\ \sqrt{2} & 0 & -\sqrt{2} \\ 1 & -\sqrt{2} & 1 \end{pmatrix} \qquad (17.10)$$

where fields are ordered by increasing values of their finite Dynkin labels: $(0), (1), (2)$, of respective conformal dimension $0, \frac{3}{16}, \frac{1}{2}$. Looking at the first and third rows of the matrix, we see that the combination $\chi_0 + \chi_2$ decouples from χ_1. (The subscript gives the value of the corresponding finite Dynkin label and parentheses are omitted.) Thus, at first sight, it seems that

$$|\chi_0 + \chi_2|^2 = |\chi_0|^2 + |\chi_2|^2 + \chi_0 \bar{\chi}_2 + \chi_2 \bar{\chi}_0 \qquad (17.11)$$

could qualify as a modular-invariant partition function. It would then describe a theory containing fields transforming differently in the holomorphic and anti-holomorphic sectors, corresponding to the last two terms in the above equation.

---

[4] Unfortunately, the precise characterization of the *physical commutant* is still an open problem.

However, it is not quite modular invariant: under $\tau \to \tau + 1$, it transforms into $|\chi_0 - \chi_2|^2$ because $h_2 - h_0 = \frac{1}{2}$.

The $\widehat{su}(2)_3$ model offers no analogous possibilities (and this generalizes to all odd values of k), so we turn to the $\widehat{su}(2)_4$ model:

$$
S = \frac{1}{2\sqrt{3}}
\begin{pmatrix}
1 & \sqrt{3} & 2 & \sqrt{3} & 1 \\
\sqrt{3} & \sqrt{3} & 0 & -\sqrt{3} & -\sqrt{3} \\
2 & 0 & -2 & 0 & 2 \\
\sqrt{3} & -\sqrt{3} & 0 & \sqrt{3} & -\sqrt{3} \\
1 & -\sqrt{3} & 2 & -\sqrt{3} & 1
\end{pmatrix}
\tag{17.12}
$$

Again fields are ordered in increasing values of the conformal dimension: $0, \frac{1}{8}, \frac{1}{3}, \frac{5}{8}, 1$. Here we see that $\chi_0 + \chi_4$ together with χ_2 decouple from χ_1 and χ_3 and so transform among themselves. Furthermore, since $h_4 - h_0 \in \mathbb{Z}$, the combination

$$
c_1 |\chi_0 + \chi_4|^2 + c_2 |\chi_2|^2
\tag{17.13}
$$

is T invariant. Under a S transformation, it is changed into

$$
\frac{c_1}{3} |\chi_0 + \chi_4 + 2\chi_2|^2 + \frac{c_2}{3} |\chi_0 + \chi_4 - \chi_2|^2
\tag{17.14}
$$

Hence, the ratio c_2/c_1 is fixed by S invariance to be 2. Finally, c_1 must be unity for the vacuum to be unique. As a result, the partition function

$$
Z = |\chi_0|^2 + |\chi_4|^2 + 2|\chi_2|^2 + \chi_0 \bar{\chi}_4 + \chi_4 \bar{\chi}_0
\tag{17.15}
$$

is modular invariant and satisfies all our physical requirements. Compared to diagonal theories, its field content differs in two respects: some fields are not spinless ($h \neq \bar{h}$), and one field appears more than once. Although it has a field content different from that of the diagonal $\widehat{su}(2)_4$ model, it has the same central charge.[5] Hence, as for the minimal models, the central charge of the WZW model is not sufficient in general to characterize a theory uniquely.

It should be clear that the fields that have been discarded in this operation, associated with the characters χ_1 and χ_3, cannot be used to form a sensible theory by themselves since there would be no vacuum.

The mechanism behind the decoupling of some of the fields and the occurrence of multiplicity-2 fields described here is identical to the one at work in the three-state Potts model. More than that: from the coset construction to be introduced in the following chapter, it turns out that the invariant (17.15) is actually a building block of the partition function for the three-state Potts model! From this point of view, the fact that the partition function (17.15) is the simplest $\widehat{su}(2)$ nondiagonal invariant shows that the Potts model is the simplest unitary minimal model that has a nondiagonal partition function.

---

[5] We recall that the central charge is read off the small q behavior of the partition function: $Z \sim (q\bar{q})^{-c/24}$; here $c = 2$. More generally, in any unitary theory the leading contribution comes from the vacuum character χ_0, always present in a physical modular invariant. Therefore, the different modular-invariant partition functions that can be built from a given set of fields with positive dimensions (and allowing repetitions) all have the same central charge.

We now analyze the structure of this new modular invariant. First we notice that "odd" fields, those with odd finite Dynkin labels, do not appear in the spectrum of the new theory. Comparing Eq. (17.15) with the $\widehat{su}(2)_4$ diagonal invariant, we see not only that "odd" fields are absent, but extra fields have been introduced: $\chi_0 \bar{\chi}_4$, $\chi_4 \bar{\chi}_0$ and $\chi_2 \bar{\chi}_2$. Since $k = 4$, we observe that the two fields that are combined in the square modulus, namely [4, 0] and [0, 4], are related to each other by a permutation of their Dynkin labels, that is, by an action of the outer automorphism a, where:

$$a\hat{\lambda} = a[k - n, n] = [n, k - n] \tag{17.16}$$

On the other hand, the field [2, 2] is neutral under the action of a. The extra fields are thus all of the form $\chi_n \bar{\chi}_{k-n}$.

These are the key features of a general method for constructing nondiagonal modular invariants out of diagonal ones:

(i) a projection onto states invariant under some discrete group H (here a projection onto even states, the discrete group being thus $\mathbb{Z}_2$);

(ii) the incorporation of additional fields transforming differently with respect to left and right transformations of G, called twisted fields.

To preserve the finite Lie group symmetry of the theory, the discrete group, under which part of the states are invariant, must be chosen among the subgroups of the center $B(G)$ of G, where G is the symmetry group of the initial diagonal theory.[6] Indeed, $B(G)$ is the largest possible *discrete* invariance group of the model. In our example, $G = SU(2)$ and $H = B(SU(2)) = \mathbb{Z}_2$.

Removing some states in a modular-invariant theory is bound to break modular invariance. To restore it, we have to introduce extra states, the twisted sector. The operators responsible for this twisting, the operators that generalize a, must be some sort of dual of the elements of the center, in order to maintain a compatibility between the projection operation and the twisting. From the above example, these are expected to be the outer automorphisms of $\hat{g}$. That these are the right objects can be guessed also from the isomorphism between $B(G)$ and the group of outer automorphisms $\mathcal{O}(\hat{g})$ described in Sect. 14.2.3, which actually makes elements of $B(G)$ and $\mathcal{O}(\hat{g})$ "S duals" of each other, as shown in Sect. 14.6.4.

Before turning to the general procedure, we point out that exactly the same construction found above for $k = 4$ is possible for all values of $k = 4\ell$, that is,

$$Z = \sum_{\substack{n=0 \\ n \in 2\mathbb{Z}}}^{2\ell-2} |\chi_n + \chi_{4\ell-n}|^2 + 2|\chi_{2\ell}|^2 \tag{17.17}$$

is modular invariant. In the terminology of Chap. 10 (cf. Sect. 10.7.3), this invariant is block-diagonal. For $k = 4\ell - 2$ there is also a nondiagonal modular invariant (which, in contrast, cannot be put in a block-diagonal form). However, in this case

---

[6] A generalization of this construction, for which the discrete group is no longer required to be a subgroup of the center, is presented in App. 17.B. In such case, the resulting partition function cannot be expressed in terms of the characters of the affine Lie algebra.

the situation is slightly more complicated, as odd fields reappear in the twisted sector. The detailed structure of these theories will be derived later.

§17.3. Modular Invariants Using Outer Automorphisms

17.3.1. The General Construction

We now present the general method underlying the $\widehat{su}(2)_4$ nondiagonal invariant, considered from the point of view of outer automorphisms. This method gives a systematic way of obtaining physical nondiagonal modular invariants from the diagonal ones.

The first step of the construction consists in projecting the Hilbert space of the diagonal theory onto the $B(G)$ invariant states. (Clearly, any discrete $H \subset B(G)$ could be considered. For definiteness, we will choose $H = B(G)$, but the possible generalizations should be kept in mind.) If we let b be a generator of $B(G)$ of order N, that is, $b^N = 1$,[7] the desired projection is

$$B = \frac{1}{N} \sum_{q=0}^{N-1} b^q \tag{17.18}$$

The action of b on characters is defined naturally by:

$$b\chi_{\hat{\lambda}} = \chi_{\hat{\lambda}} b(\hat{\lambda}) \tag{17.19}$$

where $b(\hat{\lambda})$ is the b eigenvalue on $\hat{\lambda}$. Indeed, the action of b on a highest-weight state can be lifted to the character because it commutes with the algebra generators; hence, all the states in a representation share the same b eigenvalue. As a result, the projection operation does not break up any representations; it simply excludes some. For instance, $B(SU(2)) = \{1, b\}$ with

$$b\hat{\lambda} = \hat{\lambda} e^{-\pi i \lambda_1} \tag{17.20}$$

so that

$$B\chi_{\hat{\lambda}} = \frac{1}{2}(1 + (-1)^{\lambda_1})\chi_{\hat{\lambda}} \tag{17.21}$$

Thus, from the diagonal theory, only those primary fields that are $B(G)$ invariants are preserved:

$$\sum_{\hat{\lambda} \in P_+^k} \chi_{\hat{\lambda}}(\tau) \, B \, \bar{\chi}_{\hat{\lambda}}(\bar{\tau}) = \sum_{\substack{\hat{\lambda} \in P_+^k \\ B\hat{\lambda} = \hat{\lambda}}} \chi_{\hat{\lambda}}(\tau) \, \bar{\chi}_{\hat{\lambda}}(\bar{\tau}) \tag{17.22}$$

[7] For simplicity, we restrict the analysis to the case in which the center is $\mathbb{Z}_N$. Thus, the only case untreated by the present analysis is $D_{2\ell}$.

As already mentioned, the natural operator description of the twisted sectors is provided by

$$\mathcal{A} = \sum_{p=0}^{N-1} A^p \qquad (17.23)$$

acting on the diagonal invariant projected onto $B(G)$ invariant states, with

$$A\chi_{\hat{\lambda}} = \chi_{A\hat{\lambda}} \qquad (17.24)$$

Here A is any nontrivial element of the outer-automorphism group $\mathcal{O}(\hat{g})$ (cf. Sect. 14.2). This leads to the candidate mass matrix $\mathcal{M} = \mathcal{A}\mathcal{B}$. We already know that the role of A and b get interchanged by an $\mathcal{S}$ transformation. Thus, under $\tau \to -1/\tau$, $\mathcal{M}$ will transform into $\mathcal{B}\mathcal{A}$. But the point is that $\mathcal{B}\mathcal{A} \neq \mathcal{A}\mathcal{B}$ simply because in general $Ab \neq bA$. (This noncommutativity does not show up in the simple example we have considered above.) Indeed, the b eigenvalue depends upon the finite Dynkin labels, and these are modified by the action of A. Therefore, we need to introduce some sort of symmetrized product $\circ$, such that

$$A \circ b = b \circ A \qquad (17.25)$$

We will show below how this product can be defined. Granting its existence, we now expect the following mass matrix

$$\tilde{\mathcal{M}} \equiv \mathcal{A} \circ \mathcal{B} = \frac{1}{N} \sum_{p,q=0}^{N-1} A^p \circ b^q \qquad (17.26)$$

to produce a modular-invariant theory. This is indeed the correct result.[8]

The rest of this subsection is divided into two parts. We first introduce the product $\circ$ and then we establish the modular invariance of Eq. (17.26). In the following subsection, we will verify the additional constraints $\mathcal{M}_{\hat{\lambda},\hat{\xi}} \in \mathbb{N}$, $\mathcal{M}_{00} = 1$ and derive a consistency condition for this whole procedure. Finally, these results will be used to derive nondiagonal $\widehat{su}(2)$ invariants.

A SYMMETRIZED PRODUCT

We recall that to every element $b \in B(G)$, there corresponds an element $A \in \mathcal{O}(\hat{g})$ via

$$b\hat{\lambda} = \hat{\lambda} \, b(\lambda) = \hat{\lambda} \, e^{-2\pi i (A\hat{\omega}_0, \lambda)} \qquad (17.28)$$

We consider now the commutation of b, associated with A, with another element of the group of outer automorphism, A'. For this, we evaluate the product bA' on

---

[8] The matrix notation is quite convenient but could be the source of some confusion. So we stress that the matrix element of a term of the form Ab is

$$(Ab)_{\hat{\lambda}\hat{\xi}} = \delta_{\hat{\lambda},A\hat{\xi}} \, b(\hat{\xi}) \qquad (17.27)$$

where $b(\hat{\xi})$ is the b eigenvalue of $\hat{\xi}$.

some $\hat{\lambda} \in P_+^k$, using the explicit form of $A'\hat{\lambda}$ given in Eq. (14.98):

$$bA'\hat{\lambda} = A'\hat{\lambda}\, b(A'\hat{\lambda})$$
$$= A'\hat{\lambda}\, e^{-2\pi i(A\hat{\omega}_0, A'\hat{\lambda})} \tag{17.29}$$
$$= A'\hat{\lambda}\, e^{-2\pi i k(A\hat{\omega}_0, A'\hat{\omega}_0)} e^{-2\pi i(A\hat{\omega}_0, w_{A'}\lambda)}$$

According to Eq. (14.114), in the second exponential factor of the last equality, we can simply let $w_{A'} = 1$, since λ is an integrable weight:

$$e^{-2\pi i(A\hat{\omega}_0, w_{A'}\lambda)} = e^{-2\pi i(A\hat{\omega}_0, \lambda)} \tag{17.30}$$

Since this is just the eigenvalue of b on $\hat{\lambda}$, we can write

$$bA' = A'b\, e^{-2\pi i k(A\hat{\omega}_0, A'\hat{\omega}_0)} \tag{17.31}$$

Therefore the actions of $\mathcal{O}(\hat{g})$ and $B(G)$ do not commute. But this shows precisely how the symmetrized product $\circ$ can be defined:

$$A' \circ b \equiv A'b\, e^{-ik\pi(A\hat{\omega}_0, A'\hat{\omega}_0))}$$
$$b \circ A' \equiv bA'\, e^{ik\pi(A\hat{\omega}_0, A'\hat{\omega}_0))} \tag{17.32}$$

leading to the desired result:

$$A' \circ b = b \circ A' \tag{17.33}$$

We note in particular that by using

$$(A^p\hat{\omega}_0, A^q\hat{\omega}_0) = pq|A\hat{\omega}_0|^2 \quad \text{mod } 1 \tag{17.34}$$

which follows directly from Eq. (14.118), we have

$$A^p \circ b^q = A^p b^q\, e^{-\pi i q p k|A\hat{\omega}_0|^2} \tag{17.35}$$

The particular element A appearing in the expression (17.26) is precisely the one related to b via Eq. (17.28). This does not induce any restriction because any other element A' can be written as an appropriate power of A, and the two summations in Eq. (17.26) are independent. Thus, in terms of an arbitrary nontrivial element A of the outer-automorphism group, the partition function $\tilde{Z}$ is

$$\tilde{Z} = \sum_{\hat{\lambda},\hat{\xi} \in P_+^k} \chi_{\hat{\lambda}}\, \tilde{\mathcal{M}}_{\hat{\lambda}\hat{\xi}}\, \bar{\chi}_{\hat{\xi}}$$
$$= \frac{1}{N} \sum_{\hat{\lambda},\hat{\xi} \in P_+^k} \sum_{p,q=0}^{N-1} \chi_{\hat{\lambda}}\, \bar{\chi}_{A^p\hat{\xi}}\, e^{-2\pi i q(A\hat{\omega}_0, \xi)} e^{-\pi i p q k|A\hat{\omega}_0|^2} \tag{17.36}$$

The first phase factor corresponds to the b^q eigenvalue, and the second one comes from the newly defined product $\circ$.

MODULAR INVARIANCE OF THE PARTITION FUNCTION

We first recall how A acts on the modular matrix S (cf. Eq. (14.255) of Sect. 14.6.4):

$$AS_{\hat{\lambda}\hat{\mu}} = S_{\hat{\lambda}\hat{\mu}}e^{-2\pi i(A\hat{\omega}_0, \mu)} \tag{17.37}$$

More compactly, this is just $S^\dagger A S = b$, whose transposition reads

$$SA^t S^\dagger = b \qquad \Longrightarrow \qquad S^\dagger b S = A^{-1} \tag{17.38}$$

since both S and b are symmetric matrices (thus unaffected by a transposition) and $A^t = A^{-1}$, which is always so for a permutation matrix.

We can now check the S invariance of Eq. (17.26) in a few straightforward steps:

$$
\begin{aligned}
S^\dagger \tilde{\mathcal{M}} S &= \frac{1}{N} \sum_{p,q=0}^{N-1} S^\dagger A^p S \circ S^\dagger b^q S \\
&= \frac{1}{N} \sum_{p,q=0}^{N-1} b^p \circ A^{-q} \\
&= \tilde{\mathcal{M}}
\end{aligned} \tag{17.39}
$$

Indeed, the $\circ$ product is symmetric, and $A^{-q} = A^{N-q} = A^\ell$ for $\ell = N - q$.

For the T transformation, we start from

$$(T^\dagger A T)_{\hat{\lambda}\hat{\mu}} = \delta_{\hat{\lambda},A\hat{\mu}}\, e^{-2\pi i [h_{A\hat{\mu}} - h_{\hat{\mu}}]} \tag{17.40}$$

With another application of Eq. (14.250), we see that

$$
\begin{aligned}
h_{A\hat{\mu}} &= \frac{|A(\hat{\mu} + \hat{\rho})|^2 - |\rho|^2}{2(k+g)} \\
&= h_{\hat{\mu}} + \frac{(k+g)}{2}|A\hat{\omega}_0|^2 + (A\hat{\omega}_0, w_A(\mu + \rho))
\end{aligned} \tag{17.41}
$$

Substituting Eq. (17.41) into Eq. (17.40) and again ignoring the factor w_A, we see that the identity (cf. Eq. (14.285))

$$e^{-2\pi i (A\hat{\omega}_0, \rho)} = e^{2\pi i (A\hat{\omega}_0, \rho)} = e^{-\pi i g |A\hat{\omega}_0|^2} \tag{17.42}$$

(the first equality holds because the phase is ± 1) leads to

$$(T^\dagger A T)_{\hat{\lambda}\hat{\mu}} = \delta_{\hat{\lambda},A\hat{\mu}}\, e^{-\pi i k |A\hat{\omega}_0|^2 - 2\pi i (A\hat{\omega}_0, \mu)} \tag{17.43}$$

This result can be written compactly as

$$T^\dagger A T = A \circ b \tag{17.44}$$

On the other hand, b, being a diagonal matrix with entries[9]

$$(b)_{\hat{\lambda}\hat{\lambda}} = e^{-2\pi i (A\hat{\omega}_0, \lambda)} \tag{17.45}$$

is clearly unaffected by the action of T, so that

$$T^\dagger b T = b \tag{17.46}$$

These two equations make the T invariance of $\tilde{\mathcal{M}}$ obvious.

---

[9] For the matrix elements of b, we write unnecessary parentheses to avoid any confusion with branching functions.

17.3.2. Constraints on the Partition Function

UNIQUENESS OF THE VACUUM

The vacuum state belongs to the diagonal part of the partition function, namely the untwisted sector:

$$\sum_{\hat\lambda\in P_+^k}\chi_{\hat\lambda}(\tau)\,\mathcal{B}\,\bar\chi_{\hat\lambda}(\bar\tau)=\frac{1}{N}\sum_{\hat\lambda\in P_+^k}\chi_{\hat\lambda}(\tau)\left(\sum_{q=0}^{N-1}(b^q)_{\hat\lambda,\hat\lambda}\right)\bar\chi_{\hat\lambda}(\bar\tau)\tag{17.47}$$

Since $(b^q)_{00}=1$ for all q (cf. Eq. (17.45)), the constraint $\mathcal{M}_{00}=1$ is automatically implemented by the construction.

INTEGRALITY OF THE MASS MATRIX

An explicit mass-matrix element reads

$$\tilde{\mathcal{M}}_{\hat\lambda\hat\mu}=\frac{1}{N}\sum_{p=0}^{N-1}\delta_{\hat\lambda,A^p\hat\mu}\sum_{q=0}^{N-1}e^{-2\pi iq(A\hat\omega_0,\mu)-\pi ipqk|A\hat\omega_0|^2}\tag{17.48}$$

Each term in the summation over q is an N-th root of unity. Thus, the sum over q vanishes unless

$$e^{-2\pi i(A\hat\omega_0,\mu)-\pi ipk|A\hat\omega_0|^2}=1\tag{17.49}$$

in which case the sum is equal to N. Hence $\tilde{\mathcal{M}}_{\hat\lambda\hat\mu}$ can be rewritten as

$$\tilde{\mathcal{M}}_{\hat\lambda\hat\mu}=\sum_{p=0}^{N-1}\delta_{\hat\lambda,A^p\hat\mu}\,\delta_1((A\hat\omega_0,\mu+\frac{pk}{2}A\hat\omega_0))\tag{17.50}$$

where

$$\delta_1(x)=\begin{cases}1 & \text{if }x\in\mathbb{Z}\\0 & \text{otherwise}\end{cases}\tag{17.51}$$

The form (17.50) makes manifest that $\tilde{\mathcal{M}}_{\hat\lambda\hat\mu}$ is a nonnegative integer. So the integrality of the mass matrix is also guaranteed by the construction.

A CONSISTENCY CONDITION

Since $b^N=1$, $\tilde{\mathcal{M}}_{\hat\lambda\hat\mu}$ must not be affected by the replacement $b^q\to b^{q+N}$. This clearly forces

$$N(A\hat\omega_0,\mu+\frac{pk}{2}A\hat\omega_0)=0\bmod 1\tag{17.52}$$

We have already checked that $N(A\hat\omega_0,\mu)$ must be an integer (Eq. (14.119)). We are thus left with the constraint:

$$\frac{Nk}{2}|A\hat\omega_0|^2\in\mathbb{Z}\tag{17.53}$$

This condition will place some restrictions on the level.

17.3.3. $\widehat{su}(2)$ Modular Invariants by Outer Automorphisms

We now apply the general technique developed in the previous subsection to obtain nondiagonal $\widehat{su}(2)$ modular invariants. The center of $SU(2)$ is $\mathbb{Z}_2$. To the two elements of the center, there correspond two outer automorphisms, the identity and a, defined in Eq. (17.16), which interchanges the two Dynkin labels of the $\widehat{su}(2)_k$ affine weights. Obviously, the identity $(A = I)$ is associated with the diagonal modular invariant: the conditions (17.50) and (17.53) give no constraints. We consider then the modular invariant generated by the nontrivial outer automorphism $A = a$. Since

$$|a\hat{\omega}_0|^2 = |\hat{\omega}_1|^2 = \frac{1}{2} \tag{17.54}$$

Eq. (17.53) requires k to be even. With $\hat{\mu} = [k - n, n]$, the argument of δ_1 in Eq. (17.50) is

$$\left(a\hat{\omega}_0, \mu + \frac{pk}{2}a\hat{\omega}_0\right) = \frac{n}{2} + \frac{kp}{4} \tag{17.55}$$

We will treat separately the cases $k = 4\ell$ and $k = 4\ell - 2$, with ℓ a positive integer.
a) $k = 4\ell \geq 4$
Mass matrix:

$$\tilde{M}_{n'n} = \delta_{n'n}\delta_1\left(\frac{n}{2}\right) + \delta_{n',k-n}\delta_1\left(\frac{n}{2}\right) \tag{17.56}$$

Partition function:

$$\tilde{Z} = \sum_{\substack{n=0 \\ n\in 2\mathbb{Z}}}^{4\ell} (\chi_n\bar{\chi}_n + \chi_n\bar{\chi}_{4\ell-n})$$

$$= \sum_{\substack{n=0 \\ n\in 2\mathbb{Z}}}^{2\ell-2} |\chi_n + \chi_{4\ell-n}|^2 + 2|\chi_{2\ell}|^2 \tag{17.57}$$

b) $k = 4\ell - 2 \geq 6$

Mass matrix:

$$\tilde{M}_{n'n} = \delta_{n'n}\delta_1\left(\frac{n}{2}\right) + \delta_{n',k-n}\delta_1\left(\frac{n+1}{2}\right) \tag{17.58}$$

Partition function:

$$\tilde{Z} = \sum_{\substack{n=0 \\ n\in 2\mathbb{Z}}}^{4\ell-2} \chi_n\bar{\chi}_n + \sum_{\substack{n=1 \\ n\in 2\mathbb{Z}+1}}^{4\ell-3} \chi_n\bar{\chi}_{4\ell-2-n}$$

$$= \sum_{\substack{n=0 \\ n\in 2\mathbb{Z}}}^{4\ell-2} |\chi_n|^2 + |\chi_{2\ell-1}|^2 + \sum_{\substack{n=1 \\ n\in 2\mathbb{Z}+1}}^{2\ell-3} (\chi_n\bar{\chi}_{4\ell-2-n} + \chi_{4\ell-2-n}\bar{\chi}_n) \tag{17.59}$$

Thus, we get two infinite sequences of nondiagonal modular-invariant partition functions. We note that these two sequences have a rather different structure: in the first case one field has multiplicity two and the vacuum couples to another field, whereas in the second case a spinless odd field reappears in the spectrum as a result of the twisting process. Moreover, the modular invariants of Eq. (17.57) are block-diagonal and those of Eq. (17.59) are not.

Other algebras can be treated in a similar way.

§17.4. The $\widehat{su}(2)_4$ Nondiagonal Invariant Revisited

In order to motivate the next general method for constructing nondiagonal modular invariants, we reconsider the $\widehat{su}(2)_4$ example from a new point of view. In this example, the vacuum sector is mixed with the representation of highest weight [0,4], associated with a primary field of conformal dimension 1. By $\mathcal{T}$ invariance, only fields with integer dimension can be mixed with the identity, and the lowest value of the level at which the $\widehat{su}(2)_k$ model has primary fields with integer conformal dimensions is 4.

We have already encountered a theory in which there occur extra dimension-1 fields at a particular value of the defining parameter. This is the free-boson theory compactified on a circle of radius R. At generic values of R there is only one dimension-1 field (namely $i\partial\phi$), but for $R = \sqrt{2}$ two new dimension-1 fields appear: $e^{\pm i\sqrt{2}\phi}$. These two extra fields, together with $i\partial\phi$, generate the $\widehat{su}(2)_1$ algebra. Thus, in that case, the presence of extra dimension-1 fields translates into an enhancement of the spectrum-generating algebra, from $\widehat{u}(1)$ to $\widehat{su}(2)_1$. Note that both affine algebras have the same Sugawara central charge, namely $c = 1$.

We now try a similar interpretation for the $\widehat{su}(2)_4$ case. The five states of the module $L_{[0,4]}$ at grade 0, which have conformal dimension 1, are then reinterpreted as five states occurring at grade 1 in the new vacuum module considered from the point of view of the enlarged algebra. The number of states at grade 1 in the vacuum module gives the dimension of the adjoint representation (i.e., the number of generators) of the corresponding finite algebra. Hence, these extra five states should correspond to five new generators which, together with the three $su(2)$ generators already present, form the generators of the extended algebra. That yields eight independent current-algebra generating fields. It hints that this new algebra is $\widehat{su}(3)$. But at which level? This is easily answered by comparing the central charge for $\widehat{su}(3)_k$ with that of $\widehat{su}(2)_4$:

$$\frac{8k}{k+3} = 2 \quad \Rightarrow k = 1 \tag{17.60}$$

To test the consistency of this interpretation, we look at the second part of the $\widehat{su}(2)_4$ invariant: $2|\chi_2|^2$. It naturally suggests an interpretation in terms of two conjugate $\widehat{su}(3)$ fields, which singles out [0, 1, 0] and [0, 0, 1]. This identification is supported by the equality of the involved conformal dimensions

$$h_{[2,2]} = h_{[0,1,0]} = h_{[0,0,1]} = \frac{1}{3} \qquad (17.61)$$

As a decisive confirmation, the (restricted) character identities

$$\chi_{[4,0]} + \chi_{[0,4]} = \chi_{[1,0,0]}$$
$$\chi_{[2,2]} = \chi_{[0,1,0]} = \chi_{[0,0,1]} \qquad (17.62)$$

can be checked in perturbative expansion of q. Actually, these character relations are nothing but the branching rules found in Eqs. (14.272) and (14.273), rewritten in terms of the normalized characters.

We have thus found a new interpretation of the nondiagonal $\widehat{su}(2)_4$ in terms of the $\widehat{su}(3)_1$ diagonal invariant. Finding the general mechanism to go in the reverse direction will provide us with a method for generating nondiagonal invariants from diagonal ones (and this will always produce block-diagonal invariants). Obviously, the above interpretation is possible because $su(2)$ can be embedded into $su(3)$. Moreover, the affine extension of this embedding must preserve the Sugawara central charge. This turns out to be a very restrictive condition. Embeddings having this special property are the key to the general method we are looking for: their branching rules yield directly new nondiagonal invariants out of diagonal ones. But before getting to the heart of this new construction, various aspects of such embeddings must first be described.

§17.5. Conformal Embeddings

Affine embeddings have been introduced in Sect. 14.7. Here we are interested in the subclass of affine embeddings that preserves the conformal invariance, a rather limited set, as the previous considerations suggest. In the first step, we derive a simple characterizing property for such embeddings, expressed in terms of the Sugawara central charge of the corresponding WZW models. Next, we display techniques for calculating branching rules well suited to these embeddings.

17.5.1. Conformally Invariant Embeddings

The representation space of the $\hat{g}_k$-WZW model is spanned by states of the form

$$J^a_{-n_1} J^b_{-n_2} \cdots |\lambda\rangle \qquad (17.63)$$

with $n_1 \geq n_2 \geq \ldots > 0$, and the vacuum being g invariant

$$J^a_0 |0\rangle = 0 \qquad (17.64)$$

A subalgebra truncation refers to a truncation of the space of states spanned by Eq. (17.63) to that generated by

$$\tilde{J}^{a'}_{-n_1} \tilde{J}^{b'}_{-n_2} \cdots |\mathcal{P}\lambda\rangle \qquad (17.65)$$

where $\tilde{J}^{a'}$ are the p generators (the indices $a', b' \dots$ run from 1 to dim p) and $\mathcal{P}$ is the projection matrix (i.e., $\mathcal{P}\lambda$ is the projection of $\lambda \in g$ onto p) . Clearly the g-invariance of the vacuum entails its p-invariance. However, nothing ensures that such a truncation will preserve the conformal invariance, as the consideration of the energy-momentum tensor shows. Indeed, in the Sugawara energy-momentum tensor for the $\hat{g}_k$-WZW model, we can isolate a part formed by those combinations of generators of g that are in p. But there is a remainder. Due to the latter, the application of the Virasoro modes on states of the form (17.65) creates states that generically lie outside of this set. Whenever this is the case, the truncation breaks the conformal invariance. This is in fact the generic situation.

However there are exceptions. We have already seen that for simply-laced algebras at level 1, the Sugawara energy-momentum tensor is equivalent to that obtained solely from the generators of the Cartan subalgebra. Therefore, in this particular case $T_{\hat{g}_k}$ can be reexpressed only in terms of the generators of the subalgebra, and the conformal invariance is manifestly preserved.

A truncation preserving conformal invariance must thus necessarily satisfy

$$T_{\hat{g}_k} = T_{\hat{p}_{\tilde{k}}} \tag{17.66}$$

This equality can hold only when the corresponding central charges are the same:

$$c(\hat{g}_k) = c(\hat{p}_{\tilde{k}}) \tag{17.67}$$

that is,

$$\frac{k \dim g}{k+g} = \frac{x_e k \dim p}{x_e k + p} \tag{17.68}$$

where we used the relation $\tilde{k} = x_e k$ derived in Sect. 14.7 (x_e is the embedding index defined in Eq. (13.250)) and we denoted by p the dual Coxeter number of p.

The following argument shows that Eq. (17.67) is indeed a sufficient condition for Eq. (17.66). If two theories have the same central charge, their difference (in the sense of the coset construction, to be introduced in Chap. 18) has zero central charge. Both theories under consideration are unitary by construction. Thus, their difference is also unitary and, having zero central charge, it is trivial.

Embeddings that satisfy the condition (17.68) are called *conformal embeddings*.

Quite remarkably, conformal embeddings exist only when $k = 1$. Indeed, consider the ratio:

$$f(k) \equiv \frac{c(\hat{g}_k)}{c(\hat{p}_{\tilde{k}})} = \frac{x_e k + p}{k + g} \frac{\dim g}{x_e \dim p} \tag{17.69}$$

On the one hand, $f(k)$ is a monotonically increasing function of k:

$$\frac{df}{dk} = \frac{x_e g - p}{(k+g)^2} \frac{\dim g}{x_e \dim p} > 0 \tag{17.70}$$

since for $p \subset g, p$ is smaller than g and $x_e \geq 1$. On the other hand,

$$c(\hat{g}_k) \geq c(\hat{p}_{\tilde{k}}) \quad \Rightarrow \quad f(k) \geq 1 \quad \forall k \tag{17.71}$$

Now suppose that there is an integer $k' > 1$ such that $f(k') = 1$. Then, for all $k < k'$, we would have $f(k) < f(k')$; hence $f(k) < 1$, in contradiction with the above inequality. Therefore, the only possible solutions of Eq. (17.68) are at $k = 1$. Thus, there is a finite number of possible conformal embeddings, and they have been fully classified.

According to the above criterion, the embedding $su(2) \subset su(3)$ with $x_e = 4$ considered above is conformal. Other interesting examples with future applications are:

$$\widehat{su}(2)_{10} \subset \widehat{sp}(4)_1 \quad , \quad \widehat{su}(2)_{28} \subset (\hat{G}_2)_1 \quad , \quad \widehat{su}(2)_{16} \oplus \widehat{su}(3)_6 \subset (\hat{E}_8)_1$$

$$(17.72)$$

17.5.2. Conformal Branching Rules

The problem of computing affine branching rules

$$\hat{\lambda} \mapsto \bigoplus_{\hat{\mu}} b_{\hat{\lambda}\hat{\mu}} \ \hat{\mu} \tag{17.73}$$

has been addressed in general terms in Sect. 14.7.2. For conformal embeddings, there are more efficient ways to proceed, which we now indicate.

First, we observe that the nonvanishing of $b_{\hat{\lambda}\hat{\mu}}$ means that the finite weight μ, the tip of the module $\mathsf{L}_{\hat{\mu}}$, can be found at some grade n in the infinite-dimensional highest-weight representation $\mathsf{L}_{\hat{\lambda}}$ at level 1. The assumed preservation of the conformal structure by the embedding implies that the conformal dimensions of the corresponding fields can be compared. This translates into the equality

$$h_{\hat{\lambda}} + n = h_{\hat{\mu}} \tag{17.74}$$

equivalent to

$$\frac{(\lambda, \lambda + 2\rho)}{2(1 + g)} + n = \frac{(\mu, \mu + 2\rho)}{2(x_e + p)} \tag{17.75}$$

A simple way of obtaining the branching rules is to compute the dimension spectrum of all integrable representations of the two algebras under consideration and find the triplets (λ, μ, n) satisfying Eq. (17.75). Then, we look at the decomposition of the module $\mathsf{L}_{\hat{\lambda}}$ at grade n in terms of irreducible representations of g, and write down all the finite branching rules of these irreducible g representations into irreducible representations of p. (This is a finite process, since the difference in the conformal dimensions is always bounded.) The number of times the particular representation μ appears in all these finite branching rules at grade n is precisely the coefficient $b_{\hat{\lambda}\hat{\mu}}$.

Some examples will clarify the procedure. Consider again the embedding $\widehat{su}(2)_4 \subset \widehat{su}(3)_1$. The corresponding lists of conformal dimensions are:

$$\widehat{su}(2)_4: \quad h_{[4,0]} = 0, \ h_{[3,1]} = \frac{1}{8}, \ h_{[2,2]} = \frac{1}{3}, \ h_{[1,3]} = \frac{5}{8}, \ h_{[0,4]} = 1$$

$$(17.76)$$

$$\widehat{su}(3)_1: \quad h_{[1,0,0]} = 0, \ h_{[0,1,0]} = h_{[0,0,1]} = \frac{1}{3}$$

which leads directly to

$$[1,0,0] \mapsto c_1[4,0]_0 \oplus c_2[0,4]_1$$
$$[0,1,0] \mapsto c_3[2,2]_0 \qquad\qquad (17.77)$$
$$[0,0,1] \mapsto c_4[2,2]_0$$

where the c_i's are integers to be determined; the subscript indicates the value of n. The coefficients c_1, c_3, and c_4 can thus be obtained from the finite branching rules at grade zero. The g irreducible content of $L_{\hat{\lambda}}$ at grade zero is just L_{λ}. Thus, from the finite branching rules

$$(0,0) \mapsto (0), \qquad (1,0) \mapsto (2), \qquad (0,1) \mapsto (2) \qquad (17.78)$$

we conclude that $c_1 = c_3 = c_4 = 1$. To find c_2, it is necessary to consider the vacuum representation at grade 1, whose $su(3)$ irreducible content is simply $(1,1)$ (obtained by adding to $(0,0)$, the only weight at grade 0, all the weights of the representation L_θ). The branching rule

$$(1,1) \mapsto (4) \oplus (2) \qquad (17.79)$$

shows that c_2 is also equal to unity.

A slightly more complicated example is $\widehat{su}(2)_{10} \subset \widehat{sp}(4)_1$, for which the finite projection matrix is $\mathcal{P} = (3,4)$. The conformal dimension of the integrable representations of $\widehat{sp}(4)$ are

$$\widehat{sp}(4)_1: \quad h_{[1,0,0]} = 0, \; h_{[0,1,0]} = \frac{5}{16}, \; h_{[0,0,1]} = \frac{1}{2} \qquad (17.80)$$

Comparing these with the set of dimensions of the integrable representations of $\widehat{su}(2)_{10}$:

$$\widehat{su}(2)_{10}: \quad h_{[10-\lambda_1,\lambda_1]} = \{0, \frac{1}{16}, \frac{5}{24}, \frac{5}{16}, \frac{1}{2}, \frac{35}{48}, 1, \frac{21}{16}, \frac{5}{3}, \frac{33}{16}, \frac{5}{2}\} \qquad (17.81)$$

(with the ordering $0 \le \lambda_1 \le 10$), we find

$$[1,0,0] \mapsto c_1[10,0]_0 \oplus c_2[4,6]_1$$
$$[0,1,0] \mapsto c_3[7,3]_0 \; \oplus c_4[3,7]_1 \qquad (17.82)$$
$$[0,0,1] \mapsto c_5[6,4]_0 \; \oplus c_6[0,10]_2$$

where again the c_i are positive integers. The finite branching rules

$$(0,0) \mapsto (0), \qquad (1,0) \mapsto (3), \qquad (0,1) \mapsto (4) \qquad (17.83)$$

imply that $c_1 = c_3 = c_5 = 1$. The representation $[1,0,0]$ at grade 1 contains only the finite representation $(2,0)$ (the adjoint), and the branching rule

$$(2,0) \to (6) \oplus (2) \qquad (17.84)$$

yields $c_2 = 1$. On the other hand, the weights in the finite representation $(1,0)$ are

$$\Omega_{(1,0)} = \{(1,0), (-1,1), (1,-1), (-1,0)\} \qquad (17.85)$$

and they all have multiplicity one. Since the zeroth Dynkin label of the affine extension of the last three weights is positive, we can subtract α_0 from these weights. From the resulting weights, we subtract $\alpha_{1,2}$ in all possible ways. The finite projection of all weights at grade 1 can then be reorganized in two irreducible representations, of highest weights $(1,1)$ and $(1,0)$. It may then be checked that (7) occurs with multiplicity 1 in the decomposition of $(1,1)$: Hence $c_4 = 1$. To obtain c_6, we should consider the irreducible content of $[0,0,1]$ at grade 2, which is somewhat involved. But there is a shortcut which uses the relation between the outer-automorphism groups $\mathcal{O}(\widehat{su}(2))$ and $\mathcal{O}(\widehat{sp}(4))$. Both outer-automorphism groups are isomorphic to $\mathbb{Z}_2$. For $\widehat{sp}(4)$, the action of its generator $\tilde{a}$ amounts to exchanging the zeroth and second Dynkin labels, leaving the first one unaffected. For $\widehat{su}(2)$, it interchanges the two affine Dynkin labels. It is simple to check that (cf. Eq. (14.278)):

$$sp(4): \qquad (\tilde{a}\hat{\omega}_0, \lambda) = (\omega_2, \lambda) = \frac{1}{2}\lambda_1 + \lambda_2 = \frac{1}{2}\lambda_1 \quad \text{mod } 1 \qquad (17.86)$$

while

$$su(2): \qquad (a\hat{\omega}_0, \mathcal{P}\lambda) = (\omega_1, (3\lambda_1 + 4\lambda_2)\omega_1) = \frac{3}{2}\lambda_1 + 2\lambda_2 = \frac{1}{2}\lambda_1 \quad \text{mod } 1 \qquad (17.87)$$

This shows that the two generators of the outer-automorphism groups are in correspondence, that is, $\tilde{a} \mapsto a$. At this point, we have found

$$[1,0,0] \mapsto [10,0] \oplus [4,6]$$
$$[0,1,0] \mapsto [7,3] \oplus [3,7] \qquad (17.88)$$

The second branching rule above is seen to be invariant under the action of $\tilde{a}$, but on the first one it yields

$$[0,0,1] \mapsto [0,10] \oplus [6,4] \qquad (17.89)$$

which shows that $c_6 = 1$.[10]

The branching coefficients satisfy a simple sum rule, which can also be used for their determination. We recall that to the branching rule (17.73) corresponds the (normalized) character identity:

$$\chi_{\mathcal{P}\hat{\lambda}}(\zeta; \tau; t) = \sum_{\hat{\mu} \in P_+^{k_{xe}}} \chi_{\{\hat{\lambda}; \hat{\mu}\}}(\tau) \, \chi_{\hat{\mu}}(\zeta; \tau; t) \qquad (17.90)$$

where

$$\chi_{\{\hat{\lambda}; \hat{\mu}\}}(\tau) = e^{2\pi i \tau(m_{\hat{\lambda}} - m_{\hat{\mu}})} \, b_{\hat{\lambda}\hat{\mu}}(\tau) \qquad (17.91)$$

with

$$m_{\hat{\lambda}} = h_{\hat{\lambda}} - \frac{c}{24} \qquad (17.92)$$

----

[10] Another example of the use of outer-automorphism groups to evaluate branching rules is given in App. 17.A.

(cf. Eq. (15.122)). (It should be clear that ζ in Eq. (17.90) must be a p weight.) For a conformal embedding, the central charges appearing in $m_{\hat{\lambda}}$ and $m_{\hat{\mu}}$ are the same and since the conformal dimensions are related by Eq. (17.74), we find

$$\chi_{\{\hat{\lambda};\hat{\mu}\}}(\tau) = e^{-2\pi i n t} b_{\hat{\lambda}\hat{\mu}}(\tau) \tag{17.93}$$

The sum rule we are looking for is obtained by evaluating Eq. (17.90) in the limit $\tau \to 0$, with $\zeta = t = 0$ (and when $\zeta = 0$, the projection operator on the l.h.s. of Eq. (17.90) is no longer required). For this, we use the asymptotic relation appropriate to characters of integrable representations:

$$\chi_{\hat{\lambda}}(\tau \to i0^+) \sim S_{\hat{\lambda}0} \, e^{i\pi c/12\tau} \tag{17.94}$$

The desired formula is

$$S_{\hat{\lambda}0} = \sum_{\hat{\mu}} b_{\hat{\lambda}\hat{\mu}} S_{\hat{\mu}0} \tag{17.95}$$

This can be used to fix the final form of the branching rules once a few coefficients have been found (e.g., those that can be fixed at grade zero). For instance, for $\widehat{su}(2)_4 \subset \widehat{su}(3)_1$, we found the branching rules (17.77) and the analysis at level 0 fixes $c_1 = c_3 = c_4 = 1$. Then, by comparing the first rows of the two S matrices

$$\widehat{su}(3)_1: \qquad S = \frac{1}{\sqrt{3}} \begin{pmatrix} 1 & 1 & 1 \\ 1 & -\frac{1}{2} + \frac{i\sqrt{3}}{2} & -\frac{1}{2} - \frac{i\sqrt{3}}{2} \\ 1 & -\frac{1}{2} - \frac{i\sqrt{3}}{2} & -\frac{1}{2} + \frac{i\sqrt{3}}{2} \end{pmatrix} \tag{17.96}$$

(with the ordering [1,0,0], [0,1,0] and [0,0,1]) and

$$\widehat{su}(2)_4: \qquad S = \frac{1}{2\sqrt{3}} \begin{pmatrix} 1 & \sqrt{3} & 2 & \sqrt{3} & 1 \\ \sqrt{3} & \sqrt{3} & 0 & -\sqrt{3} & -\sqrt{3} \\ 2 & 0 & -2 & 0 & 2 \\ \sqrt{3} & -\sqrt{3} & 0 & \sqrt{3} & -\sqrt{3} \\ 1 & -\sqrt{3} & 2 & -\sqrt{3} & 1 \end{pmatrix} \tag{17.97}$$

(with the ordering $0 \le \lambda_1 \le 4$), we see that in

$$[1,0,0] \to [4,0] \oplus c_2[0,4] \tag{17.98}$$

c_2 must satisfy

$$\frac{1}{\sqrt{3}} = \frac{1}{2\sqrt{3}} + c_2 \frac{1}{2\sqrt{3}} \qquad \Rightarrow c_2 = 1 \tag{17.99}$$

The presence of only a finite number of terms in Eq. (17.90) is directly linked to the conformal nature of the embedding, i.e., that the Sugawara central charge is the same in the two theories. Indeed, if

$$\Delta c \equiv c(\hat{\mathfrak{g}}_k) - c(\hat{\mathfrak{p}}_{\hat{k}}) \neq 0 \tag{17.100}$$

in which case $\Delta c > 0$,[11] then, for the following equality to hold in the limit $\tau \to i0^+$,

$$S_{\hat{\lambda}0} = \sum_{\hat{\mu}} e^{-i\pi\Delta c/12\tau} \chi_{\{\hat{\lambda};\hat{\mu}\}} S_{\hat{\mu}0} \qquad (17.101)$$

there must necessarily be an infinite number of terms in the sum. These are all the terms in the infinite series expansion of $\chi_{\{\hat{\lambda};\hat{\mu}\}}(q)$ evaluated in the limit $q \to 1^-$. This result is sometimes called the *finite reducibility theorem*.

Finally, it should be stressed that the naive methods presented above are not always sufficient to fully determine the conformal branching rules. A typical case occurs when p is not simple, for instance in the embedding $\widehat{su}(p)_q \oplus \widehat{su}(q)_p \subset \widehat{su}(pq)_1$. However, this case (and similar infinite series) can be treated by means of Young tableau techniques and the use of outer automorphisms, as illustrated in App. 17.A.

§17.6. Modular Invariants From Conformal Embeddings

We are now in a position to see how nondiagonal modular invariants can be constructed out of diagonal ones (or, more generally, any known modular invariant) through conformal embeddings. The construction is very simple: we just substitute, in the $\hat{g}_1$-WZW diagonal modular invariant, the branching rules in character form. Modular invariance is manifestly preserved in this process: the modular invariance of the nondiagonal theory is inherited from that of the diagonal theory.

For instance, the branching rules (17.82) for the embedding $\widehat{su}(2)_{10} \subset \widehat{sp}(4)_1$ lead to

$$Z = |\chi_{[10,0]} + \chi_{[4,6]}|^2 + |\chi_{[7,3]} + \chi_{[3,7]}|^2 + |\chi_{[6,4]} + \chi_{[0,10]}|^2 \qquad (17.102)$$

In contradistinction to the $\widehat{su}(2)_4 \subset \widehat{su}(3)_1$ example, this invariant cannot be obtained by the method of outer automorphisms.

Another example of interest is $\widehat{su}(2)_{28} \subset (\widehat{G}_2)_1$, for which the conformal branching rules are

$$[1,0,0] \mapsto [28,0] \oplus [18,10] \oplus [10,18] \oplus [0,28]$$
$$[0,0,1] \mapsto [22,6] \oplus [16,12] \oplus [12,16] \oplus [6,22] \qquad (17.103)$$

(the above $(\widehat{G}_2)_1$ weights are the only integrable weights at level one). The replacement of the rules (17.103) in the $(\widehat{G}_2)_1$ diagonal modular invariant yields

$$Z = |\chi_{[28,0]} + \chi_{[18,10]} + \chi_{[10,18]} + \chi_{[0,28]}|^2$$
$$+ |\chi_{[22,6]} + \chi_{[16,12]} + \chi_{[12,16]} + \chi_{[6,22]}|^2. \qquad (17.104)$$

---

[11] Quotienting a unitary theory from another unitary theory cannot give a nonunitary theory with $\Delta c < 0$.

This is another example of a partition function that cannot be interpreted in terms of outer automorphisms.

Up to now, we have considered only the case for which the embedded algebra is simple. Whenever it is semisimple (e.g., $p = p^{(1)} \oplus p^{(2)}$), another construction is possible. It consists in contracting the result obtained by substituting the branching rules for $\hat{p}^{(1)}_{k_1} \oplus \hat{p}^{(2)}_{\tilde{k}_2} \subset \hat{g}_1$ into the $\hat{g}_1$ invariant, with a known $\hat{p}^{(2)}$ modular invariant, producing then a $\hat{p}^{(1)}$ invariant. We denote the nondiagonal $\hat{p}^{(1)}_{k_1} \oplus \hat{p}^{(2)}_{\tilde{k}_2}$ mass matrix by:

$$\mathcal{M}_{\hat{\lambda}^{(1)}\hat{\xi}^{(2)},\hat{\mu}^{(1)}\hat{\nu}^{(2)}} \qquad \text{with} \quad \hat{\lambda}^{(1)}, \hat{\mu}^{(1)} \in \hat{p}^{(1)}_{k_1}, \quad \hat{\xi}^{(2)}, \hat{\nu}^{(2)} \in \hat{p}^{(2)}_{\tilde{k}_2} \qquad (17.105)$$

and the known $\hat{p}^{(2)}$ mass matrix by $\mathcal{M}^{(2)}_{\hat{\xi}^{(2)},\hat{\nu}^{(2)}}$. A new $\hat{p}^{(1)}_{k_1}$ invariant mass matrix is found by the following contraction:

$$\mathcal{M}^{(1)}_{\hat{\lambda}^{(1)},\hat{\mu}^{(1)}} = \sum_{\hat{\xi}^{(2)},\hat{\nu}^{(2)} \in P^{\tilde{k}_2}_+} \mathcal{M}_{\hat{\lambda}^{(1)}\hat{\xi}^{(2)},\hat{\mu}^{(1)}\hat{\nu}^{(2)}} \mathcal{M}^{(2)}_{\hat{\xi}^{(2)},\hat{\nu}^{(2)}} \qquad (17.106)$$

In practice, we usually choose the $\hat{p}^{(2)}$ invariant to be diagonal. In that case, the new mass matrix takes the form

$$\mathcal{M}^{(1)}_{\hat{\lambda}^{(1)},\hat{\mu}^{(1)}} = \sum_{\hat{\nu}^{(2)} \in P^{\tilde{k}_2}_+} \mathcal{M}_{\hat{\lambda}^{(1)}\hat{\nu}^{(2)},\hat{\mu}^{(1)}\hat{\nu}^{(2)}} \qquad (17.107)$$

This implies that whatever multiplies the $\hat{p}^{(2)}$ singlet part—that is, the set of all terms of the form $\chi_{\hat{\nu}^{(2)}} \bar{\chi}_{\hat{\nu}^{(2)}}$, for any integrable weights $\hat{\nu}^{(2)}$ of $\hat{p}^{(2)}_{\tilde{k}_2}$—is by construction modular invariant.

However, this construction does not guarantee the uniqueness of the vacuum, and to obtain a physically admissible modular invariant, we generally have to subtract from the result a known invariant, or divide it by a positive integer, or both.

To illustrate this method, we consider the conformal embedding

$$\widehat{su}(2)_{16} \oplus \widehat{su}(3)_6 \subset (\widehat{E}_8)_1 \qquad (17.108)$$

The unique $\widehat{E}_8$ integrable representation at level 1 is the vacuum. We take for granted that its branching rule is

$$[1,0,0,0,0,0,0,0,0] \mapsto ([16,0] \oplus [0,16]) \otimes ([6,0,0] \oplus [0,6,0] \oplus [0,0,6])$$
$$\oplus ([12,4] \oplus [4,12]) \otimes ([3,3,0] \oplus [0,3,3] \oplus [3,0,3])$$
$$\oplus ([14,2] \oplus [2,14] \oplus 2[8,8]) \otimes ([2,2,2])$$
$$\oplus ([10,6] \oplus [6,10]) \otimes ([4,1,1] \oplus [1,4,1] \oplus [1,1,4])$$
$$(17.109)$$

Extracting the $\widehat{su}(3)$ singlet part from the absolute square value of the character form of this expression yields

$$Z = 3|\chi_{[16,0]} + \chi_{[0,16]}|^2 + 3|\chi_{[10,6]} + \chi_{[6,10]}|^2$$
$$+ 3|\chi_{[12,4]} + \chi_{[4,12]}|^2 + |\chi_{[14,2]} + \chi_{[2,14]} + 2\chi_{[8,8]}|^2 \qquad (17.110)$$

At this point, the vacuum (represented by $\chi_{[16,0]}\bar{\chi}_{[16,0]}$) has multiplicity three. However, Eq. (17.110) contains the combination (17.57) for $\ell = 4$:

$$
\begin{aligned}
|\chi_{[16,0]} + \chi_{[0,16]}|^2 + |\chi_{[14,2]} + \chi_{[2,14]}|^2 \\
+ |\chi_{[12,4]} + \chi_{[4,12]}|^2 + |\chi_{[10,6]} + \chi_{[6,10]}|^2 + 2|\chi_{[8,8]}|^2
\end{aligned}
\tag{17.111}
$$

By subtracting Eq. (17.111) from Eq. (17.110), we find that a common factor of 2 can now be divided out, producing the new $\widehat{su}(2)$ modular invariant:

$$
\begin{aligned}
Z = |\chi_{[16,0]} + \chi_{[0,16]}|^2 + |\chi_{[12,4]} + \chi_{[4,12]}|^2 + |\chi_{[10,6]} + \chi_{[6,10]}|^2 \\
+ |\chi_{[8,8]}|^2 + \chi_{[8,8]}(\bar{\chi}_{[14,2]} + \bar{\chi}_{[2,14]}) + (\chi_{[14,2]} + \chi_{[2,14]})\bar{\chi}_{[8,8]}.
\end{aligned}
\tag{17.112}
$$

With this second method, it is possible to rederive the infinite sequences of modular invariants obtained by the method of outer automorphisms, although the present approach is more tedious. The difficulty with the method of conformal embeddings is rooted in the computation of branching rules. However, it is important to understand that it produces modular invariants that cannot be obtained by the method of outer automorphisms.

It might seem that all nondiagonal modular invariants can be obtained from conformal embeddings of affine Lie algebras. This is almost true, but a few counterexamples are known. One of them is the $(\widehat{F_4})_3$ invariant:

$$
\begin{aligned}
|\chi_{[3,0,0,0,0]}|^2 + |\chi_{[1,0,0,0,2]}|^2 + |\chi_{[1,0,0,1,0]}|^2 + |\chi_{[0,0,0,1,1]}|^2 + |\chi_{[0,1,0,0,1]}|^2 \\
+ \left\{ \chi_{[0,0,0,0,1]}\bar{\chi}_{[0,0,1,0,0]} + \chi_{[0,0,0,0,3]}\bar{\chi}_{[1,1,0,0,0]} + \text{c.c.} \right\}
\end{aligned}
\tag{17.113}
$$

Since it is trivial, the outer-automorphism group of $\widehat{F}_4$ cannot be obtained by the method of outer automorphisms and, being a permutation invariant, it cannot be obtained by a conformal embedding. Hence none of these methods is complete, nor is their union. A third general method will be presented in Sect. 17.9. Unfortunately, these three methods together do not explain all known invariants.

§17.7. Some Classification Results

17.7.1. The ADE Classification of the $\widehat{su}(2)$ Modular Invariants

The complete list of all distinct $\widehat{su}(2)$ modular invariants found in the previous sections is

$$
\begin{aligned}
k, \ (A_{k+1}): \quad & \sum_{\substack{n=0 \\ n\in\mathbb{Z}}}^{k} |\chi_n|^2 \\[2em]
k = 4\ell, \ (D_{2\ell+2}): \quad & \sum_{\substack{n=0 \\ n\in 2\mathbb{Z}}}^{2\ell-2} |\chi_n + \chi_{4\ell-n}|^2 + 2|\chi_{2\ell}|^2 \\[2em]
k = 4\ell - 2, \ (D_{2\ell+1}): \quad & \sum_{\substack{n=0 \\ n\in 2\mathbb{Z}}}^{4\ell-2} |\chi_n|^2 + |\chi_{2\ell-1}|^2 \\[1em]
& + \sum_{\substack{n=1 \\ n\in 2\mathbb{Z}+1}}^{2\ell-3} (\chi_n \bar{\chi}_{4\ell-2-n} + \chi_{4\ell-2-n}\bar{\chi}_n) \\[1em]
k = 10, \ (E_6): \quad & |\chi_0 + \chi_6|^2 + |\chi_3 + \chi_7|^2 + |\chi_4 + \chi_{10}|^2 \\[0.5em]
k = 16, \ (E_7): \quad & |\chi_0 + \chi_{16}|^2 + |\chi_4 + \chi_{12}|^2 + |\chi_6 + \chi_{10}|^2 + |\chi_8|^2 \\[0.5em]
& + \chi_8(\bar{\chi}_2 + \bar{\chi}_{14}) + (\chi_2 + \chi_{14})\bar{\chi}_8 \\[0.5em]
k = 28, \ (E_8): \quad & |\chi_0 + \chi_{10} + \chi_{18} + \chi_{28}|^2 \\[0.5em]
& + |\chi_6 + \chi_{12} + \chi_{16} + \chi_{22}|^2
\end{aligned}
$$

(17.114)

(Here we return to the notation where the subindex gives the finite Dynkin label.)

This list has been proven to be exhaustive. Quite remarkably, each of these partition functions can be associated with a simply-laced simple Lie algebra (as indicated above), and all simply-laced simple Lie algebras are represented in the list. The relation consists of the following: (1) $k+2$ is the dual Coxeter number of the associated Lie algebra and (2) the Dynkin labels of the diagonal fields appearing in a given modular invariant, taking their multiplicities into account, are exactly the exponents (minus one) of the associated Lie algebra (the exponents listed in App. 13.A).

Needless to say, this interrelation is purely structural. The simply-laced Lie algebra is not a symmetry of its corresponding modular invariant.

How can we understand the ADE classification underlying $\widehat{su}(2)$ modular invariants? Unfortunately, this must still be regarded as a mystery. A posteriori, some arguments justifying a connection between simply-laced algebras and $\widehat{su}(2)$ modular invariants have been found. But these arguments certainly do not capture the essence of this classification: this is evident by their inability to generalize it to other $\hat{g}$ invariants.[12]

By analogy with the $\widehat{su}(2)$ case, the diagonal invariant of a generic WZW model is said to be of A-type; nondiagonal invariants that can be obtained by the method

---

[12] A generalizable phenomenological interpretation of the ADE correspondence is presented in Sect. 17.10.

of outer automorphism are called D-type invariants; and the other invariants are called *exceptional*.

17.7.2. The Classification of the $\widehat{su}(3)$ Modular Invariants

The $\widehat{su}(3)$ modular invariants have also been the object of a classification, which we present here without proof. In addition to the A and D series, five exceptional invariants have been found. Two of them appear at level 5:

$$
\begin{aligned}
Z_{\mathcal{E}_5} = &|\chi_{(0,0)} + \chi_{(2,2)}|^2 + |\chi_{(0,2)} + \chi_{(3,2)}|^2 + |\chi_{(2,0)} + \chi_{(2,3)}|^2 \\
&+ |\chi_{(2,1)} + \chi_{(0,5)}|^2 + |\chi_{(3,0)} + \chi_{(0,3)}|^2 + |\chi_{(1,2)} + \chi_{(5,0)}|^2
\end{aligned}
\tag{17.115}
$$

and

$$
\begin{aligned}
Z_{\mathcal{E}_5}^{\,c} = &|\chi_{(0,0)} + \chi_{(2,2)}|^2 + |\chi_{(3,0)} + \chi_{(0,3)}|^2 \\
&+ \{(\chi_{(0,2)} + \chi_{(3,2)})(\bar{\chi}_{(2,0)} + \bar{\chi}_{(2,3)}) + \text{c.c.}\} \\
&+ \{(\chi_{(1,2)} + \chi_{(5,0)})(\bar{\chi}_{(2,1)} + \bar{\chi}_{(0,5)}) + \text{c.c.}\}
\end{aligned}
\tag{17.116}
$$

These two invariants are related by charge conjugation, that is, $\mathcal{M} \to \mathcal{CM}$ (cf. the analysis of the next section), and $Z_{\mathcal{E}_5}$ can be obtained from the conformal embedding $\widehat{su}(3)_5 \subset \widehat{su}(6)_1$. (The meaning of the subindex $\mathcal{E}_5$ will be clarified in Sect. 17.10). There are two additional exceptional invariants at level 9. One is given by

$$
\begin{aligned}
Z_{\mathcal{E}_9} = &|\chi_{(0,0)} + \chi_{(0,9)} + \chi_{(9,0)} + \chi_{(4,4)} + \chi_{(4,1)} + \chi_{(1,4)}|^2 \\
&+ 2|\chi_{(2,2)} + \chi_{(2,5)} + \chi_{(5,2)}|^2
\end{aligned}
\tag{17.117}
$$

which follows directly from the conformal embedding $\widehat{su}(3)_9 \subset (\widehat{E}_6)_1$. The other is

$$
\begin{aligned}
Z'_{\mathcal{E}_9} = &|\chi_{(0,0)} + \chi_{(0,9)} + \chi_{(9,0)}|^2 + |\chi_{(2,2)} + \chi_{(2,5)} + \chi_{(5,2)}|^2 \\
&+ |\chi_{(0,3)} + \chi_{(6,0)} + \chi_{(3,6)}|^2 + |\chi_{(3,0)} + \chi_{(0,6)} + \chi_{(6,3)}|^2 \\
&+ |\chi_{(4,4)} + \chi_{(4,1)} + \chi_{(1,4)}|^2 + +2|\chi_{(3,3)}|^2 \\
&+ \{(\chi_{(1,1)} + \chi_{(1,7)} + \chi_{(7,1)})\bar{\chi}_{(3,3)} + \text{c.c.}\}
\end{aligned}
\tag{17.118}
$$

and its origin will be discussed in the following section. The final exceptional invariant shows up at level 21:

$$
\begin{aligned}
Z_{\mathcal{E}_{21}} = &|\chi_{(0,0)} + \chi_{(4,4)} + \chi_{(6,6)} + \chi_{(10,10)} + \chi_{(21,0)} + \chi_{(0,21)} \\
&+ \chi_{(13,4)} + \chi_{(4,13)} + \chi_{(10,1)} + \chi_{(1,10)} + \chi_{(9,6)} + \chi_{(6,9)}|^2 \\
&+ |\chi_{(15,6)} + \chi_{(6,15)} + \chi_{(15,0)} + \chi_{(0,15)} + \chi_{(10,7)} + \chi_{(7,10)} \\
&+ \chi_{(10,4)} + \chi_{(4,10)} + \chi_{(7,4)} + \chi_{(4,7)} + \chi_{(6,0)} + \chi_{(0,6)}|^2
\end{aligned}
\tag{17.119}
$$

This invariant corresponds to the conformal embedding $\widehat{su}(3)_{21} \subset (\widehat{E}_7)_1$.

§17.8. Permutation Invariants and Extended Chiral Algebras

In view of classifying modular invariants, a useful concept is that of *permutation invariants* (already encountered in Sect. 10.7.4). These are of the form:

$$Z(\tau) = \sum_{\hat{\lambda} \in P_+^k} \chi_{\hat{\lambda}}(\tau) \, \bar{\chi}_{\Pi(\hat{\lambda})}(\bar{\tau}) \qquad (17.120)$$

where $\Pi(\hat{\lambda})$ stands for a permutation of the weights $\hat{\lambda} \in P_+^{(k)}$ satisfying

$$\mathcal{S}_{\hat{\lambda},\hat{\mu}} = \mathcal{S}_{\Pi(\hat{\lambda}),\Pi(\hat{\mu})} \qquad \mathcal{T}_{\hat{\lambda},\hat{\mu}} = \mathcal{T}_{\Pi(\hat{\lambda}),\Pi(\hat{\mu})} \qquad (17.121)$$

A diagonal invariant is thus a trivial permutation invariant. Another example of a permutation invariant is obtained by replacing the unit mass matrix in the diagonal invariant by the charge conjugation matrix. Since $C\chi_{\hat{\lambda}} = \chi_{\hat{\lambda}^*}$, this yields

$$Z^c(\tau, \bar{\tau}) = \sum_{\hat{\lambda} \in P_+^k} \chi_{\hat{\lambda}}(\tau) \, \bar{\chi}_{\hat{\lambda}^*}(\bar{\tau}) \qquad (17.122)$$

Its modular invariance is easily checked: since $h_{\hat{\lambda}} = h_{\hat{\lambda}^*}$, the invariance under $\tau \to \tau + 1$ is obvious; for $\tau \to -1/\tau$, it follows from

$$\mathcal{S}C\bar{\mathcal{S}} = \mathcal{S}^3\bar{\mathcal{S}} = \mathcal{S}^{-1}\bar{\mathcal{S}} = \mathcal{S}^{-2} = C \qquad (17.123)$$

using the unitarity of the $\mathcal{S}$ matrix and $\mathcal{S}^4 = 1$. If no "angle" is introduced to distinguish the eigenvalues of the Cartan subalgebra (i.e., no ζ dependence in the specialized characters), this partition function is numerically equal to the diagonal one, even though it has a different field content.

The $\widehat{su}(2)$ permutation invariants are $\{A_{k+1}, D_{2\ell+1}\}$. We now consider how the $D_{2\ell+1}$ series is obtained from the diagonal series by means of a permutation of the odd fields. For $\widehat{su}(2)$, a way of permuting the weights in $P_+^{(k)}$ is to act with some power of the basic outer automorphism a (defined by $a[\lambda_0, \lambda_1] = [\lambda_1, \lambda_0]$). A permutation whose action is restricted to the odd sector is:

$$\Pi(\hat{\lambda}) = a^{\lambda_1}(\hat{\lambda}) \qquad (17.124)$$

$\mathcal{T}$ invariance forces the condition

$$h_{a^{\lambda_1}\hat{\lambda}} - h_{\hat{\lambda}} \in \mathbb{Z} \qquad \text{with} \qquad h_{\hat{\lambda}} = \frac{\lambda_1(\lambda_1 + 2)}{4(k + 2)} \qquad (17.125)$$

This requires $k + 2 \in 4\mathbb{Z}_+$. The $\mathcal{S}$ invariance is also verified: a direct application of Eq. (14.255) yields

$$\mathcal{S}_{a^{\lambda_1}(\hat{\lambda}), a^{\mu_1}(\hat{\mu})} = \mathcal{S}_{\hat{\lambda}\hat{\mu}} e^{-2\pi i[(a^{\lambda_1}\hat{\omega}_0, a^{\mu_1}\mu) + (a^{\mu_1}\hat{\omega}_0, \lambda)]} \qquad (17.126)$$

and a simple analysis of the different possibilities shows that the extra phase factor is always unity. Thus, the permutation (17.124) preserves the modular transformation matrices when $k = 4\ell - 2$. Replacing the identity mass matrix in the diagonal invariant by this permutation matrix directly yields the $D_{2\ell+1}$ series.

Most permutation invariants are of this form, namely

$$\Pi(\hat{\lambda}) = a^{pv\cdot\lambda}(\hat{\lambda}) \qquad (17.127)$$

where a is the generating element of the outer-automorphism group, v is the congruence vector defined in Eq. (13.78) (see also App. 13.A), and p is some integer constrained by the conditions (17.121). All these cases are covered by our general construction based on the outer automorphism group. However, Eq. (17.127) is not exhaustive, as the following $(\widehat{G}_2)_4$ example shows:

$$|\chi_{[4,0,0]}|^2 + |\chi_{[1,0,3]}|^2 + |\chi_{[1,1,1]}|^2 + |\chi_{[2,0,2]}|^2 + |\chi_{[0,1,2]}|^2$$
$$+ \chi_{[2,1,0]}\bar{\chi}_{[0,0,4]} + \chi_{[0,0,4]}\bar{\chi}_{[2,1,0]} + \chi_{[0,2,0]}\bar{\chi}_{[2,0,1]} + \chi_{[2,0,1]}\bar{\chi}_{[0,2,0]} \qquad (17.128)$$

It is obtained from the diagonal invariant by the interchange

$$[2,1,0] \leftrightarrow [0,0,4] \quad \text{and} \quad [0,2,0] \leftrightarrow [2,0,1] \qquad (17.129)$$

in one sector. Such a permutation cannot be related to an outer automorphism since $\mathcal{O}(\widehat{G}_2)$ is trivial. Another example of the same type is the $\widehat{F}_4$ invariant (17.113).

For a permutation invariant to be physical, the simple condition

$$\Pi(0) = 0 \qquad (17.130)$$

must obviously be satisfied; this ensures the existence of a vacuum ($\mathcal{M}_{00} = 1$). But this is not a further requirement to be imposed on Π: $\mathcal{S}_{0,\hat{\mu}} = \mathcal{S}_{\Pi(0),\Pi(\hat{\mu})}$ requires $\Pi(0) = 0$ since the 0-th row of the $\mathcal{S}$ matrix is the only positive one. The condition

$$\mathcal{M}_{\hat{\lambda},0} = \mathcal{M}_{0,\hat{\lambda}} = \delta_{\hat{\lambda},0} \qquad (17.131)$$

provides a simple characterization of permutation invariants (as opposed to block-diagonal invariants, for which the vacuum couples to at least one other field).

Those invariants that are not of the type (17.120) necessarily contain more than one chiral field in the "vacuum block": they are of the form

$$Z = |\chi_0 + \chi_{n_1} + \cdots + \chi_{n_i}|^2 + \cdots \qquad (17.132)$$

Most of such invariants are in fact sums of squares (they are the invariants $D_{2\ell+2}, E_6, E_8$ for $\widehat{su}(2)$), but there are exceptions (E_7). The structure (17.132) signals a symmetry enhancement, an *extended chiral algebra*. Candidates for the extra conserved currents are exactly those fields associated with the characters χ_{n_i}, which appear in the vacuum block (and the built-in $\mathcal{T}$ invariance immediately implies that their dimensions are integers).

When the fields are reorganized with respect to this larger algebra, the invariants (17.132) also become permutation invariants. This means that every chiral primary field of the extended chiral theory (denoted by a Latin index) appears once and only once, and every holomorphic primary field is paired with exactly one antiholomorphic primary field. Setting $\ell = 0$ in the condition

$$\mathcal{S}_{i,j}\,\mathcal{M}_{j,\ell} = \mathcal{M}_{i,j}\mathcal{S}_{j,\ell} \qquad (17.133)$$

using Eq. (17.131), and dividing the result by $\mathcal{S}_{00}$ yields

$$\mathcal{D}(i) = \mathcal{M}_{i,j}\,\mathcal{D}(j) \qquad (17.134)$$

where $\mathcal{D}(i)$ denotes the quantum dimension of the field i, defined by Eq. (16.59). Since $\mathcal{D}(i) \geq 1$ (cf. Eq. (14.244) if the extended theory is a WZW model), the i-th row of the mass matrix must have at least one nonzero entry. Thus, every extended primary field ϕ_i contributes to the partition function. Moreover, requiring the inversion of Eq. (17.134) and the interchange of i and j to yield identical results forces

$$\mathcal{M}^{-1} = \mathcal{M}^t \qquad (17.135)$$

Hence, $\mathcal{M}$ must provide a one-to-one correspondence between the primary fields of the left and right sectors, that is, $\mathcal{M}$ must be a permutation matrix.

In summary, every invariant can be viewed as a permutation invariant! In particular, sums of squares become diagonal in the extended formalism.

What is the structure of this extended algebra? There is one circumstance for which it can be determined without any effort, namely when the invariant is obtained by a conformal embedding. Then, the extended algebra is simply the higher-rank affine Lie algebra in which the original theory has been embedded. These cases are easily detected by the presence of a dimension-1 field in the vacuum block. (We note that there could be additional integral spin fields in the vacuum block, as the E_8 example shows.)

However, the extended algebras are generally not affine Lie algebras. That they fall outside the field of affine Lie algebras substantially complicates their study.

We illustrate these considerations with the $\widehat{su}(2) \, D_{2\ell+2}$ series, with level $k = 4\ell$. The vacuum is coupled to the field of weight $[0, k]$. We thus add this field to the set of current generators $(J^{\pm}, J^0)$ and reorganize the fields with respect to this larger algebra. This simply amounts to combining into a single field all those fields that lie in the same block. We define

$$
\begin{aligned}
\phi_n &\equiv (2n \oplus k - 2n; 2n \oplus k - 2n) \\
&= (2n; 2n) \oplus (k - 2n; 2n) \oplus (2n; k - 2n) \oplus (k - 2n; k - 2n)
\end{aligned}
\qquad (17.136)
$$

for $n < \ell$ and

$$\phi_\ell = (2\ell; 2\ell) \qquad \phi'_\ell = (2\ell; 2\ell) \qquad (17.137)$$

In the notation $(n; m)$, n and m are, respectively, the left and right finite Dynkin labels. The two copies of $(2\ell; 2\ell)$ are related by a $\mathbb{Z}_2$ automorphism rooted in the sign ambiguity for the choice of the extra current. The ϕ_n's are invariant under this $\mathbb{Z}_2$ automorphism. With respect to the extended algebra, the $D_{2\ell+2}$ invariant takes the diagonal form

$$\sum_{i=1}^{\ell-1} |\chi_{\phi_i}|^2 + |\chi_{\phi_\ell}|^2 + |\chi_{\phi'_\ell}|^2 \qquad (17.138)$$

We now consider how the invariant E_7 could fit into this scheme. According to the above general statement, it is bound to be related to a permutation of the fields in the extended algebra. It is actually related to D_{10} in this precise way. To see it, we rewrite D_{10} in the form:

$$Z_{D_{10}} = |\chi_0 + \chi_{16}|^2 + |\chi_2 + \chi_{14}|^2 + |\chi_4 + \chi_{12}|^2 + |\chi_6 + \chi_{10}|^2 + 2|\chi_8|^2$$
$$= |\chi_{\phi_0}|^2 + |\chi_{\phi_1}|^2 + |\chi_{\phi_2}|^2 + |\chi_{\phi_3}|^2 + |\chi_{\phi_4}|^2 + |\chi_{\phi_4'}|^2$$
(17.139)

The interchange of $\chi_{\phi_4'}$ with χ_{ϕ_1} in the holomorphic sector yields

$$|\chi_{\phi_0}|^2 + |\chi_{\phi_2}|^2 + |\chi_{\phi_3}|^2 + |\chi_{\phi_4}|^2 + (\chi_{\phi_4'}\bar{\chi}_{\phi_1}) + (\chi_{\phi_1}\bar{\chi}_{\phi_4'})$$
(17.140)

which, when rewritten in terms of the $\widehat{su}(2)$ fields, becomes exactly the E_7 invariant:

$$Z_{E_7} = |\chi_0 + \chi_{16}|^2 + |\chi_4 + \chi_{12}|^2 + |\chi_6 + \chi_{10}|^2 + |\chi_8|^2 + \chi_8(\bar{\chi}_2 + \bar{\chi}_{14}) + (\chi_2 + \chi_{14})\bar{\chi}_8$$
(17.141)

That this permutation preserves the modular invariance is a direct consequence of the Macdonald identity (see Ex. 14.9 for a proof),

$$\chi_2 + \chi_{14} - \chi_8 = 3 \qquad (k = 16) \tag{17.142}$$

which ensures that $\chi_2 + \chi_{14}$ transforms in the same way as χ_8. The relation between the two partition functions is

$$Z_{D_{10}} - |\chi_2 + \chi_{14} - \chi_8|^2 = Z_{E_7} \tag{17.143}$$

Their difference is a constant—trivially modular invariant—which, at this particular level, has a rather nontrivial representation in terms of linear combination of characters.[13]

In the same vein, the $\widehat{su}(3)_9$ invariant (17.118) is related to the D invariant at level 9 through the interchange:

$$\chi_{(3,3)} \leftrightarrow \chi_{(1,7)} + \chi_{(7,1)} + \chi_{(1,1)} \tag{17.144}$$

validated by the Macdonald identity:

$$\chi_{(1,7)} + \chi_{(7,1)} + \chi_{(1,1)} - \chi_{(3,3)} = 8 \tag{17.145}$$

To find the permutations of the extended fields that preserve modular invariance is in general a difficult problem, essentially because the general structure of the extended algebras is not known. A constructive characterization of these invariants is still lacking. However, very few solutions are to be expected since we are looking for linear combinations of (nonextended) fields having identical modular transformation properties, a rather tight constraint. Consequently, no infinite series can be produced in this way: permutation invariants of extended algebras are always exceptional invariants. For $\widehat{su}(2)$, we can show that there are no solutions, other than the one already found by going through the Macdonald identities and searching for

---

[13] The simplest illustration of this trick occurs at level 6, where $\chi_1 - \chi_5 = 2$, but it does not produce a new invariant:

$$Z_{A_7} - |\chi_1 - \chi_5|^2 = Z_{D_5}$$

those relating the precise field combinations of interest, namely $\chi_n + \chi_{k-2n}$ or χ_ℓ. This procedure is clearly limited to low-rank algebras, where these identities can be made explicit. For $\widehat{su}(3)$, there is only one exceptional permutation invariant, and it can also be obtained in that way.

Given a permutation Π, the difficult step is to prove its commutativity with S. A simple necessary condition for this has already been reported in Sect. 10.8.4; such a permutation must be an automorphism of the fusion rules:

$$\mathcal{N}_{\hat{\lambda}\hat{\mu}}^{(k)\hat{\nu}} = \mathcal{N}_{\Pi(\hat{\lambda})\Pi(\hat{\mu})}^{(k)\Pi(\hat{\nu})} \tag{17.146}$$

(This is a direct consequence of the Verlinde formula.) This condition, however, is not sufficient to ensure commutativity with S. But we recall that since Π commutes with S, it preserves the quantum dimensions (cf. Eq. 17.134)). Finding permutations of fields of second-lowest quantum dimensions that yield fusion automorphisms has proven to be a powerful and reliable criterion.

We present a simple illustration of this approach for the case where the extended algebra is still a current algebra. The following result will be used: A Π-invariant permutation that leaves all the fundamental weights ω_i fixed (here ω_i is the finite part of the affine weight $k\hat{\omega}_0 + \omega_i$) must fix all the integrable weights $\hat{\lambda} \in P_+^k$. This is not trivial since $\Pi(p\omega_i) \neq p\Pi(\omega_i)$. For our illustrative example, it is sufficient to prove this for $k \geq \max\{a_i^\vee\}$ (i.e., at a level sufficiently large that the affine extension of all the fundamental weights are integrable). For any Lie algebra, there is a Giambelli-type formula (cf. Eq. (16.114) for $su(N)$) that gives the character of an arbitrary representation λ in terms of a polynomial P_λ in the characters of the fundamental representations:

$$\chi_\lambda = P_\lambda(\chi_{\omega_i}) \tag{17.147}$$

Evaluated at $\xi_\mu = -2\pi i(\mu + \rho)/k + g$, the finite characters become ratios of S matrix elements (cf. Eq. (14.247)):

$$\frac{S_{\hat{\lambda}\hat{\mu}}}{S_{0\hat{\mu}}} = P_\lambda\left(\frac{S_{\omega_i\hat{\mu}}}{S_{0\hat{\mu}}}\right) \tag{17.148}$$

Since, by assumption, Π is a permutation invariant (which thus satisfies Eq. (17.121)) that leaves all fundamental weights and the vacuum fixed (and in particular $S_{0,\Pi(\hat{\mu})} = S_{\Pi^{-1}(0),\hat{\mu}} = S_{0\hat{\mu}}$), the above relation implies the equality

$$S_{\hat{\lambda}\hat{\mu}} = S_{\hat{\lambda}\Pi(\hat{\mu})} \quad \forall \hat{\lambda}, \ \hat{\mu} \in P_+^k \quad \Rightarrow \quad \Pi(\hat{\mu}) = \hat{\mu} \tag{17.149}$$

since otherwise the S matrix would be singular.

Consider, for instance, $\widehat{G}_2$ at level 4. The vacuum is the unique state with quantum dimension 1, and the set of weights with second-lowest quantum dimension is $\{\omega_2, 2\omega_1\}$. The possible permutations are then $\Pi(\omega_2) = \omega_2$ and $\Pi'(\omega_2) = 2\omega_1$. Since Π is an automorphism of the fusion rules, it must leave the fusion

$$\omega_2 \times \omega_2 = 0 + \omega_1 + \omega_2 + 2\omega_2 \tag{17.150}$$

invariant, which means that either $\Pi(\omega_1) = \omega_1$ or $\Pi(\omega_1) = 2\omega_2$. The second possibility is ruled out by T invariance. Π thus leaves both fundamental weights unchanged, i.e., $\Pi = id$. The other permutation, when used in fusion rules, yields $\Pi'(\omega_1) = 4\omega_2$. This is exactly the permutation underlying the invariant (17.128). The present analysis shows that there can be no further permutation invariants at level 4. Another example is presented in Ex. 17.9.

§17.9. Galois Symmetry

All (but one) permutation invariants of WZW models can be obtained either from an outer automorphism—as already discussed—or from a permutation akin to the permutations of the roots of a polynomial equation (the so-called Galois transformations). The study of these permutations is the subject of the present section. Galois transformations are defined in the first subsection. In the context of modular invariant partition functions, they are used to: (i) constrain the mass matrix; (ii) build new block-diagonal invariants (not necessarily physical, however); and (iii) build new permutation invariants.

17.9.1. Galois Transformations on S Matrices

We consider a polynomial equation $f(x) = 0$ with coefficients defined in a certain field $\mathcal{K}$ but whose roots lie in an extension $\mathcal{L}$ of this field. For instance, if the polynomial coefficients are in $\mathbb{Q}$, the roots can be complex numbers or involve roots of rational numbers. The transformations σ that permute the roots of the polynomial equation but leave its coefficients fixed are called *Galois transformations*. For the equation $x^n - 1 = 0$, these are

$$\sigma : \zeta \mapsto \zeta^\ell \quad \zeta = \exp(2\pi i/n) \quad (\ell, n) = 1$$
$$\sigma : x \mapsto x \quad x \in \mathbb{Q} \tag{17.151}$$

(i.e., ℓ and n coprime). These transformations form a group, with composition as group multiplication, and preserve addition and multiplication:

$$\sigma(x + y) = \sigma(x) + \sigma(y)$$
$$\sigma(xy) = \sigma(x)\sigma(y) \tag{17.152}$$

Galois transformations are relevant in conformal field theory because the ratios of S matrix elements

$$\gamma_i^{(j)} = \frac{S_{i,j}}{S_{0,j}} \tag{17.153}$$

are the roots of the characteristic equation

$$\det(\lambda\mathbb{I} - N_i) = 0 \tag{17.154}$$

where N_i is the fusion matrix for the field i. The matrix entries of N_i being integers, the number field $\mathcal{K}$ can be chosen to be $\mathbb{Q}$. Since ratios of S matrix elements are usually not rational numbers, nontrivial Galois transformations can be associated with this equation. To a permutation of the roots $\gamma_i^{(j)} \mapsto \sigma(\gamma_i^{(j)})$, there corresponds the field permutation $j \mapsto \sigma(j) = j^\sigma$:

$$\sigma(\frac{S_{i,j}}{S_{0,j}}) = \frac{\sigma(S_{i,j})}{\sigma(S_{0,j})} = \frac{S_{i,j^\sigma}}{S_{0,j^\sigma}} \tag{17.155}$$

Actually, the following stronger result holds:[14]

$$\sigma(S_{i,j}) = \epsilon_\sigma(j) S_{i,j^\sigma} \tag{17.156}$$

with $\epsilon_\sigma(i) = \pm 1$. This is most easily proven for WZW models, in which the S matrix takes the form

$$S_{\hat{\lambda},\hat{\mu}} = K \sum_{w \in W} \epsilon(w) \exp\left\{ -\frac{2\pi i}{k+g} (w(\lambda + \rho), \mu + \rho) \right\} \tag{17.157}$$

where K is a constant. Let M be the smallest integer for which

$$M (w\lambda, \mu) \in \mathbb{Z} \qquad \forall \; \lambda, \mu \in P_+, \quad w \in W \tag{17.158}$$

For instance, $M = N$ for $su(N)$; in general[15]

$$M = |P/Q^\vee| \tag{17.161}$$

Then, up to the prefactor K, S is a linear combination of the n-th roots of unity, with

$$n = M(k + g) \tag{17.162}$$

The extended field $\mathcal{L}$ in which the S matrix elements live is thus a cyclotomic field $\mathbb{Q}(\zeta)$, that is, the field $\mathbb{Q}$ extended by the addition of a fundamental root of unity.[16] A permutation of the fields is simply related to the transformation

$$(\lambda + \rho) \mapsto \ell(\lambda + \rho), \qquad (\ell, n) = 1 \tag{17.163}$$

We denote the corresponding Galois transformation by σ_ℓ. Since the prefactor K is not a rational number (it involves square roots and a power of i), σ_ℓ acts nontrivially on it. But K^2 is rational, hence not affected by σ:

$$K^2 = \sigma(K^2) = \sigma(K)\sigma(K) \implies \sigma(K) = \eta_\ell K \tag{17.164}$$

---

[14] By shifting our attention from ratios of S matrix elements to the elements themselves, we would have to consider a further field extension $\mathcal{L}' \supset \mathcal{L}$. But this technicality is of no consequence here and is ignored throughout.

[15] We note that for nonsimply-laced algebras, M is not just the smallest integer M' for which

$$M'(\omega_i, \omega_j) \in \mathbb{Z} \quad \forall \, i, j \tag{17.159}$$

due to the Weyl reflections with respect to the short roots, e.g., :

$$(s_i\lambda, v) = (\lambda, v) - \lambda_i(\alpha_i, v) = (\lambda, v) - \frac{1}{2}\lambda_i v_i \alpha_i^2 \tag{17.160}$$

From App. 13.A, values of M' are found to be: n for A_{n-1}; 2 for B_{2n}, C_n, D_{2n}, E_7, and F_4; 3 for G_2 and E_6 and 4 for B_{2n+1} and D_{2n+1}. Actually, M' is merely $|P/Q|$.

[16] Formally, ζ must be a $4M$-th root of unity due to the prefactor K, but all field permutations can be described in the cyclotomic field based on a M-th root of unity.

with $\eta_\ell = \pm 1$. Hence, the action of σ on S takes the form

$$\sigma_\ell(S_{\hat{\lambda},\hat{\mu}}) = \eta_\ell \, K \sum_{w \in W} \epsilon(w) \exp\left\{-\frac{2\pi i}{k+g}(w\,\ell(\lambda+\rho),\mu+\rho)\right\} \qquad (17.165)$$

For $\ell \neq 1$, $\ell(\lambda+\rho) - \rho$ is not necessarily the finite part of an integrable weight at level $k+g$. If not, the affine weight

$$\hat{\xi} = \left(k - \sum_{i=1}^{r}[\ell(\lambda_i+1)-1]a_i^\vee\right)\hat{\omega}_0 + \ell(\lambda+\rho) - \rho \qquad (17.166)$$

must be reflected back into the fundamental affine Weyl chamber by means of an appropriate shifted affine Weyl reflection. Let $w_\ell^{(\lambda)}$ be the element of $\hat{W}$ such that

$$w_\ell^{(\lambda)} \cdot \hat{\xi} = \hat{\lambda}^{\sigma_\ell}, \qquad \text{with} \quad \hat{\lambda}^{\sigma_\ell} \in P_+^{(k)} \qquad (17.167)$$

(Since ℓ and $k+g$ are coprime, such a Weyl reflection always exists, i.e., the weight cannot be reflected onto the boundary of the fundamental chamber—cf. Ex. 17.11.) By a standard redefinition of the summation variable in the expression for S, we find

$$\sigma_\ell(S_{\hat{\lambda},\hat{\mu}}) = \eta_\ell \, \epsilon(w_\ell^{(\lambda)}) \, S_{\hat{\lambda}^{\sigma_\ell},\hat{\mu}} \qquad (17.168)$$

This is indeed of the form (17.156), with

$$\epsilon_{\sigma_\ell}(\lambda) = \eta_\ell \, \epsilon(w_\ell^{(\lambda)}) \qquad (17.169)$$

When $\ell = -1$, σ_ℓ acts as the usual charge conjugation. Galois transformations can thus be viewed as generalized charge conjugations. The group generated by the σ_ℓ's is denoted $\mathbb{Z}_n^*$; it is the multiplicative group comprised of the elements ℓ of $\mathbb{Z}_n$ that are coprime to n. For instance,

$$\mathbb{Z}_{24}^* = \{1, 5, 7, 11, 13, 17, 19, 23\} \qquad (17.170)$$

17.9.2. The Parity Rule

Galois transformations induce a simple and rather powerful constraint on physical mass matrices. Such a matrix must commute with S and be integer valued. Acting on both sides of the equality $\mathcal{M}S = S\mathcal{M}$ with σ (which does not affect $\mathcal{M}$) yields

$$\sum_{\hat{\mu}} \epsilon_\sigma(\mu) \, \mathcal{M}_{\hat{\lambda},\hat{\mu}} \, S_{\hat{\mu}^\sigma,\hat{\nu}} = \sum_{\hat{\mu}} \epsilon_\sigma(\lambda) \, S_{\hat{\lambda}^\sigma,\hat{\mu}} \, \mathcal{M}_{\hat{\mu},\hat{\nu}}$$

$$= \sum_{\hat{\mu}} \epsilon_\sigma(\lambda) \, \mathcal{M}_{\hat{\lambda}^\sigma,\hat{\mu}} \, S_{\hat{\mu},\hat{\nu}}$$

$$= \sum_{\hat{\mu}^\sigma} \epsilon_\sigma(\lambda) \, \mathcal{M}_{\hat{\lambda}^\sigma,\hat{\mu}^\sigma} \, S_{\hat{\mu}^\sigma,\hat{\nu}}$$

In the second line we used $S\mathcal{M} = \mathcal{M}S$, and in the third one we have simply redefined the summation variable. With $\sigma = \sigma_\ell$, this implies that

$$\mathcal{M}_{\hat{\lambda}^{\sigma_\ell},\hat{\mu}^{\sigma_\ell}} = \epsilon_{\sigma_\ell}(\lambda) \, \epsilon_{\sigma_\ell}(\mu) \, \mathcal{M}_{\hat{\lambda},\hat{\mu}} \qquad (17.171)$$

All elements $\mathcal{M}_{\hat{\lambda}^{\sigma_\ell},\hat{\mu}^{\sigma_\ell}}$ in the Galois $\mathbb{Z}_n^*$ orbit are thus related to each other. For a physical invariant, all these elements must be positive. Therefore, the product of signs must be positive for all values of ℓ:

$$\epsilon_{\sigma_\ell}(\lambda)\,\epsilon_{\sigma_\ell}(\mu) = \epsilon(w_\ell^{(\lambda)})\,\epsilon(w_\ell^{(\mu)}) = 1 \quad \forall\,\ell \in \mathbb{Z}_n^* \tag{17.172}$$

If $\epsilon(w_\ell^{(\lambda)})\epsilon(w_\ell^{(\mu)}) = -1$ for *some* $\ell \in \mathbb{Z}_n^*$, then *all* $\mathcal{M}_{\hat{\lambda}^{\sigma_\ell},\hat{\mu}^{\sigma_\ell}}$'s, and in particular $\mathcal{M}_{\hat{\lambda},\hat{\mu}}$, must vanish. The condition (17.172) will be referred to as the *parity rule*.

This constraint is particularly useful for probing nondiagonal invariants related to extensions of the chiral algebra. In that case, we look for chiral fields $\hat{\lambda}$ that could couple to the identity, that is, $\hat{\lambda}$'s such that $\mathcal{M}_{0,\hat{\lambda}} = 1$. This is illustrated in Ex. 17.12.

17.9.3. Modular Invariants From Galois Symmetry

Galois transformations not only put constraints on the mass matrices but also can be used to construct modular invariants. A candidate mass matrix is

$$\mathcal{M}_\sigma = G_\sigma + G_\sigma^t \tag{17.173}$$

where G_σ^t is the transpose of G_σ and

$$(G_\sigma)_{\hat{\lambda},\hat{\mu}} = \epsilon_\sigma(\mu)\,\delta_{\hat{\lambda},\hat{\mu}^\sigma} \tag{17.174}$$

This can also be written as

$$(G_\sigma)_{\hat{\lambda},\hat{\mu}} = \epsilon_\sigma(\lambda^{\sigma^{-1}})\,\delta_{\hat{\lambda}^{\sigma^{-1}},\hat{\mu}} = \epsilon_{\sigma^{-1}}(\lambda)\,\delta_{\hat{\lambda}^{\sigma^{-1}},\hat{\mu}} \tag{17.175}$$

For the second equality we used

$$S_{\hat{\lambda},\hat{\mu}} = \sigma\sigma^{-1}(S_{\hat{\lambda},\hat{\mu}}) = \sigma(\epsilon_{\sigma^{-1}}(\lambda)\,S_{\hat{\lambda}^{\sigma^{-1}},\hat{\mu}}) = \epsilon_{\sigma^{-1}}(\lambda)\,\epsilon_\sigma(\lambda^{\sigma^{-1}})\,S_{\hat{\lambda},\hat{\mu}} \tag{17.176}$$

which shows the equality of $\epsilon_{\sigma^{-1}}(\lambda)$ and $\epsilon_\sigma(\lambda^{\sigma^{-1}})$. It is then clear that

$$G_\sigma^{-1} = G_{\sigma^{-1}} = G_\sigma^t \tag{17.177}$$

The commutativity of $\mathcal{M}_\sigma$ with S is readily checked:

$$\sum_{\hat{\mu}}(\mathcal{M}_\sigma)_{\hat{\lambda},\hat{\mu}}\,S_{\hat{\mu},\hat{\nu}} = \sum_{\hat{\mu}}[\epsilon_\sigma(\mu)\,\delta_{\hat{\lambda},\hat{\mu}^\sigma} + \epsilon_\sigma(\lambda)\,\delta_{\hat{\lambda}^\sigma,\hat{\mu}}]\,S_{\hat{\mu},\hat{\nu}}$$

$$= \sigma^{-1}\left(\sum_{\hat{\mu}^\sigma}[\,\delta_{\hat{\lambda},\hat{\mu}^\sigma}\,S_{\hat{\mu}^\sigma,\hat{\nu}}]\right) + \epsilon_\sigma(\lambda)\,S_{\hat{\lambda}^\sigma,\hat{\nu}} \tag{17.178}$$

$$= \sigma^{-1}(S_{\hat{\lambda},\hat{\nu}}) + \sigma(S_{\hat{\lambda},\hat{\nu}})$$

(in the second line we used $\epsilon_\sigma(\mu)S_{\hat{\mu},\hat{\nu}} = \sigma^{-1}(S_{\hat{\mu}^\sigma,\hat{\nu}})$), and

$$\sum_{\hat{\mu}} S_{\hat{\lambda},\hat{\mu}} (M_\sigma)_{\hat{\mu},\hat{\nu}} = \sum_{\hat{\mu}} S_{\hat{\lambda},\hat{\mu}}[\epsilon_\sigma(\nu)\,\delta_{\hat{\mu},\hat{\nu}^\sigma} + \epsilon_\sigma(\mu)\,\delta_{\hat{\mu}^\sigma,\hat{\nu}}]$$

$$= \epsilon_\sigma(\nu)\, S_{\hat{\lambda},\hat{\nu}^\sigma} + \sigma^{-1}\left(\sum_{\hat{\mu}^\sigma} S_{\hat{\lambda},\hat{\mu}^\sigma}\,\delta_{\hat{\mu}^\sigma,\hat{\nu}}\right) \qquad (17.179)$$

$$= \sigma(S_{\hat{\lambda},\hat{\nu}}) + \sigma^{-1}(S_{\hat{\lambda},\hat{\nu}})$$

establishing thereby the desired result.

In the above derivation, the order of the transformation σ is arbitrary. But in order for Eq. (17.173) to commute with T, σ must necessarily be of order 2, that is, $\sigma^2 = 1$ (or, with $\sigma = \sigma_\ell$, $\ell^2 = 1 \bmod M(k+g)$). We now prove this statement. The starting point is the relation $(ST)^3 = S^2$, which we write in the form

$$T_{\hat{\lambda}}^{-1}\, S_{\hat{\lambda}\hat{\mu}}\, T_{\hat{\mu}}^{-1} = \sum_{\hat{\nu}} S_{\hat{\lambda}\hat{\nu}}\, T_{\hat{\nu}}\, S_{\hat{\nu}\hat{\mu}} \qquad (17.180)$$

The $T_{\hat{\lambda}}$'s stand for the diagonal elements of the T matrix. Using the equality

$$S_{\hat{\lambda}\hat{\mu}} = \sigma\,\sigma^{-1}(S_{\hat{\lambda}\hat{\mu}}) = \epsilon_\sigma(\lambda)\epsilon_{\sigma^{-1}}(\mu)S_{\hat{\lambda}^\sigma,\hat{\mu}^{\sigma^{-1}}} \qquad (17.181)$$

(in the last step, the symmetry of S is used to permute the two labels), this relation becomes

$$\epsilon_\sigma(\lambda)\epsilon_{\sigma^{-1}}(\mu)\, T_{\hat{\lambda}}^{-1}\, S_{\hat{\lambda}^\sigma\hat{\mu}^{\sigma^{-1}}}\, T_{\hat{\mu}}^{-1} = \epsilon_{\sigma^{-1}}(\mu)\epsilon_{\sigma^{-1}}(\lambda)\sum_{\hat{\nu}} S_{\hat{\lambda}^{\sigma^{-1}},\hat{\nu}^\sigma}\, T_{\hat{\nu}}\, S_{\hat{\nu}^\sigma,\hat{\mu}^{\sigma^{-1}}}$$

$$= \epsilon_{\sigma^{-1}}(\mu)\epsilon_{\sigma^{-1}}(\lambda)\, T_{\hat{\lambda}}^{-1}\, S_{\hat{\lambda}^{\sigma^{-1}},\hat{\mu}^{\sigma^{-1}}}\, T_{\hat{\mu}}^{-1} \qquad (17.182)$$

In the second equality, we used the assumed T invariance under the action of σ to write $T_{\hat{\nu}} = T_{\hat{\nu}^\sigma}$ and used again the relation $T^{-1}ST^{-1} = STS$. We conclude that

$$S_{\hat{\lambda}^\sigma,\hat{\mu}^{\sigma^{-1}}} = \epsilon_{\sigma^{-1}}(\lambda)\epsilon_\sigma(\lambda)\, S_{\hat{\lambda}^{\sigma^{-1}},\hat{\mu}^{\sigma^{-1}}} \qquad (17.183)$$

Since S is unitary, its rows must be linearly independent; this forces

$$\hat{\lambda}^\sigma = \hat{\lambda}^{\sigma^{-1}}, \qquad \text{that is,} \qquad \sigma = \sigma^{-1} \qquad (17.184)$$

or, with $\sigma = \sigma_\ell$,

$$\ell^2 = 1 \bmod M(k+g) \qquad (17.185)$$

This is a necessary condition for T invariance, but usually not sufficient. The sufficient condition for T invariance is simply

$$h_{\hat{\lambda}^\sigma} - h_{\hat{\lambda}} \in \mathbb{Z} \qquad (17.186)$$

that is,

$$(\ell^2 - 1)\frac{|\lambda + \rho|^2}{2(k+g)} \in \mathbb{Z} \qquad (17.187)$$

This requires

$$\ell^2 = 1 \bmod 2M'(k+g) \quad \text{if } M'|\lambda + \rho|^2 \text{ is odd}$$
$$= 1 \bmod M'(k+g) \quad \text{if } M'|\lambda + \rho|^2 \text{ is even} \tag{17.188}$$

with M' defined by Eq. (17.159).

The mass matrix $\mathcal{M}_{\sigma_\ell}$ is modular invariant if the condition (17.188) is satisfied, but it is not positive definite. However, it can be combined with another invariant, such as the identity mass matrix (associated with the diagonal invariant), to produce a new, and possibly physical, invariant:

$$\mathcal{M} = I + \mathcal{M}_\sigma = I + G_\sigma + G_\sigma^t \tag{17.189}$$

Such an invariant will be referred to as a *block-diagonal Galois invariant*. We note that whenever G_σ is symmetric, it is clearly sufficient to take $\mathcal{M}_\sigma = G_\sigma$. The potential generalization, in which $\mathcal{M}_\sigma$ could be combined with invariants other than the diagonal, should also be kept in mind.

We illustrate this construction by an example, the $\widehat{su}(2)_4$ nondiagonal invariant: we verify that it can be viewed as a block-diagonal Galois invariant, with $\mathcal{M} = I + G_\sigma$. For $su(2)$, $M = 2$ and therefore $n = M(k+g) = 12$. The group $\mathbb{Z}_{12}^*$ is $\{1, 5, 7, 11\}$. All its elements satisfy $\ell^2 = 1 \bmod 12$ as well as $\ell^2 = 1 \bmod 24$ (the second condition ensures the T invariance). The choices $\ell = 7, 11$ do not produce physical invariants: $\ell = 7$ yields $\mathcal{M}_{0,4} = -1$ and $\ell = 11$ yields $\mathcal{M}_{0,0} = 0$. We consider then $\ell = 5$. The different entries of G_σ are easily calculated. Denoting by $\hat{\Lambda}_{(\lambda_1)}$ the affine extension of $5(\lambda + \rho)$ at level 6, we find

$$(0)^\sigma : \hat{\Lambda}_{(0)} = [1, 5], \qquad \hat{\Lambda}_{(0)} - \hat{\rho} = [0, 4]$$
$$\Rightarrow (0)^\sigma = (4), \ \epsilon_\sigma(0) = 1$$
$$(1)^\sigma : \hat{\Lambda}_{(1)} = [-4, 10], \quad s_0 \hat{\Lambda}_{(1)} - \hat{\rho} = [3, 1]$$
$$\Rightarrow (1)^\sigma = (1), \ \epsilon_\sigma(1) = -1$$
$$(2)^\sigma : \hat{\Lambda}_{(2)} = [-9, 15], \quad s_1 s_0 \hat{\Lambda}_{(2)} - \hat{\rho} = [2, 2]$$
$$\Rightarrow (2)^\sigma = (2), \ \epsilon_\sigma(2) = 1 \tag{17.190}$$
$$(3)^\sigma : \hat{\Lambda}_{(3)} = [-14, 20], \quad s_0 s_1 s_0 \hat{\Lambda}_{(3)} - \hat{\rho} = [1, 3]$$
$$\Rightarrow (3)^\sigma = (3), \ \epsilon_\sigma(3) = -1$$
$$(4)^\sigma : \hat{\Lambda}_{(4)} = [-19, 25], \quad s_1 s_0 s_1 s_0 \hat{\Lambda}_{(4)} - \hat{\rho} = [4, 0]$$
$$\Rightarrow (4)^\sigma = (0), \ \epsilon_\sigma(4) = 1$$

and we indeed recover the invariant (17.15). Further examples are presented in the exercises.

17.9.4. Galois Permutation Invariants

In the previous construction, neither the positivity of the mass matrix nor the uniqueness of the vacuum is guaranteed. In this subsection, we show that, to any order-2 Galois transformation that leaves the vacuum fixed—that is, $0^\sigma = 0$—there corresponds a physical permutation invariant with mass matrix

$$\mathcal{M}_{\hat\lambda,\hat\mu} = \delta_{\hat\lambda,\hat\mu^\sigma} \tag{17.191}$$

provided that $T_{\hat\lambda} = T_{\hat\lambda^\sigma}$. It will be referred to as a *Galois permutation invariant*.

First, we observe that when the vacuum is a fixed point of σ, all fields have the same value of $\epsilon_\sigma(\lambda)$. This follows directly from the invariance of the quantum dimensions:

$$\sigma\left(\frac{S_{0\hat\lambda}}{S_{00}}\right) = \frac{S_{0\hat\lambda}}{S_{00}} = \frac{\epsilon_\sigma(\lambda)S_{\hat\lambda 0}}{\epsilon_\sigma(0)S_{00}} \tag{17.192}$$

Hence, $\epsilon_\sigma(\lambda) = \epsilon_\sigma(0)$ for any λ. The application of σ on the Verlinde formula shows that σ is a fusion automorphism:

$$\mathcal{N}_{\hat\lambda\hat\mu}{}^{\hat\nu} = \mathcal{N}_{\hat\lambda^\sigma\hat\mu^\sigma}{}^{\hat\nu^\sigma} \qquad (0^\sigma = 0) \tag{17.193}$$

This holds for transformations of arbitrary order. However, if the order is greater than 2, we have shown previously that σ cannot commute with T. Hence, not every fusion automorphism yields a permutation invariant.

If σ is of order 2, it necessarily commutes with S:

$$\begin{aligned} S_{\hat\lambda\hat\mu} &= \epsilon_\sigma(\lambda)\epsilon_{\sigma^{-1}}(\mu)\, S_{\hat\lambda^\sigma,\hat\mu^{\sigma-1}} \\ &= \epsilon_\sigma(0)\epsilon_\sigma(0)\, S_{\hat\lambda^\sigma,\hat\mu^\sigma} \\ &= S_{\hat\lambda^\sigma,\hat\mu^\sigma} \end{aligned} \tag{17.194}$$

(cf. Eq. (17.121)). It only remains to verify that σ also commutes with T. An example of a permutation invariant related to a Galois transformation is presented in Ex. 17.14.

A generalization of the above construction amounts to replacing the condition $0^\sigma = 0$ by the weaker one:

$$0^\sigma = A(0) \tag{17.195}$$

for some outer automorphism A. If σ commutes with T, then the permutation

$$\Pi(\hat\lambda) = \begin{cases} A(\hat\lambda^\sigma) & \text{if } \epsilon_\sigma(\lambda) = \epsilon_\sigma(0) \\ \hat\lambda^\sigma & \text{if } \epsilon_\sigma(\lambda) = -\epsilon_\sigma(0) \end{cases} \tag{17.196}$$

is modular invariant. Of course, if A is trivial, we recover the previous case. The analysis of this generalized construction is left as an exercise (see Ex. 17.16).

To summarize, Galois transformations provide a third general approach to the construction of modular invariants. Together, these three methods—outer automorphisms, conformal embeddings, and Galois transformations— explain almost all the known invariants. But there remain "very exceptional" invariants, which have not been explained yet. A famous example is the following $(\widehat{F}_4)_6$ invariant:

$$|\chi_{[6,0,0,0,0]} + \chi_{[2,0,0,0,4]} + \chi_{[1,1,1,0,0]} + \chi_{[0,0,0,3,0]}|^2$$
$$+ |\chi_{[2,0,1,0,1]} + \chi_{[0,1,0,1,2]}|^2 + |\chi_{[3,0,0,0,3]} + \chi_{[0,0,0,0,6]} \qquad (17.197)$$
$$+ \chi_{[1,0,0,2,1]} + \chi_{[0,2,0,1,0]}|^2 + |\chi_{[1,0,1,0,2]} + \chi_{[2,2,0,0,0]}|^2$$

We stress again that all physical modular invariants do not necessarily correspond to genuine conformal field theories since the underlying requirements by themselves do not guarantee positive integer-valued fusion numbers (for example).

§17.10. Modular Invariants, Generalized ADE Diagrams and Fusion Rules

In this section, we investigate in some detail the observed one-to-one relation between $\widehat{su}(2)_k$ modular invariants and ADE algebras (i.e., that the spinless $h = \bar{h}$ operators in the modular-invariant theory are indexed by the exponents of the corresponding algebra—cf. Sect. 17.7). We are motivated by the observation that a large class of fusion algebras can be coded into a single graph (this construction is reviewed below). The first natural question is: Do the graphs at hand, namely the ADE Dynkin diagrams, actually code some fusion algebra relevant to the conformal theories associated with the corresponding modular invariants? Indeed, for the block-diagonal invariants, namely the (A_n, D_{2n}, E_6, E_8) cases, this will turn out to be so: the resulting fusion algebra reproduces the fusion algebra of the extended theory. (As argued below, this construction is not suitable for the nondiagonal invariants obtained by a permutation of the fusion rules.)

This connection can be described roughly as follows. To each of these (A_n, D_{2n}, E_6, E_8) diagrams, we associate a commutative algebra (the *graph algebra*), whose generators are indexed by the vertices of the diagram. In all cases, this algebra has a nontrivial subalgebra, generated by a subset of the graph vertices, called the *marked vertices*. This subalgebra happens to be the fusion algebra of the extended theory. But this is not all: a striking duality relation allows us to translate the information contained in the fusion algebra of the extended theory into a well-defined prescription for the reconstruction of the modular invariants.

Quite remarkably, this correspondence can be generalized to $\widehat{su}(3)$. The graphs associated with the block-diagonal $\widehat{su}(3)$ modular invariants are found to be some generalizations of the Dynkin diagrams.

17.10.1. Graph Algebra

We have already seen in Sect. 10.7.6 (see also Ex. 8.19), how the fusion rules of minimal or rational conformal theories can be encoded in some graphs. In the simplest cases, the fusion rules are generated entirely by the fusions of a single (fundamental) representation (f), namely

$$(f) \times (j) = \bigoplus_k \mathcal{N}_{fj}{}^k (k) \tag{17.198}$$

To the matrix $\mathcal{N}_f$, with entries $[\mathcal{N}_f]_{jk} = \mathcal{N}_{fj}{}^k$, we associate a graph whose vertices are indexed by the representations, such that a given vertex j is linked to the vertex k by $\mathcal{N}_{fj}{}^k$ links. $\mathcal{N}_f$ is called the adjacency matrix of the corresponding graph. In particular, the identity representation (0) must correspond to an endpoint of the graph and be linked only to the vertex f. This follows from the identity

$$\mathcal{N}_{0i}{}^j = \delta_{i,j} \qquad \Longrightarrow \qquad \mathcal{N}_{0f}{}^j = \delta_{f,j} \tag{17.199}$$

which implies that there is exactly one link between the vertices 0 and f.

Conversely, we can attach a kind of fusion algebra to any connected graph $\mathcal{G}$. Let G denote the corresponding adjacency matrix, with nonnegative integer entries

$$G_{rs} = \text{number of links from vertex } r \text{ to vertex } s \tag{17.200}$$

For nonoriented graphs, G is therefore an indecomposable symmetric matrix with nonnegative integer entries. We would like to view the matrix G as the fundamental generator of some sort of fusion rules in the adjoint (matrix) representation, building up, by multiple fusions, matrices N_r with entries $[N_r]_{st} = \mathcal{N}_{rs}{}^t$. The "initial conditions" in this construction are

$$N_0 = \mathbb{I} \qquad N_f = G \tag{17.201}$$

The algebra admits thus a unit (or identity) element, indexed by a vertex 0 of $\mathcal{G}$, and a fundamental generator, indexed by a vertex f of $\mathcal{G}$, such that $G_{rs} = \mathcal{N}_{fr}{}^s$. In particular

$$G_{0r} = \mathcal{N}_{f0}{}^r = N_{0f}{}^r = \delta_{f,r} \tag{17.202}$$

Hence, the unit vertex 0 is linked to the fundamental vertex f only, with one link. This means that the unit vertex must be an endpoint of $\mathcal{G}$. We conclude that, for a graph algebra to exist, $\mathcal{G}$ must have at least one endpoint. Given such a graph, we fix the unit and fundamental vertices 0 and f.

In order to derive the full graph algebra out of the adjacency matrix G, we follow a strategy akin to the one used to obtain the Verlinde formula (10.171). Namely, we define the graph-algebra fusion coefficients $\mathcal{N}_{rs}{}^t$ in terms of the matrix S, which diagonalizes the adjacency matrix G.[17] The symmetric matrix G can be diagonalized in an orthonormal basis as

$$\Delta = S^{-1} G S \tag{17.203}$$

---

[17] Here we denote this matrix by S, to distinguish it from the S matrix which diagonalizes the fusion rules of a conformal field theory. The two notions coincide when the graph $\mathcal{G}$ encodes the fusion rules of a conformal field theory. In general, S is not a symmetric matrix: the symmetry of S_{am} would involve constructing a bijection between the vertices (labeled by a) and the eigenvalues (labeled by m).

where Δ is a diagonal matrix, and S some orthogonal matrix. This can also be written as

$$\sum_s G_{rs} S_{st} = S_{rt} \Delta_t \qquad (17.204)$$

Setting $r = 0$ in Eq. (17.204) yields

$$S_{ft} = \Delta_t S_{0t} \qquad (17.205)$$

Supposing that all $S_{0t} \neq 0$,[18] we may write

$$\Delta_t = \frac{S_{ft}}{S_{0t}} \qquad (17.206)$$

Inspired by the Verlinde formula (10.171), we now define

$$\mathcal{N}_{rs}{}^t = \sum_m \frac{S_{rm} S_{sm} (S^{-1})_{mt}}{S_{0m}} \qquad (17.207)$$

where the sum runs over all eigenvalues of G. It is a simple exercise to check that the matrices N_r, with entries $[N_r]_{st} = \mathcal{N}_{rs}{}^t$, generate a commutative algebra, called the *graph algebra*.

This definition is not unique, since it relies (i) on the choice of a unit vertex 0 on $\mathcal{G}$, and (ii) on the particular form chosen for the matrix S. Indeed, if some eigenvalue Δ_m of G is p times degenerate, there is a free $O(p)$ ambiguity in the definition of the corresponding S matrix elements $S_{am_1}, \cdots, S_{am_p}$. This is just a rotation in the dimension-p eigenspace for Δ_m. Different choices for this rotation lead to different numbers $\mathcal{N}_{rs}{}^t$ through Eq. (17.207).

For the analogy with fusion rules to be complete, we would like the numbers $\mathcal{N}_{rs}{}^t$ to be *nonnegative* integers. This turns out to be a nontrivial constraint on the possible graphs $\mathcal{G}$.

17.10.2. Positivity Constraints on Fusion Coefficients

We start from the list of ADE diagrams of Fig. 10.3.[19] A natural choice for the unit vertex is the end of the longest leg of the diagram.[20] With this choice, the graph algebras of D_{2n+1} and E_7 have some negative fusion numbers $\mathcal{N}_{rs}{}^t$ (for E_7, a proof is sketched in Ex. 17.21). We note that the corresponding modular invariants are not block-diagonal. They are obtained by an automorphism of the fusion rules of some block-diagonal theories, here A_{4n-1} and D_{10}. Since our discussion will

---

[18] This will be the case for all the ADE graphs under consideration.

[19] From now on in this section, we do not stick to our general convention of denoting the graphs by calligraphic letters. We instead reserve calligraphic symbols for the natural generalization of the ADE diagrams appearing in the classification of $\widehat{su}(2)$ modular invariants to their $\widehat{su}(3)$ relatives.

[20] Actually, this choice is not always unique: there are, for instance, two symmetric choices for the unit vertex on the A_n and E_6 graphs, but each leads to the same graph algebra. However, if the end of a short leg is chosen as the unit vertex, the corresponding graph algebra usually does not make any sense (see Ex. 17.20 for an illustration in the E_6 case).

concentrate on the fusion rules, it is natural to exclude these somewhat redundant cases.

On the other hand, the question of eigenvalue degeneracy occurs only for the D_{2n} diagrams, for which the eigenvalue 0 is twice degenerate. However, in all cases, there exists a particular rotation in the two-dimensional eigenspace for the eigenvalue 0 leading to nonnegative integer numbers $\mathcal{N}_{rs}{}^t$ (see Ex. 8.19 for the D_4 example).

17.10.3. Graph Subalgebra and Extended ADE Fusion Rules

In this subsection, we describe the precise relation between the $\widehat{su}(2)_k$ block-diagonal modular invariants and the A_n, D_{2n}, E_6, and E_8 Dynkin diagrams through their attached graph algebras.

To specify a block-diagonal modular invariant, all we need is the sets of representations that form the extended blocks. The representations occurring in the modular invariants are indexed by the exponents m_i (minus 1) of the corresponding Dynkin diagram. (As already noted, the exponents of the simply-laced Lie algebras also index the eigenvalues of the adjacency matrices of the Dynkin diagrams as $\Delta_{m_i} = 2 \cos \pi m_i / g$.) For instance, the E_6 WZW modular invariant reads

$$Z_{E_6} = |\chi_0 + \chi_6|^2 + |\chi_3 + \chi_7|^2 + |\chi_4 + \chi_{10}|^2 \qquad (17.208)$$

whereas the eigenvalues of the adjacency matrix of the E_6 Dynkin diagram of Fig. 17.1 read

$$\Delta_m = 2 \cos \pi \frac{m}{g} \quad m = 1, 4, 5, 7, 8, 11 \quad g = 12 \qquad (17.209)$$

Thus, the three blocks of the modular invariant (17.208) are specified by the pairs of eigenvalues $\{1, 7\}, \{4, 8\}$, and $\{5, 11\}$. As we now show, there exists a very simple and elegant way of encoding these data in the diagram itself.

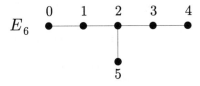

Figure 17.1. A labeling of the vertices of the E_6 diagram.

We first construct the graph algebra of E_6. We take the vertex 0 of Fig. 17.1 as the unit (hence $N_0 = \mathbb{I}$) and 1 as the fundamental vertex, so that the matrix N_1

reads

$$N_1 = G = \begin{pmatrix} 0 & 1 & 0 & 0 & 0 & 0 \\ 1 & 0 & 1 & 0 & 0 & 0 \\ 0 & 1 & 0 & 1 & 0 & 1 \\ 0 & 0 & 1 & 0 & 1 & 0 \\ 0 & 0 & 0 & 1 & 0 & 0 \\ 0 & 0 & 1 & 0 & 0 & 0 \end{pmatrix} \tag{17.210}$$

The algebra encoded in G is

$$
\begin{aligned}
G^2 &= \mathbb{I} + N_2 \\
G N_2 &= G + N_3 + N_5 \\
G N_3 &= N_2 + N_4 \\
G N_4 &= N_3 \\
G N_5 &= N_2
\end{aligned}
\tag{17.211}
$$

that is, the product GN_i is the sum of all N_j such that j is linked to i. From these relations, it is possible to express all the matrices N_i solely in terms of G:

$$
\begin{aligned}
N_2 &= G^2 - \mathbb{I} \\
N_3 &= G(G^4 - 4G^2 + 2\mathbb{I}) \\
N_4 &= (G^4 - 4G^2 + 2\mathbb{I}) \\
N_5 &= G(-G^4 + 5G^2 - 4\mathbb{I})
\end{aligned}
\tag{17.212}
$$

It is then straightforward to show that all these matrices are integral matrices with nonnegative integer entries. Notice that the matrix G is a solution of the characteristic equation

$$P(G) = (G^2 - \mathbb{I})(G^4 - 4G^2 + \mathbb{I}) = 0 \tag{17.213}$$

which simply reexpresses the relation $G\,N_5 = N_2$. Here P denotes the characteristic polynomial of G,

$$P(x) = \det(x\mathbb{I} - G) \tag{17.214}$$

The E_6 graph algebra is in fact isomorphic to the algebra $\mathbb{C}[x]/P(x)$ of polynomials modulo P.

It is easy to see that the matrices N_0, N_4, and N_5 form a subalgebra of the E_6 graph algebra. For instance,

$$
\begin{aligned}
N_4^2 - \mathbb{I} &= (N_4 + \mathbb{I})(N_4 - \mathbb{I}) \\
&= (G^4 - 4G^2 + 3\mathbb{I})(G^4 - 4G^2 + \mathbb{I}) \\
&= (G^2 - 3\mathbb{I})P(G) = 0
\end{aligned}
\tag{17.215}
$$

We find the relations

$$N_5^2 = N_0 + N_4$$
$$N_5 N_4 = N_5 \qquad\qquad (17.216)$$
$$N_4^2 = N_0$$

These are exactly the fusion rules of the extended diagonal theory associated with the invariant Z_{E_6}, namely $\widehat{sp}(4)_1$ (we recall that this invariant is obtained by the conformal embedding of $\widehat{su}(2)_{10}$ in $\widehat{sp}(4)_1$). Moreover, the fusion rules of $\widehat{sp}(4)_1$ are isomorphic to those of the Ising model. The identifications read

$$0 \;\leftrightarrow\; [10,0] \oplus [4,6] \;\leftrightarrow\; \mathbb{I}$$
$$4 \;\leftrightarrow\; [6,4] \oplus [0,10] \;\leftrightarrow\; \epsilon \qquad\qquad (17.217)$$
$$5 \;\leftrightarrow\; [7,3] \oplus [3,7] \;\leftrightarrow\; \sigma$$

In the following, we denote by $\mathcal{U} = \{0, 4, 5\}$ the set of vertices corresponding to this subalgebra.

The generalizations of this subset $\mathcal{U}$ for the other ADE algebras associated with block-diagonal invariants are indicated in Fig. 17.2, where the corresponding subsets of vertices of $\mathcal{G}$ are circled. In each case, the fusion rules associated with the graph subalgebras reproduce the *extended fusion algebra* of the corresponding WZW modular invariant theory.

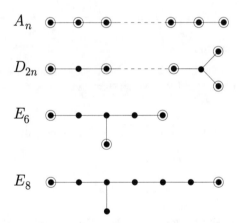

Figure 17.2. The sets $\mathcal{U}$ of circled vertices for each of the graphs A_n, D_{2n}, E_6, and E_8, giving rise to a subalgebra of the graph algebra, identical to the extended fusion algebra of the corresponding WZW model.

Therefore, by supplementing the graphs with a particular subset of their vertices, we have been able to encode more than just the set of exponents (i.e., the set of spinless operators in the theory). We will show now that, quite remarkably, this additional information actually encodes the whole structure of the corresponding modular invariant.

We need to introduce a duality relation between a graph subalgebra, with elements indexed by a subset $\mathcal{U}$ of the vertices of the graph $\mathcal{G}$, and its dual algebra, indexed by eigenvalue labels of the corresponding adjacency matrix G.[21] In concrete terms, we use the subset $\mathcal{U}$ to define an equivalence relation, denoted by $\approx$, between the eigenvalue labels of G. This equivalence relation is defined in terms of the quantities

$$f_{m,n} = \sum_{r \in \mathcal{U}} S_{rm} S_{rn} \qquad (17.218)$$

to be

$$m \approx n \text{ iff } f_{m,n} \neq 0 \qquad (17.219)$$

It can be shown that $\approx$ is indeed reflexive, symmetric, and transitive.[22] This provides an equivalence relation between the eigenvalue labels of G. The connection with the block-diagonal modular invariants is the following: all eigenvalues in a given equivalence class are the elements of a block and to each equivalence class there corresponds one block.

We now examine this equivalence relation in the E_6 case. The eigenvalues Δ_m and eigenvectors S_{rm} of the E_6 adjacency matrix are displayed in Table 17.1. With $\mathcal{U} = \{0, 4, 5\}$, the computation of the quantities $f_{m,n} = f_{n,m}$ through Eq. (17.218) is straightforward. For instance,

$$f_{1,7} = a_- a_+ + a_- a_+ + 2a_- a_+ = 4\frac{\sqrt{(3 + \sqrt{3})(3 - \sqrt{3})}}{24} = \frac{1}{\sqrt{6}}$$

For $f_{m,n}$ ($m \leq n$), we find

$$f_{1,1} = f_{11,11} = \frac{3 - \sqrt{3}}{6}$$

$$f_{5,5} = f_{7,7} = \frac{3 + \sqrt{3}}{6}$$

$$f_{1,7} = f_{5,11} = \frac{1}{\sqrt{6}} \qquad (17.220)$$

$$f_{4,4} = f_{8,8} = f_{4,8} = \frac{1}{2}$$

---

[21] The mathematical foundation of this argument is the *C-algebra duality*. A C-algebra (also called a character algebra) is simply a commutative algebra defined by some quadratic relations satisfying a number of axioms, satisfied in particular by the graph algebras for A_n, D_{2n}, E_6, and E_8. Roughly speaking, C-algebra duality generalizes the duality between classes and representations of a subgroup of a finite Abelian group.

[22] In fact, the quantities $f_{m,n}$ can be computed exactly, using the theory of C-algebra duality, and we have

$$\frac{\sum_{a \in \mathcal{U}} S_{am} S_{an}}{\sum_{a \in \mathcal{U}} S_{am}^2} = \frac{S_{0n}}{S_{0m}} \delta_{m \approx n}$$

where $\delta_{m \approx n}$ is 1 if $m \approx n$ and 0 otherwise.

and all the other f's vanish. Therefore $1 \approx 7$, $5 \approx 11$, and $4 \approx 8$, and the equivalence relation $\approx$ has three equivalence classes:

$$\{1,7\} \quad \{5,11\} \quad \{4,8\} \tag{17.221}$$

They match exactly (up to the usual shift of each exponent by -1) the three blocks of the modular invariant (17.208).

Table 17.1. Eigenvalues and eigenvectors of E_6. We display line by line the eigenvalue $2\cos\pi m/12$, the exponent m, and the six eigenvector components S_{rm}, $r = 0, \cdots, 5$, corresponding to the six vertices of E_6. Notation: $a_\pm = \sqrt{3 \pm \sqrt{3}/2\sqrt{6}}$.

Eigenvalue	Exponent	0	1	2	3	4	5
$\frac{\sqrt{3}+1}{\sqrt{2}}$	1	a_-	a_+	$\sqrt{2}a_+$	a_+	a_-	$\sqrt{2}a_-$
1	4	$\frac{1}{2}$	$\frac{1}{2}$	0	$-\frac{1}{2}$	$-\frac{1}{2}$	0
$\frac{\sqrt{3}-1}{\sqrt{2}}$	5	a_+	a_-	$\sqrt{2}a_-$	a_-	a_+	$-\sqrt{2}a_+$
$-\frac{\sqrt{3}-1}{\sqrt{2}}$	7	a_+	$-a_-$	$-\sqrt{2}a_-$	$-a_-$	a_+	$\sqrt{2}a_+$
-1	8	$\frac{1}{2}$	$-\frac{1}{2}$	0	$\frac{1}{2}$	$-\frac{1}{2}$	0
$-\frac{\sqrt{3}+1}{\sqrt{2}}$	11	a_-	$-a_+$	$\sqrt{2}a_+$	$-a_+$	a_-	$-\sqrt{2}a_-$

By repeating this calculation with A_n, D_{2n}, and E_8, we find that the equivalence classes dual to the sets $\mathcal{U}$ indicated by the circled vertices of Fig. 17.2 all reproduce the extended blocks of the corresponding WZW modular invariants. The detailed analysis of the E_8 case is presented in Ex. 17.22.

For later use in the $\widehat{su}(3)$ generalization, we stress the importance of the S matrix, which diagonalizes G. For the A_n diagrams, this matrix coincides with the S matrix of the modular transformation $\tau \rightarrow -1/\tau$ of the WZW characters at level $k = n - 1$. In particular, the eigenvalues of the A_n adjacency matrix are

$$\Delta_\mu = \gamma_f^{(\hat{\mu})} = \frac{S_{f\hat{\mu}}}{S_{0\hat{\mu}}} \tag{17.222}$$

Here f denotes the affine extension of the fundamental weight ω_1 at level k, and $\gamma_{\hat{\lambda}}^{(\hat{\mu})}$ has been defined in Eq. (14.245). It is left as an exercise to check that

$$\Delta_\mu = \frac{\sin[2\pi(\mu_1 + 1)/(k + 2)]}{\sin[\pi(\mu_1 + 1)/(k + 2)]} = 2\cos\left[\frac{\pi(\mu_1 + 1)}{(k + 2)}\right] \tag{17.223}$$

17.10.4. Generalized ADE Diagrams for $\widehat{su}(3)$

We now to generalize the above construction to the $\widehat{su}(3)$ case. In contradistinction with $\widehat{su}(2)$, here we have no diagram at hand from which we can construct the modular invariants by means of the above duality argument. The first step is to try to construct these diagrams, guided by the $\widehat{su}(2)$ results. A natural starting point is to generalize the A_n diagrams, which correspond to the diagonal invariants. We recall that for $\widehat{su}(2)_k$, all the vertices of the A_{k+1} diagram are circled. This means that the diagram itself encodes the fusions of the theory by the fundamental representation (1). This suggests defining the corresponding diagram in the $\widehat{su}(3)_k$ theory, denoted by $\mathcal{A}_k$, as the one encoding the fusions by the fundamental representation $(1,0)$. The $\mathcal{A}_k$ diagram is depicted in Fig. 17.3. Because the representation $(1,0)$ is not self-conjugate, the diagram is now *oriented*. Moreover, its vertices have a natural $\mathbb{Z}_3$-grading, namely the triality of the corresponding representation.[23] Although not symmetric, the adjacency matrix of $\mathcal{A}_k$ can be shown to be diagonalizable in an orthonormal basis. As in the $\widehat{su}(2)_k$ case, the eigenvalues and eigenvectors of this diagram are simply expressed in terms of the $\widehat{su}(3)_k$ modular matrices. The eigenvectors are the $\mathcal{S}$ matrix elements (Eq. (14.217)):

$$\mathcal{S}_{(\lambda_1,\lambda_2),(\mu_1,\mu_2)} \equiv \mathcal{S}_{[k-\lambda_1-\lambda_2,\lambda_1,\lambda_2],[k-\mu_1-\mu_2,\mu_1,\mu_2]} \tag{17.224}$$

and the eigenvalues are given by the ratios

$$\Delta_{(\mu_1,\mu_2)} = \frac{\mathcal{S}_{(1,0),(\mu_1,\mu_2)}}{\mathcal{S}_{(0,0),(\mu_1,\mu_2)}} = \sum_{j=1}^{3} e^{2i\pi(\epsilon_j,\mu+\rho)/3(k+3)} \tag{17.225}$$

where the ϵ_i are related to the fundamental weights by

$$\epsilon_1 = \omega_1 \qquad \epsilon_2 = \omega_2 - \omega_1 \qquad \epsilon_3 = -\omega_2$$

In Eqs. (17.224)–(17.225), both the vertex and eigenvalue labels run over the set P_+^k of integral weights of level k

$$P_+^k = \{(\lambda_1,\lambda_2) \,|\, \lambda_1,\, \lambda_2 \geq 0 \quad \lambda_1 + \lambda_2 \leq k\} \tag{17.226}$$

and the indices $(\lambda_1 + 1, \lambda_2 + 1) = \lambda + \rho$ now play the role of the exponents of the A_{k+1} diagram. We will still call them the *exponents* of $\mathcal{A}_k$. Equation (17.225) is actually the $\widehat{su}(3)$ generalization, for $\mathcal{A}_k$, of the formula (17.209) for the eigenvalues of A. Its proof is left as an exercise (Ex. 17.26).

Our program is now clear: reversing the logic of the $\widehat{su}(2)$ association between modular invariants and graphs, we wish to construct an oriented graph for each $\widehat{su}(3)$ block-diagonal[24] modular invariant, whose eigenvalues are indexed by the spinless fields of the theory. These fields are indexed by a certain list of integral

---

[23] Triality is the special name given to the congruence class (defined in Sect. 13.1.9) for $su(3)$; the triality of the weight $\lambda = (\lambda_1,\lambda_2)$ is $\lambda_1 + 2\lambda_2$ mod 3.

[24] We stress that the distinction between block-diagonal and nonblock-diagonal invariants is necessary only for the discussion of the extended fusion rules. In fact, generalized Dynkin diagrams can also be constructed for the other $\widehat{su}(3)_k$ invariants, but we choose not to present them here and to concentrate on the block-diagonal cases.

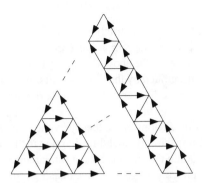

Figure 17.3. The $\mathcal{A}_k$ diagram for $su(3)$, with $(k+1)(k+2)/2$ vertices.

weights (actually a subset of Eq. (17.226) with possible repetitions), which must be the exponents of the desired graph. This can actually be done (with a little intuition and much work) and leads to the graphs of Fig. 17.4. For completeness, the list of exponents is displayed on Table 17.2.

Table 17.2. Graphs and exponents for the block-diagonal WZW $\widehat{su}(3)_k$ modular invariants.

Graph	Exponents
$\mathcal{A}_k$	$\{\lambda + \rho\},\ \lambda \in P_+^k$
$\mathcal{D}_{3k}$	$\{\lambda + \rho\},\ \lambda \in Q \cap P_+^{3k}, 2 \times (k+1, k+1)$
$\mathcal{E}_5$	$\{(1,1),(3,3),(3,2),(1,6),(2,3),(6,1),$ $(4,1),(1,4),(1,3),(4,3),(3,1),(3,4)\}$
$\mathcal{E}_9$	$\{(1,1),(10,1),(1,10),(5,5),(5,2),(2,5),$ $(3,3),(3,6),(6,3),(3,3),(3,6),(6,3)\}$
$\mathcal{E}_{21}$	$\{(1,1),(22,1),(1,22),(5,5),(5,14),(14,5),(11,11),(11,2),$ $(2,11),(7,7),(7,10),(10,7),(7,1),(16,7),(1,16),(1,7),$ $(7,16),(16,1),(5,8),(11,5),(8,11),(8,5),(5,11),(11,8)\}$

The construction presented here is not unique: several different graphs may share the same eigenvalues, hence the same set of exponents. (Some examples are worked out in Ex. 17.24.) The choice leading to Fig. 17.4 is actually justified by the relation to the extended fusion rules, which is the subject of the next section.

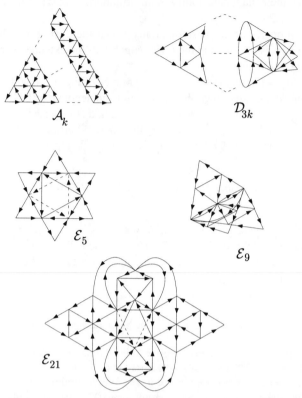

$\mathcal{A}_k$

$\mathcal{D}_{3k}$

$\mathcal{E}_5$

$\mathcal{E}_9$

$\mathcal{E}_{21}$

Figure 17.4.   The generalized Dynkin diagrams for the $\widehat{su}(3)_k$ WZW block-diagonal invariants. The names of the diagrams are indexed by the value of the level k.

17.10.5. Graph Subalgebras and Modular Invariants for $\widehat{su}(3)$

When constructing the generalized Dynkin diagrams of Fig. 17.4, an important guideline is the parallel construction of a coherent graph algebra, whose coefficients $\mathcal{N}_{rs}{}^t$ are all *nonnegative* integer numbers: this property of positivity will again restrict the result to block-diagonal modular invariants only. Coherence also forces the existence of a *unit vertex*, namely a vertex with only one link pointing out to the *fundamental* one on $\mathcal{G}$. Notice, moreover, that the graphs of Fig. 17.4 all have some particular symmetry properties. They all have the $\mathbb{Z}_3$ grading mentioned above in the $\mathcal{A}$ case. Moreover, the adjacency matrices G satisfy the *normality* condition[25]

$$[G, G^t] = 0 \qquad\qquad (17.227)$$

---

[25] For the $\mathcal{A}_k$ diagram, the matrix G codes the fusion by the fundamental representation $(1, 0)$, whereas its transpose G^t codes those of the conjugate representation $(0, 1)$. That $[G, G^t] = 0$ is a consequence of the commutativity of the fusion algebra.

which ensures their diagonalizability in an orthonormal basis. This statement is proven in Ex. 17.23.

The main property of all these diagrams is that there exists a subalgebra of the graph algebra, isomorphic to the fusion algebra of the extended version of the corresponding WZW theory.

Consider, for example, the case of the diagram $\mathcal{E}_5$ of Fig. 17.4, with 12 vertices. We choose one of the external vertices as the unit and denote it by $0 \equiv 1_0$. We label by $1_1, 1_2, 1_3, 1_4, 1_5$ the other external vertices counted clockwise from 1_0. We also label by $f \equiv 2_0$ and $2_1, 2_2, 2_3, 2_4$, and 2_5 the internal vertices, starting from the fundamental one and also counted clockwise. The graph algebra relations read

$$
\begin{aligned}
N_f\, N_{1_j} &= N_{2_j} \\
N_f\, N_{2_j} &= N_{2_{j+1}} + N_{2_{j+4}} + N_{1_{j-1}}
\end{aligned}
\tag{17.228}
$$

where the indices j are considered as elements of $\mathbb{Z}_6$. After some algebra, it can be shown that the subset of matrices corresponding to the external vertices

$$
\{N_{1_j}\}, \; j \in \mathbb{Z}_6
\tag{17.229}
$$

forms a closed subalgebra, namely (cf. Ex. 17.25)

$$
N_{1_j}\, N_{1_k} = N_{1_{j+k}}
\tag{17.230}
$$

which is isomorphic to the $\widehat{su}(6)_1$ fusion algebra. Since the $Z_{\mathcal{E}_5}$ modular invariant is obtained by the conformal embedding $\widehat{su}(3)_5 \subset \widehat{su}(6)_1$, this fusion algebra is precisely the extended fusion algebra of the $\mathcal{E}_5$ WZW model.

We have displayed, in Fig. 17.5, the generalized ADE graphs, with their circled vertices denoting the subalgebras reproducing the extended fusion rules of the block-diagonal $\widehat{su}(3)$ WZW modular-invariant theories. By means of the duality described in Sect. 17.10.3, each subalgebra can be mapped to a partition of the set of exponents (eigenvalue labels) that correspond precisely to the extended blocks of the modular invariant.

To illustrate the whole construction, we pursue the analysis of the $\mathcal{E}_5$ example and derive the corresponding partition of the set of exponents. To diagonalize the adjacency matrix G of $\mathcal{E}_5$, we use the $\mathbb{Z}_3$ grading property. With the unit $1_0 \equiv 0$ and the fundamental $2_0 \equiv f$ having triality 0 and 1, respectively, we find that all the vertices in the set $\{1_j, 1_{j+3}, 2_{j+1}, 2_{j+4}\}$ have triality j, with $j = 0, 1, 2$. (Triality is additive in tensor products, hence in fusions.) The matrix G maps the set of triality j onto that of triality $j + 1$. Therefore, in the basis

$$
\{1_j, 1_{j+3}, 2_{j+1}, 2_{j+4}\} \quad j = 0, 1, 2
\tag{17.231}
$$

the matrix G takes the block form

$$
G = \begin{pmatrix} 0 & 0 & H \\ H & 0 & 0 \\ 0 & H & 0 \end{pmatrix}
\tag{17.232}
$$

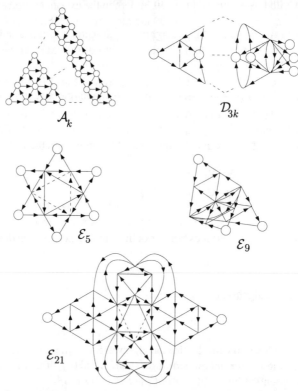

Figure 17.5. Generalized ADE graphs, with the circled vertices defining the relevant graph subalgebras, which encode the extended fusion rules of the block-diagonal $\widehat{su}(3)_k$ WZW theories.

where

$$H = \begin{pmatrix} 0 & 0 & 1 & 0 \\ 0 & 0 & 0 & 1 \\ 1 & 0 & 1 & 1 \\ 0 & 1 & 1 & 1 \end{pmatrix} \tag{17.233}$$

All the eigenvectors and eigenvalues of G may be found by diagonalizing H. We suppose that v is an eigenvector of H with eigenvalue x:

$$H v = x v \tag{17.234}$$

Then, the vector (v, v, v), with 12 components, is clearly an eigenvector of G for the same eigenvalue x. Let $\omega = e^{2i\pi/3}$ be the fundamental third root of unity. Then the vectors $(v, \omega v, \omega^2 v)$ and $(v, \omega^2 v, \omega v)$ are also eigenvectors of G with respective eigenvalues $\omega^2 x$ and ωx. This completes the diagonalization of G. The eigenvalues $\Delta_{(m,n)}$ and eigenvectors $P_{r,(m,n)}$ of H are listed in Table 17.3, with the corresponding $\mathcal{E}_5$ exponents, that is, the eigenvalue labels in Eq. (17.225). We note that if the eigenvalue x corresponds to the exponent (m, n) (i.e., is given by

Eq. (17.225) with $(\mu_1+1, \mu_2+1) = (m,n))$, then the exponents corresponding to the "rotated" eigenvalues ωx (resp. $\omega^2 x$) are also "rotated" through $\sigma(m,n)$ (resp. $\sigma^2(m,n)$), where σ is the rotation by $2\pi/3$ of the corresponding weight[26]

$$\sigma(m,n) = (8-m-n,m) \qquad (17.235)$$

For (m,n) as in Table 17.3, the eigenvalues and eigenvectors of G read, respectively,

$$\omega^j \Delta_{(m,n)}, \qquad j = 0,1,2 \qquad (17.236)$$

and

$$S_{v(r,j),\sigma^i(m,n)} = \frac{1}{\sqrt{3}}(P_{r,(m,n)}, \omega^{2j}P_{r,(m,n)}, \omega^j P_{r,(m,n)}) \qquad (17.237)$$

for $j = 0,1,2$; $r = 1,2,3,4$; and (m,n) as in Table 17.3. We have denoted by

$$\{v(r,j)\}_{1 \leq r \leq 4} = \{1_j, 1_{j+3}, 2_{j+1}, 2_{j+4}\} \qquad (17.238)$$

the corresponding vertices of $\mathcal{E}_5$.

Table 17.3. Eigenvalues and eigenvectors of H.
We display line by line the eigenvalue $\Delta_{(m,n)}$, the
corresponding exponent (m,n) of $\mathcal{E}_5$, and the four
eigenvector components, $P_{r,(m,n)}$, $r = 1,2,3,4$.
Notation: $b_\pm = \sqrt{2 \pm \sqrt{2}/2\sqrt{2}}$.

Eigenvalue	Exponent	1	2	3	4
$1+\sqrt{2}$	$(1,1)$	b_-	b_-	b_+	b_+
$1-\sqrt{2}$	$(3,3)$	b_+	b_+	$-b_-$	$-b_-$
1	$(4,1)$	$-\frac{1}{2}$	$\frac{1}{2}$	$-\frac{1}{2}$	$\frac{1}{2}$
-1	$(1,4)$	$\frac{1}{2}$	$-\frac{1}{2}$	$-\frac{1}{2}$	$\frac{1}{2}$

It is now easy to compute the quantities

$$f_{(m,n),(p,q)} = \sum_{v \in \mathcal{U}} S_{v,(m,n)} \bar{S}_{v,(p,q)} \qquad (17.239)$$

---

[26] This is merely the action of the outer automorphism a on the shifted weight $\lambda + \rho = (m,n)$ at level $k+3 = 8$.

where $\mathcal{U}$ denotes the set of vertices 1_j, $j \in \mathbb{Z}_6$. Applying the rule (17.219), which defines the equivalence relation $\approx$, we find that

$$
\begin{array}{lll}
(1,1) \approx (3,3) & (4,1) \approx (1,4) & (3,2) \approx (1,6) \\
(2,3) \approx (6,1) & (1,3) \approx (4,3) & (3,1) \approx (3,4)
\end{array}
\tag{17.240}
$$

so that the equivalence classes are

$$
\begin{array}{lll}
\{(1,1),(3,3)\} & \{(4,1),(1,4)\} & \{(3,2),(1,6)\}, \\
\{(2,3),(6,1)\} & \{(1,3),(4,3)\} & \{(3,1),(3,4)\}
\end{array}
\tag{17.241}
$$

They match exactly (up to the usual shift of all the indices by -1) the blocks of the $\mathcal{E}_5$ modular invariant (17.115)

$$
\begin{aligned}
Z_{\mathcal{E}_5} = & \ |\chi_{0,0} + \chi_{2,2}|^2 + |\chi_{3,0} + \chi_{0,3}|^2 + |\chi_{2,1} + \chi_{0,5}|^2 \\
& + |\chi_{1,2} + \chi_{5,0}|^2 + |\chi_{0,2} + \chi_{3,2}|^2 + |\chi_{2,0} + \chi_{2,3}|^2
\end{aligned}
\tag{17.242}
$$

The results presented in this section generalize to any Lie algebra. For higher-rank algebras, we have already indicated that the fusions by the fundamental representation do not generate all the fusion rules (cf. Ex. 16.11). Typically, for $\widehat{su}(N)$, we need to know the fusions by the $N/2$ first fundamental representations (the fusions by their conjugates follow by transposition of the fusion matrices N_{f_i}). Therefore, the $\mathcal{A}$ graph must be replaced by a collection of graphs, each corresponding to one of these fundamental representations. However, we believe that the above results can be adapted to these cases too. The general classification problem of modular invariants for WZW theories can therefore be rephrased in terms of a classification problem of graphs, with some particular properties relating them to the original problem. It would seem that the graph classification problem is simpler, and this is certainly the case for $\widehat{su}(2)$. However, the precise link between the two problems is still missing, and the emergence of the generalized Dynkin diagrams must still be regarded as a phenomenological observation. Nevertheless, deep relations between modular invariance, number theory, and graph theory should be expected.

We make a final remark on the generalized Dynkin diagrams for $\widehat{su}(3)$. Although they have no direct Lie-algebraic interpretation, these diagrams are the outcome of some well-defined classification problem. Just like the ADE diagrams, which have many different interpretations in mathematics and physics (e.g., Platonic solids, simply-laced Lie algebras, $\widehat{su}(2)$ modular invariants, finite subgroups of $su(2)$, catastrophes, integrable lattice models), the $\widehat{su}(3)$ diagrams may still be hiding some of their mysteries.

Appendix 17.A. $\widehat{su}(p)_q \oplus \widehat{su}(q)_p \subset \widehat{su}(pq)_1$ Branching Rules

In this appendix, a simple and powerful method is presented for the calculation of the conformal branching rules of $\widehat{su}(p)_q \oplus \widehat{su}(q)_p \subset \widehat{su}(pq)_1$. First, we explain how the corresponding finite branching rules are evaluated. Let λ be the highest weight of an irreducible representation of $su(pq)$, represented by a Young tableau of $|\lambda|$ boxes. Its decomposition into $su(p) \oplus su(q)$ is given by all pairs of $su(p)$ and $su(q)$ Young tableaux of $|\lambda|$ boxes each, which are related by transposition, that is, by interchanging rows and columns. The rule is to be understood in the sense that if the transpose of a $su(p)$ Young tableau has more than q rows, it is not a $su(q)$ Young tableau, and this pair of tableaux does not appear in the decomposition. For instance, the decomposition of ω_5 of $su(6)$ into $su(2) \oplus su(3)$ is

$$\begin{array}{c} \square \\ \to \left\{ \square \otimes \square \right\} \end{array} \qquad (17.243)$$

that is,

$$(0,0,0,0,1) \to (1) \otimes (0,1) \qquad (17.244)$$

Indeed, for the other two $su(2)$ tableaux of five boxes

$$\boxed{} \, , \quad \square \qquad (17.245)$$

the interchange of rows and columns leads to tableaux of more than three rows, that is, nonregular Young tableaux for $su(3)$.

Part of the affine branching rules can be obtained by a similar procedure, except that the level must be taken into account. This forces the transposition transformation to be slightly modified. We first describe this modification and indicate later how the complete branching rules can be found. Since we are interested only in conformal embeddings, which restricts the level of the $\widehat{su}(pq)$ algebra to one, only the branching rules for the fundamental representations are needed. The Young tableau representing $\hat{\omega}_\ell$ for $\ell \neq 0$ is a single column of ℓ boxes, and a column of pq boxes for $\hat{\omega}_0$. To obtain the decomposition of a tableau of ℓ boxes, we first look for $su(p)$ tableaux with exactly q columns (the level constraint), containing $\ell - np$ boxes, where n is a nonnegative integer. This subtraction takes into account the fact that extra columns are allowed if they have exactly p boxes each (these do not contribute to the level). Of course, the transposed tableau has the same number of boxes, namely $\ell - np$. However, we are looking for a tableau with $\ell - mq$ boxes (with m a nonnegative integer). This means that when $n, m \neq 0$, the transposition transformation must be adjusted. The required adjustment relies on the following fact: there always exists a single element of the outer-automorphism group of $\widehat{su}(q)_p$ that transforms the integrable weight associated with the Young tableau of $\ell - np$ boxes, into another integrable weight whose finite part is represented by a Young tableau of $\ell - mq$ boxes. The resulting affine weight is the second of two weights occurring in the decomposition of $\hat{\omega}_\ell$.

The above statement, concerning the existence of an element of $\mathcal{O}(\widehat{su}(q)_p)$ that can transform a Young tableau of $\ell - np$ boxes into another one of $\ell - mq$ boxes, is very simple to establish. We let $\tilde{a}$ be the basic outer automorphism of

$\widehat{su}(q)$. We recall that the action of $\tilde{a}$ on a reduced tableau (one for which columns of q boxes have been stripped off) associated with an affine weight at level p, amounts to adding on the top of the tableau a row of p boxes (cf. Eq. (16.167)). For repeated applications of $\tilde{a}$, the tableau has to be reduced at each step. Therefore, n applications of $\tilde{a}$ on a tableau of $\ell - np$ boxes transform it into a tableau of ℓ boxes, and m reductions leave a resulting tableau of $\ell - mq$ boxes, as desired.

We illustrate this procedure by the example of the branching of $\hat{\omega}_0$ of $\widehat{su}(12)_1$ into $\widehat{su}(3)_4 \oplus \widehat{su}(4)_3$, which gives:

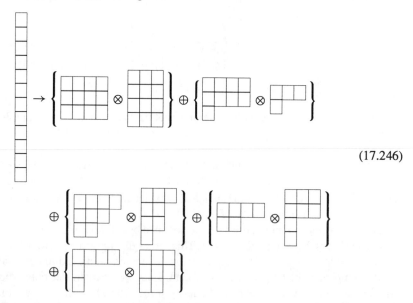

$$(17.246)$$

The above $su(3)$ tableaux (the first members of each pair) are all the possible tableaux with at most 3 rows, 4 columns, and $12 - 3n$ boxes. (Here we choose to write tableaux such that they all have exactly 4 columns, by filling with the appropriate number of columns of three boxes.) The transpose of the $su(3)$ tableau in the second pair is

$$\longleftrightarrow \quad [2,1,0,0] \qquad\qquad (17.247)$$

Since the number of boxes is 9, which cannot be written as $12 - 4m$, this is not an eligible $\widehat{su}(4)_3$ tableau in the present context. Application of the various powers of the basic element $\tilde{a}$ of the $\widehat{su}(4)$ outer automorphism group to the reduced tableau yields (performing the necessary reductions after each step):

$$\tilde{a}: \quad \square \;\rightarrow\; \; \rightarrow \; \; \rightarrow \; \qquad\qquad (17.248)$$

Among these, only the second one has the right number of boxes, hence

$$\text{(tableau)} \quad \longleftrightarrow \quad [0,2,1,0] \qquad\qquad (17.249)$$

must be chosen. Similar manipulations on the transpose of the first tableau in the third, fourth, and fifth pairs above show that eligible tableaux are obtained respectively from the action of $\tilde{a}, \tilde{a}^2$ and $\tilde{a}^2$.

The weight transcription of (17.246) is:

$$[1,0,0,0,0,0,0,0,0,0,0,0]$$
$$\mapsto \{[4,0,0] \otimes [3,0,0,0]\} \oplus \{[1,0,3] \otimes [0,2,1,0]\}$$
$$\oplus \{[2,1,1] \otimes [1,1,0,1]\} \oplus \{[0,2,2] \otimes [1,0,2,0])\}$$
$$\oplus \{[1,3,0] \otimes [0,0,1,2]\}$$
$$(17.250)$$

As previously mentioned, there is no guarantee at this point that this is the complete answer. Nevertheless, we can check, by verifying the sum rule (17.95), that for this case this is indeed the full branching rule.

To proceed, we need the branchings of the outer-automorphism groups. We denote by A, a, and $\tilde{a}$, respectively, the basic elements of the outer-automorphism group of $\widehat{su}(pq), \widehat{su}(p)$, and $\widehat{su}(q)$. The relation

$$\mathcal{P} A^{pq} = (a^p \otimes 1)\mathcal{P} = (1 \otimes \tilde{a}^q)\mathcal{P} = 1 \qquad\qquad (17.251)$$

(where $\mathcal{P}$ is the projection matrix) allows the correspondence

$$A^q \mapsto a \otimes 1, \quad A^p \mapsto 1 \otimes \tilde{a} \qquad\qquad (17.252)$$

Now, the affine branching rules are obtained from the union of the action by Eq. (17.252) in all possible ways on the partial branching rules obtained in the previous step. (We give this result without proof.) In this way, we could check that Eq. (17.250) is complete. This is left as an exercise, and a simpler example will now be considered, namely $\widehat{su}(2)_3 \oplus \widehat{su}(3)_2 \subset \widehat{su}(6)_1$. The partial branching rules obtained after the first step are:

$$\hat{\omega}_0 \mapsto \{[1,2] \otimes [0,1,1]\} \oplus \{[3,0] \otimes [2,0,0]\} \oplus \cdots$$
$$\hat{\omega}_1 \mapsto \{[2,1] \otimes [1,1,0]\} \oplus \cdots$$
$$\hat{\omega}_2 \mapsto \{[1,2] \otimes [1,0,1]\} \oplus \{[3,0] \otimes [0,2,0]\} \oplus \cdots$$
$$\hat{\omega}_3 \mapsto \{[2,1] \otimes [0,1,1]\} \oplus \{[0,3] \otimes [2,0,0]\} \oplus \cdots \qquad (17.253)$$
$$\hat{\omega}_4 \mapsto \{[1,2] \otimes [1,1,0]\} \oplus \{[3,0] \otimes [0,0,2]\} \oplus \cdots$$
$$\hat{\omega}_5 \mapsto \{[2,1] \otimes [1,0,1]\} \oplus \{[0,3] \otimes [0,2,0]\} \oplus \cdots$$

It follows from Eq. (17.252) that $A^3 \mapsto a \otimes 1$ and $A^2 \mapsto 1 \otimes \tilde{a}$. In particular, by applying A^3 on the $\hat{\omega}_4$ branching rule, we find an additional term in the $\hat{\omega}_1$ branching rule, namely $\{[3,0] \otimes [0,0,2]\}$. Further actions of A produce no additional terms.

In general, A itself has no image into $\widehat{su}(p) \oplus \widehat{su}(q)$. In fact, it is only for p and q relatively prime that A has such an image, and it is given by

$$A \mapsto a^r \otimes \tilde{a}^s \tag{17.254}$$

where the integers r and s are fixed by[27]

$$ps = 1 \bmod q \, , \qquad qr = 1 \bmod p \tag{17.255}$$

This follows from the substitution of relation (17.254) into relation (17.252). These conditions are met in the above example, for which

$$A \mapsto a \otimes \tilde{a}^2 \tag{17.256}$$

An efficient use of this relation reduces substantially the number of branching rules that have to be explicitly evaluated.

The present method provides another way of deriving $\widehat{su}(N)$ modular invariants of the outer-automorphism type. Furthermore, it can be used to find new exceptional modular invariants: the contraction on a known exceptional invariant for $\widehat{su}(p)_q$ yields an exceptional invariant for $\widehat{su}(q)_p$.

With appropriate modifications, this approach can be applied to the analysis of other infinite sequences of conformal embeddings.

Appendix 17.B. General Orbifolds: Fine Structure of the $c = 1$ Models

In this appendix, we generalize the notion of the $\mathbb{Z}_2$ orbifold encountered in Chap. 10. Whenever a conformal theory possesses some additional invariance under a finite group G, it is possible to quotient this symmetry and to produce an orbifold theory.

As an illustration, we apply the orbifold technique to the $c = 1$ bosonic theory on a circle at the self-dual point:

$$R = \frac{2}{R} = \sqrt{2} \tag{17.257}$$

at which there is an additional spectrum-generating $\widehat{su}(2)_1$ symmetry algebra. This is simply the vertex representation of the $\widehat{su}(2)_1$ model described in Sect. 15.6 (see in particular the discussion following Eq. (15.244). The $SU(2)$ symmetry of the model enables us to consider orbifolds under finite subgroups of $SU(2)$ whose conjugate action leaves the current algebra invariant (although the action on the algebra generators is nontrivial). In fact, they leave the bosonic action on the circle invariant.

---

[27] The existence of (r, s) is granted by Bezout's lemma (see Ex. 10.1), which says that for q, p relatively prime, there exist two integers (r_0, s_0) satisfying $ps_0 - qr_0 = 1$; we simply have to choose $s = s_0 \bmod q$ and $r = -r_0 \bmod p$.

17.B.1. Orbifold Based on a Group G

The $\mathbb{Z}_2$ orbifold procedure described in Sect. 10.4.3 (also applied to minimal models in Sect. 10.7.5), can be generalized to any subgroup G of the symmetry group of the initial theory. It consists of the following steps:
(i) We first consider a "twisted" partition function on the torus. This corresponds, in the Lagrangian language, to imposing boundary conditions that are not periodic, but twisted by elements a, b of G along the "space" direction 1 and the "time" direction τ of the torus, respectively.[28] The bosonic field φ then must satisfy boundary conditions of the form

$$
\begin{aligned}
\varphi(z+1) &= \mathsf{a}\,\varphi(z) \\
\varphi(z+\tau) &= \mathsf{b}\,\varphi(z)
\end{aligned}
\tag{17.258}
$$

This produces a partition function $Z_{\mathsf{a},\mathsf{b}}(\tau)$.
(ii) We next sum over all compatible boundary conditions. More precisely, if the group G is Abelian, the full orbifold partition function reads

$$
Z_{\text{orb}} = \frac{1}{|G|} \sum_{\mathsf{a},\mathsf{b}\in G} Z_{\mathsf{a},\mathsf{b}}
\tag{17.259}
$$

However, if the group G is non-Abelian, the above sum must extend only to commuting couples $(\mathsf{a},\mathsf{b}) \in G^2$.

In Eq. (17.259), the sum over a corresponds to the various twisted sectors of the Hilbert space of the theory, that is, in a fixed "time" slice, whereas the sum over b produces a group-invariant projection in each sector

$$
\mathcal{B} = \frac{1}{|G|} \sum_{\mathsf{b}\in G} \mathsf{b}
\tag{17.260}
$$

The fact that the sum (17.259) produces a modular-invariant partition function is a consequence of the T and S actions on $Z_{\mathsf{a},\mathsf{b}}$, which, from Eq. (17.258), are directly found to be

$$
\begin{aligned}
T\,Z_{\mathsf{a},\mathsf{b}} &= Z_{\mathsf{a},\mathsf{a}\mathsf{b}} \\
S\,Z_{\mathsf{a},\mathsf{b}} &= Z_{\mathsf{b},\mathsf{a}}
\end{aligned}
\tag{17.261}
$$

We see that the commutativity requirement on (a,b) for non-Abelian groups ensures that the above action of T is well-defined.

Reformulated concisely, the construction of an orbifold starts with a projection of the Hilbert space states onto a G-invariant subspace

$$
Z_{\text{proj}} = \frac{1}{|G|} \sum_{\mathsf{b}\in G} Z_{1,\mathsf{b}}
\tag{17.262}
$$

---

[28] Throughout this appendix, a or b is used to denote a group element, not to be mistaken with an outer automorphism of the affine Lie algebra or an element of the group center. Actually the choice of a and b here is rather natural in that it recalls a relation with the basic homology cycles a and b on the torus.

where the partition function remains periodic in the space direction (a twist by the identity of the group, 1, has no effect). Modular invariance is then reinforced by summing over all possible space twists a that commute with b.

We illustrate this procedure in the case $G = \mathbb{Z}_2 = \{1, -1\}$, considered as a multiplicative group. We start with the projection

$$Z_{\text{proj}} = \frac{1}{2} \left(Z_{1,1} + Z_{1,-1} \right) \tag{17.263}$$

It is transformed under the action of S and T into

$$S\, Z_{\text{proj}} = \frac{1}{2} \left(Z_{1,1} + Z_{-1,1} \right)$$
$$T\, S\, Z_{\text{proj}} = \frac{1}{2} \left(Z_{1,1} + Z_{-1,-1} \right) \tag{17.264}$$

Hence we find that the combination[29]

$$Z_{\text{orb}} = (1 + S + TS)Z_{\text{proj}} - Z \tag{17.265}$$

is modular invariant. We have subtracted the untwisted partition function $Z = Z_{1,1}$ to avoid overcounting the untwisted sector.

The $\widehat{su}(2)_1$ structure of the self-dual $c = 1$ model on a circle of radius $R = \sqrt{2}$ allows for a variety of orbifolds, based on symmetry under the action of finite subgroups of $SU(2)$.

17.B.2. Orbifolds and the Method of Outer Automorphisms

We digress briefly from our main line of argument, in order to comment on the relation between the orbifold and the outer-automorphism methods. Manifestly, the construction of nondiagonal theories using outer automorphisms, described in Sect. 17.3, is a special case of the orbifold construction. Moreover, the way modular invariance is recovered after the projection B is simpler to describe in the orbifold construction than in the one based on outer automorphisms. However, it should be stressed that in the latter method, the group that is quotiented is rather special, being the center of the symmetry group.[30] Therefore, its action commutes with the algebra generators. This has the following important implication: upon projection, the affine characters are not broken; a complete representation either survives the projection or is eliminated. As a result, the orbifold partition function can still be expressed as a bilinear combination of the affine Lie algebra characters. Thus, formulating the construction in terms of outer automorphisms keeps us closer to the affine Lie algebra structure. Moreover, generic formulas are as simple for arbitrary G as for $G = \mathbb{Z}_2$, which is not the case with the orbifold method (cf. Ex. 17.30).

Quotienting a WZW model by a group that is not in the center is thus bound to break all the affine integrable representations—and these cannot be reconstructed in the twisting process; therefore, the partition function cannot be described in

---

[29] The generalization of this formula to $\mathbb{Z}_N$ orbifolds (N prime) is given in Ex. 17.30.

[30] This is thus always an Abelian-type orbifold.

terms of affine characters. In the present case, the free-boson representation of the $\widehat{su}(2)_1$ model:

$$H = i\partial\varphi \qquad E^\pm = e^{\pm i\sqrt{2}\varphi} \tag{17.266}$$

allows us to express the orbifold partition function in terms of bosonic ones.

17.B.3. $\mathbb{Z}_2$ Orbifold of the $c = 1$ $\widehat{su}(2)_1$ Theory

The $c = 1$ theory has an obvious $\mathbb{Z}_2$ symmetry generated by the transformation

$$\mathsf{a} : \varphi \to -\varphi \tag{17.267}$$

Its action on the affine generators is

$$\mathsf{a} : \quad H \to -H \qquad E^\pm \to E^\mp \tag{17.268}$$

Equivalently, with $H = \sqrt{2}J^0 \equiv \sqrt{2}J^3$ and $E^\pm = J^1 \pm iJ^2$, it reads

$$\mathsf{a} : \quad J^1 \to J^1 \quad J^2 \to -J^2 \quad J^3 \to -J^3 \tag{17.269}$$

This transformation obviously preserves the structure of the current algebra.

We note that the currents are invariant under the transformation

$$\varphi \to \varphi + \pi\sqrt{2} = \varphi + \pi R \tag{17.270}$$

Since this $\mathbb{Z}_2$ transformation does not affect the generators, it must lie in the center of $SU(2)$.[31] Being interested in transformations that act nontrivially on the generators, we need to quotient the symmetry group by its center. This amounts to identifying the field configurations φ and $\varphi + \pi\sqrt{2}$. We now define

$$\mathsf{a}' : \varphi \to \varphi + \pi\frac{\sqrt{2}}{2} \tag{17.271}$$

whose action on the generators is

$$\mathsf{a}' : \quad J^1 \to -J^1 \quad , \quad J^2 \to -J^2 \quad , \quad J^3 \to J^3 \tag{17.272}$$

This action can be obtained from that of a by a cyclic permutation of the generators. Such a cyclic permutation can be described by a $SU(2)$ rotation. Because $SU(2)$ is a symmetry of the model, the action of a must be equivalent to that of a'. Quotienting the action of a amounts to constructing the orbifold of Sect. 10.4.3, leading to the partition function $Z_{\text{orb}}(\sqrt{2})$ of Eq. (10.84). Quotienting the action of a' simply reduces the radius of the circle by a factor of 2, leading to the partition function $Z(\sqrt{2}/2)$. This proves the equivalence

$$Z_{\text{orb}}(\sqrt{2}) = Z(\sqrt{2}/2) \tag{17.273}$$

---

[31] On the other hand, the $\mathbb{Z}_2$ transformation a does not lie in the center of the $SU(2)$ symmetry group of the theory, and this is clear from its action on the generators J^i.

In the expression Eq. (10.84), Z_{orb} appears as the half sum of $Z(R)$ and of the R–independent twisted sector contribution, namely

$$Z_{\text{orb}}(R) = \frac{1}{2}(Z(R) + Z_{\text{twist}}) \tag{17.274}$$

At $R = \sqrt{2}$, this yields, using the equivalence (17.273),

$$Z_{\text{twist}} = 2Z(\sqrt{2}/2) - Z(\sqrt{2}) \tag{17.275}$$

and hence

$$Z_{\text{orb}}(R) = \frac{1}{2}\left(Z(R) + 2Z(\sqrt{2}/2) - Z(\sqrt{2})\right) \tag{17.276}$$

This relation can also be checked directly using the expressions of $Z(R)$ and Z_{twist}, given in Eq. (10.84).

17.B.4. Quotienting by Subgroups of $SU(2)$

The above construction has a natural generalization in which $\mathbb{Z}_2$ ($\neq$ the center) is replaced by some finite subgroup $G \subset SU(2)$. G acts on $SU(2)$ through the inner automorphisms $c \rightarrow aca^{-1}$, $a \in G$. As already indicated, the center $\mathbb{Z}_2 \subset SU(2)$ always acts trivially by inner automorphisms, and only $G/\mathbb{Z}_2$ has a nontrivial effective action. We can thus restrict our study to the finite subgroups Γ of $SO(3) = SU(2)/\mathbb{Z}_2$. These are in one-to-one correspondence with the symmetry groups of regular solids in three dimensions: the cyclic group with n elements C_n; the dihedral group with $2n$ elements D_n; and the tetrahedron T, octahedron O, and icosahedron I groups, with respective numbers of elements 12, 24, and 60. Their double coverings in $SU(2)$ are known as the finite binary subgroups of $SU(2)$, namely the cyclic group C_{2n} with $2n$ elements; binary dihedral $\mathcal{D}_n$ with $4n$ elements; and tetrahedral, octahedral, and icosahedral groups $\mathcal{T}$, $\mathcal{O}$, and $\mathcal{I}$, with respective numbers of elements 24, 48, and 120.

CYCLIC GROUP C_n

The cyclic group C_n of $SO(3)$ is generated by an n–fold symmetry around a given axis. In the Cartan-Weyl basis $\{H, E^{\pm}\}$, it is generated by the element

$$a_n : \varphi \rightarrow \varphi + \pi\frac{\sqrt{2}}{n} \tag{17.277}$$

with action

$$a_n : \quad H \rightarrow H \quad E^{\pm} \rightarrow e^{\pm\frac{2i\pi}{n}} E^{\pm} \tag{17.278}$$

The translation by $1/n$ times the original period reduces the radius of the initial theory by a factor $1/n$, leading to the partition function

$$Z_n \equiv Z(\frac{\sqrt{2}}{n}) \tag{17.279}$$

DIHEDRAL GROUP D_n

The dihedral group D_n of $SO(3)$ is realized by n axes of two-fold rotation symmetry, perpendicular to one axis of n-fold rotation symmetry. It is generated by the
adjunction of the $\mathbb{Z}_2$ generator a defined in Eq. (17.267) to the $\mathbb{Z}_n$ generator a_n of
Eq. (17.277). According to the discussion of $\mathbb{Z}_2$ orbifolds, the resulting theory is
the $\mathbb{Z}_2$ orbifold of the theory at radius $\sqrt{2}/n$, with partition function $Z_{\text{orb}}(\sqrt{2}/n)$.

This result can also be obtained by following the procedure described in
Sect. 17.B.1. The problem essentially boils down to the determination of the mutually commuting elements of the group D_n. For odd n, they fall into one cyclic
subgroup C_n of order n, and n cyclic subgroups C_2 of order 2. The sum (17.259)
becomes here

$$Z_{D_n} = \frac{1}{2n}\left(nZ_n + n(2Z_2 - Z_1)\right) = \frac{1}{2}(Z_n + 2Z_2 - Z_1) \qquad (17.280)$$

The subtraction of $Z_1 = Z_{1,1}$ from the second term avoids overcounting the contribution from the identity element, which is common to all the Abelian subgroups
we summed over. For even n, the mutually commuting elements of D_n fall into
one cyclic subgroup $\mathbb{Z}_n$ and $n/2$ dihedral subgroups D_2, which in turn contain the
same C_2 as $C_n \sim C_{n/2} \times C_2$. This leads to

$$Z_{D_n} = \frac{1}{2n}\left(nZ_n + \frac{n}{2}(4Z_{D_2} - 2Z_2)\right) \qquad (17.281)$$

But since $D_2 \sim \mathbb{Z}_2 \times \mathbb{Z}_2$ is Abelian, the two $\mathbb{Z}_2$ quotients can be performed
successively: one of them reduces the radius to $\sqrt{2}/2$, and the other transforms the
partition function into

$$Z_{D_2} = Z_{\text{orb}}\left(\frac{1}{\sqrt{2}}\right) = \frac{1}{2}(3Z_2 - Z_1) \qquad (17.282)$$

Substituting into Eq. (17.281), we again find the expression (17.280). Hence,
irrespectively of the parity of n, we have

$$Z_{D_n} = \frac{1}{2}(Z_n + 2Z_2 - Z_1) = Z_{\text{orb}}\left(\frac{\sqrt{2}}{n}\right) \qquad (17.283)$$

The relation to $Z_{\text{orb}}(\sqrt{2}/n)$ follows from Eq. (17.276).)

EXCEPTIONAL GROUPS T, O, I

A thorough study of the mutually commuting elements of the three subgroups T,
O, and I leads to the following results:

$$\begin{aligned}
Z_T &= \frac{1}{12}\left(4Z_{D_2} + 4(3Z_3 - Z_1)\right) \\
&= \frac{1}{2}(2Z_3 + Z_2 - Z_1) \\
Z_O &= \frac{1}{24}\left(3(4Z_4 - 2Z_2) + 4(3Z_3 - Z_1) + 3(4Z_{D_2} - 2Z_2)\right) \qquad (17.284)
\end{aligned}$$

$$= \frac{1}{2}(Z_4 + Z_3 + Z_2 - Z_1)$$

$$Z_I = \frac{1}{60}\left(6(5Z_5 - Z_1) + 10(3Z_3 - Z_1) + 5(4Z_{D_2} - Z_1) + Z_1\right)$$

$$= \frac{1}{2}(Z_5 + Z_3 + Z_2 - Z_1)$$

where the first sum exhibits the structure of the mutually commuting elements of the various groups. These last three theories are exceptional in many respects. None of them lies on the line $Z(R)$, nor on its orbifold line $Z_{orb}(R)$. They form three isolated points in the space of $c = 1$ theories. The complete classification of modular-invariant partition functions at $c = 1$ consists simply of:[32]

$$Z(R) \quad R \in]0, \sqrt{2}]$$

$$Z_{orb}(R) \quad R \in]0, \sqrt{2}[\tag{17.285}$$

$$Z_T \quad Z_O \quad Z_I .$$

Any $c = 1$ conformal theory we can think of must thus be in this set. As a simple example, the square of the Ising model (the superposition of two noninteracting copies of the Ising model, with total central charge $c = \frac{1}{2} + \frac{1}{2} = 1$) is identified as

$$Z_{Ising}^2 = Z_{orb}(1) \tag{17.286}$$

(See Ex. 10.23 for a detailed proof.)

17.B.5. The Finite Subgroups of $SU(2)$ and $\hat{A}, \hat{D}, \hat{E}$

There is a remarkable relation between the extended Dynkin diagrams of the simply-laced affine Lie algebras (of type A, D, E) and the finite binary subgroups of $SU(2)$. Let $\ell = 1, 2, \cdots, d_G$ denote the irreducible representations of a finite binary subgroup of $SU(2)$. Fix $\ell = 1$ to be the identity representation, and $\ell = 2$ some two–dimensional faithful self-conjugate representation of G in $SU(2)$. The tensor products between representations

$$(2) \otimes (\ell) = \bigoplus_s \hat{G}_{\ell s} (s) \tag{17.287}$$

define uniquely some nonnegative integer valued matrix $\hat{G}$ of size $d_G \times d_G$. $\hat{G}$ can then be viewed as the adjacency matrix of a graph, with nodes labeled by the representations $\ell = 1, 2, \cdots, d_G$, and with $\hat{G}_{\ell s}$ links between any pair of nodes (ℓ, s). With these definitions, we have the following result, known as the *McKay correspondence*: The finite binary subgroups of $SU(2)$, namely the cyclic C_{2n}, dihedral $\mathcal{D}_{4n}$, and exceptional $\mathcal{T}, \mathcal{O}$, and $\mathcal{I}$, are in one-to-one correspondence with

---

[32] The ranges of the radius of the circle take into account the symmetry $R \leftrightarrow 2/R$. The radius $R = \sqrt{2}$ has been excluded from the range of the orbifold partition function, to avoid overcounting it (it is already given by $Z(\sqrt{2}/2)$, thanks to the equivalence (17.273)).

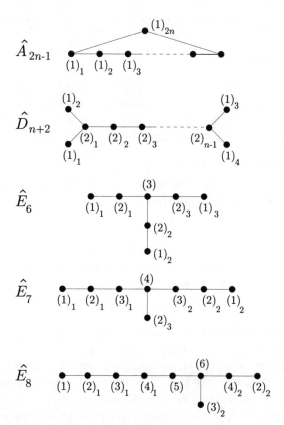

Figure 17.6. The $\hat{A}$, $\hat{D}$, $\hat{E}$ affine Dynkin diagrams corresponding to the finite binary sub-groups of $SU(2)$. The index, borrowed from the ordinary A, D, E Dynkin diagrams, denotes the number of nodes minus one. The nodes are indexed by the corresponding group representations, denoted by their dimension d as $(d)_i$.

the affine diagrams[33] $\hat{A}_{2n-1}$, $\hat{D}_n$, $\hat{E}_6$ $\hat{E}_7$, and $\hat{E}_8$, with respectively $2n$, $n + 1, 7, 8$, and 9 nodes. These diagrams are depicted on Fig. 17.6.

There is thus a close connection between the G-orbifold models constructed in this appendix and the $\hat{A}$, $\hat{D}$, $\hat{E}$ algebras. This provides a surprising remnant of the ADE classification of the modular invariants for the $\widehat{su}(2)_k$ models and the Virasoro minimal models.

We could wonder whether the affine diagrams of Fig. 17.6 have anything to do with the extended fusion rules of the rational theories obtained by the orbifold

---

[33] Note that only bicolorable affine diagrams occur here, that is, diagrams whose vertices can be consistently assigned one of two colors in such a way that any two adjacent vertices have different colors. This eliminates the $\hat{A}_{2n}$ diagrams.

method. Actually, the extended fusion numbers of the C_{2n} orbifolds, with partition function $Z(\sqrt{2}/n)$ on the torus, are easily seen to be $\mathcal{N}_{rs}{}^t = \delta_{t,r+s \bmod 2n^2}$ (see Ex. 10.21). They are encoded in the diagram $\hat{A}_{2n^2-1}$, whose adjacency matrix is $\hat{A}_{rs} = \mathcal{N}_{1r}{}^s$. There is thus a mismatch between the group fusion diagram $\hat{A}_{2n-1}$ and the corresponding RCFT fusion diagram $\hat{A}_{2n^2-1}$. There is no such relation for the other orbifold theories.[34]

The McKay correspondence is illustrated in the case of the icosahedral group $\mathcal{I}$ in Ex. 17.31.

17.B.6. Operator Content of the $c = 1$ Theories

BOSON COMPACTIFIED OPERATOR CONTENT

The torus partition function of the ordinary bosonic theories compactified on a circle of radius R is (see Sect. 12.6.2)

$$Z(R) = \frac{1}{|\eta(q)|^2} \sum_{e,m \in \mathbb{Z}} q^{h_{e,m}} \bar{q}^{\bar{h}_{e,m}} \tag{17.288}$$

It exhibits the full electromagnetic operator content: each operator $\mathcal{O}_{e,m}$ is a mixture of the purely electric vertex operator

$$\mathcal{O}_{e,0} = e^{ie\varphi/R} \tag{17.289}$$

with electric charge e, and the purely magnetic operator $\mathcal{O}_{0,m}$ which creates a line of defect along which φ has a jump discontinuity of $2\pi m$.

When R^2 is a rational number, the theory is rational, that is, the partition function can be reorganized in terms of a *finite* number of extended characters. More precisely, for

$$R = \sqrt{\frac{2p'}{p}} \tag{17.290}$$

with p, p' two positive, coprime integers, the partition function (17.288) can be rewritten as (see Ex. 10.21 for a proof)

$$Z(\sqrt{2p'/p}) = \sum_{\lambda=0}^{N-1} K_\lambda^{(N)}(\tau) \bar{K}_{\omega_0\lambda}^{(N)}(\bar{\tau}) \tag{17.291}$$

with

$$K_\lambda^{(N)}(\tau) = \frac{1}{\eta(\tau)} \sum_{n \in \mathbb{Z}} q^{(nN+\lambda)^2/2N}, \quad N = 2pp' \tag{17.292}$$

[34] Indeed, using Ex. 10.18, it can be shown that if some RCFT fusion rules are also the rules for the tensor product of the irreducible representations of a finite group, then they must also be identified with the rules for the multiplication of classes of the same group. This can be the case only for an Abelian group, so the diagrams of Fig. 17.6 other than A_{2n-1} do not encode the fusion rules of RCFTs, *a fortiori* not those of the corresponding orbifold theories at $c = 1$.

and ω_0 defined as

$$\omega_0 = pr_0 + p's_0 \mod N \tag{17.293}$$

where (r_0, s_0) is any Bezout couple for p and p' (see Ex. 10.1), that is, any couple of integers such that

$$pr_0 - p's_0 = 1 \tag{17.294}$$

The functions $K_\lambda^{(N)}(\tau)$ are the extended characters of this rational conformal theory. Note that the partition function (17.292) is nondiagonal (see Sect.10.8.7) unless $\omega_0 = 1$, which is realized only if $p = 1$ (or $p' = 1$ by the duality $R \leftrightarrow 2/R$ of Eq. (10.65)). In that case, Eq. (17.292) becomes diagonal

$$Z(\sqrt{2p'}) = \sum_{\lambda=0}^{2p'-1} |K_\lambda^{(2p')}(\tau)|^2 \tag{17.295}$$

This exhibits clearly the rational structure of the theory, with a finite collection of extended fields ϕ_λ, $\lambda = 0, 1, ..., 2p' - 1$, with conformal dimensions

$$h_\lambda = \frac{\lambda^2}{4p'} \tag{17.296}$$

Their fusion rules are easily obtained from the modular transformations of the extended characters (10.126) by applying the extended version (10.219) of the Verlinde formula. They read (see Ex. 10.21):

$$N_{\lambda\mu}{}^\nu = \delta_{\nu, \lambda+\mu \mod 2p'} \tag{17.297}$$

and therefore coincide with those of $\widehat{su}(2p')_1$.

$\mathbb{Z}_2$ ORBIFOLD OF THE BOSON COMPACTIFIED ON A CIRCLE OF RADIUS R

The $\mathbb{Z}_2$ orbifold of the bosonic theory compactified on a circle of radius R has the torus partition function (see Eq. (17.276))

$$Z_{\text{orb}}(R) = \frac{1}{2}(Z(R) + 2Z(\sqrt{2 \cdot 4}) - Z(\sqrt{2 \cdot 1})) \tag{17.298}$$

According to Eq. (17.295), we can rewrite

$$Z(\sqrt{2 \cdot 4}) = \sum_{\lambda=0}^{7} |K_\lambda^{(8)}|^2 = \sum_{\lambda=0}^{7} \left| \frac{1}{\eta} \sum_{n \in \mathbb{Z}} q^{(8n+\lambda)^2/16} \right|^2 \tag{17.299}$$

and

$$Z(\sqrt{2 \cdot 1}) = \sum_{\lambda=0}^{1} |K_\lambda^{(2)}|^2 = \sum_{\lambda=0}^{1} \left| \frac{1}{\eta} \sum_{n \in \mathbb{Z}} q^{(2n+\lambda)^2/4} \right|^2 \tag{17.300}$$

The particular combination $2Z(\sqrt{8}) - Z(\sqrt{2})$ actually subtracts some of the terms in Eq. (17.299). Indeed, we have

$$
\begin{aligned}
2\sum_{\lambda=2,6} |K_\lambda^{(8)}|^2 &= 2\sum_{\lambda=2,6} \left| \frac{1}{\eta} \sum_{n\in\mathbb{Z}} q^{(8n+\lambda)^2/16} \right|^2 = 2\sum_{\lambda=1,3} \left| \frac{1}{\eta} \sum_{n\in\mathbb{Z}} q^{(4n+\lambda)^2/4} \right|^2 \\
&= 4 \left| \frac{1}{\eta} \sum_{n\in\mathbb{Z}} q^{(4n+1)^2/4} \right|^2 = \left| \sum_{\lambda=1,3} \frac{1}{\eta} \sum_{n\in\mathbb{Z}} q^{(4n+\lambda)^2/4} \right|^2 \\
&= \left| \frac{1}{\eta} \sum_{n\in\mathbb{Z}} q^{(2n+1)^2/4} \right|^2 = |K_1^{(2)}|^2
\end{aligned}
$$

$$(17.301)$$

Analogously, the terms $\lambda = 0, 4$ in Eq. (17.299) combine with the term $\lambda = 0$ in Eq. (17.300) to yield

$$
\sum_{\lambda=0,4} |K_\lambda^{(8)}|^2 - |K_0^{(2)}|^2 = \left| \frac{1}{\eta} (\sum_{n\in\mathbb{Z}} q^{(2n)^2} - q^{(2n+1)^2}) \right|^2 \tag{17.302}
$$

Finally, the partition function (17.298) reads

$$
Z_{\text{orb}}(R) = \frac{1}{2} Z(R) + \sum_{\lambda=1,3,5,7} |K_\lambda^{(8)}|^2 + \frac{1}{2} \left| \frac{1}{\eta} \sum_{n\in\mathbb{Z}} (-1)^n q^{n^2} \right|^2 \tag{17.303}
$$

It exhibits the following operator content. The spectrum of the original bosonic theory appears with a coefficient $\frac{1}{2}$, but we note that $h_{e,m} = h_{-e,-m}$. Except for the case $e = m = 0$, the terms in Eq. (17.288) can be grouped two by two, factoring out a multiplicity 2 (which is canceled by the prefactor $\frac{1}{2}$). The corresponding operator $\mathcal{O}_{e,m}^{\text{orb}} = \frac{1}{2}(\mathcal{O}_{e,m} + \mathcal{O}_{-e,-m})$ is $\mathbb{Z}_2$-even (invariant under $\varphi \rightarrow -\varphi$). Note in particular that the electric part of the operator is $\cos(e\varphi/R)$ instead of $e^{i\varphi/R}$ in the nonorbifold model. The term $e = m = 0$ actually combines with the last term of Eq. (17.303) to give the contribution of the identity $\mathbb{I}$ and of a field Θ of dimension 1 (the appearance of this field will become clear below). In addition, the second term of Eq. (17.303) corresponds to two twice-degenerate fields $\sigma^{(i)}$ and $\tau^{(i)}$, $i = 1, 2$, with respective dimensions $\frac{1}{16}$ and $\frac{9}{16}$. These are the so-called $\mathbb{Z}_2$-twist operators of the orbifold theory. They create line discontinuities across which φ changes to $-\varphi$.

When the theory is rational and diagonal, that is, when

$$
R = \sqrt{2p'} \tag{17.304}
$$

we may write

$$
Z_{\text{orb}}(\sqrt{2p'}) = \frac{1}{2} \sum_{\lambda=0}^{2p'-1} |K_\lambda^{(2p')}|^2 + \sum_{\lambda=1,3,5,7} |K_\lambda^{(8)}|^2 + \frac{1}{2} \left| \frac{1}{\eta} \sum_{n\in\mathbb{Z}} (-1)^n q^{n^2} \right|^2 \tag{17.305}
$$

By using the symmetry $K_\lambda^{(2p')} = K_{2p'-\lambda}^{(2p')}$, we finally obtain

$$
Z_{\text{orb}}(\sqrt{2p'}) = \sum_{\lambda=1}^{p'-1} |K_\lambda^{(2p')}|^2 + 2|\tfrac{1}{2}K_{p'}^{(2p')}|^2 + 2\sum_{\lambda=1,3} |K_\lambda^{(8)}|^2
$$
$$
+ \left| \frac{1}{2\eta} \sum_{n\in\mathbb{Z}} q^{p'n^2} + (-1)^n q^{n^2} \right|^2 + \left| \frac{1}{2\eta} \sum_{n\in\mathbb{Z}} q^{p'n^2} - (-1)^n q^{n^2} \right|^2
$$

$$(17.306)$$

This exhibits the full operator content of the diagonal rational orbifold theories. More specifically, by order of appearance in Eq. (17.306), it reads:

(i) Φ_λ, $\lambda = 1, 2, ..., p' - 1$ with dimension $h_\lambda = \lambda^2/4p'$.

(ii) A twice-degenerate operator $\Phi_{p'}^{(i)}$, $i = 1, 2$, with dimension $h_{p'} = p'/4$, and character

$$
\frac{1}{2} K_{p'}^{(2p')}(\tau) = \frac{1}{2\eta(\tau)} \sum_{n\in\mathbb{Z}} q^{p'(2n+1)^2/4} = \frac{1}{\eta(\tau)} \sum_{n=0}^{\infty} q^{p'(2n+1)^2/4}
$$

where the prefactor $\frac{1}{2}$ has been canceled by grouping the n and $-n - 1$ terms in the first sum.

(iii) The twist operators $\sigma^{(i)}$ and $\tau^{(i)}$, $i = 1, 2$, with respective dimensions $\frac{1}{16}$ and $\frac{9}{16}$.

(iv) The identity operator $\mathbb{I}$ with dimension 0.

(v) The operator Θ of dimension 1.

For $p' = 2$, the orbifold model is merely the square of the Ising model

$$
Z_{\text{orb}}(2) = Z_{\text{Ising}}^2 \tag{17.307}
$$

(see Ex. 10.23 for a detailed proof). The above operator content indeed reproduces that of the square of the Ising model; the nine operators $(\mathbb{I}_1, \sigma_1, \epsilon_1) \otimes (\mathbb{I}_2, \sigma_2, \epsilon_2)$ are

$$
\Phi_1 = \sigma_1 \otimes \sigma_2 \quad (h = \frac{1}{8})
$$
$$
\Phi_2^{(i)} = \epsilon_i \otimes \mathbb{I}_{2-i} \quad (h = \frac{1}{2})
$$
$$
\sigma^{(i)} = \sigma_i \otimes \mathbb{I}_{2-i} \quad (h = \frac{1}{16})
$$
$$
\tau^{(i)} = \sigma_i \otimes \epsilon_{2-i} \quad (h = \frac{9}{16})
$$
$$
\mathbb{I} = \mathbb{I}_1 \otimes \mathbb{I}_2 \quad (h = 0)
$$
$$
\Theta = \epsilon_1 \otimes \epsilon_2 \quad (h = 1)
$$

By using the modular transformations of the extended characters appearing in (17.306), we can compute the extended fusion rules of the rational diagonal orbifold theories. For p' even, the result is (see Ex. 17.32 for a proof):

$$
\begin{aligned}
\phi_\lambda \times \phi_\mu &= \phi_{\lambda+\mu} + \phi_{\lambda-\mu} \quad \text{for } \mu \neq \lambda,\ 2p' - \lambda \\
\phi_\lambda \times \phi_\lambda &= \mathbb{I} + \Theta + \phi_{2\lambda} \\
\phi_{2p'-\lambda} \times \phi_\lambda &= \phi_{2\lambda} + \phi_{p'}^{(1)} + \phi_{p'}^{(2)} \\
\Theta \times \phi_\lambda &= \phi_\lambda \\
\Theta \times \Theta &= \mathbb{I} \\
\phi_{p'}^{(i)} \times \phi_{p'}^{(i)} &= \mathbb{I} \\
\phi_{p'}^{(1)} \times \phi_{p'}^{(2)} &= \Theta \\
\sigma^{(i)} \times \sigma^{(i)} &= \mathbb{I} + \phi_{p'}^{(i)} + \sum_{\lambda \text{ even}} \phi_\lambda \\
\sigma^{(1)} \times \sigma^{(2)} &= \sum_{\lambda \text{ odd}} \phi_\lambda \\
\Theta \times \sigma^{(i)} &= \tau^{(i)}
\end{aligned}
\tag{17.308}
$$

where the index of ϕ is defined modulo p' and takes the values $1, 2, ..., p' - 1$, and the index i takes the values $1, 2$.

When p' is odd, the fusions involving ϕ_λ (the first four lines of Eq. (17.308)) are unchanged, but those of the other fields become

$$
\begin{aligned}
\Theta \times \Theta &= \mathbb{I} \\
\phi_{p'}^{(i)} \times \phi_{p'}^{(i)} &= \Theta \\
\phi_{p'}^{(1)} \times \phi_{p'}^{(2)} &= \mathbb{I} \\
\sigma^{(i)} \times \sigma^{(i)} &= \mathbb{I} + \phi_{p'}^{(i)} + \sum_{\lambda \text{ odd}} \phi_\lambda \\
\sigma^{(1)} \times \sigma^{(2)} &= \mathbb{I} + \sum_{\lambda \text{ even}} \phi_\lambda
\end{aligned}
\tag{17.309}
$$

whereas $\Theta \times \sigma^{(i)} = \tau^{(i)}$ is unchanged.

The operator content of the exceptional theories T, O, I may also be deduced from a similar analysis of their partition functions on a torus.

Exercises

17.1 $\widehat{su}(3)_k$ invariants by outer automorphisms and permutations

a) Write down explicitly all $\widehat{su}(3)$ nondiagonal modular invariants that can be obtained by the method of outer automorphisms.

b) Use the D-invariant at level 9 and the identity (17.144) to obtain the invariant (17.119).

c) In the D-series, identify the permutation invariants and obtain the explicit form of the permutation operator.

17.2 $\widehat{su}(4)_k$ *invariants by outer automorphisms*
Obtain all nondiagonal $\widehat{su}(4)$ mass matrices that can be found by the method of outer automorphism. Note that in this case, the projection procedure can also be done with a subgroup of the center, namely $\mathbb{Z}_2$.

17.3 *Properties of $\widehat{su}(N)_k$ invariants by outer automorphisms*

a) Show that all outer-automorphism $\widehat{su}(N)_k$ invariants are permutation invariants when N and k are coprime. Give the explicit form of $\Pi(\hat{\lambda})$.

b) Show that the outer-automorphism $\widehat{su}(N)_k$ invariants are block-diagonal when N and k are not coprime. Characterize the weights that survive the projection.

17.4 $\widehat{so}(7)$ *invariants by outer automorphisms*
Write the $\widehat{so}(7)$ nondiagonal modular-invariant partition functions obtained by the method of outer automorphisms for:

a) $k = 2$;

b) $k = 3$.

17.5 *Properties of conformal embeddings*

a) Argue that for the embedding $\hat{p}_{x_e} \subset \hat{g}_1$ to be conformal, g must necessarily be simple.

b) For a chain of embeddings $\hat{p}^{(1)} \subset \hat{p}^{(2)} \cdots \subset \hat{g}$, show that all embedding indices but one, and all values of k but one, must be equal to 1.

17.6 *Application of the sum rule for conformal branching rules*
Calculate the first row of the $\widehat{sp}(4)_1$ and $\widehat{su}(2)_{10}$ modular S matrix, and use Eq. (17.95) to complete the determination of the branching rules

$$[1,0,0] \mapsto [10,0] \oplus c_2 [4,6]$$
$$[0,1,0] \mapsto [7,3] \oplus c_4 [3,7]$$
$$[0,0,1] \mapsto [6,4] \oplus c_6 [0,10]$$

which follow from an analysis of the conformal dimensions and finite branching rules at grade zero.

17.7 *Some conformal branching rules*

a) Using the following projection matrix of $so(8)$ onto $su(3)$

$$P = \begin{pmatrix} 1 & 0 & 1 & 1 \\ 1 & 3 & 1 & 1 \end{pmatrix} \tag{17.310}$$

derive an $\widehat{su}(3)$ nondiagonal invariant.

b) From the $\widehat{su}(6)$ diagonal modular invariant and the embedding $su(3) \subset su(6)$ with finite branching rules

$$(0,0,0,0,1) \mapsto (2,0)$$
$$(0,0,0,1,0) \mapsto (2,1)$$
$$(0,0,1,0,0) \mapsto (3,0) \oplus (0,3)$$

(together with their conjugates), obtain Eq. (17.115).

17.8 *Fine structure of the conformal embedding $\widehat{su}(2)_4 \subset \widehat{su}(3)_1$*

a) For the embedding $su(2) \subset su(3)$ with $x_e = 4$, express the $su(2)$ generators, in the J^a basis, in terms of those of $su(3)$. From the current extension of this equality, show directly the equivalence of the Sugawara energy-momentum tensors of the $\widehat{su}(2)_4$ and $\widehat{su}(3)_1$ theories.

b) Find the representation of those $\widehat{su}(2)_4$ fields at grade zero that transform in the adjoint representation in both sectors in terms of the three free complex fermions that provide a representation of $\widehat{su}(3)_1$.

17.9 *A $\widehat{sp}(4)_3$ permutation invariant*

In this exercise, we look for a possible $\widehat{sp}(4)_3$ permutation invariant following the strategy proposed at the end of Sect. 17.8. The set of finite weights whose affine extension have second-lowest quantum dimension is

$$\{\omega_1, 3\omega_1, \omega_1 + 2\omega_2, \omega_2, 2\omega_2\}$$

a) Show that the only permutations of this set that are compatible with T invariance are

$$\Pi(\omega_1) = \omega_1 \qquad \text{and} \qquad \Pi'(\omega_1) = \omega_1 + 2\omega_2$$

b) Derive the fusion rule:

$$\omega_1 \times \omega_1 = 0 + \omega_2 + 2\omega_1$$

Conclude that Π must leave ω_2 fixed, hence $\Pi = id$.

c) The other permutation Π' is certainly an automorphism of the fusion rules, acting as a genuine $\widehat{sp}(4)_3$ outer automorphism on ω_1: $\Pi'([2, 1, 0]) = [0, 1, 2]$. Describe the action of Π' on all the weights and relate it to the basic outer automorphism a.

17.10 *Galois symmetry and fusion rules*

a) Acting on both sides of the equation

$$\frac{S_{in}}{S_{0n}} \frac{S_{jn}}{S_{0n}} = \sum_k \mathcal{N}_{ij}^{\ k} \frac{S_{kn}}{S_{0n}} \tag{17.311}$$

with a Galois transformation σ (whose action on S is given by Eq. (17.156)), show that

$$\mathcal{N}_{i^\sigma j^\sigma}^{\ \ n^\sigma} = \sum_k \epsilon_\sigma(0) \, \epsilon_\sigma(i) \, \epsilon_\sigma(j) \, \epsilon_\sigma(k) \, \mathcal{N}_{ij}^{\ k} \, \mathcal{N}_{0^\sigma k^\sigma}^{\ \ n^\sigma}$$

(To emphasize that this result generalizes to any RCFT, we use the general field labels $i, j, \ldots$ instead of integrable weights.)

b) In terms of the matrix G_σ with components

$$(G_\sigma)_{ij} = \epsilon_\sigma(j) \, \delta_{i,j^\sigma}$$

show that the result can be written compactly as

$$\mathcal{N}_{i^\sigma} = \epsilon_\sigma(0) \, \epsilon_\sigma(i) \, G_\sigma^{-1} \, \mathcal{N}_i \, G_\sigma \, \mathcal{N}_{0^\sigma}$$

c) Illustrate this formula with the different Galois transformations of the $\widehat{su}(2)_3$ WZW model.

17.11 *Weyl reflection associated with Galois transformations*

Prove that when $(\ell, k + g) = 1$ (which is necessarily true whenever $(\ell, M(k + g)) = 1$), the affine extension of the weight $\ell(\lambda + \rho)$ can always be reflected inside the fundamental chamber.

17.12 *Parity rule*

a) Use the parity rule to probe a possible extension of the $\widehat{su}(3)_2$ chiral algebra.

b) For $\widehat{su}(3)_k$ with $k = 0 \bmod 3$, verify the compatibility of the parity rule and the outer-automorphism construction, that is, check that

$$\epsilon(w_\ell^\lambda) = \epsilon(w_\ell^{a^n\lambda})$$

where $a^n\lambda$ stands for the finite part of $a^n\hat{\lambda}$.

17.13 *Exceptional $\widehat{su}(3)_5$ invariant from Galois symmetry*
Show that the $\widehat{su}(3)_5$ exceptional invariant (17.115) can be described as a block-diagonal Galois invariant.

17.14 $\widehat{G_2}$ *Galois invariants*
Obtain the following three $\widehat{G_2}$ invariants from Galois transformations:

$$
\begin{aligned}
k = 3: \quad Z &= |\chi_{(0,0)} + \chi_{(1,1)}|^2 + 2|\chi_{(0,2)}|^2 \\
k = 4: \quad Z' &= |\chi_{(0,0)} + \chi_{(0,3)}|^2 + |\chi_{(0,4)} + \chi_{(1,0)}|^2 + 2|\chi_{(1,1)}|^2 \\
k = 4: \quad Z'' &= |\chi_{(0,0)}|^2 + |\chi_{(0,3)}|^2 + |\chi_{(1,1)}|^2 + |\chi_{(0,2)}|^2 + |\chi_{(1,2)}|^2 \\
&\quad + \chi_{(0,1)}\bar{\chi}_{(2,0)} + \chi_{(2,0)}\bar{\chi}_{(0,1)} + \chi_{(0,4)}\bar{\chi}_{(1,0)} + \chi_{(1,0)}\bar{\chi}_{(0,4)}
\end{aligned}
\tag{17.312}
$$

The first two invariants can also be obtained from conformal embeddings (in $(\widehat{E}_6)_1$ and $\widehat{so}(14)_1$ respectively), but the third one cannot be obtained from a conformal embedding nor by the outer-automorphism construction.

17.15 *Galois symmetry and the $\widehat{su}(2)$ nondiagonal invariants*

a) Show that all the $\widehat{su}(2)_{4m}$ invariants (called D_{2m+2} in Eq. (17.114)) can be obtained by an appropriate Galois transformation.

b) Show that the $\widehat{su}(2)_6$ permutation invariant cannot be described by a Galois transformation.

17.16 *Generalized Galois permutation invariants*

a) For the generalized permutation invariant (17.196), for which it is assumed only that σ commutes with T and $A(0) = 0^\sigma$, prove that A must be of order 2.

b) Verify the following consequence of the assumed commutativity of σ with T:

$$T_0 = T_{0^\sigma} = T_{A(0)} \quad \Longrightarrow \quad \frac{k}{2}|A\hat{\omega}_0|^2 = 0 \bmod 1$$

c) Derive the equality

$$S_{0,\hat{\lambda}} = e^{2\pi i(A\hat{\omega}_0,\lambda)} \, \epsilon_\sigma(0)\epsilon_\sigma(\lambda) \, S_{0,\hat{\lambda}^\sigma}$$

and conclude from this that

$$e^{-2\pi i(A\hat{\omega}_0,\lambda)} = \epsilon_\sigma(0)\epsilon_\sigma(\lambda)$$

d) With similar manipulations, show that $\epsilon_\sigma(\lambda) = \epsilon_\sigma(A(\lambda))$, where $A(\lambda)$ is the finite part of $A\hat{\lambda}$.

e) Use $S_{0\hat{\lambda}} = \sigma^2(S_{0\hat{\lambda}})$, to obtain

$$(A\hat{\omega}_0, \lambda) = (A\hat{\omega}_0, \lambda^\sigma) \bmod 1$$

f) Prove the T invariance of the generalized Galois permutation Π of Eq. (17.196), that is,
$$T_{\Pi(\hat{\lambda})} = T_{\hat{\lambda}}$$

g) Prove its S invariance: $S_{\Pi(\hat{\lambda}),\Pi(\hat{\mu})} = S_{\hat{\lambda},\hat{\mu}}$.

17.17 *An infinite sequence of permutation invariants for* $\widehat{so}(2r + 1)_2$
The integrable weights for the $\widehat{so}(2r + 1)$ affine algebra at level 2 are

$$\hat\omega_0, \ 2\hat\omega_1, \ \hat\omega_0 + \hat\omega_r, \ \hat\omega_1 + \hat\omega_r, \ \hat\omega_0 + \hat\omega_1, \ \hat\omega_j, \ 2\hat\omega_r$$

with $2 \leq j \leq r - 1$. Denote the last r weights by $\hat{v}^{(i)}$ in this order, with $1 \leq i \leq r$, that is

$$\hat{v}^{(1)} = \hat\omega_1 + \hat\omega_r$$
$$\hat{v}^{(j)} = \hat\omega_j, \qquad 2 \leq j \leq r - 1$$
$$\hat{v}^{(r)} = 2\hat\omega_r$$

All the permutation invariants of $\widehat{so}(2r + 1)_2$ can be proven to be of the form:

$$\Pi_\ell(\hat{v}^{(i)}) = \hat{v}^{([i\ell])}$$
$$\Pi_\ell(\hat\lambda) = \hat\lambda \qquad \text{if} \quad \hat\lambda \in \{2\hat\omega_0, \ 2\hat\omega_1, \ \hat\omega_0 + \hat\omega_r, \ \hat\omega_1 + \hat\omega_r\}$$

where ℓ is an integer such that $\ell^2 = 1 \bmod (2r + 1)$ and $[x]$ is the unique integer in the range $0 \leq [x] \leq r$ satisfying $x = \pm[x]$ for either sign. These are all generalized Galois permutation invariants.

a) Verify that the lowest rank at which a nontrivial invariant occurs is for $r = 7$ and find the explicit form of this new permutation invariant.

b) Show that any solution ℓ of the equation

$$\ell^2 = 1 \bmod (2r + 1)$$

may be written in the form

$$\ell + 1 = \rho\alpha$$
$$\ell - 1 = \sigma\beta$$

where α and β are two coprime integers such that $\alpha\beta = 2r + 1$, and ρ and σ are some nonnegative integers.

c) Using the Bezout lemma (Ex. 10.1), show that ρ and σ are necessarily even integers modulo $(2r + 1)$, and compute their values in terms of the Bezout couple for α and β, namely the integers r and s such that

$$r\alpha - s\beta = 1$$

Result: $\rho = 2r, \sigma = 2s$.

d) Let p denote the number of distinct prime divisors of

$$2r + 1 = \prod_{i=1}^{p-1} \rho_i^{m_i}$$

(the divisor 1 is omitted from the product). Show that the number of solutions of the equation $\ell^2 = 1 \bmod (2r + 1)$ is exactly 2^{p-1}.
Hint: There is one solution $\ell = 2r\alpha - 1$ for each divisor α of $2r + 1$ which is coprime with $(2r + 1)/\alpha$. There are exactly

$$\sum_{k=0}^{p-1} \binom{k}{p-1} = 2^{p-1}$$

such divisors. The number of permutation invariants is therefore 2^{p-1}, including the diagonal one. The next values of r where new invariants arise are then 10, 16, 17, 19, 22, 25, etc. There is a similar sequence of permutation invariants for the algebras $\widehat{so}(2r)_2$.

17.18 *Extended fusion rules*

a) Using the $\widehat{su}(2)_k$ fusion rules, derive the fusion rules for the extended fields (17.136) and (17.137). (Ambiguities related to the ϕ_ℓ, ϕ'_ℓ factors are lifted by imposing invariance under the $\mathbb{Z}_2$ automorphism $\phi_\ell \to \phi'_\ell$, which leaves the $\phi_{n<\ell}$'s unaffected.)

b) Prove that these fusion rules have a nontrivial automorphism only at $k = 16$, corresponding to the interchange

$$\chi_2 + \chi_{14} \leftrightarrow \chi_8$$

17.19 *Eigenvalues of adjacency matrices*
Diagonalize the adjacency matrices of some graphs of Fig. 17.4 and check that the eigenvalues are given by Eq. (17.225) with the sets of exponents listed in Table 17.2.

17.20 *A wrong choice for the unit vertex on E_6*
We consider the E_6 diagram of Fig. 17.1, but we choose the vertex 5 as the unit vertex.

a) What is the new fundamental vertex f?

b) Write the graph-algebra relations in terms of the matrix $G = N_f$.

c) Show that G satisfies the polynomial relation of degree 4:

$$G^4 - 5G^2 + \mathbb{I} = 0$$

d) Conclude that this choice of unit vertex cannot lead to a good graph algebra.
Hint: The dimension of the polynomial algebra generated by G is at most 4, whereas a good graph algebra should have dimension 6.

17.21 *Negative fusion numbers for the E_7 graph*
We consider the E_7 graph algebra, with unit vertex at the end of the longest leg and fundamental vertex directly linked to it. Let $G = N_f$ denote the fundamental matrix generator, also equal to the adjacency matrix of E_7.

a) Write all the generators N_r of the E_7 graph algebra as polynomials of the matrix G.

b) Compute the entries of the matrix generators N_r, and show that some of them are negative.

17.22 *E_8 graph algebra and its modular invariant*
We consider the E_8 graph algebra, with unit vertex at the end of the longest leg and fundamental vertex directly linked to it. Let $G = N_f$ denote the fundamental matrix generator, also equal to the adjacency matrix of E_8.

a) Write all the generators N_r of the E_8 graph algebra as polynomials of the matrix G.

b) Check that the consistency between the graph-algebra relations yields the vanishing of $P(G)$, where P is the characteristic polynomial of G.

c) Compute the equivalence classes for the relation $\approx$ for the choice of subalgebra indicated by the circled vertices in Fig. 17.2.

d) Compare the two classes with the blocks of the $\widehat{su}(2)_{28}$ E_8-type modular invariant given in Eq. (17.114).

17.23 *Normal matrices*

a) Let G be a *normal* matrix, that is, $[G, G^\dagger] = 0$. Show that if two eigenvalues of G are distinct, then the corresponding eigenspaces are orthogonal.

b) Deduce that a normal matrix can be diagonalized in an orthonormal basis.

17.24 *Isospectral graphs*

a) Compute the eigenvalues of the adjacency matrix of the $\mathcal{D}_3$ diagram of Fig. 17.4.

b) Diagonalize the following matrix

$$
G = \begin{pmatrix}
0 & 0 & 0 & 0 & 1 & 1 \\
0 & 0 & 0 & 0 & 1 & 1 \\
1 & 1 & 0 & 0 & 0 & 0 \\
1 & 1 & 0 & 0 & 0 & 0 \\
0 & 0 & 1 & 1 & 0 & 0 \\
0 & 0 & 1 & 1 & 0 & 0
\end{pmatrix}
$$

Deduce that the associated graph has the same exponents as $\mathcal{D}_3$.

c) Why is this graph a bad candidate for the graph-subalgebra treatment?
Hint: Look for a unit vertex.

17.25 $\mathcal{E}_5$ *graph subalgebra and extended fusion algebra*
Write explicitly all the 12×12 matrices N_{1_j} and $N_{2_j}, j \in \mathbb{Z}_6$ for the graph algebra of the $\mathcal{E}_5$ diagram and show that the N_1's are just permutation matrices, generating the multiplicative group of the 6-th roots of unity. Show also that this fusion algebra is isomorphic to the $\widehat{su}(6)_1$ fusion algebra.

17.26 *Eigenvalues of the fundamental $\widehat{su}(2)$ and $\widehat{su}(3)$ fusion matrices*
Calculate the ratio $\gamma_f^{(\lambda)}$ of S matrix elements (14.245) for the affine extension f of the fundamental representation of $\widehat{su}(2)$ and $\widehat{su}(3)$:

a) In the $\widehat{su}(2)_k$ case, show that

$$
\gamma_f^{(\lambda)} = 2 \cos \pi \frac{\lambda_1 + 1}{k + 2}
$$

b) In the $\widehat{su}(3)_k$ case, show that

$$
\gamma_f^{(\lambda)} = \sum_{j=1}^{3} e^{2i\pi(\epsilon_j, \lambda + \rho)/3(k+3)}
$$

where $\epsilon_1 = \omega_1, \epsilon_2 = \omega_2 - \omega_1$ and $\epsilon_3 = -\omega_2$, where ω_1, and ω_2 are the $su(3)$ fundamental weights.

17.27 $\mathcal{E}_9$ *and* $\mathcal{E}_{21}$ *graph algebras and their modular invariants*

a) Using a computer, diagonalize the adjacency matrices for the $\mathcal{E}_9$ and $\mathcal{E}_{21}$ diagrams of Fig. 17.4. Using the Verlinde formula, compute the matrices N_r. (A good check for the validity of the computer program is that we should get matrices with nonnegative integer entries only.)

b) Determine the classes of the equivalence relation $\approx$ by Eqs. (17.218)–(17.219). Compare the result to the blocks of the $\mathcal{E}_9$ and $\mathcal{E}_{21}$ modular invariants of Eqs. (17.117) and (17.119).

17.28 *Some $\widehat{su}(p)_q \oplus \widehat{su}(q)_p \subset \widehat{su}(pq)_1$ branching rules*

a) Calculate the embedding indices for $su(p) \oplus su(q) \subset su(pq)$ by considering the branching rule of some fundamental weight $\hat{\omega}_\ell$ with $p, q \geq \ell$, using Young tableaux.

b) Check that the branching rules obtained in App. 17.A when $(p,q) = (2,3)$ are complete by verifying the sum rule (17.95).

c) Calculate the full set of branching rules for the case $(p,q) = (2,4)$.

d) Same as (c) for the case $(p,q) = (3,4)$.

17.29 *The A-D $\widehat{su}(2)$ modular invariants from conformal embeddings*

a) Derive the complete A-D sequences of $\widehat{su}(2)$ modular invariants by an appropriate projection of the $\widehat{su}(2)_q \oplus \widehat{su}(q)_2$ invariant, obtained from the $\widehat{su}(pq)_1$ diagonal one. Hint: The required branching rules should be calculated by the method detailed in App. 17.A. The results are:

$$\hat{\omega}_\ell \mapsto \bigoplus_{\substack{n=0 \\ n\in 2\mathbb{Z}}}^{q} a^\ell \otimes \tilde{a}^{\ell(q+1)/2}\{[(q-n)\hat{\omega}_0 + n\hat{\omega}_1] \otimes [\hat{\omega}_{n/2} + \hat{\omega}_{q-n/2}]\} \quad (q \text{ odd})$$

$$\hat{\omega}_{2\ell} \mapsto \bigoplus_{\substack{n=0 \\ n\in 2\mathbb{Z}}}^{q} 1 \otimes \tilde{a}^\ell\{[n\hat{\omega}_0 + (q-n)\hat{\omega}_1] \otimes [\hat{\omega}_{(q-n)/2} + \hat{\omega}_{(q+n)/2}]\} \quad (q \text{ even})$$

$$\hat{\omega}_{2\ell+1} \mapsto \bigoplus_{\substack{n=1 \\ n\in 2\mathbb{Z}+1}}^{q} 1 \otimes \tilde{a}^\ell\{[n\hat{\omega}_0 + (q-n)\hat{\omega}_1] \otimes [\hat{\omega}_{(q-n+1)/2} + \hat{\omega}_{(q+n+1)/2}]\} \quad (q \text{ even})$$

b) Find an exceptional $\widehat{su}(10)_2$ invariant by projecting the $\widehat{su}(20)_1$ diagonal invariant onto the E_6-type $\widehat{su}(2)_{10}$ invariant.

17.30 $\mathbb{Z}_N$ *orbifold for N prime*

a) We denote by $Z_{a,b}$ the twisted partition functions, $a, b \in \mathbb{Z}_N$, and use the additive group structure of $\mathbb{Z}_N$. Write the projected partition function Z_{proj} of Eq. (17.262) in terms of $Z_{a,b}$.

b) Show that the T and S transformations on $Z_{a,b}$ act on the twists (a, b) as

$$T : (a, b) \to (a, a + b \bmod N) \qquad S : (a, b) \to (b, a)$$

c) Compute the action of $T^m S$ on the twists $(0, b)$, for any $b \in \mathbb{Z}_N$, $m = 0, 1, \cdots, N - 1$.

d) Show that if N is prime, and $b \neq 0$, then $T^m S(0, b)$ generates the twists (a, b), $a \in \mathbb{Z}_N$ *exactly once*, for $m = 0, 1, \cdots, N - 1$. Why is it crucial that N be prime? What happens to the twist $(0, 0)$?

e) Show that the form of the $\mathbb{Z}_N$ orbifold modular invariant (17.259) in terms of Z_{proj}, which generalizes Eq. (17.265) is

$$Z_{\text{orb}}^{(\mathbb{Z}_N)} = \left(1 + \sum_{m=0}^{N-1} T^m S\right) Z_{\text{proj}} - Z_{0,0}$$

17.31 *McKay correspondence for the icosahedral subgroup of $SU(2)$*
Exercise 10.18 might be a good prerequisite for generalities on group theory. The character table of the binary icosahedral subgroup of $SU(2)$, of order 120, is listed on Table 17.4. For any finite group, the group tensor-product-algebra coefficients $\mathcal{N}_{rs}{}^t$ are expressed as

$$\mathcal{N}_{ij}{}^k = \frac{1}{|G|} \sum_\alpha |C_\alpha| \chi_i(\alpha)\chi_j(\alpha)\bar{\chi}_k(\alpha)$$

(The proof of this group Verlinde formula is detailed in Ex. 10.18.) The sum extends over the classes C_α of the group, $|G|$ denotes the order of the group, $|C_\alpha|$ the order of the class C_α, and $\chi_i(\alpha)$ the value of the character of the irreducible representation i over the class C_α.

a) With the data of Table 17.4, check the orthogonality of the characters

$$\sum_j \chi_j(\alpha)\,\bar{\chi}_j(\beta) = \frac{|G|}{|C_\alpha|}\,\delta_{\alpha,\beta}$$

$$\sum_\alpha |C_\alpha|\,\chi_j(\alpha)\,\bar{\chi}_k(\alpha) = |G|\,\delta_{j,k}$$

b) Compute the tensor-product algebra of the icosahedral group $\mathcal{I}$.

c) Find a two-dimensional representation leading to the affine diagram $\hat{E}_8$ of Fig. 17.6.

Table 17.4. The character table of the binary icosahedral subgroup $\mathcal{I}$ of $SU(2)$. The columns correspond to classes (denoted by their number of elements), and the rows to representations (denoted by their dimension). We check that Σ (card. of classes) $= \Sigma$ (dim. of reps.)$^2 = 120$. Here $\alpha = (1+\sqrt{5})/2$ is the golden ratio, and $\bar{\alpha} = (1-\sqrt{5})/2$ its conjugate in $\mathbb{Q}[\sqrt{5}]$, the quadratic extension of $\mathbb{Q}$ in which the characters take their values.

Classes : Reps	[1]₁	[1]₂	[12]₁	[12]₂	[12]₃	[12]₄	[20]₁	[20]₂	[30]
(1)	1	1	1	1	1	1	1	1	1
(2)₁	2	−2	−$\bar{\alpha}$	−α	α	$\bar{\alpha}$	1	−1	0
(2)₂	2	−2	−α	−$\bar{\alpha}$	$\bar{\alpha}$	α	1	−1	0
(3)₁	3	3	$\bar{\alpha}$	α	α	$\bar{\alpha}$	0	0	−1
(3)₂	3	3	α	$\bar{\alpha}$	$\bar{\alpha}$	α	0	0	−1
(4)₁	4	−4	−1	−1	1	1	−1	1	0
(4)₂	4	4	−1	−1	−1	−1	1	1	0
(5)	5	5	0	0	0	0	−1	−1	1
(6)	6	−6	1	1	−1	−1	0	0	0

17.32 *Extended fusion rules of the rational block-diagonal $c = 1$ orbifold theory* The aim of this exercise is the derivation of Eqs. (17.308)–(17.309), by means of the extended Verlinde formula (10.219) for RCFTs.

a) For p' even, write the extended S matrix of the modular transformations of the extended characters appearing in Eq. (17.306).
Result: In the extended basis $(\mathbb{I}, \Theta, \phi_{p'}^{(i)}, \phi_\lambda, \sigma^{(i)}, \tau^{(i)})$, the S matrix reads

$$S = \frac{1}{\sqrt{8p'}}$$

$$\times \begin{pmatrix}
1 & 1 & 1 & 2 & \sqrt{p'} & \sqrt{p'} \\
1 & 1 & 1 & 2 & -\sqrt{p'} & -\sqrt{p'} \\
1 & 1 & 1 & 2(-1)^\mu & (-1)^{i-j'}\sqrt{p'} & -(-1)^{i-j''}\sqrt{p'} \\
2 & 2 & 2(-1)^\lambda & 4\cos\pi\frac{\lambda\mu}{2p'} & 0 & 0 \\
\sqrt{p'} & -\sqrt{p'} & (-1)^{i'-j}\sqrt{p'} & 0 & \delta_{i',j'}\sqrt{2p'} & -\delta_{i',j''}\sqrt{2p'} \\
\sqrt{p'} & -\sqrt{p'} & (-1)^{i''-j}\sqrt{p'} & 0 & -\delta_{i'',j'}\sqrt{2p'} & \delta_{i'',j''}\sqrt{2p'}
\end{pmatrix}$$

where the matrix elements are taken between operators $(\mathbb{I}, \Theta, \phi_{p'}^{(i)}, \phi_\lambda, \sigma^{(i')}, \tau^{(i'')})$ and $(\mathbb{I}, \Theta, \phi_{p'}^{(j)}, \phi_\mu, \sigma^{(j')}, \tau^{(j'')})$.

b) Repeat the calculation in the case p' odd. Show that the matrix elements of S differ from the even case only by some phases, which are 8-th roots of unity.

c) Deduce the extended fusion rules (17.308)–(17.309) by applying the extended Verlinde formula (10.219) for RCFTs.

Notes

The two infinite sequences of $\widehat{su}(2)$ nondiagonal modular invariants were first obtained by Gepner and Witten [172] and independently by Bernard and Thierry-Mieg [44], by projecting out odd states and reinforcing modular invariance with the introduction of extra states. The generalization of this construction, using outer automorphisms, is due to Bernard [41]. His results were further generalized by Altschuler, Lacki, and Zaugg [8]. The full list of invariants that can be obtained by the method of outer automorphisms was given in the work of Felder, Gawedzki and Kupiainen [129], using the canonical quantization of WZW models defined on nonsimply-connected manifolds. These results were confirmed by Ahn and Walton [4], who obtained them by the orbifold method; our presentation follows mainly this last reference. For WZW models, the method of outer automorphisms is essentially equivalent to that based on simple currents, advocated by Schellekens and Yankielowicz [319, 320, 321]. The latter method extends straightforwardly to any rational conformal field theory with simple currents.

Conformal embeddings first appeared in the work of Bais et al. [21]. They were completely classified by Bais and Bouwknegt [19] and Schellekens and Warner [318]. The use of conformal embeddings for the construction of nondiagonal modular invariants is originally due to Bais and Taormina [22]. It was systematized and generalized by Bouwknegt and Nahm [52]. The derivation of the $\widehat{su}(2)$ exceptional invariants from conformal embeddings was first presented in this last reference and found independently by Bernard and Thierry-Mieg (unpublished). Our presentation follows rather closely these pioneer works.

The conformal embedding $\widehat{su}(p)_q \oplus \widehat{su}(q)_p \subset \widehat{su}(pq)_1$ was first analyzed by mathematicians (Refs. [136, 210]; see also Refs. [192, 279]). The underlying level-rank duality was noted from the beginning. The same embedding was considered independently in the physics literature—without mention of level-rank duality, however—by Walton [346] (from which App. 17.A is adapted). Further developments and a more rigorous presentation can be found in the work of Altschuler, Bauer, and Itzykson [9]. Other infinite sequences of branching rules were obtained in Refs. [347, 343]. Additional results on conformal embeddings are presented in Kac and Wakimoto [218].

Conformal branching rules can easily be extracted from the existing tables: the finite branching rules can be read off (Ref. [268]), while the decomposition of a $\hat{g}$-irreducible highest-weight module into irreducible representations of the corresponding finite algebra at each grade can be found in Ref. [228].

The classification of $\widehat{su}(2)_k$ modular invariants was obtained by Cappelli, Itzykson, and Zuber [64] and independently by Kato [231]. An interesting relation between the $\widehat{su}(2)_k$ modular invariants and quaternionic coset spaces was noticed by Nahm [277]. Further mathematical curiosities related to these invariants are described in Ref. [202]. The list of $\widehat{su}(3)$ invariants was first written by Christe and Ravanini [77], except for Eq. (17.118), which was discovered by Moore and Seiberg [273]. The completeness of this list was proven by Gannon [158,159]. Partial results in this direction (Ref. [311]) displayed a curious relation between a parity symmetry in the $\widehat{su}(3)$ invariants and Fermat curves. Invariants in affine Lie algebras at level 1 were also fully classified (Refs. [201, 89, 157]).

The main result of Sect. 17.8 on the classification of modular invariants is due to Moore and Seiberg [273]. That nondiagonal invariants reveal nontrivial automorphisms of the fusion algebra was also shown by Dijkgraaf and Verlinde [102]. Fusion automorphisms were classified only for $\widehat{su}(2)$ in the above references and for $\widehat{su}(3)$ by Ruelle [309]. Permutation invariants were first introduced in Ref. [8]. Their full classification for models whose extended chiral algebra is an affine algebra was completed by Gannon, Ruelle, and Walton [161] (following the method presented at the end of Sect. 17.8; see also Ex. 17.9). Further results are presented in Ref. [160].

A powerful method for obtaining modular invariants is described in Warner [350] and Roberts and Terao [307]. It is essentially a lattice construction, in which we first translate the diagonal invariant partition function in terms of a lattice partition function and then deforms it in all possible ways that preserve the modular invariance. The completeness of this method is proven in Ref. [157], but it does not lead directly to physical models (i.e., the mass matrix entries are integers but not necessarily positive). The problem is then reduced to finding linear combinations of deformed lattice partition functions satisfying the usual physical conditions. The proof of completeness is based on the description of the commutant given in Ref. [157], generalizing the construction of the $su(N)$ commutant worked out by Bauer and Itzykson [30] (whose dimension was calculated in Ref. [310]).

The relevance of Galois symmetry in conformal field theory was first noticed by de Boer and Goeree [88] for ratios of S matrix elements. These considerations were extended to S matrix elements by Coste and Gannon [80] (from which Ex. 17.10 is taken). The parity rule has been found by Gannon [157] and independently by Ruelle, Thiran, and Weyers [311]. Galois transformations were used for constructing modular invariants by Fuchs et al. [149, 150]. Exercises 17.11, 17.13, and 17.14 are lifted from the second of these references. Quasi-Galois transformations, for which the scaling factor ℓ is not forced to be coprime with $M(k + g)$, were considered by the same group in Ref. [151]. Further results on the relation between Galois symmetry and modular invariants are presented in Ref. [161] (cf. Exs. 17.16 and 17.17). Finally, we mention the book of Stewart [331], which provides a simple and lively introduction to the Galois theory.

The ideas presented in Sect 17.10 on the interrelation between graphs, fusion of the extended algebras, and modular invariants are due to Kostov [245] and Di Francesco and Zuber [99, 98] (see also the review Ref. [91]). Recent developments can be found in the work of Petkova and Zuber [294]. The mathematics of C-algebra is presented in Ref. [24].

Further results on modular invariants can be found in Ref. [317, 6, 133]. Many invariants were obtained first by numerical methods. For instance, the $\widehat{F}_4$ invariant (17.113) and the $\widehat{G}_2$ invariant (17.128) were first found in Ref. [342]. The counterexample (17.197) was found in Ref. [319].

The classification of $c = 1$ theories was obtained by Pasquier [288] from the statistical mechanics point of view, and by Ginsparg [176] from the orbifold method. Appendix 17.B is based essentially on this latter reference. The completeness of this classification was proven by Kiritsis [238]. The orbifold construction of the $\widehat{su}(2)_1$ model was also considered by Harris [191]. The operator content of these orbifolds was studied by Dijkgraaf et al. [101]. General references on orbifolds can be found at the end of Chap 10. The McKay correspondence is described in Ref. [263, 244]. A readable exposition of the different manifestations of the ADE algebras in mathematics can be found in Ref. [327].

CHAPTER 18

Cosets

Up to this point, we have discussed two general classes of RCFTs: the minimal models, with central charge:

$$c = 1 - \frac{(p' - p)^2}{pp'} < 1 \qquad (18.1)$$

and the much larger class of models with Lie group symmetry, the $\hat{g}_k$-WZW models, with central charge:

$$c = \frac{k \dim g}{k + g} \geq 1 \qquad (18.2)$$

Actually, the second class is substantially enlarged by considering models invariant under Lie group tensor products $G_1 \otimes G_2 \otimes \cdots$, for which the spectrum-generating algebra is the direct sum of the corresponding affine algebras $(\hat{g}_1)_{k_1} \oplus (\hat{g}_2)_{k_2} \oplus \cdots$. To each affine algebra, we can assign a Sugawara energy-momentum tensor, and the total energy-momentum tensor is the sum of all these components. As a result, the central charges of all components add up to give the total central charge.

In this chapter, we introduce the coset construction, which increases tremendously the number of solvable models at hand. A coset is a quotient of two WZW models (or more generally, of direct sums of WZW models). This construction is expected to provide the framework for the complete classification of all RCFT, the WZW models being themselves represented by trivial cosets.

The description of coset conformal theories will be heavily based on the theory of WZW models developed so far. In fact, once the basic mechanism of the construction is explained, it will become evident that most results derived in previous chapters—in particular the calculation of fusion rules and the construction of modular invariants—still pertain to coset models.

The central charge of a coset is the difference of the central charges of the WZW components. This implies that models with central charge lower than one may be represented by the coset construction. On the other hand, all RCFTs with $c < 1$ are known to fall within the classification of minimal models (cf. Sect. 10.5).

Any coset with $c < 1$ must therefore provide a new representation of a minimal model. Some properties of minimal models could be established more directly from this new point of view. This is so in particular for the proof of unitarity of the minimal models with $|p' - p| = 1$. In fact, the coset construction was introduced in conformal field theory as a tool tailor-made to demonstrate this very result. In that respect, we should recall that WZW models are well-defined quantum field theories when their parameter k, which plays the role of the level in the spectrum-generating affine algebra, is a positive integer. This automatically implies unitarity, and this property is preserved by the coset construction.

Regarding unitarity, it should be stressed that minimal and WZW models differ in a fundamental way in that the former can be nonunitary. Nonunitary models cannot be represented as a quotient of unitary models. This seems to clash with the previous statement concerning the presumed completeness of coset models. As will be detailed below, a simple argument, based on the central charge, shows that a coset description of nonunitary minimal models requires WZW models with fractional values of k. Quite remarkably, for an affine Lie algebra at fractional level, there exists a finite number of so-called *admissible* representations whose characters transform among each other under modular transformations. The concept of admissible representation is the cornerstone of the coset description of nonunitary minimal models.

This chapter is organized as follows. In Sect. 18.1, the genuine conformal nature of the coset theory is established by writing the coset Virasoro algebra. Section 18.2 is concerned with the precise relation between branching functions and coset characters. This requires introducing the concept of field identification (already encountered in the minimal models) and its dual manifestation in the form of coset selection rules.

We mentioned previously that, in generic situations, the properties of the constituent WZW models can be transposed almost directly to the coset model. However, this is not always so. Problems arise when, in the WZW models, there are fixed points under the action of the outer-automorphism group. The presence of fixed points raises rather subtle issues related to the precise determination of the actual character of the coset model. This problem is an active subject of current research and is only briefly addressed in Sect. 18.2.2.

These generalities are followed, in Sect. 18.3, by a detailed presentation of the coset formulation of unitary minimal models based on a diagonal $\widehat{su}(2)_k$ coset: derivation of the primary field characters, their modular matrices and their fusion rules.

The coset description of minimal models is not unique. The complete unitary sequence can also be reproduced in two other quotienting schemes beside the diagonal $\widehat{su}(2)_k$ coset. Moreover, there are additional exceptional descriptions of the first few models. Some of these realizations are presented in Sects. 18.4 and 18.5.

The rest of the chapter is concerned with nonunitary models. In the main text, we focus on $\widehat{su}(2)$ and relegate to App. 18.B the presentation of the general case. Relevant results on the representation theory of covariant (or admissible) representations of $\widehat{su}(2)$ at fractional levels are introduced in Sect. 18.6. These results are then used in Sect. 18.7 to describe nonunitary minimal models in the coset language. More general nonunitary diagonal cosets are considered in App. 18.B.

We should stress that there are properties of the coset model that are not easily described in terms of the WZW constituents. Since the coset characters are related to the branching functions, which themselves cannot be expressed as ratios of WZW characters, the coset construction does not lead to an expression for the coset primary fields. As a result, the coset correlation functions cannot be expressed in terms of the WZW correlators in a simple product form. However, the computation of, say, the minimal model correlation functions from the coset point of view is rather academic, given that $\widehat{su}(2)_k$ correlation functions are more difficult to calculate than those of minimal models—the free-field representation requires both free bosons and ghosts in the former case, whereas only bosons are needed in the Virasoro case. The best way of calculating correlation functions of coset primary fields is to use a free-field representation of the coset model. Some representations are presented in the exercises.

In the same way, it appears difficult to derive the singular vectors of the coset model from those of the WZW theories. A general discussion of this problem is not included here. In App. 18.A, we present a Lie-algebraic transcription of the Verma submodule inclusions for the Virasoro minimal model representations.

§18.1. The Coset Construction

Consider an affine Lie algebra $\hat{g}$ and let $\hat{p}$ be a subalgebra of $\hat{g}$. We recall that if the level of $\hat{g}$ is k, that of $\hat{p}$ is given by $x_e k$, where x_e is the embedding index (cf. Eq. (14.265)). In the following, we indicate by a tilde the generators of $\hat{p}$ and denote by L_m^g and L_m^p the Virasoro modes obtained respectively from the $\hat{g}$ and $\hat{p}$ Sugawara construction.

Our primary objective is to show that the modes $L_m^g - L_m^p$ satisfy the Virasoro algebra. For this, we first recall that the $\hat{p}$ generators $\tilde{J}_n^{a'}$ are linear combinations of the generators of $\hat{g}$ (cf. Eq. (14.261)):

$$\tilde{J}_n^{a'} = \sum_a m_{a'a} J_n^a \tag{18.3}$$

It follows directly from Eq. (15.65) that

$$[L_m^g, m_{a'a} J_n^a] = -n\, m_{a'a} J_n^a \tag{18.4}$$

which, when summed over a, yields

$$[L_m^g, \tilde{J}_n^{a'}] = -n\tilde{J}_n^{a'} \tag{18.5}$$

Moreover,

$$[L_m^p, \tilde{J}_n^{a'}] = -n\tilde{J}_n^{a'} \tag{18.6}$$

and these two relations imply that the $\hat{p}$ generators commute with the difference $L_m^g - L_m^p$:

$$[L_m^g - L_m^p, \tilde{J}_n^{a'}] = 0 \tag{18.7}$$

An immediate consequence is

$$[L_m^g - L_m^p, L_n^p] = 0 \tag{18.8}$$

that is,

$$[L_m^g, L_n^p] = [L_m^p, L_n^p] \tag{18.9}$$

Defining

$$L_m^{(g/p)} \equiv L_m^g - L_m^p \tag{18.10}$$

leads to the commutation relation:

$$[L_m^{(g/p)}, L_n^{(g/p)}] = [L_m^g, L_n^g] - [L_m^p, L_n^p]$$

$$= (m - n)L_{m+n}^{g/p} + (c(\hat{g}_k) - c(\hat{p}_{x_e k}))\frac{(m^3 - m)}{12}\delta_{m+n,0} \tag{18.11}$$

Therefore $L_m^{(g/p)}$ satisfies the Virasoro algebra, and its central charge is the difference of the central charges of the constituent models:

$$\boxed{c(\hat{g}_k/\hat{p}_{x_e k}) = \frac{k \dim g}{k + g} - \frac{x_e k \dim p}{x_e k + p}} \tag{18.12}$$

where p stands for the dual Coxeter of p. This construction is often called the Goddard-Kent-Olive (GKO) construction. From now on, the quotient theory, characterized by the energy-momentum tensor $T^g - T^p$, will be referred to as the coset $\hat{g}/\hat{p}$.

In the following, we frequently encounter cosets of the form $(\hat{g} \oplus \hat{g})/\hat{g}$. They are called *diagonal coset* models. The name "diagonal" refers to the way the single copy of $\hat{g}$ is embedded into the direct sum: its generators are simply the sum of the generators of each copy of $\hat{g}$ (indexed by 1 and 2):

$$J_{\text{diag}}^a = J_{(1)}^a + J_{(2)}^a \tag{18.13}$$

Since

$$[J_{(1)}^a, J_{(2)}^a] = 0 \tag{18.14}$$

it follows that the level of the diagonal algebra is simply the sum of the other two. In other words, the embedding index is 1. Such cosets will be denoted as

$$\frac{\hat{g}_{k_1} \oplus \hat{g}_{k_2}}{\hat{g}_{k_1 + k_2}} \tag{18.15}$$

and their central charge is

$$c = \dim g \left(\frac{k_1}{k_1 + g} + \frac{k_2}{k_2 + g} - \frac{k_1 + k_2}{k_1 + k_2 + g} \right) \tag{18.16}$$

§18.2. Branching Functions and Characters

To extract the $\hat{g}/\hat{p}$ coset conformal theory from the $\hat{g}$-WZW model, we must strip off its $\hat{p}$ content. In practice, this means that we should decompose the various representations $\hat{\lambda}$ of $\hat{g}$ into a direct sum of representations $\hat{\mu}$ of $\hat{p}$:

$$\hat{\lambda} \mapsto \bigoplus_{\hat{\mu}} b_{\hat{\lambda}\hat{\mu}} \, \hat{\mu} \tag{18.17}$$

The various characters of the coset model should emerge from this decomposition. In other words, the branching functions are the natural candidates for the coset characters. However, this is not quite exact and we must first consider the precise relationship between characters and branching functions.

18.2.1. Field Identifications and Selection Rules

To the decomposition (18.17) corresponds the character identity

$$\mathrm{ch}_{\mathcal{P}\hat{\lambda}} = \sum_{\hat{\mu} \in P_+^{k x_e}} b_{\hat{\lambda}\hat{\mu}} \, \mathrm{ch}_{\hat{\mu}} \tag{18.18}$$

where $\mathcal{P}$ is the projection matrix of the embedding $p \subset g$ (cf. Sect. 13.7.1). In terms of the normalized characters evaluated at $\hat{\xi} = -2\pi i(\zeta; \tau; t)$, with ζ an arbitrary p weight, it can be written as

$$\chi_{\mathcal{P}\hat{\lambda}}(\zeta; \tau; t) = \sum_{\hat{\mu} \in P_+^{k x_e}} \chi_{\{\hat{\lambda}; \, \hat{\mu}\}}(\tau) \, \chi_{\hat{\mu}}(\zeta; \tau; t) \tag{18.19}$$

where

$$\chi_{\{\hat{\lambda}; \, \hat{\mu}\}}(\tau) = e^{2\pi i \tau (m_{\hat{\lambda}} - m_{\hat{\mu}})} \, b_{\hat{\lambda}\hat{\mu}}(\tau) \tag{18.20}$$

with $m_{\hat{\lambda}}$ defined in Eq. (14.158). The independence of the normalized branching function $\chi_{\{\hat{\lambda}; \hat{\mu}\}}(\tau)$ upon the parameters ζ and t should be clear. Indeed, since at fixed grade the g weights are reorganized into p weights, all the ζ dependence of $\chi_{\mathcal{P}\hat{\lambda}}(\zeta, \tau, t)$ is captured by the different $\chi_{\hat{\mu}}$'s. $\chi_{\{\hat{\lambda}; \hat{\mu}\}}$ is simply a multiplicity factor. Considering all grades induces a τ dependence in $\chi_{\{\hat{\lambda}; \hat{\mu}\}}$. On the other hand, t in $\chi_{\hat{\lambda}}(\zeta; \tau; t)$ appears only in the overall phase factor $e^{2\pi i k t}$ (cf. Eq. (14.161)). But since $\mathcal{P}\hat{\lambda}$ is projected onto a $\hat{p}$ weight at level $k x_e$, the t-dependent phase factors in $\chi_{\mathcal{P}\hat{\lambda}}(\zeta; \tau; t)$ and $\chi_{\hat{\mu}}(\zeta; \tau; t)$ are exactly the same. In character identities such as Eq. (18.19) the projection operator $\mathcal{P}$ is often omitted but always understood.

In the absence of nontrivial branching between elements of the outer automorphism groups of $\hat{g}$ and $\hat{p}$, $\chi_{\{\hat{\lambda}; \hat{\mu}\}}(\tau)$ is the character for the coset field $\{\hat{\lambda}; \hat{\mu}\}$. Subtleties arise when there are branchings of outer-automorphism groups as (cf. Sect. 14.7.3):

$$A \mapsto \tilde{A} \qquad \text{for} \quad A \in \mathcal{O}(\hat{g}), \quad \tilde{A} \in \mathcal{O}(\hat{p}) \tag{18.21}$$

which can be expressed equivalently under the form (14.278):

$$(A\hat{\omega}_0, \lambda) = (\tilde{A}\hat{\omega}_0, \mathcal{P}\lambda) \qquad \text{mod } 1 \qquad \forall \lambda \in g \tag{18.22}$$

For normalized branching functions, this implies

$$\chi_{\{\hat{\lambda};\,\hat{\mu}\}}(\tau) = \chi_{\{A\hat{\lambda};\,\tilde{A}\hat{\mu}\}}(\tau)$$

(18.23)

This equality is actually most directly proven at the level of modular matrices: $\chi_{\{\hat{\lambda};\hat{\mu}\}}(\tau)$ and $\chi_{\{A\hat{\lambda};\tilde{A}\hat{\mu}\}}(\tau)$ transform identically under modular transformations, which is demonstrated below, and this implies the above equality (at least up to modular invariants—e.g., constants). To these two normalized branching functions, there must correspond a single coset field $\{\hat{\lambda};\,\hat{\mu}\}$. Indeed, suppose that there are no fixed points, i.e., no coset field $\{\hat{\lambda};\,\hat{\mu}\}$ such that

$$\hat{\lambda} = A\hat{\lambda}, \qquad \text{and} \qquad \hat{\mu} = \tilde{A}\hat{\mu}$$

(18.24)

Then the number N of states in the string $A\hat{\lambda}$ generated by all allowed A is the same for all $\hat{\lambda}$. N is the number of elements of $\mathcal{O}(\hat{g})$ that branch to elements of $\mathcal{O}(\hat{p})$. Regarding branching functions as coset characters, we find that each field, and in particular the vacuum, has multiplicity N. To make the theory physical, the cure is clear: we simply divide the partition function (which codes the field content) by a factor N. In other words, we identify the fields $\{\hat{\lambda};\,\hat{\mu}\}$ and $\{A\hat{\lambda};\,\tilde{A}\hat{\mu}\}$. From now on, this will be denoted as:

$$\{\hat{\lambda};\,\hat{\mu}\} \sim \{A\hat{\lambda};\,\tilde{A}\hat{\mu}\}$$

(18.25)

An immediate consequence of the identification of characters with branching functions is that not all pairs of fields can be combined into a coset field. Indeed, for the branching function $\chi_{\{\hat{\lambda};\hat{\mu}\}}$ to be nonzero, the following condition must be satisfied:

$$\mathcal{P}\lambda - \mu \in \mathcal{P}Q$$

(18.26)

where Q is the g root lattice. This is the branching condition (13.261) already encountered.

Field identifications and selection rules are intimately related, as the comparison of Eqs. (18.26) and (18.22) shows. The branching condition requires $\mathcal{P}\lambda$ and μ to be in the same congruence class. These classes are isomorphic to the elements of the center, which are themselves S duals (in the sense of Eq. (14.256)) to the elements of the outer-automorphism group. Hence, the selection rules appear as the S duals of the field identifications.

For the diagonal cosets (18.15), the field identifications and the selection rules take a particularly simple form. Coset fields are specified in terms of three $\hat{g}$ weights, e.g., $\{\hat{\lambda},\,\hat{\mu};\,\hat{\nu}\}$, and the branching functions $\chi_{\{\hat{\lambda},\hat{\mu};\hat{\nu}\}}$, which we identify with the coset characters, are obtained from the decomposition of $\chi_{\hat{\lambda}}\chi_{\hat{\mu}}$ in terms of $\chi_{\hat{\nu}}$, where $\hat{\lambda},\,\hat{\mu},\,\hat{\nu}$ are $\hat{g}$ integrable highest weights at respective levels $k_1,\ k_2,\ k_1+k_2$. Since

$$\mathcal{P}(Q \oplus Q) = Q$$

(18.27)

with Q the root lattice of g, the selection rule is now simply

$$\lambda + \mu - \nu \in Q$$

(18.28)

Also, since for any A

$$A \otimes A \mapsto A \tag{18.29}$$

the fields are always identified according to

$$\{\hat{\lambda}, \hat{\mu}; \hat{\nu}\} \sim \{A\hat{\lambda}, A\hat{\mu}; A\hat{\nu}\} \qquad \forall A \in \mathcal{O}(\hat{g}) \tag{18.30}$$

18.2.2. Fixed Points and Their Resolutions

In the presence of fixed points, not all orbits of A have the same length (i.e., the same number of distinct elements). For instance, if $A^N = 1$, some orbits will have length $N_0 < N$, where N_0 is a divisor of N. As a result, a mere division of the partition function by a constant is no longer possible: dividing by N_0 yields a vacuum with multiplicity $N/N_0 > 1$, whereas dividing by N leads to characters with fractional coefficients in their q expansion. To cure the theory, we need to introduce extra characters—not expressible in terms of the branching functions but nevertheless compatible with modular covariance—to make the division by N meaningful. This process is called the *resolution of fixed points*.

The search for adequate and general techniques for resolving fixed points is still an active field of research and this question will not be studied further in this work. But we stress that it is a very important issue in coset conformal field theories since fixed points arise quite frequently (see, e.g., Exs. 18.10 and 18.11).

18.2.3. Maverick Cosets

Generically, Eq. (18.26) represents the only selection rules. But there are exceptional situations for which additional selection rules appear. In that case, additional field identifications also appear. The simplest of these exceptional cases are the conformal embeddings. Take, for instance, the embedding $\widehat{su}(2)_4 \subset \widehat{su}(3)_1$, for which we found (cf. Eqs. (14.272) and (14.273))

$$[1,0,0] \mapsto [4,0] \oplus q[0,4] \qquad [0,1,0] \mapsto [2,2], \qquad [0,0,1] \mapsto [2,2] \tag{18.31}$$

Hence the branching functions $\chi_{\{[1,0,0];[2,2]\}}$ and $\chi_{\{[0,1,0];[4,0]\}}$ (among others) are not ruled out by the selection rule (18.26) (we recall that here $\mathcal{P}Q_{su(3)} = Q_{su(2)}$), but they nevertheless vanish. Moreover, since the resulting coset is a unitary $c = 0$ theory, whose whole field content is the identity, all coset characters are identical and equal to 1. These are the additional field identifications announced in the introduction. Since in these examples, the numerator of the coset has level 1, we could suspect that the occurrence of extra selection rules is a low-level feature. Indeed, the only other models for which it has been found to occur are:

$$\widehat{su}(N)_2/\widehat{so}(N)_4$$
$$\widehat{so}(2N)_2/\widehat{so}(N)_2 \oplus \widehat{so}(N)_2$$
$$(\widehat{E}_6)_2/\widehat{sp}(8)_2$$

$$(\widehat{E}_7)_2/\widehat{su}(8)_2$$
$$(\widehat{E}_8)_2/\widehat{so}(16)_2$$
$$(\widehat{E}_8)_2/\widehat{su}(2)_2 \oplus (\widehat{E}_7)_2 \tag{18.32}$$

These are called *maverick cosets*. In all cases, the numerator has level 2, and, quite remarkably, it is always a simply-laced algebra. The simplest maverick is studied in Ex. 18.8. The Lie-algebraic mechanism underlying the extra selection rules and field identifications is not fully understood yet.[1]

18.2.4. Modular Transformation Properties of Coset Characters

Consider now the modular transformation properties of the branching functions. The simple modular transformation properties of the normalized affine characters lead to[2]

$$\chi_{\{\hat{\lambda};\;\hat{\mu}\}}(-1/\tau) = \sum_{\substack{\hat{\lambda}' \in P_+^k \\ \hat{\mu}' \in P_+^{k x_e}}} S^{(k)}_{\hat{\lambda}\hat{\lambda}'} S^{(k x_e)-1}_{\hat{\mu}\hat{\mu}'} \chi_{\{\hat{\lambda}';\;\hat{\mu}'\}}(\tau) \tag{18.33}$$

where $\hat{\lambda}'$ and $\hat{\mu}'$ are $\hat{g}$ and $\hat{p}$ integrable weights respectively, and

$$\chi_{\{\hat{\lambda};\;\hat{\mu}\}}(\tau+1) = e^{2\pi i(m_{\hat{\lambda}}-m_{\hat{\mu}})}\chi_{\{\hat{\lambda};\;\hat{\mu}\}}(\tau) \tag{18.34}$$

These relations are proven as follows. We first write

$$\chi_{P\hat{\lambda}}(\zeta/\tau;\; -1/\tau;\; t+|\zeta|^2/2\tau) = \sum_{\hat{\lambda}' \in P_+^k} S^{(k)}_{\hat{\lambda}\hat{\lambda}'} \chi_{P\hat{\lambda}'}(\zeta;\; \tau;\; t)$$

$$= \sum_{\substack{\hat{\lambda}' \in P_+^k \\ \hat{\mu}' \in P_+^{k x_e}}} S^{(k)}_{\hat{\lambda}\hat{\lambda}'} \chi_{\{\hat{\lambda}';\;\hat{\mu}'\}}(\tau)\, \chi_{\hat{\mu}'}(\zeta;\; \tau;\; t) \tag{18.35}$$

The projection operator has been omitted from the S matrix indices because $\chi_{P\hat{\lambda}}$ and $\chi_{\hat{\lambda}}$ have identical modular transformation properties (i.e., S does not depend upon ζ). We also have

$$\chi_{P\hat{\lambda}}(\zeta/\tau;\; -1/\tau;\; t+|\zeta|^2/2\tau) =$$

$$\sum_{\hat{\mu} \in P_+^{k x_e}} \chi_{\{\hat{\lambda};\;\hat{\mu}\}}(-1/\tau)\, \chi_{\hat{\mu}}(\zeta/\tau;\; -1/\tau;\; t+|\zeta|^2/2\tau) \tag{18.36}$$

$$= \sum_{\hat{\mu},\hat{\mu}' \in P_+^{k x_e}} \chi_{\{\hat{\lambda};\;\hat{\mu}\}}(-1/\tau)\, S^{(k x_e)}_{\hat{\mu}\hat{\mu}'}\, \chi_{\hat{\mu}'}(\zeta,\tau,t)$$

---

[1] Notice that there are extra field identifications in nonunitary models, but their Lie-algebraic interpretation is clear (cf. Sect. 18.7.2).

[2] To avoid overloading the notations, we depart from our previous convention of indicating by a prime (e.g., $\hat{\lambda}'$) a weight in the representation of highest weight $\hat{\lambda}$. In this chapter, both $\hat{\lambda}$ and $\hat{\lambda}'$ stand for integrable highest weights at level k. Our general convention is that a weight and its prime (or double prime) version are highest weights at the same level.

Since the characters $\chi_{\hat{\mu}}(\zeta, \tau, t)$ are linearly independent,[3] the comparison of the above two equations yields

$$\sum_{\hat{\lambda}' \in P_+^k} S_{\hat{\lambda}\hat{\lambda}'}^{(k)} \chi_{\{\hat{\lambda}'; \; \hat{\mu}'\}}(\tau) = \sum_{\hat{\mu} \in P_+^{kx_e}} S_{\hat{\mu}\hat{\mu}'}^{(kx_e)} \chi_{\{\hat{\lambda}; \; \hat{\mu}\}}(-1/\tau) \tag{18.37}$$

By multiplying both sides by $S_{\hat{\mu}'\hat{\sigma}}^{(kx_e)^{-1}}$, summing over $\hat{\mu}'$, and using

$$\sum_{\hat{\mu}' \in P_+^{kx_e}} S_{\hat{\mu}\hat{\mu}'}^{(kx_e)} S_{\hat{\mu}'\hat{\sigma}}^{(kx_e)^{-1}} = \delta_{\hat{\mu}\hat{\sigma}} \tag{18.38}$$

we get Eq. (18.33).

The relation (18.34) is proven in the same way. We recall that the modular anomaly is related to the conformal dimension and the central charge by Eq. (15.122); it follows that the $\mathcal{T}$ transformation matrix for $\chi_{\{\hat{\lambda}; \; \hat{\mu}\}}$ is given by

$$\chi_{\{\hat{\lambda}; \; \hat{\mu}\}}(\tau + 1) = e^{2\pi i(h_{\hat{\lambda}} - h_{\hat{\mu}} - c/24)} \chi_{\{\hat{\lambda}; \hat{\mu}\}}(\tau) \tag{18.39}$$

where c is the coset central charge. From Eq. (18.19), there actually follows a simple expression for the fractional part of the conformal dimension for the coset field $\{\hat{\lambda}; \; \hat{\mu}\}$. If the tip of the $\hat{\mu}$ representation of $\hat{p}$ lies at grade n in the $\hat{\lambda}$ representation of $\hat{g}$, then

$$h_{\{\hat{\lambda}; \; \hat{\mu}\}} = h_{\hat{\lambda}} - h_{\hat{\mu}} + n \tag{18.40}$$

(This also follows directly from Eq. (18.10).) To find n requires much work: the actual computation of the branching functions. However, the fractional part of $h_{\{\hat{\lambda}; \; \hat{\mu}\}}$ is just $h_{\hat{\lambda}} - h_{\hat{\mu}}$. Hence, as expected,

$$\chi_{\{\hat{\lambda}; \; \hat{\mu}\}}(\tau + 1) = e^{2\pi i(h_{\{\hat{\lambda};\hat{\mu}\}} - c/24)} \chi_{\{\hat{\lambda}; \; \hat{\mu}\}}(\tau) \tag{18.41}$$

We have thus shown that the transformation matrices for the $\chi_{\{\hat{\lambda};\hat{\mu}\}}$'s are simply

$$\boxed{\begin{aligned} S_{\{\hat{\lambda}; \; \hat{\mu}\},\{\hat{\lambda}'; \; \hat{\mu}'\}} &= S_{\hat{\lambda}\hat{\lambda}'}^{(k)} \, \bar{S}_{\hat{\mu}\hat{\mu}'}^{(kx_e)} \\ T_{\{\hat{\lambda}; \; \hat{\mu}\},\{\hat{\lambda}'; \; \hat{\mu}'\}} &= T_{\hat{\lambda}\hat{\lambda}'}^{(k)} \, \bar{T}_{\hat{\mu}\hat{\mu}'}^{(kx_e)} \end{aligned}} \tag{18.42}$$

where in these relations, thanks to unitarity, we have replaced inverse matrices by their complex conjugates.

Unitarity of the branching-function modular matrices is inherited from the unitarity of the WZW modular matrices. Taking field identifications into account (in the absence of fixed points) simply amounts to rescaling the coset S matrix (an example is worked out in the following section).

The study of field identifications is most conveniently formulated at the level of modular matrices: if

$$S_{ij} = S_{i'j}, \quad T_{ij} = T_{i'j} \quad \forall j \tag{18.43}$$

---

[3] This justifies our insistence on maintaining the full ζ dependence, since otherwise this statement would not be true, e.g., $\chi_{\hat{\mu}}(\tau) = \chi_{\hat{\mu}^*}(\tau)$.

then the fields ϕ_i and $\phi_{i'}$ must be identified since their characters are identical. From this point of view, the result (18.23) is easily established. The action (14.255) of the outer-automorphism group on $\mathcal{S}$,

$$\mathcal{S}^{(k)}_{A\hat{\lambda},\,\hat{\lambda}'} = \mathcal{S}^{(k)}_{\hat{\lambda}\hat{\lambda}'}\, e^{-2\pi i(A\hat{\omega}_0,\lambda')}$$
$$\bar{\mathcal{S}}^{(kx_e)}_{\tilde{A}\hat{\mu},\,\hat{\mu}'} = \bar{\mathcal{S}}^{(kx_e)}_{\hat{\mu}\hat{\mu}'}\, e^{2\pi i(\tilde{A}\hat{\omega}_0,\mu')} \tag{18.44}$$

implies that

$$\mathcal{S}_{\{A\hat{\lambda};\,\tilde{A}\hat{\mu}\},\{\hat{\lambda}';\,\hat{\mu}'\}} = \mathcal{S}_{\{\hat{\lambda};\,\hat{\mu}\},\{\hat{\lambda}';\,\hat{\mu}'\}} e^{-2\pi i[(A\hat{\omega}_0,\lambda')-(\tilde{A}\hat{\omega}_0,\mu')]} \tag{18.45}$$

Moreover, it follows from Eq. (18.22) that

$$(A\hat{\omega}_0,\lambda') - (\tilde{A}\hat{\omega}_0,\mu') = (\tilde{A}\hat{\omega}_0, \mathcal{P}\lambda' - \mu') \bmod 1 \tag{18.46}$$

By assumption, λ' and μ' are combined into a coset field, which means they must satisfy the condition (18.26). Since the comark of $\tilde{A}\hat{\omega}_0$ is necessarily 1, the above scalar product is an integer, that is,[4]

$$(\tilde{A}\hat{\omega}_0, \mathcal{P}\lambda' - \mu') = 0 \bmod 1 \tag{18.47}$$

Therefore, the phase in Eq. (18.45) is 1. It is not difficult to verify the same result at the level of the $\mathcal{T}$ matrices. This allows us to conclude directly that the fields $\{A\hat{\lambda};\,\tilde{A}\hat{\mu}\}$ and $\{\hat{\lambda};\,\hat{\mu}\}$ are identical.

By construction, coset models are thus rational conformal field theories. Indeed, since there is a finite number of primary fields in both the $\hat{g}$ and $\hat{p}$ WZW models, there is a finite number of branching rules, and thus a finite number of coset primary fields. As demonstrated in the next subsection, modular invariance is also built in, because the modular transformation properties of the branching functions are directly related to the modular transformation properties of the WZW model constituents, and also because of the dual relation between selection rules and field identifications. It is certainly quite pleasant to find that the coset and WZW modular matrices can be linked in such a simple way in spite of the rather complicated relation between their characters.

18.2.5. Modular Invariants

Given the modular transformation properties of the coset characters, the next step is to construct modular-invariant partition functions. At first sight, a straightforward way of constructing modular invariants in the coset theory is simply to take, for the coset mass matrix $\mathcal{M}$, the product

$$\mathcal{M} = \mathcal{M}^{(k)} \mathcal{M}^{(kx_e)} \tag{18.48}$$

---

[4] In more detail: with $\tilde{A}\hat{\omega}_0 = \omega_i$ and $\mathcal{P}\lambda' - \mu' = \sum n_j\alpha_j,\ n_j \in \mathbb{Z}$, we have

$$(\tilde{A}\hat{\omega}_0, \mathcal{P}\lambda' - \mu') = n_i(\omega_i, \alpha_i) = n_i(\omega_i, \alpha_i^\vee) = n_i$$

where $\mathcal{M}^{(k)}$ and $\mathcal{M}^{(kx_e)}$ denote invariant mass matrices for the $\hat{g}_k$- and $\hat{p}_{kx_e}$-WZW models, respectively, that is,

$$
\begin{aligned}
Z^{(k)} &= \sum_{\hat{\lambda},\hat{\lambda}' \in P_+^{(k)}} \chi_{\hat{\lambda}}(\tau)\, \mathcal{M}^{(k)}_{\hat{\lambda},\hat{\lambda}'}\, \bar{\chi}_{\hat{\lambda}'}(\bar{\tau}) \\
Z^{(kx_e)} &= \sum_{\hat{\mu},\hat{\mu}' \in P_+^{(kx_e)}} \chi_{\hat{\mu}}(\tau)\, \mathcal{M}^{(kx_e)}_{\hat{\mu},\hat{\mu}'}\, \bar{\chi}_{\hat{\mu}'}(\bar{\tau})
\end{aligned}
\tag{18.49}
$$

The separate modular invariance of $\mathcal{M}^{(k)}$ and $\mathcal{M}^{(kx_e)}$ ensures automatically the invariance of their product. However, this simple product mass matrix does not give the coset partition function, since the branching conditions—which impose constraints on the summations—are not taken into account. Furthermore, field identifications must be considered. If all orbits of field identifications have length N, this last point simply requires dividing the partition function by N. The candidate partition function is then

$$
Z = \frac{1}{N} \sum_{\substack{\hat{\lambda},\hat{\lambda}' \in P_+^{(k)} \ \hat{\mu},\hat{\mu}' \in P_+^{(kx_e)} \\ \mathcal{P}\lambda-\mu=\mathcal{P}\lambda'-\mu'=0 \bmod Q}} \chi_{\{\hat{\lambda};\,\hat{\mu}\}}(\tau)\, \mathcal{M}^{(k)}_{\hat{\lambda},\hat{\lambda}'}\, \mathcal{M}^{(kx_e)}_{\hat{\mu},\hat{\mu}'}\, \bar{\chi}_{\{\hat{\lambda}';\,\hat{\mu}'\}}(\bar{\tau})
\tag{18.50}
$$

Due to the constraints, this no longer has a product form, and modular invariance is not guaranteed from the onset. But this is indeed modular invariant: roughly, the S transformation changes the branching constraints into field identifications! This should be compared with the construction of nondiagonal modular invariants by outer automorphisms. We first introduced a selection rule (a projection operation) and restored modular invariance by an appropriate twisting, that is, by the insertion of outer automorphisms. What happens here is that the twisting is trivial due to the field identifications.

A large class of coset modular-invariant partition functions for the coset model can thus be obtained in this way. A posteriori, this is a further justification for the efforts devoted to the construction of modular invariants for WZW models.

§18.3. Coset Description of Unitary Minimal Models

We consider the $\widehat{su}(2)$ diagonal coset:

$$
\frac{\widehat{su}(2)_k \oplus \widehat{su}(2)_1}{\widehat{su}(2)_{k+1}}
\tag{18.51}
$$

Its central charge is

$$
c = \frac{3k}{k+2} + 1 - \frac{3(k+1)}{k+3} = 1 - \frac{6}{(k+2)(k+3)}
\tag{18.52}
$$

with

$$
k + 2 = p \geq 3
\tag{18.53}
$$

This is precisely the central charge of the unitary minimal series. This suggests that the unitary minimal models can be described by the coset (18.51). To establish this equivalence, we need to prove that, for a fixed value of k, the characters of all primary fields of the minimal model $(p+1, p)$ appear in the decomposition of the products $\chi_{\hat{\lambda}} \chi_{\hat{\mu}}$ ($\hat{\lambda}, \hat{\mu}$ at level k and 1 respectively) in terms of $\chi_{\hat{\nu}}$, with $\hat{\nu}$ at level $k+1$. This will be our first concern.

18.3.1. Character Decomposition

We then evaluate the characters $\chi_{\{\hat{\lambda}, \hat{\mu}; \hat{\nu}\}}$ defined by the decomposition

$$\chi_{\hat{\lambda}} \chi_{\hat{\mu}} = \sum_{\hat{\nu} \in P_+^{k+1}} \chi_{\{\hat{\lambda}, \hat{\mu}; \hat{\nu}\}} \chi_{\hat{\nu}} \tag{18.54}$$

where $\hat{\lambda} \in P_+^k$, $\hat{\mu} \in P_+^1$. In the character formula

$$\chi_{\hat{\lambda}} = \frac{\sum_{w \in \hat{W}} \epsilon(w) \Theta_{w(\hat{\lambda}+\hat{\rho})}}{\sum_{w \in \hat{W}} \epsilon(w) \Theta_{w\hat{\rho}}} \equiv \frac{D_{\hat{\lambda}+\hat{\rho}}}{D_{\hat{\rho}}} \tag{18.55}$$

the denominator is common to all representations, irrespective of their level. This means that Eq. (18.54) reduces to

$$D_{\hat{\lambda}+\hat{\rho}} \chi_{\hat{\mu}} = \sum_{\hat{\nu} \in P_+^{k+1}} \chi_{\{\hat{\lambda}, \hat{\mu}; \hat{\nu}\}} D_{\hat{\nu}+\hat{\rho}} \tag{18.56}$$

We now concentrate on the specialized form of this decomposition, where as usual characters are evaluated at $\hat{\xi} = -2\pi i(\zeta; \tau; t)$. Clearly, $\chi_{\{\hat{\lambda}, \hat{\mu}; \hat{\nu}\}}$ depends only upon τ. Without loss of generality, we can set $t = 0$. With $\zeta = z\omega_1$ and $p = k+2$, we have (cf. Eq. (14.176)):

$$\Theta_{\hat{\lambda}+\rho}(\zeta; \tau; 0) = \Theta_{\lambda_1+1}^{(p)}(z; \tau)$$

$$= \sum_{n \in \mathbb{Z}} e^{-2\pi i [pnz + z(\lambda_1+1)/2 - pn^2\tau - n(\lambda_1+1)\tau - \tau(\lambda_1+1)^2/4p]} \tag{18.57}$$

$$= q^{(\lambda_1+1)^2/4p} \sum_{n \in \mathbb{Z}} q^{n(\lambda_1+1)+pn^2} x^{pn+(\lambda_1+1)/2}$$

where

$$q = e^{2\pi i \tau}, \qquad x = e^{-2\pi i z} \tag{18.58}$$

so that

$$D_{\lambda_1+1}^{(p)} = \Theta_{\lambda_1+1}^{(p)} - \Theta_{-\lambda_1-1}^{(p)}$$

$$= q^{(\lambda_1+1)^2/4p} \sum_{n \in \mathbb{Z}} q^{n(\lambda_1+1)+pn^2} \left(x^{pn+(\lambda_1+1)/2} - x^{-pn-(\lambda_1+1)/2} \right) \tag{18.59}$$

On the other hand, for the character of the representations at level 1, we will use the expression derived from the vertex construction in Sect. 15.6.2, Eq. (15.244):

$$\chi_{\hat{\mu}}(z; \tau) = \eta(q)^{-1} \sum_{n \in \mathbb{Z}+\mu_1/2} q^{n^2} x^n \tag{18.60}$$

We can thus write

$$\chi_{\hat{\mu}}(z;\tau)D_{\hat{\lambda}+\hat{\rho}}(z;\tau) = \frac{q^{(\lambda_1+1)^2/4p}}{\eta(q)} \sum_{n,n'\in\mathbb{Z}} q^{n'(\lambda_1+1)+pn'^2+(n+\mu_1/2)^2}$$

$$\times (x^{pn'+(\lambda_1+1)/2+n+\mu_1/2} - x^{-pn'-(\lambda_1+1)/2-n-\mu_1/2}) \tag{18.61}$$

This expression must now be reorganized into a sum of characters for integrable representations at level $k+1$. In this view, we trade the summation index n by a new index ν_1', defined in such a way that the power of the x terms in Eq. (18.61) becomes

$$\pm[(\frac{1}{2}(\nu_1'+1)+(p+1)n'] \tag{18.62}$$

This fixes ν_1' to be

$$\nu_1' = \lambda_1 + \mu_1 + 2(n-n') \tag{18.63}$$

showing that

$$\lambda_1 + \mu_1 + \nu_1' = 0 \mod 2 \tag{18.64}$$

By replacing $n + \mu_1/2$ by $(\nu_1' - \lambda_1)/2 + n'$ in the r.h.s. of Eq. (18.61), we get

$$\chi_{\hat{\mu}}(z;\tau)D_{\hat{\lambda}+\hat{\rho}}(z;\tau) = \eta(q)^{-1} \sum_{\substack{\nu_1'\in\mathbb{Z} \\ \nu_1'+\lambda_1+\mu_1=0 \text{ mod } 2}} q^{\Delta_{\lambda_1,\nu_1'}} D_{\nu_1'+1}^{(p+1)} \tag{18.65}$$

with

$$\Delta_{\lambda_1,\nu_1'} = \left(\frac{(\lambda_1+1)^2}{4p} + \frac{(\nu_1'-\lambda_1)^2}{4} - \frac{(\nu_1'+1)^2}{4(p+1)}\right) \tag{18.66}$$

We now set

$$\nu_1' = \nu_1 + 2(p+1)t \tag{18.67}$$

with

$$t\in\mathbb{Z}, \qquad 0 \le \nu_1 \le 2p+1 \tag{18.68}$$

The condition (18.64) becomes

$$\lambda_1 + \mu_1 + \nu_1 = 0 \mod 2 \tag{18.69}$$

Since

$$D_{\nu_1'+1}^{(p+1)} = D_{\nu_1'+1+2t(p+1)}^{(p+1)} \tag{18.70}$$

we have

$$\chi_{\hat{\mu}}(z;\tau)D_{\hat{\lambda}+\hat{\rho}}(z;\tau) = \eta(q)^{-1} \sum_{\substack{\nu_1=0 \\ \nu_1+\lambda_1+\mu_1=0 \text{ mod } 2}}^{2p+1} D_{\nu_1+1}^{(p+1)} \sum_{t\in\mathbb{Z}} q^{\Delta_{\lambda_1,\nu_1}(t)} \tag{18.71}$$

where the power of q is now

$$\Lambda_{\lambda_1,\nu_1}(t) = \frac{[(\lambda_1+1)(p+1) - (\nu_1+1)p + 2tp(p+1)]^2}{4p(p+1)} \tag{18.72}$$

We are almost done: we only need to reduce by half the range of ν_1. For this, we use the relations

$$D^{(p+1)}_{-\nu_1-1} = -D^{(p+1)}_{\nu_1+1}, \quad D^{(p+1)}_{-\nu_1-1} = D^{(p+1)}_{2(p+1)-\nu_1-1} \tag{18.73}$$

With ν_1 restricted to $0 \le \nu_1 \le k+1$, the desired range, $2(p+1) - \nu_1 - 2$ covers the other values needed to fill the complete interval $0 \le \nu_1 \le 2p+1$, except for $\nu_1 = k+2$. But this value does not contribute because

$$D^{(p+1)}_{k+3} = D^{(k+3)}_{k+3} = 0 \tag{18.74}$$

The effect of replacing ν_1 by $2(p+1) - \nu_1$ in $\Lambda_{\lambda_1,\nu_1}$ is simply

$$\Lambda_{\lambda_1,2(p+1)-\nu_1}(t) = \Lambda_{\lambda_1,-\nu_1-2}(t-1) \tag{18.75}$$

We thus finally reach

$$\chi_{\hat\mu}(z;\tau)D_{\hat\lambda+\hat\rho}(z;\tau) = \sum_{\substack{\hat\nu \in P^{k+1}_+ \\ \nu_1+\lambda_1+\mu_1=0 \bmod 2}} D_{\hat\nu+\hat\rho}\left\{ \sum_{t\in\mathbb{Z}} \frac{(q^{\Lambda_{\lambda_1,\nu_1}(t)} - q^{\Lambda_{\lambda_1,-\nu_1-2}(t)})}{\eta(q)} \right\} \tag{18.76}$$

Setting

$$r = \lambda_1 + 1 \qquad s = \nu_1 + 1 \tag{18.77}$$

we have thus derived Eq. (18.54), with the following expression for the normalized branching functions $\chi_{\{\hat\lambda,\hat\mu;\hat\nu\}}$:

$$\chi_{\{\hat\lambda,\hat\mu;\hat\nu\}}(\tau) \equiv \chi_{(r,s)}(q) \tag{18.78}$$

with

$$\chi_{(r,s)}(q) = K^{(p+1,p)}_{r,s}(q) - K^{(p+1,p)}_{r,-s}(q) \tag{18.79}$$

and

$$K^{(p+1,p)}_{r,s}(q) = \eta(q)^{-1} \sum_{n\in\mathbb{Z}} q^{(2np(p+1)+r(p+1)-sp)^2/4p(p+1)}$$
$$= \frac{q^{-c/24}}{\varphi(q)} \sum_{n\in\mathbb{Z}} q^{[(2np(p+1)+r(p+1)-sp)^2-1]/4p(p+1)} \tag{18.80}$$

These are exactly the characters for the primary fields (r,s) for the unitary Virasoro minimal model $(p+1,p)$ with $c = 1 - 6/p(p+1)$ (cf. Eqs. (8.17) and (8.16)[5]).

At first sight, it is somewhat surprising to find that the coset fields actually depend upon only two $\widehat{su}(2)$ weights. But this is so because the third weight is at level 1. Given $\hat\lambda$ and $\hat\nu$, there is a unique $\hat\mu \in P^1_+$ fixed by Eq. (18.69) (which,

---

[5] In comparing with these expressions, we should let $p' \to p$, $p \to p+1$

of course, is simply Eq. (18.28)). So we can eliminate $\hat{\mu}$ and forget about the constraint (18.69): every pair $\{\lambda_1; \nu_1\}$ leads to a unique coset field $\{\hat{\lambda}, \hat{\mu}; \hat{\nu}\}$.

With λ_1 and ν_1 ranging over

$$0 \le \lambda_1 \le k, \qquad 0 \le \nu_1 \le k+1 \tag{18.81}$$

so that

$$1 \le r \le k+1 = p-1 \qquad 1 \le s \le k+2 = p \tag{18.82}$$

and with all combinations $\{\lambda_1; \nu_1\}$ allowed, we indeed find that all values $h_{r,s}$ of the $(p+1, p)$ Kac table are covered. For the minimal models, we know that the fields $\phi_{r,s}$ and $\phi_{p-r,p+1-s}$ must be identified. This is simply the field identification (18.30), whose outer-automorphism transcription is

$$\{\hat{\lambda}, \hat{\mu}; \hat{\nu}\} \sim \{a\hat{\lambda}, a\hat{\mu}; a\hat{\nu}\} \tag{18.83}$$

where a exchanges the two Dynkin labels. We note that this equivalence is compatible with Eq. (18.64) since

$$\lambda_1 + \mu_1 + \nu_1 = 0 \bmod 2 \quad \Rightarrow k - \lambda_1 + 1 - \mu_1 + k + 1 - \nu_1 = 0 \bmod 2 \tag{18.84}$$

18.3.2. Modular S Matrix

The modular S matrix for the coset character is simply

$$S_{\{\hat{\lambda}, \hat{\mu}; \hat{\nu}\}, \{\hat{\lambda}', \hat{\mu}'; \hat{\nu}'\}} = S_{\hat{\lambda}\hat{\lambda}'}^{(k)} S_{\hat{\mu}\hat{\mu}'}^{(1)} S_{\hat{\nu}\hat{\nu}'}^{(k+1)} \tag{18.85}$$

with

$$\hat{\lambda}, \hat{\lambda}' \in P_+^k, \qquad \hat{\mu}, \hat{\mu}' \in P_+^1, \qquad \hat{\nu}, \hat{\nu}' \in P_+^{(k+1)} \tag{18.86}$$

Here we used the reality of the $\widehat{su}(2)$ S matrices to omit the complex conjugation from the last S factor. We recall that

$$S_{\hat{\lambda}\hat{\lambda}'}^{(k)} = \sqrt{\frac{2}{k+2}} \sin\left(\frac{\pi(\lambda_1 + 1)(\lambda_1' + 1)}{k+2}\right)$$

which at $k = 1$ reduces to

$$S_{\hat{\mu}\hat{\mu}'}^{(1)} = \frac{1}{\sqrt{2}}(-1)^{\mu_1\mu_1'} \tag{18.87}$$

Using Eq. (18.69), this last expression can be written as

$$S_{\hat{\mu}\hat{\mu}'}^{(1)} = \frac{1}{\sqrt{2}}(-1)^{(\lambda_1+\nu_1)(\lambda_1'+\nu_1')} \tag{18.88}$$

The coset matrix is thus given by

$$S_{\{\hat{\lambda}, \hat{\mu}; \hat{\nu}\}, \{\hat{\lambda}', \hat{\mu}'; \hat{\nu}'\}} = \sqrt{\frac{2}{(k+2)(k+3)}}(-1)^{(\lambda_1+\nu_1)(\lambda_1'+\nu_1')}$$
$$\times \sin\left(\frac{\pi(\lambda_1 + 1)(\lambda_1' + 1)}{k+2}\right) \sin\left(\frac{\pi(\nu_1 + 1)(\nu_1' + 1)}{k+3}\right) \tag{18.89}$$

With

$$r = \lambda_1 + 1 \qquad r' = \lambda_1' + 1$$
$$s = \nu_1 + 1 \qquad s' = \nu_1' + 1 \qquad (18.90)$$

and $p = k + 2$, Eq. (18.89) reduces to

$$S_{(rs),(r's')} = \sqrt{\frac{2}{p(p+1)}}(-1)^{(r+s)(r'+s')} \sin\left(\frac{\pi rr'}{p}\right) \sin\left(\frac{\pi ss'}{p+1}\right) \qquad (18.91)$$

with the ranges of r and s given by Eq. (18.82). This is to be compared with the expression derived previously, namely Eq. (10.134), equivalently expressed as

$$S_{(rs),(r's')} = 2\sqrt{\frac{2}{p(p+1)}}(-1)^{(r+s)(r'+s')} \sin\left(\frac{\pi rr'}{p}\right) \sin\left(\frac{\pi ss'}{p+1}\right) \qquad (18.92)$$

which has been obtained for a fundamental range, defined by Eq. (18.82), with the additional constraint $ps < (p+1)r$. The difference between Eq. (18.91) and Eq. (18.92), a factor of 2, is to be accounted by this different range for values of r and s. Implementing the constraint $ps < (p+1)r$, we see that Eq. (18.91) requires a renormalization of the S matrix by precisely a factor of 2.

18.3.3. Fusion Rules

The substitution of Eq. (18.85) into the Verlinde formula leads directly to the following formula for the coset fusion coefficients:

$$\mathcal{N}_{\{\hat\lambda, \hat\mu; \hat\nu\},\{\hat\lambda', \hat\mu'; \hat\nu'\}}^{\{\hat\lambda'', \hat\mu''; \hat\nu''\}} = \mathcal{N}_{\hat\lambda\hat\lambda'}^{(k)\,\hat\lambda''} \mathcal{N}_{\hat\mu\hat\mu'}^{(1)\,\hat\mu''} \mathcal{N}_{\hat\nu\hat\nu'}^{(k+1)\,\hat\nu''} \qquad (18.93)$$

We now argue that the factor $\mathcal{N}_{\hat\mu\hat\mu'}^{(1)\,\hat\mu''}$ is irrelevant. To fixed values of $\hat\lambda, \hat\lambda', \hat\nu, \hat\nu'$, there correspond unique $\hat\mu$ and $\hat\mu'$. On the other hand, $\mathcal{N}_{\hat\lambda\hat\lambda'}^{(k)\,\hat\lambda''}$ and $\mathcal{N}_{\hat\nu\hat\nu'}^{(k+1)\,\hat\nu''}$ are nonzero only if

$$\lambda_1 + \lambda_1' + \lambda_1'' = 0 \bmod 2$$
$$\nu_1 + \nu_1' + \nu_1'' = 0 \bmod 2 \qquad (18.94)$$

(cf. Eq. (16.48)), which, in connection with the double prime version of Eq. (18.69), forces

$$\mu_1'' + \lambda_1 + \nu_1 + \lambda_1' + \nu_1' = \mu_1'' + \mu_1 + \mu_1' = 0 \bmod 2 \qquad (18.95)$$

Moreover, since there is only one term in the decomposition of the $\widehat{su}(2)_1$ fusion $\hat\mu \times \hat\mu'$, it follows that

$$\mathcal{N}_{\hat\mu\hat\mu'}^{(1)\,\hat\mu''} = 1 \qquad (18.96)$$

Translating the formula (16.48) for the $\widehat{su}(2)_k$ fusion coefficients in terms of the variables r, s, and $p = k + 2$, gives

$$\mathcal{N}_{(rs)(r's')}^{(r''s'')} = 1 \quad \text{if} \quad r'' \in \delta_p^{(r,r')} \quad \text{and} \quad s'' \in \delta_{p+1}^{(s,s')}$$
$$= 0 \quad \text{otherwise} \tag{18.97}$$

with

$$\delta_p^{(r,r')} = \{r'' | r + r' + r'' = 1 \text{ mod } 2,$$
$$|r - r'| + 1 \le r'' \le \min(r + r' - 1, 2p - 1 - r - r')\} \tag{18.98}$$

These are exactly the fusion rules for the unitary minimal models given by Eq. (8.131) with $p' \to p$, $p \to p + 1$, or Eq. (7.114) with $m = p$.

18.3.4. Modular Invariants

Modular invariant partition functions for the minimal models can be constructed from $\widehat{su}(2)$ invariant mass matrices as

$$\mathcal{M} = \mathcal{M}^{(1)} \mathcal{M}^{(k)} \mathcal{M}^{(k+1)} \tag{18.99}$$

taking into account the branching conditions and the field identifications. All $\widehat{su}(2)$ invariant mass matrices have been classified; the complete list is presented in Sect. 17.7. Since at level 1, as for all odd values of the level, there is only one invariant, $\mathcal{M}^{(1)} = I$. Ignoring the level 1 factor also allows us to ignore the selection rule (18.69). Field identifications are then simply accounted for by dividing the result by 2. The coset mass matrix reduces to

$$\mathcal{M} = \frac{1}{2} \mathcal{M}^{(k)} \mathcal{M}^{(k+1)} \tag{18.100}$$

Since one of k or $k+1$ is necessarily odd, an A-type (i.e., diagonal) invariant always appears in the product (18.100). For the other mass matrix, one can take either an A- or a D-type invariant (obtained by the action of the outer automorphism), or even an exceptional one if the level is 10, 16, or 28. Listing all possible pairs, we recover the invariant partition functions listed in Eqs. (10.3) and (10.4), pertaining to the case $|p - p'| = 1$ (and this justifies a posteriori their labeling in terms of two Lie algebras).

§18.4. Other Coset Representations of Minimal Models

To further illustrate the coset construction, we present other realizations of two simple minimal models (a few more examples are given in the exercises). For the character decomposition of these examples, we content ourself with a check of the first few terms in the expansion in powers of q.

18.4.1. The $\widehat{E}_8$ Formulation of the Ising Model

We consider the diagonal coset

$$\frac{(\widehat{E}_8)_1 \oplus (\widehat{E}_8)_1}{(\widehat{E}_8)_2} \tag{18.101}$$

Since dim $E_8 = 248$ and $g = 30$, its central charge is found from Eq. (18.16) to be $c = \frac{1}{2}$. It is thus bound to be another description of the Ising model. The only integrable representation at level 1 is $\hat{\omega}_0$; at level 2, these are $2\hat{\omega}_0$, $\hat{\omega}_1$, and $\hat{\omega}_7$. Their conformal dimensions are easily evaluated by means of the quadratic form matrix given in App. 13.A. For instance,

$$(\omega_1, \omega_1 + 2\rho) = 60 \qquad (\omega_7, \omega_7 + 2\rho) = 96 \tag{18.102}$$

so that

$$h_{\hat{\omega}_1} = \frac{15}{16} \qquad h_{\hat{\omega}_7} = \frac{3}{2} \tag{18.103}$$

(the dimension of the vacuum representation vanishes). Because the E_8 weight and root lattices are the same, no restriction comes from Eq. (18.26). Hence, there are three coset triplets, whose fractional dimensions are

$$h_{\{\hat{\omega}_0, \hat{\omega}_0; 2\hat{\omega}_0\}} = 0 \quad \text{mod } 1$$

$$h_{\{\hat{\omega}_0, \hat{\omega}_0; \hat{\omega}_1\}} = \frac{1}{16} \quad \text{mod } 1 \tag{18.104}$$

$$h_{\{\hat{\omega}_0, \hat{\omega}_0; \hat{\omega}_7\}} = \frac{1}{2} \quad \text{mod } 1$$

The correspondence with the Ising fields is thus:

$$\chi_{\{\hat{\omega}_0, \hat{\omega}_0; 2\hat{\omega}_0\}} = \phi_{(1,1)}$$

$$\chi_{\{\hat{\omega}_0, \hat{\omega}_0; \hat{\omega}_1\}} = \phi_{(1,2)} \tag{18.105}$$

$$\chi_{\{\hat{\omega}_0, \hat{\omega}_0; \hat{\omega}_7\}} = \phi_{(2,1)}$$

In other words, we get the following character decomposition

$$\chi_{\hat{\omega}_0} \chi_{\hat{\omega}_0} = \chi^{\text{Vir}}_{(1,1)} \chi_{2\hat{\omega}_0} + \chi^{\text{Vir}}_{(1,2)} \chi_{\hat{\omega}_1} + \chi^{\text{Vir}}_{(2,1)} \chi_{\hat{\omega}_7} \tag{18.106}$$

Because the fusion rules of $(\widehat{E}_8)_1$ are trivial (there is only $\hat{\omega}_0 \times \hat{\omega}_0 = \hat{\omega}_0$), the coset fusion rules are simply those of $(\widehat{E}_8)_2$. This explains why the $(\widehat{E}_8)_2$ WZW model has the same fusion rules as the Ising model (cf. Ex. 16.7).

18.4.2. The $\widehat{su}(3)$ Formulation of the Three-State Potts Model

We consider now the diagonal coset

$$\frac{\widehat{su}(3)_1 \oplus \widehat{su}(3)_1}{\widehat{su}(3)_2} \tag{18.107}$$

whose central charge is $c = \frac{4}{5}$. Here the central charge does not completely specify the model: at $c = \frac{4}{5}$, there are two minimal models, the diagonal theory and the Potts model.

We first identify the different coset fields. For $su(3)$, the condition $\lambda \in Q$ translates into the requirement

$$2\lambda_1 + \lambda_2 = 0 \bmod 3 \qquad (18.108)$$

For instance, the two triplets

$$\{[1,0,0], [0,1,0]; [0,0,2]\} \qquad \text{and} \qquad \{[0,1,0], [0,0,1]; [2,0,0]\} \quad (18.109)$$

are allowed. However, they are not distinct coset fields because they are related by the outer automorphism a (cf. Eq. (18.30)); they must be identified:

$$\{[1,0,0], [0,1,0]; [0,0,2]\} \sim \{[0,1,0], [0,0,1]; [2,0,0]\} \qquad (18.110)$$

The set of inequivalent fields is presented in Table 18.1.[6] The comparison of the coset-field fractional dimensions with those of the $\mathcal{M}(6,5)$ Kac table leads to the field correspondence of the last column. This correspondence is not unique, since the Kac table contains dimensions differing by an integer ($h_{(4,1)} - h_{(1,1)} = 3$ and $h_{(3,1)} - h_{(2,1)} = 1$). Notice also the double occurrence of the fields $\phi_{(3,3)}$ and $\phi_{(4,3)}$. In the coset model, the two copies of each field are not identical; they are related by charge conjugation:

$$\{[1,0,0], [0,1,0]; [1,1,0]\} = \{[1,0,0], [0,0,1]; [1,0,1]\}^* \qquad (18.111)$$

The field content appearing in Table 18.1 is exactly the one pertaining to the three-state Potts model. In addition, the multiplicities are in agreement with those required to build the nondiagonal modular invariant (A_4, D_4) of Table 10.3:

$$Z_{A_4,D_4} = |\chi_{(1,1)} + \chi_{(4,1)}|^2 + |\chi_{(2,1)} + \chi_{(3,1)}|^2 + 2|\chi_{(3,3)}|^2 + 2|\chi_{(4,3)}|^2 \quad (18.112)$$

The appropriate character decompositions are

$$\chi_{[1,0,0]}\,\chi_{[1,0,0]} = [\chi_{(1,1)}^{\text{Vir}} + \chi_{(4,1)}^{\text{Vir}}]\,\chi_{[2,0,0]} + [\chi_{(3,3)}^{\text{Vir}} + \chi_{(4,3)}^{\text{Vir}}]\,\chi_{[0,1,1]}$$
$$\chi_{[1,0,0]}\,\chi_{[0,1,0]} = \chi_{(4,3)}^{\text{Vir}}\,\chi_{[0,2,0]} + \chi_{(3,3)}^{\text{Vir}}\,\chi_{[1,1,0]} \qquad (18.113)$$

and the conjugate version of the last one. The first decomposition can be checked as follows. The $su(3)$ content at the first few grades of the modules $L_{[1,0,0]}$, $L_{[2,0,0]}$, and

---

[6] Observe that for diagonal cosets of the form $\hat{g}_k \oplus \hat{g}_k/\hat{g}_{2k}$, the interchange of the two $\hat{g}_k$ factors does not lead to field identifications. For instance, in the present case, we do not identify the two fields:

$$\{[0,1,0], [0,0,1]; [2,0,0]\} \qquad \text{and} \qquad \{[0,0,1], [0,1,0]; [2,0,0]\}$$

which are obtained from each other by the interchange of the two $\widehat{su}(3)_1$ factors. Actually, these fields are charge conjugates of each other. On the other hand, for the $\widehat{su}(2)$ coset description of the Ising model, the two fields

$$\{[0,1], [1,0]; [1,1]\} \qquad \text{and} \qquad \{[1,0], [0,1]; [1,1]\}$$

are equivalent because they are related by the action of a.

Table 18.1. Coset fields for the $\widehat{su}(3)$ diagonal coset at levels $(1,1;2)$, their fractional conformal dimensions, and their corresponding fields in the minimal model with $c = \frac{4}{5}$.

Cosetfields	h mod 1	Minimalmodelfields
$\{[1,0,0],[1,0,0];[2,0,0]\}$	0	$\phi_{(1,1)}, \phi_{(4,1)}$
$\{[1,0,0],[0,1,0];[1,1,0]\}$	$\frac{1}{15}$	$\phi_{(3,3)}$
$\{[1,0,0],[0,0,1];[1,0,1]\}$	$\frac{1}{15}$	$\phi'_{(3,3)}$
$\{[1,0,0],[1,0,0];[0,1,1]\}$	$\frac{2}{5}$	$\phi_{(2,1)}, \phi_{(3,1)}$
$\{[1,0,0],[0,1,0];[0,0,2]\}$	$\frac{2}{3}$	$\phi_{(4,3)}$
$\{[1,0,0],[0,0,1];[0,2,0]\}$	$\frac{2}{3}$	$\phi'_{(4,3)}$

$L_{[0,1,1]}$ in Tables 14.3, 18.2, and 18.3, gives the leading terms in the development of the various $\widehat{su}(3)$ characters. From Eq. (8.17) (see also Table 8.1), we have

$$
\begin{aligned}
\chi^{\text{Vir}}_{(1,1)}(q) &= q^{-1/30}(1 + q^2 + q^3 + 2q^4 + \cdots) \\
\chi^{\text{Vir}}_{(4,1)}(q) &= q^{3-1/30}(1 + q + 2q^2 + 3q^3 + 4q^4 + \cdots) \\
\chi^{\text{Vir}}_{(2,1)}(q) &= q^{2/5-1/30}(1 + q + q^2 + 2q^3 + 3q^4 + \cdots) \\
\chi^{\text{Vir}}_{(3,1)}(q) &= q^{7/5-1/30}(1 + q + 2q^2 + 2q^3 + 4q^4 + \cdots)
\end{aligned}
\tag{18.114}
$$

from which the first few terms of the decomposition can be verified. The other ones can be worked out in the same way.

Table 18.2. The $\widehat{su}(3)_2$ module $L_{[2,0,0]}$, for which $h - c/24 = -\frac{2}{15}$.

Grade	$L_{[2,0,0]} : su(3)$ content	Number of states
0	$(0,0)$	1
1	$(1,1)$	8
2	$(2,2) \oplus 2(1,1) \oplus (0,0)$	44
3	$2(2,2) \oplus 2(3,0) \oplus 2(0,3) \oplus 4(1,1) \oplus 2(0,0)$	128

Table 18.3. The $\widehat{su}(3)_2$ module $L_{[0,1,1]}$ for which $h - c/24 = \frac{3}{5} - \frac{2}{15}$.

Grade	$L_{[1,1,0]} : su(3)$ content	Number of states
0	$(1,1)$	8
1	$(3,0) \oplus (0,3) \oplus 2(1,1) \oplus (0,0)$	37
2	$2(2,2) \oplus 2(3,0) \oplus 2(0,3) \oplus 5(1,1) \oplus 2(0,0)$	136
3	$(4,1) \oplus (1,4) \oplus 5(2,2) \oplus 5(3,0)$	
	$\oplus 5(0,3) \oplus 12(1,1) \oplus 3(0,0)$	440

§18.5. The Coset $\widehat{su}(2)_k/\widehat{u}(1)$ and Parafermions

18.5.1. Character Decomposition and String Functions

A simple, yet interesting coset model is

$$\frac{\widehat{su}(2)_k}{\widehat{u}(1)} \tag{18.115}$$

with central charge

$$c = \frac{3k}{k+2} - 1 = \frac{2(k-1)}{k+2} \tag{18.116}$$

For $k = 1, 2, 3$, and 4, it describes respectively the trivial $c = 0$ theory, the Ising model, the three-state Potts model, and a rational $c = 1$ theory.

The first step in the analysis of this coset is to understand how the $su(2)$ irreducible representations decompose into $u(1)$ factors. This is rather simple: each weight in a representation of $su(2)$ is by itself a $u(1)$ representation, usually called a charge, that is,

$$(\lambda_1) \mapsto (\lambda_1)_1 \oplus (\lambda_1 - 2)_1 \oplus \cdots (-\lambda_1)_1 \tag{18.117}$$

where representations on the l.h.s. refer to $u(1)$ and the index 1 reminds us that they have dimension 1. In this section, we denote the finite Dynkin label λ_1 by ℓ (which is, in fact, the same as the partition entry ℓ_1). The above decomposition then reads

$$(\ell) \mapsto \bigoplus_{m=-\ell}^{\ell} (m)_1 \qquad (\ell + m = 0 \bmod 2) \tag{18.118}$$

The affine extension of this $u(1)$ algebra is $\widehat{u}(1)_k$. More precisely, it is an extended $\widehat{u}(1)$ theory corresponding to a free boson living on a circle of radius $\sqrt{2k}$ (or $\sqrt{2/k}$ by duality); the $u(1)$ charge is interpreted as the boson momentum. The concept of level is inherited from the covering $\widehat{su}(2)_k$ algebra. The distinct (extended) fields of $\widehat{u}(1)_k$ have charge (or finite Dynkin label) ranging over:

$$-k + 1 \leq m \leq k \tag{18.119}$$

818

18. Cosets

and dimension

$$h_m = \frac{m^2}{4k} \tag{18.120}$$

(This can be calculated from the usual dimension formula (15.104), with $\lambda = m\omega_1$, $\omega_1^2 = \frac{1}{2}$, $\rho = g = 0$, and level k.) These representations can be viewed as the integrable representations of $\widehat{u}(1)_k$, into which the $\widehat{su}(2)_k$ representations must be decomposed. That these are the only distinct representations is most directly seen from the characters, which have already been calculated in Sect. 14.4.4. To avoid notational confusion with the $\widehat{su}(2)_k$ characters $\chi_\ell^{(k)}(q)$, we will indicate the $\widehat{u}(1)_k$ characters by $K_m^{(k)}(q)$:[7]

$$K_m^{(k)}(q) = \frac{\Theta_m^{(k)}(q)}{\eta(q)} \tag{18.121}$$

where $\Theta_m^{(k)}(q)$ is the standard generalized theta function:

$$\Theta_m^{(k)}(q) = \sum_{n\in\mathbb{Z}} q^{k(n+m/2k)^2} \tag{18.122}$$

(cf. Eq. (14.176)). The invariance of the theta function under a shift of m by $2k$ implies that

$$K_m^{(k)}(q) = K_{m+2k}^{(k)}(q) \tag{18.123}$$

so that the range of m can be restricted as in Eq. (18.119). Since representations differing by a sign are charge conjugates of each other, the restricted characters also satisfy

$$K_{-m}^{(k)}(q) = K_m^{(k)}(q) \tag{18.124}$$

(but when the z dependence is reinserted, this last equality no longer holds).

The character decomposition appropriate to this coset is thus

$$\chi_\ell^{(k)}(q) = \sum_{m=-k+1}^{k} \chi_{\{\ell,m\}}(q) K_m^{(k)}(q) \tag{18.125}$$

where $\chi_{\{\ell,m\}}(q)$ stands for a coset character. The branching condition is $m + \ell = 0 \bmod 2$. As we will show, the characters for this coset turn out to be expressible directly in terms of the $\widehat{su}(2)_k$ string functions.

We first recall the relation (14.147) between (nonnormalized) characters and string functions, which we reproduce for convenience:

$$\mathrm{ch}_{\hat{\lambda}} = \sum_{\hat{\mu}\in\Omega_{\hat{\lambda}}^{\max}} \sigma_{\hat{\mu}}^{(\hat{\lambda})}(e^{-\delta}) e^{\hat{\mu}} \tag{18.126}$$

$\sigma_{\hat{\mu}}^{(\hat{\lambda})}$ is the string function of the weight $\hat{\mu}$ in the representation $\hat{\lambda}$, and $\Omega_{\hat{\lambda}}^{\max}$ is the set of the weights $\hat{\mu}$ in the representation $\hat{\lambda}$, such that $\hat{\mu} + \delta$ is not in the

---

[7] This notation was also used in App. 17.B, but there the upper index was chosen to be $N = 2k$.

representation. This set is infinite. But all weights in $\Omega_{\hat{\lambda}}^{\max}$ can be Weyl reflected—or more precisely translated—into a weight $\hat{\nu}$ such that $-k+1 \leq \nu_1 \leq k$. In other words, any $\hat{\mu} \in \Omega_{\hat{\lambda}}^{\max}$ can be written as $t_{\alpha^\vee} \hat{\nu}$ for an appropriate $\hat{\nu}$—corresponding to an integrable $\widehat{u}(1)_k$ charge—and an appropriate affine coroot. Since the string function satisfies (cf. Eq. (14.145))

$$\sigma_{\hat{\mu}}^{(\hat{\lambda})} = \sigma_{t_{\alpha^\vee} \hat{\nu}}^{(\hat{\lambda})} = \sigma_{\hat{\nu}}^{(\hat{\lambda})} \tag{18.127}$$

we can write

$$\sum_{\hat{\mu} \in \Omega_{\hat{\lambda}}^{\max}} \sigma_{\hat{\mu}}^{(\hat{\lambda})}(e^{-\delta}) \, e^{\hat{\mu}} = \sum_{-k+1 \leq \nu_1 \leq k} \sigma_{\hat{\nu}}^{(\hat{\lambda})}(e^{-\delta}) \sum_{\alpha^\vee \in Q^\vee} e^{t_{\alpha^\vee} \hat{\nu}} \tag{18.128}$$

Up to a factor, this sum is a theta function (cf. Eq. (14.154)). Its specialized version is

$$\sum_{\alpha^\vee \in Q^\vee} e^{-2\pi i (t_{\alpha^\vee} \hat{\nu}, (0;\, \tau;\, 0))} = q^{-\nu_1^2/4k} \Theta_{\nu_1}^{(k)}(q) \tag{18.129}$$

To avoid keeping track of the subscript 1, we redefine

$$\lambda_1 \equiv \ell \qquad \nu_1 \equiv m \tag{18.130}$$

and set

$$\sigma_{\hat{\nu}}^{(\hat{\lambda})} \equiv \sigma_m^\ell \tag{18.131}$$

We have thus obtained:

$$\mathrm{ch}_{\hat{\lambda}}(q) \equiv \mathrm{ch}_\ell^{(k)}(q) = \sum_{m=-k+1}^{k} \sigma_m^\ell(q) \, q^{-m^2/4k} \, \Theta_m^{(k)}(q) \tag{18.132}$$

The normalized character then reads

$$\chi_\ell^{(k)}(q) = \sum_{m=-k+1}^{k} q^{h_\ell - h_m - 3k/24(k+2)} \sigma_m^\ell(q) \, \Theta_m^{(k)}(q) \tag{18.133}$$

(with $h_\ell = h_\lambda$ and $h_m = m^2/4k$). Expressed in terms of the normalized string function, defined by Eqs. (14.223) and (14.224), that is,

$$c_m^\ell(q) = q^{h_\ell - h_m - 3k/(k+2)} \sigma_m^\ell(q) \tag{18.134}$$

Eq. (18.133) takes the form

$$\chi_\ell^{(k)}(q) = \sum_{m=-k+1}^{k} c_m^\ell(q) \, \Theta_m^{(k)}(q)$$

$$= \sum_{m=-k+1}^{k} \eta(q) \, c_m^\ell(q) \, K_m^{(k)}(q) \tag{18.135}$$

This gives exactly the character decomposition of $\widehat{su}(2)_k$ representations in terms of $\widehat{u}(1)_k$ ones. The coset characters are thus

$$\chi_{\{\ell;\, m\}}(q) = \eta(q) \, c_m^\ell(q) \tag{18.136}$$

The determination of the coset characters has been reduced to reorganizing the known $\widehat{su}(2)_k$ characters in sums of theta functions. The calculation of string functions is somewhat tedious and we will not evaluate them directly. Instead, we will take a field-theoretical point of view and use the correspondence between the coset $\widehat{su}(2)_k/\widehat{u}(1)$ and known minimal models to extract the string functions for $k \leq 3$.

Before turning to examples, we will settle the question of field identification. This amounts to finding the branching of the basic $\widehat{su}(2)$ automorphism a into an operation on the "integrable" charges. We recall that the action of a is to replace ℓ by $k - \ell$. From the finite algebra branching rules, it follows that a must branch to an operator $\tilde{a}$ that maps m to $k - m$. Under this transformation, the parity requirement $\ell - m = 0 \mod 2$ is preserved. This yields the field identification

$$\chi_{\{\ell;\, m\}}(q) = \chi_{\{k-\ell;\, k-m\}}(q) \tag{18.137}$$

Given that

$$\chi_{\{\ell;\, m\}}(q) = \chi_{\{\ell;\, -m\}}(q) = \chi_{\{\ell;\, m+2k\}}(q) \tag{18.138}$$

the character identity (18.137) takes the form

$$\chi_{\{\ell;\, m\}}(q) = \chi_{\{k-\ell;\, k+m\}}(q) = \chi_{\{k-\ell;\, m-k\}}(q) \tag{18.139}$$

Whichever relation is used is dictated by the particular value of m at hand: either $m + k$ or $m - k$ will be in the fundamental range (18.119). The net result is that half of all allowed coset pairs make distinct coset fields.

18.5.2. A Few Special Cases

EXAMPLE 1: $k = 1$

For $k = 1$, the resulting coset is a trivial unitary $c = 0$ theory, whose whole field content is the identity with no descendants. The two coset fields are

$$\{[1, 0]; (0)\} \equiv \{0; 0\} \qquad \text{and} \qquad \{[0, 1]; (1)\} \equiv \{1; 1\} \tag{18.140}$$

where the second expression uses the $\{\ell; m\}$ notation. Both have zero fractional conformal dimension. According to Eq. (18.139), these must be identified. There is therefore a single coset character, which must be equal to unity:

$$\chi_{\{0;\, 0\}}(q) = 1 \tag{18.141}$$

Equation (18.136) implies that

$$\eta(q)\, c_0^0(q) = 1 \quad \Longrightarrow \quad c_0^0(q) = \eta(q)^{-1} \tag{18.142}$$

This agrees with the result found in Sect. 15.6.2, namely that

$$\sigma_{[1,0]}^{([1,0])}(q) = \varphi(q)^{-1} \tag{18.143}$$

since in the present case the relative modular anomaly is $-\frac{1}{24}$.

This coset description based on $\widehat{su}(2)_1$ allows us to recover in a very simple way the expressions for the characters of the two integrable representations already derived from the vertex construction. Since there is only one term in each character decomposition, we have

$$
\begin{aligned}
\chi_0^{(1)}(q) &= K_0^{(1)}(q) \\
\chi_1^{(1)}(q) &= K_1^{(1)}(q)
\end{aligned}
\tag{18.144}
$$

$K_i^{(1)}$ takes a simple form in terms of Jacobi theta functions since

$$
\begin{aligned}
\Theta_0^{(1)}(q) &= \sum_{n\in\mathbb{Z}} q^{n^2} = \theta_3(q^2) \\
\Theta_1^{(1)}(q) &= \sum_{n\in\mathbb{Z}} q^{(n+\frac{1}{2})^2} = \theta_2(q^2)
\end{aligned}
\tag{18.145}
$$

This leads to

$$
\begin{aligned}
\chi_0^{(1)}(q) &= \frac{\theta_3(q^2)}{\eta(q)} = \sqrt{\frac{\theta_3^2(q) + \theta_4^2(q)}{2\eta^2(q)}} \\
\chi_1^{(1)}(q) &= \frac{\theta_2(q^2)}{\eta(q)} = \sqrt{\frac{\theta_3^2(q) - \theta_4^2(q)}{2\eta^2(q)}}
\end{aligned}
\tag{18.146}
$$

(cf. Eq. (10.233)).

EXAMPLE 2: $k = 2$

For $k = 2$, the distinct coset fields and their fractional dimension are

$$
\begin{aligned}
&\{[2,0]; (0)\} \equiv \{0; 0\} \quad && h = 0 \bmod 1 \\
&\{[2,0]; (2)\} \equiv \{0; 2\} \quad && h = \frac{1}{2} \bmod 1 \\
&\{[1,1]; (1)\} \equiv \{1; 1\} \quad && h = \frac{1}{16} \bmod 1
\end{aligned}
\tag{18.147}
$$

and they can be identified unambiguously with the Ising primary fields $\mathbb{I} = \phi_{(1,1)}$, $\epsilon = \phi_{(2,1)}$, and $\sigma = \phi_{(1,2)}$, respectively. These identifications lead to the character identities:

$$
\begin{aligned}
\chi_{(1,1)}(q) &= \frac{1}{2}\left(\sqrt{\frac{\theta_3(q)}{\eta(q)}} + \sqrt{\frac{\theta_4(q)}{\eta(q)}}\right) = \eta(q)c_0^0(q) \\
\chi_{(2,1)}(q) &= \frac{1}{2}\left(\sqrt{\frac{\theta_3(q)}{\eta(q)}} - \sqrt{\frac{\theta_4(q)}{\eta(q)}}\right) = \eta(q)c_2^0(q) \\
\chi_{(1,2)}(q) &= \sqrt{\frac{\theta_2(q)}{\eta(q)}} = \eta(q)c_1^1(q)
\end{aligned}
\tag{18.148}
$$

(Equivalent expressions for the $\widehat{su}(2)_2$ string functions are presented in Ex. 14.11.)

Again, this construction provides nice expressions for the $\widehat{su}(2)_2$ characters. The character decompositions read

$$\chi_{[2,0]} \equiv \chi_0^{(2)} = \eta \left(c_0^0 K_0^{(2)} + c_2^0 K_2^{(2)} \right)$$

$$\chi_{[0,2]} \equiv \chi_2^{(2)} = \eta \left(c_0^2 K_0^{(2)} + c_2^2 K_2^{(2)} \right) \qquad (18.149)$$

$$\chi_{[1,1]} \equiv \chi_1^{(2)} = \eta \left(c_1^1 K_1^{(2)} + c_{-1}^1 K_{-1}^{(2)} \right)$$

where these functions are understood to be evaluated at q. Using the identities

$$c_m^\ell = c_{-m}^\ell = c_{m-k}^\ell = c_{m+k}^\ell \qquad (18.150)$$

and the relation between K and the generalized theta functions, we write

$$\chi_0^{(2)} = c_0^0 \, \Theta_0^{(2)} + c_2^0 \, \Theta_2^{(2)}$$

$$\chi_2^{(2)} = c_0^0 \, \Theta_2^{(2)} + c_2^0 \, \Theta_0^{(2)} \qquad (18.151)$$

$$\chi_1^{(2)} = 2 c_1^1 \, \Theta_1^{(2)}$$

Again, the generalized theta functions are expressible in terms of the Jacobi theta functions as follows:

$$\Theta_0^{(2)} + \Theta_2^{(2)} = \sum_{n \in \mathbb{Z}} \left(q^{(2n)^2/2} + q^{(2n+1)^2/2} \right) = \theta_3(q)$$

$$\Theta_0^{(2)} - \Theta_2^{(2)} = \sum_{n \in \mathbb{Z}} \left(q^{(2n)^2/2} - q^{(2n+1)^2/2} \right) = \theta_4(q) \qquad (18.152)$$

$$\Theta_1^{(2)} = \sum_{n \in \mathbb{Z}} q^{2(n+1/4)^2} = \frac{1}{2} \sum_{n \in \mathbb{Z}} \left(q^{(2n+\frac{1}{2})^2/2} + q^{(2n+1+\frac{1}{2})^2/2} \right) = \frac{1}{2} \theta_2(q)$$

With the string functions given by Eq. (18.148), we find

$$\chi_0^{(2)} = \frac{1}{2} \left(\left[\frac{\theta_3}{\eta} \right]^{\frac{3}{2}} + \left[\frac{\theta_4}{\eta} \right]^{\frac{3}{2}} \right)$$

$$\chi_2^{(2)} = \frac{1}{2} \left(\left[\frac{\theta_3}{\eta} \right]^{\frac{3}{2}} - \left[\frac{\theta_4}{\eta} \right]^{\frac{3}{2}} \right) \qquad (18.153)$$

$$\chi_1^{(2)} = \left[\frac{\theta_2}{2\eta} \right]^{\frac{3}{2}}$$

The corresponding partition function takes the form

$$Z = \frac{1}{2|\eta|^3} \left(|\theta_2|^3 + |\theta_2|^3 + |\theta_2|^3 \right)$$

$$= 4 \sum_{\nu=2,3,4} Z_\nu^3 \qquad (18.154)$$

where

$$Z_\nu = \frac{1}{2} \left| \frac{\theta_\nu}{\eta} \right| \qquad (18.155)$$

(cf. Chap. 12). Z_ν gives the partition function of a free fermion: $\nu = 2, 3$, and 4 corresponds respectively to the (NS,R), (R,NS), and (NS,NS) boundary conditions (cf. Eq. (10.51)). The $\widehat{su}(2)_2$ theory is known to be realized in terms of three fermions (cf. Ex. 15.15). The above partition function provides another way of seeing this equivalence. It further implies that the three fermions must have the same periodicity or antiperiodicity conditions on the torus.

EXAMPLE 3: $k = 3$

For $k = 3$, the central charge is $\frac{4}{5}$, but again this does not uniquely fix the field content of the theory. The appropriate minimal model can be identified from the values of the fractional dimensions of the distinct coset fields:

$$
\begin{aligned}
&\{0; 0\} & h &= 0 \text{ mod } 1 \\
&\{0; 2\} & h &= \frac{2}{3} \text{ mod } 1 \\
&\{0; -2\} & h &= \frac{2}{3} \text{ mod } 1 \\
&\{1; 3\} & h &= \frac{2}{5} \text{ mod } 1 \\
&\{1; 1\} & h &= \frac{1}{15} \text{ mod } 1 \\
&\{1; -1\} & h &= \frac{1}{15} \text{ mod } 1
\end{aligned}
\tag{18.156}
$$

This reproduces exactly the (fractional) spectrum of the three-state Potts model, with the required multiplicities. In the coset description, fields with multiplicity larger than 1 are seen to be genuine charge conjugates of each other—here the conjugation is inherited from the $u(1)$ sector. The identification of the Virasoro and coset characters leads to the following expressions for the level-3 string functions:

$$
\begin{aligned}
\chi_{(1,1)} + \chi_{(4,1)} &= \eta c_0^0 \\
\chi_{(2,1)} + \chi_{(3,1)} &= \eta c_3^1 \\
\chi_{(4,3)} &= \eta c_2^0 \\
\chi_{(3,3)} &= \eta c_1^1
\end{aligned}
\tag{18.157}
$$

18.5.3. Parafermions

The coset (18.116) is usually called a *parafermionic model*. Without embarking on the analysis of conformal theories whose symmetry is enhanced by the conservation of parafermionic currents, we simply give the rationale for the epithet *parafermion*. The $\widehat{su}(2)_k$ WZW model can be viewed as composed of two building blocks: an $\widehat{su}(2)_k/\widehat{u}(1)$ piece, associated with parafermions, and a $\widehat{u}(1)$ factor, associated with a free boson. This must reflect itself in the composition of the $\widehat{su}(2)_k$ generators, which should then be expressible as

$$J^+(z) = \sqrt{k}\, \psi_{\mathrm{par}}(z) e^{i\sqrt{2/k}\,\varphi(z)}$$

$$J^-(z) = \sqrt{k}\, \psi^\dagger_{\mathrm{par}}(z) e^{-i\sqrt{2/k}\,\varphi(z)} \qquad (18.158)$$

$$J^0(z) = i\sqrt{2k}\, \partial_z \varphi(z)$$

where ψ_{par} and $\psi^\dagger_{\mathrm{par}}$ stand for conserved parafermionic fields, and φ is a standard free boson:

$$\varphi(z)\varphi(w) \sim -\ln(z-w) \qquad (18.159)$$

Since $e^{\pm i\sqrt{2/k}\,\varphi(z)}$ has dimension $1/k$, the parafermions must have dimension $(k-1)/k$. For $k=1$, parafermions are absent: the $\widehat{su}(2)_1$ WZW model is realized in terms of a single free boson (cf. Sect. 15.6); for $k=2$ they are genuine fermions. With $k>2$, parafermions do not have integer or half-integer fractional dimensions; they satisfy a $\mathbb{Z}_k$ parastatistics, hence their name.

To complete the study of these cosets, we mention the relation between the coset—or parafermionic—fields $f^{\ell,\bar\ell}_{m,\bar m}(z,\bar z)$ and the $\widehat{su}(2)_k$ WZW fields $\phi^{\ell,\bar\ell}_{m,\bar m}(z,\bar z)$:

$$\phi^{\ell,\bar\ell}_{m,\bar m}(z,\bar z) = f^{\ell,\bar\ell}_{m,\bar m}(z,\bar z) e^{im\varphi(z)/\sqrt{2k} + i\bar m\bar\varphi(\bar z)/\sqrt{2k}} \qquad (18.160)$$

Correlations of parafermionic fields can therefore be computed from those of the $\widehat{su}(2)_k$ theory.

For reference, we write the holomorphic conformal dimension of the parafermionic fields f^ℓ_m:

$$h^\ell_m = \frac{\ell(\ell+2)}{4(k+2)} - \frac{m^2}{4k} \qquad (18.161)$$

The fields that are primary with respect to the parafermionic algebra are the f^ℓ_ℓ's.

18.5.4. Parafermionic Formulation of the General $\widehat{su}(2)$ Diagonal Cosets

As an application of our study of the coset $\widehat{su}(2)_k/\widehat{u}(1)$, we now show that it can be viewed as the building block for the general $\widehat{su}(2)$ diagonal coset

$$\frac{\widehat{su}(2)_k \oplus \widehat{su}(2)_l}{\widehat{su}(2)_{k+l}} \qquad (18.162)$$

More precisely, the characters of the diagonal coset will be shown to have a natural decomposition in terms of the $\widehat{su}(2)$ string functions. We write the character decomposition of Eq. (18.162) in the form

$$\chi^{(k)}_\ell \chi^{(l)}_{r-1} = \sum_{s=1}^{l+k+1} \chi_{\{\ell,\, r;\, s\}} \chi^{(l+k)}_{s-1} \qquad (18.163)$$

For two characters, we choose to write the Dynkin label in a form that will facilitate the comparison with the Virasoro characters for $l=1$ and the expressions obtained in the Coulomb-gas representation for all l (Ex. 18.15). Note, in particular, that $1 \le r \le l+1$. The starting point in our quest for an explicit formula for $\chi_{\{l,r;s\}}$ consists in rewriting $\chi_\ell^{(k)}$ as a sum over string functions, as in Eq. (18.135):

$$\chi_\ell^{(k)}(x,q) = \sum_{m=-k+1}^{k} c_m^\ell(q)\, \Theta_m^{(k)}(x,q) \tag{18.164}$$

with $x = e^{-2\pi i z}$. This key step makes the following character derivation as simple as in the $l = 1$ case presented in Sect. 18.3.1. With

$$\chi_{r-1}^{(l)} = \frac{\Theta_r^{(l+2)} - \Theta_{-r}^{(l+2)}}{\Theta_1^{(2)} - \Theta_{-1}^{(2)}} = \frac{D_r^{(l+2)}}{D_1^{(2)}} \tag{18.165}$$

Eq. (18.163) becomes

$$\sum_{m=-k+1}^{k} c_m^\ell(q)\, \Theta_m^{(k)}(x,q)\, D_r^{(l+2)}(x,q) = \sum_{s=1}^{l+k+1} \chi_{\{l,r;s\}}\, D_s^{(l+k+2)}(x,q) \tag{18.166}$$

We concentrate first on the product:

$$\Theta_m^{(k)}(x,q)\, \Theta_r^{(l+2)}(x,q) = \sum_{n,n' \in \mathbb{Z}} x^{k(n+m/2k)+(l+2)(n'+r/(2l+4))}$$
$$\times q^{k(n+m/2k)^2+(l+2)(n'+r/(2l+4))^2} \tag{18.167}$$

To reexpress the exponent of x in the form $(l+k+2)n + s'/2$ for some $s' \in \mathbb{Z}$, we must set

$$s' = m + r + 2(l+2)(n'-n) \tag{18.168}$$

The integer s' is then decomposed in two parts as

$$s' = s + 2t(k+l+2) \tag{18.169}$$

with

$$1 \le s \le 2(k+l+2)+1 \qquad \text{and} \quad t \in \mathbb{Z} \tag{18.170}$$

This yields

$$\Theta_m^{(k)}(x,q)\, \Theta_r^{(l+2)}(x,q) = \sum_{s=1}^{2(k+l+1)+1} \sum_{t \in \mathbb{Z}} \Theta_s^{(l+k+2)}(x,q)\, q^{\beta_{r,s}(t)} \tag{18.171}$$

with

$$\beta_{r,s} = \frac{[(k+l+2)r - (l+2)s + 2(k+l+2)(l+2)t]^2}{4k(l+2)(l+k+2)} \tag{18.172}$$

The same result holds with the signs of m, r and s reversed; using $c_m^\ell = c_{-m}^\ell$, we obtain

$$\sum_{m=-k+1}^{k} c_m^\ell(q)\, \Theta_m^{(k)}(x,q)\, D_r^{(l+2)} = \sum_{m=-k+1}^{k} \sum_{s=1}^{2(l+k+1)+1} \sum_{t\in\mathbb{Z}} c_m^\ell(q)\, q^{\beta_{r,s}(t)}\, D_s^{(l+k+2)}$$

(18.173)

This is close to the desired result; we need only to restrict the range of s appropriately, using $D_{-s}^{(l+k+2)} = -D_s^{(l+k+2)}$. With

$$\chi_{\{\ell,\, r;\, s\}}(q) = \sum_{m=-k+1}^{k} c_m^\ell(q)\, F_m(q)$$

(18.174)

this yields

$$F_m(q) = \sum_{t\in\mathbb{Z}} \left(\delta_{m,m'} q^{\beta_{r,s}(t)} - \delta_{m,m''} q^{\beta_{r,-s}(t)} \right)$$

(18.175)

with

$$m' = |s - r - 2t(l+2)| \bmod 2k$$
$$m'' = |s + r - 2t(l+2)| \bmod 2k$$

(18.176)

The present derivation provides a rationale for the Coulomb-gas construction presented in Ex. 18.15: the coset characters are naturally decomposed into parafermionic characters, $\eta\, c_m^\ell$, and a piece that looks like a deformed minimal Virasoro character or, more properly formulated, a deformed Coulomb-gas contribution.

§18.6. Conformal Theories With Fractional $\widehat{su}(2)$ Spectrum-generating Algebra

To complete the coset description of the minimal models, we now have to consider nonunitary models. In order to reproduce

$$c = 1 - \frac{6(p - p')^2}{pp'}$$

(18.177)

with a coset of the form (18.51), we clearly need

$$k = \frac{3p' - 2p}{p - p'}$$

(18.178)

With $p - p' \neq 1$, the level is fractional!

At first sight, WZW models at fractional levels are not well-defined: the Wess-Zumino action is not single valued. In spite of this, we can set up an algebraic formulation of these WZW models starting from the current algebra. There are no immediate problems with such a formulation: the Sugawara energy-momentum tensor can be constructed exactly as before, primary fields are again associated with highest-weight states and their conformal dimension is still evaluated by means of the formula (15.104). As for the integrable case, not every highest-weight state qualifies as a state associated with a primary field. These are only those states that are the highest-weight states of the admissible representations introduced below. For a fixed value of k, there is a finite number of admissible representations. Their characterizing property is modular covariance: they all transform linearly into each other under modular transformations.

A priori, the algebraic formulation of a fractional-level WZW model is not to be viewed as a cure for the intrinsic "sickness" of the model defined in terms of an action. Although there are no problems in the mere algebraic formulation of the theory, inconsistencies could very well arise at some point. Indeed, the fusion rules of the theory calculated from the Verlinde formula are not positive![8]

Our position here with regard to these models is the following: our immediate purpose is not to make physical sense of fractional-level WZW models for themselves; they will be used as coset building blocks and it is only the final coset theory that really has to be well-defined.

Since our primary objective is to complete the coset description of the minimal models, we restrict ourself to the presentation of the $\widehat{su}(2)$ admissible representations, deferring consideration of the general case to App. 18.B.

18.6.1. Admissible Representations of $\widehat{su}(2)_k$

In this section we return to the theory of representations for affine Lie algebras, in order to characterize those $\widehat{su}(2)$ representations at fractional levels that are modular covariant.

Let the fractional level be of the form

$$k = \frac{t}{u} \tag{18.179}$$

where

$$u \in \mathbb{N}, \qquad t \in \mathbb{Z}/\{0\}, \qquad (t, u) = 1 \tag{18.180}$$

That is, u is a positive integer, t is a positive or negative (but nonzero) integer, and t and u are relatively prime. Notice that k can be negative. We define the *admissible representations* of $\widehat{su}(2)_k$ as highest-weight representations whose highest weight $\hat{\lambda}$ can be broken into two integrable weights $\hat{\lambda}^I$ and $\hat{\lambda}^F$ as[9]

$$\hat{\lambda} = \hat{\lambda}^I - (k+2)\hat{\lambda}^F \tag{18.181}$$

---

[8] Nevertheless, a deeper understanding could very well show the value of algebraic formulation of fractional-level WZW models. For instance, a naive application of the Verlinde formula might be inappropriate in this context, but the naive adaptation of the integer-level WZW action does not work.

[9] We warn the reader that such a simple characterization of the admissible representations is not typical, and is valid only for $\widehat{su}(2)_k$.

at respective levels k^I and k^F given by

$$k^I = u(k+2) - 2 \geq 0$$
$$k^F = u - 1 \geq 0 \tag{18.182}$$

(Since the level is additive upon weight addition, we verify that the level of $\hat{\lambda}$ is indeed k.) The superscript I or F refers to integer or fractional, but we stress that even though $\hat{\lambda}^F$ is responsible for the fractional part of $\hat{\lambda}$, it is itself an integrable weight, i.e., $\hat{\lambda}_i^F \in \mathbb{Z}_+$. The integrability of $\hat{\lambda}^I$ and $\hat{\lambda}^F$ forces the above inequalities on the levels. The condition $u - 1 \geq 0$ is always satisfied but the other condition provides a lower bound on the value of t:

$$t \geq 2 - 2u \tag{18.183}$$

Because t and u must be relatively prime, t can take the lower bound value only when u is odd. When $u = 1$, admissible representations reduce to integrable ones: $k^F = 0$, $\hat{\lambda}^F = 0$, $k^I = k$.

Because $\hat{\lambda}$ is built from two integrable weights at finite positive levels, there is a finite number of admissible representations at each fractional level. Actually, this number is simply $(k^I + 1)(k^F + 1)$. The list of admissible representations for $k = -\frac{4}{3}, -\frac{1}{2}$ and $-\frac{1}{3}$ is given in Tables 18.4, 18.5, and 18.6, respectively, together with their conformal dimensions:

$$h_{\hat{\lambda}} = \frac{\lambda_1(\lambda_1 + 2)}{4(k+2)} \tag{18.184}$$

We note the occurrence of negative conformal dimensions, the clearest signal of nonunitarity.

Table 18.4. Admissible
representations of $\widehat{su}(2)_{-\frac{4}{3}}$

$\hat{\lambda}$	$\hat{\lambda}^I$	$\hat{\lambda}^F$	h
$[-\frac{4}{3}, 0]$	$[0, 0]$	$[2, 0]$	0
$[-\frac{2}{3}, -\frac{2}{3}]$	$[0, 0]$	$[1, 1]$	$-\frac{1}{3}$
$[0, -\frac{4}{3}]$	$[0, 0]$	$[0, 2]$	$-\frac{1}{3}$

18.6.2. Character of Admissible Representations

The character of the $\widehat{su}(2)_k$ admissible representations, evaluated at the special point $\hat{\xi} = -2\pi i(\zeta; \tau, 0)$, with $\zeta = z\omega_1$, is

$$\chi_{\hat{\lambda}}(z; \tau) = \frac{\Theta_{b_+}^{(d)}(z/u; \tau) - \Theta_{b_-}^{(d)}(z/u; \tau)}{\Theta_1^{(2)}(z; \tau) - \Theta_{-1}^{(2)}(z; \tau)} \tag{18.185}$$

Table 18.5. Admissible representations of $\widehat{su}(2)_{-\frac{1}{2}}$

$\hat{\lambda}$	$\hat{\lambda}^I$	$\hat{\lambda}^F$	h
$[-\frac{1}{2},0]$	$[1,0]$	$[1,0]$	0
$[1,-\frac{3}{2}]$	$[1,0]$	$[0,1]$	$-\frac{1}{8}$
$[-\frac{3}{2},1]$	$[0,1]$	$[1,0]$	$\frac{1}{2}$
$[0,-\frac{1}{2}]$	$[0,1]$	$[0,1]$	$-\frac{1}{8}$

Table 18.6. Admissible representations of $\widehat{su}(2)_{-\frac{1}{3}}$

$\hat{\lambda}$	$\hat{\lambda}^I$	$\hat{\lambda}^F$	h
$[-\frac{1}{3},0]$	$[3,0]$	$[2,0]$	0
$[\frac{4}{3},-\frac{5}{3}]$	$[3,0]$	$[1,1]$	$-\frac{1}{12}$
$[3,-\frac{10}{3}]$	$[3,0]$	$[0,2]$	$\frac{2}{3}$
$[-\frac{4}{3},1]$	$[2,1]$	$[2,0]$	$\frac{9}{20}$
$[\frac{1}{3},-\frac{2}{3}]$	$[2,1]$	$[1,1]$	$-\frac{2}{15}$
$[2,-\frac{7}{3}]$	$[2,1]$	$[0,2]$	$\frac{7}{60}$
$[-\frac{7}{3},2]$	$[1,2]$	$[2,0]$	$\frac{6}{5}$
$[-\frac{2}{3},\frac{1}{3}]$	$[1,2]$	$[1,1]$	$\frac{7}{60}$
$[1,-\frac{4}{3}]$	$[1,2]$	$[0,2]$	$-\frac{2}{15}$
$[-\frac{10}{3},3]$	$[0,3]$	$[2,0]$	$\frac{9}{4}$
$[-\frac{5}{3},\frac{4}{3}]$	$[0,3]$	$[1,1]$	$\frac{2}{3}$
$[0,-\frac{1}{3}]$	$[0,3]$	$[0,2]$	$-\frac{1}{12}$

where

$$\Theta_b^{(d)}(z;\tau) = \sum_{\ell \in \mathbb{Z}+b/2d} q^{d\ell^2} x^{d\ell} \tag{18.186}$$

and $q = e^{2\pi i \tau}$, $x = e^{-2\pi i z}$.[10] The parameters d and $b_{\pm}$ are given by

$$d = u^2(k+2)$$
$$b_{\pm} = u\,[\pm(\lambda_1^I + 1) - (k+2)\lambda_1^F] \qquad (18.187)$$

We note that

$$b_{+\omega_1} = 1 \cdot (\lambda^I + \hat{\rho}) - (k+2)\lambda^F$$
$$b_{-\omega_1} = s_1 \cdot (\lambda^I + \hat{\rho}) - (k+2)\lambda^F \qquad (18.188)$$

which means that the character can be expressed as a sum over the Weyl group as in the integrable case, but with the action of the Weyl group restricted to the integral part of the weight. Actually, when $u = 1$, Eq. (18.185) reduces to Eq. (14.174).

We consider the limit $z \to 0$ of the above character. It is simple to see that the denominator vanishes linearly in z:

$$\Theta_1^{(2)}(z; \tau) - \Theta_{-1}^{(2)}(z; \tau) = -2\pi i z q^{1/8} \sum_{s \in \mathbb{Z}} q^{2s^2+s}(4s+1) + O(z^2) \qquad (18.189)$$

But this is not so for the numerator:

$$\Theta_{b_+}^{(d)}(z/u; \tau) - \Theta_{b_-}^{(d)}(z/u; \tau) = q^{b_+^2/4d} x^{b_+/2u} \sum_{s \in \mathbb{Z}} q^{s^2 d + s b_+}$$
$$\times (x^{ds/u} - q^{\lambda_1^F[2us(k+2)+\lambda_1^I+1]} x^{-\lambda_1^I-1-ds/u}) \qquad (18.190)$$

When $\lambda_1^F \neq 0$, this expression remains finite as $z \to 0$. As a result, the specialized character of admissible representations with fractional finite weight is infinite.

To understand this feature, we recall that the specialized character codes the number of states at each grade. In the integrable case, the number of states at each grade is finite and these states are organized in $su(2)$ finite dimensional representations. The specialized character is thus finite. In the fractional case, states at each grade are also organized in $su(2)$ representations; but $su(2)$ representations with fractional highest weights are infinite dimensional. Consequently, the specialized character is infinite.

18.6.3. Modular Covariance of Admissible Representations

The modular S matrix for admissible representations can be obtained as before, by means of the Poisson resummation formula. The details of the derivation are left to the reader (see Ex. 18.16); the result is

$$S_{\hat{\lambda},\hat{\mu}} = \sqrt{\frac{2}{u^2(k+2)}} (-1)^{\mu_1^F(\lambda_1^I+1)+\lambda_1^F(\mu_1^I+1)}$$
$$\times e^{-i\pi\mu_1^F\lambda_1^F(k+2)} \sin\left[\frac{\pi(\lambda_1^I+1)(\mu_1^I+1)}{k+2}\right] \qquad (18.191)$$

---

[10] As in the integral case, the character formula is given without proof (and convergence issues are ignored throughout). However, the above expression for the generalized theta function is equivalent to Eq. (14.176).

The $\mathcal{T}$ matrix is (as usual) given by

$$T_{\hat{\lambda}\hat{\mu}} = \delta_{\hat{\lambda}\hat{\mu}} e^{2\pi i(h_{\hat{\lambda}} - c/24)} \tag{18.192}$$

Both matrices are unitary.

Take, for instance, the model at level $-\frac{4}{3}$. Because the finite integer part of all the fields is zero, only the phase prefactor contributes to the $\mathcal{S}$ matrix. With the fields ordered as in Table 18.4, we find

$$S = \frac{-1}{\sqrt{3}} \begin{pmatrix} 1 & -1 & 1 \\ -1 & \epsilon^2 & -\epsilon \\ 1 & -\epsilon & \epsilon^2 \end{pmatrix}, \qquad \epsilon = e^{2\pi i/3} \tag{18.193}$$

whose unitarity is easily verified.

18.6.4. Charge Conjugation

From the list of primary fields given in Tables 18.4–18.6, we notice that there are distinct fields with the same conformal dimension. For WZW models at integer level, this is a frequent feature: primary fields associated with conjugate representations always have the same dimension, and fields with the same dimension are necessarily conjugates of each other. The conjugation of these fields is inherited from the conjugation of the Lie algebra representations. This, in turn, reflects the automorphism of the Dynkin diagram.

But here there appears to be a sort of conjugation for the fractional-level representations with no integer-level analogue ($su(2)$ representations are always self-conjugate), hence with no relation to the Dynkin diagram automorphisms.

These examples reveal that states in conjugate pairs all have nonzero fractional parts. The other states, with integral finite parts, are self-conjugate, as expected. Since the dimension formula is invariant under the transformation $\lambda \to \lambda - 2\rho$, it is natural to guess that

$$\hat{\lambda}^* = \hat{\lambda} \, \delta_{\lambda_1^F, 0} + (-\hat{\lambda} - 2\hat{\rho})(1 - \delta_{\lambda_1^F, 0}) \tag{18.194}$$

This will be confirmed below from the calculation of $\mathcal{S}^2 = \mathcal{C}$. We first prove that when $\lambda_1^F \neq 0$, $-\hat{\lambda} - 2\hat{\rho}$ is admissible if $\hat{\lambda}$ is so. This shows that conjugate states in the fractional sector always occur in pairs. When $\lambda_1^F \neq 0$, the conjugate transformation is

$$\begin{aligned} \lambda_1^I &\to k^I - \lambda_1^I \\ \lambda_1^F &\to u - \lambda_1^F \end{aligned} \tag{18.195}$$

or equivalently

$$\begin{aligned} \hat{\lambda}^I &\to a\hat{\lambda}^I \\ \hat{\lambda}^F &\to a(\hat{\lambda}^F + \hat{\omega}_0) - \hat{\omega}_0 \end{aligned} \tag{18.196}$$

where a is the $\widehat{su}(2)$ outer automorphism. If $\hat{\lambda}^I$ is integrable, $a\hat{\lambda}^I$ is also integrable. Similarly, if $\hat{\lambda}^F$ is integrable and $\lambda_1^F \geq 1$, this means that

$$1 \leq \lambda_1^F \leq u - 1 \quad \Longrightarrow \quad 1 \leq u - \lambda_1^F \leq u - 1 \tag{18.197}$$

that is, $u - \lambda_1^F$ is the nonzero finite part of an integrable weight at level $u - 1$. Finally, we show that there can be no self-conjugate state with $\lambda_1^F \neq 0$, resulting from fixed points in the I and F sector. That would correspond to states such that

$$\lambda_1^I = k^I - \lambda_1^I \quad \text{and} \quad \lambda_1^F = k^I - \lambda_1^F \tag{18.198}$$

This requires k^I and k^F both to be even. But, since $k^I = t - 2u - 2$, this contradicts the condition $(t, u) = 1$.

The transformation (18.194) implies that

$$b_\pm \to -b_\pm \tag{18.199}$$

when $\lambda_1^F \neq 0$. From the expression (18.186) for the theta function, it is readily seen that

$$\Theta_{-b}^{(d)}(z; \tau) = \Theta_b^{(d)}(-z; \tau) \tag{18.200}$$

The denominator of the character is not affected by this conjugate transformation because it involves only integer weights. But if we want to express the character at the point $-z$, it picks up a minus sign:

$$\Theta_1^{(2)}(z; \tau) - \Theta_{-1}^{(2)}(z; \tau) = -[\Theta_1^{(2)}(-z; \tau) - \Theta_{-1}^{(2)}(-z; \tau)] \tag{18.201}$$

This leads to

$$\chi_{\hat{\lambda}}(-z; \tau) = (-1)^{\delta_{\lambda_1^F, 0}} \chi_{\hat{\lambda}^*}(z; \tau) \tag{18.202}$$

Writing this relation under the form (cf. Eqs. (14.226) and (14.229))

$$\chi_{\hat{\lambda}}(-z; \tau) = C_{\hat{\lambda}, \hat{\mu}} \chi_{\hat{\mu}}(z; \tau) \tag{18.203}$$

allows us to read off the charge conjugation matrix:

$$C_{\hat{\lambda}, \hat{\mu}} = \delta_{\hat{\lambda}^*, \hat{\mu}} (-1)^{\delta_{\lambda_1^F, 0}} \tag{18.204}$$

Since $z \to -z$ can be obtained from two applications of $\tau \to -1/\tau$ (cf. Sect. 14.6), we have thus essentially checked that $\mathcal{S}^2 = C$, with C given above. With our simple example at level $-\frac{4}{3}$, we obtain

$$\mathcal{S}^2 = C = \begin{pmatrix} 1 & 0 & 0 \\ 0 & 0 & -1 \\ 0 & -1 & 0 \end{pmatrix} \tag{18.205}$$

18.6.5. Fusion Rules

We have already indicated that for WZW models at fractional levels the fusion coefficients calculated from the Verlinde formula are negative. For a simple illustration, take the model at level $-\frac{4}{3}$ whose S matrix is given in Eq. (18.193). Applying the Verlinde formula directly, we find

$$[-\frac{2}{3},-\frac{2}{3}] \times [0,-\frac{4}{3}] = -[-\frac{4}{3},0] \qquad (18.206)$$

It is not too difficult to derive, from the Verlinde formula, a closed expression for the fusion coefficients. However, in the following, we will need fusion rules only for fields with vanishing finite fractional parts. This case is easily treated because

$$\gamma_{\hat{\lambda}}^{(\hat{\sigma})} = \frac{S_{\hat{\lambda},\hat{\sigma}}}{S_{0,\hat{\sigma}}} = \chi_\lambda(\xi_\sigma) \qquad (18.207)$$

where $\chi_\lambda(\xi_\sigma)$ is the $su(2)$ finite character evaluated at the special point $\xi_\sigma = -2\pi i(\sigma + \rho)/(k + 2)$. The method of Sect. 16.2 applies and we end up with the very simple relation

$$\lambda_1^F = \mu_1^F = 0: \quad \mathcal{N}_{\hat{\lambda},\hat{\mu}}^{(k)\ \hat{\nu}} = \mathcal{N}_{\hat{\lambda},\hat{\mu}}^{(k)\ \hat{\nu}}\Big|_{\nu_1^F=0} = \mathcal{N}_{\hat{\lambda}^I,\hat{\mu}^I}^{(k^I)\ \hat{\nu}^I} \qquad (18.208)$$

The fact that the fusion rules at level k, for weights with zero finite fractional part, are given by the fusion of their integral parts at level $u(k+2) - 2$ will be used in the coset construction of nonunitary minimal models, once it will be proven, from field-identification considerations, that fields with zero finite fractional parts are suitable coset representatives.

§18.7. Coset Description of Nonunitary Minimal Models

Armed with these results on fractional-level representations of $\widehat{su}(2)$, we now turn to the coset description of nonunitary minimal models. As already pointed out, with a coset of the form (18.51), we need

$$k = \frac{3p' - 2p}{p - p'} \qquad (18.209)$$

so that p and p' are related to k by

$$p = u(k+2) + u, \qquad p' = u(k+2) \qquad (18.210)$$

Characters of coset fields are still given by normalized branching functions $\chi_{\{\hat{\lambda},\hat{\mu};\hat{\nu}\}}$ where $\hat{\lambda}$, $\hat{\mu}$ and $\hat{\nu}$ are admissible weights at levels k, 1 and $k+1$ respectively. Again, this implies a selection condition of the form Eq. (18.26) which, in the present case, takes the form

$$\lambda_1^I - (k+2)\lambda_1^F + \mu_1 - \nu_1^I + (k+3)\nu_1^F = 0 \quad \mathrm{mod}\ 2 \qquad (18.211)$$

Because k is fractional, this can be satisfied only if

$$\lambda_1^F = \nu_1^F \qquad (18.212)$$

Actually, this needs to be verified only modulo u. But since changing k by $k + 1$ does not change the value of u, the level of both $\hat{\lambda}^F$ and $\hat{\nu}^F$ is $u - 1$; since both weights are integrable, λ_1^F and ν_1^F cannot differ by a nonzero multiple of u. The I-part of the weights must then satisfy

$$\lambda_1^I + \mu_1 - \nu_1^I = 0 \quad \mathrm{mod}\ 2 \tag{18.213}$$

We note that the levels of $\hat{\lambda}^I$ and $\hat{\nu}^I$ differ by $u > 1$. The derivation of the coset characters from the character decomposition (18.54) goes through exactly as in the unitary case. However, the derivation can be simplified once the problem of field identification is under control. To appreciate the novelties brought by nonunitarity in this respect, we consider the simplest nonunitary model, the Yang-Lee singularity, with $(p, p') = (5, 2)$.

18.7.1. The Coset Description of the Yang-Lee Model

We first list all the admissible weights at levels $-\frac{4}{3}$ and $-\frac{1}{3}$. This has been done in Tables 18.4 and 18.6. Next, we look at all triplets $\hat{\lambda}, \hat{\mu}, \hat{\nu}$ at levels $-\frac{4}{3}, 1$ and $-\frac{1}{3}$ respectively that satisfy Eqs. (18.212) and (18.213), and evaluate their fractional dimension from $h_{\hat{\lambda}} + h_{\hat{\mu}} - h_{\hat{\nu}}$. Collecting fields with the same fractional dimension leads to the results presented in Table 18.7, where triplets are labeled by their finite parts.

This model has central charge $-\frac{22}{5}$ and, since there is a unique minimal model for this value of c, it is bound to be the Yang-Lee model. It must have two primary fields of dimension 0 and $-\frac{1}{5}$. Therefore, fields in Table 18.7 within each set have to be identified. Clearly, the way fields are identified goes beyond the mere action of the outer automorphism a since that can account for only two field identifications (since $a^2 = 1$), whereas six are required here.

Table 18.7. Coset fields for $k = -\frac{4}{3}$

$h = 0$	$\{0; 0; 0\}$	$\{-\frac{4}{3}, 1; -\frac{1}{3}\}$	$\{-\frac{2}{3}, 1; -\frac{5}{3}\}$
	$\{-\frac{2}{3}, 0; \frac{4}{3}\}$	$\{-\frac{4}{3}, 0; -\frac{10}{3}\}$	$\{0; 1; 3\}$
$h = -\frac{1}{5}$	$\{0; 1; 1\}$	$\{-\frac{4}{3}, 0; -\frac{4}{3}\}$	$\{-\frac{2}{3}, 0; -\frac{2}{3}\}$
	$\{-\frac{2}{3}, 1; \frac{1}{3}\}$	$\{-\frac{4}{3}, 1; -\frac{7}{3}\}$	$\{0; 0; 2\}$

18.7.2. Field Identification in the Nonunitary Case

Using the S matrix (18.191), it is easily verified that

$$S^{(k)}_{\hat{\lambda}^*,\hat{\lambda}'} = -e^{2\pi i k \lambda_1^F \lambda_1^{\prime F}} S^{(k)}_{\hat{\lambda},\hat{\lambda}'}$$
$$S^{(k)}_{a\hat{\lambda},\hat{\lambda}'} = (-1)^{\lambda_1^{\prime F}} e^{2\pi i k(\lambda_1^F + \frac{1}{2})\lambda_1^{\prime F}} S^{(k)}_{\hat{\lambda},\hat{\lambda}}$$

(18.214)

In turn, this implies

$$S_{\{\hat{\lambda}^*,\,\hat{\mu};\,\hat{\nu}^*\},\{\hat{\lambda}',\,\hat{\mu}';\,\hat{\nu}'\}} = S_{\{\hat{\lambda},\,\hat{\mu};\,\hat{\nu}\},\{\hat{\lambda}',\,\hat{\mu}';\,\hat{\nu}'\}}\, e^{2\pi i k(\lambda_1^F \lambda_1^{\prime F} - \nu_1^F \nu_1^{\prime F})}$$

(18.215)

Because $\lambda_1^F = \nu_1^F$ and $\lambda_1^{\prime F} = \nu_1^{\prime F}$, the phase factor is unity. Since the conformal dimension is not affected by conjugation, we find the field identification

$$\{\hat{\lambda}^*, \hat{\mu}; \hat{\nu}^*\} \sim \{\hat{\lambda}, \hat{\mu}; \hat{\nu}\}$$

(18.216)

This is compatible with the branching conditions (18.212)–(18.213). Similarly, since

$$S_{\{a\hat{\lambda},\,a\hat{\mu};\,a\hat{\nu}\},\{\hat{\lambda}',\,\hat{\mu}';\,\hat{\nu}'\}} = S_{\{\hat{\lambda},\,\hat{\mu};\,\hat{\nu}\},\{\hat{\lambda}',\,\hat{\mu}';\,\hat{\nu}'\}}$$

(18.217)

we conclude that the outer automorphism still yields a field identification:

$$\{a\hat{\lambda}, a\hat{\mu}; a\hat{\nu}\} \sim \{\hat{\lambda}, \hat{\mu}; \hat{\nu}\}$$

(18.218)

We now return to the Yang-Lee model. We recall that the coset fields are given in the form $\{\lambda_1, \mu_1; \nu_1\}$, which are the finite parts of affine weights at respective levels $-\frac{4}{3}$, 1 and $-\frac{1}{3}$. Therefore,

$$a\{\lambda_1, \mu_1; \nu_1\} = \{-\frac{4}{3} - \lambda_1, 1 - \mu_1; -\frac{1}{3} - \nu_1\}$$

(18.219)

On the other hand, if λ_1 is an integer, $\lambda_1^* = \lambda_1$. However, if it is fractional, $\lambda_1^* = -\lambda_1 - 2$ (cf. Eq. (18.194)). In that case, with

$$c\lambda \equiv \lambda^*$$

(18.220)

we have

$$c\{\lambda_1, \mu_1; \nu_1\} \equiv \{\lambda_1^*, \mu_1; \nu_1^*\} = \{-\lambda_1 - 2, \mu_1; -\nu_1 - 2\}$$

(18.221)

Consider the action of a and c on the fields with fractional conformal dimension 0 in Table 18.7:

$$a\{0, 0; 0\} = \{-\frac{4}{3}, 1; -\frac{1}{3}\}$$
$$c\{-\frac{4}{3}, 1; -\frac{1}{3}\} = \{-\frac{2}{3}, 1; -\frac{5}{3}\}$$
$$a\{-\frac{2}{3}, 1; -\frac{5}{3}\} = \{-\frac{2}{3}, 0; -\frac{4}{3}\}$$
$$c\{-\frac{2}{3}, 0; -\frac{4}{3}\} = \{-\frac{4}{3}, 0; -\frac{10}{3}\}$$
$$a\{-\frac{4}{3}, 0; -\frac{10}{3}\} = \{0, 0; 3\}$$

(18.222)

All these fields are then identified through the chain $acaca$. Exactly the same chain relates the six fields with $h = -\frac{1}{5}$ mod 1 in Table 18.7. Note that in this chain, we start and end with a coset field of zero fractional part.

It is not difficult to show that, for the diagonal coset

$$\frac{\widehat{su}(2)_{t/u} \oplus \widehat{su}(2)_1}{\widehat{su}(2)_{t/u+1}}$$

and starting from a coset field with $\lambda_1^F = 0$, fields can be identified according to the sequence or canonical chain:

$$(a\ c)^{u-1}\ a \tag{18.223}$$

which reduces to the previous chain when $u = 3$. The proof of this result is left as an exercise (Ex. 18.22). Since there are $2u - 1$ elements in the chain, $2u$ fields are identified. We now count the number of distinct coset fields. For λ_1^I, λ_1^F and ν_1^I, there are respectively $k^I + 1$, u, and $k^I + u + 1$ possible choices; μ_1 is uniquely fixed by λ_1^I and ν_1^I, whereas ν_1^F must be equal to λ_1^F. Dividing the total number by $2u$ yields

$$\frac{1}{2}(k^I + u + 1)(k^I + 1) = \frac{1}{2}(p - 1)(p' - 1) \tag{18.224}$$

which is precisely the number of distinct primary fields in the minimal model (p, p').

A coset field can be represented by any triplet of weights in the orbit of the canonical chain. Since every chain contains two sets of weights with zero fractional parts, any one of these provides a convenient choice of coset representatives. In the subclass of coset fields with zero fractional part—i.e., of the form $\{\hat{\lambda}^I, \hat{\mu}; \hat{\nu}^I\}$—there is now only one field identification, relating the two fields at the ends of the canonical chain:

$$\{\hat{\lambda}^I, \hat{\mu}; \hat{\nu}^I\} \sim (ac)^{u-1}a\{\hat{\lambda}^I, \hat{\mu}; \hat{\nu}^I\} \tag{18.225}$$

As far as the integer parts are concerned, we then act u times on $\hat{\mu}$ with a, but $2u + 1$ times on both $\hat{\lambda}^I$ and $\hat{\nu}^I$ (since the action of c on the I part is simply a). Because $a^2 = 1$, this produces

$$\{\hat{\lambda}^I, \hat{\mu}; \hat{\nu}^I\} \sim \{a\hat{\lambda}^I, a^u\hat{\mu}; a\hat{\nu}^I\} \tag{18.226}$$

and this is compatible with the branching condition since

$$\lambda_1^I + \mu_1 - \nu_1^I = 0 \mod 2 \Rightarrow k^I - \lambda_1^I + u(1 - \mu_1) - k^I - u + \nu_1^I = 0 \mod 2 \tag{18.227}$$

Without the factor of u in Eq. (18.226), the branching conditions are not satisfied, and to obtain this factor, we have to go through the full analysis of field identification using all admissible representations.

18.7.3. Character Decomposition, Modular Matrices, and Modular Invariants

The results of the preceding subsection indicate that the character of primary fields in nonunitary minimal models can be obtained directly from the decomposition of the character product $\chi_{\hat{\lambda}'}^{(k)} \chi_{\hat{\mu}}^{(1)}$ into a sum of characters $\chi_{\hat{\nu}'}^{(k+1)}$. Since this derivation is in all points similar to the one pertaining to the unitary case, it is left to the reader (see Ex. 18.23).

Coset fields can thus be specified by two integrable weights, $\hat{\lambda}'$ and $\hat{\nu}'$, at levels k' and $k' + u$, respectively or, equivalently, by two labels (r, s) defined by

$$
\begin{array}{llll}
r = \lambda_1' + 1 & 1 \leq r \leq p' - 1 & p' = k' + 2 \\
s = \nu_1' + 1 & 1 \leq s \leq p - 1 & p' = k' + 2 + u
\end{array}
\tag{18.228}
$$

The S matrix for Virasoro primary fields is easily recovered from that of its WZW components. This can then be used to show, following the analysis of the unitary case, that the Virasoro fusion coefficients are simply the products of two $\widehat{su}(2)$ coefficients, one at level k' and the other at level $k' + u$, in perfect agreement with the result (8.131).

Modular invariants for the nonunitary minimal models can be obtained, again exactly as in the unitary case, by the product-form mass matrix:

$$
\mathcal{M} = \frac{1}{2} \mathcal{M}^{(k')} \mathcal{M}^{(k'+u)}
\tag{18.229}
$$

Since t and u are relatively prime, so are k' and $k' + u$. Hence, one of k' or $k' + u$ must be odd, and one factor in Eq. (18.229) must be of the A-type. Listing all pairs of invariants containing at least one A-type factor leads to the full list of invariant partition functions presented in Eqs. (10.3) and (10.4).

Appendix 18.A. Lie-Algebraic Structure of the Virasoro Singular Vectors

The coset construction described in Sect. 18.3.1 yields the characters of the irreducible Virasoro modules directly. As seen in Chap. 8, these characters code the subtractions of the different singular-vector submodules from the original reducible Verma module. This plain fact naturally raises the following question: How are these singular vectors described from the Lie-algebraic point of view? This appendix addresses this question.[11]

In a finite simple Lie algebra, the integrable representation of highest weight λ is obtained from a reducible Verma module V_λ by quotienting the singular vectors

$$
(E_0^{\alpha_i})^{\lambda_i + 1} |\lambda\rangle = |s_i \cdot \lambda\rangle
\tag{18.230}
$$

---

[11] The more interesting question: "How is the structure of the coset singular vectors inferred from those of the WZW constituents?" is rather difficult. The coset construction of a given model is not well suited for the study of its singular vectors.

adding the intersection of their submodules, and so on. The Verma module V_λ is generated by the free action of the lowering generators $E^{-\alpha}$, $\alpha > 0$. The corresponding character is thus

$$\chi(V_\lambda) = \frac{e^\lambda}{\prod_{\alpha>0}(1 - e^{-\alpha})} \tag{18.231}$$

The character of the irreducible module L_λ encodes the submodule embeddings:

$$\chi(L_\lambda) \equiv \chi_\lambda = \sum_{w\in W} \epsilon(w)\chi(V_{w\cdot\lambda}) \tag{18.232}$$

Similarly, for an affine integrable representation described in terms of a reducible Verma module $V_{\hat\lambda}$,

$$(E_0^{\alpha_i})^{\lambda_i+1}|\hat\lambda\rangle = |s_i \cdot \hat\lambda\rangle \qquad (i \neq 0) \tag{18.233}$$

and

$$(E_{-1}^\theta)^{k-(\lambda,\theta)+1}|\hat\lambda\rangle = |s_0 \cdot \hat\lambda\rangle \tag{18.234}$$

are singular vectors[12] (cf. Sects. 14.3.1 and 15.3.4). The irreducible character is

$$\text{ch}(L_{\hat\lambda}) \equiv \text{ch}_{\hat\lambda} = \sum_{w\in\hat W} \epsilon(w)\text{ch}(V_{w\cdot\hat\lambda}) \tag{18.235}$$

with

$$\text{ch}(V_{\hat\lambda}) = \frac{e^{\hat\lambda}}{\prod_{\hat\alpha>0}(1 - e^{-\hat\alpha})^{\text{mult}(\hat\alpha)}} \tag{18.236}$$

Since the modular anomaly of the various terms $w \cdot \hat\lambda$ is independent of w (i.e., $h_{\hat\lambda} = h_{w\cdot\hat\lambda}$), Eq. (18.235) holds for the normalized characters also. For $\widehat{su}(2)$, the structure of module embeddings is easily described: the module $V_{w\cdot\hat\lambda}$ is embedded in the module $V_{w'\cdot\hat\lambda}$ if the (minimal) length of w is longer than that of w'.

As we will now see, the structure of $\chi(L_{\hat\lambda})$ (or any of the three weights) is directly transposed to the coset field $\{\hat\lambda, \hat\mu; \hat\nu\}$. For the present purpose, it is more convenient to label the coset field by the pair of shifted weights $\{\hat\lambda + \hat\rho; \hat\nu + \hat\rho\}$. We denote the Virasoro Verma module character by

$$\chi(V_{\{\hat\lambda+\hat\rho; \hat\nu+\hat\rho\}}) \equiv \chi(V_{(r,s)}) = \frac{q^{h_{(r,s)}}}{\eta(q)} \tag{18.237}$$

where, as usual, the labels r and s stand for the finite Dynkin labels of the weights $\hat\lambda + \hat\rho$ and $\hat\nu + \hat\rho$. More precisely, in order to take into account the nonunitary

---

[12] The expression for $s_0 \cdot \hat\lambda$ follows from Eq. (14.73). That is,

$$s_0 \cdot \hat\lambda = s_0(\hat\lambda + \hat\rho) - \hat\rho$$
$$= \hat\lambda + \hat\rho + (k+g)\hat\theta - (\lambda + \rho, \theta)\hat\theta - \hat\rho$$
$$= \hat\lambda + [k+1+(\lambda,\theta)]\hat\theta$$

where $\hat\theta$ is the level-zero affine extension of θ, and we used the relation $(\rho, \theta) = g - 1$. This is indeed equal to the weight obtained by subtracting $k - (\lambda, \theta) + 1$ times θ from $\hat\lambda$.

as well as the unitary cases, we identify r and s with finite Dynkin labels of the weights $\hat{\lambda}^I + \hat{\rho}$ and $\hat{\nu}^I + \hat{\rho}$, respectively, as in Eq. (18.228) ($u = 1$ corresponds to the unitary case). The conformal dimension $h_{r,s}$ is

$$h_{r,s} = \frac{|p(\lambda^I + \rho) - p'(\nu^I + \rho)|^2 - |p\rho - p'\rho|^2}{2pp'} = \frac{(pr - p's)^2 - (p - p')^2}{4pp'}$$

(18.238)

The character of the irreducible representation can be written as

$$\chi_{\{\hat{\lambda}+\hat{\rho};\, \hat{\nu}+\hat{\rho}\}} = \sum_{w \in \hat{W}} \epsilon(w)\chi(V_{\{w(\hat{\lambda}+\hat{\rho});\, \hat{\nu}+\hat{\rho}\}})$$

(18.239)

This gives a series of embedded submodules that reproduces exactly the tower obtained in Sect. 8.1 for the reducible Virasoro modules (cf. Fig. 8.1). This correspondence is displayed in more detail in the next paragraph.

In the following, we set

$$\{\hat{\lambda} + \hat{\rho};\, \hat{\nu} + \hat{\rho}\} \equiv \{[p' - r, r];\, [p - s, s]\} \equiv (r, s) \sim (p' - r, p - s) \quad (18.240)$$

and denote by $s_i^{(1)}$ (resp. $s_i^{(2)}$) the action of s_i on the first (resp. second) weight of the coset pair. By acting successively on the (r, s) state with $s_0^{(1)}$ and $s_1^{(1)}$, we obtain the following sequence of singular vectors:

$$\{[p' - r, r];\, [p - s, s]\} \xrightarrow{s_0^{(1)}} \{[-p' + r, 2p' - r];\, [p - s, s]\}$$

$$\xrightarrow{s_1^{(1)}} \{[3p' - r, -2p' + r];\, [p - s, s]\}$$

$$\xrightarrow{s_0^{(1)}} \{[-3p' + r, 4p' - r];\, [p - s, s]\}$$

(18.241)

$$\xrightarrow{s_1^{(1)}} \cdots$$

This reproduces the left side of the tower of Fig. 18.1. It is conventional to choose, between the zeroth and first Dynkin labels, those which are positive for the two weights. The right side of the figure is obtained by modifying the ordering of the applications of the two basic Weyl reflections:

$$\{[p' - r, r];\, [p - s, s]\} \xrightarrow{s_1^{(1)}} \{[p' + r, -r];\, [p - s, s]\}$$

$$\xrightarrow{s_0^{(1)}} \{[-p' - r, 2p' + r];\, [p - s, s]\}$$

$$\xrightarrow{s_1^{(1)}} \{[3p' + r, -2p' - r];\, [p - s, s]\}$$

(18.242)

$$\xrightarrow{s_0^{(1)}} \cdots$$

The second set of labels appearing in Fig. 18.1 is obtained by means of the identities

$$h_{r,s} = h_{-r,-s} = h_{p'+r,p+s}, \qquad h_{p'+r,s} = h_{r,-p+s}$$

(18.243)

which are trivial consequences of the formula (18.238). The embedding of the submodules on the different sides are obtained by acting with $s_0^{(2)}$ and $s_1^{(2)}$ on the second set of labels given in Fig. 18.1: $s_0^{(2)}$ acts from left to right and $s_1^{(2)}$ from right to left.[13] For example:

$$s_0^{(2)} s_1^{(1)} \{[p' - r, r]; [p - s, s]\} = s_0^{(2)} \{[p' + r, -r]; [p - s, s]\} \qquad (18.244)$$

According to our convention, the state on which $s_0^{(2)}$ acts is denoted by $(p'+r, p-s)$. An equivalent labeling, obtained by means of the identities (18.243), is

$$(p' + r, p - s) \sim (-p' - r, -p + s) \sim (p' - r, p + s) \qquad (18.245)$$

therefore

$$\{[p' + r, -r]; [p - s, s]\} \sim \{[p' - r, r]; [p + s, -s]\} \qquad (18.246)$$

It is on this last field that we must act with $s_0^{(2)}$, with the result

$$s_0^{(2)} \{[p' - r, r]; [p + s, -s]\} = \{[p' - r, r]; [-p - s, 2p + s]\} \qquad (18.247)$$

denoted by $(r, 2p + s)$. Using $s_0^{(2)} s_1^{(1)}$, we have then moved from the top of the module to the first submodule on the right, and from then, to the second submodule on the left.

Appendix 18.B. Affine Lie Algebras at Fractional Levels and General Nonunitary Coset Models

18.B.1. Admissible Representations of Affine Lie Algebras at Fractional Levels

In this section we present the general characterization of admissible representations of any affine Lie algebra at fractional levels

$$k = \frac{t}{u} \qquad (18.248)$$

where t is a nonzero integer (positive or negative), u is a positive integer, and t and u are relatively prime. To every element y of the subgroup $W/W(A)$ of the finite Weyl group W is associated a set of possible admissible highest weights $\hat{\lambda}$. $W(A)$ is the subgroup of W isomorphic to the outer-automorphism group $O(\hat{g})$ of $\hat{g}$, that is, it is generated by all the elements w_A associated with $A \in \mathcal{O}(\hat{g})$ via

---

[13] On the primary field, the action of $s_i^{(1)}$ is the same as that of $s_i^{(2)}$, but this is no longer true for the submodules, due to the noncommutativity of the Weyl reflections. For instance,

$$|ps_0(\lambda + \rho) - p'(\nu + \rho)| = |p(\lambda + \rho) - p' s_0(\nu + \rho)|$$

but

$$|ps_1 s_0(\lambda + \rho) - p'(\nu + \rho)| = |p(\lambda + \rho) - p' s_0 s_1(\nu + \rho)| \neq |p(\lambda + \rho) - p' s_1 s_0(\nu + \rho)|$$

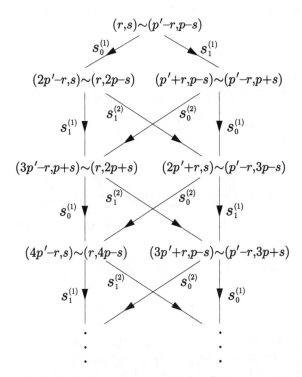

Figure 18.1. Embedding of submodules in the $V_{r,s}(p,p')$ Verma module from the coset approach.

Eq. (14.98). Furthermore, each of these weights may be broken up into an integer (I) and a fractional (F) part:

$$\hat{\lambda} = y \cdot \left(\hat{\lambda}^I - (k+g)\hat{\lambda}^{F,y}\right) \tag{18.249}$$

where $\hat{\lambda}^I$ and $\hat{\lambda}^{F,y}$ are both integral weights. The level of the integer part $\hat{\lambda}^I$ is

$$k^I = u(k+g) - g \geq 0 \tag{18.250}$$

and that of the fractional part $\hat{\lambda}^{F,y}$ is

$$k^F = u - 1 \geq 0 \tag{18.251}$$

The integer part $\hat{\lambda}^I$ is the highest weight of an integrable representation,

$$\hat{\lambda}^I \in P_+^{k^I} \tag{18.252}$$

On the other hand, the Dynkin labels of $\hat{\lambda}^{F,y}$ must satisfy the following two conditions:

$$\lambda_j^{F,y} \in \frac{a_j}{a_j^\vee} \mathbb{Z} \quad j = 0, 1, ., r \tag{18.253}$$

(we recall that $a_j/a_j^\vee$ is always an integer) and

$$\lambda_j^{F,y} \sum_{i=0}^{r} a_i^\vee \alpha_i^\vee + y(\alpha_j^\vee) \in \hat{Q}_+^\vee \qquad (18.254)$$

where $\hat{Q}_+^\vee$ is the set of real coroots of $\hat{g}$:

$$\hat{Q}_+^\vee = \sum_{i=0}^{r} n_i \alpha_i^\vee, \qquad n_i \in \mathbb{Z}_+ \quad \text{and at least one} \quad n_j \neq 0 \qquad (18.255)$$

Thus, given $y \in W$, we can determine the possible values of $\lambda_j^{F,y}$ at a given level k^F and then construct the admissible weights $\hat{\lambda}$ at level k corresponding to the choice of y. This set of admissible highest weights for a fixed y will be denoted P_y^k. The set of all admissible highest weights at level k is just the union of these:

$$P^k = \bigcup_{y \in W/W(A)} P_y^k \qquad (18.256)$$

When $u = 1$, we find $P^k = P_+^k$.

We illustrate the construction of admissible weights for $\widehat{su}(2)$, $\widehat{su}(3)$, and $\widehat{so}(5)$. For $\widehat{su}(2)$, all marks and comarks being 1, $\lambda_j^{F,y}$ is an integer. We first pick an element $y \in W/W(A)$ and derives the corresponding restrictions on the values of $\lambda_1^{F,y}$ from Eq. (18.254). Since $W = W(A)$ for $su(2)$, the only choice for y is the identity. Then Eq. (18.254) reduces to the following two requirements:

$$(\lambda_0^F + 1)\alpha_0^\vee + \lambda_0^F \alpha_1^\vee \in \hat{Q}_+^\vee$$
$$\lambda_1^F + \alpha_0^\vee + (\lambda_1^F + 1)\alpha_1^\vee \in \hat{Q}_+^\vee \qquad (18.257)$$

The coefficients of the coroots must be greater than or equal to zero, with at least one being positive. This forces

$$\lambda_0^F \geq 0 \quad \text{and} \quad \lambda_1^F \geq 0 \qquad (18.258)$$

Therefore, the two Dynkin labels must be positive definite, and any $\widehat{su}(2)$ admissible weight is of the form

$$\hat{\lambda} = \hat{\lambda}^I - (k+2)\hat{\lambda}^F \quad \text{with} \quad \hat{\lambda}^I \in P_+^{k^I}, \quad \hat{\lambda}^F \in P_+^{k^F} \qquad (18.259)$$

This is the way $\widehat{su}(2)$ admissible weights have been presented in Sect. 18.6.1.

For $\widehat{su}(3)$, the Dynkin labels of $\hat{\lambda}^F$ are still elements of $\mathbb{Z}$. The $su(3)$ finite Weyl group is

$$W = \{1, s_1, s_2, s_1 s_2, s_2 s_1, s_1 s_2 s_1\} \qquad (18.260)$$

The elements of $W(A)$ are

$$W(A) = \{1, s_1 s_2, s_2 s_1\} \qquad (18.261)$$

corresponding respectively to the outer automorphisms 1, a and a^2 (a is a cyclic permutation of the affine Dynkin labels). Therefore, we can restrict y to the set $\{1, s_1\}$. The conditions (18.254) take the form:

$$y = 1 : \quad \lambda_{0,1,2}^{F,1} \geq 0 \tag{18.262}$$
$$y = s_1 : \quad \lambda_{0,2}^{F,s_1} \geq 0, \quad \lambda_1^{F,s_1} \geq 1$$

We take, for instance, the simplest $\widehat{su}(3)$ nonunitary model, at level $-\frac{3}{2}$ ($u = 2$, $k^F = 1$ and $k^I = 0$); the allowed $\hat{\lambda}^{F,y}$ are

$$\hat{\lambda}^{F,1} : [1,0,0], \quad [0,1,0], \quad [0,0,1] \tag{18.263}$$
$$\hat{\lambda}^{F,s_1} : [0,1,0].$$

Therefore there are four admissible highest weights:

$$[-\frac{3}{2},0,0], \ [0,-\frac{3}{2},0], \ [0,0,-\frac{3}{2}], \ [-\frac{1}{2},-\frac{1}{2},-\frac{1}{2}] \tag{18.264}$$

where the first three are obtained from $y = 1$ and the last one from $y = s_1$.

We now consider $\widehat{so}(5)$. The simple roots are

$$\alpha_0 = [2,0,-2], \quad \alpha_1 = [0,2,-2], \quad \alpha_2 = [-1,-1,2] \tag{18.265}$$

the last one being the short root. All marks and comarks are equal to unity except $a_2 = 2$. This implies that $\lambda_{0,1}^{F,y} \in \mathbb{Z}$ and $\lambda_2^{F,y} \in 2\mathbb{Z}$. The sets W, $W(A)$, and $W/W(A)$ are, respectively

$$W = \{1, s_1, s_2, s_1s_2, s_2s_1, s_1s_2s_1, s_2s_1s_2, s_1s_2s_1s_2\}$$
$$W(A) = \{1, s_1s_2s_1\} \tag{18.266}$$
$$W/W(A) = \{1, s_1, s_2, s_1s_2\}$$

The constraints on the $\lambda_i^{F,y}$'s with $y \in W/W(A)$ are found to be

$$\lambda_{0,1,2}^{F,1} \geq 0,$$
$$\lambda_{0,2}^{F,s_1} \geq 0, \quad \lambda_1^{F,s_1} \geq 1$$
$$\lambda_{0,1}^{F,s_2} \geq 0, \quad \lambda_2^{F,s_2} \geq 1 \tag{18.267}$$
$$\lambda_0^{F,s_1s_2} \geq -1, \quad \lambda_1^{F,s_1s_2} \geq 0, \quad \lambda_2^{F,s_1s_2} \geq 2,$$

We note that $\lambda_2^{F,s_2} \geq 2$ and, since the other two Dynkin labels must be positive, the sector $y = s_2$ is allowed only for $u \geq 3$. For the simple case $k^I = 0$, $u = 2$, the list of admissible weights is thus

$$[-\frac{3}{2},0,0], \ [0,-\frac{3}{2},0], \ [0,-\frac{1}{2},-1], \ [-\frac{1}{2},0,-1] \tag{18.268}$$

with y respectively equal to 1, 1, s_1, and s_1s_2.

18.B.2. Modular Properties of Characters for Admissible Representations

Since we will not need the explicit expressions for the characters of admissible representations, we only present their S modular transformation matrix:[14]

$$S_{\hat{\lambda}\hat{\mu}} = i^{|\Delta_+|} |P/Q^\vee|^{-\frac{1}{2}} [u^2(k+g)]^{-r/2} \epsilon(yy')$$
$$\times \exp\{2\pi i[(\lambda^I + \rho, \mu^F) + (\lambda^F, \mu^I + \rho) - (k+g)(\lambda^F, \mu^F)]\} \quad (18.269)$$
$$\times \sum_{w\in W} \epsilon(w) e^{-2\pi i(w(\lambda^I+\rho),\mu^I+\rho)/(k+g)}$$

Here $\hat{\lambda} \in P^k_y$ and $\hat{\mu} \in P^k_{y'}$, and the fractional parts λ^F and μ^F are defined by

$$\hat{\lambda}^F = y(\hat{\lambda}^{F,y}) \quad (18.270)$$

As usual, $\epsilon(y)$ stands for the signature of the Weyl group element y. This matrix is unitary. For $\widehat{su}(2)$, it reduces to Eq. (18.191). When $k^F = 0$, we have to consider only $y = 1$, and the modular matrix S of integrable representations is recovered.

Because the summand depends only on the integer part $\hat{\lambda}^I$, the fusion rules (calculated by the Verlinde formula) are essentially determined by the integer part of the weights.

18.B.3. Charge Conjugation and the Associated Weyl Group

Given the S matrix, we can calculate the charge conjugation matrix C from S^2. But since the result is most neatly presented in terms of a particular subgroup of W, we first introduce the appropriate concept.

To the finite part λ of an admissible weight $\hat{\lambda}$, there corresponds a subgroup of the finite Weyl group, called the *associated Weyl group* W^λ. It is generated by the reflections with respect to all positive roots α such that $(\lambda, \alpha^\vee) \in \mathbb{Z}$. If all finite Dynkin labels are integers, it coincides with the full finite Weyl group. However, if some of the finite Dynkin labels are not integers, W^λ will be a proper subgroup of W. A simple $su(3)$ example is:

$$(\frac{1}{2}, -\frac{3}{2}): \quad W^\lambda = \{1, s_1 s_2 s_1 = s_\theta\} \quad (18.271)$$

As another example, consider the $su(4)$ weights and their associated Weyl group:

$$(-\frac{1}{3}, 0, -\frac{5}{3}): \quad W^\lambda = \{1, s_2, s_1 s_2 s_3 s_1 s_2 s_1\}$$
$$(-\frac{1}{3}, -\frac{2}{3}, -\frac{5}{3}): \quad W^\lambda = \{1, s_1 s_2 s_1\} \quad (18.272)$$

[14] We note that for nonsimply-laced algebras, this formula is valid only when u is not a multiple of the ratio $a_i/a_i^\vee$ corresponding to the short roots. Therefore, this formula should not be used when u is even for B_r, C_r, and F_4, and when $u \in 3\mathbb{Z}$ for G_2: the corresponding admissible sets are ill-defined from the point of view of modular transformations.

In the first case the two positive roots such that $(\lambda, \alpha^\vee) \in \mathbb{Z}$ are α_2 and θ, and in the second case, the corresponding set of positive roots is $\{\alpha_1 + \alpha_2\}$.

We denote by w_λ the longest element of W^λ and, as before, w_0 stands for the longest element of W. Then $\mathcal{S}^2 = C$ yields:

$$C_{\hat{\lambda}, \hat{\mu}} = \epsilon(w_0)\epsilon(w_\lambda)\delta_{\hat{\lambda}^*, \hat{\mu}} \tag{18.273}$$

where

$$\hat{\lambda}^* = (-w_\lambda) \cdot \hat{\lambda} \tag{18.274}$$

For $\widehat{su}(2)$, this is equivalent to Eq. (18.194). We stress that charge conjugation in the fractional case is not related to symmetries of the finite Dynkin diagram. We illustrate this formula with two $su(4)$ examples:

$$
\begin{aligned}
(-\tfrac{1}{3}, 0, -\tfrac{5}{3})^* &= (-s_1 s_2 s_3 s_1 s_2 s_1) \cdot (-\tfrac{1}{3}, 0, -\tfrac{5}{3}) = (-\tfrac{5}{3}, 0, -\tfrac{1}{3}) \\
(-\tfrac{1}{3}, -\tfrac{2}{3}, -\tfrac{5}{3})^* &= (-s_1 s_2 s_1) \cdot (-\tfrac{1}{3}, -\tfrac{2}{3}, -\tfrac{5}{3}) = (-\tfrac{1}{3}, -\tfrac{4}{3}, -\tfrac{1}{3})
\end{aligned}
\tag{18.275}
$$

18.B.4. Nonunitary Diagonal Coset Models

We now consider diagonal cosets of the form

$$\frac{\hat{g}_k \oplus \hat{g}_l}{\hat{g}_{k+l}} \tag{18.276}$$

where

$$k = t/u, \quad l \in \mathbb{N} \tag{18.277}$$

The corresponding central charge is

$$c = \frac{l \dim g}{l+g}\left\{1 - \frac{g(g+l)(p'-p)^2}{l^2 pp'}\right\} \tag{18.278}$$

where we have introduced the integers p and p' defined by

$$k + g = \frac{lp'}{(p-p')}, \quad p - p' = lu \tag{18.279}$$

Note that for $l = 1$, requiring $(t, u) = 1$ forces p and p' to be relatively prime.

In the following, we denote a diagonal coset primary field by $\{\hat{\lambda}, \hat{\mu}; \hat{\nu}\}$, with $\hat{\lambda} \in P_y^k$, $\hat{\mu} \in P_+^l$, and $\hat{\nu} \in P_{y'}^{k+l}$. Fixed points are ignored throughout. Coset characters are obtained from the decomposition

$$\chi_{\hat{\lambda}}^{(k)} \chi_{\hat{\mu}}^{(l)} = \sum_{\hat{\nu} \in P_y^{(k+l)}} \chi_{\{\hat{\lambda}, \hat{\mu}; \hat{\nu}\}} \chi_{\hat{\nu}}^{(k+l)} \tag{18.280}$$

and a necessary condition for their nonvanishing is

$$\hat{\lambda} + \hat{\mu} - \hat{\nu} \in Q, \quad y = y', \quad \hat{\lambda}^F = \hat{\nu}^F \tag{18.281}$$

where Q is the root lattice of g. Two important aspects of this decomposition must be emphasized. The first is that $\hat{\lambda}$ and $\hat{\nu}$ are associated with the same Weyl group element y. The second is the equality of the fractional parts of $\hat{\lambda}$ and $\hat{\nu}$. This identification is made possible because $\hat{\lambda}^F$ and $\hat{\nu}^F$ have the same level $(u - 1)$, even though the levels of $\hat{\lambda}$ and $\hat{\nu}$ are different. The fractional part appears then as a conserved charge under tensor-product decomposition.

The fractional part of the conformal dimension of the coset primary field, given by

$$h = h_{\hat{\lambda}} + h_{\hat{\mu}} - h_{\hat{\nu}} \quad \text{mod } 1 \tag{18.282}$$

can be written in the form

$$h = \frac{|p(\lambda + \rho) - p'(\nu + \rho)|^2 - (p' - p)^2 |\rho|^2}{2lpp'} + \frac{(\mu, \mu + 2\rho)}{2(l + g)} - \frac{|\lambda - \nu|^2}{2l} \quad \text{mod } 1 \tag{18.283}$$

in terms of the coprime numbers p and p' introduced in Eq. (18.279), or equivalently

$$p = k^l + g + lu, \qquad p' = k^l + g \tag{18.284}$$

For $l = 1$ and simply-laced algebras, it can be checked that the last two terms in Eq. (18.283) cancel.

The coset S matrix can be calculated from Eq. (18.269), and used as the starting point for the study of field identifications. This analysis proceeds as in the unitary case or $\widehat{su}(2)$ at fractional level. Accordingly, the results will be presented rather briefly. Outer automorphisms still yield field identifications:

$$\{\hat{\lambda}, \hat{\mu}; \hat{\nu}\} \sim \{A\hat{\lambda}, A\hat{\mu}; A\hat{\nu}\} \tag{18.285}$$

It turns out that the coset S matrix does not depend upon y and $\hat{\lambda}^{F,y}$. This immediately shows that the two coset fields $\{\hat{\lambda}, \hat{\mu}; \hat{\nu}\}$ and $\{\hat{\lambda}', \hat{\mu}'; \hat{\nu}'\}$ can be identified if

$$\hat{\mu} = \hat{\mu}', \quad \hat{\lambda}^l = (\hat{\lambda}')^l, \quad \hat{\nu}^l = (\hat{\nu}')^l,$$
$$\hat{\lambda}^{F,y} = (\hat{\lambda}')^{F,y} \bmod Q^\vee \tag{18.286}$$

Equality of the fractional parts modulo the coroot lattice $Q^\vee$, instead of the root lattice, is required in order to preserve the conformal dimension. This can be seen from Eq. (18.283).

A large class of field identifications can be obtained by assuming that $\hat{\lambda}^{F,y} = (\hat{\lambda}')^{F,y'}$ and $y \neq y'$. Then $\hat{\lambda}' = w \cdot \hat{\lambda}$ and $\hat{\nu}' = w \cdot \hat{\nu}$, where $w = y'y^{-1}$. Given an admissible weight $\hat{\lambda}$, it is always possible to find elements w of the Weyl group such that $w \cdot \hat{\lambda}$ is also admissible. In fact, these elements belong to the coset W/W^λ. Since $\hat{\lambda}$ and $\hat{\nu}$ have the same fractional part, they share the same associated subgroup W^λ. Thus, if $w \cdot \hat{\lambda}$ is admissible, so is $w \cdot \hat{\nu}$. Furthermore, it is clear that if $\hat{\mu}$, $\hat{\lambda}$ and $\hat{\nu}$ satisfy the branching condition (18.281), this condition is also satisfied by the weights $\hat{\mu}$, $w \cdot \hat{\lambda}$, and $w \cdot \hat{\nu}$. Finally, the conformal dimension (18.282) of the coset field is not affected by a simultaneous shifted action of the Weyl group on any weight of the coset field. Thus, we have the identification

$$\{\hat{\lambda}, \hat{\mu}; \hat{\nu}\} \sim \{w \cdot \hat{\lambda}, \hat{\mu}; \ w \cdot \hat{\nu}\} \tag{18.287}$$

We stress that such field identifications arise only in the fractional sector (i.e., the sector with noninteger finite weights) since it requires $w \in W/W^\lambda$. For integrable weights $W^\lambda = W$, and this yields no identification.

This class of field identifications, together with identifications from outer automorphisms, appears to perform all necessary field identifications. A special case of identifications by the shifted action of W is particularly useful for $\widehat{su}(N)$ cosets. We associate with each element $A \in O(\hat{g})$ an operator C_A whose action on a weight λ is defined by

$$C_A \hat{\lambda} = \begin{cases} (yw_A y^{-1}) \cdot \hat{\lambda} & \text{if the result is admissible with the same } y \\ \hat{\lambda} & \text{otherwise} \end{cases} \tag{18.288}$$

We note that

$$(yw_A y^{-1}) \cdot \hat{\lambda} = y \cdot \left(A\hat{\lambda}^I - (k+g)(A(\hat{\lambda}^{F,y} + \hat{\omega}^0) - \hat{\omega}^0) \right) \tag{18.289}$$

For $\widehat{su}(2)$, $y = 1$; this is simply the charge conjugation (18.194). This particular shifted action of W, together with outer automorphisms, is sufficient to perform all field identifications for $su(N)$ diagonal cosets.

We consider now the possibility that inequivalent coset primary fields can be chosen with a vanishing finite fractional part, i.e., can be of the form

$$\{\hat{\lambda}^I, \hat{\mu}; \hat{\nu}^I\} \qquad \text{with} \qquad \lambda^I + \mu - \nu^I = 0 \bmod Q \tag{18.290}$$

modulo the action of the outer-automorphism group. For $su(N)$, it turns out that it is always possible to choose inequivalent[15] y's such that all $\hat{\lambda}^{F,y} \in P_+^{\mu-1}$. If $y \neq 1$, some Dynkin labels must satisfy a stronger constraint than $\lambda_i^{F,y} \geq 0$. As a result, for $su(N)$, the set of admissible $\lambda^{F,y}$ for $y \neq 1$ is a proper subset of the set of admissible $\lambda^{F,1}$. From Eq. (18.286), it is therefore manifest in this case that all fields from the $y \neq 1$ sectors can be identified with fields of the $y = 1$ sector, and it is sufficient to consider this sector only. Moreover, in the $y = 1$ sector, all fields with $\lambda^{F,1} \neq 0$ can be related to fields with $\lambda^{F,1} = 0$ by using the operators A and C_A. This is illustrated in Ex. 18.27, with an $\widehat{su}(3)$ coset realization of the Yang-Lee model.

For other classical Lie algebras, the group of outer automorphisms is not sufficiently large to relate all fields with $\lambda^{F,y} \neq 0$ to those with $\lambda^{F,y} = 0$, even in the $y = 1$ sector. Furthermore, we cannot choose representative y's in $W/W(A)$ such that $\hat{\lambda}^{F,y} \in P_+^{\mu-1}$, so that a priori we have no reason to restrict ourselves to the $y = 1$ sector. Nevertheless, we can still use Eq. (18.286) to identify coset fields, and for all examples that have been considered, it is possible to pick a set of primary field representatives by restricting the search to $y = 1$ and $\lambda^{F,1} = 0$.

---

[15] y and y' are equivalent if there is no w_A such that $y = y'w_A$.

Exercises

18.1 *Field identification: T matrix part*
Check that when $A \mapsto \tilde{A}$, with $A \in \mathcal{O}(\hat{g})$ and $\tilde{A} \in \mathcal{O}(\hat{p})$,

$$T_{\{A\hat{\lambda};\ \tilde{A}\hat{\mu}\},\{\hat{\lambda}';\ \hat{\mu}'\}} = T_{\{\hat{\lambda};\ \hat{\mu}\},\{\hat{\lambda}';\ \hat{\mu}'\}}$$

for $\hat{\lambda}, \hat{\lambda}' \in P_+^k(\hat{g})$, $\hat{\mu}, \hat{\mu}' \in P_+^{(k x_e)}(\hat{p})$.

18.2 *Partition functions for minimal models*
Derive the partition function for the Ising and the three-state Potts models starting from
Eq. (18.100), using the appropriate $\widehat{su}(2)_k$ mass matrices.

18.3 *Two coset descriptions of the unitary series*
Verify that the central charges of the following two infinite sequences span the whole unitary
minimal series:

$$\frac{\widehat{su}(N+1)_2}{\widehat{su}(N)_2 \oplus \widehat{u}(1)} \qquad \text{and} \qquad \frac{\widehat{sp}(2N+2)_1}{\widehat{sp}(2N)_1 \oplus \widehat{sp}(2)_1}$$

18.4 *Some exceptional coset realizations of the minimal models*
Identify the minimal models corresponding to the following cosets:

$$\frac{(\widehat{E}_7)_1 \oplus (\widehat{E}_7)_1}{(\widehat{E}_7)_2} \ , \qquad \frac{(\widehat{E}_6)_1 \oplus (\widehat{E}_6)_1}{(\widehat{E}_6)_2} \ , \qquad \frac{(\widehat{G}_2)_1}{\widehat{su}(3)_1} \ , \qquad \frac{(\widehat{F}_4)_1}{\widehat{so}(9)_1}$$

If the central charge does not fix the theory uniquely, whether it is an $A - A$ or $A - D$
block-diagonal invariant can be inferred from the presence of Virasoro primary fields of
multiplicity 2, rooted here in a nontrivial conjugation of at least one WZW constituent. (Obviously, a safer but more tedious approach consists in calculating the fractional dimensions
of the different coset fields.)

18.5 *Some character decompositions*

a) Check the character decomposition (18.106) to order q^2, using the following $\widehat{E}_8$
characters:

$$\text{ch}_{\hat{\omega}_0}(q) = 1 + 248\, q + 4124\, q^2 + \cdots$$
$$\text{ch}_{2\hat{\omega}_0}(q) = 1 + 248\, q + 31124\, q^2 + \cdots$$
$$\text{ch}_{\hat{\omega}_1}(q) = 248 + 34504\, q + \cdots$$
$$\text{ch}_{\hat{\omega}_7}(q) = 3875 + 181753\, q + \cdots$$

b) Verify the expressions given in Eq. (18.114) and complete the calculation of the first
decomposition in Eq. (18.113), to order q^3.

c) Verify the second decomposition in Eq. (18.113). For that case, the relevant affine modules
must first be worked out.

18.6 *An E_6 nondiagonal coset*
The projection matrix and outer-automorphism branching rules appropriate to the coset

$$\frac{(\widehat{E}_6)_2}{\widehat{su}(2)_2 \oplus \widehat{su}(6)_2}$$

have been described in Ex. 14.17. List all allowed coset fields and indicate those fields that
have to be identified. Determine the corresponding minimal model.

18.7 *Quantum dimensions of coset fields*

a) Defining the coset field quantum dimension directly from a ratio of $\mathcal{S}$ matrices (cf. Eq. (16.59)):

$$\mathcal{D}_{\{\hat{\lambda};\,\hat{v}\}} = \frac{\mathcal{S}_{\{\hat{\lambda};\,\hat{v}\};\{0;\,0\}}}{\mathcal{S}_{\{0;\,0\};\{0;\,0\}}}$$

show that

$$\mathcal{D}_{\{\hat{\lambda};\,\hat{v}\}} = \mathcal{D}_{\hat{\lambda}}\,\mathcal{D}_{\hat{v}}$$

b) Prove that for the $\widehat{su}(2)$ diagonal cosets at levels $1, k$, and $k+1$, the level 1 field does not contribute to the quantum dimension. Calculate the quantum dimension of the Ising model primary fields.

c) Consider the $\widehat{su}(2)$ nondiagonal coset realization of the Potts model, where in

$$\frac{\widehat{su}(2)_3 \oplus \widehat{su}(2)_1}{\widehat{su}(2)_4}$$

the $\widehat{su}(2)_4$ factor is described by the corresponding nondiagonal invariant. Write the Virasoro primary fields in terms of the coset characters and calculate the quantum dimension of the three-states Potts primary fields.

d) Generalize the result observed in part (**c**): when the identity block contains more than one term, the quantum dimensions, as calculated from the coset characters, must be rescaled by the number of terms in the block.

e) Verify the decompositions (18.113) from the matching of the quantum dimensions.

18.8 *A maverick formulation of the three-state Potts model:* $\widehat{su}(3)_2/\widehat{su}(2)_8$

a) Consider the coset $\widehat{su}(3)_2/\widehat{su}(2)_8$, whose central charge is $\frac{4}{5}$. List all coset fields compatible with Eq. (18.26) and their fractional dimensions. Compare these dimensions with the Kac table for the minimal model $\mathcal{M}(6,5)$. Indicate the coset fields that should be absent in order that the set of coset dimensions be a subset of the Kac table.

b) Use the outer-automorphism branching rule $1 \mapsto a$ (cf. Sect. 14.7.3) to find all the fields that can be identified with the outer automorphisms. By comparing the resulting set of a priori distinct fields with the spectrum of the three-state Potts models, show that there must be further field identifications, and list them.

c) Show, up to the order q^3 (included), that

$$\chi_{\{[2,0,0];\,[8,0]\}} = \chi_{\{[2,0,0];\,[0,8]\}} = \chi_{\{[0,1,1];\,[4,4]\}}$$
$$= q^{-1/30}(1 + q^2 + 2q^3 + 3q^4 + 4q^5 + \cdots)$$
$$= \chi_{(1,1)}^{\text{Vir}} + \chi_{(4,1)}^{\text{Vir}}$$

(the additional two terms are given for reference). To proceed, decompose the $su(3)$ content of the modules $L_{[2,0,0]}$ and $L_{[0,1,1]}$ given in Tables 18.2-18.3 into $su(2)$ representations. Construct the various $\widehat{su}(2)_8$ modules with even finite parts and find the decomposition the $\widehat{su}(3)$ modules in terms of $\widehat{su}(2)$ ones. (To speed up the last step, observe that in the first few grades, in the absence of singular-vector constraints, the irreducible content at grade n is obtained from the tensor product of the irreducible content at grade $n-1$ with the adjoint representation.) As a side result, $\chi_{\{[2,0,0];\,[6,2]\}}$ and $\chi_{\{[2,0,0];\,[2,6]\}}$ vanish up to $\mathcal{O}(q^3)$. As the above relation between coset and Virasoro characters indicates, this coset description is equivalent to the diagonal $\widehat{su}(3)$ coset at levels (1,1;2). In fact, all but the last $\hat{g}_2/\hat{p}_{2x_e}$

maverick models listed in Eq. (18.32) have an equivalent description in terms of the diagonal coset $(\hat{g}_1 \oplus \hat{g}_1)/\hat{g}_2$.

18.9 Wakimoto free-field representation for the coset $\hat{g}/\hat{p}$

Find the expression for the coset energy-momentum tensor in the Wakimoto free-field representation.

Hint: Start with the expression

$$T^g = -\frac{1}{2}(\partial\varphi \cdot \partial\varphi) - i\alpha_+\rho \cdot \partial^2\varphi + \sum_{\alpha \in \Delta_+}(\gamma_\alpha \beta_\alpha)$$

for the energy-momentum tensor of the $\hat{g}_k$ WZW model, and the analogous expression for the $\hat{p}_{kx_e}$ WZW model. Distinguishing the quantities related to p by a tilde ($\tilde{\ }$), identify

$$\tilde{\beta}^\alpha = \beta^\alpha, \qquad \tilde{\gamma}^\alpha = \gamma^\alpha$$

for those α that are roots of p $\subset$ g. To find the relation between $\tilde{\varphi}^i$ and φ^i, enforce the equality

$$\tilde{h}^i = h^i \quad i = 1, \cdots \tilde{r}$$

using the general form of h^i given in Ex. 15.27.

18.10 A c = 1 coset with fixed points

a) For the diagonal coset

$$\frac{\widehat{su}(2)_2 \oplus \widehat{su}(2)_2}{\widehat{su}(2)_4}$$

write all the distinct coset fields and calculate their fractional dimension.

b) Identify the free bosonic $c = 1$ theory that has the same operator content as this parafermionic coset (cf. the last subsection of App. 17.B).

Result: $Z_{orb}(\sqrt{12})$

Observe that the operator contents of the two theories do not match exactly: the multiplicity of the coset field of conformal dimension $\frac{1}{24}$, corresponding to the triplet

$$\{\hat{\lambda}, \hat{\mu}; \hat{\nu}\} = \{[1,1], [1,1]; [2,2]\}$$

is 1 whereas the field of (fractional) dimension $\frac{1}{24}$ has multiplicity 2 in the bosonic theory (i.e., one copy has dimension $\frac{1}{24}$ and the other has dimension $\frac{25}{24}$). This reflects the fact that the coset field is a *fixed point of the outer automorphism*. To define the theory properly, the fixed point must be *resolved*, which, in the present case, amounts to replacing

$$\chi_{\frac{1}{24}} \rightarrow \frac{1}{2}\left(\chi_{\frac{1}{24}} + 1\right) + \frac{1}{2}\left(\chi_{\frac{1}{24}} - 1\right)$$

This modifies the partition function by a constant.

c) For the general $\widehat{su}(N)$ diagonal coset

$$\frac{\widehat{su}(N)_{k_1} \oplus \widehat{su}(N)_{k_2}}{\widehat{su}(N)_{k_1+k_2}}$$

find the values of k_1 and k_2 at which there are fixed points.

18.11 *Rational c* $= 1$ *models from* $\widehat{so}(N)$ *diagonal cosets*

a) Verify that the different cosets

$$\frac{\widehat{so}(N)_1 \oplus \widehat{so}(N)_1}{\widehat{so}(N)_2}$$

are rational theories at $c = 1$. Prove their equivalence with $\widehat{u}(1)$ models on an appropriate rational square radius. Observe first that, for $N = 3$, this corresponds to the previous example (cf. Ex. 18.10) since $\widehat{so}(3)_1$ is essentially the same as $\widehat{su}(2)_2$; it is equivalent to $Z_{orb}(\sqrt{12})$. For $N = 4$, the coset reduces to two copies of the Ising model since $so(4) = su(2) \oplus su(2)$, and it is thus equivalent to $Z_{orb}(2)$.

b) For $N = 5$, write the 15 distinct coset fields and calculate their fractional dimension. Find the corresponding $c = 1$ theory. Observe that the multiplicities do not match exactly; identify the source of this difficulty (cf. Ex. 18.10). Note that $so(5)$ is the same as $sp(4)$ with ω_1 and ω_2 interchanged.

c) For $N = 2r + 1$, the three finite weights corresponding to the coset field of lowest dimension are

$$\{\omega_r, \omega_r; 2\omega_r\}$$

Check that this field has dimension

$$h = \frac{1}{8(2r + 1)}$$

and argue that this fixes the value of the radius of the corresponding free-boson theory to be

$$R = 2\sqrt{2r + 1}$$

Show also that there are two coset fields of dimensions $\frac{1}{16}$, with finite weights

$$\{\omega_1, 0; \omega_1\} \qquad \{0, \omega_1; \omega_1\}$$

These data uniquely identify the equivalent $\widehat{u}(1)$ theory as $Z_{orb}(2\sqrt{2r + 1})$. (For $N = 3$, this agrees with the conclusion of part (**a**).)

d) For $N = 2r$, the three finite weights corresponding to the coset field of lowest dimension are

$$\{\omega_1, 0; \omega_1\}$$

Check that this field has dimension

$$h = \frac{1}{4r}$$

Calculate the dimension of the two fields

$$\{0, \omega_r; \omega_r\} \qquad \{0, \omega_{r-1}; \omega_{r-1}\}$$

Conclude that the coset is equivalent to $Z_{orb}(\sqrt{2r})$. (For $N = 4$, this agrees with the conclusion of part (**a**); for $N = 6$, since $su(4) \approx so(6)$, the coset is equivalent to $\widehat{su}(2)_4/\widehat{u}(1)$ (cf. Ex. 18.13), and this is shown to be equivalent to $Z_{orb}(\sqrt{6})$ in Ex. 18.12).

Remark: These results reveal the surprising $B - D$ duality:

$$\frac{\widehat{so}(2r + 1)_1 \oplus \widehat{so}(2r + 1)_1}{\widehat{so}(2r + 1)_2} \approx \frac{\widehat{so}(8r + 4)_1 \oplus \widehat{so}(8r + 4)_1}{\widehat{so}(8r + 4)_2}$$

18.12 *The parafermionic coset $\widehat{su}(2)_4/\widehat{u}(1)$*

a) Verify the equivalence

$$\frac{\widehat{su}(2)_4}{\widehat{u}(1)} \approx Z_{\text{orb}}(\sqrt{6})$$

b) By replacing the diagonal $\widehat{su}(2)_4$ modular invariant by the nondiagonal one, the coset invariant becomes:

$$|\chi_0 + \chi_1|^2 + 2|\chi_{1/12}|^2 + 2|\chi_{1/3}|^2 + |\chi_{3/4} + \chi'_{3/4}|^2$$

(the index gives the conformal dimension). The fields corresponding to each term are, respectively:

$$\{0; 0\} \oplus \{4; 0\}, \ 2\{2; 0\}, \ 2\{2; 2\}, \ \{0; 2\} \oplus \{4, 2\})$$

in the $\{\ell; m\}$ notation of Sect. 18.5. Find the equivalent $c = 1$ theory.
Result: $Z(\sqrt{6})$.

c) Using the identification obtained in part **(b)**, derive explicit expressions for the string functions $c_0^0 + c_0^4$, $c_2^0 + c_2^4$, c_0^2, and c_2^2.

18.13 *Equivalent parafermionic cosets*

a) Prove the equivalence of the following two coset models:

$$\frac{\widehat{su}(2)_k}{\widehat{u}(1)} \qquad \text{and} \qquad \frac{\widehat{su}(k)_1 \oplus \widehat{su}(k)_1}{\widehat{su}(k)_2}$$

by comparing their central charge and the number of distinct fields and charge conjugated fields in each case. To simplify the analysis of the second coset, prove first that, given two arbitrary weights $\hat{\lambda} \in P_+^1$ and $\hat{\nu} \in P_+^2$, there is a unique $\hat{\mu} \in P_+^1$ such that the three weights satisfy Eq. (18.30). Hence, the number of coset triplets is the number of independent pairs $\{\hat{\lambda}; \hat{\nu}\}$.

Remark: The exact spectra of these two coset families have already been shown to be identical for $k = 1, 2, 3$ in Sect. 18.5.2; for $k = 4$, this follows from Ex. 18.12 and Ex. 18.11.

b) The $su(N)$ generalization of the $su(2)$-type parafermionic models analyzed in Sect. 18.5 is

$$\frac{\widehat{su}(N)_k}{[\widehat{u}(1)]^{N-1}}$$

where $[\widehat{u}(1)]^{N-1} = \widehat{u}(1) \oplus \widehat{u}(1) \oplus \cdots \oplus \widehat{u}(1)$ ($N-1$ factors). Argue that an equivalent coset realization based on $\widehat{su}(k)$ algebras is

$$\frac{[\widehat{su}(k)_1]^N}{\widehat{su}(k)_N}$$

where $[\widehat{su}(k)_1]^N$ stands for the direct sum of N $\widehat{su}(k)_1$ factors, by comparing the central charge of the two cosets.

c) A simple argument, based on conformal branchings, proves coset equivalences at the level of the energy-momentum tensor. Given two distinct semisimple conformal embeddings into $\hat{g}_1$, that is,

$$\hat{g}_1 \supset \hat{p}_{k_1}^{(1)} \oplus \hat{p}_{k_2}^{(2)}$$

$$\hat{g}_1 \supset \hat{h}_{k'_1}^{(1)} \oplus \hat{h}_{k'_2}^{(2)}$$

there follows the equality

$$T_1^g = T_{k_1}^{p^{(1)}} + T_{k_2}^{p^{(2)}} = T_{k_1'}^{h^{(1)}} + T_{k_2'}^{h^{(2)}}$$

If $h^{(2)} \subset p^{(1)}$ and $p^{(2)} \subset h^{(1)}$, we obtain

$$T_{k_1}^{p^{(1)}} - T_{k_2'}^{h^{(2)}} = T_{k_1'}^{h^{(1)}} - T_{k_2}^{p^{(2)}}$$

and this proves the equality of the energy-momentum tensors for the two cosets $\hat{p}_{k_1}^{(1)}/\hat{h}_{k_2'}^{(2)}$ and $\hat{h}_{k_1'}^{(1)}/\hat{p}_{k_2}^{(2)}$. Reconsider the problem in part (**b**) using this argument and the conformal embeddings

$$\widehat{su}(pq)_1 \supset \widehat{su}(p)_q \oplus \widehat{su}(q)_p$$
$$\widehat{su}(p+q)_1 \supset \widehat{su}(p)_1 \oplus \widehat{su}(q)_1 \oplus \widehat{u}(1)$$

18.14 *Bosonization of parafermions*

From the bosonization of the $\widehat{su}(2)_k$ current generators (cf. Ex. 15.21 with $J_0 = h/2$, $J^+ = e, J^- = f$):

$$J^+(z) = i\partial v\, e^{-u+iv}$$

$$J^0(z) = \frac{i}{\sqrt{2}\alpha_+}\partial\varphi + \partial u$$

$$J^-(z) = \left[\frac{-i\sqrt{2}}{\alpha_+}\partial\varphi - (k+2)\partial u + (k+1)\partial v\right]e^{u-iv}$$

obtain the expression of the parafermions ψ_{par} and $\psi_{par}^\dagger$ defined by Eq. (18.158).
Hint: ψ_{par} and $\psi_{par}^\dagger$ must commute with the $\widehat{u}(1)$ subalgebra of $\widehat{su}(2)_k$, that is, with J^0. Express the result in terms of v and the linear combination of u and φ that is orthogonal to J^0.

a) Find the energy-momentum tensor T_{par} of the parafermionic theory by quotienting the $\widehat{u}(1)$ piece generated by J^0 from the $\widehat{su}(2)_k$ energy-momentum tensor.

b) Using the chiral version of the relation (18.160) and the bosonized version for the WZW primary fields (15.287), obtain the expression for the parafermionic fields f_m^ℓ. Calculate the OPE $T_{par}(z)f_m^\ell(w)$ and find the value of h_m^ℓ.

18.15 *Generalized $\widehat{su}(2)$ diagonal cosets and Coulomb-gas representation*

a) Show that the central charge of the general coset

$$\frac{\widehat{su}(2)_k \oplus \widehat{su}(2)_l}{\widehat{su}(2)_{k+l}}$$

factorizes as

$$c = 1 - \frac{6k}{(l+2)(k+l+2)} + \frac{2(k-1)}{k+2}$$

When $k = 1$, the third term disappears and we recover the coset realization of the unitary minimal models, which can be described by a Coulomb-gas formalism with an appropriate background charge. The third term is the central charge of the parafermionic $\widehat{su}(2)_k/\widehat{u}(1)$ theory. Together, these two pieces (boson and parafermionic coset) provide a generalized Coulomb-gas representation of the above coset model.

b) The screening charges of the theory are

$$Q_+ = \int dz\, \psi_{\text{par}}(z)\, e^{i\alpha_+ \varphi(z)}$$

$$Q_- = \int dz\, \psi_{\text{par}}^\dagger(z)\, e^{i\alpha_- \varphi(z)}$$

where ψ_{par} and $\psi_{\text{par}}^\dagger$ are the parafermionic fields of dimension $(k-1)/k$ introduced in the representation of the $\widehat{su}(2)_k$ generators of Eq. (18.158). The field φ is a standard free boson:

$$\varphi(z)\varphi(w) \sim -\ln(z-w)$$

Determine $\alpha_\pm$. Fix the indeterminate signs from the requirement that in the limit $l \to \infty$—where the coset reduces to a simple $\widehat{su}(2)_k$ theory—the screening charges become the zero modes of the $\widehat{su}(2)_k$ ladder operators:

$$Q_\pm \to J_0^\pm \quad \text{as} \quad l \to \infty$$

c) From the Coulomb-gas representation, the coset primary fields are:

$$\Phi_{r,s} = f_\ell^\ell\, e^{i\alpha_{r,s}\varphi}$$

with

$$\alpha_{r,s} = \frac{1}{2}(1-r)\alpha_+ + \frac{1}{2}(1-s)\alpha_-$$

$$\ell = |s - r \bmod 2k|$$

$$1 \le r \le l+1, \quad 1 \le s \le k+l+1, \quad 0 \le \ell \le k$$

where the f_ℓ^ℓ's are the primary fields of the parafermionic theory, of dimension

$$h_\ell^\ell = \frac{\ell(\ell+2)}{4(k+2)} - \frac{\ell^2}{4k}$$

Using this representation, calculate the dimensions of the coset primary fields.

d) Compare the result with the dimensions extracted from the character decomposition worked out in Sect. 18.5.4, Eqs. (18.174) and (18.175) (treating the factors $-c/24$ carefully).

e) The coset can equally well be described in term of an $\widehat{su}(2)_l/\widehat{u}(1)$ parafermionic theory and a modified background charge. Find the primary-field conformal dimensions in the dual formulation obtained by interchanging the values of k and l, and compare with the results of part **(c)**.

18.16 $\widehat{su}(2)_k$ modular S matrix at fractional level
From the expression (18.185) for the character of admissible representations, derive the S matrix (18.191).

18.17 $\widehat{su}(2)_k$ charge conjugation matrix at fractional level
Starting directly from the S matrix (18.191), obtain the charge conjugation matrix (18.204).

18.18 The $\widehat{su}(2)_{-\frac{1}{2}}$ model

a) Calculate explicitly the S and T matrices of the admissible representations at level $-\frac{1}{2}$ and check the relations

$$SS^\dagger = 1, \quad S^2 = (ST)^3 = C$$

b) Use the Verlinde formula to calculate the fusion rules in the theory. Verify associativity, that is, the commutativity of the fusion matrices.

18.19 $\widehat{su}(2)_k$ *modular invariants at fractional level*

a) We recall the S duality relation $AS = Sb$, where A is an element of the outer auto-morphism group (1 or a here) and b is the corresponding element of the center (not to be confused with a branching function):

$$(b)_{\mu,\mu} = e^{-2\pi i(A\hat{\omega}_0,\mu)}$$

Show that its generalization at fractional level is

$$A' S = S \, b'$$

where A' acts only on $\hat{\lambda}^I$, that is,

$$A'\hat{\lambda} = A\hat{\lambda}^I - (k+2)\hat{\lambda}^F$$

and b' is

$$(b')_{\mu,\mu} = e^{-2\pi i(A\hat{\omega}_0, u(\mu+\rho)-\rho)} = [(b)_{\mu,\mu}]^u \, [\epsilon(w_A)]^{u-1}$$

(No relations of that form are obtained by restricting the action of A to the fractional part.)

b) Show that the appropriate generalization of the symmetrized product (17.35) takes the form

$$(A')^p \circ (b')^q = (A')^p (b')^q e^{-\pi i q p k^I u |A\hat{\omega}_0|^2}$$

c) Following the analysis of Sect. 17.3, establish the modular invariance of the partition function whose mass matrix is

$$M = \frac{1}{2} \sum_{p,q=0,1} (A')^p \circ (b')^q$$

d) Show that for the vacuum to have multiplicity 1, u must be odd and for the other multiplicities to be positive integers,

$$uk^I |A\hat{\omega}_0|^2 \in \mathbb{Z}$$

must hold.

e) Derive the following sequences of modular invariants (where $\lambda_1^I = n$, $\lambda_1^F = m$ and $\chi_{\hat{\lambda}} \equiv \chi_{n,m}$): the A_{k^I+1} series for all values of k,

$$Z_{A_{k^I+1}} = \sum_{m=0}^{u-1} \sum_{n=0}^{k^I} |\chi_{n,m}|^2$$

the $D_{2\ell+2}$ series for k such that $k^I = 4\ell$,

$$Z_{D_{2\ell+2}} = \sum_{m=0}^{u-1} \sum_{\substack{n=0 \\ n\in 2\mathbb{Z}}}^{2\ell-1} |\chi_{n,m} + \chi_{4\ell-n,m}|^2 + 2|\chi_{2\ell,m}|^2$$

and the $D_{2\ell+1}$ series for $k^I = 4\ell - 2 \geq 6$,

$$Z_{D_{2\ell+1}} = \sum_{m=0}^{u-1} \left(\sum_{\substack{n=0 \\ n\in 2\mathbb{Z}}}^{4\ell-2} |\chi_{n,m}|^2 + |\chi_{2\ell-1,m}|^2 \right.$$

$$\left. + \sum_{\substack{n=1 \\ n\in 2\mathbb{Z}+1}}^{2\ell-3} [\chi_{n,m}\bar{\chi}_{4\ell-2-n,m} + \chi_{4\ell-2-n,m}\bar{\chi}_{n,m}] \right)$$

f) Modular invariants that mix the fractional parts of the weights are obtained by multiplying the above mass matrix by the charge conjugation matrix. Write the charge conjugate versions of the above series.

18.20 *A remarkable relation between $\widehat{su}(2)$ S matrices at fractional level and the minimal model S matrices*

Verify the following relation at the level of S matrices

$$\sum_{w\in W} \epsilon(w) S^{(k)}_{w^F\lambda,\hat{\mu}} = i S^{\text{Vir}\,(p,p')}_{(\lambda_1^I+1,\lambda_1^F),(\mu_1^I+1,\mu_1^F)} \tag{18.291}$$

where w^F means that the finite Weyl reflection w acts only on the fractional part:

$$w^F\hat{\lambda} = \hat{\lambda}^I - (k+2)w\hat{\lambda}^F$$

$S^{\text{Vir}\,(p,p')}$ is given in Eq. (10.134), with parameters p,p' related to the level k by

$$p' = u, \quad p = 2u + t$$

Observe that when $\lambda_1^F = 0$, the sum vanishes so that the index λ_1^F in $S^{\text{Vir}\,(p,p')}$ is actually a nonzero positive integer.

18.21 *A remarkable relation between $\widehat{su}(2)$ characters at fractional level and the minimal model characters*

Show that the residue of the character of an admissible $\widehat{su}(2)_k$ representation, multiplied by $\eta^2(q)$, that is,

$$\tilde{\chi}_{\hat{\lambda}}(\tau) \equiv \eta^2(q)\, 2\pi i \lim_{z\to 0} z\, \chi_{\hat{\lambda}}(z;\tau)$$

$$= \eta^2(q)\, \frac{q^{b_+^2/4d} \sum_{s\in\mathbb{Z}} q^{s^2d+sb+}(1-q^{\lambda_1^F[2us(k+2)+\lambda_1^I+1]})}{q^{1/8}\sum_{s\in\mathbb{Z}} q^{2s^2+s}(4s+1)} \tag{18.292}$$

is exactly a minimal Virasoro character, for a model with $p' = u$ and $p = 2u + t$.

18.22 *Canonical chain of field identifications in $\widehat{su}(2)$ diagonal nonunitary coset*

a) For the coset

$$\frac{\widehat{su}(2)_k \oplus \widehat{su}(2)_1}{\widehat{su}(2)_{k+1}}$$

with fractional k, starting from a coset field with zero fractional part, show that $2u$ fields can be identified through the canonical chain

$$(a\, c)^{u-1}a$$

Show in particular that all triplets of weights in this string are distinct and that the last triplet has a zero fractional part.

b) For $k = -\frac{1}{2}$, list all triplets of weights satisfying the branching conditions and calculate the fractional conformal dimension of the corresponding coset fields. Compare with the dimensions in the Kac table of the corresponding minimal model. Identify fields with the same fractional dimension using the operations a and c. Compare with the canonical chain obtained in part (a).

c) The presence of fixed points does not preclude fields to be identified from their S matrix. Verify that for the coset

$$\frac{\widehat{su}(2)_k \oplus \widehat{su}(2)_l}{\widehat{su}(2)_{k+l}}$$

fields can still be identified with a and c. Consider the particular case where $l = 2$ and $k = -\frac{4}{3}$. Obtain all triplets with fractional conformal dimension 0 and $-\frac{7}{32}$. Obtain the canonical chain of identification in each case. How does the presence of fixed points reflected itself in this chain?

18.23 *Character decomposition in the nonunitary case* $(\widehat{su}(2)_k)$

a) Obtain the character of primary fields in nonunitary minimal models (p, p') from the decomposition

$$\chi^{(k)}_{\hat{\lambda}^I} \chi^{(1)}_{\hat{\mu}} = \sum_{\hat{\nu}^I \in P^{kI}_+ + u} \chi_{\{\hat{\lambda}^I, \hat{\mu}; \hat{\nu}^I\}} \chi^{(k+1)}_{\hat{\nu}^I}$$

with $p = k^I + 2 + u$, $p' = k^I + 2$.
Hint: Follow closely the derivation of Sect. 18.3.1.

b) Find the corresponding S matrices.

18.24 *Examples of admissible representations*
List the admissible weights of:

a) $\widehat{G}_2$ at $k = -\frac{8}{3}$ (there are 6 of them);

b) $\widehat{so}(7)$ at $k = -\frac{5}{2}$ (there are 8 of them).

18.25 *Admissibility condition*
Prove that if $\hat{\lambda} = y \cdot (\hat{\lambda}^I - (k+g)\hat{\lambda}^F)$ is admissible, so is $v \cdot (\hat{\lambda}^I - (k+g)\hat{\lambda}^F)$ with v related to y by $y = vw_A$, for any w_A.

18.26 *The* $\widehat{su}(3)_{-\frac{3}{2}}$ *model*
Calculate the S and C matrices for the $\widehat{su}(3)_{-\frac{3}{2}}$ model. Verify that the admissible representations in the $y = 1$ sector do not transform covariantly among themselves. This illustrates the necessity of the $y \neq 1$ sectors.

18.27 *Another coset realization of the Yang-Lee model*
Consider the coset

$$\frac{\widehat{su}(3)_{-\frac{3}{2}} \oplus \widehat{su}(3)_1}{\widehat{su}(3)_{-\frac{1}{2}}}$$

Use the results of App. 18.B to obtain all the admissible weights for each WZW model, and find all triplets that satisfy the branching conditions. Organize fields in classes according to the fractional value of their conformal dimension. Show that all coset fields which involve weights in the $y \neq 1$ sector can be related to fields with $y = 1$. Compare with Eq. (18.287). In the class $y = 1$, find the canonical chain of field identification in terms of A and C_A defined in Eq. (18.288). Show that fields with zero fractional finite weights provide convenient coset representatives. In this subclass of coset fields, find the residual field identifications.

18.28 *A duality relation for cosets*
Argue, at the level of the central charge, that for a particular value of k', the following two diagonal cosets are equivalent:

$$\frac{\widehat{su}(N)_k \oplus \widehat{su}(N)_1}{\widehat{su}(N)_{k+1}} \quad \text{and} \quad \frac{\widehat{su}(u)_{k'} \oplus \widehat{su}(u)_1}{\widehat{su}(u)_{k'+1}}$$

with $k = t/u$. Characterize this $u \leftrightarrow N$ duality in terms of k'. The simplest example of this duality is furnished by the two coset representations of the Yang-Lee model, corresponding to the pair $(N, u) = (2, 3)$.

18.29 *Technical aspects of nonunitary diagonal cosets*

a) From the diagonal coset (18.276), prove that the last two terms in Eq. (18.283) cancel (mod 1) when $l = 1$ and g is simply laced.

b) For manipulations of the numerous phase factors in nonunitary WZW-model S matrices, the following result is useful: If ζ and ν are integrable weights, then for any $w \in W$ we have

$$\left((w - 1)\zeta, \nu \right) = 0 \bmod 1$$

provided

$$\zeta_i \text{ or } \nu_i \in (a_i \alpha_i^\vee) \, \mathbb{Z}$$

a condition satisfied by the fractional parts of admissible weights. Prove this statement.

c) Write the explicit form of the coset S matrix $S_{\{\hat{\lambda}, \hat{\mu}; \hat{\nu}\}, \{\hat{\lambda}', \hat{\mu}'; \hat{\nu}'\}}$, with $\hat{\lambda} \in P_y^k$, $\hat{\mu} \in P_+^l$, $\hat{\nu} \in P_y^{k+l}$ and $\hat{\lambda}^F = \hat{\nu}^F$. Verify that it depends neither upon y nor $\hat{\lambda}^{F,y}$.

d) Prove that fields related by Eq. (18.286) have the same T matrix, i.e., same value of the fractional conformal dimension.

e) Verify that Eq. (18.287) is compatible with the branching condition (18.281). The results of parts **(c)**–**(e)** establish the equivalence of fields related by Eq. (18.286).

Notes

The coset construction goes back to the work of Bardacki and Halpern [25, 189]. It appeared in conformal field theory in the seminal paper of Goddard, Kent, and Olive [178], where the general construction of the coset energy-momentum tensor was first exposed and used to prove the unitarity of the sequences $\mathcal{M}(m+1, m)$. The character decomposition presented in Sect. 18.3 follows closely the original derivation in Ref. [179]. Further properties of minimal models, viewed from the coset angle, were worked out by Gepner [165]. Although there the coset framework is not made completely manifest, it is shown that minimal models can be viewed as a product of two $\widehat{su}(2)$ theories at appropriate levels. Additional results can be found in Bowcock and Goddard [58]. This reference also contains the complete list of coset realizations of the unitary minimal models. Most of the explicit character decompositions pertaining to these exceptional representations were derived by Kac and Wakimoto [221, 218]. More decompositions are presented in Refs. [218, 221, 179, 54, 113, 198, 306]. Mavericks were discovered by Dundar and Joshi [113, 114] (Ex. 18.8 is adapted from these references). The last example in Eq. (18.32) is due to Fuchs, Schellekens, and Schweigert [152]. The calculation of correlation functions in the coset framework, mainly illustrated with the minimal models, is discussed in the work of Douglas [112]. Not all the coset modular invariants have the simple product form described in Sect. 18.2.5; further solutions are explored in Ref. [162].

The detailed analysis of the coset $\widehat{su}(2)_k/u(1)$ is due to Gepner and Qiu [170]. The related character decompositions, that is, the string functions, are worked out in Refs. [195, 215]. Zamolodchikov and Fateev [366] studied the underlying parafermionic theory, and they first noticed the relation with an $\widehat{su}(2)$ coset. Generalized parafermions, defined in terms of the coset $\hat{g}_k/u(1)^r$, were introduced in Ref. [166].

The issue of field identification, and its relation to selection rules, was addressed first by Gepner [167] and Moore and Seiberg [274]; it was further analyzed by Lerche, Vafa, and

Warner [255] and Ahn and Walton [5]. Resolution of fixed points was studied thoroughly in the works of Schellekens and Yankielowicz [321, 322] and recent progress is reported by Fuchs, Schellekens, and Schweigert [152] (it is shown, for diagonal cosets, that the WZW models do actually code all the information required for the fixed-point resolutions).

The necessity of having fractional-level affine algebra for a coset description of the nonunitary models was first noticed by Kent [234]. Admissible representations and their modular properties were discovered by Kac and Wakimoto [219, 220]. A somewhat simplified description of the admissible weights and some aspects of the nonunitary minimal models from the coset approach are given in Ref. [265]. The Fock representation of the $\widehat{su}(2)$ models at fractional level is studied in Ref. [42]. Modular invariants for $\widehat{su}(2)_k$ at fractional levels were considered by Koh and Sorba [243] and fully classified by Lu [258]. The construction based on outer automorphisms sketched in Ex. 18.19 is due to Ref. [265] (where the result is given for any affine algebra). Fusion rules in nonunitary $\widehat{su}(2)$-WZW models are discussed in Ref. [265, 243] (see also [42]) from the point of view of the Verlinde formula and in Refs. [16, 293] from an OPE approach (and with different results).

A general analysis of nonunitary diagonal coset models, with a focus on the question of field identification, can be found in Ref. [266] (from which most aspects of the nonunitary coset discussion are borrowed). The duality presented in Ex. 18.28 was found in Ref. [248] (see also Refs. [246, 10]).

There is a large literature on the relation of fractional-level Lie algebras with minimal models, based on Hamiltonian reduction (see, for instance, Refs. [46, 124, 135]). Hints of such a connection follow from Ex. 18.20 (Refs. [265, 78]) and Ex. 18.21 (Ref. [275]).

The parafermionic description of the diagonal $\widehat{su}(2)$ coset models in Ex. 18.15 is based on Refs. [230, 17, 304]; the Wakimoto free-field representation for cosets in Ex. 18.9 is described in Refs. [232, 173]; the bosonization of parafermions of Ex. 18.14 is taken from Refs. [173, 104, 280]. The general method discovered by Altschuler [7] to establishing the equivalence of some coset models is illustrated in part (c) of Ex. 18.13 (see also Ref. [18] for this exercise). Finally, Ex. 18.7 is partly based on Ref. [242].